Global Aspects of the Environment
Volume II

Environmental Analysis and Economic Policy

Series Editors: Kenneth Button
Professor of Public Policy
The Institute of Public Policy, George Mason University, USA

Peter Nijkamp
Professor in Regional, Urban and Environmental Economics
Free University, Amsterdam, The Netherlands

1. Global Aspects of the Environment (Volumes I and II)
   Robert U. Ayres, Kenneth Button and Peter Nijkamp
2. Environment, Land Use and Urban Policy
   David Banister, Kenneth Button and Peter Nijkamp
3. Environmental Valuation (Volumes I and II)
   Kenneth G. Willis, Kenneth Button and Peter Nijkamp
4. Environment and Transport
   Yoshitsugu Hayashi, Kenneth Button and Peter Nijkamp
5. Environmental Economics and Development
   J(Hans). B. Opschoor, Kenneth Button and Peter Nijkamp
6. Environmental Instruments and Institutions
   Thomas Tietenberg, Kenneth Button and Peter Nijkamp
7. Ecosystems and Nature
   R. Kerry Turner, Kenneth Button and Peter Nijkamp

Wherever possible, the articles in these volumes have been reproduced as originally published using facsimile reproduction, inclusive of footnotes and pagination to facilitate ease of reference.

For a list of all Edward Elgar published titles visit our site on the World Wide Web at http://www.e-elgar.co.uk

Global Aspects of the Environment Volume II

Edited by

Robert U. Ayres

Sandoz Professor of Environment and Management, Professor of Economics and Director of the Centre for the Management of Environmental Resources, European Business School, INSEAD, France

Kenneth Button

Professor of Public Policy, The Institute of Public Policy, George Mason University, USA

and

Peter Nijkamp

Professor in Regional, Urban and Environmental Economics, Free University, Amsterdam, The Netherlands

ENVIRONMENTAL ANALYSIS AND ECONOMIC POLICY

An Elgar Reference Collection
Cheltenham, UK • Northampton, MA, USA

Published by
Edward Elgar Publishing Ltd
Glensanda House
Montpellier Parade
Cheltenham
Glos GL50 1UA
UK

Edward Elgar Publishing, Inc.
136 West Street
Suite 202
Northampton
Massachusetts 01060
USA

A catalogue record for this book is available from the British Library.

Library of Congress Cataloguing in Publication Data

Global aspects of the environment / edited by Robert U. Ayres,
Kenneth Button and Peter Nijkamp.
(Environmental analysis and economic policy: 1)
Includes bibliographical references and index.
1. Environmental sciences. 2. Ecological engineering. 3. Human ecology. I. Ayres, Robert U. II. Button, Kenneth. III. Nijkamp, Peter. IV. Series. An Elgar reference collection.
GE105.G53 1999
333.7'2—dc21 99–29678
CIP

ISBN 1 85898 741 5 (2 volume set)

Printed and bound in Great Britain by Biddles Ltd, Guildford and King's Lynn

Contents

Acknowledgements

The editors and publishers wish to thank the authors and the following publishers who have kindly given permission for the use of copyright material.

Academic Press, Inc. for articles: Richard Grace, R. Kerry Turner and Ingo Walter (1978), 'Secondary Materials and International Trade', *Journal of Environmental Economics and Management*, **5** (2), June, 172–86; Curt L. Anderson (1987), 'The Production Process: Inputs and Wastes', *Journal of Environmental Economics and Management*, **14** (1), March, 1–12; Don Fullerton and Thomas C. Kinnaman (1995), 'Garbage, Recycling, and Illicit Burning or Dumping', *Journal of Environmental Economics and Management*, **29** (1), July, 78–91.

American Economic Association for articles: Robert U. Ayres and Allen V. Kneese (1969), 'Production, Consumption, and Externalities', *American Economic Review*, **LIX**, June, 282–97; Robert M. Solow (1974), 'The Economics of Resources or the Resources of Economics', *American Economic Review*, **64** (2), May, 1–14; Herman E. Daly (1974), 'The Economics of the Steady State', *American Economic Review*, **64** (2), May, 15–21; John M. Hartwick (1977), 'Intergenerational Equity and the Investing of Rents from Exhaustible Resources', *American Economic Review*, **67** (5), December, 972–4.

Blackwell Publishers Ltd for articles: Ralph C. d'Arge (1971), 'Essay on Economic Growth and Environmental Quality', *Swedish Journal of Economics*, March, 25–41; Udo E. Simonis (1989), 'Ecological Modernization of Industrial Society: Three Strategic Elements', *International Social Science Journal*, **121**, August, 347–61.

Cambridge University Press for article: David Pearce, Kirk Hamilton and Giles Atkinson (1996), 'Measuring Sustainable Development: Progress on Indicators', *Environment and Development Economics*, **1** (1), February, 85–101.

Elsevier Science BV for excerpt: David James (1985), 'Environmental Economics, Industrial Process Models, and Regional-Residuals Management Models', in Allen V. Kneese and James L. Sweeney (eds), *Handbook of Natural Resource and Energy Economics*, Volume 1, Chapter 7, 271–324.

Elsevier Science Ltd for articles: Bruce Hannon and John Joyce (1981), 'Energy and Technical Progress', *Energy*, **6**, 187–95; A. Rose and C.Y. Chen (1991), 'Sources of Change in Energy Use in the U.S. Economy, 1972–1982: A Structural Decomposition Analysis', *Resources and Energy*, **13** (1), April, 1–21; Jeroen C.J.M. van den Bergh and Peter Nijkamp (1994), 'Dynamic Macro Modelling and Materials Balance: Economic–Environmental Integration for Sustainable Development', *Economic Modelling*, **11** (3), July, 283–307; S.M. de Bruyn and J.B. Opschoor

(1997), 'Developments in the Throughput–Income Relationship: Theoretical and Empirical Observations', *Ecological Economics*, **20** (3), March, 255–68.

L.S. Gross and E.C.H. Veendorp for their own article: (1990), 'Growth with Exhaustible Resources and a Materials-Balance Production Function', *Natural Resource Modeling*, **4** (1), Winter, 77–94.

Martin Jänicke, Harald Mönch and Manfred Binder for their own article: (1993), 'Ecological Aspects of Structural Change', *Intereconomics*, **28** (4), July/August, 159–69.

Johns Hopkins University Press for excerpts: Kenneth E. Boulding (1966), 'The Economics of the Coming Spaceship Earth', in Henry Jarrett (ed.), *Environmental Quality in a Growing Economy*, 3–14; J.E. Stiglitz (1979), 'A Neoclassical Analysis of the Economics of Natural Resources', in V. Kerry Smith (ed.), *Scarcity and Growth Reconsidered*, Chapter 2, 36–66.

Kluwer Academic Publishers BV for article: Thomas D. Crocker and John Tschirhart (1992), 'Ecosystems, Externalities, and Economies', *Environmental and Resource Economics*, **2** (6), 551–67.

MIT Press Journals for article: Ralph C. d'Arge and Allen V. Kneese (1972), 'Environmental Quality and International Trade', *International Organization*, **26** (2), Spring, 419–65.

National Academy Press for excerpt: David T. Allen and Nasrin Behmanesh (1994), 'Wastes as Raw Materials', in Braden R. Allenby and Deanna J. Richards (eds), *The Greening of Industrial Ecosystems*, 69–89.

RAND for articles: Edward A. Hudson and Dale W. Jorgenson (1974), 'U.S. Energy Policy and Economic Growth, 1975–2000', *Bell Journal of Economics*, **5** (2), Autumn, 461–514; Kenneth C. Hoffman and Dale W. Jorgenson (1977), 'Economic and Technological Models for Evaluation of Energy Policy', *Bell Journal of Economics*, **8** (2), Autumn, 444–66.

Regional Science Association International for article: John H. Cumberland (1966), 'A Regional Interindustry Model for Analysis of Development Objectives', *Papers of the Regional Science Association*, **17**, 65–94.

Review of Economic Studies Ltd for articles: P.S. Dasgupta (1969), 'On the Concept of Optimum Population', *Review of Economic Studies*, **XXXVI** (3), July, 295–318; R.C. d'Arge and K.C. Kogiku (1973), 'Economic Growth and the Environment', *Review of Economic Studies*, **XL** (1), January, 61–77; Partha Dasgupta and Geoffrey Heal (1974), 'The Optimal Depletion of Exhaustible Resources', in *Symposium on the Economics of Exhaustible Resources, Review of Economic Studies*, 3–28.

University of Chicago Press for articles: Herman E. Daly (1968), 'On Economics as a Life Science', *Journal of Political Economy*, **76** (3), May/June, 392–406; Paul M. Romer (1990), 'Endogenous Technological Change', *Journal of Political Economy*, **98** (5, Part 2), October, S71–102.

World Bank for excerpt: Partha Dasgupta and Karl-Göran Mäler (1991), 'The Environment and Emerging Development Issues', *Proceedings of the World Bank Annual Conference on Development Economics 1990*, 101–31.

Every effort has been made to trace all the copyright holders but if any have been inadvertently overlooked the publishers will be pleased to make the necessary arrangement at the first opportunity.

In addition the publishers wish to thank the Library of the London School of Economics and Political Science, the Marshall Library of Economics, Cambridge University and B & N Microfilm, London for their assistance in obtaining these articles.

Series Preface

Kenneth Button and Peter Nijkamp

The environment has been a dominant theme for research and public policy during the latter part of the twentieth century and there is no sign that this will change as we move to the new millennium. The global interest in environmental matters in part stems from the increased pressures that a mounting population and increased production puts on the planet's natural resource base. Additionally, as personal incomes rise and leisure time becomes more freely available in the developed world, concern with more immediate human needs give way to an interest in preservation and conservation for future generations and for others. Science has also contributed by both highlighting emerging environmental problems and discovering the importance of many components of the natural environment for medical and other uses.

Historically, the interest in the environment can be traced through several phases, each with its own particular emphasis. The interest is also far from new and in antiquity there were often laws governing such things as the removal of night soil and the use of noisy wagons into towns. Traditionally, also the natural environment was seen as a source of the necessities of life and was controlled through farming techniques to ensure that food was adequate to maintain the population. Land use was controlled and water resources were often regulated at the local and regional levels.

With the Industrial Revolution came population growth, urbanization and the accelerated exploitation of natural resources. Malthus advanced ideas of over population while others, such as Jevons, became concerned about the finite nature of such resources as coal. The new technologies contaminated water courses and the widespread use of carbon fuels led to atmospheric pollution. In cities, dirt, noise and urban decay led to new ideas of town planning and ideal cities.

Much of the interest in the latter part of the last century centred around public health. As knowledge of how disease was spread – urban authorities in particular sought to improve the local environment by such measures as sewage control and clean water supply to reduce the spread of germs and infection. This trend, much later, spread to policies embodied in various pieces of clean air legislation to reduce local atmospheric pollution that causes smog and other harmful effects.

Wealthy societies, and the better off within poorer societies, with the time and resources to expand, became concerned with the built environment and in shaping nature in ways they found aesthetically pleasing. Over the centuries this has led to landscaping of the countryside and the provision of parks and gardens in urban areas.

The twentieth century has seen many of these trends continue although some of the problems, such as depletion of natural resources, initially appeared less severe as new sources were found in Africa, South America and Asia. Difficulties in feeding and otherwise sustaining the burgeoning populations of these regions, however, resurrected Malthusian concerns over whether the planet could sustain itself. Increasing reliance on fossil fuels for energy and the geographical concentration of the main reserves of oil in politically unstable areas resurrected concern over natural resource reserves in the post Second World War period.

More fundamental to modern debates has been the global orientation of recent concerns. This has been accompanied by a much more popular interest in the state of the natural environment. Kenneth Boulding's development of the idea of 'spaceship earth' attracted much attention, as did the rather doom ladened prophesies of the Club of Rome, but this was subsequently overshadowed by the publication of the Brundtland Report, *Our Common Future*, with its emphasis on the need to maintain resources for future generations. With the report came the much touted, but universally ambiguous, concept of sustainable development.

Just as the academic interest in the environment has evolved so has the institutional setting in which the topic is debated and researched. In a very general way as our understanding of the environment has progressed and as technology has developed so there has been a movement out from micro-policy making in the hands of city councils, lords of the manor and the like to macro-policy making, involving global agencies such as the United Nations and World Bank. Indeed, one of the pressing issues of contemporary environmental debate is the appropriate jurisdiction for different levels of government.

For those interested in researching environmental topics, either from the standpoint of the beginner or of that of an expert wishing to expand an established basis of knowledge, there are practical problems. Environmental analysis extends across virtually all academic disciplines although many such as economics and engineering have specific sub-areas of interest specifically devoted to environmental issues. The multidisciplinary and transdisciplinary nature of environmental research makes it difficult to keep track of the key literature. To those new to the field there is the added problem of tracing back to find classic papers or research results. There are now a large number of specialist journals that deal with environmental topics but this is often adding to the problem of collecting information rather than reducing it. While these journals and periodicals provide explicitly environmental research findings, other general journals in economics, sociology, geography, engineering, medicine, biology and so on still continue to publish important findings.

The aim of this series of edited books is to provide both a set of reference volumes for those already in the field and to facilitate newcomers with accessible core material. The aim is not to offer readings on the purely scientific issues but rather to limit the papers to those concerned with social science aspects of the environment. This does not mean that we have excluded papers with a natural science orientation but rather the focus is on the softer debates and research. There are seven volumes in the series. The papers have all been previously published but do not constitute a set of traditional classics. While many of the articles have an established pedigree, equally there are less often referenced pieces. The aim of each volume is to paint a picture of how a topic area has developed, provide an indication as to some of the key contributions to this development but also look at some of the more recent literature that covers policy considerations.

The books cover a wide range of topic areas. At one extreme there are the macro-oriented volumes edited with Bob Ayres looking at the *Global Aspects of the Environment* while, in conjunction with Kerry Turner, there is a collection concerned with *Ecosystems and Nature*. Also at the aggregate level, edited with Hans Opschoor, there is a volume which focuses on *Environmental Economics and Development*. At the other extreme there are volumes concerned with micro-issues such as that along with Ken Willis that deals with *Environmental*

Valuation and with David Banister on *The Environment, Land Use and Urban Policy*. Transport's impact on the environment is multifaceted and can be severe hence there is a volume edited with Yoshi Hayashi, *The Environment and Transport* dealing exclusively with that subject. Finally, edited with Tom Tietenberg there is a collection that looks more generally at *Environmental Instruments and Institutions*.

We accept that there are many omissions from the collections and that others may have made different selections or have structured the material differently. Space limits what can be included and there is much more that could have been justified in a larger exercise (although this would add further to deforestation). We have also had to omit work by many eminent and important scholars for the same reason. As for structuring the material, a number of possible cuts are possible, the ones selected here just seem logical to us and would seem to fit with the needs of both researchers and, possibly, teachers.

Part I
Resource/Environmental Perspective: Materials, Energy, Exergy, Entropy, Mass Balance

[1]

A REGIONAL INTERINDUSTRY MODEL FOR ANALYSIS OF DEVELOPMENT OBJECTIVES

JOHN H. CUMBERLAND, University of Maryland

Rapid increases in population, industrialization, and urbanization in recent decades have added to the urgency of understanding the processes of urban and regional growth. Our inability to cope adequately with problems of urban design, regional development, mass transportation, and the quality of the human environment suggests the need for more comprehensive and appropriate concepts of regional analysis. Regional analysts have recognized that a basic problem to date has been failure to specify clearly the objectives of regional development, and a literature is emerging in this field.[1]

A major purpose of this study is to present a model of regional and urban development that can assist in identifying some primary objectives of regional development and in analyzing the variables which affect these objectives. Another purpose of the paper is to explore the extent to which the regional development model can contribute to a location theory which goes beyond classical theory of profit maximization for the firm toward a more general theory of spatial distribution of economic activity from the viewpoint of public objectives.

The basic model used is an open static regional interindustry model that can be closed on various accounts in order to generate total flows and iterated in order to examine growth processes. Social accounts including stock and flow relationships are derived from the interindustry accounts in order to trace changes in the significant variables.

No single economic model can illuminate all of the significant aspects of a phenomenon as complex as regional economic development. The relatively simple model presented here will thus, in many respects, fall short of encompassing all of the complexities and dynamism of the development process. The model does attempt to extend conventional regional interindustry models in order to emphasize the public sector and to include some critical environmental relationships in order to examine the externalities that typically receive inadequate attention in forming development policy and objectives.

[1] For excellent discussion of regional development objectives see: Joseph L. Fisher, "Concepts in Regional Economic Development," *Papers and Proceedings of the Regional Science Association*, I (1955); Roland Artle, "Public Policy and the Space Economy of the City," in Lowdon Wingo, Jr., ed., *Cities and Space, The Future Use of Urban Land*, (Baltimore: Johns Hopkins University Press, for Resources For the Future, Inc., 1963); Charles L. Leven, "Establishing Goals for Regional Economic Development," *Journal of the American Institute of Planners*, XXX, No. 2 (May, 1964), pp. 100–10; Thomas A. Reiner, "Sub-National and National Planning: Decision Criteria," and references therein, *Papers and Proceedings of the Regional Science Association*, XIV, (1965); Wilbur R. Thompson, *A Preface to Urban Economics* (Baltimore: Johns Hopkins University Press, for Resources For the Future, Inc., 1965), p. 1.

It is the thesis of this study that fundamental objectives of regional economic development are: (1) to increase regional average per capita personal income, (2) to improve the capability of the public sector to meet the demands upon it, and (3) to maintain and improve the over-all quality of the human environment of the region. Statistical guidelines are derived from the model for determining the extent to which specific development projects satisfy each of these three objectives.

It is recognized that other economic objectives may also be significant, such as cyclical stability, steady growth, and full employment. These are related to and consistent with the three objectives which are suggested as fundamental. There are also social and political objectives of regional development that, though significant, are beyond the scope of this paper.[2]

THE TRANSACTIONS TABLE

The regional economy is represented by a conventional interindustry model, as shown in Table 1.[3] The region is defined as a political unit, such as a country, metropolitan area, or state, whose residents, acting through some level of government, can affect development decisions. The region is an open substructure of a larger national economy. Since the model is regional rather than interregional, it is assumed that indirect feedback effects upon the region resulting from development impulses originating within the region are small enough to be ignored. Table 1 shows a static snapshot of the regional economy in the initial period.

Economic activity in the region is measured by the entries that show the distribution of outputs or sales by each sector across its rows and the distribu-

[2] For an appraisal of social goals to be sought in city planning see Harvey S. Perloff, "New Directions in Social Planning," *Journal of the American Institute of Planners*, XXXI, No. 4, (November, 1965), p. 299.

[3] For discussion of interindustry models see W. W. Leontief, *The Structure of American Economy, 1919–1939* (2nd ed.; New York: Oxford University Press, 1951); Hollis B. Chenery and Paul G. Clark, *Interindustry Economics* (New York: John Wiley & Sons, Inc., 1959); W. Duane Evans and Marvin Hoffenberg, "The Interindustry Relations Study for 1947," *Review of Economics and Statistics*, XXXIV, No. 2, (May, 1952), pp. 97–142; William H. Miernyk, *The Elements of Input-Output Analysis* (New York: Random House, 1965). For discussion of regional interindustry models see Walter Isard *et al*, *Methods of Regional Analysis*, New York: John Wiley & Sons, Inc., 1960), chap. 8; Werner Hochwald, Herbert E. Striner, and Sidney Sonenblum, *Local Impact of Foreign Trade* (Washington, D.C.: National Planning Association, 1960); Roland Artle, *Studies in the Structure of the Stockholm Economy*, (Stockholm: Business Research Institute, Stockholm School of Economics, 1959); and *The Impact of Space and Space-Related Activities on a Local Economy* (Boulder: Bureau of Economic Research, Institute of Behavioral Science, University of Colorado, July, 1965); NASA Research Grant No. NSG-474, Part I, "The Input-Output Analysis," by William H. Miernyk, Ernest R. Bonner, John H. Chapman, Jr., and Kenneth Shellhammer; Part II, "The Income-Product Accounts," by Don Seastone, Ernest R. Bonner, Charles M. Franks, and William McCormick. For a critique of the interindustry approach see Charles M. Tiebout, "Regional and Interregional Input-Output Models: An Appraisal," *The Southern Economic Journal*, XXIV, No. 2, (October, 1957), pp. 140–47.

tion of inputs or purchases by each sector down the columns. Totals are shown at the end of rows and columns. Expenditure impulses originate in the final demand sectors as the result of purchase decisions by consumers, investors, and government agencies.

Entries in the household column show how consumers, using their incomes, savings, and credit, make purchases of goods and services and consume inputs from the other sectors. Similarly, the entries in the government column indicate how the regional government expends its tax and other revenue in order to provide the government services and social overhead capital required by households and businesses in the region. The capital formation column entries list all of the purchases of inputs that go into investment in new physical capital within the region.

In order to deliver goods and services to these ultimate consumers, producers in the intermediate sectors must purchase inputs, as shown in the column for each sector. Producers purchase inputs both from other intermediate producers (in the endogenous sectors) and from households and governments. Purchases from households of labor and management services result in personal income to households, and payments of taxes to governments by each sector can be thought of as payments for government services.

Total income paid to households, as shown in the personal income row, permits consumers to make purchases, whose size and composition are shown in the household column. Tax payments to state and local governments, measured in the government receipts row, provide the basis for government expenditures, as distributed in the government column. Household savings (income less expenditures) provide funds for investment, as shown in the capital formation column. Except for the entries in the capital formation column, which record all investment activity, all entries represent transactions on current account. All figures are in terms of producers' rather than purchasers' prices, with markups shown in the trade row.

Because the regional economy is an open one having trade relationship with other regions, imports are important sources of inputs. All expenditures in and payments to other regions for goods, services, tax payments to higher levels of government, and imports from abroad are included in the imports row. Similarly, all sales to and receipts from other regions for goods and services, taxes and transfers from other levels of government, and sales abroad are indicated in the exports column for each sector.

For purposes of analysis and projection, the transactions entries in Table 1 can be extended in various conventional ways. As a first step, the column cell entries in Table 1 can be divided through by the column totals in order to obtain input coefficients specifying the per cent of total inputs into each sector named at the top of the column contributed by each supplying sector named in each row.[4] This easily obtained transformation has the advantage that sector input coefficients, when multiplied by the actual or projected ac-

[4] Input coefficient $a_{ij} = \dfrac{x_{ij}}{X_j}$, see Appendix I.

TABLE 1
REGIONAL INTERINDUSTRY MODEL INCLUDING ENVIRONMENTAL FACTORS

	Purchasing Sector / Producing Sector		Intermediate Sale							Final Demand							
			Agriculture		Manufacturing	Service	Trade	Other	*Total Intermediate*	Household	Capital Formation	Government	Total Final Demand	Total Purchases by Regional Economy	Total Exports	Gross Regional Product Total Purchases From Regional Economy	Environmental Balance
			1	·	2	j	·	n	w	C	I	G	Y	D	E	X	B
Intermediate Purchases	Agriculture	1	x_{ll}	·		x_{ij}	·	x_{ln}	w	C_l	I_l	G_l	Y_l	D_l	E_l	X_l	B_l
	Manufacturing	2	·														
	Services	i	x_{il}			x_{ij}	·	x_{in}	w_i	C_i	I_i	G_i	Y_i	D_i	E_i	X_i	B_i
	Trade	·	·														
	Other	n	x_{nl}			x_{nj}		x_{nn}	w_n	C_n	I_n	G_n	Y_n	D_n	E_n	X_n	B_n
	Total Intermediate	u	u_l	·		u_j	·	u_n	u_w	C_u	I_u	G_u	Y_u	D_u	E_u	X_u	B_u
Value Added	Wages and Salaries	L	L_l			L_j		L_n	L_w	C_L	I_L	G_L	Y_L	D_L	E_L	L	B_l
	Other Household Income	R	R_l			R_j		R_n	R_w	C_r	I_r	G_r	Y_r	D_r	E_r	R	B_r
	Regional Personal Income	H	H_l			H_j		H_n	H_w	C_h	I_h	G_h	Y_h	D_h	E_h	H	B_h
	Business Savings	F	F_l			F_j		F_n	F_w	C_f	I_f	G_f	Y_f	D_f	E_f	F	B_f
	Government Receipts	T	T_l			T_j		T_n	T_w	C_t	I_t	G_t	Y_t	D_t	E_t	T	B_t
	Value Added	V	V_l			V_j		V_n	V_w	C_v	I_v	G_v	Y_v	D_v	E_v	V	B_v
	Total Sales by Regional Economy	S	S_l			S_j		S_n	S_w	C_s	I_s	G_s	Y_s	D_s	E	X	B
	Total Imports	M	M_l			M_j		M_n	M_w	C_m	I_m	G_m	Y_m	M			
	Total Sales to Regional Economy	Z	Z_l			Z_j		Z_n	Z_w	C	I	G	Y	D			
	Environmental Benefits	(+)Q															
	Environmental Costs	(−)C															
	Environmental Balance	A	A_i			A_j		A_n	A_w	A_c	A_i	A_g	A_y	A			

tivity level for that column, will indicate the actual dollar magnitude of inputs directly required from each sector of the regional economy.

A more powerful analytic device can be obtained in the form of matrix inverse, through a matrix algebra operation on the input coefficient matrix. This matrix inverse is a generalized solution to the system of simultaneous linear equations representing the regional economy. This inverse, when multiplied by any actual or projected changes in the final demand sector, yields estimates of the total direct and indirect outputs required from every regional activity in order to support the proposed bill of final demand. In order to permit focusing attention upon the important policy variables (income, public finance, and environmental factors), a set of simple regional social accounts can be derived directly from the final demand and value added rows and columns.

The Regional Development Process

Although the model is a static one, it can be used to illustrate the process of regional growth and development. For purposes of analysis it is helpful to break down artificially the complex process of economic development into processes of income generation, economic growth, and structural change. Each of these can be illustrated in Table 1.

First, in order to focus upon the process of income generation, we assume a static regional economy with a specified industrial structure and with excess but fixed capacity in each sector. As indicated in the previous section, the activity level of each local industry is determined by the expenditure decisions of final consumers within the region, households, investors, and government and by the demand for exports by purchasers from outside the region. Initially, we also assume stable price levels and linear homogeneous production functions in each sector. In order to make delivery to purchasers, regional producers must purchase inputs of materials and services, including purchases of labor and management services from regional households. The resulting payments to regional households are shown in row (*H*) of Table 1 as regional personal income.

Since the level of per capita personal income that an economy can produce is generally regarded as the single most valid indicator of efficiency and performance, the entries in row (*H*) assume a critical importance in the regional model. In Table 1, these entries have the advantage of indicating sector by sector the relative importance of various activities as generators of income within the region. It is only necessary to read across row (*H*) to determine the immediate sources of personal income within the region in the base period. For most regions, wages and salaries, row (*L*), will be the most important category of personal income. With the addition of other data, such as employment by industry, usually used in conjuction with the interindustry table but not necessarily included in it, it is easily possible to reduce the personal income or wage and salary figures to a per employee basis in order to indicate the comparative mean level of annual earnings in each industry of the regional economy.[5]

The important factor to be noted is that the level of personal income in the region, as shown by the entries in row (*H*), depends upon the activity levels of each sector of the regional economy, which, in turn, depend upon expenditure

[5] Employment data for each sector may be added conveniently as a row vector at the bottom of Table 1.

decisions by consumers, investors, and government within and outside the region. When these exogenously determined expenditures fall, because of national recession, or regional economic distress, or for other reasons, regional personal income will decline, along with other activity levels within the model. As expenditure impulses increase because of exogenous factors within the region or nation, personal income and other values in the table rise until capacity constraints are approached in the various sectors.[6] Also of major policy importance is the mean level of per capita personal income for the region, as determined by summing row (H) and adjusting for population in the region (H/P).

Economic Growth and Structural Change

Changes in activity levels discussed in the previous section are important in the short run, especially as they determine per capita personal income. But of greater importance to the region over the long run are growth and changes in the structure of the regional economy. Growth and structural change result from investment in new plant and equipment. Within the regional economy, investment can take place by enlarging the capacity of existing firms or industries, or by adding new activities. In terms of Table 1, growth can be thought of as increasing the capacity of any existing sector of the table or as adding one or more new sectors, with a row and column for each.[7] The investment decision may come from investors within the region, or outside it, or both. The type of economic growth easiest to visualize is simple addition to the capacity of a plant within an industry (X_j) already existing within the region. This investment process has two effects: first, by changing a capacity constraint it permits raising the maximum activity level (and income generated) in that sector, and, second, it creates a Keynesian investment multiplier effect. As the new capacity is utilized, added labor (and other inputs) are purchased, and aggregate regional personal income (H) increases. Whether or not per capita personal income (H/P) also increases depends upon what, if any, changes in population occur.

The Keynesian investment multiplier effect resulting from capacity change may be considerably more complex than the simple capacity income effect, depending upon the extent to which the inputs purchased in order to construct the new capacity come from within the region. If all of these are purchased from outside the region (e.g., in the form of a new machine tool installed by an outside labor force[8]), the only new income generation to occur in the region will be that in the expanded plant, plus any additional income generated in regional activities supplying other added inputs to the expanded plant. It is

[6] The expansion of existing capacity through increments of existing technological processes does not change the input coefficients in the endogenous part of the model. If expansion of existing plants and industries involves changed technologies, or if new industries are added within the region, then the input coefficient structure of the model must be modified accordingly.

[7] Capacity constraints can be indicated in Table 1 as a column vector at the far right of the table.

[8] This would correspond to import leakage in the Keynesian model.

possible, however, that even if there were a total leakage outside the region of funds spent for the capital inputs required by the expansion, once the new capacity becomes operational, the new regional demand for flow inputs required by the new productive capacity might trigger investment in other regional activity through an accelerator process.

In Table 1, the capacity expansion will be recorded in the capital formation column (*I*) during the period when the investment is taking place and will appear after its completion in the form of larger capacity in the appropriate sector. During the investment process, if no regional inputs are required, the only entry in column (*I*) will be for imports. However, if regional inputs are incorporated in the new capacity expansion, entries will appear in other cells of column (*I*).[9]

In this event, activity levels of other regional sectors will rise, and will generate additional income within the region during the investment period through a Keynesian multiplier effect. After the new capacity is completed, regional construction activity levels and income will recede to previous levels, except to the extent that new investment impulses appear, and flow input requirements rise within the region for providing inputs to the new capacity.

Thus far, the discussion has been in terms of capacity expansion in existing plants or creation of new plants in industries already represented in the region and identified in the model. Regional growth can also occur through the establishment of a new industry or activity in the region. Conceptually, this can be thought of as adding a row and column to the table (or adding a new equation to the model), or it is perhaps more easily visualized as inserting the appropriate values in a row and column already in the model but previously containing all zeros (such as the "other" row and column).[10]

As in the previous discussion of the investment process, introduction of the new capacity will first be reflected in the capital formation column, which registers purchases of inputs going into the new plant and equipment. Regional personal income will increase during the investment period, unless all investment inputs are imported into the region. After completion of the new plant and equipment, entries in the capital formation column recede to previous levels except to the extent that induced investment occurs. When the new plant and equipment go into operation, the structure of the model will be modified to record the new set of inputs to the new column and the distribution of outputs across the new row.

To the extent that the new industry purchases inputs from other regional industries and sells output to them, the structure of the regional economy will have been expanded and changed. These changes should be reflected in a re-

---

[9] Capital formation and other final demand columns are assumed to be exogenous to the model, and their input structure is free to vary. Regional economic development is assumed to take place primarily in response to investment impulses defined in the capital formation column.

[10] The analysis has been framed in terms of net addition to plants and industries in the region. The case of a declining region would be reflected initially in lower activity levels and, after a period of disinvestment, in disappearance of plant and industry capacity.

vised version of Table 1, recording the new activity and its trade relationships inside and outside of the region. Structural change in the regional economy occurs also through technological change, which should be reflected in periodic revision of input coefficients and the matrix of inverse elements.

In summary, regional economic change can be visualized in terms of Table 1 as: (1) growth, through increasing the activity levels of existing sectors of the regional economy, (2) development, through addition of new sectors (and disappearance of others), and (3) technological transformation, through change in the input structure of activities.

Table 1, by tracing the linkages between the various sectors of economic activity, emphasizes that as regional growth occurs all of the activity levels within the region tend to rise together, even though they may not all rise proportionately. With new investment and sales, regional personal incomes increase, and household consumers increase their purchases. The flow of tax revenues to local government rises, but public expenditures must also increase in order to provide public services essential to both households and businesses. And, as economic activity increases and densities increase, the regional economy draws more heavily upon its natural resources and other environmental factors.

SOME INCOME ASPECTS OF REGIONAL DEVELOPMENT

A primary reason for the almost universal acceptance of regional economic development as an objective is its association with increased income. The transaction relationships represented symbolically in Table 1, (and in their transposed form as input coefficients) indicate how increases in activity levels or in capital additions to capacity cause other inputs to rise *pari passu*. Inputs of household services (labor and management) rise with other inputs, resulting in increases in income. However, the linkage between regional development and regional welfare is not at all direct or simple. From the standpoint of the quality of economic development, it is per capita regional income, rather than aggregate income, which is important, and only under special circumstances does regional development increase per capita personal income.

Income Generation

With regional industrial expansion, additional inputs will be required, including labor and management services. To the extent that these are purchased from residents of the region, total regional personal income will rise. But the effect of new activity upon per capita regional personal income depends upon many factors, including the level of incomes in the new activity, and the source of its new labor and management services. To the extent that income payments in the new activity exceed the average of those previously earned in the region, and to the extent that these incomes go to residents of the region, regional per capita personal income will tend to rise.[11]

[11] Perloff, Dunn, Lampard, and Muth have drawn an important distinction between

If income levels in the new activity are lower than the previous average level in the region, the new activity can still raise average per capita regional income but only by hiring those already in the region previously unemployed or earning lower incomes. Movement of new labor into a region at lower average incomes than previously prevailed necessarily must depress regional average per capita income, unless significantly large indirect and induced effects result in increased earnings in other industries of the region.

If the new activity purchases its inputs from outside the region, no secondary income effects will occur in the region. If it purchases inputs from within the region, regional personal income will rise to the extent that supplying industries increase labor inputs from within the region. Whether or not these secondary income increments raise regional per capita incomes will depend upon the income-mix of the supplying industries and upon the extent to which in-migration occurs.

Table 1 is intended to emphasize that a regional economy is an open economy characterized by cross-regional flows of goods, services, and persons. Creation of new jobs in a region may defeat regional income objectives if the new jobs are filled by inmigrants from other regions and if the income level of the new residents fails to increase the existing average per capita level of regional personal income.

It may be argued on the basis of conventional economic theory that, because of competition for new workers, new economic activity is not likely to be established in a region at wage rates lower than those already prevailing. In terms of spatial analysis, this is clearly not correct. Workers can quite conceivably migrate to new job opportunities in regions paying wages higher than those in their former region but lower than average wages in their new region. Regional social accounts measuring income on a place of employment basis would indicate lower per capita regional personal income after the establishment of such new jobs. Average regional personal incomes on a place of residence basis could also be reduced as the result of new residental development in a region where a high income, low density, residential area was subdivided and developed for lower income residents.

It can also be argued that reduction of average incomes in subnational regions may be in the national interest because of mobility and growth considerations, and that restrictive attempts to maintain and raise local average incomes are a type of suboptimization which would defeat larger national goals. This proposition hinges upon relationships between regional and national objectives as well as upon criteria for optimum spatial distribution of economic activity. It will be examined in following sections.

The important implication of this analysis for the development process is that growth and change within a region may increase total and per capita personal income in a region, but they do not necessarily do so. If leakages are

"volume growth" and "welfare growth" in regional development, with welfare growth depending upon per capita income increase. See *Regions, Resources, and Economic Growth* (Baltimore: Johns Hopkins University Press, for Resources For the Future, Inc., 1960).

large enough (if inputs are purchased from outside the region, or from resource owners who reside outside the region) investment and growth within the region may have very little effect even on total regional income. If, prior to the new investment, there was unemployed labor in the region, and if this unemployed regional labor is employed in new regional plants, total and per capita regional personal income will rise as a result of the investment. If unemployed from within the region are not employed in the new plant, either because little unemployment existed in the region or because the skills of the unemployed fail to match the requirements of the new plant, recruitment of labor from outside the region will be necessary, either on a commuting basis or new resident basis. The income of the new residents will represent an addition to the total personal income of the region. To the extent that their average incomes exceed the previous average per capita income of the region, then average per capita income of the region will rise.

Conceptually, inmigration of a large labor force earning incomes lower than the previous average of the region could result in a situation in which regional growth caused lower average regional per capita income. This outcome is quite possible in a high income, sparsely populated region in which a large scale low income development takes place. But even in more typical situations, it is clear that unless new development puts to work unemployed or underemployed already resident in the region, or unless it pays its new immigrants incomes significantly above existing scales, growth and development may fail to contribute positively to regional welfare.

The regional interindustry model and its associated social accounts are designed to identify and measure changes in this income variable after all of the direct, indirect, and induced income effects resulting from any program or project have been fully worked out.

Importance of the Capital Formation Sector

The universal pressures for regional development cannot be understood fully without analysis of the investment process and of the role of the capital formation sector in the region. The input transactions shown in the interindustry table for producing industries represent purchases on current account. All capital purchases are combined and shown in the capital formation column. This procedure is followed in order to treat investment conceptually as a final demand sector, and in order that the flow input coefficients represent stable technological relationships as nearly as possible.

The investment sector (capital formation column) is likely to be the one most significantly affected by regional growth. During the investment and construction stages, regional businesses (and those outside the region) increase their total sales to regional capital formation. If construction of new plant and equipment or residential development occur, significant increases in sales are likely to be experienced by all of those activities associated with the construction industry. These include service activities such as real estate, legal services, banking, finance, insurance, surveying, architecture, and utilities. Of course most of these activities also sell outputs, even in the absence of growth, on current account to service and maintain existing regional activities. However if a large part of their sales is geared to new construction and new development, and if the development process discontinues, or if its activity level drops, or in some cases, depending upon growth patterns, if the rate of construction increase falters, then the development servicing activities may find sales falling off sharply.

Construction and development activities build up during the growth process and then decline, but the plant and equipment, once in place, continue to function over a long period. Unless continued growth takes place, growth serving activities must then recede to lower activity levels only sufficient to service and maintain the new plant and equipment.

As pointed out earlier, the entire region benefits from development over the long run on a per capita income basis only if significant amounts of inputs for the new activities are purchased on a continuing basis from within the region, and if incomes earned in the new activity are significantly higher than the previous per capita average. But in the short run it is clear that the development supporting activities benefit directly during the construction period, regardless of the eventual effect of the new construction upon the general income structure of the region. It is therefore also clear that the development supporting activity sectors have a direct interest in fostering development as a continuing process in the region.

Because the construction and development supporting sectors of the regional economy can keep their sales at high levels only if development continues as a process, these sectors have a direct interest in working actively for the adoption of regional policies which encourage regional economic development on as large and continuous a scale as possible. This type of policy does not necessarily lead to significant long-run general regional improvements in regional per capita income, and, as will be indicated subsequently, it may aggravate local public finance problems and lead to serious deterioration in the quality of the local environment. But a successful development policy tends to generate its own pressures for continuity, and it results in large sales by development related sectors, including construction, real estate, legal services, banking, finance, insurance, and utilities. When the development-oriented sectors of an urban or regional economy become a large portion of the total regional economy, then maintenance of income in the region hinges directly upon maintenance of continued growth.

For an objective evaluation of the over-all income effects of any development program, it is necessary to have a comprehensive model which measures the direct and indirect effects of growth upon income during the development period and, over the long run, for the development sectors and for the rest of the regional economy. The regional model shown in Table 1, with its extensions, represents an effort to define such a model.

Measurement of Income Change

Changes in regional personal income can be shown in Table 1, row (*H*), and in a household income account, derived from the household row, which is designed to focus attention upon this particular variable. Base period regional personal income is recorded in row (*H*), (regional personal income) of the interindustry table. Transposed into column form, this row in effect becomes the regional personal income account. The entries in this account identify the sources of regional personal income in terms of their relative significance for the region. Division of the summation figure for the column by the figure for regional population in the base period yields the regional average per capita income estimate. Other entries can be divided by appropriate employment figures in order to obtain per capita income estimates by sectors.

Base period income can be compared with income of subsequent periods arising from actual regional development or from hypothetical development programs. These comparisons can be made either by tracing them through various stages, using Table 1 and the table of input coefficients, or they can be made using computer techniques. If the regional economy is limited in size without extensive structural interrelations (i.e., with numerous zero entries in the flow table) and if the impulse whose effect is to be measured is limited, perhaps to an increase in activity levels or minor additions to capacity, several iterative multiplications of changes in the activity levels against the appropriate input coefficients and summation of results will indicate income changes between the two periods.[12]

More formal and more complex procedures are available for using the model to measure income effects and other impacts over time. Stipulated changes in patterns of government expenditures, investment programs, and consumption spending can be multiplied by the matrix inverse in order to yield estimated flows and activity levels over each time period with resulting income effects summarized in a time series of aggregate and per capita income estimates showing the effects over time of any projected or hypothetical change in expenditure patterns.[13]

Income and Employment Multipliers

The flows identified in Table 1, and the input coefficients and inverse elements which can be computed from them, reveal the structural relationships

[12] In terms of Table 1, $X_{h_{t-1}} - X_{h_t}$ measures the change in aggregate personal income for the region, and $X_{h_{t-1}}/P_{t-1} - X_{h_t}/P_t$ measures the change in average regional per capita personal income.

[13] In matrix notation, $[X] = [I - a]^{-1}$. $[Y]$, see Appendix I. Dynamic properties over time can be observed by appropriate time phasing of entries in the capital formation column. Conceptually, the model can be made dynamic through use of capital coefficients but data deficiencies and other problems have prevented the development of fully operational dynamic intersectoral models. See Richard M. Goodwin, "The Multiplier as Matrix," *Economic Journal*, LIX, No. 236 (December, 1949), pp. 237-55; Wassily W. Leontief, "Dynamic Analysis," *Studies in the Structure of the American Economy* (New York: Oxford University Press, 1953), pp. 53-90; Robert N. Grosse, "The Structure of Capital," *ibid*, pp. 185-242; R. Dorfman, P. Samuelson, and R. Solow, *Linear Programming and Economic Analysis* (New York: McGraw-Hill, 1958), chaps. 11, 12; and Hollis B. Chenery and Paul G. Clark, *Interindustry Economics* (New York; John Wiley & Sons, Inc., 1959), pp. 71-80.

which cause any increase in final demand sales (ΔY) to result in general increases in related activity levels, including, in most cases, income and employment. The ratios between these increases in final sales to consumers, investors, government, or to other regions, and increases in income and employment are defined as regional income and employment multipliers.[14]

In economic base models dealing with aggregate income, sales, and employment, it is often assumed that there is a single average regional income and employment multiplier which can be used to estimate regional income and employment increases resulting from increased sales to exports and to other final demand sectors. The disaggregated interindustry model reveals that the size of the multiplier is not independent of the composition of the final demand increase. Different patterns of final demand sales will differ in the extent to which they draw upon regional as compared with imported inputs and thus will result in different regional multipliers.[15]

Income Criterion for Regional Development

The first basic criterion for evaluating the impact of a proposed development upon a regional economy is whether or not, after all of the income multipliers and migration effects have worked themselves out, average per capita incomes in the region would rise and, if so, how much. In terms of Table 1 and the social accounts derived from it, the necessary condition is:

$$\frac{X_{h_t}}{P_t} > \frac{X_{h_{t-1}}}{P_{t-1}} .$$

Where:

X_h = regional personal income,
P = regional population, and
t = time period.

---

14 $$\frac{\Delta H}{\Delta Y} = k, \text{ or } k \cdot \Delta Y = \Delta H .$$

Where:

ΔH = increase in regional personal income,
ΔY = increase in autonomous sales to final demand, and
k = regional income multiplier.

15 Pioneering conceptual and empirical work on regional interindustry income and employment multipliers was published by Frederick T. Moore and James W. Petersen in "Regional Analysis: An Interindustry Model of Utah," *Review of Economics and Statistics*, XXXVII, No. 4 (November, 1955), pp. 368–83. In his St. Louis study, which is one of the earliest and most comprehensive interindustry studies for a metropolitan region, Werner Hirsch computed detailed income and employment multipliers for twenty-seven economic sectors and examined some of the policy implications of these results. See Werner Z. Hirsch, "Interindustry Relations of a Metropolitan Area," *The Review of Economics and Statistics*, XLI, No. 4 (November, 1959), pp. 360–69. William H. Miernyk, *op. cit.*, in his recent interindustry study of Boulder, Colorado, has significantly refined the conceptual and statistical basis for regional multipliers by disaggregating by income size class and by examining separately the consumption functions of established residents as compared with those of new residents.

This criterion is subject to the important limitation that it does not take into account changes in income distribution among size classes and among sectors. It is assumed here that problems of income distribution can more efficiently be approached directly through programs designed for that purpose in fields of education, occupational training, and welfare rather than indirectly through regional development programs.

PUBLIC FINANCE ASPECTS OF REGIONAL DEVELOPMENT

The critical decision making units in evolving regional development policies are local governments: county, metropolitan, state, and interstate. There are two main reasons for this. First, local government agencies influence or control strategic development factors such as planning and zoning, location of highways, schools, water supply, waste removal, and promotional efforts. Second, local electorates increasingly expect local government to provide a rising level of public services such as education, public safety, outdoor recreation, health services, and environmental quality control. The growth of the public sector is stimulated not only by quantitative factors, such as population growth and urbanization, but by qualitative factors, such as aspirations for improvement in the level of public services, and expansion of the range of services desired by urban dwellers. As incomes rise and tastes change, it appears likely that consumers will increasingly turn to public agencies for those community and environmental values which can be achieved through community action. Above all, sharp increases in population density and urbanization seem likely to force the citizen to turn to public agencies in the quest for orderly development.

Despite substantial grants and aid from federal and other sources, local governments are under continuous pressure to derive larger revenues from local sources. Local governments therefore have a direct interest in identifying regional development policies which can generate added revenues. Until quite recently, it was taken as self-evident that any type of local economic development contributed in a positive way to meeting public revenue needs by broadening the tax base. However, there is growing recognition that regional development and growth also generate added requirements for public expenditure so that development cannot be assumed automatically to leave a region in a better fiscal situation.[16] It is quite possible for regional development to trigger heavy requirements for new schools and other facilities which cannot

[16] Walter Isard and Robert Coughlin, *Municipal Costs and Revenues Resulting from Community Growth*, Federal Reserve Bank of Boston and American Institute of Planners, (Wellesly, Mass,: Chandler-Davis Publishing Co., 1957); Werner Z. Hirsch, "A General Structure For Regional Economic Analysis," and discussion in Werner Hochwald, ed., *The Design of Regional Accounts* (Baltimore: Johns Hopkins University Press, for Resources For the Future, Inc., 1961), pp. 1-37; Louis K. Loewenstein, "The Impact of New Industry on the Fiscal Revenues and Expenditures of Suburban Communities," *National Tax Journal*, XVI, No. 2, (June, 1963), pp. 113-36; Harold M. Groves and John Riew, "The Impact of Industry on Local Taxes, A Simple Model," *National Tax Journal*, XVI, No. 2, (June, 1963), pp. 137-45; Mabel Walker, *Business Enterprise and the City* (Princeton, New Jersey: Tax Institute, Inc.).

be financed solely from the added revenue resulting directly from this development. Local taxes must then be raised, as they frequently are, or further development ventures must be pursued, in a cycle of efforts to provide public service requirements generated by previous efforts to broaden the tax base. Communities may thus find themselves facing continuous local tax increases even though construction and business activity are expanding. The treadmill of local industrialization does not necessarily lead to tax reductions.[17]

For several reasons, local development efforts may not always achieve the net fiscal benefits anticipated in development programs. The new activity, industrial or otherwise, may simply require more direct and indirect services in the form of water supply, waste removal, and service roads, than its taxes support. Or, in an effort to compete for new activity, the community may have aggravated the situation by granting subsidies or tax relief. Also, the new activity may bring into the region new residents who require more schools and services than the associated business and personal taxes can cover; and the new activity, by increasing the activity levels of suppliers in the region, may generate indirect requirements for public expenditures in excess of tax revenue. In assessing the net fiscal impact of any development project, it is highly important to include consideration of the new labor force and residential development that will be associated with it.

To avoid these difficulties, a community could emphasize a selective development policy with the objective of attracting the type of activity that would most satisfactorily: (1) generate taxes adequate to meet the associated public service requirements, given existing tax structures, and (2) employ a labor force with income levels, residental standards, and tax payments sufficiently high to support their public service requirements.

The problem for local governments is to identify development policies that give some promise of producing new revenue in some satisfactory relationship to the new public services required by the new development. The regional development model discussed here is designed to assist in identifying activities meeting these criteria.[18]

[17] The empirical evidence on this point is inconclusive. George H. Borts and Jerome L. Stein take the position that taxes paid by immigrants usually exceed the cost of public services they generate. See *Economic Growth in a Free Market* (New York: Columbia University Press, 1964), p. 214. However, in a study of the impact of industrialization on financing education, Werner Hirsch found that while low wage industries may produce negative fiscal results, expansion of industries with high wage, low employment characteristics was most likely to improve the local fiscal status. See "Fiscal Impact of Industrialization on Local Schools," *The Review of Economics and Statistics*, XLVI, No. 2, (May, 1964), pp. 191-99. It is therefore essential to evaluate proposed programs on their own merits, using a comprehensive analytic framework.

[18] Jesse Burkhead has emphasized the importance of bringing the public sector explicitly into development planning models through construction of detailed public finance accounts. See "Public Finance as an Integral Part of Regional Accounts," in Committee on Regional Accounts, Werner Z. Hirsch, ed. *Elements of Regional Accounts*, (Baltimore: Johns Hopkins University Press, for Resources For the Future, Inc., 1964).

The critical relationships can be visualized with the assistance of Table 1. The local government receipts row (*T*) shows every sector purchasing services (paying taxes to) from the local government. Actually, for purposes of more comprehensive analysis, this row would be disaggregated into separate rows for each major type of local tax, such as income, sales, property, and others because various firms and industries pay different amounts and rate of taxes depending upon the impact of the local tax structure. Intergovernment transfers, which are becoming more important on the federal and state levels, should be separately identified.

It is the increased flow of tax revenue, indicated in row (*T*), which is usually cited as the tax base argument for local development. What is often overlooked is that as government revenues from development rise, local government expenditures necessarily rise in order to provide essential government services to the new activities, or the relative quality of local government services must fall. These local government expenditures are shown in column (*G*). Thus Table 1 and the table of input coefficients derived from it emphasize the fact that in the regional development process, all economic sectors of the regional economy are interlocked and tend to grow or contract simultaneously. Growth rates will not be precisely equal among all sectors but will generally move in the same direction. The significance of this for regional development is that there is no easy road to generation of added net local tax revenue. Local growth is accompanied by rising public expenditures unless the quality of public services deteriorates or unless there are special circumstances. However, some routes to regional development can generate a more favorable revenue to expenditure ratio than others. The purpose of this model is to permit exploration of the entire range of regional development routes and to examine the implications of each.

Impact of Development on the Public Sector

The impact of establishment of a new industrial plant upon the structure of local public finance is a complex, dynamic one. The planning, investment, and the construction phases must be distinguished from the longer run operating phase. It must also bo recognized that the direct, indirect, and induced effects, described above, will also affect the operations and tax status of other activities in the region, as well as affecting other regions.

A high tax, low wage plant which draws labor on a commuting basis from other regions might shift part of the development burden to these other regions. But any analysis of public service costs of plants which ignores the tax revenue and cost aspects of the associated labor force is partial and incomplete.[19]

---

[19] Logically, it is conceivable that a new plant or other development which did not itself pay wages or taxes high enough to improve the development status of a region could, by its indirect effects upon other regional activities (e.g. through providing markets for local products), contribute constructively to local development. This is a major reason for using a general equilibrium, interindustry model which can identify all of the indirect and induced impacts of a new development.

The complexities and interrelationships involved in systematic analysis of the fiscal aspects of regional development can be comprehensively treated by the regional interindustry model suggested in Table 1. It is necessary to postulate the nature of the new plant in the investment and operating stages, its wage structure, the sources of its labor force (local population, inmigrants, or commuters from other regions), and the local business and residential tax structure. Only when all of these factors are analyzed can a determination be made of whether or not the proposed development would result in net fiscal advantage to the community.

The construction phase of investment in the new plant could be handled analytically as an addition to regional final demand. The operating phase should be treated either as a new activity (row and column) or as an increase in activity level of an existing sector, with interindustry impacts upon other sectors. Net tax revenue and expenditures before and after development can be compared. Similar computation would be required for any residential construction, consumer purchases, tax revenue, and government services required by that part of the new labor force settling in the region. Per capita fiscal capabilities before and after the proposed development can thus be compared on a comprehensive basis. The comparisons are summarized in the social accounts under the expenditure and revenue columns. Detailed analysis would require that these accounts be disaggregated on the revenue side. Separate rows should be constructed for each type of tax, since development would affect each of these differently. On the expenditure side, the columns could be inserted on a program basis, in order to identify separately the various spending categories and their relation to development. The expenditure and revenue comparisons should be adjusted for population change in order to analyze the effects of any proposed development upon per capita local government revenue and expenditures.

The public finance criterion for judging whether or not a proposed economic change or development would contribute positively to regional economic development is whether or not the proposed development would leave the public sector of the regional economy in a better position to meet its expenditures requirements. In Table 1, local government revenues are shown in row (T), and local government expenditures in column (G). Since governments can generate deficits or surpluses, the sum of row (T) and column (G) need not be equal. Adjusting for population change and allowing for all possible income, employment, and population effects to work themselves out in the regional economy, the necessary condition that a proposed change must meet in order to contribute positively to regional economic development is:

$$\frac{G_t - T_t}{P_t} < \frac{G_{t-1} - T_{t-1}}{P_{t-1}} .$$

Where:

G = regional government expenditures,
T = regional government revenues,

P = regional population, and
t = time.

What this states is that for any given per capita level of local government expenditures (G/P) and any given tax structure, a proposed development or change in the economy should reduce the per capita deficit or increase the per capita surplus. Alternatively, the development should support, given the tax structure, a higher level of public service per capita; or, given the expenditure level, permit tax relief. This is not to suggest that maximization of the fiscal residual per se should be a major objective of development policy. The essential point is that development planners should recognize that all forms of development imply an increase in local government expenditures as well as in revenues; therefore, they should have a consistent framework for anticipating costs as well as revenues.

Of course, in the case of a development project that failed to meet these public finance criteria but was desirable in every other way, it might be possible to raise tax rates and, in effect, to pay the higher taxes out of higher per capita income, leaving higher per capita disposable income, depending upon the structural relationships involved. Or, it is conceivable that after evaluation of all the facts, a local electorate might prefer to finance local expenditure needs, such as better education, through tax increases without industrialization. In fact, given the public expenditure-generating force of industrialization, direct financing of improved public services through initial tax increases might be the least cost method of achieving better public services. It is not by any means clear that industrialization will always lower tax rates, but even if it would in a particular situation, voters should be given the option of accepting the higher tax rate without industrialization and its environmental impacts. These will be examined in the next section.

SOME ENVIRONMENTAL ASPECTS OF DEVELOPMENT

Preceding sections have emphasized the increase of per capita income and strengthening of the public sector as major goals of regional development and have explored some of the implications of these objectives. It is now postulated that a third fundamental objective to be sought in regional development—and indeed in economic development generally—is high quality in the environment. It is further postulated that this objective is not only analytically distinct from income and fiscal goals but may even conflict with them.[20]

All types of development generate some impact upon the natural environ-

[20] In a rigorous and scholarly analysis of the natural resource and conservation problem, Harold J. Barnett and Chandler Morse conclude that although resource shortage is not likely to become a serious constraint to economic development, deterioration in the quality of life and the quality of the environment as the result of population pressures and the impact of industrial society upon natural resources may become a critical problem. See *Scarcity and Growth, The Economics of Natural Resource Availability* (Baltimore: Johns Hopkins Press, for Resources For the Future, Inc., 1963), especially chap. 12, pp. 252–66.

ment, and some types of industrial development can so deteriorate the health, recreation, aesthetic, and other qualitative aspects of the regional environment that they may overshadow the income and fiscal contributions of development, even if these latter be positive.[21] Therefore, unless a development project or program meets all three of the income, fiscal, and environmental criteria, it will not, on balance, increase regional welfare.

For example, a recent Department of Health, Education, and Welfare publication concludes that "Lake Erie and its tributaries are polluted."[22] Industrial, municipal, and military installations in New York, Pennsylvania, Ohio, Indiana, Michigan, and Canada are discharging such quantities of pollutants into the water that recreation, commercial fishing, sport fishing, navigation, water supply, and esthetic values have been seriously damaged. Part of the wealth, prosperity, and industrial growth of the entire region has been based upon transfer of private industrial costs of waste disposal to the public sector by avoidance of waste treatment practices. These costs must now be faced, either through acceptance of drastically deteriorated natural resources, or through enormous costs of rehabilitation. In effect, real income has been transferred from resource-oriented activities to waste producers in the region.

Accumulating evidence indicates that man's expanding population and industrial society are measurably affecting the total environment of the earth through addition of carbon dioxide, smog, and radioactivity to the atmosphere, addition of organic and inorganic chemicals and silt to river basins, destruction of forests and open space, accumulation of spoil banks and mine seepage, rising background levels of noise and radioactivity, congestion, and environmental poisons.[23]

Current patterns of economic development push society continually into dependence upon more complex technological systems for survival, especially in urban concentrations of increased population magnitude and density. For

[21] In his pioneering efforts to develop a general structure for regional analysis, Werner Z. Hirsch formulated a remarkably comprehensive regional interaction model which explicitly considers the impact of development on the "private sector environment," the "public sector environment," and on the "amenities of life." The private and public sector environments are included as regular accounts, and effects upon amenities are identified as a final step. Conceptually, Hirsch's model is an interindustry model, even though it is not presented in typical interindustry format, and could be formulated to analyze the policy variables emphasized in this study. See Werner Z. Hirsch, "A General Structure for Regional Economic Analysis," in Werner Hochwald, ed., *Design of Regional Accounts* (Baltimore: Johns Hopkins University Press, for Resources For the Future, Inc., 1961) pp. 1–32, and comments by Roland Artle and Harold J. Barnett, pp. 33–7.

[22] United States Department of Health, Education, and Welfare, *Report on Pollution of Lake Erie and Its Tributaries*, Parts I, II, and III, (Washington, D.C.: Public Health Service, July, 1965).

[23] As an indication of accumulating environmental impacts of industrialization, see The President's Science Advisory Committee, "Restoring the Quality of Our Environment," Report of the Environmental Pollution Panel, (Washington, D.C., November, 1965). See also M. Neiburger, "Diffusion and Air Pollution," *Bulletin of the American Meteorological Society*, XXXXVI, No. 3 (March, 1963), pp. 131–34.

example, water supply and waste disposal systems become more complex and more vulnerable as dependence escalates from individual streams and wells to municipal systems, to distant water storage and delivery systems, and ultimately to nuclear desalination plants. Paul W. McGann has drawn 200 year projections of recent trends which would result in an industrial economy supporting a utopian society with world incomes of 100 to 1,000 times present levels. This could be achieved at the cost of a highly technological society in which the land surface of the earth would be covered with skyscrapers on the Manhattan pattern with population spillover dwelling in subsurface living areas and on barges over the water surface of the earth. The world material dream of opulence would have been achieved but at the cost of dependence upon a complex and highly vulnerable environment.[24]

Economic and technological advance has created wealth and affluence, but they have been achieved at great cost to the environment and are based upon technological systems whose impact upon the environment is destructive of many values. At certain minimum stages, development can improve the environment from the human point of view by providing essential transportation and communication needed to open a region to habitation. But all economic development has an impact upon the environment, and, at higher levels of population density and industrialization, congestion and pollution begin to have an adverse effect. The process of technological and economic development becomes a self-sustaining and self-reinforcing one which is not entirely responsive to human objectives and which feeds upon new technological possibilities often without consideration of effects upon resources and the human environment. Growing population pressures aggravate the problem and make solutions more difficult to find. It is for this reason that the identification of the human objectives and priorities of economic development is critically important.

The quality of the environment is significantly more difficult to define, to quantify, and to analyze than are income and fiscal relationships. Thus, although conservationists and other social critics in growing numbers have drawn attention to the heavy price that society pays for industrialization and affluence, economists have in general neglected some increasingly important aspects of environmental change which typically accompany economic development. Necessarily concerned with quantifiable, market-price dimensions of development, economists have devoted relatively little effort to the analysis of qualitative aspects of regional change. Because of rapidly increasing population densities and urbanization with their serious impacts upon water, air, open space, and other resources, it has now become much more urgent to develop methods for analyzing the qualitative aspects of development and to identify rational criteria for design of development programs, difficult as this task may be.

But as population densities increase, and qualitative deterioration of the

[24] Paul W. McGann "Technological Progress and Minerals," in Joseph J. Spengler, ed., *Natural Resources and Economic Growth* (Ann Arbor, Michigan, 1960). Conference sponsored by Resources For the Future, Inc. and Committee on Economic Growth of the Social Science Research Council, University Microfilms, Inc.

environment becomes more serious, some aspects of quality become more amenable to quantification and valuation in money terms. For example, as air pollution becomes more serious and more widespread, more attention is given to methods of air pollution control (such as smoke abatement and antismog devices) and to their costs. The same is true of water pollution, pesticide use and, in limited instances, noise control. Thus, even though impacts of development upon the environment are largely qualitative, many of them can be quantified in such terms as amounts of pollutants discharged per unit of air and water, background levels of noise and radioactivity, and parkland per household. Resources which were formerly regarded as free goods become scarce factors of production as densities increase, and price tags may then be imposed or imputed.

Air and water, under certain circumstances, can be analyzed as common property resources that are uneconomically exploited because no single user or group has an incentive to manage them efficiently.[25] It is also possible to analyze some qualitative aspects of the environment in terms of social stock accounts. It can be conceived that at relatively low levels of density and development, all residents of a region can draw freely upon stocks of pure water, pure air, quiet, privacy, open space, natural scenery, and other amenities. But with increasing densities and levels of development, these amenities are drawn down at rates greater than those at which they are replenished and per capita availability of the factors falls.[26]

Environmental factors can also be analyzed in terms of external economies and diseconomies. The use, which may appear to be a free good to the factory management, of a river to carry away industrial waste may result in costly diseconomies for downstream residents. A major objective of regional development will be to achieve patterns of spatial distributions of economic activity that explicitly recognize the central role played by externalities and to utilize these interrelationships to maximize external economies and minimize external diseconomies. The concept of externalities is fundamental in analyzing efficiency in regional development. The existence of substantial economies of scale in providing community services and social overhead capital suggests that area development efforts are more likely to be successful in assisting movement towards sustained growth if aid is concentrated at existing and potential growth points than if they are too widely dispersed.[27]

Environmental factors can also be analyzed with the advantage of insights gained from the field of ecology. The use of the cooling powers of a river to

[25] Francis T. Christy, Jr., and Anthony Scott, *The Common Wealth in Ocean Fisheries* (Baltimore: Johns Hopkins University Press, for Resources For the Future, Inc., 1965).

[26] For a treatment of stock and flow relationships in regional social accounts, see Harvey S. Perloff and Charles L. Leven, "Toward an Integrated System of Regional Accounts: Stocks, Flows, and the Analysis of the Public Sector," in *Elements of Regional Accounts, op. cit.*

[27] University of Maryland, Bureau of Business and Economic Research, "Tidewater Maryland Development Problems," *Studies in Business and Economics*, XV, No. 2 (College Park, Maryland: Regional Industrial Centers, September, 1961), pp. 19-20.

absorb the heat from power generation, although it may add to the profits of the power company, may have profound effects upon the fish and wildlife of the river basin, causing drastic changes in the entire ecology of the region.[28] In the case of water and air quality standards, the establishment of objective public health minimum purity requirements is essential.[29] Regional development objectives can thus be based not only upon economic criteria but must take into consideration ecological, public health, and other environmental factors.

Methods of Analysis

The environmental aspects of regional economic development can be introduced into the analysis by adding to the interindustry model rows and columns that attempt both to identify the environmental benefits and costs of development and to distribute these by sector. An experimental method for doing this is shown in Table 1. In this formulation, environmental balance rows and columns are used to border the conventional regional interindustry model.

As indicated above, the environmental data are more difficult to obtain than are the conventional economic statistics. They may require extensive estimation and subjective judgment. The natural environment is hardly conceivable as a conventional industry with a production function and distribution pattern. The values involved may be curvilinear or discontinuous. It is desirable, given limitations of data and knowledge, to be able to analyze these factors as a side calculation, outside the model. The environmental factors to be dealt with are primarily those which do not enter the market place and, therefore, do not ordinarily appear as priced inputs or outputs within the body of the model. Developmental impacts upon water and air resources are examples.

It should be noted that other data related to environmental impacts may already have been included in the conventional rows and columns of the table. For example, in the case of regions already enforcing public health requirements for air and water quality, the production functions of firms or the expenditure patterns of local government already include inputs purchased for waste treatment purposes. A major objective of regional development planning should be to obtain meaningful dollar values to insert in environmental balance rows and columns so that vital environmental effects of development programs can be objectively evaluated. A corresponding objective of development policy should be to build environmental quality objectives into regional planning. Requiring development projects to carry the cost of restoring resource quality,

---

[28] J. A. Mihursky, "Patuxent River Estuary Study with Special Reference to the Effects of Heated Steam Electric Station Condenser Water upon Estuarine Ecology," (College Park, Maryland: University of Maryland Natural Resources Institute, Ref. No. 63-6.).

[29] Harold A. Thomas, Jr., "The Animal Farm: A Mathematical Model for the Discussion of Social Standards for the Control of the Environment," *Quarterly Journal of Economics*, LXXVII, No. 1 (February, 1963), pp. 143–48. Thomas points out that setting a quality criterion is the equivalent of imputing a cost-benefit ratio.

through measures such as waste treatment, would have the effect of building some of the environmental balance rows and columns into the internal structure of the model.

One purpose of adding the environmental balance rows and columns to the model is to permit explicit evaluation of external economies and diseconomies, which are often ignored in partial analysis, but which are the key factors in regional and urban development. The general equilibrium regional interindustry model provides a comprehensive framework for the examination of these externalities, explicitly identifying them sector by sector. Another purpose of these rows and columns is to provide an objective, quantitative basis for analyzing the resource quality issues which are generally considered under the broad heading of conservation.

Row (A) in Table 1 is intended to measure the environmental effects of any development project or program. It consists of estimated dollar measurements of any environmental benefits by sector, as shown in row (Q), and environmental costs by sector, as shown in row (C). The entries in the environmental balance column, (B), are defined as the costs which would have to be incurred by the public or private sectors of the regional economy in order to neutralize any adverse environmental effects and to restore the environment to quality levels acceptable by public health, ecological, aesthetic, or other relevant standards. These rows and columns should be disaggregated into subsectors for as many environmental and resource factors as were relevant to the region and for which data or estimates could be obtained. They would include, at a minimum, water, air, and open space, with possible additions of less measurable factors such as aesthetics, personal safety, absence of noise, freedom from congestion, ecological balance, etc.[30]

For any proposed new development project, row (A) includes estimates of the dollar value per year, at the margin, of specific benefits and costs to each sector resulting from the inputs of environmental factors not priced in the market economy. The environmental balance row and column entries and totals are proposed in order to provide guidelines, indicators, or criteria for shaping regional development objectives and programs. As a hypothetical case we can consider a proposed new regional water-cooled thermal power plant. The probable effects of this plant upon regional per capita personal income and upon the public sector of the region could first be evaluated in the model according to the income and public finance criteria suggested above. If construction of the plant were found to promise a positive contribution to development in the

[30] For an excellent discussion of environmental factors see Orris Herfindahl and Allan V. Kneese, *The Quality of the Environment* (Baltimore: Johns Hopkins University Press, for Resources For the Future, Inc., 1965). The environmental factor rows and columns in the interindustry model presented here can be regarded as summarizing all of the data which would be generated from comprehensive analysis of environmental factors and externalities as discussed by Herfindahl and Kneese considered in a framework of general equilibrium analysis of the total development process. See also Allen V. Kneese, *The Economics of Regional Water Quality Management* (Baltimore: Johns Hopkins University Press, for Resources For the Future, Inc., 1964).

region with respect to per capita income and public finance, it should then be evaluated with respect to the environmental costs and benefits it would generate.

For example, the total environmental benefits to the plant might be $750,000 annually as compared with those of the next most favorable location available. However, the existence of a large power plant imposes environmental costs upon other sectors, especially upon the household sectors. The resulting thermal pollution might damage a vital link in the life cycle of local sport fisheries for an estimated loss annually of $600,000 to sport fishermen. Estimated losses to households from aesthetic deterioration and air pollution might total $300,000. The environmental gain to the power company of $750,000 is greater than the environmental loss to any sector taken individually. However, environmental losses for all sectors exceed the gains. This is illustrated in Table 2 in the environmental balance sectors. The environmental balance column

TABLE 2

	Utilities (Electric)	Households (Aesthetic and Health)	Households (Recreation, Fishing)	Total	Environmental Balance B
Utilities (Electric)					$ 700,000
Total					$ 700,000
Environmental Benefits +Q	+750,000			+750,000	
Environmental Costs −C		−300,000	−600,000	−900,000	
Environmental Balance A	+750,000	−300,000	−600,000	−150,000	

shows a cost of $700,000, which is the cost that would be required to erect cooling towers, to install smoke precipitators, and to take other measures to neutralize the environmental effects of the power plant and avoid damage to the local environment. The power company could incur these costs of neutralizing its effects upon the environment and still enjoy net environmental benefits of $50,000 annually. If the plant were unable to cover the costs of environmental protection in this region, it should be placed in another location where its environmental impact would be less severe, i.e. where the external diseconomies were lower and external economies greater.

Various decision rules concerning environmental criteria could be established in a region to reflect its development objectives. It is open to question whether a region should accept net environmental damage in order to achieve economic development. Where environmental benefits from development greatly exceed environmental losses, it may be possible to tax some of these gains in order to provide for restoration of the environmental balance or for compensation of the sectors suffering environmental damage.

With respect to the environmental criteria, a proposed development project can potentially be capable of making a positive contribution to the development and welfare of a region only if the total environmental gains (Q) from the project to one or more sectors exceed the total environmental losses (C) of all sectors. If the gains exceed the losses, then the project can be justified if equitable ways can be found either to assess the benefiting sectors an amount equivalent to part of their benefits and the proceeds used either to compensate fully the losing sectors, or to apply measures for restoring the damaged resources to at least the quality level existing before the project was undertaken, whichever costs less.

Thus a development project may contribute positively to regional economic development and welfare even though the cost of environmental quality restoration (B) exceeds the net environmental benefits from that project (A), but only if the gross environmental benefits (Q) exceed gross environmental costs (C) and part of these benefits are used to compensate fully the losing sectors.

Formally, in terms of the model suggested in Tables 1 and 2, the necessary environmental condition that a proposed project or program must meet to make a positive contribution to the economic development and welfare of a region is:

$$Q > C, \text{ or } Q > B.$$

Sufficient conditions are either that if

$$B > C,$$

compensation of the disadvantaged sectors must be achieved through assessment of the benefiting sectors, or if

$$C > B,$$

restoration of the quality of the environment to a level at least as high as that prevailing before the development project be financed from assessments against the benefiting sectors.

SUMMARY

There are three major criteria which determine whether or not an economic development program or any of its elements (such as a new plant) can make a positive contribution to the economic development and welfare of a city or region. These criteria, relating to income, public finance, and environmental quality, can be presented symbolically in terms of the model presented. The necessary conditions are:

$$\frac{X_{H_{t-1}}}{P_t} > \frac{X_{H_{t-1}}}{P_{t-1}}, \qquad (1)$$

$$\frac{G_t - T_t}{P_t} < \frac{G_{t-1} - T_{t-1}}{P_{t-1}}. \qquad (2)$$

$$Q_t - C_t > Q_{t-1} - C_{t-1} \text{ or } Q_t - B_t > Q_{t-1} - B_{t-1}. \qquad (3)$$

Where:

X_H = regional personal incomes,
G = regional government expenditures,
T = regional government revenues,
P = regional population,
Q = environmental benefits,
C = environmental costs,
B = cost of environmental restoration, and
t = time period.

The values for these variables are determined after all of the direct, indirect, and induced effects have been worked out through the general equilibrium model for any proposed development program or project.

IMPLICATIONS FOR URBAN AND REGIONAL DEVELOPMENT

This paper has emphasized the point that regional and urban development processes should be analyzed and planned in terms of major objectives to be sought. The goals given primary attention here have been raising the level of per capita personal income in the private sector, improving the capability of the public sector to meet the growing demands upon it, and improving the quality of environment in which both the private and public sectors must function. The analytic model proposed here has been designed primarily for relevance to these objectives but, because of its general equilibrium characteristics, could be used for the analysis of a wide range of alternative objectives. The income, public finance, and environmental objectives emphasize the quality aspects of development and, to the extent they are significant, have numerous implications for development planning.

A major implication is that the real problem of the future for regions within the United States and other high income societies will not be how to achieve maximum growth rates but how to enjoy the benefits of industrialization while minimizing deterioration in the environment and in the quality of life

It is not likely that industrialization can be or should be retarded or reversed. The operational question is how to design and manage a technological society in order to obtain the productivity needed for the achievement of social objectives, affluence, and security without creating intolerable damage to the environment. This task is made both more difficult and more important as population grows. Economic growth in itself is necessarily linked with environmental impacts, whose nature will be a function of the type of economic activities involved, and of the density and rate of development. Rapid growth of an area can result in the emergence of local development-oriented sectors, such as construction, whose continued existence depends upon further expansion. When these development-linked sectors are a significant portion of the regional economy and of the local leadership structure, the pressures for seeking continued expansion may dominate local objectives, regardless of effects on the local fisc and regardless of impacts upon the quality of the local environment. This may help explain the active role played in local government and local development associations by representatives of the development-linked sectors of the local economy. It may become recognized that an important socio-economic problem is that of finding acceptable ways of actually reducing growth rates in cities and regions where development-linked sectors have succeeded in generating self-sustaining growth at rates which excessively threaten the quality of the environment.

The use of a general equilibrium model for the analysis of regional growth reveals that the pursuit of conventional development objectives (creation of new jobs, broadening the tax base, developing local resources) does not necessarily improve the welfare of a region. Job creation is usually linked with inmigration of new residents and does not necessarily raise per capita incomes. Industrialization is accompanied by new demands upon the public sectors, which may or may not be covered by any new taxes. All growth affects the quality of air, water, and other environmental characteristics. It is therefore questionable whether it is in the interest of any city or region to pursue growth and development without major attention to selectivity and quality of new activities sought in the light of local conditions, needs, and objectives. It is also questionable whether the use of public funds to provide subsidies for industrialization can be justified in terms of development objectives consistent with the public interest. The debate on this issue, which appears to have been based thus far upon the implicit objective of maximizing local industrialization, could be conducted in more operational terms by reference to specific sets of objectives.[31]

Regions and cities already having high income levels, strong public sectors, and high quality environments may find it in their interest not only to avoid subsidizing more industrialization but to give serious consideration to careful

[31] See John E. Moes, *Local Subsidies for Industry* (Chapel Hill: University of North Carolina Press, 1962), and the following articles in the *Southern Economic Journal*, XXIV, No. 2 (October, 1962): W. David Maxwell, " 'Review of Local Subsidies for Industry,' by John E. Moes," pp. 137–9; Irving J. Goffman, "Local Subsidies for Industry," pp. 111–4; James H. Thompson, "Local Subsidies for Industry, Comment," pp. 114–9; and John E. Moes, "Local Subsidies to Industry, Reply," pp. 119–26. See also Andre Simmons, "Review of Local Subsidies for Industry," *The American Economic Review*, LII, No. 4 (September, 1962) pp. 881–3; and Douglas Jay, "Review of Local Subsidies for Industry," *The Economic Journal*, LXXII, No. 290, (June, 1963), pp. 315–7.

screening of any new activity in order to make certain that it meets their development objectives. Conceivably, actual removal of offensive activities, with compensation, could result in movement towards development objectives. This might be true in the case of an activity with low employment, low tax payments, and serious environmental effects. In urban renewal programs, the principle has been accepted of removing undersirable housing in order to replace it with improved housing or other uses on the assumption that total benefits, direct and indirect, exceed costs, including costs of demolition and compensation. On precisely the same principle, programs might be considered to remove and replace offensive industrial and commercial activity in the interest of over-all urban and regional objectives. This approach could involve construction activity and might enlist the support of the development-oriented sectors of the economy. Or, exploration of all the realistic alternatives may suggest that the least-cost, most efficient route to achievement of local objectives, e.g. for better education and public safety, might be simply to change the tax structure or even to raise taxes rather than to pursue industrialization in the hope that new revenues will rise faster than new expenditures.

This is not an argument for exclusion of low income residents from particular areas or for creating mercantilistic areas of affluence. It is an argument for efforts to increase per capita income in every region, for attacking the root causes of poverty, and for avoiding indiscriminate geographic spread of industry, particularly in areas where its environmental impact would be objectionable. The logic of this argument is that every region should seek its best development opportunities, given local economic structure, resources, and objectives, and that each region should seek as high a level of environmental quality as it can achieve. Economic activities which fall at the low end of the scale vis-à-vis income payments, tax payments, and environmental demand will then be located in regions which are least selective, and which presumably are most in need of development, and where the development opportunity cost is lowest.

To the extent that regional development and location theory come to be analyzed from the viewpoint of public objectives and to the extent that more regions raise the entrance requirements for economic activity, all industrial sectors hopefully will be encouraged to raise productivity in order to be able to pay higher wages and taxes and to carry the costs of reducing unfavorable environmental effects. In designing development strategy, each region should explore the entire range of alternative growth patterns open to it and study the implications of each.

Among the alternatives to industrialization which should be considered as carriers for development are the highly income-elastic activities such as education, recreation, tourism, urban redevelopment, historic preservation,[32] research

[32] On the economics of historic preservation and tourism see Tony P. Wrenn, "The Tourist Industry and Promotional Publications," *Historic Preservation*, XVI, No. 3 (1964), pp. 111–8, which suggests that Virginia could have justified the restoration of Colonial Williamsburg on the basis that the added revenues from the gas tax alone exceeded the $60 million spent on restoration. See also, Robert L. Montague, III and Tony P. Wrenn, *Planning for Preservation* (Chicago: American Society of Planning Officials, 1964).

and development, and other environment-sensitive activities. The public sector should be explored as a development-determining activity which can, by positive investment in human ecapital and in various forms of social overhead capital, as well by exclusion, guide regional and urban development into preferred channels.

There are a number of factors operating today that appear likely to confer very strong locational advantages upon those urban areas and regions most successful in preserving environmental quality and amenities. Rising incomes and leisure time in which to enjoy them have put a high premium on outdoor recreation opportunities. Technological increases in efficiency along with relatively decreased cost of transportation and communication have reduced locational constraints and made it easier to locate economic activity wherever decision makers wish to locate. It is therefore highly probable that the regions best able to attract high income, highly mobile activities in the future will be those which emphasize the quality of their environment and public sectors.

However, even if decision makers in particular regions introduce selectivity constraints on the types of economic activity permitted in their regions, and environment-damaging activity is then channelled into regions where environmental quality is of less concern, the resulting spatial distribution of economic activity may not be optimal from a national point of view. National guidelines to regional development are becoming and will become more essential, especially where interstate river basins and unique national assets such as shorelines, scenic resources, and historic areas are concerned.[33] The recent emergence of significant federal programs affecting urban and regional development, resource management, and environmental quality makes even more apparent the urgent need for a consistent, constructive national policy on regional development.

Competitive regional and urban development are clearly suboptimal. They may involve regions in a competitive race to offer up for private exploitation their air and water quality. The resulting resource deterioration involves transfer of income from local residents to business firms. Competitive tax concessions to attract development may also result in relative weakening of the public sector. Competitive regional development may involve serious external diseconomies resulting from failure to treat environmental units, such as river basins, as planning units. The larger the planning region, the more adequately externalities can be assessed.

The logic of this argument and the implications of increasing population pressures, urbanization, and technological advance suggest the growing importance of regional planning, zoning, and land use projection. These have now been formulated for many local areas, but the need for national land use planning of priorities and objectives is urgent. Growing problems of population densities and environmental deterioration can be expected eventually to make international land use and regional development planning essential.

The implications of designs for alternative patterns of regional development

---

[33] The need for national perspective on regional development has been noted by John Friedman in "Regional Development in Post-Industrial Society," *Journal of the American Institute of Planners*, XXX, No. 2, (May, 1964), pp. 84-90.

can be explored using models such as the one here suggested or improved variants of it. In research and planning for regional development, this suggests a change in emphasis away from promotion and expansion toward quality and selectivity.

If the quality of economic development is to be accepted as an important objective, much more research is needed on the nature of externalities and upon the costs and benefits of alternative development patterns. Traditional location theory must be broadened to account for the public interest through location of economic activity in patterns that do least violence to the quality of the environment.

Continued enjoyment of the fruits of economic development on a long run viable basis, in the face of expanding population and urbanization and without intolerable damage to the environment, will require much more sensitivity in the future to the role of environmental factors, the space variable, and regional factors in economic analysis.

Copies of Appendix I and II may be obtained upon request from The Bureau of Business and Economic Research, University of Maryland, College Park, Maryland 20742.

Appendix I, "The Regional Interindustry Model and Equation System."

Appendix II, "The Regional Social Accounts."

[2]

Kenneth E. Boulding

THE ECONOMICS OF THE COMING SPACESHIP EARTH

We are now in the middle of a long process of transition in the nature of the image which man has of himself and his environment. Primitive men, and to a large extent also men of the early civilizations, imagined themselves to be living on a virtually illimitable plane. There was almost always somewhere beyond the known limits of human habitation, and over a very large part of the time that man has been on earth, there has been something like a frontier. That is, there was always some place else to go when things got too difficult, either by reason of the deterioration of the natural environment or a deterioration of the social structure in places where people happened to live. The image of the frontier is probably one of the oldest images of mankind, and it is not surprising that we find it hard to get rid of.

Gradually, however, man has been accustoming himself to the notion of the spherical earth and a closed sphere of human activity. A few unusual spirits among the ancient Greeks perceived that the earth was a sphere. It was only with the circumnavigations and the geographical explorations of the fifteenth and sixteenth centuries, however, that the fact that the earth was a sphere became at all widely known and accepted. Even in the nineteenth century, the commonest map was Mercator's projection, which visualizes the earth as an illimitable cylinder, essentially a plane wrapped around the globe, and it was not until the Second World War and the development of the air age that the global

---

Kenneth E. Boulding *is professor of economics at the University of Michigan and author of numerous books, pamphlets, and articles. His most recent book,* The Meaning of the Twentieth Century, *was published in 1964. A fourth edition of his* Economic Analysis, *first published in 1941, came out in 1965. Mr. Boulding was born in Liverpool, England, in 1910 and received his B.A. and M.A. degrees from Oxford University. He became a United States citizen in 1948. Before going to the University of Michigan in 1949, he taught at universities in Scotland, Canada, and the United States.*

nature of the planet really entered the popular imagination. Even now we are very far from having made the moral, political, and psychological adjustments which are implied in this transition from the illimitable plane to the closed sphere.

Economists in particular, for the most part, have failed to come to grips with the ultimate consequences of the transition from the open to the closed earth. One hesitates to use the terms "open" and "closed" in this connection, as they have been used with so many different shades of meaning. Nevertheless, it is hard to find equivalents. The open system, indeed, has some similarities to the open system of von Bertalanffy,[1] in that it implies that some kind of a structure is maintained in the midst of a throughput from inputs to outputs. In a closed system, the outputs of all parts of the system are linked to the inputs of other parts. There are no inputs from outside and no outputs to the outside; indeed, there is no outside at all. Closed systems, in fact, are very rare in human experience, in fact almost by definition unknowable, for if there are genuinely closed systems around us, we have no way of getting information into them or out of them; and hence if they are really closed, we would be quite unaware of their existence. We can only find out about a closed system if we participate in it. Some isolated primitive societies may have approximated to this, but even these had to take inputs from the environment and give outputs to it. All living organisms, including man himself, are open systems. They have to receive inputs in the shape of air, food, water, and give off outputs in the form of effluvia and excrement. Deprivation of input of air, even for a few minutes, is fatal. Deprivation of the ability to obtain any input or to dispose of any output is fatal in a relatively short time. All human societies have likewise been open systems. They receive inputs from the earth, the atmosphere, and the waters, and they give outputs into these reservoirs; they also produce inputs internally in the shape of babies and outputs in the shape of corpses. Given a capacity to draw upon inputs and to get rid of outputs, an open system of this kind can persist indefinitely.

There are some systems—such as the biological phenotype, for instance the human body—which cannot maintain themselves indefinitely by inputs and outputs because of the phenomenon of aging. This process is very little understood. It occurs, evidently, because there are some outputs which cannot be replaced by any known input. There is not the same necessity for aging in organizations and in societies, although an analogous phenomenon may take place. The structure and composition

[1] Ludwig von Bertalanffy, *Problems of Life* (New York: John Wiley and Sons, 1952).

of an organization or society, however, can be maintained by inputs of fresh personnel from birth and education as the existing personnel ages and eventually dies. Here we have an interesting example of a system which seems to maintain itself by the self-generation of inputs, and in this sense is moving towards closure. The input of people (that is, babies) is also an output of people (that is, parents).

Systems may be open or closed in respect to a number of classes of inputs and outputs. Three important classes are matter, energy, and information. The present world economy is open in regard to all three. We can think of the world economy or "econosphere" as a subset of the "world set," which is the set of all objects of possible discourse in the world. We then think of the state of the econosphere at any one moment as being the total capital stock, that is, the set of all objects, people, organizations, and so on, which are interesting from the point of view of the system of exchange. This total stock of capital is clearly an open system in the sense that it has inputs and outputs, inputs being production which adds to the capital stock, outputs being consumption which subtracts from it. From a material point of view, we see objects passing from the noneconomic into the economic set in the process of production, and we similarly see products passing out of the economic set as their value becomes zero. Thus we see the econosphere as a material process involving the discovery and mining of fossil fuels, ores, etc., and at the other end a process by which the effluents of the system are passed out into noneconomic reservoirs—for instance, the atmosphere and the oceans—which are not appropriated and do not enter into the exchange system.

From the point of view of the energy system, the econosphere involves inputs of available energy in the form, say, of water power, fossil fuels, or sunlight, which are necessary in order to create the material throughput and to move matter from the noneconomic set into the economic set or even out of it again; and energy itself is given off by the system in a less available form, mostly in the form of heat. These inputs of available energy must come either from the sun (the energy supplied by other stars being assumed to be negligible) or it may come from the earth itself, either through its internal heat or through its energy of rotation or other motions, which generate, for instance, the energy of the tides. Agriculture, a few solar machines, and water power use the current available energy income. In advanced societies this is supplemented very extensively by the use of fossil fuels, which represent as it were a capital stock of stored-up sunshine. Because of this capital stock of energy, we have been able to maintain an energy input into the system, particularly over the last two centuries, much larger than we would have been able to do with existing techniques if we had had to rely on the current input of available energy from the sun or the earth itself. This supplementary input, however, is by its very nature exhaustible.

The inputs and outputs of information are more subtle and harder to trace, but also represent an open system, related to, but not wholly dependent on, the transformations of matter and energy. By far the larger amount of information and knowledge is self-generated by the human society, though a certain amount of information comes into the sociosphere in the form of light from the universe outside. The information that comes from the universe has certainly affected man's image of himself and of his environment, as we can easily visualize if we suppose that we lived on a planet with a total cloud-cover that kept out all information from the exterior universe. It is only in very recent times, of course, that the information coming in from the universe has been captured and coded into the form of a complex image of what the universe is like outside the earth; but even in primitive times, man's perception of the heavenly bodies has always profoundly affected his image of earth and of himself. It is the information generated within the planet, however, and particularly that generated by man himself, which forms by far the larger part of the information system. We can think of the stock of knowledge, or as Teilhard de Chardin called it, the "noosphere," and consider this as an open system, losing knowledge through aging and death and gaining it through birth and education and the ordinary experience of life.

From the human point of view, knowledge or information is by far the most important of the three systems. Matter only acquires significance and only enters the sociosphere or the econosphere insofar as it becomes an object of human knowledge. We can think of capital, indeed, as frozen knowledge or knowledge imposed on the material world in the form of improbable arrangements. A machine, for instance, originated in the mind of man, and both its construction and its use involve information processes imposed on the material world by man himself. The cumulation of knowledge, that is, the excess of its production over its consumption, is the key to human development of all kinds, especially to economic development. We can see this pre-eminence of knowledge very clearly in the experiences of countries where the material capital has been destroyed by a war, as in Japan and Germany. The knowledge of the people was not destroyed, and it did not take long, therefore, certainly not more than ten years, for most of the material capital to be reestablished again. In a country such as Indonesia, however, where the knowledge did not exist, the material capital did not come into being either. By "knowledge" here I mean, of course, the whole cognitive structure, which includes valuations and motivations as well as images of the factual world.

The concept of entropy, used in a somewhat loose sense, can be applied to all three of these open systems. In the case of material systems, we can distinguish between entropic processes, which take concentrated materials and diffuse them through the oceans or over the earth's surface or into the atmosphere, and anti-entropic processes, which take diffuse materials and concentrate them. Material entropy can be taken as a measure of the uniformity of the distribution of elements and, more uncertainly, compounds and other structures on the earth's surface. There is, fortunately, no law of increasing material entropy, as there is in the corresponding case of energy, as it is quite possible to concentrate diffused materials if energy inputs are allowed. Thus the processes for fixation of nitrogen from the air, processes for the extraction of magnesium or other elements from the sea, and processes for the desalinization of sea water are anti-entropic in the material sense, though the reduction of material entropy has to be paid for by inputs of energy and also inputs of information, or at least a stock of information in the system. In regard to matter, therefore, a closed system is conceivable, that is, a system in which there is neither increase nor decrease in material entropy. In such a system all outputs from consumption would constantly be recycled to become inputs for production, as for instance, nitrogen in the nitrogen cycle of the natural ecosystem.

In regard to the energy system there is, unfortunately, no escape from the grim Second Law of Thermodynamics; and if there were no energy inputs into the earth, any evolutionary or developmental process would be impossible. The large energy inputs which we have obtained from fossil fuels are strictly temporary. Even the most optimistic predictions would expect the easily available supply of fossil fuels to be exhausted in a mere matter of centuries at present rates of use. If the rest of the world were to rise to American standards of power consumption, and still more if world population continues to increase, the exhaustion of fossil fuels would be even more rapid. The development of nuclear energy has improved this picture, but has not fundamentally altered it, at least in present technologies, for fissionable material is still relatively scarce. If we should achieve the economic use of energy through fusion, of course, a much larger source of energy materials would be available, which would expand the time horizons of supplementary energy input into an open social system by perhaps tens to hundreds of thousands of years. Failing this, however, the time is not very far distant, historically speaking, when man will once more have to retreat to his current energy input from the sun, even though this could be used much more effectively than in the past with increased knowledge. Up to now, certainly, we have not got very far with the technology of using current solar energy, but the possibility of substantial improvements in the future is certainly high. It may be, indeed, that the biological revolution which is just beginning will produce a solution to this problem, as we develop artificial organisms which are capable of much more efficient transformation of solar energy into easily available forms than any that we now have. As Richard Meier has suggested, we may run our machines in the future with methane-producing algae.[2]

The question of whether there is anything corresponding to entropy in the information system is a puzzling one, though of great interest. There are certainly many examples of social systems and cultures which have lost knowledge, especially in transition from one generation to the next, and in which the culture has therefore degenerated. One only has to look at the folk culture of Appalachian migrants to American cities to see a culture which started out as a fairly rich European folk culture in Elizabethan times and which seems to have lost both skills, adaptability, folk tales, songs, and almost everything that goes up to make richness and complexity in a culture, in the course of about ten generations. The American Indians on reservations provide another example of such degradation of the information and knowledge system. On the other hand, over a great part of human history, the growth of knowledge in the earth as a whole seems to have been almost continuous, even though there have been times of relatively slow growth and times of rapid growth. As it is knowledge of certain kinds that produces the growth of knowledge in general, we have here a very subtle and complicated system, and it is hard to put one's finger on the particular elements in a culture which make knowledge grow more or less rapidly, or even which make it decline. One of the great puzzles in this connection, for instance, is why the take-off into science, which represents an "acceleration," or an increase in the rate of growth of knowledge in European society in the sixteenth century, did not take place in China, which at that time (about 1600) was unquestionably ahead of Europe, and one would think even more ready for the breakthrough. This is perhaps the most crucial question in the theory of social development, yet we must confess that it is very little understood. Perhaps the most significant factor in this connection is the existence of "slack" in the culture, which permits a divergence from established patterns and activity which is not merely

[2] Richard L. Meier, *Science and Economic Development* (New York: John Wiley and Sons, 1956).

devoted to reproducing the existing society but is devoted to changing it. China was perhaps too well-organized and had too little slack in its society to produce the kind of acceleration which we find in the somewhat poorer and less well-organized but more diverse societies of Europe.

The closed earth of the future requires economic principles which are somewhat different from those of the open earth of the past. For the sake of picturesqueness, I am tempted to call the open economy the "cowboy economy," the cowboy being symbolic of the illimitable plains and also associated with reckless, exploitative, romantic, and violent behavior, which is characteristic of open societies. The closed economy of the future might similarly be called the "spaceman" economy, in which the earth has become a single spaceship, without unlimited reservoirs of anything, either for extraction or for pollution, and in which, therefore, man must find his place in a cyclical ecological system which is capable of continuous reproduction of material form even though it cannot escape having inputs of energy. The difference between the two types of economy becomes most apparent in the attitude towards consumption. In the cowboy economy, consumption is regarded as a good thing and production likewise; and the success of the economy is measured by the amount of the throughput from the "factors of production," a part of which, at any rate, is extracted from the reservoirs of raw materials and noneconomic objects, and another part of which is output into the reservoirs of pollution. If there are infinite reservoirs from which material can be obtained and into which effluvia can be deposited, then the throughput is at least a plausible measure of the success of the economy. The gross national product is a rough measure of this total throughput. It should be possible, however, to distinguish that part of the GNP which is derived from exhaustible and that which is derived from reproducible resources, as well as that part of consumption which represents effluvia and that which represents input into the productive system again. Nobody, as far as I know, has ever attempted to break down the GNP in this way, although it would be an interesting and extremely important exercise, which is unfortunately beyond the scope of this paper.

By contrast, in the spaceman economy, throughput is by no means a desideratum, and is indeed to be regarded as something to be minimized rather than maximized. The essential measure of the success of the economy is not production and consumption at all, but the nature, extent, quality, and complexity of the total capital stock, including in this the state of the human bodies and minds included in the system. In the spaceman economy, what we are primarily concerned with is stock maintenance, and any technological change which results in the maintenance of a given total stock with a lessened throughput (that is, less production and consumption) is clearly a gain. This idea that both production and consumption are bad things rather than good things is very strange to economists, who have been obsessed with the income-flow concepts to the exclusion, almost, of capital-stock concepts.

There are actually some very tricky and unsolved problems involved in the questions as to whether human welfare or well-being is to be regarded as a stock or a flow. Something of both these elements seems actually to be involved in it, and as far as I know there have been practically no studies directed towards identifying these two dimensions of human satisfaction. Is it, for instance, eating that is a good thing, or is it being well fed? Does economic welfare involve having nice clothes, fine houses, good equipment, and so on, or is it to be measured by the depreciation and the wearing out of these things? I am inclined myself to regard the stock concept as most fundamental, that is, to think of being well fed as more important than eating, and to think even of so-called services as essentially involving the restoration of a depleting psychic capital. Thus I have argued that we go to a concert in order to restore a psychic condition which might be called "just having gone to a concert," which, once established, tends to depreciate. When it depreciates beyond a certain point, we go to another concert in order to restore it. If it depreciates rapidly, we go to a lot of concerts; if it depreciates slowly, we go to few. On this view, similarly, we eat primarily to restore bodily homeostasis, that is, to maintain a condition of being well fed, and so on. On this view, there is nothing desirable in consumption at all. The less consumption we can maintain a given state with, the better off we are. If we had clothes that did not wear out, houses that did not depreciate, and even if we could maintain our bodily condition without eating, we would clearly be much better off.

It is this last consideration, perhaps, which makes one pause. Would we, for instance, really want an operation that would enable us to restore all our bodily tissues by intravenous feeding while we slept? Is there not, that is to say, a certain virtue in throughput itself, in activity itself, in production and consumption itself, in raising food and in eating it? It would certainly be rash to exclude this possibility. Further interesting problems are raised by the demand for variety. We certainly do not want a constant state to be maintained; we want fluctuations in the state. Otherwise there would be no demand for variety in food, for variety in scene, as in travel, for variety in social contact, and so on. The demand for variety can, of course, be costly, and sometimes it seems to be too costly to be tolerated or at least legitimated, as in the case of marital partners, where the maintenance of a homeostatic state in the family is usually regarded as much more desirable than the variety and excessive throughput of the libertine. There are problems here which the economics profession has neglected with astonishing singlemindedness. My own attempts to call attention to some of them, for instance, in two articles,[3] as far as I can judge, produced no response whatever; and economists continue to think and act as if production, consumption, throughput, and the GNP were the sufficient and adequate measure of economic success.

It may be said, of course, why worry about all this when the spaceman economy is still a good way off (at least beyond the lifetimes of any now living), so let us eat, drink, spend, extract and pollute, and be as merry as we can, and let posterity worry about the spaceship earth. It is always a little hard to find a convincing answer to the man who says, "What has posterity ever done for me?" and the conservationist has always had to fall back on rather vague ethical principles postulating identity of the individual with some human community or society which extends not only back into the past but forward into the future. Unless the individual identifies with some community of this kind, conservation is obviously "irrational." Why should we not maximize the welfare of this generation at the cost of posterity? *"Après nous, le déluge"* has been the motto of not insignificant numbers of human societies. The only answer to this, as far as I can see, is to point out that the welfare of the individual depends on the extent to which he can identify himself with others, and that the most satisfactory individual identity is that which identifies not only with a community in space but also with a community extending over time from the past into the future. If this kind of identity is recognized as desirable, then posterity has a voice, even if it does not have a vote; and in a sense, if its voice can influence votes, it has votes too. This whole problem is linked up with the much larger one of the determinants of the morale, legitimacy, and "nerve" of a society, and there is a great deal of historical evidence to suggest that a society which loses its identity with posterity and which loses its positive image of the future loses also its capacity to deal with present problems, and soon falls apart.[4]

Even if we concede that posterity is relevant to our present problems, we still face the question of time-discounting and the closely related

[3] K. E. Boulding, "The Consumption Concept in Economic Theory," *American Economic Review,* 35:2 (May 1945), pp. 1–14; and "Income or Welfare?," *Review of Economic Studies,* 17 (1949–50), pp. 77–86.

[4] Fred L. Polak, *The Image of the Future,* Vols. I and II, translated by Elise Boulding (New York: Sythoff, Leyden and Oceana, 1961).

question of uncertainty-discounting. It is a well-known phenomenon that individuals discount the future, even in their own lives. The very existence of a positive rate of interest may be taken as at least strong supporting evidence of this hypothesis. If we discount our own future, it is certainly not unreasonable to discount posterity's future even more, even if we do give posterity a vote. If we discount this at 5 per cent per annum, posterity's vote or dollar halves every fourteen years as we look into the future, and after even a mere hundred years it is pretty small—only about 1½ cents on the dollar. If we add another 5 per cent for uncertainty, even the vote of our grandchildren reduces almost to insignificance. We can argue, of course, that the ethical thing to do is not to discount the future at all, that time-discounting is mainly the result of myopia and perspective, and hence is an illusion which the moral man should not tolerate. It is a very popular illusion, however, and one that must certainly be taken into consideration in the formulation of policies. It explains, perhaps, why conservationist policies almost have to be sold under some other excuse which seems more urgent, and why, indeed, necessities which are visualized as urgent, such as defense, always seem to hold priority over those which involve the future.

All these considerations add some credence to the point of view which says that we should not worry about the spaceman economy at all, and that we should just go on increasing the GNP and indeed the gross world product, or GWP, in the expectation that the problems of the future can be left to the future, that when scarcities arise, whether this is of raw materials or of pollutable reservoirs, the needs of the then present will determine the solutions of the then present, and there is no use giving ourselves ulcers by worrying about problems that we really do not have to solve. There is even high ethical authority for this point of view in the New Testament, which advocates that we should take no thought for tomorrow and let the dead bury their dead. There has always been something rather refreshing in the view that we should live like the birds, and perhaps posterity is for the birds in more senses than one; so perhaps we should all call it a day and go out and pollute something cheerfully. As an old taker of thought for the morrow, however, I cannot quite accept this solution; and I would argue, furthermore, that tomorrow is not only very close, but in many respects it is already here. The shadow of the future spaceship, indeed, is already falling over our spendthrift merriment. Oddly enough, it seems to be in pollution rather than in exhaustion that the problem is first becoming salient. Los Angeles has run out of air, Lake Erie has become a cesspool, the oceans are getting full of lead and DDT, and the atmosphere may become man's major problem in another generation, at the rate at which we are filling it up with gunk. It is, of course, true that at least on a microscale, things have been worse at times in the past. The cities of today, with all their foul air and polluted waterways, are probably not as bad as the filthy cities of the pretechnical age. Nevertheless, that fouling of the nest which has been typical of man's activity in the past on a local scale now seems to be extending to the whole world society; and one certainly cannot view with equanimity the present rate of pollution of any of the natural reservoirs, whether the atmosphere, the lakes, or even the oceans.

I would argue strongly also that our obsession with production and consumption to the exclusion of the "state" aspects of human welfare distorts the process of technological change in a most undesirable way. We are all familiar, of course, with the wastes involved in planned obsolescence, in competitive advertising, and in poor quality of consumer goods. These problems may not be so important as the "view with alarm" school indicates, and indeed the evidence at many points is conflicting. New materials especially seem to edge towards the side of improved durability, such as, for instance, neolite soles for footwear, nylon socks, wash and wear shirts, and so on. The case of household equipment and automobiles is a little less clear. Housing and building construction generally almost certainly has declined in durability since the Middle Ages, but this decline also reflects a change in tastes towards flexibility and fashion and a need for novelty, so that it is not easy to assess. What is clear is that no serious attempt has been made to assess the impact over the whole of economic life of changes in durability, that is, in the ratio of capital in the widest possible sense to income. I suspect that we have underestimated, even in our spendthrift society, the gains from increased durability, and that this might very well be one of the places where the price system needs correction through government-sponsored research and development. The problems which the spaceship earth is going to present, therefore, are not all in the future by any means, and a strong case can be made for paying much more attention to them in the present than we now do.

It may be complained that the considerations I have been putting forth relate only to the very long run, and they do not much concern our immediate problems. There may be some justice in this criticism, and my main excuse is that other writers have dealt adequately with the more immediate problems of deterioration in the quality of the environment. It is true, for instance, that many of the immediate problems of pollution of the atmosphere or of bodies of water arise because of the failure of the price system, and many of them could be solved by corrective taxation. If people had to pay the losses due to the nuisances which they create, a good deal more resources would go into the prevention of nuisances. These arguments involving external economies and diseconomies are familiar to economists, and there is no need to recapitulate them. The law of torts is quite inadequate to provide for the correction of the price system which is required, simply because where damages are widespread and their incidence on any particular person is small, the ordinary remedies of the civil law are quite inadequate and inappropriate. There needs, therefore, to be special legislation to cover these cases, and though such legislation seems hard to get in practice, mainly because of the widespread and small personal incidence of the injuries, the technical problems involved are not insuperable. If we were to adopt in principle a law for tax penalties for social damages, with an apparatus for making assessments under it, a very large proportion of current pollution and deterioration of the environment would be prevented. There are tricky problems of equity involved, particularly where old established nuisances create a kind of "right by purchase" to perpetuate themselves, but these are problems again which a few rather arbitrary decisions can bring to some kind of solution.

The problems which I have been raising in this paper are of larger scale and perhaps much harder to solve than the more practical and immediate problems of the above paragraph. Our success in dealing with the larger problems, however, is not unrelated to the development of skill in the solution of the more immediate and perhaps less difficult problems. One can hope, therefore, that as a succession of mounting crises, especially in pollution, arouse public opinion and mobilize support for the solution of the immediate problems, a learning process will be set in motion which will eventually lead to an appreciation of and perhaps solutions for the larger ones. My neglect of the immediate problems, therefore, is in no way intended to deny their importance, for unless we at least make a beginning on a process for solving the immediate problems we will not have much chance of solving the larger ones. On the other hand, it may also be true that a long-run vision, as it were, of the deep crisis which faces mankind may predispose people to taking more interest in the immediate problems and to devote more effort for their solution. This may sound like a rather modest optimism, but perhaps a modest optimism is better than no optimism at all.

[3]

On Economics as a Life Science

Herman E. Daly

Federal University of Ceará, and Louisiana State University

There is no wealth but life.—JOHN RUSKIN
All flesh is grass.—Isa. 40:6

I. Introduction

The purpose of this essay is to bring together some of the more salient similarities between biology and economics and to argue that, far from being superficial, these analogies are profoundly rooted in the fact that the ultimate subject matter of biology and economics is *one*, viz., *the life process*. Most of biology concentrates on the "within skin" life process, the exception being ecology, which focuses on the "outside skin" life process (Bates, 1960, pp. 12–13). Economics is the part of ecology which studies the outside-skin life process insofar as it is dominated by *commodities* and their interrelations. In what follows the traditional economic (outside skin) and the traditional biological (within skin) views of the total life process will be considered, both in their *steady-state* aspect and in their *evolutionary* aspect. Finally an approach to a more general "general equilibrium" model will be suggested by considering the human economy from an ecological perspective.

II. Biological Analogies in Economics

Analogy is so fundamental to our way of thinking that the ability to recognize analogies is generally considered one of the criteria of intelligence. While there is a vast difference between analogy on the one hand and logical proof and empirical verification on the other, it by no means follows that the former belongs only to poetry and not to science. Analogy is the essence of the inductive side of science. Furthermore, the dominant mode of thought in economics today is the "analytical simile" (Georgescu-Roegen, 1966, pp. 114–24), the mathematical or geometric model based on a Pythagorean analogy between fuzzy, dialectical reality and well-defined, analytic number. The fruitfulness of this analogy for all science is obvious —but it is an analogy nonetheless, with its roots in the same insight which inspired the mystical Pythagorean brotherhood. That economists have also found biological analogies useful is only slightly less obvious. The circular flow of blood and the circular flow of money, the many parallel phenomena of specialization, exchange, interdependence, homoeostasis, and evolution are well known. In the opposite direction, economic analogies in biology are also common, as witnessed by Malthus' influence on Darwin and by the very etymology of the word "ecology." Finally, an ultimately central place for biological analogies in economics has been claimed by no less an authority than Alfred Marshall in this famous statement, "The Mecca of the economist lies in economic biology rather than in economic dynamics" (Marshall, 1920, Preface, p. 14), and in his further statement that "in the later stages of economics, when we are approaching nearly to the conditions of life, biological analogies are to be preferred to mechanical" (Marshall, 1925, p. 317). Among current economic theorists it would appear that only the works of Kenneth Boulding (1950, 1958, 1966) and Nicholas Georgescu-Roegen (1966) (both freely drawn upon here) reveal a disposition to take Marshall seriously on this point.

Perhaps the intellectual genealogy of the ideas to be developed in this paper can be more specifically indicated by a pair of quotations from two seminal thinkers of the early part of this century—one a biologist (A. J. Lotka) and the other an economist (J. A. Hobson).

Lotka (1956) informs us that "underlying our economic manifestations are biological phenomena which we share in common with other species; and...the laying bare and clearly formulating of the relations thus involved—in other words the analysis of the biophysical foundations of economics—is one of the problems coming within the program of physical biology."

Just what these "biophysical foundations" are, and how they support the economic superstructure, is in large part the subject of Section V.

From Hobson (1929) we learn that

> all serviceable organic activities consume tissue and expend energy, the biological costs of the services they render. Though this economy may not correspond in close quantitative fashion to a pleasure and pain economy or to any conscious valuation, it must be taken as the groundwork for that conscious valuation. For most economic purposes we are well-advised to prefer the organic test to any other test of welfare, bearing in mind that many organic costs do not register themselves easily or adequately in terms of conscious pain or disutility, while organic gains are not always interpretable in conscious enjoyment.

The "groundwork for conscious valuation" and the "organic test of welfare" are ideas with close counterparts in Section III, to which we now turn.

III. The Steady-State Analogy

The close similarity of the basic within-skin life process of metabolism (anabolism and catabolism) with the outside-skin life process of economics (production and consumption) is evident from Figure 1.

In either process the only *material* output is *waste*. The purpose (value produced) of the metabolic process is the maintenance of life. The purpose (value produced) of the economic process is the maintenance and enjoyment of life. An accounting balance equation of the life process in value terms would state that the value of life enjoyment plus the value of material waste (zero) equals the sum of the values of all the matter and energy upon which the total life process is based. The total value of life (our subjective estimate thereof) is *imputed* to the total quantity of things necessary for its enjoyable maintenance.[1] The Austrian economists have taught us that this imputation also determines the *relative* values (prices) of individual things according to the principle of diminishing marginal utility, which for Böhm-Bawerk was "the key-stone of all economic theory" (1891, p. 149). Since commodities are priced according to their diminishing marginal utilities, the sum of all goods in the economy valued at their marginal utilities (or prices) would be very small relative to the total utility of all goods (total life value), which is probably infinite.[2] The infinite difference between the finite sum of prices of all goods and the infinite sum of total utility of all goods is an infinite "global consumers' surplus." Hence, insofar as economics concentrates on value in exchange (marginal utility) to the exclusion of value in use (total utility)—to that extent it is concerning itself with only an infinitesimal portion of total life value. This is not meant to minimize the importance of exchange values, since it is precisely by considering margins that we maximize totals. The point is that, while margins are reliable means for *maximizing* totals, they are very treacherous means for *evaluating* totals, as any student who has pondered the diamonds-water paradox must realize. Any sort of economic numerol-

[1] Value is not permanently imputed to the (non-material) technology within which matter and energy are used, unless that technology is made artificially scarce by patents. Following Schumpeter we can say that a new technology, while it is temporarily scarce by virtue of its novelty, will earn a temporary profit but will not receive a permanent imputed share of total value produced.

[2] To say that "total life value" is infinite is not to say that it is *ultimate*—"For whosoever will save his life shall lose it: and whosoever will lose his life for My sake shall find it. For what is a man profited, if he shall gain the whole world and lose his own soul? Or what shall a man give in exchange for his soul?" (Matt. 16:25, 26). On the commonsense infinitude of total utility, see Böhm-Bawerk (1891, Book III, pp. 147–53).

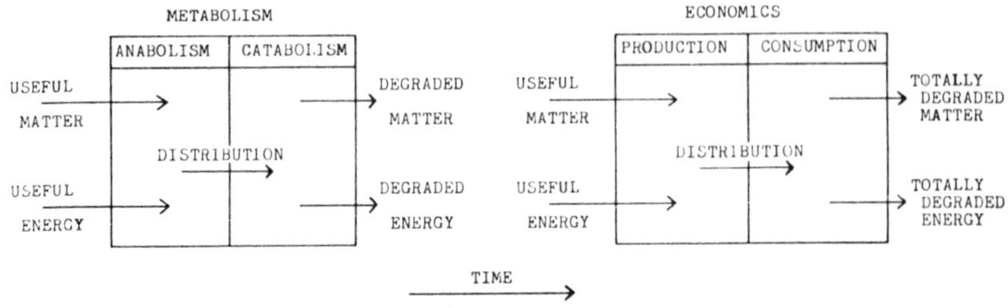

FIG. 1

ogy which, with one-eyed devotion to Pythagoras, insists on glossing over this treachery deserves a thorough dunking in the satirical acid of Jonathan Swift's *A Modest Proposal*.[3] Perhaps Hobson's "organic test of welfare" is simply the idea that it is better to make imprecise statements about unmeasurable but relevant magnitudes (use value, total utility) than to make more precise statements about the measurable but irrelevant magnitude (for evaluating total welfare) of exchange value. Economists shy away from thinking too much about total utility mainly because it is unmeasurable and dependent on value judgments—both embarrassing for a "positive science." But perhaps, as Joan Robinson suggests (1962, p. 54), this aversion to total utility also stems from its tendency to make one question "an economic system in which so much of the good juice of utility is allowed to evaporate out of commodities by distributing them unequally"; furthermore "this egalitarian element in the doctrine was sterilized mainly by slipping from utility to physical output as the object to be maximized." But as we have seen, the ultimate *physical* output of the economic process is waste, and there is no sense in maximizing that!

There is also a balance equation of the life process in physical units, based on the law of conservation of matter-energy. But more significant than the physical balance, from an economic viewpoint, is the one-way, non-circular, irreversible nature of the flow of matter-energy through all divisions of the life process. Since useful (low entropy) matter-energy is apparently finite, the total life process could be brought to a halt by what Boulding has called "the entropy trap." Thus one of the ultimate natural sources of scarcity, and hence of economic activity, is the second law of thermodynamics (Georgescu-Roegen, 1966, pp. 66–82). Indeed, if one were perversely to insist on a real-cost theory of value, it would seem that entropy, rather than labor or energy, should be the source of value. Even in the subjective theory of value, however, entropy, the common denominator of all forms of scarcity, determines the locations of the margins and

[3] In which, using exchange-value calculations, Swift logically demonstrates the "economic desirability" of eating children!

hence enters into the determination of marginal utilities and exchange values.

Erwin Schroedinger (1945) has described life as a system in steady-state thermodynamic disequilibrium which maintains its constant distance from equilibrium (death) by feeding on low entropy from its environment—that is, by exchanging high-entropy outputs for low-entropy inputs. *The same statement would hold verbatim as a physical description of the economic process.* A corollary of this statement is that an organism cannot live in a medium of its own waste products. With this principle in mind, one can better appreciate the significance of the following recent observation by J. J. Spengler (1966) in his presidential address to the American Economic Association, "Witness here in America the endless dumping of trash (four pounds per person per day)....Indeed, some hold, J. K. Galbraith had better labeled ours an effluent society than an affluent one." This four pounds per person per day does not disappear—it becomes a part of the physical environment in which we must live. Great stress has been put on the reciprocal nature of the relation of fitness between organism and environment by L. J. Henderson (1958). If the organism fits the environment, then it is also the case that the environment is fit for the organism. Henderson argues that there must have been some not-yet-understood process of physical evolution prior to the emergence of life in order for the environment to attain the rather exacting preconditions for supporting life. Thus man's newly acquired ability to degrade his material environment at the rate of four pounds per person per day is likely to be even more dangerous than commonly realized, in view of our ignorance of ecological relations.

How do the economic and metabolic processes fit together? Clearly metabolism is partly contained within the economic subprocess of consumption. Many of the material inputs into metabolism are economic products, and some outputs of metabolism are generally not totally degraded and thus can be further consumed—for example, manure fertilizer and carbon dioxide. But the ultimate physical output of the economic process *is* totally degraded matter-energy, in Marx's term, "devil's dust." Continuing in Chinese-box fashion, the total economic process is itself a subprocess on the consuming side of the total ecological life process, the producing side of the latter consisting mainly of photosynthesis carried on by green plants, which draw their inputs from the physical environment of air, soil, water, and sunlight.

Both the within-skin and outside-skin life processes have a permanently maintained physical basis which undergoes continual replacement over relatively short time periods (steady-state aspect) and which is capable of qualitative change and reorganization over long periods (evolutionary aspect). In other words "capital" represents "exosomatic organs" and biological organs represent "endosomatic capital." In each case, we observe both short-term depreciation and replacement and long-term technological change. Physical capital is essentially matter that is capable of trapping energy and channeling it to human purposes. Hence, in a very real sense the entire physical environment is capital, since it is only through the agency of air, soil, and water that plant life is able to capture the solar energy upon which the whole hierarchy of life (and value) depends. Should not these elements receive the same care we bestow upon our other machines? And is not any theory of value that leaves them out rather like a theory of icebergs that fails to consider the submerged 90 per cent?

IV. The Evolutionary Analogy

The material basis of the life process grows when the rate of production (anabolism) exceeds the rate of consumption (catabolism). Growth merges into development as alterations in the rates of increase of different parts give rise to new proportions, new qualitative relations, and new technologies. Although development is not well understood by either science, the subtle influence of size on organization has led both biologists and economists to the concept of a proper or optimum scale for a given organizational plan. That Marx, who emphasized this dialectic interplay of quantity and quality, also tended to view economics as a part of natural history is evident in the following quotation (1967, I, 372):

"Darwin has interested us in the history of Nature's Technology, that is, in the formation of the organs of plants and animals, which organs serve as instruments of production for sustaining life. Does not the history of the productive organs of man, of organs that are the material basis of all social organization, deserve equal attention?"

The same idea has been expressed by Lotka (1956, p. 208), viz., "Man's industrial activities are merely a highly specialized and greatly developed form of the general biological struggle for existence," and further in a passage (1956, p. 369) that would have pleased Marx:

> The most singular feature of the artificial extensions of our natural body is that they are shared in common by a number of individuals. When the sick man consults the physician, who, we will say, makes a microscopic examination, for example, the patient is virtually hiring a pair of high power eyes. When you drop a nickel into a telephone box, you are hiring the use of an ear to listen to your friend's voice five or ten miles distant. When the workingman accepts a wage of forty dollars for his weekly labor, he is in fact paying to his employers an undetermined amount for the privilege of using his machines as artificial members to manufacture marketable wares.
>
> The modern development of artificial aids to our organs and faculties has exerted two opposing influences.

> On the one hand, it has in a most real way bound men together into one body: so very real and material is the bond that society might aptly be described as one huge multiple Siamese twin.
>
> On the other hand, since the control over certain portions of this common body is unevenly distributed among the separate individuals, certain of them may be said in a measure to own parts of the bodies of others, holding them a species of refined slavery, and though neither of the two parties concerned may be clearly conscious of the fact, it is often resented in a more or less vague way by the one less favored.

In biological evolution genes transmit the "knowledge" of organic forms over time, and gene mutations introduce occasional modifications, resulting in the success of the forms best suited to the environment. In economic evolution, culture transmits knowledge over time, and new ideas produce mutant organizations from which competition again determines the fittest. Indeed, Teilhard de Chardin (1959) argues that "cultural evolution" is simply a new evolutionary mechanism that superseded the old mechanism in importance.

A natural history of economic evolution might be built around the theme of "economic surplus" and its progressive growth and cultivation. The original surplus was produced by plants, since they capture more solar energy than that necessary for their own maintenance. Animal life depends on this surplus, and perhaps man's greatest discovery was that he could cultivate and expand that upon which his existence depended, thus "exploiting niggardly nature."[4]

As soon as this primary activity became efficient enough to produce a surplus above the maintenance needs of those engaged in primary production, it became possible to evolve secondary economic activities, etc. Although economic activity moves far away from direct contact with nature, the "biophysical foundations of economics" remain ever present in the background, and it is to these foundations that we now direct our attention.

V. The Human Economy in Ecological Perspective

Although the life process is essentially one, it seems that for many analytical purposes the most convenient boundary by which to divide the process is the natural boundary of skin. The outside-skin life process is the subject of ecology, but ecologists abstract from the human economy and study only natural interdependences, while economists abstract from

[4] And, Marx would argue, man even discovered that he could "cultivate and extract" an analogous surplus from other men in the factory "hothouse."

nature and consider only interdependences among commodities and man. But what discipline systematically studies the interdependences which clearly exist between the natural and human parts of the outside-skin life process? Marston Bates, a biologist, addresses himself to this point in the following quotation (1960, p. 247):

> Then we come to man and his place in the system of life. We could have left man out, playing the ecological game of "let's pretend man doesn't exist." But this seems as unfair as the corresponding game of the economists, "let's pretend that nature doesn't exist." The economy of nature and ecology of man are inseparable and attempts to separate them are more than misleading, they are dangerous. Man's destiny is tied to nature's destiny and the arrogance of the engineering mind does not change this. Man may be a very peculiar animal, but he is still a part of the system of nature.

Any attempt to isolate a segment of reality is always somewhat misleading, but not for that reason less necessary. Our purposes dictate the manner in which we abstract from reality, and as economists well know, many useful purposes can be served by partial analysis—that is, studying one industry in abstraction from its matrix of interconnections with the rest of the economy. While this is a useful procedure for studying the peanut industry, no economist would want to study the automobile industry under such limitations. Too many important feedbacks from the rest of the economy would be left out. Until recently the economy of man was "peanuts" in the total economy of nature. Now it is more like the automobile industry, and to continue *ceteris paribus* treatment of nature (even in general-equilibrium analysis) is indeed dangerous to our purpose if that purpose is to say something about how human wants can best be served.

A rather dramatic example of this kind of danger has been indicated by Dr. Edward Teller (1965), who pointed out that since the Industrial Revolution the tremendous consumption of carbon fuels has resulted in an increased concentration of carbon dioxide in the atmosphere. Since this gas increases the heat retention of the atmosphere, thus raising the average temperature, it may well be that the ultimate effect of the Industrial Revolution will be the melting of the polar ice cap and the inundation of large parts of the world. The more concrete case of the unintentional destruction wrought on the environment by chemical insecticides has been forcefully documented by Rachel Carson (1962). Also, we know that the entire chain of life depends heavily on bacteria—for example, nitrogen fixation and decomposition of dead organisms. Is it not possible that some export from the human economy (for example, detergents) could prove lethal to certain of these organisms? Conversely, might not some human exports be highly beneficial to the propagation of particular disease-causing bacteria? And one need only mention the problem of radioactive fallout. At a less dramatic but increasingly serious level, we have ubiquitous instances of air and water pollution plaguing the world's cities, not to mention the problems of deforestation, soil erosion, and noise.

Such phenomena have long been recognized (grudgingly) in economic theory under the heading of *externalities*—that is, interrelations whose connecting links are external to the economists' abstract world of commodities but very much internal to the world in which we live, move, and have our being. Perhaps "non-market interdependence" is a more descriptive term.

It would be easy to liken this concept to a *deus ex machina* lowered into the scene by our theoretical playwrights to save an awkward plot, but it is by no means easy to suggest a better treatment. A better treatment is called for, however, since externalities are spending more time on center stage and less time in the wings than previously. Or, changing the metaphor, to continue theoretical development via continued ad hoc introduction of externalities is reminiscent of adding epicycles and in the long run will lead only to Ptolemaic complications in economic theory. Our economic cosmos is not one of uniform circular motion of commodities among men but one of eliptical orbits through interdependent ecological sectors.

How does one integrate the world of commodities into the larger economy of nature? Perhaps this is a problem in which economics can provide a useful analogy. Leontief's input-output model has proved useful in dealing with phenomena of interdependence, and it may offer the most promising analytical framework within which to consider the above question.[5] Just as the annual flow of gross national product, or *final* commodities, requires a supporting matrix of flows of *intermediate* commodities, so does the annual flow of all economic commodities (final and intermediate) require a supporting matrix of flows of physical things which carry no price tag but nonetheless are necessary complements to the flows of those things which do carry price tags.

In its simplest input-output representation the total economy can be divided into its human and non-human sectors, as in Table 1.

Cell or quadrant (2) is the domain of traditional economics, that is, the study of inputs and outputs to and from various subsectors within the human-to-human box. Cell (4) represents the traditional area of concern of ecology, the inputs to and outputs from subsectors in the non-human–

[5] The Leontief input-output model derives from a line of thought beginning with Francois Quesnay's "tableau economique," which was described by Mirabeau as "the great discovery which glorifies our century and will yield posterity its fruits." (For an exposition see Leontief, 1966.) It is more than coincidental that we should find the input-output model relevant to economics considered as a life science, since Quesnay (a physician) and the physiocrats emphasized the supremacy of nature and the biological analogy.

TABLE 1

FROM	To: Human	To: Non-Human
Human	(2)	(1)
Non-human . . .	(3)	(4)

to–non-human box. Cells (1) and (3), respectively, contain the flows of inputs from human subsectors to non-human subsectors and from non-human subsectors to human subsectors. All of the items exchanged in (2) are *economic commodities*, by which we mean that they have positive prices. All items of exchange in cells (1), (3), and (4) may by contrast be labeled *ecological commodities*, which consist of free goods (zero price) and "bads" (negative price). The negative price on bads is not generally observed, since there usually exists the alternative of exporting the bad to the non-human economy, which cannot pay the negative price (that is, charge us a positive price for the service of taking the "bad" off our hands, as would be the case if it were transferred to another sector of the human economy). Ecological commodities that are bads are bad in relation to man, not necessarily to the non-human world. The difficulty, however, is that these more than gratuitous exports from the human economy in cell (1) are simultaneously inputs to the non-human economy and as such strongly influence the outputs from the non-human back to the human sector—that is, cell (1) is connected to cell (3) via cell (4), and cell (3) directly influences human welfare.[6] These relationships will perhaps be more evident in Table 2, which is an expansion of Table 1, with the four quadrants corresponding to the quadrants of Table 1. Note that in both tables the basic vision is still a "world of commodities," although a bigger world that now includes both economic commodities (the q_{ij} in quadrant [2]) and ecological commodities (the q_{ij} in quadrants [1], [3], and [4]). The q_{ij} in quadrants (1), (3), and (4) are the "biophysical foundations of economics."

In Table 2, quadrant (2) is the simplest form of the usual Leontief input-output table, with two transforming sectors (agriculture and industry) and one primary sector (households). Agriculture consists of living transformers of matter-energy, and industry consists of non-living transformers of matter-energy. The non-human economy has likewise been divided into the "transforming sectors" of animal, plant, and bacteria (living sectors)

[6] If the reader will pardon the liberties taken with Luke 11:24–26 we may say that sometimes a bad cast out of cell (2) wanders through the waterless places of cells (1) and (4) seeking rest. And finding none it gathers seven new bads, which then descend upon the well-garnished human household through the back door of cell (3). And the last state of that household is worse than the first.

TABLE 2

Output from	Input to: Agri-culture (1)	Industry (2)	Households (Final Consump-tion) (3)	Ani-mal (4)	Plant (5)	Bac-teria (6)	Atmosphere (8)	Hydro-sphere (8)	Lytho-sphere (9)	Sink (Final Consump-tion) (10)	Total
	Quadrant (2)			Quadrant (1)							
1. Agriculture	...	q_{12}	...	...	...	...	q_{17}	...	...	...	Q_1
2. Industry	q_{21}	(q_{22})	q_{23}	...	...	...	q_{27}	...	...	...	Q_2
3. Households (primary services)	...	q_{32}	...	...	...	...	q_{37}	...	...	...	Q_3
	Quadrant (3)			Quadrant (4)							
4. Animal	...	...	...	...	...	...	q_{47}	...	...	...	
5. Plant	...	...	...	...	...	...	q_{57}	...	...	...	
6. Bacteria	...	...	...	...	...	...	q_{67}	...	...	...	
7. Atmosphere	q_{71}	q_{72}	q_{73}	q_{74}	q_{75}	q_{76}	(q_{77})	q_{78}	q_{79}	$q_{7,10}$	Q_7
8. Hydrosphere	...	...	...	...	...	...	q_{87}	...	...	...	
9. Lithosphere	...	...	...	...	...	...	q_{97}	...	...	...	
10. Sun (primary services)	...	...	...	...	...	...	$q_{10,7}$	...	...	...	

and of atmosphere, hydrosphere, and lithosphere (non-living sectors). In addition, in row 10 we have a primary-service sector providing the ultimate source of low-entropy matter-energy, the sun, and, in column (10), the great thermodynamic sink into which finally consumed high-entropy matter-energy goes, forever degraded as devil's dust.[7] The annual flow of low entropy consists of direct solar energy currently received, plus a running down of the stock of low entropy that came from the sun in the distant past. The table records the passage of low-entropy matter-energy through its life-supporting input-output transformations into high-entropy waste. These transformations are not all known or understood, but certainly the scope they offer for non-market interdependence far exceeds the standard examples of externalities in the literature, "somewhat bucolic in nature, having to do with bees, orchards and woods" (Scitovsky, 1954).

Table 2 has thus far been considered only as a descriptive catalogue for economically filing vast amounts of information about the exchanges of economic and ecological commodities making up the total economy of life. Any realistic table would probably have to have at least one hundred sectors, and the resulting ten thousand cells would be pigeonholes for storing measured data about the ten thousand most important exchanges in the total economy of life. Would it be possible to convert the table from a descriptive and heuristic device to a statistical tool, a matrix of technical coefficients useful for planning and prediction—that is, could one do with the whole table what Leontief has done with quadrant (2)?

Each row of Table 2 can be stated as a physical balance equation, thus:

$$\sum_{j=1}^{n} q_{ij} = Q_i; \qquad i = 1, \ldots, n,$$

where i = row and j = column.

Technical coefficients could be defined as $a_{ij} = q_{ij}/Q_j$.

The a_{ij} in quadrant (2) are the usual technical coefficients of the Leontief system, and the a_{ij} in the remaining quadrants are natural technical coefficients. For example, if i is water and j is alfalfa, then a_{ij} would be nine hundred, since it takes nine hundred pounds of water to produce one pound of dried alfalfa (Storer, 1954, p. 96). Assuming all a_{ij} are known, and noting that $q_{ij} = a_{ij}Q_j$, we have the following n equation in n unknowns:[8]

$$\sum_{j=1}^{n} a_{ij}Q_j = Q_i; \qquad i = 1, \ldots, n.$$

[7] Cf. Lotka's (1956, chap. xxiv) concept of the "world engine."

[8] If we separate out household consumption as having no meaningful "technical" coefficients, then we would have n equations in $2n$ unknowns (n of the Q_j and n of the q_{ik}, where k is the household sector). Arbitrarily setting any n of these magnitudes determines the remaining n unknowns. This corresponds to the "open" Leontief model. The assumption of technical coefficients for the household sector would give the "closed" model.

These equations are formally identical to Leontief's quantity table, in which we can sum across rows but not down columns. The assumptions by which Leontief breathes usefulness into this formalism are discussed below and are shown to present no greater theoretical problems for the whole Table 2 than for quadrant (2). To begin, Leontief's basic assumption of constant (slowly changing) technology over time seems to be much closer to the facts for Table 2, since in the non-human economy technical change (evolution) is much slower than in the human economy. Linearity or constant-costs assumptions (a_{ij} constant with respect to Q_j) would seem to be at least equally appropriate as a first approximation. Perhaps this assumption, too, is closer to reality for Table 2, since biological populations grow by adding identical units—hence input-output relations of biological populations are more likely to be proportional to scale (linear) than are such relations for populations of firms (that is, industries) in which new members are never such close replicas of old members. The assumption of single production processes with no joint products appears, at first sight, to be less true for nature than for the human economy. However, this is not all clear, especially if we include bads and free goods as outputs in our traditional production functions. In general, aggregation and classification criteria used in input-output models (similarity of input structure and fixity of proportions among outputs) would remain applicable in the larger table. Certainly no single classification would give a complete representation of the exquisitely tangled web of physical life relations—but then the usual input-output model is also a very incomplete picture of economic relations. Different classifications can be used to serve different limited purposes.

Although there appear to be no theoretical problems in extending the input-output model in this way, there is the obvious practical difficulty that most of the q_{ij} and a_{ij} in quadrants (1), (3), and (4) have never been measured. Nevertheless they all seem to be measurable or at least subject to indirect calculation. Probably the major reason this information has not been acquired is that we have not had many theoretical pigeonholes into which it would fit. Also, the model does not really require a Laplacian knowledge of the universe, as it may appear from the presentation. Application can be confined to a given spatial or conceptual region, with an export row and an import column summarizing relations with the "rest of the world." In any case, application appears rather less utopian than "cost-benefit analysis," which on the slender reed of exchange-value calculations attempts to "maximize the present value of all benefits less all costs, subject to specified restraints" (Prest and Turvey, 1965, p. 4). In fact, something like Table 2 would be necessary for indentifying "all" costs and benefits in the organic sense of Hobson. The construction of such a table would require the co-operation of many disciplines—which may be a point in its favor.

In conclusion, to summarize and support the point of view taken here, I can do no better than to remind the reader of the introductory aphorisms from Ruskin and Isaiah and to quote Lotka (1956, p. 183) one last time:

> For the drama of life is like a puppet show in which stage, scenery, actors and all are made of the same stuff. The players indeed, "have their exits and their entrances," but the exit is by way of translation into the substance of the stage; and each entrance is a transformation scene. So stage and players are bound together in the close partnership of an intimate comedy; and if we would catch the spirit of the piece our attention must not all be absorbed in the characters alone, but must be extended also to the scene, of which they are born, on which they play their part, and with which, in a little while, they merge again.

References

Bates, Marston. *The Forest and the Sea.* New York: Random House, 1960.

Böhm-Bawerk, E. *The Positive Theory of Capital.* Translated by William Smart. New York: G. E. Stechert & Co., 1891.

Boulding, K. E. *A Reconstruction of Economics.* New York: John Wiley & Sons, 1950.

———. *The Skills of the Economist.* Toronto: Clarke, Irwin & Co., 1958.

———. "Economics and Ecology," in F. Frazer Darling and John P. Milton (eds.). *Future Environments of North America.* New York: Natural Hist. Press, 1966.

Carson, Rachel. *Silent Spring.* Boston: Houghton Mifflin Co., 1962.

Georgescu-Roegen, Nicholas. *Analytical Economics.* Cambridge, Mass.: Harvard Univ. Press, 1966.

Henderson, L. J. *The Fitness of the Environment.* Boston: Beacon Press, 1958. Originally published in 1913.

Hobson, J. A. *Economics and Ethics.* Boston: D.C. Heath & Co., 1929.

Leontief, Wassily. *Input-Output Economics.* New York: Oxford Univ. Press, 1966.

Lotka, A. J. *Elements of Mathematical Biology.* New York: Dover Publications, 1956. Previously published under the title *Elements of Physical Biology.*

Marshall, Alfred. *Principles of Economics.* London: Macmillan & Co., 1920.

———. *Memorials of Alfred Marshall.* Edited by A. C. Pigou. London: Macmillan & Co., 1925.

Marx, Karl. *Capital.* Edited by Friedrich Engels. New York: Internat. Publishers, 1967. Reproduction of the English edition of 1887.

Prest, A. R., and Turvey, R. "Cost Benefit Analysis; A Survey," *Econ. J.* (December, 1965), pp. 1–49.

Robinson, Joan. *Economic Philosophy.* London: C. A. Watts & Co., 1962.

Schroedinger, Erwin, *What is Life?* New York: Macmillan Co., 1945.

Scitovsky, Tibor. "Two Concepts of External Economies," *J.P.E.* (April, 1954).

Spengler, J. J. "The Economist and the Population Question," *A.E.R.* (March, 1966).

Storer, John H. *The Web of Life*. New York: Devin-Adair Co., 1954.

Teilhard de Chardin, Pierre. *The Phenomenon of Man*. New York: Harper & Row, 1959.

Teller, Edward. Public address at Louisiana State Univ., Baton Rouge, 1965.

[4]

Production, Consumption, and Externalities

By Robert U. Ayres and Allen V. Kneese*

> "For all that, welfare economics can no more reach conclusions applicable to the real world without some knowledge of the real world than can positive economics" [21].

Despite tremendous public and governmental concern with problems such as environmental pollution, there has been a tendency in the economics literature to view externalities as exceptional cases. They may distort the allocation of resources but can be dealt with adequately through simple *ad hoc* arrangements. To quote Pigou:

> When it was urged above, that in certain industries a wrong amount of resources is being invested because the value of the marginal social net product there differs from the value of the marginal private net product, it was tacitly assumed that in the main body of industries these two values are equal [22][1].

And Scitovsky, after having described his cases two and four which deal with technological externalities affecting consumers and producers respectively, says:

> The second case seems exceptional, because most instances of it can be and usually are eliminated by zoning ordinances and industrial regulations concerned with public health and safety. The fourth case seems unimportant, simply because examples of it seem to be few and exceptional [25].

We believe that at least one class of externalities—those associated with the disposal of residuals resulting from the consumption and production process—must be viewed quite differently.[2] They are a normal, indeed, inevitable part of these processes. Their economic significance tends to increase as economic development proceeds, and the ability of the ambient environment to receive and assimilate them is an important natural resource of increasing value.[3] We will argue below that

* The authors are respectively visiting scholar and director, Quality of the Environment Program, Resources for the Future, Inc. We are indebted to our colleagues Blair Bower, Orris Herfindahl, Charles Howe, John Krutilla, and Robert Steinberg for comments on an earlier draft. We have also benefited from comments by James Buchanan, Paul Davidson, Robert Dorfman, Otto Eckstein, Myrick Freeman, Mason Gaffney, Lester Lave, Herbert Mohring, and Gordon Tullock.

[1] Even Baumol who saw externalities as a rather pervasive feature of the economy tends to discuss external diseconomies like "smoke nuisance" entirely in terms of particular examples [3]. A perspective more like that of the present paper is found in Kapp [16].

[2] We by no means wish to imply that this is the only important class of externalities associated with production and consumption. Also, we do not wish to imply that there has been a lack of theoretical attention to the externalities problem. In fact, the past few years have seen the publication of several excellent articles which have gone far toward systematizing definitions and illuminating certain policy issues. Of special note are Coase [9], Davis and Whinston [12], Buchanan and Stubblebine [6], and Turvey [27]. However, all these contributions deal with externality as a comparatively minor aberration from Pareto optimality in competitive markets and focus upon externalities between two parties. Mishan, after a careful review of the literature, has commented on this as follows: "The form in which external effects have been presented in the literature is that of partial equilibrium analysis; a situation in which a single industry produces an equilibrium output, usually under conditions of perfect competition, some form of intervention being required in order to induce the industry to produce an "ideal" or "optimal" output. If the point is not made explicitly, it is tacitly understood that unless the rest of the economy remains organized in conformity with optimum conditions, one runs smack into Second Best problems" [21].

[3] That external diseconomies are integrally related to economic development and increasing congestion has been noted in passing in the literature. Mishan has commented: "The attention given to external effects in

the common failure to recognize these facts may result from viewing the production and consumption processes in a manner that is somewhat at variance with the fundamental law of conservation of mass.

Modern welfare economics concludes that if (1) preference orderings of consumers and production functions of producers are independent and their shapes appropriately constrained, (2) consumers maximize utility subject to given income and price parameters, and (3) producers maximize profits subject to the price parameters; a set of prices exists such that no individual can be made better off without making some other individual worse off. For a given distribution of income this is an efficient state. Given certain further assumptions concerning the structure of markets, this "Pareto optimum" can be achieved via a pricing mechanism and voluntary decentralized exchange.

If waste assimilative capacity of the environment is scarce, the decentralized voluntary exchange process cannot be free of uncompensated technological external diseconomies unless (1) all inputs are fully converted into outputs, with no unwanted material residuals along the way,[4] and all final outputs are utterly destroyed in the process of consumption, or (2) property rights are so arranged that all relevant environmental attributes are in private ownership and these rights are exchanged in competitive markets. Neither of these conditions can be expected to hold in an actual economy and they do not.

Nature does not permit the destruction of matter except by annihilation with antimatter, and the means of disposal of unwanted residuals which maximizes the internal return of decentralized decision units is by discharge to the environment, principally, watercourses and the atmosphere. Water and air are traditionally examples of free goods in economics. But in reality, in developed economies they are common property resources of great and increasing value presenting society with important and difficult allocation problems which exchange in private markets cannot resolve. These problems loom larger as increased population and industrial production put more pressure on the environment's ability to dilute and chemically degrade waste products. Only the crudest estimates of present external costs associated with residuals discharge exist but it would not be surprising if these costs were in the tens of billions of dollars annually.[5] Moreover, as we shall emphasize again, technological means for processing or purifying one or another type of waste discharge do not destroy the residuals but only alter their form. Thus, given the level, patterns, and technology of production and consumption, recycle of materials into productive uses or discharge into an alternative medium are the only general options for protecting a particular environmental medium such as water. Residual problems must be seen in a broad regional or economy-wide context rather

the recent literature is, I think, fully justified by the unfortunate, albeit inescapable, fact that as societies grow in material wealth, the incidence of these effects grows rapidly . . . " [21]; and Buchanan and Tullock have stated that as economic development proceeds, "congestion" tends to replace "co-operation" as the underlying motive force behind collective action, i.e., controlling external diseconomies tends to become more important than cooperation to realize external economies [7].

[4] Or any residuals which occur must be stored on the producer's premises.

[5] It is interesting to compare this with estimates of the cost of another well known misallocation of resources that has occupied a central place in economic theory and research. In 1954, Harberger published an estimate of the welfare cost of monopoly which indicated that it amounted to about .07 percent of GNP [15]. In a later study, Schwartzman calculated the allocative cost at only .01 percent of GNP [24]. Leibenstein generalized studies such as these to the statement that " . . . in a great many instances the amount to be gained by increasing allocative efficiency is trivial . . . " [19]. But Leibenstein did not consider the allocative costs associated with environmental pollution.

than as separate and isolated problems of disposal of gas, liquid, and solid wastes.

Frank Knight perhaps provides a key to why these elementary facts have played so small a role in economic theorizing and empirical research.

> The next heading to be mentioned ties up with the question of dimensions from another angle, and relates to the second main error mentioned earlier as connected with taking food and eating as the type of economic activity. The basic economic magnitude (value or utility) is service, not good. It is inherently a stream or flow in time . . . [18].[6]

Almost all of standard economic theory is in reality concerned with services. Material objects are merely the vehicles which carry some of these services, and they are exchanged because of consumer preferences for the services associated with their use or because they can help to add value in the manufacturing process. Yet we persist in referring to the "final consumption" of goods as though material objects such as fuels, materials, and finished goods somehow disappeared into the void—a practice which was comparatively harmless so long as air and water were almost literally free goods.[7] Of course, residuals from both the production and consumption processes remain and they usually render disservices (like killing fish, increasing the difficulty of water treatment, reducing public health, soiling and deteriorating buildings, etc.) rather than services. Control efforts are aimed at eliminating or reducing those disservices which flow to consumers and producers whether they want them or not and which, except in unusual cases, they cannot control by engaging in individual exchanges.[8]

I. *The Flow of Materials*

To elaborate on these points, we find it useful initially to view environmental pollution and its control as a materials balance problem for the entire economy.[9] The inputs to the system are fuels, foods, and raw materials which are partly converted into final goods and partly become waste residuals. Except for increases in inventory, final goods also ultimately enter the waste stream. Thus goods which are "consumed" really only render certain services. Their material substance remains in existence and must either be reused or discharged to the ambient environment.

In an economy which is closed (no imports or exports) and where there is no net accumulation of stocks (plant, equipment, inventories, consumer durables, or residential buildings), the amount of residuals inserted into the natural environment must be approximately equal to the weight of basic fuels, food, and raw materials entering the processing and production system, plus oxygen taken from the atmosphere.[10] This result, while obvious

[6] The point was also clearly made by Fisher: "The only true method, in our view, is to regard uniformly as income the *service* of a dwelling to its owner (shelter or money rental), the *service* of a piano (music), and the *service* of food (nourishment) . . . " (emphasis in original) [14].

[7] We are tempted to suggest that the word consumption be dropped entirely from the economist's vocabulary as being basically deceptive. It is difficult to think of a suitable substitute, however. At least, the word consumption should not be used in connection with goods, but only with regard to services or flows of "utility."

[8] There is a substantial literature dealing with the question of under what conditions individual exchanges can optimally control technological external diseconomies. A discussion of this literature, as it relates to waterborne residuals, is found in Kneese and Bower [17].

[9] As far as we know, the idea of applying materials balance concepts to waste disposal problems was first expressed by Smith [26]. We also benefitted from an unpublished paper by Joseph Headley in which a pollution "matrix" is suggested. We have also found references by Boulding to a "spaceship economy" suggestive [4]. One of the authors has previously used a similar approach in ecological studies of nutrient interchange among plants and animals; see [1].

[10] To simplify our language, we will not repeat this essential qualification at each opportunity, but assume it applies throughout the following discussion. In addition, we must include residuals such as NO and NO_2 arising from reactions between components of the air itself but occurring as combustion by-products.

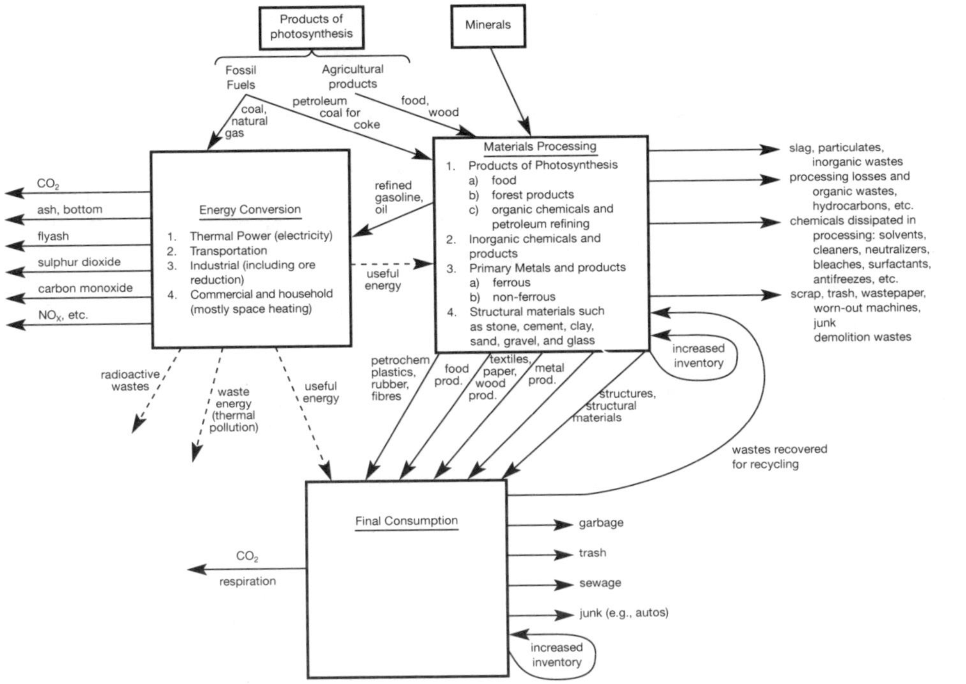

FIGURE 1.—MATERIALS FLOW

upon reflection, leads to the, at first rather surprising, corollary that residuals disposal involves a greater tonnage of materials than basic materials processing, although many of the residuals, being gaseous, require no physical "handling."

Figure 1 shows a materials flow of the type we have in mind in greater detail and relates it to a broad classification of economic sectors for convenience in our later discussion, and for general consistency with the Standard Industrial Classification. In an open (regional or national) economy, it would be necessary to add flows representing imports and exports. In an economy undergoing stock or capital accumulation, the production of residuals in any given year would be less by that amount than the basic inputs. In the entire U.S. economy, accumulation accounts for about 10–15 percent of basic annual inputs, mostly in the form of construction materials, and there is some net importation of raw and partially processed materials amounting to 4 or 5 percent of domestic production. Table 1 shows estimates of the weight of raw materials produced in the United States in several recent years, plus net imports of raw and partially processed materials.

Of the active inputs,[11] perhaps three-quarters of the overall weight is eventually discharged to the atmosphere as carbon (combined with atmospheric oxygen in the form of CO or CO_2) and hydrogen (combined with atmospheric oxygen as H_2O) under current conditions. This results from combustion of fossil fuels and from animal respiration. Discharge of carbon dioxide can be considered harmless in the short run. There are large "sinks" (in the form of vegetation and large water bodies,

[11] See footnote to Table 1.

TABLE 1—WEIGHT OF BASIC MATERIALS PRODUCTION IN THE UNITED STATES PLUS NET IMPORTS, 1963 (10^6 tons)

	1963	1964	1965
Agricultural (incl. fishery and wildlife and forest) products			
Food: Crops (excl. livestock feed)	125	128	130
Food: Livestock	100	103	102
Other products	5	6	6
Fishery	3	3	3
Forestry products (85 per cent dry wt. basis)			
Sawlogs	53	55	56
Pulpwood	107	116	120
Other	41	41	42
Total	434	452	459
Mineral fuels	1,337	1,399	1,448
Other minerals			
Iron ore	204	237	245
Other metal ores	161	171	191
Other nonmetals	125	133	149
Total	490	541	585
Grand total[a]	2,261	2,392	2,492

[a] Excluding construction materials, stone, sand, gravel, and other minerals used for structural purposes, ballast, fillers, insulation, etc. Gangue and mine tailings are also excluded from this total. These materials account for enormous tonnages but undergo essentially no chemical change. Hence, their use is more or less tantamount to physically moving them from one location to another. If this were to be included, there is no logical reason to exclude material shifted in highway cut and fill operations, harbor dredging, land-fill, plowing, and even silt moved by rivers. Since a line must be drawn somewhere, we chose to draw it as indicated above.

Source: R. U. Ayres and A. V. Kneese [2, p. 630].

mainly the oceans) which reabsorb this gas, although there is evidence of net accumulation of CO_2 in the atmosphere. Some experts believe that the latter is likely to show a large relative increase, as much as 50 per cent by the end of the century, possibly giving rise to significant—and probably, on balance, adverse—weather changes.[12] Thus continued combustion of fossil fuels at a high rate could produce externalities affecting the entire world. The effects associated with most residuals will normally be more confined, however, usually limited to regional air and water sheds.

The remaining residuals are either gases (like carbon monoxide, nitrogen dioxide, and sulfur dioxide—all potentially harmful even in the short run), dry solids (like rubbish and scrap), or wet solids (like garbage, sewage, and industrial wastes suspended or dissolved in water). In a sense, the dry solids are an irreducible, limiting form of waste. By the application of appropriate equipment and energy, most undesirable substances can, in principle, be removed from water and air streams[13]—but what is left must be disposed of in solid form, transformed, or reused. Looking at the matter in this way clearly reveals a primary interdependence between the various waste streams which casts into doubt the traditional classification of air, water, and land pollution as individual categories for purposes of planning and control policy.

Residuals do not necessarily have to be discharged to the environment. In many instances, it is possible to recycle them back into the productive system. The materials balance view underlines the fact that the throughput of new materials necessary to maintain a given level of production and consumption decreases as the technical efficiency of energy conversion and materials utilization increases. Similarly, other things being equal, the longer that cars, buildings, machinery, and other durables remain in service, the fewer new materials are required to compensate for loss, wear, and obsolescence—although the use of old or worn machinery (e.g., automobiles) tends to increase other residuals problems. Technically efficient combustion of (desulfurized) fossil fuels

[12] See [30]. There is strong evidence that discharge of residuals has already affected the climate of individual cities; see Lowry [20].

[13] Except CO_2, which may be harmful in the long run, as noted.

would leave only water, ash, and carbon dioxide as residuals, while nuclear energy conversion need leave only negligible quantities of material residuals (although thermal pollution and radiation hazards cannot be dismissed by any means).

Given the population, industrial production, and transport service in an economy (a regional rather than a national economy would normally be the relevant unit), it is possible to visualize combinations of social policy which could lead to quite different relative burdens placed on the various residuals-receiving environmental media; or, given the possibilities for recycle and less residual-generating production processes, the overall burden to be placed upon the environment as a whole. To take one extreme, a region which went in heavily for electric space heating and wet scrubbing of stack gases (from steam plants and industries), which ground up its garbage and delivered it to the sewers and then discharged the raw sewage to watercourses, would protect its air resources to an exceptional degree. But this would come at the sacrifice of placing a heavy residuals load upon water resources. On the other hand, a region which treated municipal and industrial waste water streams to a high level and relied heavily on the incineration of sludges and solid wastes would protect its water and land resources at the expense of discharging waste residuals predominantly to the air. Finally, a region which practiced high level recovery and recycle of waste materials and fostered low residual production processes to a far reaching extent in each of the economic sectors might discharge very little residual waste to any of the environmental media.

Further complexities are added by the fact that sometimes it is possible to modify an environmental medium through investment in control facilities so as to improve its assimilative capacity. The clearest, but far from only, example is with respect to watercourses where reservoir storage can be used to augment low river flows that ordinarily are associated with critical pollution (high external cost situations).[14] Thus internalization of external costs associated with particular discharges, by means of taxes or other restrictions, even if done perfectly, cannot guarantee Pareto optimality. Investments involving public good aspects must enter into an optimal solution.[15]

To recapitulate our main points briefly: (1) Technological external diseconomies are not freakish anomalies in the processes of production and consumption but an inherent and normal part of them. (2) These external diseconomies are quantitatively negligible in a low-population or economically undeveloped setting, but they become progressively (nonlinearly) more important as the population rises and the level of output increases (i.e., as the natural reservoirs of dilution and assimilative capacity become exhausted).[16] (3) They cannot be properly dealt with by considering environmental media such as air and water in isolation. (4) Isolated and *ad hoc* taxes and other restrictions are not sufficient for their optimum control, although they are essential elements in a more systematic and coherent program of environmental quality management. (5) Public investment programs, particularly including transportation systems, sewage disposal, and river flow regulation, are intimately related to the amounts and

[14] Careful empirical work has shown that this technique can fit efficiently into water quality management systems. See Davis [11].

[15] A discussion of the theory of such public investments with respect to water quality management is found in Boyd [5].

[16] Externalities associated with residuals discharge may appear only at certain threshold values which are relevant only at some stage of economic development and industrial and population concentrations. This may account for their general treatment as "exceptional" cases in the economics literature. These threshold values truly were exceptional cases for less developed economies.

effects of residuals and must be planned in light of them.

It is important to develop not only improved measures of the external costs resulting from differing concentrations and duration of residuals in the environment but more systematic methods for forecasting emissions of external-cost-producing residuals, technical and economic trade-offs between them, and the effects of recycle on environmental quality.

In the hope of contributing to this effort and of revealing more clearly the types of information which would be needed to implement such a program, we set forth a more formal model of the materials balance approach in the following sections and relate it to some conventional economic models of production and consumption. The main objective is to make some progress toward defining a system in which flows of services and materials are simultaneously accounted for and related to welfare.

II. *Basic Model*

The take off point for our discussion is the Walras-Cassel general equilibrium model,[17] extended to include intermediate consumption, which involve the following quantities:

resources and services

$$\overbrace{r_1, \cdots\cdots\cdots, r_M}$$

products or commodities

$$\overbrace{X_1, \cdots\cdots\cdots, X_N}$$

resource prices

$$\overbrace{v_1, \cdots\cdots\cdots, v_M}$$

product or commodity prices

$$\overbrace{p_1, \cdots\cdots\cdots, p_N}$$

final demands

$$\overbrace{Y_1, \cdots\cdots\cdots, Y_N}$$

[17] The original references are Walras [28] and Cassel [8]. Our own treatment is largely based on Dorfman *et al.* [13].

The M basic resources are allocated among the N sectors as follows:

$$\begin{aligned} r_1 &= a_{11}X_1 + a_{12}X_2 + \cdots + a_{1N}X_N \\ r_2 &= a_{21}X_1 + a_{22}X_2 + \cdots + a_{2N}X_N \\ &\vdots \\ r_M &= a_{M1}X_1 + a_{M2}X_2 + \cdots + a_{MN}X_N \end{aligned}$$

(1a) or

$$r_j = \sum_{k=1}^{N} a_{jk}X_k \qquad j = 1, \cdots, M$$

In (1a) we have implicitly assumed that there is no possibility of factor or process substitution and no joint production. These conditions will be discussed later. In matrix notation we can write:

$$[r_{j1}]_{M,1} = [a_{jk}]_{M,N} \cdot [X_{k1}]_{N,1} \tag{1b}$$

where $[a]$ is an $M \times N$ matrix.

A similar set of equations describes the relations between commodity production and final demand:

$$X_k = \sum_{l=1}^{N} A_{kl}Y_l \qquad k = 1, \cdots, N \tag{2a}$$

$$[X_{k1}]_{N,1} = [A_{kl}]_{N,N} \cdot [Y_{l1}]_{N,1} \tag{2b}$$

and the matrix $[A]$ is given by

$$[A] = [I - C]^{-1} \tag{3}$$

where $[I]$ is the unit diagonal matrix and the elements C_{ij} of the matrix $[C]$ are essentially the well known Leontief input coefficients. In principle these are functions of the existing technology and, therefore, are fixed for any given situation.

By combining (1) and (2), we obtain a set of equations relating resource inputs directly to final demand, viz.,

$$\begin{aligned} r_j &= \sum_{k=1}^{N} a_{jk} \sum_{l=1}^{N} A_{kl}Y_l = \sum_{k,l=1}^{N} a_{jk}A_{kl}Y_l \\ &= \sum_{l=1}^{N} b_{jl}Y_l \qquad j = 1, \cdots, M \end{aligned} \tag{4a}$$

or, of course, in matrix notation (4b).

$$\text{(4b)}\qquad [r_{j1}]_{M,1} = [a_{jk}]_{M,N}\cdot[A_{kl}]_{N,N}\cdot[Y_{l1}]_{N,1} = [b_{jl}]_{M,N}\cdot[Y_{l1}]_{N,1}$$

We can also impute the prices of N intermediate goods and commodities to the prices of the M basic resources, as follows:

$$\text{(5a)}\qquad p_k = \sum_{j=1}^{M} v_j b_{jk} \qquad k = 1, \cdots, N$$

$$\text{(5b)}\qquad [p_{1k}]_{1,N} = [v_{1j}]_{1,M}\cdot[b_{jk}]_{M,N}$$

To complete the system, it may be supposed that demand and supply relationships are given, a priori, by Pareto-type preference functions:

$$\text{(6)}\quad \text{Demand:}\quad Y_k = F_k(p_1, \cdots, p_N) \qquad k = 1, \cdots, N$$

$$\text{(7)}\quad \text{Supply:}\quad r_k = G_k(v_1, \cdots, v_M) \qquad k = 1, \cdots, M$$

where, of course, the p_j are functions of the v_j as in (5b).

In order to interpret the X's as physical production, it is necessary for the sake of consistency to arrange that outputs and inputs always balance, which implies that the C_{ij} must comprise *all* materials exchanges including residuals. To complete the system so that there is no net gain or loss of physical substances, it is also convenient to introduce two additional sectors, viz., an "environmental" sector whose (physical) output is X_0 and a "final consumption" sector whose output is denoted X_f. The system is then easily balanced by explicitly including flows both to and from these sectors.

To implement this further modification of the Walras-Cassel model, it is convenient to subdivide and relabel the resource category into tangible raw materials $\{r^m\}$ and services $\{r^s\}$:

$$\begin{bmatrix} r_1 \\ r_2 \\ \vdots \\ r_L \end{bmatrix} \text{ becomes } \left.\begin{bmatrix} r_1^m \\ r_2^m \\ \vdots \\ r_L^m \end{bmatrix}\right\} \text{raw materials (units)}$$

$$\begin{bmatrix} r_{L+1} \\ \vdots \\ r_M \end{bmatrix} \text{ becomes } \left.\begin{bmatrix} r_1^s \\ \vdots \\ r_P^s \end{bmatrix}\right\} \text{service (units)}$$

where, of course,

$$\text{(8)}\qquad L + P = M$$

It is understood that services, while not counted in tons, can be measured in meaningful units, such as man-days, with well defined prices. Thus, we similarly relabel the price variables as follows:

$$\begin{bmatrix} V_1 \\ \vdots \\ V_L \end{bmatrix} \text{ becomes } \left.\begin{bmatrix} V_1^m \\ \vdots \\ V_L^m \end{bmatrix}\right\} \text{raw material (prices)}$$

$$\begin{bmatrix} V_{L+1} \\ \vdots \\ V_M \end{bmatrix} \qquad \left.\begin{bmatrix} V_1^s \\ \vdots \\ V_P^s \end{bmatrix}\right\} \text{labor and service (prices)}$$

The coefficients $\{a_{ij}\}$, $\{b_{ij}\}$ are similarly partitioned into two groups,

$$\text{e.g.,}\qquad \begin{matrix} b_{1j} \\ \vdots \\ b_{Lj} \\ b_{L+1,j} \\ \vdots \\ b_{Mj} \end{matrix} \quad \text{becomes} \quad \begin{matrix} b_{1j}^m \\ \vdots \\ b_{Lj}^m \\ b_{1j}^s \\ \vdots \\ b_{Pj}^s \end{matrix}$$

These notational changes have no effect whatever on the substance of the model, although the equations become somewhat more cumbersome. The partitioned matrix notation simplifies the restatement of the basic equations. Thus (1b) becomes (9), while (5b) becomes (10).

$$(9)\qquad M\left\{\begin{bmatrix}\vdots\\ r\\ \vdots\end{bmatrix}\right. = \begin{bmatrix} r^m \\ \cdots \\ r^s\end{bmatrix}\begin{matrix}\}L\\ \\ \}P\end{matrix} = M\left\{\overbrace{\begin{bmatrix}\}L & b^m\\ & \cdots \\ \}P & b^s\end{bmatrix}}^{N}\begin{bmatrix} \\ Y \\ \\ \end{bmatrix}\right\}N$$

$$(10)\qquad \underset{N}{[p_1,\cdots,p_N]} = \underbrace{\underset{L\quad P}{[v^m : v^s]}}_{M}\underbrace{\begin{bmatrix} b^m \\ \cdots \\ b^s\end{bmatrix}}_{N}\Bigg\}M$$

$$= [\cdots v^m \cdots]\begin{bmatrix}\vdots & b^m & \vdots\\ \cdot & \cdots & \cdot\end{bmatrix} + [\cdots v^s \cdots]\begin{bmatrix}\vdots & b^s & \vdots\\ \cdot & \cdots & \cdot\end{bmatrix}$$

The equivalent of (5a) is:

$$(11)\qquad p_k = \underbrace{\sum_{j=1}^{L} b^m_{jk} v^m_j}_{\substack{\text{prices imputed}\\ \text{to cost of raw}\\ \text{materials}}} + \underbrace{\sum_{j=1}^{P} b^s_{jk} v^s_j}_{\substack{\text{prices imputed}\\ \text{to cost of}\\ \text{services}}}$$

where $k = 1, \cdots, N$

We wish to focus attention explicitly on the flow of materials through the economy. By definition of the Leontief input coefficients (now related to materials flow), we have:

$C_{kj}X_j$ (physical) quantity transferred from k to j

$C_{jk}X_k$ quantity transferred from j to k

Hence, material flows *from* the environment to all other sectors are given by:

$$(12)\qquad \sum_{k=1}^{N} C_{0k}X_k = \sum_{j=1}^{L} r^m_j = \sum_{j=1}^{L}\sum_{k=1}^{N} a^m_{jk}X_k = \sum_{j=1}^{L}\sum_{k=1}^{N} b^m_{jk}Y_k$$

using equation (1), as modified.[18] Obviously, comparing the first and third terms,

$$(13)\qquad \underbrace{C_{0k}}_{\substack{\text{total material}\\ \text{flow (0 to }k)}} = \underbrace{\sum_{j=1}^{L} a^m_{jk}}_{\substack{\text{all raw materials}\\ \text{(0 to }k)}}$$

Flows into and out of the environmental sector must be in balance:

$$(14)\qquad \underbrace{\sum_{k=1}^{N} C_{0k}X_k}_{\substack{\text{sum of all raw}\\ \text{material flows}}} = \underbrace{\sum_{k=1}^{N} C_{k0}X_0 + C_{f0}X_0}_{\substack{\text{sum of all return}\\ \text{(waste) flows}}}$$

Material flows to and from the final sector must also balance:

$$(15)\qquad \underbrace{\sum_{k=1}^{N} C_{kf}X_f}_{\substack{\text{sum of all}\\ \text{final goods}}} = \underbrace{\sum_{k=1}^{N} C_{fk}X_k}_{\substack{\text{sum of all}\\ \text{materials}\\ \text{recycled}}} + \underbrace{C_{f0}X_0}_{\substack{\text{waste residuals}\\ \text{(plus accumulation}^{19})}}$$

[18] Ignoring, for convenience, any materials flow from the environment *directly* to the final consumption sector.

[19] For convenience, we can treat accumulation in the final sector as a return flow to the environment. In truth, structures actually *become* part of our environment, although certain disposal costs may be deferred.

Of course, by definition, X_f is the sum of the final demands:

$$(16) \qquad X_f = \sum_{j=1}^{N} Y_j$$

Substituting (16) into the left side of (15) and (2a) into the right side of (15), we obtain an expression for the waste flow in terms of final demands:

$$(17) \quad C_{f0}X_0 = \sum_{j=1}^{N}\sum_{k=1}^{N}(C_{jf} - C_{fj}A_{jk})Y_k$$

The treatment could be simplified slightly if we assumed that there is no recycling per se. Thus, in the context of the model, we could suppose that all residuals return to the environmental sector,[20] where some of them (e.g., waste paper) become "raw materials." They would then be indistinguishable from new raw materials, however, and price differentials between the two would be washed out. In principle, this is an important distinction to retain.

III. *Inclusion of Externalities*

The physical flow of materials between various intermediate (production) sectors and the final (consumption) sector tends to be accompanied by, and correlated with, a (reverse) flow of dollars.[21] However, the physical flow of materials from and back to the environment is only partly reflected by actual dollar flows, namely, land rents and payments for raw materials. There are three classes of physical exchange for which there exist no counterpart economic transactions. These are: (1) private use for production inputs of "common property" resources, notably air, streams, lakes, and the ocean; (2) private use of the assimilative capacity of the environment to "dispose of" or dilute wastes and residuals; (3) inadvertent or unwanted material inputs to productive processes—diluents and pollutants.

All these goods (or "bads") are physically transferred at zero price, not because they are not scarce relative to demand—they often are in developed economies—or because they confer no service or disservice on the user—since they demonstrably do so—but because there exist no social institutions that permit the resources in question to be "owned," and exchanged in the market.

The allocation of resources corresponding to a Pareto optimum cannot be attained without subjecting the above-mentioned nonmarket and involuntary exchanges to the moderation of a market or a surrogate thereof. In principle, the influence of a market might be simulated, to a first approximation, by introducing a set of shadow (or virtual) prices.[22] These may well be zero, where supply truly exceeds demand, or negative (i.e., costs) in some instances; they will be positive in others. The exchanges are, of course, real.

The Walras-Cassel model can be generalized to handle these effects in the following way:

1. One can introduce a set of R common-property resources or services of raw materials $\{r_1^{cp}, \cdots, r_R^{cp}\}$ as a subset of the set $\{r_j\}$; these will have corresponding virtual prices $\{v_j^{cp}\}$, which would constitute an "income" from the environment. Such resources include the atmosphere; streams, lakes, and oceans; landscape; wildlife and biological diversity; and the indispensable assimilative capacity of the environment (its ability to accept and neutralize or recycle residuals).[23]

[20] In calculating actual quantities, we would (by convention) ignore the weight of oxygen taken free from the atmosphere in combustion and return as CO^2. However, such inputs will be treated explicitly later.

[21] To be precise, the dollar flow governs and is governed by a combined flow of materials and services (value added).

[22] A similar concept exists in mechanics where the forces producing "reaction" (to balance action and reaction) are commonly described as "virtual forces."

[23] Economists have previously suggested generalization of the Walras-Cassel model to take account of public goods. One of the earliest appears to be Schles-

2. One can introduce a set of S environmental *disservices* imposed on consumers of material resources, by forcing them to accept unwanted inputs $\{\overset{u}{r_1}, \cdots, \overset{u}{r_s}\}$ (pollutants, contaminants, etc.); these disservices would have negative value, giving rise to *negative* virtual prices $\{u_j\}$.[24]

The matrix coefficients $\{a_{ij}\}$ and $\{b_{ij}\}$ can be further partitioned to take account of this additional refinement, and equations analogous to (9), (10), and (11) can be generalized in the obvious way. Equation (6) carries over unchanged, but (7) must be appropriately generalized to take account of the altered situation. Actually, (7) breaks up into several groups of equations:

$$\overset{m}{r_k} = \overset{m}{G_k}(p_1, \cdots, p_N) \qquad k = 1, \cdots, L \tag{18}$$

$$\overset{s}{r_k} = \overset{s}{G_k}(p_1, \cdots, p_N) \qquad k = 1, \cdots, P \tag{19}$$

However, as we have noted at the outset, the supplies of common-property resources and environmental services or disservices are *not* regulated directly by market prices of other goods and services. In the case of common-property resources, the supplies are simply constants fixed by nature or otherwise determined by accident or noneconomic factors.

The total value of these services performed by the environment cannot be calculated but it is suggestive to consider the situation if the natural reservoir of air, water, minerals, etc., were very much smaller, as if the earth were a submarine or "spaceship" (i.e., a vehicle with no assimilative and/or regenerative capacity). In such a case, all material resources would have to be recycled,[25] and the cost of all goods would necessarily reflect the cost of reprocessing residuals and wastes for reuse. In this case, incidentally, the ambient level of unrecovered wastes continuously circulating through the resource inventory of the system (i.e., the spaceship) would in general be nonzero because of the difficulty of 100 percent efficient waste-removal of air and water. However, although the quantitity of waste products in constant circulation may fluctuate within limits, it cannot be allowed to increase monotonically with time, which means that as much material must be recycled, on the average, as is discarded. The value of common resources plus the assimilation services performed by the environment, then, is only indirectly a function of the ambient level of untreated residuals per se, or the disutility caused thereby, which depend on the cost efficiency of the available treatment technology. Be this as it may, of course, the bill of goods produced in a spaceship economy would certainly be radically different from that we are familiar with. For this reason, no standard economic comparison between the two situations is meaningful. The measure of worth we are seeking is actually the difference between the total welfare "produced" by a spaceship economy, where 100 percent of all residuals are promptly recycled, vis-à-vis the existing welfare output on earth, where resource inventories are substantial and

inger [23]. We are indebted to Otto Eckstein for calling our attention to this key reference.

[24] The notion of introducing the possibility of negative prices in general equilibrium theory has apparently been discussed before, although we are not aware of any systematic development of the idea in the published literature. In this connection, it is worth pointing out the underlying similarity of negative prices and effluent taxes—which have been, and still are being considered as an attractive alternative to subsidies and federal standard-setting as a means of controlling air and water pollution. Such taxes would, of course, be an explicit attempt to rectify an imbalance caused by a market failure.

[25] Any consistent deviation from this 100 per cent rule implies an accumulation of waste products, on the average, which, by definition, is inconsistent with maintaining an equilibrium.

complete recycling need not be contemplated for a very long time to come.

This welfare difference might well be very large, although we possess no methodological tools for quantifying it. In any case, the resource inventory and assimilative capacity of the environment probably contribute very considerably to our standard of living.

If these environmental contributions were paid for, the overall effect on prices would presumably to be push them generally upward. However, the major *differential* effect of undervaluing the environmental contribution is that goods produced by high residual-producing processes, such as papermaking, are substantially underpriced vis-à-vis goods which involve more economical uses of basic resources. This is, however, not socially disadvantageous per se: that is, it causes no misallocation of resources unless, or until, the large resource inventory and/or the assimilative capacity of the environment are used up. When this happens, however, as it now has in most highly industrialized regions, either a market must be allowed to operate or some other form of decision rule must be introduced to permit a rational choice to be made, e.g., between curtailing or controlling the production of residuals or tolerating the effects (disservice) thereof.

It appears that the natural inventory of most common resources used as inputs (e.g., air as an input to combustion and respiratory processes) is still ample,[26] but the assimilative capacity of the environment has already been exceeded in many areas, with important external costs resulting. This suggests a compromise treatment. If an appropriate price could be charged to the producers of the residuals and used to compensate the inadvertent recipients—with the price determined by appropriate Pareto preference criteria—there would be no particular analytic purpose in keeping books on the exchange of the other environmental benefits mentioned, although they are quantitatively massive. We will, therefore, in the remainder of the discussion omit the common-property variables $\{r_j^{cp}\}$ and the corresponding virtual-price variables $\{v_j^{cp}\}$ defined previously, retaining only the terms $\{r_j^u\}$ and $\{u_{jk}\}$. The variable $\{r_j^u\}$ represents a physical quantity of the jth unwanted input. There are S such terms, by assumption, whose magnitudes are proportional to the levels of consumption of basic raw materials, subject to the existing technology. However, residuals production is not immutable: it can be increased or decreased by investment, changes in materials processing technology, raw material substitutions, and so forth.

At first glance it might seem entirely reasonable to assert that the *supplies* of unwanted residuals received will be functions of the (negative) prices (i.e., compensation) paid for them, in analogy with (7). Unfortunately, this assertion immediately introduces a theoretical difficulty, since the assumption of unique coefficients $\{a_{ij}\}$ and $\{C_{ij}\}$[27] is not consistent with the possibility of factor or process substitution or joint-production, as stated earlier. To permit such substitutions, one would have to envision a very large collection of alternative sets of coefficients: one complete set of a's and C's for each specific combination of factors and processes. Maximization of any objective function (such as GNP) would involve solving the entire system of equations as many times as there are combinations of factors and pro-

[26] Water is an exception in arid regions; in humid regions, however, water "shortages" are misnomers: they are really consequences of excessive use of watercourses as cheap means of waste disposal. But some ecologists have claimed that oxygen depletion may be a very serious long-run problem; see Cole [10].

[27] Or $\{b_{ij}\}$ and $\{A_{ij}\}$.

$$[r] = \begin{bmatrix} r^m \\ r^s \\ r^u \end{bmatrix} = M \left\{ \underbrace{\begin{bmatrix} a^m \\ \cdots \\ a^s \\ \cdots \\ a^u \end{bmatrix}}_{N} \begin{bmatrix} X \end{bmatrix} \right\} N = \begin{bmatrix} b^m \\ \cdots \\ b^s \\ \cdots \\ b^u \end{bmatrix} \begin{bmatrix} Y \end{bmatrix} \tag{21}$$

cesses, and picking out that set of solutions which yields the largest value. Alternatively, if the a's and C's are assumed to be continuously variable functions (of each other), the objective function could also, presumably, be parameterized. However, as long as the a's and C's are uniquely given, the supply of the kth unwanted residual is only marginally under the control of the producer, since it will be produced in strict relationship to the composition of the bill of final goods $\{Y_j\}$.

Hence, for the present model it is only correct to assume

$$\overset{u}{r}_k = \overset{u}{G}_k(Y_1, \cdots, Y_N) \tag{20}$$

This limitation does not affect the existence of an equilibrium solution for the system of equations; it merely means that the shadow prices $\{u_{jk}\}$ which would emerge from such a solution for given coefficients $\{a_{ij}\}$, $\{b_{ij}\}$, and $\{C_{ij}\}$ might be considerably higher than the real economic optimum, since the latter could only be achieved by introducing factor and process changes.

Of course, the physical inputs are also related to the physical outputs of goods, as in (21).

Written out in full detail (21) is equivalent to:

(22) raw materials $$\overset{m}{r}_k = \sum_{j=1}^{N} \overset{m}{a}_{kj} X_j = \sum_{j=1}^{N} \overset{m}{b}_{kj} Y_j \qquad k = 1, \cdots, L$$

(23) labor and technical services $$\overset{s}{r}_k = \sum_{j=1}^{N} \overset{s}{a}_{kj} X_j = \sum_{j=1}^{N} \overset{s}{b}_{kj} Y_j \qquad k = 1, \cdots, P$$

(24) unwanted inputs $$\overset{u}{r}_k = \sum_{j=1}^{N} \overset{u}{a}_{kj} X_j = \sum_{j=1}^{N} \overset{u}{b}_{kj} Y_j \qquad k = 1, \cdots, S$$

where, of course,

$$L + P + S = M \tag{25}$$

The corresponding matrix equation for the prices of goods, in terms of production costs, is

$$[p_1, \cdots, p_N] = [v^m \vdots v^s \vdots u] \begin{bmatrix} b^m \\ \cdots \\ b^s \\ \cdots \\ b^u \end{bmatrix} \tag{26}$$

Written out in the standard form, we obtain

$$p_k = \underbrace{\sum_{j=1}^{L} \overset{m}{b}_{jk} \overset{m}{v}_j}_{\text{cost of raw materials}} + \underbrace{\sum_{j=1}^{P} \overset{s}{b}_{jk} \overset{s}{v}_j}_{\text{cost of labor and technical services}} + \underbrace{\sum_{j=1}^{S} \overset{u}{b}_{jk} \overset{u}{v}_j}_{\text{cost (compensation) for providing environmental disservices}} \qquad k = 1, \cdots, N \tag{27}$$

Evidently, the coefficients b_{jk}^{u} are empirically determined by the structure of the regional economy and its geography. It is assumed that a single overall (negative) price for each residual has meaning, even though each productive sector—and even each consumer—has his own individual utility function. Much the same assumption is conventionally made, and accepted, in the case of positive real prices.

All of the additional variables now fit into the general framework of the original Walras-Cassel analysis. Indeed, we have $2N+2M-1$ variables (r_i, Y_i, p_i, v_i) (allowing an arbitrary normalization factor for the price level) and $2N+2M-1$ independent equations.[28] If solutions exist for the Walras-Cassel system of equations, the arguments presumably continue to hold true for the generalized model. In any case, a discussion of such mathematical questions would carry us too far from our main theme.

IV. *Concluding Comments*

The limited economics literature currently available which is devoted to environmental pollution problems has generally taken a partial equilibrium view of the matter, as well as treated the pollution of particular environmental media, such as air and water, as separate problems.[29] This no doubt reflects the propensity of the theoretical literature to view externalities as exceptional and minor. Clearly, the partial equilibrium approach in particular is very convenient theoretically and empirically for it permits external damage and control cost functions to be defined for each particular case without reference to broader interrelationships and adjustments in the economy.

We have argued in this paper that the production of residuals is an inherent and general part of the production and consumption process and, moreover, that there are important trade-offs between the gaseous, liquid, or solid forms that these residuals may take. Further, we have argued that under conditions of intensive economic and population development the environmental media which can receive and assimilate residual wastes are not free goods but natural resources of great value with respect to which voluntary exchange cannot operate because of their common property characteristics. We have also noted, in passing, that the assimilative capacity of environmental media can sometimes be altered and that therefore the problem of achieving Pareto optimality reaches beyond devising appropriate shadow prices and involves the planning and execution of investments with public goods aspects.

We have exhibited a formal mathematical framework for tracing residuals flows in the economy and related it to the general equilibrium model of resources allocation, altered to accommodate recycle and containing unpriced sectors to represent the environment. This formulation, in contrast to the usual partial equilibrium treatments, implies knowledge of all preference and production functions including relations between residuals discharge and external cost and all possible factor and process substitutions. While we feel that it represents reality with greater fidelity than the usual view, it also implies a central planning problem of impossible difficulty, both from the standpoint of data collection and computation.

What, if any, help can the general interdependency approach we have outlined offer in dealing with pollution problems effectively and reasonably efficiently? A minimal contribution is its warning that partial equilibrium approaches, while more

[28] There is one redundant equation in the system, which expresses the identity between gross product and gross income for the system as a whole (sometimes called "Walras law").

[29] See, for example, the essays in Wolozin [29].

tractable, may lead to serious errors. Second, in projecting waste residuals for an economy—a regional economy would usually be the most relevant for planning and control—the inter-industry materials flow model can provide a much more conceptually satisfying and accurate tool for projecting future residuals production than the normal aggregative extrapolations.[30] The latter not only treat gaseous, liquid, and solid wastes separately, but do not take account of input-output relations and the fact that the materials account for the region must balance.

We think that in the next few years it will be possible to make improved regional projections of residuals along the lines sketched above. Undoubtedly, there will also be further progress in empirically estimating external costs associated with residuals discharge and in estimating control costs via various alternative measures. On the basis of this kind of information, a control policy can be devised. However, this approach will still be partial. Interrelations between the regional and national economy must be treated simplistically and to be manageable, the analysis must confine itself to a specific projected bill of goods.

The basic practical question which remains to be answered is whether an iterated series of partial equilibrium treatments—e.g., focusing on one industry or region at a time, *ceteris paribus*—would converge toward the general equilibrium solution, or not. We know of no theoretical test of convergence which would be applicable in this case but, in the absence of such a criterion, would be willing to admit the possible relevance of an empirical sensitivity test more or less along the following lines: take a major residuals-producing industry (such as electric power) and parametrize its cost structure in terms of emission control levels, allowing all technically feasible permutations of factor (fuel) inputs and processes. It would be a straightforward, but complicated, operations research problem to determine the minimum cost solution as a function of the assumed (negative) price of the residuals produced. If possible industry patterns—factor and process combinations—exist which would permit a high level of emission control at only a small increase in power production cost, then it might be possible to conclude that for a significant range of (negative) residuals prices the effect on power prices—and therefore on the rest of the economy—would not be great. Such a conclusion would support the convergence hypothesis. If, on the other hand, electric power prices are very sensitive to residuals prices, then one would at least have to undertake a deeper study of consumer preference functions to try to determine what residuals prices would actually be if a market mechanism existed. If people prove to have a strong antipathy to soot and sulfur dioxide, for instance, resulting in a high (negative) price for these unwanted inputs, then one would be forced to suspect that the partial equilibrium approach is probably not convergent to the general equilibrium solution and that much more elaborate forms of analysis will be required.

[30] Some efforts to implement these concepts are already underway. Walter Isard and his associates have prepared an input-output table for Philadelphia which includes coefficients representing waterborne wastes (unpublished). The recent study of waste management in the New York Metropolitan region by the Regional Plan Association took a relatively broad view of the waste residuals problem [31]. Relevant data on several industries are being gathered. Richard Frankel's not yet published study of thermal power in which the range of technical options for controlling residuals, and their costs, is being explored is notable in this regard. His and other salient studies are described in Ayres and Kneese [2].

References

1. R. U. Ayres, "Stability of Biosystems in Sea Water," Tech. Rept. No. 142, Hudson Laboratories, Columbia University, New York 1967.

2. ——— AND A. V. KNEESE, "Environmental Pollution," in U.S. Congress, Joint Economic Committee, *Federal Programs for the Development of Human Resources*, Vol. 2, Washington 1968.
3. W. J. BAUMOL, *Welfare Economics and the Theory of the State*. Cambridge 1967.
4. K. E. BOULDING, "The Economics of the Coming Spaceship Earth," in H. Jarrett, ed., *Environmental Quality in a Growing Economy*, Baltimore 1966, pp. 3–14.
5. J. H. BOYD, "Collective Facilities in Water Quality Management," Appendix to Kneese and Bower [17].
6. J. W. BUCHANAN AND WM. C. STUBBLEBINE, "Externality," *Economica*, Nov. 1962, *29*, 371–84.
7. ——— AND G. TULLOCK, "Public and Private Interaction under Reciprocal Externality," in J. Margolis, ed., *The Public Economy of Urban Communities*, Baltimore 1965, pp. 52–73.
8. G. CASSEL, *The Theory of Social Economy*. New York 1932.
9. R. H. COASE, "The Problem of Social Cost," *Jour. Law and Econ.*, Oct. 1960, *3*, 1–44.
10. L. COLE, "Can the World be Saved?" Paper presented at the 134th Meeting of the American Association for the Advancement of Science, December 27, 1967.
11. R. K. DAVIS, *The Range of Choice in Water Management*. Baltimore 1968.
12. O. A. DAVIS AND A. WHINSTON, "Externalities, Welfare, and the Theory of Games," *Jour. Pol. Econ.*, June 1962, *70*, 241–62.
13. R. DORFMAN, P. SAMUELSON AND R. M. SOLOW, *Linear Programming and Economic Analysis*. New York 1958.
14. I. FISHER, *Nature of Capital and Income*. New York 1906.
15. A. C. HARBERGER, "Monopoly and Resources Allocation," *Am. Econ. Rev.*, Proc., May 1954, *44*, 77–87.
16. K. W. KAPP, *The Social Costs of Private Enterprise*. Cambridge 1950.
17. A. V. KNEESE AND B. T. BOWER, *Managing Water Quality: Economics, Technology, Institutions*. Baltimore 1968.
18. F. H. KNIGHT, *Risk, Uncertainty, and Profit*. Boston and New York 1921.
19. H. LEIBENSTEIN, "Allocative Efficiency vs. 'X-Efficiency,'" *Am. Econ. Rev.*, June 1966, *56*, 392–415.
20. W. P. LOWRY, "The Climate of Cities," *Sci. Am.*, Aug. 1967, *217*, 15–23.
21. E. J. MISHAN, "Reflections on Recent Developments in the Concept of External Effects," *Canadian Jour. Econ. Pol. Sci.*, Feb. 1965, *31*, 1–34.
22. A. C. PIGOU, *Economics of Welfare*. London 1952.
23. K. SCHLESINGER, "Über die Produktionsgleichungen der ökonomischen Wertlehre," *Ergebnisse eines mathematischen Kolloquiums*, No. 6. Vienna, F. Denticke, 1933.
24. D. SCHWARTZMAN, "The Burden of Monopoly," *Jour. Pol. Econ.*, Dec. 1960, *68*, 627–30.
25. T. SCITOVSKY, "Two Concepts of External Economies," *Jour. Pol. Econ.*, Apr. 1954, *62*, 143–51.
26. F. SMITH, *The Economic Theory of Industrial Waste Production and Disposal*, draft of a doctoral dissertation, Northwestern Univ. 1967.
27. R. TURVEY, "On Divergencies between Social Cost and Private Cost," *Economica*, Nov. 1962, *30*, 309–13.
28. L. WALRAS, *Elements d'economie politique pure*, Jaffé translation. London 1954.
29. H. WOLOZIN, ed., *The Economics of Air Pollution*. New York 1966.
30. CONSERVATION FOUNDATION, *Implications of Rising Carbon Dioxide Content of the Atmosphere*. New York 1963.
31. REGIONAL PLAN ASSOCIATION, *Waste Management*, a Report of the Second Regional Plan. New York 1968.

[5]

ENVIRONMENTAL QUALITY AND INTERNATIONAL TRADE

Ralph C. d'Arge and Allen V. Kneese

RALPH C. D'ARGE is associate professor of economics, University of California, Riverside, and ALLEN V. KNEESE is director, Quality of the Environment Program, Resources for the Future, Inc., Washington. The authors would like to express their indebtedness to E. Brook, O. Bubik, S. Follmer, T. Munnecke, and W. Schulze.

I. INTERNATIONAL ASPECTS OF ENVIRONMENTAL PROBLEMS

The Nature of International Environmental Problems

Environmental problems are currently a matter of international interest and concern for a wide variety of reasons ranging from those which are rather general and even vague to others which are concrete and sometimes very pressing.[1] There are, hopefully, more than a few people who are concerned about the welfare of their fellow man wherever he may be; moreover, there is an evident feeling of "one-worldness" resulting from an increasing degree of interdependence in several spheres, including the economic and the cultural. Further, the developed countries feel that they can learn useful lessons from one another about how to cope with environmental problems which are quite similar from country to country. These problems usually result from high population densities combined with large and rapidly growing per capita production and consumption which are often brought about by the use of technologies that generate large amounts of destructive residual materials. There is also an almost universal absence of satisfactory institutions for collective management of a number of "common property resources" including the air mantle, watercourses, and other large ecological systems. Interest in study-

[1] The matters passed over lightly in this section are discussed in some detail in Clifford S. Russell and Hans H. Landsberg, "International Environmental Problems—A Taxonomy," *Science,* June 25, 1971 (Vol. 172, No. 3990), pp. 1307-1314; Marion Clawson, "Economic Development and Environmental Impact: International Aspects" (Paper prepared for the Symposium on Political Economy of Environment—Problems of Method, Ecole Pratique des Hautes Etudes, Paris, July 1971); and William Baumol, *Environmental Protection, International Spillovers, and Trade* (Wicksell Lectures 1971) (Stockholm: Almquist and Wiksell, 1971).

ing these problems in a cooperative manner and exchanging information about them is evident in much of the current activity of the World Health Organization (WHO), the Economic Commission for Europe (ECE), the Organization for Economic Co-operation and Development (OECD), the North Atlantic Treaty Organization (NATO), and other international institutions, and, no doubt, the 1972 United Nations Conference on the Human Environment will occupy itself considerably with them.

Another primary source of concern is the pressingly evident fact that economic activities in one country are producing direct "spillover" or, in economic terms, "external" effects on one or more other countries. Examples are legion: the salt pollution of the Rio Grande in the United States which adversely affects agriculture in Mexico; the virtually unrestricted discharge by many countries of residuals into the Baltic Sea which is already in a most precarious ecological state; the progressive and accelerating degradation of the Rhine; and the discharges of sulfur oxides in the Ruhr area and elsewhere in northern Europe which are causing "acid rain" in Scandinavia. One of the most distressing features of the present world condition is that although many of these are clear and pressing problems, advanced countries seem totally unable to develop effective international measures to deal with them. This itself should not be very surprising, however, since many of those advanced countries appear incapable of coping even with their internal environmental problems.

While many specific regions of the world are afflicted with more or less uncontrolled pollution problems, there are some truly global problems which exist either in fact or at least as potential problems. The possible climatic effects of man's already detectable alteration of the atmosphere's chemical properties have been amply publicized although they are not well understood scientifically. More pressing are the disturbances associated with the introduction of "exotic" and very persistent substances into large ecosystems: DDT, mercury, and other substances have already aroused concern.

These are all salient and important reasons for trying to achieve international monitoring and information exchange and for working toward agreements on the management of substances actually or potentially deleterious to both regional and global natural environments.

Other more subtle international problems related to the natural environment result from demands (or preferences) for goods produced in one country on the part of another country's citizens. Sometimes these linkages are not expressible through markets; for example, substantial numbers of people throughout the world may feel quite unhappy if a unique natural feature like Lake Baikal is irreversibly changed or if an animal species is driven to extinction, even though these events occur wholly within another country. In instances like these discontent is not individually expressible through markets. If people's preferences are to count at all in these cases, they must be expressed through collective action and international agreement. In a vast number of other cases the preferences for goods produced in one country on the part of citizens in another, and vice versa, are both expressible and expressed through markets. These preferences may result from qualitative or price differences with respect to domestically produced goods, and the result of the expression of those preferences is, of course, international trade.

This discussion shows that both environmental preferences and dispersal of residuals into the environment may occur on an international scale. We do not intend to treat transnational and global residuals problems in any depth here. Rather, we will deal rigorously and empirically only with those problems (or apparent problems) associated with the influence on international trade of environmental controls instituted within a country. Also, we limit our consideration to pollution control or "residuals management" problems although our general approach and conclusions apply to other types of environmental disruption. We appreciate not only that this is but a part of the problem but also that many countries and industries regard it as an extremely important one. We comment about other aspects of environmental controls only in general terms.

Impact of Environmental Controls on International Trade of Major Trading States

The major trading states are clearly concerned about the effect that environmental controls in their own countries or in other states may have on their international trade position, on their national income, and on the prosperity of specific industries in their economies. This set of issues was at the center of discussion during the January 1971 Atlantic Council meetings.[2]

One fear is that cost increases which home industries may have to sustain because of environmental controls will adversely affect both the industry's and the state's international trade position, the level of real national income, and long-term comparative advantage. This fear becomes particularly acute when it is suggested that one trading country move to curb environmental degradation without coordinate action by others or when it is thought that other countries will subsidize the environmental controls instituted by industry while

[2] See Allen V. Kneese, Sidney E. Rolfe, and Joseph W. Harned, eds., *Managing the Environment: International Economic Cooperation for Pollution Control* (New York: Praeger, 1971).

the country in question will not subsidize but perhaps even charge or tax them for any environmental disruptions.

The latter possibility is rapidly becoming a genuine problem in the United States and a few other countries. There is now a substantial theoretical and empirical literature in the United States which concludes that from the domestic economy's point of view "effluent charges" or taxes are the most effective and efficient means for achieving control of industrial residuals discharges.[3] Moreover, and politically much more important, the "conservation" groups, which previously opposed the effluent charges technique, are now solidly behind it.[4] There are also proposals to tax particular commodities which cause notable environmental problems; a tax on containers which are particularly resistant to reuse is an example.

Other impacts on international trade may result from direct controls imposed in the interest of environmental protection on either emissions of residuals or on the qualitative characteristics of commodities. For example, if the conventional internal combustion engine remains the power source for new automobiles, it appears that foreign manufacturers, who typically use small engines in their cars, will have a disproportionately difficult time meeting increasingly strict emissions standards relative to the problems associated with large understressed engines in American cars. Moreover, they may be faced with the problem of producing different cars for different markets with relatively short production runs of each type. Such disproportionate effects raise both the possibility and the fear that alleged environmental controls may be used as restrictive trade practices. A related example is provided by quality control on pharmaceuticals in France. The French require inspection of production processes and by this means have effectively frozen foreign manufacturers out of the market. We discuss the various policy options which a country might adopt in domestic environmental protection further in section II.

Some Special Concerns of the Developing Countries

Developing countries may also have some special cause for concern. Due to the costs of residuals disposal in developed countries pressures for more efficient materials use through increased recycling and other means may affect the developing countries' economic development adversely if they are still heavily dependent on the export of

[3] See Allen V. Kneese and Blair T. Bower, *Managing Water Quality: Economics, Technology, Institutions* (Baltimore, Md: Johns Hopkins Press [for Resources for the Future], 1969), chapters 5, 6.

[4] United States Congress, Subcommittee on Priorities and Economy in Government, Joint Economic Committee, *Hearings, Economic Incentives to Control Environmental Pollution,* 92nd Cong., 1st sess., July 12, 1971.

basic resource commodities. It appears somewhat doubtful, however, that the efficiency of materials used in developed countries could grow to such an extent that the international demand for these commodities would actually be reduced, at least in the next few decades. But it is quite plausible that the demand might grow at a lesser rate than now anticipated.

There are other fears in developing countries that environmental policies in developed countries will adversely affect their trade positions, international assistance, and the transfer of technology.[5] One aspect of this thinking is a fear that cost increases for environmental controls in developed economies will give rise to "neoprotectionism" as a means to protect home industries from those in developing countries whose environmental standards are lower. Agricultural products might be especially vulnerable. For example, nontariff barriers may be erected against developing countries' products carrying some small environmental hazard which is unacceptable to the developed country; the banning in some European countries of fruits and vegetables carrying traces of DDT is one such case. Another aspect is the thought that government subsidies for pollution control in developed countries will reduce funds for foreign aid. Further, there is concern that in providing aid the developed countries will unfairly or unthinkingly press their own environmental standards, instituted for domestic environmental protection, onto developing countries. Finally, there is concern that the kind of technology transferred from the developed to developing countries will be effected in a manner adverse to development in more economically primitive circumstances. For example, technology might develop in the direction of engine types in which considerable cost is incurred to keep emissions low for domestic purposes but lower than developing countries would find optimal.

There are, of course, possible influences of environmental concerns in developed countries favorable to developing countries. For instance, the concern for global pollution may help to stimulate a "one-world" feeling which could extend to a concern for poverty all over the world and which might lend support to projects and programs in the human resources sector which have been rather neglected in aid programs. Suspicion or demonstration of the pollution effects of synthetics may rebound to the benefit of natural resources products of some developing countries. Perhaps most important of all, controls on high residuals

[5] These issues are examined to some degree in a draft report *Development and Environment,* submitted by a panel of experts convened by the secretary-general of the United Nations Conference on the Human Environment, Founex, Switz., June 4-12, 1971. (Mimeographed.) Our summary here is based largely on this report. A discussion of the impacts of environmental controls on the balance of payments and the national income of developing countries is offered in section IV.

industries in developed countries will tend to stimulate their growth in developing countries where controls are less strict. This relates to basic materials as well as processing industries. For example, present technologies used in the production of basic copper generate sulfur oxides almost pound for pound with copper, and control technologies are costly. Emissions controls or taxes in the developed countries would tend to stimulate the copper market in the less developed countries. This raises the issue of "exporting pollution," discussion of which has so far generated a lot more heat than light. Many of the fears, concerns, and positive possibilities surrounding relationships between environmental policies in developed countries and rates of development in less developed areas are so ill formed and vague that little more can be said about them. Our further analysis will have some bearing, however, on the influence of environmental controls on trade patterns, and we will have more to say on the "exporting pollution" issues.[6]

Analysis of International Trade Impacts at the Macroeconomic and the Microeconomic Levels

The two levels at which international trade impacts of environmental controls may appropriately be considered have already been implied by the above discussion. On the one hand, it is pertinent to consider the "macroeconomic" effects on the balance of payments, national income, and employment. On the other, impacts on specific industries are of concern both to the industries themselves and to the government of the states in which they manufacture. It should be pointed out that there is no sharp distinction between macro and micro since what occurs cumulatively at the micro level determines the extent of macro adjustments. Our primary focus in this article is on the macroeconomic level, but we will also have some comments to make about the microeconomic level and comparative advantage in the next

[6] Another broad question, of course, is whether environmental protection really deserves the low status the developing countries are giving it in their domestic priorities. Our view is that in many instances it does not. Developing countries, especially in the western hemisphere, exhibit great diversity internally from region to region. In addition to economically primitive backcountry with poor environmental conditions related to extreme poverty and often ignorance of even basic sanitation they also have the dubious distinction of being centers of some of the worst environmental pollution in the world: Sao Paulo and Mexico City are excellent examples. Moreover, one of the most important characteristics of these situations is the rate at which they are deteriorating. Some measure of this is given by rates of population growth for these cities which range up to 500,000 people, net, per year. It seems that these facts are neglected by those persons in developing countries who claim that "this is not a problem for us." On the contrary, inclusion of environmental considerations in development planning appears to be a matter of some urgency.

section.

In section III we turn to 1) a discussion of an econometric model which has been constructed to assist in the analysis of macroeconomic effects for several countries and 2) a presentation of some numerical results obtained from using the model in connection with various assumptions concerning domestic environmental policy. Measures of macroeconomic impacts are made when unilateral action on environmental controls is undertaken by a particular country (which raises costs in its export industries) while no other countries take action. We also experiment with various combinations of multilateral action. We conclude with discussions of ways to mitigate adverse effects on domestic industries resulting from losses in competitive international advantage and problems of separating legitimate governmental actions directed toward improving environmental quality from those that are taken for other objectives.

II. NATURE AND MAGNITUDE OF IMPACTS AT THE MICRO LEVEL: POSSIBLE EFFECTS ON SPECIFIC INDUSTRIES AND COMPARATIVE ADVANTAGE

Domestic Civil Costs of Residuals Discharge

In the introductory section we indicated, in general terms, the relationship between the common property characteristic of resources like the air mantle, watercourses, and ecological systems and the problem of environmental pollution. These resources have not been and in many cases could not be converted to private property which could be exchanged in markets to establish its value in alternative uses. Accordingly, unless some sort of public policy prevents it (and we consider later what sorts of public policies might do so) these resources are the recipients of residuals from extractive, processing, and consumption activities that are discharged without regard to their destructive effects. A residual's discharge may have a highly degrading effect on the ability of a watercourse to support a valuable ecological system or on the air's life-support capacity. But since these services of the common property resource are not priced, the residuals discharger does not have to pay for his discharge; in fact, he has no incentive to even take them into account. Thus, a kind of real cost goes unmet. It is called variously, in the literature, an external diseconomy, an external cost, or a social cost. Several results follow from this situation: Too much residual is discharged, too much residual is generated which is either treated or controlled too little, and the price of the good which is finally sold to consumers is too low. A public policy must be devised to cancel these effects if we are to have an efficient economy in which all social costs are met when a decision is made to undertake an extractive, processing, or consumption activity.

The distinction between extractive and processing industries, on the one hand, and consumption activities, on the other, is important in connection with international trade issues. If a consumption activity gives rise to external costs, it is the external cost-generating aspect which must be controlled or managed, and it is a matter of indifference whether the good whose services are being consumed is of foreign or domestic origin. For example, restrictions on automobile exhausts imposed by a given country or region within a country would apply equally to automobiles of foreign or domestic manufacture. Another example is a "disposal charge" levied on automobiles. Again, any distinction made would not be contingent on whether the automobile is foreign or domestic. Moreover, if an automobile is exported, domestic emission standards should not govern, nor should the disposal fee be imposed on it for, in this case, the external cost associated with its use will occur in the destination country which may or may not wish to control it. Thus, controls on consumption-related externalities are simliar to excise taxes levied for domestic purposes. They apply equally to foreign and domestically produced items and should not be levied on exported items. Only when an externality-producing good is imported but not produced domestically would a tax or other control on it come to resemble a tariff in the trade-restrictive sense. But, as long as this restriction reflects a genuine domestic environmental objective, the resemblance is only superficial for what is involved is simply an effort to equate domestic social and private costs.

The situation is quite different in the case of an external cost associated with an extractive or processing activity. When these costs are made internal, they become "embodied" in the price of goods to which they furnish material or component inputs, and an effort to isolate those costs which are "attributable" to environmental protection resembles unscrambling an omelet. But the difference is deeper than the mere difficulty of isolating a cost component. The incidence of the external or social costs in this case is where the goods are produced and not where they are consumed. Thus, for example, external costs associated with the discharge of biologically active or chemical residuals from a pulp and paper plant or costs associated with reducing such discharges are social costs in the country where the paper is produced whether or not it is consumed domestically.

From this discussion we see among other things that a tariff-like, excise-type levy on a foreign-produced good as it enters the country (one not applicable to a similar domestically produced good) would be an appropriate environmental protective device only in rare instances. This would only be true if a similar good were not produced domestically, if it were not possible to tax or otherwise appropriately restrict the externality-generating activity itself (e.g., tax auto emissions rather than autos), and if the appropriate tax were uniform nationally. Thus, tariffs imposed with the nominal purpose of environmental protection should be viewed with suspicion as in fact being trade restrictions.

Effluent Charges, Standards, and Subsidies within a Domestic and an International Context

We have spoken in general terms about internalizing external costs, but now we must confront the question of how this might be done more directly. In doing this we neglect the matter of consumption-related externalities because, insofar as relationships to international trade issues are concerned, they have been adequately described above.

There are really two steps involved in approaching this question. The first involves the matter of how much environmental capacity is to be allotted to residuals discharge, and the second is the question of how this available supply is to be distributed among residuals dischargers. The process whereby the first is determined involves some sort of weighting of the costs of controlling residuals discharge against the external cost or damage which can be avoided by reducing it. In practice this process is usually crude, with a large intuitive or judgmental element.

Once a permissible amount of environmental capacity is allotted to residuals discharge, its distribution among actual and potential residuals dischargers must be determined. In general, economic efficiency demands that this capacity be distributed in such a way that those potential users to which it is most valuable are able to use it most. In other words, those that find it least costly (everything considered) to reduce their waste discharge should reduce it most and those who find it most costly should reduce it least. If we put aside many qualifications, the economic criterion is that the incremental costs of reducing residuals discharge must be equated for all dischargers.

It follows from this discussion that two major types of errors may occur in residuals management policy within a domestic or an international context: 1) Too much or too little environmental capacity may be allotted to residuals discharge; and 2) the available capacity may be ill distributed within and among countries. An extreme case of the former occurs when no environmental control is exercised. In this case all costs remain external and most of these are visited directly upon the populations as consumers — in the form of poor health, reduced aesthetic value of the environment, or lower productivity. The second type of error involves needlessly high costs of attaining the environmental standards and consequent unnecessary restrictions on the production of other goods and services. The international trade implications of these errors are discussed below.

As a means of diminishing the second type of error and at the same time reflecting the external costs of production in the various salient decisions economists have long advocated the use of publicly administered charges and taxes on external cost producing activities. The level of the charge would be set so that waste dischargers would have an economic incentive to cut back just enough so that the allotted environmental capacity would not be exceeded. This would have the effect of providing an incentive to those that can control residuals most cheaply to do so.[7]

Of course, other techniques can be used to limit the use of environmental capacity. In principle, a payment could be made to a waste discharger per unit of discharge withheld. This is a rather idealized subsidy scheme. The general conclusion is that informational and administrative problems make it virtually impossible to obtain an efficient result with this type of payment.[8] Omitting for the moment considerations of long-run adjustments and information-administrative costs associated with subsidies, we can make several statements with regard to effluent charges and subsidies within an international context. First, if subsidies are offered strictly for pollutant emission reductions, effluent charges and effluent subsidies or payments will achieve the same degree of short-run domestic efficiency in reduction of emissions.[9] The question of whether this result still holds when international trade is introduced is answered by a qualified yes. Since the firm is presumed to view the subsidy as an opportunity cost of production, it will adjust production downward so that the marginal loss associated with waste residuals reduction is equated with marginal revenue gains from the subsidy for the reduction. If all firms within a country respond in this manner, however, the domestic industry's output will be reduced and prices will thereby rise. That is, prices will rise provided that the domestic industry produced more than an insignificant share of world consumption of the commodity. Thus, if the commodity is being produced domestically and exported, an outcome of increased domestic and export prices will occur which is identical in immediate impact to an efficient effluent charges system. The same result holds for domestic substitutes for imported commodities. Effluent subsidies will also increase domestic

[7] There is considerable economic literature on this. See, for example, Allen V. Kneese, "The Political Economy of Pollution," *American Economic Review*, May 1971 (Vol. 61, No. 2), pp. 153-166.

[8] See the discussion in Kneese and Bower.

[9] This has been demonstrated in ibid., chapter 5.

opportunity costs of producing imports with the same immediate impact on comparative advantage of producing the imported good domestically as with implementation of effluent charges. However, while the immediate theoretical impact is identical, long-run impacts are likely to be extremely different. This is due to changes in relative profitability of a particular industry between countries which will induce shifts in international investment and perhaps even international relocation of existing capital and labor.[10]

The effect of such differences in profitability that emerge, in the long run, from the use of domestic effluent subsidies rather than charges may be a serious misallocation of international investment, particularly if all countries did not adopt the subsidies approach. International investment would probably accumulate in those countries willing to continuously maintain a subsidy for such investment which may or may not have any direct relationship to productivity of capital, at the margin, between countries. Thus, long-run efficiency from the global viewpoint, by utilizing domestic subsidies which are geared to emission reductions, is conceptually infeasible.

A subsidy device frequently used is to meet part of the cost of constructing (and possibly operating) waste treatment plants from general public funds. This procedure has several severe drawbacks, including the fact that it tends to bias the discharge-reducing technology selected in favor of waste treatment plants even though other methods for reducing residuals generation would often be cheaper. More generally it makes it impossible to equate social and private costs, which we have already argued is necessary for efficiency in the domestic economy. Moreover, this form of subsidy does not by itself provide an incentive to reduce waste discharge. Even though the government pays part (say half) of the cost of a waste-treatment plant, it is still cheaper from the private viewpoint of the residuals generator to dump them into the "commons" without reductions. Thus, this system undoubtedly reinforces the inefficiencies inherent in schemes to enforce direct restriction on residuals discharge. Nevertheless, the combination of treatment subsidies and enforcement is now in (highly imperfect) use in the United States and is probably the most common single approach used elsewhere.

Enforcement of specified residuals discharge controls has suffered from great difficulties in practice. The legal processes used have been

[10] There is some evidence, though meager, that long-term direct investment flows are quite responsive to short-term changes in relative profitability. See R. C. d'Arge, "Customs Unions and Direct Foreign Investments: A Correction and Further Thoughts," *Economic Journal*, June 1971 (Vol. 81, No. 322), pp. 352-355.

cumbersome, expensive, and very time-consuming.[11] Even if these problems could be largely overcome, efficient allocation of permissible environmental capacity would tend to be precluded by other factors. Problems of information, administrative complexities, and equity usually push this approach in the direction of applying uniform discharge standards to all dischargers. This effectively forecloses an efficient allocation of permissible capacity for reasons we have discussed above. We briefly consider the international trade implication of this problem in the succeeding sections.

Some Observations on Externalities, Welfare, and International Trade

One of the most important welfare propositions in the international trade literature is that given a world composed of sovereign states independent of each other except through international market exchanges, world welfare will be maximized if domestic and international prices fully reflect the social costs of production and consumption for all states.[12] This proposition is central to the often-stated belief that tariffs are undesirable. Tariffs have the effect of distorting the relationship between prices and costs of production for individual countries. Such distortions do not allow either domestic or international markets to establish prices which reflect production costs. Consequently, resources such as labor, capital, and land may be used in the domestic production of commodities that are protected by tariffs and which could be produced more cheaply abroad. With full employment of all resources worldwide, domestic resources, if reemployed in producing nonprotected domestic commodities or exports, could yield a net increment to world output. That is, without tariffs foreign resources would be committed to producing those commodities for which they have a comparative advantage while domestic resources would be employed in producing commodities for export and domestic consumption which can be produced at least cost domestically. For example, if country x can produce commodity A at half the cost of resources compared with country y, but country x can only produce commodity B with equal resource costs as country y, country x has a comparative advantage in producing commodity A. Without tariffs x will obviously import commodity B and export commodity A. If x institutes a tariff rate of above 100 percent on commodity B and places no restrictions on domestic producers as to the mix of domestic production between A and B, and y does not retaliate with its own

[11] See *Water Wasteland* (Report of Ralph Nader's Task Force on Water Pollution) (Washington: Center for the Study of Responsive Law, 1971).

[12] See P. A. Samuelson, "The Gains from International Trade Once Again," *Economic Journal,* December 1962 (Vol. 72, No. 288), pp. 820-829.

tariffs, then producers in x will shift resources to production and export of commodity B and attempt to import commodity A. The net result will be that the combined output of x and y in producing A and B will fall. In the extreme case in which all resources in country x are transferred to the production of B and all resources in country y are utilized in producing commodity A, and each country contains an equal endowment of resources, world production of B will remain constant but production of A will decline by 50 percent. While extremely simple, this example illustrates how tariffs distort the global allocation of resources and thereby reduce the level of global output.[13] In reducing global output and efficiency of resource use tariffs are presumed to also reduce world welfare since world output could be increased and distributed by tariff removal in such a way that no country would be made worse off while at least one country would be made better off.

External diseconomies associated with waste residuals are like tariffs in creating distortions in international markets and thereby in affecting world output and welfare. A domestic external diseconomy which has not been internalized yields a separation between private production costs and social costs meaning that the real production costs are not reflected in domestic or export prices. The effect of such a separation on world welfare is precisely the same as that of tariffs. Since resources are inefficiently allocated both domestically and internationally, at least one country can be made better off without harming others (given suitable redistribution of goods) by correcting for the external diseconomy. Note, however, that external diseconomies involve the production of "bads" which are unpriced by domestic or international markets. Therefore, unlike tariff removal, internalizing external diseconomies may result in reduced production and trade of commodities and therefore a *reduction* in world output. But this reduction is only in output of commodities already traded and not in output of such untraded goods as domestic air or water quality.

Tariff and Externality Distortions

Import tariffs or exports subsidies cause distortions in two ways:

[13] While tariffs tend to reduce global output and efficient utilization of resources viewed globally, tariffs may improve the welfare position of any individual country, especially if there is little or no tariff retaliation by other countries. As a simple example, if country x placed a tariff on comodity B and this action caused y to demand less of commodity A but pay more for it in terms of commodity B, i.e., inelastic demand for commodity B in country y, then country x may be made better off. This occurs because x receives in addition to the tariff revenue on B a greater amount of B per unit of A exported which is large enough to offset the loss in revenue from reduced exports of A.

1) They drive a wedge between domestic and international costs of production at the margin; and 2) they induce a distortion between domestic valuation of commodities and their foreign valuation by consumers. Domestic prices do not adequately reflect costs of production *or* consumer valuations abroad. Unlike monopolistic factor or product markets which only distort the production side of domestic and international markets, tariffs induce distortions from both production and consumption forces in domestic and international markets. International differences in productivity, factor availability, and technology are not allowed to be fully reflected in cost differences between states when tariffs are introduced. Consequently, the correct price signals are not available to domestic producers. The result is that domestic producers tend to produce smaller quantities of exports and larger quantities of commodities competing with imports than they would if tariffs were not imposed. Also, consumers responding to relatively lower prices of commodities produced domestically consume too small a quantity of imports and excessive amounts of the import-competing commodity produced domestically.

When external diseconomies arise domestically in the production of goods normally imported, the distortion in domestic production and consumption patterns is similar to that of tariffs. The major difference between tariffs and external diseconomies in domestic production is that tariffs affect price signals while external diseconomies affect cost signals to domestic producers. Domestic production costs of import-type commodities are lower since environmental services are unpriced domestically. This induces too high a rate of domestic production of commodities normally imported and too low a level of production of exports, which is identical in direction of effect to import tariffs. Likewise, because of lower domestic prices for imports domestically produced, domestic demand is too large. Further, because the external diseconomy results in unpriced damages to consumers, consumer patterns of consumption are distorted in the domestic economy. Thus, import tariffs and external diseconomies in domestic production of import-type commodities are similar in their effect in distorting both production and consumption patterns.

External diseconomies in the domestic production of exports (and not in imports) are similar to import subsidies or export tariffs in impact on production and consumption. Export prices due to non-priced environmental services used in their production will be lower than they should be both in domestic and in international markets. The result will be excessive production of exports and excessive domestic consumption of commodities which should be exported. In the case in which both imports and exports have identical external diseconomies associated with their domestic production, domestic production and consumption of both exports and imports will be too large. Finally, when external diseconomies arise in both domestic and foreign production processes, little can be said regarding excessive domestic production of export *or* import type of commodities without detailed information on the relative magnitude of the external diseconomies encountered both domestically and abroad.

The foregoing statements lead us to the conclusion that external diseconomies are structurally more like tariff distortions than other types of domestic distortions (e.g., the presence of monopolies in product or factor markets). This is because uncorrected external diseconomies, like tariffs, distort price and cost signals to both domestic producers and consumers whereas domestic distortions in factor and/or product markets distort domestic production patterns but not domestic consumer responses to competitively established international prices.[14] In addition, and perhaps most importantly, the distortions are likely to be more pervasive on the consumption side with external diseconomies than with tariffs. The reason is simply that receptors or parties damaged by waste residuals stemming from production are generally consumers, particularly in highly urbanized locations. In the absence of a suitable negotiating mechanism consumers are unable to register their preferences for less waste residuals through payments or other means. In consequence public "bads" go unpriced and thereby encourage excessive expenditure on other commodities. On the production side external diseconomies also may distort productive efficiency since, with one factor price (for environmental assimilative capacity) constrained to zero, factors of production will not be allocated efficiently. Unlike tariffs, unregulated external diseconomies induce distortions in the economy's efficiency beyond distortions in the amount of exports and imports produced.

Long-run Comparative Advantage, Externalities, and Cost Structure

Efficiency in the production of commodities by a group of trading

[14] In the economist's language there is distortion between domestic consumers' marginal rates of substitution between environmental goods and other goods which thus distorts marginal rates of substitution between other goods and domestic prices. Likewise, the existence of the externality in production means that producers' marginal rates of transformation are equalized with international price ratios but not with the social marginal rates of transformation for the domestic economy. Therefore, external diseconomies not only induce discrepancies in equality of marginal rates of transformation between producers in different countries but also marginal rates of substitution between consumers in different countries on traded commodities. The one case in which such consumer preference distortions would not occur is when no substitution existed between environmental "bads" and internationally traded commodities.

states requires that each country utilize its unique attributes in terms of productive resources to their best advantage. Unique attributes may include such things as special labor skills, relative capital abundance, soil fertility, and, of increasing importance, the environment's capacity to assimilate and neutralize waste residuals. Countries with relatively large environmental assimilative capacities (EAC) should, other things being equal, produce commodities with relatively high wasteloads per unit of production for export and import commodities with relatively low wasteloads in their production. Also, such countries would consume relatively more commodities with greater residuals generated in consumer use (or in final disposal). Of course, these statements are predicated on the assumption that the citizens of countries with relatively high environmental capacity do not have a greater preference for environmental quality compared to countries relatively less well endowed. The introduction of EAC as another determinative factor for patterns of trade is likely in the long run to: 1) reduce the flow of trade in commodities with relatively high residuals generation in their production *and* consumption; 2) increase trade in commodities with relatively high residuals generation in production; and 3) reduce trade in commodities with relatively high residuals generation in their final disposal and use. Given variations in preferences of different countries for improved environmental quality, the resulting patterns of trade are not clearly identifiable without knowledge of each country's preferences and EACs.

There have been assertions that developing countries will prefer to adopt a lower level of environmental quality in order to achieve a higher and more rapidly growing domestic income. If this assertion is true, then in the long run there will be economic pressures for high residuals generating industries to relocate in developing countries with substantial natural EACs. Such relocations are desirable from a global efficiency viewpoint, since comparative advantage and differences in preferences are reflected, but may be harmful for some industrialized countries.[15] A fear has emerged that some developing countries may become "pollution havens" and thereby achieve a trade position superior to industrialized states with assumed higher emission or environmental quality standards. While in the short run particular industries in the developed countries may be placed at a competitive disadvantage, in the long run domestic factors of production will be shifted to more internationally productive uses pro-

[15] In this and later sections we have assumed that developing countries can and do develop adequate analytical procedures for assessing damages and monitoring waste flow as well as reasonable enforcement provisions. If this is not true, then the statements on global efficiency and comparative advantage may well be invalid.

vided domestic and international prices are allowed to reflect all costs including waste disposal costs. Several policy tools to mitigate the impact of loss in comparative advantage in the short run by particular domestic industries are discussed in section IV.

The shift in domestic and export prices accompanying implementation of domestic environmental controls depends crucially on the magnitude of control costs to domestic industries and on the structure of all production costs. Several estimates have been developed as to the likely extent of changes in production costs resulting from domestic environmental control programs. These estimates, although not strictly comparable, indicate that pollution control costs for the United States will increase total costs for producing exports in the approximate range of 2 to 8 percent.[16] However, for specific industries the *full* effect of changes in costs including the industry's own pollution control costs as well as pollution control costs for other industries which supply inputs to the industry may be much larger. For example, Wassily Leontief estimates that the range of increase in costs necessary for air pollution control (for products that enter the export sector) spreads from 0.2 percent for petroleum refining to 16.8 percent for primary nonferrous metals. Changes in domestic costs of these probable magnitudes may have a substantial impact on export prices and thereby comparative advantage, but the exact magnitude depends on the structure of all production costs.

The first question that may be posed regarding the connection between changes in domestic costs and changes in export prices is whether the export industry produces its products under conditions of increasing, decreasing, or constant costs per unit of product.[17] If the domestic pollution industry produced under conditions of constant costs per unit of product, then export prices will be shifted upward by an amount equal to pollution control costs (including damages not ameliorated by pollution abatement if effluent charges are imposed) per unit of product. However, for industries with increasing costs per unit of production, as production expands a proportion of the pollution control costs will be paid for via lower profits

[16] See W. Leontief and W. Ford, "Air Pollution and the Economic Structure: Empirical Results of Input-Output Computations" (mimeographed), referenced in the General Agreement on Tariffs and Trade study cited below; Ralph C. d'Arge, "International Trade and Domestic Environmental Control: Some Empirical Estimates," in Kneese, Rolfe, and Harned, appendix F; and "Industrial Pollution Control and International Trade," *GATT Studies in International Trade* (Geneva: General Agreement on Tariffs and Trade, July 1971), p. 26.

[17] In this discussion we implicitly presume that pollution control costs or damages are not so pervasive as to cause a basically decreasing-cost industry to become an increasing-cost industry or vice versa.

to the industry, and export prices will increase by less than the domestic pollution abatement costs and damages, when these are added to other production costs which increase prices but in so doing induce a reduction in sales. The reduction in domestic and export sales raises the average cost per unit of production in decreasing-cost industries, creating a greater adjustment in prices than the cost of pollution including abatement and damages. Thus, countries with decreasing-cost industries in the export sector will tend to have relatively larger adjustments in export prices stemming from pollution abatement costs.

In the United States a substantial proportion of exports (very crudely more than 50 percent) are produced by industries with estimated increasing returns to scale which implies decreasing costs. Perhaps it can be inferred that this is true for other developed countries. Alternatively, agriculture and raw materials industries tend to be subject to constant or decreasing returns which on balance is suggestive that developing countries' exports may be produced under conditions of constant or increasing costs. In consequence it can be expected that developed countries will undergo export price changes in excess of domestic pollution control costs while developing countries' export prices, if these states institute domestic controls which are reflected in export prices and costs, will shift by less than pollution control costs. Of course, the above statements may not be applicable to any particular industry or country since scale effects or differences in technology could make an industry subject to increasing costs in one country and decreasing costs in another. What this brief discussion does illustrate, however, is that a simple one-to-one correspondence between pollution control costs and changes in export prices undoubtedly does not exist.[18]

Some Observations on Domestic External Diseconomies and Protection

So far, our remarks on comparative advantage have been directed

[18] In addition there is some slight evidence that decreasing-cost industries in the United States will have higher costs for pollution abatement. A correlation coefficient of .58 was obtained between measured returns to scale and air pollution abatement costs as estimated by Leontief and Ford, where both variables were transformed into logarithms. From this one might even infer that a proportion of the observed increasing returns to scale is"explained" by excessive utilization of the natural environment although the meager evidence at this point does not allow more than a casual hypothesis. Data on increasing returns to scale for the United States by industry was obtained from: George H. Hildebrand and Ta-Chung Liu, *Manufacturing Production Functions in the United States, 1957: An Inter-Industry and Interstate Comparison of Productivity* (Ithaca, N.Y: New York State School of Industrial and Labor Relations, Cornell University, 1965).

toward the effects on comparative advantage of effluent charges, emission standards, or any other control strategy in which costs are wholly reflected in domestic prices. We have also briefly discussed the connection between changes in domestic costs and changes in domestic and thereby international prices. However, questions regarding the use of domestic environmental control strategies which have no impact or a substantially different impact on the domestic and international price structure have not been posed or answered. Also, we have as yet not considered the effect of using tariffs or nontariff barriers to reduce the domestic impact of environmental control programs.

As was mentioned earlier, a trend appears to have emerged in developed countries toward the utilization of emission, effluent, and/or quality standards coupled with subsidies for the construction of particular types of waste treatment facilities. Of course, strictly enforced emission standards with no compensating treatment cost subsidies will have a similar impact on domestic costs and therefore export prices as effluent charges since domestic firms must incur costs to meet such standards. However, partial or complete subsidization through such means as investment tax writeoffs, special leasing arrangements, relocation costs of pollution-affected groups, or any payments strategy which is not reflected in higher production costs will cause domestic and international prices to be distorted. If all countries adopt a standards-subsidies approach, then only by chance will the resulting international trade patterns be efficient. In addition there will be pressures for countries, especially developing ones, to lower standards and pressures in the developed countries to increase subsidies. Consequently, little or no relationship will emerge between the types of goods a country exports and the country's natural endowment of environmental assimilative capacity or preferences for environmental quality. Pressures may also develop to equalize standards *and* subsidies between countries so that no particular country would be given a net advantage for location of relatively high polluting industries. But such an outcome would only be globally efficient if all countries had identical preferences, technologies, and natural endowments including waste residuals assimilation, a most unrealistic hypothesis since in that case there is no economic reason for trade. In consequence, in order to avoid such distortions and politically expedient but economically unsound retaliation between countries most, if not all, pollution control costs should be reflected in export prices. The possibility of pressures for individual countries to take remedial action in order to protect domestic industries would still be present but might be countered by the threat of retaliatory action by others. Trade patterns would at least be partially, if not completely, determined by comparative advantage.

If higher import tariff rates are adopted in conjunction with domestic environmental control programs which affect costs and prices, then the economy is simultaneously undergoing a reduction in one distortion while being confronted by an increase in another. The net effect may or may not be advantageous to the country. In the long run it is not likely to be advantageous if tariff imposition induces other countries to also implement protective tariffs. This is because the country will not be utilizing its domestic resources efficiently, *and* its trade will be reduced substantially by multilateral adoption of tariffs.

Nontariff barriers such as artificially high quality restrictions, quantitative restrictions or quotas, and quarantine requirements have the same impact as tariffs in distorting comparative advantage. However, in many cases regarding environmental protection it will be difficult to discover whether nontariff barriers are being used to protect the domestic environment or are subtle ways of achieving protection for domestic firms confronted with rising domestic costs (and thereby loss in competitive international advantage). As we have already argued, nontariff barriers are inappropriate mechanisms for domestic control of waste residuals resulting from consumption. Trade in intermediate products such as petroleum which yields residuals in the production process also should not be subject to tariff or nontariff barriers but be taxed or subject to standards only when used. However, if nontariff barriers are instituted by certain countries, then an attempt should be made to convince them to institute the least restrictive import barriers on commodities that these countries have a comparative disadvantage in producing. Also, if possible, fixed time limits on tariff or nontariff barriers should be encouraged.

Even though traditional tariff or nontariff barriers are not utilized by countries to mitigate the impact of domestic environmental controls, other more subtle forms of barriers to trade will undoubtedly arise. An example is the recent law in the Federal Republic of Germany (West Germany) restricting the lead content of gasoline to 0.15 grams per liter.[19] Ostensibly, the law was passed to reduce lead emissions from automobiles in West Germany although it has the simultaneous impact of restricting sales of other countries' automobiles in West Germany, which is clearly a nontariff trade barrier. West German vehicles can be easily adjusted to low lead gasolines, but several major Italian and French automobile makers' high compression engines cannot be so easily adjusted. Is this a nontariff barrier? We do not know, but the example indicates the potential complexity of deciding

[19] See Claire Sterling, "Depolluting Autos is Complex Issue," *Washington Post*, November 9, 1971, p. A18.

which environmental controls are barriers and which are not. We would recommend that the General Agreement on Tariffs and Trade (GATT) or a GATT-type of organization be used to analyze and arbitrate such barriers so that if possible, environmental control does not become the instrument of selective nontariff barriers.

Environmental Controls and the Developing Countries

Perhaps one of the most discouraging aspects of international environmental agreements is the feeling that such agreements between developed countries will have a negative economic impact on developing countries. There are several dimensions to the presumed negative economic impact. The first deals with the implicit assumption by developing countries that governmental expenditure of all types is relatively fixed in the developed countries and that environmental control costs in these countries will come, at least partially, at the expense of aid funds for developing countries. Thus, environmental control is viewed as a potential substitute for a use of expenditure normally earmarked for foreign aid by developed countries. Whether budgets are relatively fixed and the developed countries will actually view domestic environmental control as a higher-priority use of funds than foreign aid cannot be foreseen. However, a few very crude indicators can potentially be developed for the United States. A rough estimate of investment and other costs for a "substantial" reduction in environmental pollution in the United States is $95 billion for the next six years, or about $16 billion per year.[20]

This amount, if viewed as a percentage of the expected increase in gross national product (GNP) for the United States during the next six years, is substantial — about 35 percent.[21] Also, if one presumes that the entire environmental pollution program is going to be financed via subsidies by the United States government, then a $16 billion annual increase constitutes a sizeable expansion of about 8 percent in federal expenditure. Such a budgetary increase may very well induce some net reductions in foreign aid, but it is more likely to reduce or eliminate any net increases. What is overlooked is that it is doubtful that a substantial proportion of the total costs for environmental pollution reductions will be borne by the United States government. If the "polluter must pay" principle is adopted extensively, then direct and indirect subsidization for pollution control can be expected to be substantially less than 50 percent of total pollu-

[20] Estimate is taken from Allen V. Kneese, "The Economics of Environmental Pollution in the United States," in Kneese, Rolfe, and Harned, chapter 1.

[21] Ibid.

tion costs, or roughly $7 to $8 billion per year.[22] However, this amounts to only about a 3 to 4 percent increase in governmental expenditures per year which is unlikely to have a very pronounced negative impact on the level of foreign aid funds. As regards foreign aid, at least for the United States, the impact on foreign aid funds to developing countries will likely be smaller if an extensive set of emission standards and/or effluent charges system is adopted rather than a substantial subsidization system.

A second dimension of the potential negative economic impact of environmental control programs of developed economies on developing economies is due to the implied change in terms of trade between these two blocs of countries. If the developed economies adopt simultaneously a domestic charges or standards system for waste emissions control, this implies that some increase will occur in their export prices and some decrease in the profitability of investment relative to the developing economies. The normally expected impact would be a deterioration of the balance of payments in both current and capital accounts for the developed economies and improvement for the developing economies. Higher export prices would be expected to reduce exports, and lower profit rates would divert international capital flows toward more profitable ventures in developing countries. Alternatively, extensive governmental financing of subsidies programs in developed economies is likely to have the opposite effect unless subsidies are tied explicitly to waste emissions damages, according to the arguments we presented in section II. In the subsidies case profit rates will not decrease but will more than likely increase. On balance, subsidies related to reductions in effluent discharges will have an impact on export prices similar to but probably smaller than that of effluent charges but a different impact on international capital flows. Certainly, from the viewpoint of developing countries, emission standards or effluent charges adopted by the developed economies would appear to offer the fewest disadvantages in terms of both aid funds and trade repercussions. However, as will be discussed in section III, when export earnings adjustments and international income repercussions are considered, this tentative conclusion may or may not be correct.

Risking gross and potentially error-prone simplification, let us assume for the moment that there are three types of countries: developed, developing, and a third category containing countries which

[22] This calculation is made by presuming the United States government subsidies 50 percent of the change in costs for treatment of municipal sewage and 100 percent of the costs associated with separation of storm and sanitary sewers. All other industrial, municipal, and vehicular waste emissions control costs are presumed to be paid for by the emitter or some other nongovernmental entity.

are assumed not to participate substantially in international pollution agreements or to undertake very extensive domestic environmental control programs but which export commodities which are close substitutes for the developed economies' exports. Also, we assume that international competition for markets exists: between developed countries; to a lesser degree between developed and industrialized developing countries; and between developing countries. Competition for international markets between developing and developed or developing and industrialized developing is presumed to be relatively small. Thus, it is envisioned that there is a substantial degree of horizontal competition but not vertical competition between countries when vertical classification is specified by degree of development, i.e., level of income or income per capita. In this synthetic trading world it is assumed that export price elasticities of demand (a measure of the percentage change in exports demanded given a one percent change in the ratio of import prices) for particular countries increase (in absolute value) as degree of development increases.[23]

Finally, to complete this system of assumptions it will be taken as given that only the developed countries, and in certain instances the industrialized developing countries, impose substantial domestic pollution control programs which raise domestic and trading prices.

The first case to be considered is that in which all developed countries simultaneously institute waste emission control programs of varying degrees utilizing emission standards or charges. Clearly, all will undergo some reduction in exports since these countries' export prices will increase relative to their semi-competitors, the industrialized developing countries. Since export elasticities of demand are assumed to be elastic for the developed countries, export earnings will decrease for them. Likewise, for the developing countries with presumed inelastic demands for exports and declining relative prices, exports will increase but export earnings will also decrease. Alternatively, for the industrialized developing economies, rising prices of their competitors, the developed countries, may or may not lead to changes in export prices. However, it could be expected that these countries would have increased export earnings since export prices may be relatively stable but demand for their exports would increase.

The negative impact on export earnings of both developed and developing countries might be reduced or ameliorated if the industrialized developing countries were to adopt the strategem of standards and/or charges but implement them more slowly than did the developed countries. In the extreme case in which both developed and indus-

[23] For some evidence on this assertion see H. S. Houthakker and Stephen P. Magee, "Income and Price Elasticities in World Trade," *Review of Economics and Statistics,* May 1969 (Vol. 51, No. 2), pp. 111-125.

trialized developing countries simultaneously adopted relatively similar levels of nonsubsidized environmental controls, export earnings of all countries might fall.[24] However, such a decrease in export earnings in isolation tells us little about the impact on domestic incomes or the balance of payments of each group of countries. It only indicates that the amount of foreign exchange exchanged is reduced.

In the next section balance-of-payments and income impacts of domestic environmental controls are analyzed. However, before proceeding to impacts we should mention a third dimension of the special problems encountered by developing countries. This dimension arises with regard to threats of retaliation by developed countries to induce either industrialized developing or developing countries to impose more rigid environmental controls than they otherwise would consider. Retaliatory action such as trade barriers, which markedly influence international prices and thereby comparative advantage, should be discouraged for the many reasons outlined above. Only in cases of transnational pollution of a significant degree might there be some justification for potential threats of retaliation.[25]

III. NATURE AND MAGNITUDE OF MACROECONOMIC IMPACTS: BALANCE OF PAYMENTS AND NATIONAL INCOME

Introduction

In section II the long-term ramification for comparative advantage and competitive trading position of countries resulting from domestic environmental control programs were discussed. In this section we turn to a shorter time horizon and concentrate on the immediate or short-term effect of domestic environmental controls on two important indicators of social welfare for individual countries, national income or product and the balance of payments. Empirical estimates of the impact of environmental control programs on these two indicators are presented in this section. We omit explicit consideration of retaliatory actions by countries and concentrate explicitly on changes in national income and the balance of payments of selected countries resulting from the unilateral *and* multilateral imposition of environmental controls. Only those environmental controls such as standards and effluent charges which may have a substantial impact on domestic and export prices are considered in terms of empirically estimating impacts.

[24] That is, provided our assumptions on demand elasticities of the different blocs are valid.

[25] On this point see the discussion by Baumol.

An upward shift in domestic and export prices caused by one country unilaterally adopting environmental controls will induce adjustments in the balance of payments and national income emanating from at least two sources. First, substitution of relatively cheaper imports for domestic commodities will occur if: a) import and domestic-export commodities are substitutes; and b) if environmental controls on domestic firms result in higher domestic-export prices. Second, higher domestic and export prices through shifts in demand for imports and domestic commodities will influence the level of domestic production and thus domestic incomes. These initial shifts in domestic demand and income through multiplier-type repercussions will induce further adjustments in domestic income and imports.

The chain of economic events might be described as follows. A set of unilateral domestic environmental controls (related to effluent reductions) is imposed. This action increases export prices. The potential rise in export prices changes demands for exports, imports, and domestic products in domestic markets *simultaneously* via changes in the terms of trade. These changes in both foreign and domestic demand alter the equilibrium level of domestic expenditure and income. As both foreign and domestic incomes adjust to the changes in demand, if full employment is not achieved, multiplier-type repercussions will occur thus further altering the level of income and employment. A price-income adjustment system with foreign repercussions is thus hypothesized for measuring changes in the balance of payments and national income when there is less than full employment and in the short run no governmental actions are taken to remove the impact of unilateral domestic pollution controls on domestic or export prices.

The repercussions as structured are only intermediate in focus since the impact originates from an exogenous shift in export prices and concludes with the resulting adjustments in the balance of payments and national income. However, provided the government did not take remedial action on income, balance of payments, or the terms of trade via barriers to trade, a second adjustment would occur. The second adjustment would originate from the highly probable balance-of-payments disequilibrium caused by the initial shift in domestic and export prices. If the balance of payments was in deficit due to the price shift, normally we would expect a negative adjustment in the terms of trade so that balance-of-payments equilibrium (no deficit or surplus) would be achieved. The resulting movement in the terms of trade would not only influence the balance of payments in the direction of an equilibrium but also would induce further multiplier- and price-related adjustments to the level of domestic income via the mechanism described earlier. Our empirical results do not contain an estimate of this "second-round" effect which brings about balance-of-payments equilibrium. Thus, these estimates are of a short-run or impact nature and do not indicate the complete equilibrium movement of domestic income. What they do offer is an estimate of the degree of potential change in the balance of payments and national income that arises without considering the "feedback" of balance-of-payments disequilibrium on the terms of trade. Such a set of estimates, while not yielding reliable measures of national income or balance-of-payments adjustments, does indicate the potential direction of movement and the potential severity of environmental controls as a domestic policy problem. For example, if country y's estimated adjustment in national income is twice that of x's in percentage terms, then we can conclude that unilaterally imposed environmental controls will be a much more difficult policy question and require more extensive remedial actions on the part of policymakers in country y.

Assumptions and Basic Structure of the Empirical Model

In order to clarify the meaning and indicate the extreme qualifications of the empirical results given later in this section we present a very precise list of underlying assumptions used to develop the empirical model. Some, if not most, are fairly standard assumptions in the international trade literature for balance-of-payments adjustment models.[26] However, we present them hoping that the reader will become more fully aware of the basic structure and potential biases of the empirical model.

1) The trading world can be reasonably divided into six major trading blocs or countries: France, Japan, the United Kingdom, the United States, West Germany, and the rest of the world aggregated into one bloc.

2) It is assumed that a meaningful aggregation of different types of export, import, and domestic commodities is obtained by simply comparing the aggregate money value in real terms of all exports, all imports, or domestic consumption.

3) Adjustment of domestic and export commodity costs due to domestically imposed environmental controls is transmitted primarily through changes in prices and is not to any significant extent absorbed by losses in profits. If polluting firms tended to absorb these costs in profits and prices shifted very little, the impact on comparative advantage would be similar to the impact of subsidies discussed in earlier sections. However, an impact may result on international capital flows, and thus on the balance of payments, if domestic

[26] See, for example, Jaroslav Vanek, *International Trade: Theory and Economic Policy* (Homewood, Ill: Richard Irwin, 1962).

profit rates are substantially reduced.

4) Exchange rates are pegged, and domestic monetary and fiscal policies are implicitly assumed to be operating to maintain constant interest rates and exchange-rate stability.

5) There is less than full employment in the domestic economy, and goods are assumed to be produced under conditions of constant marginal costs. These assumptions make the empirical model "demand-oriented" in that shifts in demand for domestic, export, and import commodities *completely* determine changes in the level of domestic output and income.

6) Domestic and export prices of the other countries and rest of the world bloc are constant in the unilateral case of imposing domestic environmental controls. In the case of multilateral adoption of environmental controls it is assumed that the net impact on any one country is the cumulative sum of the effects on that country of each other country imposing domestic environmental controls.

7) Estimated behavioral coefficients including marginal propensities to absorb domestic product, to export, to import, and to save and all import price elasticities and cross price elasticities of demand are constant or do not markedly change over the range of estimated domestic and export price changes.

8) The trading world is inherently stable in that the sum of the marginal propensities to absorb domestic product, import, and save equal one. If this were not the case, the empirical estimates of domestic income and balance-of-payments changes would not necessarily be toward a new equilibrium.

9) There are no compensating actions initiated by the countries selected or the rest of the world bloc when unilateral or multilateral environmental controls are instituted. Thus, such actions as tariffs or nontariff trade barriers which may be instituted unilaterally to protect domestic industries from foreign competition are not considered. Also, barriers placed on imported goods which have relatively high residuals generation in their use or final disposal are not taken into account. In consequence the estimates we derive later must be viewed as the probable *extreme* changes in income and balance of payments.

10) The base year selected for empirical estimation was 1968. Thus, all weights as regards export and import composition by commodity type, domestic and export price levels, and aggregate indices of the value of domestic consumption, production, exports, and imports relate to 1968. All estimates of coefficients were derived from annual data by country for the period 1952 through 1968. It appeared that 1968 was not a year of serious aggregate distortions in normal trade patterns.

11) All demand relationships for imports and domestically produced goods of all countries are assumed to be additively separable, i.e., a shift in one variable affecting demand will not influence the magnitude of the effect of any other variable. (This is partially implied in assumption 7. It is restated here to emphasize the specific assumptions associated with demand.)

12) Demand for imports by country y from country x is assumed to be primarily determined by: the level of gross national income in country y; the terms of trade between country x and country y, i.e., domestic prices in y divided by export prices of x; and the terms of trade between country y and the rest of the world bloc defined as the import prices of y divided by y's export prices. All prices were expressed in index number form with quantity weights for 1968 by commodity type (the United Nations classification was utilized for identifying commodity types). Demand for domestic goods was presumed to be determined by gross national income and the terms of trade between the domestic economy and all other countries. A mathematical exposition of the demand relationships and other aspects of the impact model is presented in appendix 1.

Given this set of assumptions, an empirical model can be derived which will yield estimates of the change in the balance of payments and national income of the five countries and the rest of the world. Before proceeding to a discussion of the actual estimates we should make several statements with regard to the assumptions. First, if monetary and fiscal policies are operated so as to insure full employment and, perhaps, stable domestic interest rates, then the impact on real income from a shift in terms of trade due to environmental controls can be expected to be small, if not zero. Thus, the assumption that the government will take no unilateral countervailing actions (and proceed as just another producer and consumer in the economy) is perhaps the most restrictive one. However, unless the pattern of governmental actions is known in advance, no empirical estimation can be undertaken which accounts for this. And unless the government involved knows in advance the severity of the problem, it is doubtful if it can decide on the extent of countervailing action. Second, adding together into one dollar sum such diverse commodity types as agricultural products, electrical machinery, and petroleum does not allow one to discover which types of products will be affected most or least from changes in comparative international advantage. Also, such aggregations over commodity types for individual countries probably bias upward the estimated magnitude of change in balance of payments and national income but conceal partially the potential magnitude of change in the commodity composition of trade. This is due to the omission of considerations on adjustments within the country to changes in the terms of trade except at the aggregate demand level. Third, the division of the trading world into five major countries and the rest of the world introduced some inherent biases. For example, the European Economic Community (EEC) or European Free Trade Association (EFTA) trading blocs are not explicitly recognized. Also, preferential agreements or other forms of particular trading arrangement between these countries or blocs are not recognized by the division of trading blocs selected. In partial defense of the division it should be recognized that domestic environmental programs are likely to be decided on without substantial international negotiation. In light of this consideration and the fact that the countries selected represent a very large proportion of free-world exports the division appeared to be appropriate at least for an experimental model of the type described here.

Finally, it should be stressed again that given all of the qualifications set forth above, the resulting empirical estimates must be viewed as basically experimental at least at this stage of model development and testing.

Preliminary Estimates of Balance-of-Payments and National Income Impacts

In tables 1 and 2 the estimates of income and balance-of-payments changes in response to increased export prices are given. The percentage changes in export prices were calculated by taking estimated domestic increases in costs due to a substantial environmental control program by commodity type and weighting these cost changes by the dollar value of the commodity exported in 1968. Cost estimates and the weighting procedure are described in an earlier article.[27] These estimated export price changes are recorded in column 1 of tables 1 and 2 by country. It should be noted that we assumed all cost increases are completely reflected in export price changes. This is consistent with the assumption of the model that all commodities are produced under conditions of constant costs but requires severe qualification as was pointed out in section II.

Income and balance-of-payments adjustments were only calculated for the case in which the five countries unilaterally or, in selected combinations, multilaterally instituted "substantial" domestic environmental control programs while the "rest of the world" bloc as a whole had *no perceptible* change in export prices. Thus, the "rest of the world" sector taken as an aggregate was presumed not to participate significantly in domestic environmental programs which would raise export prices. This assumption may not be valid for certain relatively "developed" countries in Western Europe and elsewhere.

[27] See d'Arge, in Kneese, Rolfe, and Harned, appendix F.

However, given the degree of aggregation implied in using a "rest of the world" trading bloc, it appears reasonable to make this assumption.

A rather startling result emerged for estimated gross national income adjustments. For the United States, Japan, and France a unilateral export price increase resulted in a positive change in income for them as well as for the "rest of the world" bloc. Alternatively, export price increases (unilaterally adjusted) in West Germany and the United Kingdom resulted in a negative change in their domestic income. Also, if the United States, Japan, or France institutes unilateral environmental controls, the changes in income of all other countries as well as their own are positive. This occurs because of the multiplier nature, and the less than full employment assumption, of the model. An increase in export prices stimulates substitution to other countries' exports and away from the country with the price rise. But, expanded export sales and income of the other countries leads to *increased* exports sales by the price-raising country. This stimulation of export sales from increased incomes of other countries is greater than the reduction in income due to export price increases, resulting in a *net* positive change in domestic income. If full employment already exists, these results suggest that domestic and export prices will increase and thereby create domestic inflationary pressures. Note also that implicit in the model is the assumption that the balance of payments adjusts to a disequilibrium level which is consistent with the change in domestic income. The estimated negative change in balance of payments for all countries with unilateral export price changes (see table 2) indicates that the change in income is accompanied by and sustained by this deficit.

Given the assumptions underlying the model, we are able to calculate the cumulative effect on income of unilateral export price adjustments by all five countries. This process yields an estimate of the net change in income of each country if all five countries adopted price-changing environmental controls simultaneously. An estimate of this net change in income is recorded in row seven of table 1. Rows eight, nine, and ten contain the estimates of net change in income for multilateral adoption of environmental controls with Japan, France, and Japan plus France omitted from the multilateral decision, respectively. In the case of multilateral decisions involving all five of the countries or the subsets just specified, net income changes for all countries and the "rest of the world" were estimated to be positive. This result is surprising and should be taken at least initially as severely qualifying interpretations of the model's estimates. However, these very preliminary results raise some issues regarding the usually expected effect on incomes resulting from increasing export prices, or more importantly, from domestic changes in production costs. They

TABLE 1. ESTIMATED CHANGE IN INCOME RESULTING FROM THE IMPOSITION OF ENVIRONMENTAL CONTROLS, FIVE COUNTRIES, 1968 (MILLIONS OF UNITED STATES DOLLARS)

Export Price	Income Changes in Millions of Dollars and as Percentage of 1968 Gross National Income by Country											
	United States		Japan		West Germany		United Kingdom		France		Rest of the World	
Percentage Increase by Country	Dollars (Millions)	Percentage	Dollars (Millions)	Percentage	Dollars (Millions)	Percentage	Dollars (Millions)	Percentage	Dollars (Millions)	Percentage	Dollars (Millions)	Percentage
(1) 3.6 percent by the United States	13,869	1.80	3,410	2.98	3,843	3.29	1,686	1.97	2,091	1.95	3,868	1.48
(2) 4.0 percent by Japan	1,606	0.21	1,594	1.39	394	0.34	225	0.26	235	0.22	411	0.16
(3) 4.0 percent by West Germany	—496	—0.06	—383	—0.33	—3,047	—2.61	—263	—0.31	—707	—0.66	—949	—0.36
(4) 4.0 percent by United Kingdom	—560	—0.07	—294	—0.26	—519	—0.45	—1,236	—1.45	—322	—0.30	—930	—0.36
(5) 3.7 percent by France	5,494	0.71	2,922	2.55	8,110	6.95	2,282	2.68	6,956	6.50	5,426	2.07
Sum of (1), (2), (3), (4), and (5)	19,914	2.58	7,249	6.33	8,782	7.52	2,695	3.15	8,254	6.77	7,826	2.99
Sum of (1), (3), (4), and (5)	18,307	2.37	5,655	4.94	8,387	7.19	2,470	2.89	8,019	7.49	7,415	3.55
Sum of (1), (2), (3), and (4)	14,420	1.87	4,320	3.77	671	0.57	412	0.48	1,298	1.21	2,400	0.92
Sum of (1), (4), and (5)	18,804	2.43	6,038	5.27	11,434	9.80	2,732	3.20	8,726	8.15	8,364	3.91

TABLE 2. ESTIMATED CHANGE IN BALANCE OF PAYMENTS RESULTING FROM THE IMPOSITION OF ENVIRONMENTAL CONTROLS, FIVE COUNTRIES, 1968 (MILLIONS OF UNITED STATES DOLLARS)

	Balance-of-Payments Changes in Millions of Dollars and as Percentage of 1968 Exports by Country											
Export Price	United States		Japan		West Germany		United Kingdom		France		Rest of the World	
Percentage Change by Countries	Dollars (Millions)	Percentage	Dollars (Millions)	Percentage	Dollars (Millions)	Percentage	Dollars (Millions)	Percentage	Dollars (Millions)	Percentage	Dollars (Millions)	Percentage
(1) 3.6 percent by the United States	—1,222	—2.66	454	3.13	292	0.94	152	0.76	233	1.37	91	0.13
(2) 4.0 percent by Japan	435	0.95	—274	—1.89	37	0.12	34	0.17	19	0.11	—251	—0.36
(3) 4.0 percent by West Germany	183	0.40	—35	—0.24	—970	—3.12	26	0.12	955	5.60	—157	—0.23
(4) 4.0 percent by the United Kingdom	63	0.12	160	1.08	—24	—0.08	—228	—1.12	162	0.96	—133	—0.19
(5) 3.7 percent by France	858	1.89	155	1.07	477	1.55	129	0.63	—1,393	—8.18	—225	—0.32
Sum of (1), (2), (3), (4), and (5)	317	0.69	459	3.16	—189	—0.61	112	0.55	—24	—0.14	—675	—0.97
Sum of (1), (3), (4), and (5)	—119	—0.26	734	5.06	—226	—0.73	78	0.38	—43	—0.25	—425	—0.61
Sum of (1), (2), (3), and (4)	—541	—1.18	304	2.09	—665	—2.15	—17	—0.08	1,369	8.03	—450	—0.64
Sum of (1), (4), and (5)	—301	—0.66	769	5.30	745	2.40	52	0.26	—997	5.85	—268	—0.38

certainly indicate that substantial research must be undertaken by any country before an adequate prediction can be made of changes in domestic income resulting from environmental controls.

In table 2 estimates of balance-of-payments adjustments are recorded for the five countries and the "rest of the world" bloc. In the unilateral case all countries had an estimated negative change in their balance of payments resulting from a positive change in export prices although the magnitude of change varied markedly between countries. With unilateral imposition of environmental controls by West Germany and the United Kingdom there were calculated balance-of-payments deficits for other countries as well. If West Germany unilaterally adopted domestic environmental controls which raised export prices, West Germany, Japan, and the "rest of the world" bloc would undergo a balance of payments deficit; if the United Kingdom unilaterally initiated such controls, the United Kingdom, West Germany, and the "rest of the world" bloc would probably undergo a balance of payments deficit. Only in the case of unilateral controls by the United States would the "rest of the world" bloc not be estimated to have a balance-of-payments deficit.

In the multilateral case in which all five countries adopt environmental controls simultaneously the *net* effect on the balance of payments is positive for the United States, the United Kingdom, and Japan while it is negative for France, West Germany, and the "rest of the world." For the various combinations of multilateral action the net effect can be either positive or negative depending on which countries are omitted except for Japan (with all positive changes) and the rest of the world (with all negative changes). It should be emphasized that there is a substantial variability in the estimated change in balance of payments among the countries with unilateral or multilateral shifts in export prices. However, for the multilateral case involving all countries none of the estimated negative changes in balance of payments exceeded one percent of 1968 exports.

Comparison with Other Estimates and Further Qualifications

In an earlier study one of the present authors reported estimates of income and balance-of-payments adjustments for the five countries.[28] The same export price change estimates were utilized in that study as are applied here. However, the previous model was much more aggregative in that it contained only one country and a "rest of the

[28] See d'Arge, in Kneese, Rolfe, and Harned, appendix F. Some of the results of the earlier aggregated model were also presented in: R. C. d'Arge, "Essay on Economic Growth and Environmental Quality," *Swedish Journal of Economics*, March 1971 (Vol. 73, No. 1), pp. 25-41.

world" bloc embodying all other countries. The estimates of income changes derived from the earlier model were larger in magnitude (except for the United States) and mostly negative in sign and with balance-of-payments changes positive. Disaggregation from a two- to a six-sector model had the impact of reducing the magnitude and changing the signs of the estimates. However, the sensitivity analyses reported on in the earlier study contained a range of estimates broad enough to include almost all of the estimates given in tables 1 and 2. Also, in the earlier study empirical estimates of elasticities and marginal propensities were collected from secondary sources which were not estimated using precisely identical data. In addition, estimates of elasticities and marginal propensities were not done simultaneously. Finally, and most importantly, the earlier study did not contain interdependencies between countries which have the impact of altering international shifts in demand. Thus, as was stipulated in the previous study, the earlier estimates could not be accepted without severe qualification.

In appendix 1 three tables list the set of estimated coefficients for each country including all marginal propensities to import, marginal propensities to absorb domestic product, price elasticities of demand for imports and exports, and cross price elasticities. Cross price elasticities reflect the impact of changing terms of trade between a country and the rest of the world on imports of that country from some other country or the impact on demand for domestically produced goods resulting from a change in the terms of trade with the "rest of the world." The estimates in some cases are not statistically significant from zero at reasonable test levels for econometric models. Thus, the empirical estimates in tables 1 and 2 are partially derived from a model which contains relationships not adequately observed in the historical past. In order to see how important such nonobserved but imputed relationships were the model was recalculated with all cross price elasticities constrained to zero. This resulted in very minor changes in magnitude of the estimated income and balance-of-payments adjustments and no significant changes in signs. Also, instead of constraining the model so that the sum of all marginal propensities associated with expenditure equaled one we used the original statistical estimates. This resulted in only minor changes in magnitude and sign. This would indicate, at least partially, that the estimates contained in tables 1 and 2 are not markedly affected by changes in magnitude of the estimates of coefficients (unlike the previous model). However, a relatively complete sensitivity analysis has not been worked out as yet for the model developed here.

The model described here probably yields a more accurate (though not necessarily highly accurate) portrayal of the unilateral and multilateral consequences of environmental control programs than the earlier model. The empirical estimates of coefficients are more compatible, the data analyzed is more homogeneous, and the structural characteristics of the model are much more realistic. More work needs to be completed before the empirical estimations and the model outlined here can be reasonably used as a basis of prediction and, most importantly, as a starting point for international assessment of the desirability of multilateral agreements on environmental control strategies. The model's results do indicate two relatively important conclusions: First, the impact of unilateral and multilateral environmental control programs is unlikely to be neutral in effect between countries, but it is crucially dependent on which countries decide to initiate environmental control programs; second, completely multilateral programs are likely to result in relatively small balance-of-payments problems between countries accepting them, at least in the short run.

IV. MITIGATING IMPACTS WITHOUT TRADE BARRIERS: IS IT A GOOD IDEA?

Introduction

We direct the present discussion primarily at the United States situation, but what is described could be utilized by other developed countries (or even developing countries) to mitigate partially the trade impact of environmental controls. Also, we focus on those situations in which the impact of a cost increase is initially on industry and secondarily on consumers. As we explained in section II, in the case of taxes levied on consumer goods (because of an external cost associated with their use or disposal), such as taxes on nonreturnable containers or automobiles, the situation is quite simple: These should be handled as excise taxes presently are. They should not be imposed on exported goods and should be imposed on imported goods when consumed domestically.

Emission controls or taxes imposed at the extraction or fabricating stage will, as we also discussed in section II, affect the basic price structure to one degree or another. Therefore, these cost increases cannot readily be excluded from the price of exported goods, nor should they be. There is no rationale for doing so since the external costs are associated with the production of the article rather than with its consumption.

As we have seen in the above analysis, if the major trading states moved multilaterally to impose controls on a somewhat similar scale, immediate or near-term effects on the balance of payments would be largely canceled. Aside from the adjustment costs which might be avoided by such action, there is good reason to favor more or less simultaneous action by the major industrial economies and to an extent by the industrialized developing economies since, in its absence, or even though all might favor vigorous action in their own countries, each may fear to act individually. But the United States, or any other country, can neither depend on this happening nor delay its own urgently needed programs in anticipation of it. Furthermore, some countries may well decide to provide subsidies to their industries to reduce the generation of residuals.[29]

The question that must be addressed is: "Should the United States move unilaterally to internalize the external cost of production?" We believe that this question must be answered affirmatively if we are to have a socially efficient economy. This general position should not be substantially affected by considerations of possible international trade effects. From the point of view of the domestic economy the case for making industry bear the external costs of production is compelling. At the level of national policy we can largely avoid the international question because, as our analysis above shows, the effects are likely to be minor (probably insignificant) relative to the many other influences which play on international trade.

This is not to say that impacts on particular industries, and especially on some specific plants, might not be substantial—indeed they probably will be. It is quite likely, for example, that some copper mines and pulp mills would close. But these are likely to be marginally profitable operations which are unable to meet the social costs of production even aside from international trade considerations. We favor a program of limited special adjustment assistance (adjustment to a situation in which external costs of production are internalized) not contingent on whether or not the problem for the firm results from its inability to meet either foreign or domestic competition.

Adjustment Assistance

While one can be confident that the adjustments resulting from a well-structured and administered program of internalizing external costs will improve the overall social efficiency of the domestic economy, there may be instances in which there is good reason to publicly assist

[29] Despite the present United States administration's position against such action (see *Environmental Quality: The Second Annual Report of the Council on Environmental Quality together with the President's Message to Congress* [Washington: Government Printing Office, August 1971]) it is already embodied in American policy to some extent through tax writeoff provisions. There seems to be a consensus developing among the major industrial powers (at least at the international conference level) that industry should in general bear the cost of residuals control.

the adjustment process. Basic to any such effort are, of course, policies which sustain a high level of employment generally in the economy. But even when this is the case, pockets of unemployment may persist because of geographical remoteness from employment opportunities or obsolete skills. This situation can be alleviated by policies which are directly targeted on the worker and by those which assist selected firms in adapting to the new situations. An appropriate way to pursue these ends would be through expanding and improving the program of adjustment assistance now available under the Trade Expansion Act. The act's adjustment provisions should be extended to cover all severe adjustment problems (in the sense that substantive long-term unemployment is likely to result) associated with environmental pollution control. A good summary statement concerning the present program is found in the recent report by the Commission on International Trade and Investment:

> Adjustment assistance . . . was introduced in the Trade Expansion Act of 1962 (TEA) to accomplish two specific goals: to alleviate injury stemming from increased competition from imports; and to facilitate the process of domestic adjustment by bringing about a more effective utilization of our manpower and capital. Workers were to be assisted in their transition to new jobs through allowances covering a limited period of retraining or relocation. Firms injured by import competition were to be aided in modernizing their plants and production methods and in switching lines of production through technical, financial, and tax assistance.
>
> Attainment of the first goal would ensure that particular groups in the economy do not suffer undue harm as a result of policies that increase the general well-being, and attainment of the second would improve the productivity of economic resources by enhancing their mobility and upgrading their quality.
>
> Unfortunately, these objectives have rarely been achieved by the current program of adjustment assistance because the program is too narrow, inadequate emphasis is placed on anticipating the need for adjustment, the criteria for eligibility have proved too restrictive, and time-consuming procedures have caused lengthy delays in delivery of benefits. There is little question that the extensive procedural and substantive detail enacted by Congress to limit the scope and prevent misapplication of the adjustment assistance program has unintentionally prevented its effective development.[30]

[30] *United States International Economic Policy in an Interdependent World; Report to the President, Submitted by the Commission on International Trade and Investment Policy* (Government Printing Office: Washington, July 1971).

Because the present transition to environmental controls will no doubt require some substantial adjustments with consequent possible pockets of unemployment we recommend that the Trade Expansion Act adjustment provisions (except escape clause relief) be expanded to apply to adjustment problems associated with environmental controls whether these are due to domestic or foreign competition. Furthermore, we suggest that these provisions be strengthened along the lines recently recommended by the Commission on International Trade and Investment Policy. Since the extension of these provisions is intended to apply to what we regard as essentially a transitional problem, it should be limited to a specified period—say, ten years. The commission's recommendations are aimed at making adjustment relief more prompt, effective, and equitable. We reproduce a brief summary of them in appendix 2.

There are other programs in existence which can be helpful to workers and firms during the transitional phases even though they are not specifically directed toward problems associated with environmental controls. The Department of Labor administers such programs under the Manpower Development Act. These are aimed at retaining and reemploying workers and are currently reaching about one million persons. In addition, the Economic Development Administration can provide loans, grants, and technical assistance to local governments and industries in economically depressed areas. Finally, the Small Business Administration (SBA) can make, participate in, or guarantee loans of up to $350,000 for the SBA share. These could be used by small businesses to modernize plants and otherwise control residuals. In view of the added adjustments which will have to be made in the economy during the transition period it would appear to be desirable to strengthen present retraining, reemployment, and small business assistance programs.

If possible, adjustment assistance and retraining programs should be geared to relocation in industries for which the United States (or for that matter any country) will have a long-run export advantage *when* residuals generation, assimilative capacity, and preferences for improved environmental quality are adequately taken into account. Such assistance would then tend to accentuate comparative advantage rather than retard it between countries, as might happen with continuously increasing subsidization for industrial pollution control regardless of the degree of competitiveness of the industry.

V. CONCLUSIONS AND SOME THOUGHTS ON THE "NEW PROTECTIONISM"

It is currently a matter of some debate whether the major trading states are poised on a course leading toward a new protectionist period. During the past decade tariffs, and to some degree nontariff barriers, have exhibited a downward trend. Concurrently, however, specific trading blocs have developed such as the EEC and EFTA, which tend to expand trade within the bloc but diminish it with other countries. In recent years balance-of-payments deficits have induced some developed economies to adopt protectionist policies, at least for short-term intervals. Whether these and other signals indicate movement toward or away from a new degree of protectionism is a matter of discussion much beyond the scope of this article. What is pertinent is the possibility that domestic environmental protection will become the rationalization for greater protection. In sections II and III of this article we have attempted to demonstrate that such a rationalization cannot be substantiated on the basis of either long-term comparative advantage and efficiency or on likely short-term impacts on the balance of payments and domestic incomes of selected countries.

On efficiency premises there is little if any justification for tariffs or nontariff barriers applied to imports destined for consumption in the importing country. A much more efficient taxation device is through excise or other forms of taxes applied on the import *and* on domestically produced commodities of a similar waste-generating nature at the point of consumption or waste generation. If the waste residuals generated by the production of commodities create external costs and since these external costs are imposed on the producing country, they should be taxed or otherwise regulated by that country. In the spirit of no distortions in world trade, such as tariffs or other barriers, external costs should be fully reflected in export prices. We have argued that the decision by an individual country not to reflect such external costs in export prices is even more pervasive in distorting trade patterns than the imposition of tariffs.

From the standpoint of developing countries (largely the rest of the world sector in the model discussed in section III) the case for external costs being reflected in export prices for developed economies is particularly strong. Our calculations indicate that the terms-of-trade effect and the potential income and balance-of-payments adjustments are likely to be small in the short run with the added benefit of lessened pressures on channeling foreign aid to domestic environmental control in the developing economies. With reference to the United States experience in trade adjustment programs we have attempted to outline approaches to temper the impact, in the short run, of domestic environmental control programs with their costs reflected in export prices. A near-term loss may be encountered with such assistance programs. But this relatively small loss may be more than offset with long-run gains in comparative advantage adjusted for citizens' desires for environmental amenities.

We hope that international negotiation and the harmonizing of environmental control programs can proceed unhindered by unjustified unilateral decisions for protection of domestic industries which will result in inefficiencies in production that are international in scope.

APPENDIX 1: THE EMPIRICAL MODEL

In section III a verbal description of the empirical model used to calculate the shifts in domestic income and the balance of payments resulting from domestic and export price changes was given. Here, a highly simplified mathematical derivation of the model is developed along with three tables containing the marginal propensities and elasticities utilized in all calculations. The model involves two sets of equations in which all but price ratio variables are expressed in constant dollars, i.e., adjusted for secular drifts in prices. The first is for domestic demand for domestically produced commodities:

1) $M_{ii} = a_0 + a_1 (P_w/P_i) + a_2 Y_i$

where:

- M_{ii} denotes gross domestic absorption of country i defined as consumption expenditure, investment, and government expenditure;
- P_w/P_i denotes an index of import prices from the rest of the world divided by an index of export prices for country i;
- Y_i represents gross national income in country i;

and a_0, a_1, a_2 are constants obtained by regression analysis utilizing data for each country for the period 1952-1968.[1] The second demand equation estimated was for imports of each country where the subscripts j and i denote imports of country i originating in country j:

2) $M_{ji} = b_0 + b_1 (P_j/P_i) + b_2 (P_w/P_i) + b_3 Y_i$

with:

- M_{ji} representing imports of country i originating in country j;
- P_j/P_i representing domestic or export price indices in country j divided by domestic or export price indices in country i;
- P_w/P_i identical to the ratio defined for equation (1);
- Y_i identical to the definition given for equation (1);

In addition to the demand equations specified a set of basic identities were postulated as follows:

[1] The estimated regression equations are not reproduced here, and the set of data utilized is not documented in detail. The authors will gladly make this information available upon request.

$$3)\ Y_i - \sum_{j=1}^{6} M_{ij} = 0 \qquad i = 1, \ldots, 6$$

and

$$4)\ \sum_{j=1}^{6} M_{ij} - \sum_{i=1}^{6} \frac{rP_j}{P_i} M_{ji} = B_i \qquad i = 1, \ldots, 6$$

where:

B_i denotes the current balance-of-payments deficit or surplus in real terms of country i;

r denotes an exchange rate expressing the quantity of country j's currency it takes to purchase one unit of i's currency.

Substituting the demand relationships postulated in (1) and (2) into the set of equations (3) and (4) and differentiating the resulting system of equations with respect to P_i, we obtain a linear system. Before describing the result it should be noted that in a trading world by definition $\sum_{i=1}^{6} B_i = 0$. That is, the balance-of-payments deficit for all countries taken together must be zero. Thus, one of the equations causes the system of equations to be linearly dependent and therefore must be omitted in order to obtain an independent set of relationships. In the computational system the equation specifying B_i where i denotes the "rest of the world" was arbitrarily excluded. To illustrate the resulting system of equations we present it for five countries and the "rest of the world" where the only export price change is assumed to occur in the first country, i.e., $P_i = P_1$.

In matrix notation:

$$5)\qquad \left[\begin{array}{c|c} M & O \\ \hline N & -1 \end{array}\right] \cdot V = Z$$

where:

M = a six by six matrix of one minus the marginal propensity to absorb domestic product for each country on the main diagonal and marginal propensities to import (with negative sign) country i's product by country j composing the remainder of the elements where i denotes row number and j column number, i.e., j = 1,2, . . . 6.

N = a five by six matrix with each row containing the sum of marginal propensities to export of country i (row number) to all other countries composing the diagonal elements and marginal propensities to import (with positive sign) of country i's products by country j composing the off-diagonal elements of each row where i again denotes row number and j column number. Note: The row for the rest of the world is deleted.

O = a six by five matrix of zeroes.

−1 = a five by five diagonal matrix with all diagonal elements equal to minus one.

V = a column vector (of dimension 1 × 11) with dY_i/dP_1 for i = 1, . . . , 6 as the first six elements and dB_i/dP_1 for i = 1, . . . , 5 as the next five elements where the change in prices is for country 1. An analogous vector is obtained if the system is differentiated with respect to any other countries' price levels.

Z = a column vector (of dimension 1 × 11) which is different whenever the system is differentiated with respect to one of the five countries or bloc prices. We only discuss the vector when the first country's domestic and export prices are presumed to be exogenously changed. The first row of the vector contains the sum of all export elasticities of demand for country 1's products. The second through sixth rows contain the import elasticities of demand by country 1 for the other countries' products where row number indicates the export country; added to this is the weighted cross price elasticity between imports to country 1 and the terms of trade between the rest of the world and country 1. Row seven contains the weighted sum of all export, import, and cross price elasticities of demand for country 1's products. Rows eight through eleven contain the sum of import, export, and cross price elasticities of demand derived from the b_2 coefficient in equation (2) between country 1 and country i with the eighth row containing i = 2, ninth row i = 3, tenth row i = 4, and eleventh row i = 5. This sum was omitted for i = 6 since no balance-of-payments equation was included for the "rest of the world."

Taking the estimates of all marginal propensities and elasticities recorded in tables 1, 2, and 3 and inserting these into the M and N matrices and into the Z vector in the system specified by (5) yields a linear system of eleven equations and eleven endogenous variables. The empirical estimates of the magnitudes of the eleven endogenous variables including changes in income of the five countries and the rest of the world and changes in the balance of payments of the five countries, derived by solving this system for each country, are presented in tables 1 and 2 in section III of the text. It should be noted that the system of equations was recalculated for unilateral domestic and export price changes in each country and the rest of the world. Given the presumed linearity of the system, the total effect on incomes and balance of payments of all countries imposing environmental controls would equal, by definition, the sum of the effects from unilateral action by all countries.

TABLE 1. ESTIMATES OF THE MARGINAL PROPENSITIES TO ABSORB DOMESTIC PRODUCT AND TO IMPORT, SELECTED COUNTRIES, 1968

	Importing Country					
Exporting Country[a]	United States	Japan	West Germany	United Kingdom	France	Rest of World
United States	0.2200[b]	0.0283**	0.0193**	0.0276*	0.0192**	0.1250**
Japan	0.0101**[c]	0.1790	0.0032**	0.0026*	0.0016**	0.0732**
West Germany	0.0087**	0.0047**	0.2750	0.0214**	0.0430**	0.1445**
United Kingdom	0.0065**	0.0041**	0.0105**	0.3150	0.0087**	0.0598**
France	0.0016**	0.0014**	0.0347**	0.0173**	0.2070	0.0571
Rest of World	0.0353**	0.0846**	0.1511**	0.1887**	0.0800**	0.5400

[a] Each row contains one minus the marginal propensity to absorb domestic product (excluding saving) with $i = j$ and marginal propensity to export from country i to country j (or import to j from i), with $i \neq j$, where i indicates row and j column.

[b] Marginal propensities to absorb domestic product were calculated by subtracting the estimated marginal propensities to import and 0.05, to account for savings propensities, from one. Thus, for each country the marginal propensity to absorb is calculated as a residual. This procedure was necessary since the sum of the marginal propensities to import and consume domestic product derived from the regression estimates exceeded one, thereby implying instability in world trade.

[c] * indicates marginal propensity was statistically significant at the 10 percent level and ** significant at the 5 percent level using a two-tailed "t" test.

TABLE 2. IMPORT ELASTICITIES OF DEMAND[a]
SELECTED COUNTRIES, 1968

	Importing Country					
Exporting Country	United States	Japan	West Germany	United Kingdom	France[b]	Rest of World
United States	—	—2.18**	—1.17*	—0.70	—2.60	0.29
Japan	—1.12*[c]	—	—1.42	—1.98**	—1.30	0.77*
West Germany	—2.40**	—1.29**	—	—0.94**	—8.30	0.01
United Kingdom	—1.85*	—2.14*	—0.29	—	—6.80	—0.19
France	0.30	—0.68*	—0.59	—1.12**	—	2.03*
Rest of the World	—0.29	0.26	—0.18	0.21	—0.84	—

[a] Import elasticity of demand is a measure of the percentage change in imports resulting from a one percent change in the terms of trade between the importing country and the exporting country. The terms of trade are defined as a composite price index of the importing country's domestic prices divided by a price index of exports of the exporting country.

[b] The empirical estimates for France were derived using non-least-squares estimating techniques. In consequence these estimates are not strictly comparable to the others reported on in this table.

[c] * Statistically significant at the 20 percent level. ** Statistically significant at the 5 percent level.

TABLE 3. ESTIMATES OF CROSS PRICE ELASTICITIES[a] OF DEMAND SELECTED COUNTRIES, 1968

Exporting Country	Importing Country: United States	Japan	West Germany	United Kingdom	France
United States	—0.099**	0.490	0.020	0.190	0.660
Japan	—0.124[b]	—0.053*	—0.090	—0.420	—4.740**
West Germany	—0.003	0.500	—0.189*	0.030	—1.080*
United Kingdom	—0.008**[c]	—0.110	—0.220	—0.054*	—1.450*
France	—1.040**	0.140	0.390	0.210	—0.036

[a] Cross price elasticity of demand measures the percentage change in imports or in domestic absorption resulting from a one percent change in the terms of trade between the "rest of the world" and the exporting country. The terms of trade were estimated by taking the ratio of an index of export prices of the "rest of the world" and dividing it by an index of export prices for the exporting country. Generally, if the "rest of the world" exports are substitutes for the exporting country's, the sign of the cross price elasticity should be negative.

[b] Most calculated cross price elasticities of demand were not statistically significant at the 5 percent or even the 20 percent level. In consequence the empirical model was also estimated constraining cross price elasticities to zero. The results were not markedly different from those reported on in section III, tables 1 and 2. Research is continuing in an attempt to obtain more precise cross elasticity estimates.

[c] * Statistically significant at the 20 percent level. ** Statistically significant at the 5 percent level.

APPENDIX 2: BRIEF SUMMARY OF THE UNITED STATES ADJUSTMENT ASSISTANCE PROGRAM AND RECOMMENDATIONS OF THE COMMISSION ON INTERNATIONAL TRADE AND INVESTMENT POLICY

The Commission on International Trade and Investment Policy identified two cases under which adjustment assistance can be obtained. We quote from their summary document:

> First, a firm or a group of workers may apply to the Tariff Commission for certification of eligibility to apply for adjustment assistance. If the Tariff Commission finds that the injury requirements of the law are met, it so reports to the President, who may then certify that the applicant is eligible to apply for relief.
>
> At that point, a firm seeking relief submits a proposal to the Secretary of Commerce, who certifies its eligibility to receive assistance upon determination that increased imports have caused or threaten to cause serious injury to the firm. Groups of workers apply for assistance to the Secretary of Labor, who must satisfy himself that increased imports have caused or threaten to cause serious injury to the firm. Groups of workers apply for assistance to the Secretary of Labor, who must satisfy himself that increased imports have caused or threaten to cause unemployment or underemployment of a significant number of workers in an afflicted firm or in one of its subdivisions. Individual workers must apply separately for benefits, and if they meet the strict criteria of the Act, they will be certified as eligible to receive those benefits.
>
> Second, a trade association, firm, labor union, or other representative of an industry may apply for "tariff adjustment" (escape-clause relief). If the Tariff Commission finds that the industry represented by the applicant meets the injury test of the law, the President may raise or impose import restrictions through the escape clause; may initiate adjustment assistance procedures as an alternative; or he may do both.[1]

The Commission on International Trade recommended that time delays in handling applications be reduced by eliminating Tariff Commission investigation and findings except for escape clause action. The commission thought it best to allow the executive branch to make decisions about whether firms or workers have met the injury criteria. We strongly concur since reducing delays under this assistance program would undoubtedly reduce pressures for trade barriers. For improved assistance to workers the Commission on International Trade recommends: a) making adequate amounts of assistance available as quickly as possible after eligibility has been established; b) providing

[1] *United States International Economic Policy in an Interdependent World.*

greater incentives, including wider benefits, to accept training or relocation assistance; c) making allowances available for the full period of retraining, in addition to retroactive allowances paid during a period of unemployment while awaiting retraining; d) giving qualified workers the opportunity to pursue technical, professional, or academic as well as vocational training; e) relaxing the requirement concerning previous work and earnings, thereby enabling recent entrants into the labor force to qualify for the program; f) providing family health benefits for workers in the adjustment assistance program; and g) providing subsidies to allow older workers, who are not yet eligible for full social security benefits, to enter the social security program without reduction of benefits.

Adjustment assistance to firms can be provided under the Trade Expansion Act for "modernizing" existing plants and "developing new lines of more competitive products." The Commission on International Trade recommends that: a) operating responsibility be centralized to expedite the delivery of benefits; b) more attractive terms of financial assistance and tax benefits be provided; c) interim financing be made available between approval and delivery of assistance; d) adjustment assistance to firms be normally available only to small businesses; and e) United States antitrust legislation be altered to permit mergers of firms encountering difficulty because of competition of imports. We concur with these recommendations except for restricting adjustment assistance to small businesses in the case of loss in comparative advantage due to emission standards and/or taxes since these controls are likely to affect the competitive position of domestic industries characterized by high degrees of concentration and size as much or more than small businesses.

[6]

Ecosystems, Externalities, and Economies

THOMAS D. CROCKER[1] and JOHN TSCHIRHART[2]
[1] *Department of Economics & Finance, University of Wyoming, P.O. Box 3985, Laramie, WY 82071-3985, U.S.A.;* [2] *Department of Economics & Finance, Box 3985 University Station, Laramie, WY 82071-3985, U.S.A.*

Abstract. This paper incorporates an ecosystem model into a model of a simple economy. The decisionmaking agents in the ecosystem are individual organisms aggregated to the species level. A species may provide utility directly to humans, or it may provide utility indirectly because it is used either as a raw material in goods fabrication or as sustenance for other species. We describe a comparative static equilibrium of the ecosystem where species' demands for other species are equal to the supplies of those other species, and energy is conserved. The ecosystem is then embedded in the economy so that the effects of human intervention can be traced through both the ecosystem and the economy. Human intervention creates ecosystem externalities such that ecosystem equilibria are shifted and the new equilibria affect the utility or the production processes of other humans. This framework allows us to describe in principle *which* ecosystem services can be efficiently usurped by humans, *which* waste flows can be efficiently allowed into ecosystems, and *which* ecosystem organisms and physical attributes can be efficiently maintained.

Key words. Ecosystems, externalities, economies, sustainable development.

1. Introduction

Ecosystems are valuable in their natural states and as sources of raw materials but their markets are often incomplete. One might value ecosystems by treating their individual species as economic goods. An ecosystem may be composed of thousands of interacting species, each one filling a niche and contributing to its ecosystem's health and stability. Some species, or goods, may provide obvious and measurable benefits (e.g., trout), others may go unnoticed (e.g., soil microbes), and still others may be considered pests with negative benefits (e.g., mosquitos). But if each species contributes to its ecosystem, then it must have a positive imputed value based on the value of the ecosystem and the niche that the species occupies. Thus if a pest species provides nourishment to a beneficial species, the beneficial species' derived demand for the pest can be used to determine the pest's value.

A problem with this approach is that ecosystems exhibit very complex interactions and dynamics. But so do economic systems, and they have been modelled with varying degrees of success. If the complexities of an economic system are similar to the complexities of an ecosystem, then the techniques used to model the former can be transferred to the latter. The similarities between the two systems are striking.[1] Demand and supply, competition and

Environmental and Resource Economics **2**: 551–567, 1992.

marginal rates of substitution are just some of the economic concepts that we apply to ecosystems in the following pages.

Our approach allows us to incorporate an ecosystem model into a model of a simple economy. The decision making agents in the ecosystem are the individual organisms aggregated to the species level. A species may provide utility directly to humans, or it may provide utility indirectly because it is used either as raw material in fabricating a good, or as sustenance for other species. We describe an equilibrium of the ecosystem wherein species' demands for other species are equal to the supplies of those other species, and energy is conserved. The ecosystem is then embedded into the economy so that the effects of human intervention can be traced through both the ecosystem and the economy.

To motivate our approach further, we introduce the notion of an ecosystem externality in Section II. The idea is that the action of one economic agent shifts ecosystem equilibria, and the new equilibria affect the utility or the production process of a second agent. In Section III, the behavior and actions of individual organisms are described. The concept of an ecosystem equilibrium is developed in Section IV. The ecosystem is then embedded into an economy and the effects of human intervention are examined in Section V. Section VI is a conclusion.

II. Ecosystem Externalities

Around the turn of this century, the citizens of Kern County, California, a rural area of farms and small towns, decided to eliminate various predators that killed domestic animals and frightened children.[2] Armed with shotguns, traps, and strychnine, they decimated populations of skunks, foxes, badgers, weasels, snakes, owls, ravens, and hawks. In addition, they hired a team from the U.S. Department of Agriculture to exterminate all coyotes in the county. The agents were effective. Both campaigns took place over two decades. Then, during a year when farmers were enjoying a bumper crop, hordes of mice appeared. Their vast numbers, unchecked by natural predators, were too much for the usual wild grains to feed, and the mice began to encroach on silos, barns, and homes. In places, the mice were ankle-deep, and people were killing them by the thousands. Distribution of poisoned grain subdued the mice for a few months, but they eventually regrouped and descended on the villages. U.S. Highway 399 became so slippery from squashed mice that cars ran into ditches and "go slow" signs were erected. One particular school had mice in every classroom, in all the waste paper baskets, and in some desks. The affair was labelled by the U.S. Biological Survey as the greatest rodent infestation in U.S. history.

When the Kern County residents decimated predator populations, they generated an ecosystem externality. Their action was an exogenous shock that changed the ecosystem's equilibrium values. As the ecosystem moved to a new equilibrium, it had feedback effects on the economy. To formalize this notion, we model the Kern County episode in a one consumer—one producer world. The consumer has preferences over a single final good and leisure. The final good is produced from labor and an intermediate good. The intermediate good is produced from labor, and it provides no direct utility to the consumer.[3] The consumer is endowed with L units of labor/leisure. The intermediate good will be called grain which is combined with labor to produce the final good called bread. The ecosystem is comprised of three species in a simple food chain. Predators feed on mice which feed on grain.[4] Again, the grain must be combined with labor before providing utility to the consumer. Predators provide disutility to the consumer as observed in Kern County.

The consumer's utility is given by the function

$$U(b, p, l), \tag{1}$$

where b is bread, p is a measure of the predator which will be formalized in the next section, and l is leisure. Partial derivatives are $U_b > 0$, $U_p < 0$ and $U_l > 0$. The consumer sacrifices L_p units of leisure to eliminate the predators so that

$$p = p(L_p), \tag{2}$$

with $p' < 0$. Grain is produced according to the function

$$g(L_g, p), \tag{3}$$

where L_g is the labor used in grain production. Note that grain will also depend on the predators. This dependency is the source of the ecosystem externality. A formal description of the dependency is derived in Part IV. Bread is produced according to the function

$$b(g, L_b), \tag{4}$$

where L_b is the labor used in bread production. Finally, the market clearing condition for labor is

$$L = L_b + L_g + L_p + l. \tag{5}$$

A state of the economy is Pareto optimal if it maximizes (1) subject to (2)—(5). The first-order necessary conditions to the maximization problem yield the condition

$$U_p/U_b = \left[\frac{\partial b}{\partial L_b} - \frac{\partial b}{\partial g} \cdot \frac{\partial g}{\partial L_p}\right] \Big/ \frac{\partial p}{\partial L_p}. \tag{6}$$

The left-side term is the consumer's marginal rate of substitution between grain and predator elimination, and the right-side term is the rate of product transformation between grain and predator elimination. Condition (6) characterizes the Pareto optimum allocation of labor between grain production and predator elimination.

To characterize a competitive equilibrium for this economy, let π and w be the price of bread and labor, respectively. For brevity, all production takes place in one vertically integrated firm with profit given by

$$v = \pi b - wL_g - wL_b, \tag{7}$$

where b is given in (4). The firm chooses L_g and L_b to maximize (7). The consumer chooses L_g, L_p and leisure to maximize utility given by (1) subject to the budget constraint

$$\pi b = w(L - l - L_p). \tag{8}$$

The solutions to the producer's and consumer's problems yield the condition

$$U_p/U_b = \frac{\partial b}{\partial L_b} \bigg/ \frac{\partial p}{\partial L_p}. \tag{9}$$

The competitive solution for labor allocation does not satisfy the necessary condition for Pareto optimality, because the second term in the numerator on the right side of (6) does not appear in (9). This second term is the ecosystem externality; it represents the feedback effect of predator elimination on the production of grain. Assuming well behaved utility and production functions, the competitive equilibrium results in too little bread and too much labor devoted to predator elimination because the consumer does not account for the negative effect predator elimination has on the production of grain.

What makes this externality particularly vexing is its indirectness. The traditional externality is formulated with one agent's decision variable having a direct impact on another agent's objective function. However, in the ecosystem case, the consumer's choice to eliminate predators impacts the firm's production of grain via a series of adjustments internal to the ecosystem but external to the economy. Formally, the term $\partial g/\partial L_p$ in (6) represents this series of adjustments. Although in this example the adjustments involve only three species, in practice hundreds may be involved. To be complete, a model of the ecosystem that identifies the adjustments must be embedded into this simple economy.

The indirectness also complicates the use of taxes or subsidies to remedy the situation. If a market for predators was established, a tax on the consumer could be levied on each unit of labor used to eliminate predators. Since the consumer makes labor supply decisions based on parametric prices, he must be given an incentive to reduce the quantity of L_p that he supplies at a zero price. Thus, if t is the tax/subsidy, tL_p would be added to the left side of (8). Both the consumer and producer would choose L_p, and it is easy to show that the competitive equilibrium would then yield a Pareto optimum. The solution for the optimal tax on the consumer is

$$t = \pi \frac{\partial b}{\partial g} \frac{\partial g}{\partial p} \frac{\partial p}{\partial L_p}. \tag{10}$$

The complication lies in knowing what to tax or subsidize. If the agents are unaware of the link between predators and grain (i.e., they do not account for p in (3)), then they will not know where to introduce taxes and subsidies. Generally, as information about ecosystem interactions expands, so would the opportunities to use taxes and subsidies or other remedies for ecosystem externalities. Currently available information about the link between predators and rodents presumably would prevent a recurrence of the Kern County experience. But examples of uninformed choices still abound. The pesticide DDT was widely used in the U.S.A. until studies revealed that it was responsible for significant declines in bird of prey populations. Although adult birds were not seriously affected, their eggshells were too thin to support the adult weight. The link between DDT and eggshells is obscure. Prior to the research that uncovered the link, no one suggested a tax on the use of DDT to reduce the number of deficient eggshells.

Ecosystem externalities present a twofold problem. First, there must be adequate biological information to understand ecosystem interactions, and second, there must be a framework for using this information in economic analyses. We concentrate on the second aspect by formulating models of species bahavior that can be incorporated into standard economic analyses.

III. Optimization by Individual Organisms

Initially, we develop a model of an ecosystem sans humans. The model can be considered a depiction of prehistoric times or of very primitive areas in modern times. In this world, energy is the limiting resouce and all energy is derived from the sun. Organisms may use this energy directly, in the case of plants, or indirectly, in the case of carnivores and other animals. Each organism is a member of a particular species, and the members of a species feed from the same set of sources and do not feed upon each other. In essence, one or more species can be thought of as a stratum in a food pyramid, and the objective is to link the species mathematically to provide a framework for discussing ecosystem equilibria.

Before deriving the links, however, the actions of the individual organisms must be described. In a general equilibrium model of an economy, individual consumers and firms are described as maximizers. But in an ecosystem, do nonhuman organisms maximize? Most people do not credit a weasel with thoughtful preference revelation when it raids the chicken coop instead of ferreting out a mouse or two. ". . . men consciously optimize, animals do not — they survive by adopting successful strategies 'as if' conscious optimization takes place" (Hirshleifer, 1977, p. 4). This "as if" assumption is sufficient to capture much of the behavior of nonhuman organisms and thereby establish a fruitful model.

Various suggestions have been made as to what it is that nonhuman organisms behave as if they are maximizing. Lotka (1925) developed a model where the maximand is the rate of increase of the species. This rate is a function of food capture, shelter, and other physical demands which require energy expenditure. If a species is to be successful, then the energy expended must at least be equal to the energy acquired. Lotka characterizes a maximum in this system with a set of equations where the marginal productivity (i.e., an increase in the species) of an energy expenditure equals the marginal loss (i.e., a decrease in the species) of that energy expenditure. Modern work has emphasized the role of energy more directly in the search for a maximand. Odum (1971, p. 90) points out that life requires power and "... the maximum and most economical collection, transmission, and utilization of power must be one of the principal selective criteria ... ." Finally, Hannon (1976) develops a model using stored energy as the maximand. Stored energy is simply the energy acquired by the organism less the energy needed to maintain itself. Hannon (1976) and Caraco and Lima (1985) argue for the reasonableness of this objective based on general observation, and on the increased organism stability it provides during periods of fluctuating inputs.

The stored energy approach is used here. Though this approach may be an open question to many ecologists, it is consistent with the pioneering work of Lotka and frequently finds expression in modern ecological literature.

IV. Natural Equilibrium of an Ecosystem

For brevity, we omit modelling the individual organism's allocation problem and move directly to the species. We employ the convention that an entire species is a component that acts as if it maximizes its collective stored energy, the sum of the energies stored by the assumed identical individuals in the species. For each species, dynamic considerations are put aside. This is equivalent to assuming that the resources each species uses are completely regenerated in each period.

Each species is said to have a physiology function which specifies how it transforms energy inputs into energy outputs. The species has some discretion over the outputs actually realized from a set of inputs, but the physiology function places biochemical bounds on this discretion. For example, a predator species cannot continually capture prey without also producing output in the form of heat expenditures. The bounds that the physiology function sets will depend on ambient temperature, season, and other environmental conditions, all of which are assumed to be parametric to the behavior of the species.

Given the above simplifying assumptions, the allocation problem of the ith species can be written as:

$$\text{Maximize } R_i = \sum_{j=0}^{n} E_{ji}X_{ji} - \sum_{j=1}^{n} E_{ij}\overline{X}_{ij} - E_{i,n+1}X_{i,n+1} \tag{11}$$

$$\text{subject to } F^i(X_i, \overline{X}_i, X_{i,n+1}) = 0, \tag{12}$$

where R_i is the ith species' stored energy, n is the number of species, the X_{ji} $(i, j = 1, \ldots, n)$ are inputs defined in terms of mass flows from the jth species to the ith species, the $\overline{X}_{ij}$ are outputs defined in terms of mass flows from the ith species to the jth species, and the E's represent the energy content or calorific content per unit of mass flow. F^i is the physiology function which is assumed to be twice differentiable with X_i the vector of X_{ji} inputs, and $\overline{X}_i$ the vector of $\overline{X}_{ij}$ outputs. For the ith species, its output, $\overline{X}_{ij}$, to other species is assumed to be invariant; the ith species cannot prevent being harvested by other species. Note, however that even though the $\overline{X}_{ij}$ enter (11) and (12) parametrically and are therefore exogenous for the ith species, they are endogenous for the entire ecosystem since they are inputs, and are therefore decision variables, for other species. Energy used as waste heat, metabolism, growth, etc. that is neither stored nor used by other organisms, is combined into the single term $E_{i,n+1}X_{i,n+1}$.

Subscript $j = 0$ refers to the sun so that E_0X_{0i} is species i's intake of solar energy. Thus, X_{0i} can be thought of as exposure to the sun and E_{0i} the energy absorbed per unit of exposure. For a species that derives no energy from the sun, $X_{0i} = 0$. Solar energy is the only input not derived from other species, and ultimately, it is the supply of all energy in the system. Each species receives this solar energy either directly, or indirectly through other species, and then transforms it into biomass which supplies other species.

When individuals belonging to a predator species, i, capture and consume members of a prey species, j, a total of $E_{ji}X_{ji}$ energy units is assimilated by the predator. Because this predator must expend energy units in the capture and consumption, $E_{ji}X_{ji}$ actually represents net energy assimilation:

$$E_{ji}X_{ji} \equiv (E'_{ji} - E''_{ji})X_{ji}, \quad i, j = 1, \ldots, n. \tag{13}$$

In (3), E'_{ji} is the calorific content per unit of prey biomass, and E''_{ji} is the energy expended in capturing and consuming this unit. In accordance with Ulanowicz (1972), each prey species has a constant E'_{ji}, and we assume that E''_{ji} is independent of the quantity of X_{ji}.

The solution to the constrained maximization problem given by (11)–(13) is characterized more easily if $F^i(\cdot)$ is explicitly solved for $X_{i,n+1}$:

$$X_{i,n+1} = G^i(X_i, \overline{X}_i). \tag{14}$$

Thus, for given levels of X_i and $\overline{X}_i$, $G^i(\cdot)$ can be thought of as the minimum possible biomass expended on metabolism, detritus, and mechanical activities, which together will be referred to as maintenance.

If equation (14) is substituted for $X_{i,n+1}$ in (11), the first-order conditions for a maximum are

$$E_{ji} - E_{i,n+1} \frac{\partial G^i(\cdot)}{\partial X_{ji}} = 0, \quad j = 0, \ldots, b, \ldots, \beta, \ldots, n. \tag{15}$$

When any two (say the bth and βth) of the conditions in (15) are divided by one another, we obtain

$$\frac{\partial G^i(\cdot)/\partial X_{bi}}{\partial G^i(\cdot)/\partial X_{\beta i}} = \frac{E_{bi}}{E_{\beta i}}. \tag{16}$$

For the ith species, the rate at which it can trade biomass from species b for biomass from species β at the maximum, while achieving the same maintenance level, $X_{i,n+1}$, must equal the ratio of the net energy contents from the biomass of the two prey species.

Assuming the sufficient second-order conditions for a maximum are satisfied implying that $G^i(\cdot)$ is strictly convex, the implicit function theorem can be used to invert the first-order conditions in order to obtain continuous, differentiable functions representing species i's demand for the biomass of species j.

$$X_{ji}(\underline{E}, \overline{X}) \quad i, j = 1, \ldots, n, i \neq j. \tag{17}$$

$\underline{E}$ represents a vector of net energy contents. Using these demands, we refer to a natural equilibrium of the ecosystem, where "natural" indicates no human intervention, as a state where all species are maximizing their stored energy, all biomass markets are cleared:

$$\sum_{j=1}^{n} E_{ij} X_{ij} = \sum_{j=1}^{n} E_{ji} \overline{X}_{ji} \quad i = 1, \ldots, n \tag{18}$$

and the energy budget equation is satisfied:

$$\sum_{j=1}^{n} E_{0j} X_{0j} = \sum_{j=1}^{n} [E_{j,n+1} \overline{X}_{j,n+1} + R_j]. \tag{19}$$

The natural equilibrium is provided by the solution of all species' first-order conditions. Each species provides at most $n + 1$ equations from its $n + 1$ input variables $(X_{0i}, X_{1i}, \ldots, X_{ni})$. There may be fewer equations if there is no direct interaction with some species. In total there are at most $n(n + 1)$ equations and variables. We do not explore the question of existence in this paper. In the long-run, all species have zero assimilated energy. The driving force for this outcome is the change in net energy contents that occur when stored energies are nonzero. For example, suppose

$R_i > 0$. The E_{ij} for $j = 1, \ldots, n$ will decrease, because species i is bountiful and more easily preyed upon. As more predation occurs, R_i will tend towards zero. A similar but reverse story can be told for $R_i < 0$.

A comparative static analysis can be used to derive testable hypotheses about (17). For example, it can be shown that

$$\frac{\partial X_{ji}(\underline{E}, \overline{X})}{\partial E_{ji}} > 0, \quad j = 0, 1, \ldots, n. \tag{20}$$

Recall that for $j = 1, \ldots, n$, $E_{ji} = E'_{ji} - E''_{ji}$. Expression (20) thus implies that an increase (E'_{ji}) in the unit energy content of j's biomass, or a decrease (E''_{ji}) in the energy that species i expends to obtain a unit of j's biomass will result in an increase in demand for species j by species i.[5] The signs of cross-partials such as

$$\frac{\partial X_{ji}(\underline{E}, \overline{X})}{\partial E_{ji}} \tag{21}$$

are ambiguous unless prior information is available on the signs and/or the magnitudes of the cross-partial terms for $G^i(X_i, \overline{X}_i)$.

Finally, note that the demands in (17) can be substituted into the objective function to obtain an indirect stored energy function. Thus, for species i, $i = 1, \ldots, n$,

$$R_i(E, \overline{X}) = \sum_{j=0}^{n} E_{ji} X_{ji}(\underline{E}, \overline{X}) - \sum_{j=1}^{n} E_{ij} \overline{X}_{ij}$$
$$- E_{i,n+1} G^i(X_i(\underline{E}, \overline{X}), \overline{X}). \tag{22}$$

The function $R_i(\cdot)$ gives the maximum stored energy possible for species i over all energy content values and biomass and energy outputs. The envelope theorem can be used to show that for $i, j = 1, \ldots, n$, $i \neq j$

$$\frac{\partial R_i}{\partial E'_{ji}} = -\frac{\partial R_i}{\partial E''_{ji}} = X_{ji} > 0, \tag{23}$$

$$\frac{\partial R_i}{\partial E_{ij}} = -\overline{X}_{ji} < 0. \tag{24}$$

Increased energy content per unit of biomass of input species j, or decreased energy expenditures for obtaining units of biomass from species j, or decreased energy content of biomass yield to species j will all result in reduced stored energy for species i. However, the sign of

$$\frac{\partial R_i}{\partial \overline{X}_{ij}} = E_{ij} - E_{i,n+1} \partial G^i / \partial \overline{X}_{ji} \tag{25}$$

is ambiguous because changes in exogenous outputs to predators have opposing effects on stored energy. On the one hand, if individuals are lost to predators, this immediately lowers stored energy of the species through the first term on the right side of (25). On the other hand, fewer remaining individuals in the species implies fewer losses to waste heat, predators, metabolism, etc., and this is captured by the second term.

V. The Economic Problem

HUMAN INTERVENTIONS

Key features of the natural ecosystem are the inputs, energy contents of biomass, and physiology functions. Humans intervene in the ecosystem by directly or indirectly affecting all of these features. By introducing fossil-fuels, humans also may alter the overall energy equation. By augmenting the model of the previous sections, we wish to show one way in which these interventions can be captured. In particular, we consider the effects of humans cropping nonhuman species.

Humans have an initial endowment of labor that can be used to crop the n ecosystem species.[6] The cropped ecosystem species can be used in the production of m manufactured goods. Let $b = (b_1, \ldots, b_m)$ be the manufactured goods, and let L_q^b be the amount of labor used in the production of the qth manufactured good. Furthermore, let L_i be the amount of labor used to crop the ith species. Then if L is the initial endowment of labor and l is leisure, it follows that

$$\sum_{q=1}^{m} L_q^b + \sum_{i=1}^{n} L_i = L - l. \tag{26}$$

The problem for the ith species can be rewritten as

$$\text{maximize } R_i^h = [1 - c_i]R_i, \tag{27}$$

where $G^i(X_i, \overline{X}_i)$ becomes $G^i(X_i, \overline{X}_i, \overline{L}_i)$ in R_i. Following Hannon (1976), the cropping is accomplished by removing a fraction, c_i, of the energy the ecosystem would otherwise put into stored energy. We assume that $0 \leqslant c_i < 1$, that c_i does not exceed a level which would threaten the viability of the species, and that species have no control over the human intervention. This inability of species to prevent human intervention accounts for the bar notation on $\overline{L}_i$.[7] By including $\overline{L}_i$ in the physiology function, we can account for either positive or negative impacts of human intervention on the species. For positive intervention, we assume

$$\frac{\partial G^i(\cdot)}{\partial \overline{L}_i} < 0 \quad \text{and} \tag{28}$$

$$\frac{\partial^2 G^i(\cdot)}{\partial X_{ji} \partial \overline{L}_i} < 0, \quad j = 0, \ldots, n, \; j \neq 1. \tag{29}$$

Ceteris paribus, human intervention decreases the energy loss to maintenance, and decreases the marginal energy loss to maintenance for inputs from other species or the sun. Positive intervention may occur, for example, by genetic improvements in domestic animals or development of disease resistant crops. Thus, an increase in positive intervention: (1) increases stored energy.

$$\frac{\partial R_i^h}{\partial L_i} = [1 - c_i] \frac{\partial R_i}{\partial \bar{L}_i} = -[1 - c_i] E_{i,n+1} \frac{\partial G^i}{\partial \bar{L}_i} > 0 \tag{30}$$

by (28) and the envelope theorem applied to the ith species, maximization problem; and (2) increases the ith species' use of other inputs

$$\frac{\partial X_{ij}}{\partial \bar{L}_i} > 0, \quad j = 0, \ldots, n, \tag{31}$$

which follows from (29) and comparative statics applied to the maximization problem.

For negative intervention, the signs on (28)—(31) are reversed. Examples of negative intervention are the effects of pesticides on the pests, or the effect of oil spills that coat marine mammals and waterfowl and lower their resistance to cold. Agriculture is a good example where the cropping from the ecosystem is used as an input into the production of a manufactured good. A manufactured good may be a tomato *in the supermarket*. The ecosystem provides the tomato which is combined with other resources (only labor in our simplified model) to produce the manufactured good.

Production of the qth good is given by the production function

$$b_q = b_q(L_q^b, c_1 R_1, \ldots, c_n R_n). \tag{32}$$

If biomass from species i is not used to produce b_q, then $c_i R_i = 0$ in (28).

Maximizing (27) and solving the first-order conditions for the species' demands reveals that these demands for biomass are now dependent on the human inputs as well. Thus, (17), the demand of the ith species for the jth species, becomes

$$X_{ji}(\underline{E}, \bar{X}, \bar{L}_i) \tag{33}$$

and the indirect stored energy function of species i which replaces (22) is

$$R_i^h(\underline{E}, \bar{X}, \bar{L}_i) \tag{34}$$

where the superscript refers to human intervention.

PARETO OPTIMALITY

We now have a framework for incorporating ecosystem interventions into the economy. To illustrate its use, we return to the Kern County example and derive necessary conditions for a Pareto optimal allocation of inputs. The example is expanded to include a second consumer and a second manufactured good. Again, there are three species comprising a simple food chain with species 3 feeding on species 2 only, and 2 feeding on 1 only. Although this masks much of the richness of ecosystem interactions, it suffices to show how human intervention can reverberate through the ecosystem. Humans crop species 1 (grain) for use as an input in the production of manufactured good 1, and they intervene in species 3 (predators) which is not an input into any production process.

The utility function for consumer p is given by

$$U^p(b_1^p, b_2^p, R_3^h, l^p), \quad p = 1, 2, \tag{35}$$

where

$$b_1^1 + b_1^2 = b_1(L_1^b, c_1 R_1), \tag{36}$$

$$b_2^1 + b_2^2 = b_2(L_2^b), \tag{37}$$

$$\begin{aligned} R_1^h &= [1 - c_1]R_1 \\ &= [1 - c_1]\,[E_{01}X_{01} - E_{12}\overline{X}_{12} - E_{14}G^1(X_{01}, \overline{X}_{12}, \overline{L}_1)], \end{aligned} \tag{38}$$

$$R_2^h = E_{12}X_{12} - E_{23}\overline{X}_{23} - E_{24}G^2(X_{12}, \overline{X}_{23}), \tag{39}$$

$$R_3^h = E_{23}X_{23} - E_{34}G^3(X_{23}, \overline{L}_3), \tag{40}$$

$$L_1^b + L_2^b + \overline{L}_1 + \overline{L}_3 = L - l^1 - l^2. \tag{41}$$

Expressions (36) and (37) are the manufactured goods that are divided between the two consumers. The first manufactured good requires an ecosystem input from species 1. Expressions (38)—(40) describe the stored energy of the three species in the ecosystem which is assumed to be in equilibrium. Note that in equilibrium we have the species' demand functions

$$X_{01} = X_{01}(E_{01}, E_{12}, E_{14}, X_{12}, \overline{L}_1) \tag{42}$$

$$X_{12} = X_{12}(E_{12}, E_{23}, E_{24}, X_{23}) \tag{43}$$

$$X_{23} = X_{23}(E_{23}, E_{34}, \overline{L}_3). \tag{44}$$

Expression (41) accounts for the total available labor.

Humans derive positive utility in (35) from manufactured goods and leisure. They derive negative utility from species 3, therefore, the stored energy of species 3 is included in the utility functions. Stored energy can be thought of as a proxy for the prevalence of a species, and we assume it is this prevalence that humans respond to. Let $p = 1, 2$ throughout, and let $U_q^p = \partial U^p / \partial b_q^p > 0$ for $q = 1, 2$ and $U_{hi}^p = \partial U^p / \partial R_3^h < 0$.

Necessary conditions for a Pareto optimum allocation of labor can be derived by maximizing

$$U^1(b_1^1, b_2^1, R_3^h, l^1) + \lambda[U^2(b_1^2, b_2^2, R_3^h, l^2) - \overline{U}^2] \tag{45}$$

with respect to the manufactured goods allocated to each consumer and the labor/leisure inputs, and the utility of consumer two fixed at $\overline{U}^2$ with λ as a multiplier. Maximization is subject to the constraints and functions given by (36)–(44). Details on the derivation of the first-order conditions are available from the authors; here, the results obtained from those conditions are summarized.

First, a Pareto optimum requires that both consumers' marginal rates of substitution (MRS) between the manufactured goods be equal to the marginal rate of product transformation (MRPT) for the goods:

$$U_2^1/U_1^1 = U_2^2/U_1^2 = \frac{\partial b_1/\partial L_1^b}{\partial b_2/\partial L_2^b}. \tag{46}$$

Second, the allocation of labor between manufactured good one and cropping species one satisfies the condition for production efficiency:

$$\frac{\partial b_1}{\partial L_1^b} = \frac{\partial b_1}{\partial c_1 R_1} \frac{\partial c_1 R_1}{\partial L_1}. \tag{47}$$

The marginal product of labor (MPL) used directly in the production of b_1 equals the MPL used indirectly via species one.

We can generalize (47) by assuming that species one provides utility directly to humans so that R_1^h enters both utility functions. For instance, instead of grain, suppose species one is Douglas fir trees which can be cropped to make the manufactured good lumber, or which can be left standing for aesthetics or biological diversity. Then (47) would become

$$\frac{U_{h1}^1}{U_1^1} + \frac{U_{h1}^2}{U_1^2} = \frac{\dfrac{\partial b_1}{\partial L_1^b} - \dfrac{\partial b_1}{\partial c_1 R_1} \dfrac{\partial c_1 R_1}{\partial L_1}}{\partial R_1^h/\partial L_1} \tag{48}$$

Equation (48) requires the sum of the MRS's between species one and manufactured good one over both consumers to equal the MRPT between species one and manufactured good one. That the MRS's are summed follows from the public good nature of species one in its nonconsumptive uses. The MRPT accounts for the use of labor as input into the manufactured good in two distinct ways: directly and indirectly through species one. Increasing direct labor to manufacturing good one implies, *ceteris paribus*, that less labor is available to crop species one, therefore, less of species one is available as an input to this manufactured good.

Examining (48) further, if $U_{h1}^1 = U_{h2}^2 = 0$, which is the case where species one is grain, then the numerator on the right side is zero. This case corresponds to (47). Alternatively, if U_{h1}, $U_{h2} > 0$, as would be the case if species one were Douglas fir trees, then the numerator is negative. (Recall, the denominator is negative from (30).) This implies that L_1 is not employed to the point where the MPL's are equal, because using labor to crop species one creates losses in nonconsumptive utility. The greater the direct marginal utility of species one, the more that direct labor is substituted for species one (and, therefore, indirect labor) in the production of manufactured good one. The familiar trade off between development and allowing ecosystems to remain in their natural state is exemplified.

Lastly, returning to the case of Kern County, where species one is grain, we have equation (49) which is analogous to (6).

$$\frac{U_{h3}^{1}}{U_{1}^{1}} + \frac{U_{h3}^{2}}{U_{1}^{2}} = \frac{\dfrac{\partial b_1}{\partial L_1^h} - \dfrac{\partial b_1}{\partial c_1 R_1}\dfrac{\partial c_1 R_1}{\partial L_3}}{\partial R_3^h / \partial L_3} \tag{49}$$

Again there is the public good aspect where MRSs are summed and equal to an MRPT. For Kern County, the MRPT is between grain and predators. The second term in the numerator on the right side is the ecosystem externality corresponding to $(\partial b/\partial g)\,(\partial g/\partial L_p)$ in (6). However, now we can be more precise about the nature of this term because the interactions within the ecosystem are included in the problem. Accounting for the functional relationships in (36)–(44), the ecosystem externality is

$$\frac{\partial c_1 R_1}{\partial L_3} = c_1(L_1)\frac{\partial X_{23}}{\partial L_3}\left\{\frac{\partial X_{12}}{\partial X_{23}}\left[E_{01}\frac{\partial X_{01}}{\partial X_{12}} - E_{12}\right.\right.$$

$$\left.\left. - E_{14}\left[\frac{\partial G^1}{\partial X_{01}}\frac{\partial X_{01}}{\partial X_{12}} - \frac{\partial G^1}{\partial X_{12}}\right]\right]\right\}. \tag{50}$$

The externality represents how the grain cropped for human consumption is affected by predator elimination. The effect occurs via linkages between predators, mice and grain. The first partial on the right-side of (50) shows how human intervention in the predator species reduces predator consumption of mice. The second partial shows how more mice changes the available grain. The bracketed term shows how the change in mice's grain consumption affects the solar absorption and maintenance activities of grain.

Returning to (49), the left-side is negative in the Kern County example owing to the human distate for predators. The denominator on the right-side is negative, because of the negative intervention. (See the discussion for (30).) Therefore, the numerator on the right-side is positive, and because the first two partials are positive, the sign of the ecosystem externality is ambiguous. This is consistent with (50) where the sign of the term in braces is ambiguous. However, as an empirical test, we know that (50) was negative in Kern County.

VI. Conclusions

What is academically important in natural resource and environmental questions has become progressively less distinguishable from what is academically and societally urgent. For several academic disciplines, including ecology and economics, analytical constructs and empirical methods that took 100 years or more to establish have been drawn tight in one-tenth the time. With its insistence that one not confuse the depletion of valuable environmental assets with the generation of income, and its skepticism of temporary improvements in human consumption purchased by permanent losses in environmetal wealth and productivity, the sustainable development literature has made vivid the inadequacies of many academic formulations and policy discussions of environmental protection. Practical implementation of the common sense of sustainable development insights requires a common or at least conjugate understanding of environment-economy interactions, however. This paper is an attempt to lay a foundation for building that understanding.

Current understandings of environment-economy interactions crudely distinguish only between the fabricated good inputs and the life support and amenity services that ecosystems offer. They arbitrarily lump together the disparate organisms found in the lower trophic levels of ecosystems in a fashion that hinders predictions and evaluations of *which* ecosystem services are to be usurped by humans, *which* waste flows are to be allowed into ecosystems, and *which* ecosystem organisms and physical attributes are to be maintained. Arguably the contributions of natural environments to human well-being depend as much upon the details of ecosystem and economic states as upon the gross relationship between ecological systems and economic systems. The formulation in this paper provides one unified but limited framework, consistent nevertheless with broadly accepted paradigms in ecological and economic theory, within which these systematic insights about details may evolve.

If sustainable development ideas are to be operational in economic policy, a means must be found to integrate national environmental accounts such as those in Japan, Canada, France, and the Netherlands with standard national economic accounts. These environmental accounts are stated in physical units. The start toward reconciliation of ecological and economic paradigms that this paper offers provides a means, in principle, of incorporating environmental protection and natural resource management based on physical units into economic policymaking. The frequent tensions and occasional bizarre anomalies that now arise between environmental and economic perspectives do not allow the researcher to account systematically for the economic consequences of externalities that impinge upon remote activities through the trophic levels of ecosystems. Many symptoms of damage to an ecosystem and to the human activities that employ ecosystem services appear far from the biological point of entry of a toxin to the system. Any unified and robust framework for ascertaining economically efficient environmental management programs must admit ecosystem process descriptions and be able to incorporate all of the detail that the ecosystem and the economic models provide. Research designs for ecosystem management must simultaneously be narrower and broader than current efforts: narrower in the sense that less spatial, temporal, and species aggregation is involved; broader in that a wide range of conditions can be treated.

We submit that the key to a more robust perspective on sustainable development is explicit recognition in the fashion of this paper that some human decision variables simultaneously influence and are influenced by the detailed behaviors of ecosystems. Acknowledgement of the parallels between ecological systems and economic systems implies that economic activity has physical and biological foundations. Though such acknowledgement can lead to praiseworthy attempts to incorporate restrictions about structure, functional forms, and parameter values from other disciplines into the behavioral postulates of economics, it must be inadequate if it does no more than treat physical foundations as exogenous or predetermined. Our premise has been that these foundations are endogenous, not wholly because of constraints on invention and innovation. They are endogenous because human-induced ecosystem adaptations place limits upon economic activity. This paper represents our attmept to frame this point.

Acknowledgements

This research was funded in part by U.S. Environmental Protection Agency Grant R-806972010. Very helpful comments on various segments of this research were provided by Shmuel Amir, Gardner Brown, Ronald Cummings, A. M. Freeman III, Bruce Forster, Bruce Hannon, Susan Kask, Robert Kendall, Jon Miller, James Moore and three anonymous referees.

Notes

[1] See Hirshleifer (1977) for a more lengthy coverage of this point.

[2] See Muize (1977).

[3] The following newspaper story from the *Laramie (WY) Daily Boomerang*, Feb. 8, 1988, provides an example of economic-ecosystem interactions when a species directly enters the utility function.

"A great horned owl, not a bizarre cult, mutilated more than 50 rabbits in a Teaneck park, New Jersey wildlife officials have concluded."

"A $1,000 reward was offered by the Fund for Animals. . . . It was thought a cult might be responsible because several of the carcasses were lined up in a row."

". . . the owl, . . . was eating only the heart and lungs of the rabbits — the blood rich parts — because there is an abundance of rabbits in the park."

When the wildlife officials tried to collect the reward, the Fund said the owl theory was " 'inconclusive'."

[4] We include grain in the ecosystem rather than as a manufactured good because there are significant interactions between it and other species. Pimm (1982, p. 186) notes that food chains are short, with the most common chains having only three or four trophic levels.
[5] There are many field and laboratory studies that support these comparative statics. For example, Menge (1972) observed the foraging strategy of starfish in the Sunloun Islands. Starfish would consume more of a particular barnacle species in the summer and autumn when the species was relatively more abundant. Greater relative abundancy implies less energy expended to assimilate a given amount of energy. Also, prey species with relatively less effective escape responses were consumed more frequently. In the laboratory, Rapport (1978) observed the foraging strategy of protozoans. He found that increases in the relative abundance of any prey species led to greater consumption of that species. See also Hof *et al.* (1990) and the papers in Staddon (1980).
[6] For simplicity, the labor endowment is taken as exogenous. Ultimately, labor depends on cropping ecosystem outputs from agriculture, and this could be incorporated in the model.
[7] Species having no control is a simplification because mammals will avoid areas where cohorts are killed, and even plant species may experience shifts in gene frequencies such that less desirable genotypes may come to dominate and dissuade the cropper.

References

Caraco, T. and S. L. Lima (1985), 'Survivorship, Energy Budgets, and Foraging Risk', in M. Commons, S. J. Shettleworth, and J. Nevin, eds., *Quantitative Analyses of Behavior*, Vol. 6, Cambridge, MA: Ballinger.

Hannon, B. (1976), 'Marginal Product Pricing in the Ecosystem', *Journal of Theoretical Biology* **56**, 253—267.

Hirshleifer, J. (1977), 'Economics from a Biological Standpoint', *The Journal of Law and Economics* **20**, 1—52.

Hof, J., D. Rideout, and D. Binkley (1990), 'Carbon Fixation in Trees as a Micro Optimization Process: An Example of Combining Ecology and Economics', *Ecological Economics* **2**, 286—294.

Lotka, A. J. (1925), *Elements of Physical Biology*, Williams and Wilkins, Baltimore, MD.

Menge, B. (1972), 'Foraging Strategy of a Starfish in Relation to Actual Prey Availability and Environmental Predictability', *Ecological Monographs* **42**, 25—50.

Muize, K. P. (1977), 'The Great Kern County Mouse War', *Audobon* **79**, 159.

Odum, E. P. (1971), *Fundamentals of Ecology*, Philadelphia: W.B. Sanders Company.

Pimm, S. L. (1982), *Food Webs*, New York, NY: Chapman and Hall.

Rapport, D. J. (1978), 'Foraging Behavior of *Stentor Coeruleus*: A Microeconomic Interpretation', in A. C. Hamil and T. D. Sargent, eds., *Foraging Behavior, Ecological, Ethological, and Psychological Approaches*, Garland STPM Press, New York, pp. 77—93.

Staddon, J. E. R. (1980), *Limits to Action: The Allocation of Individual Behavior*, New York, NY: Academic Press.

Ulanowicz, R. E. (1972), 'Mass and Energy Flow in Closed Ecosystems', *Journal of Theoretical Biology* **34**, 239—253.

Zucahetto, J. (1975), 'Energy-Economic Theory and Mathematical Models for Combining the Systems of Man and Nature', *Ecological Modelling* **1**, 241—268.

[7]

Economic and technological models for evaluation of energy policy

Kenneth C. Hoffman

Brookhaven National Laboratory

and

Dale W. Jorgenson

Harvard University

Models for energy policy assessment have been developed using both process analysis and econometrics. The process approach provides for the incorporation of information on future technological and structural changes based on detailed engineering studies. The econometric approach is well adapted to the description of aggregative consumer behavior and economic activity. This paper presents a new approach for policy assessment, integrating process analysis and econometric models that have been used extensively in energy policy analysis and technology assessment. We illustrate the application of this approach by an analysis of a national research, development, and demonstration plan for the United States.

1. Introduction

■ The Arab oil embargo of 1973 and the extraordinarily cold winter of 1977 are part of a series of events that have elevated energy policy in the United States to the highest level of social priority. A new Federal government agency, the Federal Energy Administration, was established to administer price controls and associated allocation mechanisms adopted in the face of higher world oil prices. A second agency, the Energy Research and Development Administration, was established to coordinate research, development, and demonstration projects for the energy sector. These agencies have now been incorporated into a new cabinet-level Department of Energy. Energy research and development in the private sector has expanded with increased financial support from both government and private sources.

The guidelines of a national energy policy for the United States have emerged; detailed programs are being developed and implemented. To preserve flexibility in the face of uncertainty a continuous assessment of existing programs is required. Analysis is needed for likely impacts of such policy measures as price controls, taxes, and regulation to stimulate energy conservation in the private sector, government support to generate additional conventional energy supplies, and government sponsored research and development programs, directed toward providing new technology for energy production, conversion, and utilization. The evaluation of new and existing energy policies must incorporate information from detailed engineering studies of specific technologies emerging from research and development programs and must include the assessment of policy impacts on the structure of the energy sector and on the overall level and composition of economic activity.

Alternative models for energy policy assessment have been developed on the basis of both process analysis and econometrics. In the process analysis approach energy flows and energy conversion processes are described in physical terms. The description need not be limited to a particular technology, but can encompass the entire system for the production and utilization of energy. In the econometric approach the representation of technology is based on behavioral and technical responses of production patterns to alternative prices; a similar approach can be employed for the representation of consumer preferences. Flows of economic activity, including energy flows, are described in terms of economic accounts in current and constant prices.

The process approach provides for the incorporation of explicit information on anticipated technological change in the energy system and in the economy based on detailed engineering studies, including studies of technologies that are under development and consideration for future implementation. This approach is well adapted to the description of the energy sector; however, the representation of aggregate consumer behavior and economic activity by means of process analysis is infeasible. The econometric approach including appropriate exogenous variables to represent new policy initiatives is well adapted to the description of aggregate economic activity in summary form and provides for the analysis of policy impacts on the overall level of economic activity and its distribution among industry groups or groups of consumers. However, this approach is infeasible for the study of technologies that are not already in use or for the study of consumer preferences for commodities not already in existence.

A satisfactory framework for the assessment of the full range of alternative energy policies requires an approach that encompasses both process analysis and econometrics. Since the output of the energy producing industries is largely consumed by other industries rather than by final consumers such as households, governments, and the rest of the world, a natural focal point for the study of the impact of energy policy is the matrix of interindustry transactions, representing flows of commodities, including energy, among industrial sectors. For the energy sector these transactions can be expressed in economic terms, in current and constant prices, to provide a link with econometric models. These energy sector transactions can also be expressed in physical terms, in British thermal units, to provide a link with process analysis models. By using both forms for the expression of energy flows, process analysis and econometric modeling can be combined.

The purpose of this paper is to present a new approach for the assessment of energy research, development, and demonstration policy, integrating process analysis and econometric models that have already been used extensively in energy policy analysis and technology assessment. The first component of the approach is an econometric model of interindustry transactions together with a macroeconomic model, presented in Section 2, developed for the Energy

FIGURE 1

COMBINED ECONOMETRIC AND PROCESS ANALYSIS MODEL

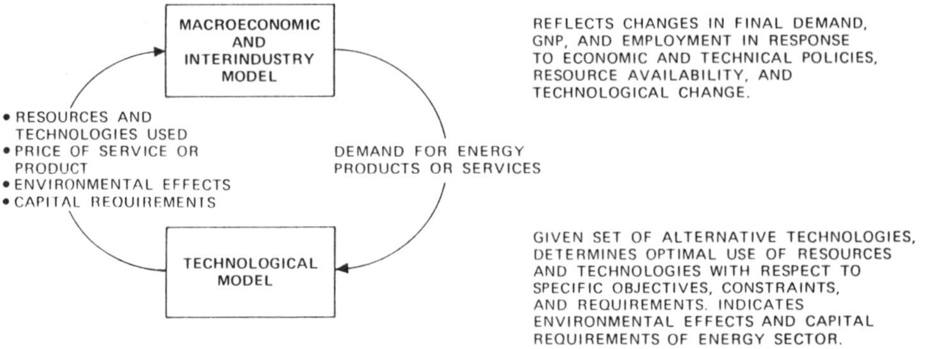

Policy Project. The second component is a process analysis model of the energy sector, also presented in Section 2, developed for the Energy Research and Development Administration. We present the combined econometric and process analysis approach in Section 3; a preliminary model based on this approach has been employed by the Energy Research and Development Administration in the construction and analysis of a national research, development, and demonstration plan. In Section 4 we employ the results of a policy analysis prepared for the national plan to illustrate the application of this approach to energy policy assessment.

The methodology for our combined econometric and process analysis model is illustrated in Figure 1. The econometric model reflects economic impacts at the aggregate level, including changes in final demand and employment, that result from changes in energy policy. The process analysis model determines the optimal use of resources for a given energy policy and a given economic environment. Our methodology can be applied to a wide range of national policy questions where the technological component is significant. In any policy area involving technology it is necessary to make explicit the relationships between choice of technology and the economic and social environment. Process analysis is the most appropriate methodology for describing alternative technologies; econometrics provides a basis for describing consumer behavior in the economy as a whole. With this combined approach, provision is made to allow for structural changes in the energy system and economy arising from technological change and specific policy actions. Our model represents the first attempt to implement a model that combines the advantages of both methodologies.

2. Econometric and process analysis models

■ The first component of our model for the analysis of energy research, development, and demonstration policy is an econometric model of interindustry transactions, developed by Hudson and Jorgenson (1974a). This model is based on a system of accounts for the private domestic sector of the U.S. economy, including final demand, primary input, and interindustry transactions in current and constant prices. By means of this accounting system we can trace the process of production for energy and nonenergy products from the purchase of primary inputs through all stages of intermediate processing to deliveries to final demand. The accounts in constant prices correspond to commodity flows in physical terms. For energy sectors (or industries) these flows can be measured in physical units such as tons of coal, barrels of petroleum, and thousands of cubic feet of natural gas, or, alternatively, in energy units such as British thermal units (Btu's).[1] The accounts in current prices correspond to flows in financial terms and can be used to generate financial accounts for each industry group included in the model. For energy and nonenergy sectors the prices can be expressed as index numbers; for energy sectors the prices can also be given in terms of physical or energy units.

In our system of accounts the private domestic sector of the U.S. economy is divided among nine industry groups, including five groups within the energy sector—coal mining, crude petroleum and natural gas, petroleum refining, electric utilities, and gas utilities. Our representation of the energy sector provides for an analysis of the impact of energy research, development, and demonstration policy on the industrial sectors directly affected by changes in energy technology. By incorporating final demand and four industry groups making up the nonenergy sector we can assess the impact of changes in energy technology on the sectors that consume energy products. Our complete system of accounts is represented in diagrammatic form in Figure 2. The nine industry groups included in the accounting system are listed in Figure 2. In this Figure we also list three categories of primary inputs—capital services, labor services, and imports—and four categories of final demand—consumption, investment, government purchases, and exports.

In our system of accounts for interindustry transactions, each industry group purchases primary inputs and intermediate inputs produced in each of the nine industrial sectors. These purchases are represented as columns of the matrix of interindustry transactions in Figure 2. Intermediate inputs include five types of energy—coal, crude petroleum and natural gas, refined petroleum, refined natural gas, and electricity—and four types of nonenergy products. The output of each industry is distributed to final demand and to intermediate demand by each of the nine industrial sectors. These deliveries are represented as rows of the matrix of interindustry transactions in Figure 2. The rows corresponding to the five industries that make up the energy sector include deliveries of energy products to energy and nonenergy sectors and to final demand. Similarly, the rows corresponding to the four industries of the nonenergy sector include deliveries of nonenergy products.

Our econometric model of interindustry transactions includes balance equations between supply and demand for the products of each

[1] The data are described in more detail in a report by Jack Faucett Associates (1973). Interindustry accounts for the year 1967 have been compiled for a more detailed industry breakdown by Bullard and Herendeen (1973a, b). The corresponding interindustry model has been linked to the Brookhaven Energy System Optimization Model discussed below by Hoffman, Palmedo, Marcuse, and Goldberg (1973) and by Behling, Marcuse, Swift, and Tessmer (1975). Energy flows for six years have been compiled for the Federal Energy Administration by Jack Faucett Associates (1975).

FIGURE 2

INTERINDUSTRY TRANSACTIONS IN THE ECONOMETRIC MODEL

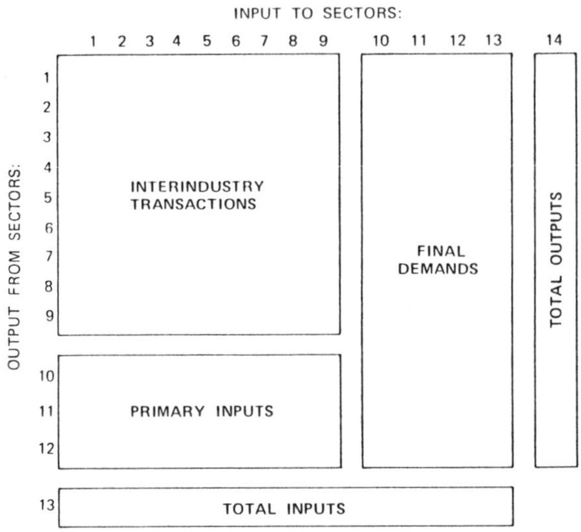

INDUSTRY SECTORS:

1. AGRICULTURE, NONFUEL MINING, AND CONSTRUCTION.
2. MANUFACTURING, EXCLUDING PETROLEUM REFINING.
3. TRANSPORTATION.
4. COMMUNICATIONS, TRADE, AND SERVICES.
5. COAL MINING.
6. CRUDE PETROLEUM AND NATURAL GAS.
7. PETROLEUM REFINING.
8. ELECTRIC UTILITIES.
9. GAS UTILITIES.

PRIMARY INPUTS:

10. IMPORTS.
11. CAPITAL SERVICES.
12. LABOR SERVICES.

FINAL DEMANDS:

10. PERSONAL CONSUMPTION EXPENDITURES.
11. GROSS PRIVATE DOMESTIC INVESTMENT.
12. GOVERNMENT PURCHASES OF GOODS AND SERVICES.
13. EXPORTS.

of the nine industrial sectors included in the model.[2] These balance equations state that the output of each sector in constant prices must be equal to deliveries of this output to all nine industrial sectors and to all four categories of final demand. For energy products the balance equations assure that for each form of energy, the energy units produced must be equal to the energy units consumed by all industrial groups and by final demand. Similarly, our econometric model includes balance equations stating that the output of each sector in current prices must be equal to the value of deliveries of this output to all nine industrial sectors and to final demand. These equations assure that differences between prices received by producers and prices paid by consumers reflect excise and sales taxes paid on the value of each product.

Our econometric model of interindustry transactions includes models of producer behavior for each industrial group included in the model.[3] Producer behavior in each industrial sector can be characterized by a system of technical coefficients, giving primary and intermediate inputs per unit of output of the sector. The model of producer behavior gives the technical coefficients as functions of the prices of output and of primary and intermediate input. For each sector the technical coefficients as functions of the prices are generated from the price possibility frontier, giving the minimum price of output of the sector attainable for given prices of primary and inter-

[2] The model of interindustry transactions is described by Hudson and Jorgenson (1974a), especially pp. 467–474.

[3] The model of producer behavior is described by Berndt and Jorgenson (1973). See also Christensen, Jorgenson, and Lau (1973) and Berndt and Wood (1976).

mediate inputs and for a given level of productivity of the sector. The minimum price of output depends on the technological possibilities for substitution among primary and intermediate inputs, including the substitution between energy and nonenergy inputs and the substitution among different forms of energy. The price possibility frontier for each sector provides a representation of the technology of that sector. This representation assures that the value of output of the sector is equal to the sum of the values of all primary and intermediate inputs into the sector.

Finally, our econometric model of interindustry transactions includes a model of consumer behavior that allocates personal consumption expenditures among the commodity groups included in final demand.[4] Consumer behavior can be characterized by a system of quantities purchased per capita. The model of consumer behavior gives the quantities purchased as functions of total personal consumption expenditures per capita, prices of the products of the nine industrial sectors, and prices of capital services and noncompetitive imports. The quantities purchased as functions of total expenditure and the prices can be generated from the indirect utility function, giving the maximum level of utility attainable for given total expenditure and given prices. The maximum level of utility depends on the substitutability of alternative goods and services in consumption, so that the indirect utility function provides a representation of consumer preferences. This representation assures that the sum of the values of all quantities purchased is equal to total personal consumption expenditures.

Starting with prices of primary inputs—capital services, labor services, and imports—and levels of productivity in each of the nine industrial sectors, the prices of both energy and nonenergy products are determined by the nine price possibility frontiers. With prices of primary inputs and prices of energy and nonenergy products determined from our model of production, we can generate the matrix of technical coefficients, giving primary and intermediate inputs per unit of the output of each of the nine industrial sectors. Similarly, with total personal consumption expenditures, the prices of capital services and noncompetitive imports, and the prices of energy and nonenergy products, we can generate the quantities purchased per capita of the products of the nine industrial sectors, capital services, and noncompetitive imports. Given the level of population, we can convert these quantities per capita to quantities of personal consumption expenditures as a component of final demand. To obtain final demand for the output of each of the nine industrial sectors we add personal consumption expenditures to gross private domestic investment, government purchases of goods and services, and exports.

From the quantities of final demand for the output of each of the nine industrial sectors and the matrix of technical coefficients, providing intermediate input per unit of output of each sector, we can determine the quantities of output of both energy and nonenergy sectors. We can also determine the distribution of the output of each sector between intermediate and final demand and the distribution of intermediate demand among intermediate inputs to each of the nine

[4] The model of consumer behavior is described by Jorgenson (1975). See also Christensen, Jorgenson, and Lau (1975) and Jorgenson and Lau (1975).

industrial sectors. The output of energy sectors and its distribution can be expressed in constant prices, physical units such as tons of coal or barrels of petroleum, or energy units such as Btu's. From the matrix of technical coefficients of primary input per unit of output, we can determine the quantities of primary input into each sector. Finally, given the nine industrial prices and the prices of primary inputs, we can express the flow of primary input, interindustry transactions, and final demand in current prices. We can generate the complete system of interindustry accounts in current and constant prices from the prices of primary inputs, the levels of productivity in each interindustry sector, total personal consumption expenditures, and the quantities of final demand for the output of each sector for investment, government purchases, and exports.[5]

The second component of our model for the analysis of energy policy is a process analysis model of the energy sector, developed at the Brookhaven National Laboratory (1973). This model is based on the Reference Energy System presented in Figure 3. This description

FIGURE 3

REFERENCE ENERGY SYSTEM

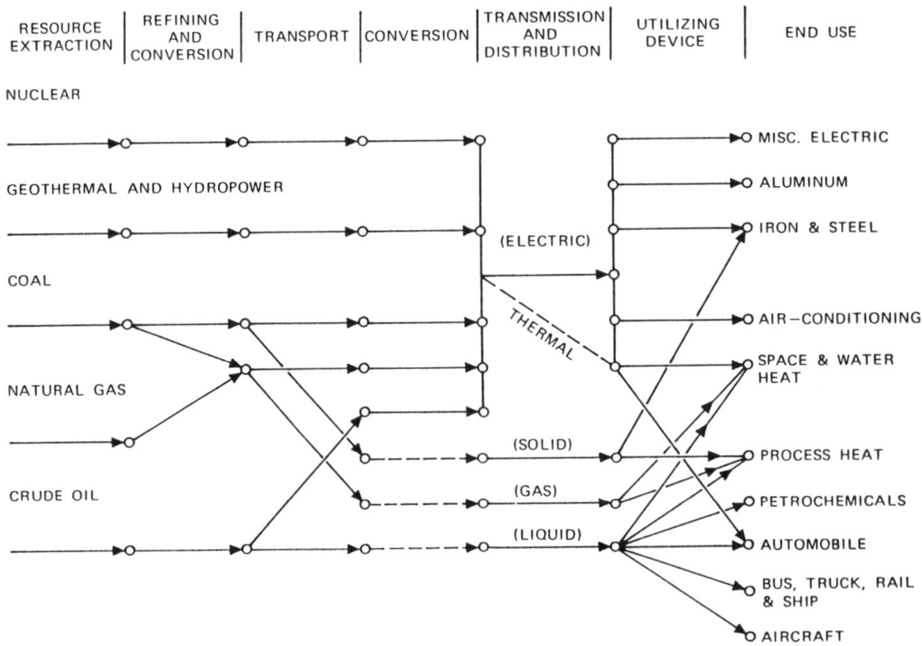

of the U.S. energy system provides a complete physical representation of the technologies, energy flows, and conversion efficiencies from extraction of primary energy sources through refining and various

[5] Applications of the econometric model to policy analysis are given by Hudson and Jorgenson (1974 a, b, c; 1975 a, b) and Jorgenson and Wright (1975).

stages of conversion from one energy form to another, and through transportation, distribution, and storage of energy.[6] In the Reference Energy System energy supplies such as nuclear fuels, fossil fuels, and hydropower are allocated to energy demands defined on a functional basis, such as space heating, industrial process heat, and automotive transportation. The characteristics of utilizing technologies, which are important in the identification of conservation and fuel substitution options, are included at the same level of detail as supply technologies. The allocation of energy resources to specific demands depends on the energy technologies that are available for the production, transportation, distribution, and storage of energy and on the cost and efficiency of these technologies. The allocations may be determined by a judgmental or optimization approach. Conversion losses are represented by the efficiency of each conversion process in physical terms. In the Reference Energy System all energy flows are measured in British thermal units (Btu's).

In the Reference Energy System energy supplies and demands are linked by energy conversion processes, such as steam generation of electricity from coal. This process converts a primary energy supply, coal, into an intermediate form of energy, electricity. Electricity can be used to satisfy demands for a variety of energy products, such as base, intermediate, and peak load electricity, space heat, air conditioning, and water heat. For each process we can specify the efficiency of conversion of primary energy supplies into intermediate forms of energy and the efficiency of conversion of the intermediate forms into final energy services or products. For the coal steam electric process the conversion loss from the primary to the intermediate form of energy is associated with the generation of electricity. Similarly, the conversion loss from intermediate to final form of energy is associated with transmission and distribution losses for electric energy and the conversion efficiency of the end use device. The supply efficiency for a given energy source is defined as the product of the supply efficiencies on a path from the primary resource to the intermediate form of energy. Similarly, the demand efficiency is defined as the product of the demand efficiencies on a path from the intermediate form to the final energy product.

The Brookhaven Energy System Optimization Model (BESOM) is based on the allocation of energy supplies to energy demands to minimize cost.[7] The minimization of cost can be formulated as a linear programming problem of the transportation type. Sources in the transportation problem can be identified with energy supplies; in the optimization model there are eleven types of energy supplies, including underground and strip-mined coal, domestic, shale, and imported oil, domestic and imported natural gas, and hydro electric, nuclear, geothermal, and solar energy. Uses can be identified with energy demands, including base, intermediate, and peak load electricity, low, intermediate, and high temperature thermal, ore reduction, petrochemicals, space heat (including heat pumps as well as electric resistance heat), air conditioning, and water heat, and air, truck and bus, rail and automobile transportation. Energy storage and synthetic fuels including hydrogen are also incorporated in the model.

[6] The Reference Energy System is described in Beller *et al.* (1975).

[7] The Brookhaven Energy System Optimization model is described by Hoffman (1973) and Cherniavsky (1974).

The optimization model is designed around the Reference Energy System. Each trajectory through the system from a resource to a specific end use is represented by a single activity. The input data for the model are:

(1) The level of demand for energy services for energy products, consistent with those determined in the integrated model.
(2) Annual production constraints on supply of energy resources and availability constraints on new technologies.[8]
(3) Characterization of energy supply and use technologies in terms of conversion efficiency, capital, and operating cost, and emissions to the environment (air, water, and land).
(4) Definition of objective function.[9] This can be based on annual cost, including the cost of energy resources, or on alternative objectives that include utilization of primary resources, capital requirements, environmental impact, and dependence on imports.
(5) Definition of any special constraints required to reflect policies or market forces that result in departures from an unconstrained optimum.

Given these data, the optimization determines resource utilization, technology, and fuel mix employed to satisfy the energy product requirements. Activity levels are given in terms of the quantity of fuel or energy delivered from a supply trajectory to the end use.

In BESOM each energy supply-demand combination is associated with costs of extraction, refining and conversion, transportation and storage, and final utilization. Annual costs per unit of operation of an energy conversion process include both capital costs and operating costs. Capital costs are converted into annual form and are conceptually equivalent to the capital service prices that enter into the econometric model of interindustry transactions. Constraints on the supply of energy resources and the degree of implementation of new technologies are based on geological information, market surveys, and engineering judgment. A version of BESOM has been developed which incorporates supply elasticities to relate annual resource production levels to the shadow prices on resources determined in the model. Each energy conversion process produces environmental pollutants as well as intermediate forms of energy. Constraints can be imposed on the level of environmental pollution as well as on the level of energy demand and supply. Capacity limitations on energy conversion processes can be included as separate constraints in the optimization model; additional constraints corresponding to balance requirements between peak and off-peak electricity generation are also included.

The energy sector optimization model determines a set of energy conversion levels that minimizes the cost of satisfying energy product demands from energy resource supplies. The dual to this linear programming problem is to maximize the value of energy products less the value of primary energy supplies by choosing a set of energy

[8] The constraints of the optimization model are described by Cherniavsky (1974), pp. 9–18.

[9] The objective function of the optimization model is described by Cherniavsky (1974), especially pp. 18–23.

product and energy resource shadow prices. These shadow prices assure that the value of the output of each conversion process in actual use is equal to the value of input, including the input cost of primary energy supplies and any scarcity shadow prices and costs of extraction, conversion, and transportation, just as in the econometric model of interindustry transactions. Any energy resource with a positive price is fully utilized; similarly, the demand for any energy product with a positive price is exactly satisfied. For energy products and resources with positive shadow prices, supplies and demands are balanced in both physical terms and in current prices, as in our econometric model of interindustry transactions.

The solution of the dual to the energy sector optimization model determines the shadow prices associated with energy products and energy resources. The assignment of energy supplies to energy demands through energy conversion processes determined by the model can be represented in physical terms in the Reference Energy System format. Given the prices of resources and the costs associated with energy conversion processes, BESOM also provides a complete description of the energy sector in financial terms. We can generate the resultant energy system scenario in the format of the Reference Energy System in both physical and financial terms from the costs of energy conversion processes, the availability of energy resources, the requirements for energy products, and any additional constraints associated with conversion capacities and environmental restrictions.

In appraising alternative energy research, development, and demonstration policies we first associate with each policy the resulting technology for the energy sector; changes in energy research and development policy are associated with changes in dates of commercial implementation, costs, and technical characteristics of specific energy technologies.[10] We can introduce the corresponding changes in energy technology into the Reference Energy System in two ways. First, the introduction of new technologies provides new energy conversion processes in addition to those that already exist. Accelerated research, development, and demonstration programs may make it possible to accelerate the introduction of new technologies. Second, the improvement of existing technologies may increase the efficiency of energy conversion or may reduce the costs of extraction, conversion, or transportation. More extensive research, development, and demonstration may speed the increase in efficiency or the reduction in cost. The introduction of new energy technology or the improvement of existing technology may reduce the costs associated with meeting given demands for energy products from given energy resource supplies. For any change in the technology options resulting from new policy initiatives we can assess the effects on the energy sector in both physical and financial terms, using the energy sector optimization model. We can also assess the environmental impact of the changes, using the environmental impact associated with alternative energy conversion processes.

---

[10] Methodology for application of the optimization model for assessment of alternative energy research, development, and demonstration policies is described by Hoffman and Cherniavsky (1974) and Cherniavsky (1975). This methodology was applied in a series of twelve scenarios for 1985 and 2000 in *A National Plan for Energy Research, Development, and Demonstration* (1975).

3. Model integration

■ We have presented the two models utilized for the analysis of energy research, development, and demonstration policy. The first is an econometric model of interindustry transactions providing a representation of the technology of the energy sector through models of producer behavior for five industrial groups that make up that sector. In addition, the econometric model provides a representation of the nonenergy sector through models of producer behavior for four industrial groups making up that sector and a model of consumer behavior for the personal consumption expenditures component of final demand. The second model for the analysis of energy policy is an optimization model for the energy sector providing a much more detailed representation of the technology of that sector through the specification of characteristics of energy conversion processes linking energy resource supplies with energy resource demands. The energy sector optimization model includes existing technologies, such as steam generation of electricity from coal, and technologies that can be developed through research, development, and demonstration programs, such as the liquid metal fast breeder reactor for the generation of electricity.

Although both components of our energy policy model can be used to generate a description of the energy sector in physical and financial terms, the energy sector optimization model provides a far more detailed characterization of technology and permits the analysis of the effects of introducing new technologies. The econometric model also provides a description of the nonenergy sector and generates a complete description of the U.S. economy, including flows of primary input, interindustry transactions, and final demand in current and constant prices. The energy sector optimization model is especially well suited to the assessment of the impacts of alternative research, development, and demonstration policies on the energy sector. The econometric interindustry model is well suited to the assessment of the impact of these policies on the economy as a whole. By integrating the two models we can combine the detailed characterization of technology available from the energy sector optimization model with the complete representation of the economy, including energy and nonenergy sectors, available from the econometric interindustry model. This section describes the conceptual basis for the integration of the models. A preliminary version of the integrated model has been implemented and applied by Behling, Dullien, and Hudson (1976).

Our integrated model is based on an expanded system of interindustry accounts for the private domestic sector of the U.S. economy. In our expanded system of interindustry accounts the energy sector is divided into energy resource sectors, energy conversion processes, and energy product sectors. The remaining components of our original system of interindustry accounts—interindustry transactions in nonenergy products, primary inputs, and final demands—are also included in the expanded system. The expanded system of interindustry accounts is presented in diagrammatic form in Figure 4.[11] The

[11] The sectors in this expanded system of accounts include:

(a) energy resource sectors: (1) underground coal, (2) strip-mined coal, (3) domestic oil, (4) shale oil, (5) imported oil, (6) domestic natural gas, (7) imported natural gas, (8) hydro energy, (9) nuclear energy, (10) geothermal energy, and (11) solar energy.

FIGURE 4

INTERINDUSTRY TRANSACTIONS IN THE INTEGRATED MODEL

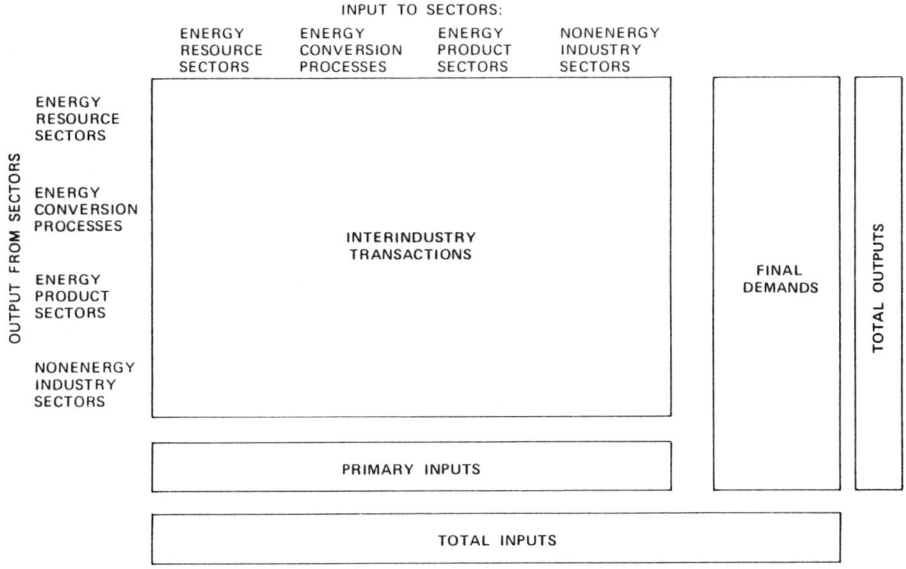

nonenergy industry sectors of our expanded system correspond to industry groups that can be found in a conventional interindustry accounting system. Similarly, primary inputs such as capital and labor services and final demands such as personal consumption expenditures and gross private domestic investment occur in a conventional system.

To incorporate a detailed physical representation of the energy sector we have introduced categories of transactions involving energy resources, energy conversion processes, and energy products that do not correspond to industry sectors in a conventional interindustry accounting system. The activities of the five industry groups compris-

(b) energy conversion processes: (12) coal steam electric, (13) coal steam combined cycle, (14) oil steam electric, (15) oil steam combined cycle, (16) gas turbines, (17) gas steam electric, (18) total energy systems, (19) LWR electric, (20) LMFBR electric, (21) HTGR electric, (22) hydro electric, (23) geothermal electric, (24) solar electric, (25) pumped storage, (26) synthetic gas from oil, (27) synthetic gas from coal, (28) electrolytic hydrogen, (29) methanol, (30) hydrogen from coal, and (31) synthetic oil from coal.

(c) secondary energy forms and energy product sectors: (32) base load miscellaneous, electric, (33) intermediate load miscellaneous electric, (34) peak load miscellaneous electric, (35) storage and synthetic fuel, (36) miscellaneous thermal, low temperature, (37) miscellaneous thermal, intermediate temperature, (38) miscellaneous thermal, high temperature, (39) ore reduction (iron), (40) petrochemicals, (41) space heat, (42) air conditioning, (43) water heat, (44) air transport, (45) truck, bus, (46) rail, and (47) automobile.

(d) nonenergy industry sectors: (48) agriculture, nonfuel mining, and construction, (49) manufacturing, excluding petroleum refining, (50) transportation, and (51) communications, trade, and services.

(e) primary inputs: (52) imports, (53) capital services, and (54) labor services.

(f) final demands: (55) personal consumption expenditures, (56) gross private domestic investment, (57) government purchases of goods and services, and (58) exports.

ing the energy sector in our econometric model are allocated among energy resources, energy conversion processes, and energy products in the integrated model. For example, the electric utility sector in the original system converts fossil fuels and other energy resources into base load, intermediate load, and peak load electricity and into other energy products such as space heat and air conditioning. In our integrated model energy flows are represented in energy units (Btu's) as in the Reference Energy System. Nonenergy flows are represented in constant dollars, as in our econometric model of interindustry transactions. Given a set of energy product and energy resource prices and prices for nonenergy products and primary inputs, our expanded system of interindustry accounts can also be represented in current prices.

In the integrated model energy resources are delivered to energy conversion processes and to final demand for inventory accumulation and for exports. Energy resources are also used in the production of energy resources; for example, coal is used as a fuel in the coal mining industry. However, energy resources are not delivered to energy products, to nonenergy sectors of our integrated model, to personal consumption expenditures, or to government purchases of goods and services. The corresponding entries in our expanded system of interindustry accounts are equal to zero. Similarly, in the integrated model energy conversion processes deliver their output to energy product categories such as miscellaneous electric demand and rail transportation. The outputs of energy conversion processes are used in energy conversion; for example, production and distribution of electricity require electric energy. There are no direct deliveries to energy resource sectors, to nonenergy sectors, or to final demands in our integrated model; the corresponding interindustry accounting entries are zero. Finally, energy products are delivered to nonenergy industrial sectors and to final demand. There are no deliveries of energy products to energy resource, energy conversion, or energy product sectors, so that the corresponding interindustry accounting transactions are zero.

All final demands for energy in our integrated model, except for inventory accumulation and exports of energy resources, are supplied by deliveries of energy products. Similarly, all demands for energy by nonenergy producing sectors are supplied by deliveries of energy products. The first step in the construction of our expanded system of interindustry accounts is to disaggregate flows of energy from the five industry groups that make up the energy sector in our econometric interindustry model. Flows from these groups to the four nonenergy industry groups and the four categories of final demand are distributed among energy product categories in our integrated model on the basis of historical data. For example, deliveries from the electric utility sector to nonenergy industry groups and final demand categories are divided among base load, intermediate load, and peak load miscellaneous electric and among the other energy products that can be supplied by electric energy. Not all energy products can be supplied by means of electricity. Miscellaneous thermal, air transportation, and truck and bus transportation are examples of energy products supplied by nonelectric energy sectors in our econometric model of interindustry transactions.

The energy sectors in our integrated model employ inputs of capital and labor services, nonenergy intermediate goods, and energy. The energy inputs are represented in the same way as in the Reference Energy System. Energy resources are delivered to energy conversion processes and energy products receive deliveries from energy conversion processes. Inputs of labor services and nonenergy intermediate goods are components of operating cost in the objective function of the energy sector optimization model. To obtain total operating cost in current prices we evaluate labor services and nonenergy intermediate goods given a system of prices for primary inputs and nonenergy products. Similarly, inputs of capital services are a component of capital cost in the objective function of the optimization model. Capital cost in current prices is expressed in annual form. The second step in construction of our expanded system of interindustry accounts is to disaggregate flows of nonenergy products and primary inputs in the energy sector optimization model. For each energy resource, energy conversion process, and energy product these flows must be distributed among the four nonenergy industry groups, capital services, and labor services.

Our integrated model is based on an expanded system of accounts for interindustry transactions. The integrated model includes balance equations between supply and demand for products of each of the fifty-one sectors included in the model—eleven energy resource sectors, twenty energy conversion processes, sixteen energy products, and four nonenergy industrial sectors. The model incorporates a process analysis representation of the technology of the energy sectors of the model and an econometric representation of the technology of the nonenergy industry groups. It also incorporates an econometric model for personal consumption expenditures. The integrated model can be used to generate a complete system of interindustry accounts in current and constant prices. The integrated model can also be used to generate energy flows in physical terms for the forty-seven categories of energy included in the model.

A flow chart of the integrated model is presented in Figure 5. The first step in solving the model is to solve the econometric model of interindustry transactions and the energy sector optimization model separately. From these initial solutions we determine the final demands for the nonenergy industrial sectors and the demands for the five energy industrial sectors in the econometric model. Using a fixed distribution of final demands for the products of the five energy sectors to energy product categories, we can determine final demands for sixteen energy products and for inventory accumulation and exports of energy resources. We obtain technical coefficients for the four nonenergy industrial sectors from the econometric model. Using a fixed distribution of intermediate demands for the products of the five energy sectors to energy product categories, we allocate demands for energy by the four nonenergy sectors among the sixteen energy products. We obtain technical coefficients for the forty-seven energy sectors of the integrated model from the energy sector optimization model. Given the technical coefficients and the final demands, we can determine levels of output for all fifty-one sectors of our integrated model. Given the prices of primary inputs and nonenergy industrial products from the econometric model, energy resource prices, energy

FIGURE 5

FLOW CHART OF THE INTEGRATED MODEL (STARTING WITH INPUTS TO INTERINDUSTRY MODEL)

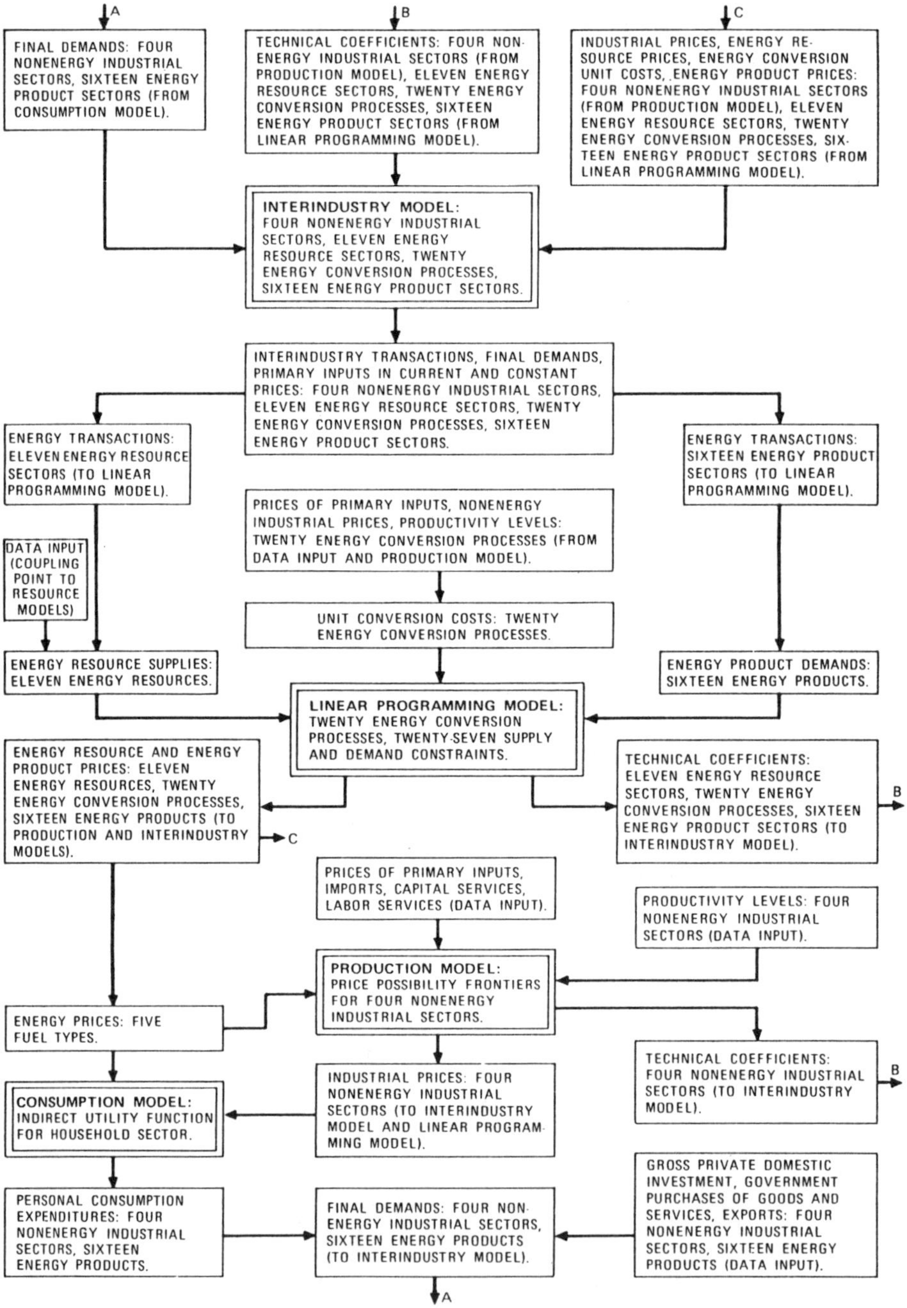

conversion costs, and energy product prices from the energy sector optimization model, we can convert the array of interindustry transactions, final demands, and primary inputs into current prices.

The second step in solving the integrated model is to generate input data for the energy sector optimization model. The eleven energy resource supplies correspond to levels of energy resource output in the integrated model. The sixteen energy product demands correspond to levels of energy product output in the model. Unit conversion costs for the twenty energy conversion processes are the sums of unit operating and capital costs for the conversion processes of the model. Operating costs include costs of labor services and costs of intermediate goods employed in the conversion processes.[12] Capital costs correspond to costs of capital services in the model. These data define a linear programming model that is identical to the energy sector optimization model. Given the unit costs, the energy product demands, and the energy resource supplies, the energy sector optimization model generates cost minimizing levels for the energy conversion processes and value maximizing prices associated with energy supplies and demands.

The third step in solving the integrated model is to determine prices for products of each of the five energy sectors that appear in the econometric interindustry model. Prices of domestic petroleum, natural gas, and coal correspond to energy resource prices from the energy sector optimization model. The prices of gas delivered by gas utilities, electricity, and refined petroleum products are determined as a weighted average of the corresponding energy product prices. The weights are based on the proportions of deliveries of each fuel to each energy product in the total of all deliveries of the fuel. Given the prices of the products of the five energy sectors that appear in the econometric model, the prices of primary inputs, and the levels of productivity in each of the four nonenergy industrial sectors, we can determine prices for the products of the four nonenergy sectors from the four price possibility frontiers for these sectors from the econometric model. From energy and nonenergy prices and the prices of primary inputs, we can generate the technical coefficients for the nonenergy industrial sectors of the integrated model. On the basis of these prices we can allocate total personal consumption expenditures among the products of the nine sectors of the econometric model and primary inputs.

We have outlined three steps in the solution of the integrated model. At the completion of these steps we have generated a new set of data to initiate the process of solution. We repeat the sequence of three steps until the data employed to initiate the process are generated as a solution of the integrated model. The integrated model includes technical coefficients for the energy sector from the energy sector optimization model, technical coefficients for the nonenergy sector from the production models for the four nonenergy industrial sectors from the econometric interindustry model, and final demands for energy and nonenergy products from the econometric interindustry model, allocated among the energy products of the energy sector optimization model. The expanded interindustry model assures that

[12] This link between prices and costs has not been included in the initial implementation of our model. Research is now underway to complete this linkage.

supplies and demands are balanced for energy resources, energy conversion processes, energy products, and nonenergy products.

To summarize: a solution of the integrated model consists of a solution of the expanded interindustry model for which the following conditions hold:

(1) The energy conversion levels minimize cost for the corresponding levels of energy demands and supplies and for the corresponding unit costs of the energy conversion processes.
(2) The prices of energy products and energy resources maximize the value of the products less the value of the resources.
(3) The prices of the five fuel types in the econometric interindustry model are generated by dual solution of the LP model.
(4) The prices of the four nonenergy products are consistent with the energy product and fuel prices and the exogenously given prices of primary inputs.
(5) The unit costs of the energy conversion processes are consistent with nonenergy product prices and the given prices of primary inputs.
(6) The technical coefficients for the energy sectors are those associated with the cost minimizing solutions of the energy sector optimization model; the technical coefficients for the nonenergy sectors are those associated with the prices for primary inputs, the four nonenergy products, and the five fuel types.
(7) The final demands for energy and nonenergy products are those associated with the prices for these products.

Under these conditions the value of the output of each sector of the expanded interindustry model is equal to the value of the input of that sector.

4. Application

■ We next present an application of the preliminary version of the model to the analysis of energy research, development, and demonstration policy.[13] Our first step is to establish a base case, representing a projection of the U.S. economy and the energy sector through 1985, that meets the following specifications:

(1) Real gross national product will grow at 4 percent per year from 1975 to 1985.
(2) Energy prices will grow relative to the implicit deflator of the gross national product at 1.3 percent per year over the same period.
(3) Energy supplies for 1985 are similar to the "calibration case" presented in *1976 National Energy Outlook* (1976), as modified for the purposes of this analysis.
(4) The availability of new energy technology for 1985 is given by the "combination scenario," in *A National Plan for Energy Research, Development and Demonstration* (1975).

Under these assumptions imports rise from 8.6 percent of total U.S. energy supply in 1967 to 18.4 percent of total supply in 1985.

The dramatic increase in the proportion of energy resource supplies that must be imported under the base case assumptions for 1985 suggests that energy policy may have to be modified to meet national

[13] This application is based on the research of Behling, Dullien, and Hudson (1976).

security objectives. Accordingly, we have examined the implications of a reduction of imports from 18.4 percent of total energy supply in 1985 to no more than 10 percent of total supply in that year. This objective requires the introduction of a tariff on imported petroleum at the rate of 51.8 percent; we have also introduced taxes on domestic supplies of oil and gas so as to leave the prices received by domestic suppliers unchanged. Energy product and conversion levels under the base case assumptions and under our alternative case for 1985 are presented in Table 1. The results of the proposed change in policy are

TABLE 1

ENERGY PRODUCTS AND CONVERSION LEVELS, BASE CASE AND ALTERNATIVE CASE 1985 (QUADRILLION BTU'S)

	BASE CASE	ALTERNATIVE CASE	PERCENTAGE CHANGE
1. ENERGY PRODUCTS			
AIR CONDITIONING	2.14	2.01	− 6
PRIVATE GROUND TRANSPORTATION	3.04	2.75	−10
AIR TRANSPORTATION	0.85	0.78	− 8
TRUCK, BUS, & DIESEL RR	0.95	0.84	−12
SPACE HEAT	12.73	11.96	− 6
WATER HEAT	1.79	1.67	− 7
PROCESS HEAT	10.91	10.64	− 2
MISC. ELECTRIC (INCL. ELEC. RR)	5.39	5.09	− 6
COKE FOR IRON PRODUCTION	1.64	1.56	− 5
PETROCHEMICALS	4.83	4.70	− 3
2. ELECTRICITY GENERATION			
COAL STEAM	10.71	11.53	+ 8
COAL STEAM COMB. CYCLE	0.00	0.00	0
SOLVENT REFINED COAL STEAM	0.00	0.00	0
OIL STEAM	3.71	2.21	−40
GAS TURBINE	1.57	1.73	+10
GAS STEAM	4.39	2.70	−38
TOTAL ENERGY SYSTEMS	0.00	0.02	——
LWR	9.55	10.51	+10
HTGR	0.00	0.25	---
HYDROELECTRIC	3.38	3.38	0
GEOTHERMAL	0.69	1.66	+141
SOLAR	0.00	0.00	0
TOTAL ELECTRICITY INPUTS	34.00	34.00	0
OTHER INPUTS	66.00	62.70	− 5

SOURCE: BEHLING, DULLIEN, HUDSON (1976), TABLE V.4, PAGE 84; TABLE V.8, PAGE 96.

presented in Table 2. Energy prices rise by an average of 11.6 percent with the greatest increase in the price of refined petroleum products. New technologies such as oil from shale, direct use of solar energy, generation of electricity from geothermal sources, and coal gasification and liquefaction are introduced in greater quantities under the alternative energy policy. In addition, there is greater reliance on nuclear energy through use of both light water reactors and high temperature gas-cooled reactors for generation of electricity. Finally, electric automobiles are introduced on a modest scale to conserve gasoline.

TABLE 2

ALTERNATIVE CASE 1985

1. ENERGY PRICES (PERCENTAGE CHANGE FROM BASE CASE 1985).	
COAL	0.0
REFINED PETROLEUM	35.3
REFINED GAS	− 4.6
ELECTRICITY	2.4
ENERGY	11.6
2. NEW TECHNOLOGY (CHANGE FROM BASE CASE 1985 IN QUADRILLION BTU'S).	
OIL SHALE	1.00
DIRECT SOLAR	0.20
GEOTHERMAL ELECTRIC	0.97
COAL GASIFICATION AND LIQUEFACTION	1.42
LWR	0.96
HTGR	0.25
ELECTRIC CAR	0.09

SOURCE: BEHLING, DULLIEN, AND HUDSON (1976), TABLE V.3, PAGE 81; TABLE V.1, PAGE 77.

The reduction in oil and gas imports resulting from the energy policy underlying our Alternative Case for 1985 is analyzed in Table 3. Of the total reduction in imports of 8.42 quadrillion Btu's in 1985, reduced demand accounts for 3.30 quadrillion Btu's; increased supplies account for an additional 2.17 quadrillion Btu's; the substitution of alternative forms of energy for oil and gas accounts for the remaining 2.72 quadrillion Btu's of import reduction. We recall that domestic supplies of oil and gas remain unchanged under the assumptions of our alternative case. The economic impact of the alternative case for 1985 is analyzed in Table 4. Higher prices of imported oil result in higher prices for the output of the agricultural, manufacturing, transportation, and service sectors of the U.S. economy with the greatest increase in transportation prices. In Table 4 we also present

TABLE 3

REDUCTION IN OIL AND GAS IMPORTS, ALTERNATIVE CASE 1985 (QUADRILLION BTU'S)

1. REDUCED DEMAND	3.30
2. INCREASED SUPPLIES	
OIL SHALE	1.00
DIRECT SOLAR	0.20
GEOTHERMAL ELECTRIC	0.97
3. SUBSTITUTION	
COAL GASIFICATION AND LIQUEFACTION	1.42
LWR	0.96
HTGR	0.25
ELECTRIC CAR	0.09
OTHER	0.23
4. TOTAL REDUCTION	8.42

SOURCE: BEHLING, DULLIEN, HUDSON (1976), TABLE V.2, PAGE 78.

TABLE 4

ECONOMIC IMPACT, ALTERNATIVE CASE 1985

	BASE CASE	ALTERNATIVE CASE	PERCENTAGE CHANGE
1. NONENERGY PRICES (PERCENTAGE CHANGE FROM BASE CASE 1985).			
AGRICULTURE			1.10
MANUFACTURING			0.62
TRANSPORTATION			1.33
SERVICES			0.26
2. TECHNICAL COEFFICIENTS.			
AGRICULTURE			
CAPITAL	0.1755	0.1753	
LABOR	0.2515	0.2542	
ENERGY	0.0148	0.0125	
NONENERGY	0.5582	0.5580	
MANUFACTURING			
CAPITAL	0.1143	0.1140	
LABOR	0.2959	0.2977	
ENERGY	0.0194	0.0189	
NONENERGY	0.5705	0.5694	
TRANSPORTATION			
CAPITAL	0.1831	0.1814	
LABOR	0.4074	0.4082	
ENERGY	0.0390	0.0344	
NONENERGY	0.3706	0.3761	
SERVICES			
CAPITAL	0.3197	0.3201	
LABOR	0.4025	0.4036	
ENERGY	0.0166	0.0161	
NONENERGY	0.2612	0.2603	
3. FINAL DEMAND (PERCENTAGE COMPOSITION).			
AGRICULTURE	11.66	11.62	
MANUFACTURING	33.58	33.61	
TRANSPORTATION	2.48	2.46	
SERVICES	49.19	49.41	
ENERGY	3.09	2.91	
4. OUTPUT (PERCENTAGE COMPOSITION).			
AGRICULTURE	11.94	11.91	
MANUFACTURING	38.40	38.44	
TRANSPORTATION	4.50	4.44	
SERVICES	40.44	40.78	
ENERGY	4.73	4.43	

SOURCE: BEHLING, DULLIEN, HUDSON (1976), TABLE V.14 PAGE 110; TABLE V.17, PAGE 118; TABLE V.15, PAGE 111; TABLE V.16, PAGE 115.

the resulting changes in technical coefficients for each of the four nonenergy sectors and changes in the composition of final demand and total output for these sectors and for the energy sector as a whole.

5. Conclusion

HOFFMAN AND JORGENSON / 463

■ We have presented an assessment of the impact of energy research, development, and demonstration policies in combination with tax and tariff policies to reduce imports of energy resources for the year 1985. A complete evaluation of alternative energy policies requires assessments for a wider range of policies over a longer period of time. In addition to the impact of energy policy on the structure of the energy sector and the level and composition of overall economic activity, assessments must also be made of the impact on levels of well-being, life styles, and environmental pollution in the United States; international repercussions of alternative policies must also be considered. These assessments must be combined into an overall evaluation of energy policy.[14]

The application of the preliminary version of the model for the analysis of energy research, development, and demonstration policy involves the creation of base case projections for the U.S. economy for additional years along the lines of our base case for 1985. The assessment of alternative energy policies requires the development of alternative case projections for each policy. Our model can be used to analyze the impact of energy policy on the energy sector and on overall economic activity. In addition, the energy system optimization model and other closely related models can be used to assess the environmental impact of alternative energy policies, to estimate capital requirements associated with the implementation of new technologies, and to evaluate the effect of changes in energy prices on the domestic supply of primary energy resources such as uranium, coal, oil, and gas.[15]

While the successful integration of process analysis and econometric models of energy policy is an important step in the development of a framework for the evaluation of energy policy, we must emphasize the limitations of our current approach. A fully satisfactory data base requires the development of interindustry transactions accounts for energy flows in physical terms as well as current and constant prices. It would be very useful to incorporate these accounts into the U.S. national income and product accounts on an annual basis. It would also be useful to disaggregate the nonenergy sectors of our model and to extend our modeling effort to incorporate primary factor input supplies and supplies of primary energy resources. Finally, additional research is required on the most efficient techniques for solution of our model.

Ultimately, projections for different years over a planning horizon could be developed within a dynamic version of our integrated process analysis and econometric model. In a dynamic model capital requirements and investment costs, together with the prices of capital services and labor services, would be generated endogenously by incorporating the supply of primary factors of production along with the demands included in our existing framework. Such a dynamic model could be extended to encompass the development of reserves of primary energy resources, production from reserves, and the pricing of current supplies of these resources for each period of time. Research is now under way that will enable us to extend our existing

---

[14] An overall evaluation of U.S. energy research, development, and demonstration policy, incorporating assessments of policy impacts based on our integrated econometric and process analysis model is given by the Energy Research and Development Administration (1976).

[15] For a detailed discussion of the application of our model, see Behling, Dullien, and Hudson (1976).

model in the direction of a dynamic model for the assessment of alternative energy policies.[16]

Our objective has been to present a model that integrates a process analysis model of the energy sector with an econometric model of interindustry transactions for application to policy analyses where the technological component is significant. We have illustrated the use of this model for assessment of the impact of energy policy. We have not attempted a comprehensive evaluation of alternative energy policies in order to focus on the methodology we have developed for model integration. Integration of process analysis and econometric models, using an extended accounting framework for interindustry transactions, has proved to be feasible. A great deal of additional research will be required to develop the most appropriate framework for evaluation of alternative energy policies.

References

Behling, D. J., Dullien, R., and Hudson, E. A. *The Relationship of Energy Growth to Economic Growth under Alternative Energy Policies*, BNL 50500. Upton, N.Y.: Brookhaven National Laboratory, 1976.

Behling, D. J., Marcuse, W., Swift, M., and Tessmer, R. G. *A Two-Level Iterative Model for Estimating Inter-Fuel Substitution Effects*, BNL 19863. Upton, N.Y.: Brookhaven National Laboratory, 1975.

Beller, M., ed. *Sourcebook for Energy Assessment*, BNL 50483. Upton, N.Y.: Brookhaven National Laboratory, 1975.

Bernanke, B. and Jorgenson, D. W. "The Integration of Energy Policy Models." *Computers and Operations Research*, Vol. 2, No. 3 (September 1975).

Berndt, E. R. and Jorgenson, D. W. "Production Structure," Chapter 3 in D. W. Jorgenson *et al.*, *U.S. Energy Resources and Economic Growth*, Final Report to the Energy Policy Project, Washington, D.C., 1973.

Berndt, E. R. and Wood, D. O. "Technology, Prices, and the Derived Demand for Energy." *Review of Economics and Statistics*, Vol. 58, No. 1 (February 1976), pp. 1–10.

Bullard, C. W. and Herendeen, R. A. *Energy Cost of Consumption Decisions*, Document 135. Center for Advanced Computation, Urbana, Illinois, University of Illinois at Urbana-Champaign, 1973a.

———. *Energy Cost of Consumer Goods 1963/67*, Document 140. Center for Advanced Computation, Urbana, Illinois, University of Illinois at Urbana-Champaign, 1973b.

Cherniavsky, E. A. *Brookhaven Energy System Optimization Model*, BNL 19569. Upton, N.Y.: Brookhaven National Laboratory, 1974.

———. *Linear Programming and Technology Assessment*, BNL 20053. Upton, N.Y.: Brookhaven National Laboratory, 1975.

Christensen, L. R., Jorgenson, D. W., and Lau, L. J. "Transcendental Logarithmic Production Frontiers." *Review of Economics and Statistics*, Vol. 55, No. 1 (February 1973), pp. 28–45.

———. "Transcendental Logarithmic Utility Functions." *The American Economic Review*, Vol. 65, No. 3 (June 1975), pp. 367–383.

Dullien, R., ed. *User's Guide to the DRI Long-Term Interindustry Transactions Model*. Interim Report to the U.S. Department of the Interior. Washington, D.C.: U.S. Government Printing Office, 1976.

Energy Research and Development Administration. *A National Plan for Energy Research and Development, and Demonstration: Creating Energy Choices for the Future*, ERDA 48. Washington, D.C.: U.S. Government Printing Office, 1975.

———. *A National Plan for Energy Research Development, and Demonstration:*

[16] A dynamic version of the Brookhaven Energy Systems Optimization Model has been developed by Marcuse, Bodin, Cherniavsky, and Sanborn (1975). A dynamic version of our econometric interindustry model has been developed by Dullien *et al.* (1976). Incorporation of models of primary resource supply into our econometric interindustry model is discussed by Bernanke and Jorgenson (1975).

Creating Energy Choices for the Future, 1976, ERDA 76–1. Washington, D.C.: U.S. Government Printing Office, 1976.

JACK FAUCETT ASSOCIATES. *Data Development for the Input-Output Energy Model*. Final Report to the Energy Policy Project. Washington, D.C., 1973.

———. *Historical Energy Flow Accounts*. Final Report to the Federal Energy Administration. Washington, D.C., 1975.

FEDERAL ENERGY ADMINISTRATION. *1976 National Energy Outlook*. Washington, D.C.: U.S. Government Printing Office, 1976.

HOFFMAN, K. C. "A Unified Framework for Energy System Planning," in M. F. Searl, ed., *Energy Modeling*, Washington, D.C.: Resources for the Future, 1973.

——— AND CHERNIAVSKY, E. A. *Interfuel Substitution and Technological Change*, BNL 18919. Upton, N.Y.: Brookhaven National Laboratory, 1974.

HOFFMAN, K. C., PALMEDO, P. F., MARCUSE, W., AND GOLDMAN, M. D. *Coupled Energy Systems—Economic Models*, BNL 19293. Upton, N.Y.: Brookhaven National Laboratory, 1974.

HUDSON, E. A. AND JORGENSON, D. W. "U.S. Energy Policy and Economic Growth, 1975–2000". *The Bell Journal of Economics and Management Science*, Vol. 5, No. 2 (Autumn 1974a), pp. 461–514.

———. "Tax Policy and Energy Use," in Committee on Finance, United States Senate, *Fiscal Policy and the Energy Crisis*, Washington, Ninety-Third Congress, First and Second Sessions, 1974b, pp. 1681–1694.

———. "Economic Analysis of Alternative Energy Growth Patterns." Report to the Energy Policy Project, Ford Foundation; published in D. Freeman *et al.*, *A Time to Choose*, Cambridge: Ballinger, 1974c, pp. 493–511.

———. "Tax Policy and Energy Conservation," in D. W. Jorgenson, ed., *Econometric Studies of U.S. Energy Policy*, Amsterdam: North-Holland, 1975a.

———. "Projections of U.S. Economic Growth and Energy Demand." Report to the Edison Electric Institute, September; published in condensed form in R. W. Greenleaf, ed., *Structural Change and Current Problems Facing Regulated Public Utilities*, Indianapolis, Graduate School of Business, Indiana University, 1975b, pp. 75–128.

JORGENSON, D. W. "Consumer Demand for Energy," in *Proceedings of the Workshop on Energy Demand*, W. D. Nordhaus, ed., Laxenburg, Austria, IIASA, 1975, pp. 765–802.

——— AND LAU, L. J. "The Structure of Consumer Preferences." *Annals of Social and Economic Measurement*, Vol. 4, No. 1 (January 1975), pp. 49–101.

JORGENSON, D. W. AND WRIGHT, B. D. "The Impact of Alternative Policies to Reduce Oil Imports." Report to the Office of the Secretary, U.S. Department of the Treasury; published in *Data Resources Review*, Vol. 4, No. 6 (June 1975).

MARCUSE, W., BODIN, L., CHERNIAVSKY, E. A., AND SANBORN, Y. *A Dynamic Time Dependent Model for the Analysis of Alternative Energy Policies*, BNL 19406. Upton, N.Y.: Brookhaven National Laboratory, 1975.

[8]

JOURNAL OF ENVIRONMENTAL ECONOMICS AND MANAGEMENT 5, 172–186 (1978)

Secondary Materials and International Trade

RICHARD GRACE

University of East Anglia, England

R. KERRY TURNER

University of East Anglia, England

AND

INGO WALTER

New York University, New York, New York 10006

Received March 4, 1977; revised June 12, 1977

The paper develops the international trade dimensions of recycling. It begins by outlining the sources of demand and supply that determine the volume, composition and direction of trade in waste materials. The model is intended to describe factors affecting international flows of any type of secondary material, although its usefulness in an empirical context is restricted in many cases by data limitations and complex lag structures on the supply side. Parts of the model are then applied to the case of waste paper, focusing on observed differences in national recycling rates and the role of trade in reconciling those differences.

Each nation has a unique materials flow configuration consisting of (a) domestic sources of virgin materials, (b) imports and exports of materials, (c) generation of production waste, (d) generation of consumption waste, and (e) basic materials demand.[1] Each nation also has a unique environmental profile involving population density, industrial agglomeration, environmental assimilative capacity, and social demand for environmental quality. There is evidence that both the materials flow and environmental configurations differ significantly between nations—that the existence of an international market for recycled materials can perform an important role in increasing the economic incentive to recycle in some countries, thereby reducing environmental damage, while providing an additional source of basic materials supplies in others.[2] International trade can thus perform much the same function in the secondary materials sector as it does in other industries, by reconciling intercountry variations in supply and demand and permitting nations to move to higher levels of social welfare than they could otherwise attain.

[1] One of the early studies is [2]. For a survey, see [4]; recent and comprehensive account is presented in [3].

[2] The economics of recycling and their international and comparative dimensions are discussed in detail in [5].

0095-0696/78/0052-0172$02.00/0

SECONDARY MATERIALS EXCESS SUPPLY AND DEMAND

The relationship of international trade to recycling of waste materials is rather straightforward. At the national level, the supply of recycled materials derives from *stock* accumulations of past wastes and *flows* of current wastes. The higher are national secondary materials prices, the larger is the proportions of the current waste stream that tends to be recovered and the higher the rate of depletion of accumulated secondary materials stocks. The supply of secondary materials will also be higher, the greater is the cost of waste disposal in ways *other than* recycling, e.g., through disposal charges, enforcement of environmental standards, or prohibition of certain disposal options. Last, the available supply of secondary materials depends on the cost of collection and reprocessing to the extent required for reuse in production.

Secondary materials flows are exceedingly complex. Once virgin materials have been derived from renewable or nonrenewable natural resources, they normally go through an initial refining or processing stage where a certain amount of waste is generated, some of which is immediately reused within the plant itself—generally called "home" scrap in the metals industries. This material never enters commercial channels and can essentially be ignored from an international trade standpoint. Refined or processed basic materials then enter the manufacturing stage, where again production waste is generated, e.g., metal filings, textile trimmings, and the like. Recovered materials at this stage are generally sold to intermediaries in the secondary materials industry or directly back to the refiner/processor and therefore are commercially traded—often called "prompt" or "industrial" scrap. Such wastes are relevant from an international trade standpoint.

Manufactured goods enter distribution channels, where once more a certain amount of waste is generated through accidental breakage, obsolete goods, and discarded packaging materials. There are also considerable flows of waste from the services sector (e.g., banking, government, education) largely in the form of paper. Retired capital equipment and construction waste constitute another important source. And finally, there are consumer goods that have reached the end of their useful lives and generally wind up in municipal garbage.[3] To the extent that end-product wastes from the distribution, services, industrial, and consumer sectors are in fact recovered, collected, reprocessed, and sold, usually by firms in the secondary materials sector, they have demonstrated commercial value. Secondary materials at this stage is often called "old" scrap, and is again relevant from the standpoint of international trade. Secondary materials that are internationally traded thus comprise principally "commercial" wastes consisting of prompt, or industrial, and old scrap.

As prices of secondary materials rise, increasing quantities will tend to be recovered beginning with concentrated, homogeneous flows and stocks, and proceeding to more dispersed and heterogeneous sources where the incremental costs of collection, transport, and separation are successively higher. We thus have a hypothesized domestic supply function for a given type of secondary material that is upward-sloped over the relevant range of prices, but may be

[3] It is perhaps useful to note that the composition of municipal waste in the U.S. is estimated to be about 30% paper, 10% glass, 10% metals (largely ferrous), roughly 3% each of rubber and plastic, 4% wood, and 1.5% textiles, the remainder being sand, yard waste and miscellaneous organics. See [1]. The proportions in British household waste are very similiar, as indicated in [6].

more elastic at very low prices and less elastic at very high prices, at least in the short run. Its position in part depends on the cost of alternative disposal in the environment imposed on the consumer or producer by pollution control measures. Price depends largely on the cost of available competitive virgin materials.

On the domestic demand side, the price of substitute virgin resources presumably sets an upper limit to the price at which there will be some demand for recycling—although in certain industrial processes (e.g., steelmaking and glassmaking) technical factors may mandate a certain proportion of scrap. Depending on the quality of secondary materials involved, they may be subject to a "discount" from the counterpart virgin materials if additional processing is required, if the end product may turn out inferior in quality, or if production processes run less efficiently e.g., due to greater use of energy or greater pollution that has to be eliminated. Beyond this, the demand for secondary materials can be expected to behave normally as a negative function of price, with its elasticity some composite of (a) the substitutability of the secondary material in question for the counterpart virgin material; (b) its substitutability for competitive materials in the same applications; and (c) the elasticity of demand for the products to be manufactured.

Two additional points should be noted. First, public policy plays an important role in both the domestic demand for and supply of secondary materials. Subsidies for the extraction of virgin materials, tax depletion allowances, or direct grants tend to depress demand for recycled materials. Pollution-control costs imposed upon producers of consumers will tend to stimulate supply of secondary materials. Pollution abatement measures will also tend to generate a greater stream of potentially recoverable materials. Pollution control in the extractive industries will tend to stimulate secondary demand. Government may also stimulate demand directly by specifying recycled materials in products acquired in its own procurement activities. And because of the low unit value of secondary materials, the relative cost of transportation is very high, so that rate structures in transport industries may affect demand for recycled materials. Finally, technical advances developed by or with the assistance of government may affect both the supply and demand sides.

The second point concerns the volatility that apparently characterizes the markets for most kinds of secondary materials. If there is high substitutability and cross price elasticity between a particular type of virgin and secondary material, then even a small change in the price of the former may trigger a very large change in demand for the latter. Moreover, during periods of economic expansion periodic shortages of basic materials tend to develop, which lead to burgeoning secondary demand. As the boom subsides and virgin materials supplies catch up, secondary materials are often the first to feel the impact and be displaced. Such volatility makes some of the secondary materials industries extremely risky and poor prospects for large-scale investment. Consequently, with some notable exceptions, firms tend to be small, poorly capitalized, and labor intensive, often with substandard technology and management skills, and themselves beset with environmental problems. And while primary industries tend to be organized along oligopolistic lines with some price and output stability and vertical integration, secondary materials firms tend to operate in a much more competitively structured environment.[4]

[4] For a comprehensive discussion, see [10].

Secondary materials export supply and import demand clearly depend on the balance between domestic supply and demand at prevailing prices. Current availability of exportable secondary materials in a particular country is a function of the level of past domestic production minus past exports plus past imports, subject to three kinds of lags: (a) a "distribution lag" for old scrap, between the end of the manufacturing process (or date of importation), and procurement by final consumers and users—highly variable and product-specific, involving transport, warehousing, inventories, and the like; (2) a "consumption lag" between the initial purchase and the end of a product's useful life, again highly variable; and (3) a "recovery lag" between the end of a product's useful service and completion of the recycling process for the recovered component of the waste stream. For any given category of secondary material (e.g. copper scrap) the lag structure is thus highly complex. Furthermore, it applies only to "old" scrap, as noted, since "prompt" and "industrial" scrap are subject only to the "recovery lag." The higher past production or import levels of goods containing a given recyclable material, all else equal, the higher the current export level is likely to be.

Also on the export-supply side, current secondary materials shipments are inversely related to the domestic costs of collection, separation, and reprocessing, including internalized environmental costs which raise secondary materials prices. The lower the domestic price of substitutable virgin materials, the more secondary materials will tend to be available for export at higher foreign prices. Exports are also likely to be greater, the higher the domestic nonrecycling disposal costs into the environment. On the foreign demand side, exports of secondary materials would seem to be a function of current production levels in the rest of the world (or major target markets). The higher the foreign price of substitutable virgin materials, furthermore, the greater will be the foreign demand for secondary material exports from the home country. And the more rigorous are antipollution policies abroad and the higher the effective non-recycling disposal costs, the greater the import competing supplies of secondary materials and the smaller the potential market for exports from the home country. Last, we have already noted that transport costs play a major role in determining the competitiveness of recycled materials—due to generally low value per unit weight or volume—so that exports would tend to be an negative function of transport charges from reprocesssing facility to target markets.

Such behavioral relationships may be specified and tested empirically using time series data, wherein the lag structure as well as the size of the nonrecycling waste disposal cost and recycling cost variables would appear to be most problematic in terms of data availability. The data problems are often exceedingly difficult to manage, and some of the parameters and variables—such as substitution elasticities between virgin and secondary materials, and environmental disposal costs—are often impossible to establish at this stage in any meaningful way.[5] It is instructive, nevertheless, to examine the relationship of trade to recycling in the case of a single secondary material, waste paper.

[5] If there is free trade in virgin materials, prices in exporting and importing countries will of course be the same, or will differ only by the amount of transport costs. Recycling costs may differ internationally if recycling technologies or factor costs differ between countries—in effect, these determine the cost of recycled materials if the waste itself is assumed to be a free good.

THE CASE OF WASTE PAPER

As we have seen, the demand for a particular secondary material is principally a function of the price of the substitute primary material (*Piv*), and normally the price of the primary material will set an upper limit on the price of the secondary material. Where a secondary material is a perfect substitute for the primary material the demand curve will be perfectly elastic, although this is not usually the case. Waste paper, for example, is generally an inferior product to pulp because contaminants may be present that increase risks during production and may lower quality, and because the fiber length of repulped paper is shorter and hence weaker than virgin pulp and results in an inferior final product and a lower price (*Pis*). A combination of waste paper and pulp is normally used in production to arrive at a desired quality of final product. Hence the demand for waste paper is not only a function of primary product price (*Piv*) but also related to the technical ability of the producer to mix these inputs (TS_i). It may therefore be assumed that the demand curve for waste paper is downward sloping. This is shown in Fig. 1b, where DP_t represents the demand for pulp, SP_t the supply of pulp. Equilibrium in the pulp market will set price at P_2 and this becomes the upper limit for the demand curve of waste paper DWP_t in Fig. 1a, hence $P_2 = P_M$.

The supply of a secondary material is a function of the cost of collection (C_i) and the quantity or stock of final product waste (*Qmi*). If quantity supplied is to be increased, collectors will have to search more diverse and composite sources for waste, which will increase costs and therefore the supply function will be upward sloping. In the case of waste paper, the stock available for collection will be given by the relationship

$$Qmi = CP_{t-1} - \alpha_{t-1},$$

where *Qmi* represents the stock of waste paper in period *t*, CP_{t-1} represents consumption of paper in period (*t*-1), and α_{t-1} denotes losses during consumption (cigarette paper, toilet tissue, and the like). Unlike other secondary materials, the single-period lag is quite representative in the case of waste paper due to

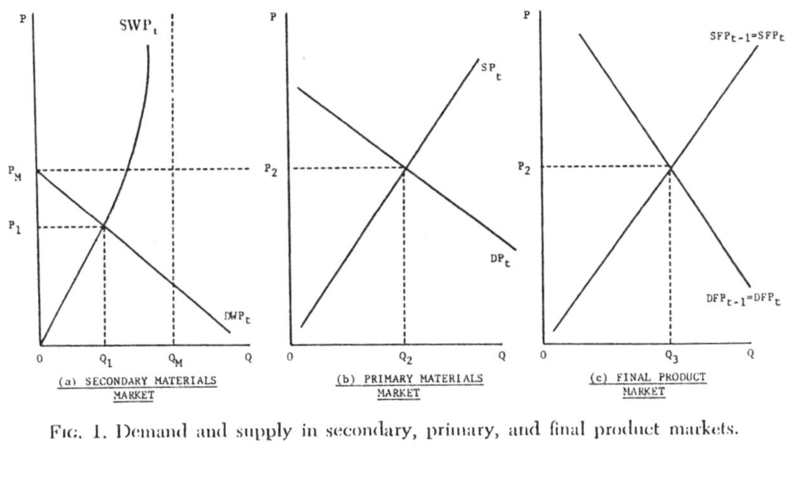

FIG. 1. Demand and supply in secondary, primary, and final product markets.

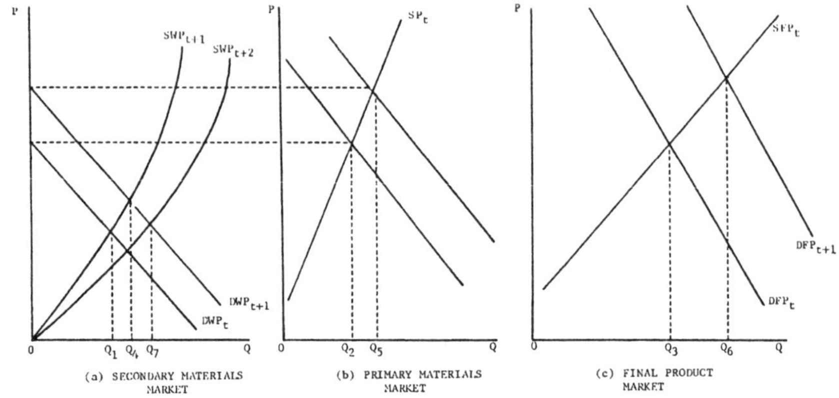

FIG. 2. Effects of shifting demand for paper.

the short product life. The supply curve for waste paper is shown as SWP_t in Fig. 1a. P_1 is the equilibrium price of waste paper, and Q_1 is the quantity of waste paper generated. Q_2 represents the quantity of pulp generated and, if it is assumed that the consumption of paper in period t is the same as consumption in period t-1, then $Q_1 + Q_2 \equiv Q_3$ in Fig. 1c, the equilibrium quantity of paper produced and consumed, assuming no losses during production.

In the event that the equilibrium quantity in the final product market is not constant over time, then there are important repercussions in each of the two input markets. Consider an increase in the demand for paper in period $t + 1$ causing a shift from DFP_t to DFP_{t+1} in Fig. 2c, such that the new equilibrium price generates a quantity Q_5 of pulp, and also causes a shift in the demand function of waste paper from DWP_t to DWP_{t+1}, such that a quantity Q_4 is generated. $Q_4 + Q_5 = Q_6$, the new level of consumption in the final product market. However, this is not the final equilibrium because, as noted, the supply of waste paper is a *lagged* function of consumption of the final product. Hence, in period $t + 2$ the supply of waste paper SWP_{t+1} will shift to SWP_{t+2} in Fig. 2a. Quantity Q_7 of waste paper will be supplied, but $Q_7 + Q_5 \neq Q_6$ so that the demand for pulp will shift to the left until equilibrium is once again found in all three markets. The decline in demand for pulp will be determined by the elasticity of substitution between pulp and waste paper. It should be noted that any such change in the pulp market equilibrium will in turn shift the waste paper demand function, and hence this final adjustment will affect both of the input markets.

INTERNATIONAL TRADE IN WASTE PAPER

Most countries, even the major pulp producers, have been attempting to intensify the collection and utilization of waste paper as a raw material for the manufacture of paper and board. Two key descriptive measures are the so called "recovery" and "utilization" rates. The *recovery rate* denotes the supply of waste paper from domestic sources and is defined as domestically recovered waste paper and board relative to apparent consumption of paper and board, or

$$R = \frac{W_D}{Q_{DB} + M_B - X_B},$$

where W_D is domestically recovered waste, Q_{DB} is domestically produced paper and board, X_B is exports of paper and board and M_B is imports of paper and board. The *utilization* rate denotes the demand for waste paper as a fiber input in the paper and board industry, and is defined as total waste paper relative to total fiber usage, or

$$U = \frac{W_D - W_X + W_M}{TFU},$$

where W_M and W_X are waste paper imports and exports, respectively, and TFU is total fiber usage in the manufacturing of paper and board. For our purposes, we can let TFU be equal to Q_{DB}—although there are clearly losses in the conversion process such that, strictly $Q_{DB} = \beta\, TFU$, where β is less than 1.

Therefore, we obtain

$$DWP_t = U_t Q_{DB_t}$$

and

$$SWP_t + R_t(Q_{DB} - X_B + M_B)_{t-1},$$

where t denotes the current time period. Note that the supply of waste paper is a fraction of the apparent consumption of paper and board in the previous time period $(t-1)$. Therefore,

$$U_t = \frac{DWP_t}{Q_{DB_t}},$$

and similarly

$$R_t = \frac{SWP_t}{(Q_{DB} - X_B + M_B)_{t-1}}.$$

Thus it is clear that U and R are composite measures encompassing the individual variables contained in the conceptual discussion presented earlier.

In a closed economy, without international trade, we have $R = W_D/Q_{DB}$ and $U = W_D/Q_{DB}$, so that $R = U$. It follows that the role of trade in both paper and board and in waste paper is one of "absorbing" the actual disparities between R and U that are observable. If there is trade in paper and board, but not in waste paper, then a net paper and board *importing* country will have a utilization rate above its recovery rate, while a net paper and board *exporting* country will have a utilization rate below its recovery rate. Without W_X and W_M we have $R = W_D/(Q_{DB} + \bar{M}_{PB})$ and $U = W_D/Q_{DB}$, where $\bar{M}_{PB}$, represents net imports of paper and board. Hence

$$W_D = R(Q_{DB} + \bar{M}_{PB})$$

and

$$U = \frac{R(Q_{DB} + \bar{M}_{PB})}{Q_{DB}} = \frac{R(\bar{C}_{PB})}{Q_{DB}},$$

where $\bar{C}_{PB}$ represents apparent consumption of paper and board. For a net paper and board importing nation, $\bar{C}_{PB} > Q_{DB}$, and hence $U > R$. For a net paper and board exporting nation, $C_{PB} < Q_{DB}$, so that $U < R$. Sweden for example, as a net exporter of paper and board had a utilization rate of only 7.3% in 1974 and a recovery rate of 28%. The United Kingdom, a net importer of paper and board, had a utilization rate of 45.8% and a recovery rate of 27.6%.

TABLE I

International Trade in Waste Paper, 1973[a]

Importing country	West Germany	Belgium/ Luxembourg	Denmark	France	Eire	Italy	Netherlands	U.K.	EEC	Iceland	Austria	Spain	Finland	Greece	Norway	Sweden	Switzerland	Turkey	Canada	U.S.A.	North America	Japan	Australia	New Zealand	Total OECD	Non OECD	Total based on importing Country figure[a]
	Exporting country																										
West Germany	—	74.3	23.4	33.5	0.9	0.1	184.9	15.9	333.0	—	1.3	—	4.6	—	1.6	37.1	8.3	—	0.1	2.3	2.4	—	—	—	388.3	76.5[b]	464.8
Belgium/ Luxembourg	1.6	—	—	4.7	—	—	12.2	2.9	21.4	—	—	—	—	—	0.2	—	—	—	—	0.2	0.2	—	—	—	21.8	—	21.8
Denmark	2.1	—	—	—	—	—	—	0.5	2.6	0.3	—	—	—	—	0.5	8.4	—	—	—	0.1	0.1	—	—	—	11.9	—	11.9
France	30.6	89.5	—	—	—	—	34.4	15.7	170.2	—	—	0.2	0.2	—	0.1	—	10.3	—	—	—	—	—	—	—	181.0	1.3	182.3
Eire	—	—	—	—	—	—	—	22.8	—	—	—	—	0.6	—	—	—	—	—	—	0.4	0.4	—	—	—	23.8	—	23.8
Italy	61.7	6.2	1.7	69.6	—	—	20.2	10.2	169.6	—	15.2	1.5	0.8	0.1	0.6	27.8	161.1	0.1	0.1	36.1	36.2	—	—	—	368.0	75.2[b]	443.2
Netherlands	48.1	72.0	—	2.5	0.1	—	—	15.0	137.7	—	—	—	0.6	0.1	0.2	0.2	—	—	0.2	0.1	0.3	—	—	—	139.1	1.2	140.3
United Kingdom	0.1	2.8	—	0.2	5.6	—	8.6	—	17.3	—	—	—	0.7	—	0.1	0.6	—	—	—	—	—	—	—	—	18.7	9.3	28.0
Total EEC	144.2	244.8	25.1	10.5	6.6	0.1	260.3	83.0	874.6	0.3	16.5	1.7	7.5	0.2	3.3	74.1	134.7	0.1	0.4	39.2	39.6	—	—	—	1152.6	163.5	1316.1
Iceland	—	—	—	—	—	—	—	—	—	—	—	—	—	—	—	—	—	—	—	—	—	—	—	—	—	—	—
Austria	35.0	—	3.7	—	—	—	0.6	0.4	39.7	—	—	—	0.1	—	0.1	10.5	4.0	—	—	—	—	—	—	—	54.4	74.9[b]	129.3
Spain	0.1	1.9	0.2	26.3	—	—	0.5	9.4	38.4	—	0.2	—	0.0	—	—	1.0	0.3	—	0.5	2.4	2.9	—	—	—	43.4	—	43.4
Finland	—	—	—	—	—	—	—	—	—	—	—	—	—	—	—	—	—	—	—	—	0.1	—	—	—	—	—	—
Greece	—	—	—	—	—	—	—	4.6	4.6	—	0.4	—	0.5	—	—	—	—	—	—	0.1	—	—	—	—	5.6	4.5	10.1
Norway	—	—	0.9	—	—	—	—	—	0.9	—	—	—	—	—	—	22.7	—	—	—	—	—	—	—	—	23.6	—	23.6
Sweden	14.8	—	14.1	—	—	—	7.2	1.1	37.2	—	—	—	0.1	—	5.9	—	—	—	—	—	—	—	—	—	43.2	0.1	43.3
Switzerland	19.0	—	2.0	18.0	—	—	2.0	—	41.0	—	1.0	—	1.0	—	—	2.0	—	—	—	—	—	—	—	—	45.0	—	45.0
Turkey	—	—	—	—	—	—	—	—	—	—	—	—	—	—	—	0.5	—	—	—	—	—	—	—	—	0.5	—	0.5
Canada	—	—	—	—	—	—	—	—	—	—	—	—	—	—	—	—	—	—	—	141.1	141.1	—	—	—	141.1	—	141.1
U.S.A.	—	0.1	—	—	—	—	—	0.1	0.2	—	—	—	—	—	—	—	—	—	78.2	—	78.2	—	—	—	78.4	0.3	78.7
Total North America	—	0.1	—	—	—	—	—	0.1	0.2	—	—	—	—	—	—	—	—	—	78.2	141.1	219.3	—	—	—	219.5	0.3	219.8
Japan	—	—	—	—	—	—	—	—	—	—	—	—	—	—	—	—	—	—	—	—	—	—	—	—	—	—	—
Australia	—	—	—	—	—	—	—	—	—	—	—	—	—	—	—	—	—	—	—	3.9	3.9	—	—	—	10.5	0.8	11.3
New Zealand	—	—	—	—	—	—	—	—	—	—	—	—	—	—	—	—	—	—	—	—	—	—	—	—	—	—	—
Total OECD	213.1	246.8	46.0	154.8	6.6	0.1	270.6	98.6	1036.6	—	18.1	1.7	9.8	0.2	9.3	110.8	139.0	0.1	79.1	186.7	265.8	—	—	6.6	1599.3	244.1	1842.4
Non-OECD	3.0	3.5	10.4	—	—	—	1.5	2.8	21.2	—	—	—	—	—	0.3	7.2	—	—	1.0	265.4	266.4	—	—	0.8	—	—	295.4
Total based on importing country figures	216.1	250.3	56.4	154.8	6.6	0.1	272.1	101.4	1057.8	—	18.1	1.7	9.8	0.2	9.6	118.0	139.0	0.1	80.1	270.9	371.0	—	—	—	1599.3	—	—
Total based on exporting country figures	208.0	251.6	54.8	—	6.6	0.1	269.8	102.2	893.1	—	16.5	3.1	—	—	13.0	121.1	136.0	—	92.7	619.9	712.0	—	—	—	1899.8	—	—

[a] Source: Computed from Organization for Economic Cooperation and Development, *Pulp and Paper Industry, 1973/1974*, OECD, Paris, 1974.

Dominated by Eastern European Sources.

However, countries do engage in international trade in waste paper as well as paper and board. The total volume and direction of the world trade in waste paper in 1973 is presented in the trade matrix in Table I, showing the quantity flows of waste paper among the EEC countries and four other country-groups: (a) non-EEC Western Europe, (b) North America, (c) the developed countries of the Pacific Basin, and (d) non-OECD countries. The pattern of trade that emerges is one that is dominated by a network of fairly close-knit trading ties between neighboring countries. Most European countries trade with each other and only some 15% of U.S. exports reached OECD Europe in 1973. Italy, the Netherlands, France, and Germany get the bulk of their waste paper imports through intra EEC trade. This picture tends to confirm that transport costs represent a very significant variable in secondary materials trade.

Table II summarizes trade, recovery and utilization of waste paper for the major trading nations. Several countries have very high export ratios compared to the level of domestic recovery of waste paper, Belgium-Luxembourg (65%), Denmark, Netherlands, and Switzerland (about 35% each), and Sweden (30%). Substantially lower rates between 10 and 20% exist for West Germany, France, Norway, and Canada, with the United States, Finland, Spain, Austria, and the United Kingdom exporting less than 10% of domestic waste paper recovery. Countries relatively dependent on imports for their waste paper consumption are Eire, Italy, and Austria, with Belgium–Luxembourg, France, Spain, Sweden, and Denmark at the low end of the scale. Neither the United Kingdom nor the United States is significantly reliant on waste paper imports.

The difference between the utilization and recovery rates for any country is determined by the effect of trade in both paper and board and waste paper. The observed disparity between the two sets of ratios is thus a composite measure of the effect of both types of trade on recycling activity. Figure 3a illustrates these effects for sixteen OECD countries. We have seen that in a closed economy the waste paper utilization rate would be equal to the recovery rate, and points representing this situation would lie on the 45° line. The level of recycling that takes place in each country is indicated very generally by the distance a point lies away from the origin coordinates (0, 0) indicating no recycling, and the top right-hand corner of the graph (1, 1) defines 100% recycling. Hence points along the 45° line represent the level of recycling. Interpretation of points lying off this line is more problematic. Both the recovery and utilization rates are indicators of recycling activity, therefore any movement in a NE direction in Fig. 3a indicates recycling activity. Where trade exists, the point may be shifted to the left or right of the 45° line—the direction and magnitude of the shift depending upon the composite effect of trade both in paper and board and in waste paper. Canada, Norway, and Austria, for example, are net importers of waste paper and net exporters of paper and board, both of which will increase the relevant utilization and recovery rates but it is the magnitude of the trade that will determine whether the point will be the left or the right of the 45° line.

Figure 3b shows how each type of trade affects the recovery and utilization rates. Each country is again represented by a point whose position is determined by the proportion of trade to domestic production of paper and board and recovery of waste paper. Each quadrant represents one of four combinations of trade flow (net imports or exports of waste paper or paper and board) in relation to production of paper and board or recovery of waste paper. Using Fig. 3b

TABLE II

Importance of Waste Paper Trade 1973 and 1974[a]

	Exports as a percent of domestic recovery rate of waste paper		Imports as a percent of apparent consumption of waste paper	
	1973	1974	1973	1974
Austria	7.3	6.6	37.8	52.5
Belgium–Luxembourg	65.8	68.5	14.2	16.2
Canada	11.8	13.7	17.0	22.3
Denmark	31.0	37.1	8.9	8.3
Eire	10.0	9.4	28.7	27.6
Finland	2.5	4.4	0	0
France	10.1	11.0	10.7	12.8
Italy	0	0	32.6	28.9
Netherlands	32.6	36.2	20.1	23.1
Norway	12.9	17.5	21.1	28.0
Spain	0.5	1.4	6.7	11.7
Sweden	29.2	28.1	12.8	8.8
Switzerland	37.7	37.1	16.7	21.6
U.K.	4.9	6.5	1.4	4.8
U.S.A.	5.3	9.6	0.7	0.7
West Germany	8.2	10.0	16.7	19.1

[a] Source: Computed from OECD, *The Pulp and Paper Industry, 1973/1974*, OECD Paris, 1974 and 1976.

in conjunction with Fig. 3a, it is possible to establish a detailed picture of the trade-effect for each country. Austria (point 10) provides an interesting example. Figure 3a shows Austria lying close but to the right of the 45° line, suggesting that the level of trade is small. However, Fig. 3b shows that Austria's position is due to the very large net imports of waste paper causing the utilization rate to be high. At the same time, Austria's large net exports of paper and board push its recovery rate up. The combined effect of the two types of trade flows is that

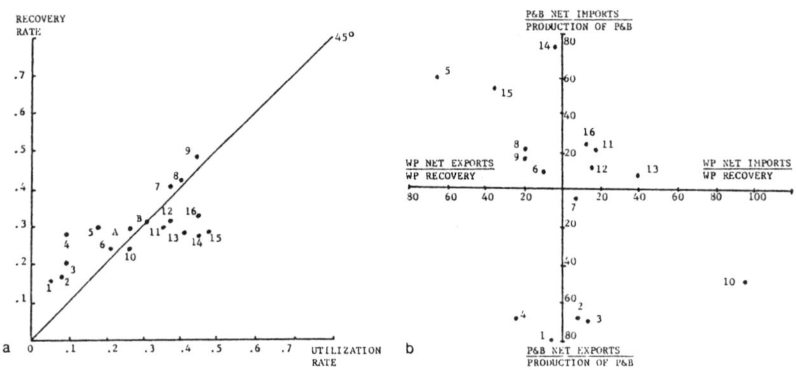

A=OECD total, B=OECD Europe, 1=Finland, 2=Canada, 3=Norway, 4=Sweden, 5=Belgium-Luxembourg, 6=USA, 7=Japan, 8=Switzerland, 9=Netherlands, 10=Austria, 11=France, 12=Spain, 13=Italy, 14=United Kingdom, 15=Denmark, 16=F.R. Germany.
Source: Computed from OECD, The Pulp and Paper Industry, 1974/75 (Paris: OECD, 1075).

FIG. 3(a). Impact of trade on recovery and utilization rates of waste paper, 1974. (b) Direction of trade, 1974, waste paper, and paper and board.

Austria is represented by a point close to the 45° line—the trade in waste paper marginally outweighing the impact of the net export trade in paper and board.

The pattern that emerges from Fig. 3a is also related to the domestic availability of pulp. The major pulp producing countries such as Finland, Canada, Norway, and Sweden can easily be identified as a group shown by the cluster of points 1, 2, 3, and 4. The cluster is rather distant from and to the left of the 45° line but close to the origin. Fig. 3b confirms that each of these countries is a major net exporter of paper and board, the effect of which outweighs the waste paper trade effect which is negligible.[6] This indicates the importance of the relative price variables *Piv/Pis* and *Pjs/Pis*. A second distinct group of countries, those with small indigenous supplies of pulp, can also be identified in Fig. 3a. This second group comprises the European countries numbered 11 to 16. The location of this second group in Fig. 3a results from the combined effects of net import trade in paper and board and (with two exceptions) net import trade in waste paper as shown in Fig. 3b. Both factors lead to high utilization rates and low recovery rates, and hence a grouping to the right of the 45° line. Two exceptions are the United Kingdom and Denmark (points 14 and 15) which are net exporters of waste paper, the effects of which are negated by the large net import trade in paper and board products.

A third group of countries (Points 5 to 10) are generally scattered along and reasonably close to the 45° line. Figure 3b shows that this is because these countries fall in the two quadrants where the impacts of trade work against each other and the magnitudes of the trade flows are of the same order, thus bringing the utilization and recovery rates close together. Belgium–Luxembourg (point 5) is the only exception because the two trade flows—although yielding opposite effects—do not cancel each other out. It is pushed further away and to the left of the 45° line by the impact of the large net export trade in waste paper, which tends to depress its utilization rate.

We noted earlier that the secondary materials industry is subject to cyclical instability. To illustrate this, Fig. 4a shows U.K. waste paper consumption, adjusted for trend, during the period 1962–1975. A clear 4- to 4½-year consumption cycle emerges that is largely accounted for by variations in the level of economic activity (industrial production).[7]

This consumption cycle, in turn, generates a price cycle—although the exact relationship between the two variables is complicated in the U.K. case by the existence of vertically integrated mill dealer organizations and by lagged responses by mills to changes in demand pressure. The issue is also complicated by stock adjustment behavior. As a result, prices appear to "follow" the consumption cycle. It can also be shown that these price cycles tend to move together internationally. Market prices of waste paper, undeflated by any other price index, are shown for several countries in Fig. 4b.[8] West German prices (not shown) behave in much the same manner—see [9].

[6] Figure 3a is not meant to be used in isolation. The impact of international trade on recovery and utilization rates is in fact described by the use of both Figs. 3a and b. Neither diagram in isolation tells very much about this due to their two-dimensional nature.

[7] For a detailed discussion, see [7] and [8].

[8] The waste paper trade figures presented so far are overall totals as no breakdown according to grades of waste paper traded are available. However, it can be shown that prices of different grades in any one country do move together, and that prices of the same grade move together between countries.

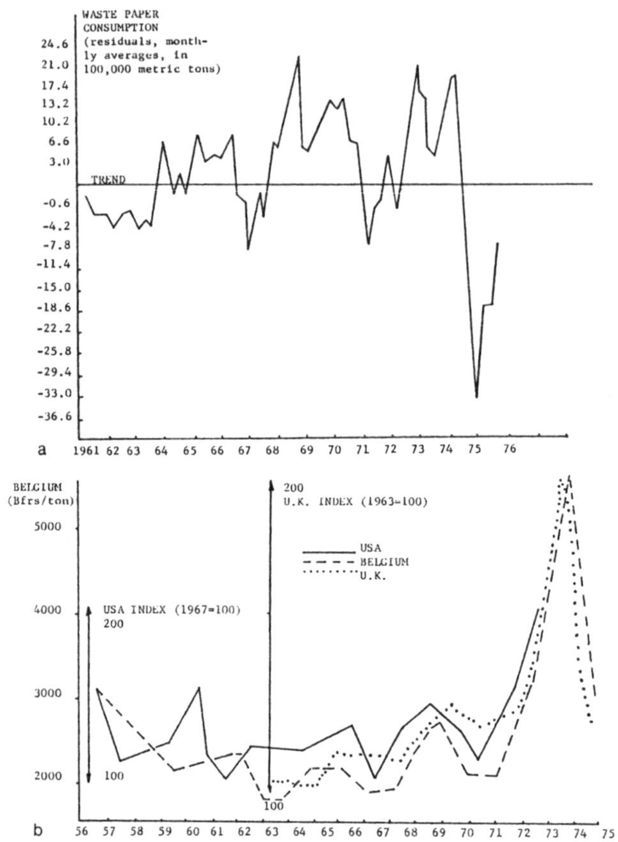

FIG. 4(a). Waste paper consumption cycles in the United Kingdom, 1961–75. (b) Waste paper price movements in U.K., U.S.A., and Belgium, 1956–1975.

Thus there would seem to be a comparatively free trade in all grades of waste paper, with exports and imports responding to prices *Pis*/*Pjs*. In 1969, for example, mixed waste paper in the U.K. sold at £10.50 per ton f.o.b., as compared with £13 per ton f.o.b. in the Netherlands, i.e., $Pis < Pjs$. This premium, however, proved too small to generate substantial additional trade between the two markets, presumably because it was not large enough to offset transport costs. Unfortunately, trade statistics covering waste paper are not broken down by grade. Nevertheless, the total export trade flow between the U.K. and OECD Europe in 1969 was 99,100 metric tons. By contrast in 1974, with price controls in the U.K. home market and strong demand on the Continent, mixed papers selling at £20 per ton in the U.K. sold for £50 per ton in Europe and exports were stimulated. Total U.K. exports to OECD Europe in 1974 were $136,600 metric tons, a 38% increase over 1969 levels. The average annual waste paper export figure to OECD Europe in the intervening years was 83,000 metric tons.

In this respect then, trade became a mechanism for reducing price fluctuations at the national level.

SOME POLICY ASPECTS

We have attempted to establish a link between national and international markets for secondary materials, both conceptually and to a limited extent emprically using waste paper as a case in point. If there is free international trade in waste products, and if barriers to knowledge and economic distance can be overcome, then prices on international markets will determine the economics of recycling within a particular country. We have seen that international prices will tend to encourage recycling in some countries and discourage it in others. but that the *overall* volume of recycling will tend to be greater with international trade than without it. Moreover, the stability of markets for recycled materials is expected to be greater under conditions of international supply and demand diversification than in its absence. In short, international trade appears to play much the same role with respect to secondary materials as it does in other goods and services sectors—it allocates them where their marginal product (and hence price) is highest, and thereby provides gains to both users and suppliers. In both cases, these gains are economic as well as environmental. There are, however, a number of additional dimensions that might be touched upon briefly here.

If market forces, environmental policy, and secondary materials policy succeed in bringing about higher levels of recycling that are considered beneficial from a national welfare standpoint, the international financial implications can be beneficial as well. If the country in question imports virgin raw materials, recycling can serve to displace purchases from abroad, with a positive effect on the current account of its balance of payments. If it is self-sufficient in raw materials, recycling will generate exports of secondary or primary materials, again having a positive effect on international payments.[9] If—as is currently the case—there exists a system of floating exchange rates, increased recycling will tend to cause the country's currency to appreciate on foreign exchange markets and improve its commodity terms of trade by reducing the cost of imports and increasing export prices. In either case, the implications of increased recycling are likely to be beneficial.

Second, we noted at the beginning of this paper a tendency for increased governmental interference in basic materials markets—interference designed to displace volumes of production, consumption and trade, as well as prices, from their market-determined equilibria. This position is advocated by developing countries to redress what they allege has been a long-term deterioration in the terms of trade of primary for manufactured goods, with a concurrent transfer of real income from the poor to the rich countries. Whether this has indeed happened in recent decades is arguable, although it is accepted as fact by the developing nations in stating their case. If market rigging in primary commodities is successful, it will substantially increase price incentives for recycling, especially in the nonrenewables sector. This has already made itself felt in waste oil, and may affect ferrous and nonferrous metals sectors next. Countries with major recycling capability may be spared some of the associated economic dislocations.

[9] For a preliminary discussion of the magnitude involved, see [11].

There is also a movement underway for international commodity price stabilization. It is recognized that international primary commodity prices tend to be more unstable than the prices of manufactured goods through time, and that this makes coherent economic policy difficult particularly for developing countries specialized in basic materials and agricultural products. Instability engenders costs, and these costs may fall disproportionately on the poor countries. To remedy this, both the advanced and developing countries have agreed to investigate the feasibility of price stabilization schemes, including buffer stocks, for individual commodities. If such stabilization does indeed materialize, it would benefit the secondary materials sector as well by reducing price fluctuations of substitute primary materials. This could affect the economic structure and efficiency of the recycling industry and ultimately read to higher rates of recycling—and reduced environmental damage.

Third, there is the problem of secure access to raw materials supplies. As the 1973–1974 oil crisis demonstrated, damage to national economies caused by embargoes or precipitous price rises of basic materials may be substantial indeed. It makes sense to insure against such risk by diversifying sources of basic materials supplies even if these alternatives involve higher costs. Aside from domestic and third country supplies of such materials, one obvious possibility is recycling. This could justifiy higher domestic materials prices through import levies on virgin resources (and hence increased incentives to recycle), direct subsidies, or other incentives to the recycling industry itself.

Finally, the principal trade policy issue directly relating to the secondary materials sector is export control. In times of rising materials prices and possible domestic shortages, exporters are tempted to restrict sales abroad. The incentive to interference with exports of secondary materials will grow as supplies of substitute primary materials become more costly, and particularly if they become uncertain as a result of export policies pursued by primary suppliers. The result of export control on the domestic side are predictable. Prices of secondary materials will be lower than if they had free access to export markets. Consequently the incentive structure that determines the volume of secondary materials recovery is negatively affected, and the actual recovery rate is likely to decline. Unless the initial collection phase is still undertaken and the wastes are stockpiled, environmental damage may result. Perhaps more important, export controls may disprupt lasting and orderly supply relationships between buyers and sellers of secondary materials, and promote market disintegration and fragmentation that may be difficult to reverse later on. In short, export controls may indeed yield short-term economic gains, but only at potentially significant long-term economic and environmental cost.

ACKNOWLEDGEMENTS

For research support, the authors are grateful to the Social Science Research Council (U.K.) and the Organization for Economic Cooperation and Development, Paris. The authors are indebted for helpful comments on earlier drafts to David W. Pearce, Michel Potier, and several anonymous referees.

REFERENCES

1. Environmental Protection Agency, "Second Report to the Congress: Resource Recovery and Source Reduction," U.S. Government Printing Office, Washington, D.C. (1974).

2. A. V. Kneese, R. U. Ayres, and R. C. d'Arge, "Economics and the Environment," Johns Hopkins Press, Baltimore (1960).
3. T. Page, "Conservation and Economic Efficiency," John Hopkins Press, Baltimore (1977).
4. D. W. Pearce, "Environmental Economics," Longmans, Green, London (1976).
5. D. W. Pearce and I. Walter, Eds., "Resource Conservation: Social and Economic Dimensions of Recycling," New York Univ. Press, New York (1977).
6. Royal Commission on Environmental Pollution, "Fourth Report," Her Majesty's Stationery Office, London (1975).
7. R. K. Turner, R. Grace, and D. W. Pearce "Forecasting the Demand for Waste Paper," PSERC/MRP Discussion Paper No. 76-04, Leicester University (1976).
8. R. K. Turner and R. Grace, "Forecasting the Demand for Secondary Materials," PSERC/MRP Discussion Paper No. 76-07, Leicester University (1976).
9. Umweltbundesamt, "Materialien zum Abfallwirtschaftsprogranm 75:II-Papierabfälle," Federal Ministry of the Interior, Bonn (1976).
10. I. Walter, "A Discussion of the International Economic Dimensions of Secondary Materials Recovery," Institute for International Economic Policy, Washington, D.C. (1975).
11. I. Walter and S. P. Maltezou, International trade and secondary materials recovery: The case of waste oil, *Environ. Affairs* 3, No. 3 (1974).

[9]

ENVIRONMENTAL ECONOMICS, INDUSTRIAL PROCESS MODELS, AND REGIONAL-RESIDUALS MANAGEMENT MODELS

DAVID JAMES

Macquarie University, Australia

1. Conceptual framework for industrial activity analysis and regional environmental quality management

1.1. Materials balance, residuals discharges, and environmental external effects

The concept of externalities associated with residuals in the environment has long standing within the discipline of economics ever since Pigou (1932) discussed the welfare implications of smoke from a railroad. But it was not until the environmental movement of the later 1960s, together with new insights gained from industrial and water research, especially in the United States, that economists were forced to consider the full significance of economic activity in terms of its effects on environmental quality. The work of Ayres and Kneese (1969) and Kneese, Ayres and d'Arge (1970) from Resources for the Future (RFF) represented a turning point in the literature. These writers applied the laws of mass balance (or conservation of mass), which were well known to chemical engineers in their design and analysis of industrial plants, to the economic system as a whole.

Within the framework of a Walras–Cassel general equilibrium model (specifically a physical Leontief system), Ayres and Kneese formulated a set of identities relating to the use and disposal of materials by the various production and consumption sectors of the economy. Without going into mathematical proofs, which can be found in the references cited, the main balances can be summarized as: the mass of all materials inputs from the environment to the economy, ignoring flows from the environment directly to the final consumption sector, equals the mass of inputs to the intermediate product sectors; the mass of inputs to the intermediate product sectors equals the mass of products supplied to the final consumption sector plus the mass of residuals discharged to the environment minus the mass of materials recycled; and the mass of all final products equals the

Handbook of Natural Resource and Energy Economics, vol. I, edited by A.V. Kneese and J.L. Sweeney

mass of materials recycled plus residuals generated by the final consumption sector. Assuming no accumulation or recycling, the mass of all inputs from the environment must equal the mass of all residuals discharged to the environment.

Several strands of research arose from this work. Elaborations and criticisms of the model came from Noll and Trijonis (1971), Converse (1971), and Victor (1972), among others. The welfare implications were profound. Ayres and Kneese demonstrated that, far from constituting a theoretical curiosum, environmental externalities associated with the discharge of residuals to natural systems are a normal and inevitable outcome of economic activity. Research on pollution and environmental damage functions soon followed [OECD (1974, 1976)]. At the national level, input–output models were extended to operational simulation tools capable of projecting discharges of liquids, solids, and gases to air, water, and land, under various economic growth scenarios [Ridker (1972)], and efforts were made to identify the industrial sectors most responsible for particular types of residuals discharges and environmental damage [Victor (1972), Ayres and Gutmanis (1972)].

The outcome of primary interest here is the development of multidisciplinary quantitative models designed to analyze industrial activities, the generation and discharge of residuals by activities to the environment, the effects of residuals on ambient environmental quality (AEQ) and receptors, and the actions that can be taken to manage AEQ at the regional level. Such models are described in the literature as residuals–environmental quality management (REQM) models.

The plan of the chapter is as follows. First, a set of operational concepts and definitions is provided. The construction of REQM models is then discussed in general terms. The relationship between economic activity and residuals flows is analyzed in greater detail, emphasizing the distinction between residuals generation and discharges to the environment. The behavior of residuals in the environment is dealt with next, demonstrating the use of air and water dispersion models and the role of dose–response functions. Approaches to residuals management are then presented, with an essentially theoretical treatment of alternative modeling techniques. The chapter concludes with a series of actual case studies demonstrating application of REQM models to the management of individual industrial plants, air and water quality, and integrated regional economic–environmental systems.

1.2. Concepts and definitions

Appropriate terminology for REQM modeling was established by RFF researchers in conjunction with their work on materials balance and industrial process analysis [Ayres (1972), Russell and Spofford (1972), Bower (1977), Bower et al. (1977), Kneese and Bower (1979)]. The same terminology is followed here.

Regional-residuals management begins with an analysis of industrial activities. The term "industrial" refers to the transformation of materials (or energy) to desired commodities or "product outputs". An activity is defined as a decision-making establishment (mine, plant, mill) consisting of a set of unit processes or operations with given site boundaries. This definition of an industrial activity draws heavily on the concepts of chemical engineering. As described by Ayres (1972), a unit operation is a physical transformation of materials within a plant, and a unit process involves chemical reactions. Each operation or process is distinct, and has a specific set of technological characteristics, production inputs, and product outputs. The industrial process approach to production analysis thus involves a detailed study of everything that takes place within an industrial activity. At a higher level of aggregation, activities of a similar kind may be classified as a regional production sector, although considerable variation could exist in product mixes, input requirements, and production technologies. Whether such variation is important or not will be discussed at a later stage.

Flows of materials and energy that remain from the production of product outputs are referred to as nonproduct outputs. Some nonproduct outputs have further economic use, either as marketable by-products or as recovered materials (or energy) that can be used in production processes in the same plant. In sugar refining, for example, sugarcane fiber (bagasse) can be converted to building boards or placed in combustion chambers to raise steam for process heat or electricity generation.

Recovery of nonproduct outputs is affected by economic factors. Costs usually are involved in recovery operations, and in an efficient industrial plant, recovery will proceed to the level at which the marginal recovery cost equals the economic value of the recovered materials or energy. Among other things, prices of product outputs, by-products and production inputs to an activity will govern recovery operations.

Remaining nonproduct materials and energy are defined as residuals. Materials residuals take the form of gases, liquids, and solids. Energy residuals comprise noise, heat, and vibration. Radioactive residuals are a third category with their own special characteristics. The term "residual" is clearly economic. Whether or not a nonproduct output becomes a residual depends on prevailing cost and price conditions and on process technologies. Government regulations and charges may also be important determinants. It is worth noting, finally, that residuals are not defined as "wastes" or "pollutants", since at their point of generation their subsequent economic or environmental significance is unspecified.

Residuals can be left untreated or subjected to modification processes. On-site modification of residuals may entail add-on processes (e.g. flue-gas scrubbing) or constitute an integral part of the activity's production processes. In either event, modification processes require their own materials and energy inputs and may lead to by-products, secondary materials and energy recovery, and further residuals generation, depending on particular circumstances. Disposal alternatives include discharge to the environment at site boundaries, storage on site, and conveyance to other locations for off-site storage or discharge.

The environment consists of natural systems outside economic activity boundaries. For the purpose of residuals management, AEQ can be measured in terms of the concentrations of residuals at receptor points within a region. Linkages between discharges of residuals and ambient concentrations can be quite complicated, as natural systems processes and effects are involved. For nonreactive residuals, this refers to the physical dispersion of residuals in air, water, and land systems. Critical factors are the locations of discharge points, the mass or volume of residuals discharged, the type of residuals, temporal release patterns, and physical processes in the environment, such as diffusion, sedimentation, fallout, and accumulation. Reactions among residuals such as oxidation (and reduction), biological assimilation, or transformation may, however, occur in the environment. For policy reasons, it may be necessary to consider the exposure of receptors (plants, humans, animals, materials, and structures) to ambient concentrations and assess the ultimate environmental impacts through dose–response relationships.

Management of environmental quality involves physical measures to alter the release or functions of residuals in the environment, and implementation incentives (inducements or penalties) introduced by government to control the disposal of residuals by activities. Various institutional arrangements may be established to facilitate this task. Each particular set of physical measures, implementation incentives, and institutional arrangements to control AEQ is defined as an environmental management strategy. The conceptual framework for a residuals-environmental quality management system is presented in Figure 7.1. In any situation, many alternatives usually exist as a means of reaching AEQ goals.

1.3. Components of a regional environmental quality model

Quantitative models are available to deal with all phases of the analysis and management of residuals in a regional environment. The researcher has a choice of constructing a comprehensive modeling system incorporating all possible components, or of carefully selecting the most appropriate models to tackle specific management problems.

The conceptual framework of Figure 7.1, for example, has been translated into an operational regional model by Russell and Spofford (1972) and applied to the Delaware Estuary, as discussed in Section 5.4.1. Under the Russell–Spofford approach, production decisions within an activity are simulated by means of a linear programming model that incorporates different technologies, by-products, materials recovery, and residuals modification processes. Technologies and

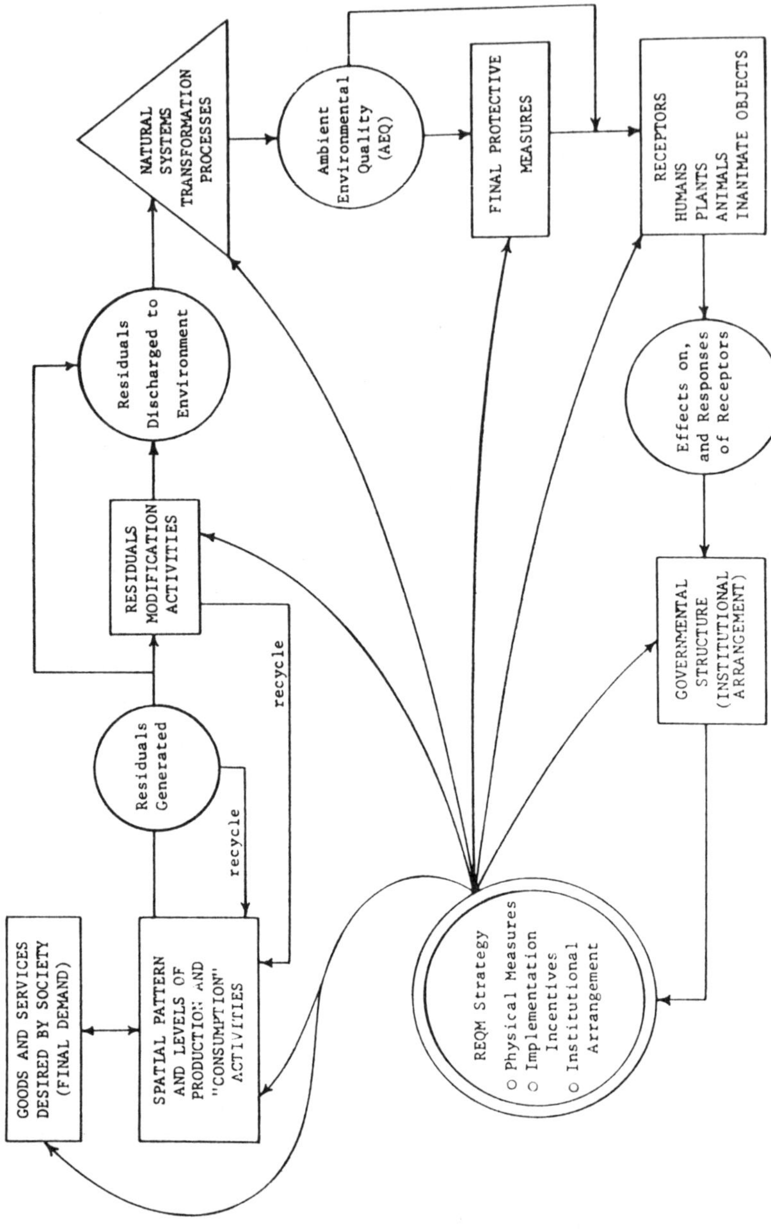

Figure 7.1. Concept of a residuals–environmental quality management (REQM) system. *Source*: Bower (1977, p. 8).

residuals handling respond to cost parameters in the objective function or to constraints imposed on residuals discharge rates.

The behavior of residuals released to the environment is analyzed by means of natural systems models, which have been surveyed by Bower (1983) and Basta and Bower (1982), among others (and briefly in Chapter 3 of this Handbook). Ambient concentrations of nonreactive residuals in air and water systems can be estimated with physical dispersion models. Reactive residuals require more elaborate analysis taking physical and chemical processes into account. Biological models combine residuals movement in the environment with ecological processes such as the transmission of toxic substances in food chains and changes in the population levels of organisms of terrestrial or aquatic ecosystems.

Environmental damage functions may play an important role in an environmental quality management model. Estimates of exposure of receptors to ambient conditions in the environment are required, and physical receptor effects determined through dose–response functions or models. Monetary estimates can, in principle, be made to obtain monetary representations of economic welfare changes (see the discussion in Chapter 6 of this Handbook).

Whether damage functions are needed in a model depends largely on the kind of management objectives pursued. As pointed out by Lesuis et al. (1980), objectives often conflict, and compromise solutions must be found. Typical goals for a regional system are the maximization of output or income, minimization of environmental damage costs and minimization of production and residuals management costs. Quantitative models, designed to assess tradeoffs between economic and environmental goals, are described later in Section 4.4. Many of these models incorporate environmental standards rather than environmental damage (or improvement) functions.

1.4. Approaches to model construction

Construction of REQM models can be approached in a number of different ways. Bower (1983) classifies approaches according to: mathematical rigor (simple empirical functions, two-variable correlation models, multivariate regression models, and analytical models); natural systems media (air, water, land, and ecosystems); and phenomena under study (natural systems or receptor effects).

Models may also be classified in terms of their mathematical form. Simulation models, static or dynamic, are useful when decisionmakers' priorities are not clearly specified. Alternative outcomes of different environmental management strategies can be presented to decisionmakers for their consideration. A drawback with simulation modeling is that the full range of management options may not be properly identified, and the results of simulations may be difficult to interpret. Optimization models, especially those based on mathematical programming techniques, offer a more systematic means of meeting management goals.

Regardless of the kind of model proposed, certain criteria should be met when setting up the analytical framework. The first is that due attention should be given to the spatial, temporal, physical, chemical, ecological, and receptor-effects aspects of the model. Second, flexibility should be maintained so that the model can be adapted to a variety of policy applications. Third, the overall framework should be decomposable so that research skills can be directed to specific modeling tasks and provision is made for the monitoring of outputs from each component of the model. Fourth, the problem of uncertainty must be addressed. One solution is to build a stochastic model. Other approaches include sensitivity analysis and the estimation of probability distributions for key explanatory variables in the model. Finally, certain logistical difficulties in constructing, operating, and applying the model need to be overcome.

1.5. Model accuracy, cost, and applications

Costs of an REQM model can be divided into several categories. Data acquisition costs comprise the first. The data requirements for a comprehensive model incorporating natural systems and dose–response effects are considerable. Unless data can be provided by existing research institutions, private industry, and government agencies, primary data collection has to be carried out, and this is a very costly operation. Models based on Leontief input–output systems may be quite expensive if a set of regional input–output accounts has to be specially compiled. Costs of research personnel may also be high. Environmental modeling is a multidisciplinary endeavor and the necessary skills, particularly those of coordination and direction, may be difficult to locate and buy. A third cost category is that of computing time (where needed) which usually rises rapidly as model complexity increases.

Problems of accuracy may create serious difficulties in implementing a REQM model. Data may be unreliable or outdated. The model itself may not be properly specified. Problems arise, for example, if an attempt is made to use a steady-state model to predict a dynamic situation, or if a linear model does not appropriately represent actual relationships. Verification may be difficult to achieve in practice; expensive simulation and monitoring may be required. Natural systems models, especially those of ecosystem behavior, are notorious for their unreliability. Statistical analyses of dose–response relationships have similarly been heavily criticized. With all of these problems, tradeoffs typically must be made between model accuracy and cost.

Communication of the results of an REQM model to users is the final logistical problem to overcome. Unless users are frequently consulted in the early stages of model construction, the wrong issues may be analyzed and irrelevant conclusions drawn. It is often necessary to simplify the results to convey the findings to environmental managers. In this respect, benefit–cost analysis is a useful approach, as a wide spectrum of effects can be condensed to a single indicator. Cost-effectiveness analysis, with environmental standards set by scientific and medical experts, is another realistic route to follow. Problems of this kind are dealt with extensively in the general economic literature, but references that are specifically concerned with REQM models are Spofford et al. (1975), EPRI (1979a), Bower (1983), Kneese (1977), and Kneese and Bower (1979).

2. Economic activity and the determination of residuals flows

2.1. Methods of determining residuals discharges

Approaches to the estimation of residuals discharges from economic activities should be guided by the kind of environmental planning or management problem to be tackled. In some situations, the aim is to find the best set of residuals management practices and technologies that meets AEQ goals while maintaining existing levels of product outputs for an established regional industrial framework. Controls may simply be required for a single major plant within a region, so that an in-depth study of a particular industrial activity is called for. In other situations, the objective might be to obtain "broad-brush" scenarios of future economic development patterns, with associated residuals discharge and AEQ effects, to identify general problem areas that subsequently can be examined in greater detail.

In principle, industrial process models of the kind developed by Resources for the Future (RFF) should yield the most accurate estimates of residuals discharges at the individual plant level and at the regional level, and the relevant methodologies are highly commended in this review. It is worth noting, nevertheless, that the RFF approach is suitable only in certain contexts. Activity levels are assumed to already exist, and an optimal REQM strategy is sought in an essentially static production situation. Future projections could of course be made, and either a comparative statics or dynamic programming approach adopted, but a commensurate increase in data requirements and modeling effort must also be expected. The costs of complex dynamic optimization models are invariably high, although they have been borne by various member countries of the OECD in their development of national energy system assessment models [International Energy Agency (1980)].

For relatively low-cost projections of regional-residuals discharges, multisectoral simulation models – in particular, generalized regional input–output models – have been a popular choice. Industrial activity levels in such models are usually quantified in terms of sectoral monetary output values. Important detail can easily be hidden on input patterns, production technologies, product mixes, discharge rates, and relative sizes of individual establishments in the same regional industrial category, but the technique nevertheless generates information on "average" conditions, which is undoubtedly better than having no information at all. Furthermore, if desired, multisectoral models featuring mixed indicators of industrial output – both monetary and physical – can be constructed, and engineering detail pursued in the industrial sectors that are the most significant environmentally. Residuals recovery and modification processes can fairly easily be incorporated. Multisectoral simulation models thus do have a role to play in REQM programs, their main contribution being to provide rough estimates of residuals discharges under different technological and economic growth scenarios.

The level of aggregation of industrial activities within an REQM model is clearly a major factor governing the approach to residuals discharge determination. This point having been made, it is appropriate to consider the methods that have been applied to individual industrial plants, linked industrial activities, and multisectoral regional production systems.

2.2. *Residuals from individual plants*

Residuals discharged by industrial activities can be determined by applying the laws of conservation of mass to individual plants. The mass of all inputs to a plant, including environmental inputs such as oxygen used in combustion or chemical processes, must equal the mass of all final products and by-products, materials accumulated on-site, and residuals directly discharged to the environment or transported to other locations.

To estimate residuals discharges from a given activity, one approach is to establish residuals discharge coefficients and apply these to output levels of the activity. As demonstrated by Kneese and Bower (1979), such coefficients may be obtained in two stages: first, by ascertaining the net generation of residuals within the activity, and second, by determining the actual discharge of residuals into the environment. The two concepts are quite distinct. In the absence of controls over residuals discharges, generation coefficients may be written as a simple function:

$$R_g = f(M, TP, PO, POS), \tag{1}$$

where R_g is a vector of residuals generated per unit of output or raw material input, M is a vector of raw materials used, TP is a set of process technologies, PO is product mix, and POS a set of product specifications.

It is preferable to calculate generation rates in terms of physical output levels or in relation to materials or energy inputs. As an example of the input approach, combustion emission factors can be combined with fuel-use patterns in a plant to estimate airborne emissions [U.S. EPA (1973), Stern (1976), and Tomany (1975)].

Residuals discharge coefficients depend on the same factors as generation coefficients, but new determinants must be taken into account. An appropriate expression is

$$R_d = F(M, TP, PO, POS, EC, TR), \tag{2}$$

where R_d is a vector of residuals discharged per unit of output or raw material input, EC reflects a set of controls, standards or charges imposed on specific residuals, and TR indicates residuals modification technology. Changes in any of these factors may result in new discharge coefficients. Other methods of estimating discharge coefficients include the use of chemical engineering analysis, plant monitoring data, plant performance specifications submitted to government authorities, international studies, or industrial surveys.

As pointed out in Chapter 3 of this Handbook, the spatial characteristics of residuals discharge are important from an environmental management viewpoint. Within a plant, discharges may occur at many points. Discharge standards imposed under the "bubble concept" [Deland (1979)] can be expected to result in engineering solutions quite different from those obtained when restrictions are applied to specific on-site locations or unit processes. As previously noted, discharges may occur at locations other than site boundaries, as with flyash from power stations or with coal washery rejects that can be discharged off-site through slurry piplelines.

Discharges also have a time dimension. Seasonal variations can occur because of fluctuating demand patterns or the effect of climate on raw material supplies. Short-term variations can be caused by peak-load demands as in electric power generation, maintenance operations, work shifts, accidents, and even the deliberate release of residuals at night to avoid detection and prosecution. Long-term variations in discharges will tend to arise from changes in the scale of production, physical depreciation of equipment, the introduction of different process technologies, and alterations in cost and price conditions.

To handle the complexities of process technologies, residuals generation and residuals discharge for individual industrial plants, mathematical programming can be applied. Programming has been commonly used in industry for management decisions, especially petroleum refining [Manne (1963)], but it was not until the work of Russell (1971, 1973) and Russell and Vaughan (1974, 1976) that specific recognition was given to the role of residuals. Russell's model offers a powerful method of assessing the effects of different economic, technological, and regulatory factors on plant operations. It is a linear programming model of the form

$$\max_{\text{s.t.}} -c_x X - c_b B - c_w W - C_t T - c_v V + c_y Y - c_d D$$

$$\begin{bmatrix} A_{11} & A_{12} & & & & A_{16} & \\ A_{21} & & A_{23} & & & & \\ A_{31} & & & & & & \\ A_{41} & A_{42} & A_{43} & A_{44} & A_{45} & & A_{47} \\ & A_{52} & A_{53} & A_{54} & A_{55} & & A_{57} \\ & & & & & & A_{67} \end{bmatrix} \begin{bmatrix} X \\ B \\ W \\ T \\ V \\ Y \\ D \end{bmatrix} \gtreqless \begin{bmatrix} 0 \\ b_2 \\ b_3 \\ 0 \\ 0 \\ b_6 \end{bmatrix} \qquad (3)$$

$$X, B, W, T, V, Y, D \geq 0.$$

Vectors representing process levels in the model are: X, a set of production alternatives; B, production of by-products; W, materials recovery from residuals;

T, treatment of residuals; V, transport of residuals; D, discharge of residuals; and Y, sale of products. The objective function maximizes revenue from the sale of all products ($c_y Y$) where c_y is a vector of output prices, and minimizes the total costs of production ($c_x X$), by-product production ($c_b B$), raw material recovery ($c_w W$), residuals treatment ($c_t T$), and residuals transport ($c_v V$), where the subscripted c's are row vectors indicating corresponding unit costs. If charges are imposed on residuals discharges, then the vector c_d in the objective function takes on positive values. The model treats charges as costs and minimizes them together with all other production costs.

The rows of the activity matrix indicate the model's constraints. Row set 1 ensures that all products sold are actually produced. Row set 2 constrains the supplies of inputs to production activities, including those from materials recovery processes. Row set 3 controls output quality. Row set 4 is a mass balance equation for primary residuals generated by all activities. Row set 5 represents the generation and handling of secondary residuals, which can take the form of primary residuals remaining from treatment processes, new types of residuals created by a treatment process, and primary residuals at different locations. Row sets 4 and 5 ensure that all residuals generated are accounted for by means of materials recovery, production of by-products, treatment, transport or discharge. The last set of rows introduces discharge constraints. A maximum limit can be imposed on the total discharge from all activities of any particular residual. These constraints can of course be disregarded if the model is used to simulate effluent or emission charges instead of discharge standards. In a completely uncontrolled situation, the pattern of residuals discharges can be determined by setting the "costs" of discharges in the objective function equal to zero and deleting the residuals discharge constraints.

Russell's model has certain limitations because of its linear properties. It does not allow for economies of scale in production or for possible nonlinearities in residuals handling functions. The time dimension is also greatly simplified, as the model is steady state. Such models may, furthermore, be very expensive to build.

One of the earliest studies undertaken by RFF – a beet sugar production plant – is an excellent example of the principle of mass balance applied to an individual industrial plant [Löf and Kneese (1968)]. Details of the production process, with associated product outputs and residuals generated in the absence of materials recovery, are given in Figure 7.2. The process is comparatively simple, and permitted RFF to gain experience in developing a methodology that later could be applied to other industrial activities. The residual of greatest concern is biochemical oxygen demand (BOD), but considerable amounts of flue gas and lime cake slurry are also generated. The process was analyzed again with extensive recycling that removes most of the waterborne residuals. Mass balance for the improved process indicate that only a small increase of gaseous and solid residuals must be incurred to obtain a large reduction in BOD discharges. Other individual industry studies undertaken by RFF include a dairy, livestock

High Residual Beet Sugar Production Process—No Recirculation

(In pounds per ton of beets processed except where otherwise stated)
Intake: 5250 gallons/ton sliced beets—regular; 175 gallons/ton sliced beets—Steffens additional.

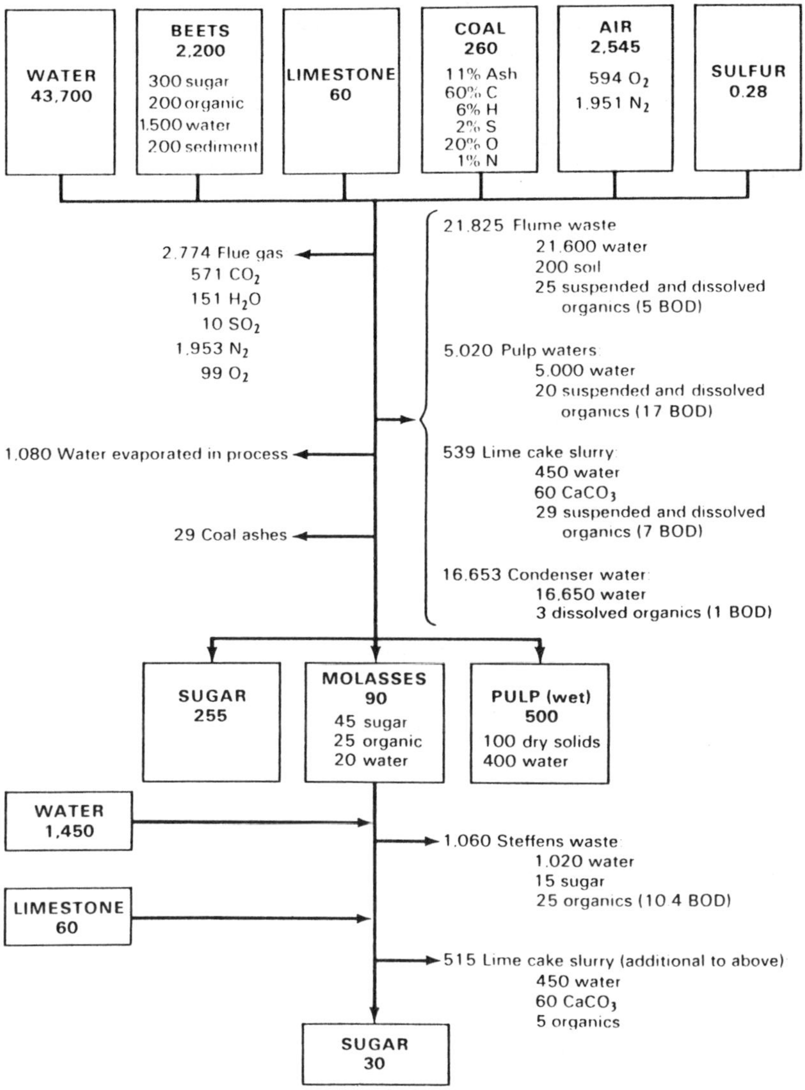

Source: Kneese et al. (1970, p. 50).

production and food processing [Kneese et al. (1970)], pulp and paper manufacture [Bower et al. (1972), Löf et al. (1973)], a brewery and lignite-fired electric power plant [Basta et al. (1978)], petroleum refining [Russell (1973)], and a steel mill [Russell and Vaughan (1976)]. Residuals generation coefficients for a wide range of industrial activities are presented in Basta et al. (1978).

Although a great deal of pioneering work was done by RFF on industrial activity analysis and residuals management, other research covered similar ground, for example, the engineering approaches to residuals discharges from energy systems followed by Brookhaven National Laboratory [Beller (1976) and Foell (1979a)].

A voluminous literature also appeared throughout the 1970s on residuals generation in input–output models, at national and regional levels [James et al. (1978)]. Most of these models use fairly aggregated economic sectors and measure output in monetary rather than physical units, thus losing much of the accuracy and detail of the RFF approach.

2.3. Regional production systems

At the regional level, residual discharges need to be identified by type, by activity or industry source, by location, and over time. The first step in the analysis is to specify a set of regional economy activity levels. Various approaches may be used. The simplest involves an exogenous estimation of output for each activity in the region for a specified time period. Regional activity levels can also be determined by "driving" the regional economic system with a national model, as in the Brookhaven model suite [Marcuse (1979)], the SEAS/RFF model [Ridker and Watson (1980)], and the Hunter Valley model [James et al. (1982)]. Other possibilities are regional, interregional, and multiregional models.

In a regional economic activity model, discharges of residuals are estimated as

$$R = P\hat{q}, \tag{4}$$

where R is a matrix of residuals discharges (residual type by industrial activity or sector), P is matrix of discharge coefficients, and $\hat{q}$ is a diagonal matrix of activity levels. Each discharge coefficient p_{kj} indicters the mass of residual k discharged per unit of sectoral output j measured in monetary or physical units. Total discharges are simply

$$z = Ri = Pq, \tag{5}$$

where z is a vector of residuals by type, discharged by all activities in the region, and i is a sum vector.

If the economic activity model includes a regional input-output matrix, sectoral outputs are found by using the Leontief inverse matrix:

$$q = (I - A)^{-1} f, \tag{6}$$

where A is the regional input–output coefficient matrix, and f is a vector of regional final demands.

Regional discharges can thus be written as a function of final demands:

$$z = P(I - A)^{-1} f. \tag{7}$$

The matrix in the above equation indicates the discharge of residuals resulting directly and indirectly from the supply of final products by economic sectors.

The basic model can be converted to a comparative statics system by adding a time dimension to the simple Leontief model:

$$z_t = P(I - A)^{-1} f_t, \tag{8}$$

where t represents time periods. If a regional capital formation matrix is available, a dynamic Leontief model can be developed, of the form

$$z_t = P(I - A - B)^{-1} f_t, \tag{9}$$

where B is a capital matrix with typical element b_{ij} indicating the output of sector i directly required to permit a one-unit production capacity expansion in sector j. Miernyk and Sears (1974) have constructed a model of this kind to analyze discharges of airborne residuals.

If a matrix of fuel use by sectors is available, regional discharges of airborne residuals can be estimated by applying combustion emission factors as in the equation

$$z = FEq, \tag{10}$$

where E is an energy-use coefficient matrix (fuel type by input–output sector) and F is a matrix of combustion emission factors (residual type by fuel type). This approach assumes that combustion processes for each fuel type are the same across all sectors and that no residuals recovery or modification occurs.

Regional input–output models have been developed by Shefer (1973), Hite and Laurent (1972), Miernyk and Sears (1974), Kohn (1975), Muller (1973), and Howe (1977), among others. Various national models of the Netherlands could also be considered as regional, such as those constructed by the Free University of Amsterdam [Jansen et al. (1978)] and DenHartog and Houweling (1974). This last reference follows the theoretical method of Leontief (1970) and incorporates residuals activities (with secondary residuals discharges) within the generalized input–output framework.

Regardless of the approach used to determine regional production levels and of the degree of aggregation that is chosen, economic activities must be allocated spatially if dispersion modeling is to be carried out. This topic is taken up in the next sections. An introductory discussion of dispersion models is found in Chapter 3 of this Handbook.

3. Environmental systems models

3.1. Air quality models

The relationship between regional air quality and the discharge of airborne residuals by industrial activities is complex. Important factors that govern ambient concentrations of residuals include classes of sources, spatial and temporal release patterns, and composition of discharges, meteorological conditions, chemical reactions in the atmosphere, the type of ground surface, and background concentrations. The airborne residuals appearing most commonly in REQM studies are sulfur oxides, carbon monoxide, hydrocarbons, nitrogen oxides, and particulates.

Because of the three-dimensional nature of air dispersion, each class of emission sources has to be modeled separately. The main classes are high-level point sources, for example, power plant stacks; medium-level point sources, such as industrial stacks; low-level area sources, such as industrial zones in which petroleum fuels are widely used; and line-source emissions that usally apply to traffic flows but may also refer to industrial activities.

The most commonly used model for high- and medium-level point sources is the Gaussian plume equation [Bach (1972)], which is of the form:

$$C(x, y, z, H) = \frac{Q}{2\pi S\sigma_y\sigma_z}\left[\exp - \frac{y^2}{2\sigma_{y^2}}\right]\left[\exp - \frac{(z-H)^2}{2\sigma_{x^2}} + \exp - \frac{(z+H)^2}{2\sigma_{z^2}}\right], \tag{11}$$

where C is the ambient concentration of the residual (g/m^3), x is the downwind coordinate (m), y is a crosswind coordinate (m), z is the vertical coordinate (m), Q is the emission rate (g/sec), S is the average windspeed (m/sec), and H is the effective stack height (m). The model assumes that, downwind from the stack, the residual will be normally distributed (in the statistical sense) in the crosswind and vertical directions. The crosswind vertical standard deviations of the plume are given (in meters) by σ_y and σ_z, respectively.

For a receptor point at a distance y from the centerline of the plume, the ground level concentration is estimated as

$$C(x, y, 0, H) = \frac{Q}{\pi S\sigma_y\sigma_z}\exp\left[-\frac{y^2}{2\sigma_{y^2}} - \frac{H^2}{2\sigma_{x^2}}\right]. \tag{12}$$

The values of σ_y and σ_z depend on atmospheric stability conditions, as shown by Pasquill (1962) and Gifford (1961). Wind speed and direction must also be specified.

The effective stack height is given by the equation

$$H = h + \Delta h, \tag{13}$$

where h is the height of the stack and Δh is plume rise (m).

For medium-level point sources, the most commonly used formula for plume rise is that of Carson and Moses (1969) which takes the form:

$$\Delta h = A\left[-0.029\frac{V \cdot D}{S} + 5.35\frac{F^{1/2}}{S}\right], \tag{14}$$

where V is the stack exist velocity (m/sec), D is the stack diameter (m), F is heat flux (kcal/sec), S is average wind speed (m/sec), and A is an atmospheric parameter with values of 2.65, 1.08, and 0.68 for unstable, neutral, and stable conditions, respectively.

For high-level stacks, the formula of Briggs (1969) is preferred. With neutral and unstable atmosphere, the formula is

$$\Delta h = 2.5(F^{1/3}h^{2/3}), \tag{15}$$

and with stable atmosphere:

$$\Delta h = 2.96\left(\frac{F}{0.0277}^{1/3}\right). \tag{16}$$

The heat flux is calculated from the formula

$$F = R\frac{(T_e - T_a)}{T_a} \cdot K, \tag{17}$$

where R is the volume flow rate (m^3/sec), T_e is the exit temperature (degrees Kelvin), T_a is ambient temperature, and K is a constant (84.88 kcal/m^3). When $F < 5000$ kcal/sec, the Carson–Moses formula should be used, and when $F >$ 5000, the Briggs formula is more appropriate.

The Gaussian plume model has been applied to electric power plants [Mendlesohn (1979, 1980)]. Under steady-state conditions, concentrations at receptor points are proportional to discharge rates. With multiple source and receptor points, a matrix of fixed transfer coefficients can be used. For a given airborne residual, ambient concentrations can be estimated as

$$c = DR + b, \tag{18}$$

where c is a vector of ambient concentrations at different receptor points, r is a vector of discharges by source points, D is the matrix of transfer coefficients, and b is a vector of background concentrations at receptor points.

Dispersion of emissions from low-level area sources can be handled by means of a simple model developed by Gifford (1973) and Hanna (1973) of the form:

$$C = KQ, \tag{19}$$

where C is the concentration of an airborne residual in a particular grid square, Q is the level of discharges, and K is a constant that depends on meteorological factors and the size of the grid squares in the model. Howe (1977) reports use of this model in a study of the Upper Colorado River Basin.

An effective model combining Gaussian plume equations for medium- and high-level sources with a model of low-level area sources is the Smeared Concentration Approximation (SCA) model developed by Dennis (1978). The SCA model is a simplified version of the more elaborate models and can be applied using only a hand calculator. It has be incorporated in energy–environment studies [Foell (1979a)] and has been modified for the analysis of industrial emissions in the Hunter Valley [Chambers (1983)].

Gaussian models rely on analytical solution procedures to determine ambient concentrations of residuals, but other modeling approaches can be taken. In a "mixing bowl" situation, ambient concentrations depend solely on total discharges within a region, and the locations of emission sources can be ignored. This situation, however, does not occur frequently in practice. Dispersion models can be built using statistical techniques. In the Ljubljana regional study [Basta et al. (1978)], regression equations were fitted to observed values of industrial emissions and ambient concentrations. The method is straightforward, but does have obvious statistical traps. Another approach to dispersion modeling is the use of mass conservation equations, normally specified as partial differential or difference equations and solved through numerical integration techniques. The method is discussed and compared with Gaussian models for line source emissions in papers by Sistla et al. (1979) and Rao et al. (1980). Other applications are described by MacCracken and Sauters (1975) and in the U.S. Nuclear Regulatory Commission Guide 1.iii (1977).

The preceding analysis deals with conservative (nonreactive) residuals. Chemical reactions in the atmosphere may also be significant for regional air quality. The most notable case is photochemical smog, which is formed by reactive hydrocarbons, nitrogen oxides, and sunlight. Although primarily attributable to motor vehicle emissions, photochemical smog can be affected by supplementary industrial discharges, and air quality models should take both sources into account. Models used to study air pollution in Sydney, Australia, a city badly affected by photochemical smog, are described by the NSW State Pollution Control Commission (1977). Industrial emissions were handled with a Gaussian model, and the Gifford-Hanna model was applied to motor vehicle emissions.

3.2. Water quality models

Various factors need to be taken into account when modeling the effects of water borne discharges by industrial activities, the most obvious being whether the receiving water system is a stream, estuary, lake, ocean, or groundwater table. In the case of freshwater streams, flow conditions are affected by precipitation, economic activities, and reservoir operations [Howe (1979), Maass et al. (1962), Fiering (1967), Spofford (1980)]. Other systems depend also on natural conditions and economic activities. Dispersion of residuals is closely linked with the hydrologic characteristics of the system under study. Physical, chemical, and biological reactions taking place within water systems are, however, also of significance. The kinds of discharges and the spatial and temporal patterns of release are further important determinants of ambient water quality.

Some of the key concerns in regional water quality management are shown in Table 7.1.

Attempts have been made to derive general water quality indices [Brown et al. (1975), Luken et al. (1976), Jansen et al. (1978)], but for industrial discharges, the most commonly used indicators are BOD_5, TDS, and TSS [OECD (1980)].

BOD discharges involve reactive dispersion processes, as recognized by Streeter and Phelps (1925) in their study of the Ohio River. Dissolved oxygen in a stream is affected by two processes: a deoxygenation process in which oxygen is consumed by bacteria and micro-organisms as they degrade the released organic material, and a reaeration process that raises the dissolved oxygen level through surface/air interchange. Both processes are time-dependent, but a spatial determinant can be introduced by translating time into downstream distances from a given discharge point to different receptor points.

The two processes can be described mathematically as

$$\frac{\mathrm{d}D}{\mathrm{d}t} = k_1 L_t - k_2 Dt, \tag{20}$$

Table 7.1
Water quality indicators

Oxygen	*Pathogens*
Dissolved oxygen (DO)	Fecal coliforms
5-Day biological oxygen demand (BOD_5)	
Chemical oxygen demand (COD)	
	Solids
Salinity	Total suspended solids (TSS)
Total dissolved solids (TDS)	
	Water properties
Nutrients	Temperature
	pH
Nitrates	Turbidity
Phosphates	Hardness
	Color
Toxics	Odor
	Taste
Pesticides	Electrical conductance
Heavy metals	

where D_t is the dissolved oxygen deficit below saturation concentration, L_t is BOD remaining in the flow, k_1 is a deoxygenation rate constant, k_2 is a reaeration rate constant, and t is time. L_t depends on the initial discharge of BOD (L_a) as in the equation

$$L_t = L_a e^{-k_1 t}. \tag{21}$$

Substituting for L_t and integrating the differential equation yields the classical Streeter–Phelps oxygen-sag equation:

$$D = \frac{k_1}{k_2 - k_1} L_a (e^{-k_1 t} - e^{-k_2 t}) + D_a e^{-k_2 t}, \tag{22}$$

where D_a is the initial dissolved oxygen deficit when BOD is released to the stream (that is, when $t = 0$).

In converting a steady-state discharge of BOD to DO levels at different receptor points, fixed transfer coefficients can be used. More complex versions of the Streeter–Phelps model are described by Kneese and Bower (1968). These include multiple discharge systems that can be modeled, under steady-state assumptions, by additive linear relationships, and systems that simulate non-steady-state conditions. Camp (1965) has specified a model in which dissolved oxygen depends on sedimentation or scour of organic material from the stream bed, on photosynthesis by floating plants and algae, and on deoxygenation caused by respiration of plants and animals. Camp's model has been applied by Moodie (1979). Other recent work on *DO* modeling has been carried out by Wang et al. (1979) and Wen and Kao (1980, 1982), among others.

Diffusion models for TDS can be of many different forms. The simplest case is that of a stream with constant release rates by dischargers, constant flow conditions, complete mixing of TDS, and an absence of sinks or contributing sources in the system itself. This is a case of a "conservative pollutant", in which a simple additive approach is taken to determine total mass or ambient concentrations of TDS at receptor points. The same kind of model as specified for airborne residuals in eq. (18) is appropriate.

More complex models are needed once the above assumptions are relaxed, for example, episodic release of residuals, variations in stream flow because of precipitation patterns or the operation of reservoirs, incomplete mixing, and the role of sinks [Gosz (1980)]. Industrial discharges of TDS may be supplemented by those from watercourse runoff, as modeled by Moodie (1979). Mixing zones are discussed by Neely (1982) who shows that the concentrations of a chemical in a stream can be modeled by means of a Gaussian equation, constrained by physical stream parameters such as river width, depth, and shear velocity. It is difficult to generalize about the effects of sinks on the dispersion of TDS. The type of chemical can be critical. Heavy metals, for example, usually have strong affinity for clay because of high rates of adsorption. A net loss of heavy metal salts can be expected initially through chemical reactions with stream-bed material, but over time, releases of heavy metals from saturated sinks may occur. Biological uptake by aquatic organisms can add further complications. The dynamics of such processes are likely to be quite complex, and must be modeled according to each specific set of circumstances.

Dispersion of TDS in other water systems can be mentioned only briefly here. For a discussion of estuaries, see Gameson (1973). An excellent survey of aquatic dispersion models is the U.S. Nuclear Regulatory Commission Regulatory Guide 1.113 (1976). Although designed for estimating dispersion of radioactive effluents, the models apply also to TDS discharges.

3.3. Solid residuals

Solid residuals are classified by Kneese (1977) according to the type of material and by source. Those associated with industrial activities include combustibles, noncombustibles, ashes, bulky items, hazardous materials, construction and demolition residuals, obsolete vehicles, and tailings and overburden from mining. Solid residuals are generated also by other activities, and include food scraps, garden debris, street sweepings, dead animals, animal residues, crop residues, slush and logging residues, and sewage treatment residues. Kneese produces statistics indicating that mixed solid residuals generated by industrial processes in the United States in 1971 comprised about 2.5 percent of all solid waste (excluding mine overburden).

Mixed solid residuals generation coefficients appear in the Ljubljana regional study by Basta et al. (1978). Much of the information was obtained from salvage firms that collected waste and sold it for reuse. Nonsalable solids were estimated from municipal collection sites, but only on a volume basis. More detailed estimates of solid residuals discharges may be obtained from process analysis at different industrial activity sites. Recycling of solid residuals is discussed by Kneese, Ayres, and d'Arge (1970) and by Schlottmann (1977). Automobile scrap is dealt with by Sawyer (1974).

4. Control of regional environmental quality

4.1. Mechanisms for regional environmental quality management

The various components of an REQM system have been dealt with in the preceding sections. To manage environmental quality, three steps are required: first, an analysis of alternative physical measures or combinations of those measures that can control residuals discharge or dispersion in the environment; second, evaluation of the effects of alternative physical controls to determine the "best" or preferred plan of action; and third, selection of appropriate instruments or implementation incentives that will enable the preferred strategy to be successfully put into operation. As will be shown, management actions should ideally be directed toward the control of ambient environmental quality rather than the control of residuals discharges themselves.

4.2. *Physical measures of control*

At the regional level, various kinds of physical measures may be taken to control ambient environmental quality. Kneese and Bower (1979) classify them as: regulation of output; changing process technologies; modification of residuals on-site; making better use of the environment's assimilative capacity; and increasing the environment's assimilative capacity.

Output can be controlled by altering product mix or the regional economic structure. At the individual plant level, the product mix can be altered, or different product specifications introduced. Input–output models are capable of demonstrating the direct and indirect regional effects of changes in the final demand vector on total discharge of specific types of residuals. Measures can be taken to regulate categories of final demand that are responsible for residuals of major concern, as in the model of Norway developed by Førsund and Strøm (1974) (see Chapter 9 in this Handbook) a model of the Netherlands economy [Jansen et al. (1977)], and studies of the Rijnmond region [Muller (1973, 1979)].

Changes in process technologies may involve altering the input mix, the timing of production, or the introduction of completely new processes. Such measures affect the level and composition of residuals generated within industrial plants. Other measures modify residuals after they have been generated, for example, materials or energy recovery processes, production of by-products, or add-on treatment such as filtration, scrubbing, neutralization, sedimentation, and landfill. In some cases, as with fluidized bed combustion in electric power stations, changes in residuals generation and modification take place simultaneously.

The list of residuals reduction processes in industry shown in Table 7.2 is based on Kneese, Ayres and d'Arge (1970) and Luken et al. (1976), with some additions. Making better use of the environment's assimilative capacity may be achieved by controlling the spatial and temporal distribution of discharges and activities. Examples of spatial control of residuals discharges are the use of high chimney stacks or slurry pipelines. Temporal control of residuals discharges is illustrated by regulation of BOD discharges from a pulp and paper mill to maintain dissolved oxygen in receiving waters at an acceptable level. Activities may be controlled in temporal terms by appropriate production scheduling. Location of activities to regulate regional environmental quality has received a great deal of attention. The literature includes studies of power plant siting [Mendelsohn (1979,

1980), EPRI (1979b), Guldman and Shefer (1977)] and mathematical programming models of spatial planning [Nijkamp (1977), Mastenbroek and Nijkamp (1974)].

The assimilative capacity of the environment can be increased in many ways. One is the addition of water to effluents to increase the dilution of residuals in the environment. Artificial mixing can be carried out in water bodies to improve dispersion and aeration. Direct oxygenation of water is also possible through air pumps or by churning in air with ship propellers. In some cases, as with freshwater lakes affected by acid rain, chemical buffers can be added to raise the pH level.

From a modeling viewpoint, restrictions on physical alternatives for AEQ management are of major significance in setting up an appropriate mathematical structure. If the temporal aspects of a problem are emphasized, a comparative static or dynamic model needs to be built. Even for a static optimization model, difficulties may be encountered. For example, if all physical measures can coexist

Table 7.2
Residuals treatment processes

Waterborne residuals	
Sedimentation	Stripping
Screening	Biological filtration
Filtration	Chemical coagulation
Flotation	Emulsion breaking
Separation	Neutralization
Centrifuging	Equalization
Holding pond	Carbon adsorption
Spray irrigation	Chemical oxidation
Landfill	Chemical reduction
Drying beds	Wet oxidation
Incineration	Fermentation
Evaporation	Activated sludge
Distillation	Anaerobic digestion
Airborne residuals	
Two-stage combustion	Recycling
Fluidized bed combustion	Electrostatic precipitation
Afterburning	Cyclones
Condensation	Fabric filters
Absorption	Wet scrubbing
Adsorption	Dry scrubbing
Solid residuals	
Incineration	Ocean dumping
Pyrolysis	Recycling
Landfill	Compaction
Road ballast	

and be operated at nonnegative levels, linear programming is an acceptable method for optimizing residuals management alternatives. Frequently, however, only one or a limited number of activities within a given set of alternatives is permissible, and integer programming may have to be used. In siting an electric power plant, for example, only one location can usually be chosen. Indivisibilities can also occur within industrial plants. One might choose a fluidized bed combustion process or a conventional boiler with a wet scrubber, but not both. Some of the economic implications of such restrictions are discussed in the next section.

4.3. Evaluation criteria for environmental quality goals

Baseline simulation modeling of residuals discharges, dispersion, and dose-response effects produces a given set of physical, economic, and environmental characteristics. Changes in the system, such as would occur with implementation of physical measures to control environmental quality, will result in new sets of characteristics. Rational environmental quality management implies selection of the "best" outcome.

Evaluation criteria are clearly necessary for the best outcome to be determined. Bower et al. (1977) take a general view of evaluation criteria for environmental quality management and list the main considerations as: physical effects; economic effects, including distributional aspects, administrative simplicity and flexibility; timing; political preferences and institutional factors; intermedia and resource use effects; and accuracy of estimates.

4.4. Operational REQM models

Analytical methods can be applied with some, but not all, evaluation approaches. The most widely used modeling techniques can be listed as: simulation modeling combined with multiattribute utility analysis; simulation modeling supplemented by full benefit–cost analysis, including monetary damage functions for adverse environmental effects (for a discussion of benefit–cost applications, see Chapter 6 of this Handbook); cost-minimization models incorporating penalty functions if prescribed AEQ standards are transgressed; cost-minimization models designed to fulfill prescribed output constraints and environmental standards for either emissions or ambient concentrations of residuals; and income maximization models with resource restrictions and ambient concentration or emission standards. All these options are available in static or dynamic modes.

Multiattribute utility analysis [Keeney and Raiffa (1976)] has been advocated by the Wisconsin energy modeling group in their work on energy–environment management [Buehring et al. (1978), Foell (1979b)]. The technique is designed to elicit decisionmakers' preferences and assist them to choose the scenario that generates the highest level of expected utility. The method does have limitations. It may be difficult for decisionmakers to reach consistent preferences if a large spectrum of effects needs to be considered. The process can, furthermore, be very time-consuming. Finally, decisionmakers often prefer to assess outcomes in terms of community values rather than their own subjective values, and require information obtained by more objective methods. Multiattribute utility analysis is therefore likely to be useful as an evaluation technique only in rather rare situations.

Benefit–cost analysis has recently received much attention as a means of evaluating environmental impacts and guiding environmental quality management [Hufschmidt and Hyman (1982), Hufschmidt et al. (1983)]. Early work in this area concentrated on environmental damage functions [OECD (1974, 1976)]. Further work on the economic valuation of environmental effects has been carried out in the United Kingdom [Pearce (1978)] and in the United States [Freeman (1979), Sinden and Worrell (1979)]. It is inappropriate to enter a full discussion of benefit–cost analysis in this chapter, but several difficulties should be briefly mentioned. Aside from awkward value judgements underlying the method (for example, discounting long-run irreversible ecological damage costs or placing monetary values on human life and health), many practical problems need to be overcome in applying it in a REQM context. Bower and Brady (1981) report numerous difficulties in quantifying damages caused by air pollution. Another problem is that, with so many alternatives available for measures to control AEQ, estimation of all associated damage functions could be an almost unmanageable task. The net benefit response surface may contain nonlinearities and the global optimum may be difficult to find, although mathematical techniques have been applied to this problem for water resource systems [Hufschmidt (1962)] and hydroelectric power schemes [Holling (1978)].

A more realistic REQM approach is cost-effectiveness analysis combined with standard-setting. As a practical technique, cost-effectiveness has several advantages. The procedure forces decisionmakers to carefully specify the objectives they are trying to reach, for example, control of discharges at source, ambient concentrations (by residual type or location), indoor or outdoor concentrations, health levels, etc. Standards must be properly specified also with respect to time – daily, weekly, monthly, or annual averages, or perhaps short-term peak levels.

Regional cost-minimization models range from simple discharge control models to those involving ambient concentrations standards and spatial planning of activities. The Russell–Spofford model incorporates penalty functions if AEQ standards are exceeded. This approach is discussed in detail in the Delaware Valley case study (Section 5.4.4).

The model developed by Kohn (1971) features regional discharge constraints. The theoretical form of the model is presented here, and its empirical applications in Section 5.2. The model can be written as

$$\min_{\text{s.t.}} c'x \tag{23i}$$

$$Px \leq z^*, \tag{23ii}$$

$$Ux = q, \tag{23iii}$$

$$x \geq 0, \tag{23iv}$$

where x is a vector of unit processes for all activities in the region; P is a residuals discharge matrix; c is a vector of associated (incremental) costs; z^* is a vector of permissible regional discharges, by residual type; q is a vector of prescribed output targets; and U is an elementary matrix.

Each product output volume, say q_h, is producible with several processes. Restrictions on the relevant process levels, $x_{h1}, x_{h2}, \ldots x_{hm}$ can be written

$$q_h = \sum_j u_{hj} x_{hj}. \tag{24}$$

The typical element of the matrix U is u_{hj}, which takes a value of one if the jth activity can produce the hth product, and otherwise a value of zero.

Kohn later incorporated an input–output matrix in the model [Kohn (1975)]. The dual problem generates shadow prices for the discharge standards, so that tradeoffs can be assessed between discharge constraints on specific residuals and regional environmental control costs.

Discharge constraint models do not in general lead to control of AEQ. (An exception is a mixing bowl situation when location of dischargers and receptors does not matter.) Dispersion models must be added to estimate ambient concentrations. If a set of fixed transfer coefficients is appropriate, ambient concentrations at receptor points can be found by means of a diffusion matrix. Ignoring background concentrations, a separate matrix can be used for each residual, as follows:

$$c^k = D^k r^k, \tag{25}$$

where c^k is a vector of ambient concentrations of the kth residual at different receptor points, D^k is the diffusion matrix, with typical element d^k_{jn} indicating transfer of the kth residual from activity j to receptor n, and r^k is a vector of discharges from activities.

As discharges depend on activity levels,

$$r^k = \hat{p}^k q, \tag{26}$$

where $\hat{p}^k$ is a diagonal matrix of discharge coefficients for the kth residual. AEQ constraints enter the model as

$$D^k \hat{p}^k q \leq c^{k*}, \tag{27}$$

where c^{k*} is a vector of permissible ambient concentrations (which may, of course, be uniform) at receptor points.

The air quality management model developed by Muller (1973, 1979) places environmental restrictions on ambient concentrations of airborne residuals. Unlike Kohn's model, which uses a cost-minimization approach, Muller's model maximizes regional income. The objective function is

$$\max v'q, \tag{28}$$

where v is a vector of value-added coefficients and q is a vector of sectoral activity levels. In the absence of environmental constraints, output levels are restricted by maximum labor use and prescribed minimum supplies of final products:

$$(I - A - B - M)q \geq f^*, \tag{29i}$$

$$n'q \leq w^*, \tag{29ii}$$

$$q \geq 0, \tag{29iii}$$

where I is an identity matrix, A is the regional input–output coefficient matrix, B is a capital formation matrix, M is a matrix of inputs to antipollution investments, and f^* is a vector of prescribed final supplies. The region's workforce is represented by the scalar w^*, and n is a vector of labor input coefficients.

Muller uses the same diffusion matrix for all residuals. Discharge points, furthermore, are allocated to grid positions by subdividing regional sector outputs. Ambient concentrations are estimated as

$$C = P\hat{q}SD, \tag{30}$$

where $\hat{q}$ is a diagonal matrix of sector outputs, P is a residuals discharge coefficient matrix (residual type by sectoral output), S is a matrix that distributes outputs to grid positions, D is the diffusion matrix transferring residuals from discharge to receptor points, and C is the required matrix, indicating concentrations of residuals, by type, at different receptor points. AEQ restrictions take the form:

$$P\hat{q}SD \leq c^*, \tag{31}$$

where c^* is a vector of ambient concentration standards for different residuals, applied at a particular receptor point.

The models discussed so far permit all activities, even unit processes within activities, to enter the optimal solution. If the number of unit processes has to be limited, integer variables must be introduced. For example, in the Kohn model, if x_1, x_2, and x_3 are three alternative methods of producing q_1, and only one of the three is permissible, an extra set of constraints must be added to the model, for example

$$x_1 - v_1 x_1 \geq 0, \tag{32i}$$

$$x_2 - v_2 x_2 \geq 0, \tag{32ii}$$

$$x_3 - v_2 x_3 \geq 0, \tag{32iii}$$

$$v_1 + v_2 + v_3 = 1, \tag{32iv}$$

$$0 \leq v_j \leq 1, \quad v_j \text{ an integer,} \tag{32v}$$

The overall result is a mixed linear-integer programming model. Burton and Sanjour (1970) used integer programming methods to model air pollution control. Optimal siting of electric power stations could be assessed with this method, for the formulation is similar to that for the selection of unit processes within plants. One disadvantage of the method is that shadow prices cannot be obtained.

Dynamic optimization models of regional-residuals management are described by Guldmann (1978) and Parvin and Grammas (1976), among others. Such models have not been used as frequently as static optimization models.

4.5. Implementation methods

An extensive literature has appeared on alternative institutional and administrative approaches to implementation of regional environmental quality management strategies. Bower et al. (1977) list the main alternatives as: regulatory, comprising laws, ordinances, and permits controlling production processes, performance, location, timing, and procedures; economic, applied to residuals, inputs, product outputs, and residuals modification processes; administrative; judicial; and educational/informational.

Effluent (or emission) charges have been strongly advocated on economic efficiency grounds. The OECD (1980) refers to "incentive" charges that are high enough to induce dischargers to introduce residuals modification schemes, and "redistributive" charges that raise revenue – without necessarily altering discharge patterns – that can be diverted to municipal sewerage works, research on pollution abatement, and subsidization of new investment in residuals modification technologies in private industry.

Uniform incentive charges can be shown to achieve a given reduction of discharges at minimum direct economic cost to the region. Consider a situation involving only one residual and many dischargers. Initially, the jth activity discharges an amount of r_j^*. The amount of residual removed by the jth activity is y_j. Actual discharge by the jth activity is thus

$$r_j = r_j^* - y_j. \tag{33}$$

Suppose the abatement cost function for the jth discharger is

$$TC_j = f_j(y_j). \tag{34}$$

Substitution of eq. (33) in (34) leads to

$$TC_j = f_j(r_j^* - r_j). \tag{35}$$

Finally, suppose that, for all dischargers together, the permissible total discharge rate is r^*.

The problem takes the form:

$$\min \sum TC_j, \quad \text{for all } j, \tag{36i}$$

s.t.

$$\sum r_j = r^*, \quad \text{for all } j. \tag{36ii}$$

The optimal conditions can be found by minimizing the Lagrangian function:

$$L = \sum f_j(r_j^* - r_j) + \lambda\left(\sum r_j - r^*\right). \tag{37}$$

The first-order conditions of direct interest are

$$\partial L/\partial r_j = -f_j' + \lambda = 0, \tag{38}$$

or that

$$f_j' = f_i', \quad \text{for all } i, j, i \neq j. \tag{39}$$

Equation (39) implies equalization of marginal treatment costs, which can be achieved by imposing a uniform charge. Each discharger will carry out abatement up to the point where the marginal abatement cost equals the charge, so a uniform charge will, in principle, achieve a given total reduction at minimum cost.

This argument is valid for total discharges, but does not necessarily hold true for control of ambient concentrations. To demonstrate this, a linear dispersion model can be introduced. The same objective function is minimized, but the new constraint, which is assumed to apply at only one receptor point, becomes

$$c_n^* = \sum d_{nj} r_j, \tag{40}$$

where c_n^* is the prescribed ambient concentration of the residual at receptor point n, and d_{nj} is a transfer coefficient from source point j to receptor point n.

A new Lagrangian function should now be minimized:

$$L_0 = \sum f_j(r_j^* - r_j) + \lambda_0\left(\sum d_{nj} r_j - c_n^*\right), \tag{41}$$

which leads, among other results, to

$$\frac{\partial L_0}{\partial r_j} = -f_j' + \lambda_0 d_{nj} = 0 \tag{42}$$

or

$$f_j' = f_i' \frac{d_{nj}}{d_{ni}}, \quad \text{for all } i, j, i \neq j. \tag{43}$$

Only if the transfer coefficients are the same for all dischargers ($d_{nj} = d_{ni}$) should marginal abatement costs be equalized and a uniform change imposed; otherwise a set of differential charges is required. If ambient concentrations are imposed at several receptor points, the Kuhn–Tucker theorem is needed to determine the optimal conditions.

Kneese (1977) cites the Delaware estuary water quality study to show that in a real situation a system of uniform charges would produce a result very close to the minimum cost solution. However, completely opposite conclusions are drawn in a study of nitrogen oxide control in the Chicago air quality control region [U.S. Council on Environmental Quality (1979)] which revealed that a uniform charges approach to meeting an ambient concentration standard of 250 micrograms per cubic meter could involve a total cost twelve times that of the least-cost strategy. Differential charges to bring about the least-cost solution, furthermore, ranged from $60 per pound per hour to $40 000 per pound per hour. These studies indicate that it is quite dangerous to generalize about the most efficacious method of policy implementation. Each case should be considered separately.

The above exercises deal only with the direct costs of residuals management. The total costs of any implementation scheme include costs of acquiring information, enforcement, monitoring, and administration. Charges supposedly offer a low-cost approach, which may be true for uniform charges. However, if differential charges are needed, the same amount of administrative effort will probably be involved as any other method. The OECD (1980) points out that, in practice, charges have never been used on their own, but always in conjunction with direct regulations. Very little attempt has been made to manage air quality by means of charges. The main application has been in the area of water quality management. Actual cases are covered by the OECD (1981), Harrington (1981), and Bower et al. (1981), among others (for further discussion of these policy issues, see Chapter 10 of this Handbook).

5. Case studies of model applications

5.1. Control of individual industrial plants

Results of the petroleum refinery study by Russell (1973) are too extensive to be reported fully. A sample is sufficient, however, to illustrate the application of linear programming to problems of residuals management in an industrial plant. The basic refinery modeled by Russell is a plant handling 150 000 barrels of crude

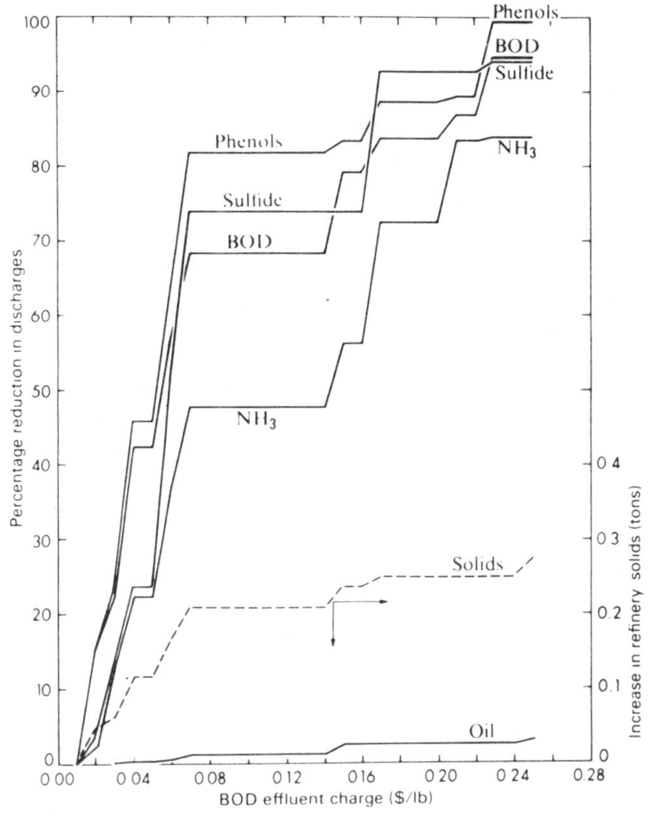

Figure 7.3. Response to BOD effluent charge, 150 000 barrels per day petroleum refinery. *Source*: Russell (1973, p. 139).

oil per day. In the uncontrolled situation, the only residuals management processes consist of oil-water separation and sour-water scrubbing.

The effects of an effluent charge on BOD discharges are shown in Figure 7.3. Because of the properties of a linear programming model, the relationship between the charge and the response is not continuous. The charge needs to increase over a range before a jump from one process technology to another occurs. Figure 7.3 reveals that as the BOD charge rises from $0.01 to $0.07, there is rapid reduction of BOD discharges, almost 70 percent. Raising the charge produces no further response until $0.16 per pound, where 80 percent BOD removal occurs. A 95 percent reduction is achieved when the charge is $0.25. The power of the model in demonstrating the interdependence of residuals is evident in the response curves for residuals other than BOD. Even though the charge is imposed on BOD itself, the consequent shifts in process technologies lead to significant reductions in ammonia, sulfides, and phenols. There is, however, a large increase in the discharge of solids, occurring principally as sludge from BOD treatment, but including particulates from incineration.

Equal moderate charges on other residuals yield disparate results. For sulfur dioxide, oil, sulfides, and residual heat, the charge has virtually no effect; positive responses are obtained for phenols, BOD, and ammonia; and dramatic reductions occur for particulates.

Other technological options for the plant include hydrocracking and *H*-oil (both processes involving introduction of excess hydrogen into the reaction vessel). A new model is developed for an "advanced" refinery using these processes. In both the basic and advanced plants, the total and marginal costs of BOD reduction rise dramatically when high levels of control are sought. These costs are, however, quite small in comparison with total refining costs (approximately 0.045 percent when a 75 percent reduction of BOD is achieved). Similar results are obtained for other residuals.

The methodological approach applied by Russell and Vaughan (1976) to residuals management in a steel production plant is the same as in Russell's model of a petroleum refinery. The main processes in a steel mill are: coking; sintering; ironmaking; steelmaking; and rolling and finishing. Steelmaking itself can be carried out with an open hearth furnace, basic oxygen furnace, or electric arc. Residuals from iron and steel production occur in large quantities, especially particulate emissions and wastewater discharges.

In the base case, the only gas cleaning controls assumed, with removal rates for particulates shown in brackets, are: dry cyclone on the sinter plant (90 percent); dry cyclone plus wet scrubber and electrostatic precipitator on the blast furnace (99 percent); electrostatic precipitator on an open hearth furnace (97 percent); electrostatic precipitator on a basic oxygen furnace (94 percent); and fabric filters on an electric arc furnace (97 percent).

Changes in the specification of the kind of steel produced have little effect on residuals discharges. A more significant factor is the type of furnace used, and the technically feasible level of added scrap metal. The electric arc and open hearth furnaces can use up to 100 percent scrap, although in practice, only the electric arc fully utilizes scrap capacity; blast furnaces have been operated on about 50 percent scrap. The basic oxygen furnace can accept only 30 percent scrap and thus is responsible for the highest levels of residuals discharges – approximately 45 percent more than an open hearth as a result of higher coke consumption. Although the electric arc typically uses 100 percent scrap, it may generate the same particulate load as the basic oxygen furnace and more sulfur dioxide than the other two kinds. The reason for this in the Russell–Vaughan model is that electricity for the furnace is generated on-site from residual fuel oil. Only if electricity is brought in is the electric arc furnace the cleanest, but then, of course, discharges occur off site.

The price of scrap metal clearly has a major influence on the choice of process technology and hence on the pattern of residuals generation. If the price of scrap is doubled, there is a 45 percent increase in iron ore used by blast furnaces, and much higher levels of discharge occur for all residuals: particulates 24 percent; sulfur dioxide 22 percent; solids 29 percent; and 1.5–25 percent for various waterborne residuals (BOD, oil, phenols, and ammonia). Residual waterborne heat increases by 5.5 percent.

Other changes analyzed with the model include the introduction of continuous casting, which changes raw material costs; variation in the cost of water; changes in by-product prices; the use of cooling towers to dissipate waste heat; the introduction of modification and recovery processes for waterborne residuals, which results in only modest increases in costs; and adoption of more effective particulate control technologies, which adds substantially to total costs.

5.2. *Air quality models*

A good example of a model dealing with control of regional discharges of airborne residuals is Kohn's model of the St. Louis airshed in Missouri [Kohn (1971, 1975)]. The model has been constructed and applied in several versions. The first version has been described in Section 4.4. It minimizes the total incremental regional cost of process technologies that meet output targets and discharge standards. It uses linear programming and treats sectoral activities as economically independent. The vector of prescribed product outputs has 94 elements. Outputs are set for the year 1975. The vector of possible unit processes covers the 94 already in existence plus another 215 processes, given 309 altogether. Control costs are measured in 1968 prices. The five residual types in the model are carbon monoxide, hydrocarbons, nitrogen oxides, sulfur dioxide, and particulates. Permissible discharges are calculated from ambient concentration standards, but the model described assumes that ambient concentrations are proportional to discharge, so it is relatively easy to estimate permissible discharge levels. The results of this model showed that an extra $35.3 million would need to be spent on abatement to meet regional air quality goals.

A later version of Kohn's model includes an input–output matrix based on an existing 23-sector input–output model of the region developed by Liu (1968). Outputs are supplied to activities to meet ordinary production requirements and also to implement residuals reduction measures, thus indirect sectoral discharges attributable to residuals control measures are accounted for in the model. Regional final output levels are specified so that both types of final demand must be met. Incremental abatement costs are then minimized subject to output and emission constraints. The input–output system is simplified to identify sectoral inputs to antipollution activities and to estimate the elements of the cost function. Liu's model is reduced to a rectangular system with six rows and seventeen columns. The first six sectors interact in usual input–output fashion, but the remaining eleven draw direct inputs only from the first six.

Only some of the results of the second version of the model are presented here. By adding an input–output matrix and capturing indirect residuals control costs, the total incremental cost to the region of meeting the same output and emission goals as in the first version is $36.1 million, an increase of only 2.3 percent. Unit processes do not change equiproportionally. One major change is a 15 percent overall increase in the use of natural gas to control emissions. A large substitution of gas for coal occurs in grate stokers fitted with cyclone collectors. As regards activity levels, the largest response is from the electricity generation sector which has to supply extra power for residuals-control processes, but this is only a 1.4 percent change. These results suggest that inclusion of an input–output model does not radically change the outcome obtained with an ordinary linear programming model. Given the relatively small numbers of interacting sectors, however, this finding is not altogether surprising. Finally, it is worth noting the shadow prices obtained for the different residuals discharged, which turn out to be: carbon monoxide, 0.4 cents per pound; hydrocarbons, 2.5 cents per pound; nitrogen oxides, 33.3 cents per pound; sulfur dioxide, 2.2 cents per pound; and particulates, 7.9 cents per pound.

In yet another version of the model, Kohn derives emission constraints from ambient concentrations by means of Gaussian dispersion formulae [Kohn (1974)].

The air quality model developed by Muller (1973, 1976) was designed for management of ambient concentrations of airborne residuals in the Rijnmond region in the Netherlands. The model, without environmental restrictions, is described by the linear programming problem in (28) and (29). The model has ten sectors, and uses 1965 as a reference year for all economic data. The optimal solution to the problem yields a regional income of 16 184 million guilders (Dfl), higher, in fact, than the actual 1965 income of Dfl 7614m.

Discharge coefficients for each of the ten sectors are introduced for sulfur dioxide, nitrogen oxides, carbon monoxide, hydrogen fluoride, and small particulates. Dispersion is simulated with a Gaussian model which has 23 emission points, 5 receptor points, and 16 wind directions, giving 1840 diffusion coefficients. Ambient concentrations are determined according to eq. (30). With no controls, the model indicates that standards would be exceeded for nitrogen oxides, hydrogen fluorides, and sulfur oxides at all locations, and particulates at two locations.

Ambient concentration standards are added to the model as in eq. (31), and initially it is assumed that the only method of reducing discharges is by adjusting output levels. Full employment can still be maintained, but regional income falls to Dfl 7889m. The "cost" of reaching environmental quality goals is thus Dfl 8295m. In the environmentally constrained solution, the model shows also that the highest shadow price occurs with the standard for nitrogen oxides at Rotterdam.

The model is then extended by achieving residuals control through antipollution investments. The pollution abatement investment cost function has a rectangular hyperbola form, which converts the model to a nonlinear system. With antipollution investments, discharges are reduced, environmental targets and full employment are reached, and regional income is Dfl 13 113m. This solution is attainable if environmental investments in the oil and chemical industries are 16 times the 1965 levels, 5 times greater for public utilities (mainly electric power stations), and 2 times greater for real estate and other industries. A six-fold increase in output of the oil and chemicals sector is indicated. The shadow price for nitrogen oxides at Rotterdam falls to almost zero. Price increases generated by the antipollution investments are simulated through the input-output model. The largest is a 7.8 percent rise for the public utilities sector, but increases of 3.7 and 3.1 percent occurs for the oil and chemicals and real estate sectors, respectively.

5.3. Water quality modeling

A good example of reactive residuals management in a stream can be found in the work carried out by the U.S. Army Corps of Engineers and by RFF on dissolved oxygen in the Potomac River. This research is described by Kneese and Bower (1968), Kneese (1977), and Davis (1968).

The Potomac River Basin has a total area of 14 000 square miles and is situated in the states of Pennsylvania, West Virginia, Maryland, Virginia, and in the District of Columbia. In the 1960s, population in the watershed was 3 million, most of which was located in the Washington metropolitan area. Dissolved oxygen levels in the river frequently fell to low levels, to a large extent attributable to urban sewage releases, but also upstream BOD discharges.

In 1963, the Corps of Engineers proposed a water management plan for the river after projecting water demands and waste discharges up to the year 2010. A standard of 4 parts per million (ppm) for DO was established. The recommended means of achieving this standard was 90 percent removal of BOD from discharge points through secondary sewage treatment and increased flushing in low flow conditions. To provide the necessary water, the Corps suggested construction of sixteen major dams and 400 headwater structures. Although the primary aim was to control dissolved oxygen levels, secondary objectives related to water supply for municipal and industrial use, flood mitigation, and regulation of water quality for recreational activities.

The plan, unfortunately, had several serious deficiencies. The 4 ppm standard was arbitrarily selected so there was no indication of whether an alternative standard might have yielded higher net benefits or if indeed the project should have been undertaken. Other alternatives were not investigated. Questions could be asked about the discount rate employed. Perhaps the most critical aspect of the plan was that only a fairly crude method (the mass curve approach) was used to estimate the storage capacity required to meet the minimal flow standard. The mass curve or Rippl diagram approach is a simple deterministic hydrologic model that indicates minimum reservoir capacity in relation to cumulative flow over a preceding historical period; in the case of the Potomac study, the preceding thirty years.

It was found that the low-flow augmentation method of achieving a 4 ppm dissolved oxygen standard would cost $115 million over a fifty year period (in 1965 prices) using a four percent discount rate. Some waste treatment methods, such as carbon adsorption could cost $127 million, but other single process approaches were much less expensive, for example, $29 million for reoxygenation. The lowest attainable cost was $22 million, entailing combinations of low-flow augmentation, reoxygenation, polymer precipitation, and step aeration.

The minimal cost could be reduced to only $8 million by lowering the standard to 2 ppm. Even with the economically inefficient method (low-flow augmentation) advocated by the Corps of Engineers, dramatic cost savings could be achieved by lowering the standard. The total cost fell to $27 million with a 3 ppm standard, and $8 million with 2 ppm. The plan proposed by the Corps of Engineers was rejected partly as a result of the RFF study, and a more flexible water quality management approach was subsequently adopted.

5.4. *Integrated models*

5.4.1. *The Lower Delaware Valley*

Few regions have received as much attention as the Delaware estuary in the states of Pennsylvania, Delaware, and New Jersey. Management of water quality began with formation of the Delaware River Basin Commission in 1936 to tackle the problems of residuals discharges and low levels of dissolved oxygen. A study of the region was undertaken in 1957–58 by the U.S. Public Health Service following a request by the U.S. Army Corps of Engineers. A more detailed analysis of the estuary was initiated in 1961 by state and interstate agencies, leading to a report released by the Federal Water Pollution Control Administration in 1966. The contents of this report are discussed by Kneese and Bower (1968) and Kneese (1974, 1977). Briefly, the study used a linear programming model to find the least-cost method of meeting prescribed water quality objectives at different points on the estuary, examined alternative systems of charges and regulations, and attempted to assess the relevant economic benefits of the water quality goals.

Development and applications of an integrated model of the Lower Delaware Valley, probably the most detailed and comprehensive REQM model ever built, took place in the 1970s at RFF, representing the culmination of years of research on industrial activities, residuals generation, and management of the environment. Similar studies have since been carried out in other parts of the world, although few have reached the complexity of the RFF model itself.

The most complete description of the RFF model can be found in Spofford et al. (1976). Briefer accounts have been written by Spofford, et al. (1975, 1977) and Kneese and Bower (1979). The conceptual basis of the model appears in a noteworthy paper by Russell and Spofford (1972). Voting mechanisms were considered in a didactic version of the model. [Russell et al. (1974)], but were not actually incorporated in the final formulation.

The Lower Delaware Valley region consists of the counties: Bucks, Montgomery, Chester, Delaware, and Philadelphia in Pennsylvania; Mercer, Burlington, Camden, Gloucester, and Salam in New Jersey; and New Castle in Delaware. The region covers an area of 4700 square miles, and in 1970, had a population of 5.5 million concentrated mostly in the urban areas of Philadelphia, Trenton, Camden, and Wilmington. The region is heavily industrialized, with 7 petroleum refineries, 5 large steel mills and many smaller mills, 13 large pulp and paper mills and numerous smaller ones, 15 large and 2 small thermal power stations, and other industrial plants including chemicals and petrochemicals, foundries, automobile manufacture, and electronic assembly. Sewage is discharged into the system by 140 treatment plants. There are also 17 municipal incinerators and numerous landfill sites.

The model was built with several objectives in mind. The first was to establish a framework for the analysis of solid, gaseous, and liquid residuals, their interrelationships, and their disposal in environmental media of air, water, and land. A second was to gain experience in the collection of data, the construction of models, computational problems, and derivation of policy-relevant conclusions and recommendations. A third objective was to discover whether an optimization model, incorporating a natural ecosystem model, could be developed to determine least-cost ways of meeting AEQ goals instead of the more commonly used recursive simulation models. A fourth was to explore ways of incorporating noneconomic constraints (distributional goals or political factors) in a model based essentially on an economic efficiency (cost-minimization) criterion. A final objective was to derive policy prescriptions on regional residuals management for the Lower Delaware Valley. It will be noted that four of the five research objectives are methodological rather than case-specific.

The model is a mathematical optimization model with deterministic variables. A unique feature is that it contains a nonlinear dynamic ecosystem model. Steady state solutions from the ecosystem model are fed back into the main system. Thus, the model as a whole is basically a static-equilibrium nonlinear system, which is solved through iterative procedures.

The general structure of the model is shown in Figure 7.4. The key components of the system are: an economic activity model, with residuals handling functions, including inputs, outputs, costs, and environmental modifications; natural sys-

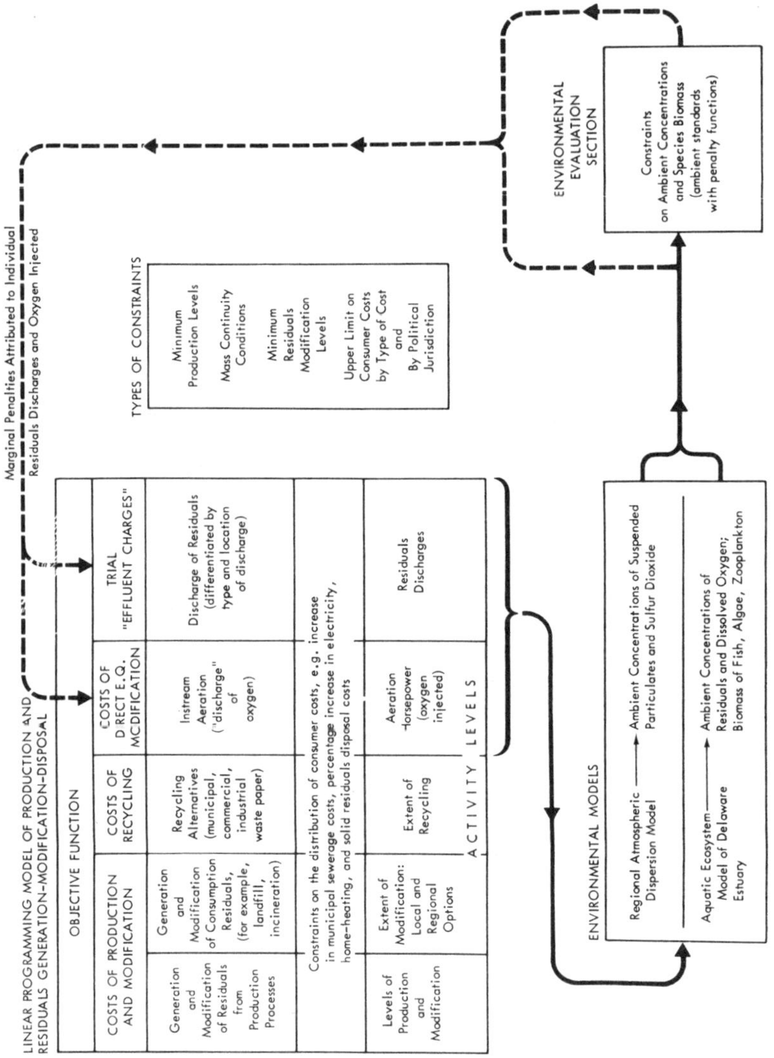

Figure 7.4. Schematic diagram of the Lower Delaware Valley residuals–environmental quality management model. *Source*: Spofford et al. (1976, p. 22).

tems models, comprising an air quality model and an aquatic ecosystem model; and an environmental evaluation module which compares environmental quality variables with predetermined standards and generates penalty functions for particular discharges in the activity model. The system also has various additional constraints (minimum product output levels, cost distribution constraints) and an objective function. These components can now be discussed in greater depth.

The main "driver" of the system is the linear programming model of activities in the region, with residuals generation, physical measures to alter discharges, recycling possibilities, and alternative methods of altering the environment to improve assimilative capacities. The objective function for the activities model includes the social costs of residuals discharged to environment, and the resource costs of alternative process technologies and environmental management options. Individual activity models are derived from separate studies carried out by RFF for different industries, as described in Section 2.2.

The air dispersion model is based on a model developed for the U.S. Environmental Protection Agency. It used the Gaussian plume approach and is a steady-state, deterministic model. Transfer matrices are obtained for 240 dischargers and 57 receptors for SO_2 and suspended particulates. Ambient concentrations are measured at ground level in average annual values. Necessary point-source data comprise stack coordinates, discharge rates, stack height, stack diameter, gas exit temperature, and gas exit velocity. To simplify the analysis, multiple stacks in a specific location were combined to obtain an "average" stack, to which the Gaussian plume equation was applied. Meteorological data required are mean annual temperature and pressure, mean annual atmospheric mixing depth, and frequency distributions for wind speed, wind direction, and atmospheric stabilities.

The aquatic ecosystem module is described in detail by Kelly (1976) and Kelly and Spofford (1977). This part of the model represents an attempt to go beyond the usual analysis of water quality in terms of dissolved oxygen and ambient concentrations of salts and solids, to simulate the effects of ambient conditions on interacting populations of aquatic organisms. Endogenous variables in the model comprise algae, zooplankton, bacteria, fish, dissolved oxygen, BOD, nitrogen, phosphorous, toxics, suspended solids, and temperature. Discharges affecting the ecosystems are BOD, nitrogen, phosphorous, toxics, suspended solids, and heat. Management goals are specified for algae, fish, and dissolved oxygen. The ecosystem model is nonstochastic and nonlinear and generates a set of results following a dynamic simulation process.

The objective function minimizes the total cost of meeting environmental quality targets (ecosystem variables and ambient concentration standards). Costs include annualized capital and operating costs of different process technologies, recycling and environmental modification, and social costs reflecting damages of residuals in the environment. This last category of costs is derived from a set of penalty functions, such that when ambient concentration standards are exceeded, values of the cost parameters for the relevant residuals in the objective function are raised. Minimum levels of product outputs have to be met, thus the model optimizes the quantities of residuals discharged, the physical control measures used, and the allocation of abatement burdens to dischargers in the region.

The solution procedure is described by Spofford et al. (1975). The penalty function is quadratic, necessitating nonlinear solution techniques. An iterative gradient method is utilized. An additional term is included in the penalty function, however, that results in a cut-off for computations when the solution is close to the full optimum. For each iteration, an ordinary linear programming algorithm is employed. The model is quite expensive to operate. Bower and Kneese (1979) comment that a typical 30-iteration run of the model would cost, at 1975 rates, $1220. After 30 iterations, ambient standards would be met and a fairly constant pattern of total discharges would be established, but the optimal levels of residuals control by individual dischargers would still not be completely determined.

The findings of the study can be evaluated in two ways: first, as regards the appropriateness of the methodology for REQM modeling in general, and second, in terms of the results on the Delaware region itself.

Several conclusions were drawn for the management of environmental quality in the Lower Delaware Region. First, it appears that air quality raises more problems than water quality because of technological difficulties and the high associated costs. Second, by using combinations of on-site residuals discharge reduction, sewage treatment and in-stream aeration, water quality standards could be met. Third, solid wastes can most effectively be disposed of by landfill in the region, but in the future, recycling and export of residuals by rail will probably be required. Municipal incinerators are not a cost-effective method of dealing with solid wastes. Finally, a wide variety of physical measures to handle residuals in the region could be used without much variation in total cost, but different combinations selected involved substantially different patterns in the incidence of management costs.

The conclusions on methodology should be of wider interest to the modeling community. Only a brief summary can be given here. There is no doubt that the research resources, data requirements, and computer time for the RFF model presented considerable difficulties for those engaged in the project. Nevertheless, if construction of such a model is contemplated, it is worthwhile identifying all the linkages among solid, liquid, and gaseous residuals rather than considering each in isolation. Computations with the nonlinear programming algorithm proved to be very costly and probably did not satisfactorily indicate the true optimum. This arose because an iterative solution method was employed and computations had to be stopped (for cost reasons) well before the optimum was reached. RFF has converted from the aquatic ecosystem model to a simple, linear Streeter–Phelps model to simplify the computations. Finally, the model was difficult to operate when distributional constraints on the incidence of costs were imposed. This, however, is a common problem in mathematical programming. If the model is overconstrained, feasible solutions may not be found.

5.4.2. The Upper Colorado River Basin

The Upper Colorado River Basin has been the subject of several regional environmental quality studies. It consists of the Green, Upper Main Stem, and San Juan sub-basins of the main Colorado River system, covers an area of 102 000 square miles, and contains sections of five states: Wyoming, Colorado, Utah, New Mexico, and Arizona. Precipitation reaches 50 inches per year in the headwaters and is below 6 inches in downstream desert areas. The outflow per unit area of 60 acre-feet per square mile is the lowest of any river basin in the United States. The region nevertheless is an important ecological resource and contains a large number of rare and endangered fish and wildlife species.

Human activities have already had major environmental impacts on the region. The population in 1970 was 346 000, engaged mostly in agriculture, grazing, and metal mining. About 70 percent of the area of the basin is owned by the federal government and used for a wide variety of recreational pursuits. Water supply and quality present difficult management problems. The river has many dams, reservoirs, and aqueducts, and water flows are carefully regulated. Downstream flows are fully utilized; typically, no water reaches the Colorado's end point in the Gulf of California. Salinity is of major concern. Natural processes (release of salts from shales, evapotranspiration, concentration of salts by vegetation) produce high background levels of dissolved solids, and this is exacerbated by irrigation practices, acid drainage from mines, use of cooling water by power stations, and discharge of sewage into the river system.

The newest threat to the basin's environment is the proposed development of the extensive deposits of coal and oil shale that underlie the region. This will result in large strip mines, oil shale refining operations, and the construction of electric power stations. Air quality in the region has already deteriorated from power station activity. The proposed energy developments will contribute further airborne residuals and place added stresses on water resources, raising the consumptive use of water and increasing salinity loadings. Uranium mining and milling in the basin will add to these problems.

The study reported by Howe (1977) represents one attempt to model the Upper Colorado River Basin, undertaken by the University of Colorado under sponsorship of the Economic Development Administration, U.S. Department of Commerce. The model has three main components: a set of regional input–output tables for the Upper Basin; a hydrologic model that simulates surface and groundwater flows and their uses in the basin; and an air dispersion model that determines ambient concentration of airborne residuals discharged from point and area sources. The model is a classic example of a generalized input–output model applied at the regional level.

The University of Colorado model was designed for several applications. The first was analysis of the effects of a large reduction in the use of water for irrigation in the Upper Main Stem sub-basin on the regional economy, hydrology, and water quality. The second application was to assess the effects of two possible coal development patterns in the Gunnison River sub-basin – either increased coal mining for export purposes, or extended coal mining accompanied by power station construction and operation. A third aim was to estimate the damage costs of uncontrolled salinity and examine the regional significance of alternative irrigation techniques and agricultural practices. Finally, the model was designed to determine whether energy developments in the Upper Basin might be constrained by environmental factors and what the tradeoffs are between environmental quality and different patterns of economic activity.

Economic scenarios for the Upper Basin are generated with input–output tables for the three subregions: Upper Main Stem (33 sectors); Green River (24 sectors); and San Juan River (30 sectors). The input-output models are not formally linked in an interregional system. Howe comments that the data were over ten years old so that the reliability of the model was somewhat in doubt. It is interesting to note also that the tables took eighteen months to compile and cost $500 000.

The hydrosalinity model uses monthly data and is driven by precipitation, temperature, and water inflows. Incoming water and total dissolved solids entering the basin establish the initial conditions. Flows within the basin are affected by tributary inflows, reservoir storage, water releases, water use by industry and municipal activities, agricultural practices, evaporation, imports and exports of water, groundwater use, and reservoir operation within the basin. Water flows and salinity are calibrated against data for a 25-year period. The model generates not only average values, resulting from hydrometeorological data driving the system, but frequency distributions of the key variables.

Air quality is analyzed by means of APGDM, a computerized air pollution generation and discharge model. The model determines ground level concentrations of airborne residuals on a quarterly basis, given the levels of economic activity indicated by the input–output model or from independent surveys. Each of the sub-basins is fairly self-contained as regards airborne residuals because of the high mountain ridges in the region; thus the sub-basin input–output models can be used separately to determine residuals discharge and dispersion. Area and point sources of discharges are distinguished. Ambient concentrations are reported on a grid system or as an isopleth mapping. An accuracy of ± 20 percent for the air dispersion model is claimed.

One of the frustrations of the University of Colorado's exercise was the failure to establish close working relationships with state and federal government agencies responsible for development decisions in the region. According to Howe, because of vested interests of the individual states involved, little progress toward efficient management of air and water quality was apparent.

A second study of the Upper Colorado River Basin was conducted by the University of New Mexico and RFF at the request of the Office of Biological Services of the U.S. Fish and Wildlife Service as part of the Southwest Region Under Stress Project. The results of the study, a collection of papers based on an RFF forum held in Albuquerque, New Mexico, 1976, are reported by Spofford et al. (1980). The interesting feature of this study is that, even though a formal REQM model was not developed from the inputs available, the papers do interrelate closely and provide an ideal basis for constructing a simulation model at some future stage. Moreover, the sequence in which the separate analyses are presented in the final report offers a valuable guide to the methodology that can be followed in gathering and organizing the information required to establish an operational, analytical structure. The study is noteworthy also in pushing beyond a consideration of ambient air and water quality to dose–response relationships for animal and fish species in the region.

Energy developments in the region are assessed in two ways: an autonomous study of possible energy expansion, emphasizing coal mining, power generation, synthetic natural gas, synthetic crude oil, and uranium mining; and with the aid of a 40-sector interregional input–output model driven by a demographic module. Feedbacks from the economic model determine employment and immigration. Environmental impacts resulting from energy developments, industrial production, and recreational activities are identified.

The impacts are assessed initially in terms of water quantity and quality. Particular properties of water systems, of significance to aquatic organisms, are shown to be: concentrations of dissolved gases; levels of toxic compounds; supply of nutrients; turbidity; physical characteristics of the river system; and temperature. The effects of changes in these characteristics are then traced for fish populations in the Colorado River. Impacts of energy developments on big game in the region are dealt with, although this analysis is not integrated with the economic and water resource scenarios. Separate papers are also included on energy developments in the Yampa Basin, with effects on water quality and trout populations. The study concludes with a discussion of institutional approaches to water quality management, with implications for fish and wildlife populations.

5.4.3. Hunter Valley

The Hunter Valley is situated in southeast Australia about 100 miles north of Sydney. It is approximately 130 miles from east to west and 75 miles north to south. The Hunter River rises in the north and reaches the coast at Newcastle, the major city in the region. The Hunter is joined by its western tributary, the Goulburn River, in the center of the valley.

The main population centers are situated in the city of Newcastle, and in surrounding towns and smaller villages. Lake Macquarie is located south of Newcastle and forms an extension of the metropolitan area.

The main activities in the region are black coal mining, electric power generation, iron and steel production, aluminum smelting, and metal fabrication. There are 52 active coal mines, producing about 35 million tonnes each year, or roughly one-third of Australia's total black coal output. Electricity generation is around 22 000 gigawatt hours.

Environmental concerns were expressed in the later 1970s with the impending natural resource boom, as projections indicated a trebling of coal output from the Hunter Valley, especially from new open cut mines. Plans were also made for an expansion of electricity generation and aluminum smelting. Much of the growth impetus has now slowed as a result of the world recession, but a further expansionary phase is very likely in the late 1980s and many of the same problems will need to be addressed once again.

A model is currently being constructed within the Centre for Environmental Studies at Macquarie University under sponsorship of the National Energy Research Development and Demonstration Council [James et al. (1982)]. The model combines a national energy-economic input–output model with an environmental input–output model for the Hunter region based on a set of transactions tables prepared by Garlick (1979). Aims of the exercise include: integration with MARKAL, an engineering-economic model of energy use in Australia developed by the Commonwealth Scientific and Industrial Research Organisation [Musgrove et al. (1983)]; simulation of national economic and energy scenarios, tracing the implications for coal-related developments in the Hunter region; simulation of the air and water quality effects of economic scenarios in the Hunter region; and investigation of possible regional environmental constraints on regional or national development.

The regional input–output model has 23 sectors, with all transactions flows measured in dollars, except for raw coal, washed coal, and electricity, which are measured in physical units. Regional economic scenarios will be driven by the national energy model or obtained from autonomous studies of the regional economy and its future prospects. The main airborne residuals under examination are sulfur oxides and fluorides. Sulfur oxides are a reasonable general indicator of air quality, and fluorides are important because of their potential damaging effects on vineyards in the region. An improved version of the Dennis model is being used to simulate average monthly concentrations of these residuals, and to determine the meteorological conditions under which peak concentrations can arise [Chambers (1983)].

A water supply model is being constructed, based on data supplied by the NSW Water Resources Commission. Water demands will be projected by means of the input–output model, together with appropriate assumptions about irrigation practices in the agricultural sector. A major problem is salinity, as the coal fields have been formed from ancient marine sediments. Salinity occurs naturally through groundwater extrusion, but as yet the impact of mining on salinity levels, especially during times of low river flow, is not properly known. These problems are being investigated by the Macquarie group.

Preliminary results suggest that the water supply/demand pattern and salinity levels will be the main environmental quality problems to tackle. When this is established with greater certainty, more detailed analysis of environmental management alternatives, together with their economic costs, can be undertaken.

5.4.4. RIM

RIM is the acronym for "Reken-en Informatiesysteem Milieuhygiene", a multi-sectoral economic–environmental model of the Netherlands currently under development at the Institute for Environmental Studies, Free University, Amsterdam. The Institute's English language version is ECIS (Environmental Computation and Information System). RIM is discussed here as an example of an REQM model, as it contains all the elements of a regional environmental model, and even though RIM is designed to provide information for the management of residuals at the national level, the Netherlands itself is so small and compact geographically that the entire country can be treated as a single region.

RIM has evolved from several lines of research over the last ten years, but the common thread has been a residuals-generation and modification input–output model of the Netherlands economy that has been successively improved and expanded to its present form. Much of this research, including current development of the model, has been sponsored by the Ministry of Environment and Public Health. Very little documentation is available, unfortunately, in the English language.

The origins of RIM can be traced to early work on combustion emissions associated with energy use in a paper written by Jansen and Stapel (1974) that made use of a 23-sector input–output model. By 1977, the system had been enlarged to a 60-sector model, and linear programming was applied to assess tradeoffs between economic variables and energy-related residuals discharges [Jansen et al. (1977)]. A brief description of this work can be found in James et al. (1978).

In the next stage of development, process residuals were added to the model [Jansen et al. (1978)] and, for each of the 60 sectors, discharge coefficients were calculated for four environmental indicators: an air quality index obtained by weighting different airborne residuals and aggregating them, a water quality index reflecting heavy metals discharges, a BOD water quality index, and chemical wastes measured by mass. The model was used to create alternative scenarios for 1985, analyzing policies emphasizing economic goals, such as income and employment, or environmental quality as measured by the four indicators just described.

The feasibility study for RIM was carried out in 1980 [Hordijk et al. (1980)] after a research tour of the United States where detailed discussions were held on the U.S. EPA SEAS (Strategic Environmental Assessment System) model. The conclusion was reached that a much simpler model would need to be constructed for the Netherlands given the proposed manpower and funding. The most recent

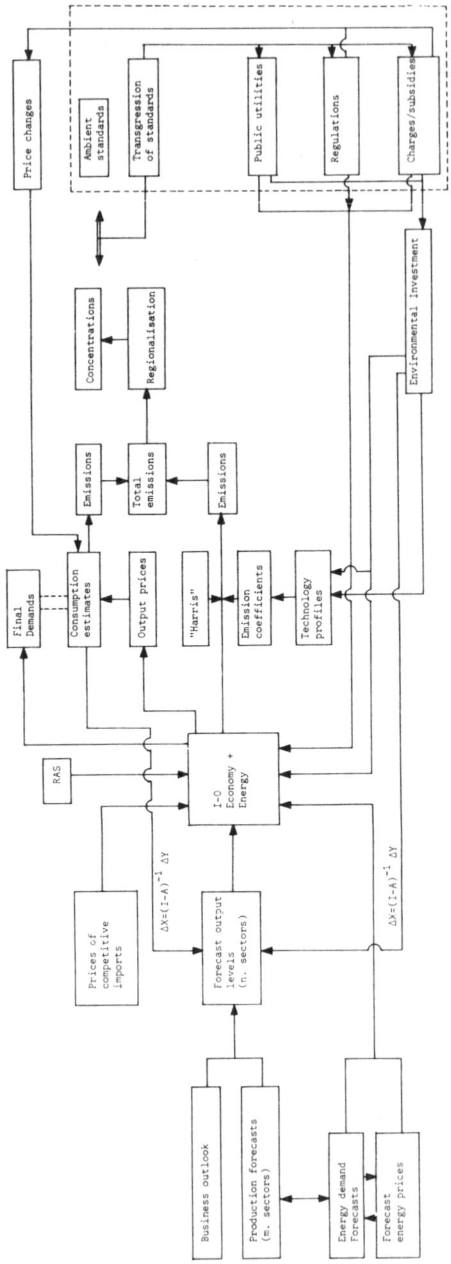

Figure 7.5. RIM. *Source*: Hordijk et al. (1980, p. 13).

description of the model appears in Olsthoorn and Jansen (1982). The first phase of the project, producing an inventory of nitrogen oxide emissions in the Netherlands and possible emission patterns for the period 1978–2000 under different technologies for NO_x control has now been completed [Olsthoorn et al. (1982)].

The general structure of RIM appears in Figure 7.5. Economic forecasts from the Central Planning Bureau and the Ministry of Economic Affairs, together with assumptions about energy prices and energy demands, are translated into a set of final demands that drive the 60-sector input–output model. The RAS method is proposed for updating and forecasting the input–output coefficients. It is possible also that these coefficients will be modified according to the results of econometric studies already underway in the Netherlands on energy use and technical change [Lesuis et al. (1980)].

Emission coefficients in the model apply to specific processes within each sector, as far as possible from physical emission factors rather than from monetary data. NO_x emissions, for example, are estimated for stationary and mobile sources. The stationary source emissions are based on fuel use patterns, including different combustion processes and emission control technologies, and on process emissions from the outputs of products such as nitric acid and fertilizers. With mobile sources, emissions are estimated separately for private automobiles driven in different modes and using different fuels, commercial vehicles, taxis, motorcycles, rail transport, and special vehicles. The calculations reflect earlier work carried out on NO_x emissions in the Netherlands [METRA (1977)]. Future research on sulfur oxides emissions will likewise be assisted by existing studies [Ministerie van Volksgezondheid en Milieuhygiene (1980)].

Emission coefficients are affected in the model by new technological profiles created through environmental investment. This part of the model follows from the implementation of environmental quality standards. National emissions, once determined, are translated into ambient concentrations at different receptor points through a regionalization program. Ambient concentrations are then compared with selected standards. If the standards are transgressed, management actions are simulated. These may take the form of residuals discharge controls for public utilities such as water and sewage plants, or regulations and subsidies/charges to control residuals discharges from the private sector. Price effects from these measures lead to changes in final demands, requiring an iterative adjustment of the model. In addition, environmental investment may take place, affecting sectoral outputs and emission coefficients as previously mentioned.

RIM will perform a number of functions in the Netherlands. It will be used to obtain advance warnings of environmental pollution by comparing predicted ambient concentrations of residuals with existing standards. In the longer term, it will simulate patterns of economic growth and energy use, and indicate the requirements for new standards before environmental damage actually occurs. A time span of twenty years is proposed. RIM will assist general evaluation of government policies and will be particularly useful in integrating environmental goals with other socioeconomic goals. RIM will also be a valuable information-gathering system. Data will be collected and analyzed as on ongoing process. Another function is publicity. Regular reports can be issued to advise the community of trends in the economy and environmental quality, encouraging critical public discussion of environmental policies and establishing a rational basis for future action. Finally, RIM will help to coordinate institutions involved in managing the economy, energy use, and environmental quality. All participants in the planning and management process will have common use of a detailed data and modeling base.

6. Concluding remarks

Sufficient experience has now been gained in the industrial process/REQM area for a number of conclusions to be drawn, and for various challenges to be identified in future research. Perhaps the most important fact to emerge is that if REQM modeling is to have practical relevance – in terms of analyzing alternative industrial process technologies, assessing their effects on ambient environmental quality, and indicating guidelines for environmental managers to follow – economists need access to detailed physical and engineering models and data. Is multidisciplinary teamwork the best way to achieve this goal? Should economists endeavor to take over relevant branches of engineering and environmental science? Can the physical systems analysts provide reference guides with appropriate model descriptions and areas of application that economists and environmental managers can incorporate in an REQM framework? Answers to these questions must be found if REQM modeling is to maintain credibility in future years.

The relationship between modelers and users is crucial. Above all, modelers must realize that the object of analysis is not the construction of theoretically elaborate mathematical systems, but the solution of specific problems. Methodology is, to a certain extent, transferrable from one set of problems (or geographic region) to another, but continued sponsorship by policymakers can be expected only if practical results are achieved. Modelers accordingly need to make careful judgments about the directions of research before data collection and analysis begin. For example, what is the scope of the problem? Is pollution in a region associated with all activities, or only a few? Is is necessary to study all residuals discharged, or are only certain residuals of prime significance? How many classes of receptors must be analyzed? Should dose–response relationships be considered, or are widely used ambient concentration standards acceptable proxies for environmental goals? If institutional and environmental boundaries (airsheds, watersheds, ecosystems) do not coincide, how can policymakers be shown the value of a broad-ranging modeling exercise? Institutional constraints on management strategies may be just as important as economic, environmental and technological parameters.

For future research on industrial activities, physical measures of inputs, commodity outputs and residuals discharges associated with specific process technologies are clearly superior to monetary measures. Compilation of commonly used industrial technologies and residuals modification processes would be extremely useful to REQM modelers. Cost functions specific to countries and regions could be readily constructed by economists by applying financial data to engineering process models, permitting tradeoff analysis of the costs of residuals discharge management and levels of control. Where multiactivity REQM models are involved, as in linked activity models or input–output systems, the prospects of identifying intersectoral or interactivity flows measured in physical units should be carefully explored, as this approach lends itself readily to mass balance and energy balance analysis. Less significant regional production flows may be modeled in monetary values if other data are unavailable.

Nonreactive dispersion modeling appears to have reached a satisfactory stage of development, especially for steady-state calculations. Further work remains to be done, however, in several areas. So far, only a limited number of residuals have been studied. Much more needs to be known about types of residuals discharged from industrial activities, particularly toxic substances. Although acid rain and photochemical smog have recieved much attention, reactive residuals modeling is still in a state of infancy. The dynamics of residuals behavior in the environment is a topic of emerging interest. Very little is understood about the accumulation of substances in physical (and biological) systems. Compilation of emission inventories, increased field monitoring, and model verification are important requirements in future REQM work.

Probably the weakest link in the entire chain of analysis is dose–response modeling and the construction of damage functions. According to recent studies a large component of damage resulting from air pollution may be human health, yet estimates vary by orders of magnitude. Much better data on dose–response relationships are urgently needed. Concern is also expressed by medical researchers about the paucity of reliable data on long-term exposure of humans to low-level concentrations of various substances. In this respect uncertainty exists even about the more commonly studied air-borne residuals such as sulphur oxides and carbon monoxide. Effects of residuals on ecosystems is another area which, as yet, remains largely unexplored. Work on the economic valuation of environmental effects has progressed rapidly in the last few years. Delays in specifying damage functions are the result of inadequate physical modeling rather than valuation methodologies. Deficiencies in damage function estimation should not preclude further REQM research. What can be expected in the foreseeable future is an emphasis on standard-setting and cost-effectiveness analysis. Mathematical programming and simple tradeoff models can make important contributions in this area. When more experience in environmental systems modeling has been gained, full benefit–cost analyses of regional industrial development programs and environmental quality management may evolve.

References

Ayres, Robert U. (1972) "A Materials-Process-Product Model", in: Allen V. Kneese and Blair T. Bower (eds.), *Environmental Quality Analysis: Theory and Method in the Social Sciences* (Johns Hopkins University Press, Baltimore) pp. 35–67.

Ayres, Robert U. and Ivars Gutmanis (1972) "Technological Change, Pollution and Treatment Costs in Input–Output Analysis", in: Ronald G. Ridker (ed.), *Population, Resources and the Environment*, U.S. Commission on Population Growth and the American Future (U.S. Government Printing Office, Washington) pp. 313–337.

Ayres, Robert U. and A.V. Kneese (1969) "Production, Consumption and Externalities", *American Economic Review* 59, 282–297.

Bach, Wilfred (1972) *Atmospheric Pollution* (McGraw-Hill, New York).

Basta, Daniel J. and Blair T. Bower (eds.) (1982) "Analyzing Natural Systems: Analysis for Regional Residuals-Environmental Quality Management", Research Paper (Resources for the Future, Washington).

Basta, Daniel J., James L. Lounsbury and Blair T. Bower (1978) "Analysis for Residuals – Environmental Quality Management", RFF Research Paper R-11 (Resources for the Future, Washington).

Beller, M. (ed.) (1976) "Energy Systems Studies Program Annual Report", BNL 50539, Brookhaven National Laboratory, Upton, New York.

Bower, Blair T. (1967) "The Economics of Industrial Water Utilization", in: Allen V. Kneese and Stephen C. Smith (eds.), *Water Research* (Johns Hopkins University Press, Baltimore).

Bower, Blair T. (1975) "Studies of Residuals Management in Industry", in: Edwin S. Mills (ed.), *Economic Analysis of Environmental Problems*, National Bureau of Economic Research (Columbia University Press, New York) pp. 275–324.

Bower, Blair T. (1977) "The Why and What of Regional Residuals Environmental Quality Management Modeling", in: Blair T. Bower (ed.), *Regional Residuals Environmental Quality Management Modeling*, Research Paper R-7 (Resources for the Future, Washington), pp. 1–32.

Bower, Blair T. (1983) "Analyzing Effects on Natural Systems and Receptors", in: Maynard M. Hufschmidt, David E. James, Anton D. Meister, Blair T. Bower and John A. Dixon (eds.), *Environment, Natural Systems and Development: An Economic Valuation Guide* (Johns Hopkins University Press, Baltimore).

Bower, Blair T., Remi Barre, Jochen Kuhner and Clifford S. Russell (1981) "Incentives in Water Quality Management: France and the Ruhr Area", Research Paper R-24 (Resources for the Future, Washington).

Bower, Blair T., Charles N. Ehler and Allen V. Kneese (1977) "Incentives for Managing the Environment", *Environmental Science and Technology* 11, 250–254.

Bower, Blair T., George O.G. Lof and W.M. Hearon (1972) "Residuals Management in the Pulp and Paper Industry", *National Resources Journal* 11, 605–623.

Brady, Gordon L. and Blair T. Bower (1981) "Air Quality Management: Quantifying Benefits", Research Report no. 7, East–West Environment and Policy Institute, Honolulu.

Briggs, G.A. (1969) "Plume Rise", Atomic Energy Commission Critical Review Series, TID 25075.

Brown, F. Lee and A.O. Lebeck (1976) *Cars, Cans and Dumps: Solutions for Rural Residuals* (Johns Hopkins University Press, Baltimore).

Brown, Robert M., Nina I. McClelland, Rolf A. Deininger and Michael F. O'Connor (1975) "A Water Quality Index—Crashing the Psychological Barrier", in: William A. Thomas (ed.), *Indicators of Environmental Quality* (Plenum/Rosetta, New York) pp. 173–182.

Buehring, William A. and Wesley K. Foell (1976) *Environmental Impact of Electrical Generation: A Systemwide Approach* (International Institute for Applied Systems Analysis, Laxenburg, Austria).

Burton, Ellison S. and William Sanjour (1970) "A Simulation Approach to Air Pollution Abatement Program Planning", *Socio-Economic Planning Sciences* 4, 147–159.

Camp, T.R. (1965) "Field Estimates of Oxygen Balance Parameters", *Journal of Sanitary Engineering Division*, American Society of Civil Engineers, 91, 1–16.
Carson, J.E. and H. Moses (1969) "The Validity of Several Plume Rise Formulas", *Journal of the Air Pollution Control Association* 19, 862–966.
Chambers, John A. (1983) "A Dispersion Model for the Prediction of Ambient Air Pollution Concentrations in the Hunter Region, NSW", Center for Environmental and Urban Studies, Macquarie University, North Ryde, NSW.
Cherniavski, E.A. (1974) "Brookhaven Energy System Optimization Model", Report no. BNL 19569 Brookhaven National Laboratory, Upton, New York.
Converse, A.O. (1971) "On the Extension of Input–output Analysis to Account for Environmental Externalities", *American Economic Review* 61, 197–198.
Davis, Robert K. (1968) *The Range of Choice in Water Management* (Johns Hopkins University Press, Baltimore).
Deland, Michael R. (1979) "The Bubble Concept", *Environmental Science and Technology* 13, 277.
Den Hartog, H. and A. Houweling (1974) "Pollution, Pollution Abatement and the Economic Structure: Empirical Results of Input–output Computations for the Netherlands", Occasional Papers no. 1/1974 Central Planning Bureau, The Hague.
Dennis, R.L. (1978) "The Smeared Concentration Approximation Method: A Simplified Air Pollution Dispersion Methodology for Regional Analysis", IIASA Research Paper RR-78-9 International Institute for Applied Systems Analysis, Laxenburg, Austria.
EPRI (1979a) "Workshop Proceedings: Integration of Environmental Considerations Into Energy-Economic System Models", Report no. WS-78-95 Electric Power Research Institute, Palo Alto.
EPRI (1979b) "Modeling Future Power Plant Location Location Patterns", Report no. EA-1063 Electric Power Research Institute, Palo Alto.
Fiering, Myron B. (1967) *Streamflow Synthesis* (Harvard University Press, Cambridge).
Foell, Wesley K. (1979a) *Management of Energy/Environment Systems: Methods and Case Studies* (Wiley, New York).
Foell, Wesley K. (1979b) "The Wisconsin/IIASA Energy/Environment Models: An Overview and Recent Applications", in: Workshop Proceedings: Integration of Environmental Considerations Into Energy-Economic System Models, Report no. WS-78-95 (Electric Power Research Institute, Palo Alto) pp. 4.37–4.50.
Foell, W.K., J.W. Mitchell and J.L. Pappas (1975) "The Wisconsin Regional Energy Model: A Systems Approach to Regional Analysis", Report no. 56 (Energy Systems and Policy Research Group, University of Wisconsin, Madison).
Forsund, Finn R. and Steinar Strom (1976) "The Generation of Residual Flows in Norway: An Input–output Approach", *Journal of Environmental Economics and Management* 3, 129–141.
Freeman, A. Myrick (1979) *The Benefits of Environmental Improvement: Theory and Practice* (Johns Hopkins University Press, Baltimore).
Gameson, A.L.H. (1973) *Mathematical and Hydraulic Modelling of Estuarine Pollution*, Water Pollution Research Technical Paper no. 13 (Her Majesty's Stationery Office, London).
Garlick, S. (1979) "New Input–output Tables for the Hunter Region", (Hunter Valley Research Foundation, Tighes Hill, N.S.W.).
Gifford, Jr., F.A. (1961) "Use of Routine Meteorological Observations for Estimating Atmospheric Dispersion", *Nuclear Safety* 2, 47–51.
Gifford, Jr., F.A. (1973) "The Simple ATDL Air Pollution Model", rev., ATDL Contribution File no, 78 (Oak Ridge, NOAA Atmospheric and Diffusion Laboratory).
Gosz, James R. (1980) "The Influence of Reduced Stream Flows on Water Quality", in: Walter O. Spofford, Jr., Alfred L. Parker and Allen V. Kneese, (eds.), *Energy Developments in the Southwest*, RFF Research Paper R-18 (Resources for the Future, Washington).
Green, A.E.S. (ed.) (1980) *Coal Burning Issues* (University of Florida Press, Gainsville).
Guldmann, Jean-Michel (1978) "Industrial Location, Air Pollution Control and Meteorological Variability: A Dynamic Optimization Approach", *Socio-Economic Planning Sciences* 12, 197–214.
Guldmann, Jean-Michel and Daniel Shefer (1977) "Optimal Plant Location and Air Quality Management Under Indivisibilities and Economies of Scale", *Socio-Economic Planning Sciences* 11, 77–93.
Hafkamp, Wim and Peter Nijkamp (1980) "An Integrated Interregional Model for Pollution Control", in: T.R. Lakshmanan and P. Nijkamp (eds.), *Economic-Environmental-Energy Interactions: Modeling and Policy Analysis* (Martinus Nijhoff, The Hague).
Hanna, Steven R. (1973) "A Simple Dispersion Model for the Analysis of Chemically Reactive

Pollutants", *Atmospheric Environment* 7, 803–817.

Harrington, Winston (1981) "The Regulatory Approach to Air Quality Management: A Case Study of New Mexico", Research Paper R-25 (Resources for the Future, Washington).

Hite, James C. and Eugene A. Laurent (1972) *Environmental Planning: An Economic Analysis* (Praeger, New York).

Holling, C.S. (1978) *Adaptive Environmental Assessment and Management* (Wiley, New York).

Hordijk, L., H.M.A. Jansen, A.A. Olsthoorn and J.B. Vos (1980) "Reken-en Informatiesysteem Milieuhygiene, Een Studie Naar Haalbaarheid" (Institute for Environmental Studies, Free University, Amsterdam).

Howe, Charles W. (1977) "A Coordinated Set of Economic, Hydro-salinity and Air Quality Models of the Upper Colorado River Basin with Applications to Current Problems", in: Blair T. Bower (ed.), *Regional Residuals Environmental Quality Management Modeling*, RFF Research Paper R-7 (Resources for the Future, Washington) pp. 108–138.

Howe, Charles W. (1979) *Natural Resource Economics* (Wiley, New York).

Hufschmidt, Manyard, M. (1962) "Analysis by Simulation: Examination of Response Surface", in: Arthur Maass, Maynard M. Hufschmidt, Robert Dorfman, Harold A. Thomas, Jr., Stephen A. Marglin and Gordon Maskew Fair (eds.), *Design of Water Resource Systems* (Macmillan, London).

Hufschmidt, Maynard M. and Eric L. Hyman (eds.) (1982) *Economic Approaches to Natural Resource and Environmental Quality Analysis* (Tycooly, Dublin).

Hufschmidt, Maynard M., David E. James, Anton D. Meister, Blair T. Bower and John A. Dixon (1983) *Environment, Natural Systems and Development: An Economic Valuation Guide* (Johns Hopkins University Press, Baltimore).

International Energy Agency (1980) *A Group Strategy for Energy Research, Development and Demonstration* (Organisation for Economic Cooperation and Development, Paris).

James, David, Alison Gilbert and John Chambers (1982) "An Integrated Energy-Economic-Environmental Model for Australia: Background, Conceptual Framework and Research Design" (Centre for Environmental Studies, Macquarie University, North Ryde, N.S.W.).

James, D.E., H.M.A. Jansen and J.B. Opschoor (1978) *Economic Approaches to Environmental Problems* (Elsevier, Amsterdam).

Jansen, H.M.A. et al. (1978) "Milieuverontreiniging en Productiestructuur in Nederland" (Institute for Environmental Studies, Free University, Amsterdam).

Jansen, H.M.A., J.B. Opschoor, J.H.A. Stapel, A.A. Olsthoorn, J.B. Vos and J.L. ten Broek (1977) "Luchtverontreiniging en Economische Struktuur" (Institute for Environmental Studies, Free University, Amsterdam).

Jansen, H.M.A. and J.H.A. Stapel (1974) "Vuil uit Nederlands Pijpen" Werknota no. 40 (Institute for Environmental Studies, Amsterdam).

Keeney, Ralph L. and Howard Raiffa (1976) *Decisions with Multiple Objectives* (Wiley, New York).

Kelly, Robert A. (1976) "Conceptual Ecological Model of the Delaware Estuary", in: B.C. Patten (ed.), *Systems Analysis and Simulation in Ecology* vol. IV (Academic Press, New York).

Kelly, Robert A. and Walter O. Spofford, Jr. (1977) "Application of an Ecosystem Model to Water Quality Management: The Delaware Estuary", in: C.A.S. Hall and J.W. Day, Jr. (eds.), *Ecosystem Modeling in Theory and Practice: An Introduction with Case Histories* (Wiley, New York).

Kneese, Allen V. (1974) "The Application of Economic Analysis to the Management of Water Quality: Some Case Studies", in: J. Rothenberg and Ian G. Heggie (eds.), *The Management of Water Quality and the Environment* (Macmillan, London) pp. 73–103.

Kneese, Allen V. (1977) *Economics and the Environment* (Penguin, Harmondsworth, Middlesex) pp. 151–194.

Kneese, Allen V., Robert U. Ayres and Ralph C. d'Arge (1970) *Economics and the Environment: A Materials Balance Approach* (Resources for the Future, Washington).

Kneese, Allen V. and Blair T. Bower (1968) *Managing Water Quality: Economics Technology, Institutions* (Johns Hopkins University Press, Baltimore).

Kneese, Allen V. (1977) "Issues Surrounding Regional Residuals-Environmental Quality Management Modelling", in: Blair T. Bower, (ed.), *Regional Residuals Environmental Quality Management Modeling*, Research Paper R-7 (Resources for the Future, Washington) pp. 33–51.

Kneese, Allen V. and Blair T. Bower (1979) *Environmental Quality and Residuals Management* (Johns Hopkins University Press, Baltimore).

Kohn, Robert E. (1971) "Optimal Air Quality Standards", *Econometrica* 39, 983–995.

Kohn, Robert E. (1974) "Industrial Location and Air Pollution Abatement", *Journal of Regional Science* 14, 55–63.

Kohn, Robert E. (1975) "Input–output Analysis and Air Pollution Control", in: Edwin S. Mills (ed.), *Economic Analysis of Environmental Problems*, National Bureau of Economic Research (Columbia University Press, New York) pp. 239–274.

Leontief, Wassily W. (1970) "Environmental Repercussions and the Economic Structure", *Review of Economics and Statistics* 52, 262–271.

Lesuis, P.J.J., F. Muller and P. Nijkamp (1980) "Operational Methods for Strategic Environmental and Energy Policies", in: T.R. Lakshmanan and P. Nijkamp (eds.), *Economic-Environmental-Energy Interactions: Modeling and Policy Analysis* (Martinus Nijhoff, The Hague).

Liu, Ben-chieh (1968) *Interindustrial Structure of the St. Louis Region, 1967* (St. Louis Regional Industrial Development Corporation).

Lof, George O.G., W.M. Hearon and Blair T. Bower (1973) "Residuals Management in Pulp and Paper Manufacture", in: Walter S. Kaghan (ed.), *Forest Products and the Environment*, American Institute of Chemical Engineers Symposium Series, vol. 69, no. 133.

Lof, George O.G. and Allen V. Kneese (1968) *The Economics of Water Utilization in the Beet Sugar Industry* (Johns Hopkins University Press, Baltimore).

Luken, Ralph A., Daniel J. Basta and Edward H. Pechan (1976) *The National Residuals Discharge Inventory* (National Research Council, Washington).

Maass, Arthur, Maynard M. Hufschmidt, Robert Dorfman, Harold Thomas, Jr., Stephen A. Marglin and Gordon Maskew Fair (1962) *Design of Water-resource Systems* (Macmillan, London).

MacCracken, M.C. and G.D. Sauters (eds.) (1975) *Development of an Air Pollution Model for the San Francisco Bay Area: Final Report to the National Science Foundation* (Lawrence Livermore Laboratory, Livermore, California).

Manne, Alan S. (1963) *Scheduling of Petroleum Refinery Operations* (Harvard University Press).

Marcuse, William (1979) "A Hierarchial Decomposition Analysis to Environmental Policy Analysis", in: Workshop Proceedings: Integration of Environmental Considerations Into Energy-Economic Systems Models, Report no. WS-78-79 (Electric Power Research Institute, Palo Alto) pp. 4.77–4.86.

Mastenbroek, A.P. and P. Nijkamp (1974) "Een Multi-Regional Milieumodel Voor Een Ruimtelijke Allocatie Van Investeringen", in: P. Nijkamp (ed.), *Milieu En Economie* (Universitaire Pers, Rotterdam).

Mendelsohn, Robert O. (1979) *Towards Efficient Regulation of Air Pollution from Coal-Fired Power Plants* (Garland, New York).

Mendelsohn, Robert O. (1980) "An Economic Analysis of Air Pollution from Coal-Fired Power Plants", *Journal of Environmental Economics and Management*, 7, 30–43.

Mendelsohn, Robert O. and Guy Orcutt (1979) "An Empirical Analysis of Air Pollution Dose Response Curves", *Journal of Environmental Economics and Management* 6, 85–106.

Metra Consulting Group (1977) "Long-term Consequences of NO_x and SO_2 Abatement", Report to Ministerie van Volksgezondheid en Milieuhygiene, Metra Consulting Group, London.

Miernyk, William H. and John T. Sears (1974) *Air Pollution Abatement and Regional Economic Development* (D.C. Heath, Lexington).

Ministerie van Volksgezondheid en Milieuhygiene (1980) *SO_2 Beleidskaderplan* (Zitting 1979–80, 15834, Nrs. 1-2, Tweede Kamer).

Moodie, A.R. (1979) "Modeling of Water Quality and Hydrology in an Urban Watercourse", Australian Water Resources Council, Technical Paper no. 45 (Australian Government Publishing Service, Canberra).

Muller, Frederik (1973) "An Operational Mathematical Programming Model for the Planning of Economic Activities in Relation to the Environment", *Socio-Economic Planning Sciences* 7, 123–138.

Muller, Frederik (1976) "An Integrated Regional Environmental Economic Model", in: P. Nijkamp (ed.), *Environmental Economics*, vol. 2, Methods (Martinus Nijhoff, The Hague).

Muller, Frederik (1979) *Energy and Environment in Interregional Input–output Models* (Martinus Nijhoff, The Hague).

Musgrove, A.R. de L., K.J. Stocks, P. Essam, D. Le and J.V. Hoetzl (1983) "Exploring Some Australian Energy Alternatives Using MARKAL", Technical Report TR-2 (Division of Energy Technology, CSIRO, Lucas Heights, N.S.W.).

Neely, W. Brock (1982) "The Definition and Use of Mixing Zones", *Environmental Science and Technology* 16, 518–521.

New South Wales State Pollution Control Commission (1977) "Report on Air Pollution Constraints in the Sydney Region" (State Pollution Control Commision, Sydney).

Nijkamp, P. (1977) *Theory and Application of Environmental Economics* (North-Holland, Amsterdam).

Noll, Roger G. and John Trijonis (1971) "Mass Balance, General Equilibrium and Environmental Externalities", *American Economic Review* 61, 730–735.

OECD (1974) *Environmental Damage Costs* (Organisation for Economic Cooperation and Development, Paris).

OECD (1976) *Economic Measurement of Environmental Damage* (Organisation for Economic Cooperation and Development, Paris).

OECD (1980) *Pollution Charges in Practice* (Organisation for Economic Cooperation and Development, Paris).

OECD (1981) *The Costs and Benefits of Sulphur Oxide Control: A Methodological Study* (Organisation for Economic Cooperation and Development, Paris).

Olsthoorn, A.A. and H.M.A. Jansen (1982) "RIM: Reken-en Informatiesysteem Milieuhygiene", in: H. Aiking, P. Ester, L. Hordijk and H.E. van de Veen (eds.), *Mozaiek van de Mileuproblematiek* (VU Boekhandel/Uitgeverij, Amsterdam) pp. 65–79.

Olsthoorn, A.A., R. Thomas, J.B. Vos, J.P. Hettelingh, L. Hordijk and H.M.A. Jansen (1982) *Drie Scenarios Voor de Ontwikkeling van de NO_x Emissies in Nederland 1978–2000* (Institute for Environmental Studies, Free University, Amsterdam).

Pasquill, F. (1962) *Atmospheric Diffusion* (Van Nostrand, New York).

Parvin, Manoucher and Gus W. Grammas (1976) "Optimization Models for Environmental Pollution Control: A Synthesis", *Journal of Environmental Economics and Management* 3, 113–128.

Pearce, David W. (ed.) (1978) *The Valuation of Social Cost* (Allen and Unwin, London).

Pearce, David W. (1978) "Air Pollution", in: David W. Pearce (ed.), *The Valuation of Social Cost* (Allen and Unwin, London) pp. 54–67.

Pigou, A.C. (1932) *The Economics of Welfare*, 4th Edition (Macmillan, London).

Rao, S. Trivkrama, Gopal Sistla, Michael T. Keenan and John S. Wilson (1980) "An Evaluation of Some Commonly Used Highway Dispersion Models", *Journal of the Air Pollution Control Association* 30, 239–246.

Ridker, Ronald G. (ed.) (1972) *Population, Resources and the Environment*, U.S. Commission on Population Growth and the American Future (U.S. Government Printing Office, Washington).

Ridker, Ronald G. and William D. Watson (1980) *To Choose a Future* (Johns Hopkins University Press, Baltimore).

Russell, Clifford S. (1971) "Models for the Investigation of Industrial Response to Residuals Management Actions", *Swedish Journal of Economics* 73, 134–156.

Russell, Clifford S. (1973) *Residuals Management in Industry: A Case Study of Petroleum Refining* (Johns Hopkins University Press for Resources for the Future, Baltimore).

Russell, Clifford S. and Walter O. Spofford, Jr. (1972) "A Quantitative Framework for Residuals Management Decisions", in: Allen V. Kneese and Blair T. Bower (eds.), *Environmental Quality Analysis: Theory and Method in the Social Sciences* (Johns Hopkins University Press, Baltimore) pp. 115–179.

Russell, Clifford S., Walter O. Spofford, Jr. and Edwin T. Haefele (1974) "The Management of the Quality of the Environment", in: J. Rothenberg and Ian G. Heggie (eds.), *The Management of Water Quality and the Environment* (Macmillan, London) pp. 224–277.

Russell, Clifford S. and William J. Vaughan (1974) "A Linear Programming Model of Residuals Management for Integrated Iron and Steel Production", *Journal of Environmental Economics and Management* 1, 17–42.

Russell, Clifford S. and William J. Vaughan (1976) *Steel Production: Processes, Products and Residuals* (Johns Hopkins University Press, Baltimore).

Sawyer, James W., Jr. (1974) *Automotive Scrap Recycling: Processes, Prices and Prospects* (Johns Hopkins University Press, Baltimore).

Schlottmann, Alan (1977) "New Life for Old Garbage: Resource and Energy Recovery from Solid Wastes", *Journal of Environmental Economics and Management* 4, 57–67.

Shefer, Daniel (1973) "Forecasting Industrial Air Pollution in the Haifa Bay Area with an Input–output Model", *Socio-Economic Planning Sciences* 7, 397–406.

Sinden, John A. and Albert C. Worrell (1979) *Unpriced Values: Decisions Without Market Prices* (Wiley Interscience, New York).

Sistla, Gopal, Perry Samson, Michael Keenan and S. Trivikrama Rao (1979) "A Study of Pollutant Dispersion Near Highways", *Atmospheric Environment* 13, 669–685.

Spofford, Jr., Walter O. (1980) "Potential Impacts of Energy Development on Streamflows in the Upper Colorado River Basin", in: Walter O. Spofford, Jr., Alfred L. Parker and Allen V. Kneese (eds.), *Energy Development in the Southwest*, RFF Research Paper R-18 (Resources for the Future, Washington) pp. 351–429.

Spofford, Jr., Walter O., Alfred L. Parker and Allen V. Kneese (eds.) (1980) *Energy Development in the Southwest*, RFF Research Paper R-18 (Resources for the Future, Washington).

Spofford, Jr., Walter O., Clifford S. Russell and Robert A. Kelly (1975) "Operational Problems in Large Scale Residuals Management Models", in: Edwin S. Mills (ed.), *Economic Analysis of Environmental Problems*, National Bureau of Economic Research (Columbia University Press, New York) pp. 171–238.

Spofford, Jr., Walter O., Clifford S. Russell and Robert A. Kelly (1976) *Environmental Quality Management: An Application to the Lower Delaware Valley*, RFF Research Paper R-1 (Resources for the Future, Washington).

Spofford, Jr., Walter O., Clifford S. Russell and Robert A. Kelly (1977) "The Lower Delaware Valley Integrated Residuals Management Model: A Summary", in: Blair T. Bower (ed.), *Regional Residuals Environmental Quality Management Modeling* RFF Research Paper R-7 (Resources for the Future, Washington) pp. 52–106.

Stern, Arthur C. (1976) *Air Pollution*, vol. III (Academic Press, New York).

Streeter, H.W. and E.B. Phelps (1925) "A Study of the Pollution and Natural Purification of the Ohio River—III: Factors Concerned in the Phenomena of Oxidation and Reaeration", Public Health Bulletin no. 146.

Tomany, James P. (1975) *Air Pollution: The Emissions, the Regulations and the Controls* (American Elsevier, New York).

Turner, R.K. (1978) "Water Pollution", in: David W. Pearce (ed.), *The Valuation of Social Cost* (Allen and Unwin, London) pp. 97–119.

U.S. Council on Environmental Quality (1979) *Environmental Quality: Tenth Annual Report* (Council on Environmental Quality, Washington).

U.S. Environmental Protection Agency (1973) "Compilation of Air Pollutant Emission Factors", Publication no. AP-42 (Environmental Protection Agency, Washington).

U.S. Environmental Protection Agency (1979) *Energy from the West: Impact Analysis Report* (U.S. Environmental Protection Agency, Washington).

U.S. Nuclear Regulatory Commission (1976) *Estimating Aquatic Dispersion of Effluents from Accidental and Routine Reactor Releases for the Purpose of Implementing Appendix I*, Regulatory Guide 1.113 (U.S. Regulatory Commission, Washington).

U.S. Nuclear Regulatory Commission (1977) *Regulatory Guide* 1.111: *Methods for Estimating Atmospheric Transport and Dispersion of Gaseous Effluents in Routine Releases from Light-Water Cooled Reactors* (U.S. Nuclear Regulatory Commission, Washington).

U.S. Office of Technology Assessment (1979) *The Direct Use of Coal* (Government Printing Office, Washington).

Victor, Peter A. (1972) *Pollution: Economy and Environment* (Allen and Unwin, London).

Wang, Mu Hao, Lawrence K. Wang, Jao-Fuan Kao, Ching-Gung Wen and David Vielkind (1979) "Computer-aided Stream Pollution Control and Management", Part I, *Journal of Environmental Management* 9, 165–183.

Wang, Lawrence K., Mu Hao Wang, Jao-Fuan Kao and Ching-Gung Wen (1979) "Computer-aided Stream Pollution Control and Management", Part II, *Journal of Environmental Management* 9, 185–204.

Wen, Ching-Gun and Jao-Faun Kao (1980) "Mathematical Modeling of Stream Water Quality by a New Moment Method: Theoretical Development", *Journal of Environmental Management* 10, 1–11.

Wen, Ching-Gun and Jao-Fuan Kao (1982) "Determination of Sensitivity of Water Quality Parameters for Stream Pollution Control", *Journal of Environmental Management* 74, 17–34.

232-43
[87]

JOURNAL OF ENVIRONMENTAL ECONOMICS AND MANAGEMENT **14**, 1–12 (1987)

The Production Process: Inputs and Wastes[1]

CURT L. ANDERSON

Department of Economics, University of Minnesota–Duluth, Duluth, Minnesota 55812

0223
7220
U.S.

Received July 5, 1982; revised December 5, 1983

A simple model of the production process is developed which blends the traditional theory of productive services with the concepts of material processing, energy use, and waste generation. Taxonomies for inputs by function and wastes by origin are offered which more clearly define the nature of substitution possibilities between inputs and illustrate trade-offs which exist between different types of production wastes. Implications of the model for production function specification, economic growth, econometric estimation, and emissions control policy are discussed. © 1987 Academic Press, Inc.

Q20
D24

This paper develops a production model which blends the traditional theory of productive services with the concepts of material processsing, energy use, and waste generation. Previous work in this area [1, 2, 4, 5, 7, 8, 11, 13, 17] has provided excellent treatments of the technical aspects of various parts of this blending. This paper attempts to draw these together to produce a simple, stylized representation of the role of all resources (particularly natural resources) in the production process. Section 1 provides a description of the nature of production. Sections 2 and 3 formalize input and waste relationships for increasingly realistic assumptions regarding waste generation. Finally, Section 4 considers some of the more important applications and implications of the production model presented, while Section 5 summarizes.

1. A DESCRIPTION OF THE PRODUCTION PROCESS[2]

For many productive processes it appears that basically three types of inputs, distinguished by the roles they play in the production process, may be identified: agent, material, and energy.[3] The agent inputs are the actual performers of the production, or the "actors" in the process. They are funds which provide productive services.[4] The material inputs are those inputs upon which the agent inputs act.

[1] The author wishes to acknowledge invaluable comments from the editor and numerous anonymous referees. Any remaining oversights or errors are merely violations of the assumption that the author possesses perfect information.

[2] The majority of the paper considers production in a microcontext but most of what follows may be applied to aggregate production as well.

[3] This terminology was first suggested by Georgescu-Roegen [7]. In addition he identifies a fourth type of input—maintenance inputs. These are input flows required to maintain the agent inputs at a given productive level. This might include, for example, crankcase oil for an engine, chemical solvents, or gloves for a laborer. This type of input will not be considered here. Also not considered explicitly is yet a fifth type of input—land. Land may be seen as the input providing "space" for production to occur.

[4] The distinction between a "stock" and a "fund" is made clearly by Georgescu-Roegen [8, Chap. 5.] The distinction is important in what follows in that use (depletion) of a fund (or more precisely, its services) requires time while a stock may be used (depleted) instantaneously.

0095-0696/87 $3.00
Copyright © 1987 by Academic Press, Inc.
All rights of reproduction in any form reserved

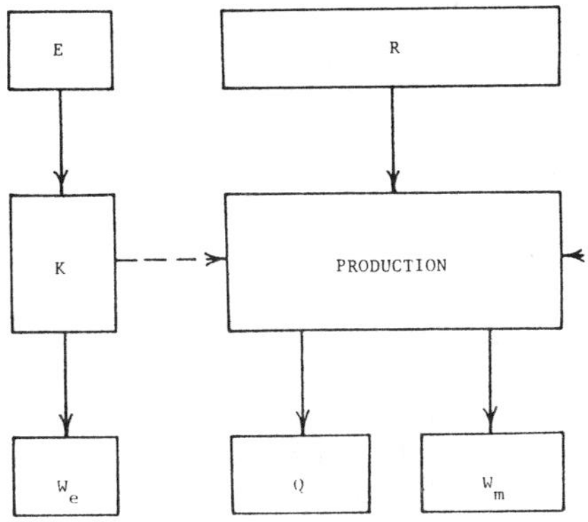

FIG. 1. A schematic of the production process. K = fund of capital; L = fund of labor; E = stock of energy outputs; R = stock of material inputs; Q = stock of output; W_e = stock of wastes from the energy inputs; W_m = stock of wastes from the material inputs.

They are transformed from their original states by the agent inputs into output and possibly waste flows. Finally, the energy inputs are those inputs required to enable the agent inputs to act. Since the agent inputs perform work in the production process a conversion of energy is required.

Given these rough distinctions, a typical production process may be viewed (as is shown in Fig. 1) as a fund of agent inputs (K and L), acquiring their ability to act from energy inputs (E), whose productive services (represented by the dashed lines) transform some stock of material inputs (R) into a stock of material outputs (Q and W_m). Material flows are illustrated with the solid lines.

Several points about this representation are noteworthy. Three basic flows of materials out of the process may be identified: the output flow (Q), waste from the incomplete use of the material inputs (W_m), and the flow of materials from the energy conversion component of the process (W_e). Given the necessity of materials balance, R must be equal to $Q + W_m$ and E must equal W_e when all are measured in mass units and no accumulation within the process is assumed.[5] Besides materials balance the first law of thermodynamics (neglecting nuclear processes) also requires that there be energy balance throughout the process (not shown in Fig. 1). In particular, the total energy embodied in the energy inputs must equal the work energy expended by the agents plus any energy which may become embodied in the material flows leaving the process plus any "waste" energy released (usually in the form of heat). Thus, as is well established (see especially [1, 2, 4, 11]), there are mass and energy constraints on the production process.

[5] Note that W_e is *not* energy waste (i.e., available energy of the energy inputs not used in the production process—see footnote 8) but instead a flow of materials. Measuring the energy inputs (coal, gas, oil, etc.) in mass terms is important in determining the mass of materials (steam, hydrocarbons, sulfur oxides, particulates, etc.) leaving the process after energy conversion. While the *composition* of these "wastes" may be changed with improved efficiency, the *mass* of the materials generated from burning, for example, one ton of coal remains one ton.

Further, the amount of material inputs a given fund of agent inputs is capable of processing clearly depends on how long it is allowed to act and the efficiency of the agent inputs in converting the energy inputs into work. So, time and thermodynamic considerations will also act as constraints on production [5, 7, 8, 17].

Finally, note in Fig. 1 that while capital is shown to have a servicing energy input, none is shown for labor. This is to emphasize that the energy input for labor, while required, is supplied externally to the firm by the laborer and further that the purchase of labor services is in reality the simultaneous purchase of an agent input *and* its required energy input. In any case, to use an agent input for an hour necessarily requires an energy input flow sufficient to run it for an hour. This suggests that a complementary relationship likely exists between these input classes.[6]

The production process, thus, is essentially made up of two relationships: one describing the material requirements of the output and another describing the ability of the agents, with sufficient energy inputs, to act upon these materials to produce the output desired. The material inputs may be seen as "supplying" mass, while the agent inputs "supply" productive services. Viewed in this manner, the direct substitution between these two classes of inputs appears not to be possible.

However, there is yet a third relationship that must be considered. This is the relationship that exists between the amount of agent inputs used and the degree to which the material inputs exit the production process as waste rather than output. Large amounts of material wastes in the form of unwanted by-products are generated by many production processes, most noticeably perhaps in the paper and mining industries. Other material wastes may result from the incomplete utilization of useable material inputs. For example, every possible ounce of sugar is not likely to be extracted from a load of raw sugar beets.

In this latter case it appears reasonable to argue that as more agent inputs are added, better utilization of a given amount of material inputs is possible (further processing could extract more of the previously unextracted sugar).[7] This implies less material inputs would be required to produce a given output (up to the limit imposed by mass balance). Thus, some substitution possibilities are likely to exist between material and agent inputs.

2. PRODUCTION WHEN $W_m = 0$

Suppose that the output of the firm has already been decided upon in regard to its technical specifications and material requirements. Second, assume that there is but one agent input (with its necessary energy inputs) and one material input, K and R, respectively, where K is the total quantity of agent inputs and R represents the total mass of the material input. Finally, suppose $W_m = 0$.

The above assumptions yield the production relationship

$$Q' = R/m, \tag{1}$$

[6] This relationship is even more apparent when one considers that many agent inputs are energy-input specific—a diesel truck requires diesel fuel, a power saw requires electricity, a gas furnace requires natural gas, and a hammer requires food energy (here the agent is best viewed as a carpenter-with-a-hammer).

[7] In a recent Carnegie-Mellon University Study, "Impacts of Robotics on the Workforce and Workplace," "reduced materials waste" was listed as one of the principle factors influencing firms to decide to install robots (agent inputs).

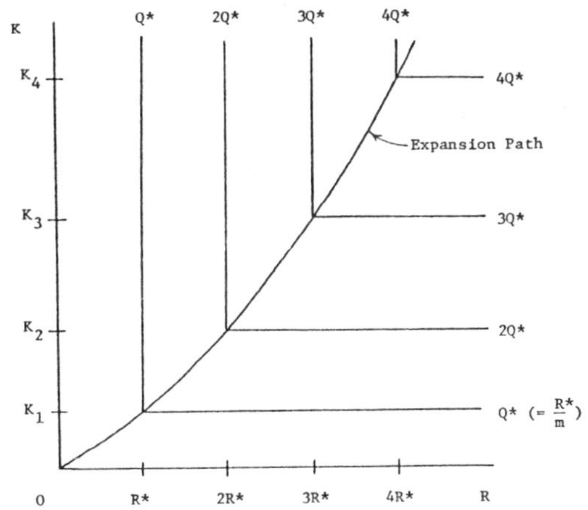

FIG. 2. Production without material waste.

where Q' is the maximum amount of output (measured in units of output) that could be produced given R units of the material input (measured in mass terms) and a mass per unit of output of m. This clearly is just the materials balance condition in the absence of material input wastes.

The relationship between the agent inputs and output must take explicit recognition of time and the thermodynamic efficiency of the agent input in converting the available energy of the energy inputs into work. To do this, assume the production period is of length T^* and e^* $(0 < e^* < 1)$ is the coefficient of thermodynamic efficiency of K[8]. Now consider the question of how much of the material input a given amount of agent inputs could process in this time. Let this be determined by

$$Q'' = f(K; T^*, e^*), \tag{2}$$

where K is the quantity of agent inputs, and Q'' is the maximum number of units of the output that could be produced. Letting $f_K > 0$ and $f_{KK} < 0$, (2) merely represents traditional productivity relationships.[9]

[8]Let e represent the "second law" efficiency measure for energy conversion given by e = available work of outputs/available work of inputs [1, 12, 15]. The second law of thermodynamics, the entropy law, precludes the possibility that $e = 1$, essentially stating that each time energy is transferred or transformed from one form or system to another, its quality (our ability to extract work from it) is degraded. While the maximum allowable efficiency for a particular energy transfer process ($e = \bar{e} < 1$) is determined by the second law of thermodynamics, the actual working efficiency ($e = e^*$) will likely be below this limit ($e^* < \bar{e}$) since higher efficiency levels (approaching $\bar{e}$) would in general require greater flows of maintenance inputs (footnote 1). Thus, the firm will likely trade some energy efficiency for lower maintenance costs. Finally note that even if the firm producing where $e = \bar{e}$, a loss of available work would still occur.

[9]At the microlevel in the short run some factors, most often land and/or capital, are fixed; thus diminishing marginal productivity is likely (and generally assumed) for any variable agent inputs. Viewed as an aggregate production model for an economy as a whole, land plays the implicit fixed factor leading again to diminishing marginal productivity for the agent inputs.

Clearly, (1) and (2) together imply

$$Q = \min(Q', Q''), \tag{3}$$

where Q is the actual output that can be produced given different values of K, R, T^*, and e^*. This relationship is summarized in Fig. 2.

The right-angle nature of the isoquants indicates the expected lack of substitution possibilities between the material input and the agent input. Further, there are constant returns to scale with respect to the material input but not with respect to all inputs given the assumptions above about f.[10] Finally, although not illustrated in Fig. 2, it is important to recognize that associated with each higher level of K is use of greater amounts of some energy input and the resulting greater flows of W_e out of the process.

3. PRODUCTION WHEN $W_m > 0$

Often in production processes, the firm cannot obtain material inputs with the exact mass characteristics desired. For example, in the making of potato chips, the principal material input is potatoes. However, attached to these potatoes are "skins," which are usually not desired and are peeled off early in the production process. They are that part of the material input which is not useable. Similarly, in the production of aluminum from bauxite ore, the ore is the material input but only a fraction of it is actually useable in the production of aluminum. These examples suggest that one source of W_m in a production process is the unuseable part of the material inputs (the eventual by-product). Let the term "useable" be loosely defined as "able to physically become a part of the produced output." If it is assumed that roughly a constant fraction of any material resource is useable in a particular production process, then the existence of unuseable portions of the material input requires that (1) be rewritten as

$$Q' = vR/m, \tag{4}$$

where v is the proportion of the mass of the material input which is useable, $0 \leq v \leq 1$.

This suggests waste production of $(1 - v)R$, or waste generation proportional to the amount of material inputs used. This is noteworthy in that this type of assumption is prevalent in the literature of waste accumulation models (e.g., Plourde [14]). This component of W_m will be referred to as "by-product" waste.

The effect of by-product waste on the isoquant mapping of Fig. 2 is fairly straightforward. The mass balance constraint implies that given R^*, Q' equals vR^*/m instead of R^*/m. If v is less than one, then this implies that less output is possible from the same amount of material input as before.

[10] Georgescu-Roegen [7] notes that production from agent inputs is subject to CRS with respect to changes in T^* but not necessarily with respect to changes in the quantity of the agents.

It is also likely that

$$Q'' = f(K; T^*, e^*, v) \tag{5}$$

since the value of v is likely to influence the amount of agent inputs required to produce a given output. For example, if a mining company wished to produce one ton of aluminum, it would have to process four tons of 25% pure ore or only two tons of 50% pure ore. In the former case, it must process twice as much ore, which, given production periods of equal length, and no change in thermodynamic efficiency, would require additional agent inputs. Hence, it appears reasonable to assume that $f_v > 0$.

The exact relationship between v and the amount of K required, however, may be difficult to specify in that f is likely to be structurally dependent on v (that is, v may not merely act as a shift parameter but could alter marginal relationships). For example, if v is less than one, extra steps in the production process will likely be required simply to separate the useable from the unuseable material input. This additional processing is likely to be of a different nature than the actual production of the output, so the production process viewed as a whole has been changed.

Thus, assuming 100% separation, the isoquants appear much as they did in Fig. 2 except that with $v < 1$, the minimum requirements of material *and* agent inputs to produce a given level of Q will be greater than before.

The amount of by-product waste is not dependent on how the output is to be produced but solely on the mass specifications of the output and physical characteristics of the material input. Thus, the only ways to reduce this type of waste flow (other than reducing output) are to change the nature of the output or purchase a material input more closely matching what is desired.[11]

As discussed earlier, another potential source of W_m comes from the incomplete utilization of the useable portion of the material input (usually because total separation from the by-product and/or complete utilization would be prohibitably expensive). This component of W_m will be called economic waste. The extent of this waste depends on the production technique chosen. It appears reasonable to assume that in the production of a given level of output the marginal rate of technical substitution of an agent input for a material input diminishes. In other words, while additional units of the agent input would likely lead to a more complete separation of the useable portion of the material input and/or better utilization of already separated useable portions, each additional unit reduces the amount of material input required by less and less (and only up to the constraint imposed by materials balance).

Figure 3 represents the nature of the production relationship when both by-product and economic wastes are present. Note that these isoquants do not approach *either* axis in the limit but instead are still constrained on the one hand by the materials balance and on the other by time and thermodynamic efficiency. They

[11]Some potato chip manufacturers are now marketing "old-fashion" style potato chips, wherein the skin is left on. This reduces their by-product waste and its associated disposal problems and allows better utilization of their material inputs. This points out how the choice of the product may affect and be affected by waste considerations.

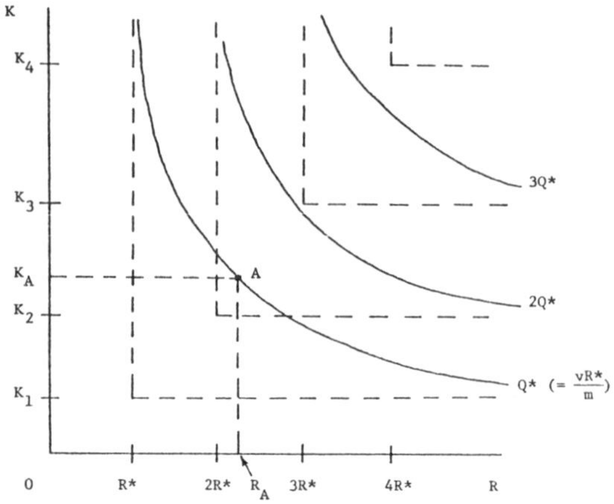

FIG. 3. Production with material waste.

illustrate the very simple and often overlooked proposition that the production of a given amount of a material output requires at least a certain amount of material inputs and (*not* and/or) a certain amount of productive work (provided by the agent inputs with their associated energy inputs).

It should be clear from the discussion above that as the level of output increases so do the minimum requirements of each the material, agent, and energy inputs. In fact, the minimum requirement of the material input rises proportionately with the output level. Thus, each isoquant approaches asymptotically different and higher levels of the material and agent inputs. So, while the isoquants resemble those of traditional production economics, it is important to recognize that there are constraints on their shape and position relative to each other due to materials balance, time, and thermodynamic considerations.

4. IMPLICATIONS OF THE MODEL

a. An Example

Since by physical necessity any production process bearing a material output must possess the properties described above, it is useful to ask what production function specifications satisfy these requirements. To a certain extent, this question has been addressed by Ayres [1], Dasgupta and Heal [5], and Winston [17]. Unfortunately, only some of the constraints are considered by each. Dasgupta and Heal deal only with the thermodynamic constraint placed on production. Ayres deals with this only indirectly and concentrates primarily on the materials balance constraint while ignoring the time constraint. Winston provides a most encompassing account of the role of time but ignores the other physical constraints.

It is clear that linear and Cobb–Douglas specifications do not have the necessary asymptotic properties. Fixed proportion production functions and activity–analysis production functions have these properties; however, the former implicitly rules out the possibility of economic waste while both ignore the effect diminishing marginal productivity may have on the relationship between isoquants. Dasgupta and Heal suggest that CES production functions with an elasticity of substitution less than one have the required properties.

To investigate this proposition, consider the class of CES production functions represented by

$$Q = A\{aK^{(b-1)/b} + (1-a)R^{(b-1)/b}\}^{b/(b-1)}, \tag{6}$$

where b is the elasticity of substitution coefficient and $b > 0$, $A > 0$, $0 < a < 1$. Rearranging, this becomes

$$A(1-a)R^{(b-1)/b} = Q^{(b-1)/b} - AaK^{(b-1)/b}. \tag{7}$$

Consider now a movement along the Q^*-isoquant. If $b < 1$, then $(b-1)/b < 0$. As K tends toward infinity, this implies that the term in K vanishes and the limiting value of R becomes

$$\lim_{K \to \infty} R = Q^*/(A - Aa)^{b/(b-1)} = nQ^*, \tag{8}$$

where n is a constant equal to $1/(A - Aa)^{b/(b-1)}$. Clearly, this does satisfy the mass balance constraint in that the isoquant approaches a positive value of R asymptotically; and further, as Q increases the value of R approached asymptotically increases proportionately.[12] In fact, it is easy to confirm that the isoquants of a CES production function with $b < 1$ also approach an asymptotic value of K in a similar manner. Dasgupta and Heal did not explicitly acknowledge that these asymptotic properties were required with respect to *both* arguments, so that these particular CES functions are even better representations than they suggest.

However, given diminishing marginal productivity it is not the case that the K-asymptotes would expand proportionately with changes in Q. A general class of production functions which exhibits this type of behavior may be given as

$$Q = A\{aK^{z(b-1)/b} + (1-a)R^{(b-1)/b}\}^{b/(b-1)}, \tag{9}$$

where A, a, and b are defined as before and $b < 1$, while $0 < z < 1$.

The R-asymptote for the Q^*-isoquant of this function is shown earlier in (8). The K-asymptote, however, is given by

$$\lim_{R \to \infty} K = Q^{*(1/z)}\left(1/Aa^{b/(zb-z)}\right) = n'Q^{*(1/z)}. \tag{10}$$

[12] Dasgupta and Heal present a more generalized CES production function in which they express the limiting value of R as $R(Q^*)$. While they demonstrate that indeed there exist positive R-asymptotes, it is *not* the case for their generalized function that $R(2Q^*) = 2R(Q^*)$, which is a necessary property implied by materials balance.

Now as Q increases, the value of the relevant K-asymptote increases at a greater than proportionate rate (since $0 < z < 1$).[13]

b. Long-Run Growth Models

The results of models which address the possibility of maintaining aggregate consumption forever or perhaps even having it grow in the face of finite resource stocks depend heavily on the properties of the production function at the extreme values of its arguments (Dasgupta and Heal [6] and Solow [16] are noteworthy examples). Considering the model presented here as a representation of an aggregate production function it is clear that with a fixed state of technology (and so fixed e^*'s for all agent inputs) and no change in the nature of the material output, there are definite minimum requirements of K (and so, E) and R to produce a given output level (i.e., the natural resource inputs are "essential" in that $g(K, E, 0) = g(K, 0, R) = 0$). This can lead to nothing but pessimistic growth implications à la Dasgupta and Heal.

Optimistic growth possibilities are generated primarily only when K is seen as being "infinitely substitutable" for R and E. For the former substitution (K for R) this assumption clearly violates materials balance given that the nature of the material output is fixed. For the latter substitution (K for E) this assumption violates the complementary relationship that must exist between an agent input and its associated energy input (assuming e^* remains constant and the type of agent input is unchanged). *Agent inputs are seldom substituted for energy inputs but instead one type of agent input with its accompanying energy input is substituted for another type of agent input with its accompanying energy input.* While such substitutions may indeed lower energy requirements by changing to agent inputs with higher thermodynamic efficiencies (i.e., $\bar{e}$'s closer to unity), it is clear that the transformation of a given amount of material inputs into output flows still ultimately requires some minimum amount of work so that even substitutions of this type do not alleviate the necessity of E.

Finally, note that not even technological growth will save us from this pessimistic scenario but will merely delay the inevitable. If the nature of the material output is fixed there will always be minimum requirements of R and even if the $\bar{e}$ of all agent inputs used approach unity there will still be a minimum requirement of E associated with each output level. Sustained growth thus depends more on the extent to which these minimum R and E flows can be supplied by renewable and/or nondepletable natural resources.

c. Empirical Estimation

Econometric studies at the microlevel frequently indicate that Cobb–Douglas and CES specifications, as well as the more general translog specification of the production process, fit actual data quite well even though they violate the con-

[13]Although E, the quantity of energy inputs, does not appear explicitly in this specification, it should be evident that given a fixed e^* for K and fixed T^*, there is a unique amount of E associated with each level of K which expands proportionately with K.

TABLE I

Partial Elasticities of Substitution among Capital (K), Labor (L), Energy (E), and Other Intermediate Materials (R): U.S. Manufacturing 1947–1971

σ_{KE}	σ_{LE}	σ_{KR}	σ_{LR}	σ_{ER}
− 3.2	0.64	0.55	0.57	0.76

Source: Berndt and Wood [2], Table 4.

straints described in this paper.[14] Since these functions are typically estimated from nonextreme observations where these constraints are not binding this is not surprising and so the specification offered here has no real advantage.

The taxonomy of inputs and wastes presented, however, may be most useful in addressing aggregation questions. For example, it points out the difficulty of interpreting elasticity of substitution coefficients when all natural resource inputs are aggregated together while in reality some are energy inputs, others material inputs, and still others maintenance inputs (see footnote 1). Also of note is the implicit aggregation in most studies of labor and its energy input (i.e., food) together when this is not done for other agent inputs like capital.

In one of the few studies which break down inputs into the classes discussed in this paper, Berndt and Wood [3] considered substitution possibilities in the U.S. manufacturing sector. Average values for the partial elasticities of substitution are shown in Table I. Howe [10] comments that these "estimates seem plausible with the exception of the high degree of complementarity indicated for capital and energy." This complementarity, however, is the expected result given the earlier discussion and supports the formulation of the production process presented here. Further, while the value of σ_{LE} appears to contradict the complementarity expected between an agent and energy input, it is merely reflecting the fact that purchases of L are in reality purchases of the agent labor *and* its required energy input simultaneously. Thus, σ_{LE} is in reality a measure of the ease of substitution between *an agent and its associated energy input* and some other energy input. Given this it is not surprising that a degree of substitutability is evidenced. σ_{KR} and σ_{LR} reflect the expected substitution possibilities due to economic wastes. Finally, σ_{ER} is not suggesting that energy inputs are directly substitutable for material inputs (or vice versa) but instead that an increased use of E is associated with increased use of K which again may be substituted for R to lower economic wastes.

d. The True Cost of an Agent Input

Because agent inputs (particularly capital) are generally energy-input specific, the choice of an agent is actually the choice of an energy–agent combination. Thus, the real cost of K may be given by $w + ps$, where w is the price of the agent's services, p is the price per mass unit of its required energy input, and s is the amount of this energy input (measured in mass terms) required to keep one unit of K "working" for a specified amount of time at a given thermal efficiency of e^*. Thus, the demand for energy inputs depends directly upon the demand for the agents which utilize it

[14]See, for example, the recent work of Kopp and Smith [9].

while the demand for an agent input is clearly influenced by the price of its associated energy input. In particular, this implies that high energy prices induce firms not only to seek more thermally efficient agent–energy combinations but also to consider more labor inputs since their associated energy inputs are free (to the extent that food costs are not reflected in wage rates).

e. Optimal Waste Generation

Given the curvature assumptions with respect to f and positive prices for K, R, and E, optimality will generally imply the generation of some W_m and W_e.[15] Clearly, as one moves along any isoquant in Fig. 3 toward its vertical asymptote, less W_m and more W_e are generated and vice versa. To calculate the waste generated suppose the firm cost-minimizes by producing at the combination depicted by point A in Fig. 3, where it uses K_A units of agent input and R_A units of the material input. Since Q^* units of the output are produced and this physically requires only vR^* amount of material mass, the total amount of W_m generated must be $R_A - vR^*$. Of this, the by-product waste is given by $(1 - v)R_A$, while economic waste is given by $v(R_A - R^*)$. If s is as defined above the amount of W_e is sK since the total mass of energy inputs exit as waste.

f. Emissions Control Policy

It must be recognized that W_e is an inevitable flow from any production process so that policy directed at the elimination of this flow is doomed to failure. Incentives are instead required to induce firms to use agent–energy combinations which have high thermal efficiencies (thus, lowering the required amount of E and reducing unnecessary flows of W_e) and/or result in flows of a less harmful nature (i.e., more steam, less sulfur oxides). Second, it must be recognized that for by-product wastes waste reduction can only be achieved by a reduction of the output produced (assuming the nature of the output and material input are to remain unchanged). Finally, while policies directed at reducing economic waste may be effective in doing so they are likely to result in increased flows of W_e if output levels are to be maintained. Thus, the proper choice of environmental policy tool depends on an identification of the type of waste flow which is to be managed and must consider the possible ramifications of the policy on other waste flows generated in the process.

5. SUMMARY

Natural resources enter the production process through several avenues the principle being as material and energy inputs. Previous production models have separately considered the materials balance constraint that material inputs place on production, the thermodynamic constraint energy inputs place on production, and

[15] In addition, wastes from the use of maintenance inputs (see footnote 1) would be generated. Although not emphasized in this paper, this may be the primary source of most so-called hazardous wastes (waste oil, chemical solvents, greases, etc.). Also note that, like W_e, these wastes tend to increase proportionately with increased use of K.

the constraint time places on production. The model presented here considers all these constraints simultaneously while further dealing with the flows of the natural resources out of the process as output and waste. It is intended to serve as a general framework from which more technical, more process-specific representations may be developed.

A functional taxonomy of the principal input classes—agent, material, and energy—clarifies the roles played by various inputs in the production process and highlights substitution linkages and limitations. Pessimistic implications for growth given finite resource stocks result. Further, the taxonomy serves as a guideline for aggregating inputs in empirical studies, as well as for interpreting elasticity of substitution coefficients.

A taxonomy of wastes by their source of origin—by-product, economic, energy, and maintenance—clarifies waste trade-offs that exist within the production process and reemphasizes the need for a general equilibrium approach to environmental policy.

Only when there is a more complete understanding of the roles played by natural resources in the production process and the physical laws which govern them can we even begin to believe we can "manage the environment."

REFERENCES

1. R. U. Ayres, "Resources, Environment, and Economics," Wiley, New York, (1978).
2. R. U. Ayres and A. V. Kneese, Production, consumption and externalities, *Amer. Econ. Rev.* 282–297 (June 1969).
3. E. R. Berndt and D. O. Wood, Technology, prices, and the derived demand for energy, *Rev. Econ. Statist.* **52**, No. 3, 259–268 (August 1975).
4. R. Cummings and W. Schulze, Ramsey, resources and the conservation of mass energy, *J. Environ. Econ. Manage.*, in press.
5. P. S. Dasgupta and G. M. Heal, "Economic Theory and Exhaustible Resources," Cambridge Economics Handbook Series, Cambridge Univ. Press, London/New York (1980).
6. P. S. Dasgupta and G. M. Heal, The optimal depletion of exhaustible resources, in "Review of Economic Studies Symposium," pp. 3–28 (1974).
7. N. Georgescu-Roegen, The economics of production, Papers and Proceedings of the American Economics Association, *Amer. Econ. Rev.* (1969).
8. N. Georgescu-Roegen, "The Entropy Law and the Economic Process," Harvard Univ. Press, Cambridge, Mass. (1971).
9. R. J. Kopp and V. K. Smith, Measuring factor substitution with neoclassical models: An experimental evaluation, *Bell J. Econ.* (Autumn 1980).
10. C. W. Howe, "Natural Resource Economics," Wiley, New York (1979).
11. K.-G. Mäler, "Environmental Economics: A Theoretical Inquiry," Johns Hopkins Press for Resources for the Future, Inc., Baltimore (1974).
12. G. T. Miller, Jr., "Energy and the Environment: The Four Energy Crises," Wadsworth, Belmont, Calif. (1975).
13. E. S. Mills, "The Economics of Environmental Quality," Norton, New York (1978).
14. C. B. Plourde, A model of waste accumulation and disposal, *Canad. J. Econo.* **5**(1), 119–125 (February 1972).
15. M. Ross, R. Socolow *et al.*, "Effective Use of Energy: A Physics Perspective," Report of the Summer Study on Technical Aspects of Efficient Energy Utilization, Amer. Phys. Soc. New York, (January 1975).
16. R. M. Solow, Intergenerational equity and exhaustible resources, in "Review of Economic Studies: Symposium on the Economics of Exhaustible Resources," pp. 29–46 (1974).
17. G. C. Winston, "The Timing of Economic Activities," Cambridge Univ. Press, London/New York (1982).

Martin Jänicke, Harald Mönch and Manfred Binder*

Ecological Aspects of Structural Change

Structural change in advanced industrialized countries has followed a very ambivalent course, with the resulting reductions in environmental stress in one area often being cancelled out by increases in another. What role has been played in this to date by industrial and economic policy? What options are open to policy-makers for the future?

In environmental policy two fundamental strategies can be distinguished: a remedial "end-of-pipe" strategy that does not revise existing problematic technologies, and a preventive strategy[1] that alters production and consumption ex ante towards ecologically better adapted forms. A similar distinction can be drawn between "additive" and "integrated" environmental protection. The term preventive (or integrated) environmental protection policy as we understand it refers to all forms of environmental policy or ecologically motivated economic policy that envisage structural change of the economy. Technological progress in the sense of "ecological modernization" forms part and parcel of this concept of structural change. It is after all a matter of the notion, covered nowadays by a variety of terms, of the transition to "sustainable development" or "eco-restructuring". We refer to the policy pursuing such aims as ecological structural policy. Integrated environmental protection of this sort requires not only an integrated technical solution but also an integrated political solution in the sense of well-developed cooperation between policy fields.

We call the subject of ecological structural policy "structural environmental stress". The extent of structural environmental stress does not depend on end-of-pipe environmental protection measures. Every economic act causes environmental stress in so far as, without additional cleaning technology, it would cause actual environmental damage. As a rule, however, structural environmental stress goes beyond this, characteristically also involving considerable residual pollution, even where remedial environmental protection measures are taken. In the case of a coal-fired power station, for example, end-of-pipe cleaning technologies do nothing to eliminate subsequent environmental stress (partly even caused by cleaning measures) from:

□ CO_2 emissions,

□ residual emissions, which can increase again as a result of growth processes,

□ extremely high water consumption,

□ heavy transport commitment (fuels, limestone, gypsum and other waste),

□ high waste production,

□ waste water production from cleaning plants,

□ detrimental impact on the landscape from coal extraction, limestone quarrying or power lines,

□ considerable consumption of electricity by cleaning plants,

□ the consumption of resources.

Nuclear power generation, even with a high level of environmental protection and safety, is also a large consumer of water and involves considerable risk in the operation of power stations and the transport and disposal of radioactive materials.

Existing pollution is, furthermore, often simply shifted from one region or medium to another by end-of-pipe

* Free University, Berlin, Germany.

[1] Cf. Jürgen Gerau: Zur politischen Ökologie der Industrialisierung des Umweltschutzes, in: Martin Jänicke (ed.): Umweltpolitik, Beiträge zur Politologie des Umweltschutzes, Opladen 1978, pp. 114-149; Martin Jänicke: Wie das Industriesystem von seinen Mißständen profitiert, Opladen 1979; Udo Ernst Simonis (ed.): Präventive Umweltpolitik, Frankfurt am Main, New York 1989.

environmental protection measures. Moreover, remedial environmental protection presupposes goods production that is detrimental to the environment.

Environmental protection, especially in the long term, will therefore necessarily be conceived in terms of structural policy. This is inevitable, not least of all because residual problems can rapidly return to their initial levels due to growth processes. This may also be formulated as an ecological/economic imperative: in a growing economy the ecologically detrimental impact of growth must constantly be compensated. Initially, remedial environmental technology appears appropriate. When this possibility has been exhausted, stress-reducing technological and structural change become essential.

At this point it is also decisive whether environmental policy overcomes its status as mere additive policy, to be integrated into other policy fields. At this stage the point of gravity in environmental policy thus shifts to other policy areas: energy policy, transport or industrial policy, farming policy etc.

This paper deals with industrial policy[2] as ecological structural policy. This is defined in the first place as the sum of all measures designed to reduce structural environmental stress by industrial production.

Environmental protection as sectoral structural policy may seek structural change in the sense of a change in the relative status of individual branches (intersectoral structural change). The change may be tackled politically by means of ecologically motivated shut-downs, branch agreements, environmental charges and so on. It may, however, also be the result of changing demands of all types, whether as a consequence of changes in values or of modernisation processes among industrial customers. It may arise because investors switch to other branches or relocate production (which amounts to shifting ecological problems). The shift of freight transportation from road haulage to rail transport can also be regarded as sectoral structural policy.

Ecological structural policy may, however, also envisage intrasectoral change within branches. In this case it is generally a matter of technological innovation in the sense of "ecological modernisation".[3] Ecological modernisation as a programme recognises that, in view of the problem of long-term destruction of the environment, technological progress should not be retarded but rather accelerated in order to achieve the switch to ecologically more appropriate production and products. Ecological modernisation involves both innovation in production processes and innovation in products. For example, changes in products and processes under the heading of "soft chemistry" have been discussed and implemented. In the field of electricity supplies, there have been relative reductions in environmental stress due to high efficiency power stations and combined heat and power production, and there has been transition to cleaner energy sources (water, gas, solar energy etc.) In the building industry, the transition to ecological building materials and building forms (e.g. zero-energy buildings with low space and water needs) represent a similar possibility. Intrasectorally oriented industrial policy will generally be aimed at lowering the intensity (per unit of value added) of specific parameters of production[4], in particular:

- ☐ material intensity (especially that of non-renewable resources),
- ☐ energy intensity,
- ☐ water intensity,
- ☐ land use intensity,
- ☐ transport intensity,
- ☐ emission intensity,
- ☐ waste intensity and
- ☐ risk intensity.

In the case of intrasectoral structural improvement, the organizational framework and the social reality of the branch remain essentially unchanged. Change affects the form and content of production. For employment policy or regional policy reasons, it will thus frequently seem best to practise environmental structural policy initially in this innovation variant.

Much can be said in favour of inducing the process of ecologically beneficial intrasectoral change primarily by means of technological progress within sectors of the economy. On the one hand, ecological modernisation results in a high demand for tertiary pre-production activities (research, development, consultancy, qualification etc.), thus increasing the importance of services with little impact on the environment; on the other hand, production which is sparing in energy and materials

[2] Cf. Chalmers Johnson (ed.): The Industrial Policy Debate, San Francisco 1984.

[3] Cf. Martin Jänicke: Umweltpolitische Prävention als ökologische Modernisierung und Strukturpolitik, Wissenschaftszentrum Berlin für Sozialforschung, IIUG discussion papers 84-1; Klaus Zimmermann, Volkmar J. Hartje, Andreas Ryll: Ökologische Modernisierung der Produktion, Strukturen und Trends, Berlin 1990.

[4] Cf. Martin Jänicke, ibid.: Marina Fischer-Kowalski et al.: Verursacherbezogene Umweltindikatoren. Abstract, Vienna 1991.

Figure 1

International Change Profile of Cement Production 1970/90

(Level and change 1970/1990 calculated as linear trend)

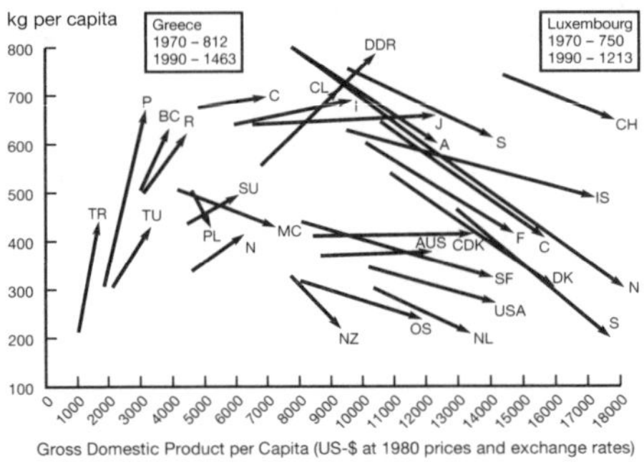

Figure 3

International Change Profile of Paper and Paperboard Production 1970/87

(Level and change 1970/1990 calculated as linear trend)

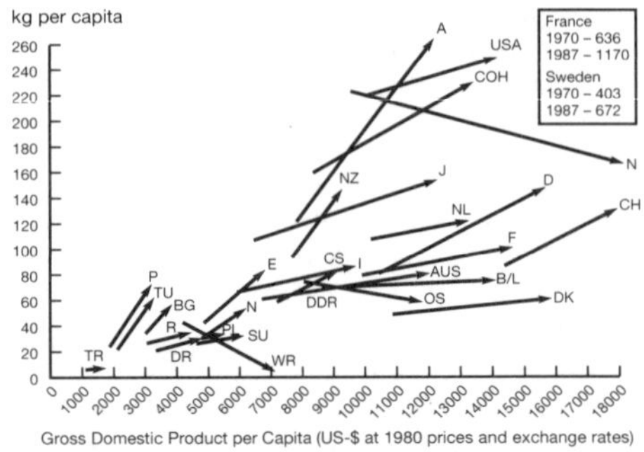

Figure 2

International Change Profile of Crude Steel Production 1970/90

(Level and change 1970/1990 calculated as linear trend)

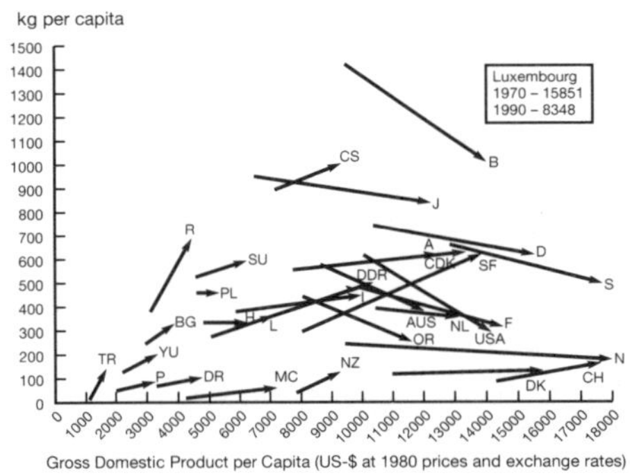

Figure 4

International Change Profile of Freight Movement 1970/88

(Level and change 1970/1990 calculated as linear trend)

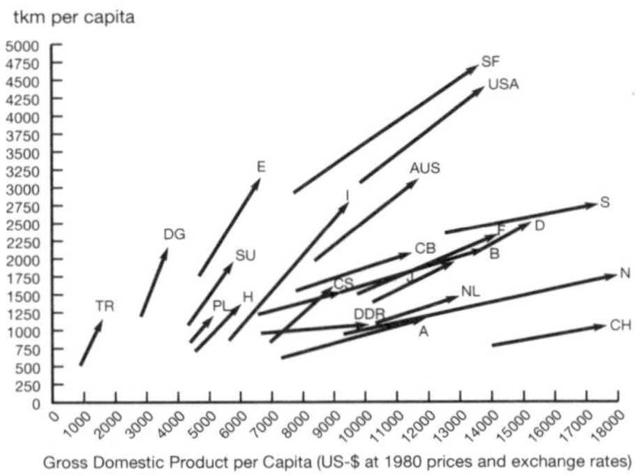

leads to an (at least relative) decline in heavy industries which cause environmental stress.

Moreover, modern service industries generally imply production which is intensive in know-how and services. Sectoral structural change–in the form of "tertiarization"–may thus be ecologically significant. This is also true for the encouragement of corresponding shifts in demand from extensive goods consumption to intangible "goods" or services.

The overall impact of shifts between branches (as intersectoral change) may be to reduce environmental stress, but they may also disguise problem relocation in the form of adjustments in the division of labour in the world market. Technological change as ecological modernisation, by contrast, tackles the problem directly. It improves the environmental situation not least by avoiding problematic substitutions or the relocation of problematic sectors of production.

Changes in Environmental Stress

How has structural environmental stress as we have defined it changed in the industrialized countries since 1970? The following treatment of this topic is based on an international comparative research project[5] at the Free University of Berlin sponsored by the Volkswagen Foundation. A total of 32 industrialized countries were

[5] Cf. Martin Jänicke, Harald Mönch, Manfred Binder et al.: Umweltentlastung durch industriellen Strukturwandel? Eine explorative Studie über 32 Industrieländer (1970-90), Berlin 1992.

examined for the period 1970 to 1990 with regard to seven production sectors evidencing a particularly high degree of structural environmental stress: crude steel, primary aluminium, cement, chlorine, pesticides, fertilizers, and paper and paperboard. These branches are representatives of the basic industries most detrimental to the environment. To these were added two branches from outside the manufacturing sector: electricity generation and freight transport, which can also serve as background variables for structural environmental stress (another important background variable dealt with is energy requirement). Selection of the nine branches causing structural environmental stress was made not only on an objective basis, but was also determined for practical research reasons by the availability of time-series permitting an additional comparison with eastern European industrial countries.

It proved that advanced industrialized countries in traditional sectors such as cement, crude steel and fertilizers evidenced at least a relative reduction in environmental stress in the form of the delinkage of these production sectors from overall economic growth. A trend towards delinking in energy requirements and in the weight of freight moved by road and rail was also observed. The less advanced industrial countries in eastern Europe showed an absolute deterioration in the sectors mentioned at an already high level. The greatest degree of structural deterioration, albeit at a low initial level, was evidenced by southern European countries (cf. Figs. 1 and 2).

By contrast, structural environmental stress related to electricity generation and freight movement by road (measured in tonne-kilometres) is on average proportionate to economic growth in most of the advanced industrialized countries. No delinking has occurred in this respect, but rather a frequent increase at a high level. In a number of advanced industrialized countries this is also true for primary aluminium production, chlorine, paper and paperboard and (where figures are available) pesticides (cf. Figs. 3 and 4).

The significance of the reduction in environmental stress in traditional production sectors and the increase in more modern sectors is well illustrated by the example of the Federal Republic of Germany (cf. Fig. 5). The two peaks of the first and second oil crises (1973 and 1979) are particularly apparent.

Four aspects of this change are dealt with below in greater detail:

☐ the significance of the energy prices, particularly industrial electricity prices;

Figure 5

National Change Profile FRG 1960-1990

(1960 = 100)

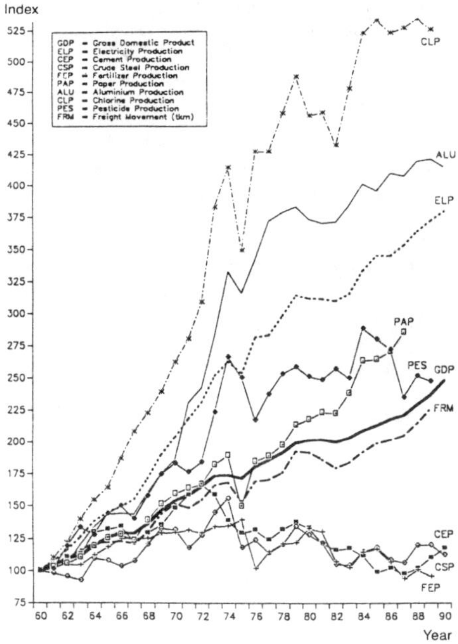

Figure 6

Electricity Price and Electricity Intensity 1989

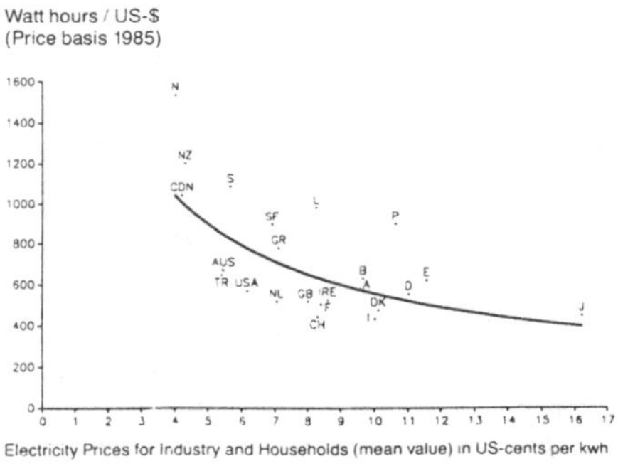

□ the contrast between technological capability and reality, especially in relation to energy and transport;

□ the considerable significance of technological change and the low significance of intersectoral change;

□ the apparently low significance of sectoral structural policy for ecologically beneficial development.

Significance of Energy Prices

In northern Europe (Scandinavia and Iceland), North America (USA and Canada) and Oceania (Australia and New Zealand) – countries with particularly low electricity prices – energy requirements developed on a high level with atypical dynamism. This was especially true for electricity consumption: where a country charged electricity prices twice those prevailing in a second country, the former generally evidenced an electricity intensity rate a third lower (cf. Fig. 6). These countries also have industries causing a particularly high degree of environmental stress (aluminium, chlorine, paper and paperboard) – a doubly negative structural impact. It is thus hardly by chance that Japan, the country with the highest industrial electricity prices, also shows the highest degree of structural change.

Especially with regard to the discussion on the ecological impact of energy taxes, however, the massive price differentials necessary to obtain this result should not be overlooked: if the long-term relation between electricity price and electricity consumption established in international comparative terms were to exist also for short-term elasticity within individual economies, a country with an average annual rate of growth of 3% would need to double relative electricity prices every 16 years simply in order to keep electricity consumption constant. Persistent political intervention in price structures on this scale would doubtless be a politically risky undertaking.

Unexploited Technological Possibilities

The delinkage of energy requirement from the economic growth trend is common in the more advanced industrialized countries. Some of these countries evidence stagnation in per capita consumption of primary energy; the United States, Denmark, Britain and Luxembourg even experienced a decrease.

The majority of more advanced industrialized countries has at least exhausted some of the conservation potential which today can be regarded as the state of the art in energy research. But there is no development towards stagnation in electricity consumption parallel to that registered for energy requirement; in the course of time and with the rise in the standard of living there has hitherto always been an increase in electricity generation. Some affluent industrialized countries (Norway, Canada, Iceland, Sweden) even exhibited an accelerated increase. As mentioned above, the countries concerned charge low electricity prices. In Sweden, for example, the advance of electricity into the heating market was questionable from the energy policy point of view. Because of the particularly large volume of waste heat produced by electricity generation, this strong trend towards electricity use is in conspicuous contradiction to the objective of conserving primary energy. In other words: if the "away from oil" policy had been less a policy of "towards electricity", a great deal more primary energy would have been saved. However, the discovery of significant technological conservation capacities in final electricity consumption entered public awareness only in the eighties.

In the field of freight movement the contrast between technological possibilities and reality was no less marked. If we consider the weight of cargo moved by rail and road, we find a trend towards stagnation in a series of advanced industrialized countries (in contrast to eastern Europe with its then prevailing "tonnage ideology"). In Sweden, Japan, France, Britain, Belgium and Luxembourg there has even been a reduction (per capita). This remarkable result seems to indicate that growth has incorporated a more distinctly "qualitative" component. The fact that the consumption of raw materials in the Federal Republic of Germany in 1990 was slightly below the level for 1978 (see below) is fully compatible with stagnation in the weight of cargo moved.

This is, however, in strong contrast to the marked increase in transportation activity in road freight (cf. Fig. 4). An apparently hardly increasing volume of goods was thus moved ever farther and ever more frequently by road. Increasing integration in the world market has certainly made an essential contribution to an ever larger proportion of goods produced being exchanged over ever greater distances. Precisely this development *could* have favoured rail transport as the cleaner alternative – but this did not occur.

Significance of Intrasectoral Change

We have hitherto dealt with intersectoral structural change and its implicit environmental impact. In brief, it brought no reduction in environmental stress in the southern and eastern European industrial countries. On the contrary: the industrial environmental situation in these areas deteriorated. In the highly advanced industrialized countries, too, there was no reduction in environmental stress in the above sense, despite a slight decline in traditional heavy industries (mineral products, crude steel, fertilizers), because modern industries with high environmental consumption – especially the chemical industry – have experienced strong growth. This is also true for countries with strong structural change such as Japan.

If we also consider change within industrial sectors and enterprises, a somewhat different picture emerges. At least in the advanced industrialized countries, this intrasectoral change has hitherto been the most significant factor from the ecological point of view. We have – unfortunately only incomplete – figures for the highly advanced industrialized countries Japan, the Federal Republic of Germany and Sweden. A comparison will be drawn with the less advanced industrial country Portugal.

With regard to intrasectoral change, it can be assumed that technological change predominates. In fact, however, there are also changes – not technological in nature – in the relative weighting of product groups within industries that have to be taken into consideration. If, for example, products in the chemical industry that have always been relatively cleaner to produce than others gain in importance, this can no more be considered modernization than can the relocation of problematic preliminary production phases. We thus refer to intrasectoral change.

We measure intrasectoral change as an adjustment in a number of the above factors in industrial ecological structures, primarily changes in energy and water intensity. Figures are also provided on the usually marked, but differentiated, trend towards electricity use (electricity use is included at the same time in data on energy consumption). Data are also available on raw materials consumption and on waste production by the manufacturing sector in the Federal Republic of Germany and Japan; in addition there are industrial land use figures for Japan. We will proceed to describe intersectoral and intrasectoral change in the four countries mentioned, showing developments in Japan and the Federal Republic of Germany in Figs. 7 and 8.

Japan experienced considerable intersectoral change and incomparably greater intrasectoral change. The relative decrease in significance of mineral products, basic metal production and the paper industry, and an absolute downturn in special basic industries (aluminium,

Figure 7

Percentage Change by Indicator and Branch 1971 - 1987: Japan

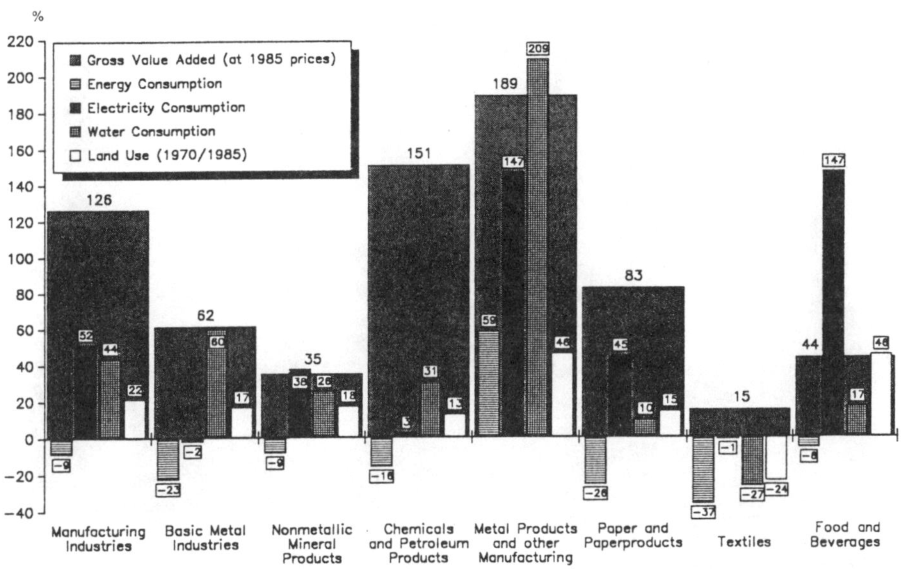

Sources: OECD: National Accounts Statistics; OECD: Energy Balances Statistics; Japan Statistical Yearbook: Kogyo tokei-hyo (Industrial Statistics).

fertilizers) was accompanied by radical change in individual branches. Although the chemical industry experienced above-average growth, it nevertheless reduced energy consumption in absolute terms by 16 per cent. Metal production developed similarly. In the textile industry there were improvements throughout.

For the period 1977 to 1987 the picture is even more favourable: stagnation or an absolute drop in energy and water consumption and land use by metal production, the chemical industry (with the exception of a 4% increase in land use), the paper industry and the textile industry.

Intrasectoral change in Japan proved to be little less than dramatic: industrial final energy consumption in 1989 was nc less than 58.6 per cent lower than it would have been without change – primarily technological – within branches since 1970. In the Federal Republic this stress-reducing effect of intrasectoral change made a difference (up to 1989) of 30.4 per cent, and in Sweden (1973-1988) of 27.6 per cent. In Portugal change within industrial branches tended to result in structural deterioration.

By contrast, the reduced significance of energy-intensive branches (intersectoral structural change) lowered energy consumption in Japan and West Germany from 1970 on balance only marginally (in each case by a good 13 per cent). In Sweden and Portugal this effect was quite unimportant.

In the case of industrial water consumption, too, technological change within branches was very much to the fore. Japan consumed 29.5 per cent less water, and the Federal Republic 36.9 per cent less (1971 to 1987) than they would have without intrasectoral change. In Sweden, where precise branch data on water consumption are not available, a marked reduction was also only possible because of conservation within industries. The intersectoral components in Japan, by contrast, produced a difference of only 8.2 per cent (1970-1987). In the Federal Republic it was of as little significance as it apparently was in Sweden.

With regard to waste produced by the manufacturing sector, intersectoral change brought a reduction in the Federal Republic (there was 10.4 per cent less waste than there would otherwise have been for the period 1977-1987). Intrasectoral change, however, resulted in a 19.8 per cent deterioration. The growth-related rise in industrial

Figure 8

Percentage Change by Indicator and Branch 1971-1987: Federal Republic of Germany

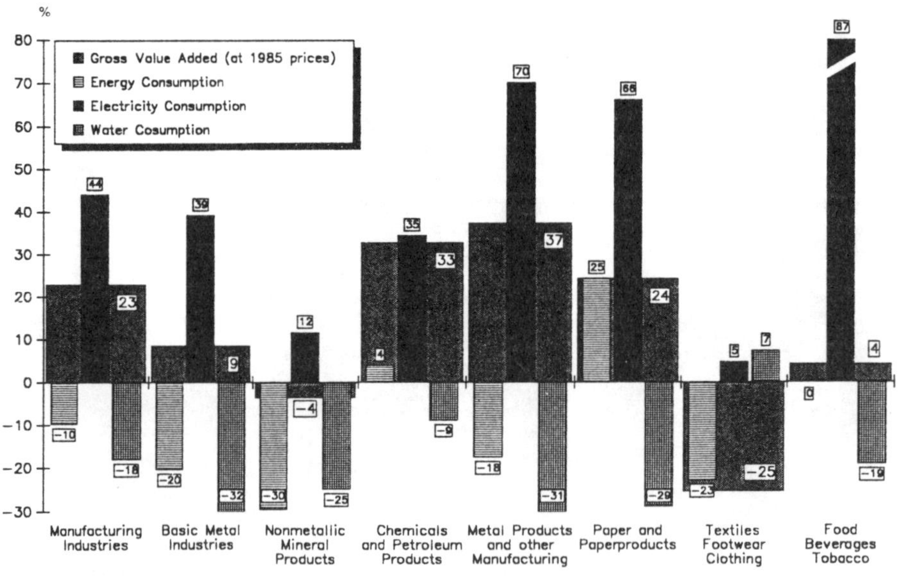

Sources: OECD: National Accounts Statistics; OECD: Energy Balances Statistics; Statistical Yearbook FRG.

waste production was thus not compensated, but – also as a consequence of environmental protection measures – aggravated, producing a 23.5 per cent increase. In Japan, too, industrial waste increased (1975-1985: 32 per cent) in absolute terms, but in contrast to the Federal Republic the rise was slower than that in industrial value added.

Reduction in Environmental Stress

What, then, was the impact of structural change on the four countries we have been looking at in detail in relation to ecologically significant criteria? The indicators used are industrial energy, water and raw materials consumption, production of waste and the volume of freight movement by rail and road (the latter concerned not only, but primarily, with industrial goods).

The greatest reduction was in industrial final energy consumption. The three highly advanced industrialized countries Japan, Sweden and the Federal Republic experienced an absolute reduction. Industrial water consumption also showed a relatively favourable trend. In the Federal Republic and Sweden there was also a drop in absolute terms. In respect of both indicators there was thus a real reduction in environmental stress. In Japan the specific reduction in both fields was much greater. Nevertheless, strong industrial growth led to an absolute increase in water consumption; from 1979 onwards, however, this stagnated.

The relative or absolute reduction in industrial energy and water consumption contrasts with the absolute deterioration in waste production (Japan, Federal Republic). Nearly the same is true with regard to freight movement (three countries), with the exception of Sweden, where transport activity was nearly stabilized after 1974. Japan evidenced at least a drop in intensity for all four indicators, thus achieving delinkage from economic growth. This was so substantial that for a long period there was no deterioration in the growth process. This observation also applies with regard to a further indicator: land use.

Data are also available on raw materials consumption in the Federal Republic of Germany for the period 1978 to 1990, according to which consumption – at constant prices – dropped slightly in absolute terms.[6] In Japan there was an increase of about 50 per cent from 1970 to 1990[7] but near stagnation between 1975 and 1985.

According to the five indicators selected, structural change in advanced industrialized countries has thus followed a very ambivalent course. This also became

---

[6] Statistisches Bundesamt: Statistisches Jahrbuch für die Bundesrepublik Deutschland, Stuttgart 1992, p. 713.

[7] Environmental Agency: Quality of the Environment in Japan, Tokyo 1992, p. 156.

apparent from the rough classification used for the nine polluting branches of the economy (see above). On balance we cannot speak of a reduction in environmental stress. However, without structural change this would have deteriorated dramatically from the ecological point of view. This is probably the most important result to be attributed to industrial change since the oil crisis.

The – relative – benefits of the alteration in the growth pattern after 1973 only become clear when the situation in Portugal is examined. Even with the unsatisfactory data situation there it is obvious that industrial energy and water consumption rose not only in absolute terms, but also relative to economic performance. This also applies with regard to freight movement.

For Japan the development of industrial land use could also be taken into consideration. After 1975 it hardly increased at all. Industrial land in Tokyo between 1975 and 1985 even decreased in area by 23 per cent.[8] On the other hand, changes in Japan, constituting the largest-scale reduction in environmental stress worldwide, were more strongly counteracted by more rapid industrial growth. From 1986 onwards energy consumption (especially in the form of electricity) once again increased conspicuously. The same is true for raw materials consumption and road and rail freight, which until 1985 had hardly risen above the 1972 level (cf. Fig. 9). The best-practice case Japan thus demonstrates two things:

☐ Firstly, it becomes clear how extensive reductions in important input factors can be effected primarily by means of adjustments within branches. For a period Japan was almost a country with qualitative growth, sufficiently delinked from ecologically relevant input factors.

☐ Secondly, however, it became apparent how difficult it is to sustain such stress-reducing delinking effects in the long-term growth process. The formulas of qualitative growth and sustainable development had hitherto hardly been associated with the notion that the reduction of environmental stress in the growth process has constantly to be renewed and extended.

The Portuguese growth pattern, representative for a large number of less advanced industrialized countries, but also for the growth pattern eastern Europe has hitherto experienced, would, if universalized, prove ecologically disastrous. But the Japanese growth pattern – with its partial relocation or export of environmental problems – does not yet offer alternative, environmentally appropriate goods production that is globally generalizable.

Table 1

Industrial Change by Ecologically Relevant Indicators in Selected Countries 1970-1987/89

(Intensity = relation to industrial value added, in brackets)

	Japan[1]	FRG[2]	Sweden[3]	Portugal[4]
Energy consumption	decrease(–)	decrease(–)	decrease(–)	increase(+)
Water consumption	increase(–)	decrease(–)	decrease(–)	increase(+)
Raw materials consumption	increase(–)	decrease(–)	–	–
Freight movement	increase(–)	increase(+)	stability(–)	increase(+)
Waste	increase(–)	increase(+)	–	–
Land use	increase(–)	–	–	–

[1]Waste: 1975-85; water consumption constant since 1980; land use: 1973/85; raw materials consumption: 1970-90.
[2]Waste: 1977/87; raw materials consumption: since 1978.
[3]Water consumption: official estimate.
[4]Water consumption: 1980/84; energy consumption: 1977/87.

\+ = increase
– = decrease

Figure 9

Resources Use by Manufacturing in Japan 1970-90

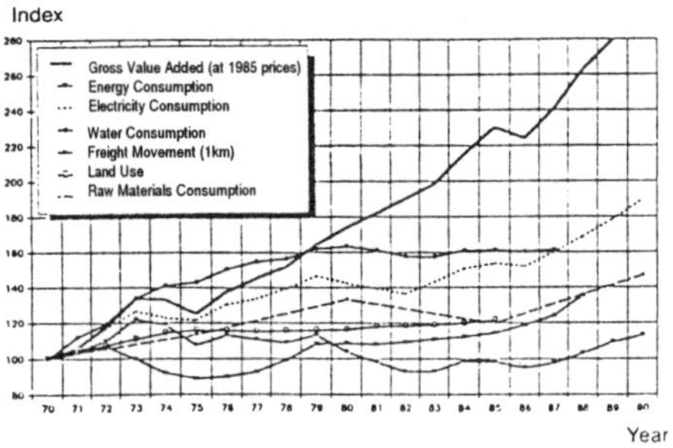

Sources: OECD: National Accounts Statistics; OECD: Energy Balances Statistics; IRF: World Road Statistics; Japan Statistical Yearbook; Environmental Agency.

Industrial Policy

We also examined the role played by industrial and economic policy in possible ecological structural improvements in the four industrialized countries described. (We did not examine the part played by energy, transport and technology policy, to which structural policy effects could also be attributed in our context.)

We found that intersectoral change had been purposively influenced so far only by Japan. It is true that,

[8] Tokyo Metropolitan Government: Second Long-Term Plan for the Tokyo Metropolis, Tokyo 1987, p. 283.

after 1973, energy policy in other industrialized countries also brought about factual reductions in the mineral oil industry. But ecological considerations generally played at best a subordinate role. Moreover, in a considerable number of countries (the United States, Britain, Canada etc.) any type of industrial policy is officially strictly rejected.

By contrast, Japan, the country which has experienced the farthest-reaching structural change, not only espoused industrial structural policy, but also gave it an ecological bias.[9] The 1971 MITI concept of a "know-how intensive production structure", with its call for the rejection of material and energy intensive production, although primarily motivated by economic policy considerations, had distinct ecological implications – accompanied by forced-pace environmental policy. After the oil crisis, the MITI in 1974 called for the traditional growth model based on branches with a high material consumption to be abandoned. Increasing attention was paid to the environmental question. Finally, in 1978 legislation provided concrete measures dealing with economically/ecologically problematic economic sectors. Subsequently the production of primary aluminium was drastically reduced (aluminium was, however, then largely imported – with a high recycling rate). Fertilizer production continued to decrease in orderly fashion. Further regulation affected synthetic fibres, petro-chemicals, electro-steel, shipbuilding, paper and textiles – all branches with high rates of material, energy and environmental consumption. Overall priority was given to reducing dependence on imports. But the environmental question was not without significance as a further motive.

At the beginning of the seventies (even before the oil crisis) there was a certain amount of discussion in the Federal Republic of Germany on the relation between industrial structure and environmental stress. The 1971 Annual Economic Report issued by the Federal Government stated, "In its structural policy the Federal Government will seek improvements in environmental conditions to a greater extent than before".[10] But a turnabout occurred at the latest in the period following 1975. And actual economic policy, in particular subsidisation policy, took the opposite direction under the banner of structural crisis: maintenance subsidisation became quite clearly predominant. In the course of time it even increased (1987: one third of all federal subsidies;[11] 1992: more than 40 per cent[12]). In the late eighties, however, reporting on economic structural matters espoused the topic of ecology.[13]

Sweden, with its pronounced industrial and environmental policy,[14] permitted the assumption to be made that a structural policy motivated, among other things, by ecological considerations, would be taken up at least within the field of economic policy. Initially this was indeed the case – at least as far as general statements of intention were concerned. But we find a cycle of discussion on ecological questions strikingly similar to that prevailing in the Federal Republic of Germany: in the early seventies long-term economic planning took the relation between environmental stress and industrial development into account. As in the Federal Republic, however, more than discussion was not on the agenda. After the oil crisis the question dropped out of sight in both countries. And only in recent years has the question of an ecologically more appropriate industrial structure received greater attention in economic structural policy.

The ecological effects of Swedish industrial development have thus been almost as ambivalent as they have been in the Federal Republic of Germany. Figures issued by the Central Office for Statistics show no favourable ecological balance for intersectoral structural change. But here, too, intrasectoral change (not least of all technological change) in polluting branches such as the paper industry has been of greater significance. Altogether in Sweden it was slow industrial growth that caused better ecological effects than in Japan, though the industrial change was less radical.

The question of ecologically influenced industrial policy in Portugal is not precluded from the outset, since this country on the southern periphery of Europe established environmental policy institutions at a relatively early date (protection of the environment was entrenched in the constitution in 1976, and as early as 1974 a ministry for social affairs and the environment was set up). Ecologically orientated structural policy was, however, not

[9] Gesine Foljanty-Jost: Industriepolitik in Japan – Ansätze für einen strukturpolitischen Umweltschutz?, Berlin: Forschungsstelle für Umweltpolitik der FU Berlin 1990, FFU rep 90-6.

[10] Jahreswirtschaftsbericht der Bundesregierung, Bonn 1971, section No. 77.

[11] Cf. Frank Stille: Umweltpolitische Auswirkungen staatlicher Subventionspolitik, in: Ulrich Petschow, Eberhard Schmidt (eds.): Staatliche Politik als Umweltzerstörung, Berlin 1990, p. 15.

[12] Deutsches Institut für Wirtschaftsforschung (DIW): DIW-Wochenbericht No. 46/92, Berlin 1992, p. 617.

[13] Rheinisch-Westfälisches Institut für Wirtschaftsforschung (RWI): Analyse der strukturellen Entwicklung der deutschen Wirtschaft (Strukturberichterstattung 1987), Schwerpunktthema: Strukturwandel und Umweltschutz, Essen 1987; HWWA Institut für Wirtschaftsforschung: Zusammenhang zwischen Strukturwandel und Umwelt, Spezialuntersuchung 2 im Rahmen der HWWA-Strukturberichterstattung 1987, Hamburg 1987.

[14] Martin Jänicke: Conditions for Environmental Policy Success: An International Comparison, in: The Environmentalist, Vol. 12 (1992), No. 1, pp. 47-58.

on the government's agenda. The EC Regional Fund thus tended to have a structurally unfavourable impact (support for the paper industry). It was only in the environmental basic law of 1987 that ecological interests in industrial development were given a sort of general-clause mention. Nevertheless, in the 1987 energy plan lower growth rates in energy-intensive branches were scheduled. In 1988 economic structure reports gave programmatic status to the restructuring of the Portuguese economy to conserve resources and energy. Up to the present, however, Portuguese industry did not develop accordingly.

Conclusions

Although the topic of the relationship between environmental policy and industrial policy had been taken up in a number of advanced industrialized countries (Japan, Sweden, the Federal Republic of Germany) as early as the beginning of the seventies, this bore little real fruit except in Japan. Until well into the late eighties environmental policy was essentially special policy in the sense of the end-of-pipe variant. Only in Japan was attention given to the deliberate and socially cushioned reduction of polluting industrial sectors. And there, too – as elsewhere with the mineral oil industry – energy policy considerations were to the fore.

By contrast, autonomous structural change in the traditional heavy and basic industries in advanced industrialized countries (crude steel, cement, fertilizers) led to reductions in environmental stress. But from the point of view of environmental policy these were to be regarded more as chance bonus effects. And they were cancelled out – just as much by chance – by disproportionately high growth in industries causing high environmental stress such as chlorine, aluminium or paper and paperboard. However, in many places these industries have been under fire from public opinion, which does not preclude a change in demand behaviour. But this, too, would constitute a form of autonomous structural change.

The greatest environmental stress reduction effect in the advanced industrialized countries was caused by – primarily technological – changes taking place within branches and enterprises (intrasectoral change). But such changes, too, seem for the most part to be autonomous, triggered especially by alterations in price patterns. In this respect Japan is also the only country to show a certain degree of regularity in (relative) environmental stress reduction for parameters such as energy, raw materials and water consumption, land use and freight movement. However, Japan's high industrial growth rate appears to be the cause for the reductions achieved being cancelled out in the course of time.

Whatever general problems ecologically motivated industrial structural policy may present, the following conclusions are, in our opinion, justified:

□ *Energy taxes:* Our investigations lend support to demands that energy be made significantly dearer by imposing taxes (or in the case of innovation processes by levying charges to be ploughed back as fully as possible). This not only brings reductions in ecologically sensitive energy requirements; it also has structural effects that are to be classified as ecologically beneficial. A concentration of such taxes/charges on electricity seems to be particularly appropriate, since there is unexploited conservation potential in this field.

□ *Renouncing ecologically detrimental maintenance subsidisation:* This leads to industrial structural policy becoming taboo. The minimum recommendation would be that governments obeying this taboo ought also to waive every type of ecologically detrimental structural maintenance. "Ecological subsidy reduction" would probably be equivalent to an environmental stress reduction effect in many industrialized countries – not to mention fiscal relief.

The ecological restructuring of industrial societies will not be possible without structural policy. This, too, is a considerable pensum in cooperation and integration for environmental policy and economic policy.

□ *Renouncing active promotion of growth:* The example of Japan shows that ecologically beneficial economic change tends to be neutralized by high growth. For growth rates are themselves an environmental problem. It is apparent that qualitative growth can in the long term only be limited growth, if ecologically negative growth effects are to be compensated by technological and structural change. If with an annual growth rate of one per cent we have a doubling in 70 years, with 5 per cent growth this factor would be attained in a mere 14 years; and in 70 years it would be thirtyfold. Such growth clearly cannot continue to be compensated by structural effects; except by relocation of production, which is no global or long-term solution.

For this reason the state must rethink its economic policy role as the engine of economic growth. The industrialized countries will not be able to afford the luxury of high growth rates for much longer. They will have to become accustomed to the universal problem-solver "economic growth" being increasingly superseded by political action (e.g. in matters of distribution). The transition from quantity (financial flow) to quality (political structuring) may be also a task for the state.

[12]

Ecological modernization of industrial society: three strategic elements

Udo E. Simonis

Introduction

At a Forum on Industry and the Environment in New Delhi, Stephan Paulus gave the following definition of ecological modernization: 'Ecological modernization focuses on prevention, on innovation and structural change towards ecologically sound industrial development. . . . It relies on clean technology, recycling, and renewable resources. . . . To introduce such a concept into the economy, it is necessary to co-ordinate various policy areas, such as industrial, fiscal, energy, transport and environmental policies'.[1]

This is in fact a rather broad and demanding definition of a concept, proposed to achieve better harmony between economy and ecology. In this article I will, therefore, concentrate on only certain aspects of such a concept. First, I will present some empirical evidence on the relationship between economic structure and environmental impacts; second, I shall point to some deficiencies of environmental policy, and third, I shall put forward some ideas on how to integrate ecological dimensions into economic policy.

Udo E. Simonis is professor of environmental policy at the Wissenschaftszentrum Berlin (Science Centre Berlin), Reich-pietschufer 50, D-1000 Berlin (West) 30, FRG, and chairman of the board of the Deutsche Umweltstiftung (German Environmental Foundation). He has published extensively on a wide range of developmental and environmental issues.

Ecological structural change of the economy

In both the East and the West, economists, planners and engineers are seeking a solution to the problem of how to change the traditional patterns of resource use. *Perestroika* and 'modernization' are two current catchwords for this process, and new environmental priorities play a part in the envisaged conversion of the economy.[2] Harmonizing ecology and economy in a specific sense relies on the premise that a reduction in the resource input of production (structural change) will lead to an *ex ante* reduction of emissions and wastes that have a negative impact on the natural environment (ecological structural change).

In order to clarify the relationship between economic structure, structural change, and environmental impacts, one needs suitable information concerning the material side of production, for environmental protection and resource conservation by the economy – and thus its long-term sustainability – cannot be described appropriately in such terms as income, investment and consumption. One possibility is to select and compare certain indicators describing the environmentally relevant features of the production process. The availability of environmental indicators such as emission data relating to 'representative' pollutants – like sulphur dioxide (SO_2) and nitrogen oxide (NO_x) – has grown recently (e.g. the Annual Reports on the Environment by several industrial nations, by UNEP and the OECD). These indicators concern certain negative environmental effects of production. Less is known about the environmental relevance of the input factors in industrial production, or on the question of which indicators provide environmentally significant information about the structure of the economy. Given the present state of statistics, only a few such indicators can be tested in a cross-national comparison of Eastern and Western countries.

De-linking economic growth from environmentally relevant inputs

Using a set of four indicators (input factors), Jänicke et al. have studied 31 member countries of both COMECON and OECD with reference to the relationship between economic structure and environmental impacts.[3] The four factors whose direct and indirect environmental significance is thought to be self-evident were: energy, steel, cement, and freight transport. Regarding their patterns of production and consumption these are environmentally 'hard' factors, characteristic of a certain structure of the economy or stage of economic development.

The main hypothesis of the research was a simple one and reads like this: Positive environmental effects of structural change of the economy are to be expected by actively de-linking economic growth from the use of environmentally relevant inputs (resources). Such active de-linking (or ecological structural policy) would

- result in a decrease of resource depletion and/ or environmental pollution;
- mean *ex ante* instead of *ex post* (end-of-pipe technology) environmental protection;
- promote those integrated technologies which touch upon several environmental effects (pollutants) at one and the same time.

Structural change as a shift of input factors to more intelligent uses can thus be conceived as a process of successive de-linking: The contribution of traditional (hard) input factors to the national product decreases, i.e. they change or lose their function in the development process.

Examples of successful and deficient de-linking

Taking the Federal Republic of Germany as an example, Figure 1 illustrates a five-fold de-linking from the growth of the Gross Domestic Product (GDP): the de-linking of energy and cement consumption and weight of freight transport from the GDP became apparent during the 1970s; regarding steel, the de-linking process already began in the 1960s. In this way, the structural change of the economy generated environmental gratis effects of various kinds:

- The stagnating consumption of primary energy led to a reduction of harmful emissions (pollutants).
- The relative decline in the weight of freight transport indicates that the volume of materials employed was reduced rather than increased.
- The fall in the use of cement represents a direct gratis effect as far as the emissions from cement factories are concerned; this decrease coincided with the trend towards labour-intensive renovation of the housing stock, as compared to new construction.
- The decrease in steel consumption accounts for a considerable reduction in harmful emissions as far as processing is concerned; this drop was strongly marked and partly due to increased recycling activities.

Environmental gratis effects occur when the rate of usage of the input factors (resources) having a (strong) negative impact on the environment remains below the growth rate of the GDP. Comparing the rates of usage of the four selected input factors with the growth rate of the GDP, Jänicke et al. discovered three different development patterns:[4]

- The factors having impacts on the environment decline absolutely; i.e. absolute structural improvements are induced, corresponding to absolute environmental gratis effects.
- The factors having impacts on the environment remain constant, or increase, but with a lower growth rate than the GDP; i.e. relative structural improvements, corresponding to relative environmental gratis effects.
- The factors having impacts on the environment increase at a higher growth rate than the GDP; i.e. structural deterioration, corresponding to absolute negative environmental effects of economic growth.

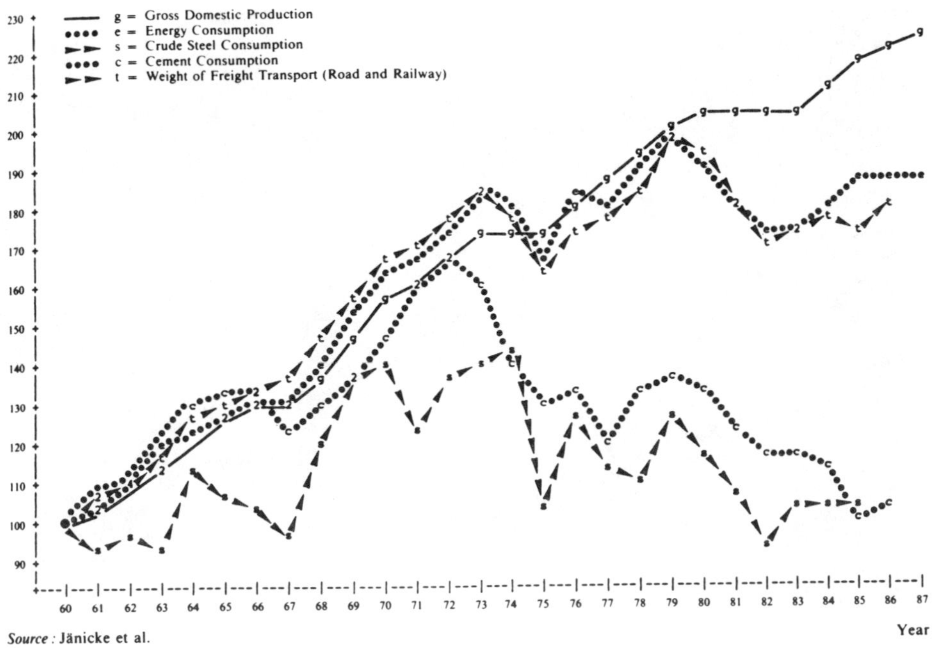

Source: Jänicke et al.

FIGURE 1. Structural economic change in the Federal Republic of Germany, 1960–1987 (1960 = 100)

In Table 1, 16 out of the 31 countries studied are grouped according to these three development patterns.

Of all the industrial countries investigated by Jänicke et al., Sweden went through the most rapid structural change. The drastic reduction in cement production (−41 per cent), the decreasing use of crude steel (−38 per cent), and the decrease in the weight of freight transport (−21 per cent) add up to notable environmental gratis effects ('absolute structural improvement').

In Japan, the de-linking process was partly neutralized by the rapid growth in industrial production and thus only resulted in 'relative structural improvement'.

In Czechoslovakia, no significant de-linking of economic growth from the four input factors took place (see Figure 2). The development profile of this country, with sluggish structural change, is to some extent representative of the other economies of Eastern Europe.

Trends towards industrial restructuring

Despite certain analytical limitations of such empirical research (as e.g. the selection of only four input factors), several conclusions can be drawn from this international comparison as regards trends of industrial restructuring:[5]

Structural change in the form of de-linking economic growth from environmentally relevant inputs was evident in most but not all of the countries studied.

Several countries enjoyed environmental gratis effects as a result of active structural change. In some cases, especially for Sweden, these effects were quite considerable.

In other countries, the possibly beneficial environmental effects of structural change were levelled off by the rapid industrial growth pursued. This was especially true for Japan.

Ecology and economy: 'In the Equatorial Bush', by Raffy le Persan. Musée d'art moderne de la Ville de Paris. Bulloz.

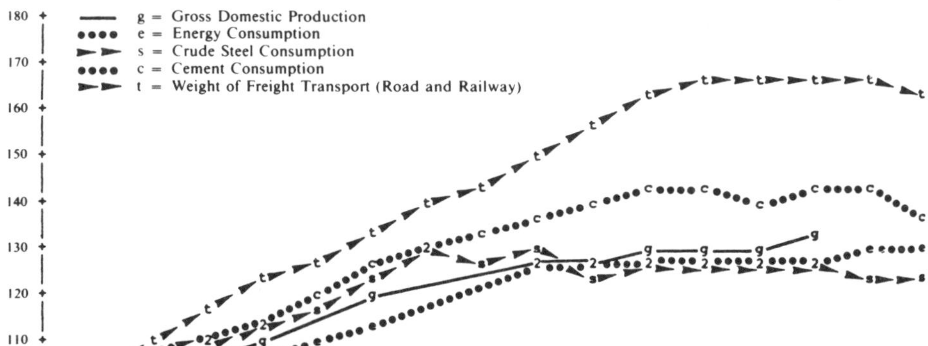

Source: Jänicke et al.

FIGURE 2. Structural economic change in Czechoslovakia, 1970–1985 (1970 = 100)

TABLE 1. Environmental impacts deriving from structural change, percentage changes 1970/1985

Country	Consumption of primary energy	Consumption of crude steel	Cement production	Weight of freight transport	GDP*
	Group 1: Absolute structural improvement				
Belgium	7.1	−24.5	−17.6	−2.2	42.7
Denmark	−2.7	−15.6	−33.2	20.1	40.8
France	30.3	−34.8	−23.4	−14.5	51.6
FRG	13.4	−26.3	−32.8	4.4	38.4
Sweden	26.4	−37.9	−41.2	−21.4	32.7
United Kingdom	−2.3	−43.5	−28.7	−18.2	32.4
	Group 2: Relative structural improvement				
Austria	32.1	−33.9	−6.0	21.3	54.3
Finland	39.6	14.8	−11.2	12.2	65.7
Japan	37.3	−2.3	27.4	7.5	90.2
Norway	51.1	−21.6	−40.3	34.7	87.5
	Group 3: Structural deterioration				
Bulgaria	74.9	24.9	42.3	77.5	37.3
Czechoslovakia	31.5	22.5	37.3	62.9	33.9
Greece**	119.3	67.3	162.9	43.1	69.1
Portugal**	89.0	34.2	133.1	27.4	69.0
Soviet Union	76.3	33.4	35.9	70.2	47.7
Turkey	218.8	184.4	173.2	118.6	118.2

Source: Jänicke et al.
* Calculation of the Gross Domestic Product percentage changes on the basis of constant (1980) $US. Bulgarian, Czechoslovak and Soviet Union data refer to percentage changes 1970/83 of the Gross National Product.
** Transport data only take railway transport into account.

The strong correlation between the level of production (GDP) and environmental impacts, still evident in the 1970s, had dissolved in the 1980s. The high-income countries exhibited fairly rapid structural change. In the medium-income countries, a distinct pattern emerged in that there were cases of rapid quantitative growth and cases of qualitative growth, i.e. economic growth with constant or decreasing resource input.

All in all, it is therefore not yet possible to speak of a single dominant trend towards industrial restructuring. However, the environmental gratis effects of active structural change are highly evident and thus provide one strategic element of ecological modernization of industrial society.

Preventive environmental policy

Theoretically speaking, environmental policy could be defined as '. . . the sum of objectives and measures designed to regulate society's interaction with the environment as a natural system; it comprises aspects of rehabilitation, conservation, and structural adjustment'.[6] Practice, however, does not conform to such a broad definition. Only parts of the interaction between society and environment become the subject of policy. So far, environmental policy has mostly been designed as react-and-cure strategies concerning air and water pollution, noise, and waste, with emphasis on the rehabilitation aspect.

For a variety of reasons, this conventional environmental policy was, and still is, meaningful and very necessary. It has, however, a number of defects, some of which are discussed below, along with some suggestions for overcoming them through preventive environmental policy, i.e. anticipate-and-prevent strategies.

Environmental expenditures – environmental damages

Since the beginning of the 1970s, when systematic records first began to keep track of the funds allocated for environmental protection the sum of public and private investments in the industrialized countries has reached large proportions. Industrial society thus appears to be paying through the nose – back payments for the negative environmental costs of production accumulated in the past.

In the Federal Republic of Germany, for instance, this sum has come to the handsome total of over 250 billion Deutschmarks (or about $140 billion). In a detailed study, Leipert et al. of the International Institute for Environment and Society (IIES) have computed and classified all existing data on investments and expenditures aimed at repairing and protecting the environment.[7]

Table 2 shows the total and sectoral environmental protection investments for the manufacturing sector of the German economy for the years 1975 to 1985.

Table 3 shows the total costs of environmental protection (current expenditures and depreciations) for both industry and government for the years 1975 to 1985.

Figures like these, however, are ambivalent. On the one hand, they give cause for proud political statements about the successes of environmental protection, according to the motto 'the more, the better'. On the other hand, they are – presumably – the absolute minimum of what is necessary to secure the very basis for society's sustainability. At the same time, they symbolize a serious structural deficit of industrial society: environmental protection expenditures are spent when damage to the natural environment has occurred and can no longer be denied. Belated, they are repairs to the process of economic growth, signs of a '*post-hoc*' policy that reacts to damages (and must react to them)

TABLE 2. Environmental protection investments, manufacturing sector, Federal Republic of Germany, 1975–1985

	Total investments		Waste disposal		Water pollution control		Noise abatement		Air pollution control	
Year	current prices	1980 prices	current prices	1980 prices	current prices	1980 prices	current prices	1980 prices	current prices	1980 prices
	in millions of DM									
1975	2,480	3,090	170	210	900	1,110	200	240	1,210	1,530
1976	2,390	2,830	200	230	820	960	220	260	1,150	1,380
1977	2,250	2,560	200	230	740	850	210	230	1,100	1,250
1978	2,150	2,370	170	180	680	750	200	220	1,100	1,220
1979	2,080	2,190	260	160	760	800	200	210	960	1,020
1980	2,650	2,650	210	210	910	910	240	240	1,290	1,290
1981	2,940	2,810	250	240	950	910	210	200	1,530	1,460
1982	3,560	3,250	390	360	1,130	1,030	230	210	1,810	1,650
1983	3,690	3,270	290	260	1,100	990	230	200	2,070	1,820
1984	3,500	3,100	270	240	1,040	920	230	190	1,960	1,750
1985	5,620	4,940	330	280	1,060	910	260	220	3,970	3,530
	Average annual change in %									
1975 84	+3.9	−0.0	+5.3	+1.5	+1.6	−2.1	+1.6	−2.6	+5.5	+1.5
1975 79	−4.3	−8.2	−1.5	−6.6	−4.1	−7.9	0.0	−3.3	−5.6	−9.6
1979 84	+11.0	+7.2	+11.0	+8.4	+6.5	+2.8	+2.8	−2.0	+15.3	+11.4

Source: IIES research project

A sea of garbage, near New York: not exactly a 'low waste economy'. B. Davidson Magnum.

TABLE 3. Total costs of environmental protection, Federal Republic of Germany, in millions of DM, 1975–1985

	Industry			Government			Industry and government		
Year	current expen-ditures	depreci-ations	total costs	current expen-ditures	depreci-ations	total costs	current expen-ditures	depreci-ations	total costs
				At current prices					
1975	3,200	1,520	4,720	3,000	1,920	4,920	6,200	3,440	9,640
1980	5,160	2,250	7,410	4,690	3,390	8,080	9,850	5,640	15,490
1985	7,930	3,160	11,090	6,430	4,340	10,770	14,360	7,500	21,860
				At 1980 prices					
1975	4,050	1,870	5,920	3,790	2,570	6,360	7,840	4,440	12,280
1980	5,160	2,250	7,410	4,690	3,390	8,080	9,850	5,640	15,490
1985	6,230	2,640	8,870	5,340	4,030	9,370	11,570	6,670	18,240

Source: IIES research project

but does not, or cannot, prevent them. Therefore, it is necessary to confront the 'success stories' of environmental protection expenditures with figures on the environmental damages themselves.

Again taking the Federal Republic of Germany as an example, a recent estimation by Lutz Wicke of the Federal Environmental Protection Agency, showed that the annual damage to the natural environment is above 103 billiion DM, or in the order of 6 per cent of the GNP, and not 3 per cent, as the OECD had estimated for the industrial countries as a whole some years ago.[8]

Table 4 is based on different estimation methods, using data on actual damage costs and findings from willingness-to-pay studies. Though the results must be interpreted with some care, the table illustrates that, despite high annual environmental protection expenditures, enormously high environmental damages still occur annually, which of course, may apply to many other countries as well.

There are more shortcomings to conventional environmental policy.[9] To name a few: environmental policy usually identifies the given problem too late, so that the ecosystems affected cannot survive. As it is pursued as a media-specific policy, i.e. separately regulating air and water quality, noise or waste, it also runs the risk of lacking co-ordination between its specific goals, measures and institutions. And this then may result in shifting a problem from one environmental medium to another, e.g. from air to water or soil, or from one place to another, as is the case with long-range, trans-boundary pollution. In addition, environmental policy often becomes entangled in a debate on principles. If measures must be taken quickly, the argument gets shifted from the 'polluter-pays-principle' – which is advocated in general – to the 'taxpayer-pays-principle', thus switching the burden of environmental protection from the individual polluter to the community, to government or to society at large.

Thus, innovations in planning and implementation are needed. Preventive environmental policy, it seems, can counter the shortcomings of conventional environmental policy. But in order to switch to preventive policy,

TABLE 4. Environmental damage in the Federal Republic of Germany ('Measurable Damage' in DM bn per year)

Environmental sectors	Environmental damage
Air pollution	ca. 48.0
Health hazards	between 2.3–5.8
Material damage	over 2.3
Degradation of vegetation	over 1.0
Forest blight, etc.	between 5.5–8.8
Water pollution	far over 17.6
Damage to rivers and lakes	over 14.3
Damage to the North Sea and Baltic Sea	over 0.3
Contamination of ground water, etc.	over 3.0
Soil contamination	far over 5.2
Coasts of Chernobyl disaster	over 2.4
Rehabilitation of 'yesterday's waste'	over 1.7
Costs of preserving biotopes and species	over 1.0
Other soil contamination, etc.	far over 0.1
Noise	over 32.7
Degradation of residential amenities	over 29.3
Productivity losses	over 3.0
'Noise rents', etc.	over 0.4
Grand total of damage	over 103.5

several conceptual as well as practical constraints have to be overcome.[10]

One constraint has to do with the particular history of an environmental impact. In cases of yesterday's wastes, when damage has already occurred, a curative strategy is probably the only conceivable option. In cases where no damage has occurred as yet but where damage is expected in the future, the choice between a preventive and a curative strategy is basically open. In such a situation, the anticipatory principle leads to encouraging the first option. As practice often is a mixture between the existing and the new, most policies will actually also include a mixture of prevention and cure. Demanding preventive environmental policy will then mean seeking and at last finding a better balance between the anticipatory and the reactive component within the policy action.

Basic conditions for preventive environmental policy

According to Scimemi and Winsemius one can conceive three factors as concomitant policy relevant processes in time: the accumulation of environmental damage; the acquisition of technical knowledge, and the rise of public awareness. The time sequence of these processes, especially the relative timing of their critical level, is decisive for the whole issue of preventive environmental policy.

To illustrate the relationship between these three factors, Scimemi has redrawn a diagram suggested by Winsemius, using three separate functions: Level of Damage, Level of Technical Knowledge, Level of Public Awareness.[11] The relative position and the shape of these functions depends, of course, on the specific circumstances (country, environmental sector, historical phase) under consideration; the common case is illustrated in Figure 3.

Line D_1–D_2 indicates the accumulation of environmental damage over time. The accumulation of damage starts at a given point in history (in the diagram somewhere between time 0 and time 10). At that point, neither the scientific community nor the general public is yet aware that anything of importance is happening. Line T_1–T_2 indicates the process of gathering technical knowledge. This process may not start until some time after damage has begun to accumulate (in the diagram somewhere between time 10 and 20), and proceeds gradually. During that phase the public may still be unaware of the hazard. Somewhere between time 20 and 30, public awareness starts to rise, as indicated by the line P_1–P_2.

Within these concomitant processes, a certain stage becomes important (critical level). As

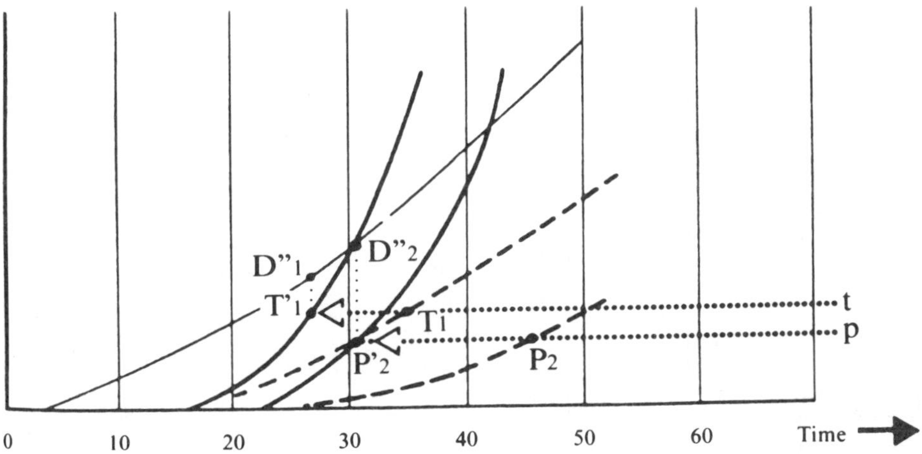

Source: Scimeni

FIGURE 3. Factors in the environmental policy cycle: damage, technical knowledge, public awareness

illustrated, the technical understanding of the issue reaches a critical level, *t*, thus ensuring the first of two conditions required for effective policy action, i.e., technical rationality. Public awareness also reaches a critical level, *p*; at that time the second condition for effective decision-making, i.e. political viability, is fulfilled. It is at this stage that action will be undertaken to avoid the occurrence of further damage.

Recalling past developments in environmental policy at the national or the international level, it is easy to recognize that the processes evolved very much in conformity with Scimemi's theoretical interpretation. Thus, the diagram may be considered to be a true representation of real events. (There may be cases, however, where public awareness – line P_1 or P_2 – started much earlier than technical knowledge – line T_1–T_2.)

What are now the opportunities to influence these basic conditions of policy action in favour of preventive environmental policy? Figure 3 helps to formulate three general and two specific options:[12]

General options:

Retarding damage accumulation (i.e. sliding the D_1–D_2 curve towards the right)

Accelerating technical knowledge (i.e. sliding the T_1–T_2 curve towards the left and/or raising its slope)

Increasing public awareness (i.e. sliding the P_1–P_2 curve towards the left and/or raising its slope).

Specific options:

Dynamic environmental standard setting (i.e. lowering threshold *t*)

Dynamic public participation (i.e. lowering threshold *p*).

All these various options make policy decisions possible at a stage when the level of environmental damage is still relatively low.[13] One of these options is illustrated in Figure 4.

Environmental impact assessment as part of preventive policy

Acceleration of knowledge and awareness can, of course, be promoted through a variety of approaches and methods and depends a great deal on the specific environmental issue at hand. *Environmental Impact Assessments* (EIA) are increasingly being applied, not only for public but also for private investment projects. They entail efforts to learn more about possible environmental impacts, and are intended to allow appropriate action to be taken before damage has occurred. In that sense, environmental impact assessments can be classified as

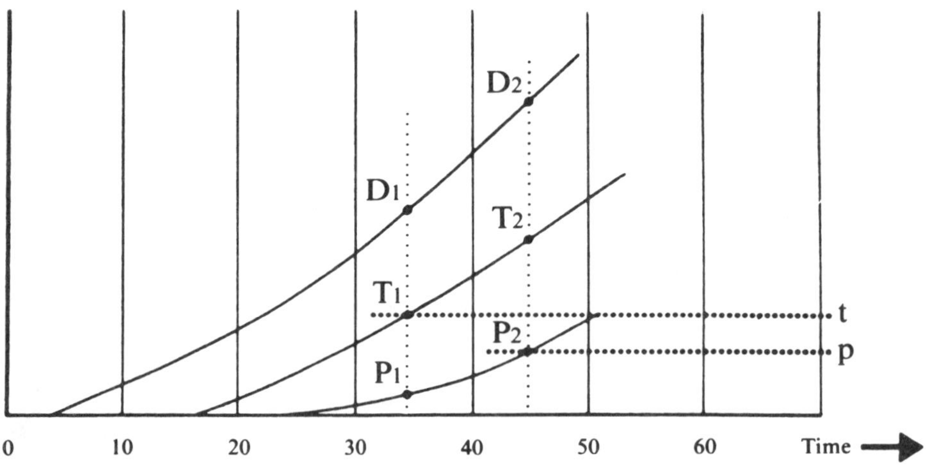

Source **: Scimeni**

FIGURE 4. Accelerating the generation of technical knowledge and/or the development of public awareness

The impact of acid rain in Czechoslovakia. Bossu/Sygma.

part and parcel of preventive environmental policy.[14]

During the last years, some headway has been made to institutionalize and standardize EIA procedures, nationally and to a minor extent also internationally. As the EIA procedure is particularly used for specific investment projects, it allows for the 'accelerating effort' to be targeted, and generally also permits the burden of such efforts to be imposed upon the project initiator himself, thus conforming to a precondition of preventive environmental policy, i.e. the polluter-pays-principle. A big problem, however, remains in how to implement EIA as a preventive procedure in cases of global change, like climate warming or ocean pollution.

The required levels (thresholds) regarding technical knowledge or public participation in environmental decision-making differ widely from one environmental medium and country to the other.[15] The question of how much knowledge/awareness is enough normally falls upon the political decision-maker (the government, the environmental protection agency, the institution in charge of the problem), even if the scientific community (or parts of it) is ready to say 'we know enough' and the public (or parts of it) demands that 'something must be done'. Therefore, stalemates in decision-making on environmental issues are quite frequent.

What is 'enough knowledge/awareness' for one country (government, institution) may not be enough for another. The normal outcome of such a situation is a compromise over the emission standards to be implemented. They will be weaker than technically/politically feasible because knowledge/awareness on cause-effect-relationships or social priorities are said to be insufficient. Eminent cases in point are the emission standards for SO_2 and NO_x in the air pollution field, and the nitrate standard in the water pollution field. Thus, the dilemma of setting stricter emission standards is serious. Meanwhile the forests may continue to die, the ozone layer may continue to be affected, and the water may continue to be contaminated.

The conclusion therefore is, that environmental standard setting must be conceived as a continuous process. With growing knowledge/awareness on actual and probable environmental damages the thresholds for action must be progressively lowered, i.e. standard setting must be dynamized to make industrial restructuring rapidly possible.

This need to come to terms with the future is not unique to environmental policy, as Scimemi rightly observed. Implementing the prevention principle is especially necessary in all other domains of policy where collective interests are at stake. One such major domain we have to address when discussing the possibilities and impediments of ecological modernization is, of course, economic policy.

Ecology and economic policy

Conflicts between economy and ecology

'Ecology in essence means the necessary and feasible harmony between man and nature, society and environment' (C. F. von Weizsäcker). Economy, however, in general means disharmony with nature. Use is made of nature both directly and indirectly when raw materials are processed into products, and nature is polluted by the emissions and wastes generated by industrial production. These are, then, the two processes in which nature remains the loser. She exchanges natural raw materials for produced waste materials. Apart from labour and capital, nature is the truly quiescent and exploited third production factor. How then can nature's position in the 'economic game' be strengthened?[16]

The use of raw materials and the generation of emissions and wastes are of course old issues. Scientific and technological development, however, has made it possible to exploit increasingly the depletable resources, and has led to an ever increasing accumulation of harmful emissions and non-decomposable wastes. Nature is no longer able to absorb all of these substances, many of which are not only toxic for flora and fauna but for human beings as well.

Efforts to hide harmful emissions and toxic wastes – in land-fill sites, in transfer stations or permanent depositions, to spread them – through high smokestacks and incinerators, or to dump them – into the water bodies and abroad, have at best been temporarily successful because many emissions and wastes are 'mobile poisons' or reappear in different form. These activities lead to what Johan Galtung called 'linearization of ecological cycles,' i.e. natural diversity is reduced, the robustness of ecosystems declines, ecological symbioses and equilibria break down. As a consequence, environmental degradation increases and the absorption capacity of the natural environment decreases.

Accordingly, the conflict between ecology and economy can be attributed to two (actually or possibly) incompatible basic principles: the ecological principle of 'stability', as a precondition for the sustainability of ecological systems, and the economic princple of 'growth', as the inherent logic of economic systems – more precisely: the principles of business profitability, national economic growth, and world market expansion.

Given the current and pending ecological crisis, the question on whether these economic principles can be changed, reshaped and finally brought into harmony with ecological principles, at which level, in what way, and at what time, is, of course, controversial both in theory and in practice, and a specific challenge to the social sciences. The answer not only depends upon respective individual and societal constellation of interests. The answer particularly depends upon the ability and willingness to engender social innovations, i.e. on (a) whether the potential for ecological self-regulation of the economy is applied, and (b) how the option of an ecological re-orientation of economic policy is implemented.

Ecological self-regulation of the economy?

To start with a general statement: most certainly, only a small fraction of current environmental problems would exist if economic contexts had remained sufficiently comprehensible, for producers and consumers personally to be able and liable to recognize and perceive the consequences of their own decisions in depleting resources and polluting nature. Or, if commercial profitability, national economic growth, and the expansion of world markets could not be increased by externalizing portions of the ensuing costs. This is the old but still relevant – because unresolved – problem of the external effects of production. Scientific and technological development has been, and still is, coupled with negative external effects, i.e. the shifting of costs to society, future generations, and nature. With respect to the environmental problem, all these cost components are relevant. Let us take, as an example, acid rain and the ensuing damage to the forests:

First, this example shows the shifting of a part of the costs of industrial production, i.e. insufficiently reduced air pollutants, onto nature, which is resistant only up to certain levels: the forests are dying.

Second, this example shows the shifting of costs onto succeeding generations, i.e. a future with fewer or regionally even no forests at all.

Third, this example shows the shifting of costs onto third parties (i.e. partial expropriation of the forest owners) and onto society, in the sense that economic and technical decisions of individual polluters (especially emissions from power plants, cars, trans-boundary pollution) impair the well-being and the physical health of the population.

The economic system thus evidently makes incorrect calculations with respect to the 'ecosystem forest'. Both business accounting and national accounting do not provide adequate signals which may prevent pollution levels that are not tolerable for the ecological system. Conventional accounting shows favourable balances for the production of energy, for the automobile manufacturers, and for the exporters of pollutants (to remain with only the three sources of pollution mentioned above), although the 'ecosystem forest' is definitely being damaged by emissions from these economic sectors. Losses here – profits there: compensation does not take place nor is liability provided for.

One of the pending tasks both for theory and practice thus can easily be prescribed: internalize the external effects of production, shift the costs back to the economic units that cause the environmental problem, include the ecological perspective into all investment decision-making. Drastically reducing the external effects of production on society, nature, and future generations seems to be the necessary step towards regaining harmony between economy and ecology. But how to proceed in practice and determine priorities?

To re-organize the economy towards a materially integrated cycle would, first of all, mean to reduce systematically the use of depletable resources and the generation of polluting emissions and wastes – which contradicts the prevailing 'throughput economy' (K. Boulding). In practice, recycling and clean technology are still at an incipient stage and not systematic economic undertakings. Especially, the step from simply disposing of wastes towards avoiding wastes ('low waste economy') has not been taken.

Certainly, this is in part because many waste products can only be recycled at high cost, if at all. But it is also true because adequate price and cost signals have not been set. Preventing waste generation and conserving depletable resources are still not sufficiently promoted. This state of affairs, however, has also to do with the above mentioned structural deficiencies of economic accounting procedures which do not measure adequately the diminishing stocks. Therefore, two contradictory trends can be observed: increasing monetary income and decreasing natural stock.

Proposals for ecological accounting at the factory level and in the national accounts, however, are promising. With ecological accounting the amount of energy, materials, wastes, land use, etc. are being computed and, by simulating the resulting shortage, accounting units are determined which then enter the accounts. Thus a measure is developed which may not only guide private investment decision-making, but at the same time will provide a public information instrument for promoting qualitative economic processes.

In industrial society another ecological principle is no longer adhered to, that of the sustainability of resource use. Traditionally, forest owners for instance have followed the rule 'Do not cut down more wood than you regrow'. This rule is being undermined: externally produced acid rain collides with internal resource conservation; and accumulated external debt led to over-exploitation of national resources. Sustaining the yield of the forest capital stock is being replaced by indirect expropriation and resource depletion.

One basic principle to be re-established in the economy is that of responsibility or liability . With respect to environmental problems, the legal system, and also economic behaviour, in most countries is marked by the strict proof of causality. Only when an injured (damaged party) can prove who caused the damage (polluting party), is the polluter held liable for compensation. In some countries, however (for some cases), statistical probability is sufficient to oblige industry to compensate for damages (collective liability). Once this principle was established by the courts and through legislation, it quickly helped to improve environmental quality through the ecological self-regulation of business activities.

In general, the liability principle would strengthen the anticipate-and-prevent strategy in environmental policy, and shift the technical solutions for environmental problems from *ex-post* to *ex-ante* approaches, i.e. from controlling or end-of-pipe technology towards low emission or integrated technology. To implement the principle in practice, small steps or big leaps could be taken: from continuous reporting on wastes or automatic monitoring of emissions, to collective funds and strict environmental liability.

Ecological re-orientation of economic policy?

Confronted with serious environmental damage, conventional economic policy is increasingly being challenged. Its guiding principles, goals, instruments, and institutions are being questioned, and a new concept is emerging: ecological economic policy.

Conventional economic policy is based on the guiding principle of maximizing flows – volume of production, income, profits, turnover (Kenneth Boulding seventeen years ago called this the 'throughput economy'). Instead, the 'ecological economy' is based on a different guiding principle, i.e. 'increasing efficiency and maintaining substance'! Aspects such as environmental compatibility and resource conservation becomes important, and structural adjustment of products and technologies according to ecological considerations becomes the task.

Regarding goals, it seems necessary to redefine and supplement the conventional economic policy goals, especially to re-assess the growth target and to include 'environmental stability' into the catalogue of economic policy goals. The conventional policy goal indicators were developed at a time when environmental pollution was already a problem but not yet a public issue, and since then they have not really been readjusted. Economic growth is still being measured in terms of goods and income categories only (Gross National Product), while effects on the stock and quality of the resources (natural capital) are not considered adequately. In the conventional concept of economic growth, all monetary transactions are summed up independently of their specific function; also, increasing expenditures are included which are solely due to the (necessary) compensation for damage originally caused by the production process ('compensatory or defensive expenditures').

Qualified goal indicators for economic policy can be defined in various ways: computations of compensatory expenditures, i.e. assessment of an environmentally related net product (ENP – Eco National Product); combined growth and distribution indices (RWG – Redistribution With Growth); an integrated system of economic and environmental indicators, or attached 'Satellite System', etc.

Regarding instruments, conventional economic policy relies mainly on two instruments, i.e. variations of interest rates and of tax rates. From an ecological point of view, new taxes and charges are required which, to some extent, should replace traditional taxes. In a situation of structural unemployment and environmental pollution the introduction of resource taxes (as e.g. an energy tax) and emission charges (as e.g. charges on SO_2, NO_x, and CO_2 emissions), and a decisive decrease of income taxes is called for. Such a structural tax reform would change the existing incentives in the economy towards accelerating resource efficiency and increasing employment opportunities.

Economic policy manifests itself in and works through particular institutions. Therefore, an ecological orientation of economic policy also requires creating new institutions, and abolishing or redefining old ones. (The current debate on the negative environmental effects of decisions by the World Bank and the IMF is just a case in point.) The actual and the pending environmental crisis requires structural institutional reforms by which economic institutions would have to incorporate the ecological perspective, and environmental institutions would have to improve their competence, and by which environmental impact assessments (EIA) would be integrated into all major economic decision-making.

Conclusions

According to the above deliberations, industrial restructuring for sustainable development, or 'ecological modernization', obviously is a demanding concept, both methodologically and practically. Its implementation requires a far-reaching conversion of the economy, a re-orientation of environmental policy, and a replenishment of economic policy: 'ecological structural change of the economy', 'preventive environmental policy', and 'ecological orientation of economic policy' seem to me to be the three main strategic elements – or points of departure – to reconcile the interests of man and nature, society and environment. The social sciences – economics, sociology, jurisprudence, political science, psychology – have to develop further the methodological foundations and to improve the institutional arrangements for a successful practical implementation of such a concept.

Notes

1. S. Paulus 'Economic Concepts for Industry Related Environmental Policies'. In: *Proceedings – Forum on Industry and Environment*. New Delhi: Friedrich Ebert Foundation, 1986.

2. Cf. G. Enyedi et al. *Environmental Policies in East and West*. London: Taylor Graham, 1987.

3. The following data and arguments rely on M. Jänicke, H. Mönch, T. Ranneberg, U. E. Simonis. 'Economic Structure and Environmental Impacts'. In *Environmental Monitoring and Assessment*, 1988.

4. Cf. Jänicke et al., op. cit.

5. ibid.

6. International Institute for Environment and Society. *Research Program 1983–1987*. Berlin: IIES, 1982, p. 6.

7. The following data are from C. Leipert, U. E. Simonis. Environmental Damage – Environmental Protection. Empirical Evidence on the Federal Republic of Germany. In *International Journal of Social Economics*, Vol. 15, 1988, No. 7, pp. 37–52.

8. L. Wicke. *Die ökologischen Milliarden* (The Ecological Billions). Munich: Kösel, 1986, p. 123.

9. U. E. Simonis. Preventive Environmental Policy. Prerequisites, Trends and Prospects. In *Ekistics*, 313, July/August 1985, pp. 368–72.

10. In the following I rely on the arguments put forward by G. Scimemi: Environmental Policies and Anticipatory Strategies. In: U. E. Simonis (Ed.), *Präventive Umweltpolitik*, Frankfurt, New York: Campus, 1988.

11. Cf. Scimemi, op. cit.

12. ibid.

13. For a detailed illustration cf. ibid.

14. Cf. P. Wathern (Ed.), *Environmental Impact Assessment. Theory and Practice*, London: Unwin Hyman, 1988.

15. Cf. Scimemi, op. cit.

16. Cf. U. E. Simonis: Ecology and Economic Policy. In A. Vlavianos-Arvanitis (Ed.), *Biopolitics*, Athens: Biopolitics International Organisation, 1988, pp. 163–7.

[13]

Resources and Energy 13 (1991) 1–21. North-Holland

Sources of change in energy use in the U.S. economy, 1972–1982

A structural decomposition analysis

A. Rose*

Pennsylvania State University, University Park, PA 16802, USA

C.Y. Chen

National Cheng Kung University, Tainan, Taiwan

Final version received May 1990

Energy use patterns in industrial nations have gone through dramatic changes over the past 20 years. This paper offers a major advance in the methodology of structural decomposition analysis (SDA) and applies it to the determination of the relative prominence of various 'sources' of change in energy use in intermediate sectors of the U.S. economy between 1972–1982. These sources correspond to the pure form of input changes in the context of production theory. We derive a mutually exclusive and completely exhaustive set of estimating equations at the level of detail of the two-tier KLEM production function. It is our contention that these equations can yield as much insight as more elaborate and data-intensive approaches utilizing flexible functional forms. Our results indicate that the major sources of upward pressure on energy use were economic growth, KLEM substitution, and the joint effects of technological change, while the major sources of downward pressure were energy conservation and technological change in materials.

1. Introduction

The pattern of energy use has changed dramatically over the past dozen years. Several reasons for this phenomenon have been advanced, including: the Arab oil embargo, OPEC price hikes, development of more energy-efficient technology, and structural shifts in the economy away from more energy-intensive heavy industry. A few studies have authoritatively explained the impact of a small subset of stimuli [see, e.g., Mork and Hall (1980), Hirst et al. (1983), Jorgenson (1984)], but none have been able to determine the

*The authors wish to acknowledge the assistance of Shih-Mo Lin in helping to compile the necessary data. Peter Blair, Andrew Wyckoff, David Wood, and Timothy Considine provided helpful comments on earlier versions of the manuscript; however, the authors alone are responsible for any remaining errors or omissions.

0165–0572/91/$03.50

relative importance of the various causative factors in a comprehensive study. This is due, in great part, to the complexity of the cause–effect relationships involved. For example, an OPEC price increase is likely first to lead to interfuel substitution, then to spur the development and subsequent adoption of energy-saving technology, and finally to accelerate the eventual decline of heavy industry.

The purpose of this paper is to take a major step toward explaining the causes of change in energy use in the U.S. between 1972 and 1982 by determining the relative prominence of various 'sources' of change. These sources are the 'pure' form that changes take in the context of production theory, as opposed to the sociopolitical, engineering, or behavioral nature of the original stimuli. They include aspects of demand shifts, technical substitution, and technological change. The results enable us to make important inferences regarding the relative size of the original stimuli, even though there is not a perfect one-to-one correspondence between them and our sources of change.

Another objective of this study is to advance the state-of-the-art of input–output 'structural decomposition analysis'. We offer a formal derivation of estimating equations at a fine level of detail. The ensuing I–O computations are remarkably straightforward, but it is our hypothesis that they can yield as much insight as more elaborate and data-intensive econometric models of production technologies typically used in this area of research.

2. Conceptual framework

Since the classic work on intermediate energy demand analysis by Hudson and Jorgenson (1974), it has been popular to employ flexible functional forms, inclusive of all inputs, for the production function modeling used in energy analysis. The most typical example is the translog production function specified in terms of capital, labor, energy, and material aggregates, often denoted by the acronym 'KLEM'. Often, the production function has two tiers, with subaggregates within some of the KLEM components (e.g., coal, oil, gas and nuclear components of the energy aggregate).

Overall, the neoclassical KLEM modeling approach has proven to be a powerful and versatile tool in analytical studies and forecasting models. Hudson and Jorgenson (1974), Berndt and Wood (1975, 1979), and others have used a time series of input–output table data to estimate the parameters of the dual price frontier of a translog specification of the KLEM model. The resulting parameters yield elasticity measures that are invaluable in policy simulation modeling, but are of less importance in structural analysis, or the study of the broad set of considerations that reflect more permanent changes in the production system. These include, for example, changes in the mix of outputs, changes in the geographic origins and destinations of goods, and changes in technology.

We choose as our model framework input–output analysis[1] applied in the manner of what has recently been referred to as *structural decomposition analysis*, or 'the analysis of economic change by means of a set of comparative static changes in key parameters of an input–output table' [Rose and Miernyk (1989)]. This approach dates from the work of Leontief (1941), Chenery et al. (1962) and Strout (1967) in analyzing changing economic patterns over time. More recently, research in this area has reached a high level of sophistication with the inclusion of adjustments to trade patterns and prices [see, e.g., Skolka (1984), Kanemitsu and Ohnishi (1989)]. Moreover, there have been several valuable studies applying this methodology in some form to energy studies [see Lager et al. (1983), Gould and Kulshreshtha (1986), Richter (1987), U.S. Office of Technology Assessment (1990)].[2]

There are still two major shortcomings of these studies, which we intend to remedy in this paper. The first is most prevalent in the case of energy analysis and relates to the level of detail at which structural decompositions are performed. Most formulations merely distinguish between effects attributable to final demand changes and an undisaggregated set of technical coefficient changes [see also the more recent general applications by Feldman et al. (1987), Blair and Wyckoff (1989)].[3] The energy studies have thus failed to distinguish the influence of changes in energy coefficients themselves from the influence of other inputs, either in terms of aggregate or subaggregates.[4] To remedy this we will extend the analysis to a two-tier KLEM framework, which yields 11 separate sources of change plus three 'interactive' effects.

Second, all of the previous models, except the most basic [see, e.g., Feldman et al. (1987)] have been specified in an ad hoc manner. The more basic versions are intuitively obvious, and simply require the comparative static adjustments of the entire set of technical coefficients and final demand level or mix between two sets of I–O tables. This method of sequential, single element adjustment has proven useful in deriving more complex versions, but there is no guarantee that this approach will yield a set of estimating equations

[1]Many of the flexible functional form models have successfully been used to adjust the coefficients of an I–O model in response to price stimuli, so there is an inherent compatibility between the two formulations [see, e.g., Hudson and Jorgenson (1974)].

[2]The first significant 'decomposition' of energy use (though not in an I–O framework) was undertaken by Schurr and Netschert (1960).

[3]A noteworthy exception is Carter (1970), who decomposed structural change into separate material input and value-added components. Furthermore, Carter extends the analysis to a dynamic level by formulating the process of coefficient change as the allocation of investment for the purpose of replacement of less productive processes by more productive ones.

[4]A very recent study of changes in energy use by U.S. Office of Technology Assessment (1990) has distinguished energy inputs from the more general category of inputs.

that are mutually exclusive and completely exhaustive. We formally derive a set of distinct estimating equations in the following section, and point to the complete accounting for energy change (no residual) in the empirical application below as evidence that the equations are comprehensive.

3. The formal model

The basic two-tier KLEM model is typically expressed as

$$Q = F[K, L, E(E_1 \dots E_g), M(M_1 \dots M_h)].$$

In the neoclassical version, the production function is assumed to be a positive, twice differentiable, strictly quasi-concave function. Also, the convention is to assume the production function is homothetically weakly separable, meaning that the optimal mix of components in each aggregate is independent of the optimal mix of the four aggregate groups. In regard to our paper, the explicit inclusion of all inputs is inherently consistent with an input–output framework. Moreover, the linearity feature of the most basic version of the I–O model encompasses homothetic weak separability, a feature which facilitates the derivation of our theoretical model.

3.1. Derivation of the estimating equations

The total energy use in an economy in a given year can be expressed as

$$TE^t = B^t(I-A)^{-1}Y^t, \tag{1}$$

where

TE = a vector energy use by fuel type f,
B = the matrix of physical energy coefficients expressed in terms of BTUs of fuel type f used to produce 1 dollar of gross output of sector j,[5]
$(I-A)^{-1}$ = the matrix of total input requirements expressed as the fraction of a dollar of good i neded to produce 1 dollar of gross output of good j,
Y = the vector of final demand for each good i,
t = a given time period, i.e., year.

The change in total energy use over time is

$$\Delta TE = TE^t - TE^{t-1}. \tag{2}$$

[5]These energy coefficients are used to convert the energy coefficients (per dollar basis) of a standard I–O table into physical flows. This approach yields energy flows equivalent to those derived from a hybrid energy/I–O table.

Simplifying the notation by letting $G=(I-A)^{-1}$,

$$\Delta TE = B^t G^t Y^t - B^{t-1} B^{t-1} Y^{t-1}. \tag{3}$$

Therefore, in the most general terms, changes in energy use can be attributed to changes in input coefficients and in final demand. Our structural decomposition begins by attempting to separate the two major categories of sources of change in energy use. They can be derived from eq. (3) by first adding and subtracting and then rearranging like terms, thus:

$$\begin{aligned} \Delta TE = {} & B^{t-1} G^{t-1} Y^t - B^{t-1} G^{t-1} Y^{t-1} \\ & + B^t G^t Y^{t-1} - B^{t-1} G^{t-1} Y^{t-1} \\ & + (B^t G^t - B^{t-1} G^{t-1})(Y^t - Y^{t-1}). \end{aligned} \tag{4}$$

The first term of eq. (4) represents a final demand effect, while the second represents an input requirement (coefficient) effect. Interestingly, the separation cannot be made clearly, as the third term represents a 'joint effect' that captures interactions between the first two terms.[6]

Each of the first two categories can be further decomposed into an effect relating to changes in the relative proportions of variables, and an effect relating to the absolute level of each variable. For the case of the change in total energy use due to a change in final demand, from eq. (4),

$$\Delta TE_Y = B^{t-1} G^{t-1} \Delta Y^t, \tag{5}$$

further decomposition is the most straightforward. First, decomposing the final demand term itself yields

$$Y^t = Y^t + \left[Y^t \frac{\sum Y_i^{t-1}}{\sum Y_i^t} - Y^t \frac{\sum Y_i^{t-1}}{\sum Y_i^t} \right] - Y^{t-1} \tag{6a}$$

$$= \left[Y^t - Y^t \frac{\sum Y_i^{t-1}}{\sum Y_i^t} \right] + \left[Y^t \frac{\sum Y_i^{t-1}}{\sum Y_i^t} - Y^{t-1} \right]. \tag{6b}$$

Further simplifying the notation we let

[6]This term has been labeled an 'interaction effect' in the decomposition studies of Casler et al. (1990) and U.S. Office of Technology Assessment (1990). A similar term has simply been interpreted by Skolka (1989) as a way of eliminating the index number problem inherent in an analysis of changes over time. For now, we simply take the term to reflect any effects of the two major categories of sources of change occurring simultaneously.

$$Y_i^{t(t-1)} \equiv Y_i^t \frac{\sum Y_i^{t-1}}{\sum Y_i^t}, \qquad i=1,\ldots,n. \tag{7}$$

Thus, in the notation $Y^{t(t-1)}$, the first superscript refers to the year for which coefficient values (i.e., proportions, or the mix) are set, while the superscript in parentheses refers to the year for which control totals (levels) are set. We can now decompose eq. (5) into two terms:[7]

$$\begin{aligned} \Delta TE_Y = {} & B^{t-1}G^{t-1}Y^t - B^{t-1}G^{t-1}Y^{t(t-1)} \\ & \text{(final demand level effect)} \\ & + B^{t-1}G^{t-1}Y^{t(t-1)} - B^{t-1}G^{t-1}Y^{t-1} \\ & \text{(final demand mix effect).} \end{aligned} \tag{8}$$

In decomposing input requirements, we distinguish each of the KLEM components of the A matrix. We begin by isolating the second term of eq. (4), representing changes in input requirements in general:[8]

$$\Delta TE_A = B^t G^t Y^{t-1} - B^{t-1}G^{t-1}Y^{t-1}. \tag{9}$$

Recall that the standard practice in the two-stage KLEM model approach is to assume the production function is homothetically weakly separable. Thus, eq. (9) can be decomposed as follows (where the notation can be simplified if we explicitly denote only those input categories whose time values vary from the remainder of the coefficients of the A matrix or its Leontief inverse, G):[9]

$$\begin{aligned} \Delta TE_A = {} & (B^{t(t-1)}G^{t-1}_{Et(t-1)Mt(t-1)}Y^{t-1} - B^{t-1}G^{t-1}Y^{t-1}) \\ & + (B^{t(t-1)}G_{(KLEM)t(t-1)}Y^{t-1} - B^{t(t-1)}G^{t-1}_{Et(t-1)Mt(t-1)}Y^{t-1}) \\ & + (B^t G^t Y^{t-1} - B^{t(t-1)}G_{(KLEM)t(t-1)}Y^{t-1}). \end{aligned} \tag{10}$$

[7]We could further decompose the final demand component along the lines of personal consumption, capital formation, and government along similar lines [see also Richter (1987)]. However, because our main focus is KLEM modeling we will analyze the former two components in conjunction with labor and capital inputs, respectively in a closed inverse (see also footnote 6). A further decomposition of exports and imports is more complicated, and will not be presented here. The reader is referred to a separate study by Rose et al. (1991) in which trade-related sources of change in a open economy extension of this model were found to be relatively small despite some significant changes in the import dependency of the U.S.

[8]In eq. (9), and in subsequent equations the Leontief inverse, G, is closed with respect to capital and labor inputs (energy and material inputs are already incorporated in the open inverse). This also means that personal consumption and capital formation elements are removed from final demand from eq. (9) on. We have chosen to cue the reader of these facts in this way rather than introduce new symbols that would unnecessarily clutter the presentation.

[9]That is, for example, in eq. (10):

$$G^{t-1}_{Et(t-1)Mt(t-1)} \equiv G^{t-1}_{Kt-1Lt-1Et(t-1)Mt(t-1)}.$$

The first term of eq. (10) refers to the bottom-tier component substitution effects within an aggregate. The second term refers to the top-tier substitution effect of the four aggregate groups, while the third term refers to the technological change effects associated with the four aggregate groups. Where, for any aggregate (K, L, E, M) within the input coefficient matrix,

$$A_i^{t(t-1)} \equiv A_i^t \frac{\sum A_i^{t-1}}{\sum A_i^t}, \qquad i = 1, \ldots, n, \tag{11a}$$

and

$$(KLEM)t(t-1) \equiv (KLEM)t \frac{(KLEM)t-1}{(KLEM)t}. \tag{11b}$$

Analogous to eq. (4), we decompose the first term (bottom-tier substitution) of eq. (10), thus:

$$
\begin{aligned}
TE_{A1} = {} & B^{t(t-1)}G^{t-1}Y^{t-1} - B^{t-1}G^{t-1}Y^{t-1} \\
& \text{(interfuel substitution, direct)} \\
& + B^{t(t-1)}G^{t-1}_{Et(t-1)}Y^{t-1} - B^{t(t-1)}G^{t-1}Y^{t-1} \\
& \text{(interfuel substitution, linkage)} \\
& + B^{t-1}G^{t-1}_{Mt(t-1)}Y^{t-1} - B^{t-1}G^{t-1}Y^{t-1} \\
& \text{(material substitution)} \\
& + [B^{t(t-1)}G^{t-1}_{Mt(t-1)Et(t-1)}Y^{t-1} - B^{t(t-1)}G^{t-1}_{Et(t-1)}Y^{t-1}] \\
& - [B^{t-1}G^{t-1}_{Mt(t-1)}Y^{t-1} - B^{t-1}G^{t-1}Y^{t-1}] \\
& \text{(substitution joint effect).}
\end{aligned} \tag{12}
$$

There are two types of interfuel substitution effects. The first involves just the direct effect of changing energy coefficients and the second pertains to changes in energy use throughout the rest of the economy as a result of second-order, or 'linkage', effects. Note also that there is a joint effect of the three substitution terms. It does not violate the assumption of weak separability because it takes place within a single tier of the production function.

The second term of eq. (10) requires no further decomposition, and we present it here in isolation as depicting substitution among aggregate input categories:

$$B^{t(t-1)}G_{(KLEM)t(t-1)}Y^{t-1} - B^{t(t-1)}G^{t-1}_{Mt(t-1)Et(t-1)}Y^{t-1} \tag{13}$$
(KLEM substitution).

In a manner similar to eq. (12), we can decompose the third term of eq. (10), thus:

$$TE_{A3} = B^{t(t-1)}G_{Kt(LEM)t(t-1)}Y^{t-1} - B^{t(t-1)}G_{(KLEM)t(t-1)}Y^{t-1}$$
(technological change in capital)

$$+ B^{t(t-1)}G_{Lt(KEM)t(t-1)}Y^{t-1} - B^{t(t-1)}G_{(KLEM)t(t-1)}Y^{t-1}$$
(technological change in labor)

$$+ B^{t}G_{(KLEM)t(t-1)}Y^{t-1} - B^{t(t-1)}G_{(KLEM)t(t-1)}Y^{t-1}$$
(technological change in energy, direct)

$$+ B^{t}G_{Et(KLM)t(t-1)}Y^{t-1} - B^{t}G_{(KLEM)t(t-1)}Y^{t-1}$$
(technological change in energy, linkage)

$$+ B^{t(t-1)}G^{t-1}_{Mt(KLE)t(t-1)}Y^{t-1} - B^{t(t-1)}G_{(KLEM)t(t-1)}Y^{t-1}$$
(technological change in materials)

$$+ [B^{t}G^{t}Y^{t-1} - B^{t(t-1)}G_{Mt(KLE)t(t-1)}Y^{t-1}]$$

$$- [B^{t(t-1)}G_{Kt(LEM)t(t-1)}Y^{t-1} - B^{t(t-1)}G_{(KLEM)t(t-1)}Y^{t-1}]$$

$$- [B^{t(t-1)}G_{Lt(KEM)t(t-1)}Y^{t-1} - B^{t(t-1)}G_{(KLEM)t(t-1)}Y^{t-1}]$$

$$- [B^{t(t-1)}G_{Lt(KEM)t(t-1)}Y^{t-1} - B^{t(t-1)}G_{(KLEM)t(t-1)}Y^{t-1}]$$

$$- [B^{t}G_{Et(KLM)t(t-1)}Y^{t-1} - B^{t(t-1)}G_{(KLEM)t(t-1)}Y^{t-1}]$$
(technological change joint effect). (14)

Analogous to the interfuel substitution effect, the technological change in energy (i.e., energy conservation) is split into direct and linkage components. Also, analogously there is a joint effect term for the technological change category, again taking place within a single tier.

We can summarize our derivation by referring first to eq. (4). The final demand effect of that equation was decomposed in eq. (8). The input acquirement effect was decomposed in terms of bottom-tier substitution [eq. (12)], top-tier substitution [eq. (13)], and technological change [eq. (14)]. This leaves the joint effect between final demand input change, the third term in eq. (4), which we repeat below to complete our equation set:

$$(B^{t}G^{t} - B^{t-1}G^{t-1})(Y^{t} - Y^{t-1})$$
(joint effect between final demand and input change). (15)

Table 1

Decomposition of change in energy demand.

Source	Estimating equation[a]
1. Level of final demand	$B^{72}G^{72}Y^{82} - B^{72}G^{72}Y^{82(72)}$
2. Mix of final demand	$B^{72}G^{72}Y^{82(72)} - B^{72}G^{72}Y^{72}$
3. Interfuel substitution (direct)	$B^{82(72)}G^{72}Y^{72} - B^{72}G^{72}Y^{72}$
4. Interfuel substitution (linkage)	$B^{82(72)}G^{72}_{E82(72)}Y^{72} - B^{82(72)}G^{72}Y^{72}$
5. Material substitution	$B^{72}G^{72}_{M82(72)}Y^{72} - B^{72}G^{72}Y^{72}$
6. Substitution joint effect	$B^{82(72)}G^{72}_{M82(72)}Y^{72} - B^{82(72)}G^{72}Y^{72} - B^{72}G^{72}_{M82(72)}Y^{72} - B^{72}G^{72}Y^{72}$
7. KLEM substitution	$B^{82(72)}G^{82(72)}Y^{72} - B^{82(72)}G^{72}Y^{72}$
8. Tech. change in capital	$B^{82(72)}G^{82(72)}_{K82}Y^{72} - B^{82(72)}G^{82(72)}Y^{72}$
9. Tech. change in labor	$B^{82(72)}G^{82(72)}_{L82}Y^{72} - B^{82(72)}G^{82(72)}Y^{72}$
10. Tech. change in energy (direct)	$B^{82}G^{82(72)}Y^{72} - B^{82(72)}G^{82(72)}Y^{72}$
11. Tech. change in energy (linkage)	$B^{82}G^{82(72)}_{E82}Y^{72} - B^{82}G^{82(72)}Y^{72}$
12. Tech. change in materials	$B^{82(72)}G^{82(72)}_{M82}Y^{72} - B^{82(72)}G^{82(72)}Y^{72}$
13. Tech. change in joint effect	$B^{82}G^{82}Y^{72} - B^{82(72)}G^{82(72)}_{M82}Y^{72} - B^{82(72)}G^{82(72)}_{K82}Y^{72} - B^{82(72)}G^{82(72)}_{L82}Y^{72} - B^{82}G^{82(72)}_{E82}Y^{72} + 3(B^{82(72)}G^{82(72)}Y^{72})$
14. Joint effect between final demand and input change	$(B^{82}G^{82} - B^{72}G^{72})(Y^{82} - Y^{72})$

[a]See text for explanation of notation.

3.2. *Presentation of the estimating equations*

The equations of our I–O structural decomposition approach are presented in summary form in table 1. Note that we have further simplified the notation by dropping subscripts between the terms of any of our equations, thus isolating the comparative static change the equation represents,[10] and, by definition, invoking notational substitutions such as $G^{t(t-1)} \equiv G_{(KLEM)t(t-1)}$. For example,

$$[B^{t(t-1)}G^{t(t-1)}_{Kt}Y^{t-1} - B^{t(t-1)}G^{t(t-1)}Y^{t-1}]$$

$$\equiv [B^{t(t-1)}G_{Kt(LEM)t(t-1)}Y^{t-1} - B^{t(t-1)}G_{(KLEM)t(t-1)}Y^{t-1}].$$

The first two equations decompose final demand changes into 'level' and 'mix' components.[11] For example, in eq. (2), the effect of a change in the

[10]We emphasize that the equations in table 1 are stylized versions of the equations derived above, intended for the purpose of simplifying the notation. We utilized the more complex equations for our empirical analysis. Strictly speaking the application of the equations in table 1 would yield slightly different results.

[11]Eqs. (1) and (2) are computed with an open I–O inverse, while eqs. (3) through (14) are computed with an I–O inverse closed to both labor and capital.

level of economic activity is modeled by utilizing 1972 energy and other technical coefficients, as well as 1972 final demands, throughout, but with a juxtaposition of 1972 and 1982 final demand mixes. This means subtracting actual energy use in 1972 from the calculation of the energy use in the U.S. economy that would have been required given a 1982 mix of final demand, set at its 1972 level, in order to determine the change due simply to a change in the mix. In the notation $Y^{82(72)}$, the first superscript refers to the year coefficient values (i.e., proportions, or the mix) are set, while the superscript in parentheses refers to the year that serves as the control total (level) for final demand, or short hand for $Y_i^{82}(\sum Y_i^{72}/\sum Y_i^{82})$.[12]

Eqs. (3) through (6) refer to substitution within the bottom-tier of the energy and material aggregates. In this case, 1982 coefficients of individual components of E and M, respectively, are utilized as weights for the corresponding aggregates, set at 1972 levels, to capture substitution effects, e.g., for the set of fuels, f, $E^{82(72)}=E^{82}(\sum E_f^{72}/\sum E_f^{82})$.[13] Again the energy-related estimation is split into two components. The first refers to the pure interfuel substitution effect, which should sum to zero if equivalent fuel efficiency requirements are assumed. The second refers to the higher-order repercussions of this substitution.

Eq. (7) determines the effects of substitution within the top-tier of the production function. Here aggregate coefficients are set at their 1982 proportions, though the components of the energy and material aggregates maintain their 1972 proportions. At the same time, the total of the aggregates is adjusted to its 1972 coefficient level, i.e. $KLEM^{82(72)}=KLEM^{82}(\sum(KLEM)^{72}/\sum(KLEM))^{82}$.

Eqs. (8) through (13) isolate productivity changes in each of the four KLEM components. Here, the subscript symbols refer to the elements of the extended A matrix under scrutiny, all other elements being held constant. In the case of capital and labor there is only a single variable each to be set at its 1982 level. Note, that there are a pair of equations relating to technological change in energy (energy conservation). Eq. (10) refers to the direct conservation effects operating through the matrix of physical unit energy coefficients, B. Eq. (11) covers the indirect and induced (linkage) effects of energy conservation via a change in the energy coefficients of the A matrix and $G=(I-A)^{-1}$. That is, a reduction in energy use per unit of output will lead to a reduction in the demand for energy products and their various direct and indirect suppliers, which in turn, sets off more rounds of energy production (level but not coefficient) reductions. In the case of materials, which is also an aggregate of inputs, 1982 coefficients are juxtaposed to

[12]All control totals are expressed in constant dollar terms.

[13]The summation of fuel types within each sector refers to coefficient values and not just flows in relation to the column total.

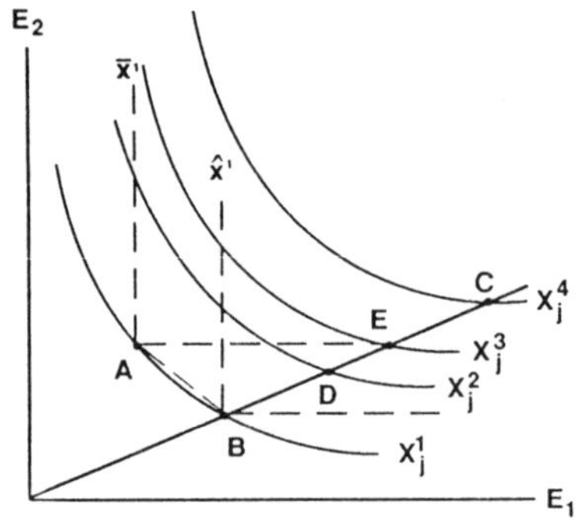

Fig. 1(a). Bottom-tier effects within an I–O framework.

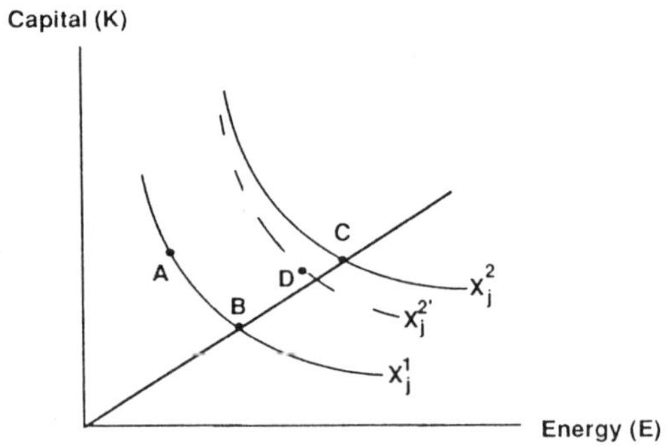

Fig. 1(b). Top-tier effects within an I–O framework.

relative proportions of these component inputs, adjusted to their overall 1982 levels.

4. Comparison with the neoclassical approach

The system of eqs. (1) through (14) are mutually exclusive and completely exhaustive of pure production theory sources of change in energy demand within an I–O framework. We can further distinguish these types of change, and compare the I–O framework to the more conventional neoclassical framework with a diagrammatic presentation. Figs. 1(a) and 1(b) illustrate the assumptions and implications of an I–O framework for the case of two energy components and two aggregate inputs, respectively. At first glance, the reader may question the presence of a set of neoclassical (convex and differentiable) isoquants in the figure, but we reiterate our intent to use the

I–O framework to analyze the economy at two different points in time, and thus to be able to capture many neoclassical properties by coefficient change. Our two-dimensional representation is consistent with the Samuelson Substitution Theorem, which in effect enables us to consider an equilibrium point in the figure as the vertex of a right-angle isoquant, more typically associated with I–O, or alternatively as a piecewise linear isoquant characteristic of a more general activity analysis problem.

In fig. 1(a) the move from point A to point B reflects the substitution of E^1 for E^2 in a given sector in response to a change in the relative price of the two energy inputs in favor of the former (technical substitution).[14] The move from point B to point C reflects substitution among aggregates in favor of energy, with the fact that both points are on a straight line emanating from the origin reflecting the homothetic weak separability assumption of the modified I–O approach, i.e., any increase in the energy aggregate due to substitution with other aggregates does not alter the relative proportion of energy subaggregates.[15] The move from C to D would be representative of (non-price-induced) energy conservation, but again the model is subject to the weak homothetic separability assumption that does not enable us to distinguish relative differences in conservation between the two fuels. The move from point D to point E is intended to correspond to an increase in the derived demand for the two fuels as a result of an increase in final demand. The linearity property of estimating eqs. (1) and (2) carries with it the assumption of strong homotheticity.

Fig. 1(b) depicts the implications of these same phenomena for energy and capital aggregates. The lowering of the price of one of the fuels lowers the price of the energy aggregate relative to capital, and results in the substitution from point A to point B [which corresponds to the move from B to C in fig. 1(a)]. The move from point B to point C reflects the final demand increase for the output of the given sector and the linearity property of the production function. The move from C to D corresponds to energy conservation, depicted as a form of technological change, by the shifting isoquant from X_j^2 to $X_j^{2\prime}$. The requirement of homotheticity is not binding in this case, hence the final equilibrium point need not be on the original ray from the origin.

While the assumption of homotheticity, equivalent to the standard

[14]Note, so as not to clutter the diagram, we have not included the conventional rotating relative price lines underlying the shift from points such as A to points such as B. Also, in both the I–O and neoclassical cases, we do not distinguish between direct (partial equilibrium) or total (general equilibrium) effects, but simply intend the diagrams to convey the latter situation. This has no bearing on the outcome. Finally, note that the technical substitution in fig. A.1, can also represent the movement from one fixed requirement process to another over time ($\bar{x}^1$ to $\hat{x}^1$).

[15]The fixed coefficient assumption of the standard I–O model implies strong homotheticity. In all but one case, our modified I–O approach is less restrictive because it only implies weak homothetic separability.

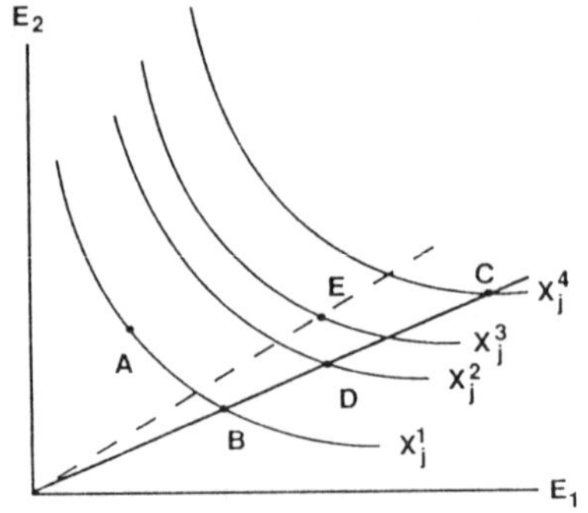

Fig. 2(a). Bottom-tier of a neoclassical production function.

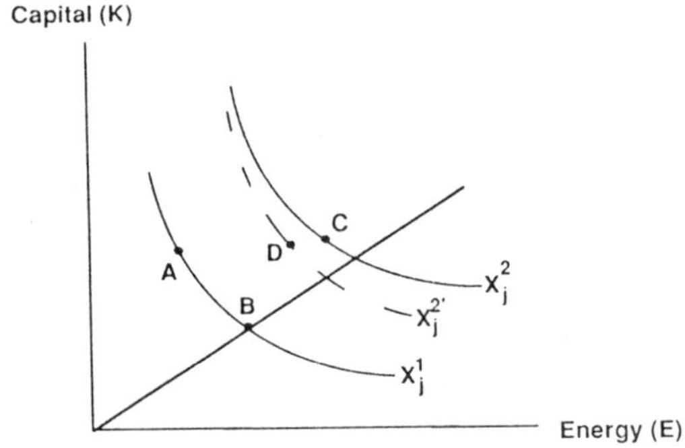

Fig. 2(b). Top-tier of a neoclassical production function.

assumption of fixed input proportions in the standard I–O model would appear to be rather restrictive, it is much less so in the time-based comparison of our modified I–O framework. Input substitution can be incorporated within the aggregate tier and within the sub-aggregate tier, as can technological change in the aggregate tier. Full homotheticity is not imposed in any of these cases and only weak homothetic separability is in effect in the substitution relations among sub-aggregates stemming from the substitution of aggregates [the move from point *B* to point *C* in fig. 1(a)]. We will show below that even neoclassical flexible functional forms are restrictive to the same degree in all of these instances. The only difference between our framework and the neoclassical one is our more restrictive result pertaining to the response to output change.

The paradigm neoclassical model is illustrated in figs. 2(a) and 2(b) for cases analogous to figs. 1(a) and 1(b). The sequences of stimuli are the same between the two sets of figures as well. Note that in fig. 2(a) the response is identical to fig. 1(a) in the move from points A to B to C to D. The neoclassical model does not, however, involve a strong or weakly separable homothetic property associated with output change, and thus point E is not restricted to the original ray from the origin. The results are similar for the case of input aggregates depicted in fig. 2(b). They differ from fig. 1(b) only for the case of the output effect.

Overall, we have shown that the modified I–O model, which is easier to estimate empirically, can yield results as extensive as a neoclassical flexible functional form, and with only one additional restriction. The restriction could result in a confusion of some substitution effects for output effects in the I–O approach. The implications for the accuracy of the results depends on the size of output change. For modest changes, the I–O approximation is likely to yield results whose accuracy can rival the neoclassical model. In cases of constant-returns-to-scale, the two methodologies would yield equivalent results.

5. Application

The decomposition equations presented in section 3 were applied to the analysis of the change in energy use in the intermediate sectors of the U.S. economy between 1972 and 1982. This period was chosen partly because the end-points correspond to benchmark years for two U.S. I–O tables, but primarily because this period coincides with one of the most dramatic periods of change in energy use in the nation's history.

5.1. *The empirical model*

The I–O tables in the analysis consist of the official 1972 U.S. BEA input–output table (U.S. Bureau of Economic Analysis (1979a)] and the U.S. Forest Services update to 1982 of the BEAs 1977 I–O table [U.S. Forest Service (1986)].[16] The 1972 I–O table was transformed to 1982 constant dollar terms with the use of sectoral deflators obtained from U.S. Office of Technology Assessment (1989). While a decomposition analysis is best performed at the finest level of detail possible, the preliminary nature of our inquiry lends itself to a more aggregated, and hence more manageable, two-digit BEA level classification consisting of 80 sectors. At this level there are four energy sectors: Coal mining (BEA Industry 7), Crude petroleum and natural gas (8), Petroleum refining (31), and Utilities (68). The latter is composed primarily of electric and gas utilities, the effects of which can be

[16]The BEA benchmark I–O table for 1982 has been delayed for some time, and thus was not available for our study.

Table 2

Intermediate sector energy use, 1972 and 1982 (in trillion BTUs).

Year	Coal	Petroleum	Natural gas	Electricity	Total[a]
1972	11,908	29,590	16,609	3,311	60,005
1982	15,233	28,533	13,359	4,221	60,591

[a]The total represents only primary energy use, and hence does not equal the sum of the elements of each row (see text).

Source: Calculated from U.S. I–O tables and data from U.S. Energy Information Administration (1987a, b).

separated by a variety of methods. We chose the most straightforward – using separate physical energy coefficients for electricity and gas.

Fuel use by sector in BTUs was calculated with a combination of actual data and average (implicit) price data. The energy input for each sector in 1972 and 1982 derived from the I–O 'use' matrix by first dividing energy coefficients (in dollar value terms) by energy prices for major sector groupings – industrial, commercial, electric utility, transportation, residential [U.S. Energy Information Administration (1987b)] – and then scaling the results to conform to energy consumption control totals [U.S. Energy Information Administration (1987a)].[17] Adjustments for energy transformation losses and primary vs. secondary energy distinctions are discussed below [see also Lager et al. (1983), Miller and Blair (1985) and Richter (1987) for a discussion of the calculations of energy use in an I–O framework].

The results of our calculation of energy use by fuel type for the intermediate sectors of the U.S. economy are presented in table 2. Despite the Arab oil embargo and OPEC price escalations, petroleum use decreased only slightly during the study period and continued to be the major fuel used in the intermediate sectors. Natural gas, which was characterized first by regulatory-induced shortages and then by price increases during phased deregulation, suffered a sizeable decline in use over the period.

The closed I–O inverse used in eqs. (3) through (13), in order to determine induced effects as well as direct and indirect ones. The employee compensation row of the tables was included in the inverse and also served as the basis for the labor input of the KLEM formulation. An estimate of personal consumption from wages in the two tables served as the spending counterparts of the income rows. On the capital side, we used a gross returns measure combining property-type income [U.S. Bureau of Economic Analysis (1979a), U.S. Forest Service (1986)] and depreciation [U.S. Bureau of Labor Statistics (1990)] on a sectoral basis. The row totals for both

[17]Distinct data were obtained for the electric and gas utility sectors for 1972 from the BEA four-digit 'use' matrix [U.S. Bureau of Economic Analysis (1979b)] and for 1982 from the four-digit IMPLAN use table [U.S. Forest Service (1989)].

Table 3

Sources of change in intermediate sector energy demand, 1972–1982 (in trillion BTUs).

Source	Coal	Petroleum	Natural gas	Electricity	Weighted average
Level of final demand	2,004	5,495	2,993	640	10,859
Mix of final demand	−1,695	−1,581	−1,357	−49	−4,661
Interfuel substitution (direct)	2,354	−1,394	−2,729	1,156	−0
Interfuel substitution (linkage)	624	131	166	−1,920	931
Material substitution	−835	−1,178	−1,177	−126	3,262
Joint effect of substitution	−35	−37	139	1,833	−46
KLEM substitution	1,187	4,114	1,398	570	7,168
Tech. change in capital	−432	−869	−428	−149	−1,852
Tech. change in labor	−650	−1,253	−608	−204	−2,678
Tech. change in energy (direct)	−395	−5,072	−1,619	−851	−7,785
Tech. change in energy (linkage)	−17	22	22	5	30
Tech. change in materials	85	−4,244	−1,195	−685	−5,917
Tech. change in joint effect	311	5,222	1,633	879	7,888
Joint effect of FD & input change	821	−414	−487	−187	−88
Total	3,325	−1,057	−3,249	910	587

conformed to the National Income and Product Accounts for 1972 and 1982. The counterpart columns represent spending out of property-related income and purchases of replacement investment goods from individual supplying sectors. In the absence of explicit sectoral data for the latter, we used the sectoral proportions of the Total Capital Formation columns of the 1972 and 1982 tables. This will result in biased results to the extent that the sectoral composition of plant and equipment purchases for replacement purposes differs from the sectoral composition of total investment.

5.2. *Results*

The results of our decomposition analysis of the structure of change in energy use are presented in tables 3 and 4. Table 3 presents levels of change in energy use of each fuel and a weighted average of fuels[18] attributable to each of our 14 estimating equations. Note that, aside from a very minor rounding error in some cases, the sum of the entries in each column equals the column total. That is, our analysis is able to account for 100 percent of the change. We summarize them in relation to the individual sources of change in energy use as follows:

(1) The change in the level of net domestic final demand is one of the most significant influences on energy use. The total 10,859 trillion BTU demand increase (table 3) attributable to this source is an 18 percent increase (table 4)

[18]Note that the weighted average includes only primary energy to avoid double-counting of, say, coal utilized directly by industry and coal used to produce electricity. For this purpose, we relied on energy conversion factors obtained from U.S. Office of Technology Assessment (1990).

Table 4

Percentage change in intermediate sector energy demand, 1972–1982.

Source	Coal	Petroleum	Natural gas	Electricity	Weighted average
Level of final demand	17	19	18	19	18
Mix of final demand	−14	−5	−8	−1	−8
Interfuel substitution (direct)	20	−5	−16	35	0
Interfuel substitution (linkage)	5	[a]	1	−58	2
Material substitution	−7	−4	−7	−4	−5
Substitution joint effect	[a]	[a]	1	55	[a]
KLEM substitution	10	14	8	17	12
Tech. change in capital	−4	−3	−3	−5	−3
Tech. change in labor	−5	−4	−4	−6	−4
Tech. change in energy (direct)	−3	−17	−10	−26	−13
Tech. change in energy (linkage)	[a]	[a]	[a]	[a]	[a]
Tech. change in materials	1	−14	−7	−21	−10
Tech. change in joint effect	3	18	10	27	13
Joint effect of FD & input change	7	−1	−3	−6	[a]
Total	28	−4	−20	27	1

[a]Less than 0.5 percent.

over 1972 total intermediate energy use levels. It is significantly less than the approximately 28 percent real increase in GNP between 1972 and 1982, because the growth of final purchases of goods and services (which give rise to intermediate energy demand) was significantly lower than the growth of total final demand (which also included especially fast-growing import and government payroll components).

(2) The mix effect is negative as expected. That is, the relatively large increase in energy prices caused a shift to less (directly and indirectly) energy-intensive final goods and services.

(3) Direct interfuel substitution is not an overall major source of change, because of the offsetting increases and decreases between the various fuel types in each sector and for the economy as a whole. Gas and petroleum demand are reduced primarily in favor of coal and electricity in that order in absolute terms (table 3) though percentage for electricity is greater than for coal (table 4). The results might at first appear somewhat surprising in terms of the large changes associated with gas and electricity: however, they can be explained by the actual changes in relative prices of the two fuels.[19] The increase in coal demand is not surprising in terms of relative prices and also suggests the lack of any significant environmental control cost premium.

(4) Indirect effects of interfuel substitution are a very modest positive

[19]In real terms, natural gas prices rose six-fold between 1972 and 1982 despite being regulated, while the price of electricity rose slightly more than three-fold. (Interestingly, despite all the attention paid to machinations in oil prices during the study period, they rose less than five-fold).

amount. We can offer no explanation of the reason for the significant negative second-order effect for electricity at this time.

(5) Material substitution (substitution between intermediate inputs other than energy) is a minor negative source of change in energy demand over the study period. It appears that the relative energy intensity of materials might have been a consideration.

(6) The joint effect of input substitution does not have any effect on the weighted average of energy use, nor any significant effect on coal, oil or gas. The large effect for electricity, however, is probably due to its secondary energy nature (i.e. the overwhelming majority of electricity is produced from coal, oil and gas).

(7) KLEM substitution was overall the third most significant positive source. A major reason for this result was the decrease in value added as a proportion of total gross output between 1972 and 1982 (the latter was a recession year). This means that energy and materials aggregates increased relative to capital and labor, thus resulting in a positive net direct and indirect stimulus to energy demand. Note also that, while the changes in subaggregate fuels in any given sector have to be equivalent because of the assumption of weak homothetic separability, because of differing energy (and individual fuel) intensities, these changes need not be a constant proportion over 1972 base year values for the economy as a whole. Since the mix of sectors changed significantly as a result of the second-order effects of KLEM substitution, and were sizeable, our results are plausible.

(8) Technological change, or productivity improvement, in capital represents one of the more minor sources of change. The increase in energy use stems from the fact that the production of the mix of capital goods became directly and indirectly less energy-intensive, thus offsetting any energy decreases stemming from an increase in capital per unit of output. Note that this is the pure productivity effect; the effect of increased capital intensity is included in estimating equation 7 of table 1.

(9) Technological change relating to labor represents another small negative source of change in energy demand. This could stem from the fact that wage increases lagged behind labor productivity increases. The decrease in wages per unit of output translates into relatively lower income, then lower consumption, and eventually to a lower derived demand for energy.

(10) Direct technological change in energy, essentially energy conservation, was the strongest negative influence on energy use. This indicates that policies promoting energy conservation in response to the energy crisis were somewhat effective [recall price induced conservation was picked up in eq. (7)]. Again, as in eq. (7), the non-uniform effect on individual fuel demand is still consistent with the assumption of weak homothetic separability, since it pertains only to direct effects in any given sector. Thus, we have a significant difference in the effectiveness of conservation across fuel types.

(11) The linkage effect of technological change in energy is very small but positive. We would, however, expect second-order effects of conservation to be negative. Several studies have found the implementation of energy saving equipment, whose production generates a derived demand for energy [see, e.g., Rose (1985)], to yield positive linkage effects. That phenomenon is really an example of KLEM substitution, and we have relegated such effects to eq. (7). We can offer no further explanation of our result at this time.
(12) Technological change in materials had a significant negative effect on energy use. To some extent, this reflects the decrease in the ratio of all intermediate goods to economy-wide gross output (47.4 percent in 1972 to 46.0 percent in 1982). If we further adjust for the displacement of intermediate goods (e.g., steel) by imports between the 2 years, we see that the 10 percent decrease in energy use is reasonable.
(13) The technological change joint effect is the only one of our interaction effects of any overall significance, accounting for a 13 percent increase in energy use. If it is a true interaction effect and not just an index number phenomenon, it would reflect the not so surprising interdependence of factor productivity. However, because the effect is positive we hesitate to draw such a conclusion without further study.
(14) Note that the joint effect of final demand and input change has a rather small impact on any single fuel and that these impacts cancel each other out to a zero overall net effect. This result may be idiosyncratic to our analysis and the reader should not conclude that this joint effect is not potentially important.[20]

6. Conclusion

This paper has presented an alternative to conventional approaches to analyzing energy use patterns. A formal input–output structural decomposition model was derived, and then applied to changing energy demand in the U.S. economy. The results are indicative of the relative prominence of the major sources of change in this demand between 1972 and 1982.

Although we have termed our framework of analysis 'structural', its potential applications extend further than typical analyses of this kind. Still, it should be viewed as complementing rather than competing with more conventional production function and macro-econometric models of energy demand. In conjunction with one or the other of these two model forms, it can yield even more extensive insights. Basic stimuli such as embargoes,

[20]Note that there are some significant differences in the results presented in this version of the paper in relation to an earlier, widely circulated draft. They are: a relative decrease in the prominence of conservation and KLEM substitution, as well as a reversal of the sign of the final demand mix effect from positive to negative. These changes are due to a combination of an respecification of the model to include joint effects and an improvement in the data base.

changes in technology, etc., can be mapped into the mechanisms, or sources, of change in energy demand designated in this paper. For example, government mandated energy conservation is likely to manifest itself in what we have designated as 'technological change in energy'. Additional implications of the stimulus might be explored through the use of more conventional models or constructs, such as demand elasticities, to translate reductions in the cost of producing intermediate goods into changes in intermediate and final demand. Another example would be the impact of an oil embargo stimulus, which corresponds most to our equations for the impact of the 'level of final demand', 'interfuel substitution', and 'KLEM substitution'. The portion of the change in final demand attributable to the embargo can be calculated with the use of a macro-econometric model, and the I–O model can be used to further partition sources of change that transmit the stimulus and allocate the impacts to individual sectors.

Input–output models have proven useful in energy analysis for years, but primarily as sources of data or for uncomplicated demand or impact studies. This paper has shown that I–O also has the potential to serve as a valuable analytical tool to be applied to complex events relating to energy and the economy.

References

Berndt, E. and D. Wood, 1975, Technology, prices and the derived demand for energy, Review of Economics and Statistics 57, 259–268.

Berndt, E. and D. Wood, 1979, Engineering and economic interpretations of energy–capital complementarity, American Economic Review 69, 342–354.

Blair, P. and A. Wyckoff, 1989, The changing structure of the U.S. economy, in: R. Miller, K. Polenske and A. Rose, eds., Frontiers of input–output analysis (Oxford University Press, New York).

Carter, A., 1970, Structural change in the American economy (Harvard University Press, Cambridge, MA).

Casler, S., A. Afriasabi and M. McCauley, 1990, Decomposing change in energy input–output coefficients, Resources and Energy 13, this issue.

Chenery, H., S. Shishido and T. Watanabe, 1962, The pattern of Japanese growth, 1914–1954, Econometrica 30, 98–139.

Feldman, S., D. McClain and K. Palmer, 1987, Sources of structural change in the United States, 1963–978: An input–output perspective, Review of Economics and Statistics 69, 461–514.

Gould, B. and S. Kulshreshtha, 1986, An interindustry analysis of structural change and energy use linkages in the Saskatchewan economy, Energy Economics 8, 186–196.

Hirst, E., R. Marlay, D. Greene and R. Barnes, 1983, Recent changes in U.S. energy consumption, Annual Review of Energy 8, 193–246.

Hudson, E. and D. Jorgenson, 1974, U.S. energy policy and economic growth, 1975–2000, Bell Journal of Economics 5, 461–514.

Jorgenson, D., 1984, The role of energy in productivity growth, Energy Journal 5, 11–25.

Kanemitsu, H. and H. Ohnishi, 1989, An input–output analysis of technological change in the Japanese economy: 1970–1980, in: R. Miller, K. Polenske and A. Rose, eds., Frontiers of input–output analysis (Oxford University Press, New York).

Lager, C., K. Musil and J. Skolka, 1983, Input–output analysis of energy conversion in Austria,

1955–1980, Proceedings of the Fourteenth IIASA Task Force Meeting on Input–Output Modeling, Laxenburg, Austria.

Leontief, W., 1941, Structure of the American economy (Oxford University Press, New York).

Miller, R. and P. Blair, 1985, Input–output analysis: Foundations and extensions (Prentice-Hall, Englewood Cliffs, NJ).

Mork, K. and R.E. Hall, 1980, Energy prices, inflation and recession, 1974–75, Energy Journal 1, 31–63.

Richter, J., 1987, Rise in energy prices and industrial response: The Austrian case, Paper presented at the IIASA Workshop on Energy, the Economy, and the Environment, Laxenburg, Austria.

Rose, A., 1985, Modeling energy conservation programs, Energy Journal 6, 87–103.

Rose, A. and C.Y. Chen, 1987, Sources of change in energy use in the U.S. economy, 1972–1982, in: J. Rowse, ed., World energy markets: Coping with instability (Friesen, Calgary, Alb.).

Rose, A. and W. Miernyk, 1989, Input–output analysis: The first fifty years, Economic Systems Research 1, 229–271.

Rose, A. and D. Kolk, 1987, Forecasting the demand for natural gas in a changing world (JAI Press, Greenwich, CT).

Rose, A., C.Y. Chen and S.M. Lin, 1991, The role of international competition in changing energy use patterns in the U.S., in: O. Guvenen, J.B. Lesourd and W. Labys, eds., Politiques économiques internationaux de matières premières: Analyses Econométriques, Economica, Paris.

Schurr, S.H. and B.C. Netschert, 1960, Energy in the American economy, 1850–1975 (Johns Hopkins University Press, Baltimore, MD).

Skolka, J., 1984, Input–output anatomy of changes in employment structure in Austria between 1964 and 1976, Austrian Economic Papers 2, 205–233.

Skolka, J., 1989, Input–output structural decomposition analysis for Austria, Journal of Policy Modeling 11, 45–66.

Strout, A., 1967, Technological change in the United States energy consumption, 1939–54, Doctoral dissertation (University of Chicago, Chicago, IL) (Reprinted by Arno Press, New York, 1979).

U.S. Bureau of Economic Analysis, 1979a, U.S. input–output table, 1972, Survey of Current Business 57, 34–72.

U.S. Bureau of Economic Analysis, 1979b, The detailed input–output structure of the U.S. economy, 1972 (USGPO, Washington, DC).

U.S. Bureau of Labor Statistics, 1990, Fixed private capital, by industry and legal form of organization, Computer printout (U.S. Bureau of Labor Statistics, Washington, DC).

U.S. Energy Information Administration, 1987a, Annual review of energy (USGPO, Washington, DC).

U.S. Energy Information Administration, 1987b, State energy price and expenditure report (USGPO, Washington, DC).

U.S. Forest Service, 1986, U.S. input–output table, 1982, Computer tape (Rocky Mountain Experiment Station, Ft. Collins, CO).

U.S. Office of Technology Assessment, 1989, ECON I-O (diskette).

U.S. Office of Technology Assessment, 1990, Energy use and the economy (USGPO, Washington, DC) forthcoming.

[14]

The Greening of Industrial Ecosystems. 1994.
Pp. 69–89. Washington, DC:
National Academy Press.

Wastes as Raw Materials

DAVID T. ALLEN and NASRIN BEHMANESH

Postconsumer waste, industrial scrap, and unwanted by-products from manufacturing operations should not be viewed as wastes. Rather, they are raw materials that are often significantly underused. One of the research challenges of the emerging discipline of industrial ecology will be to identify productive uses for materials that are currently regarded as wastes, and one of the first steps in meeting this challenge will be to understand the nature of industrial and postconsumer wastes.

More than 12 billion tons of industrial waste (wet basis) are generated annually in the United States (Allen and Jain, 1992; U.S. Environmental Protection Agency [EPA], 1988a,b). Municipal solid waste, which includes postconsumer wastes, is generated at a rate of 0.2 billion tons per year (EPA, 1990). When these material flows are compared with the annual output of 0.3 billion tons per year of the top 50 commodity chemicals (*Chemical and Engineering News*, June 28, 1993), it is apparent that wastes should not be ignored as a potential resource. While these comparisons between waste mass and the mass of commodity products make apparent the magnitude of industrial wastes, considering mass flows alone can be somewhat misleading. The extent to which industrial wastes could serve as raw materials depends not only on the mass of the waste stream, but also on the concentration of resources in the wastes. As shown in Figure 1, the value of a resource is proportional to the level of dilution at which it is present in the raw material. Resources that are present at very low concentration can be recovered only at high cost, while resources present at high concentration can be recovered economically. The primary goals of this paper will be to evaluate the flow rates and concentrations of valuable resources in waste streams and to determine the

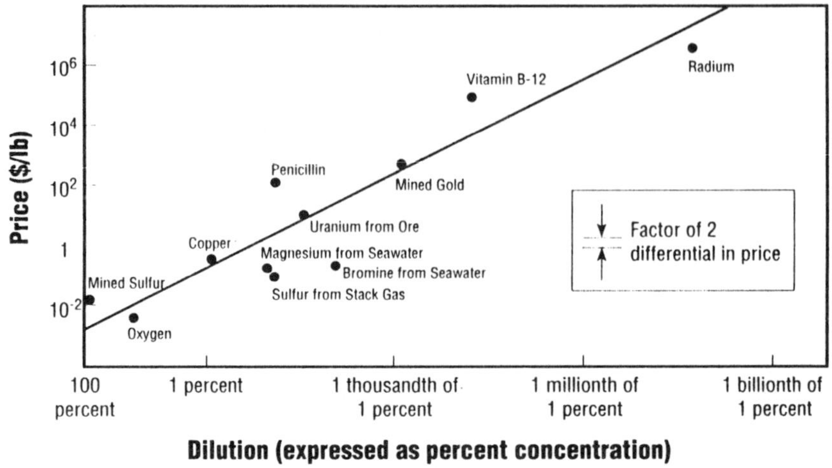

FIGURE 1 The Sherwood plot: Selling prices of materials correlate with their degree of dilution in the initial matrix from which they are being separated. Note that the horizontal axis shows increasing dilution, or decreasing concentration, in the initial matrix. SOURCE: National Research Council (1987).

extent to which materials currently regarded as wastes might be used as raw materials.

We will begin with a brief examination of the total quantities of material circulating through the waste cycles. We will then focus on a series of metals, tracking their flows as wastes and comparing the waste mass and concentration level of recycled wastes, discarded wastes, and virgin raw materials.

OVERVIEW OF INDUSTRIAL AND POSTCONSUMER WASTE GENERATION AND MANAGEMENT

Mapping the flows of more than 12 billion tons of industrial and postconsumer waste is a challenging task. Part of the challenge is integrating information from many diverse sources of data. For example, more than a dozen national sources of data on industrial wastes are available (Eisenhauer and Cordes, 1992), but each covers only a portion of industrial waste generation. Each was collected over a different time period and each considers a different subset of waste generators. Despite these difficulties, each of the data sources provides a unique perspective on industrial waste streams and can be useful. The main focus in this paper will be on just a few of these inventories, most notably the National Hazardous Waste Survey and, to a much lesser extent, the Toxic Release Inventory and the biennial survey of waste generators collected under the Resource Conservation and Recovery Act (RCRA).

The Toxic Release Inventory (TRI) (EPA, 1991), collected annually under the Superfund Amendments and Reauthorization Act (SARA), Title III, provides data on the emission profiles of more than 300 chemical species. While the TRI data are useful for profiling the releases of specific chemicals, they provide little information on the total waste stream. In contrast, the biennial survey of generators and the biennial survey of treatment, storage, and disposal facilities collected under the RCRA provide data on total waste mass but little data beyond loosely defined waste categories on the composition of the waste streams. A hybrid data base of information collected by Research Triangle Institute for the Environmental Protection Agency under RCRA (EPA, 1988c) combines some of the best features of both the TRI and the RCRA biennial survey. This data base, which is called the National Hazardous Waste Survey, combines detailed data on waste composition and data on bulk waste stream properties. It has two basic components, a generator survey focusing on waste characterization and a survey of treatment, storage, disposal, and recycling (TSDR) facilities focusing on waste treatment and disposal. The flow diagram of Figure 2 was prepared using the TSDR section of the survey and some data on air emissions from the TRI. It shows the flow patterns and approximate flow rates of industrial hazardous waste streams. All waste streams regulated under RCRA are included; also included are some wastes exempt from the RCRA regulations and some hazardous wastes managed in units exempt from RCRA permitting requirements. The total mass flow rate of all streams represented in the National Hazardous Waste Survey is approximately 0.75 billion tons per year and the total mass of releases and off-site transfers reported through the TRI totals 0.003 billion tons per year. Therefore the data represent only about 5–10 percent of the total flow rate of industrial wastes. Even though just a small fraction of industrial wastes is represented, the excluded wastes are primarily from a limited group of industries: mining, pulp and paper manufacturing, electrical power generation, and petroleum production. So, the National Hazardous Waste Survey can begin to provide a picture of the flow rates and compositions of waste streams. It is far from comprehensive and omits some major sectors of the economy that generate substantial wastes, but it represents some of the best information available on waste stream composition.

Figure 2 reports the flow rates of hazardous waste streams generated by U.S. industry in 1986, the only year for which the survey data are available. As indicated in Figure 2, a small fraction of solvent, metal, and other wastes, less than 1 percent of total waste mass generated, flows through recycling loops. The total mass involved in recycling is about 5 million tons per year (mt/yr). The largest single stream in terms of total mass flow, nearly 720 mt/yr (more than 90 percent of the total waste flow), is hazardous wastewater. Most of this stream is water with a small percentage of nonaqueous contaminants; hence the mass of the chemically hazardous component of this stream is within an order of magnitude of the

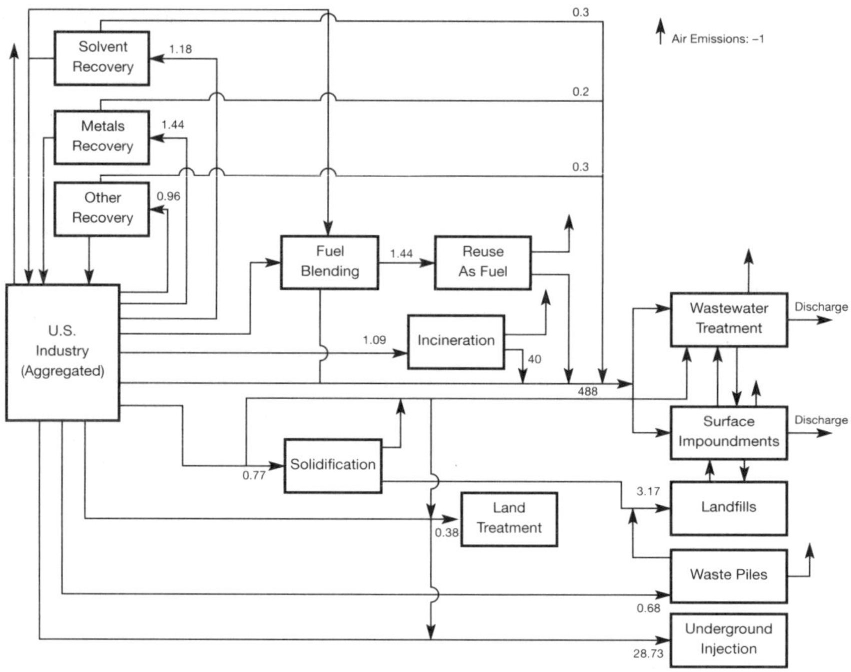

FIGURE 2 Flow of industrial hazardous wastes in treatment operations (1986 data in millions of tons per year).

components being recycled. A third set of waste streams, about 4 mt/yr, is sent to various thermal treatment technologies which include direct incineration, fuel blending, and reuse as fuel.

While an examination of total waste flows is a necessary first step in assessing the use of waste streams as raw materials, total mass is not a good indicator of the potential value of waste streams. Instead, the concentration and mass flow rates of valuable resources in the waste streams will be the most important evaluation criteria. Unfortunately, reliable composition data are not generally available for waste streams, so it is not always possible to evaluate the potential for recycling. One set of materials for which waste composition data are available is metals. The National Hazardous Waste Survey contains data on the flow rates of the 16 metals listed in CFR 261, Appendix VIII (Resource Conservation and Recovery Act, Subtitle C, Sections 3001–3013, 42 U.S.C., Sections 6921–6934 [1976] and Supplement IV [1980]). The next two sections will review the mass flow rates and concentration distributions of selected metals in industrial waste streams.

FLOWS OF SELECTED METALS IN INDUSTRIAL WASTES

As a first step in evaluating the potential of industrial wastes for use as raw materials, we will consider the flows of three metals—cadmium, chromium, and lead. Figures 3, 4, and 5 report the amount of cadmium, chromium, and lead sent to major industrial waste management operations. For cadmium and chromium, only a small fraction of the material is recovered. In the case of cadmium, approximately 1,300 out of a possible 16,000 tons were sent to recovery operations in 1986. In contrast, a major portion of lead generated as industrial wastes (106,000 out of 189,000 tons) is sent to metal recovery. Recycling is feasible for many lead-containing streams because an efficient collection and reprocessing system is in place for used automotive storage batteries. With such extensive recycling of lead, it should come as no surprise that secondary nonferrous metal processing (Standard Industrial Classification [SIC] code 3341), which is largely lead battery recycling, is the dominant source of lead wastes (Figure 6).

The flows of cadmium, chromium, and lead in industrial hazardous wastes,

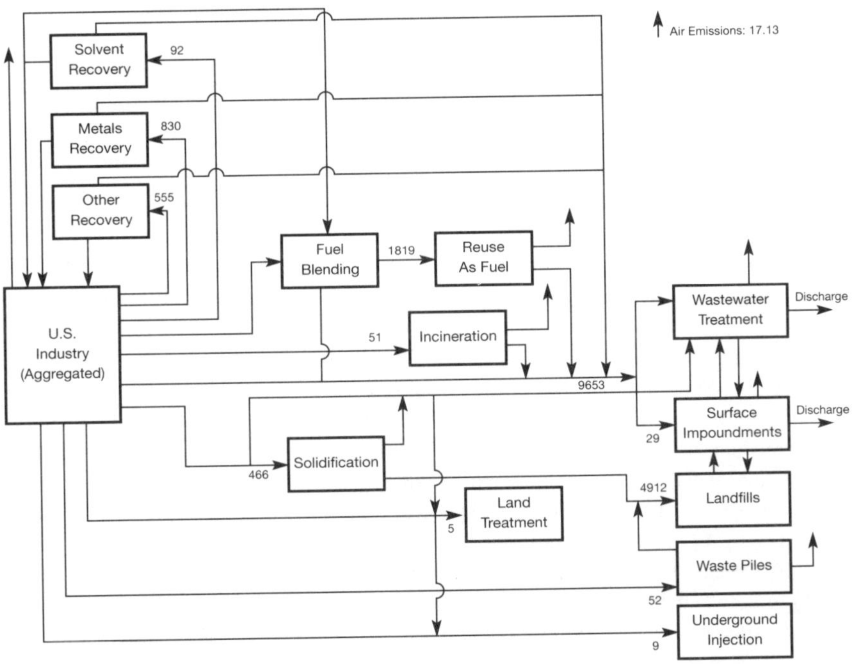

FIGURE 3 · Flow of cadmium in hazardous waste treatment operations in 1986 (tons per year).

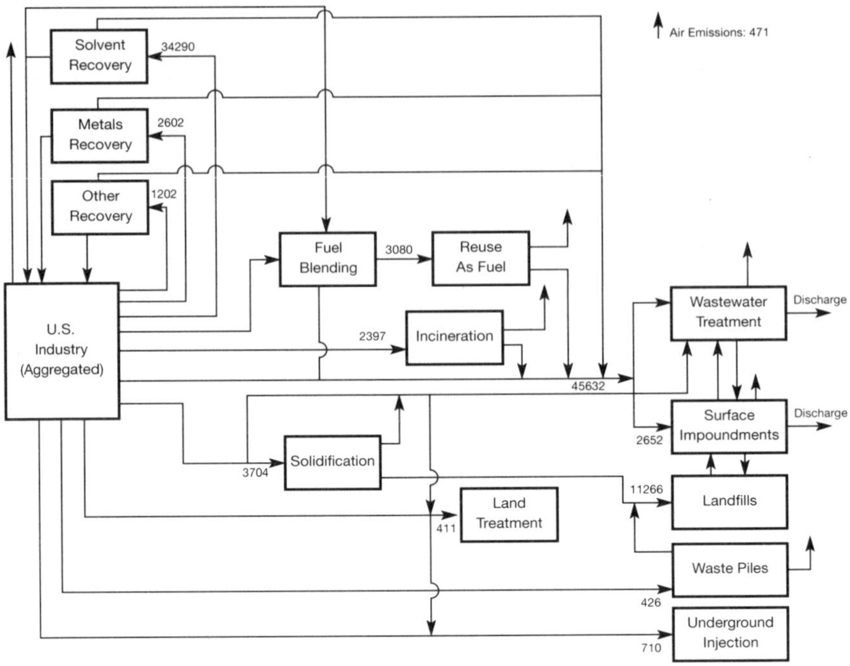

FIGURE 4 Flow of chromium in hazardous waste treatment operations in 1986 (tons per year).

illustrated in Figures 2–5, cannot be considered in isolation. They are merely a part, albeit a major part, of the total waste stream flow. A second major component of waste stream flows is municipal solid waste. The total mass flow of municipal solid waste is approximately 0.2 billion tons per year, which is considerably less than the total flow of industrial hazardous wastes (0.75 billion tons per year) and total industrial wastes (12 billion tons per year). Simple tonnage comparisons can be misleading, however. To assess the potential value of industrial hazardous waste and municipal solid wastes as sources of raw materials, it is necessary to compare the flows of specific materials. For example, according to the U.S. EPA (1989), the dominant contributor to lead in the municipal solid waste (MSW) stream is storage batteries. Depending on rates of recycling, batteries contribute between 100,000 and 150,000 tons per year of lead to MSW. Recent data (EPA, 1989) indicate a 90 percent rate of recycling, resulting in 100,000 tons of waste out of the 800,000 tons of lead in used batteries. Other sources of lead in the MSW stream include consumer electronics (estimated to be 60,000 tons), glass and ceramics (8,000 tons), plastics (4,000 tons), metals such as soldered cans

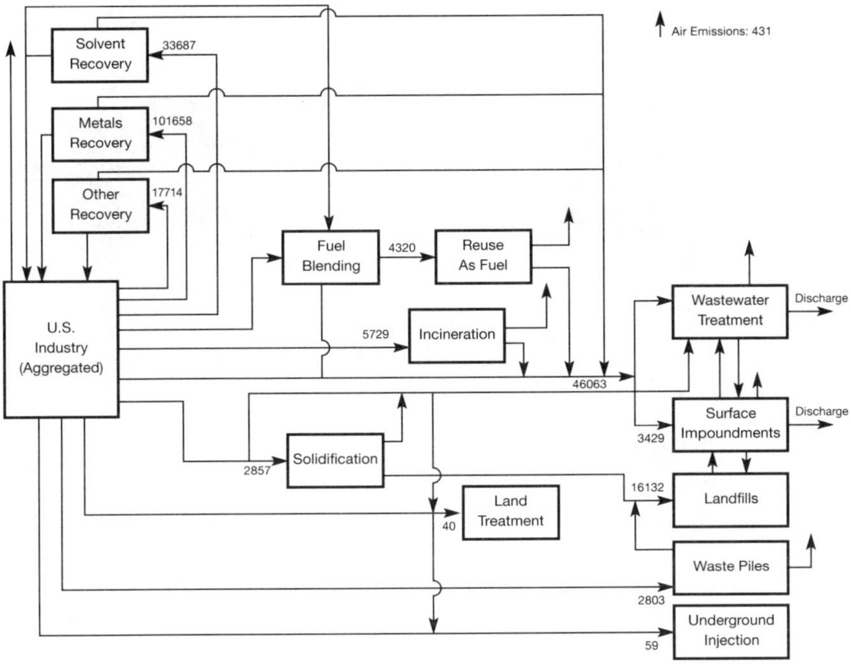

FIGURE 5 Flow of lead in hazardous waste treatment operations in 1986 (tons per year).

(1,000 tons) and pigments (1,000 tons). None of these other sources of waste result in any significant degree of lead recycling, largely because of the low concentrations of lead in the products.

Analysis of these waste management data represents a first step in performing studies in industrial ecology. The next step is to integrate waste generation data with production data. Coupling the waste flow data for lead presented above with the data on lead production and consumption presented in Table 1 yields the lead flow diagram shown in Figure 7. Lead is used at a rate of roughly 1.2 million tons per year. Most is consumed in the production of lead storage batteries and these batteries are eventually retired. Roughly 90 percent of these used batteries are recycled, so the net loss of lead through battery disposal is about 100,000 tons. Other sources of lead in MSW total roughly 70,000 tons per year. Industrial hazardous wastes are another sink for lead. Of the 189,000 tons of lead in hazardous wastes, roughly 53,000 tons are sent to disposal. The remainder is recycled, but the recycling of both hazardous and battery wastes generates roughly 100,000 tons of lead waste. Comparing the total flow of lead with the amount of lead

CADMIUM

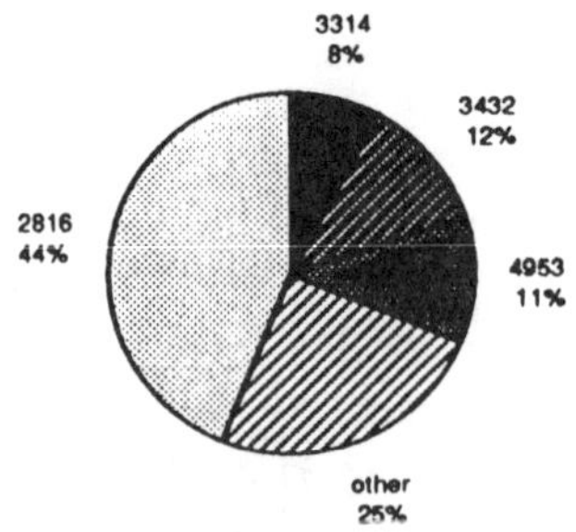

SIC	INDUSTRY
2816:	Inorganic pigments
3314:	Primary metals industries
3432:	Plumbing, fitting, & brass goods
4953:	Refuse systems

Total: 16,100 tons/year

CHROMIUM

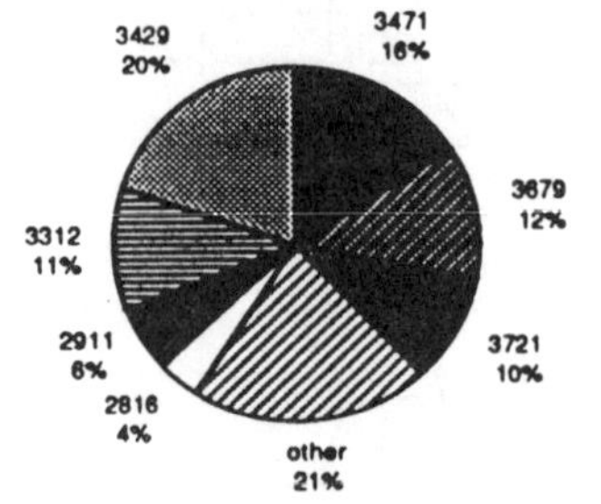

SIC	INDUSTRY
2816:	Inorganic pigments
2911:	Petroleum refining
3312:	Blast furnace & steel mills
3429:	Hardware, nec
3471:	Plating & polishing
3679:	Electronic components, nec
3721:	Aircraft

Total: 89,900 tons/year

LEAD

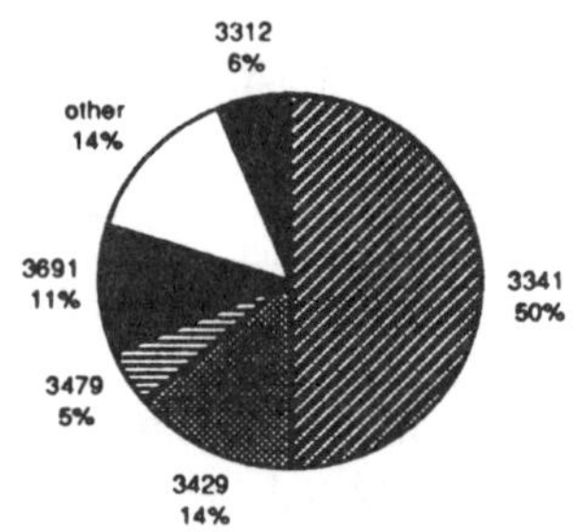

SIC	INDUSTRY
3312:	Blast furnace & steel mills
3341:	Secondary nonferrous metals
3429:	Hardware, nec
3479:	Metal coating & allied services
3691:	Storage batteries

Total: 188,600 tons/year

FIGURE 6 Industrial sources of cadmium, chromium, and lead wastes.

TABLE 1 Production and Consumption of Lead, 1986 and 1989

	Amount, 1986[a] (metric tons)	Percent, 1986[a]	Amount, 1989[b] (metric tons)	Percent, 1989[b]
Production				
Mine production	340,000		411,000	
Secondary lead	615,000		809,000	
Consumption				
Metal products	146,000	12.7	170,000	13.3
Storage batteries	854,000	77.9	1,012,000	78.9
Other oxides	69,000	5.0	58,000	4.5
Miscellaneous (including gasoline additives)	55,000	4.4	43,000	3.3
TOTAL	1,124,000	100.0	1,283,000	100.00

[a]U.S. Department of the Interior (1988).
[b]U.S. Department of the Interior (1991).

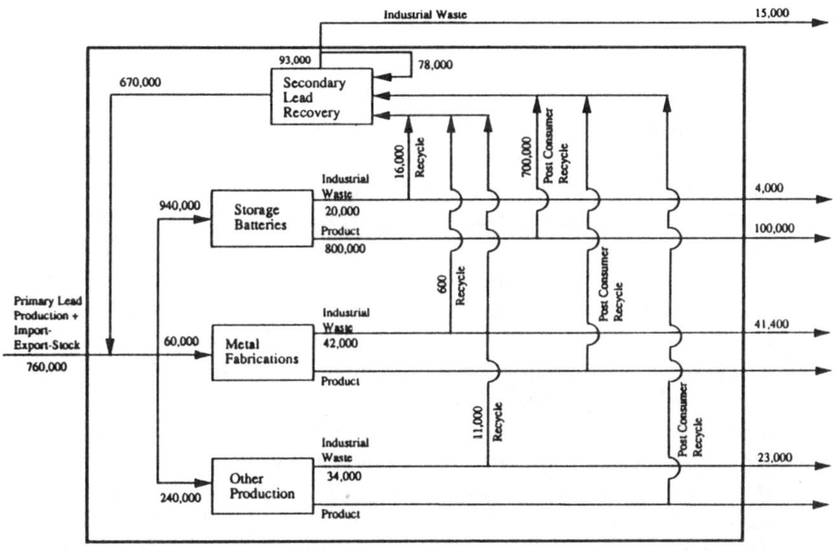

FIGURE 7 Simplified model of the industrial ecology of lead (amounts in tons per year).

eventually requiring disposal, we see that lead is reprocessed at about 75 percent efficiency:

$$\begin{aligned}\% \text{ Efficiency} &= 100 \times 1 - (300{,}000 \text{ tons discarded} \div 1{,}200{,}000 \text{ tons consumed}) \\ &= 75\%\end{aligned}$$

If this efficiency is to be improved, then the streams that are currently reaching disposal must find productive use. Overall recycling efficiency could be improved by increasing the collection of lead batteries above 90 percent, by improving the efficiency of secondary lead smelting, and by targeting for recycling industrial waste streams from nonbattery operations. Figure 8 compares some of the waste streams currently requiring disposal with those currently being recycled. Examination of Figure 8 reveals that more concentrated waste streams are more likely to be recycled than waste streams with low lead concentration. Although the decision whether or not to reclaim a metal from a waste stream is complex, it is in essence an economic question. Accordingly, it depends not only on the value of the recycled material, but to a significant extent on concentration.

DILUTION DETERMINES RECYCLABILITY

As in the case of lead, the concentration at which recycling of other materials in the waste stream becomes cost-effective depends on the value of the raw material (see Figure 9). The Sherwood diagram (Figure 1) showed that whereas materials such as gold and radium can be recovered from raw materials that are quite dilute in the resource, materials such as copper can be recovered economically only from relatively rich ores.

Given the price, it is therefore possible to estimate the concentration at which materials can be recovered. By comparing metal prices, minimum economically recoverable concentration (from the Sherwood diagram), and data on the concentration distributions of metals in waste streams (Figure 9), it is possible to estimate what fraction of metals in hazardous waste streams can be recycled. These estimates are reported in Table 2 and indicate that metals in hazardous wastes are underutilized. This could be because only waste streams with very high metal concentrations are recovered or because only a small fraction of potential recyclers at all feasible concentration levels recover metals. Figure 10 is an attempt to differentiate between these two cases.

To develop Figure 10, the concentration distribution of metals in recycled waste streams was examined. The concentration below which only 10 percent of the metal recycling took place was assumed to be a lower bound for economic metal recovery from the waste. This concentration was then plotted, together with the 1986 metal price (recall that the waste data are from 1986) to generate Figure 10. Also plotted on Figure 10 is the Sherwood diagram for virgin materials. Comparison of the Sherwood plot for virgin materials and the waste concentration data

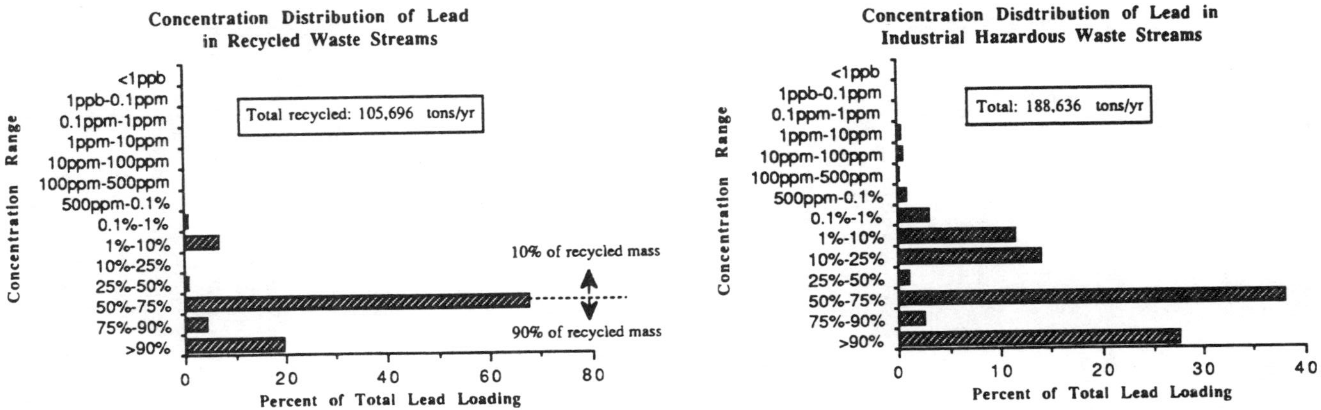

FIGURE 8 Concentration distributions of lead in waste streams undergoing recycling and concentration distributions of lead in all industrial hazardous waste streams (1986). The concentration below which only 10 percent of recyling takes place is noted.

TABLE 2 Percentage of Total Metal Loadings That Can Be Recovered Economically (derived from Sherwood Plot) from Industrial Hazardous Waste Streams

Metal	Minimum Concentration Recoverable, from Sherwood Plot (mass fraction)	Percent of Metal Theoretically Recoverable (%)	Percent Recycled in 1986 (%)
Sb	0.00405	74-87	32
As	0.00015	98-99	3
Ba	0.0015	95-98	4
Be	0.012	54-84	31
Cd	0.0048	82-97	7
Cr	0.0012	68-89	8
Cu	0.0022	85-92	10
Pb	0.074	84-95	56
Hg	0.00012	99	41
Ni	0.0066	100	0.1
Se	0.0002	93-95	16
Ag	0.000035	99-100	1
Tl	0.00004	97-99	1
V	0.0002	74-98	1
Zn	0.0012	96-98	13

reveals that most metals in waste streams are recycled only at high concentrations. The concentration of resources in recycled wastes is generally higher than for virgin materials, indicating significant disincentives to make use of waste. Figures 9 and 10 demonstrate that there are many opportunities for increased recycling.

SUMMARY

Compositions and sources of recycled waste streams can be examined and opportunities for improving recycling efficiencies can be explored. Unfortunately, these analyses rely extensively on a single data base of industrial waste, the National Hazardous Waste Survey. This collection of data, which is the only comprehensive and detailed source of information on the composition of industrial hazardous waste streams, was based on wastes generated during 1986 and is already somewhat outdated. The data are also restricted to wastes classified as hazardous under the provisions of the Resource Conservation and Recovery Act. The lack of current, comprehensive, and reliable data on waste composition remains a serious barrier to studies in industrial ecology.

If the limitations of available data are understood, however, it is possible to

81

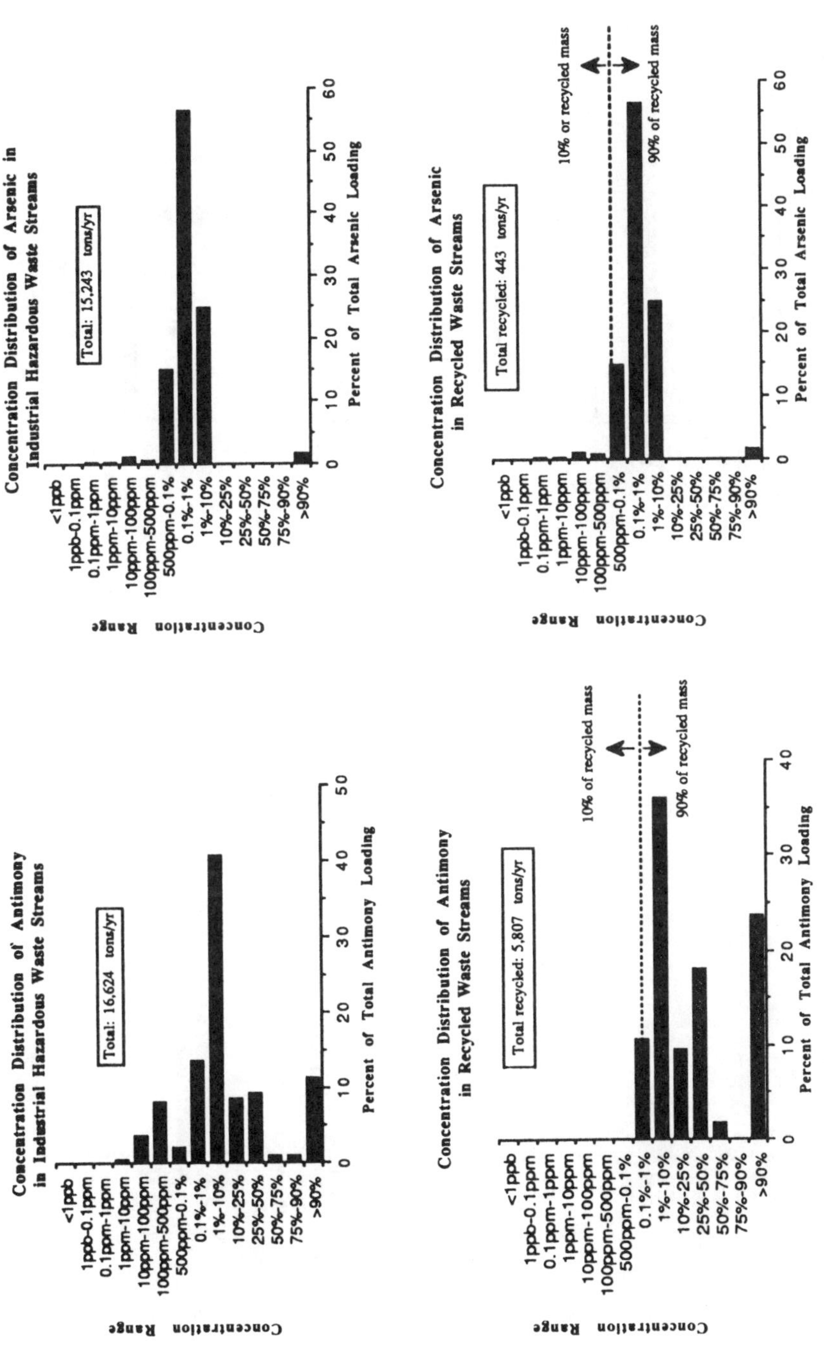

FIGURE 9 Concentration distribution of recycled metals and the concentration distributions of metals in all industrial hazardous waste streams (1986 data). The concentration below which only 10 percent of recycling takes place is noted for each metal.

82

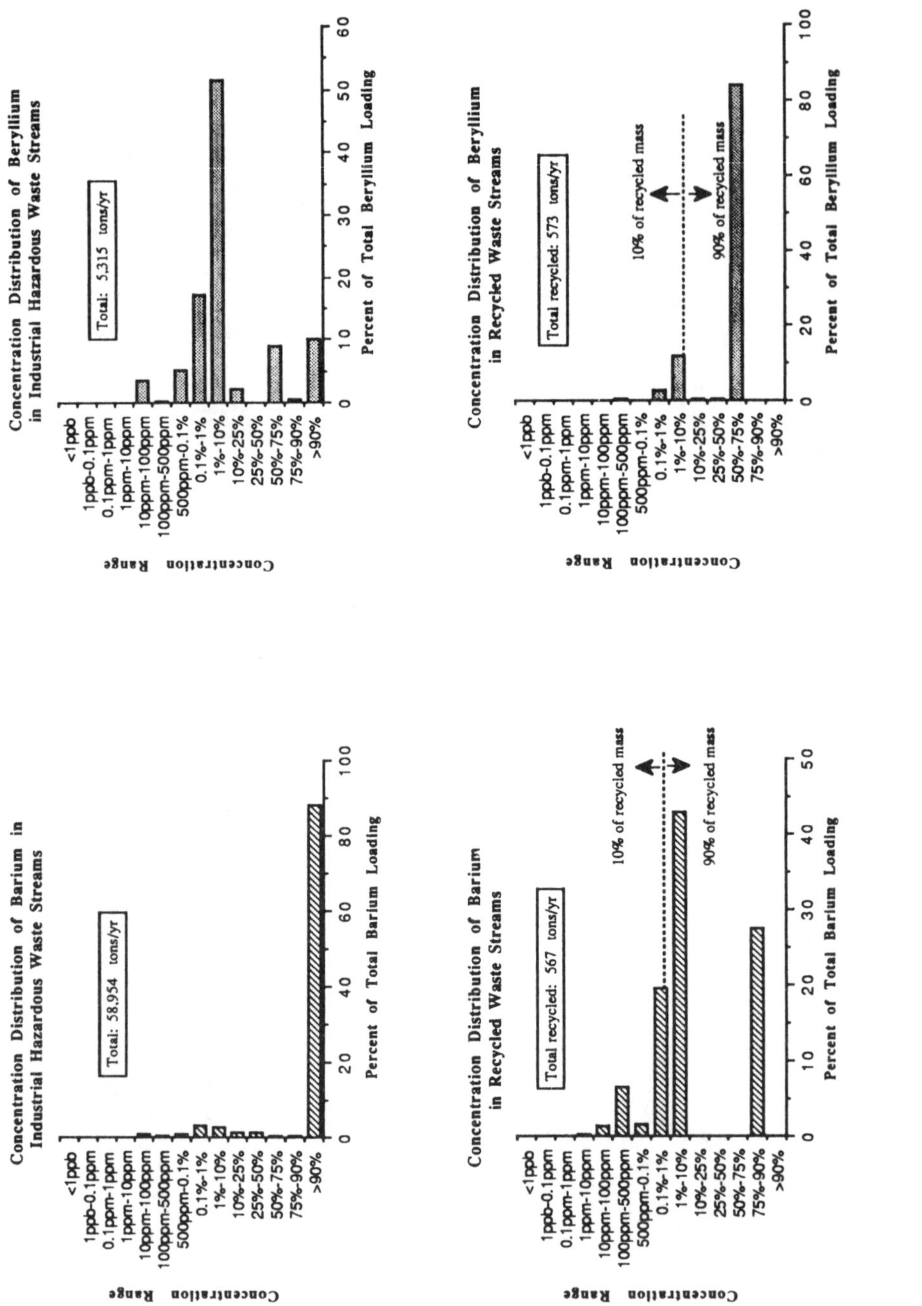
Concentration Distribution of Beryllium in Industrial Hazardous Waste Streams
Total: 5,315 tons/yr
Concentration Range
Percent of Total Beryllium Loading
Concentration Distribution of Beryllium in Recycled Waste Streams
Total recycled: 573 tons/yr
10% of recycled mass
90% of recycled mass
Concentration Range
Percent of Total Beryllium Loading
Concentration Distribution of Barium in Industrial Hazardous Waste Streams
Total: 58,954 tons/yr
Concentration Range
Percent of Total Barium Loading
Concentration Distribution of Barium in Recycled Waste Streams
Total recycled: 567 tons/yr
10% of recycled mass
90% of recycled mass
Concentration Range
Percent of Total Barium Loading
<1ppb
1ppb-0.1ppm
0.1ppm-1ppm
1ppm-10ppm
10ppm-100ppm
100ppm-500ppm
500ppm-0.1%
0.1%-1%
1%-10%
10%-25%
25%-50%
50%-75%
75%-90%
>90%

83

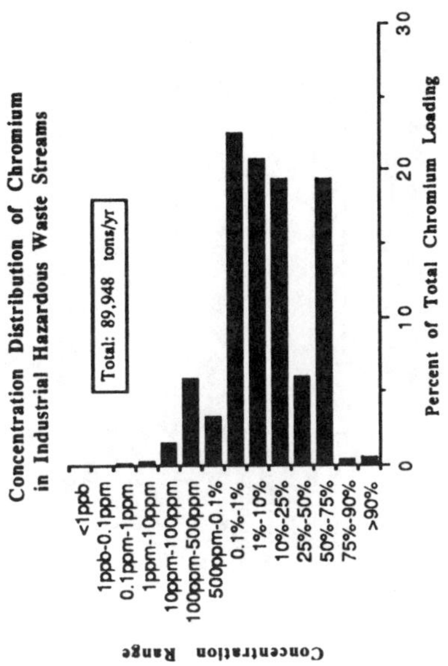
Concentration Distribution of Chromium
in Industrial Hazardous Waste Streams
Total: 89,948 tons/yr
Concentration Range
<1ppb
1ppb-0.1ppm
0.1ppm-1ppm
1ppm-10ppm
10ppm-100ppm
100ppm-500ppm
500ppm-0.1%
0.1%-1%
1%-10%
10%-25%
25%-50%
50%-75%
75%-90%
>90%
0
10
20
30
Percent of Total Chromium Loading

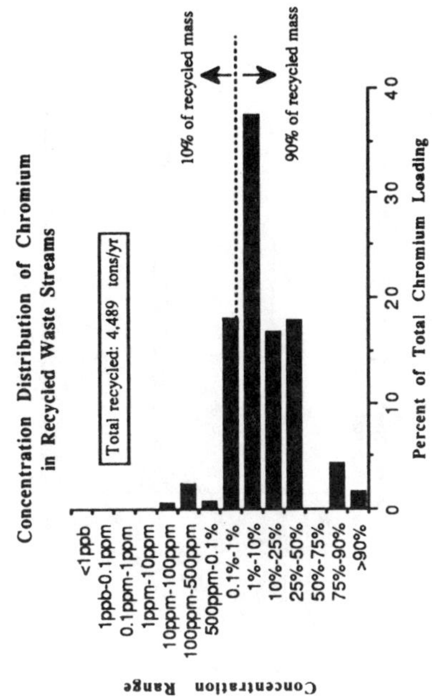
Concentration Distribution of Chromium
in Recycled Waste Streams
Total recycled: 4,489 tons/yr
10% of recycled mass
90% of recycled mass
Concentration Range
<1ppb
1ppb-0.1ppm
0.1ppm-1ppm
1ppm-10ppm
10ppm-100ppm
100ppm-500ppm
500ppm-0.1%
0.1%-1%
1%-10%
10%-25%
25%-50%
50%-75%
75%-90%
>90%
0
10
20
30
40
Percent of Total Chromium Loading

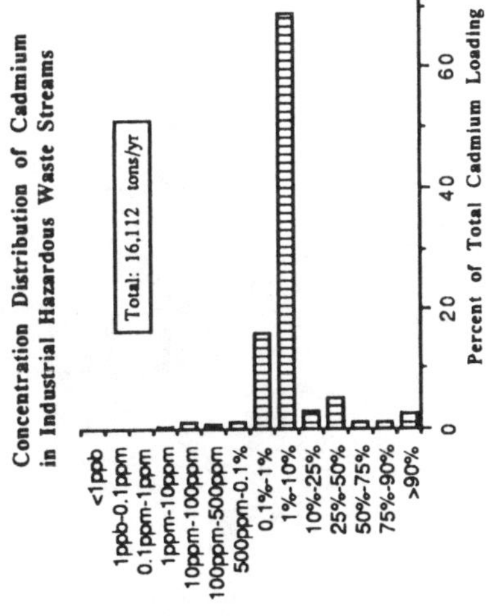
Concentration Distribution of Cadmium
in Industrial Hazardous Waste Streams
Total: 16,112 tons/yr
Concentration Range
<1ppb
1ppb-0.1ppm
0.1ppm-1ppm
1ppm-10ppm
10ppm-100ppm
100ppm-500ppm
500ppm-0.1%
0.1%-1%
1%-10%
10%-25%
25%-50%
50%-75%
75%-90%
>90%
0
20
40
60
80
Percent of Total Cadmium Loading

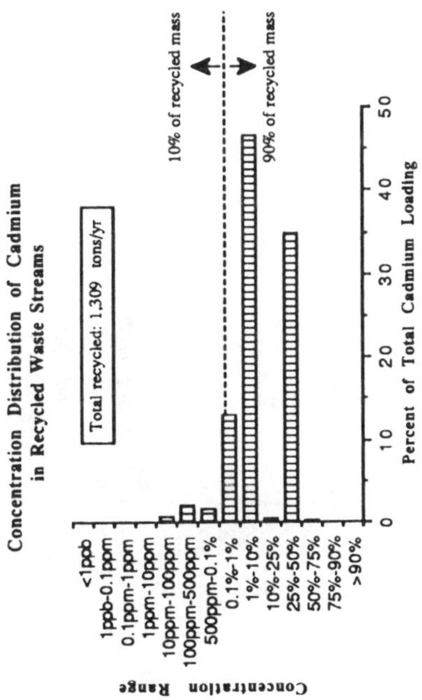
Concentration Distribution of Cadmium
in Recycled Waste Streams
Total recycled: 1,309 tons/yr
10% of recycled mass
90% of recycled mass
Concentration Range
<1ppb
1ppb-0.1ppm
0.1ppm-1ppm
1ppm-10ppm
10ppm-100ppm
100ppm-500ppm
500ppm-0.1%
0.1%-1%
1%-10%
10%-25%
25%-50%
50%-75%
75%-90%
>90%
0
10
20
30
40
50
Percent of Total Cadmium Loading

84

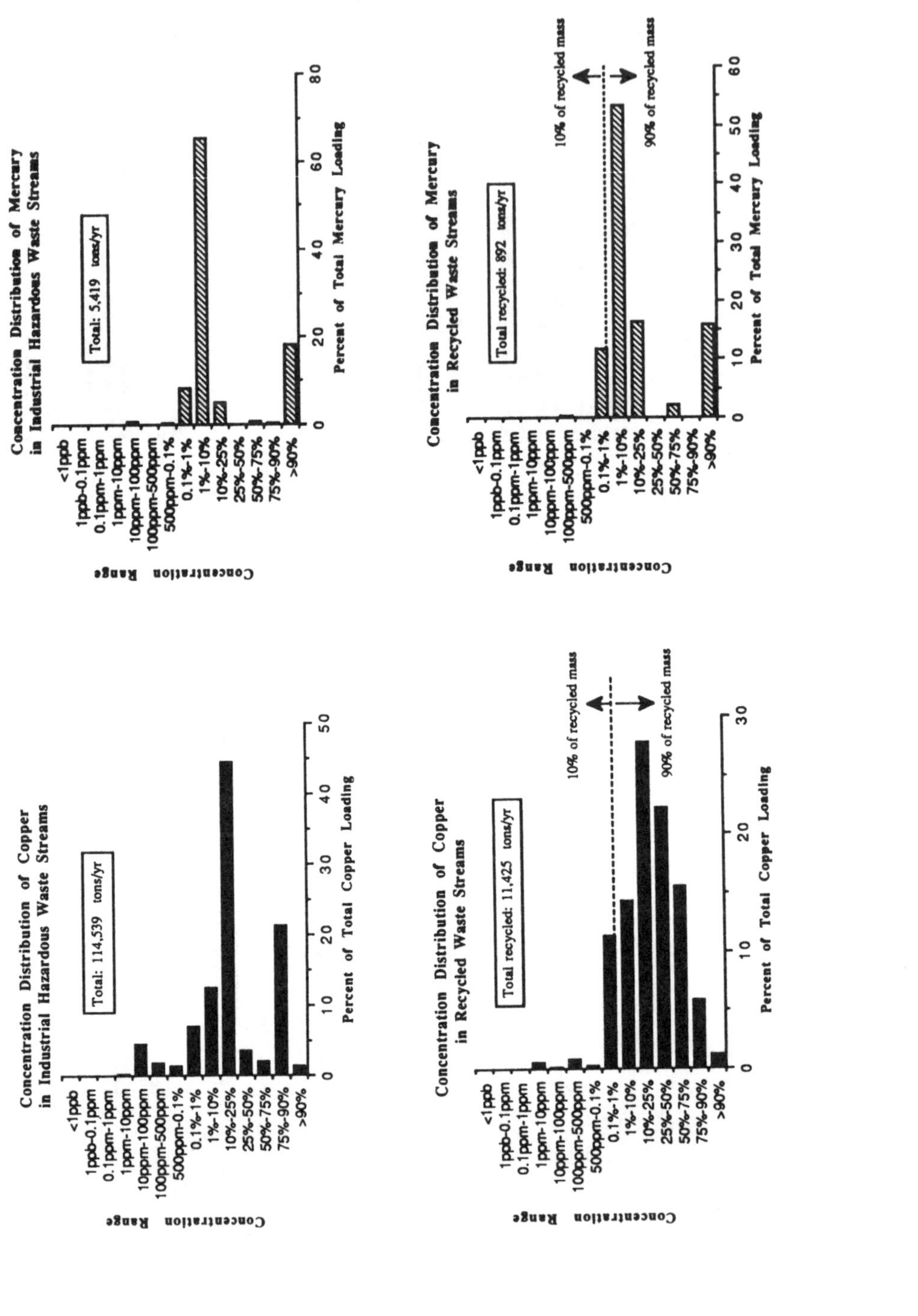
Concentration Distribution of Mercury in Industrial Hazardous Waste Streams
Total: 5,419 tons/yr
Percent of Total Mercury Loading
Concentration Range
Concentration Distribution of Mercury in Recycled Waste Streams
Total recycled: 892 tons/yr
10% of recycled mass
90% of recycled mass
Concentration Distribution of Copper in Industrial Hazardous Waste Streams
Total: 114,539 tons/yr
Percent of Total Copper Loading
Concentration Distribution of Copper in Recycled Waste Streams
Total recycled: 11,425 tons/yr
<1ppb
1ppb-0.1ppm
0.1ppm-1ppm
1ppm-10ppm
10ppm-100ppm
100ppm-500ppm
500ppm-0.1%
0.1%-1%
1%-10%
10%-25%
25%-50%
50%-75%
75%-90%
>90%

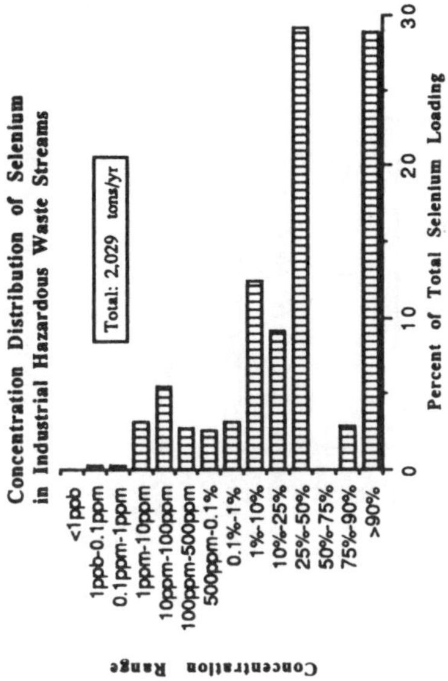
Concentration Distribution of Selenium in Industrial Hazardous Waste Streams
Total: 2,029 tons/yr
Concentration Range
<1ppb
1ppb-0.1ppm
0.1ppm-1ppm
1ppm-10ppm
10ppm-100ppm
100ppm-500ppm
500ppm-0.1%
0.1%-1%
1%-10%
10%-25%
25%-50%
50%-75%
75%-90%
>90%
0
10
20
30
Percent of Total Selenium Loading

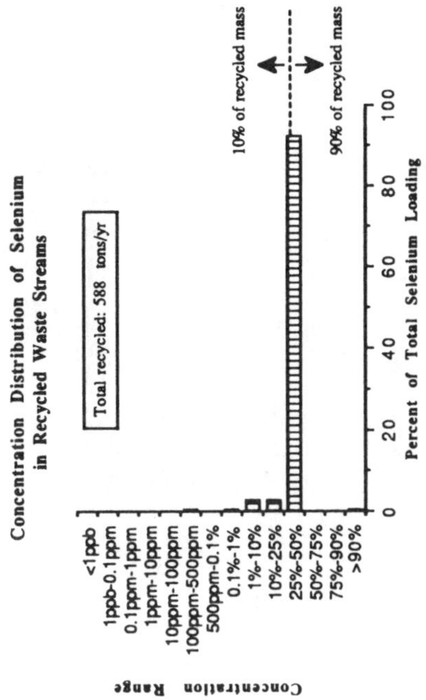
Concentration Distribution of Selenium in Recycled Waste Streams
Total recycled: 588 tons/yr
10% of recycled mass
90% of recycled mass
Concentration Range
<1ppb
1ppb-0.1ppm
0.1ppm-1ppm
1ppm-10ppm
10ppm-100ppm
100ppm-500ppm
500ppm-0.1%
0.1%-1%
1%-10%
10%-25%
25%-50%
50%-75%
75%-90%
>90%
0
20
40
60
80
100
Percent of Total Selenium Loading

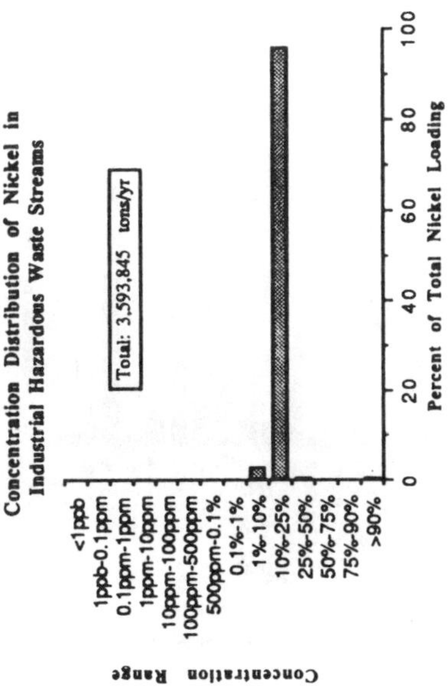
Concentration Distribution of Nickel in Industrial Hazardous Waste Streams
Total: 3,593,845 tons/yr
Concentration Range
<1ppb
1ppb-0.1ppm
0.1ppm-1ppm
1ppm-10ppm
10ppm-100ppm
100ppm-500ppm
500ppm-0.1%
0.1%-1%
1%-10%
10%-25%
25%-50%
50%-75%
75%-90%
>90%
0
20
40
60
80
100
Percent of Total Nickel Loading

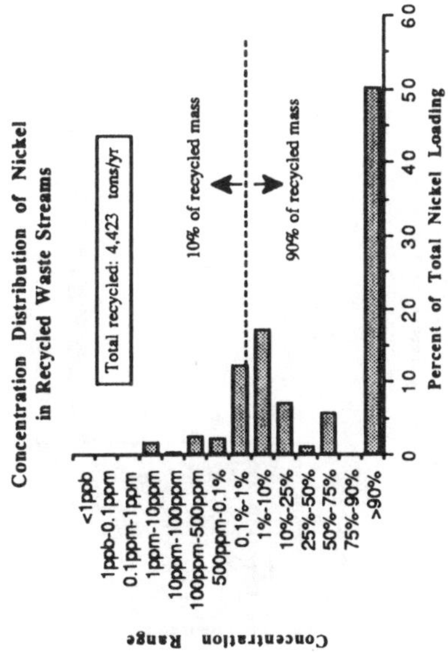
Concentration Distribution of Nickel in Recycled Waste Streams
Total recycled: 4,423 tons/yr
10% of recycled mass
90% of recycled mass
Concentration Range
<1ppb
1ppb-0.1ppm
0.1ppm-1ppm
1ppm-10ppm
10ppm-100ppm
100ppm-500ppm
500ppm-0.1%
0.1%-1%
1%-10%
10%-25%
25%-50%
50%-75%
75%-90%
>90%
0
10
20
30
40
50
60
Percent of Total Nickel Loading

86

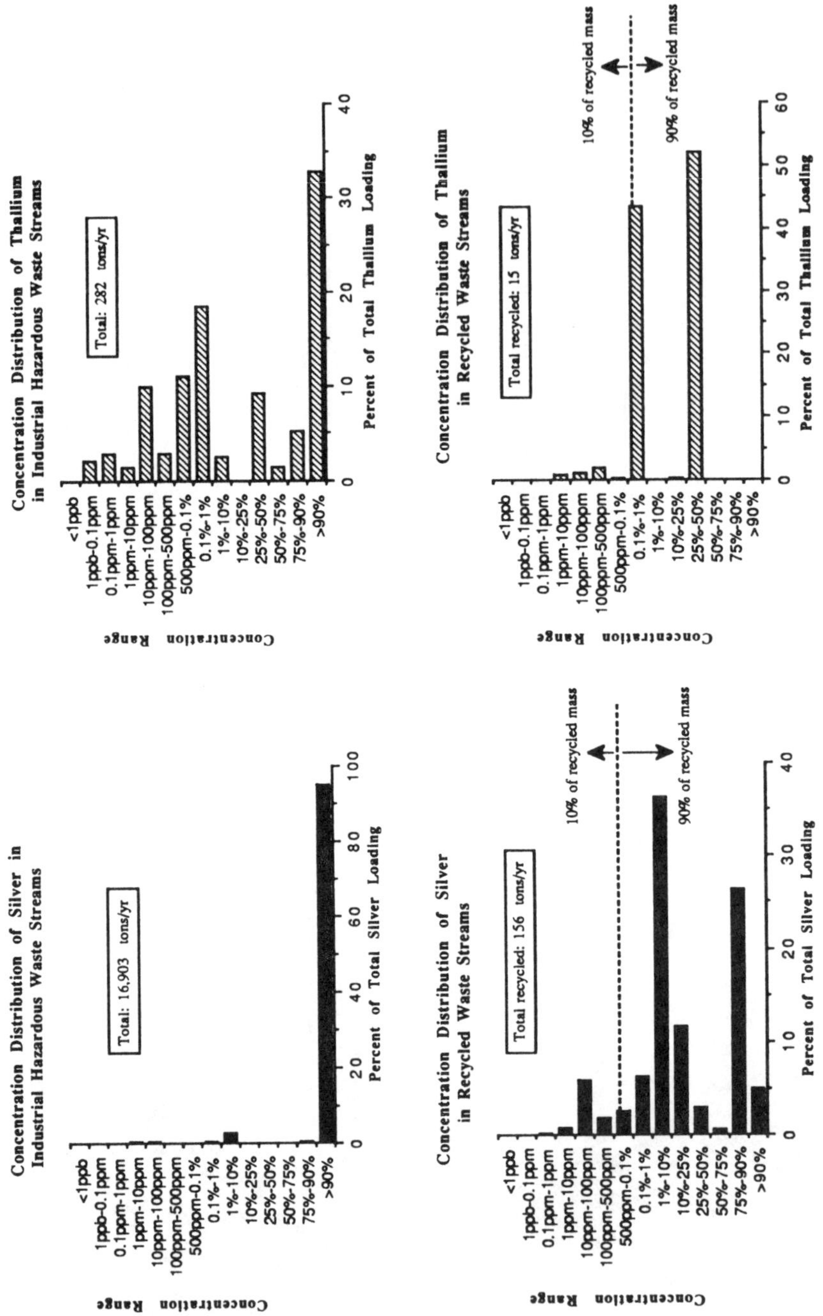

Concentration Distribution of Thallium in Industrial Hazardous Waste Streams
Total: 282 tons/yr
Percent of Total Thallium Loading
Concentration Distribution of Thallium in Recycled Waste Streams
Total recycled: 15 tons/yr
10% of recycled mass
90% of recycled mass
Concentration Distribution of Silver in Industrial Hazardous Waste Streams
Total: 16,903 tons/yr
Percent of Total Silver Loading
Concentration Distribution of Silver in Recycled Waste Streams
Total recycled: 156 tons/yr
Concentration Range

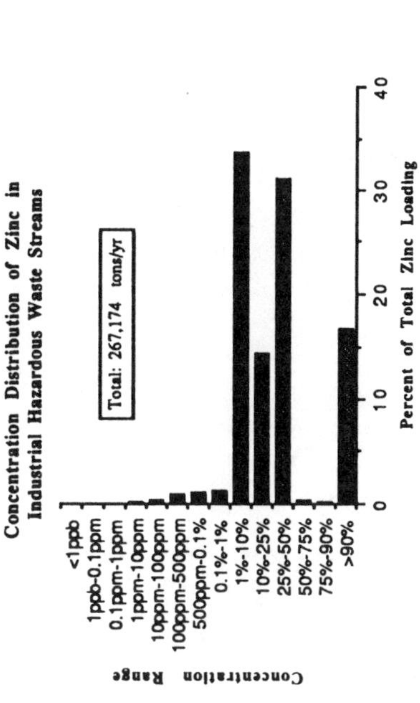
Concentration Distribution of Zinc in Industrial Hazardous Waste Streams
Total: 267,174 tons/yr
Concentration Range
<1ppb
1ppb-0.1ppm
0.1ppm-1ppm
1ppm-10ppm
10ppm-100ppm
100ppm-500ppm
500ppm-0.1%
0.1%-1%
1%-10%
10%-25%
25%-50%
50%-75%
75%-90%
>90%
Percent of Total Zinc Loading
0
10
20
30
40

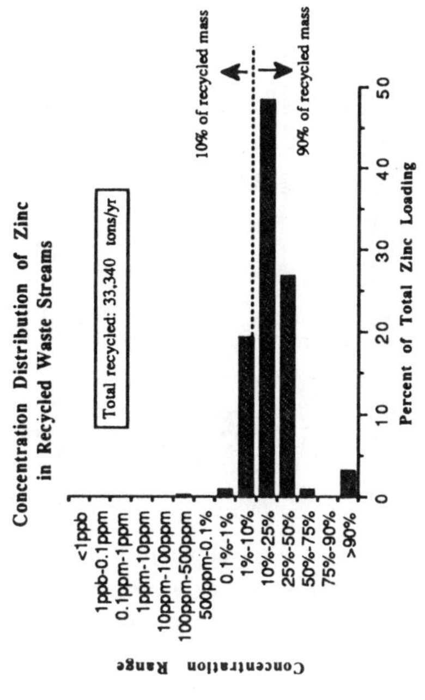
Concentration Distribution of Zinc in Recycled Waste Streams
Total recycled: 33,340 tons/yr
10% of recycled mass
90% of recycled mass
Concentration Range
<1ppb
1ppb-0.1ppm
0.1ppm-1ppm
1ppm-10ppm
10ppm-100ppm
100ppm-500ppm
500ppm-0.1%
0.1%-1%
1%-10%
10%-25%
25%-50%
50%-75%
75%-90%
>90%
Percent of Total Zinc Loading
0
10
20
30
40
50

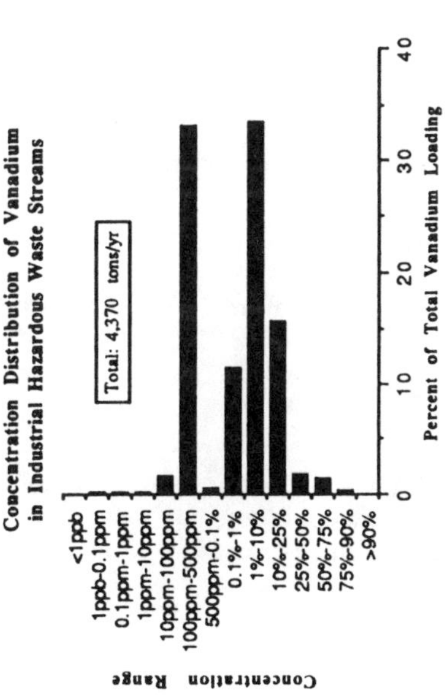
Concentration Distribution of Vanadium in Industrial Hazardous Waste Streams
Total: 4,370 tons/yr
Concentration Range
<1ppb
1ppb-0.1ppm
0.1ppm-1ppm
1ppm-10ppm
10ppm-100ppm
100ppm-500ppm
500ppm-0.1%
0.1%-1%
1%-10%
10%-25%
25%-50%
50%-75%
75%-90%
>90%
Percent of Total Vanadium Loading
0
10
20
30
40

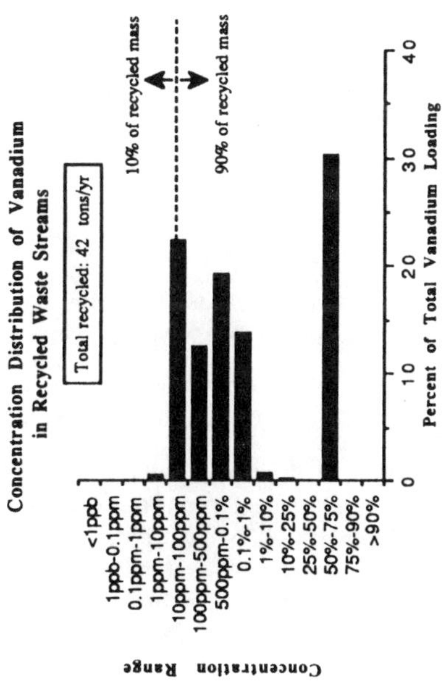
Concentration Distribution of Vanadium in Recycled Waste Streams
Total recycled: 42 tons/yr
10% of recycled mass
90% of recycled mass
Concentration Range
<1ppb
1ppb-0.1ppm
0.1ppm-1ppm
1ppm-10ppm
10ppm-100ppm
100ppm-500ppm
500ppm-0.1%
0.1%-1%
1%-10%
10%-25%
25%-50%
50%-75%
75%-90%
>90%
Percent of Total Vanadium Loading
0
10
20
30
40

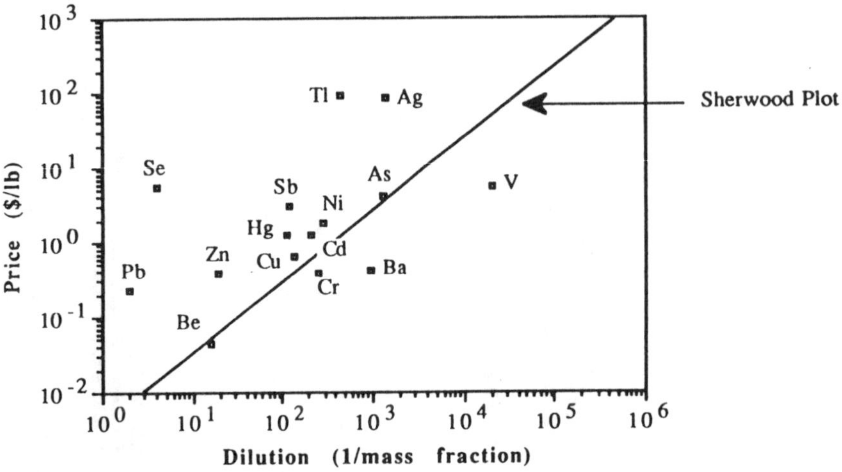

FIGURE 10 The Sherwood plot for waste streams. The minimum concentration of metal wastes undergoing recycling (see Figures 8 and 9) is plotted against metal price. The Sherwood plot for virgin materials is provided by comparison. Points lying above the Sherwood plot indicate that the metals in the waste streams are underused, that is, waste streams undergoing disposal are richer than typical virgin materials. Points lying below the Sherwood plot indicate that the waste streams are vigorously recycled.

examine the industrial ecology of some metals. The results reveal that the concentrations of metal resources in many waste streams that are currently undergoing disposal are higher than for typical virgin resources. Thus, extensive waste trading could significantly reduce the quantity of waste requiring disposal.

ACKNOWLEDGMENT

This work was supported by the University of California Toxic Substances Research and Teaching Program.

REFERENCES

Allen, D. T., and R. Jain, eds. 1992. Special issue on industrial waste generation and management. Hazardous Waste and Hazardous Materials 9(1):1–111.

Eisenhauer, J., and Cordes, R. 1992. Industrial waste databases: A simple roadmap. Hazardous Waste and Hazardous Materials 9(1):1–19.

National Research Council. 1987. Separation and Purification: Critical Needs and Opportunities. Washington, D.C.: National Academy Press.

U.S. Department of the Interior. 1988. Minerals Yearbook: 1986, Metals and Minerals, Volume 1. Washington, D.C.: U.S. Government Printing Office.

U.S. Department of the Interior. 1991. Minerals Yearbook: 1989, Metals and Minerals, Volume 1. Washington, D.C.: U.S. Government Printing Office.

U.S. Environmental Protection Agency. 1988a. Report to Congress: Solid Waste Disposal in the United States, Volume 1, EPA 530-SW-88-011.

U.S. Environmental Protection Agency. 1988b. Report to Congress: Solid Waste Disposal in the United States, Volume 2, EPA 530-SW-88-011B.

U.S. Environmental Protection Agency. 1988c. 1986 National Survey of Hazardous Waste Treatment, Storage, Disposal and Recycling Facilities, EPA/530-SW-88-035.

U.S. Environmental Protection Agency. 1989. Characterization of Products Containing Lead and Cadmium in Municipal Solid Waste in the United States, 1970 to 2000, EPA/530-SW-89-015A.

U.S. Environmental Protection Agency. 1990. Characterization of Municipal Solid Waste in the United States: 1990 Update, EPA 530-SW-90-042.

U.S. Environmental Protection Agency. 1991. Toxics in the Community, EPA 560/4-91/014.

[15]

JOURNAL OF ENVIRONMENTAL ECONOMICS AND MANAGEMENT 29, 78–91 (1995)

Garbage, Recycling, and Illicit Burning or Dumping

DON FULLERTON

Department of Economics, University of Texas at Austin, Austin, Texas 78712

AND

THOMAS C. KINNAMAN

Department of Economics, Bucknell University, Lewisburg, Pennsylvania 17837

Received July 15, 1993; revised February 18, 1994

With garbage and recycling as the only two disposal options, we confirm prior results that the optimal curbside fee for garbage collection equals the direct resource cost plus external environmental cost. When illicit burning or dumping is a third disposal option that cannot be taxed directly, the optimal curbside tax on garbage changes sign. The optimal fee structure is a deposit–refund system: a tax on all output plus a rebate on proper disposal through either recycling or garbage collection. The output tax helps achieve the first-best allocation even though it affects the choice between consumption and untaxed leisure.

1. INTRODUCTION

Solid waste disposal has become more expensive recently due to rising land prices, strict environmental regulations, and host fees paid to localities to accept new landfills or incinerators. Tipping fees in the northeastern U.S. approach $125 per ton. Most towns still pay for garbage collection and disposal using general revenues, however, with no price per bag. Thus the resident views it as free.

As an alternative, more towns are beginning to sell special bags or stickers necessary for curbside collection of each bag or can of garbage (U.S. EPA [20]). These per-unit charges can help defray the cost of collection, and they help discourage waste. Two major recent studies describe the advantages of such charges. Project 88—Round II [14], sponsored by then Senators Timothy Wirth of Colorado and John Heinz of Pennsylvania, says that unit pricing "creates strong incentives for households to reduce the quantities of waste they generate, whether through changes in their purchasing patterns, reuse of products and containers, or composting of yard wastes" (pp. 49–50). The World Resources Institute (WRI, Repetto *et al*, [15]) further extols the virtues of "pay-by-the-bag." For densely populated areas, they estimate that each 32-gallon bag of garbage costs $1.12 in direct payments to waste haulers and landfill operators, and $1.83 including external costs to others near the landfill who may suffer from noise, odor, litter, and extra traffic. They go on to measure welfare gains from charging such a price.

In response to unit charges, however, households might not just recycle, compost, and adjust purchasing habits. They might also burn paper in fireplaces and carry trash to commercial dumpsters, back woods, and vacant lots. If New York City were to sell stickers for $1.83 each, and pick up only bags with stickers, we believe that revenue would be small and piles of unidentified garbage would be large. Welfare gains would be negative.

0095-0696/95 $12.00

Should garbage be taxed to reflect its negative externality, or subsidized to avert illicit dumping? Maybe recycling should be subsidized. If not, could the same effect be achieved by a tax on virgin materials? What is the role for deposits on purchases with refunds on returns.

To address these questions, we build a simple theoretical general equilibrium model of household choice between consumption and leisure, and among three disposal options: garbage, recycling, and illicit burning or dumping. A single consumption good is produced using a single primary factor, recycled input, and virgin materials such as timber or minerals. We later consider disaggregate goods and services. The model also includes three externalities. First, municipal garbage collection and disposal may impose aesthetic and health costs on those who live near the landfill or incinerator. Second, improper burning and dumping may impose even higher costs on others. Third, the extraction of virgin materials involves clear-cutting or strip-mining that may adversely affect not only the landowner who sells timber or mineral rights, but others who enjoy wilderness and wildlife.

Others have addressed some of the above questions, usually with partial equilibrium models and only two disposal options. First, Sullivan [18] compares legal and illegal disposal. He finds both an optimal subsidy on legal disposal and an optimal degree of enforcement against illegal disposal. These are second-best policies since he does not allow for a tax on consumption (or, equivalently, a "deposit" upon purchase). Also, Dobbs [5] and Project 88—Round II [14] have discussed the problem of litter as a reason for deposits and refunds on particular commodities. Copeland [3] builds a general equilibrium model with international trade and waste from production, but no recycling or waste from consumption. Jenkins [6] and Sigman [16] consider waste and recycling, but not illegal dumping. Several other models of household decisionmaking deal with various important aspects of the solid waste problem, but also ignore illicit dumping (e.g., [4, 10, 15, 21]). Our study is unique in that (1) we address all of the above questions in (2) a single general equilibrium framework with (3) enough instruments to solve for first-best policies, and when (4) households can choose among all three disposal options: garbage, recycling, and illicit burning or dumping. Another advantage, we believe, is that our solution can be replicated easily on the back of an envelope.

To introduce our notation, and to characterize prior results, we start in Section 2 with a simple model in which garbage and recycling are the only two disposal choices. In this case the WRI study [15] is correct that the optimal garbage collection fee includes not only the direct resource cost ($1.12 per bag) but also the external cost (for a total of $1.83 per bag, in their study).

In Section 3, we add illicit dumping as a third disposal option. When all tax instruments are available, the first-best solution can be achieved by waste-end taxes on garbage *and* on illicit dumping. Not surprisingly, these two tax rates reflect only the corresponding externalities. Suppose, however, that a tax on illicit dumping is difficult or impossible to enforce. Does this mean that the first-best can no longer be supported? No. In general equilibrium, only relative prices matter—and therefore any tax can be set equal to zero as long as taxes on all other relevant activities are adjusted so as to induce first-best relative prices. Several points follow directly from this general equilibrium insight.

First, with the tax on illegal dumping set equal to zero, the proper relative price between legal and illegal disposal can be induced by subsidizing legal disposal (both garbage and recycling). That is, the optimal tax on garbage switches to a subsidy. If this subsidy for garbage is close to the direct resource cost ($1.12), then free collection of garbage is quite sensible.

Second, this subsidy for legal garbage disposal tends to subsidize consumption since it lowers the cost of waste disposal. Therefore, to restore the proper relative prices in general equilibrium, a tax on consumption is required. This result addresses a debate in the literature about whether optimal fees would be imposed "upstream" at the point of production, or "downstream" at the point of disposal.[1] In our model, if the downstream tax on illicit dumping cannot be enforced, the same first-best can be achieved by using an upstream tax instead. Consumption should be taxed at a rate that reflects not the good's disposal cost, but its possible externality from illicit burning or dumping. This tax is then returned as a subsidy on recycling and on proper disposal of garbage, leaving an implicit tax on burning or dumping. The result is a deposit–refund system, as in Bohm [2], but it applies to all consumption goods rather than just bottles or lead–acid batteries.

Third, many have suggested that recycling be encouraged by a tax on virgin materials (e.g., [9, 14, 16, and 19]). No such role arises in our model. Virgin materials are only taxed if their extraction has a negative externality (e.g., strip mining). Section 4 shows that if recycling *cannot* be subsidized for some reason, then all of the proper relative prices can still be achieved with a tax on virgin materials and on *all* other inputs except recycling.

Fourth, existing public finance literature generally finds that a consumption tax distorts the choice between taxed consumption and untaxed leisure. Here, we find that a consumption tax is part of a first-best policy, even though leisure is still untaxed. The reason is simply that consumption leads to disposal problems while leisure does not.

Section 5 considers disaggregate consumption goods, and Section 6 briefly discusses population density, multiple levels of government, and administrative cost.

2. JUST GARBAGE AND RECYCLING

Most models of household solid waste behavior ignore the possibility of illicit burning or dumping (e.g., [4, 6, 10, 15, 16, 21]). We therefore start with garbage and recycling as the only two disposal choices. The purpose of this section is not to provide any new results, but to set up our notation and to characterize prior results. Since none of the the discussion above involves distributional issues, we consider a single jurisdiction with n identical individuals or households. Each buys a single composite consumption good c, and each generates waste in two forms. We use g for garbage collection, and r for recycling and subsequent reuse in production. These alternatives are substitutes in the "technology" of household

[1] Wertz [21] finds that a (downstream) per-unit garbage fee raises the effective purchase price of goods with high disposal content. Menell [8] suggests retail disposal-content charges that reflect the subsequent disposal cost of each item. Porter [13] analyzes a deposit–refund system for bottles. Sigman [16] finds that a tax on virgin lead is equivalent to a deposit–refund system, when virgin lead and recycled lead are perfect substitutes in production. The pros and cons of alternative policies are nicely described in some of these papers, as well as in Miedema [9] and Project 88—Round II [14].

consumption,[2]

$$c = c(g, r), \tag{1}$$

where $c(\cdot, \cdot)$ is continuous and quasi-concave, and has positive first derivatives c_g and c_r. This relationship captures the degree to which the household is able to shift between disposal methods. With a given amount of consumption, the household may be able to reduce g and increase r by recycling newspapers, composting food waste, purchasing bottles in glass instead of plastic, collecting aluminum, and buying goods with less packaging. For simplicity, we specify each form of disposal as a single continuous variable. As a special case of (1), we will later discuss a "mass balance" example where $c = g + r$ (and $c_g = c_r = 1$).

Utility depends not only on household consumption, c, but also on home production, h, and the total amount of garbage, $G = ng$,

$$u = u[c(g, r), h, G], \tag{2}$$

where the first derivatives are $u_c > 0$, $u_h > 0$, and $u_G \leq 0$. For practical purposes, think of h as leisure use of time and resources. We use lower case letters to denote values per household and upper case letters for aggregates. Total garbage G may impose aesthetic and health costs, even if it is regulated in a "sanitary" landfill.[3]

On the production side of the model, output c may be produced using the constant returns to scale production function

$$c = f(k_c, r), \tag{3}$$

with input of resources k_c and recycled materials r from used consumption goods. Any reprocessing of r is folded into the production function. Also, just as we ignore transaction costs in the sale of c or k_c, we ignore the cost of collecting and trading r.

Provision of garbage collection services requires only one input, k_g, so constant returns to scale implies that production is linear. Home production uses resources k_h,

$$g = \gamma k_g, \quad h = k_h. \tag{4}$$

Finally, the model is closed by the resource constraint

$$k = k_c + k_g + k_h, \tag{5}$$

where k denotes a fixed total resource such as labor or capital. The results below do not require any distinction between labor and capital.

In this model, the social planner's problem is to maximize the utility of the representative household subject to the resource constraint, production constraints,

[2] Some readers may prefer to think of g and r as outputs of a function with input c, but Eq. (1) simply inverts that function. We think of g and r as amounts necessary to support c.

[3] In this static model of annual flows, G is total garbage per year. See V. L. Smith [17] for a dynamic treatment of waste flows into a landfill with a stock externality.

and $c(g, r) = f(k_c, r)$. The resource and production constraints can be substituted directly, to maximize

$$\mathscr{L} = u\big[c(\gamma k_g, r), k_h, n\gamma k_g\big] + \delta\big[f(k - k_g - k_h, r) - c(\gamma k_g, r)\big] \quad (6)$$

with respect to k_g, r, and k_h. This optimization recognizes that every individual imposes costs on others through the use of garbage collection services.[4] The first-order conditions are

$$u_c c_g + u_G n = \delta(c_g + f_{kc}/\gamma) \quad (7a)$$

$$u_c c_r = \delta(c_r - f_r) \quad (7b)$$

$$u_h = \delta f_{kc}, \quad (7c)$$

where f_{kc} is $\partial f/\partial k_c$, the marginal product of k used in the production of c. These equations will be employed shortly. They just indicate that the marginal utility made possible through additional g, r, or h must equal the marginal social cost.

For the case of private markets, individuals maximize utility in Eq. (2) subject to a budget constraint that may be affected by a tax or subsidy on each good,[5]

$$p_k k = (1 + t_c)c + (p_g + t_g)g + (p_r + t_r)r + p_k k_h, \quad (8)$$

where p_k is the price earned on resources, the price of consumption equals one since c is numeraire, t_c is the tax per unit of consumption, p_g is the price paid for garbage collection, t_g is the tax per unit of garbage, p_r is the price paid by the consumer for recycling (which may be positive or negative), and t_r is the tax per unit of recycling. Note that consumers pays prices gross of tax, but producers receive prices net of tax. Here, however, households ignore the effect of their own activities on the total externality. Tax and subsidy rates can simply be set to zero for the case of private markets with no government interference.

Producers receive a price ($= 1$) for selling c, and they receive the price paid by consumers for recycling, p_r, which we said could be positive or negative. They maximize profits ($c + p_r r - p_k k_c$) under perfect competition with constant returns to scale. Thus $f_{kc} = p_k$ and $f_r = -p_r$. Producers of garbage collection services similarly maximize ($p_g g - p_k k_g$), so $p_g = p_k/\gamma$.

In this decentralized economy, the consumer chooses g, r, and h to maximize utility in (2) subject to the budget in (8). The resulting first-order conditions involve prices (p_k, p_r, and p_g), but we replace those with marginal products ($f_{kc}, -f_r, f_{kc}/\gamma$) to obtain

$$u_c c_g = \lambda\big[(1 + t_c)c_g + f_{kc}/\gamma + t_g\big] \quad (9a)$$

$$u_c c_r = \lambda\big[(1 + t_c)c_r - f_r + t_r\big] \quad (9b)$$

$$u_h = \lambda f_{kc}, \quad (9c)$$

[4] We assume second-order conditions hold, solutions are internal, and a unique solution exists (see Baumol and Oates [1, pp. 37–38]).

[5] We ignore the government revenue requirement, assuming implicitly that lump-sum taxes are available to finance spending and to pay for necessary subsidies.

where λ is the marginal utility of income. These first-order conditions indicate that private marginal utility matches the cost to individuals of each activity. With tax rates of zero, it is easy to see that the outcome is not optimal. The right-hand sides of (7a) and (9a) would be similar, but only the left side of (7a) would account for the external cost of garbage, u_G.

With Pigouvian tax rates, however, private behavior in (9) can be induced to match the unique social optimum in (7). In this case, (7c) and (9c) indicate that $\delta = \lambda$. Since both problems maximize utility subject to a resource constraint, and both attain the same optimum, the social marginal utility equals the private marginal utility of the resource.

Next compare (7b) and (9b). By inspection, and using $\delta = \lambda$, these two equations will both hold when $t_c = t_r = 0$. In this model, no tax or subsidy is required for private behavior to yield this first-order condition of the social optimum.[6]

Finally, we compare (7a) and (9a). When $t_c = 0$, these equations both hold so long as $t_g = -nu_G/\lambda$. Since $u_G \le 0$, this Pigouvian tax is ≥ 0. It increases with the size of the externality (u_G) and with the number of people adversely affected (n). To convert the tax into dollars, the marginal effect on utility, u_G, is divided by the marginal utility of income, λ.

Additional garbage collection is not a pure public good but uses scarce resources such as labor, capital, and landfill. Consumers should pay $(p_g + t_g)$, equal to \$1.83 per bag in the WRI study [15], enough to cover both the resource cost and the negative externality from garbage.

3. VIRGIN MATERIALS AND ILLICIT BURNING OR DUMPING

This section makes two modifications to the model. First, for each individual, we allow a third disposal alternative,

$$c = c(g, r, b), \tag{1'}$$

where b stands for burning and other improper disposal, such as dumping by the side of the road. Again b is a single continuous variable, and $c_b > 0$. With given consumption, the household may reduce g and raise b by burning cardboard boxes in the fireplace, carrying trash to commercial dumpsters, or leaving it out in the woods. Total $B = nb$ may reduce the utility of others. In the "mass balance" example, $c = g + r + b$.

Second, we consider virgin materials, with per capita use v. Aggregate use $V = nv$ also may reduce the utility of others. Implicitly, this cost may represent the shadow price of over-using scarce minerals in a more complicated dynamic model.[7] More explicitly, in our model, total V may reduce the public enjoyment of natural areas through clear-cutting or strip-mining. Cutting timber may reduce biodiversity

[6] Actually, this condition allows for any tax on consumption t_c, as long as it is returned as a subsidy for both garbage and recycling such that the net tax is still zero. We use this property below.

[7] Neher [11, Chap. 13] shows conditions under which (1) economies systematically underprice environmental resources (p. 238), and (2) the static optimizing solution is the steady state solution of a corresponding dynamic problem (p. 243).

and increase global warming. The new utility function is

$$u = u[c(g,r,b),h,G,B,V], \tag{2'}$$

where $u_G \leq 0$, $u_B \leq 0$, and $u_V \leq 0$. In addition, we assume that $u_B \leq u_G$. In other words, garbage is bad enough in the landfill but even worse if thrown by the side of the road.[8]

Production of c is modified to use not only labor and capital resources, k, and recycling input, r, but also virgin materials, v,

$$c = f(k_c, r, v). \tag{3'}$$

These virgin materials are produced or extracted using a simple linear function,

$$v = \alpha k_v. \tag{4'}$$

Each of these goods can be provided for a market price, but improper burning and dumping cannot. Yet illegal burning does involve time, psychic costs, and perhaps risk of getting caught. We assume that burning uses private resources $k_b = \beta(b)$, with marginal costs that are positive ($\beta_b > 0$) and rising ($\beta_{bb} > 0$). We also assume that any direct tax or penalty on burning or dumping would be difficult or impossible to enforce.[9] Finally, the resource constraint becomes

$$k = k_c + k_g + k_h + k_v + k_b. \tag{5'}$$

The social planner maximizes consumer utility (2′) in a Lagrangean like (6), subject to additional constraints (3′) through (5′), with respect to k_g, r, k_h, b, and k_v.[10] First-order conditions (7a, b, c) are unchanged, and

$$u_c c_b + u_B n = \delta(c_b + f_{kc}\beta_b) \tag{7d}$$

$$u_V n = \delta(f_{kc}/\alpha - f_v). \tag{7e}$$

For the case of private markets, competitive firms set the marginal product f_v equal to their cost in terms of market price plus tax ($p_v + t_v$). Other firms produce v and maximize profits $[p_v(\alpha k_v) - p_k k_v]$, so $p_v = p_k/\alpha$.

[8] We are grateful to a referee for pointing out that it is not sufficient to assume the u_B schedule lies below the u_G schedule. Both schedules may be decreasing (with B or G). We assume that the marginal unit of B entails more externality than the marginal unit of G, even though the amount of burning B may be much less than the amount of regular garbage G.

[9] See Lee [7] for a full treatment of enforcement costs and taxpayer avoidance costs. A simple enforcement model might suggest arbitrarily high penalties, in order to save real police resources. Our model avoids this problem in two ways. First, an internal solution for the social optimum implies that a certain amount of burning may be less costly in social terms than more landfill or recycling. The optimal tax on b is finite. Second, in our model, no direct tax on b is required or even desirable. As long as the government can enforce taxes or pay subsidies on market transactions (of c, g, r, and v), we show that the first-best allocation is attainable.

[10] The Lagrangean is

$$\mathcal{L} = u\left[c(\gamma k_g, r, b), k_h, n\gamma k_g, nb, n\alpha k_v\right] + \delta\left[f(k - k_g - k_h - k_v - \beta(b), r, \alpha k_v) - c(\gamma k_g, r, b)\right].$$

The consumer's budget constraint in (8) must now include the cost of burning, $p_k\,\beta(b)$, which is not taxed for reasons cited above. First-order conditions (9a, b, c) are unchanged, and[11]

$$u_c c_b = \lambda[(1 + t_c)c_b + f_{kc}\,\beta_b] \tag{9d}$$

$$f_{kc} = (f_v - t_v)\alpha. \tag{9e}$$

Again we solve for the Pigouvian tax rates that induce private behavior in Eqs. (9) to match the social optimum in (7). Again (7c) and (9c) imply that $\delta = \lambda$. This time, however, (7d) and (9d) can be solved for a particular value of t_c that is not zero. Then (7b) and (9b) require that $t_r = -t_c c_r$. The full set of optimizing tax rates are

$$t_c^* = -nu_B/\lambda c_b \tag{10a}$$

$$t_r^* = nu_B c_r/\lambda c_b \tag{10b}$$

$$t_g^* = n[u_B c_g - u_G c_b]/\lambda c_b \tag{10c}$$

$$t_v^* = -nu_V/\lambda. \tag{10d}$$

If illicit burning or dumping has no external effect ($u_B = 0$), then this solution reduces to the previous solution, with $t_c = t_r = 0$ and $t_g = -nu_G/\lambda$. With $u_B < 0$, however, consumption must be taxed at $t_c^* > 0$ to attain the first-best conditions, even if leisure is still untaxed. This tax is returned on goods that are properly collected as garbage or recycling. The net effect is a tax on illicit burning, circumventing the problem that b could not be taxed directly.

Garbage receives the rebate of t_c^*, but it also receives a tax that depends on its own externality. In Section 2, above, the tax was positive. Here, the net tax t_g^* is likely to be negative. In the "mass-balance" case where $c_g = c_b = 1$, garbage receives a net subsidy because u_B is more negative than u_G. In general, the tax depends on the relative ease of burning versus garbage collection (c_b versus c_g). It may also depend on the consumer's willingness to break the law. If the optimal price $p_g + t_g^*$ is *near* zero, then the city or county can save administrative and billing costs by providing free garbage collection.

Finally, the tax on virgin materials is not part of any deposit–refund system. It is not used to encourage recycled input or to discourage the generation of garbage. Instead, the tax on virgin materials should only correct negative externalities from the use of virgin materials.

4. THE UPSTREAM VS DOWNSTREAM DEBATE

This model can be used to reconcile various policy recommendations, that is, to show the conditions under which different policies can each attain the first-best allocation of resources.

[11]The consumer maximizes utility (2′) subject to the new budget constraint by choosing g, r, h, and b, which yields four first-order conditions in prices. We replace those prices with marginal products as before, to get (9a) through (9d). The fifth condition (9e) results directly from producer behavior; the producer of v chooses input k_v to maximize profits $p_v(\alpha k_v) - p_k k_v$, which yields $p_v = p_k/\alpha = f_{kc}/\alpha$. Then the producer of c chooses inputs k_c, r, and v to maximize profits $f(k_c, r, v) + p_r r - p_k k_c - (p_v + t_v)v$, which yields $f_v = p_v + t_v$. Thus $p_v = f_v - t_v = f_{kc}/\alpha$, to get (9e).

TABLE I
Comparison of Various Tax Schemes to Achieve the First-Best

Case 1: illicit burning or dumping cannot be taxed	Case 2: all forms of household disposal can be taxed	Case 3: no recycling subsidy (requires tax on other inputs)	Case 4: disaggregate goods (illicit dumping cannot be taxed)
$t_c^* = -nu_B/\lambda c_b$	$t_c^* = 0$	$t_c^* = nu_B c_r/\lambda c_b f_r - nu_B/\lambda c_b$	$t_{ci}^* = -nu_{Bi}/\lambda c_{bi}$
$t_r^* = nu_B c_r/\lambda c_b$	$t_r^* = 0$	$t_r^* = 0$	$t_{ri}^* = nu_{Bi} c_{ri}/\lambda c_{bi}$
$t_g^* = n(u_B c_g - u_G c_b)/\lambda c_b$	$t_g^* = -nu_G/\lambda$	$t_g^* = n(u_B c_g - u_G c_b)/\lambda c_b$	$t_{gi}^* = n(u_{Bi} c_{gi} - u_{Gi} c_{bi})/\lambda c_{bi}$
$t_v^* = -nu_V/\lambda$	$t_v^* = -nu_V/\lambda$	$t_v^* = -nu_B c_r f_{kc}/\lambda c_b f_r \alpha - nu_V/\delta$	$t_v^* = -nu_V/\lambda$
	$t_b^* = -nu_B/\lambda$	$t_{kc}^* = -nu_B c_r f_{kc}/\lambda c_b f_r$	

Note. t_c^*, t_r^*, t_g^*, and t_v^* are, respectively, tax rates on consumption, recycling, garbage, and virgin materials. t_b^* is a tax on illicit burning or dumping (only in Case 2). t_{kc}^* is a tax on resources (labor or capital) used in production of the consumption good (only in Case 3). u_G, u_B, and u_V are the (negative) externalities from garbage collection, illicit burning, and virgin material extraction. c_g, c_r, and c_b are the (positive) derivatives of $c = c(g, r, b)$, the extra consumption enabled by more g, r, or b. λ is the private marginal utility of income, and δ is the social marginal utility of income. n is the number of individuals, and i is the index for disaggregate goods and services.

The first column of Table 1 repeats the deposit–refund system just described, and the second column shows an equivalent downstream tax on wastes. In this case we constrain t_c to zero, but we allow a tax on illicit burning or dumping at rate t_b. Equations (7b) and (9b) then require $t_r = 0$, while (7d) and (9d) require the enforcement of $t_b = -nu_B/\lambda$.

Clearly the downstream tax system in column 2 can solve the problem in theory. It is the most direct policy, since the tax on each activity reflects its own externality. All tax rates are positive, but $u_B < u_G$ implies $t_b > t_g$. However, t_b would require enforcement of a tax or penalty upon actions that are easy to hide, such as burning trash in a fireplace or leaving it along a deserted road. Fortunately, the upstream tax t_c (in column 1) is easier to implement. It requires no litter penalties, and it still achieves the same first-best outcome in this model. We therefore return to the case where $t_b = 0$.

In addition, the CBO [19] echos others (e.g., [9, 14, 16]) in suggesting that a tax on virgin materials "could bring about an increase in the recycled content of some products and an overall decrease in the amount of waste disposal" [19, p. 41]. However, our equations (7e) and (9e) above show that $t_v = -nu_V/\lambda$ (as shown in columns 1 and 2 of the table). This simple Pigouvian tax only corrects for negative externalities from the extraction of virgin materials (e.g., strip mining). Some may believe that $u_V = 0$, in which case $t_v = 0$. The point is that t_v is *not* used to correct any problem related to garbage or recycling.

A reason for this nonresult is that recycling can be subsidized directly. In addition, production uses three inputs. A tax on v would encourage not only the use of r, but also the use of k_c. If recycling *cannot* be subsidized for some reason, and if k_c can be taxed at rate t_{kc}, then we can again appeal to multiple solutions. Since only relative prices matter, column 3 shows how the same first-best can be achieved using both an extra tax on virgin materials *and* a tax on k_c. In fact, the table shows that t_v and t_{kc} contain the same unambiguously positive term.[12] This term appears with the opposite sign in t_c. In other words, a subsidy for recycling in $f(k_c, r, v)$ is equivalent to a tax on v and k_c, which is then returned on output. The tax on g is unaffected.[13]

Although we show the equivalence of this result to the deposit–refund system, it is much more complicated. The tail wags the dog, since illicit dumping is corrected by multiple taxes on all inputs other than recycling, combined with a subsidy on all output.

5. DISAGGREGATE GOODS AND SERVICES

We can further modify the model to include different consumption goods c_i, where i is an index. These goods may have different technologies $c_i(g_i, r_i, b_i)$ and production functions $f_i(k_{ci}, r_i, v_i)$. Subscripts also are required for

[12] Slight differences appear in these terms. The tax on k_c uses $f_{kc} = \partial c/\partial k_c$ to convert to changes in c, whereas the tax on v uses $f_{kc}/\alpha = \partial c/\partial v$ to convert to changes in c.

[13] The derivation of Case 3 is a bit more difficult than the earlier cases, since λ is no longer equal to δ. We assume that p_k is the net return to individuals, so producers of c pay $p_k + t_{kc}$. Still the approach is to solve all equations (7) and (9) for t_c, t_g, t_v, and t_{kc}. Also, t_c can be manipulated to show a term $(c_r - f_r)$ which is unambiguously positive, from (7b), so t_c must be negative; the rebate of t_{kc} and t_v is within t_c, and this term exceeds the previous tax paid on c.

$u_G, u_B, \delta, t_c, t_g, t_r$, and the price of output (p). The results look exactly the same, except for the subscripts. In other words, the waste-end tax rates in column 2 must be modified to collect a specific tax rate on each good $t_{bi} = -nu_{Bi}/\lambda$, which reflects the social cost of burning or dumping that particular item. Alternatively, the deposit–refund system would collect $t_{ci} = -nu_{Bi}/\lambda c_{bi}$ on each purchase and rebate the corresponding amount upon proper disposal of that item (column 4 of Table I).

Differences arise because some goods are more toxic, more unsightly, or more easily burned. To get the details exactly right, the first-best policy would have to place a different tax rate on each item. With millions of goods and services, either of these first-best systems would be a nightmare to administer. Perhaps only some items could be targeted, or all items could be placed in just a few categories. Note, however, that this diversity among commodities is no reason to prefer waste-end taxes to the deposit–refund system: for any given commodity or category, the difficulty of enforcing a tax on illegal midnight dumping can still be avoided by taxing legal daytime purchases and subsidizing legal daytime disposal.

Consider three categories, for example. The purchase of "services" might be left untaxed, since this purchase requires no subsequent disposal.[14] A second category could receive a moderate rate, and a third, "hazardous" category would receive a higher rate. If goods with different-sized externalities were thrown into the same tax category, perhaps for reasons of administrative simplicity, then consumers would be encouraged to buy relatively too much of the goods that entailed relatively more costly disposal. Even if all goods were thrown in the same category, however, at least consumers would have some incentive to dispose of those goods properly. Results here can be taken simply to help explain existing policies that tax purchases and subsidize proper garbage collection.

6. LIMITATIONS, EXTENSIONS, AND DISCUSSION

Using broad brush strokes, this paper characterizes the optimal taxation of garbage, recycling, and general consumption. These broad strokes may miss some important details, however. One extension might consider how these optimizing fees are related to population density. We hypothesize that garbage fees would work best in suburban areas or small towns where the charges can be enforced. In densely populated urban areas, any price for garbage collection may be greeted by huge piles of unidentified garbage on the streets or vacant lots. In very rural areas, similarly, dumps may appear on back roads. Thus, these results may help explain differences in actual municipal pricing mechanisms.[15]

Another problem arises with different levels of government. States traditionally set sales tax rates, but municipal governments subsidize garbage collection. Perhaps some revenue should be transferred from one to the other. Goods may be purchased in one state, however, and then traded or carried across state lines before disposal. Various spillover effects might justify a national-level tax and

[14] On the other hand, the production of services such as legal and medical services generates plenty of paper and medical wastes. The model could be expanded to consider waste by-products from production.

[15] The U.S. EPA [10] lists sixteen towns that have unit fees for garbage collection, and we have extended that list for empirical work. All are small or moderate in size, with the exception of Seattle.

rebate system, but spillovers may still cross national boundaries. This issue deserves further scrutiny (as in Copeland [3]). Also, consideration of distributional issues might affect the optimal tax and refund system.

Finally, our model ignores some compliance costs and market imperfections. With regard to recycling, Nestor [12] points out that subsidies to households may generate supplies of recycled goods when the industrial capacity to make use of them does not exist. With regard to a per-unit garbage fee, the town would have to sell special bags or stickers. This administrative cost might further justify the subsidy inherent in free garbage collection.

None of these considerations alter our four main points, however. First, existing studies find that a negative externality from garbage can be corrected by a tax on garbage. When we add illicit dumping as a third disposal option, and assume that it cannot be taxed directly, then the tax on garbage may turn negative. Garbage collection may optimally subsidized to help prevent the worse environmental costs of improper disposal. Second, to restore all the correct relative prices, output would be taxed. The result is a deposit–refund system that is equivalent to the unenforceable waste-end tax. Third, existing literature suggests that a tax on virgin materials may help encourage recycling. In our basic model, a tax on virgin materials is not useful for that purpose. It only corrects directly for the ill effects of using virgin materials. Fourth, in existing public finance literature, a consumption tax distorts the choice between labor and leisure. This paper provides a first-best case for a consumption tax, however, even when leisure is untaxed. No tax on leisure is required, because leisure does not cause disposal problems.

APPENDIX: NOMENCLATURE

n	number of identical individuals in the single jurisdiction
c	a single composite commodity consumption good
$c(\ )$	consumption function (technology of consumption)
g	quantity of garbage collection
r	quantity of recycling
c_g	partial derivative of consumption with respect to garbage
c_r	partial derivative of consumption with respect to recycling
h	quantity of home production
G	total amount of garbage
$u(\ )$	utility function of the individual
u_c	marginal utility of consumption
u_h	marginal utility of home production
u_G	marginal utility of total garbage
$f(\ )$	production function
k_c	resources used in production of the consumption good
k_g	resources used in garbage collection
k_h	resources used in home production
γ	marginal product of resources in the collection of garbage
k	fixed total resources in the economy
$\mathscr{L}$	the Lagrange function
δ	social marginal utility of the resource (k)
f_{kc}	marginal product of the resource input (k_c)
f_r	marginal product of the recycled input
p_k	price earned on resources
t_c	tax per unit of consumption
p_g	price paid for garbage collection
t_g	tax per unit of garbage

p_r	price paid by the consumer for recycling
t_r	tax per unit of recycling
λ	private marginal utility of income
b	burning and other improper disposal
c_b	partial derivative of consumption with respect to burning
B	the total amount of burning in the economy
u_B	marginal utility of total burning
V	total amount of virgin material used in production
v	per capita virgin material used in production
u_V	marginal utility of total virgin material used in production
k_v	resources used in the extraction of virgin material
α	marginal product of resources in the extraction of virgin material
k_b	resources used in burning garbage
$\beta(\ )$	a function relating resources used in burning to the quantity of burning
β_b	the first derivative of $\beta(\ )$ with respect to burning
β_{bb}	the second derivative of $\beta(\ )$ with respect to burning
f_v	the marginal product of virgin material
p_v	price of virgin material
t_v	tax per unit of virgin material
t_b	tax per unit of burning
t_c^*	the optimal tax on consumption
t_r^*	the optimal tax on recycling
t_g^*	the optimal tax on garbage
t_v^*	the optimal tax on virgin material
t_{kc}	tax on the use of resources in the production of the consumption good
i	an index
c_i	consumption of good i
$c_i(\ )$	the consumption function for good i
g_i	garbage associated with good i
r_i	recycling associated with good i
b_i	burning associated with good i
$f_i(\ )$	the production function of good i
k_{ci}	the use of resources in the production of good i
r_i	the use of recyclables in the production of good i
v_i	the use of virgin materials in the production of good i
t_{bi}	tax on dumping for good i
u_{Bi}	the marginal utility from burning waste from good i
t_{ci}	tax on the consumption of good i
c_{bi}	the partial derivative of consumption with respect to burning

ACKNOWLEDGMENTS

This research began while both authors were at the University of Virginia. We are grateful for suggestions from Debbie Nestor, Ed Olsen, Jon Skinner, Margaret Walls, anonymous referees, and seminar participants at NBER, Carnegie Mellon, Georgia, Vanderbilt, and Virginia. We are also grateful for funding from NSF Grant SES91-22785, and from the Bankard Fund at the University of Virginia. This paper is part of NBER's research program in Public Economics. Any opinions expressed are those of the authors and not those of the National Science Foundation or the National Bureau of Economic Research.

REFERENCES

1. W. J. Baumol and W. E. Oates, "The Theory of Environmental Policy," 2nd ed., Cambridge Univ. Press, New York (1988).
2. P. Bohm, "Deposit–Refund Systems: Theory and Applications to Environmental, Conservation, and Consumer Policy," John Hopkins Univ. Press, Baltimore (1981).
3. B. R. Copeland, International trade in waste products in the presence of illegal disposal, *J. Environ. Econom. Management* **20**, March, 143–62 (1991).

4. T. M. Dinan, Economic efficiency effects of alternative policies for reducing waste disposal, *J. Environ. Econom. Management* **25**, 242–56 (1993).
5. I. M. Dobbs, Litter and waste management: Disposal taxes versus user charges, *Canad. J. Econom.* **24**, February, 221–7 (1991).
6. R. R. Jenkins, Municipal demand for solid waste disposal services: The impact of user fees, dissertation for the University of Maryland Department of Economics (1991).
7. D. R. Lee, The economics of enforcing pollution taxation, *J. Environ. Econom. Management* **11**, 147–60 (1984).
8. P. S. Menell, Beyond the throwaway society: An incentive approach to regulating municipal solid waste, *Ecology Law Quart.* **17**, 655–739 (1990).
9. A. K. Miedema, Fundamental economic comparisons of solid waste policy options, *Resour. Energy* **5**, 21–43 (1983).
10. G. E. Morris and D. M. Holthausen, The economics of household solid waste generation and disposal, *J. Environ. Econom. Management* **26,** 215–34 (1994).
11. P. A. Neher, "Natural Resources Economics: Conservation and Exploitation," Cambridge Univ. Press, New York (1990).
12. D. V. Nestor, Partial static equilibrium model of newsprint recycling, *Appl. Econom.* **24**, April, 411–7 (1992).
13. R. C. Porter, A social benefit–cost analysis of mandatory deposits on beverage containers, *J. Environ. Econom. Management* **5**, 351–75 (1978).
14. "Project 88—Round II, A Public Policy Study Sponsored by Senator T. E. Wirth and Senator J. Heinz, Directed by R. N. Stavins," Washington, DC (1991).
15. R. Repetto, R. C. Dower, R. Jenkins, and J. Geoghegan, "Green Fees: How a Tax Shift Can Work for the Environment and the Economy," World Resources Institute, Washington, DC (1992).
16. H. Sigman, A comparison of public policies for lead recycling, mimeo, UCLA Department of Economics (1991).
17. V. L. Smith, Dynamics of waste accumulation: Disposal versus recycling, *Quart. J. Econom.* **86**, November, 600–16 (1972).
18. A. M. Sullivan, Policy options for toxic disposal: Laissez-faire, subsidization, and enforcement, *J. Environ. Econom. Management* **14**, 58–71 (1987).
19. U.S. Congress, Congressional Budget Office, "Federal Options for Reducing Waste Disposal," U.S. Government Printing Office, Washington, DC (1991).
20. U.S. Environmental Protection Agency, "Charging Households for Waste Collection and Disposals: The Effects of Weight or Volume-Based Pricing on Solid Waste Management," EPA 530-SW-90-047, Washington, DC (1990).
21. K. L. Wertz, Economic factors influencing households' production of refuse, *J. Environ. Econom. Management* **2,** 263–72 (1976).

[16]

ELSEVIER

Ecological Economics 20 (1997) 255–268

ECOLOGICAL ECONOMICS

ANALYSIS

Developments in the throughput–income relationship: theoretical and empirical observations

S.M. de Bruyn *, J.B. Opschoor

Free University Amsterdam, Department Spatial and Environmental Economics, De Boelelaan 1105, 1081 HV Amsterdam, The Netherlands

Received 15 May 1995; accepted 26 June 1996

Abstract

It has been suggested that the relationship between income and the associated environmental pressure in terms of derived materials inputs and pollution levels may take an inverted-U shape, indicating a 'delinking' of environmental pressure from economic growth in relation to rising per capita incomes. The likelihood of such a relationship being persistent is discussed in the context of a simple macro model of industrial metabolism, and the possibility of 'relinking' clearly emerges. Data on specific indicators of environmental pressure (i.e., the throughput of materials, energy and the volume of transport) in 19 countries have been used to investigate patterns of delinking. The results suggest that delinking may not be persistent; already some advanced economies may be entering a new period of relinking. The relationships between environmental pressure and welfare in the medium long term may be N-shaped rather than inverted-U-shaped.

Keywords: Economic growth; Materials consumption; Energy consumption; Sustainable development; Dematerialisation

1. Introduction

In recent years much attention has been devoted to the development of what has been called by Herman Daly 'environmental macro economics': the branch of environmental or ecological economics that considers its subject matter to be the physical exchanges crossing the boundary between the total ecological system and the economic system (Daly, 1991, p. 35). Theoretical research as well as empirical studies have provided inputs into the scientific and societal debates about actual and desirable material interactions between the economy and the environment. One of the positions put forward is that in the process of ongoing economic growth the economy would 'delink' itself from its environmental base, so that rising per capita income levels would be associated with declining levels of pollution and use of resources (e.g., World Bank, 1992). The relationship between environmental pressure (defined here as the aggregate of pollution, depletion and other human activity related threats to environmental quality) and income is sometimes assumed to take on an inverted-U shape referred to as the 'Environmental Kuznets Curve' (Selden and Song, 1994). Focusing on an aggregated proxy for environmentally relevant material throughput, Jänicke et al. (1989) report a tendency of declining levels of throughput beyond certain income levels, also suggesting inverted-U curves. These ideas are obviously important in the debate on the relationship between economic growth

* Corresponding author. Tel.: +31-20-4446095; fax: +31-20-4446005; e-mail: sbruyn@econ.vu.nl.

PII S0921-8009(96)00086-9

and environmental quality: if such relationships are persistent then eventually economic growth might be compatible with ever declining levels of environmental pressure. However, as several authors in a recent issue of this journal (Vol. 15, No. 2, 1995) have pointed out, theoretical considerations cast some shadows over such bright prospects.

In this paper, first an overview will be presented of some theoretical and empirical issues in Section 2. Then in Section 3 we will present some empirical work to validate and extend one particular approach: the analysis of aggregated throughput at the level of national economies. The theoretical parts will deal with: (i) a conceptual macro model of the physical exchange (cf. Daly) between the economy and the ecological system in which it is embedded, (ii) notions relevant to the description of the development of this exchange, such as metabolism, delinking and dematerialisation. The empirical part of the paper analyses changes in the consumption of selected materials, energy and transport for 19 countries over the period of 1966–90. The objective is to search for patterns of delinking that are persistent and to compare the results with other empirical studies.

2. Conceptual framework and empirical background

2.1. Environmental macro economics

Daly (1991) has addressed physical exchanges or material flows between the economy and the environment by considering them not only in terms of equity and efficiency, but also in terms of their scale or total volume relative to the ecosystem. The environmentally relevant scale of the economy (S) in year t has been defined as the product of the level of population (P_t) and per capita environmental claim or use (U_t)

$$S_t = P_t \cdot U_t. \qquad (1)$$

Daly refers to S as 'throughput', defined as the entropic physical flow of matter–energy from nature's sources, through the human economy and back to nature's sinks (e.g., Daly, 1991, p. 36). S_t thus comprises material inputs into the economy (M) and outputs of the economy into the environment (pollution including waste, W). One can distinguish between 'environmental pressure' and subsequent modifications of the 'state' of the environment (to use standard OECD jargon; OECD, 1993) and differentiate between types of environmental pressure accordingly. M and W refer to two distinct types of environmental pressure: depletion and pollution. The changes in the state of the environment they give rise to may differ according to variations in local circumstances concerning the prior state of the environment and its associated assimilative capacities. In what follows we will discuss S_t as an indicator of total environmental pressure through materials flows, that is, some aggregate of M and W. Or

$$S_t(M, W) = P_t \cdot U_t(M, W), \qquad (1.1)$$

U_t is an aggregate of per capita claims in the domains of M and W. It can be considered as being decomposable in a per capita material welfare or income term, y_t, and a variable reflecting the environmental pressure per unit of welfare, E_t. Elaborating Eq. (1.1) using the decomposition in y_t and E_t, we obtain

$$S_t(M, W) = P_t \cdot y_t \cdot E_t(M, W). \qquad (2)$$

This equation shows that Daly's environmental macroeconomics is related conceptually to several other perspectives that have emerged in the environmental and economic literature. First, Eq. (2) is equivalent to the famous '$I = PAT$' — formula that has emerged from the debate between Ehrlich, Holdren and Commoner in the early 1970s about the determinants of environmental degradation (Ehrlich and Holdren, 1971). [1] Secondly, Daly's environmental macroeconomics is embedded in the notion of industrial metabolism. Industrial metabolism (IM) has been defined as: the set of physico-chemical

[1] If we equate S with I for environmental impact, and replace y by A (for 'affluence') and E for T (technology), Eq. (2) turns into $I = PAT$, which is an identity since I/Y is defined as T and Y/P is defined as A. This formula is often used in extrapolations of the increased pressure on the environment in alternative scenarios for economic and population growth and technological development (e.g., Weterings and Opschoor, 1992; Vonkeman and Maxson, 1994).

transformations that convert raw materials and energy, plus labour, into finished products and wastes in a more or less steady state condition (Ayres, 1994), i.e., it is a metaphor for a set of processes driving materials flows on the interface between the economy and the environment. [2] S_t can be regarded as a variable indicating the environmentally relevant levels at which these processes operate, or the level of metabolism. As is evident from the physical basis of the IM, environmental macro economics operates within constraints set by the Law of Conservation of Matter: matter can be transformed, transported or stored, but cannot be created or destroyed. That is, a materials balance approach should be taken at least in principle (see e.g., Ayres and Kneese, 1969). If we define Y_t as the total income (the product of per capita income y_t and population P_t), the materials balance can be captured by

$$M_t - W_t = f(Y_t) - \sum_{j=0}^{t} g_j(Y_j), \tag{3}$$

where M and W are the values in period t of materials used and emissions/wastes released, $f(Y_t)$ is an expression of the volume of M_t embodied in the production of Y in period t, while $g(Y)$ are the wastes released to the environment from materials embodied in previously produced products. Or: materials input flows equal output flows, corrected for changes in materials stocks. Theoretically, such changes in stocks are levelled out in the long run so that the inputs and outputs must equal

$$\int_0^{\infty} F(M_t - W_t)\mathrm{d}t = 0. \tag{4}$$

Compared to neoclassical economics, Daly adds a concern for optimal or 'proper scale' to other, more conventional economic objectives such as efficiency. This proper scale can be found by reference to the preferences of those agents that are thought to be relevant constituents of society today as well as in the future; it is now quite common to consider this proper scale to be at least not beyond the carrying capacity (C) of the total ecological system that the economy operates within (Daly, 1991)

$$S_t(M, W) \leq = C \tag{5}$$

Economies operating at or within C are called (ecologically) 'sustainable' (Daly, 1991, p. 44). This notion of carrying capacity has to be developed in a dynamic sense in order to correspond better to the ecological complexities of the relationships between economy and environment. One may conceive of an 'environmental utilisation space' as a set of steady states in terms of levels of M_t and W_t that are sustainable, i.e., compatible with the ecological processes (such as regeneration and absorption processes) and life-support systems underlying the economy (Siebert, 1982; Opschoor, 1992; Opschoor, 1995). This environmental utilisation space can be regarded as a multidimensional sustainability frontier analogous to the familiar production possibilities frontiers of neoclassical welfare economics. The notion of environmental space as a sustainability frontier implies that an initial carrying capacity C_0, if exceeded by $S_t(W, M)$ at some point in time, will lead to a subsequent capacity $C_t < C_0$, which could in turn be sustained, but with lower yields in terms of sustainable environmental pressure than in the initial situation. Thus

$$C_t = h(M_t, W_t, C_{t-1}), \tag{6}$$

which makes the environmental utilisation space at the end of period t a function of its initial value C_{t-1} and the level of metabolism (M_t and W_t) during the period.

We can now put forward the ideas of Daly c.s. in more precise terms. They imply that effectively some value C^* (e.g., C_0, if the current level of environmental potential is to be maintained) is chosen on the sustainability frontier, which is then imposed on the economic process as an environmental constraint

$$S_t(M_t, W_t) \leq C^*(M_t, W_t, C_{t-1}). \tag{5.1}$$

Eqs. (1) and (2) are ways of linking environmental pressure to driving forces such as the variation in income, population and technology. Eq. (3) connects this with claims in the past. Eq. (6), together with Eqs. (2) and (3), describes the pattern of environmental change over time and provides a link between

[2] The notion does not relate so much to the metabolism of industry alone (as opposed to, e.g., agriculture), but in fact refers to the metabolism induced or entailed by all economic activity. What is described here could have also been labelled 'economic metabolism' or 'societal metabolism' (Moll, 1993), but we will use the more familiar notion of industrial metabolism.

environmental pressure and subsequent modifications of the state of the environment in terms of reduced carrying capacities. Incorporating (5.1) would turn the model into a prescriptive one on sustainable economic development

2.2. Dematerialisation and delinking

Having thus described some 'first principles' of environmental macro economics in one conceptual structure, we now proceed a few steps. If C^* can be perceived as being or becoming a binding sustainability constraint on economic development, then a crucial question is whether the economy will (or can be made to) move to a position compatible with it, or whether the economy will proceed ignoring the constraint, thereby advancing along unsustainable paths. Given the high preference for rising material welfare, this may lead to questions on the possibility for economic growth to continue and on the factual behaviour of S as y rises. The latter question is normally addressed by looking at changes in the level of 'throughput' S per unit of Y and will be elaborated below; the former gives rise to a search for ways to change throughput rates and is beyond this paper's scope.

Ratios M_t/Y_t and W_t/Y_t can be labelled the materials intensity and the pollution intensity of production; their inverses are the productivity coefficients of materials and pollution. A more efficient IM would then be characterised by lower levels of materials and pollution intensities, or by raised productivities. Reduction of the materials intensities over time can be referred to as 'dematerialisation' (cf. Herman et al., 1989) and, analogously, a reduction in the pollution intensity could be labelled 'depollution'. Given the fact that data on materials and energy use are more abundant than data on emissions and wastes, some researchers take the rate of change in total materials inputs as a measure of potential environmental impact, or throughput S, and hence focus on materials intensity and dematerialisation (Ayres, 1989; Jänicke et al., 1989; Simonis, 1989; Bringezu et al., 1994). We will follow that line for a moment for the sake of the gains in transparency of the argument.

If environmental pressure S is a function of M (M being a proxy of U), and if we define m_t as the materials intensity M/Y in period t, then Eq. (1) can be rewritten as

$$S_t = m_t \cdot Y_t \tag{1.2}$$

and

$$\frac{dS}{dt} = Y_t \cdot \frac{dm}{dt} + m_t \cdot \frac{dY}{dt}. \tag{7}$$

We can now proceed by distinguishing two forms of dematerialisation in a growing economy: weak dematerialisation and strong dematerialisation. For dematerialisation to be called weak, the materials intensity must fall. Hence, a sufficient condition is

$$\frac{dm}{dt} < 0. \tag{8}$$

For dematerialisation to be called strong, total materials consumption must become smaller over time. In view of Eq. (7) this may be the case if

$$\frac{-dm/dt}{m} > \frac{dY/dt}{Y}. \tag{9}$$

Or, there will be strong dematerialisation if the rate of change of the materials intensity has an absolute value higher than the growth of output (or economic growth).

The concept of 'delinking' (World Bank, 1992) embraces both dematerialisation and depollution; it refers to the process whereby aggregate economic activity gives rise to reduced environmental impacts (either in the form of materials and energy inputs, or its emissions and wastes). Delinking can be described as strong and weak analogous to Eqs. (8) and (9). Delinking may result from different sets of developments (or combinations thereof):

1. changes in production processes and product design;
2. changes in the composition of production or consumption.

Hence, delinking may result from technological change leading to increased M- and W-productivity and changes in prevailing production patterns due to changes in consumption patterns or life styles. Some refer to these processes as 'structural change' (Simonis, 1989; Jänicke et al., 1989), indicating that the reductions in pollution and materials consumption are not due to economic fluctuations, but to more structural changes underlying the economic process. Such a process of structural change may be the result of endogenous processes operating within the economy (e.g., due to differences in income elasticities, innovation, etc.) or they may be exogenous (e.g., changes in other, related, economies, or policy interventions). To the extent that prices reflect environmental costs, markets may be assumed to generate signals (in the form of price changes) supporting the processes of substitution mentioned above, and providing incentives to innovate in terms of the development of new products, processes or inputs. If prices do adequately capture these costs, and if market processes are flexible enough, then delinking may be a (near-) spontaneous, endogenous development within the economic process. Theoretical as well as empirical investigations, however, point at rather rigid market processes for materials as input in production processes, at least in the short run (see Malenbaum, 1978; Auty, 1985; Holmes, 1990).

If the rates of dematerialisation correspond to the rates of depollution, changes over time in environmental pressure could be related adequately to dematerialisation only. In fact, Eq. (4) suggests that this will be the case in the very long run at least, but in empirical applications on politically relevant time scales, Eq. (4) may be of limited value. Moreover, owing to information constraints, normally only subsets of the total material flows M and total pollution W will be considered. For these reasons factual analysis on subsets of M and W may exhibit different trends. This need not cause a problem for empirical applications, since depletion and pollution also differ in their impacts for the relevant scale of economic activities. Relatively abundant resources may be substituted for relatively scarce resources, but such substitutions may be harmful if ecosystem resilience is at stake. These remarks must be kept in mind when discussing empirical studies on dematerialisation below.

2.3. Some empirical evidence on delinking

The data on pollution that are available generally show reductions of pollution per unit of production/income, as income levels rise, and often even reductions in absolute levels of pollution. Such strong delinking has been reported for the Netherlands for emissions (1985–1990) such as CFC's (−46%), NH_3 (−16%), SO_2 (−20%), whereas NO_x remained constant and CO_2 and waste flows still increased (13% and 7%, respectively) (RIVM, 1993). The World Bank has reported strong delinking for emissions of SO_x, lead and particulates for all OECD countries since 1970, and for NO_x since 1980 (World Bank, 1992). Several other references confirm, at least for certain types of pollutants, such trends (e.g., Shafik and Bandyopadhyay, 1992; Selden and Song, 1994). The more extensive research on dematerialisation to a certain degree discloses similar tendencies. Several studies reveal decreasing material and energy intensities in a range of OECD countries, especially since 1970 (see for example for energy Bossanyi, 1979; Chesshire, 1986 and for materials Larson et al., 1986; Tilton, 1990).

It is worth observing that each of these studies refers to single pollutants and single materials, but the aggregated developments of several pollutants or materials together are not analyzed. It might be the case that materials substitution fully accounts for the perceived dematerialisation and hence one should speak of 'transmaterialisation' only: a shift from materials that the study focuses on to unobserved ones. Wadell and Labys (1988) present results that indicate that such material shifts have taken place during this century. To exclude the possibility of transmaterialisation, it is important to include a more complete aggregate of materials consumption. Only few empirical studies analyze such aggregates over time: Ayres (1989) presents figures for the US economy for four years between 1960–1975, and Bringezu et al. (1994) presents total material consumption (including removal of earth for mining) for Western Germany for five years between 1980–1989. Both studies aggregate on the basis of mass and give a comprehensive estimation of the total materials flows as inputs in an economy, albeit the results are too fragmentary to compare developments over time and among countries.

Jänicke et al. (1989) have used an aggregated indicator approximating the volume of throughput. Comparing 1970 and 1985 for a set of 31 COMECON and OECD countries, they concluded that economic growth seems to be delinked from this aggregate for the more developed economies. We have taken our cue from that study and hence will describe it in some more detail. Jänicke et al. (1989) have used four types of proxies for throughput:

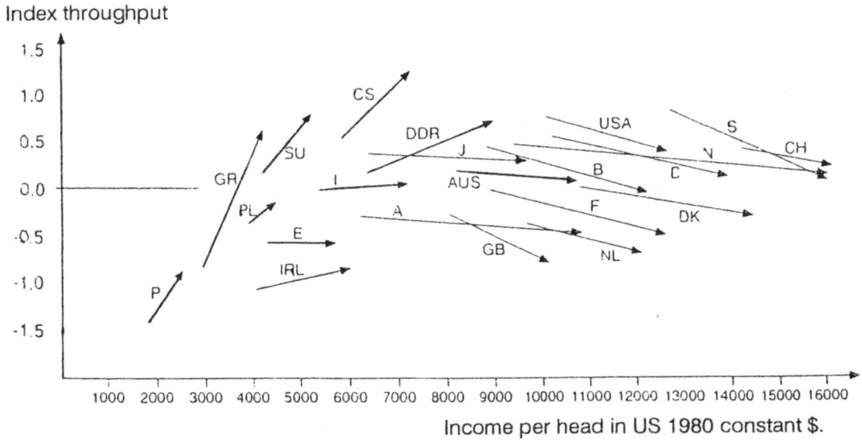

Fig. 1. Delinking income from the EI for some countries (Mean EI 1970 = 1985 = 0). Cross-section analysis (1970) EI = − 1.2362 + 0.17Y, $R^2 = 0.57$; (1985) EI = −0.3951 + 0.046Y, $R^2 = 0.09$. Source: Weizsäcker and Schmidt-Bleek (1994), after Jänicke et al. (1989).

energy consumption, steel consumption, cement production and weight of freight transport on rail and road (as a general measure of the volume aspect of an economy). These proxies for throughput capture to a large extent the environmentally relevant physical realities of the industrial metabolism, according to the authors, and have been aggregated into a single indicator, which we shall refer to as *throughput indicator* (TI, see Section 3 for calculation). The results of this analysis are summarised in Fig. 1. The arrows give the linearised developments between 1970 and 1985 for various countries. They suggest a clear tendency of delinking.

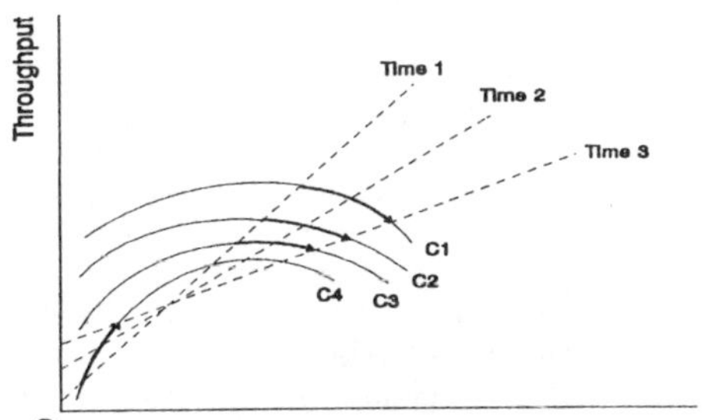

Fig. 2. Graphical illustration that cross-section analyses that measure decreasing slopes and increasing intercepts assume declining intensities of use for developed economies.

If we assume that TI captures the developments of dematerialisation satisfactorily, then strong delinking with associated absolute reductions in the level of approximated throughput can be found in e.g., France, Sweden, the UK and Western Germany, whereas other countries were undergoing weak delinking with the associated decline in the throughput intensities; some Southern European economies, such as Greece and Portugal (and all COMECON economies), would still have increasing levels of throughput.

Jänicke et al. (1989) also found that the correlations between per capita GDP and the TI for the whole sample was much less significant and showed a smaller slope and higher intercept in 1985 than in 1970; this they interpreted as another sign of the process of delinking through structural change. This can be illustrated by using Fig. 2, which gives the stylized inverted-U relationship between environmental pressure and income for four countries ($C1 \ldots C4$) with a shifting inward for poorer countries, resulting from the diffusion of technology. [3]

[3] As remarked by Malenbaum (1978), diffusion of technology prevents economic late-comers from experiencing the same levels of materials consumption per unit of GDP than older industrialized countries have experienced in the past.

The bold line in each curve represents the period of measurement (time 1, 2, 3) for each country. If these countries are taken together in linear cross-section regression analysis, period 1 will show a higher slope and a lower intercept than period 2 and 3. So the cross-section results of Jänicke et al. confirm inverted-U curves (or Environmental Kuznets Curves) for individual countries. Interpreted in this way, the results can indeed be seen as a suggestion for a process of delinking materials consumption from economic growth for the more developed countries in the period considered.

2.4. The 'relinking hypothesis'

As we have seen, empirical evidence goes far in suggesting that delinking has occurred in the domain of M as well as in that of W for developed economies. Given the likelihood that both globally and regionally economic growth will continue to be pursued, delinking environmental pressure from economic growth is clearly an important issue. In terms of our model, delinking will continue if dm is a negative function of income or time. Eq. (7) suggests that the manifestation as well as the magnitude of delinking depend on the relative values of the rate of economic growth (dY/Y) and the rate of change of the materials intensity coefficient (dm/m). As income grows, consumption patterns may change in less materials intensive directions and means may become mobilised towards technological innovations conducive to $(dm/m) < 0$. However, the question is whether such trends can be easily extrapolated. There may come a time, or income level, at which conditions Eqs. (8) and (9) no longer hold, simply because the possibility of increasing energy and materials efficiency may have a technological or even economic upper bound. From that point onward, the growth component in Eq. (7) may become more dominant and the economy and environmental pressure S will be relinked (Opschoor, 1990), at least until further breakthroughs in research and development occur or a more intensive application of environmental policy checks is implemented. This prediction we call the 'relinking hypothesis', and we take it that empirical manifestations of $dS/dt > 0$ in specific areas of environmental concern (M and/or W) are validating it at least in part.

3. An empirical investigation of changes in throughput, 1966–1990

3.1. Calculating throughput

The calculation of any indicator for S involves several steps. First, proxies have to be identified that reflect an adequate subset of aspects of S and that allow for data collection. Secondly, these proxies need to be integrated in a throughput indicator, TI, that is supposed to give a comprehensive picture of the developments of S. Finally, the developments in this indicator have to be linked to economic developments in order to assess the impacts of economic activity on the TI.

The choice of proxies that are indicative for the developments of S may include input (M) as well as output factors (W). The choice made by Jänicke et al. (1989) was to look at a small set of variables indicative of materials flows in a range of industrialised economies for which data are widely available: the consumption of steel and energy, the production of cement and the volume of transport by rail and road, which were supposed to capture roughly the throughput aspects of a country. These proxies reflect M-related aspects of S and capture a substantial part of the energy and mass-related aspects of production. For example, cement consumption accounts for 18% and steel consumption for 22% of all industrial minerals [4] and metals consumed in 1990 in the United States (Berry, 1994). For the same reason, the weight of freight transport in tonnes (and not in ton-kilometres) is used as a general indicator of the volume aspect of production. Finally, the addition of energy consumption results in an integrated proxy of material and energy flows.

The decision to focus on M-related aspects of S only is mainly driven by data limitations: there are actually no data on emissions and wastes over a long time period for a reasonable number of countries that would allow cross-sectional analysis as depicted in Fig. 2. Therefore, we will below also focus on aggregated materials/energy consumption and hence follow the dematerialisation argument using the same

[4] Construction materials like crushed stone, sand and gravel are not taken into account.

kind of indicators as Jänicke et al. (1989). Contrary to Jänicke et al. (1989) we will use cement consumption as an indicator instead of cement production in order to reduce the sensitivity of the TI for import substitution. Opschoor (1990); Stern et al. (1994) have suggested that the observed inverted-U curves may be the result of changes in international specialisation: poor countries may attract 'dirty' and material intensive production while richer countries specialize in 'clean' and material extensive production, without altering the consumption patterns. In fact, import substitution related reductions in environmental pressure should not be counted as an environmental gain (Opschoor and Reijnders, 1991).

In a next step, the proxies chosen can be integrated in a throughput indicator, TI, that is supposed to give a comprehensive picture of the developments of M. This causes a problem, since materials consumption and energy consumption are calculated in different units, respectively, tonnes and Joules. The approach that has been put forward by Jänicke et al. (1989) is to add up the relative deviations from the mean per capita consumption of each proxy with an equal weighting. We will follow that approach below in order to make our results comparable. First, per capita sample means are calculated for each proxy made up from all countries and all years. The throughput index (TI) for each country c and for each year t is then calculated by adding the relative deviations from the mean per capita consumption of each indicator ($M_{i,c,t}$) and dividing by 4 which results in an equal weighting of each proxy [5]

$$\mathrm{TI}_{c,t} = \tfrac{1}{4} \sum_{i=1}^{4} \left(M_{i,c,t} - \overline{M}_i \right) / \overline{M}_i \qquad (10)$$

where i = proxy, c = country $(1 \ldots n)$, t = year $(0 \ldots T)$, M_i = per capita use of proxy i.

The equal weighting of cement consumption, steel consumption, freight transport and energy consumption may be unsatisfactory from an environmental point of view, but this point will not be elaborated here — dealing with it would either imply the application of some set of weights, or multicriteria analysis. We would like to emphasize that virtually every aggregation is arbitrary from the environmental point of view: aggregation by mass also assumes that every material is equally important for either depletion or pollution.

Finally, and more importantly, the methods that have been used to analyze the relationship between the TI and income need to be discussed. Jänicke et al. (1989) apply cross-section analysis to two distinct years (1970 and 1985). This ignores the possibility that the patterns of M/Y over time may take on several shapes. It is possible that the aggregate materials consumption declined from 1970 to 1980 as a result of the two oil crises, but increased again after 1980. Delinking may not be a stable or persistent process. By extending the range of sub-periods, more insight can be gained in the stability of the observed delinking. We will focus below explicitly on periods instead of single years, due to the measurement of the data. Steel and cement consumption are being measured as apparent consumption (production + import–export). As a result of changes in stocks, data for one single year might be unreliable.

In the next sections, then, we shall incorporate into the analysis the points raised above. In Section 3.2 we will extend the analysis by reviewing the performance of a set of countries for a longer time span (1966–1990), in order to capture a wider range of possible forces operating on the M-Y-relationship. We will cut down that period into four sub-periods to investigate the relationship over time, rather than make a simple comparison between the initial and the final year. We will use cement consumption as an indicator instead of production in order to diminish the sensitivity of the TI for import substitution. In Section 3.3 we will take a closer look at especially the last decade in order to dig slightly deeper into the phenomenon of the emerging upswing in throughput per unit of Y (or 'relinking') in some countries.

3.2. The development of throughput as a function of income, 1966–1990

As a first step we will explore the development of throughput in several countries, using TI as a proxy

[5] It appears that the throughput index is on a per capita base, which contrasts with our model in Section 2.2, where environmental pressure and throughput are defined in terms of their absolute levels. However, if we want to relate the throughput index to per capita income, population growth merely acts as a scaling factor for both throughput and income. The basic relationship between throughput and income, therefore, is not affected.

for M. A sample of 19 countries has been used, containing Belgium–Luxembourg, Denmark, Finland, France, Greece, Hungary, Italy, Japan, the Netherlands, Norway, Poland, Spain, Sweden, Switzerland, Turkey, United Kingdom, the United States, Western Germany and (former) Yugoslavia. The TI has been calculated as the aggregation of the deviations from the mean per capita consumption of the selected proxies: steel consumption, cement consumption, energy consumption and tonnes freight transport by rail and road (data sources are listed in the Appendix). Using Eq. (10) we have calculated the mean for all the countries together over the entire period 1966–1990. To obtain an overall view of the relationship between the TI and GDP, we divide the period under study into four sub periods of seven years: 1966–72, 1972–78, 1978–84 and 1984–90. The values for TI and Y for each of these periods are calculated as the average value in each period of seven years. In this way incidental shocks in the economies or data are levelled out.

The four periods distinguished each represent a different stage of economic development in most of the market-economies. In the first period, economic growth was high and not disturbed by external shocks. The second period was marked by the first oil crisis in 1973, which influenced several economic parameters. The third period was marked by the second oil crisis and the subsequent negative growth rates in the beginning of the 1980s. The last period is characterized by a recovery in Western economies from the crisis in the beginning of the 1980s while the economy in the formerly socialist countries collapsed.

In a cross-section analysis the computed per capita TI is regressed on per capita income over all four periods. This analysis leads to some observations on the strength of the M-Y-relationship and on the development of its slope (see Table 1). The measure of strength of the relationship drops considerably after the first period, to drop to a low in the third period, and to pick up again in the final one. The slope has decreased until 1984, but increased slightly in the last period. The intercept increased until 1984, but decreased in the last period.

In accordance with Fig. 2, the interpretation of the diminishing slope over time is that countries converge in their approximated throughput. It explicitly does not indicate that a one unit increase in GDP will result in an increase of the TI by 0.76 in the first period and 0.41 in the third. The interpretation must be that over time, aggregate materials and energy consumption in the poorer countries rose faster than the increase in GDP, while the consumption in the richer countries declined in both relative and absolute levels. This development of convergence is explicitly made clear by the diminishing standard deviation of the TI given in Table 1. All this is still fully compatible with the hypothesis that there exists a parabolic relationship between income and environmental pressure. Over time, richer countries tend to consume less materials and energy while poorer countries tend to consume more. The most interesting, and in fact, novel feature of Table 1 is the suggestion that this development has come to an end in the last period. Interpreted in terms of Fig. 2, it may be the case that some of the developed economies in our sample are entering a period of

Table 1
Relationship between per capita TI and per capita GDP (measured in 1000 1985 US$) for 20 countries: descriptive statistics for four periods[a]

	Years			
	1966–1972	1972–1978	1978–1984	1984–1990
Slope β (T-value)	0.0765 (5.830)	0.0552 (3.926)	0.0409 (3.381)	0.0435 (4.711)
Intercept α (T-value)	−0.5892 (−6.725)	−0.4188 (−3.800)	−0.3938 (−3.734)	−0.4624 (−5.061)
Measure of strength R^2	0.666	0.476	0.402	0.566
St. deviation of TI	0.330	0.311	0.275	0.283

[a] Regression equation: $TI = \alpha + \beta GDP + e$

relinking, rather than continuing on the delinking path. This development is made more explicitly in Fig. 3 where the developments for some individual countries from period to period are shown. Each dot indicates one period; the lines connect the dots for each country. Country symbols refer to international automobile numberplates, except for Belgium–Luxembourg (BL), Spain (SP) and the United Kingdom (UK).

As can be seen from this figure, three groups of countries can be distinguished. First, there may be relatively poor countries (e.g., Turkey) which display a persistent positive linkage between growth and environmental pressure as measured by the throughput index. Secondly, a range of countries (Greece, Hungary, Poland, Spain and former Yugoslavia) have experienced absolute reductions in their levels of throughput in the last period (1984–1990) compared to the third period (1978–1984). The former communist economies are in a period of structural change at least since 1989, but drastic changes occurred even prior to 1989. The group of richer countries showed in general absolute reductions in their levels of throughput in the third period, between 1978 and 1984. It is noteworthy that this development of declining levels of throughput continued in the last period only for France and the United States. All the other countries showed a rise in their levels of throughput for that period. For Italy and Japan this has been the case in each period, but the growth in their levels of TI has been slackening. Switzerland is the only exception to this generalized pattern: decreases in the second period and increases in the third and fourth period as well.

3.3. Developments in 1980–1990: relinking in advanced economies?

Most of the countries under study have experienced absolute reductions in their levels of aggregate material throughput in one or more periods, with rising per capita incomes, i.e., strong dematerialisation as defined in Eq. (9). However, our analysis suggests that for the most developed economies, this degree of delinking may have come to an end in the 1980s. We now compare the end of the 1980s (1989) with the beginning of the 1980s (1983) by recalculating the TI as a moving three years' average value for these years and the 1986 value (Table 2).

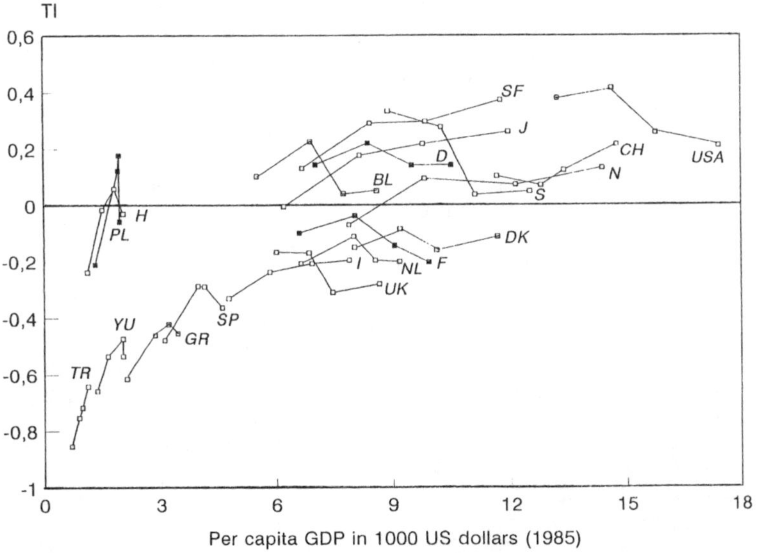

Fig. 3. Developments of selected countries in levels of TI and per capita GDP, 1966–1990.

S.M. de Bruyn, J.B. Opschoor / Ecological Economics 20 (1997) 255–268 265

Table 2
Developments of throughput intensities and their components (1983–1989)

Countries	Intensities (1983 = 100)		Per capita changes 1983–1989 (%)					
	1986	1989	TI	GDP	steel	cement	energy	transport
Poland	88.4	78.3	−10.6	14.1	−19.1	−8.7	−6.5	−8.7
Hungary	90.9	80.2	−13.4	8.0	−30.8	−12.8	4.1	−13.1
Spain	81.2	83.6	2.3	22.4	32.8 *	−8.6	20.7	−15.3
Norway	97.7	85.3	−1.5	15.4	−29.7	−19.9	10.8	16.0 *
Finland	91.5	87.2	5.3	20.8	−15.0	2.1	17.2	9.2
Japan	88.3	89.7	12.6	25.6	28.9 *	−2.6	16.4	9.5
Yugoslavia	93.8	90.5	−14.7	−5.7	−26.8	−16.1	12.0 *	−24.7
Greece	90.1	91.6	1.3	10.6	10.8 *	−9.1	25.6 *	13.5 *
Sweden	94.9	92.7	5.0	13.3	3.2	18.1 *	13.1	−5.5
USA	97.2	94.1	7.8	14.6	6.5	15.9 *	7.2	5.1
France	94.7	94.1	6.1	12.7	1.2	2.1	12.3	8.7
West-Germany	94.1	94.1	8.3	15.1	16.8 *	−5.6	6.5	13.3
Italy	92.8	95.0	11.5	17.4	34.6 *	−0.9	12.6	14.7
Denmark	103.3	96.6	10.7	14.6	12.3	4.4	6.7	16.3 *
United Kingdom	93.5	97.8	18.7	21.4	18.2	37.5 *	7.6	18.3
Switzerland	101.1	97.9	10.9	13.3	−7.1	19.0 *	10.6	12.3
Belgium	96.4	98.5	13.2	14.9	8.0	19.9 *	14.0	11.4
Netherlands	98.2	100.1	12.4 *	12.3	24.0 *	10.1	9.8	9.0
Turkey	113.0	112.8	36.4 *	20.9	62.8 *	34.2 *	21.9 *	25.9 *

All figures based on three-year moving averages 1983–1989.
* Figures with an asterisk indicate that this indicator rose more than proportional to the increase in GDP.

The second and third column of this table show the developments of the throughput intensities (dm/m) in 1986 and 1989, respectively, compared to 1983. Countries are ranked according to their success in reducing their throughput intensities over this time span. It appears that the formerly communist countries together with the northern Scandinavian countries were quite successful in reducing their throughput intensities. Among these countries Hungary, Norway and Poland experienced strong delinking; i.e., absolute reductions in their levels of throughput while their economy was growing. A second subset is that of countries with weak delinking, that is, with a growing environmental pressure as measured by our throughput-proxies, but with dropping throughput intensities. For Finland, Greece and Spain, this increase in throughput is relatively low compared to their growth of GDP. In other countries the throughput rose at a rate sometimes close to that of the increase in GDP. However, if the throughput intensities of 1989 are compared with those of 1986, it can be seen that several of these countries (Belgium–Luxembourg, Italy, Japan, Spain, Western Germany and the United Kingdom) showed an upswing in their throughput intensities. That is, their (approximated) throughput rose more than their increase in GDP, which is indicative of a period of relinking. For the Netherlands and Turkey this rise in throughput intensities was persistent over the whole period. For Turkey, this was the case for all materials, a typical feature of a rapidly developing country. For the Netherlands this was due to strongly increased consumption of steel and a relatively low increase of GDP.

If the individual proxies are considered, we see some similarities and some differences between countries. On average, the growth in energy consumption was relatively low compared to the growth in GDP. Most countries experienced increases of steel or cement consumption, but rarely of both. The wide range of developments of the individual proxies justifies the choice to work with aggregated indicators that approximate throughput, since single indicators may not be representative.

We conclude that the observed absolute improvements in aggregate material consumption for developed economies as observed by Jänicke et al. in 1985 as compared to 1970 did not continue in the late 1980s (with the exception of Norway). This confirms our observations of Section 3.2.: a tendency of the throughput intensities of relatively poor countries of our sample to drop, and a tendency of the throughput intensities of the more developed countries to level off and sometimes even increase. In terms of our theoretical framework, this may indeed be indicative of a development where technological and/or economic constraints prevent subsequent delinking rates in terms of (dm/dt) as were experienced during the 1970s. Once the technological options for further dematerialisation have been exhausted, the growth component in Eq. (7) (dY/dt) may become a more persistent determinant to explain developments in throughput, in spite of arguments that have been made by other authors.

4. Discussion and conclusions

Most studies on material and energy consumption calculate single material or energy intensities. These studies often report substantial decreases in the consumption of materials and energy compared to GDP for OECD countries over the last twenty years. Jänicke et al. have integrated consumption of materials, energy and transport into one single indicator which they then calculated for a range of OECD as well as COMECON countries, for 1970 and 1985. This too suggested a trend of developed economies to 'delink'. However, as our analysis shows, this does not prove to be persistent. There are strong reasons to believe that this development is coming to an end and that we may now enter a phase of relinking, at least as far as materials inputs are concerned.

The levels of aggregate materials consumption through time may show an 'N-shape' rather than an inverted-U curve, with several Western European countries now finding their aggregate material consumption again increasing faster than GDP. As was to be expected on the qualitative, theoretical grounds presented earlier on, delinking does not appear as a process which prevails or persists under conditions of sustained economic growth. If this conclusion holds, and presupposing an ongoing tendency of economic growth, then there is no reason to feel assured that this process of growth will induce developments such that the economic process will adhere to environmental constraints. Sustained growth is not necessarily sustainable. This is intuitively even more appealing if we link the observed patterns of environmental pressure to changes in the state of the environment. It can be expected that developed economies in the past already have exhausted part of their assimilative capacities and thereby more easily expand carrying capacities than less developed economies.

As has been noted above, these conclusions may critically hinge on the choice of parameters or proxies for environmental pressure. This is an aggregate of environmental inputs such as materials and energy into the economic process and environmentally relevant outputs of that process, such as pollution and waste flows. Here we have used a specific subset of proxies for throughput. They may incompletely reflect materials inputs (though in terms of mass they capture most of the relevant materials flows in developed economies) and they do inadequately capture pollution. Therefore, one obvious area for further research is that of improved indicators for environmentally relevant industrial (or economic) metabolism. This holds for materials-related indicators and even more so for pollution and waste-related indicators. The latter area of research is especially interesting as other studies indicate continuously declining patterns of emissions for most pollutants, thus perhaps neutralising or contradicting some of the tendencies reported on in this paper.

Also the aggregation scheme that has been chosen in this paper may be too simplistic and can be refined in future work using multicriteria analysis or factor analysis with weights attached to the policy relevance of various indicators. The lack of a uniform aggregation scheme for proxies of environmental pressure is still hampering empirical research.

More research will also be needed to appropriately decompose the changes in materials, energy and pollution intensities and to disclose the driving forces behind these changes (such as structural changes in the patterns of production and final demand, technological change, environmental and other policies) in order to develop a strategy for maintaining and accelerating delinking where possible, and to discern areas where complementary scale-related approaches may be required.

Acknowledgements

The authors would like to thank Jeroen van den Bergh from our department, Paul Ekins of the Department of Economics, University of London and the anonymous referees for useful comments on earlier drafts.

Appendix A. Data sources

Our data were taken from a number of international sources as listed below.

A.1. Cement consumption

Economic Commission for Europe of the United Nations: Annual Report on Housing and Building. United States CRB: Commodity Yearbook. For Italy, Japan, Spain and Turkey cement production has been taken as a proxy for consumption. Production data from Statistical Office of the United Nations: Statistical Yearbook.

A.2. Energy consumption

Measured as total primary energy supply (TPES). International Energy Agency (IEA): Energy Balances of OECD Countries, Energy Balances of Non-OECD countries, World Energy Statistics and Balances. For Hungary, Poland and (former) Yugoslavia before 1971 growth rates extracted from: United Nations: World Energy Supplies (total final consumption).

A.3. Steel consumption

Until 1987. United Nations: Statistical Yearbook. After 1987. Eurostat: Eisen und Stahl for EC members; Economic Commissions for Europe of the United Nations: Annual Steel Statistics for Europe for non-EC members. Series have been standardized on each other using three-year averages.

A.4. Freight transport

Economic Commission for Europe of the United Nations: Annual Bulletin of Transport Statistics for Europe. International Road Federation: World Road Statistics. European Conference of Ministers of Transport (ECMT): Transport Statistics. Especially for road transport series show differences. Series have been standardized. Rail transport has been used for all countries, road transport for all countries except United States, Spain, Italy and Greece. For these countries the quantity of freight transport was measured by rail only. This is the same routine as Jänicke et al. took in their investigations in 1989. The choice of only taking into account rail and road transport can be justified by pointing out that these modes of transport represent the final stages of a transport network and the magnitude of transported tonnes gives a good approximation of the total transport chain which also includes international marine, inland waterways and air transport.

A.5. Population

World Bank: Worldtables.

A.6. Gross domestic product

World Bank: Worldtables, 1992. For (former) Yugoslavia, Poland and Hungary before 1971 growth rates have been extracted from: Summers, R. and A. Heston: A new set of international comparisons of real product and price levels estimates for 130 countries, 1950–1985, in: Income and Wealth, 1988.

References

Auty, R., 1985. Materials intensity of GDP: Research issues on the measurement and explanation of change. Resour. Pol., 11: 275–283.

Ayres, R.U., 1989. Industrial Metabolism. In: J.H. Ausubel & H.E. Sladovich (Editors), Technology and Environment. National Acadamy Press. Washington, D.C., pp. 23–50.

Ayres, R.U., 1994. Industrial metabolism: Theory and policy. In: R.U. Ayres and U.E. Simonis (Editors), Industrial Metabolism: Restructuring for Sustainable Development. UNU Press, New York, pp. 3–21.

Ayres, R.U. and Kneese, A.V., 1969. Production, Consumption and Externalities. Am. Econ. Rev., LIX: 282–297.

Berry, D., 1994. Information needs for sustainable resource policy. In: F.J. Duinhouwer, G.J. van der Meer and H. Verbruggen (Editors), Sustainable Resource Management and Resource Use: Policy Questions and Research Needs. RMNO publ., No. 97, Rijswijk, pp. 175–190.

Bossanyi, E., 1979. UK primary energy consumption and the changing structure of final demand. Energy Pol., 7: 253–258.

Bringezu, S., Hinterberger, F. and Schütz, H., 1994. Integrating Sustainability into the System of National Accounts: The Case of Interregional Material Flows. Paper prepared for the international AFCET symposium on Models of Sustainable Development, Paris, March 1994.

Chesshire, J., 1986. An energy-efficient future: a strategy for the UK. Energy Pol., 14: 395–412.

Daly, H.E., 1991. Steady State Economics: Second Edition with New Essays. Island Press, Washington, DC, 297 pp.

Ehrlich P.R. and Holdren, J.P., 1971. Impact of population growth. Science 171: 1212–1217.

Herman, R., Ardekani, S.A. and Ausubel, J.H., 1989. Dematerialization. In: J.H. Ausubel and H.E. Sladovich (Editors), Technology and Environment. National Academy Press, Washington, DC, pp. 50–69.

Holmes, M.J., 1990. Material substitution in the UK window industry, 1983–1987. Resour. Pol., 16: 128–142.

Jänicke M., Monch, H., Ranneberg, T. and Simonis, U.E., 1989. Economic structure and environmental impacts: East–West comparisons. Environmentalist, 9: 171–182.

Larson, E.D., Ross, M.H. and Williams, R.H., 1986. Beyond the era of materials. Sci. Am., 254: 34–41.

Malenbaum, W., 1978. World Demand for Raw Materials in 1985 and 2000. McGraw-Hill, New York, 126 pp.

Moll, H.C., 1993. Energy Counts and Materials Matter in Models for Sustainable Development: Dynamic Lifecycle Modelling as a Tool for Design and Evaluation of Long-Term Environmental Strategies. Styx, Groningen, 396 pp.

Organisation for Economic Co-operation and Development (OECD), 1993. OECD Core Set of Indicators for Environmental Performance Reviews: A synthesis report by the Group on the State of the Environment. Environ. Monogr., No. 83, OECD, Paris.

Opschoor, J.B., 1990. Ecologische duurzame economische ontwikkeling: Een theoretisch idee en een weerbarstige praktijk. In: P. Nijkamp and H. Verbruggen (Editors), Het Nederlands Milieu in de Europese Ruimte: Preadviezen van de Koninklijke Vereniging voor Staathuishoudkunde. Stenfert Kroese, Leiden, pp. 77–126.

Opschoor, J.B., 1992. Sustainable development, the economic process and economic analysis. In: J.B. Opschoor (Editor), Environment, Economy and Sustainable Development. Wolters-Noordhoff, Groningen, pp. 25–53.

Opschoor, J.B., 1995. Ecospace and the fall and rise of throughput intensity. Ecol. Econ., 15: 137–141.

Opschoor, J.B. and Reijnders, L., 1991. Towards sustainable development indicators. In: H. Verbruggen and O. Kuik (Editors), In Search of Indicators for Sustainable Development. Kluwer Academic Press, Dordrecht, pp. 7–29.

Rijksinstituut voor Volksgezondheid en Milieuhygiëne (RIVM), Nationale Milieuverkenning 1993–2010. Samson H.D. Tjeenk Willink, Alphen aan de Rijn, 167 pp.

Selden, T.M. and Song, D.S., 1994. Environmental quality and development: Is there a Kuznets Curve for air pollution emissions? J. Environ. Econ. Manage., 27: 147–162.

Shafik, N. and Bandyopadhyay, S., 1992. Economic Growth and Environmental Quality: Time-Series and Cross-Country Evidence. World Bank Working Papers, WPS 904, Washington, 52 pp.

Siebert, H., 1982. Nature as a life support system: Renewable resources and environmental disruption. J. Econ., 42: 133–142.

Simonis, U.E., 1989. Industrial restructuring for sustainable development: Three points of departure. Science Centre Berlin FS II 89–401, Berlin.

Stern, D.I., Common, M.S. and Barbier, E.B., 1994. Economic Growth and Environmental Degradation: A Critique of the Environmental Kuznets Curve. Discussion Papers in Environmental Economics and Environmental Management, EEEM 9409, University of York.

Tilton, J.E., 1990. The OECD countries: Demand trend setters. In: J.E. Tilton (Editor), World Metal Demand: Trends and prospects. Resources for the Future, Washington, DC, pp. 35–76.

Vonkeman, G.H. and Maxson, P., 1994. International views on long-term environmental policy. In: Dutch Committee for Long-Term Environmental Policy, The Environment: Towards a Sustainable Future. Kluwer Academic Publishers, Dordrecht, pp. 219–245.

Wadell, L.M. and Labys, W.C., 1988. Transmaterialization: Technology and materials demand cycles. Mater. Soc., 12: 59–85.

von Weizsäcker, E.U. and Schmidt-Bleek, F., 1994. Signs of hope for the 21st century? In: Dutch Committee for Long-Term Environmental Policy, The Environment: Towards a Sustainable Future. Kluwer Academic Publishers, Dordrecht, pp. 21–45.

Weterings R. and Opschoor, J.B., 1992. The Ecocapacity as a Challenge to Technological Development. RMNO Publ., No. 74a, Rijswijk.

World Bank, 1992. Development and the Environment: World Development Report 1992. Oxford University Press, Oxford, New York, 308 pp.

Part II
Sustainability and Growth Perspective: Limits, Factors …

[17]

On the Concept of Optimum Population[1]

1. INTRODUCTION

1.1. Concern about the optimum size of population dates back at least to Wicksell. Later economists like Cannan, Dalton, Wolfe and Robbins were all primarily concerned with the problem that Wicksell originally posed, namely: what density of population under given circumstances is most advantageous? The view held by all those I have mentioned above was roughly the following: the optimum population is that population which, with given natural resources, state of arts, and standard of working time, would secure the largest *per capita* output.

Population policies pose difficulties that are both practical and conceptual. However, it is the purely conceptual problems that concern us here and I shall in this paper pose some of these problems in the context of a number of highly stylized models that will allow the different considerations that are thought to be relevant to be expressed in a manner precise enough to obtain definite answers.

1.2. The economic welfare of a community is affected by policies that determine (1) the rate of capital accumulation; and (2) the rate of growth of population. At any moment of time the optimum size of population will depend on the size of the existing capital stock and the optimum rate of savings will depend on the existing number of people. In this sense a population policy cannot be formulated without a concurrent savings policy. The two must be considered together. This, then, is the approach that I shall take.[2]

The criterion of optimality that I shall be adopting is the maximization of the total discounted welfare of all generations from now to infinity. The problem will be to choose that rate of savings and that size of population at every moment which will attain this maximum welfare, if indeed, a maximum exists. Much of the time I shall inquire into the existence of an optimum policy under various circumstances.

1.3. I have heard it argued that the welfare of each generation ought to be reflected by its standard of living and thence to the argument that it is the discounted sum of welfares per head that ought to be maximized. Quite apart from the *a priori* ethical objections that one may have to this valuation it is interesting to note one consequence of this criterion. Unless there are strong economies of scale (reflected in an increasing average product of labour with the number of workers) given a policy there is always a better policy in which population is run down faster. Such a consequence, it seems to me, can be used as a compelling reason for our rejecting this criterion for optimality. Modifying one's valuations in the light of their consequences is often a legitimate procedure and the purpose of much that is involved in analyses such as this paper attempts is to evaluate the consequences of various ethical beliefs.

Maximizing discounted total welfare will require of us to compare the ultimate value of having one more person in the world with the ultimate value of present people having

[1] For their many incisive comments regarding the problems discussed in this paper I am deeply indebted to Professor J. E. Meade, and Professor J. A. Mirrlees.

[2] For a masterly presentation of the problem of optimum population policies, see Meade [4], Chap. 6.

a bit more to consume. In this sense we shall have to agree on the exact origin of the welfare index. That is, we shall have to agree on the consumption rate at which individual utility is judged to be zero. The fact that this is a difficult judgement to make is not, of course, an argument against the choice of our welfare criterion. The choice of the welfare criterion ought not be made on the basis of simplicity, but on ultimate values.

It is as well to emphasize at this point that for the most part I shall be concerned with the optimum size of population at every moment—including the present. Quite apart from the analytical advantages this will afford the discussion it will enable us to see quite sharply the actual interplay of the two variables which, in our model, affect welfare. The only initial condition, therefore, that we shall be given will be the size of the community's capital stock. This assumption will be changed in section 7, where I shall discuss the problems raised by the imposition of constraints on the rate of growth of population.

I shall assume that the welfare of every individual is reflected by a utility function (u) which is an increasing, twice differentiable, and strictly concave function of his consumption rate. It follows that at any moment a perfectly egalitarian distribution of the generation's consumption allotment is the best distribution. It must be admitted that there is a strong source of externality that affects the economic welfare of an individual: I refer to the degree of congestion in his community. Except in section 6 I shall assume an absence of this externality. Granted that a powerful argument in favour of restricting births is the fear of congestion. It must also be granted that for many regions the argument has been the current unavailability of resources and it is the latter that I shall, for the most part, explore.

In order to avoid referring explicitly to a physical subsistence level I assume that the marginal utility of consumption tends to infinity as consumption tends to zero. When the consumption rate of an individual is " low " his welfare is assumed to be " negative "; when it is " high " life is enjoyable and his welfare is " positive ". It follows that there exists a consumption rate (C_0) at which life is just about enjoyable. Following Professor Meade ([4], Chap. 6), I shall refer to C_0 as the " welfare subsistence level ". Welfare of an individual is, therefore, represented by the utility function $u(C)$ which satisfies the following conditions:

$$u'(C)>0,\ u''(C)<0,\ \lim_{C\to 0} u'(C)=\infty,$$

$$u(C_0)=0, \text{ and } \int_{C_0}^{\infty} u'(C)dC = B \leq \infty.^1 \qquad \ldots(1.1)$$

Much of the time I shall work with utility functions that are:

$$u(C) = B - C^{-v}\ (v>0), \text{ and } B = C_0^{-v}, \qquad \ldots(1.2)$$

i.e. marginal utility is iso-elastic.

1.4. Turning to the technological aspects of the economy under review I shall be concerned with a one-good economy that has a smooth production function. Two kinds of production functions will be discussed: (1) where it is constant returns to scale and (2) where it is Cobb-Douglas of degree less than unity.

One's interest in a constant returns to scale technology is immediate. It is interesting to inquire into the conditions under which an optimum policy exists—as well as the character of such a policy if and when it exists—in a model in which the only limitations to production are capital and labour, both of which can be increased indefinitely. For a model of population growth it is important also to investigate a world that has a fixed factor in production. There is force in the argument that an ultimate constraint on

[1] I have friends who feel that " $\lim_{C\to 0} u'(C)=\infty$ " is too mild a condition and that the stronger condition, $\lim_{C\to 0} u(C) = -\infty$, ought to be imposed on the individual utility function. I am inclined to agree, but I have retained only the former because of its generality.

indefinite population growth is the presence of only a finite quantity of natural resources.

For the most part I shall assume that there is no technical progress. This assumption will be relaxed in section 5. No explicit cost of controlling population will be taken into account. That is, irrespective of the rate of population growth all investment goes to the increase in the capital stock. This is perhaps not an unduly strong assumption. Rough calculations indicate that the costs of population control are quite low in comparison to the overall rates of investment.[1]

I assume that people begin work on being born and remain workers till they die. I apologise for this assumption, which is bad—particularly since the structure of population changes with time when its rate of growth is not constant. I have retained it because of its obvious analytical advantage.

The problem, then, is to determine the optimum consumption and population profile through time given that total consumption $C(t)$ can be predicted by the relation

$$C(t) = F[K(t), L(t)] - \dot{K}(t), \qquad \ldots(1.3)$$

where $K(t)$ is the community's non-deteriorating capital stock and $L(t)$ is the total population (labour force) at time t. $K(0)$ is given as an initial condition. For feasibility we have

$$C(0) \geqq 0,\ K(0) \geqq 0,\ L(0) \geqq 0.$$

As I have mentioned earlier we shall be comparing alternative policies by the sum of the discounted total welfares of all generations from now to infinity. But infinite integrals need not converge and one needs a different ordering. Following Professors Koopmans [3], and von Weizsacker [8], I shall say that a given policy $\{\underline{C}(t), \underline{L}(t)\}$ is better than a policy $\{C(t), L(t)\}$ if there exists a $T>0$ such that for all $t \geqq T$

$$\int_0^t e^{-\rho\tau}\underline{L}(\tau)u[\underline{C}(\tau)/\underline{L}(\tau)]d\tau \geqq \int_0^t e^{-\rho\tau}L(\tau)u[C(\tau)/L(\tau)]d\tau.$$

Furthermore, the policy $\{\underline{C}(t), \underline{L}(t)\}$ will be said to be optimal if it is better than any other feasible policy.

In the following two sections I shall investigate the existence and character of optimum policies under the assumptions that (1) F is linear and homogeneous; and (2) F is Cobb-Douglas of degree less than unity. In section 4 I shall inquire into the consequences of giving equal weight to all individuals at all times ($\rho = 0$). Section 5 will contain a discussion of the case when F is an explicit function of time, the technical change being Harrod neutral. In section 6 I shall explicitly take into account the disutility of congestion. Section 7 will contain a discussion of the optimum path when the rate of population growth is restricted to a given range, and finally in section 8 I shall go into some of the wider issues pertaining to the problems posed in this paper.

I am afraid the presentation might appear to some as unduly taxonomic, but in a sense it is a useful approach. For it is important to inquire as to how the different features affect the existence and character of optimum policies. It is all the more important in this area where what appears to be obvious is by no means necessarily true.

2. CONSTANT RETURNS TO SCALE TECHNOLOGY

2.1. In this section I shall examine the problems of optimum population growth as posed earlier within the context of an economy that has a constant returns to scale production function. That is, F is linear homogeneous.

We then have $F(K, L) = LF(K/L, 1) \equiv Lf(K/L)$ writing $c = C/L$, $k = K/L$, $x = \dot{L}/L$, (1.3) reduces to

$$\dot{k}(t) = f\{k(t)\} - c(t) - x(t)k(t). \qquad \ldots(2.1)$$

[1] For an interesting summary of such calculations see Ohlin [6], Chap. 8.

As to f, I assume that it is increasing, strictly concave, twice differentiable, so that

$$\left.\begin{array}{l} f(0)=0,\ (2.2a);\ f'(0)=\infty,\ (2.2b); \\ f'(\infty)=0,\ (2.2c);\ f(\infty)=\infty,\ (2.2d). \end{array}\right\} \quad \ldots(2.2)$$

(2.2b) is perhaps overly strong but there seems little purpose in weakening it in this section. Recall that $K(0)$ is given initially, but there is choice over the values of $c(0)$ and $k(0)$.

The theorem that explores the questions of existence in this model is:

Theorem 2.1. *An economic policy described by* $\underline{C}(t)$, $\underline{K}(t)$, $\underline{L}(t)$, $\underline{c}(t)$, $\underline{k}(t)$ *is optimal if, writing* $p(t)=e^{-\rho t}u'\{\underline{c}(t)\}$.[1]

$$-\frac{\dot{p}}{p}=f'(\underline{k}), \quad \ldots(2.3)$$

$$u(\underline{c})=u'(\underline{c})[\underline{c}-\{f(\underline{k})-\underline{k}f'(\underline{k})\}], \quad \ldots(2.4)$$

$$\text{and} \quad p\underline{K}\rightarrow 0 \text{ as } t\rightarrow\infty. \quad \ldots(2.5)$$

Proof. It is simple to assure oneself that $Lu(C/L)$ is concave in L and C. Consider now some feasible development policy C, K, L, etc. other than the given policy. Then,

$$\int_0^T\{\underline{L}u(\underline{c})-Lu(c)\}e^{-\rho t}dt \geqq \int_0^T e^{-\rho t}[\{u(\underline{c})-\underline{c}u'(\underline{c})\}(\underline{L}-L)+u'(\underline{c})\{\underline{C}-C\}]dt. \quad \ldots(2.6)$$

We want to show that the right hand side of (2.6) is positive for large enough T. To demonstrate this notice that the R.H.S. is equal to

$$\int_0^T e^{-\rho t}[\underline{L}u(\underline{c})-Lu(\underline{c})+L\underline{c}u'(\underline{c})-Lc\,u'(\underline{c})]dt$$

$$=\int_0^T e^{-\rho t}[(\underline{L}-L)u'(\underline{c})\{\underline{c}-f(\underline{k})+\underline{k}f'(\underline{k})\}+u'(\underline{c})\{L\underline{c}-Lc\}]dt$$

$$=\int_0^T p_t[(L-\underline{L})\{f(\underline{k})-\underline{k}f'(\underline{k})\}+\underline{c}\underline{L}-cL]dt,$$

which, on using (2.1)

$$=\int_0^T p[(L-\underline{L})\{f(\underline{k})-\underline{k}f'(\underline{k})\}+\underline{L}\{f(\underline{k})-\underline{k}-(\dot{\underline{L}}/\underline{L})\underline{k}\}-L\{f(k)-k-(\dot{L}/L)k\}]dt$$

$$=\int_0^T p[L\{f(\underline{k})-\underline{k}f'(\underline{k})\}+\underline{L}\underline{k}f'(\underline{k})-(\dot{\underline{L}}\underline{k}+\underline{L}\dot{\underline{k}})-Lf(k)+(\dot{L}k+L\dot{k})]dt$$

$$=\int_0^T p[L\{f(\underline{k})-\underline{k}f'(\underline{k})\}+\underline{L}\underline{k}f'(\underline{k})-Lf(k)]dt-\int_0^T p\{\dot{\underline{k}\underline{L}}\}dt+\int_0^T p\{\dot{kL}\}dt$$

$$=\int_0^T pL[\{f(\underline{k})-\underline{k}f'(\underline{k})\}-\{f(k)-kf'(\underline{k})\}]dt$$

$$+p(0)\{\underline{K}(0)-K(0)\}+p(T)\{K(T)-\underline{K}(T)\}. \quad \ldots(2.7)$$

$\{f(k)-kf'(\underline{k})\}$ attains its maximum at $k=\underline{k}$. Therefore the integral in (2.7) is positive. Now, $\underline{K}(0)=K(0)$. Finally, recalling (2.5) we note that given any negative number $-\varepsilon$ there exists a T_0 such that for $T\geqq T_0$

$$p(T)\{K(T)-\underline{K}(T)\}\geqq -\varepsilon.$$

It follows that for T large enough the expression in (2.7) is positive. This proves that the given programme is optimal.

[1] Where there is no risk of confusion I shall normally drop the suffix t from the variables.

It is worthwhile pausing over the economic content of the optimality conditions. Equation (2.3) is immediate. It says that on the optimal path the percentage rate of fall of discounted marginal utility (the rate of time preference, if one likes) is equal at every moment to the marginal product of capital. This is the Ramsey Rule. Equation (2.4) has an equally simple common-sense appeal. Suppose the optimum time profile of capital has been obtained. We are then considering the advisability of introducing a new member into society. If we introduce a new member with an equal distribution of consumption then the *gain* in social welfare is the discounted utility of this new member's consumption (i.e. discounted utility per head). There is, however, a *loss* in total welfare by existing members of society having to make up this new member's consumption to the average. This loss is clearly the discounted marginal utility of consumption multiplied by the excess of the new member's consumption over his marginal product. It follows that the absolute level of population is optimal only when the gain and the loss to total welfare on introducing this last member are equal. Equation (2.4) asserts this.[1]

Theorem (2.1) states that a development programme satisfying equations (2.3) and (2.4) is optimal if the present value of capital in the very distant future tends to zero on this path (condition (2.5)). This is an appealing condition. It is, however, natural to ask if this is not excessively strong. Certainly in the analysis of the optimum rate of savings under an exogenously given population growth rate borderline cases arise where on the optimum path the present value of capital tends to a finite positive number.[2] I shall argue later that this will not carry over to our present problem.

2.2. If we want sharp answers we must specify our functions and so for a while I shall assume that u satisfies (1.2).

Equations (2.3) and (2.4) reduce to

$$\frac{\dot{c}}{c} = \{f'(k)-\rho\}/(1+v), \qquad \ldots(2.8)$$

$$B-c^{-v} = vc^{-(1+v)}[c-f(k)+kf'(k)]. \qquad \ldots(2.9)$$

It is simplest to consider first the stationary values of the system given by equations (2.1), (2.8) and (2.9).

Let $\bar{k}$ be the solution of $f'(k) = \rho$. Rearranging (2.9) we have, with $k = \bar{k}$

$$B = c^{-v}-vc^{-(1+v)}[f(\bar{k})-\bar{k}f'(\bar{k})-c] = \chi(c). \qquad \ldots(2.10)$$

Notice that $\chi(c)\to-\infty$ as $c\to 0$ and that $\chi(c)\to 0$ as $c\to\infty$. Write

$$c^* = f(\bar{k})-\bar{k}f'(\bar{k}), \qquad \ldots(2.11a)$$

and

$$\bar{B} = c^{*-v}. \qquad \ldots(2.11b)$$

Then $\chi(c^*)>0$ and χ attains a maximum at $c = c^*$. The general shape of χ is drawn in Fig. 1.

It is immediate that equation (2.10) has a solution in c only when $B \leqq \bar{B}$. For $0<B<\bar{B}$ there are two solutions in c, one less than, and the other greater than c^*. Anticipating matters somewhat I merely glance at the smaller of these two values and concentrate on the larger of the two. Call the smaller value $\underline{c}$. Since $\underline{c}<c^*$ we have $B-\underline{c}^{-v}<0$.

Label the larger solution $\bar{c}$. Then $\bar{c}>c^*$. Therefore $B-c^{-v}>0$. Turning to the stationary value of equation (2.1) we have

$$\bar{x} = (f(\bar{k})-\bar{c})/\bar{k}. \qquad \ldots(2.12)$$

[1] Professor Meade gives the same argument for establishing this formula. See [4], p. 91. I shall subsequently refer to this formula as the Meade Rule.

[2] T. C. Koopmans [3].

Now when $B = \bar{B}$ we have $\underline{c} = \bar{c} = c^*$ and so $\bar{x} = f'(\bar{k}) = \rho$. It follows that when $B < \bar{B}$, $\bar{c} > c^*$ and so $\bar{x} < \rho$. Finally when $\bar{c} = f(\bar{k})$ we have

$$B = \underline{B} = \{f(\bar{k})\}^{-v}\{1 + [v\bar{k}f'(\bar{k})/f(\bar{k})]\}. \quad \ldots(2.13)$$

Therefore when $B < \underline{B}$ we have $\bar{x} < 0$.

To summarise, stationary values of the optimal equations exist only when $B \leqq \bar{B}$. For $B < \bar{B}$, concentrating on the larger of the two solutions of (2.10) it has been shown that $B - \bar{c}^{-v} > 0$ and that $\bar{x} < \rho$. Also when $\underline{B} \leqq B < \bar{B}$ we have $0 \leqq \bar{x} < \rho$.

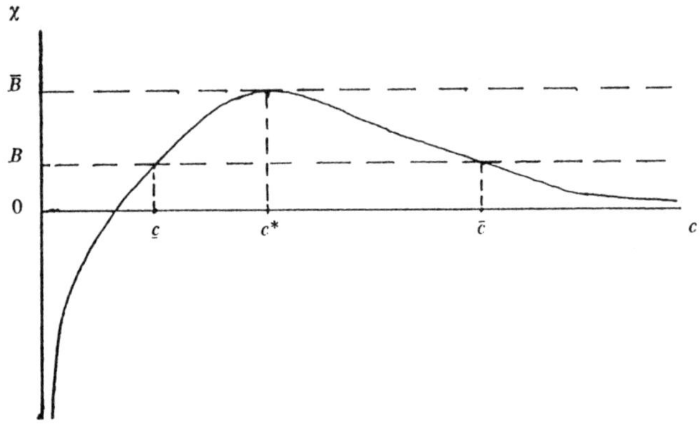

FIGURE 1

The general behaviour of the system (2.1), (2.8) and (2.9) is now worked out. Rearranging equation (2.9) and writing

$$w(k) = f(k) - kf'(k)$$

we have

$$vw(k) = (1+v)c - Bc^{1+v}. \quad \ldots(2.14)$$

Now $w(k) = 0$ when $c = 0$ and when $c = [(1+v)/B]^{1/v}$. Furthermore, $w(k)$ attains a maximum at $c = (1/B)^{1/v} = c_0$. Fig. 2 below represents equation (2.14) in its essentials.

Three paths have been drawn, each with a different value of B. The path with $B = \bar{B}$ is tangent to the horizontal line $vw(\bar{k})$. The path with $B = B_2 > \bar{B}$ does not intersect this line. On this path $k < \bar{k}$ and so $\dot{c} > 0$ (see (2.8)).

It follows that an economy obeying equations (2.1), (2.8) and (2.9) will exhaust its capital in finite time if $B > \bar{B}$. But these are necessary conditions for optimality. Therefore there is no optimal policy if $B > \bar{B}$. For $0 < B < \bar{B}$ the analysis is once again simple. Essentially, the rate of savings and the rate of population growth have to be so controlled as to force the economy to follow this path. The two stationary points are $\underline{c}$ and $\bar{c}$. Using (2.8) it is clear that movement along the path is away from $\bar{c}$ on both sides and towards $\underline{c}$ from both sides. Now given $K(0)$ consider the population level $L(0)$ so that $K(0)/L(0) = \bar{k}$. Next obtain $\bar{c}$ from (2.10). Then obtain $\bar{x}$ from (2.12). It has been demonstrated that $\bar{x} < \rho$. Therefore, when $0 < B < \bar{B}$, this *steady state* $(\bar{k}, \bar{c}, \bar{x})$ satisfies conditions (2.3), (2.4) and (2.5). It follows that this steady state policy is the optimal policy.

It is interesting to inquire about existence of an optimum policy when $B = \bar{B}$. Notice that in this case $\underline{c} = \bar{c} = c^* = (1/B)^{1/v} = c_0$. On this steady state (which is meta-stable, as Fig. 2 shows) $\bar{x} = \rho = f'(\bar{k})$, and it is characterized by the Golden-Rule savings ratio. On this path consumption per head is equal to the welfare subsistence level. Furthermore the value of capital tends to a finite positive level on this path. It is simple to show that this steady state policy, which is the only possible candidate for optimality, is not optimal. For, ignoring equation (2.9) we can consider an alternative steady state policy which has a constant rate of population growth less than $f'(\bar{k})$. Consumption per head on this path will exceed c^*. Therefore on this path welfare is positive. Consequently

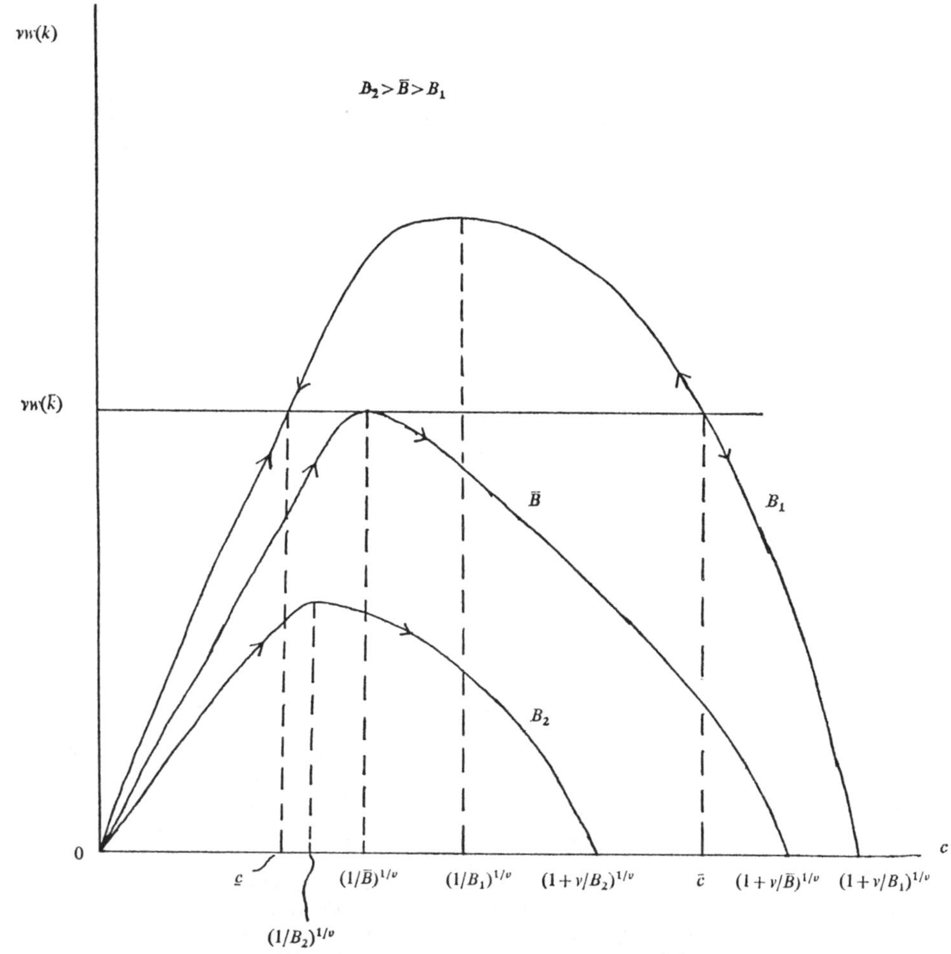

FIGURE 2

total welfare with this policy will be greater than that reached with our candidate. But this alternate policy does not satisfy (2.9). It follows that with $B = \bar{B}$ there is no optimum policy.

The results of this section can be summarized as

Theorem 2.2. *If F is constant returns to scale satisfying* (2.2) *and u satisfies* (1.2) *and $\rho > 0$ then an optimal policy exists if and only if $B < \bar{B}$.*

Theorem 2.3. *Under the technological and welfare assumptions of theorem 2.2, the optimal policy, when it exists, is characterized by a steady growth state* $(\bar{k}, \bar{c}, \bar{x})$ *where* $f'(\bar{k}) = \rho$ *and* $\bar{c}$ *is the larger of the two solutions of* $B - c^{-v} = vc^{-1+v}\{c - [f(\bar{k}) - \bar{k}f'(\bar{k})]\}$ *and* $\bar{x} = [f(\bar{k}) - \bar{c}]/\bar{k}$. *When* $\bar{B} > B > \underline{B}$ *(see equation* (2.13)) *we have* $\rho > \bar{x} > 0$. *(i.e. the savings ratio is positive). If* $B = \underline{B}$ *then* $\bar{x} = 0$ *(i.e. the savings ratio is zero). If* $B < \underline{B}$, *then* $\bar{x} < 0$ *(i.e. the savings ratio is negative).*

2.3. Theorem 2.2 possesses an amiable simplicity. Essentially it says that an optimum exists if and only if B is not too large. Since $B = c_0^{-v}$, this is another way of saying that an optimum exists if and only if the welfare subsistence rate is not too small. This is really expected. For consider the set of all steady state paths. If c_0 is " small " one can maintain a " high " rate of population growth and enjoy, at the same time, positive utility. Given the production function f and the rate of pure time discount ρ, if c_0 is small enough the

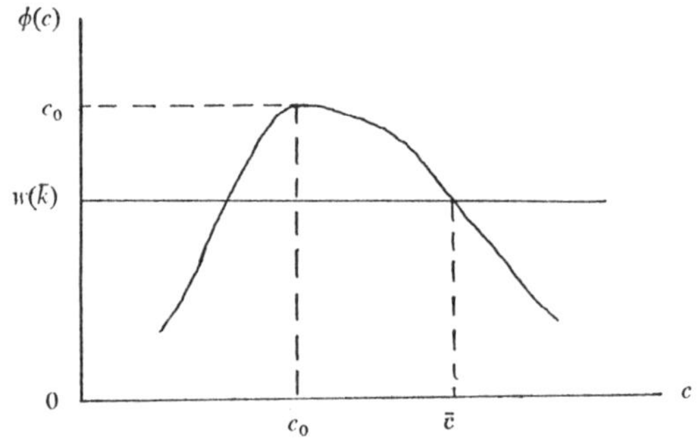

FIGURE 3

economy can enjoy positive welfare throughout and simultaneously follow a policy that has a constant rate of growth of population exceeding ρ. Once this is possible it is clear that our welfare criterion will not yield an optimum. For given any policy we can better it by keeping consumption low (and keeping utility non-negative) for a finite initial period and then enjoying a splurge in the distant future, when there will be a lot more souls to enjoy the splurge. But social welfare can always be increased by postponing the splurge. The claims of the future are much too large. An optimum policy, in such cases, does not exist.

On the other hand when the welfare subsistence rate is large enough one can have high rates of population growth only at the expence of positive utility. It is meaningful in such cases to strike a balance, so to speak, between the number of individuals enjoying life, and consumption rates at which life is indeed deemed enjoyable.

The borderline between existence and non-existence of an optimum policy is given by $B = \bar{B}$. It was shown that the obvious candidate for optimality here is the Golden Rule of accumulation. Along this path the value of capital tends to a finite positive level. Welfare along this path, however, is zero and so it is not surprising that this borderline case yields no optimum.

2.4. More general utility functions deserve a brief mention. There is really no difficulty in handling them. Let u, therefore, simply satisfy 1.1. Then the optimality equations are

$$-[u''(c)/u'(c)]\dot{c} = f'(k)-\rho, \qquad \text{...(2.15)}$$

$$u(c) = u'(c)[c-f(k)+kf'(k)]. \qquad \text{...(2.16)}$$

The optimal policy, when it exists is, as before, a steady state. Let $\bar{k}$ be the solution of $f'(k) = \rho$. Using (2.16) we have

$$\omega(\bar{k}) = c-[u(c)/u'(c)] \equiv \phi(c). \qquad \text{...(2.17)}$$

Now $\phi(c_0) = c_0$. Also $\phi'(c)<0$ when $c>c_0$ and $\phi'(c)>0$ when $c<c_0$. The general shape of ϕ is drawn in Fig. 3. An optimum exists if and only if $\omega(\bar{k})<\phi(c_0)$, and the optimum rate of consumption per head is given by the larger solution ($\bar{c}$) of equation (2.17) and the optimum rate of growth of population is derived from equation (2.12).

It ought to be noticed that unbounded utility functions are not ruled out. e.g. $u(c) = \log c$. The welfare subsistence level for this utility function is unity. So long as $\omega(\bar{k})<1$, an optimum exists.

3. TECHNOLOGY WITH A FIXED FACTOR

3.1. The economy that I shall discuss in this section contains a fixed factor of production which causes diseconomies of scale to appear. It is natural to capture this feature by a production function that is homogeneous of degree less than unity and it is caught sharply by a Cobb-Douglas function. Most of the following analysis will clearly carry over to the case of a technology that is not Cobb-Douglas in quality. The argument in favour of working with so highly special a technology is that it will enable us to assess rather easily the influence of the relevant economic parameters on our optimum policy. I therefore assume that

$$F(K, L) = K^{\alpha}L^{\beta}, \qquad \text{...(3.1)}$$

where α, $\beta>0$ and $\alpha+\beta<1$. Clearly $\partial F/\partial K = \alpha K^{\alpha-1}L^{\beta}$ and $\partial F/\partial L = \beta K^{\alpha}L^{\beta-1}$. Writing $c = C/L$ equation (1.3) becomes

$$\dot{K} = K^{\alpha}L^{\beta}-cL. \qquad \text{...(3.2)}$$

The central theorem of this section can be written as

Theorem 3.1. *An economic policy described by* $\underline{K}(t)$, $\underline{L}(t)$, $\underline{C}(t)$, $\underline{c}(t)$ *is optimal, if writing* $p = e^{-\rho t}u'\{\underline{c}(t)\}$

$$-\dot{p}/p = \alpha\underline{K}^{\alpha-1}\underline{L}^{\beta}, \qquad \text{...(3.3)}$$

$$u(\underline{c}) = u'(\underline{c})[\underline{c}-\beta\underline{K}^{\alpha}\underline{L}^{\beta-1}], \qquad \text{...(3.4)}$$

$$p\underline{K}\to 0 \text{ as } t\to\infty. \qquad \text{...(3.5)}$$

Proof. Consider some feasible development programme C, K, L, c other than the one given. In view of the concavity of $Lu(C/L)$ as a function of L and C we have

$$\int_0^T \{\underline{L}u(\underline{C}/\underline{L})-Lu(C/L)\}e^{-\rho t}dt \geqq \int_0^T e^{-\rho t}[[u(\underline{c})-\underline{c}u'(\underline{c})](\underline{L}-L)+u'(\underline{c})[\underline{L}\underline{c}-Lc]]dt. \qquad \text{...(3.6)}$$

We wish to show that for large enough T the right hand side of (3.6) is positive. To see this notice that the R.H.S. is

$$= \int_0^T e^{-\rho t}[(\underline{L}-L)u(\underline{c})+u'(\underline{c})\{L\underline{c}-Lc\}]dt$$

$$= \int_0^T p[(\underline{L}-L)\{\underline{c}-\beta \underline{K}^{\alpha}\underline{L}^{\beta-1}\}+\{L\underline{c}-Lc\}]dt$$

$$= \int_0^T p[(L-\underline{L})\beta K^{\alpha}\underline{L}^{\beta-1}+\underline{L}\underline{c}-Lc]dt$$

$$= \int_0^T p[(L-\underline{L})\beta \underline{K}^{\alpha}\underline{L}^{\beta-1}+\underline{K}^{\alpha}\underline{L}^{\beta}-K^{\alpha}L^{\beta}]dt-\int_0^T p\dot{\underline{K}}dt+\int_0^T p\dot{K}dt$$

$$= \int_0^T p[(L-\underline{L})\beta \underline{K}^{\alpha}\underline{L}^{\beta-1}+\underline{K}^{\alpha}\underline{L}^{\beta}-K^{\alpha}L^{\beta}]dt-\int_0^T p\alpha \underline{K}^{\alpha}\underline{L}^{\beta}dt$$

$$+p(0)\{\underline{K}(0)-K(0)\}+p(T)\{K(T)-\underline{K}(T)\}+\int_0^T p\alpha \underline{K}^{\alpha-1}\underline{L}^{\beta}Kdt$$

$$= \int_0^T p[[1-(\alpha+\beta)]\underline{K}^{\alpha}\underline{L}^{\beta}-[K^{\alpha}L^{\beta}-\alpha \underline{K}^{\alpha-1}K\underline{L}^{\beta}-\beta \underline{K}^{\alpha}\underline{L}^{\beta-1}L]]dt+p(T)[K(T)-\underline{K}(T)]. \quad \ldots(3.7)$$

The expression $[K^{\alpha}L^{\beta}-\alpha K^{\alpha-1}\underline{L}^{\beta}K-\beta K^{\alpha}\underline{L}^{\beta-1}L]$ attains its maximum when $K=\underline{K}$ and $L=\underline{L}$. Consequently, the integral in (3.7) is positive. Finally (3.5) implies that given any number $-\varepsilon$ there exists a T_0 such that for $T \geqq T_0$

$$p(T)[K(T)-\underline{K}(T)] \geqq -\varepsilon.$$

It follows that the given policy is optimal.

3.2. Restricting ourselves to the case when u satisfies (1.2) we find that (3.3) and (3.4) reduce to

$$\dot{c} = c(\alpha K^{\alpha-1}L^{\beta}-\rho)/(1+v), \quad \ldots(3.8)$$

$$B-c^{-v} = vc^{-(1+v)}(c-\beta K^{\alpha}L^{\beta-1}). \quad \ldots(3.9)$$

We are given $K(0)$ as an initial condition. The point will again be to choose the " correct " values for $C(0)$ and $L(0)$. It is useful to determine the stationary values of the system of equations (3.2), (3.8) and (3.9). Denoting these values as $\underline{K}$, $\underline{c}$, $\underline{L}$ we find

$$\underline{c} = \underline{K}^{\alpha}\underline{L}^{\beta-1}, \quad \ldots(3.10)$$

$$\alpha \underline{K}^{\alpha-1}\underline{L}^{\beta} = \rho, \quad \ldots(3.11)$$

$$B-\underline{c}^{-v} = (1-\beta)v\underline{c}^{-v}. \quad \ldots(3.12)$$

Solving the last three equations one gets

$$\underline{c} = [[1+v(1-\beta)]/B]^{1/v}, \quad \ldots(3.13)$$

$$K = (\alpha/\rho)^{(1-\beta)/[1-(\alpha+\beta)]}[[1+v(1-\beta)]/B]^{-\beta/[v[1-(\alpha+\beta)]]}, \quad \ldots(3.14)$$

$$\underline{L} = (\alpha/\rho)^{\alpha/[1-(\alpha+\beta)]}[[1+v(1-\beta)]/B]^{-(1-\alpha)/[v[1-(\alpha+\beta)]]}, \quad \ldots(3.15)$$

It is natural to expect this stationary set to be the long run values of the optimal policy. Several features of these values merit attention at this stage. In contrast to the results of the last section a stationary solution exists for *all* positive values of B. Next, the stationary values are gross terms. We expect, therefore, that the optimal policy tends in the long run to a stationary state (i.e. with zero savings and constant population). This is anticipated, in view of the diminishing returns in this economy.

It is interesting also to note that in the long run the standard of living, as caught by consumption per head, is independent of the rate of discount. This differs sharply from the analysis in section 2 where a higher rate of pure time discount *cet. par.* entails a higher consumption per head.

It is clearly expected that the larger the value of B (i.e. the smaller the value of the welfare subsistence level) the smaller is the long run consumption per head. This is indeed the case.

I shall now consider the general behaviour of the system of equations (3.2), (3.8) and (2.9).

Re-arranging (3.9) we have

$$L^{\beta} = \left[\frac{v\beta K^{\alpha}}{c\{(1+v)-Bc^{v}\}}\right]^{\beta/(1-\beta)}. \qquad \ldots(3.16)$$

This implies that for feasibility $c<[(1+v)/B]^{1/v}$. Using (3.16) in (3.2) we have

$$\dot{K} = \left[\frac{v\beta}{c(1+v)-Bc^{1+v}}\right]^{\beta/1-\beta} K^{\alpha/(1-\beta)}\left[1-\frac{v\beta}{\{(1+v)-Bc^{v}\}}\right]. \qquad \ldots(3.17)$$

Moreover, using (3.16) in (3.8) we have

$$\dot{c} = \frac{c}{(1+v)[c(1+v)-Bc^{1+v}]^{\beta/(1-\beta)}}\left[\alpha(v\beta)^{\beta/(1-\beta)}K^{-[1-(\alpha+\beta)]/(1-\beta)} - \rho[c(1+v)-Bc^{1+v}]^{\beta/(1-\beta)}\right]. \qquad \ldots(3.18)$$

(3.17) and (3.18) therefore, are a pair of first order differential equations in c and K and I shall draw their solutions in the (c, K) space, (Fig. 4). The optimum policy, if it exists, is one of these paths.

Now $\dot{K} = 0$ when $c = [[1+v(1-\beta)]/B]^{1/v}$.

Also $\dot{c} = 0$ on the curve

$$\alpha(v\beta)^{\beta/(1-\beta)}K^{-[1-(\alpha+\beta)]/(1-\beta)} = \rho[c(1+v)-Bc^{1+v}]^{\beta/(1-\beta)},$$

which reduces to

$$K = A[c(1+v)-Bc^{1+v}]^{-q} = \phi(c). \qquad \ldots(3.19)$$

where

$$A = (\alpha/\rho)^{(1-\beta)/[1-(\alpha+\beta)]}(v\beta)^{\beta/[1-(\alpha+\beta)]}>0,$$

$$q = \beta/[1-(\alpha+\beta)]>0.$$

Looking at (3.19) we see that $\phi(c)\to\infty$ when $c\to 0$ and $c\to[(1+v)/B]^{1/v}$. Moreover ϕ is convex in c and attains its minimum at $c = (1/B)^{1/v}$. Turning to (3.17) and (3.18) notice that apart from the stationary point $(\underline{c}, \underline{K})$ given by (3.13) and (3.14), the origin and the point $[(1+v)/B]^{1/v}, 0]$ are singular points of the system.

M is the point $(\underline{c}, \underline{K})$. I shall now argue that there is a curve $OQMN$ passing through O and M which represents two solutions of the system (3.17) and (3.18) both terminating at M. It is this curve that is the obvious candidate for optimality.

To see this, consider a path passing through $[c(0), K(0)]$ that lies under the curve PM but to the left of SM. Since $\dot{c}>0$ and $\dot{K}>0$ in this region the path must move towards the north-east. It must eventually cross PM or cross SM or go towards M. We know also that through each point one and only one path must pass. Some paths pass through PM and some through SM. There must be at least one point $[c(0), K(0)]$ such that the path through this must tend to M. But given $K(0)$ there can be only one such point $c(0)$, since it is easy to see that for any other value of c, say c', the path through $[c', K(0)]$ will diverge from OM. A similar argument applies to the branch MN.

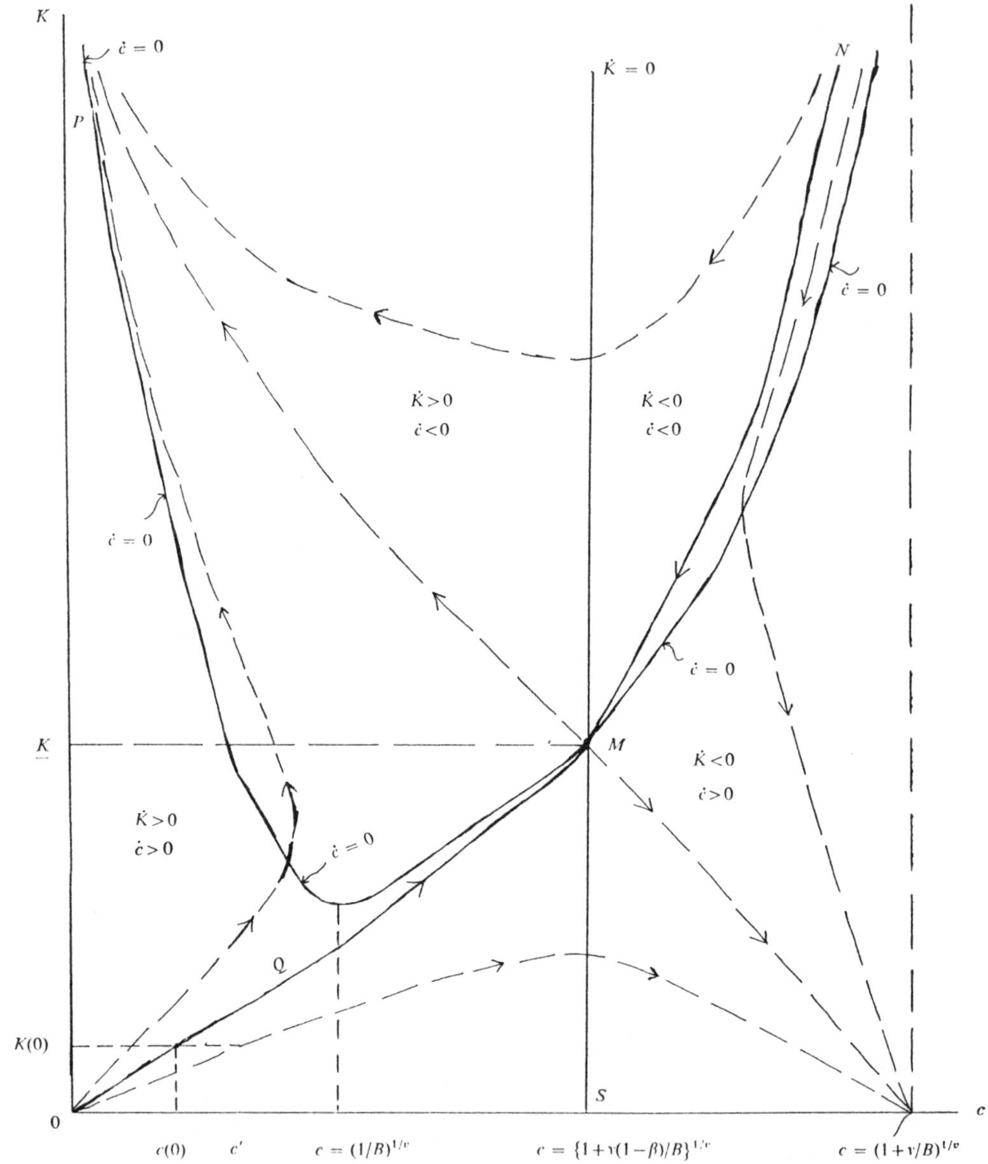

K
ċ = 0
P
K̇ = 0
N
ċ = 0
K̇>0
ċ<0
K̇<0
ċ<0
ċ = 0
ċ = 0
K
M
K̇<0
ċ>0
K̇>0
ċ>0
ċ = 0
Q
K(0)
S
0
c
c(0)
c'
c = (1/B)^{1/v}
c = {1+v(1−β)/B}^{1/v}
c = (1+v/B)^{1/v}

FIGURE 4

I have, therefore, argued that M is a saddle point and that the curve $OQMN$ forms the two stable branches. All other points either collapse onto the point $[[(1+v)/B]^{1/v}, 0]$ or are asymptotic to the curve MP. Given a path from the set that collapses onto

$$[[(1+v)/B]^{1/v}, 0]$$

it is simple to see that it does so in finite time.[1] Moreover, paths that are asymptotic to the curve MP accumulate capital unnecessarily. Following such a path would be inefficient, in a Pareto sense, as compared to following $OQMN$.

The economy travelling along OM or NM has in the long run $K(t) \to \underline{K}$ and $c(t) \to \underline{c}$. The value of capital, $e^{-\rho t}u'\{\underline{c}(t)\}\underline{K}(t)$ along this path, therefore, tends to zero as t goes to infinity. It follows that the stable branches $OQMN$ represent the unique optimum policy. Given $K(0)$ the correct value of $c(0)$ is read off from the curve $OQMN$ and the correct value of $L(0)$ is obtained from (3.16). Differentiating L with respect to time in expression (3.16) one gets the optimum rate of population growth. It is simple to check that $\dot{L}<0$ on the branch MN and that $\dot{L}>0$ on the branch OQM. Finally, as M is approached, $\dot{L} \to 0$.

The results of this section can be summarized as:

Theorem 3.2. *If F satisfies* (3.1) *and u satisfies* (1.2) *and* $\rho>0$, *then a unique optimum policy exists for all* $B>0$.

Theorem 3.3. *Under the technological and welfare assumptions of Theorem* 3.2, *the long run steady state to which the optimum policy tends is characterized by the values* $\underline{c}$, $\underline{K}$, *and L given by* (3.13), (3.14) *and* (3.15). *If* $K(0)<\underline{K}$, *then on the optimum programme c, K, and L monotonically increase with time and tend to the stationary values. If* $K(0)>K$ *then on the optimum programme c, K, and L monotonically decrease with time and tend to the stationary values. If* $K(0) = \underline{K}$ *the optimum policy is the stationary state given by* $\underline{c}$, $\underline{K}$, $\underline{L}$.

3.3. Notice that if $K(0)$ is sufficiently low there is a finite period of time during which welfare is negative along the optimum path. Welfare remains negative until per head consumption rises to the level $(1/B)^{1/v}$, from which instant welfare becomes positive and increases asymptotically to its long run value. Theorem 3.3 asserts that population is increasing all the while in this case. It appears, therefore, that even when per head consumption is below welfare subsistence people are being " brought in ". This may seem paradoxical but a little reflection shows that it is not. For it is the over-all welfare of all generations that is being maximized. When the initial capital stock is low we expect some generations to experience a tightening of the belt. During the interval when welfare is negative people are alloted consumption rates less than their marginal products. They are being brought in so as to accumulate capital in the interest of posterity.

It is tempting to make numerical assessments. I am impressed that

$$c = [[1+v(1-\beta)]/B]^{1/v} < (1+v)^{1/v}c_0 < ec_0.$$

That is, the ratio of the long run consumption per head and the welfare subsistence rate is less than 2·74.... Taking exact values; with $\beta = 0{\cdot}5$ and $v = 1$, we have $c = (3/2)c_0$. It is curious as I have remarked earlier, that the value of ρ does not appear in this ratio. The larger is the elasticity of marginal utility the smaller is the ratio. This is understandable. In any case, it seems that in a world that has a fixed factor of production the long run standard of living on the optimum path is not much larger than the welfare subsistence rate—no matter how the latter is defined.

[1] It might be of interest to notice that as $[[(1+v)/B]^{1/v}, 0]$ is approached $\dot{K} \to -\infty$ and $\dot{c} \to \infty$. (See equations (3.17) and (3.18)).

3.4. One ought, perhaps, to glance at unbounded utility functions. For example, if $u(c) = \log(c)$, equations (3.10)-(3.12) reduce to

$$\underline{c} = \underline{K}^{\alpha}\underline{L}^{\beta-1}, \quad \text{...(3.20)}$$

$$\alpha\underline{K}^{\alpha-1}\underline{L}^{\beta} = \rho, \quad \text{...(3.21)}$$

$$\log \underline{c} = \frac{1}{\underline{c}}[\underline{c} - \beta\underline{K}^{\alpha}\underline{L}^{\beta-1}]. \quad \text{...(3.22)}$$

Using equations (3.20) and (3.22) one obtains

$$\underline{c} = e^{(1-\beta)} > 1.$$

Similarly $\underline{K}$ and $\underline{L}$ may be evaluated. The unique optimum policy tends, in the long run, to the stationary state $\{\underline{c}, \underline{K}, \underline{L}\}$.

4. ZERO RATE OF PURE TIME PREFERENCE

4.1. Opinions are bound to vary as regards what the rate of pure time preference ought to be. Certainly much has been written on it. Ramsey found discounting the future at a positive rate, " . . . a practice which is ethically indefensible ", and that it, " . . . arises merely from the weakness of the imagination . . . ", [7]. Harrod finds it, " . . . a polite expression for rapacity and the conquest of reason by passion ". ([2], p. 40).

There is force in these sentiments. The point really is, one can argue that it is not *our* valuation of future generations' welfare that ought to be considered but future generations' valuation of their own welfare. To put it another way, suppose that we were to look back at the rates of accumulation of past generations with a view of discovering if they have been in some sense optimal. It is most unlikely that in such an evaluation we shall find it reasonable to give larger weight to past generations than we allot ourselves.

Nevertheless, many economists today would probably not regard a small positive discount of the future as unethical. One would not like to weight generations in the very distant future the same as the present simply because one is not sure if the world will not cease to exist in the distant future, and a positive discount rate captures in some (albeit, a special) sense the possibility of extinction.

At any rate, quite independent of where one's morals lead one so far as the rate of pure time preference is concerned it is interesting to see what theoretical features arise when $\rho = 0$. In this section I briefly survey this question.

4.2. Consider first a constant returns to scale technology, with F satisfying (2.2). It is simple to check that for all $B > 0$, a policy satisfying (2.8) and (2.9) exhausts its capital in finite time. It is clear that unless there is production saturation no optimum exists.

Take next the technology of section 3. Consider the optimality equations (3.8) and (3.9). If $\rho = 0$ then $\dot{c} \to 0$ when $K \to \infty$ and/or $L \to 0$, Re-arranging (3.9) we have

$$Bc^{1+v} - (1+v)c = -v\beta K^{\alpha}L^{\beta-1}. \quad \text{...(4.1)}$$

If either $K \to \infty$ or $L \to 0$, the R.H.S. of (4.1) tends to $-\infty$ which will be inconsistent with (4.1). The optimality conditions are inconsistent with any steady state. It seems safe to reject the possibility of an optimum policy that oscillates permanently with finite amplitude. It looks as though no optimum policy exists.

4.3. It is interesting to look into the consequences of introducing a capital saturation level. I shall now assume that up to a capital labour ratio $\hat{k}$ constant returns to scale prevails but near $\hat{k}$, strong diminishing returns set in. That is, for all $k < \hat{k}$, per head output $f(k)$ is increasing, strictly concave and twice differentiable and that $f(0) = 0$ and $f'(0) = \infty$, but that for $k \geq \hat{k}$, $f'(k) = 0$.

The optimality equations (2.8) and (2.9) read

$$\dot{c}/c = f'(k)/(1+v) \qquad \text{...(4.2)}$$

$$B - c^{-v} = vc^{-(1+v)}[c - [f(k) - kf'(k)]]. \qquad \text{...(4.3)}$$

Now $\dot{c} = 0$ for $k \geqq \hat{k}$. Write $\hat{c} = f(\hat{k})$. Then if $B < \hat{c}^{-v}$ the curve defined by (4.3) is discontinuous, as shown in Fig. 5, where, once again, $w(k) = f(k) - kf'(k)$.

As before, no optimum exists if $B \geqq \hat{c}^{-v}$. If $B < \hat{c}^{-v}$, a unique optimum policy exists and it is characterized by the steady state values at G. At G, $\bar{c} > \hat{c}$ and so $\bar{c} > f(\hat{k})$. It follows that the steady state at G is always characterized by a negative savings ratio and therefore a negative rate of growth of population. This is not surprising. If B is small enough (i.e. $< \hat{c}^{-v}$) a rate of consumption that yields positive welfare is large. But there is production saturation in the economy at $\hat{k}$. Therefore such a high rate of consumption cannot be maintained unless capital is regularly eaten into.

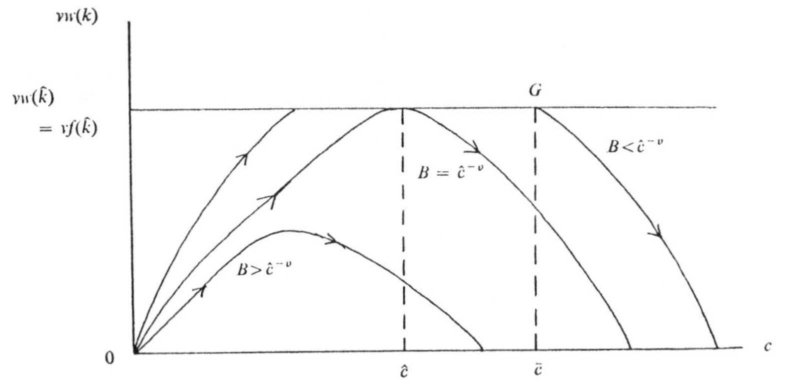

FIGURE 5

5. HARROD-NEUTRAL TECHNICAL PROGRESS

5.1. When population is growing at an exogenously given rate in a constant returns to scale technology with technical change the conditions under which an optimum accumulation policy exists are weaker than are the conditions for existence in an economy with no technical change.[1] Just the opposite is the case when the rate of population growth is a policy variable. It is the purpose of this section to sketch the features that arise when an economy experiences Harrod-neutral technical progress.

I shall first survey the question for an economy with a fixed factor and then glance at the problems raised by a constant returns to scale economy. As usual, u satisfies (1.2).

5.2. We now have $F = e^{mt}K^{\alpha}L^{\beta}$ where m, α, $\beta > 0$ and $\alpha + \beta < 1$ ...(5.1)
write

$$y = F/L,\ k = K/L,\ c = C/L,\ x = \dot{L}/L.$$

Then it is simple to establish that such an economy can experience steady growth (i.e. $\dot{k} = \dot{y} = 0$) if and only if the rate of growth of population is $\bar{x}$ where

$$\bar{x} = m/[1 - (\alpha + \beta)]. \qquad \text{...(5.2)}$$

The point is that at this rate the effect of technical progress just cancels off the diminishing returns due to the fixed factor.

[1] Mirrlees [5].

The production relation (1.3) now becomes

$$\dot{k} = e^{mt}k^{\alpha}L^{\alpha+\beta-1} - c - xk. \quad \text{...(5.3)}$$

The necessary conditions for optimality are

$$\dot{c}/c = [\alpha e^{mt}k^{\alpha-1}L^{\alpha+\beta-1} - \rho]/(1+v), \quad \text{...(5.4)}$$

$$B - c^{-v} = vc^{-(1+v)}[c - \beta e^{mt}k^{\alpha}L^{\alpha+\beta-1}]. \quad \text{...(5.5)}$$

The difficulty about the system (5.3), (5.4) and (5.5) is its non-autonomous nature. For purely illustrative reasons, suppose $L = \underline{L}e^{\bar{x}t}$. Then the stationary values $\underline{c}$ and $\underline{k}$ of this system are obtained by solving

$$\begin{aligned} \underline{c} &= \underline{k}^{\alpha}\underline{L}^{\alpha+\beta-1} - \bar{x}\underline{k}, \\ \alpha\underline{k}^{\alpha-1}\underline{L}^{\alpha+\beta-1} &= \rho, \\ B - \underline{c}^{-v} &= v\underline{c}^{-(1+v)}[\underline{c} - \beta\underline{k}^{\alpha}\underline{L}^{\alpha+\beta-1}]. \end{aligned} \quad \text{...(5.6)}$$

It is clear that (5.6) is consistent for all $B>0$. We now want to determine the condition under which $B-\underline{c}^{-v}>0$. This is simple. The first two equations in (5.6) give us

$$\underline{k} = \alpha\underline{c}/(\rho - \alpha\bar{x}).$$

Using this and the second equation of (5.6) the third equation in (5.6) gives us as the condition for $B-\underline{c}^{-v}>0$ as

$$[1-(\alpha+\beta)] > \rho[1-(\alpha+\beta)] - \alpha m. \quad \text{...(5.7)}$$

On the face of it (5.7) looks innocent enough. However, for the value of capital to tend to zero on this balanced path we need the condition

$$\rho > \bar{x} = m/[1-(\alpha+\beta)]. \quad \text{...(5.8)}$$

It is clear, then, that if by chance $K(0) = \underline{L}\underline{k}$ and if (5.7) and (5.8) are satisfied the balanced growth path with $c = \underline{c}$, $k = \underline{k}$ and $L = \underline{L}e^{\bar{x}t}$ is indeed the optimal path. My purpose in considering this highly unlikely case is that it is tractable and it yields readily the difficulties that arise as regards the existence of optimum policies when technical progress is present. For notice that *cet. par.* if $(\alpha+\beta)$ is too close to unity conditions (5.7) and (5.8) will be inconsistent with each other. At any event, (5.8) demands much too high a value for ρ. It is difficult to judge what values of m are regarded as reasonable approximations to reality. I have heard it claimed that m is as large as 3·5 per cent. There are again others who are convinced that $m = 0$. For reasons of illustration let us assume that $m = 0{\cdot}01$. (This is taking $\beta = \frac{1}{2}$ and a 2 per cent. Harrod neutral technical change.) With $\alpha+\beta = 0{\cdot}9$ we find that ρ should be greater than 10 per cent.

It seems clear that one will have to discount at a high rate to obtain an optimal policy. It looks also that the more serious the diminishing returns are (i.e. the smaller is the value of $\alpha+\beta$) the lower is the lower bound for ρ that will yield an optimum.

If $K(0) \neq \underline{k}\underline{L}$ the analysis becomes difficult and for k and c to asymptote to stationary values we must have

$$e^{mt}\left[e^{\int_0^t x d\tau}\right]^{\alpha+\beta-1} \to d < \infty.$$

It seems clear that $x \to \bar{x}$. One conjectures that given *any* $K(0)$ a policy exists which takes the economy to the steady values for c, k and x. Even here, for the value of capital on such a path to tend to zero, the rate of discount will need to be large.

5.3. The discussion in paragraph (5.2) implied that if the technology has near constant returns to scale, the discount rate required is very large. In the following remarks I shall consider only the necessary conditions for optimality for a technology that has constant returns to scale.

Let the rate of Harrod-neutral technical progress be λ. F is linear homogeneous and f satisfies three of the conditions in (2.2) but not (2.2*b*). Assume that $f'(0) = T<\infty$. Write

$$z = c/(e^{\lambda t}L),\ y = k/(e^{\lambda t}L),\ \text{and}\ f = F/(e^{\lambda t}L).$$

Equation (1.3) now becomes

$$\dot{y} = f(y)-z-(\lambda+x)y. \qquad \ldots(5.9)$$

The Ramsey and the Meade Rules are now

$$\dot{z}/z = [f'(y)-(1+v)\lambda-\rho]/(1+v). \qquad \ldots(5.10)$$

$$Be^{v\lambda t}+vz^{-(1+v)}[f(y)-yf'(y)-z]-z^{-v} = 0. \qquad \ldots(5.11)$$

It follows from (5.11) that

$$vz^{-(1+v)}[f(y)-yf'(y)-z]-z^{-v} \xrightarrow[t\to\infty]{} -\infty,$$

i.e.

$$vz^{-(1+v)}[f(y)-yf'(y)]-(1+v)z^{-v} \xrightarrow[t\to\infty]{} -\infty. \qquad \ldots(5.12)$$

Since $vz^{-(1+v)}[f(y)-yf'(y)]>0$ it follows that

$$z\to 0 \text{ as } t\to\infty.$$

Re-arranging (5.12) we get that for large enough t

$$[(1+v)/v]z>[f(y)-yf'(y)].$$

Since $z\to 0$ as $t\to\infty$ it follows that $\dot{z}<0$ for large enough t and that $y\to 0$ as $t\to\infty$. It follows that $\rho \geqq f'(0)-(1+v)\lambda$. This is a fairly strong condition.

It is not difficult to see why technical progress causes difficulties. Essentially, the point is that with no technical progress maintaining too high a population growth rate would keep low the consumption rates accrued to individuals. With technical progress going on, later generations are assured high enough consumption rates even if population growth rates are high. Consequently large discount rates are required to ensure the existence of an optimum.

6. EXTERNALITIES

6.1. So far I have ignored an important source of externality—I refer to the effect of congestion on an individual's welfare. In what follows I shall sketch the broad features that appear to arise. The analysis will be brief since it is fairly simple to tidy up the arguments that I shall present with the results obtained in sections 2 and 3.

I shall, as usual, work with utility functions that are iso-elastic in consumption and, as in the earlier sections, I assume that there is no technical progress and that we are discounting the future at a positive rate.

Essentially we want an individual's welfare (u) to depend not merely on consumption per head, but also on the number of people in his society. It is difficult to judge how an individual's marginal utility of consumption is affected by the number of people around him. It is simple to think that they are independent of each other. I take it then,

$$u = B(L)-c^{-v} \quad (v>0)$$

and that $\qquad \ldots(6.1)$

$$B'(L)<0,\ B''(L)<0, \text{ and for some } L_0>0,\ B(L_0) = 0.$$

That is, B is a strictly decreasing and strictly concave function of L and attains the value of zero at some level of population.

It can be argued that B as a function of L ought to rise in value initially and then fall. I do not suppose there is much point in introducing this complication. The externality that I am trying to capture here is that due to congestion rather than that due to isolation and the main features of this seem to be caught quite sharply by the assumptions that I have made, viz. (6.1). For example, the concavity of B is clearly right.

6.2. I begin with the technology of section 3. A separate existence theorem is needed since utility per head depends not only on consumption per head but also on the level of population. I shall not go into it here, since it is fairly simple to see that the conditions for existence are the same as in Theorem 3.1.

The long run stationary values of the optimal policy are given, as before, by equations (3.10), (3.11), and (3.12). I rewrite them as

$$\underline{c} = \underline{K}^{\alpha}\underline{L}^{\beta-1},$$
$$\alpha\underline{K}^{\alpha-1}\underline{L}^{\beta} = \rho, \qquad \text{...(6.2)}$$
$$B(\underline{L}) - \underline{c}^{-v} = v\underline{c}^{-(1+v)}[\underline{c} - \beta\underline{K}^{\alpha}\underline{L}^{\beta-1}].$$

Re-arranging (6.2) we get

$$\underline{c} = (\rho/\alpha)^{\alpha/(\alpha-1)}\underline{L}^{-[1-(\alpha+\beta)]/(1-\alpha)}, \qquad \text{...(6.3)}$$

$$B(\underline{L}) = [1+v(1-\beta)]\underline{c}^{-v} = [1+v(1-\beta)](\rho/\alpha)^{\alpha v/(1-\alpha)}\underline{L}^{v[1-(\alpha+\beta)]/(1-\alpha)} = \Pi(\underline{L}), \qquad \text{...(6.4)}$$

$$\underline{K} = (\alpha/\rho)^{1/(1-\alpha)}\underline{L}^{\beta/(1-\alpha)}. \qquad \text{...(6.5)}$$

Fig. 6 obtains the solution to equation (6.4).

It is clear that the long run population level is smaller than what it would have been had B been a constant (at $B(0)$). Long run consumption per head is, therefore, higher. Finally the long run capital stock is lower. Just as in section 3, the stationary point given by (6.2) is a saddle point and there is a unique optimum path leading to this point.

6.3. It is interesting to ask how the results of section 2 (constant returns to scale technology) are affected by congestion. For convenience I consider only a Cobb-Douglas

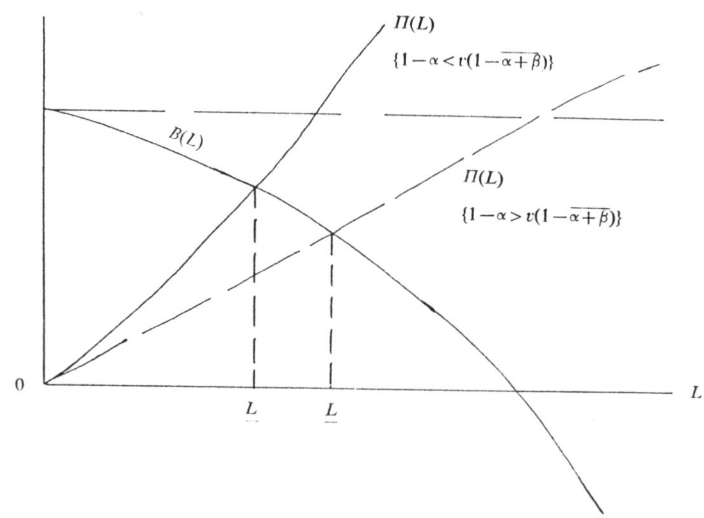

FIGURE 6

economy. Recall that in section 2, when $B = \underline{B}$ (2.13) the optimum policy is a steady state with zero growth. Suppose now u satisfies conditions (5.1) and that $\lim_{L\to 0} B(L) > \underline{B}$. Next put $\alpha+\beta = 1$ in (6.2). Then (6.4) becomes

$$B(\underline{L}) = [1+v(1-\beta)](\rho/\alpha)^{\alpha v/(1-\alpha)}. \qquad \ldots(6.6)$$

But with $F = K^\alpha L^\beta$ $(\alpha+\beta = 1)$ it is simple to confirm (2.13), that

$$\underline{B} = [1+v(1-\beta)](\rho/\alpha)^{\alpha v/(1-\alpha)}.$$

It is now obvious as to what is happening. If $B(L)$, for some value of L, is large enough (i.e. greater than $\underline{B}$) the optimum policy tends in the long run to a steady state with zero growth, and L, c, and K tend to the solution of system (6.2). (With $\alpha+\beta = 1$.)

I should emphasise that the optimum policy is, in general, not a steady state. Notice that only $K(0)$ is given. The solution of (6.6) is unique. Now, except by a fluke $\underline{K}(0)/L$ will not be equal to the per head capital characterizing the steady state optimum path with $B = \underline{B}$ in section 2. Therefore, in the present case, if $K(0) < \underline{K}$ the optimum policy will consist of an increasing L, K and c, tending in the long run, to the solution of (6.2).

If, for all L, $B(L) < \underline{B}$, there is trouble. (6.2) is now incompatible with a steady state solution. The point is that population cannot be run down fast enough. No optimum exists.

7. CONSTRAINTS ON POPULATION CHANGE

7.1. The analysis so far has carried no constraints on the rate of growth of population. The stock of the community's capital has been the only initial condition. The analysis has been carried out with a view to determining the optimum size of population at all times. Such a formulation, apart from its obvious academic interest is important for governments that seriously consider policies for migration. Granted that in many counties the level of population is much too large by any reasonable welfare criterion. It must also be granted that there are large regions that are sparsely populated in this sense. Anyhow it is not clear, I think, whether the population of the world as a whole is above the optimum level as determined by some reasonably acceptable form of the welfare function. Nor is it clear that costs incurred in migration are large enough to rule out such policies as unworkable.

It has to be admitted, however, that for closed economies it is important to impose constraints on the rate of growth of population. Given an initial capital stock it is not much good calculating what the optimum initial population is since there is no means of attaining it—unless by a fluke the initial population level is equal to the optimum. In what follows I shall sketch the arguments that are relevant for the analysis of optimum policies when the rate of growth of population is constrained to lie between two bounds. We are, therefore, given both $K(0)$ and $L(0)$.

It is as well to discuss what entails the bounds on the rate of growth of population. Given a constant mortality rate it is natural that the percentage rate of change of population cannot be below this rate. It will also be generally agreed that it is impossible to push the fertility rate down to zero. For reasons of illustration—it seems not unreasonable to consider a fertility rate as low as 1·5 per cent. Add to this the fact that in many parts of the world the rate of mortality is around 2 per cent and one expects that it is fair to regard the lower bound to the rate of population growth as a negative number $(-m)$. As regards the upper bound (M) it is natural to refer to it as the " biological maximum ". It is difficult to judge what its value is. Clearly it varies from region to region. It is also not a very precise concept. But a figure of around 6 per cent is probably not implausible. I shall assume that $\rho < M$. This is reasonable. It is unlikely that many would advocate a very large value for ρ. Even at 2 per cent the generation following the present one is assigned about half the weight assigned to the present.

7.2. I shall sketch the argument for only the constant returns to scale technology where F satisfies (2.2). It will be convenient to work with the savings ratio (s) initially. So I write $c = (1-s)f(k)$. (The notation is just as in section 2.)

The optimization problem is to maximize (in our overtaking sense)

$$\int_0^\infty Lu[(1-s)f(k)]e^{-\rho t}dt$$

subject to

$$\dot{k} = sf(k)-xk, \quad \text{...(7.1)}$$

$$\dot{L} = xL, \quad \text{...(7.2)}$$

and it is required that $s \leqq 1$ and $-m \leqq x \leqq M$. We are given $k(0)$ and $L(0)$.

To investigate this problem I appeal to Pontriyagin's maximum Principle and write down the Hamiltonian of the system as

$$H = Lu[(1-s)f(k)]e^{-\rho t}+\psi e^{-\rho t}xL+\xi e^{-\rho t}[sf-xk] \quad \text{...(7.3)}$$

where $\psi e^{-\rho t}$ and $\xi e^{-\rho t}$ have their usual price interpretations. These prices must satisfy the following equations:

$$\frac{d}{dt}[\psi e^{-\rho t}] = -(\partial H/\partial L) = -ue^{-\rho t}-\psi e^{-\rho t}x,$$

i.e.

$$\dot{\psi} = \psi(\rho-x)-u, \quad \text{...(7.4)}$$

and

$$\frac{d}{dt}[\xi e^{-\rho t}] = -(\partial H/\partial k) = -Le^{-\rho t}(1-s)f'(k)u'-\xi e^{-\rho t}[sf'(k)-x],$$

i.e.

$$\dot{\xi} = -[sf'(k)-(x+\rho)]\xi-L(1-s)f'(k)u'. \quad \text{...(7.5)}$$

Write $\zeta = \frac{\xi}{L}$; then (7.5) becomes

$$\dot{\zeta} = -[sf'(k)-\rho]\zeta-(1-s)f'(k)u'. \quad \text{...(7.6)}$$

Next, H must be maximized with respect to s and x. The maximization with respect to s is clearly an interior one since $u'\to\infty$ as $s\to 1$. The necessary and sufficient condition for maximization, $\partial H/\partial s = 0$, yields

$$\zeta = u'. \quad \text{...(7.7)}$$

Maximization of H with respect to x gives us the following rules

$$\psi = k\zeta \Rightarrow -m \leqq x \leqq M, \quad \text{...(7.8}a\text{)}$$

$$\psi > k\zeta \Rightarrow x = M, \quad \text{...(7.8}b\text{)}$$

$$\psi < k\zeta \Rightarrow x = -m. \quad \text{...(7.8}c\text{)}$$

Using (7.7), equation (7.6) becomes

$$(d/dt)u' = -[f'(k)-\rho]u', \quad \text{...(7.9)}$$

which is the Ramsey Rule.

It is now convenient to work in terms of consumption per head. The optimum policy therefore must satisfy the system

$$\dot{\psi} = (\rho-x)\psi-u(c), \quad \text{...(7.10}a\text{)}$$

$$\dot{c} = -[u'(c)/u''(c)][f'(k)-\rho], \quad \text{...(7.10}b\text{)}$$

$$\dot{k} = f(k)-c-xk, \quad \text{...(7.10}c\text{)}$$

subject to the rules (7.8).

As would be expected, the conditions $\psi = ku'$ and $(d/dt)\psi = (d/dt)(ku')$ combined with (7.10*a*) and (7.10*b*) will yield the Meade Rule (2.4). That is, the Meade Rule is operative so long as x does not assume one of its extreme values and keep to that value.

Assume now that u satisfies (1.2). In order to focus on the essentials let us also assume that $\underline{B} \leqq B < \bar{B}$ ($\underline{B}$ and $\bar{B}$ are defined by (2.13) and (2.11*b*)).

Then the system of equations (7.10) becomes

$$\begin{aligned} \dot{\psi} &= (\rho - x)\psi - (B - c^{-v}), \\ \dot{c} &= c[f'(k) - \rho]/(1+v), \\ \dot{k} &= f(k) - c - xk. \end{aligned} \qquad \ldots(7.11)$$

Recall that in the unconstrained problem, as dealt with in section 2, the optimum policy is a steady state policy $\bar{c}$, $\bar{k}$ and $\bar{x}$ where $0 \leqq \bar{x} < \rho$. The constrained problem, as found in this section, therefore, reduces to this: what is the optimal method of correcting this difference when $k(0) \neq \bar{k}$?

We expect that in the long run the system (7.11) will tend to the optimum policy $\bar{c}$, $\bar{k}$, $\bar{x}$. Consider now the stationary values $\bar{\psi}$, $\bar{c}$, $\bar{k}$ and $\bar{x}$ of (7.11). They are given by

$$\bar{\psi} = (B - \bar{c}^{-v})/(\rho - \bar{x}), \qquad \ldots(7.12)$$

$$f'(\bar{k}) = \rho, \qquad \ldots(7.13)$$

$$f(\bar{k}) - \bar{c} - \bar{x}\bar{k} = 0. \qquad \ldots(7.14)$$

The system is not under-determined, since, because we expect that $0 \leqq \bar{x} < \rho$ and therefore $-m < \bar{x} < M$, we must have, by (7.8*a*) the condition

$$\bar{\psi} = v\bar{k}\bar{c}^{-(1+v)}. \qquad \ldots(7.15)$$

Suppose now, in our constrained problem, $k(0) < \bar{k}$. Set $\psi(0) < k(0)\zeta(0)$ where $\zeta(0) = v[c(0)]^{-(1+v)}$.

Rule (7.8*c*) instructs us to put $x = -m$. The system (7.11), therefore, takes the form:

$$\dot{\psi} = (\rho + m)\psi - (B - c^{-v}), \qquad \ldots(7.15a)$$

$$\dot{c} = c[f'(k) - \rho]/(1+v), \qquad \ldots(7.15b)$$

$$\dot{k} = f(k) - c + mk; \qquad \ldots(7.15c)$$

(7.15*b*) and (7.15*c*) are independent of ψ. This is advantageous. Because, given $k(0)$, where is a unique $c(0)$ for which the solution of (7.15*b*) and (7.15*c*) will pass through the point $(\bar{c}, \bar{k})$. Let T_0 be the time at which the solution starting at $[c(0), k(0)]$ attains $(\bar{c}, \bar{k})$. We now possess the profile of c and k for the interval $(0, T_0)$. We still have to choose $\psi(0)$ correctly. It is safe to conjecture that we want ψ to attain the value $\bar{\psi}$ at T_0. This requirement gives us $\psi(0)$. One supposes that these are consistent with the demand that $\psi \leqq k\zeta$, all t.

Essentially then one expects that if $k(0) < \bar{k}$, the optimum policy is to put $x = -m$ until a time T_0. In this interval the Ramsey Rule is operative but not the Meade Rule. Both consumption per head and capital per head are increasing. When, at T_0, consumption per head and capital per head reach their long run values $\bar{c}$, $\bar{k}$ (as determined by the unconstrained problem (theorem 2.3), a jump is made in the value of x and it is given its long run value $\bar{x}$. From this moment the policy is to keep to the steady state as given in theorem 2.3). If, on the other hand, $k(0) > \bar{k}$, initially $x = M$ and along the Ramsey Rule consumption per head and capital per head fall until they reach the long run values $\bar{c}$, $\bar{k}$ at T_0. At this moment a jump in the value of x is made and it is given the value $\bar{x}$.

7.3. The intuitive simplicity of the previous argument is, I think, lost if the economy has a fixed factor. My arguments for supposing this are as follows: the unconstrained problem (section 3) yielded an optimum policy that is not a steady state. Along this optimum policy it is not clear if the constraint on the population rate of growth may not be violated at some intervals followed by intervals when the constraint is satisfied. If this were indeed the case then the optimum policy of the unconstrained model would be unfeasible at non-overlapping intervals of time. This presumably implies that the optimum policy of the constrained model has a number of jumps in the value of x, rather than the unique jump that the constant returns to scale model appears to yield in the last paragraph.[1] Whether this is indeed the case will depend largely, I think, on the degree of variation of the elasticity of the utility function at different rates of consumption. The question remains an open one.

8. CONCLUDING REMARKS

8.1. There are at least two immediate reasons why one might look askance at the analysis of this paper. The assumption that people work for every moment of their lives is clearly unsound. After all, in the real world an individual spends a good part of his life to mature into a productive being, and so, to play around haphazardly with the rate of birth can have momentous consequences on the productive capability of a future generation. It is relatively simple to construct a formal model of individual life cycles—one which has a person facing an initial unproductive phase, followed by a phase when he can work and procreate and this, followed (hopefully) by a period of retirement. But it seems to be rather intricate to incorporate such a model into a theory of optimum growth and make it answer the questions we would really like to ask.

Next, the assumption that the cost to population control is negligible is perhaps not too bad for small adjustments to the growth rate but is clearly questionable for non-negligible variations from the natural rate of growth. Essentially one would like to capture the fact that a planner wants not merely to evaluate the optimum rate of savings, but wants also to allocate the savings optimally between the formation of capital and expenditure on population control. We might, then, want to assume that there is a natural rate of growth of population μ and that *per capita* expenditure in obtaining a rate x is given by a function

$$g = g(x-\mu),$$

where $g(0) = 0$. It is natural to assume $dg/dx \gtrless 0$ as $x \gtrless \mu$. Total investment, I, must now be equal to $\dot{K}+Lg(x-\mu)$ and our basic technological relation (7.1) will now read

$$\dot{k} = sf(k)-xk-g(x-\mu).$$

The optimization problem with this added feature is, I am inclined to believe, not unduly complicated. Lack of space prevents me from attempting to deal with it here.

8.2. It will not have escaped the reader of the immense difficulty in judging what ought to be the welfare subsistence rate. There is indeed a basic difficulty in interpreting it. To take the two most immediate interpretations, ought we to regard it as the rate of consumption at which one is indifferent between being alive and not having been born or ought we to regard it as the minimum rate of consumption which an individual, as a parent, would like to ensure for his child before conceiving it? The former possesses the difficulty of our having to deal with a counter factual. But this in itself ought not to be too worrying. The real difficulty, it seems to me, lies in that a person has to compare the state of being alive at a given consumption rate with the " state " of not having been born.

[1] Formally the problem posed in this section resembles that of optimizing the rate of savings in a model where there is a lower bound to the rate at which consumption and investment can be made to fall. It can be shown that for a wide class of utility functions the optimum policy for the unconstrained (no lower bound to the rate of fall of consumption and investment) model is unfeasible for the constrained model at non-overlapping intervals of time. The consequence is that the optimum policy for the constrained model consists of any number of switches between the planning phase during which the rate of fall of investment is kept at its lower bound and the planning phase where the Ramsey Rule (along which the rate of change of investment is above the lower bound) is followed. See [1].

It is difficult to judge whether the two alternatives are always comparable. In extreme situations such a comparison may not pose a problem. In the face of abject poverty a comparison between these two " states " is understandable in its poignancy. Alternatively if the rate of consumption is extremely high one might be confident in asserting that one is glad to have been born. This does not, however, imply that it is possible to make such a comparison at intermediate standards of living and in particular to adjudge a specific rate of consumption as one at which one is indifferent between the two alternatives.

The latter interpretation appears to be more appealing in that it is readily operational but this too poses problems. To begin with, one might argue that in actual fact parents do occasionally consider, prior to conceiving, the well-being of their children. This, however, is not a compelling argument because in considering this problem parents ought really to assess their children's welfare functions rather than their own. To the extent that there is a certain constancy in tastes and values within two generations a parent may happily use his own welfare function as a reference to judge the welfare functions of his children. But this is a contingent matter and does not resolve the analyatical problem that we are faced with; and so, to regard the welfare subsistence rate as a minimum rate of consumption that a parent would like to ensure for his children is to abide by a kind of intergenerational buck-passing. And this runs counter to the very commendable moral that in making a decision about savings and population growth the well-being of an individual ought to be judged from his own welfare function and not have it assessed from the welfare function of someone else.

There is, of course, a third way of viewing the welfare subsistence rate which, prima-facie, appears much the same as the first interpretation but is not really so. This is to regard utility in the classical utilitarian sense of Mill and Sidgwick and this, for our problem, requires of us to introspect the quantity of joy in an economic state. This is not the place to discuss whether happiness or joy is in any sense measurable as a cardinal magnitude or indeed if there are any compelling moral grounds for maximizing total happiness even if it *were* measurable. But it appears to be generally accepted that when later utilitarians, like Marshall, thought about utility they viewed it as a magnitude that can be measured uniquely up to a proportional transformation—a degree of measurability that we have in this paper taken for granted. No doubt, then, there are difficulties in viewing the problem of population policies through utilitarian spectacles. That is, however, not to say that we could think more clearly about these issues if we were to abandon it, and it must be granted that this approach provides a convenient setting for us to ask with incisiveness the questions that one feels it is appropriate to ask.[1]

In one sense at least the welfare criterion used in this paper is defective. Individual utility, it is judged, depends only on the rate of consumption. One can persuasively argue that a person's utility depends also on the number of children he has and consequently the utility function ought to be a function of not only *per capita* consumption but also of, say, the rate of growth of population. And indeed it ought. It would not be correct to argue that in many circumstances (e.g. in section 2, when $\bar{B}>B>\underline{B}$) population grows along the optimal policy even when we ignore this added feature and so it should make no difference. It *will* make a difference since ignoring the added feature would allow the economy to grow at the wrong rate. It is, however, difficult to judge how the rate of growth of population ought to enter the individual utility function and it is for this reason I have neglected it.

Certain features of our welfare functional are clearly right. For example, the inclusion of all time for the evaluation of the social optimum appears to be inescapable. The

[1] Recently two versions of utilitarianism have been distinguished as (i) act utilitarianism and (ii) rule utilitarianism. There are situations where it is not entirely clear if the two versions are really distinct but I would judge that it is the latter that best fits the utilitarian approach that I have persued in this paper—the rule being to maximize total welfare, and we follow this even if it entails that certain acts yield negative utility, e.g. in section 3, where people may be " brought in " even with negative utility for the sake of over-all welfare.

agreeable moral that we ought to regard the born and the unborn in essentially the same way lies behind our adding all individuals' utility functions to arrive at the social welfare functional. Finally, it seems to me that the appeal of the welfare criterion followed in this paper lies in our being able to make the moral judgment that if an individual can be introduced into the world at a desirable level of well-being, and this without affecting the welfare of others in any way then it is desirable to have him. But this poses the serious difficulty of the welfare criterion not necessarily yielding an optimum policy. Suppose, then, that in the world described by the model in section 2 the welfare subsistence level is indeed very low. What are the policy makers to do? They are faced with the uncomfortable fact that no matter how " good " a development policy they can think of there is a yet better one. Ought they to revise their judgment and raise the welfare subsistence level to the region where an optimum exists? This, to my mind at least, is surely illegitimate.

Trinity Hall, Cambridge P. S. DASGUPTA.

First version received 15.5.68; *final version received* 15.1.69

REFERENCES

[1] Dasgupta, P. S. " Optimum Growth when Capital is Non-transferable ", *Review of Economic Studies*, **36** (January 1969).

[2] Harrod, R. F. *Towards a Dynamic Economics* (London, Macmillan & Co., 1948).

[3] Koopmans, T. C. *On the Concept of Optimal Economic Growth*, Cowles Foundation Discussion Paper 238 and 238A.

[4] Meade, J. E. *Trade and Welfare* (Oxford University Press, 1955).

[5] Mirrlees, J. A. " Optimum Growth when Technology is Changing ", *Review of Economic Studies*, **34** (January 1967).

[6] Ohlin, G. *Population Control and Economic Development* (O.E.C.D. Development Centre, 1966).

[7] Ramsey, F. " A Mathematical Theory of Saving ", *Economic Journal* (1928).

[8] von Weizsäcker, C. C. " Existence of Optimal Programs of Accumulation for an Infinite Time Horizon ", *Review of Economic Studies*, **32** (April 1965).

[18]

ESSAY ON ECONOMIC GROWTH AND ENVIRONMENTAL QUALITY*

*Ralph C. d'Arge***

University of California, Riverside, California, USA

Summary

The major premise of this essay is that production-consumption and waste emissions tend to be joint products of the human species. Given this premise, a parable of an astronaut irretrievably lost in space is discussed in order to deduce several propositions on the astronaut's optimal rate of consumption. A Harrod type of model is also analyzed with regard to the rate of consumption over time where the model includes a simplified depiction of the interaction between the economy and natural environment. Empirical estimates of the impact of effluent charges on comparative international advantage of selected countries are also presented. The major conclusion is that national and international economic policies and national environmental policies are not separable.

I

Pervasive environmental changes accompanying the development of modern societies have wrought doubts regarding the insignificance of uncontrolled interdependencies between man and the natural environment. The economic acts of production and consumption not only benefit man through his provision but simultaneously involve the creation of wastes with the potential of degrading his environment. It can be asserted that production-consumption and waste emission are joint products of the human species ... they certainly are for all other animal species and plant life. A conservative estimate of waste flow in the United States currently would be 6–7 pounds of "active" waste produced per dollar of gross national product, and there is some evidence that waste flow increases at a faster rate than income.[1] This would

---

* A portion of the research reported on here was financially supported by Resources For The Future, Inc., Washington, D.C. with no responsibility for the inferences or results contained herein.

** I wish to acknowledge the comments and criticisms of the following individuals without committing them to agreement with any part of this essay: K. C. Kogiku, T. Crocker, K. Oddson, E. Brook, O. Bubik, T. Clark, and H. Lawton.

[1] See R. U. Ayres and A. V. Kneese, "Environmental Pollution" in United States Congress, Joint Economic Committee, *Federal Programs for the Development of Human Resources*, Vol. 2, Washington, D. C. (1968); P. Leighton, "The Air Resource", in proceedings of Statewide Conference on: *Man in California—1980's*, Sacramento, California (January 1964); or T. Scitovsky, "External Diseconomies in the Modern Economy", *Western Economic Journal*, Vol. IV (summer 1966).

undoubtedly be true for less developed economies attempting to industrialize with a very small services sector. In highly developed economies the interdependencies between the economy and natural environment are accentuated because of vast agglomerations of people and high waste generating industries in locations with only limited waste assimilative capability. The natural environment containing highly industrialized and densely grouped populations can no longer be viewed as a bathtub by economists, where once each month the ring needs to be cleaned, but the drain never clogs. As an example, the somewhat isolated and rare disasters in the Meuse Valley, Donora, Pennsylvania, and London due to "smogs" are rapidly becoming seasonal threats in such cities as Los Angeles and Tokyo.

The purpose of this essay is to analyze some of the basic issues involved in the dichotomy between economic growth and environmental degradation and conceivably, decimation. We shall start by discussing an extremely simple parable about an economic spaceman and then attempt to relate it to global decisions with respect to the environment. I believe several enlightening propositions can be obtained from such an extreme parable. Next, a brief exposition of the interconnection between macro growth models and the natural environment will be made, again to attempt to find some useful propositions. Our third task will be to discuss briefly and present tentative empirical evidence on some of the major difficulties in developing international agreements for the regulation of waste flows.

Before presenting the spaceman parable, a comment is deserved on the connections between highly localized and interregional-international problems of environmental pollution. One factor, in my opinion, which has led economists to ignore the general interdependency questions of man-environment themes has been the apparent implicit belief that externalities associated with pollution were essentially two-party problems ... the emittor versus the receptor.[1] If emittor and receptor are identified, Coase, Buchanan, Turvey, and others have pointed out that through private exchange, regardless of who owns the pollutant receiving medium, efficiency will result through the equating of marginal pollutant damages and marginal waste treatment costs.[2] These cases, if they existed at all, have already been resolved under the aegis of private property rights and legal adversary proceedings, at least in the western countries. However, the less mundane but pervasive cases where receptors are only vaguely identified and (or) damages induced by emittors are synergistic in their effects on receptors cannot be resolved with the application of two-party principles of property rights.

[1] For an extremely lucid discussion of the problems encountered utilizing the "two-party" mentality, see A. V. Kneese, "Environmental Pollution and Economic Policy", to be presented to the American Economic Association, Detroit, Michigan (December, 1970).
[2] See the definitive articles by: R. H. Coase, "The Problem of Social Cost", *Journal of Law and Economics* (October, 1960), J. M. Buchanan and W. C. Stubblebine, "Externality", *Economica* (November, 1962), and R. Turvey, "On Divergencies Between Social Cost and Private Cost", *Economica* (August, 1963).

As Allen Kneese has frequently argued, the receptor in most pervasive cases of environmental pollution is or cannot be well defined ... usually being a group of peeple, both present and future generations, without a viable organization to represent their interests.[1] Recipients of the effects of long-term accumulations of DDT, mercury, or CO_2 in the atmosphere have little or no voice in the decision on emission rates of these potentially dangerous pollutants with nearly global dispersion. River basin commissions such as are found in the Ruhr industrial basin or for the Delaware in the United States may improvise sub-optimal plans for maintaining water quality within a basin, but how cognizant are these commissions of pollutant loads at the mouth of the river? Relatively high water quality standards or effluent charges will undoubtedly induce shifts in production processes toward a greater pollutant load to the atmosphere via burning or industrial sludge, or toward creating solid waste disposal problems, unless the air and land resources are also protected by standards or effluent charges.

It seems to me that in concentrating on two-party or localized effects, economists have not taken heed of the warnings of ecologists and have ignored one of the major dictums of economic thought: "that everything depends on everything else".[2] In theoretical discussions of environmental control, I would take the side of the ecologists who not only echo the general equilibrium principle of economics, but demand that it be applied in analyzing issues of environmental pollution. A general equilibrium stance should be adopted which attempts to recognize at least the major interdependencies between man's economic behavior and the effect and interaction of such behavior on the environment in which he resides. What follows must be viewed in this light even though at most the concepts and parables discussed are no more than suggestive of extremely feeble beginnings.

---

[1] A. V. Kneese and B. T. Bower, *Managing Water Quality: Economics, Technology, and Institutions*, Baltimore: Johns Hopkins Press (1968).

[2] In several recent instances, economists have attempted to integrate some rather simple concepts of the man-environment interaction. For an integration of economic equilibrium and materials balance, see R. U. Ayres and A. V. Kneese, "Production, Consumption, and Externalities", *American Economic Review* (June, 1969); and for a less rigorous but philosophical categorization on a similar theme, H. Daley, "Economics as a Life Science", *Journal of Political Economy* (April, 1968). For a discussion of Pareto optimality, second best, and environmental taxation within the context of materials balance, see A. V. Kneese and R. C. d'Arge, "Pervasive External Costs and the Response of Society", United States Congress, Joint Economic Committee, *Economic Analysis of the P.P.B.S. System* ..., Washington, D.C. (May, 1969). On production and renewable resources, see V. L. Smith, "The Economics of Production from Natural Resources", *American Economic Review*, Vol. LVIII (June, 1968), and R. G. Cummings and O. R. Burt, "The Economics of Production from Natural Resources: Note", *American Economic Review*, Vol. LIX (December, 1969). For an integration of pollution and utilization of exhaustible resources concepts, see R. C. d'Arge, "Economic Growth, Recycling, and the Natural Environment", in A. V. Kneese, editor, *Environmental Quality Analysis: Research in the Social Sciences*, Baltimore: Johns Hopkins Press (forthcoming, 1970).

II

The analogy applied here is with spacemen since astronauts have replaced cowboys and Vikings as recipients and victims of the laurels of contemporary hero worship. What is deduced in the following paragraphs is equally applicable in many instances to the former hero's "rational" decisions to "break camp" or depart from Jutland.

Let us assume that there is a lone astronaut within a sealed space capsule who is suddenly confronted with the realization that he will not be able to return to earth. He is irretrievably lost in space. Within his capsule are stored enough provisions to sustain life for a period of time shorter than the astronaut's "normal life" span even if he consumes at minimal subsistence in terms of caloric intake. Clearly if the spaceman's criterion of welfare is the singular objective of maximizing the length of existence, he will not consume above the minimum subsistence level. Alternatively, he may be more or less myopic, depending on one's philosophy, and decide to consume at a faster rate, particularly if he evaluates an additional year of existence at near starvation levels to be of less value than an increment of present consumption. The rational spaceman would consume at a rate during each period of consumption such that he equated the marginal utility of a unit of consumption in that period with the foregone opportunity, in terms of utility, of consuming that unit in any future consumption period including the period following the one in which he runs out of provisions. The price of this foregone opportunity will also include a positive shadow price for not being able to consume below a minimal subsistence level if such a level is constraining to the astronaut's desired rate of consumption. These consequences are obvious and are related to a negative analogue of Ramsey's theory of saving.[1] However, here the astronaut dissaves and approaches a finite anti-bliss point. The conclusions stipulated are also nearly identical to recent models of mine exploitation.[2]

From this extremely naive model we are able to make several additional inferences. First, if the astronaut has a positive finite rate of time preference he will consider consuming at a higher rate than minimal subsistence only if his evaluation of an additional time interval of survival is not infinite. Second, if his preferences and capacity are such that he exhibits diminishing marginal utility of consumption, he is likely to spread consumption over a longer time span than if he achieved bliss through an immediate orgy of satiation.

Next, let us revise our assumptions and introduce the idea that since the astronaut resides in a sealed capsule he must endure the effects of his own waste generation. The astronaut is now confronted with the problem that higher immediate rates of consumption will mean a higher density of wastes

---

[1] F. Ramsey, "A Mathematical Theory of Saving", *Economic Journal*, Vol. 38 (1928).
[2] See R. G. Cummings and O. K. Burt, op. cit.

to live with during future time intervals. In order to analyze the question of rational planning by the astronaut, let us stipulate the following assumptions: (*a*) The astronaut's preferences are such that he has diminishing marginal utility of consumption and increasing marginal disutility induced by expanding waste density; (*b*) There is some waste density or concentration which is lethal to the astronaut and he is aware of this; (*c*) There is very little or no assimilation of wastes or recycling within the sealed capsule; (*d*) The astronaut has enough provisions to last his normal lifetime and this lifetime is a very large number such that his only binding constraint is his generation of wastes; (*e*) Consumption and waste emissions are joint products, i.e., the principles of conservation of matter-energy is operative; and (*f*) The astronaut is not decreasing in size and does not discount future consumption at a negative or too high a positive rate. Given these assumptions, the astronaut following a rational policy of maximizing utility over time will follow one of two rules:

Proposition 1. If maximizing the length of existence is of primary concern, the astronaut will immediately reduce consumption to the minimal subsistence level and remain there until decimation occurs.
Proposition 2. If continued existence is not of overriding importance, the astronaut will immediately reduce consumption upon discovering that he is lost in space but increase it thereafter.

The formal proof of Proposition 2 is given elsewhere[1] and the logic of Proposition 1 is clearly seen from the earlier discussion of the astronaut's dilemma on dissaving over time. Clearly, the mining and dissaving problem is identical to the pollution problem in many ways but different in one vital respect ... dissaving to the astronaut *per se* does not cause disutility whereas waste generation is usually assumed to.

Professor Robert Strotz came up with what I believe to be an ingenious and simple explanation for Proposition 2. He stated the problem as one of campers entering a campground and planning to stay for some finite interval of time. The campers are confronted with the problem that if they consume their provisions and discard waste at a high rate initially they will have to suffer the "disutilities" of this waste during the remainder of their stay. However, if the campers consume at a low rate initially and build up consumption over time, they need not suffer too much from waste generated in early intervals and yet increasing consumption allows them to compensate for the increasing waste density. At some point they will depart leaving a "debacle of debris" following a "grand orgy of consumption".

Like the factor price equalization theorem of international trade, these propositions at best have very tenuous connections with reality. However,

---

[1] See R. C. d'Arge, "Economic Growth, Recycling ...", op. cit. or R. C. d'Arge and K. C. Kogiku, "Economic Growth and the Natural Environment", paper to be given at the Econometric Society Meetings, Detroit, Michigan (December, 1970).

I believe these propositions lead to some important thoughts on very long-run environmental quality problems. If the earth is truly a spaceship or closed resource system characterized by finite amounts of energy and natural waste assimilative capability, energy dispersion coupled with environmental degradation will, at some finite point in time, severly limit man's proclivities for reproduction and greater material abundance. And even if the finite planning interval is assumed to be extremely large, say the length of time before the sun's gases undergo complete combustion, we are confronted with the fact that optimization, under a set of environmental constraints, generally leads to minimal subsistence or at least, retarded current consumption.[1] This is particularly true if population is expected to continue growing at positive rates for the indefinite future. Not only is the environmental quality management problem one of applying effluent charges, standards, and (or) payments which reflect current damages to the population, but these charges should also include a substantial adjustment for the effects on future generations, especially for reducing opportunities of a higher standard of living which includes less environmental degradation. Thus, it appears that environmental quality management must involve the analysis of what might be termed dynamic or time related externalities ... the impact of one generation's ignoring discrepancies between private and social costs on subsequent generations' economic choices. Perhaps there is an emerging rationale in these simple propositions for placing a ceiling on per capita consumption or even considering the possibility of zero growth of material goods in more advanced economies.

The spaceship analogies discussed here contain several major weaknesses. Always implicit within the previous discussion of spaceships are the very long-run facets of survival conditioned by limited energy reserves, energy dispersion, an expanding or constant population, and no compensating investments in recycling technologies or in augmenting the natural environment's capacity to assimilate wastes. Of course, these assumptions yield the most cataclysmic outcomes. Next, we turn to a discussion within the context of

---

[1] It should be noted that a finite planning interval coupled with a finite environmental capacity for wastes are extremely rigid assumptions. Of course, infinite planning intervals in conjunction with an unlimited natural environmental assimilative capacity reduce to trivial nonexistent scarcity cases. Alternatively, an infinite planning interval in conjunction with a finite environmental waste assimilative capability appears to be philosophically inconsistent provided population is increasing over time, consumption per capita has a positive lower bound, rates of environmental waste assimilation are very low or zero, waste generating technology is completely static in character, and wastes at some level of concentration become lethal to the human species. These conditions quite obviously insure that at some moment in the finite future decimination will occur. The alternative case of a finite planning interval and infinite waste assimilative capability is also philosophically inconsistent. Specifying an infinite planning interval for mankind, of course, implies the imposition of a constraint on the decision process, namely that regardless of how degraded man's habitat becomes, he must continue to exist. A more general and encompassing decision process would stipulate that the survival period be optimally chosen with infinity as one possible choice. I have chosen the course of assuming a finite planning interval, thinking that finite numbers can be very large. However, such a decision precludes consideration of an infinite time horizon.

neo-classical economic growth in which the outcomes are not so bleak by introducing the possibilities of recycling investments and investments which increase the natural environment's capability to alter, assimilate, or otherwise neutralize wastes.

III

Saving or the augmentation of the existing capital stock plays a key role in contemporary thought on economic growth. In the simple Harrod types of growth, the greater the proportion of current output saved and invested the faster the economy is warranted to grow. For economies with the option to invest in augmenting waste assimilative capability through such investments as reservoir construction to provide stabilized river flows, waste treatment plants, or catalytic mufflers on internal combustion engines, saving becomes the source of funds for such investments. But saving also has an additional effect on the environment in that the higher the saving rate in relation to depreciation or decay rates the smaller the current flow of wastes. Thus, saving can have the impact of redistributing waste flows or residuals accruing to the environment over more extended intervals of time.

Investment directed toward reducing environmental degradation appears to be able to take three rather general forms: recycling, augmentation of the natural environment's assimilative capability, or investments which reduce the need for, or can compensate for, increased consumption of material goods. Recycling investments reduce waste loads by channeling wastes of the production and consumption processes back into these processes. Investments in augmenting assimilative capacity reduce the effects of waste loads by altering or augmenting the natural environment through such activities as reservoir construction, stream reaeration, and chemical and biological waste treatment. The third category of investments are directed toward inducing the populace toward consuming less time devoted to material consumption and more time toward services, solitude, or recreational-aesthetic pursuits involving little or no waste generating goods. In the model which follows, we will concentrate only on one category of such investments, namely those which augment environmental assimilative capability and (or) the special category of recycling investments which do not increase or decrease productive output.

In order to describe briefly some of the ramifications of growth and environmental quality, we shall adopt an extremely simple model of the Harrod type. Let W denote a homogeneous waste flow, F the current production of material goods, K_r a measure of the capital stock committed to augmenting the assimilative capability of the environment with $dK_r/dt = I_r$, and S saving. Then, in the spirit of the concept that production-consumption and waste generation are joint products, the following equation is postulated:

$$W = g_c(F-S) + g_f F \qquad (1)$$

where g_c and g_f are coefficients relating the amount of waste flow generated from current consumption and production, respectively. Next, it is postulated that density of wastes is the proper measure of environmental quality. Let D denote average waste density, V a volume measure of the natural environment, for example the size of the global natural life zone which is assumed fixed, $\dot{D}$ definitionally equal to dD/dt, and δ a decay rate in waste density which can be considered as symbolic of the natural assimilative capability of the environment:

$$\dot{D} = \frac{1}{V} W - hI_r - \delta \tag{2}$$

where h is a coefficient reflecting the rate of augmentation of the natural assimilative capability of the environment by capital investments as previously described.[1] For now, we shall assume K_r does not depreciate though the model could easily be made to accommodate instantaneous and constant decay rates in K_r. From equations (1) and (2) we obtain a simple expression relating changes in waste density to investment in assimilative capability, current saving and output.

$$\dot{D} = \frac{1}{V}(g_c + g_f) F - \frac{1}{V} g_c S - hI_r - \delta \tag{3}$$

Note from equation (3) that even if current saving and investment in augmenting the environment's assimilative capacity are both zero, there is a positive rate of output which would yield no increase or decrease in waste density. In ecologist's terms, this is a natural rate of production which hypothetically at least, balances man's proclivity to produce waste with nature's ability to absorb it ... a biological equilibrium.

To complete this simple model, define K_f as the non-depreciating capital stock devoted to producing material output, s as the marginal and average propensity to save from output, and σ as the incremental output–capital ratio. Then:

$$sF = S = I_f + I_r \tag{4}$$

$$\dot{F} = \sigma I_f \tag{5}$$

with $dK_f/dt = I_f$, and $dF/dt = \dot{F}$. Substituting equation (5) into (4), the connection between this model and the Harrod-Domar model is observed:

$$\dot{F}/F = G_f = s\sigma - \sigma(I_r/F) \tag{6}$$

Thus, if investment in non-productive waste assimilation is not undertaken, one ends up with a warranted rate of growth (G_f) equivalent to $s\sigma$. Such a

[1] Specifying h to be a constant in equation (2) carries with it the implicit assumption that the *level* of waste density at any point in time is linearly determined by past levels of production, assimilative investment and the cumulative neutralizing effects of the natural environment on waste density.

rate would be warranted if the natural environment could be characterized as limitless in capacity or size and in regenerative capability. Substituting (6) and (4) into (3) with rearrangement of terms, one obtains:

$$\left[\frac{g_c}{V}(1-s)+\frac{g_f}{V}-hs\right]F+\frac{h}{\sigma}\dot{F}-\delta=\dot{D} \tag{7}$$

In order to obtain a warranted rate of growth, the constraint is postulated of no change in average waste density, or $\dot{D}=0$. Such a constraint might not be desirable in that variations in waste density over time may yield greater welfare and (or) less disutility associated with waste flows. Such variations have been briefly mentioned earlier in conjunction with the spaceman concept. However, in terms of the model proposed here it seems preferable to presume a "steady state" in terms of environmental degradation via waste flows. Such a steady state could proceed without the hindrance of a finite planning interval for man's survival, or the philosophical contradiction between finite resources and an infinite planning horizon. It should be pointed out, however, that the following references to a warranted rate of growth presume not only a non-deteriorating natural environment but also one that is not allowed to improve either. With the assumption $\dot{D}=0$, equation (7) becomes a first order non-homogeneous linear differential equation of the form:

$$\dot{F}+\alpha F-\eta=0 \tag{8}$$

and

$$\alpha=\sigma\left[\frac{g_c}{hV}(1-s)+\frac{g_f}{hV}-s\right]$$

$$\eta=\frac{\sigma\delta}{h}$$

which has a solution:

$$F(t)=He^{-\alpha t}+\eta/\alpha \tag{9}$$

where H is a constant determined by initial conditions. Given the assumption that $H>0$, for output to increase over time, $\alpha<0$ or $sh>g_c/V(1-s)+g_f/V$. This simply indicates saving rates and the efficiency of investment to reduce waste density must be high enough to compensate for the wastes generated through consumption and production.[1] As would be expected, the greater the

[1] Take $F(0)=F_0>0$; then at $t=0$, $H=F_0-\eta/\alpha$ and therefore:
$F(t)=F_0e^{-\alpha t}+\eta/\alpha(1-e^{-\alpha t})$
Thus, $\dot{F}>0$, provided $\alpha<0$, or $\alpha>0$ and $F_0<\eta/\alpha$. In words, if initial output is less than the "biological equilibrium", production is warranted to increase even though the economy cannot sustain $\dot{D}=0$ above the "biological equilibrium" without governmental controls. In this discussion we shall presume $F_0>\eta/\alpha$.

savings rate, efficiency of investment in the natural environment's assimilative capability, size or extent of the natural environment, and the smaller the amount of waste generated per unit of output and consumption, the more likely a sustainable positive rate of growth would be which would not simultaneously involve an increasing waste density. The incremental output-capital ratio alternatively is neutral, either accelerating the rate of growth in output if the warranted rate is positive or increasing the negative change in output toward the biological equilibrium.

In order to give some empirical content to the meaning of a warranted positive or negative change in output, let us redefine the h coefficient, which connects reduction in waste density to assimilative investment, in terms of waste magnitude, i.e., $h' = h \cdot V$ where h' denotes the reduction in waste tonnage polluting the environment per dollar of assimilative investment. Next, let the savings rate equal 0.20 and the sum of g_c and g_f equal 6 pounds. The latter is consistent with the estimate for the United States reported earlier. Then the criterion for a positive rate of growth in output is $h' > 30\text{-}g_c$. In other words, if pounds of waste generated per dollar of consumption expenditure equal two, then an average dollar of investment in waste assimilative capacity must yield at minimum a reduction of twenty-eight pounds in terms of waste density. The efficiency of waste assimilative investments must be substantially greater for lower savings rates. Lowering the rate to .10 requires investments in waste assimilation via augmenting the natural environment to rise to fifty-eight pounds per dollar of investment.

This model is perhaps the most simple that can be constructed which includes the major basic elements of both a growing economy and waste generation-assimilation processes. It embodies many extreme simplifications which undoubtedly will not withstand the probing of empirical verification.[1] In its defense, one can indicate that waste flows and increasing waste density relationships are not tied to any specific environmental phenomenon. The model is equally applicable to problems of mercury, lead, DDT, or BOD concentrations where assimilative investment may be chemical or physical processes of neutralization, constructing reservoirs to increase river flows during seasonal water shortages, or mechanical stream reaeration which raises

[1] The constant incremental output–capital ratio can be generalized by including a constant returns to scale production function, with the usual restriction on convexity with labor services as an added argument. In addition, labor force can be assumed to be a constant proportion of population, and a population growth equation with constant percentage growth rate can be included. These complications add the additional restriction that population cannot be increasing at too rapid a rate if output per capita is to be increasing and with $\dot{D} = 0$. But a compensating force arises in that labor can be substituted for the production augmenting capital stock, thus allowing a larger proportion of saving to be allocated to investments which increase the natural environment's assimilative capability. Likewise, logarithmic technical change can be assumed for either environmental investments or investments in production of goods and services. Such complications affect the interpretation of feasible positive and negative rates of growth, but add only a marginal increment to the viewpoints expressed here.

dissolved oxygen concentrations. Thus, the model is nonspecific regarding processes to augment the environment's assimilative capability.

The model suffers, however, in requiring extreme aggregation assumptions on different types of waste flow and in specifying a linear coefficient connecting assimilative investment to reductions in waste density. The latter is very likely to be not only non-linear, but non-additive in terms of different types of wastes. Also, no consideration was given to the possibility of recycling investments which not only reduce current waste flow but may also contribute to production as well.[1] However, the model has the advantage of being highly illustrative of the policy arguments that I now wish to consider. Many additional minor conclusions can be squeezed out, but this would induce additional pollution via the paper and pulp industry so I shall leave them for the reader.

What we have enumerated from this model is the fact that unregulated behavioral conditions such as the marginal propensity to save influence in the long-run whether a positive rate of growth in output is warranted when a constraint is imposed on the utilization of environmental assimilative capacity. Certainly in a private market economy, the common property character of almost all environmental resources and failure of legal systems to adequately define common property rights means that there is little or no individual incentive to invest in augmenting the environment's waste assimilative capability. In consequence, such economies will generally be plagued with overutilization of the natural environment, unless these economies are blessed with exceedingly large natural environments relative to the magnitude of wastes generated from production-consumption. But what is most important in terms of mixed public-private economies is the realization of the degree of interdependence between decisions on economic growth and the environment. Most certainly production-consumption and waste generation need not be strictly joint products provided the natural environment is treated by public authorities as an asset which needs to be considered in *all* other decisions with respect to the economy's development and stability. There are no neutral fiscal or monetary policies when environmental assets are considered. Either growth in consumption as well as rates of investment which augment the capacity for increasing physical product are encouraged or discouraged. In either case, there is a direct effect on the utilization of the natural environment's assimilative capability.

What the simple model emphasizes is the types of policy instruments the government has at its disposal to influence the natural environment beyond effluent charges, subsidies, or direct standards on waste emissions of industries and municipalities. First, not only are shifts in propensities to save likely to increase potential growth rates but also provide the necessary investment for

[1] One should note the possibility that certain types of recycling investment may cause very high waste loads per dollar of production, thus partially or completely offsetting recycling gains.

increasing the environment's capacity to assimilate wastes. Secondly, higher saving increases the time span between production and waste generation at least for some types of capital equipment investment. However, if assimilative investments are not undertaken, higher savings rates only aggravate the environmental problem in the long run. For a warranted positive rate of growth in output to be commensurate with no or little change in waste densities, a reasonably high efficiency must be obtained in assimilative investments.

A nationally consistent environmental policy means the integration and coordination of economic growth and stability with specific decisions on pollution. Not only are relative prices not reflecting discrepancies between marginal social and private costs due to environmental pollution, but governmental policies directed toward "pure growthmanship" are also inducing externalities by accelerating the generation and emission of wastes.

A fiscal policy that emphasizes maintenance of high levels of aggregate demand and employment quite innocently also accelerates rates of waste generation and accumulation if it is not coupled with an environmental investment program. Monetary policies aimed at easing credit have the simultaneous impact of increasing immediate waste flows via a positive shift in consumption expenditure but a negative impact through encouragement of investment and thus retardation in the flows of goods immediately consumed. Selective instruments such as investment tax credits, depletion allowances, and capital gains provisions also have direct and indirect effects on the rate of consumption and the distribution of production between current consumption and savings. Therefore these instruments also influence the potential rate of environmental deterioration.[1]

This discussion is not meant to be an analysis or even a complete statement on the effects of economic policies on the generation of wastes. However, I believe this most limited exposition suggests that macroeconomic policy instruments and decisions on economic development tactics have a definite and often unrecognized impact on the natural environment through their effects on rates of resource extraction, growth in consumption per capita, and disruption of natural food and energy chains.

Policies directed toward accelerating depreciation and even policies aimed at improving balance of payments influence rates of waste generation. In recent decades, there is some evidence the United States and several other developed nations have become or are becoming net importers of natural resources and net exporters of consumer durables which is indicative, on

[1] It is interesting to speculate whether the opponents of the oil depletion allowance in the United States considered the question whether such an allowance actually increased the rate of waste flows, particularly automotive emissions of carbon monoxide and oxides of nitrogen? The proponents for the Aswan Dam project most certainly neglected to analyze the problems resulting from nutrient imbalances downriver, with the consequent reduction in both soil fertility and fish yield.

balance, they may very well have become net importer's of waste emissions.[1] Most certainly the current terms of trade *do not* reflect long-term social costs (or disutilities) associated with the wastes emitted.

IV

One of the most perplexing difficulties confronting a rational utilization of the world environment is the effect environmental policies will have on the growth of developing and developed countries. It is difficult to even conceptualize the waste loads that would be exerted on the world environment if all countries had standards of living approaching those of the mature nations. Here, only one dimension of this problem will be briefly examined: namely, the effect on international comparative advantage of selected developed countries if these countries undertake a control program directed toward reducing or neutralizing their pollution loads.

Specifically, the establishment of effluent charges or standards to reduce industrial and municipal waste emissions will raise domestic costs of production, provided in-plant processing changes, substitution between factor inputs, recycling of coolants, and other changes induce additional expenditure by polluters. In part, such standards or charges should have the impact of shifting investment from augmenting production to directly or indirectly augmenting the natural environment's assimilative capability.[2] The expected domestic effect however will be reduced profits to relatively high waste load emitting industries and increased domestic and export prices. The extent of the division between higher prices and lower profits depends on such factors as market and cost structures.

If a nation *unilaterally* imposes higher waste emission standards on its domestic industries, a rise in domestic and export prices will usually result in placing the nation at an international disadvantage. For relatively open economies exporting highly competitive commodities internationally, the impact of such standards could spell a significant reduction in domestic income. This income reduction, of course, would reduce further the amount of waste flow generated by decreasing the rate of consumption. It may also have the impact of reducing the availability of investment for augmenting assimilative capability via the results of the growth model discussed earlier. Thus, a set of effluent charges which reflect current domestic social damages and are optimal from a domestic standpoint may have international repercussions which reduce waste flows to less than optimal levels domestically.[3]

---

[1] See for example, J. Vanek, *The Natural Resource Content of United States Foreign Trade: 1870–1955*, Cambridge, Massachusetts: MIT Press (1963).

[2] The effects of such shifts can be readily seen from the model discussed in Section III.

[3] In effect, the old transfer problem of international trade on over and under compensation has reared its ugly head in terms of the newer problem of environmental control. See J. M. Keynes, "The German Transfer Problem", *Economic Journal*, Vol. XXXIX (March, 1929); and B. Ohlin, "The Reparation Problem: A Discussion", *Economic Journal*, Vol. XXIX (June, 1929); or the modern synthesis in J. Vanek, *International Trade: Theory and Economic Policy*, Homewood, Illinois: Richard Irwin (1962).

Table 1. *Estimated change in domestic income resulting from the imposition of environmental controls, five countries, 1968 (billions of dollars)*[b]

Country	% change in domestic and export prices[a]	Projected negative change in domestic income	Projected % negative change in domestic income
United States	1.0–3.6	1.0– 3.6	0.1– 0.4
Japan	1.0–4.0	7.0–28.1	4.9–20.0
West Germany	1.0–4.0	0.9– 3.7	0.7– 2.7
United Kingdom	1.0–4.0	6.5–25.8	6.3–25.0
France	1.0–3.7	9.6–35.3	7.5–28.0

[a] Estimated range of changes in domestic and export prices resulting from this imposition of effluent charges or standards.
[b] These estimates must be viewed as purely experimental and preliminary as their magnitude and sign are extremely sensitive to slight changes in empirical specifications.

Alternatively, augmenting waste assimilative capability through general taxes by subsidizing industrial and municipal pollution control is likely to shift relative prices less but place the impact directly on lowering domestic real incomes. However, it is difficult to assess whether effluent charges or subsidies will have the greater impact via international advantage.

In Table 1 are recorded empirical estimates of the impact a significant shift in domestic and export prices induced by unilaterally imposed effluent charges and/or standards would have on domestic incomes for five selected countries. Many simplifying assumptions were necessary to make these calculations and they will not be documented in detail here. Briefly, a three equation model was utilized connecting each country with the "rest of the world" where import demand and domestic absorption were determined by domestic output and domestic terms of trade.[1] The model utilized was one which expli-

[1] The three equation model utilized to calculate the changes in income was of the form:

$$Y_i - M_{ii}(Y_i, T) - M_{ij}\left(Y_j, \frac{1}{T}\right) = 0$$

$$Y_j - M_{ji}(Y_i, T) - M_{jj}\left(Y_j, \frac{1}{T}\right) = 0$$

$$M_{ij} - \frac{1}{T}M_{ji} = B_i$$

where Y_i, M_{ii}, M_{ij}, and B_i are respectively: domestic output, domestic absorption, exports to the rest of the world, which is denoted as j, and balance of payments on current account of country i. T denotes the gross barter terms of trade of country i. Differentiating these three equations with respect to a change in domestic prices; solving for the change in income of country i; and inserting empirical estimates of import elasticities, domestic cross price elasticities and marginal propensities to import and consume, yielded estimates in Table 1. Published estimates were used for all elasticities and propensities except the cross elasticities of demand which were derived from our own single equation regression estimates. Weighted elasticities and propensities were used for the "rest of the world"

citly omitted the governmental sector, and therefore monetary policy is implicitly assumed to be operating in such a way that interest rates are stabilized and exchange rates are pegged. In essence then, the model does not consider *any* type of governmental policy which could be directed toward alleviating or eliminating the effect of domestic effluent charges on international advantage. Consequently, the crude estimates in Table 1 should be viewed as the most extreme effect on domestic incomes resulting from the imposition of effluent charges or standards.

The shift in domestic and export prices from a given change in environmental control which attempts to equate marginal private and social costs is difficult if not impossible to measure at the level of aggregation applied here. For reference purposes, I have utilized the range of from one percent to a percentage subjectively determined which would yield an "appreciable" reduction in waste load for individual industries. These percentages were then weighted according to the export composition for each of the countries listed in Table 1. Therefore, these ranges must be considered as pure judgments from limited experience on United States industrial pollution loads. The estimated percentage changes appear, however, to be biased downward for any relatively large scale national effort to reduce pollution loads.

The estimates in Table 1, although qualified with a long list of assumptions, lead to several interesting conjectures on the establishment of national and/or international pollutant emission controls. First, the effect of establishing environmental controls is very unlikely to be internationally neutral ... some countries, due to the type of product exported and variations in demand elasticities and marginal propensities, will undoubtedly suffer a greater "burden". This burden in terms of reduced domestic incomes will probably result in too large an adjustment to waste induced damages domestically. Second, countries which are relatively "open", in that exports account for a sizable proportion of domestic output, will tend to be affected more by impositions of environmental controls than economies, such as the United States, which are relatively closed. Economies confronted with elastic (price) demands for export commodities with relatively high pollution loads per unit of product produced and consumed will be subject to falling export revenues and reductions in domestic demand. In consequence, such economies undoubtedly face a greater burden than economies with less elastic demands and greater environmental assimilative capabilities such that effluent charges have less impact on export prices.

It should be recognized that economies may (and probably will) place high tariffs or quotas on imports which induce a relatively high waste load

where the weights corresponded to the volume of imports to country i from other countries. Also, application of the three equation model above implies that *all* commodities are produced under conditions of constant marginal cost. For a complete discussion of the procedure applied see, R. C. d'Arge "Environmental Standards and Comparative Advantage", unpublished manuscript (November, 1970).

during or following their "consumption" even if defensive action via tariffs directed toward alleviating the effect of domestic effluent charges is not undertaken. Such a set of tariffs is not only necessary for protecting domestic environmental quality but also for free international trade to result in global type of efficiencies.[1]

The final consequences of nationally imposed controls on each country cannot be foretold since they depend not only on price elasticities, income propensities, and domestic substitutions in production, but also on publically induced tariff, quota, and indirect measures to offset environmental controls. What I have attempted to illustrate is not the particular policy apparatus which will affect comparative advantage but only that pollution control efforts can potentially have a marked effect on comparative international advantage of particular countries. If we accept these rather fragmentary bits of evidence, then the conclusion follows again that governmental decisions at the national and international level are influencing and therefore ought to be influenced by considerations of the interaction of man and the environment.

V

Several threads of thought have permeated each section of this essay and it seems appropriate to draw them together here. Production, consumption, and waste generation are joint products except for the activities of man toward recycling his own or the environment's wastes, his actions toward augmenting the natural environment's assimilative capability through investments, or his actions toward reducing his own demands for material goods. In the very long run, mankind may be faced with a finite amount of resources and an infinite planning horizon ... Nietzsche confronted this problem in the doctrine of Eternal Recurrence.[2] If a finite planning interval is assumed along with some unsevere assumptions regarding utility, disutility, and population change, minimum current consumption appears to be appropriate.

A declining rate of current consumption and output was also found to be consistent with the steady state assumption on environmental pollution, provided the saving and assimilative investment coefficients of an economy were not large enough to compensate for the economy's rate of waste generation. Economic growth and environmental quality are only compatible in the long run provided that as growth in output occurs, a significant proportion

[1] For a discussion of the assumptions necessary for world efficiency in terms of free trade, see: P. A. Samuelson, "The Gains from International Trade", *Canadian Journal of Economics and Political Science*, Vol. V (May, 1939); P. A. Samuelson, "The Gains from International Trade Once Again", *Economic Journal*, Vol. LXXII (December, 1962); or the excellent discussion in M. C. Kemp, *The Pure Theory of International Trade and Investment*, Englewood Cliffs, New Jersey: Prentice Hall (1969), pp. 253–285.

[2] See, A. C. Danto, *Nietzsche as Philosopher*, New York: Macmillan (1965) pp, 205–206.

of investment is directed toward recycling wastes or increasing the assimilative capability of the environment. These findings indicate to me the necessity of analyzing very carefully policies directed toward stabilizing and even reducing per capita incomes in the "mature" economies.

The effects of environmental controls on international advantage of particular economies appear to be quite pervasive, even if one heavily discounts for all of the qualifications necessary to justify the empirical methods employed in estimation. This is especially true for relatively trade dependent nations. Economic policies directed toward the traditional targets of growth in output, stability, full employment, and the balance of payments all have seemingly unrelated but potentially destructive interactions with the natural environment and therefore cannot be separated from *any* national (or international) environmental policy.

[19]

Economic Growth and the Environment [1, 2, 3]

R. C. D'ARGE and K. C. KOGIKU

University of California, Riverside

I

The premise that man resides within a bounded and closed natural environment, an environment with relatively fixed dimensions, has been discussed in many forms for at least three centuries. The idea of a " spaceship earth " has recently caught the imagination and interest of both economists and environmentalists in particular, and the public in general [3], [5], [6] and [17]. A basic tenet of this viewpoint is that whatever has been produced, consumed, not recycled and discarded within the sealed " spaceship earth " is still here and will continue to be. Of course, energy dispersion may be offset by nuclear fusion or breeder reactors so that what was previously discarded may be completely reused without seriously impinging on the earth's natural energy reservoirs. Economically and technically-feasible extraterrestrial transportation systems, destruction of wastes by symmetrical anti-wastes, or increased efficiencies in solar energy conversion also may become realities. However, such energy-augmenting technological changes, at least at this time, appear to be in the very distant future. What is of pressing current importance, if we accept the premise of a nearly sealed capsule earth with relatively fixed dimensions in terms of mass-energy and waste assimilative capability, is how rapidly economic growth should proceed.

Rather than attempting to dissect the " spaceship earth " premises regarding limited energy reserves and assimilative capability, we take these as given in this paper. We then construct several simple paradigms of an economy confronted with a bounded, closed natural environment and then we analyze how such an economy should optimally produce and consume over time. The next section of the paper deals with an extremely simple model of waste generation, based on the conservation of matter-energy principle, and with the consumption behaviour of the economy's inhabitants assumed to be predetermined. In subsequent sections, the model is generalized to an optimal control problem where consumption and waste generation are allowed to be regulated, and an attempt is made to integrate the non-mutually exclusive processes of resource extraction and waste generation.

II

The materials balance view of a closed resource system indicates that tonnages of raw materials extraction utilized by an economy are approximately equal to tonnages of waste products generated by the economy in the long-run [3]. A basic identity derivable from the principle of conservation of matter-energy, given the assumption of materials balance and no recycling is:

$$\begin{aligned} R &= W_i + W_f = W \\ F &= W_f \end{aligned} \qquad \ldots(1)$$

1 *First version received December* 1971; *final version received April* 1972 (*Eds.*).

2 A portion of the research reported here was financially supported by Resources for the Future, Inc., with no responsibility for results or interpretation. A previous draft of this paper was presented at the 1970 Econometric Society Meetings in Detroit, Michigan.

3 The authors wish to acknowledge the very helpful comments and suggestions of K. Oddson, K. Hamada, G. Heal, and two referees.

where R, F, and W denote material extraction, production consumed by consumers, and total waste flows, respectively. W_i and W_f are the amounts of waste flow originating in the production and consumption sectors.

Given these fundamental identities and assumptions we are able to write an additional identity provided the flow (F) of material is assumed homogeneous. Basically, we assume that there is only one type of consumer good. Then:

$$F = y.N \qquad \ldots(2)$$

where y is *per capita* income in units of material flow *per capita* and N is total population. Finally, if production and consumption are joint products of waste emissions, total waste flows can be assumed to be proportional to final product [3], [5]:

$$W = gF. \qquad \ldots(3)$$

Then, by assuming that each variable previously defined refers to time (t), waste flows are linearly related to total income measured in material units:

$$W(t) = g.y(t).N(t). \qquad \ldots(4)$$

Thus, we obtain a relation between waste flows and output *per capita*. Actually, the assumption of proportionality is even more restrictive than one might at first presume, since it completely specifies an implicit technology relating output to raw materials. By definition, $W(t) = R(t)$, so $W(t) = g.y(t).N(t)$ yields $y(t) = \frac{1}{g}\frac{R(t)}{N(t)}$. Thus, in this most simple case, production results only from the magnitude of raw materials and there is no substitution between labour and raw materials in the production process. In terms of crude empirical approximation the number $g_f.y(t)$ for the consumption sector i.e., $W_f(t)/N(t)$ is estimated to be greater than one ton *per capita* per year for the United States at current production levels.

Our next task is to " model " the environment in some semi-realistic and yet manageable way. Environmental pollution, at least in its quantitative dimensions, is usually expressed in terms of concentrations, i.e., parts per million (ppm) of dissolved solids or DDT, ppm (suitably indexed) of carbon monoxide or oxides of nitrogen concentrations in the air, or tons per cubic acre of solid wastes. Thus, a natural single unit of waste concentration appears to be average density. Of course, concern with densities only may be misleading, particularly with such pollutants as carbon monoxide or methane gas generated from solid wastes. For now, however, we will assume that density is a reasonable abstract measure of waste concentration.

Let $D(t)$ and v denote waste density at time t and total environmental waste holding capacity, respectively. We, in effect, identify the closed resource system by a simple fixed volumetric magnitude, v. Then using the definition that waste divided by volume equals waste density and applying it to (4) yields:

$$D(t) = D(0) + \frac{g}{v}\int_0^t y(t)N(t)dt. \qquad \ldots(5)$$

Given exogenously determined percentage rates of growth in population of $100n$ and income *per capita* of 100β, waste density can be easily related to initial population levels and material flow *per capita*:

$$D(t) = D(0) + \frac{g}{v}\frac{e^{(\beta+n)t}}{\beta+n}y(0)N(0). \qquad \ldots(6)$$

Increasing density of wastes within the closed resource system is not necessarily bad *per se*. However, if the natural environment is relatively small in comparison to current and expected future sums of waste flows, people may suffer physiological, psychic, or other damages. As an example, we could presume for each individual a monotonic, continuous increasing equation relating *per capita* damages and waste densities, where aggregate damages are in terms of some common unit measure such as dollars or utils. Since waste density is increasing more rapidly than either population or material flow *per capita*, damage costs *per capita* must also be increasing. If the *per capita* damage function is linear then *per capita* damage costs increase at the same proportionate rate as waste density. Also, since total damage costs are a constant multiple of population and damage costs *per capita*, total damage costs due to waste accumulation must be increasing at an even *faster rate* than damage costs *per capita*. Thus, we obtain a spiralling rise in damage costs associated with waste accumulation, resulting from an ever increasing population, and rising material flow *per capita*. Each individual's waste from production and consumption activities affects a toll of damages on themselves and all other individuals including future generations unless the natural environment is able to assimilate a substantial proportion of the waste generated.

Without controls, this very simplified depiction of an extraction-consumption-waste economy yields a Malthusian type of scenario: population and material flow *per capita* increase until some threshold waste density for viable human populations is reached. At that point, instead of subsistence level existence, either material output *per capita* must fall to zero or the population is decimated.[1] Alternatively, we may contemplate other variants of this scenario where population, for the most part, is decimated but a viable few continue to survive. If the natural environment is able, over long periods of time, to regenerate and assimilate these wastes, then a cyclical pattern is obtained for the human species. The pattern is characterized by periods of rapid growth in population and " consumption ", decimation thresholds, and then very small populations at minimum subsistence consumption until environmental regeneration allows another phase of " economic development " to proceed. Clearly, from the biologist's viewpoint, this population-material flow model suggests the possibility of an almost continuous interspecies-disequilibrium [11], [17].

The previously developed model of the economy's interaction with the natural environment cast man in the role of a rabbit (or lemming). Growth rates of population and material flow were assumed to be exogenously determined, and the natural environment was viewed as an unregenerative, non-assimilative volume. In addition, the possibilities of investment in recycling or technical progress toward reducing waste flow per unit product, reduction in rates of material flow through saving and increasing durability, and altering rates of growth in material flow and population were not considered. Here we shall start by analyzing a hypothetical economy for which all of the previous paradigm's assumptions are retained except for exogenously determined material flows *per capita*. Thus, we wish to inquire into the possibilities for and behaviour of an economy characterized as a purely extraction-consumption-waste process.

In order to examine a time-related optimization within such an economy, in addition to equations (1), (2), (4), and (5) a time dependent utility function, relating material flow *per capita* to utility *per capita*, must be specified along with a disutility function for cumulative waste flows. For expositional purposes, we shall also assume that all variables including time are continuous, and that the optimization planning interval is finite τ.

The following *per capita* utility function will be assumed.

$$u(y, D, \tau) = u(y) - u(D) + u(\tau), \qquad \ldots(7)$$

which prevails for each member of society. Several conditions are presumed with regard

[1] There might be other biological thresholds before catastrophic decimation is reached. For example, increasing concentrations of waste may adversely affect reproduction rates so that population growth is reduced as waste density builds up. Potentially, then, we may have an " iron law " of pollution: the " natural " negative rate of growth in population is just sufficient to offset the positive rate of growth in density of wastes necessary to support population at some minimal subsistence level. Alternatively, if the natural environment could assimilate a certain amount of wastes, population may reach some " steady state " where wastes generated are exactly offset by environmental assimilation or regeneration and material flow *per capita* is at some minimum subsistence level. Several simple models of a " steady state " have been recently described [17], [18].

to the form of the utility function in addition to separability which sidesteps the problem of interpreting cross partial derivatives of the utility function:

$$u_y>0 \quad u_D>0 \quad u_\tau>0$$
$$u_{yy}<0 \quad u_{DD}>0 \quad u_{\tau\tau}<0 \qquad \ldots(8)$$

In essence, these conditions stipulate that *per capita* utility of material flow is increasing at a decreasing rate, while *per capita* disutility generated by increasing waste densities is increasing at an increasing rate. As the finite planning interval τ is increased, it is assumed that *per capita* utility at any time t is increased, but at a decreasing rate. This is perhaps one of our most tenuous assumptions since it implies that, regardless of how close t is to τ, individuals will be precommitted to applying a non-shifting evaluation to an increment in τ.[1] *Consequently*, the optimal finite planning interval is deduced with no adjustment occurring over time to account for potential "future disobedience" in the Strotz [1] sense, as t approaches τ(1).

From equation (5), we are able to write a relation between waste flow and waste densities:

$$\dot{D} = \frac{1}{v} W \qquad \ldots(9.1)$$

$$\dot{D} = \frac{g}{v} yN. \qquad \ldots(9.2)$$

Equation (9.2) is, in effect, a description of the changing state of the natural environment as a result of the amount of material consumption at time t. Equations (7) and (9.2) yield the " pollution process " model except for specification of initial and terminal conditions and inclusion of a social rate of time preference factor. The terminal time τ is conceptualized to be that period of survival which is optimum for the maximization to be stated below. Given a presumed decimation density level (for example, 2,000~4,000 ppm of carbon monoxide), $\bar{D}$, then $D(\tau) = \bar{D}$.

To complete this exposition, the initial density is specified as $D(0) = \underline{D}$, and population at t as:

$$N(t) = N(0)e^{nt} \qquad \ldots(10)$$

and the utility function is discounted with a discount factor a.[2]

[1] It should be noted that the assumptions of a finite planning interval coupled with a finite environmental capacity for wastes are extremely rigid. Of course, infinite planning intervals in conjunction with an unlimited natural environment assimilative capacity reduces to a trivial non-existent scarcity case unless scarcity is introduced in some other way, i.e., available technology, knowledge, or capital. Alternatively, an infinite planning interval in conjunction with a finite environmental waste assimilative capability appears to be philosophically inconsistent, provided population is constant or increasing through time, material flow *per capita* has a positive lower bound, rates of environmental waste assimilation are very low or zero, waste generating technology is completely static in character, and wastes at some level of concentration become lethal to the human species. These conditions quite obviously ensure that at some moment in the finite future, decimation will occur. The alternative case of a finite planning interval and infinite waste assimilative capability also appears to be philosophically inconsistent. Specifying an infinite planning interval for mankind, of course, implies the imposition of a constraint on the decision process, namely that regardless of how degraded man's habitat becomes, he must continue to exist. A more general and encompassing decision process would stipulate that the survival period be optimally chosen with infinity as one possible choice. We have chosen the course of specifying a finite planning interval and making it subject to choice by current generations. However, such a decision precludes consideration of an infinite time horizon. The marginal utility of an additional unit of time may approach or be at infinity for those concerned only with preservation of the human species. In such a case, the length of τ would dominate all other decisions in a trivial manner since minimal consumption at each instant of time must prevail with finite resources, specifically waste assimilative capability.

[2] Although we shall apply a discount factor which reflects a positive or zero rate of time preference for material-consumption flows, several potentially explosive ethical questions are brushed aside. If the human race faces the possibility of extinction by pollution (or other means) over a finite interval, are inter-generational utility comparisons defensible? Within a different context, Dasgupta recently suggested that a " small positive discount of the future " may be accepted as " ethical " [12]. However, if we assume that each

Given the above definitions, an optimal control problem can be formulated as follows; maximize

$$J = \int_0^\tau [u(y)-u(D)+u(\tau)]e^{-at}dt = \int_0^\tau \left\{[u(y)-u(D)]e^{-at} + \frac{d}{dt}\frac{1-e^{-at}}{a}u(t)\right\} dt \qquad \ldots(11.1)$$

where y is the control variable and D is a state variable, whose equation of motion is

$$\dot{D} = \frac{g}{v}yN = bye^{nt} \quad (b \equiv gN(0)/v). \qquad \ldots(11.2)$$

The problem is to select the admissable control y so that a maximum integral of discounted individual utility is achieved over the interval $0 \leqq t \leqq \tau$ where τ is free to vary.

The Hamiltonian function for the problem can be written:

$$H = [u(y)-u(D)]e^{-at} + \frac{d}{dt}\frac{1-e^{-at}}{a}u(t) + pbe^{nt}y$$

$$= [u(y)-u(D)+pbye^{(n+a)t}]e^{-at} + \frac{d}{dt}\frac{1-e^{-at}}{a}u(t) \qquad \ldots(12)$$

where p is the costate variable corresponding to D.

According to the theory of optimal control, optimality calls for the following conditions[1]:

$$\dot{D} = bye^{nt} \quad D(0) = \underline{D} \quad D(\tau) = \bar{D} \qquad \ldots(13.1)$$

$$\dot{p} = -\frac{\partial H}{\partial D} = u_D(D)e^{-at} \qquad \ldots(13.2)$$

$$y^* \text{ maximizes } u(y) + pbye^{(n+a)t} \qquad \ldots(13.3)$$

$$[u(y)-u(D)+pbye^{(n+a)t}]e^{-at} = \int_t^\tau \{a[u(y(s))-u(D(s))] - npbye^{(n+a)s}\}e^{-as}ds - \frac{d}{dt}\frac{1-e^{-at}}{a}u(t)\Big|_\tau \qquad \ldots(13.4)$$

$$[u(y(\tau))-u(D(\tau))+pbye^{(n+a)\tau}]e^{-a\tau} + \frac{d}{dt}\frac{1-e^{-at}}{a}u(t)\Big|_\tau = 0. \qquad \ldots(13.5)$$

If we define a new variable

$$z = -bpe^{(n+a)t}$$

generation can be exactly separated from others, then current generations with higher rates of time preference may actually eliminate the existence of some distant future generation. The faster we consume in a closed resource system, the more rapidly extinction occurs. While the utility of distant generations may seem valueless now, if we were that distant generation, we may very well value our continued existence at or approaching infinity. Even very high rates of time preference over a finite interval would not make it worthwhile to consume currently above some basic subsistence level, provided the last future generation's utilities were considered. In any case, we shall assume that such inter-generational utility comparisons can be contrived so that each generation would be equitably considered, including potential but excluded future generations, and in the remainder of the discussion emphasize the results where all generations are treated as equals, i.e., $a = 0$.

[1] See, e.g., Michael Athans and Peter L. Falb, *Optimal Control*, New York, N.Y.: McGraw-Hill, 1966, Theorem 5-7, p. 293.

then (12) and (13.1)-(13.5) become

$$H = [u(y) - u(D) - zy]e^{-at} + \frac{d}{dt}\frac{1-e^{-at}}{a}u(t) \qquad \ldots(12')$$

$$dD/dt = be^{nt}y \quad D(0) = \underline{D} \quad D(\tau) = \bar{D} \qquad \ldots(13.1')$$

$$dz/dt = (n+a)z - be^{nt}u_D \qquad \ldots(13.2')$$

$$u(y) - zy \geqq u(w) - zw \text{ for all } w \text{ such that } 0 \leqq v_0 \leqq w \leqq v_1 < \infty \qquad \ldots(13.3')$$

where v_0 is a lower bound and v_1 an upper bound for consumption;

$$u_y - z \leqq 0$$

$$[u(y(t) - u(D(t)) - z(t)y(t)]e^{-at} = \int_t^\tau e^{-as}\{a[u(y(s)) - u(D(s))] + nz(s)y(s)\}ds - \frac{d}{dt}u(t)\frac{1-e^{-at}}{a}\Big|_\tau \qquad \ldots(13.4')$$

$$[u(y(\tau)) - u(D(\tau) - z(\tau)y(\tau)]e^{-a\tau} + \frac{d}{dt}\frac{1-e^{-at}}{a}u(t)\Big|_\tau = 0. \qquad \ldots(13.5')$$

For (13.3') the control should be:

$$\begin{aligned} y &= v_0 && \text{if } z \geqq u_y(v_0) \\ &= [u_y]^{-1}(z) && \text{if } u_y(v_1) \leqq z \leqq u_y(v_0) \\ &= v_1 && \text{if } z \leqq u_y(v_1). \end{aligned} \qquad \ldots(13.6)$$

That is, $y = F(z)$, which is Lipschitz continuous for all z and non-increasing, should have the graph shown in Figure 1. Using $F(z)$ the system (13.1')-(13.2') can be replaced with

$$dD/dt = be^{nt}F(z) \quad D(0) = \underline{D} \qquad \ldots(13.1'')$$

$$dz/dt = (n+a)z - be^{nt}u_D \quad z(0) = z_0 \qquad \ldots(13.2'')$$

where z_0 is free. A solution exists and is unique for all z_0. Several possible timepaths are shown in Figure 2.

From the preceding two equations:

$$\frac{dz}{dD} = \frac{(n+a)z - be^{nt}u_D}{be^{nt}F(z)} \gtreqless 0 \qquad \ldots(14.1)$$

depending on

$$z \gtreqless \frac{be^{nt}u_D}{n+a}.$$

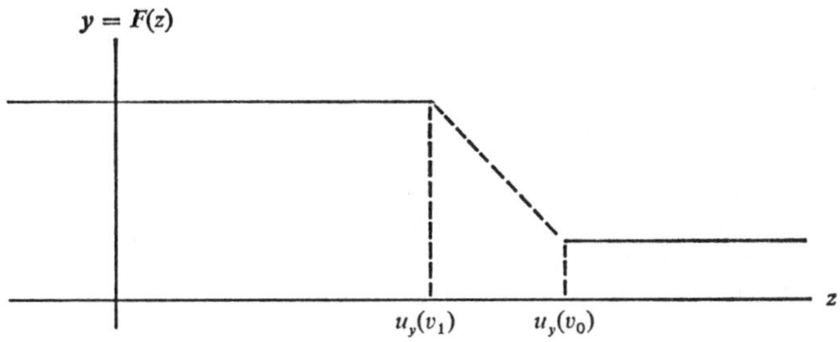

FIG. 1

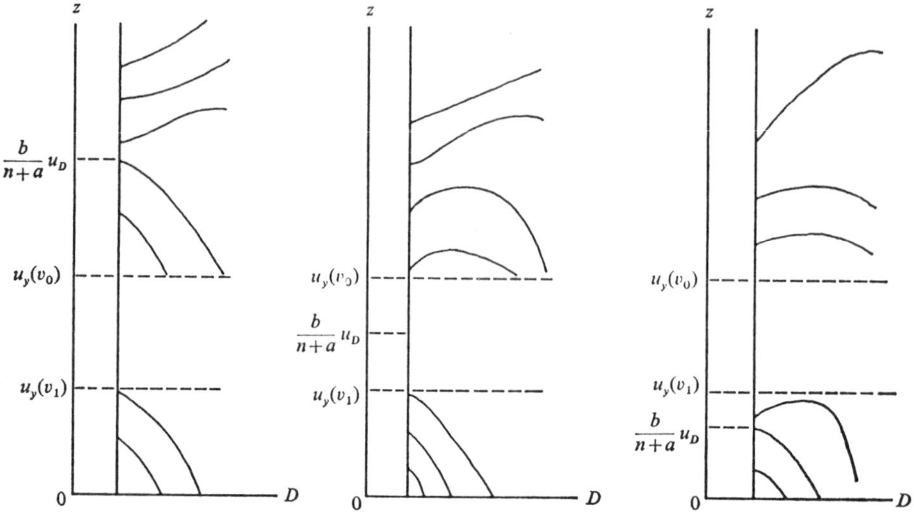

FIG. 2

At $t = 0$

$$\frac{dz}{dD} \gtreqless 0 \qquad \ldots(14.2)$$

depending on

$$z \gtreqless \frac{bu_D}{n+a}.$$

Whether z will start out increasing or decreasing can be determined by (14.2). As t and D increase, z will increase (remain constant, or decrease) depending on whether z is greater than (equal to, or less than) $be^{nt}u_D/(n+a)$. The control will shift according to (13.6).

Since we assumed that τ satisfying (13.5′) is positive and finite, the interpretation of the above conditions are as follows. Equation (13.3′) indicates that the variable z is positive (and the costate variable p is negative). Equations (13.6) indicate that optimal consumption y is a non-increasing function of z; that consumption should be kept at the minimum consumption level v_0 if the variable z (a " cost " of utilizing an increment of waste density resulting from consumption where the cost implicitly reflects a negative value of p associated with waste density) exceeds the *per capita* marginal utility of subsistence consumption $u_y(v_0)$; that consumption should be kept at the maximum level v_1 if z falls short of the marginal utility of the maximum consumption $u_y(v_1)$; and that consumption should be sufficient to equate the marginal utility of consumption to z when z is between $u_y(v_0)$ and $u_y(v_1)$. Equation (13.2″) indicates that the shadow price, z, of waste density is changing in response to the marginal disutility of the individual due to the increment in waste density, and that the shadow price z should fall when t and D are large enough because D increases by (13.1) and u_D increases by (8).

At first glance this result may appear incorrect since, with increasing total waste flows over time, a greater degree of " control " (compared with no regulation) on pollution loads would seem imperative. Note, however, that condition (13.2″) refers to optimal utilization of waste storage through time and not to unregulated utilization patterns. Accordingly, optimal utilization commencing from an unregulated state requires that in the initial period a substantial undiscounted shadow price should be applied to waste density. Following the initial large-scale shift the shadow price decreases over time.

Re-writing (13.3′) for an interior solution

$$u_y(y) = z. \qquad \text{...(15)}$$

Hence in this case

$$\dot{y} = \dot{z}/u_{yy}(y) \qquad \text{...(16)}$$

and y is increasing when z is decreasing. Equation (13.2″) shows that z will be falling when

$$(n+a)z < be^{nt}u_D$$

that is,

$$z = u_y(y) < \frac{be^{nt}}{n+a} u_D. \qquad \text{...(17)}$$

Therefore, y will be increasing when the marginal utility of consumption is small relative to the marginal disutility associated with rising waste density. Combining (16) and (13.2″)

$$\dot{y} = \frac{(n+a)z - be^{nt}u_D}{u_{yy}(y)} \qquad \text{...(18)}$$

when t is sufficiently large,

$$\dot{y} = \frac{-be^{nt}u_D}{u_{yy}}. \qquad \text{...(19)}$$

If the discount factor a is zero, implying equal valuation of all future generations that do exist, and the population growth rate n is also zero, then (13.2″) implies that z falls from the beginning and by (13.6), consumption, if not at the maximum level initially, will eventually begin to increase. Thus, if these sufficient but not necessary conditions (see Figure 2) are fulfilled, such an economy will increase consumption over time to compensate for the disutility associated with rising waste concentrations.

However, this model is much too simple in terms of at least two groupings of characteristics. First, extractive and renewable resources are assumed to be quite literally free. Second, a non-progressive economy facing an inert natural environment is assumed, where waste recycling, reduction in obsolescence rates, technological change, savings rates, the environment's assimilative capacity and other factors are not allowed to continuously alter over time the magnitudes and thereby disutilities of waste flows. In what follows, we will attempt to remove some of these limitations. But before proceeding, one important tentative conclusion can be inferred: if our assumptions are not without some element of validity, waste generating processes in both production and consumption sectors need to be taxed at relatively high rates in the near future, particularly if social rates of time preference are close to or are, by design, constrained to be zero.

Issues emerging from environmental management arise not only from the optimal rate of waste generation in a closed medium but also with regard to the rate of extractive (or renewable) resource exploitation. If extractive resources are finite in magnitude and can for all practical purposes be exhausted, then optimal environmental management involves a " conjunctive use " type of allocation problem where one must consider rates of extraction *and* rates of waste generation. Thus, the " pure " mining problem must be coupled with the " pure " pollution problem and questions like these become relevant: which should we run out of first, air to breathe or fossil fuels to pollute the air we breathe?

For expositional purposes, we shall assume an initial finite stock of extractive and non-renewable resources such that:

$$\dot{Q} = -R. \qquad \text{...(20.1)}$$

where R denotes, as previously, the amount of resources extracted. Also boundary conditions will be stipulated for Q:

$$Q(\tau) = \underline{Q} \qquad \ldots(20.2)$$

$$Q(0) = \bar{Q} \qquad \ldots(20.3)$$

with $\underline{Q}$ and $\bar{Q}$ constants equal to the terminal and initial stock of extractive resource.[1]

In order to consider the possibilities of exhausting extractive resources and substitution between labour and resources in producing consumable goods, excluding services, we will introduce several changes into the model given earlier. Thus:

$$R(t) = W(t) + \text{residual determined by (20.5) (21.1) (22.1)} \qquad \ldots(20.4)$$

defines the materials balance relationship at any instant of time.[2] Also, we shall assume that total population is employed in each period, and that population equals the labour force. A homogeneous production function of degree one is assumed in which substitution exists between raw materials and labour:

$$F = F(R, N) = f(r)N \qquad \ldots(20.5)$$

with r denoting extractive resources *per capita*. It is also assumed that this technological relationship fulfils the usual first and second order diminishing marginal factor productivity conditions, namely $f_r > 0, f_{rr} < 0$. It is assumed that waste flows resulting from the production process are directly related to the quantity of extractive resources:

$$W_i = H(R) \qquad \ldots(21.1)$$

or

$$W_i = h(r)N \quad h_r > 0 \quad h_{rr} > 0. \qquad \ldots(21.2)$$

Finally, waste flows originating in the consumption sector are presumed to be related to the amount of goods produced and consumed[3]:

$$W_f = G(F) \qquad \ldots(22.1)$$

or

$$W_f = g(y)N \quad g_y > 0 \quad g_{yy} > 0. \qquad \ldots(22.2)$$

[1] The case depicted here is one where the change in stock of resources equals flow: a purely extractive process. Alternatively, if the resource is renewable, i.e., animal or plant populations (20.1) would need to include a relationship for biological growth. It has become common practice to relate fish population growth rates to the initial mass of fish [11], [12], [15] and [16]. In terms of our notation and variables, and provided $E(Q)$ denotes the growth-bio-mass relationship, (20.1) would then be altered to:

$$\dot{Q} = E(Q) - R$$

For simplicity (20.1) will be applied, though it is to be noted that the model, with some changes in conclusions and inclusion of the possibility that $\tau = \infty$ or a terminal value of $E(Q)$, will also accommodate biological growth relationships, of the above type.

[2] As long as there is a "mass balance" in that raw materials extraction equals waste flows in each instant of time or even over short gestation-production lags, saving, recycling, or other long term activities which markedly reduce current waste flows per unit of production, must by definition be zero. Thus, in order to accommodate such waste reducing possibilities it is necessary to adjust the mass balance relationship [5]. However, the theoretical possibility of continuous recycling within a production cycle does exist, though we shall not treat this case except via a constraint on recycling in a variant of the model which is developed later.

[3] Savings hypotheses in most neoclassical models of growth indicate that a certain proportion of production is not consumed currently, and thus in our terminology does not contribute to current waste flow. However, obsolescence or decay rates in these same models indicate the proportions of accumulated saving-investment which do enter as waste currently. Thus, the relationship $g(y)$ (relating waste flow from the consumption sector to the amount of goods produced) can be thought of as embodying both accumulation rates and depreciation rates for durable goods within the consumption sector. Additionally, in the simplified model the $g(y)$ relationship implicitly contains recycling rates from the consumption sector back into the consumption sector. The processes of depreciation, consumption-saving, and recycling (within the consumption sector) are therefore all submerged in $g(y)$, and certainly some of the most important policy questions on environmental pollution require that these processes be carefully separated.

The $h(r)$ function links rates of raw materials extraction directly to the magnitude of waste flow from the production sector. Thus, this relationship embodies such important determinants of pollutant loads as rates of obsolescence and decay of machinery and equipment, and efficiency of raw materials conversion.

Relation (22.1) yields a one to one correspondence between waste flows and final goods produced if no saving or recycling occurs, and $W_i = 0$. However, in what follows, such an equality will not be assumed. Given (21.2) and (22.2), total waste flows can be defined in terms of r and y:

$$W = [h(r)+g(y)]N \qquad ...(23)$$

and thus from (9.1):

$$\dot{D} = \frac{1}{v}[h(r)+g(y)]N. \qquad ...(24)$$

Extracting resources such as harvesting forests, oil spills from off-shore oil derricks, strip mining, reductions in river flow, and soil erosion resulting from removal of natural vegetative cover for crop production may impose social costs (or disutilities) emanating from psychic and aesthetic values associated with undisturbed environments. It therefore seems appropriate, within the context of the model studied, to include a *per capita* loss function which reflects disturbance of the natural environment through extraction (or distortions in interspecies equilibrium). Let us incorporate such a *per capita* loss function of the following form:

$$u(Q) \qquad ...(25)$$

with $u_Q>0$, $u_{QQ}<0$. These conditions imply the stock of natural resources offers some pristine enjoyment such that a reduction in their availability (or their disturbance) causes increasing *per capita* marginal disutility. The *per capita* utility function that emerges is:

$$u(y, D, Q, \tau) = u(y)-u(D)+u(Q)+u(\tau).$$

Then the new control problem is to maximize

$$J = \int_0^\tau u(y, D, Q, \tau)e^{-at}dt = \int_0^\tau \left\{[u(y)-u(D)+u(Q)]e^{-at}+\frac{d}{dt}\frac{1-e^{-at}}{a}u(t)\right\}dt \qquad ...(26)$$

where

$$y = f(r). \qquad ...(27)$$

The control variable is r, and the state variables are D and Q, with

$$\dot{D} = \frac{1}{v}[h(r)+g(y)]N \qquad ...(28)$$

$$\dot{Q} = -R = -rN \qquad ...(29)$$

$$D(0) = \underline{D} \quad D(\tau) = \bar{D} \qquad ...(30)$$

$$Q(0) = \bar{Q} \quad Q(\tau) = \underline{Q}. \qquad ...(31)$$

The Hamiltonian function for this problem is

$$H = [u(f(r))-u(D)+u(Q)]e^{-at}+\frac{d}{dt}\frac{1-e^{-at}}{a}u(t)+\frac{N(0)}{v}e^{nt}p[h(r)+g(f(r))]-N(0)e^{nt}qr$$

$$= \{u(f(r))-u(D)+u(Q)-z_1[h(r)+g(f(r))]-z_2r\}e^{-at}+\frac{d}{dt}\frac{1-e^{-at}}{a}u(t) \qquad ...(32)$$

where p and q are the costate variables for D and Q, respectively, and z_1 and z_2 are their transformations:

$$z_1 = -\frac{N(0)}{v}e^{(n+a)t}p$$

$$z_2 = N(0)e^{(n+a)t}q.$$

The rest of the analysis is analogous to the case analyzed earlier without a constraint on extractable resources, Q. The condition corresponding to (13.3′) is

$$r^* \text{ maximizes } u(f(r)) - z_1[h(r)+g(f(r))] - z_2 r \qquad \ldots(33.1)$$

$$u_y f_r - z_1(h_r + g_y f_r) - z_2 \leqq 0$$

The control corresponding to (13.6) is:

$$\begin{aligned} r &= r_0 \text{ if } z_1(h_r+g_yf_r)+z_2 \geqq u_yf_r \,|_{r_0}, \qquad \ldots(33.2)\\ &= [u_yf_r]^{-1}(z_1h_r+g_yf_r)+z_2)\\ &\quad \text{if } u_yf_r\,|_{r_1} \leqq z_1(h_r+g_yf_r)+z_2 \leqq u_yf_r\,|_{r_0},\\ &= r_1, \text{ if } z_1(h_r+g_yf_r)+z_2 \leqq u_yf_r\,|_{r_1} \end{aligned}$$

where r_0 and r_1 correspond to v_0 and v_1 for the case without Q, in defining the relationship between resource extraction and production.

Define:

$$G(r) = h(r)+g(f(r)) \text{ then } G_r = h_r + g_y f_r \qquad \ldots(33.3)$$

and

$$z \equiv z_1(h_r+g_yf_r)+z_2 = z_1G_r+z_2 \qquad \ldots(33.4)$$

then (33.2) defines $r = F(z)$, which should have a graph as shown in Figure 3. Using $F(z)$, equations (28) and (29) can be rewritten as:

$$\frac{dD}{dt} = \frac{G}{v}N(0)e^{nt} = \frac{N(0)}{v}e^{nt}G(r) \quad D(0) = \underline{D} \qquad \ldots(33.5)$$

$$\frac{dQ}{dt} = -N(0)e^{nt}F(z) \quad Q(0) = \bar{Q}. \qquad \ldots(33.6)$$

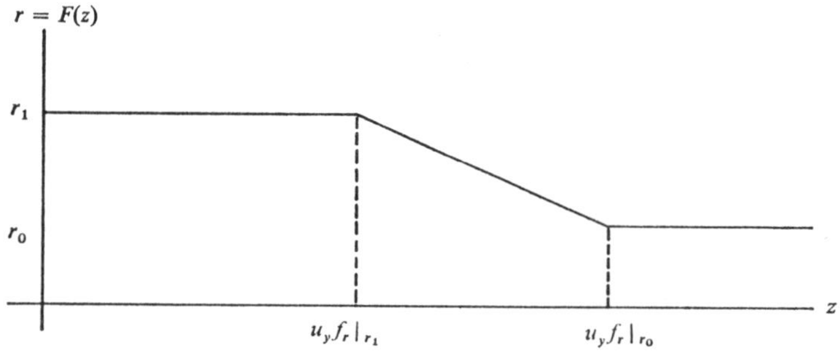

FIG. 3

The canonical equations for the costate variables are:

$$\frac{dp}{dt} = -\frac{\partial H}{\partial D} = u_D(D)e^{-at} \qquad \ldots(33.7)$$

$$\frac{dq}{dt} = -\frac{\partial H}{\partial Q} = -u_Q(Q)e^{-at}. \qquad \ldots(33.8)$$

Hence

$$\frac{dz_1}{dt} = (n+a)z_1 - \frac{N(0)}{v} u_D e^{nt} \qquad \text{...(33.9)}$$

$$\frac{dz_2}{dt} = (n+a)z_2 - N(0)u_Q e^{nt} \qquad \text{...(33.10)}$$

$$\frac{dz}{dt} = G_r \frac{dz_1}{dt} + \frac{dz_2}{dt} = (n+a)z - \left[\frac{G_r}{v} u_D + u_Q\right] N(0)e^{nt}. \qquad \text{...(33.11)}$$

Hence:

$$\frac{dz}{dD} = \frac{(n+a)z - \left[\frac{G_r}{v} u_D + u_Q\right] N(0)e^{nt}}{\frac{N(0)}{v} e^{nt} G(r)} \qquad \text{...(33.12)}$$

and

$$\frac{dz}{dQ} = \frac{(n+a)z - \left[\frac{G_r}{v} u_D + u_Q\right] N(0)e^{nt}}{-N(0)e^{nt} F(z)}. \qquad \text{...(33.13)}$$

It is obvious from (33.12) and (33.13) that as t and D increase and Q decreases, z will increase (remain constant, or decrease) depending on whether z is greater than (equal to, or less than):

$$\frac{\left[\frac{G_r}{v} u_D + u_Q\right] N(0)e^{nt}}{n+a}. \qquad \text{...(33.14)}$$

The transformed variable z will start out increasing provided (33.14) is less than z when evaluated at $t = 0$.

Given the assumption that τ is positive and finite, the interpretation of the above conditions is as follows: (33.1) indicates that the transformed variable z is positive. Equations (33.2) indicate that the optimal resource extraction rate r is a non-increasing function of z; that r should be kept at the minimum level r_0 if the variable z exceeds the *per capita* marginal utility product of subsistence resource extraction $u_y(v_0)f_r(r_0)$; that r should be kept at the maximum level r_1 if z falls short of the *per capita* marginal utility product at the maximum rate of resource extraction $u_y(v_1)f_r(r_1)$; and that r should be selected to equate the marginal utility product of resource extraction to z when z is between $u_y(v_0)f_r(r_0)$ and $u_y(v_1)f_r(r_1)$.

Equations (33.9)-(33.11) show that the transformed shadow price z is changing in response to the marginal disutilities of the population associated with the increment in waste density and decrement in extractable resources and that the shadow price z should fall when t and D are sufficiently large and Q is sufficiently small. Hence, r and therefore consumption *per capita* (y) will eventually begin to rise if both are not already at a maximum.

If the discount factor a is zero and population growth is zero, (33.11) implies that z falls from the beginning and by (33.2), resource extraction, if not at a maximum or minimum level initially, will immediately begin to rise. The reason for this result is obvious upon reflection. Extracting resources and rising waste densities are both associated with loss in utility so it is optimal to postpone high consumption rates until they are needed to compensate for the increasing disutility associated with rising waste density and depletion of the environment.

In the simplified paradigms examined thus far, the possibility of recycling from the consumption sector back into the production sector has not been examined. One possibility of including such inter-sector recycling in our last model would be by defining a new decision variable as representing the amount of recycled materials from consumption to production sectors. Hypothetically, we should also include it as a separate argument of the production function (20.5) since it might be expected *a priori* that recycled material would be less productive than "original" raw materials and contribute differently to total waste flows. Also, there must be some upper limit established on the magnitude of recycling in each instant of time, or there is no "scarcity" of new and recycled extractive resources.

Coupling these modifications for recycling between production and consumption sectors to the earlier model given by (26) through (31) a new set of first order conditions are derivable. These conditions indicate that recycling between production and consumption should be enough so that the marginal utility product of recycled material is equal to (the case of an interior maximum) or does not exceed (the case of a boundary maximum) the marginal disutility associated with waste density which is not increased as a result of recycling, plus the shadow price of the upper-limit constraint on number of recycles per instant of time.[1] Conditions for rising *per capita* consumption can be derived as in the preceding control problem. The only difference is that, with recycling, two controls need to be selected rather than one.

V

In order to briefly explore several facets of investment in recycling machinery and equipment, we shall propose a simple investment model where, for convenience, waste flows from the production sector are assumed to be channelled through the consumption sector, i.e., $W_i = 0$ and $W = W_f$. Also, to reduce complications, we will also assume extractable resources are not scarce. Then let k denote the *per capita* amount of capital committed to recycling, s the *per capita* amount of recycled material, c *per capita* consumption, δ an instantaneous decay rate of recycling capital, and α the ratio of recycled material to recycling capital. Then, the problem becomes:

maximize
$$J = \int_0^\tau u(c, D, \tau)e^{-at}dt \qquad \ldots(34.1)$$

subject to

$$\dot{D} = \left[\frac{g(r, s)}{v} + \delta k\right] N \qquad w = W/N = g(r, s) > 0,\ g_r > 0,\ g_{rr} < 0,\ g_s < 0,\ g_{ss} > 0 \qquad \ldots(34.2)$$

$$\dot{k} = y - \delta k - c \qquad \ldots(34.3)$$

$$y = f(r, s) \qquad \ldots(34.4)$$

$$s = \alpha k \qquad \ldots(34.5)$$

where the rate of resource extraction r and consumption c are the control variables. In the utility function in (34.1), *per capita* consumption c appears rather than *per capita* output y. In (34.2), the rate of change of waste density is affected not only by the rate of production and recycling, but also by the depreciation of the recycling capital stock whose life is $1/\delta$. Equation (34.3) postulates that the rate of accumulation of recycling capital stock is net output minus consumption. Equation (34.4) is the production function for output, while equation (34.5) is the assumed constant ratio between recycled material and capital stock.

[1] A more complete description of recycling results are contained in: R. C. d'Arge and K. C. Kogiku, "Economic Growth and the Natural Environment," Program in Environmental Economics, University of California, Riverside, *Working Paper Series*, #1, (April 1971).

If p is the costate variable for D as before, and q is now the costate variable for k, then the Hamiltonian function for this problem is

$$H = [u(c)-u(D)]e^{-at}+N(0)e^{nt}p\left[\frac{g(r, \alpha k)}{v}+\delta k\right] +q[f(r, \alpha k)-\delta k-c]+\frac{d}{dt}\frac{1-e^{-at}}{a}u(t). \quad \ldots(34.6)$$

Introducing the transformed costate variables

$$z_1 = -N(0)e^{(n+a)t}p \quad \ldots(34.7)$$

$$z_2 = e^{at}q \quad \ldots(34.8)$$

the Hamiltonian function becomes

$$H = \left\{u(c)-u(D)-z_1\left[\frac{g(r, \alpha k)}{v}+\delta k\right]+z_2[f(r, \alpha k) -\delta k-c]\right\}e^{-at}+\frac{d}{dt}\frac{1-e^{-at}}{a}u(t) \quad \ldots(34.9)$$

and

$$\frac{\partial H}{\partial r} = \left(-z_1\frac{g_r}{v}+z_2f_r\right)e^{-at} \lesseqq 0 \quad \ldots(35.1)$$

$$\frac{\partial H}{\partial c} = [u_c-z_2]e^{-at} \lesseqq 0. \quad \ldots(35.2)$$

Hence for a maximum of the Hamiltonian function $r = r_0$ and $c = c_0$, i.e., r and c are at minimum levels, if

$$v\frac{z_2}{z_1} < \left.\frac{g_r}{f_r}\right|_{r0} \qquad z_2 > u_c\,|_{c0}, \quad \ldots(35.3)$$

$$r = \left(\frac{g_r}{f_r}\right)^{-1}\left(v\frac{z_2}{z_1}\right) \text{ and } c = (u_c)^{-1}(z_2) \quad \ldots(35.4)$$

when

$$\left.\frac{g_r}{f_r}\right|_{r0} \leqq v\frac{z_2}{z_1} \leqq \left.\frac{g_r}{f_r}\right|_{r1} \quad \ldots(35.5)$$

and

$$u_c\,|_{c_1} \leqq z_2 \leqq u_c\,|_{c0}, \quad \ldots(35.6)$$

and $r = r_1$, and $c = c_1$ (r and c take on maximum levels) when

$$v\frac{z_2}{z_1} > \left.\frac{g_r}{f_r}\right|_{r1} \qquad z_2 < u_c\,|_{c_1}. \quad \ldots(35.7)$$

In the case of an interior maximum

$$z_1\frac{g_r}{v} = u_cf_r. \quad \ldots(35.8)$$

The relationship between controls and costate variables are depicted in Figures 4 and 5. The canonical equations are (34.2)-(34.3), and

$$\dot{z}_1 = (n+a)z_1-u_De^{-at}N(0) \quad \ldots(36.1)$$

$$\dot{z}_2 = az_2+z_1\left(\frac{g_s\alpha}{v}+\delta\right)-z_2(f_s\alpha-\delta). \quad \ldots(36.2)$$

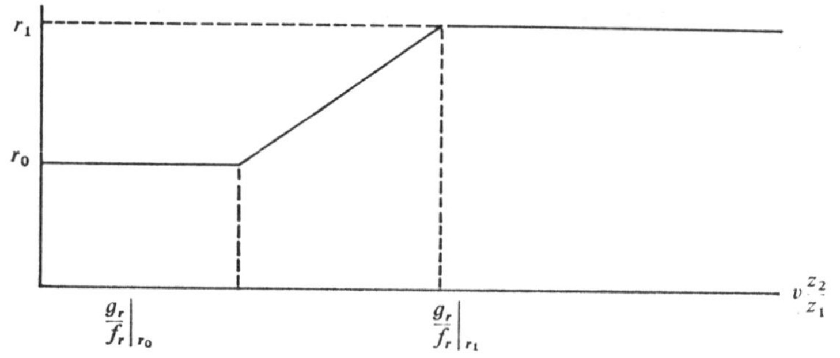

FIG. 4

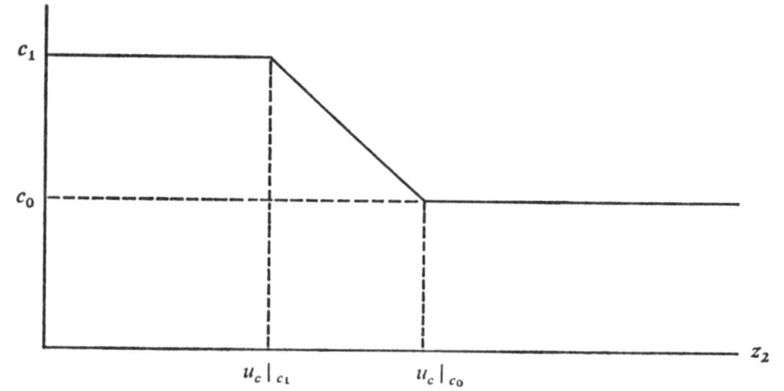

FIG. 5

That is, the shadow price of waste density changes according to the marginal disutility of waste density, while the shadow price of investment (which equals or does not exceed the marginal utility of consumption) changes according to the difference between the marginal valuation of output due to an additional unit of capital stock and the marginal valuation of waste density due to an additional unit of capital stock.

If $a = n = 0$, z_1 will fall. z_2 will fall and therefore consumption *per capita* will rise through time if:

$$z_1\left(\frac{g_s\alpha}{v}+\delta\right) < z_2(f_s\alpha-\delta). \qquad \ldots(37)$$

The left-hand side can be interpreted as the shadow price of net waste reduction due to an additional unit of recycling investment while the right-hand side is a shadow price of net output resulting from an additional unit of recycling investment. By algebraic manipulation of (37), (35.2), and (35.8), assuming an interior solution and $f_s\alpha > \delta$, one can deduce that the inequality (37) is more likely the smaller the rate of depreciation on recycling investment δ, and the smaller the magnitude of v, total waste holding capacity.

If z_2 falls faster than z_1, $v\frac{z_2}{z_1}$ falls with the final result that consumption *per capita*, c, increases while resource extraction *per capita*, r, decreases over time. This can be verified by a quick glance at Figures 4 and 5. Also, if z_2 falls more slowly than z_1, $v\frac{z_2}{z_1}$ rises such that both c and r increase over time. Finally, if z_2 rises, i.e., the inequality in (37) is reversed, then r rises and c falls. Which case is the more likely? We have already seen that the smaller v and δ, the more likely it is that z_2 will decrease over time. We can also assert that in the case of an interior optimum, the greater the *per capita* marginal disutility of rising waste density, the more rapidly z_1 will fall and the greater the marginal utility of consumption *per capita*, the faster z_2 will fall if the inequality (37) holds. Therefore, the more distasteful a degraded environment is compared with higher current consumption, and the smaller the natural environment the more likely that z_1 will fall faster than z_2, and both consumption *per capita* and resource extraction *per capita* will rise through time.

Throughout the brief discussion of consumption behaviour defining the optimum in a closed resource system, we have omitted comparisons with what might result in a non-regulated economy. Yet it is obvious that if the closed medium analogy contains some degree of validity, and there are disutilities associated with increasing waste densities and disturbance of the natural environment, a hypothetical economy characterized by individual, unconstrained decisions will be inefficient in its utilization of the environment. Not only will the closed medium tend to be used up too rapidly, but the economy's uncoordinated choices or piece-meal planning with regard to both resource extraction and pollution rates will generally result in an inefficient consumption path toward decimation, or the " anti-bliss " point.

To end on a rather facetious remark, it is interesting to contemplate whether the global economy is, in fact, near the optimal consumption path given the increasingly degraded environment, rapid exploitation of energy reserves, and long-run rising *per capita* consumption level.

REFERENCES

[1] Strotz, R. H. " Myopia and Inconsistency in Dynamic Utility Maximization ", *Review of Economic Studies*, **23** (1957), 165-180.

[2] Scitovsky, T. " External Diseconomies in the Modern Economy ", *Western Economic Journal*, **4**, No. 3 (Summer 1966).

[3] Ayres, R. U. and Kneese, A. V. " Production, Consumption and Externalities ", *American Economic Review*, **19** (June 1969).

[4] Barnett, H. J. and Morse, C. *Scarcity and Growth . . .*, Resources for the Future, Inc. (Baltimore: Johns Hopkins Press, 1963), 10.

[5] Kneese, A. V., Ayres, R. U. and d'Arge, R. C. *Economics and the Environment: A Materials Balance Approach*, Resources for the Future, Inc., (Baltimore: Johns Hopkins Press, 1970), Ch. III.

[6] Boulding, K. " The Economics of the Coming Spaceship Earth ", in H. Jarrett, *Environmental Quality in a Growing Economy*, Resources for the Future, Inc., (Baltimore: Johns Hopkins Press, 1966).

[7] Cummings, R. G. and Burt, O. R. " The Economics of Production from Natural Resources: Note ", *American Economic Review*, **59** (December 1969).

[8] Dorfman, R. " An Economic Interpretation of Optimal Control Theory ", *American Economic Review*, **59** (December 1969).

[9] Brown, G. " An Optimal Program for Managing Common Property Resources with Congestion Externalities ", unpublished manuscript (1969).

[10] Smith, V. L. " The Economics of Production from Natural Resources ", *American Economic Review*, **58** (June 1968).

[11] Lotka, A. J. *Elements of Mathematical Biology* (New York: Dover Publications, 1956).

[12] Dasgupta, P. S. " On the Concept of Optimum Population ", *Review of Economic Studies*, **36** (July 1969).

[13] Smith, V. L. " On Models of Commercial Fishing ", *Journal of Political Economy*, **77** (March-April 1969).

[14] Beverton, R. J. and Holt, S. V. *On the Dynamics of Exploited Fish Populations*, **19** (London: Ministry of Agriculture . . ., 1957).

[15] Daly, H. " Economics as a Life Science ", *Journal of Political Economy*, **76**, No. 3 (May-June 1968).

[16] Pielou, E. C. *An Introduction to Mathematical Ecology* (New York: John Wiley, 1969).

[17] d'Arge, R. C. " Essay on Economic Growth and Environmental Quality ", *Swedish Journal of Economics* (March 1971).

[18] Keeler, E., Spence, M. and Zeckhauser, R. " The Optimal Control of Pollution ", *Journal of Economic Theory*, **4**, No. 1 (February 1972).

401-07 [74]

1110

040

[20]

The Economics of the Steady State

By HERMAN E. DALY*

> But if your theory is found to be against the second law of thermodynamics, I can give you no hope; there is nothing for it but to collapse in deepest humiliation.
>
> *Sir Arthur Eddington*

My title is somewhat pretentious since at present this "new economics" consists only of a definition of a steady-state economy, some arguments for its necessity and desirability, and some disciplined speculations on its appropriate institutions and the problem of transition, each of which will be briefly discussed below.

I. What is a Steady-State Economy?

A steady-state economy is defined by constant stocks of physical wealth (artifacts) and a constant population, each maintained at some chosen, desirable level by a low rate of throughput—i.e., by low birth rates equal to low death rates and by low physical production rates equal to low physical depreciation rates, so that longevity of people and durability of physical stocks are high. The throughput flow, viewed as the cost of maintaining the stocks, begins with the extraction (depletion) of low entropy resources at the input end, and terminates with an equal quantity of high entropy waste (pollution) at the output end. The throughput is the inevitable cost of maintaining the stocks of people and artifacts and should be minimized subject to the maintenance of a chosen level of stocks (Kenneth E. Boulding).

The services (want satisfaction) yielded by the stocks of artifacts (and people) are the ultimate benefit of economic activity, and the throughput is the ultimate cost. The stock of physical wealth is an accumulated flow of throughput, and thus in the final analysis is a cost. Ultimate efficiency is the ratio of service to throughput. But to yield a service, the throughput flow must be first accumulated into stocks even if of short duration. It is the existence of a table or a doctor at a point in time that yields services, not their gradual depreciation nor the productive process by which they are replaced. Stocks are intermediate magnitudes that yield services and require throughput for maintenance and replacement. This may be expressed in the equation:

$$(1)\quad \frac{\text{Ultimate}}{\text{Efficiency}} = \overset{(1)}{\frac{\text{Service}}{\text{Throughput}}} = \overset{(2)}{\frac{\text{Service}}{\text{Stock}}} \times \overset{(3)}{\frac{\text{Stock}}{\text{Throughput}}}$$

Since by definition stocks are constant at a level corresponding to some concept of sufficiency or maturity, progress in the steady state consists in increasing ultimate efficiency (ratio 1) in two ways: by maintaining the stock with less throughput (increase ratio 3 or "maintenance efficiency") and by getting more service per unit of time from the same stock (increase ratio 2 or "service efficiency"). The laws of thermodynamics provide a theoretical limit to the improvement of maintenance efficiency. Whether there is any theoret-

* Professor of economics, Louisiana State University.

ical limit to increase in service efficiency resulting from the limits of the human stomach and nervous system is less clear, but in my opinion likely.

Over short periods of time the throughput cost of maintaining the constant stock may decrease due to improvements in maintenance efficiency, but over the long run it must increase because as better grade (lower entropy) sources of raw materials are used up, it will be necessary to process ever larger amounts of materials using ever more energy and capital equipment to get the same quantity of needed mineral. Thus a steady-state economy, as here defined, does not imply constant throughput, much less static technology, nor does it imply eternal life for the economic system. It is simply a strategy for good stewardship, for maintaining our spaceship and permitting it to die of old age rather than from the cancer of growthmania. It is basically an extension of the demographers' model of a stationary population to include the populations of physical artifacts, and the fundamental idea is found in John Stuart Mill's discussion of the stationary state of classical economics.

The term "economic growth" conventionally refers to an increase in the flow of "real *GNP*," which is a value *index* of the *physical* flow of throughput. The (measurable) throughput is, in turn, an index of (unmeasurable) service only if ratios 2 and 3 (or their product) are constant or increasing. This may have been the case in the past, but for the future it is doubtful. As the growing throughput pushes against biophysical limits, it provokes a decline in service efficiency (more of the stock must be devoted to the defensive use of repairing life support systems that formerly provided free services). Also, since our institutions are geared to a continually increasing throughput, we may willingly lower maintenance efficiency for the sake of permitting a larger throughput (e.g. planned obsolescence andfashion). If someone wants to redefine "economic growth" as an increase in nonmaterial services and then argue that it can and should grow forever, he is free to do so. But this hardly constitutes a refutation of the steady-state economy, which is defined in terms of measurable physical stocks, not unmeasurable psychic fluxes.

Nor are the levels at which the stocks of people and artifacts are maintained necessarily frozen for all eternity. As a result of technical and moral evolution it may become both possible and desirable to grow or to decline to a different level. But then growth or decline would be seen as a temporary transition from one steady state to another and not as the norm for a healthy economy. Technical and moral change would lead growth rather than being blindly pushed down the path of least resistance by the growth juggernaut.

At what point should growth in stocks and maximization of production flow give way to stock maintenance and the minimization of the production flow? There are a large number of steady-state levels of stocks to choose from, and such a choice is a difficult problem of ecology and ethics. But our inability to define the optimum level does not mean that we will not someday discover that we have grown beyond it. It is more important to learn to be stable at existing or nearby levels than to know in advance which level is optimal. Knowledge of the latter without the former merely allows us to recognize and wave goodbye to the optimum as we grow through it. Besides, the optimum may well be a broad plateau within which one place is as good as another as long as we don't go too near the edge.

The radical change implied by a steady state is evident from the foregoing, and from W. W. Rostow's characterization of our present economy of high mass consumption, to which all countries unrealistically aspire, as one "in which compound interest becomes built, as it were, into our habits and institutions" (p. 7). This built-in exponential growth and its unfortunate consequences constitute the theme of the much-maligned little book, *Limits to Growth*, by D. H. Meadows et al. Before discussing the radical departure of "deinstitutionalizing" compound interest or at least uncoupling it from all physical dimensions, we must consider whether such a change is really necessary and/or desirable.

II. The Necessity and Desirability of the Steady State

Our economy is a subsystem of the earth, and the earth is apparently a steady-state open system. The subsystem cannot grow beyond the frontiers of the total system and, if it is not to disrupt the functioning of the latter, must at some much earlier point conform to the steady-state mode. The technocratic project of redesigning the world (substituting technosphere for ecosphere) so as to allow for indefinite economic growth is a bit of hubris that has received the insufficiently pejorative label of "growthmania."

The conceptual roots of growthmania are to be found in the orthodox doctrines of "relative scarcity" and "absolute wants." Relative (or "Ricardian") scarcity refers to the scarcity of a particular resource relative to another resource or to a lower quality of the same resource. Absolute (or "Malthusian") scarcity refers to the scarcity of all resources in general, relative to population and per capita consumption levels. The solution to relative scarcity is substitution. Absolute scarcity assumes that all economical substitutions are made so that the total burden of absolute scarcity is minimized but still exists and may still increase. Even an efficiently borne burden can become too heavy. Substitution is always of one form of low entropy matter-energy for another. There is no substitute for low entropy itself, and low entropy is scarce, both in its terrestrial source (finite stocks of concentrated fossil fuels and minerals) and in its solar source (a fixed rate of inflow of solar energy). (See Georgescu-Roegen.) Both the human economy and the nonhuman part of the biosphere depend on the same limited budget of low entropy and on the allocative pattern which that budget has evolved over millennia. The entropy of the human sector is reduced and kept low by the continual importation of low entropy from, and exportation of high entropy to, the nonhuman sector (Daly, 1968). If too much low entropy is diverted to economic growth in the human sector, or if too many evolutionary allocative patterns are disrupted in the process of diversion, then the complex life support systems of the biosphere will begin to fail. Growth in population and per capita consumption result in increasing absolute scarcity, which is manifested in the increasing prevalence of "external costs"—i.e., the system becomes more generally sensitive to particular interferences as the web of general interdependence is stretched ever tighter by growth in the populations of people and artifacts and the resulting stress on the entropy budget.

Orthodox economic theory has assumed that all scarcity is relative: "Nature imposes particular scarcities, not an inescapable general scarcity" (Harold J. Barnett and Chandler Morse, p. 11). Therefore the answer to scarcity is always substitution, and since relative price changes induce substitution, the policy recommendation is "internalization of externalities," usually via pollution taxes. The following statement is representative of orthodox complacency: " . . . the problem of environmental pollution is a simple matter of correcting a minor resource misallocation by means of pollution charges . . . " (Wilfred Beckerman, p. 327). But price rigging by itself is ineffective in coping with increasing absolute scarcity since its mode of operation is only to induce substitution. What substitute is there for resources in general, for low entropy? How is it possible to raise the *relative* price of *all* resources? Attempts to do so result in inflation rather than substitution.

A similar distinction between absolute and relative wants has also been obscured by orthodox economics. Following Keynes we may define absolute wants as those we feel independent of the situation of our fellow human beings. Relative wants are those that we feel only if their satisfaction makes us feel superior to our fellows. The importance of this distinction is that only relative wants are infinite and that relative wants cannot be universally satisfied by growth because the relative satisfactions of the elite are cancelled as growth raises the general level. This effect can be avoided, and often is, by allowing growth to increase inequality so that the relatively well off become relatively better off. But it is quite impossible for everyone to become better off relative to everyone else. In spite of this extremely important distinction, orthodox theory assumes that wants *in general* are insatiable and extends to all wants the dignity of absolute status—i.e., the satisfaction of relative and absolute wants is considered equally legitimate and equally capable of satisfaction in the aggregate by means of economic growth. The assumption of equal legitimacy is a value judgment (though it is treated by many economists as the avoidance of a value judgment), and the assumption of equal capability of satisfaction is either a logical error or an implicit acceptance of a value judgment in favor of increasing inequality.

The implication of the dogmas of the relativity of all scarcity and the absoluteness of all wants is growthmania. If there is no absolute scarcity to limit the *possibility* of growth (infinite substitutability of relatively abundant for relatively scarce resources) and no merely relative wants to limit the *desirability* or efficacy of growth (wants in general are infinite and all wants are equally worthy and capable of satisfaction by growth), then "growth forever and the more the better" is the logical consequence. It is also the *reductio ad absurdum* that exposes the growth orthodoxy as a rigorous exercise in wishful thinking, as a theory that is against the second law of thermodynamics as well as against common sense. It is simply a brute fact that there is such a thing as absolute scarcity and such a thing as relative wants. Furthermore, these latter categories eventually become dominant at the margin as growth continues. The implication of absolute scarcity and relative wants is the opposite of growthmania, namely, the steady state.

At this point the growthmaniacs usually make a burnt offering to the god of technology: surely economic growth can continue indefinitely because technology will continue to "grow exponentially" as it has in the past. This elaborately misses the point. The alleged "exponential growth" of technology is not directly measurable and is only inferred from the permissive role that it has played in making possible the measured exponential growth in the physical magnitudes of production, depletion and pollution (i.e., the throughput). Such technical progress is more a part of the problem than the solution. What must be appealed to is a *qualitative change* in the direction of technical progress, not a continuation of alleged quantitative trends. The institutions to be discussed in the next section seek to induce just such a change toward resource-saving technology and patterns of living, and to a greater reliance on solar energy and renewable resources. But we can be fairly certain that no new technology will abolish absolute scarcity because the laws of thermodynamics apply to all possible technologies. No one can be absolutely certain that we will not some day discover perpetual motion and how to create and destroy matter and energy. But the reasonable assumption for economists is that this is an unlikely prospect and that while technology will continue to pull rabbits out of hats, it will not pull an elephant out of a hat—much less an infinite series of ever-larger elephants!

But the ideology of growth continues to transcend the ordinary logic of elementary economics. Growth is the basis of national power and prestige. Growth offers the prospect of prosperity for all with sacrifice by none. It is a substitute for redistribution. The present sins of poverty and injustice will be washed away in a future sea of abundance, vouchsafed by the amazing grace of compound interest. This evasion, common to both capitalism and communism, was never totally honest. It is now increasingly exposed as absurd.

III. Speculations on the Steady State

The first design principle disciplining our speculations on institutions is to provide the necessary social control with a minimum sacrifice of personal freedom, to provide macro stability while allowing for micro variability, to combine the macro static with the micro dynamic. A second design principle, closely related to the first, is to maintain considerable slack between the actual environmental load and the maximum carrying capacity. The closer the actual approaches the maximum, the more rigorous, finely tuned, and micro oriented our controls will have to be. We lack the knowledge and ability to assume detailed central control of the spaceship, even if such were desirable, so therefore we should leave it on "automatic pilot" as it has been for eons. But the automatic pilot only works if the actual load is small relative to the maximum. A third design principle, important for making the transition, is to start from existing initial conditions rather than an imaginary "clean slate," and a fourth is to build in the ability to tighten constraints gradually. Minimum faith is placed in our ability to plan a detailed blueprint for a new society. Maximum faith is placed in the basic regenerative powers of life and in the possibility of moral growth, once the root physical process of degeneration (unlimited growth) is arrested.

The kinds of institutions required follow directly from the definition. We need (1) an institution for stabilizing population, (2) an institution for stabilizing physical wealth and keeping throughput below ecological limits, and, less obviously but most importantly, (3) an institution limiting the degree of inequality in the distribution of the constant stocks among the constant population since growth can no longer be appealed to as the answer to poverty.

What specific institutions can perform these functions and are most in harmony with the general design principles discussed above? Elsewhere I have outlined a model and can here only briefly describe it (Daly, 1973, 1974a). The model builds on the existing institutions of private property and the price system and is thus fundamentally conservative. But it extends these institutions to areas previously not included: control of aggregate births (marketable birth license plan as first proposed by Boulding) and control of aggregate depletion of basic resources (depletion quotas auctioned by the government). Extending the market, under the discipline of aggregate quotas, to these vital areas is necessary to deal with increasing absolute scarcity since, as argued above, price controls deal only with relative scarcity. Quantitative limits are set with reference to ecological and ethical criteria, and the price system is then allowed, by auction and exchange, to allocate depletion quotas and birth quotas efficiently. The throughput is controlled at its input (depletion) rather than at the pollution end because physical control is easier at the point of lower entropy. Orthodox economics suggests price controls at the output end (pollution taxes), while steady-state economics suggests quantitative controls at the input end (depletion quotas).

With more vital areas of life officially subject to the discipline of the price system, it will become more urgent to establish the institutional preconditions of free and mutually beneficial exchange, namely, to limit the degree of inequality in the distribution of income and wealth and to limit the monopoly power of corporations. A distributist institution establishing a minimum income and a maximum income and wealth would go a long way toward achieving that end, while leaving room for differential reward and incentives within reasonable limits. There might be one set of limits for individuals, one for families, and one for corporations. Natural monopolies should be publicly owned and operated.

Birth quotas, depletion quotas, and distributive limits can all be varied continuously and applied with any degree of gradualism desired. Moreover, all three control points are price system parameters, and altering them does not interfere with the static allocative efficiency of the market. Externalities involving ecological, demographic, and distributive issues are "externalized" by means of quotas rather than "internalized" in rigged market prices. Yet the effect is much the same in that prices rise to reflect previously unaccounted dimensions of scarcity, and prices become a safer guide to market decisions. The net advantage of the quota scheme is that it limits aggregate through-put, whereas price controls merely alter through-put composition, providing a useful fine-tuning supplement to quotas, but not a substitute. The higher resource prices resulting from limited depletion would have the dynamic effect of inducing resource-saving technology and a shift to greater dependence on solar energy and renewable resources. The receipts of the depletion quota auction could help finance the minimum income. The marketable birth license plan would also have an equalizing effect on per capita income distribution.

Such institutional change is obviously not on the political agenda for 1974. Nor should it be since it is speculative, has not had the benefit of widespread professional criticism, and thus may contain terrible mistakes. But mistakes will not be discovered and better ideas will not be offered unless economists awake from the dogmatic slumber of growthmania induced by the soporific doctrines of relative scarcity and absolute wants and put the steady-state paradigm on the agenda for academic debate.

REFERENCES

H. J. Barnett and C. Morse, *Scarcity and Growth,* Baltimore 1963.

W. Beckerman, "Economists, Scientists, and Environmental Catastrophe," *Oxford Econ. Papers,* Nov. 1972, *24,* 327.

F. H. Bormann and W. R. Burch, eds., *Growth, Limits and the Quality of Life,* San Francisco 1974 (forthcoming).

K. E. Boulding, *Economics as a Science,* New York 1970.

J. Culbertson, *Economic Development: An Ecological Approach,* New York 1971.

H. E. Daly, "On Economics as a Life Science," *J. Polit. Econ.,* May/June 1968, *76,* 392–406.

———, "In Defense of a Steady-State Economy," *Amer. J. Agri. Econ.,* Dec. 1972, *54,* 945–954.

———, ed., *Toward a Steady-State Economy,* San Francisco 1973.

———, "Long Run Environmental Constraints and Trade-Offs Between Human and Artifact Populations," *Internat. Populat. Conf.*, IUSSP, Liege 1973, *3,* 453–460.

———, "A Model for a Steady-State Economy" in Bormann and Burch, 1974a.

———, "Steady-State Economics Versus Growthmania: A Critique of the Orthodox Conceptions of Growth, Wants, Scarcity, and Efficiency," *Policy Scien.* (forthcoming) Summer 1974b.

Editors of *The Ecologist,* "A Blueprint for Survival," *The Ecolo.*, Jan. 1972.

N. Georgescu-Roegen, *The Entropy Law and the Economic Process,* Cambridge, Mass. 1971.

D. H. Meadows et al., *The Limits to Growth,* New York 1972.

D. L. Meadows and D. H. Meadows, eds., *Toward Global Equilibrium: Collected Papers,* Cambridge, Mass., 1973.

W. Ophuls, *Prologue to a Political Theory of the Steady State,* (unpublished) Ph.D. dissertation in political science, Yale University, 1973.

W. W. Rostow, *The Stages of Economic Growth,* New York 1960.

"The No-Growth Society," *Daedalus,* American Academy of Arts and Sciences Proceedings, *102,* Fall 1973.

[21]

The Optimal Depletion of Exhaustible Resources[1,2]

PARTHA DASGUPTA
LSE and Stanford University

and

GEOFFREY HEAL
University of Sussex

0. INTRODUCTION

0.1. In any given economy there must be a number of commodities that enter into production and which are used up, and whose available stock cannot be increased. Examples like fossil fuels come readily to mind. It would seem reasonable to argue that in the long run the limited availability of these commodities, together with their technological importance, would begin to act as a constraint on the economy's growth potential. In fact several recent studies have laid great emphasis on this possibility.[3] Although the point is an obvious one, most economic studies of the properties of long-run plans neglect it.[4] In this paper, therefore, we explore in a rather preliminary way the problems that appear to arise naturally when the existence of exhaustible resources is incorporated into the study of intertemporal plans.

It appears that questions in this area are hard to answer. For even in the simplest of environments where it is supposed that there is perfect foresight, one is concerned not merely with the optimal depletion of exhaustible resources but with the optimal rate of investment as well. The two must plainly be interrelated. The latter problem on its own is hard, and the combined problem is very complex. In fact it will appear in the course of the arguments that follow that intuition is not a very good guide for this joint problem. In any case, the assumption of perfect foresight is particularly dubious here since it would appear almost immediate that an investigation of intertemporal plans in the presence of exhaustible resources readily invites consideration of the possibilities of large-scale alterations in technology at dates in the future that are inherently uncertain. Moreover, it is clear that such extensive technological changes would not be achievable costlessly. In this paper, therefore, we attempt to demonstrate how one might, in a relatively simple manner, bring such considerations as these to bear on a set of questions that have generated a considerable amount of interest in recent years.

0.2. It is plain of course that the mere *existence* of a resource that is exhaustible is not a

1 *First version received April* 1973; *final version accepted February* 1974 (*Eds*.).

2 Among the many to whom acknowledgement is due for their comments we would like to mention Christopher Bliss, Steve Glaister, Terence Gorman, Frank Hahn, Tjalling Koopmans, Bill Nordhaus, Robert Solow, Joseph Stiglitz, Niel Vousden and, in particular, Harl Ryder.

3 We have in mind the works of Forrester [8] and Meadows *et al.* [20]. But see also Cole *et al.* [4] for a penetrating commentary on the Forrester analysis; as well as the lively comments of Beckerman [3].

4 Thus the extensive literature on optimal planning is clearly concerned with horizons long enough for such constraints to become effective, and yet it rarely mentions them. Exceptions are Anderson [1], Ingham and Simmonds [15], Vousden [27], and the seminal and rather neglected work of Hotelling [14]. Professors Koopmans [18], Solow [25], and Stiglitz [26] have recently, and independently of us, explored different aspects of the question under review here. Their findings are complementary to ours and, in what follows, we shall have occasion to refer to the tasks that they have undertaken in their contributions. For a detailed survey of recent modelling experiences in the area of exhaustible resources, see Dasgupta [7].

sufficient basis for apocalyptic visions. Even if we assumed away the possibility of technological change, exhaustible resources would pose a " problem " only if they are, in some sense, essential in production. Intuition suggests that one regards a resource as being *essential* if output of final consumption goods is nil in the absence of the resource. Otherwise, call the resource inessential.[1] It is then natural to ask whether feasible output must eventually decline to zero in an economy that possesses an essential exhaustible resource. Rather surprisingly, perhaps, the answer turns out to be " no ", and this even if there is no technological progress.[2] The point, of course, is that the possibilities open to such an economy depend crucially on the ease with which reproducible inputs can be substituted for the exhaustible resources. It follows, then, that the effect that the existence of an exhaustible resource will have on the characteristics of an *optimal* plan will depend on the extent of such substitution possibilities. In particular we demonstrate below that the *elasticity of substitution* between reproducible inputs and exhaustible resources plays an important and a rather direct role in the properties of an optimal plan. A related question that arises is whether one should, as intuition suggests, deplete an *essential* exhaustible resource slowly over the planning horizon, or whether one should exhaust it in finite time, closing down and living off the existing capital stock. In the same way one wants to know whether it is optimal to deplete an *inessential* exhaustible resource in finite time and, subsequently, rely solely on reproducible capital to continue production. In Section 1 we investigate this range of issues under the assumption that future technology is known with certainty to be the same as that available at present. As one would expect, the analysis will give an indication of the structure of relative prices that would sustain an optimal policy.

While intuition suggests that the elasticity of substitution will be an important parameter, it also suggests that it would be inappropriate to regard the technology of the economy as unchanging over time. Even if under present technological knowledge an exhaustible resource is judged essential it would be unwise to rule out the possibility of substitutes being discovered in the future. In one sense this is at the core of the controversy between Forrester and his critics. In their critiques of the results published by Forrester [8], both Maddox [19] and Nordhaus [21] argue that as an essential exhaustible resource is depleted its market price will rise, forcing entrepreneurs to search for cheaper substitutes. This argument clearly has force, though in the absence of a well-articulated intertemporal plan or a satisfactory set of forward markets it is not at all plain that market prices will be providing the correct signals.[3] Moreover, there are some additional problems that arise in this context, for many of the difficulties that are involved in making policy recommendations about the rate of depletion of exhaustible resources stem from the fact that crucial aspects of this problem are inherently uncertain, and it is not clear that an adequate class of contingent markets exists. There is, of course, some uncertainty about the amounts of resources actually available at any given date, but one would be inclined to feel that this is not the most interesting source of uncertainty. Rather, it would seem plausible that the really important source of uncertainty is connected with future technology. There is, for example, a chance that the discovery of substitutes will render previously essential resources inessential. This has happened a number of times in the past, and one would expect that the best policy towards resource depletion would depend on the probability of such an occurrence. In Section 2 we attempt to analyse this aspect of the problem. We assume in particular that technical change is costlessly attained and that we know in advance the form of the technical change (i.e. the nature of future technology and resource availability) but that the date of its appearance is random.

1 Later we shall note that this method of partitioning resources that are exhaustible is not as helpful as might appear at first sight.

2 See Solow [25] for a demonstration of this.

3 Stiglitz [26] and Dasgupta [7] have investigated the properties of a competitive system where futures markets do not exist and where expectations are not necessarily fulfilled. As would be suspected, it is very easy in such an economy to have a systematic over or under utilization of the resource: for an analysis of the manner in which they have actually allocated resources over time see Heal [10, 12].

In point of fact, of course, substitutes are not obtained at zero cost. One would like (finally) to analyse a problem where resources are deployed in the search for substitutes and where, at the same time, there is no guarantee that the search will be successful. In such a situation the problem is not merely one of obtaining the optimal depletion rate for the exhaustible resources; it also involves finding the correct allocation of the resources used between the production of goods and expenditure on research. In a subsequent paper we shall analyse this aspect of the problem.

The problems that we are considering in this paper arise from the implications that a finite earth has for intertemporal planning. In this context the treatment of population poses considerable difficulties. It would clearly be unreasonable to consider an exogenously given rate of population increase, because the very factors that we are attempting to allow for must eventually force a voluntary or involuntary reduction in rates of population growth on society. There is thus, in principle, a clear need for an optimal population policy, together with an optimum consumption and depletion policy. Professor Koopmans [18] has recently investigated a special form of this problem which consists in assuming a *constant* and *given* population size, with the interesting twist that the planning horizon is regarded as choice variable. The Koopmans problem in effect consists of locating an optimal depletion policy and the optimal survival period for an economy that contains a given and fixed population. It looks as though the population problem in its general setting is particularly complex and we have, so far, been able to obtain some preliminary results only.[1] In this paper, therefore, we assume a constant level of population through time. We are, then, analysing a world in which the objective of zero population growth has been acccepted and achieved. Such a device poses its own interpretive problems, since one has to assume away the existence of a level of consumption below which life cannot be sustained. We can, of course, by suitable assumptions, avoid *zero* consumption along the optimum programme. But in effect we shall be supposing that the subsistence level of consumption is nil.

0.3. In order to clarify ideas it will prove convenient to review briefly the simplest of the models that attempt to capture the presence of exhaustible resources. What we review here is a slight generalization of the well-known " cake-eating " problem analysed first by Hotelling [14] and later by Gale [9]. It would seem plausible that the " cake-eating " problem would be a basic building block of any production model that is to catch the issues that we are concerned with in this paper. In subsequent sections of this paper we shall establish the precise sense in which this claim is true.

But for the moment assume that there is no production. The economy possesses, to begin with, a finite stock (inventory), S_0, of a homogeneous consumption good. By an absence of production in this economy we mean that the social rate of return to investment is zero. Our simple generalization consists of the supposition that a substitute for the resource is already available, in the sense that there is a steady flow of the consumption good that is fed into the economy at the rate M. Denote by C_t the rate of consumption at time t, and by $U(C_t)$ the instantaneous utility of consuming C_t. We assume, as is usual, that $U(\cdot)$ is monotonically increasing, strictly concave, and twice differentiable everywhere. We also suppose, as seems natural, that

$$\operatorname*{lt}_{C\to 0} U'(C) = +\infty.^{2} \qquad \ldots(0.1)$$

Write $\eta(C) = -CU''(C)/U'(C)$, for the *elasticity of marginal utility*. We take it that

$$+\infty > \operatorname*{lt}_{C\to 0} \eta(C) = \eta > 0. \qquad \ldots(0.2)$$

[1] For an analysis of optimum population policies in the presence of a fixed factor (land), and external diseconomies through large population sizes, see Dasgupta [5].

[2] We follow the by now standard notation: $\dot{x}_t = dx_t/dt$ and $f'(x) = df/dx$. In what follows we shall often drop the time subscript. This should not cause any confusion.

Social welfare over the interval $[0, T]$ will be taken to be of a utilitarian form, yielding an amount

$$\int_0^T e^{-\delta t}U(C_t)dt, \qquad \text{...(0.3)}$$

where δ is strictly positive.

One is therefore concerned with obtaining that consumption profile, C_t, which will maximize

$$\left.\begin{aligned} &\int_0^\infty e^{-\delta t}U(C_t)dt \text{ subject to} \\ &\dot{S}_t = M - C_t, \\ &\quad\text{where } C_t,\ S_t \geqq 0,\ M \geqq 0, \text{ and where } S_0 \text{ is given} \end{aligned}\right\}. \qquad \text{...(0.4)}$$

In exploring the problem as presented in (0.4) the natural thing to do is to introduce multipliers for the various constraints and write down the Hamiltonian of the system as

$$\mu = e^{-\delta t}U(C_t) + e^{-\delta t}p_t(M - C_t) + e^{-\delta t}q_tS_t \qquad \text{...(0.5)}$$

where

$$q_t \geqq 0 \text{ and } q_tS_t = 0. \qquad \text{...(0.6)}$$

We have not explicitly introduced the multiplier associated with the non-negativity of C_t simply because we know in advance that given (0·1) the multiplier will throughout be zero. It is then immediate that for a programme to be optimal it is necessary that

$$p_t = U'(C_t) \qquad \text{...(0.7)}$$

and also that p_t, the spot price of consumption, should satisfy the differential equation

$$\dot{p}_t = -q_t + \delta p_t. \qquad \text{...(0.8)}$$

Using (0.7) in (0.8) one obtains

$$\frac{\dot{C}}{C} = -\frac{\delta}{\eta(C)} + \frac{q_t}{\eta(C)U'(C)}. \qquad \text{...(0.9)}$$

From (0.6) and (0.9) it is clear that there are two (possibly repeated) phases, namely: *Phase A*. During which $S_t > 0$ and hence

$$\frac{\dot{C}}{C} = -\frac{\delta}{\eta(C)} < 0 \qquad \text{...(0.10)}$$

and *Phase B*. During which equation (0.9) holds with $q_t > 0$, so that $S_t = 0$.

The solution to problem (0.4) for the case $M = 0$ is well known (see e.g. Heal [11]). Consequently we sketch the argument for the case $M > 0$. One begins by noting from equation (0.9) that phase A cannot continue for all $t \geqq 0$ along an optimal policy. For then by (0.2) and (0.10), given any $\varepsilon > 0$ and C_0 there will exist a T such that $C_t \leqq \varepsilon$ for $t \geqq T$, and this would imply that $S_t \to \infty$, which is plainly inefficient. From (0.4) it is clearly feasible to have $C_t \geqq M$ for $t \geqq 0$. This suggests a policy of following phase A for an initial period $[0, T]$ followed by phase B for $t \geqq T$. During phase A equation (0.10) holds. Both C_0 and T are determined by the requirements

$$\int_0^T C_tdt = S_0 + MT \text{ and } \lim_{t \to T^-} C_t = C_T = M.$$

For $t \geqq T$ (i.e. during phase B) we have $C_t = M$. Thus one sets $q_t = 0$ for $0 \leqq t < T$ and

$$q_t = \frac{\delta}{U'(M)} \text{ for } t \geqq T.$$

The proposal satisfies the necessary conditions for an optimal policy. The Hamiltonian (0.5) is concave. It is also the case that along the proposed policy

$$\lim_{t \to \infty} e^{-\delta t} p_t S_t = 0.$$

It follows that we have

Proposition 1. *An optimal policy for the problem* (0.4) *exists and it is uniquely given.*

Proposition 2. *If $M > 0$ the optimal policy consists of precisely two phases. It consists initially of phase A until a time T at which the stock, S_t, is nil. From T onwards the policy consists of phase B during which $C_t = M$.* (Figure 1 illustrates the policy described in Proposition 2.)

The following proposition is well known.

Proposition 3. *If $M = 0$, the optimal policy consists of only one phase, namely phase A. The initial consumption rate, C_0, is so chosen that*

$$\lim_{t \to \infty} S_t = 0. \quad \text{(See Figure 1.)}$$

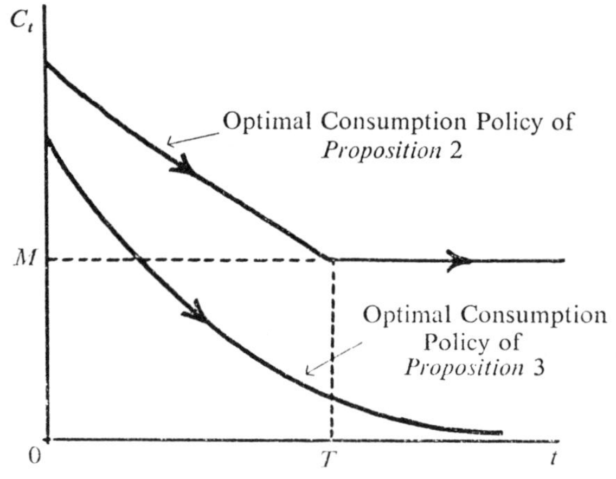

FIGURE 1

It is an implication of both Propositions 2 and 3 that the consumption rate is a non-increasing function of time. Furthermore, Proposition 2 might suggest that if a resource is inessential the economy should not carry a positive inventory of it indefinitely. It turns out that neither of these implications necessarily carries over into a model with production.

1. PRODUCTION

1.1. The economy introduced in Section 0.3 was instrumental merely in fixing ideas and was in itself of limited interest, as there was no production. Consequently we introduce production now. But before introducing exhaustible resources into a production model in a fully-fledged way we present, in this sub-section, a quick review of a production model that incorporates one aspect of the economy discussed in the previous section—namely that a steady stream of a resource that enters in production is fed into the economy from outside. By this device we are, of course, assuming away the essence of the problem.

There are, nevertheless, two reasons why we shall wish to review such a model here. First, it will allow one to see the sharp differences in the properties of an optimal path in an economy that has no exhaustible resources, and one in which such a resource constraint bites in the long run. Second, we shall use the model presented in this sub-section in our discussion of the specific issues that arise when one considers technological change.

We suppose now that there is a single non-deteriorating consumption good which, in conjunction with the service provided by a perfectly durable commodity (e.g. an energy source), can reproduce itself. The quantity of the durable commodity providing service is given and cannot be augmented. It is assumed to provide service at the constant rate M. We can, if we like, assume that the flow of service from the durable good can be stored costlessly, but we shall wish to suppose here that at the start of the planning period there is no inventory of the service.

Denote by K_t the stock of the reproducible composite commodity and by Z_t the rate of utilization of the service at time t. Efficient output possibilities are represented by the production function $G(K, Z)$. It is supposed that G is increasing, twice differentiable, strictly concave and that

$$G(0, Z) \geqq 0 \qquad \text{...(1.1}a\text{)}$$

$$\operatorname*{lt}_{K\to\infty} \frac{\partial G}{\partial K}(K, M)<\delta \qquad \text{...(1.1}b\text{)}$$

$$\operatorname*{lt}_{K\to 0} \frac{\partial G}{\partial K}(K, M)>\delta. \qquad \text{...(1.1}c\text{)}$$

It follows that consumption possibilities can be predicted by the equation

$$\dot{K} = G(K, Z)-C. \qquad \text{...(1.2)}$$

Denote by V_t the inventory of the service of the durable commodity at t. It follows that the rate, Z_t, at which one utilizes the service can be predicted by the equation

$$\dot{V}_t = M-Z_t, \text{ where } V_t \geqq 0 \text{ all } t \geqq 0. \qquad \text{...(1.3)}$$

The problem we are interested in is one of obtaining time profiles of $\tilde{C}_t$ and $\tilde{Z}_t$ which will maximize

$$\left.\begin{array}{l}\displaystyle\int_0^\infty e^{-\delta t}U(C_t)dt \text{ subject to equations (1.2), (1.3) and the constraints} \\ C_t,\ Z_t,\ V_t,\ K_y \geqq 0, \text{ and where } K_0(>0) \text{ is given and } V_0 = 0.\end{array}\right\}. \qquad \text{...(1.4)}$$

The problem, though cast here in a rather unusual form, is a familiar one. Consequently we merely state the results and furnish no proofs. Write $g(K) = G(K, M)$ and define $\bar{K}$ as the solution of $g'(K) = \delta$. One has then

Proposition 4. *Problem* (1.4) *possesses a unique solution. Along the optimal policy* $Z_t = M$ *for all* $t \geqq 0$ (*i.e.* $V_t = 0$ *for all* $t \geqq 0$), *and the economy tends in the long run to the stationary-state consumption rate* $\bar{C} = g(\bar{K})$, *and therefore the capital stock level* $\bar{K}$. *If* $K_0<\bar{K}$, *then along the optimal path* $\dot{C}>0$ *and* $\dot{K}>0$. *If* $K_0>\bar{K}$ *then along the optimal path* $\dot{C}<0$ *and* $\dot{K}<0$.

(Figure 2 presents a typical consumption profile of Proposition 4 for the more plausible case of $K_0<\bar{K}$.)

Consider now a special case of the model just discussed. Assume that the production function G satisfies (1.1b)-(1.1c) but now strengthen (1.1a) to the form

$$G(0, Z)>0 \text{ for } Z>0. \qquad \ldots(1.1a')$$

Consider problem (1.4) for this economy, but with the difference that now $K_0 = 0$ as well. That is, the economy begins with no capital stock and no inventory of the service. For completeness we express this special case of Proposition 4 as

Proposition 4′. *If G satisfies* (1.1a') *in addition to its other properties and if $K_0 = V_0 = 0$, then Problem* (1.4) *possesses a unique solution. Along the optimal path $Z_t = M$ for all $t \geqq 0$,* (*i.e. $V_t = 0$ for all $t \geqq 0$*), *and the economy tends, in the long run, to the stationary state consumption rate $\bar{C} = g(\bar{K})$, and therefore, to the capital stock level $\bar{K}$. Along the optimal path $\dot{C}>0$ and $\dot{K}>0$.*

(Figure 2 presents a typical consumption profile of Proposition 4′.)

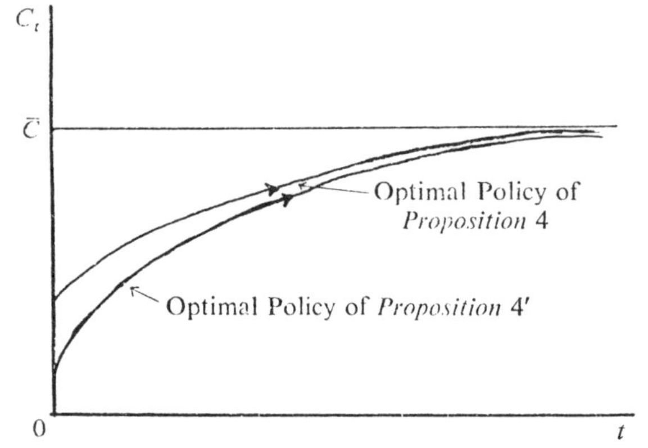

FIGURE 2

1.2. The economy discussed in the previous sub-section possessed no resource that was exhaustible. But it was suggestive, in that the technical possibilities in it were, albeit in a pristine form, the kind that one might be inclined to contemplate when reflecting on the circumstance that would prevail if, to take an example, some totally different source of energy (e.g. the sun) is harnessed. But presumably these are not the conditions that prevail as yet. Consequently we introduce exhaustible resources explicitly into production.

We suppose now that there is a non-deteriorating composite consumption good which, in conjunction with an exhaustible resource in production, can reproduce itself. We continue to denote by K_t the stock of the reproducible composite commodity at time t, but by R_t the flow of the exhaustible resource into production at t. Efficient output possibilities are, therefore, represented by a production function $F(K, R)$. It will be supposed that F is increasing, *strictly* concave, twice differentiable, and homogeneous of degree unity. Consumption possibilities are thus predicted by the condition

$$\dot{K}_t = F(K_t, R_t) - C_t. \qquad \ldots(1.5)$$

One assumes that the economy begins with a stock, S_0, of the exhaustible resource, and a stock, K_0, of the composite commodity. The planning problem then can be represented as being one of

$$\left.\begin{aligned}&\text{maximizing } \int_0^\infty e^{-\delta t}U(C_t)dt \text{ subject to equation (1.5) and the constraint [1]}\\&\int_0^\infty R_t dt \leqq S_0, \text{ where}\\&C_t,\ K_t,\ R_t \geqq 0 \text{ and } K_0(>0) \text{ is given}\end{aligned}\right\}. \quad \ldots(1.6)$$

As before, it would be pointless to introduce the multipliers associated with the non-negativity of C and K, since we have supposed (0.1). Consequently we express the Hamiltonian associated with problem (1.6) as

$$H = e^{-\delta t}U(C_t) + e^{-\delta t}p_t(F(K_t, R_t) - C_t) - \lambda R_t + e^{-\delta t}\mu_t R_t, \quad \ldots(1.7)$$

where

$$\mu_t \geqq 0 \text{ and } \mu_t R_t = 0 \quad \ldots(1.8)$$

and

$$\lambda \geqq 0 \text{ and } \lambda\left(S_0 - \int_0^\infty R_t dt\right) = 0. \quad \ldots(1.9)$$

One expects (and this will be confirmed) that $\lambda > 0$. It should be noted that λ is independent of time. Maximizing (1.7) with respect to C yields the condition

$$p_t = U'(C_t). \quad \ldots(1.10)$$

Likewise, maximizing (1.7) with respect to R_t yields the condition

$$\lambda = e^{-\delta t}\left(\mu_t + p_t \frac{\partial F}{\partial R_t}\right). \quad \ldots(1.11)$$

Write $F_R = \partial F/\partial R$, $F_K = \partial F/\partial K$, etc. Equations (1.8) and (1.11) imply that when $R_t > 0$ along an optimal policy

$$\lambda = e^{-\delta t}p_t F_{R_t}. \quad \ldots(1.12)$$

But when $R_t > 0$, λ denotes the present value shadow price of the exhaustible resource in utility numeraire. It follows that during an interval along an optimal path when $R_t > 0$ this present value shadow price remains constant. However, during this interval the price of the resource relative to the composite commodity numeraire is not, of course, constant, but equal to the marginal product of the resource. A question of importance is the behaviour of this relative price along an optimal programme.[2]

[1] We are thus supposing that extraction of the exhaustible resource is costless. Extraction costs do not appear to introduce any great problem, provided that we assume away non-convexities. One might for example, wish to assume that to extract the resource at a rate R when the remaining stock is of size S requires $E(R, S)$ units of the composite good. One might suppose that E is a convex function and that, in particular, $E(0, S) = 0$, $\partial E/\partial R > 0$ and $\partial E/\partial S \leqq 0$, the last inequality indicating that one is, as it were, digging deeper when the stock has shrunk. To have an interesting problem one would, in addition, need to assume that $\operatorname{lt}_{R\to 0} \partial F/\partial R > \operatorname{lt}_{R\to 0} \partial E/\partial R$. The planning problem (1.6) would in this event be modified only to the extent that equation (1.5) would be replaced by the equation,

$$\dot{K} = F(K, R) - C - E(R, S).$$

[2] If extraction costs were introduced (see previous footnote) equation (1.12) would be replaced by the condition

$$\lambda = e^{-\delta t}p_t[F_R - E_R].$$

Thus, in fact, during an interval when $R_t > 0$ the shadow price of the unextracted exhaustible resource relative to the composite commodity is the *difference* between the marginal product of the resource and its marginal extraction cost. But the price of the extracted resource relative to the composite commodity is of course its marginal product. The extraction cost could be regarded as a pure transport cost. We have found that even amongst professional economists there is often the belief that the price of an exhaustible resource should be its marginal extraction cost. It is, of course, rather obvious why this belief is false.

A related question is whether or not the exhaustible resource ought to be depleted in *finite* time. Towards this consider an economy for which

$$\lim_{R \to 0} F_R = \infty. \qquad ...(1.13)$$

It is an immediate implication of equation (1.11) that we shall have

Proposition 5. *If* $F(K, R)$ *satisfies* (1.13), *then along an optimal policy, if one exists,* $R_t > 0$ *for all* $t \geqq 0$.

The sufficient condition (1.13) in Proposition 5 is, of course, unduly restrictive. One can relatively easily, as we shall presently, weaken it. But it is suggestive in that a condition that prohibits the depletion of the stock of the exhaustible resource in finite time rests not so much on whether the resource is essential (in the sense of output being nil in its absence), but rather on the behaviour of the marginal product of the resource for low rates of usage of the resource. To an economist this is very intuitive.

Proposition 5 is analogous to Proposition 3 of the " cake-eating " problem. It is tempting to construct an analogue of Proposition 2. Towards this we complete the set of conditions necessary for optimality by noting that one must have as well that

$$\frac{d}{dt}(e^{-\delta t}p) = -e^{-\delta t}pF_K, \qquad ...(1.14)$$

or

$$-\dot{p}/p = F_K - \delta. \qquad ...(1.15)$$

Now, equation (1.15), which is the familiar Ramsey condition can, in turn, be re-expressed on using (1.10) as

$$\frac{\dot{C}}{C} = (F_K - \delta)/\eta(C). \qquad ...(1.16)$$

Consider now a time interval of positive measure, if one exists, during which it is optimal to set $R_t > 0$. During this interval equation (1.12) is operative. It follows that on differentiating (1.12) with respect to time one obtains the condition

$$\frac{\partial(F_R)}{\partial t}\frac{1}{F_R} = \frac{-\dot{p}}{p} + \delta,$$

which, on using equation (1.15) reduces to the form

$$\frac{\partial(F_R)}{\partial t}\frac{1}{F_R} = F_K. \qquad ...(1.17)$$

Condition (1.17) is really rather obvious, for it is a statement concerning the equality of the rates of return on the two assets (the exhaustible resource and reproducible capital).

Write $x \equiv K/R$ and $f(x) \equiv F(K/R, 1)$. It follows from our assumptions on F that $F_K = f'(x) > 0$, $F_R = f(x) - xf'(x) < 0$, and $f''(x) < 0$. Thus one has as well that $d(F_R)/dx > 0$, for $x > 0$. Write

$$\sigma = \frac{-f'(x)(f(x) - xf'(x))}{xf(x)f''(x)}, \quad (\infty > \sigma > 0)$$

for the elasticity of substitution between K and R. It is then immediate that equation (1.17) reduces to the form

$$\frac{\dot{x}}{x} = \sigma\frac{f(x)}{x}. \qquad ...(1.18)$$

Equations (1.16) and (1.18) govern an optimal programme, if one exists, during an interval when $R_t>0$. Moreover, we have already established (Proposition 5) that if F satisfies (1.13) then equations (1.16) and (1.18) govern an optimal programme for all $t \geqq 0$.

It is remarkable that the conditions governing an optimal programme of aggregate consumption and resource depletion should be so simple in form. In particular, equation (1.18) makes the role of the elasticity of substitution very clear. The capital-resource ratio changes at a percentage rate equal to the product of the elasticity of substitution and the average product per unit of fixed capital. The former gives an indication of the ease with which substitution can be carried out, and the latter can be regarded as an index of the importance of fixed capital in production. Thus, the easier it is to substitute, and the more important is the reproducible input, the more one wants to substitute the reproducible resource for the exhaustible one.

We have already established that if F satisfies (1.13) then along an optimal policy, if one exists, $R_t>0$ for all $t \geqq 0$. The following Lemma yields considerable information regarding the broad characteristics of an optimal depletion policy.

Lemma. *Along an optimal programme, if one exists, an interval during which $R = 0$ cannot be followed by an interval during which $R>0$.*

Proof. Recall that $d(F_R)/dx>0$. Assume that $\sup(f(x)-xf'(x))$ is finite (for if not, the Lemma is trivially true in view of Proposition 5). Suppose then the contrary and let there be a change in phase at T. Assume first that the interval during which $R = 0$ is non-degenerate and that $R_T>0$. Then since K_t is continuous one has

$$\lim_{t\to T^-} (f(x_t)-x_tf'(x_t)) = \sup(f(x)-xf'(x))>(f(x_T)-x_Tf'(x_T)).$$

But $\lim_{t\to T^-} e^{-\delta t}p_t = e^{-\delta T}p_T$. Such a phase change, therefore, is impossible since, given equations (1.11) and (1.12), one must have

$$\lim_{t\to T^-} e^{-\delta t}(\mu_t+p_t(f(x_t)-x_tf'(x_t))) = e^{-\delta T}p_T(f(x_T)-x_Tf'(x_T)),$$

with $\mu_t \geqq 0$.

The other possibility is that $R_T = 0$ but that during an interval (T, T_1) one sets $R_t>0$. But if this is so then during (T, T_1) equation (1.18) holds, and in order for (1.18) to hold it must be the case that $\lim_{t\to T^+} R_t>0$. In other words, there must be a discontinuity in R at T. Thus $f(x_T)-x_Tf'(x_T)> \lim_{t\to T^+} (f(x_t)-x_tf'(x_t))$. One argues now, as earlier, that such a phase change is impossible since equations (1.11) and (1.12) dictate that

$$e^{-\delta T}(\mu_T+p_T(f(x_T)-x_Tf'(x_T))) = \lim_{t\to T^+} e^{-\delta t}p_t(f(x_t)-x_tf'(x_t))$$

and $\mu_T \geqq 0$. ‖

The Lemma is powerful in its implications, for we can now assert

Proposition 6. *Along an optimal policy, if one exists, R_t is a continuous function of time. Furthermore, either $R_t>0$ for all $t \geqq 0$ or there exists a finite $T(>0)$ such that $R_t>0$ for $0 \leqq t<T$ and $R_t = 0$ for $t \geqq T$.*

Proof. The second part (with the exception of the demonstration that $T>0$) of the proposition is a direct implication of the Lemma, and a simple argument (as in the proof of the Lemma) showing that we cannot have $R_t=0$ for $0 \leqq t \leqq T$ and $R_t > 0$ for $t>T$. Therefore we begin by proving that R_t must be continuous if $T>0$. Thus if $R_t>0$ for all $t \geqq 0$ then the continuity of R_t is trivially true since x_t must satisfy (1.18) for all $t \geqq 0$ and since K_t must be a continuous function. Suppose then that there exists a finite $T(>0)$ at which there is a phase change from $R>0$ to $R = 0$. During the interval $[0, T)$ equation (1.18) holds. Since $\dot{x}>0$ during $[0, T)$, $\lim_{t\to T^-} x_t$ exists. But it cannot be finite. For suppose it is. By hypothesis $x_T = \infty$. This implies that we would violate the condition necessary at the switch point, namely that

$$\operatorname*{lt}_{t\to T^-} e^{-\delta t}p_t(f(x_t)-x_tf'(x_t)) = e^{-\delta T}\{\mu_T+p_T[f(x_T)-x_Tf'(x_T)]\},$$

where $\mu_T \geqq 0$.

It follows that during $[0, T)$, x_t increases continuously to ∞. That is $\operatorname*{lt}_{t\to T^-} x_t = x_T = \infty$. Since $K_t > 0$ for all $t \geqq 0$ this implies that $\operatorname*{lt}_{t\to T^-} R_t = R_T = 0$.

What remains to be proved is that $T>0$. So suppose on the contrary that the phase change occurs at $t = 0$. That is, assume $T = 0$ (i.e. that R_t is a δ-function). Once again we appeal to the assumption that $\frac{d}{dx}(f(x)-xf'(x))>0$ and note that if there is a singularity at $t = 0$ one would violate a condition necessary at this switch point, which is

$$p_0\{f(x_0)-x_0f'(x_0)\} = \operatorname*{lt}_{t\to 0^+} e^{-\delta t}p_t\{f(x_t)-x_tf'(x_t)\}+\mu_t,$$

where $\mu_t \geqq 0$, $x_0 = 0$ and $\operatorname*{lt}_{t\to 0^+} x_t = \infty$.

We are now in a position to construct a proposition analogous to Proposition 2 of the previous section. Towards this consider an economy for which

$$\operatorname*{lt}_{x\to\infty}(f(x)-xf'(x)) = \gamma < \infty$$

and ...(1.19)

$$F(K, 0)>0 \text{ for } K>0.$$

It is then simple to confirm that it cannot be optimal to maintain $R_t>0$ for all $t \geqq 0$ in an economy that satisfies (1.19). For suppose the contrary. From (1.18) one has $x_t \to \infty$. Furthermore, from the second part of (1.19) it follows that

$$\operatorname*{lt}_{x\to\infty} f(x)/x = \operatorname*{lt}_{x\to\infty} f'(x) = \rho > 0.$$

Therefore, from (1.14) one has that the present value price of consumption $(e^{-\delta t}p_t)$ goes to zero, and from the first part of (1.19) the fact that the marginal product of the exhaustible resource is bounded above by γ. But these would contradict the requirement that the economy satisfies (1.12). Thus one has

Proposition 7. *If $F(K, R)$ satisfies (1.19) then along an optimal policy, if one exists, $R_t>0$ for $0 \leqq t<T$ where T is finite and $R_t = 0$ for $t \geqq T$. Moreover R_t is continuous for all $t \geqq 0$.*[1]

1.3. The arguments of the previous sub-section were fairly general, and were concerned with glancing at the conditions that an optimal programme must necessarily satisfy. In particular, we noted the rather direct way in which the elasticity of substitution enters into the characterization of an optimal programme during a phase when $R_t>0$. In this section we parametrize the economy somewhat further to obtain some definite answers to questions that one would like to ask. One would, for example, like to know under what circumstances an optimal programme exists, and whether the rate of consumption along an optimal policy necessarily falls if the resource is essential and indeed, whether the rate of consumption necessarily rises if the resource is inessential. Furthermore, one

[1] An example of an economy, satisfying (1.19) is one where $f(x) = \rho(1+x+\phi(x))$ with $\rho>0$, $\phi(0) = 0$, $\phi'(x)>0$, $\phi''(x)<0$ and $\operatorname*{lt}_{x\to\infty}\phi(x)<\infty$. One should note that we have avoided the case $f(x) = \rho(1+x)$ (i.e. $F(K, R) = (K+R)\,\rho$, $\sigma = \infty$) by our assumption of strict concavity of $F(K, R)$. We shall have occasion to comment on this case in Section 1.4.

would like to know if the optimal utilization rate of the resource is non-increasing over time, as well as conditions under which the capital stock ought to be built up over time.

Towards this we turn to the simplest laboratory that might provide answers, namely the class of production functions for which the elasticity of substitution is constant. It might seem plausible that the class of CES production functions captures the variety of the issues under review. In fact, the CES production functions (with the exception of those for which $\sigma = 1$ and ∞) have rather unusual properties at the " corners ". Consequently some of the characteristics of an optimal policy in a CES world are, at first blush, counter-intuitive. It will be plain that one can establish rather easily conditions under which an optimal programme exists for a much larger environment than the CES world provides. Since the arguments turn out to be rather similar to the ones that we shall provide for the CES case, we do not elaborate on them here.

Recall that the CES production functions is of the form

$$F(K, R) = [\beta K^{(\sigma-1)/\sigma} + (1-\beta)R^{(\sigma-1)/\sigma}]^{\sigma/(\sigma-1)},$$

where $0<\beta<1$ and $\infty \geqq \sigma \geqq 0$.

The analysis of the previous sub-section implied that one is particularly interested in the properties of the production function for $R = 0$. It is convenient, then, to catalogue the following properties:

If $\sigma = 1$ (i.e. the Cobb-Douglas form)

$$F(K, 0) = 0 \qquad \ldots(1.20a)$$

$$\lim_{x\to\infty} f'(x) = \lim_{x\to\infty} \frac{f(x)}{x} = 0 \qquad \ldots(1.20b)$$

$$\lim_{x\to\infty} (f(x) - xf'(x)) = \lim_{x\to\infty} f(x) = \infty. \qquad \ldots(1.20c)$$

If $0 \leqq \sigma < 1$ then

$$F(K, 0) = 0 \qquad \ldots(1.2a)$$

$$\lim_{x\to\infty} f'(x) = \lim_{x\to\infty} \frac{f(x)}{x} = 0 \qquad \ldots(1.21b)$$

$$\lim_{x\to\infty} (f(x) - xf'(x)) = \lim_{x\to\infty} f(x) = (1-\beta)^{\sigma/(\sigma-1)}. \qquad \ldots(1.21c)$$

If $\infty > \sigma > 1$ then

$$F(K, 0) = \beta^{\sigma/(\sigma-1)}K \qquad \ldots(1.22a)$$

$$\lim_{x\to\infty} f'(x) = \lim_{x\to\infty} \frac{f(x)}{x} = \beta^{\sigma/(\sigma-1)} > 0 \qquad \ldots(1.22b)$$

$$\lim_{x\to\infty} (f(x) - xf'(x)) = \lim_{x\to\infty} f(x) = \infty. \qquad \ldots(1.22c)$$

An exhaustible resource is essential to production only when $\sigma \leqq 1$. Otherwise it is inessential. The pair (1.22c) and (1.21c) are curious: an inessential resource is infinitely valuable and an essential resource is finitely valuable at the margin when the rate of utilization of the resource is zero. Only the Cobb-Douglas form may be said to have properties that are reasonable at the corner.

From (1.20c) and (1.22c) and Proposition 5 it is immediate that when $\infty > \sigma \geqq 1$, along an optimal policy, if one exists, $R_t > 0$ for all $t \geqq 0$. We demonstrate now that this must be true for the case $0 < \sigma < 1$ as well. Recall Proposition 6. Assume that there exists a finite $T(>0)$ such that $R_t = 0$ for $t \geqq T$. During the interval $[0, T)$ equation (1.18) must be satisfied. Moreover R_t must be continuous at T. But this is impossible, for the constancy of σ and condition (1.21c) imply that x cannot increase to infinity via equation (1.18) in finite time.[1]

We have yet to demonstrate that an optimal programme exists. To keep the *presentation* simple we restrict ourselves to the case of utility functions that are iso-elastic. That is, assume $\eta(c) = \eta > 0$. Write

$$\rho = \operatorname*{lt}_{x\to\infty} f'(x) = \operatorname*{lt}_{x\to\infty} \frac{f(x)}{x} \text{ and } S_t = S_0 - \int_0^t R_t d\tau.$$

We can then establish

Proposition 8. *If $\sigma(\infty > \sigma > 0)$ is constant, $\eta(>0)$ is constant, and $\delta > \rho(1-\eta)$, a unique optimal policy exists and it satisfies equations* (1.5), (1.16) *and* (1.18)*for all $t \geqq 0$. Moreover, the optimal programme $(\tilde{C}_t, \tilde{R}_t, \tilde{K}_t, \tilde{x}_t, \tilde{S}_t)$ satisfies the following asymptotic properties:*

$$\operatorname*{lt}_{t\to\infty} \tilde{x} = \infty \qquad \text{...(1.23)}$$

$$\operatorname*{lt}_{t\to\infty} \tilde{C} = \operatorname*{lt}_{t\to\infty} \tilde{K} = \infty \text{ if } \rho > \delta \qquad \text{...(1.24)}$$

$$\operatorname*{lt}_{t\to\infty} \tilde{C} = \operatorname*{lt}_{\to\infty} \tilde{K} = 0 \text{ if } \rho < \delta \qquad \text{...(1.25)}$$

$$\operatorname*{lt}_{t\to\infty} \frac{\dot{\tilde{C}}}{\tilde{C}} = \operatorname*{lt}_{t\to\infty} \frac{\dot{\tilde{K}}}{\tilde{K}} = \frac{\rho-\delta}{\eta} \qquad \text{...(1.26)}$$

$$\operatorname*{lt}_{t\to\infty} \frac{\tilde{C}}{\tilde{K}} = \rho + \frac{\delta-\rho}{\eta} \qquad \text{...(1.27)}$$

$$\operatorname*{lt}_{t\to\infty} \frac{\dot{\tilde{R}}}{\tilde{R}} = \operatorname*{lt}_{t\to\infty} \frac{\dot{\tilde{S}}}{\tilde{S}} = \frac{\rho-\delta}{\eta} - \sigma\rho < 0 \qquad \text{...(1.28)}$$

$$\operatorname*{lt}_{\to\infty} \frac{\tilde{R}}{\tilde{S}} = \sigma\rho + \frac{\delta-\rho}{\eta} > 0. \qquad \text{...(1.29)}$$

Proof. It has already been established that the conditions necessary for optimality are that equations (1.16) and (1.18) must hold for all $t \geqq 0$. Condition (1.23) is then immediate from (1.18). From equation (1.18) one recognizes as well that for any pair of paths $x_t^{(1)}$ and $x_t^{(2)}$ one has $x_t^{(1)} > x_t^{(2)}$ for all $t > 0$ if $x_0^{(1)} > x_0^{(2)}$. From (1.16) it follows that $C_0^{(1)} > C_0^{(2)}$ if, and only if, $C_t^{(1)}(C_0^{(1)}, x_0) > C_t^{(2)}(C_0^{(2)}, x_0)$. Now write equation (1.5) as

$$\frac{\dot{K}}{K} = \frac{f(x)}{x} - \frac{C}{K}. \qquad \text{...(1.30)}$$

It follows from (1.30) that $K_t^{(1)}(C_0^{(1)}, x_0) < K_t^{(2)}(C_0^{(2)}, x_0)$ if, and only if, $C_0^{(1)} > C_0^{(2)}$.

Write $y = C/K$, and use equations (1.16) and (1.30) to obtain

$$\frac{\dot{y}}{y} = y - \frac{\delta}{\eta} + \frac{f'(x)}{\eta} - \frac{f(x)}{x}. \qquad \text{...(1.31)}$$

Recall that K_0 and S_0 are given and that we are to choose optimally C_0 (i.e. y_0) and R_0(i.e. x_0). The point is to observe that these choices can, in principle, be separated out

[1] It is now immediate that Proposition 5 can be generalized to

Proposition 5′. *Along an optimal policy, if one exists, $R_t > 0$ for all $t \geqq 0$ if either $F(K, R)$ satisfies* (1.13) *or $F(K, 0) = 0$.*

Since the sufficiency conditions in Propositions 5′ and 7 are not merely mutually exclusive but exhaustive as well, it follows that these two Propositions cover all the cases.

conveniently. We have, of course, assumed $\sigma>0$. Then, using (1.23) it follows from (1.31) that for large t

$$\frac{\dot{y}}{y} \simeq y - \frac{(\delta-\rho)}{\eta} - \rho. \qquad \text{...(1.32)}$$

It follows from (1.31) and (1.32) that *given* x_0, if C_0 (i.e. y_0) is chosen large enough, then $y_t(x_0)\to\infty$ as $t\to\infty$. But this entails from (1.30) that the feasibility requirement, $K_t \geqq 0$, is violated in finite time. It follows as well from (1.31) and (1.32) that *given* x_0, if C_0 (i.e. y_0) is chosen positive, but low enough, then $y_t(x_0)\to 0$ as $t\to\infty$.

Thus let $\tilde{C}_0(x_0) = \sup[C_0 \mid y_t(C_0, x_0) \geqq 0 \text{ for } t \geqq 0]$. From (1.32) it is immediate that

$$\operatorname*{lt}_{t\to\infty} y_t(\tilde{C}_0(x_0), x_0) = \rho+(\delta-\rho)/\eta>0,$$

confirming (1.27). From (1.16) and (1.30) condition (1.26) follows directly.

One can now verify that

$$\operatorname*{lt}_{t\to\infty} \frac{d(e^{-\delta t}p_t\tilde{K}_t)/dt}{e^{-\delta t}p_t\tilde{K}_t} = \frac{\rho-\delta}{\eta} - \rho<0, \text{ and thus that } \operatorname*{lt}_{t\to\infty} e^{-\delta t}p_tK_t = 0. \qquad \text{...(1.33)}$$

But $\dot{R}/R = \dot{K}/K - \dot{x}/x$. On using (1.26) and (1.18) it is immediate that

$$\operatorname*{lt}_{t\to\infty} \frac{\dot{R}}{R} = \frac{\rho-\delta}{\eta} - \sigma\rho<0. \qquad \text{...(1.34)}$$

It follows from (1.34) that $S(x_0) \equiv \int_0^\infty R_t(\tilde{C}_0(x_0), x_0)dt$ is well defined. It follows as well that we can choose x_0 large enough (i.e. R_0 small enough) so that $S(x_0)<S_0$, and x_0 small enough (i.e. R_0 large enough), so that $S(x_0)>S_0$. The former would be inefficient and the latter choice unfeasible. Thus write

$$\tilde{x}_0 \equiv \sup\{x_0 \mid S(x_0) \leqq S_0\}.$$

Then

$$S_t = \int_t^\infty \tilde{R}_\tau(\tilde{C}_0(\tilde{x}_0), \tilde{x}_0)d\tau \text{ and } \operatorname*{lt}_{t\to\infty} \tilde{S}_t = 0.$$

But $\dot{S}/S = -R/S$. Hence

$$\operatorname*{lt}_{t\to\infty} \dot{\tilde{S}}/\tilde{S} = \operatorname*{lt}_{t\to\infty} -\tilde{R}/\tilde{S} = \operatorname*{lt}_{t\to\infty} \dot{\tilde{R}}/\tilde{R} = (\rho-\delta)\eta-\sigma\rho<0,$$

and conditions (1.28) and (1.29) are confirmed. Plainly

$$\operatorname*{lt}_{t\to\infty} \lambda\tilde{S}_t = 0. \qquad \text{...(1.35)}$$

Conditions (1.24) and (1.25) are implied immediately by (1.26).

Finally, the Hamiltonian (1.7) is strictly concave. The two transversality conditions (1.33) and (1.35) have been verified. The proposition is thus established. ‖.

Having established the main proposition of this section we present some specific commentaries in the following sub-section.

1.4. The case $0<\sigma<1$ is at once the simplest and the most pessimistic of all.[1] As total output is bounded if $\sigma<1$, it is plain that *feasibility* dictates that $C_t\to 0$ as $t\to\infty$. Since $\rho = 0$ it is clear that an optimum exists if $\delta>0$. This is so, even if $U(\cdot)$ is unbounded above (i.e. $0<\eta<1$). Given (1.23), the shadow price of the exhaustible resource relative to the composite is monotonically increasing. Given (1.21*c*) in addition, it is clear that

[1] We have, of course, assumed away the case $\sigma = 0$ by our assumptions on F. It is not a case that presents any special problems of analysis but as its analysis calls for some extra notation, we do not discuss it here.

this shadow price tends in the long run to the value $(1-\rho)^{\sigma/(\sigma-1)}$ (see equation (1.12)). This is mildly surprising, given that the resource is essential.

The Cobb-Douglas case is particularly interesting since the analysis can relatively easily be taken further. To begin with, since $\rho = 0$, an optimum exists if $\delta > 0$. Furthermore, given (1.20*c*) and (1.23), the price of the exhaustible resource relative to fixed capital is increasing and tends to infinity with time. Let $f(x) = x^{\alpha}$ with $0 < \alpha < 1$. Professor Solow [25] has demonstrated that consumption can be unbounded if $\alpha > 1/2$, but that feasible consumption must tend to zero if $\alpha < 1/2$. But from (1.25) we have that for all α, $0 < \alpha < 1$, $\tilde{C}_t \to 0$.

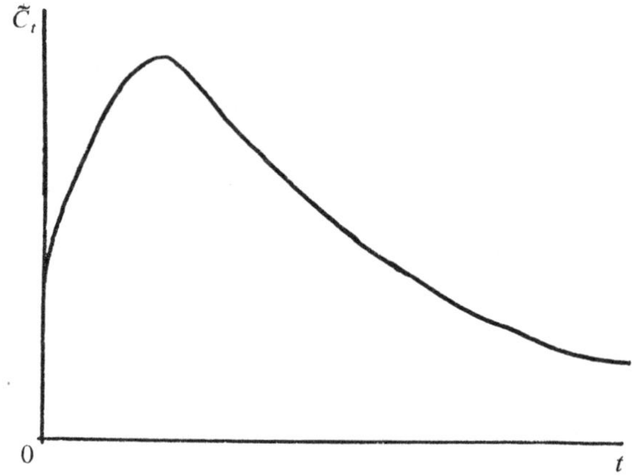

FIGURE 3

Now integrate (1.18) to obtain

$$\tilde{x}_t = [(1-\alpha)t + x_0^{(1-\alpha)}]^{1/(1-\alpha)}, \qquad \text{...(1.36)}$$

and thus that in the long run the price of the exhaustible resource relative to the composite commodity should be approximately

$$(1-\alpha)^{1/(1-\alpha)} t^{\alpha/(1-\alpha)}.$$

It is also the case that in the long run the percentage rate of change in the rate of depletion of the exhaustible is approximately equal to $-\delta/\eta$ (see (1.28)) and, in particular, the rate of depletion, as a fraction of the then existing stock of the resource, is δ/η (see (1.29)).

Using (1.36) in equation (1.16) and integrating, the optimal consumption profile is

$$\tilde{C}_t = \tilde{C}_0 \tilde{x}_0^{-\alpha/\eta} [(1-\alpha)t + \tilde{x}_0^{(1-\alpha)}]^{\alpha/\eta(1-\alpha)} e^{(-\delta/\eta)t}. \qquad \text{(...1.37)}$$

From (1.37) it would appear that if the initial stocks K_0 and S_0 are " large " the optimal rate of consumption will, during an initial period, be rising, and will in fact be single peaked, with a maximum at T where [1]

$$T = \frac{\alpha}{\delta(1-\alpha)} - \frac{\tilde{x}_0^{\,1-\alpha}}{(1-\alpha)}.$$

A typical consumption profile is presented in Figure 3. It is routine to show that

$$\dot{\tilde{R}}_t = \tilde{C}_0 \tilde{x}_0^{-\alpha/\eta} [(1-\alpha)t + \tilde{x}_0^{1-\alpha}]^{\alpha/\eta} e^{(-\delta/\eta)t},$$

[1] It should, of course, be evident that T is not independent of η, since $\tilde{X}_0$ is dependent on η. Professor Solow ([25], appendix) considers the case where $\alpha > 1/2$ and $\delta = 0$. This last ensures that our existence theorem is not applicable to his case. He shows, in effect, that if η is large enough, then $T = \infty$. For this limiting case of $\delta = 0$ he assumes, therefore, that $\alpha > 1/2$ to ensure that consumption can be unbounded and large enough η to ensure the existence of an optimal policy.

from which

$$\text{sign}\,(\ddot{\tilde{R}}_t) = \text{sign}\left[\delta - \frac{\alpha(1-\alpha)}{(1-\alpha)t+\tilde{x}_0^{1-\alpha}}\right].$$

It follows that while $\dot{\tilde{R}}<0$ for all $t \geqq 0$ and $\ddot{\tilde{R}}_t>0$ for large enough t, one would have $\ddot{\tilde{R}}_t<0$ for an initial period for large values of K_0 and S_0, yielding a time profile of resource utilization as in Figure 4.

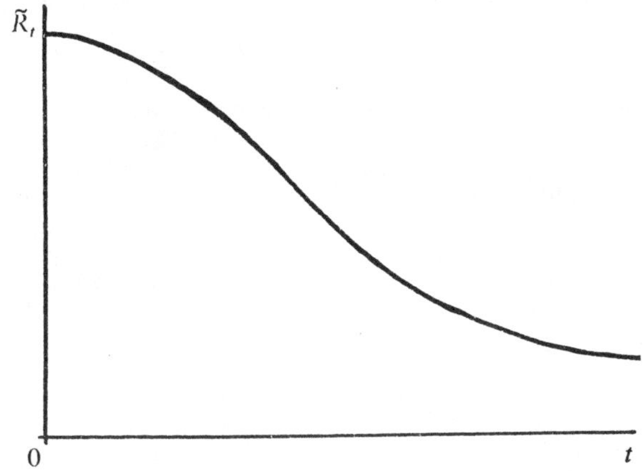

FIGURE 4

The case $\sigma>1$ is one in which exhaustible resources are not a fundamental problem, since they are inessential in production. Given (1.23) the shadow price of the exhaustible resource relative to fixed capital is monotonically increasing, but given (1.22*c*) one does have the surprising result that this price tends to infinity with time. Since $\rho>0$ it is plain that $\delta>0$ is not sufficient for the existence of an optimum policy. But if $U(\cdot)$ is bounded above (i.e. $\eta>1$) an optimum exists for $\delta \geqq 0$. As conditions (1.24) and (1.25) make clear, whether or not it is optimal to have a growing economy in the long run depends on whether ρ is greater than, or less than δ, a condition that is intuitively plain.

The case, $\sigma = \infty$, though silly, merits a brief glance, though we have so far assumed it away. Assume then that $f(x) = \rho(1+x)$, where $\rho>0$. It satisfies (1.19). But since f is linear, Proposition 7 is not strictly operative. Nevertheless, it is an easy matter to confirm that if $\delta>(1-\eta)\rho$ an optimal programme exists and that the optimal depletion rate, $\tilde{R}_t$, is a Dirac-δ-function with the singularity at $t = 0$. The point is that since the marginal product of the resource is constant, it does not pay to leave the resource lying idle. As we have imposed no costs of extraction one can, in principle, use up the entire initial stock, S_0, instantaneously to build up the capital stock, which is, in fact, the optimal strategy.[1]

2. TECHNICAL CHANGE AND UNCERTAINTY

2.1. Technological progress is, as usual, hard to model. We are concerned now with those situations where technical change causes a resource that was previously essential to become inessential. In practice this is most likely to be due to the discovery of a synthetic substitute (for example, being able to harness solar energy or nuclear fusion), and it is this type of alteration in the technological possibilities that provides the motivation

[1] The reader can easily check that if $\delta>(1-\eta)\rho$ then a unique optimum exists if the production function $f(\cdot)$ is of the form given in the footnote to Proposition 7. Given Proposition 7, the optimal depletion function, $\tilde{R}_t$, is continuous for this case, but the entire stock is exhausted in *finite* time.

for the discussion that follows. In such situations, although one may not know in advance exactly what the substitute will be like, there will probably be general information about its likely characteristics. Industrial research directed towards innovation is, after all, not carried out aimlessly. A more important source of uncertainty seems to be the exact *timing* of the availability of the substitute. It would seem reasonable to argue that we are much more uncertain about when, if ever, such an event will occur, than we are about the characteristics it will have if it does occur, and we shall focus attention on this aspect of the problem. Consequently we suppose that we know exactly the nature of the technical change that will occur, but we treat the date at which the event occurs as a random variable. In this paper, however, we suppose that the distribution of the random variable is exogenously given.

The basic framework that we shall present appears to us to be fairly robust in that it can accommodate quite diverse characterizations of the technological breakthrough. Furthermore, it would seem that one could, in principle, extend the analysis quite directly to allow for the possibility of a *finite* sequence of such events in the future. But it seems most interesting to conceive of the event as being the discovery of a substitute for an exhaustible resource which releases the economy from the resource constraints. Moreover, for the sake of simplicity we take it that only one such event is to be considered. Towards this we amalgamate the production model considered in Sections 1.1 and 1.2 in a sequential manner.

Assume that the economy is endowed initially with a stock, K_0, of the composite consumption good, and a stock, S_0, of an exhaustible resource. Production possibilities until the date of the technological breakthrough are represented, as in Section 1.2, by the production function $F(K_t, R_t)$, where, as before, R_t is the flow of resources into production at t. Suppose the technological breakthrough occurs at date $T>0$. Then the constraints that limit the choice of consumption till T can be denoted by the conditions

$$\dot{K}_t = F(K_t, R_t) - C_t,\ 0 \leqq t \leqq T;\ \int_0^T R_t dt \leqq S_0 \text{ and } K_0 \text{ and } S_0 \text{ are given.} \qquad \text{...(2.1)}$$

We characterize the technological innovation as being the discovery of a perfectly durable commodity which provides a flow of service at the constant rate M. But it would seem hard to think that the production possibilities after the event would remain the same as those before. It is here that we revert to the economy discussed in Section 1.1, and suppose that the production function after the innovation can be regarded as being $G(K_t, Z_t)$, where Z_t is the rate of utilization of the service at date t. Now, there is of course nothing to suggest that the service provided by the durable commodity is a perfect substitute for the exhaustible resource in the production function, G. In fact there is every reason to suppose that it is not. Nevertheless, for the moment, to keep the presentation simple we suppose that they are perfect substitutes. It follows that the constraints that limit the choice of a consumption profile for the period *beyond* T are represented by the conditions

$$\left.\begin{aligned} &\dot{K} = G(K, Z) - C \\ &\dot{V} = M - Z \\ &\text{where } Z_t, K_t, C_t, V_t \geqq 0 \text{ and } V_T = S_0 - \int_0^T R_t dt \\ &\text{and } K_T \text{ given} \end{aligned}\right\}. \qquad \text{...(2.2)}$$

The date T, is assumed unknown. In fact suppose that T is random number with a probability density function ω_t. Thus we suppose that

$$\omega_t > 0 \text{ and } \int_0^\infty \omega_t dt = 1. \qquad \text{...(2.3)}$$

The second condition in (2.3) supposes that the substitute source will certainly be discovered *sometime* in the future. One might wish to allow for the possibility of it *never* being discovered. Nothing substantial seems to emerge by supposing so. Therefore we suppose (2.3).[1]

We take it that the valuation function for planning is $E\int_0^\infty e^{-\delta t}U(C)dt$, where E is the expectation operator. An important question then arises as to what penalty one would wish to attach to depleting completely the exhaustible resource *before* the substitute is discovered. It would seem most interesting to formulate the problem in such a way as to make the penalty infinite. Thus we impose the restriction that along a feasible strategy the probability of being caught short with none of the exhaustible resource before the substitute is discovered, is nil.

Define $W(K_T, V_T)$ = maximum value of $\int_T^\infty e^{-\delta(t-T)}U(C_t)dt$ subject to (2.2). It follows that

$$E\int_0^\infty e^{-\delta t}U(C)dt = \int_0^\infty \omega_T\left\{\int_0^T e^{-\delta t}U(C)dt + W(K_T, V_T)e^{-\delta t}\right\}dT. \qquad ...(2.4)$$

Write $\Omega_t = \int_t^\infty \omega_\tau d\tau$, and integrate the RHS of (2.4) by parts to obtain

$$E\int_0^\infty e^{-\delta t}U(C)dt = \int_0^\infty e^{-\delta t}\{U(C)\Omega_t + \omega_t W(K_t, V_t)\}dt. \qquad ...(2.5)$$

The problem then reduces to one of

$$\left.\begin{array}{l} \text{maximizing expression (2.5) subject to the constraints} \\ \dot{K} = F(K, R) - C \\ \dot{S} = -R \\ \text{where } V_t = S_t \text{ and } K_t, C_t, R_t, S_t \geqq 0 \end{array}\right\}. \qquad ...(2.6)$$

It is as well to point out that the optimal policy is to pursue the solution path of problem (2.6) only until the substitute is discovered. At the date the substitute is discovered there is a switch in regime and the optimal policy from then on is to follow the policy which yields $W(K_T, V_T)$.[2]

In order to analyse (2.6) it is convenient to express the Hamiltonian of the system as

$$H = e^{-\delta t}\{U(C)\Omega_t + \omega_t W(K_t, V_t)\} + e^{-\delta t}p_t\{F(K, R) - C\} - e^{-\delta t}q_t R_t + e^{-\delta t}\pi_t R_t + e^{-\delta t}\gamma_t S_t, \qquad ...(2.7)$$

where
$$\left.\begin{array}{l} \pi_t \geqq 0 \text{ and } \pi_t R_t = 0 \\ \gamma_t \geqq 0 \text{ and } \gamma_t S_t = 0 \end{array}\right\}, \qquad ...(2.8)$$
and

We shall assume $W(K_t, V_t)$ to be bounded and differentiable. The former can be justified if we assume that $U(\cdot)$ is bounded above, and we shall assume that this is so. The concavity of W follows from assumptions that we have already made. From (2.1) it is plain that $e^{-\delta t}p_t$ is the present value price of the composite consumption good and that $e^{-\delta t}q_t$ is the present value price of the exhaustible resource. One notes as well from (2.8) that $\gamma_t > 0$ implies that $\pi_t > 0$.

1 For simplicity of exposition we shall take it, in fact, that $\omega_t > 0$ for $t \geqq 0$, though this is not at all essential.

2 The solution to (2.6) is thus a conditional strategy, conditional upon the substitute not having been discovered.

For a programme to be optimal it is then necessary that it satisfies the following conditions

$$p = \Omega_t U'(C) \qquad \text{...(2.9)}$$

$$pF_R = q - \pi_t \qquad \text{...(2.10)}$$

$$\dot{p} = \delta p - \omega_t W_K - pF_K \qquad \text{...(2.11)}$$

and

$$\dot{q} = \delta q - \omega_t W_V - \gamma_t, \qquad \text{...(2.12)}$$

where

$$W_K = \frac{\partial W}{\partial K}, \text{ etc.}$$

Now $W_V \geqq 0$ and $W_K \geqq 0$. It follows from (2.10) to (2.12) that if F satisfies (1.13) then we shall have $R_t > 0$ for all $t \geqq 0$ along an optimal policy. We suppose, therefore, that this is so. Consequently $\gamma_t = \pi_t = 0$ for all $t \geqq 0$.

Write $\Psi_t = \omega_t/\Omega_t$, for the conditional probability of the substitute being discovered at t given that it has not been discovered earlier. From equations (2.9) and (2.11) one has [1]

$$\frac{\dot{C}}{C} = \frac{F_K - \delta + \Psi_t\{(W_K - U'(C))/U'(C)\}}{\eta(C)}. \qquad \text{...(2.13)}$$

Differentiate (2.10) with respect to time and use equations (2.12), (2.13) and the production constraint in (2.6) to obtain

$$\dot{x} = \sigma f(x)\left\{1 + \frac{W_K \Psi_t}{U'(C)f'(x)}\right\} + \frac{W_V \Psi_t}{x f''(x) U'(C)}, \qquad \text{...(2.14)}$$

where x, σ and $f(\cdot)$ are as in Section 1.2.[2] Equations (2.13) and (2.14) govern the solution to the problem expressed in (2.6). The nature of the path they define is far from obvious except in some special cases. We turn to these. Now the technology G subsequent to the technological breakthrough is likely to be considerably different from that prior to T. The simplification that we introduce lies in the supposition that at T (the date of the discovery), the then existing stocks of capital and the exhaustible resource come to have no economic value. This is, of course, a terribly strong assumption, but not totally unreasonable. It seems, for example, very likely that if fusion reactors ever become a commercial proposition, turning water into an abundantly available substitute for fossil fuel, then power stations generating electricity from fossil fuels will be rapidly phased out. Both the capital equipment that they comprise and the remaining fossil fuel stocks will have little economic value as sources of energy, although of course fuels would have value as sources of organic chemicals. This assumption implies

$$W_K = W_V = 0. \qquad \text{...(2.15)}$$

Given (2.15) we shall certainly have to suppose that $G(0, Z) > 0$ for $Z > 0$ (see condition (1.1a')). Using (2.15) we note that equations (2.13) and (2.14). reduce to the forms

$$\frac{\dot{C}}{C} = \frac{f'(x) - (\delta + \Psi)}{\eta(C)} \qquad \text{...(2.16)}$$

and

$$\dot{x} = \sigma f(x). \qquad \text{...(2.17)}$$

Assume that the utility function is iso-elastic. Having already assumed that $U(\cdot)$ is bounded above, this implies that $\eta > 1$. But the conditions (2.16) and (2.17) which an

[1] As one would expect, equation (2.13) reduces to equation (1.16) when $\Psi_t = 0$.
[2] As would be expected, (2.14) reduces to (1.18) if $\Psi_t = 0$.

optimal policy must satisfy differ from (1.16) and (1.24) only by the addition of an amount Ψ_t to the discount rate. Thus the optimal policy is to suppose that the substitute will *never* be discovered and to pursue the solution described in Section 1.4 with the single proviso that it be assumed that the utility rate of discount is not merely δ but $\delta+\Psi_t$. Since $\Psi_t>0$, one is in effect discounting at a higher rate in this uncertain problem than in the problem of Section 1.3. This is intuitively very reasonable. If, at T, the substitute appears, the economy switches regime and the optimal policy is to pursue the programme

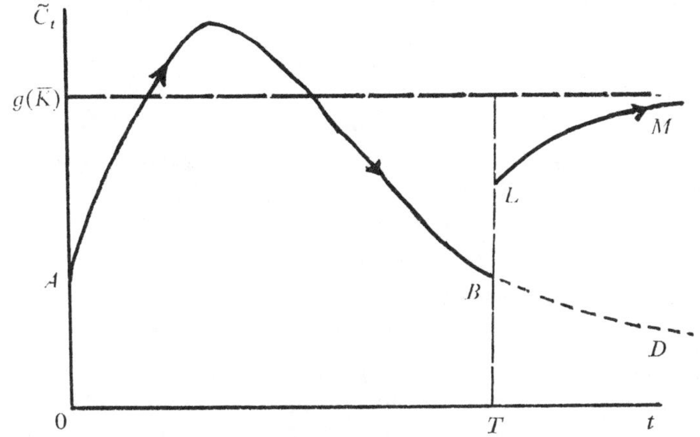

FIGURE 5

described in Proposition 4′. In fact Proposition 4′ is precisely what is required since assumption (2.15) entails a total break with the past at T. At T the economy is totally new, with no existing capital stock and no inventory of the resource. The break with the past is complete.[1] In Figure 5 we present the optimal consumption profile for this economy. *ABD* describes in broad terms the consumption profile of Figure 3. Throughout, however, the rate of discount is higher than δ by an amount Ψ_t. *LM* is the optimal consumption policy described in Proposition 4′.[2] We have then

Proposition 9. *Consider the two following problems:*

$$\left.\begin{aligned}&\textit{maximize} && \int_0^\infty [U(C)\Omega_t+\omega_t W(K_t, S_t)]e^{-\delta t}dt\\ &\textit{subject to} && \dot{K}=F(K, R)-C\\ & && \dot{S}=-R\\ & && C_t, K_t, R_t, S_t \geq 0 \textit{ and } K_0, S_0 \textit{ given}\end{aligned}\right\} \quad ...(2.18)$$

and

$$\left.\begin{aligned}&\textit{maximize} && \int_0^\infty e^{-\gamma_t}U(C)dt\\ &\textit{subject to the same constraints as in (2.18),}\end{aligned}\right\} \quad ...(2.19)$$

where γ is a discount rate which is independent of the path followed by the economy.

[1] In effect, then, condition (2.15) reduces the model to one where the economy faces an uncertain terminal date, a case that has been explored in another context by Yaari [28].

[2] As would be expected, there is generally a discontinuity in consumption at T. By assumption (2.15) the height TL is independent of the value taken by T.

Then, necessary and sufficient conditions for Problems (2.18) *and* (2.19) *to have identical solutions are*

(i) $W_K = W_S = 0$ *for all* K, S.

(ii) $\gamma_t = \delta t + \int_0^t \Psi_\tau d\tau.$

Proof. Sufficiency is obvious. To establish necessity note that the conditions that a solution of problem (2.19) must satisfy are

$$\frac{\dot{C}}{C} = \frac{(F_K - \gamma)}{\eta(C)} \qquad \text{...(2.20)}$$

and that

$$\dot{x} = \sigma f(x). \qquad \text{...(2.21)}$$

It follows that if equation (2.13) is identical to equation (2.20) we must have $W_K = 0$. But if (2.14) is identical to (2.21) we must have

$$\frac{\sigma f(x) W_K \Psi}{U'(C)f'(x)} + \frac{W_V \Psi}{x f''(x) U'(C)} = 0.$$

But since $W_K = 0$ it follows that $W_V = 0$. ||

What is interesting about Proposition 9 is that it gives a precise statement of a set of conditions under which we can allow for uncertainty merely by solving a certain problem with a suitably chosen discount factor depending only on the nature of the uncertainty (a kind of certainty equivalent result), *and* where the equivalent certain problem is independent of the expected supply of the substitute. The amount Ψ_t that has to be added to the utility discount rate, δ, is of course generally time variant and equals the conditional probability of the substitute arriving at t, given no previous occurrence. We need hardly point out that non-constancy of Ψ_t, does *not* imply that the optimal policy is intertemporally inconsistent. In general one suspects that the conditional probability distribution is complex, but possibly single peaked. For example, it has been suggested that the conditional probability of harnessing nuclear fusion will rise to a peak at about the end of this century but will then fall, the argument being that expectations, though rising over the near future, will diminish if the discovery is not made by the turn of the century. A conditional distribution having such attributes would result if ω_t were log normally distributed. For the simple case where $\omega_t = \pi e^{-\pi t}$ (i.e. ω_t, is a Poisson distribution) Ψ_t is constant and equal to π and the planning problem becomes still easier.

Proposition 9 yielded a sufficient *and* necessary set of conditions under which the optimal depletion problem could be solved by pretending that the substitute will never appear and solving for the optimal depletion problem by a simple increase in the utility discount rate. Raising the utility discount rate to handle uncertainty has much intuitive appeal. Proposition 9 implies that an error will be committed in doing this if the conditions do not hold. It is then, naturally interesting to ask whether we have any reasons for believing that the error is as likely to go one way as the other. In fact, it is possible to establish the following proposition about the nature of the errors introduced if the " certainty equivalent " problem is used as a guide to decision-making when it should not be.

Proposition 10. *If we suppose* $W_K = W_V = 0$ *and solve the certainty equivalent problem, when in fact* $W_K = 0$ *but* $W_V > 0$, *then the values of* R_0 *and* C_0 *thus chosen will be greater than or equal to their correct values, and at least one will be strictly greater.*

2.3. The results established in the previous sub-section may seem to be dependent upon some of the unpalatable simplifying assumptions used there—a one-sector economy, a homogeneous capital stock, etc. The point of this section is to observe that in fact similar results hold in a far more general situation. In order to see this point, it is helpful to consider an economy whose evolution is characterized by three functions of time, $\underline{C}$, $\underline{K}$ and $\underline{S}$, which are respectively mappings to R^1, R^m and R^n: $\underline{C}_t$ is the consumption vector at t; $\underline{K}_t$ the vector of capital stocks, and $\underline{S}_t$ the vector of remaining resource stocks. Over any finite period $[0, T]$, and given the initial values $\underline{K}_0$ and $\underline{S}_0$, there exists a set $\mathscr{E}_T$ of triples $(\underline{C}_t, \underline{K}_t, \underline{S}_t)$ such that a triple is in $\mathscr{E}_T$ if and only if it describes an evolution of the economy from 0 to T which is feasible given the technology and initial conditions. Associated with each such path over $[0, T]$ is another triple $(\underline{C}_T, \underline{K}_T, \underline{S}_T)$ giving the terminal values of consumption, stocks of capital and resources. For such an economy, we consider the following finite-horizon problem:

$$\text{maximize} \qquad \int_0^T U(\underline{C})e^{-\delta t} + e^{-\delta T}W(\underline{K}_T, \underline{S}_T)$$

$$\text{subject to} \qquad (\underline{C}_t, \underline{K}_t, \underline{S}_t) \in \mathscr{E}_T.$$

This involves maximizing a utility integral and a terminal stock valuation function, subject to the usual technological constraints. It is assumed that a solution to this problem exists for each T.

Now suppose that the horizon T is a random variable with density function ω_T, and that planners seek a path of the economy which will maximize the expected value of the utility integral plus stock valuation function over this random horizon—i.e. they seek to

$$\text{maximize} \qquad \int_0^\infty \omega_T \left\{ \int_0^T U(\underline{C})e^{-\delta t}dt + e^{-\delta T}W(\underline{K}_T, \underline{S}_T) \right\} dT.$$

To make this a well-defined maximization problem, it is necessary to specify the operative constraints. A reasonable formulation would be to require the programme adopted to be feasible whatever the horizon should turn out to be: if $\omega_T > 0$ for all t, this gives as a constraint

$$(\underline{C}_T, \underline{K}_T, \underline{S}_T) \in \mathscr{E}_\infty$$

where $\mathscr{E}_\infty$ is the set of programmes feasible *ad infinitum*, given the technology and initial conditions. We can now prove the following simple but useful result:

Proposition 11. *The solution to the problem.*

$$\textit{Maximize} \qquad \int_0^\infty \omega_T \left\{ \int_0^T U(C)e^{-\delta t}dt + e^{-\delta T}W(K_T, S_T) \right\} dT$$

$$\textit{subject to} \qquad (C_t, K_t, S_t) \in \mathscr{E}_\infty$$

is identical to the solution to

$$\textit{maximize} \qquad \int_0^\infty U(C)e^{-\delta t - \int_0^t \Psi_\tau d\tau} dt + \int_0^\infty \omega_t e^{-\delta t}W(K_t, S_t)dt$$

subject to the same constraint, where

$$\Psi_\tau = \omega_\tau \Big/ \int_\tau^\infty \omega_t dt.$$

Proof. Consider the term

$$\int_0^\infty \omega_T \left(\int_0^T U(C)e^{-\delta t}dt \right) dT.$$

Defining $\Omega_t = \int_t^\infty \omega_\tau d\tau$ and integrating by parts, this equals

$$\int_0^\infty U(C)\Omega_t e^{-\delta t}dt.$$

Now $\dot{\Omega}_t = -\omega_t$ so that $\dot{\Omega}_t/\Omega_t = -\omega_t/\Omega_t = -\Psi_t$. Hence on integration

$$\Omega_t = e^{-\int_0^t \Psi_\tau d\tau}. \|$$

An immediate corollary of this is the following

Proposition 12. *Consider the problems*

maximize $\int_0^\infty \omega_T \left\{ \int_0^T U(C)e^{-\delta t}dt + e^{-\delta T}W(K_T, S_T) \right\} dT$

subject to $(C_t, K_t, S_t) \in \mathscr{E}_\infty$ *and*

maximize $\int_0^\infty U(C)e^{-\alpha_t}dt$ *subject to the same constraint.*

Then a sufficient condition for these two problems to have the same solution is that

(i) $W(K_T, S_T)$ *should be independent of* K_T *and* S_T *for all values of these variables, and*

(ii) $\alpha_t = \delta t + \int_0^t \Psi_\tau d\tau$.

Proposition 12 is clearly a generalisation of Proposition 9, in that it states a set of conditions under which it is possible to replace an uncertain problem (which is clearly a generalisation of the problem of the previous sub-section), by a certain problem derived from it by a modification of the discount rate that makes use solely of the probability function.[1]

3. CONCLUDING REMARKS

In this paper we have attempted to answer some questions that appear to arise rather naturally when one thinks of intertemporal planning in the presence of exhaustible resources. Since many of these questions are, in fact, independent of one's exact notion of intergenerational equity we have posed them within the context of the, by now standard, utilitarian framework.[2] In Section 1.2 we presented a simple production model with no uncertainty. While it is plain that the characteristics of an optimal *depletion policy* depend crucially on whether or not the resource is essential to the production of final goods, it is not *a priori* plain as to what constitutes essentiality. The analysis led rather naturally to a simple articulation of this notion, namely the sufficiency conditions in Proposition 5 (see footnote 1, p. 15): that is, that *either* the marginal product of the resource is unbounded *or* that ouput of final goods is nil in the absence of the resource. It is intuitively clear that the elasticity of substitution between reproducible capital and exhaustible resources is an important determinant of the characteristics of an optimal

[1] The " certainty equivalent " problem that we have discussed here should not be confused with the problem of determining the optimal depletion rate where it is supposed that the substitute *will* become available at its *expected date* of arrival. That the latter problem may, in some case, yield a bias towards excessive depletion rates is intuitively clear, and it is this issue that is discussed by Henry [13]. In spite of its shortcomings the certainty equivalent approach (of either variety) is adopted explicitly or implicitly in much of the more empirically oriented literature in this type of problem. On this see Posner [24].

[2] Dasgupta [6] and Solow [25] have explored some of these questions within the context of alternative notions of intergenerational equity.

policy. It turned out that the parameter enters in a natural and in fact strikingly simple way in the intertemporal plan (viz. equation (1.18)). One would judge that for the near future at least, one will have to resort to simple production models (such as, say, the class of CES functions) to obtain numerical results. It is partly for this reason (and partly also for pedagogic ones) that we have presented our existence theorem (Proposition 8) for a world that has a production function that is of the CES form, and a utility function that is iso-elastic. It is in fact relatively simple to extend Proposition 8 to cover the case of production functions that are not of the CES variety as well as utility functions that are not iso-elastic. We need hardly point out that one would need to make appropriate assumptions (such as (0.2)) about the behaviour of the elasticity of marginal utility for low and high rates of consumption and that it is these asymptotic values of the elasticity of marginal utility that would enter into the condition that guarantees the existence of an optimal policy, namely that $\delta > \rho\ (1-\eta)$. If, in particular $\delta > \rho$, then we would be interested in the value of the elasticity of marginal utility as consumption tends to zero. Likewise, if $\rho > \delta$ then we would be interested in the value of the elasticity of marginal utility as consumption tends to infinity.

The main novelty that exhaustible resources introduce in a planning exercise is that one has to be particularly conscious about the properties of production functions at the " corners ". The banality of this observation is matched only by the problems this poses in obtaining empirical estimates. Certainly it is possible that we live in a world where for " moderate " values of the capital-resource ratio, the elasticity of substitution between capital and the resource exceeds one.[1] The point of concern, of course, is its behaviour for large values of the capital-resource ratio, given that large values cannot be avoided in the long run.[2] For the purposes of planning one is particularly interested in the conditions under which it is optimal to spread the exhaustible resource thinly over the distant future and thereby never to exhaust it completely. In a *very* general sense the analysis of Section 1.2 has been re-assuring to intutition in that the conditions under which it is optimal to exhaust the resource in finite time are really rather stringent (see Proposition 7). Certainly, if we wish to rely on the working hypothesis that the elasticity of substitution is constant then irrespective of its value (except for the case where it is infinite, which is, of course, just silly) we should not exhaust the resource in finite time. But the analysis does suggest that if $\sigma \geqq 1$ but σ is not unduly large, the price of the exhaustible resource relative to output ought to be rising rather rapidly. Experience does not suggest that this has been the case.[3]

In Section 2 we presented a model incorporating technical progress under uncertainty. It seems entirely reasonable to suppose that the important uncertainty concerns the date of arrival of new knowledge. Furthermore, it seemed to us that for the kind of problem that we have in mind here, it would not be very appropriate to make use of the law of large numbers (as say, would be implicitly the case for models such as those of Arrow [2] and Kaldor and Mirrlees [16]) to generate technical progress in a continuous fashion. Rather, we are here trying to envisage it as coming in a discrete manner. The event (or perhaps we should say " Event "!) can, of course, be quite anything. It can, for instance, be the discovery of a new stock of the same resource (e.g. new beds of oil or natural gas). But in this event exhaustible resources would remain a fundamental problem if they were initially so. Certainly the discussions that are currently conducted amongst energy experts suggest that what they have in mind is the possibility of discovering a source that will provide, for all practical purposes, an unlimited flow of energy. It is for such reasons that we have explored the problems that arise if technological change consists in the economy being provided by a steady stream of an alternative resource which can replace

1 The important work by Nordhaus and Tobin [22] suggests that this may be true.

2 Either this or one runs down capital sufficiently rapidly as well, in which event consumption trivially goes to zero in the absence of technical progress.

3 For an analysis of the behaviour of oil prices in particular, see Heal [12].

a resource that was previously essential. The date of the discovery we have supposed uncertain, but in this paper we have supposed that the random variable is uninfluenced by policy; that is, that the acquisition of knowledge is costless.

While it appears to be customary in practice to cope with uncertainty by raising the rate of discount by an amount that merely reflects the uncertainty, it is generally recognized that it leads to errors. In Section 2.2 we explored the conditions under which no such error will in fact be committed. It emerged that this is so only when both the existing stock of capital and the stock of the exhaustible resource is judged totally devoid of value at the date of the technical change (Proposition 9). It appeared, not surprisingly perhaps, that in this instance the model bears a strong resemblance to one where the economy faces an uncertain terminal date; a case that has been explored at some length by Yaari [28]. That one will, in general, commit an error by this simple method of coping with uncertainty is not surprising. But it is not plain that there will be a bias. In fact, one can say something specific about such biases, and we have presented the result in the form of Proposition 10.

Finally, and at a more primitive level, while one can take the view (reflected in the model economy of Section 2) that even though exhaustible resources may be essential currently they will not remain so over the indefinite future, it is not plain that it is a particularly comforting view. The model of uncertainty that we have presented here certainly articulates the view that the probability of the new source being discovered *sometime* over the indefinite future is precisely one. But it may well be a long while coming. If in fact major discoveries occur only at a great distance in the future, the intervening generations will naturally be that much worse off. In particular, one can argue that one's conception of intergenerational justice ought to be influenced by the likelihood of the arrival of technological change. This may well be an appropriate view. In this paper we have not attempted to face this set of difficult questions. Our purpose has been to explore some of the immediate implications of the existence of exhaustible resources.

REFERENCES

[1] Anderson, K. P. " Optimal Growth when the Stock of Resources is Finite and Depletable ", *Journal of Economic Theory* (1972).

[2] Arrow, K. J. " The Economic Implications of Learning by Doing ", *Review of Economic Studies* (1962).

[3] Beckerman, W. " Economists, Scientists and Environmental Catastrophe ", *Oxford Economic Papers* (1972).

[4] Cole, H. S. D. *et al.* *Thinking about the Future: A Critique of the Limits to Growth* (Chatto and Windus, for Sussex University Press, 1973).

[5] Dasgupta, P. " On the Concept of Optimum Population ", *Review of Economic Studies* (1969).

[6] Dasgupta, P. " Some Alternative Criteria for Justice Between Generations ", Paper presented at the ISPE Conference on the Demand for Public Goods, Siena (1973).

[7] Dasgupta, P. " Some Theoretical Explorations in the Economics of Exhaustible Resources ", Paper presented at the Bayerisch Academy Symposium on Systems Analysis, Reisenburg (1973).

[8] Forrester, J. W. *World Dynamics* (Wright-Allen Press Inc., Cambridge, Mass., 1971).

[9] Gale, D. " Optimal Development in a Multi-Sector Economy ", *Review of Economic Studies* (1967).

[10] Heal, G. M. " The Depletion of Exhaustible Resources: Some Theoretical Issues ", in M. Parkin and R. Nobay (eds.) *Issues in Contemporary Economics*, (Manchester University Press, 1974).

[11] Heal, G. M. *The Theory of Planning* (North Holland, 1973).

[12] Heal, G. M. " Economic Aspects of Natural Resource Depletion ", in D. Pearce (ed.) *Economics of Natural Resource Depletion*, (Macmillan, London, 1975).

[13] Henry, C. " Irreversible Investment Decisions Under Uncertainty ", *American Economic Review*, forthcoming.

[14] Hotelling, H. " The Economics of Exhaustible Resources ", *Journal of Political Economy* (1931).

[15] Ingham, A. and Simmonds, P. " The Optimal Use of Natural Resources ", *Review of Economic Studies*, forthcoming.

[16] Kaldor, N. and Mirrlees, J. A. " A New Model of Economic Growth ", *Review of Economic Studies* (1962).

[17] Koopmans, T. C. " A Proof for the Case where Discounting Advances the Doomsday". *Review of Economic Studies* (this issue).

[18] Koopmans, T. C. " Some Observations on ' Optimal ' Economic Growth and Exhaustible Resources ", Cowles Foundation Discussion Paper, No. 356 (1973).

[19] Maddox, J. *The Doomsday Syndrome* (Macmillan, London, 1971).

[20] Meadows, D. H. *et al.* *The Limits to Growth* (Universe Books, New York, 1972).

[21] Nordhaus, W. D. " World Dynamics: Measurement without Data ", Cowles Foundation Discussion Paper (1972).

[22] Nordhaus, W. D. and Tobin, J. " Is Economic Growth Obsolete ", in *Economic Growth* (Fiftieth Anniversary Colloquium, V), National Bureau of Economic Research (1972).

[23] Pontryagin, L. S. *et al.* *The Mathematical Theory of Optimal Processes* (Interscience, New York, 1972).

[24] Posner, M. V. *Fuel Policy: A Study in Applied Economics* (MacMillan, London, 1972).

[25] Solow, R. M. " Intergenerational Equity and Exhaustible Resources ", *Review of Economic Studies* (this issue).

[26] Stiglitz, J. E. " Growth with Exhaustible Natural Resources ", *Review of Economic Studies* (this issue).

[27] Vousden, N, " Basic Theoretical Issues in Resource Depletion ", *Journal of Economic Theory* (1973).

[28] Yaari, M. E. " Uncertain Lifetime, Life Insurance, and the Theory of the Consumer ", *Review of Economic Studies* (1965).

[22]

U. S. energy policy and economic growth, 1975-2000

Edward A. Hudson

Senior Economist
Data Resources, Incorporated

and

Dale W. Jorgenson

Professor of Economics
Harvard University

This paper presents a new approach to the quantitative analysis of U.S. energy policy, based on an integration of econometric modeling and input-output analysis. It incorporates a new methodology for assessing the impact of economic policy on both demand and supply for energy within a complete econometric model of the U.S. economy. The model consists of production models for nine industrial sectors, a model of consumer demand, and a macro-econometric growth model for the U.S. economy. The model is first used to project economic activity and energy utilization for the period 1975 to 2000 under the assumption of no change in energy policy. The model is then employed to design a tax program for stimulating energy conservation and reducing dependence on imported sources of energy. The overall conclusion of the analysis of tax policy is that substantial reductions in energy use can be achieved without major economic cost.

1. Introduction

■ The dramatic increase in world petroleum prices associated with the Arab oil embargo of October 1973 has highlighted the need for a new approach to the quantitative analysis of economic policy. Econometric models in the Tinbergen-Klein mold have proved to be very useful in studying the impact of economic policy on aggregate demand.[1] At the same time these models do not provide an adequate basis for assessing the impact of economic policy on supply. Input-output analysis in the form originated by Leontief is useful for a very detailed analysis of supply, predicated on a fixed technology at any point of time.[2] Input-output analysis does not provide

Dale W. Jorgenson received the B.A. from Reed College (1955), and the A.M. and Ph.D. from Harvard University (1957 and 1959, respectively). He was the 1971 recipient of the John Bates Clark Medal of the American Economic Association. His interests include the econometrics of production and investment.

Edward A. Hudson received the B.A. from Victoria University, New Zealand, in 1968 and the Ph.D. from Harvard University in 1973. His research interests include energy economics and public finance.

[1] The seminal contribution to macro-econometric modeling of the U. S. economy is the Klein-Goldberger model in [27]. For a recent review of macro-econometric models of the United States, see Hickman [23].

[2] For the original development of input-output analysis, see Leontief [30].

a means of assessing the impact of changes in technology induced by price variations associated with changes in economic policy.

The purpose of this paper is to present a new approach to the quantitative analysis of U. S. energy policy.[3] This approach is based on an integration of econometric modeling and input-output analysis and incorporates an entirely new methodology for assessing the impact of economic policy on supply. We combine the determinants of energy demand and supply within the same framework and relate patterns of U. S. economic growth to both demand and supply. Our approach can be used to project U. S. economic growth and energy utilization for any proposed U. S. energy policy. It can be employed to study the impact of specific policy changes on energy demand and supply, energy price and cost, energy imports and exports, and on U. S. economic growth.

The first component of our framework for energy policy analysis is an econometric model of interindustry transactions for nine domestic industries. We have subdivided the business sector of the U. S. economy into nine industrial groups in order to provide for the detailed analysis of the impact of U. S. energy policy on the sectors most directly affected by policy changes. The nine sectors included in the model are:

(1) agriculture, nonfuel mining, and construction
(2) manufacturing, excluding petroleum refining
(3) transportation
(4) communications, trade, and services
(5) coal mining
(6) crude petroleum and natural gas
(7) petroleum refining
(8) electric utilities
(9) gas utilities.

Our interindustry model includes a model of demand for inputs and supply of output for each of the nine industrial sectors. The model is closed by balance equations between demand and supply for the products of each of the nine sectors.

The principal innovation of our interindustry model is that the input-output coefficients are treated as endogenous variables rather than exogenously given parameters. Our model for producer behavior determines the input-output coefficients for each of the nine sectors listed above as functions of the prices of products of all sectors, the prices of labor and capital services, and the prices of competing imports. We determine the prices of all nine products and the matrix of input-output coefficients simultaneously. In conventional input-output analysis the technology of each sector is taken as fixed at any point of time. Prices are determined as functions of the input-output coefficients, but the input-output coefficients themselves are treated as exogenously given parameters. Our approach integrates conventional input-output analysis with a deter-

A recent compendium of research on input-output analysis is Carter and Brody [10].

[3] A more detailed presentation of our approach is contained in Jorgenson, Berndt, Christensen, and Hudson [26].

mination of the structure of technology through models of supply for each industrial sector.

The second component of our framework for energy policy analysis is a macro-econometric growth model. The complete model consists of endogenous business and household sectors and exogenous foreign and government sectors. The chief novelty of our growth model is the integration of demand and supply conditions for consumption, investment, capital, and labor. The model is made dynamic by links between investment and changes in capital stock and between capital service prices and changes in investment goods prices. The model determines both components of gross national product in real terms, as generated by conventional macro-econometric models, and relative prices of labor and capital services required by our econometric model of interindustry transactions.

Our approach to the analysis of macro-economic activity can be contrasted with the analysis that underlies macro-econometric models used for short-term forecasting. Short-term forecasting is based on the projection of demand by foreign and government sectors and the determination of the responses of households and businesses in the form of demands for consumption and investment goods. The underlying economic theory is essentially the Keynesian multiplier, made dynamic by introducing lags in the responses of households and businesses to changes in income. In short-term macro-econometric models the supply side is frequently absent or present in only rudimentary form.[4] Our approach integrates the determinants of demand employed in conventional macro-econometric models with the determinants of supply; this integration is essential to the successful implementation of our interindustry model for longer run analyses.

Given a framework that incorporates the determinants of demand and supply for energy in the U. S. economy, our first objective is to provide a reference point for the analysis of energy policy by establishing detailed projections of demand and supply, price and cost, and imports and exports for each of the nine industrial sectors included in our model. For this purpose we project the level of activity in each industrial sector and relative prices for the products of all sectors for the years 1975 through 2000. Our projections include the level of macro-economic activity in the U. S. economy and the matrix of input-output coefficients for each year. Projections for the five industrial sectors that form the energy sector of the U. S. economy provide the basis for translating our detailed projections into the energy balance framework that has become conventional in the analysis of patterns of energy utilization.[5]

Our interindustry approach imposes the same consistency requirements as the energy balance approach, namely, that demand is equal to supply in physical terms for each type of energy. In addition, our approach requires that demand and supply be consistent with the same structure of energy prices. This additional consistency

---

[4] In the Klein-Goldberger model the determination of prices can be completely suppressed with a resulting improvement in forecasting accuracy for real magnitudes. See Suits [38] and Goldberger [22].

[5] The energy balance framework has been employed by Dupree and West [20] and the National Petroleum Council [35, 36].

requirement is absent from energy balance projections and requires the integration of energy balance projections with projections of energy prices. Our interindustry model provides a means of combining these projections within a framework that also includes prices and interindustry transactions for the sectors that consume but do not produce energy.

To illustrate the application of our model to the analysis of U. S. energy policy we have analyzed the effects of tax policies to stimulate energy conservation on the future pattern of energy utilization. Our methodology for policy analysis begins with a set of projections that assume no major new departures in energy policy. We then prepare an alternative set of projections incorporating the proposed change in policy. In analyzing the impact of tax policy we have incorporated the effect of energy taxes on demand and supply for energy. We find that price increases provide the economic incentive for the adoption of energy conservation measures that will result in considerable savings of energy. Tax policies or other measures to increase the price of energy could result in U. S. independence from energy imports by 1985.

We present our model for interindustry transactions in Section 2 of the paper. In Section 3 we present econometric models of producer behavior for each of the nine sectors included in our interindustry model. We then outline our macro-econometric growth model in Section 4. In Section 5 we present projections of economic activity and energy utilization for the period 1975 through 2000. In Section 6 we discuss a tax program for stimulating energy conservation and eliminating reliance of the U. S. economy on energy imports.

2. Interindustry model

■ The first component of our framework for energy policy analysis is a model of interindustry transactions for the United States. Rather than analyzing energy utilization in isolation, we begin with an analysis of the entire U. S. economy and then proceed to a detailed examination of the energy sector as one among many interdependent components of the economy. This perspective is necessarily more complex and more detailed than traditional perspectives on the analysis of the energy sector, but is indispensable to the study of the interaction of energy resources and the growth of the U. S. economy.

Our interindustry model permits the analysis of the entire chain of production from the purchase of primary inputs through the various intermediate stages of production to the emergence of final products to be absorbed in consumption, investment, government, or export final demand. The structure of production includes all of U. S. domestic supply of goods and services, but our specification provides for detailed analysis of the impact of U. S. energy policy on the sectors most directly affected by policy changes. We have classified production into nine industrial sectors, each of which purchases primary inputs, makes purchases from and sales to the other producing sectors, and sells finished output to final users. The flow of interindustry transactions is represented in diagrammatic form in Figure 1.

FIGURE 1

INTERINDUSTRY TRANSACTIONS: DIAGRAMMATIC REPRESENTATION

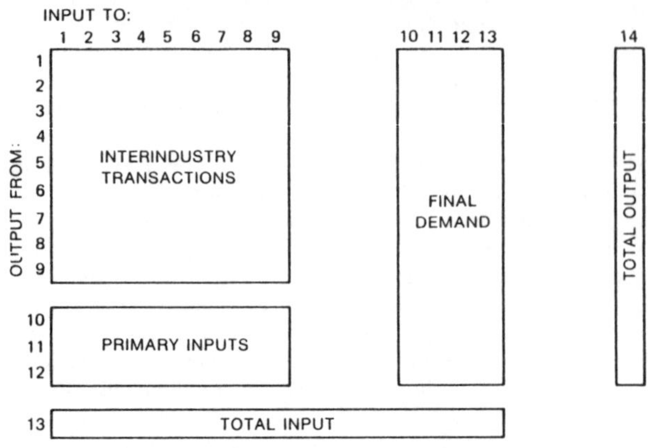

INTERMEDIATE SECTORS:
1. AGRICULTURE, NONFUEL MINING AND CONSTRUCTION.
2. MANUFACTURING, EXCLUDING PETROLEUM REFINING.
3. TRANSPORT.
4. COMMUNICATIONS, TRADE, SERVICES.
5. COAL MINING.
6. CRUDE PETROLEUM AND NATURAL GAS.
7. PETROLEUM REFINING.
8. ELECTRIC UTILITIES.
9. GAS UTILITIES.

PRIMARY INPUTS, ROWS:
10. IMPORTS.
11. CAPITAL SERVICES.
12. LABOR SERVICES.

FINAL DEMAND, COLUMNS:
10. PERSONAL CONSUMPTION EXPENDITURES.
11. GROSS DOMESTIC PRIVATE INVESTMENT.
12. GOVERNMENT PURCHASES OF GOODS AND SERVICES.
13. EXPORTS.

Our interindustry model consists of balance equations between supply and demand for the products of each of the nine sectors included in the model. The model also includes accounting identities between the value of domestic availability of these products and the sum of values of intermediate input into each industry, value added in the industry, and imports of competing products. Demands for the products include demands for use as inputs by each of the nine sectors included in the model. The rest of domestic availability is allocated among four categories of final demand: personal consumption expenditures, gross private domestic investment, government expenditures, and exports.

In the model for projecting energy demand and supply we take the levels of final demand for all industries from the macro-econometric model presented in Section 4, below. Second, for the five energy sectors of the model we take the price and quantity of imports to be exogenous. For the four nonenergy sectors we take the prices of imports as exogenous and determine import quantities along with the quantities of capital and labor services in each industry.[6] The prices of capital and labor services are determined within the macro-econometric model. We take the quantities of exports and government purchases of the output of each industry as exogenous. We also take the allocation of investment among the industries of origin to be exogenous.

Our interindustry model consists of models of producer behavior for each of the nine industries included in the model. Producer behavior in each industry can be characterized by input-output coefficients for the input of products of each of the nine sectors, inputs of capital and labor services, and, for the four nonenergy sectors, the level of competitive imports. An interindustry approach to the study of energy resources is essential, since most energy is consumed as an intermediate rather than a final product of the economy. Examples of intermediate products would be fossil fuels consumed by the electric generating sector. Examples of final products would be gasoline and heating oil consumed by the household and government sectors. Energy balance models provide projections of the levels of both intermediate and final demand.

Given the prices of domestic availability of the output of each sector included in our model, we determine the allocation of personal consumption expenditures among commodity groups distinguished in the model, using our model of consumer behavior. Personal consumption expenditures include deliveries to the household sector by eight of the nine sectors included in our interindustry model of the producing sector. There are no direct deliveries of crude petroleum and natural gas to personal consumption expenditures. These products are delivered first to the petroleum refining and gas utility sectors and then to personal consumption expenditures and to other categories of intermediate and final demand.

Personal consumption expenditures also include noncompetitive imports and the services of dwellings and consumers' durables. The levels of personal consumption expenditures on each of the eleven commodity groups included in our model of the household sector are determined from the projected level of personal consumption expenditures from the macro-econometric model, from the prices of domestic availability of the output of each sector included in the interindustry model, and from the prices of noncompetitive imports, consumers' durables services, and housing services. The price of noncompetitive imports is taken to be exogenous. The capital service prices for consumers' durables services and housing services are determined from the price of capital services determined in the macro-econometric model.

The equations representing the balance of demand and supply for each of the nine sectors of the interindustry model set domestic availability equal to the sum of intermediate demands and final demand. Intermediate demands are determined simultaneously with the levels of output of each industry, given input-output coefficients determined in the model of producer behavior. The input-output

[6] Energy imports are significant only for crude and refined petroleum products and natural gas. For the period 1958 to 1972 petroleum imports were subject to a system of quotas. Natural gas imports are subject to regulation by the Federal Power Commission. For a discussion of the petroleum import quota system, see Burrows and Domencich [8].

coefficients are determined simultaneously with the prices of domestic availability of the output of each industry. Finally, levels of capital and labor services for all sectors and competitive imports for the four nonenergy sectors are determined from the levels of domestic availability and the corresponding input-output coefficients. These levels can be compared with the levels projected in the macro-econometric model.

☐ **Interindustry transactions.** We first describe our model of interindustry transactions and then outline the application of this model to the projection of energy demand and supply. Our notation is as follows:

XIJ = intermediate demand for the output of industry I by industry J;
YI = final demand for output of industry I;
XI = domestic availability of the output of industry I;
PI = price of the output of industry I.

To simplify the notation we take the price of the output of each industry to be the same in all uses. The deflators for each category of intermediate and final demand can differ. In projecting energy demand and supply we take the ratios of the deflator for the individual categories of demand to the deflator for domestic availability of output of the industry to be exogenous.

The interindustry model consists of equality between demand and supply for each of the nine sectors included in the model. The balance equations for the nine sectors are:

$$XI = \sum_{J=1}^{9} XIJ + YI, \qquad (I = 1, 2 \dots 9). \tag{1}$$

In addition, the model includes accounting identities between the value of domestic availability and the sum of values of intermediate input into the industry, value added in the industry, and, for the four-nonenergy sectors, the imports of competing products:

$$PI^*XI = \sum_{J=1}^{9} PJ^*XJI + PK^*KI + PL^*LI + PRI^*RI, \qquad (I = 1, 2 \dots 9), \tag{2}$$

where:

KI = quantity of capital services in industry I;
LI = quantity of labor services in industry I;
RI = competitive imports of the output of industry I;
PK = price of capital services;
PL = price of labor services;
PRI = price of competitive imports to industry I.

Again, prices of capital and labor services can differ among industries. To simplify notation we take the prices of these productive factors to be the same in all industries. In projecting energy demand and supply we take the ratios of service prices for each industry to the corresponding prices from the macro-econometric model to be exogenous.

Our interindustry model includes models of producer behavior for each of the nine industrial sectors included in the model. These models of producer behavior can be derived from price possibility frontiers for the nine sectors:

$$AI*PI = GI(P1, P2, \ldots P9; PK, PL, PRI), \quad (3)$$
$$(I = 1, 2 \ldots 9),$$

where $AI(I = 1, 2 \ldots 9)$ is an index of the level of Hicks-neutral technology in industry I. The price possibility frontier for each sector can be derived from price possibility frontiers for each of the three submodels employed in our analysis of production structure:[7]

(1) a model giving the price of output as a function of prices of four aggregate inputs in each sector—capital (K), labor (L), energy (E), and materials (M);

(2) a model giving the price of aggregate energy input in each sector as a function of the prices of the five types of energy included in the model—coal, crude petroleum and natural gas, refined petroleum products, electricity, and gas as a product of gas utilities;

(3) a model giving the price of aggregate nonenergy input in each sector as a function of the prices of the five types of nonenergy input into each sector—agriculture, manufacturing, transportation, communications, and, for the four nonenergy sectors, competitive imports.

Given the prices of capital services, labor services, and competitive imports in each of the four nonenergy sectors, we can determine the prices of domestic availability of output PI ($I = 1, 2 \ldots 9$) for all nine sectors. To determine these prices we solve twenty-seven equations for prices of domestic availability, prices of aggregate energy input, and prices of aggregate nonenergy input into all nine sectors. This system of twenty-seven equations consists of three equations for each sector. These three equations correspond to production possibility frontiers for each of the three submodels for each sector. In these computations we are making use of a nonsubstitution theorem of the type first discussed by Samuelson.[8] This theorem states that for given prices of the factors of production and competitive imports, the prices of domestic availability of the output of each sector are independent of the composition of final demand.

The second step in our analysis of interindustry transactions is to derive input-output coefficients for each of the nine industrial sectors included in our interindustry model. The input-output coefficients can be expressed as functions of the prices. First, the relative share of the jth intermediate input can be determined from the identity:

$$\frac{\partial \ln PI}{\partial \ln PJ} = \frac{PJ*XJI}{PI*XI} = \frac{PJ}{PI}*AJI, \quad (I, J = 1, 2 \ldots 9), \quad (4)$$

where AJI is the input-output coefficient corresponding to XJI; it represents the input of the output of industry J per unit of output of

[7] For a detailed interpretation of the price possibility frontier, see Christensen, Jorgenson, and Lau [18], especially pp. 32–33.

[8] See Samuelson [37].

industry I. Similar identities determine the relative shares of capital and labor services and competitive imports.[9]

Second, we can divide the relative shares by the ratio of the price of domestic availability of the output of the Jth industry, PJ to the price for the Ith industry, PI, to obtain the input-output coefficients:

$$\frac{XJI}{XI} = AJI(P1, P2 \ldots P9; PK, PL, PRI), \qquad (I, J = 1, 2 \ldots 9); \tag{5}$$

and

$$\begin{aligned} \frac{KI}{XI} &= AKI(P1, P2 \ldots P9; PK, PL, PRI), \\ \frac{LI}{XI} &= ALI(P1, P2 \ldots P9; PK, PL, PRI), \\ \frac{RI}{XI} &= ARI(P1, P2 \ldots P9; PK, PL, PRI), \\ &(I = 1, 2 \ldots 9). \end{aligned} \tag{6}$$

For each industry we derive the input-output coefficients in two steps. First, we determine the input-output coefficients for the aggregate inputs—capital (K), labor (L), energy (E) and materials (M). Second, we determine the input-output coefficients for the input of each type of energy input per unit of total energy input and the input of each type of nonenergy input per unit of total nonenergy input. To obtain the input-output coefficients required for our interindustry model we multiply the input-output coefficients for each type of energy by the input-output coefficient for total energy. Similarly, we multiply the input-output coefficients for each type of nonenergy input by the input-output coefficient for total nonenergy input. We obtain input-output coefficients for capital services, labor services, five types of energy inputs into each sector, and five types of nonenergy inputs into each sector.

The input-output coefficients for each of the nine industrial sectors included in our model of interindustry transactions are functions of the prices of capital services, labor services, and competitive imports for the four nonenergy sectors and the prices of domestic availability of the output of each of the nine sectors. The prices of domestic availability are functions of the prices of capital services, labor services, and competitive imports for the four nonenergy sectors. By the nonsubstitution theorem both prices of domestic availability and input-output coefficients are independent of the composition of final demand.[10]

---

[9] For further discussion of the model of producer behavior, see Section 3, below.

[10] The idea of treating input-output coefficients as functions of prices can be traced to Walras [41], especially pp. 382–392; this approach has been extensively discussed by Samuelson [37], pp. 513–536, and Morishima [33], pp. 54–92. A more influential idea is to model trends in input-output coefficients without treating them as part of a model of producer behavior. This alternative approach has been employed by Leontief [31], Carter [9] and Almon, *et al.* [1]. Comparisons of input-output coefficients for 1947, 1958, and 1961 are given by Carter [9] and Vaccara [40].

☐ **Final demand.** Final demand for domestic availability of the output of each of the nine sectors included in our interindustry model is allocated among personal consumption expenditures, gross private domestic investment, government expenditures, and exports. In projecting energy demand and supply we take aggregate levels of each category of final demand from our macro-economic projections. We allocate personal consumption expenditures among the nine sectors included in our model, employing aggregate personal consumption expenditures as total expenditures, on the basis of the prices of domestic availability of the output of all nine sectors. Government expenditures and exports of the output of each sector are exogenous. Imports of the output of the five energy sectors are also exogenous so that we include only exports net of imports in final demand for these sectors. We take aggregate private domestic investment from our macro-economic projections. We take the relative proportion of investment in the output of each industrial sector included in our interindustry model to be exogenous.

The final step in determining the level and composition of interindustry transactions is to determine the levels of output, employment, and utilization of capital for each of the nine industrial sectors included in our model and competitive imports for the four nonenergy sectors included in the model. This part of our model coincides with conventional input-output analysis. Given the input-output coefficients for all nine sectors, we can determine the level of output for each sector for any given levels of final demand for the output of all nine sectors. We present projected matrices of interindustry transactions in energy for the years 1975, 1980, 1985, and 2000 in Section 5 below. We also present projections of energy prices for each of these years.

Final demand for domestic availability of the output of each of the nine sectors included in the model is allocated among consumption, investment, government expenditures, and exports:

$$YI = CI + II + GI + ZI, \qquad (I = 1, 2 \ldots 9), \qquad (7)$$

where:

CI = personal consumption expenditures on the output of industry I;
II = gross private domestic investment in the output of industry I (the sum of gross private fixed investment and net inventory change);
GI = government expenditure on the output of industry I;
ZI = exports of the output of industry I (exports less imports for the five energy sectors).

In our model for projecting energy demand and supply we take the levels of final demand for all industries from the macro-economic projections. We link our interindustry model to our macro-econometric model through the identities:

$$PC^*C = \sum_{I=1}^{9} PI^*CI,$$

$$PI^*I = \sum_{I=1}^{9} PI^*II, \tag{8}$$

$$PG^*G = \sum_{I=1}^{9} PI^*GI,$$

$$PZ^*Z = \sum_{I=1}^{9} PI^*ZI.$$

The values of personal consumption expenditures PC^*C, gross private domestic investment PI^*I, government expenditure PG^*G, and exports PZ^*Z from the macro-econometric model are set equal to the sums of each of these categories of expenditures over all nine industries included in the interindustry model.

In the macro-econometric model government expenditures on goods and services are divided into two parts:

$$PG^*G = PIG^*IG + PCG^*CG, \tag{9}$$

where

IG = quantity of government expenditures on investment goods;
CG = quantity of government expenditures on consumption goods;
PIG = price of government expenditures on investment goods;
PCG = price of government expenditures on consumption goods.

In making projections of energy demand and supply we project total government expenditures in current and constant prices. Government expenditures on goods and services are exogenous to the macro-econometric model. We then allocate total government expenditures among the nine industry groups included in the interindustry model.

In our macro-econometric model net exports of goods and services are taken to be exogenous. For the purposes of projecting energy demand and supply we divide net exports between imports and exports, allocate exports among the nine industry groups included in the model, and project the prices of competitive imports for each of the nine sectors of the model. Since net exports in current and constant prices are exogenous to the macro-econometric model, this change in the treatment of net exports does not alter the structure of the complete model. In projecting energy demand and supply we take the prices of imports for each industrial sector together with levels of capital and labor services for each sector.

We project gross private domestic investment in current and constant prices in our macro-econometric model. To project energy demand and supply we allocate gross private domestic investment among the nine industry groups included in the model. The relative proportions of investment originating in each sector are taken to be exogenous. In a completely dynamic model the allocation of investment by sector of origin and sector of destination would be endogenous. Our macro-econometric model incorporates the dynamics of saving and investment only in the projection of total investment. The allocation of capital by sector of destination is endogenously determined, but the allocation of investment by sector of origin is exogenous.

The final step in determining final demand for the domestic availability of the output of each sector included in our interindustry model is to allocate personal consumption expenditures among the products of the nine sectors included in the model and expenditures on noncompetitive imports and the services of consumers' durables, which are not included among the products of the nine sectors. For this purpose we employ an econometric model of consumer behavior. This model is based on an indirect utility function that can be represented in the form:[11]

$$\ln V = \ln V\left(\frac{P1}{PC{*}C}, \frac{P2}{PC{*}C} \cdots \frac{P11}{PC{*}C}\right), \tag{10}$$

where V is the level of utility, PI is the price of the Ith commodity, and $PC{*}C$ is total personal consumption expenditures. There are eleven commodity groups included in our model of consumer behavior: one for each of the nine industrial sectors, excluding crude petroleum and natural gas, and one each for housing, services of consumers' durables, and noncompetitive imports.

For each commodity group we take the budget share to be fixed. This assumption corresponds to a linear logarithmic indirect utility function. The quantity demanded for each of the eleven commodity groups included in our model of consumer behavior is a function of the prices of the corresponding commodity group and the level of total personal consumption expenditures. We determine the level of total personal consumption expenditures in our macroeconometric model. We determine the prices of domestic availability of the output of each sector in the interindustry model from our models of producer behavior. Given total expenditures and the prices, we can determine the quantities demanded of the output of each sector of the model for personal consumption expenditures.

Final demand for all nine industrial sectors included in our interindustry model is the sum of final demands for personal consumption expenditures, gross private domestic investment, government expenditures, and exports. We add personal consumption expenditures for the output of communications, trade, and services, less housing, to personal consumption expenditures for housing to obtain personal consumption expenditures on communications, trade, and services. Otherwise, there is a direct correspondence between commodity groups in our model of consumer behavior and the sectors included in our interindustry model. Given all final demands, the prices of domestic availability of the output of each sector and the matrix of input-output coefficients, we can determine the matrix of interindustry transactions in both current and constant prices.

The equations representing the balance of demand and supply for each of the nine industrial sectors included in our interindustry

[11] For a detailed discussion of the indirect utility function, see Christensen, Jorgenson, and Lau [19].

model can be presented in the form:

$$XJ = \sum_{J=1}^{9} XIJ + YI, \tag{11}$$
$$= \sum_{J=1}^{9} AIJ{*}XJ + CI + II + GI + ZI, \quad (I = 1, 2 \ldots 9).$$

The input-output coefficients $\{AIJ\}$ are determined together with the prices of domestic availability of the output of each industry $\{PI\}$. Given prices and the level of aggregate personal consumption expenditures, the levels of personal consumption expenditures $\{CI\}$ are determined. The remaining components of final demand $\{II, GI, ZI\}$ are projected by industry of origin. These projections are consistent with levels of gross private domestic investment, government expenditures and exports from our macro-econometric model.

In matrix form the demand and supply balance equation for the model can be represented as:

$$x = Ax + y, \tag{12}$$

where x and y are vectors of outputs and final demands:

$$x = \begin{vmatrix} X1 \\ X2 \\ \cdot \\ \cdot \\ \cdot \\ X9 \end{vmatrix}, \qquad y = \begin{vmatrix} Y1 \\ Y2 \\ \cdot \\ \cdot \\ \cdot \\ Y9 \end{vmatrix},$$

and A is the matrix of input-output coefficients:

$$A = \begin{vmatrix} A11 & A12 & \ldots & A19 \\ A21 & A22 & \ldots & A29 \\ \cdot & \cdot & & \cdot \\ \cdot & \cdot & & \cdot \\ \cdot & \cdot & & \cdot \\ A91 & A92 & \ldots & A99 \end{vmatrix}.$$

Levels of domestic availability of the output of each vector are obtained by solving this system of equations:

$$x = (I - A)^{-1}y. \tag{13}$$

Levels of capital and labor services and competitive imports are determined from the levels of domestic availability and the corresponding input-output coefficients:

$$\begin{aligned} KI &= AKI{*}XI, \\ LI &= ALI{*}XI, \\ RI &= ARI{*}XI, \\ (I &= 1, 2 \ldots 9). \end{aligned} \tag{14}$$

The input-output coefficients for capital and labor services and competitive imports are functions of the prices of the outputs of the nine sectors included in our interindustry model, the prices of capital and labor services and the prices of competitive imports for the four nonenergy sectors of the model. Our complete econometric model for interindustry transactions is presented in diagrammatic form in Figure 2.

FIGURE 2

INTERINDUSTRY ECONOMETRIC MODEL: DIAGRAMMATIC REPRESENTATION

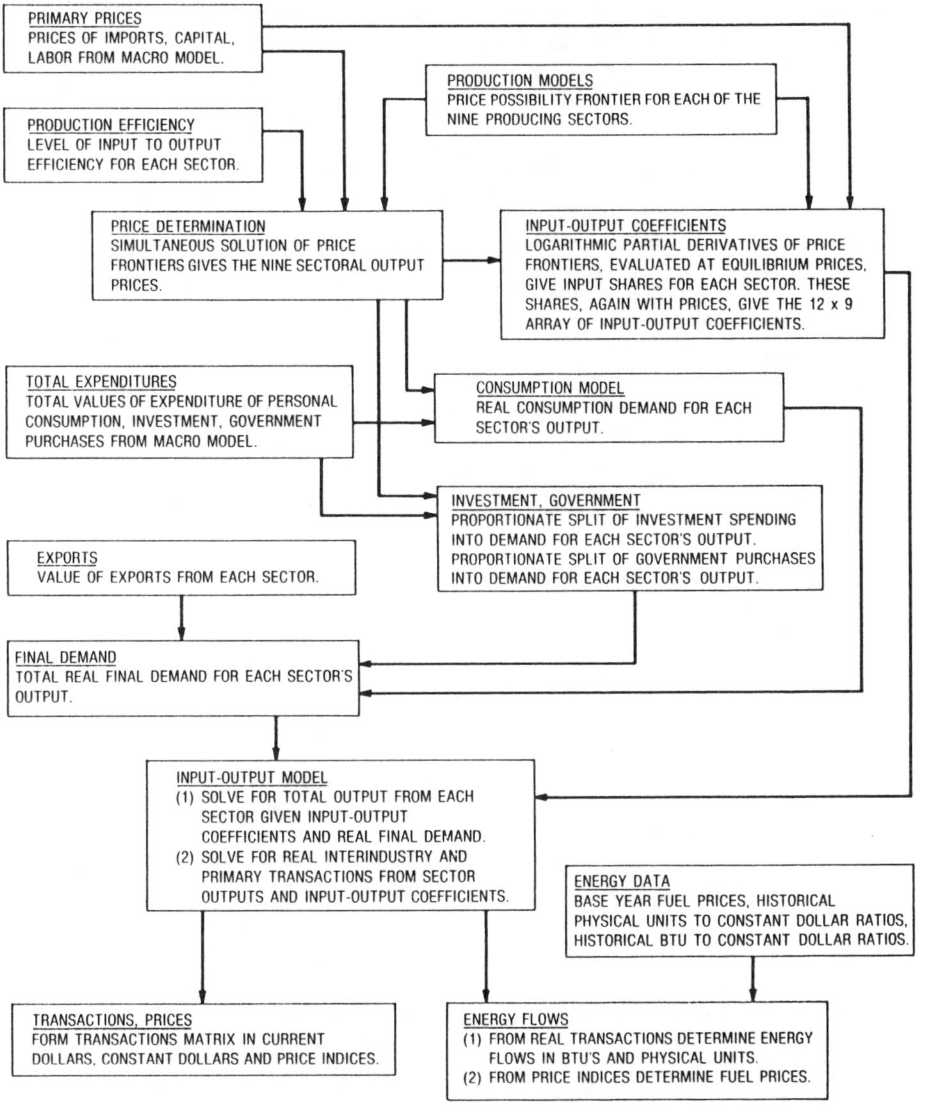

3. Producer behavior

■ Our interindustry model includes econometric models of producer behavior for each of the nine industrial sectors included in the model.[12] In implementing an econometric model of producer behavior for each sector our primary objective is to explore the

EDWARD A. HUDSON AND
474 / DALE W. JORGENSON

[12] This Section is based on Berndt and Jorgenson [5].

interrelationships between relative demand for energy and relative demand for capital services, labor services, and nonenergy inputs. Similarly, we wish to explore the interrelationships among relative demands for the five types of energy included in our model—coal, crude petroleum and natural gas, refined petroleum products, electricity, and gas as a product of gas utilities. We have imposed a structure on the price possibility frontier that permits us to deal with relative demand for energy as a whole and relative demands for the five types of energy included in our model as two separate problems.

To construct a model of the interrelationship of relative demands for energy, capital services, labor services, and nonenergy inputs, we first define groups of inputs that are aggregates of the twelve inputs included in our model of interindustry production structure. These commodity groups are:

(1) Capital (K).
(2) Labor (L).
(3) Energy (E). This group consists of inputs of coal, crude petroleum and natural gas, refined petroleum products, electricity, and gas as a product of gas utilities.
(4) Materials (M). This group consists of inputs of agriculture, manufacturing, transportation, communications, trade, and services, and competitive imports for the nonenergy sectors.

We first construct a model for producer behavior in terms of the four aggregates—capital, labor, energy, and materials. We represent the price of domestic availability of the output of each sector as a function of the prices of each of the aggregates. A sufficient condition for the price possibility frontier to be defined on the prices of the four aggregates is that the overall price possibility frontier is separable and homogeneous in the inputs within each aggregate.[13] The price possibility frontier is separable in the commodities within an aggregate if and only if the ratio of the relative shares of any two commodities within an aggregate is independent of the prices of commodities outside the aggregate.[14] For example, the five types of energy make up an appropriate aggregate if the relative value shares of any two types of energy depend only on the prices of energy and not on the prices of nonenergy intermediate inputs or the prices of capital and labor services.

The second step in constructing a model of producer behavior is to represent the price possibility frontier for the energy and materials aggregates as functions of the prices of inputs that make up each of the aggregates. For the energy aggregate the price of energy is represented as a function of the prices of the five types of energy that make up the aggregate—coal, crude petroleum and natural gas, refined petroleum products, electricity, and gas as a product of gas utilities. For the materials aggregate the price of materials is represented as a function of the five types of inputs that make up the aggregate—agriculture, manufacturing, transportation, communications, trade, and services, and competitive imports for the nonenergy sectors.

[13] See Christensen, Jorgenson, and Lau [18], pp. 29–32.
[14] See Leontief [29].

☐ **Econometric specification.** The system of relative demand functions employed in our econometric model of producer behavior for each of the nine industrial sectors of our model is generated from the price possibility frontier for the corresponding sector. For each of the three submodels that make up our model of producer behavior we represent the price possibility frontier by a function that is quadratic in the logarithms of the prices of the inputs into the sector. The resulting price possibility frontier provides a local second-order approximation to any price possibility frontier. We refer to our representation as the transcendental logarithmic price possibility frontier or, more simply, the *translog price possibility frontier*. The price possibility frontier is a transcendental function of the logarithms of the prices of inputs. The translog price possibility frontier was introduced by Christensen, Jorgenson, and Lau.[15]

As an example, the price possibility frontier for the aggregate (*KLEM*) submodel takes the form:

$$\ln AI + \ln PI = \alpha^I_0 + \alpha^I_K \ln PK + \alpha^I_L \ln PL + \alpha^I_E \ln PE + \alpha^I_M \ln PM + \frac{1}{2}[\beta^I_{KK}(\ln PK)^2 + \beta^I_{KL} \ln PK \ln PL + \ldots], \quad (15)$$

where PK is the price of capital services, PL the price of labor services, PE the price of energy, and PM the price of materials. For this form of the price possibility frontier, the equations for the relative shares of the four input aggregates take the form:

$$\begin{aligned}
\frac{PK^*KI}{PI^*XI} &= \alpha^I_K + \beta^I_{KK}\ln PK + \beta^I_{KL}\ln PL + \beta^I_{KE}\ln PE + \beta^I_{KM}\ln PM,\\
\frac{PL^*LI}{PI^*XI} &= \alpha^I_L + \beta^I_{LK}\ln PK + \beta^I_{LL}\ln PL + \beta^I_{LE}\ln PE + \beta^I_{LM}\ln PM,\\
\frac{PE^*EI}{PI^*XI} &= \alpha^I_E + \beta^I_{EK}\ln PK + \beta^I_{EL}\ln PL + \beta^I_{EE}\ln PE + \beta^I_{EM}\ln PM, \qquad (16)\\
\frac{PM^*MI}{PI^*XI} &= \alpha^I_M + \beta^I_{MK}\ln PK + \beta^I_{ML}\ln PL + \beta^I_{ME}\ln PE + \beta^I_{MM}\ln PM,\\
&\qquad (I = 1, 2 \ldots 9),
\end{aligned}$$

where KI is the quantity of capital services in the Ith sector, LI the quantity of labor services, EI the quantity of energy input, and MI the quantity of materials input.[16]

The dependent variable in each of the four functions generated from the translog price possibility frontier is the relative share of the corresponding input. To derive the input-output coefficient for that input, we divide the relative share by the ratio of the price of the input to the price of the output of the sector. For example, the

[15] See Christensen, Jorgenson, and Lau [17].

[16] A *KLEM* model for total U. S. manufacturing based on the translog price possibility frontier has been developed by Berndt and Wood [6]. Berndt and Christensen [2, 3, 4] have developed models of capital-labor substitution for U. S. manufacturing based on the translog production function, which is dual to the translog price possibility frontier.

input-output coefficients for capital services are:

$$AKI = \frac{KI}{XI} = (\alpha^I_K + \beta^I_{KK} \ln PK + \beta^I_{KL} \ln PL$$
$$+ \beta^I_{KE} \ln PE + \beta^I_{KM} \ln PM)/(PK/PI), \ (I = 1, 2 \ldots 9).$$

Similar expressions can be obtained for the input-output coefficients for labor services, energy, and materials.

The value of domestic availability of the output of each sector is equal to the sum of the values of capital and labor services in that sector and the value of energy and nonenergy inputs into the sector:

$$PI*XI = PK*KI + PL*LI + PE*EI + PM*MI, \quad (17)$$
$$(I = 1, 2 \ldots 9).$$

Given this accounting identity, the relative shares of the four aggregate inputs into each sector add to unity. The parameters of the four relative demand functions for capital and labor services and energy and nonenergy inputs must satisfy the restrictions:

$$\begin{aligned}
&\alpha^I_K + \alpha^I_L + \alpha^I_E + \alpha^I_M = 1,\\
&\beta^I_{KK} + \beta^I_{LK} + \beta^I_{EK} + \beta^I_{MK} = 0,\\
&\beta^I_{KL} + \beta^I_{LL} + \beta^I_{EL} + \beta^I_{ML} = 0, \quad (18)\\
&\beta^I_{KE} + \beta^I_{LE} + \beta^I_{EE} + \beta^I_{ME} = 0,\\
&\beta^I_{KM} + \beta^I_{LM} + \beta^I_{EM} + \beta^I_{MM} = 0,\\
&(I = 1, 2 \ldots 9).
\end{aligned}$$

Given estimates of the parameters of any three equations for the relative shares, estimates of the parameters of the fourth equation can be determined from these restrictions.

The logarithm of the price possibility frontier for each sector is twice differentiable in the logarithms of the prices of inputs, so that the Hessian of this function is symmetric. This gives rise to a set of restrictions relating the parameters of cross partial derivatives. For the aggregate (*KLEM*) submodel three of these restrictions are explicit in the three equations we estimate directly, namely:

$$\begin{aligned}
&\beta^I_{KL} = \beta^I_{LK},\\
&\beta^I_{KE} = \beta^I_{EK}, \quad (19)\\
&\beta^I_{LE} = \beta^I_{EL},\\
&(I = 1, 2 \ldots 9).
\end{aligned}$$

In addition, we estimate the parameters β^I_{MK}, β^I_{ML}, and β^I_{ME} $(I = 1, 2 \ldots 9)$ from the equations:

$$\begin{aligned}
&\beta^I_{MK} = -\beta^I_{KK} - \beta^I_{LK} - \beta^I_{EK},\\
&\beta^I_{ML} = -\beta^I_{KL} - \beta^I_{LL} - \beta^I_{EL},\\
&\beta^I_{ME} = -\beta^I_{KE} - \beta^I_{LE} - \beta^I_{EE},\\
&(I = 1, 2, \ldots 9),
\end{aligned}$$

so that three additional symmetry restrictions are implicit in the equations we estimate, namely:

$$\begin{aligned}
&\beta^I_{KM} = \beta^I_{MK},\\
&\beta^I_{LM} = \beta^I_{ML}, \quad (20)\\
&\beta^I_{EM} = \beta^I_{ME},\\
&(I = 1, 2 \ldots 9).
\end{aligned}$$

For each of the nine industrial sectors, the aggregate (*KLEM*) submodel involves six symmetry restrictions.

The price possibility frontier for each sector is homogeneous of degree one; proportional changes in the prices of all inputs result in a proportional change in the price of output. Homogeneity of the price possibility frontier is implied by the symmetry restrictions outlined above and the restrictions implied by the accounting identity between the value of output and the value of input. In the absence of the symmetry restrictions and the restrictions implied by the accounting identity between the value of output and the value of input, the aggregate (*KLEM*) submodel involves twenty unknown parameters. Taking these restrictions into account, we reduce the number of unknown parameters to nine.

We have presented the aggregate (*KLEM*) submodel of our model of producer behavior in detail. The forms of the energy (*E*) and materials (*M*) submodels are analogous to the form of the aggregate submodel. For the energy submodel we can write the translog price possibility frontier in the form:

$$\begin{aligned} \ln PE = \alpha_0^{EI} + \alpha_1^{EI} \ln PE1 + \alpha_2^{EI} \ln PE2 + \alpha_3^{EI} \ln PE3 \\ + \alpha_4^{EI} \ln PE4 + \alpha_5^{EI} \ln PE5 \qquad (21) \\ + \frac{1}{2} [\beta_{11}^{EI} (\ln PE1)^2 + \beta_{12}^{EI} \ln PE1 \ln PE2 + \ldots], \\ (I = 1, 2 \ldots 9), \end{aligned}$$

where $PE1$ is the price of coal, $PE2$ the price of crude petroleum and natural gas, $PE3$ the price of refined petroleum products, $PE4$ the price of electricity, and $PE5$ the price of gas as a product of gas utilities. Similarly, we can write the translog price possibility frontier for the materials submodel in the form:

$$\begin{aligned} \ln PM = \alpha_0^{MI} + \alpha_1^{MI} \ln PM1 + \alpha_2^{MI} \ln PM2 + \alpha_3^{MI} \ln PM3 \\ + \alpha_4^{MI} \ln PM4 + \alpha_5^{MI} \ln PM5 \qquad (22) \\ + \frac{1}{2} [\beta_{11}^{EI} (\ln PM1)^2 + \beta_{12}^{EI} \ln PM1 \ln PM2 + \ldots], \\ (I = 1, 2 \ldots 9), \end{aligned}$$

where $PM1$ is the price of agriculture, nonfuel mining, and construction, $PM2$ the price of manufacturing, excluding petroleum refining, $PM3$ the price of transportation, $PM4$ the price of communications, trade, and services, and $PM5$ the price of competitive imports.

For both energy (*E*) and materials (*M*) submodels we can derive a system of five equations for determining the relative shares of the five commodity groups making up each submodel. Each equation gives the relative share of one of the commodity groups as a function of the prices of all five groups included in the submodel. We can derive the relative demand functions for each commodity group by dividing the relative value share of the group by the ratio of the price of that group to the price of the corresponding aggregate. For example, to derive the demand for coal relative to total energy we divide the relative value share of coal by the ratio of the price of coal to the price of total energy. We can derive the input-output coefficient for coal by multiplying the demand for coal relative to total energy by the demand for energy relative to the output of the corresponding industrial sector.

The value of each aggregate is equal to the sum of the values of the commodity groups that make up that aggregate. For example, the value of energy is equal to the sum of the values of the five types of energy:

$$PE*EI = PE1*E1I + PE2*E2I + PE3*E3I + PE4*E4I + PE5*E5I, \quad (I = 1, 2 \ldots 9), \tag{23}$$

where $E1I$ is the quantity of coal, $E2I$ the quantity of crude petroleum and natural gas, $E3I$ the quantity of refined petroleum products, $E4I$ the quantity of electricity, and $E5I$ the quantity of gas as a product of gas utilities. As before, the relative shares of the five energy inputs add to unity, so that the parameters of the five relative demand functions for these inputs must satisfy restrictions analogous to the restrictions given above for the parameters of the aggregate ($KLEM$) submodel. Similar restrictions hold for the five relative demand functions for nonenergy inputs.

☐ **Parameter estimation.** For each of the nine industrial sectors included in our interindustry model of the production structure the aggregate ($KLEM$) submodel consists of four equations. We fit the three equations for relative shares of capital (K), labor (L), and energy (E). The relative shares of materials (M) can be determined from these three equations and the accounting identity between the value of output and the value of input. Taking into account the symmetry restrictions on the parameters of the three equations of the aggregate ($KLEM$) submodel, the number of unknown parameters to be estimated is reduced to nine. Taking convexity restrictions into account where appropriate, we further reduce the number of unknown parameters.[17]

For four of the industrial sectors included in our interindustry model the energy (E) submodel consists of five equations for the relative shares of coal, crude petroleum and natural gas, refined petroleum products, electricity, and gas as a product of gas utilities. These four sectors are:

(1) agriculture, nonfuel mining, and construction
(2) manufacturing, excluding petroleum refining
(4) communication, trade, and services
(7) petroleum refining.

For these industrial sectors we fit the four equations for relative shares of coal, crude petroleum and natural gas, refined petroleum products, and electricity. The relative share of gas as a product of gas utilities can be determined from these four equations and the accounting identity between the total value of energy and the sum of the values of the five types of energy.

For the four industrial sectors listed above the energy (E) submodel involves six symmetry restrictions in the four equations we

[17] Methods for imposing convexity restrictions have been developed by Lau [28].

estimate directly and four additional restrictions that are implicit in these equations and the accounting identity for the total value of energy. In the absence of these restrictions and the restrictions implied by the accounting identity, the energy (E) submodel for the four industrial sectors involves unknown parameters. Taking these restrictions into account, we reduce the number of unknown parameters to fourteen.

For four of the industrial sectors included in our interindustry model the energy (E) submodel consists of four equations for relative shares of four types of energy. For transportation, coal mining, and electric utilities the relative share of crude petroleum and natural gas is zero. For gas utilities the relative share of electricity is zero. For these four sectors the form of the energy (E) submodel is analogous to the form of the aggregate ($KLEM$) submodel. We fit three equations for the relative shares of three types of energy, excluding the equation for the relative share of gas as a product of gas utilities. The relative share of gas can be determined from the accounting identity for the total value of energy. We fit these three equations subject to symmetry restrictions so that the number of unknown parameters is reduced to nine.

For the crude petroleum and natural gas sector of our interindustry model the energy (E) submodel consists of three equations for relative shares of three types of energy. For this sector the relative shares of coal and gas as an output of gas utilities are equal to zero. We fit two equations for the relative shares of crude petroleum and natural gas and refined petroleum products. The relative share of electricity can be determined from the accounting identity for the total value of energy. Fitting these equations subject to symmetry restrictions, we reduce the number of unknown parameters to five.

For the four nonenergy sectors included in our interindustry model the materials (M) submodel consists of five equations for the relative shares of agriculture, manufacturing, transportation, communications, and competitive imports. For the five energy sectors the materials submodel consists of four equations for the relative shares of agriculture, manufacturing, transportation, and communications. The form of the materials (M) submodel for the four nonenergy sectors is analogous to the form of the energy (E) submodel with five equations for the relative shares of five types of energy. For these sectors we fit four equations for the relative shares of nonenergy inputs, excluding competitive imports. The relative shares of competitive imports can be determined from these four equations and the accounting identity between the total value of materials and the sum of the values of the five types of nonenergy inputs. Each of these submodels involves fourteen unknown parameters.

For the five nonenergy sectors we fit three equations for the relative shares of nonenergy inputs, excluding inputs of communications, trade, and services. The materials (M) submodels for these sectors are analogous to the aggregate ($KLEM$) submodel and involve nine unknown parameters.

For each of the nine industrial sectors included in our interindustry model of production all three submodels—aggregate

(*KLEM*), energy (*E*), and materials (*M*)—have been fitted to annual data on interindustry transactions, capital and labor services, and competitive imports for the period 1947 to 1971.[18] Our method to estimation is the minimum distance estimator for nonlinear simultaneous equations, treating the prices of competitive imports as exogenous variables[19] For each of these submodels the system of equations is nonlinear in the variables but linear in the parameters.

In Tables 1, 2, and 3 we present estimates of the parameters of the translog price possibility frontier for each of the three submodels of our econometric model of producer behavior for all nine industrial sectors. For convenience of the reader we list again the nine sectors:

(1) agriculture, nonfuel mining, and construction
(2) manufacturing, excluding petroleum refining
(3) transportation
(4) communications, trade, and services
(5) coal mining
(6) crude petroleum and natural gas
(7) petroleum refining
(8) electric utilities
(9) gas utilities.

Table 1 contains estimates of the parameters of the translog

TABLE 1

ESTIMATES OF THE PARAMETERS OF THE TRANSLOG PRICE POSSIBILITY FRONTIER FOR THE AGGREGATE (KLEM) SUBMODEL FOR NINE INDUSTRIAL SECTORS OF THE U.S. ECONOMY, 1947-1971

	SECTORS								
PARAMETER	1	2	3	4	5	6	7	8	9
α_K^I	0.1785	0.1149	0.1799	0.2994	0.1277	0.4272	0.1373	0.3458	0.2165
α_L^I	0.2354	0.2940	0.4096	0.4171	0.4139	0.0987	0.0978	0.1925	0.1085
α_E^I	0.0244	0.0202	0.0380	0.0182	0.1857	0.1101	0.4553	0.2120	0.5547
α_M^I	0.5616	0.5708	0.3726	0.2653	0.2727	0.3640	0.3096	0.2496	0.1204
β_{KK}^I	0.0851	0.0590	0.1018	0.0595	0.0280	0.2447	0.0849	0.1330	0.0749
β_{KL}^I	-0.0366	0.0030	-0.0601	0.0114	-0.0357	-0.0422	-0.0242	0.0288	-0.1181
β_{KE}^I	-0.0052	-0.0055	-0.0137	0.0011	0.0099	-0.0470	-0.0772	-0.1682	-0.0941
β_{KM}^I	-0.0434	-0.0565	-0.0280	-0.0719	-0.0022	-0.1555	0.0165	0.0064	0.1373
β_{LL}^I	0.0287	0.0737	0.0582	0.0848	-0.0751	-0.0131	-0.0174	-0.0968	-0.2773
β_{LE}^I	0.0023	0.0054	0.0180	0.0098	0.1145	0.0459	-0.0122	0.0239	0.1318
β_{LM}^I	0.0056	-0.0821	0.0199	-0.1059	-0.0037	0.0093	0.0538	0.0441	0.2636
β_{EE}^I	0.0072	0.0188	0.0198	0.0020	0.0087	0.0184	0.2282	0.0638	0.1413
β_{EM}^I	-0.0044	-0.0187	0.0119	-0.0129	-0.1332	-0.0173	-0.1388	0.0805	-0.1790
β_{MM}^I	0.0422	0.1573	-0.0038	0.1907	0.1392	0.1635	0.0685	-0.1311	-0.2219

[18] These data were compiled by Jack Faucett Associates [21].

[19] The minimum distance estimator for nonlinear simultaneous equations is discussed by Malinvaud [32], pp. 325–373.

price possibility frontier for the aggregate submodel. The aggregate price possibility frontier is defined on the prices of capital (K), labor (L), energy (E), and materials (M). For each of the nine industrial sectors the parameters of the aggregate submodel are estimated from a system of three equations. The dependent variables in these equations are the relative shares of capital, labor, and energy in the value of total output. Parameters in the equation for the relative share of materials are estimated from the restrictions implied by the accounting identity between the value of output and the value of input. We employ constraints across the equations arising from symmetry restrictions on the price possibility frontier for each sector, reducing the number of parameters to be estimated to nine. We also employ convexity restrictions where appropriate to further reduce the number of parameters to be estimated.

Table 2 contains estimates of the parameters of the translog price possibility frontier for the energy (E) submodel. The energy price possibility frontier gives the price of energy for each sector as a function of the prices of five types of energy inputs. The five types of energy are:

(1) coal
(2) crude petroleum and natural gas
(3) refined petroleum products
(4) electricity
(5) gas as a product of gas utilities.

For each of the nine industrial sectors the parameters of the energy submodel are estimated from a system containing as many as four equations. The dependent variables in these equations are the relative shares of each type of energy in the total value of energy. We employ the restrictions implied by the accounting identity between the total value of energy and the sum of the values of all types of energy, the symmetry restrictions, and, where appropriate, convexity restrictions in reducing the number of parameters to be estimated.

Finally, Table 3 contains estimates of the parameters of the translog price possibility frontier for the materials (M) submodel. For the four nonenergy sectors the materials price possibility frontier is defined on the prices of the five types of nonenergy inputs. These are:

(1) agriculture, nonfuel mining, and construction
(2) manufacturing, excluding petroleum refining
(3) transportation
(4) communications, trade, and services
(5) competitive imports.

For the five energy sectors the materials price possibility frontier is defined on the prices of four types of nonenergy inputs, excluding competitive imports. The dependent variables are relative shares of each type of nonenergy input. We employ restrictions on the parameters that are analogous to the restrictions used for the energy submodel.

TABLE 2

ESTIMATES OF THE PARAMETERS OF THE TRANSLOG PRICE POSSIBLITY FRONTIER FOR THE ENERGY (E) SUBMODEL FOR NINE INDUSTRIAL SECTORS OF THE U.S. ECONOMY, 1947-1971

	SECTORS								
PARAMETER	1	2	3	4	5	6	7	8	9
α_1^{EI}	0.0053	0.2040	0.0799	0.1142	0.8510	0.0738	0.0448	0.3165	0.0909
α_2^{EI}	0.0021	0.0002	0.0000	0.0111	0.0000	0.1464	0.2131	0.0000	0.1408
α_3^{EI}	0.8389	0.3384	0.8107	0.3520	0.3997	0.0894	0.2289	0.1173	0.0940
α_4^{EI}	0.1212	0.2858	0.0406	0.4136	0.1062	0.6904	0.5132	0.3829	0.6743
α_5^{EI}	0.0325	0.1716	0.0688	0.1091	0.0029	0.0000	0.0000	0.1829	0.0000
β_{11}^{EI}	0.0052	0.1624	0.0735	0.1011	-0.0118	-0.0557	-0.0165	0.0762	-0.0752
β_{12}^{EI}	0.0000	0.0000	0.0000	-0.0013	0.0000	0.1098	0.0483	0.0000	0.1850
β_{13}^{EI}	-0.0068	-0.0690	-0.0648	-0.0402	0.0220	-0.0255	0.0005	0.0833	-0.0192
β_{14}^{EI}	0.0011	-0.0583	-0.0032	-0.0472	-0.0100	-0.0286	-0.0323	-0.0578	-0.0906
β_{15}^{EI}	0.0005	-0.0350	-0.0055	-0.0125	-0.0002	0.0000	0.0000	-0.1017	0.0000
β_{22}^{EI}	0.0010	-0.0029	0.0000	0.0110	0.0000	0.0077	0.1113	0.0000	-0.1416
β_{23}^{EI}	0.0007	-0.0001	0.0000	-0.0039	0.0000	0.0053	-0.0592	0.0000	0.0455
β_{24}^{EI}	-0.0007	0.0073	0.0000	-0.0046	0.0000	-0.1228	-0.1003	0.0000	-0.0889
β_{25}^{EI}	-0.0010	-0.0043	0.0000	-0.0012	0.0000	0.0000	0.0000	0.0000	0.0000
β_{33}^{EI}	-0.0252	0.2239	0.1534	0.2281	-0.0287	-0.0418	0.0680	0.0001	-0.0553
β_{34}^{EI}	0.0128	-0.0967	-0.0329	-0.1456	0.0064	0.0620	-0.0092	-0.0966	0.0289
β_{35}^{EI}	0.1854	-0.0581	-0.0557	-0.0384	0.0003	0.0000	0.0000	0.0162	0.0000
β_{44}^{EI}	-0.0410	0.1868	0.0389	0.2425	0.0055	0.0894	0.1418	0.2077	0.1505
β_{45}^{EI}	0.0278	-0.0390	-0.0028	-0.0451	-0.0019	0.0000	0.0000	-0.0502	0.0000
β_{55}^{EI}	-0.0458	0.1364	0.0640	0.0972	0.0018	0.0000	0.0000	0.1357	0.0000

4. Growth model

■ The second component of our framework for energy policy analysis is a model of long-term U. S. economic growth. A model of U. S. economic growth is required to provide totals for consumption, investment, government and export final demand in our econometric model of interindustry transactions. For this purpose conventional econometric models could be used as an alternative to our macro-econometric growth model. However, a growth model is also required to provide the relative prices of capital and labor services, which enter as primary inputs into the nine sectors of our interindustry model. Prices of primary inputs must be generated within a framework that also incorporates primary input quantities and the prices and quantities of final products. For this purpose a new approach to macro-econometric model-building is required.

Our approach to the explanation of economic growth is closely related to the neoclassical theory of economic growth.[20] The build-

[20] See, for example, the discussion of the neoclassical two-sector growth

TABLE 3

ESTIMATES OF THE PARAMETERS OF THE TRANSLOG PRICE POSSIBILITY FRONTIER FOR THE MATERIALS (M) SUBMODEL FOR NINE INDUSTRIAL SECTORS OF THE U.S. ECONOMY, 1947-1971

	SECTORS								
PARAMETER	1	2	3	4	5	6	7	8	9
α^{MI}_{1}	0.2578	0.1348	0.1221	0.0819	0.0193	0.0779	0.0562	0.1134	0.1759
α^{MI}_{2}	0.3777	0.5933	0.1373	0.2548	0.4270	0.0869	0.1656	0.1046	0.1546
α^{MI}_{3}	0.0653	0.0472	0.1932	0.0532	0.0675	0.0501	0.1736	0.1200	0.1453
α^{MI}_{4}	0.2674	0.1643	0.4382	0.5774	0.4839	0.5517	0.4659	0.6620	0.4426
α^{MI}_{5}	0.0318	0.0603	0.1091	0.0327	0.0023	0.2334	0.1388	0.0000	0.0815
β^{MI}_{11}	0.0799	0.0376	0.1072	-0.0454	0.0190	0.0718	0.0530	0.0170	-0.0407
β^{MI}_{12}	-0.1012	-0.0200	-0.0168	0.0848	-0.0083	-0.0068	-0.0093	0.0755	0.0845
β^{MI}_{13}	0.0629	0.0043	-0.2360	-0.0094	-0.0013	-0.0039	-0.0097	-0.0292	-0.0703
β^{MI}_{14}	-0.0672	-0.0571	-0.0535	-0.0806	-0.0094	-0.0430	-0.0262	-0.0633	-0.0899
β^{MI}_{15}	0.0256	0.0352	-0.0133	0.0506	-0.0000	-0.0182	-0.0078	0.0000	0.1164
β^{MI}_{22}	0.2349	0.1958	0.1185	0.0973	0.2447	0.0794	0.1382	0.0023	-0.0070
β^{MI}_{23}	-0.0219	-0.0361	-0.0265	-0.0091	-0.0288	-0.0044	-0.0288	0.0037	-0.0123
β^{MI}_{24}	-0.1009	-0.0710	-0.0602	-0.1179	-0.2066	-0.0480	-0.0772	-0.0815	-0.0895
β^{MI}_{25}	-0.0109	-0.0687	-0.0150	-0.0550	-0.0010	-0.0203	-0.0230	0.0000	0.0243
β^{MI}_{33}	0.0039	0.0435	0.1559	0.0502	0.0629	0.0476	0.1434	0.1027	0.1095
β^{MI}_{34}	-0.0187	-0.0030	-0.0847	-0.0321	-0.0327	-0.0276	-0.0809	-0.0773	-0.0739
β^{MI}_{35}	-0.0263	-0.0087	-0.0211	0.0005	-0.0002	-0.0117	-0.0241	0.0000	0.0471
β^{MI}_{44}	0.1959	0.1218	0.2462	0.2348	0.2497	0.2473	0.2488	0.2221	0.2346
β^{MI}_{45}	-0.0090	0.0093	-0.0478	-0.0042	-0.0011	-0.1287	-0.0646	0.0000	0.0187
β^{MI}_{55}	0.0206	0.0329	0.0972	0.0081	0.0023	0.1789	0.1195	0.0000	-0.2065

ing blocks of our model are submodels of household and production sectors and submodels of foreign and government sectors. Economic growth results from the link between current capital formation and future productive capacity. In our model this link is provided by a macro-econometric production function, relating the output of consumption and investment goods to the input of capital and labor services. Preferences between present and future consumption, which determine the allocation of income between saving and consumption, complete our model of economic growth.

Our macro-econometric growth model of the U. S. economy provides for the simultaneous determination of the values of products and factors of production in both current and constant prices. The model links demand for capital formation by savers to the

model by Burmeister and Dobell [7] and the references given there. A more detailed discussion of our model is presented in Hudson and Jorgenson [25]; see also, Jorgenson, Berndt, Christensen, and Hudson [26], Chapter 2.

supply of investment goods by producers. Similarly, the model links demand for consumption goods and supply of labor services by households to supply of consumption goods and demand for labor by producers. Finally, given the supply of capital stock, the demand for capital services determines the overall rate of return to capital.

The theory of U. S. economic growth that underlies our macroeconometric growth model is a theory of the behavior of the private sector of the U. S. economy. The behavior of the foreign and government sectors is taken to be exogenous. Demographic trends—the growth of population, labor force, and unemployment—are also exogenous to the model. The main determinant of growth in productivity is capital formation. Growth in productivity over and above growth due to capital formation is exogenous to the model. We have projected demographic trends and trends in productivity growth on the basis of post-war experience in the United States.[21]

☐ **Variables.** Our econometric growth model is summarized in the following series of tables. In Table 4 we present our notation for the variables that appear in the model. The first group of variables converts aggregates from one basis of classification to another. For example, the index of total factor productivity *A* converts input to output. The index of *AW* converts investment weights for capital formation to the weights appropriate for the measurement of wealth. All of the aggregation variables are taken to be exogenous.

The second group of variables appearing in Table 4 comprises the quantities of products and factors of production, broken down by sector of origin and destination. Variables beginning with *C* are quantities of consumption goods. Similarly, variables beginning with *I* are quantities of investment goods. Variables beginning with *L* are quantities of labor services, while variables beginning with *K* are quantities of capital services. The third group of variables includes prices corresponding to the quantities of products and factors of production. Each price begins with *P* and continues with the corresponding quantity. For example, the variable *C* is personal consumption expenditures and the variable *PC* is the price of personal consumption expenditures.

The fourth group of variables consists of financial variables: rates of depreciation and replacement, the nominal rate of return, gross private national saving, and private national wealth. Finally, the fifth group of variables comprises tax and transfer variables. The variable *EL* represents government transfer payments to persons other than social insurance funds, an expenditure category. The variables beginning with *T* are tax rates. Each of the products and factors included in the model—consumption goods, investment goods, capital services, and labor services—is associated with an effective tax rate. The variable *TP* is the effective tax rate for capital stock.

☐ **Equations.** In Table 5 we present the equations for our macroeconometric growth model. The model includes five behavioral equations, describing the behavior of household and business sec-

21 Detailed projections are presented by Hudson and Jorgenson [25].

TABLE 4

MACRO-ECONOMIC GROWTH MODEL: NOTATION

1. AGGREGATION VARIABLES			
A	TOTAL FACTOR PRODUCTIVITY (INPUT TO OUTPUT).	AK	CAPITAL STOCK, LAGGED TO CAPITAL SERVICE.
ACI	INVESTMENT TO CHANGE IN BUSINESS INVENTORIES, CONSUMPTION GOODS.	APC	IMPLICIT DEFLATOR OF CONSUMPTION GOODS TO IMPLICIT DEFLATOR OF CHANGE IN BUSINESS INVENTORIES, CONSUMPTION GOODS.
AI	INVESTMENT TO CAPITAL STOCK.	AW	INVESTMENT TO WEALTH.
AL	INVESTMENT TO CAPITAL STOCK, LAGGED.		
2. QUANTITIES			
C	PERSONAL CONSUMPTION EXPENDITURES, INCLUDING SERVICES OF CONSUMERS' DURABLES.	IG	GOVERNMENT PURCHASES OF INVESTMENT GOODS.
CE	SUPPLY OF CONSUMPTION GOODS BY GOVERNMENT ENTERPRISES.	IR	NET EXPORTS OF INVESTMENT GOODS.
CG	GOVERNMENT PURCHASES OF CONSUMPTION GOODS.	IS	SUPPLY OF INVESTMENT GOODS BY PRIVATE ENTERPRISES.
CI	CHANGE IN BUSINESS INVENTORIES OF CONSUMPTION GOODS.	L	SUPPLY OF LABOR SERVICES.
CR	NET EXPORTS OF CONSUMPTION GOODS, LESS INCOME ORIGINATING, REST OF THE WORLD.	LD	PRIVATE PURCHASES OF LABOR SERVICES.
CS	SUPPLY OF CONSUMPTION GOODS BY PRIVATE ENTERPRISES.	LGE	GOVERNMENT ENTERPRISES PURCHASES OF LABOR SERVICES.
G	NET CLAIMS ON GOVERNMENT.	LGG	GENERAL GOVERNMENT PURCHASES OF LABOR SERVICES.
R	NET CLAIMS ON REST OF THE WORLD.	LH	TIME AVAILABLE.
I	GROSS PRIVATE DOMESTIC INVESTMENT, INCLUDING PURCHASES OF CONSUMERS' DURABLES.	LJ	LEISURE TIME.
		LR	NET EXPORTS OF LABOR SERVICES.
		LU	UNEMPLOYMENT
		K	CAPITAL STOCK.
		KD	CAPITAL SERVICES.
3. PRICES			
PC	IMPLICIT DEFLATOR, PERSONAL CONSUMPTION EXPENDITURES, INCLUDING SERVICES OF CONSUMERS' DURABLES.	PIG	IMPLICIT DEFLATOR, GOVERNMENT PURCHASES OF INVESTMENT GOODS.
PCE	IMPLICIT DEFLATOR, SUPPLY OF CONSUMPTION GOODS BY GOVERNMENT ENTERPRISES.	PIR	IMPLICIT DEFLATOR, NET EXPORTS OF INVESTMENT GOODS
PCG	IMPLICIT DEFLATOR, GOVERNMENT PURCHASES OF CONSUMPTION GOODS.	PIS	IMPLICIT DEFLATOR, SUPPLY OF INVESTMENT GOODS BY PRIVATE ENTERPRISES.
PCI	IMPLICIT DEFLATOR, CHANGE IN BUSINESS INVENTORIES OF CONSUMPTION GOODS.	PL	IMPLICIT DEFLATOR, SUPPLY OF LABOR SERVICES.
PCR	IMPLICIT DEFLATOR, NET EXPORTS OF CONSUMPTION GOODS, LESS INCOME ORIGINATING, REST OF THE WORLD.	PLD	IMPLICIT DEFLATOR, PRIVATE PURCHASES OF LABOR SERVICES.
PCS	IMPLICIT DEFLATOR, SUPPLY OF CONSUMPTION GOODS BY PRIVATE ENTERPRISES.	PLGE	IMPLICIT DEFLATOR, GOVERNMENT ENTERPRISES PURCHASES OF LABOR PURCHASES.
PG	IMPLICIT DEFLATOR, NET CLAIMS ON GOVERNMENT.	PLGG	IMPLICIT DEFLATOR, GENERAL GOVERNMENT PURCHASES OF LABOR SERVICES.
PR	IMPLICIT DEFLATOR, NET CLAIMS ON REST OF THE WORLD.	PLR	IMPLICIT DEFLATOR, NET EXPORTS OF LABOR SERVICES.
PI	IMPLICIT DEFLATOR, GROSS PRIVATE DOMESTIC INVESTMENT, INCLUDING PURCHASES OF CONSUMERS' DURABLES.	PKD	IMPLICIT DEFLATOR, CAPITAL SERVICES.
4. FINANCIAL VARIABLES			
D	RATE OF DEPRECIATION, PRIVATE DOMESTIC TANGIBLE ASSETS.	N	NOMINAL RATE OF RETURN, PRIVATE DOMESTIC TANGIBLE ASSETS.
M	RATE OF REPLACEMENT, PRIVATE DOMESTIC TANGIBLE ASSETS.	S	GROSS PRIVATE NATIONAL SAVING.
		W	PRIVATE NATIONAL WEALTH.
5. TAX AND TRANSFER VARIABLES			
EL	GOVERNMENT TRANSFER PAYMENTS TO PERSONS, OTHER THAN FROM SOCIAL INSURANCE FUNDS.	TK	EFFECTIVE TAX RATE, CAPITAL SERVICES.
TC	EFFECTIVE TAX RATE, CONSUMPTION GOODS.	TL	EFFECTIVE TAX RATE, LABOR SERVICES.
TI	EFFECTIVE TAX RATE, INVESTMENT GOODS.	TP	EFFECTIVE TAX RATE, CAPITAL STOCK.

tors.[22] The demand for labor is determined by the total level of production, the amount of capital services available, and the relative

[22] Our model of the household sector was originated by Christensen and Jorgenson [12]. Our model of the business sector was originated by Christensen, Jorgenson, and Lau [18].

TABLE 5

MACRO-ECONOMIC GROWTH MODEL: EQUATIONS

1. BEHAVIORAL EQUATIONS

INVESTMENT SUPPLY:

$$\frac{PIS*IS}{PKD*KD} = 1.1717 - 0.5006 * (LOG\ CS - LOG\ IS).$$

LABOR DEMAND;

$$\frac{PLD*LD}{PKD*KD} = 1.5655.$$

PRODUCTION POSSIBILITY FRONTIER:

0= - LOG KD-1.5655*LOG LD + 1.3938 * LOG CS - 2.5655 * LOG A
+ 1.1717 * LOG IS + 0.2503 * (LOG CS - LOG IS) ** 2.

CONSUMPTION DEMAND:

PC * C = 0.0034 * W(-1) + 0.1469 * (PL * LH + EL).

LEISURE DEMAND:

PL * LJ=0.0196 * W(-1) + 0.8403 * (PL *LH + EL).

2. ACCOUNTING IDENTITIES

CAPITAL STOCK AND INVESTMENT:

K=AI * I + (1-M) * K(-1).

CAPITAL SERVICE AND CAPITAL STOCK:

KD = AK * K(-1).

VALUE OF OUTPUT AND INPUT:

PIS * IS + PCS * CS = PKD * KD + PLD * LD.

VALUE OF CONSUMPTION GOODS:

(1 + TC) * PCS * CS + PCE * CE
= PC * C + PCG * CG + PCI * CI + PCR * CR.

VALUE OF INVESTMENT GOODS:

(1 + TI) * PIS * IS + PCI * CI = PI * I + PIG * IG + PIR * IR.

VALUE OF CAPITAL SERVICES:

(1 - TK) * (PKD * KD - TP * PI[-1] * AW [-1] * K[-1])
= N * PI(-1) * AW(-1) * K(-1) + D * PI * AL * K(-1)
+ PI(-1) * AW(-1) * K(-1) - PI * AL * K(-1).

VALUE OF LABOR SERVICES:

(1 - TL) * (PLD * LD + PLGE * LGE + PLGG * LGG + PLR * LR) = PL * L.

SAVING:

S = PI * I + PG * (G - G[-1]) + PR * (R - R[-1]).

WEALTH:

W = PI * AW * K + PG * G + PR * R.

3. BALANCE EQUATIONS

CONSUMPTION:

CS + CE = C + CG + CI + CR.

INVESTMENT:

IS + CI = I + IG + IR.

TIME:

LH = L + LJ.

LABOR:

L = LD + LGE + LGG + LR + LU.

4. AGGREGATION EQUATIONS

IMPLICIT DEFLATOR, CHANGE IN BUSINESS INVENTORIES, CONSUMPTION GOODS:

PCI = PC * APC.

CHANGE IN BUSINESS INVENTORIES, CONSUMPTION GOODS:

PCI * CI = PI * I * ACI.

prices of capital and labor services. Output of investment goods is determined by the price of investment goods, the prices of capital services and the available supply of capital services, and the amount of productive capacity being devoted to the output of consumer goods and services. The production that takes place in the U. S. private sector, whether of consumption or of investment goods, is limited by the total productive capacity, which in turn depends on available supplies of capital and labor services as well as on the level of technology.

The level of household expenditure on consumer goods and services is determined by the wealth and resources held by the household sector, including the time endowment of the sector. The desired amount of work input provided by the household sector is determined by the total amount of time available, the wage rate, and the extent of other resources available to the household sector in the form of wealth and transfer payments.

The behavioral equations of our macro-econometric growth model have been estimated from historical data for the United States for the period 1929 to 1971.[23] In addition to the five behavioral equations the model includes accounting identities for capital stock, investment and capital services, for the value of input and output, for saving and wealth, and for the value of consumption goods, investment goods, capital services, and labor services. These accounting identities incorporate the budget constraints for household and business sectors and the flow of each product and factor of production in current prices.

The model is completed by balance between demand and supply of products and factors of production in constant prices and by aggregation equations that determine inventory accumulation of consumption goods. Although gross private domestic investment is determined in the model, the allocation of investment between fixed investment and inventory accumulation is not determined in the model. An allocation between inventory accumulation in the form of consumption goods and other components of gross private domestic investment is required for the balance between demand and supply of consumption and investment goods. In Figure 3 we outline the working of our macro-econometric growth model by means of a diagram.

5. Energy projections

■ Our next objective is to provide a reference point for the analysis of energy policy by establishing detailed projections of energy demand and supply, energy price and cost, and energy imports and exports. Our projections cover the years 1975, 1980, 1985, and 2000 and are based on the assumption that there are no major changes in energy policy, either by U. S. or foreign governments, over the forecast period. The projections include the entire matrix of interindustry transactions in current dollar and in constant dollar flow. We translate this information into physical terms by converting the constant dollar energy transactions into British thermal units (Btu) for each fuel.[24] We also convert the price indices into dollars per physical unit. This transformation permits the expression

---

[23] The data are presented in a series of articles by Christensen and Jorgenson [13, 14, 15, 16].

[24] The conversion process is discussed in the following section.

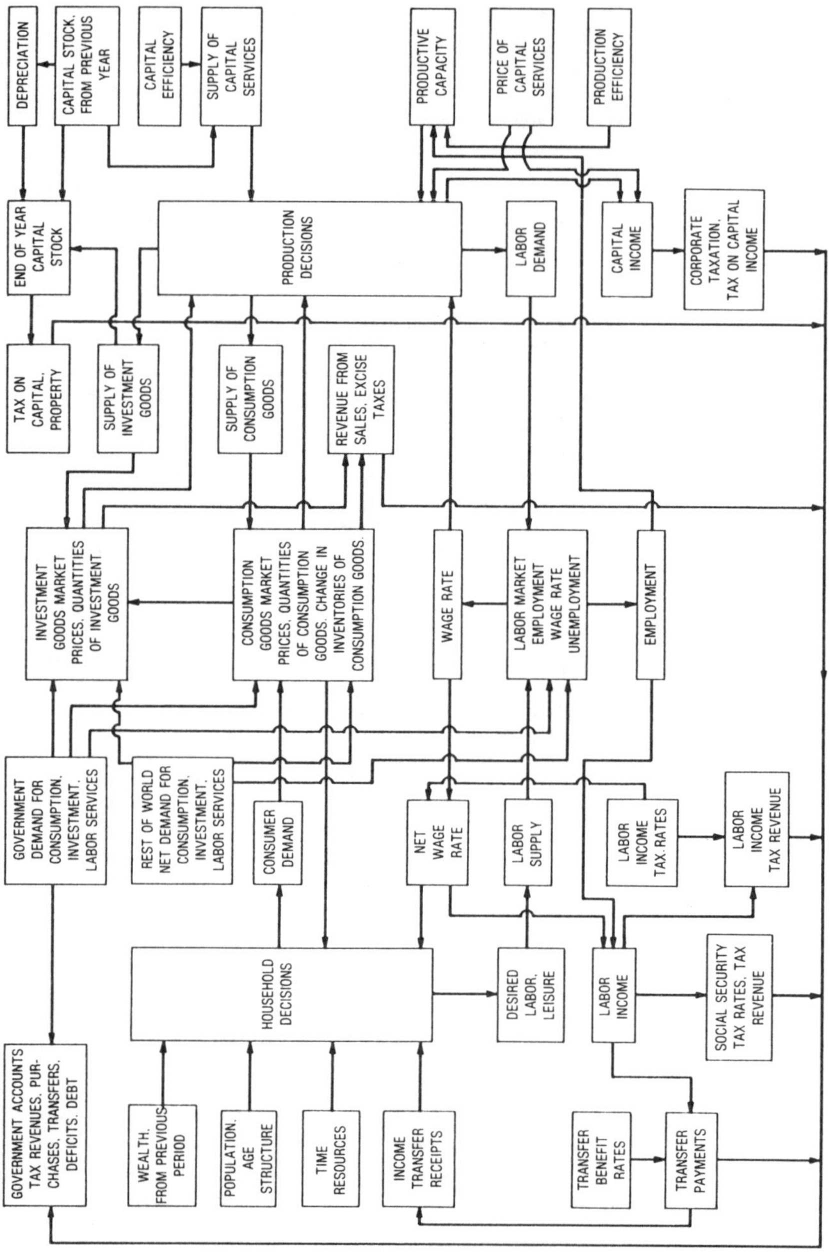

FIGURE 3
MACRO-ECONOMETRIC GROWTH MODEL: DIAGRAMMATIC REPRESENTATION

of our detailed interindustry projections in the energy balance framework that is conventional in energy analysis.[25]

A summary of the composition and growth of the interindustry transactions for the U. S. economy is given in Table 6. This table

TABLE 6

U.S. INDUSTRY OUTPUT AND PRICES, 1975-2000 (AVERAGE ANNUAL GROWTH RATES, PERCENT)

OUTPUT OF SECTOR	REAL DOLLARS	PRICES	CURRENT DOLLARS
1	2.82	4.97	7.93
2	3.92	3.49	7.55
3	4.30	2.33	6.73
4	3.51	4.15	7.81
5	2.73	6.45	9.36
6	1.64	4.46	6.17
7	2.15	5.80	8.06
8	5.25	3.54	8.07
9	1.05	6.68	8.98
CONSUMPTION	3.82	3.64	7.80
INVESTMENT	3.80	3.93	7.88
GOVERNMENT	3.78	4.11	8.05
NET EXPORTS	6.29		
GNP	3.85	3.76	7.75

presents information on the gross output of each producing sector, together with information on the disposition side of GNP. The rate of growth of real GNP is expected to slow somewhat from recent levels, in large part because of the expected decline in the rate of increase of the labor force, but only to around 3.85 percent a year. Inflation also is expected to slow from rates experienced in the recent past but, at 3.76 percent a year, to remain above typical historical rates.

The composition of GNP is expected to change gradually. Net exports absorb an increasing fraction of GNP as the terms of trade, particularly relating to raw materials, continue to move against the United States. Real government purchases fall in relation to total output, although the rapid rate of increase in the price of government purchases, primarily of its purchases of labor and services, offsets this and results in a small increase in the current dollar share of government purchases of GNP. Personal consumption in real terms increases in line with GNP but, because of the slower than average increase in consumption prices, the current dollar share of consumption in GNP falls. Private investment in real terms increases more slowly than GNP but, in current dollars, it increases more rapidly.

The composition of production changes more noticeably. Agriculture, nonfuel mining, and construction output increases relatively slowly, as the output is income inelastic, while the expected productivity advance in manufacturing and transport permits a comparatively rapid increase in output from these sectors with a less than average increase in prices. Communications, trade, and services output continues to increase but less rapidly than real GNP and, due to the slow productivity advance in this sector and its

[25] See note 5, above.

relative intensity in an increasing cost input, labor, its prices increase comparatively rapidly.

☐ **Energy balance.** Our projection of total U. S. energy utilization for the period 1975 to 2000 is summarized in Table 7. Total

TABLE 7

U.S. ENERGY INPUT, 1970-2000

	1970	1975	1980	1985	2000
1. U.S. TOTAL ENERGY INPUT (QUADRILLION BTU)					
COAL	12.922	13.260	14.304	16.538	26.891
PETROLEUM	29.614	33.696	39.176	44.547	58.148
NATURAL GAS (EXC.SYN.GAS)	22.029	24.727	26.888	28.145	28.681
HYDRO, NUCLEAR, OTHER	2.879	5.416	10.115	16.096	47.521
TOTAL	67.444	77.099	90.484	105.326	161.241
2. TOTAL ENERGY INPUT, GROWTH RATES (AVERAGE PERCENT PER ANNUM)					
COAL		0.52	1.52	2.94	3.29
PETROLEUM		2.62	3.06	2.60	1.79
NATURAL GAS		2.34	1.69	0.92	0.13
HYDRO, NUCLEAR, OTHER		13.47	13.31	9.74	7.48
TOTAL		2.71	3.25	3.08	2.88
3. TOTAL ENERGY INPUT, COMPOSITION (PERCENT OF TOTAL)					
COAL	19.16	17.20	15.81	15.70	16.68
PETROLEUM	43.91	43.70	43.30	42.29	36.06
NATURAL GAS	32.66	32.07	29.72	26.72	17.79
HYDRO, NUCLEAR, OTHER	4.27	7.02	11.18	15.28	29.47
TOTAL	100.00	100.00	100.00	100.00	100.00

energy input is forecast to increase from 77.099 quadrillion Btu in 1975 to 161.241 quadrillion Btu in 2000, an average annual growth rate of 3.0 percent. The rate of increase of this total varies over time. In the past the rate of increase of total input has varied from 3.7 percent a year over the 1960 to 1965 period, to 4.8 percent over 1965 to 1970, to 3.9 percent for 1970 to 1973. Our forecast 1975 total corresponds to a 2.7 percent increase for the 1970 to 1975 period.

The continued increase in fuel prices and the continued presence of some general interest in energy conservation, together with the changing structure of the economy, prevent the rate of increase in energy use from returning to historical trends. These forces lead to several avenues of energy conservation. Wasteful energy use is reduced, existing capital is gradually replaced with more energy efficient capital, and output shifts towards less energy-intensive forms, leading to a steady decline in the projected rate of increase of total energy input. Thus, projected U. S. gross energy input increases at 3.2 percent over the 1975 to 1980 period, at 3.3 percent over 1980 to 1985, and at 2.9 percent over 1985 to 2000.

The composition of total energy input is also expected to change markedly. Coal is projected to decline slightly in relative importance until 1985 but then to increase until 2000 as new demand for coal for synthetic gas production is superimposed on the continuing demand for coal for electricity generation. The net result is that the share of coal in total input is almost the same in 2000 as in 1975. Petroleum shows a continuing decline in relative importance. Use of petroleum continues to increase at around 2.8 percent a year until 1985 but this increase gradually slows as the introduction of more energy efficient capital and change in economic patterns have their principal effects in reducing demand for petroleum. For example, use of more efficient automobiles, more use of public transport, better building insulation, use of heat pumps in heating and cooling all have a major effect in slowing the increase in petroleum use. Thus, by 2000, petroleum is projected to form 36 percent of total energy input, compared to 44 percent in the 1970s.

Natural gas is predicted to decline dramatically in relative importance. This is due primarily to expected supply limitations which prevent its use from keeping pace with the other fuels. The share of natural gas in total energy falls from 32 percent in 1975 to 18 percent in 2000 (although, in terms of consumption, synthetic gas supplements the availability of gaseous fuels). Finally, the hydro, nuclear, and other share in total input is expected to increase dramatically. This is due to rapidly increasing use of electricity and to the steadily increasing importance of nuclear generation within the electricity sector. The 9-percent average annual increase of hydro and nuclear input over the 1975 to 2000 period results in its share in total input's increasing from 7 percent to 29 percent.

Consumption of each type of energy is shown in Table 8. Coal

TABLE 8

U.S. ENERGY CONSUMPTION, 1970-2000*

	1970	1975	1980	1985	2000
1. U.S. ENERGY CONSUMPTION (QUADRILLION BTU)					
COAL	12.922	13.260	14.304	16.538	26.891
PETROLEUM	29.614	33.696	39.176	44.547	58.148
ELECTRICITY	5.218	6.646	8.854	11.499	23.873
GAS	22.029	24.727	26.888	28.835	32.321
2. ENERGY CONSUMPTION, GROWTH RATES (AVERAGE PERCENT PER ANNUM)					
COAL		0.52	1.52	2.94	3.29
PETROLEUM		2.62	3.06	2.60	1.79
ELECTRICITY		4.96	5.90	5.37	4.99
GAS		2.34	1.69	1.41	0.76
3. ENERGY CONSUMPTION, PHYSICAL UNITS					
COAL (MILLION SHORT TONS)	525.00	573.00	618.00	715.00	1163.00
PETROLEUM (MILLION BARRELS A DAY)	14.70	16.72	19.44	22.11	28.86
ELECTRICITY (BILLION KILOWATT HOURS)	1530.00	1938.00	2590.00	3363.00	6981.00
GAS (TRILLION CUBIC FEET)	21.37	23.97	25.90	27.76	31.10

*COAL CONSUMPTION INCLUDES COAL USED IN ELECTRICITY GENERATION AND IN THE PRODUCTION OF SYNTHETIC GAS. PETROLEUM CONSUMPTION INCLUDES PETROLEUM INPUTS INTO ELECTRICITY GENERATION AND SYNTHETIC GAS. GAS CONSUMPTIONS INCLUDES BOTH NATURAL AND SYNTHETIC GAS.

consumption increases at an average rate of 2.7 percent over the 1975 to 2000 period, petroleum consumption at a rate of 2.2 percent, electricity at 5.3 percent and gas (natural plus synthetic) at 1.1 percent. These growth rates vary within the period, typically increasing over the 1975 to 1980 period, compared to rates for

1970 to 1975, which are reduced as a result of the 1973 to 1974 energy crisis. After 1980 the growth rates steadily decline over the remainder of the century. The dominant trend in these aggregate consumption figures is the continued rapid increase in electricity consumption.

The rapid increase in electricity use has several implications. First, since electricity is a secondary form of energy suffering large energy losses in the conversion from primary fuels, it is very expensive in terms of energy input and its rapid growth produces a rapid growth in the use of primary fuels. In other words, the cost of the electricity growth in energy terms is the absorption of ever increasing proportions of total energy input in electricity conversion loss. For example, in 1975 electricity conversion losses are projected to amount to 13.9 quadrillion Btu, or 18.1 percent of total energy input, but in 2000 these losses are projected to be about 50.0 quadrillion Btu, which is 31.0 percent of total energy.

The second implication of electricity use is derived from the fact that any fuel can be used as the primary input to electricity generation. Uranium and hydro resources, at present, have no alternative use, although uranium enrichment for present generation nuclear reactors is a heavy user of electricity. Also, the ability to use any fossil fuel gives electricity generation some degree of comparative advantage over final use of these fuels. Electricity generation can exploit coal, which is in relative abundance in the United States, and residual oil, which, for technical reasons, has only limited value as a fuel in other uses. In short, electricity has a property that partially offsets its large energy requirements, for the cheapest and most abundant fuels can be used in its generation, leaving scarce oil and gas supplies available for direct use.

Projected fuel prices are shown in Table 9. All prices are

TABLE 9

U.S. ENERGY PRICES, 1970-2000

	1970	1975	1980	1985	2000
1. FUEL PRICES, CURRENT DOLLARS					
COAL ($/SHORT TON)	7.59	11.48	16.38	22.84	54.83
CRUDE PETROLEUM ($/BARREL)	3.39	7.12	8.64	10.97	21.20
ELECTRICITY (c/KILOWATT HR.)	1.61	2.11	2.56	3.01	5.04
GAS ($/THOUSAND CUBIC FT.)	0.83	1.05	1.48	1.97	5.03
2. FUEL PRICES, GROWTH RATES (AVERAGE PERCENT PER ANNUM)					
COAL		8.63	7.36	6.88	6.01
CRUDE PETROLEUM		16.00	3.95	4.89	4.49
ELECTRICITY		5.56	3.94	3.29	3.50
GAS		4.81	7.11	5.89	6.45
GNP PRICE DEFLATOR		5.63	4.11	3.70	3.66
3. FUEL PRICES, CONSTANT DOLLARS					
COAL ($1975/SHORT TON)	5.77	11.48	13.39	15.57	21.80
CRUDE PETROLEUM ($1975/BARREL)	4.46	7.12	7.06	7.48	8.42
ELECTRICITY (c1975/KILOWATT HR.)	2.12	2.11	2.09	2.06	2.00
GAS($1975/THOUSAND CUBIC FT.)	1.09	1.05	1.21	1.34	2.00

expected to increase in current dollars, coal by an average of 6.5 percent a year over the 1975 to 2000 period, crude petroleum by

4.5 percent, refined petroleum products by 5.8 percent, electricity by 3.5 percent, and gas by 6.5 percent. These rates of increase reflect both supply and demand factors. The rapid increase in coal prices is due largely to the slow productivity advance expected in the coal mining industry; increasing difficulty and expense in securing crude oil both domestically and from foreign sources, combined with demand conditions in petroleum product markets, operate to produce continuing rapid increases in oil prices; electricity prices increase but, due to continued productivity advance in the electric utilities sector, at a much slower rate than other prices; gas prices rise rapidly as an increasing demand faces a relatively inelastic supply (this presupposes a relaxation of price controls on natural gas).

When the rates of increase of fuel prices are compared to general inflation, which averages 3.8 percent over the 1975 to 2000 period, a somewhat different picture emerges. First, coal, petroleum, and gas prices all show a significant increase in real terms, which is in marked contrast to historical experience, e.g. between 1951 and 1971 real coal prices fell 15 percent and real prices for refined petroleum products fell 17 percent. This rising price trend provides a strong incentive for economy in the use of fossil fuels. The second price feature is that electricity prices, in real terms, fall slightly, although the 5-percent fall between 1975 and 2000 is again much less than the 43-percent fall that occurred between 1951 and 1971. Compared to past experience, the slower decline in real electricity prices exerts some pressure to slow the historical rate of increase in electricity use, but the main effect of the real electricity price decline is to continue to promote the substitution of electricity for other fuels.

The fuel supply sections of the energy forecasts are shown in Table 10. Although our interindustry model takes account of production and supply possibilities after the primary extractive stage, it does not include any detailed information on the supply characteristics of the U. S. fuel extracting industries. Indeed, given current knowledge, any supply side predictions, particularly to 2000, are hazardous. The supply figures shown in Table 10 represent, essentially, estimates appended to the model. Coal has no supply problems concerning its existence; the only problems concern the prices at which mining and land reclamation become economic. Our coal supply information represents an extrapolation of past data and is embodied in the coal sector production structure of the model. The U. S. oil output figures are based on an *ad hoc* combination of past price elasticities and current information concerning future oil fields, such as the North Slope and other Alaskan fields. Imports make up the difference between U. S. demand and supply at each oil price. Our estimates for U. S. output of natural gas are based upon estimates made by various government and private agencies, as are our estimates for the output of synthetic gas and of shale oil.

The energy flows forecast for the 1975 to 2000 period are presented in detail in Table 11 and in summary form in Table 12. The interindustry energy flows are shown in trillions of Btu for each of the four fuels—coal, petroleum, electricity, and gas—into

TABLE 10

U.S. ENERGY SUPPLY, 1970-2000

	1970	1975	1980	1985	2000
COAL (MILLION SHORT TONS)					
PRODUCTION	598	658	707	819	1291
EXPORTS	72	85	89	104	129
U.S. CONSUMPTION	525	573	618	715	1163
PETROLEUM (MILLION BARRELS A DAY)					
U.S. CRUDE OUTPUT (INCLUDING GAS LIQUIDS)	11.54	10.61	13.91	17.82	21.41
U.S. OUTPUT OF SHALE OIL	0.00	0.00	0.00	0.00	4.29
IMPORTS	3.42	6.40	5.78	4.51	3.28
U.S. CONSUMPTION	14.70	16.72	19.44	22.11	28.86
EXPORTS	0.26	0.29	0.24	0.23	0.12
GAS (BILLION CUBIC FEET)					
U.S. OUTPUT OF NATURAL GAS	20616	21838	22510	23620	23812
U.S. OUTPUT OF SYNTHETIC GAS	0	0	0	664	3502
IMPORTS	821	2210	3443	3525	3813
U.S. CONSUMPTION	21367	23966	25898	27762	31097
EXPORTS	70	81	55	47	30
TOTAL (QUADRILLION BTU)					
U.S. ENERGY INPUT	67.444	77.099	90.484	105.326	161.241
EXPORTS	2.610	2.604	2.601	2.905	3.256
TOTAL DEMAND	70.054	79.703	93.085	108.231	164.497
IMPORTS	8.235	15.880	15.886	13.273	11.040
SUPPLEMENTAL (SYNTHETIC GAS, SHALE OIL)	0.000	0.000	0.000	0.690	12.281
IMPORTS AS PERCENTAGE OF TOTAL DEMAND	11.76	19.92	17.07	12.26	6.71
SUPPLEMENTAL AS PERCENTAGE OF TOTAL DEMAND	0.00	0.00	0.00	0.64	7.47

each of the nine intermediate and four final use sectors. These Btu flows are obtained by applying exogenous Btu/constant dollar ratios to the fuel entries in the constant dollar transaction matrices. Since we are operating at a greater level of disaggregation of fuel uses than is given in energy data, it was necessary to make some simplifying assumptions to obtain the translation of dollars to Btu's.

Information is available for the Btu/constant dollar ratio for each fuel and this ratio for fuel inputs to electricity generation can also be accurately derived from published data. Thus, there is no problem in deriving the figures for total use of each fuel or for fuel inputs into electricity. To fill in the remaining Btu entries it was assumed that, within each fuel, the same Btu/constant dollar ratio applied to each remaining entry with certain exceptions. The exceptions, which embody what other information is available, are that sales of electricity to households and to agriculture take place at a price 67 percent above that charged other users, that petroleum products to households and to agriculture have, because of the product mix, a price 50 percent and 25 percent, respectively, above the average price charged other users, and that the price of gas to households in 1971 was $1.48/million Btu, the price to manufacturing was $0.56, and that gas prices to other users are equal.

Coal is used primarily in manufacturing, coal mining, electricity generation, and for export. The first two uses are relatively unchanging; the projected growth in coal consumption comes in its use in electricity generation, its use for export, and, in the latter

TABLE 11

COMPOSITION OF U.S. ENERGY USE, 1970-2000, INTERMEDIATE AND FINAL DEMAND BY SECTOR (ALL FIGURES IN TRILLION BTU)

	SUPPLY		TOTAL DEMAND, SUPPLY	DEMAND ENERGY INPUTS TO SECTOR												
	U.S. OUTPUT	IMPORTS		1	2	3	4	5	6	7	8	9	10	11	12	13
COAL																
1970	14911	2	14913	27	3342	15	440	1223	0	17	7483	6	173	159	36	1991
1975	15208	9	15217	0	3225	0	147	1228	0	0	8522	0	102	4	33	1955
1980	16356	9	16365	0	3040	0	74	853	0	0	10227	0	74	4	33	2061
1985	18929	10	18939	0	2833	0	66	863	0	0	11818	844	75	4	34	2401
2000	29850	16	29866	0	3904	0	90	1316	0	0	15985	5504	32	7	43	2975
PETROLEUM																
1970	22773	7388	30161	3830	4747	2614	3903	32	83	150	2087	148	10523	126	1360	547
1975	20677	13592	34269	4349	5587	2931	4383	34	51	127	2375	165	12115	79	1500	573
1980	27330	12327	39658	4897	6476	3696	5057	59	44	138	2567	197	14261	107	1677	482
1985	35212	9628	45001	5586	7575	4480	5931	94	29	160	2830	318	15555	125	1865	453
2000	51252	7092	58395	8244	11602	6210	9340	165	0	181	3918	625	15366	119	2379	247
ELECTRICITY																
1970	5220	0	5220	101	1388	19	1459	21	44	43	596	0	1396	0	149	4
1975	6646	0	6646	122	1796	22	1753	29	46	31	671	0	1987	0	184	4
1980	8857	0	8857	173	2261	30	2248	40	73	14	835	0	2950	0	228	4
1985	11502	0	11502	150	2709	36	2910	61	119	0	928	0	4254	0	331	4
2000	23876	0	23876	245	4771	55	5487	180	227	0	1613	0	10697	0	597	4
GAS																
1970	21255	846	22101	171	8884	281	1804	7	0	968	4015	595	4722	0	582	72
1975	22654	2279	24933	204	10215	347	2024	7	0	817	4245	599	5509	0	889	76
1980	23389	3550	26943	235	11389	414	1931	9	0	1231	4505	589	5716	0	869	55
1985	25276	3635	28882	278	13643	431	1742	13	0	1160	4856	568	5321	0	798	47
2000	28587	3932	32351	334	17425	450	1461	26	0	586	6472	475	4274	0	809	30

part of the forecast period, as an input to synthetic gas. The summary table brings out the fact that coal is entirely used as an intermediate fuel and that its use is projected to increase at an increasing rate.

Petroleum products are used heavily in all sectors. Here, growth rates are projected to differ sharply, for the demand for petroleum in final use increases only slowly, while intermediate demand continues to increase rapidly. Use of petroleum products in all intermediate sectors continues to increase at similar rates, with the average increase being 3.0 percent in the 1975 to 1985 period, then falling to 2.7 percent from 1985 to 2000. Final demand, particularly for personal consumption use, grows much more slowly, rising to a peak in the mid-1980s and then remaining steady over the rest of the forecast period.

The slow growth of final demand is due to various factors—higher petroleum prices and the conservation ethic's stimulating economy in use of petroleum, the incorporation of energy-saving devices such as more building insulation, heat pumps, and more efficient heating units into the capital stock of buildings, the replacing of large cars by more energy efficient vehicles, both smaller cars and more public transport, the substitution of electronic communication for some purposes presently achieved by moving people, and the increasing use of electricity in the home's partially replacing petroleum. The time involved in these changes is rather long, for

TABLE 12

COMPOSITION OF U.S. ENERGY USE, 1970-2000, SUMMARY

		USE (QUADRILLION BTU)			GROWTH RATES (AVERAGE PERCENT PER ANNUM)		
		FINAL	INTERMEDIATE	U.S. TOTAL	FINAL	INTERMEDIATE	U.S. TOTAL
COAL	1970	0.368	12.554	12.922			
	1975	0.139	13.123	13.260	-17.69	0.89	0.52
	1980	0.111	14.193	14.304	- 4.40	1.58	1.52
	1985	0.113	16.424	16.538	0.36	2.96	2.94
	2000	0.092	26.799	26.891	- 1.36	3.32	3.29
PETROLEUM	1970	12.019	17.595	29.614			
	1975	13.694	20.002	33.696	2.64	2.60	2.62
	1980	16.045	23.131	39.128	3.22	2.95	3.06
	1985	17.545	27.002	44.547	1.80	3.14	2.60
	2000	17,864	40.284	58.148	0.12	2.70	1.79
ELECTRICITY	1970	1.545	3.673	5.218			
	1975	2.171	4.471	6.642	7.04	4.01	4.96
	1980	3.178	5.676	8.854	7.92	4.89	5.90
	1985	4.585	6.914	11.499	7.61	4.02	5.37
	2000	11.294	12.579	23.873	6.19	4.07	4.99
GAS	1970	5.304	16.725	22.029			
	1975	6.398	18.459	24.857	3.82	1.99	2.34
	1980	6.585	20.303	26.888	0.58	1.92	1.69
	1985	6.119	22.716	28.835	— 1.46	2.27	1.41
	2000	5.083	27.238	32.321	— 1.23	1.22	0.76

first attitudes must be changed, by economic, regulatory, and other pressures, and then these attitudes must be translated into new capital stock which, typically, must wait upon the depreciation and replacement of the existing stock, or in some cases, such as building insulation, can be incorporated into existing, highly durable stock. In any event, a long time lag is predicted before the full effects of this conservation become felt in terms of reduced energy input but ultimately, in the 1980s and 1990s, the scope for conservation is large enough to permit increasing levels of effective services to be sustained from a constant input of petroleum energy.

Electricity consumption increases at a rapid rate—5.3 percent a year between 1975 and 2000. The overall rate of increase steadily decreases, after an increase in the 1975 to 1980 period compared to the previous five years, due to the 1975 consumption figure's being depressed by the reaction to the current energy crisis. As in petroleum use, final and intermediate demand components grow at different rates but, with electricity, it is final demand that grows more rapidly. Intermediate use of electricity increases at an average of 4.2 percent a year with use in the services sector increasing the most rapidly, at 4.7 percent, and manufacturing use increasing at 4.0 percent. Final use of electricity increases at 6.8 percent a year with the most rapid increase taking place in personal consumption which increases at 6.9 percent a year over the 1975 to 2000 period. This use of electricity in personal consumption forms the most rapid increase of any use of any fuel and, in this, continues the same trend that has been observed in the past. The rate of increase is slower than that of the past due to the different price behavior, the new awareness in energy conservation, the introduction of more energy efficient capital such as heat pumps and thermal insulation, and the approach to saturation of such heavy electricity users as home air conditioning, dishwashers, laundries, ranges and so on.

Gas consumption is forecast to increase comparatively slowly and at a declining rate. Increasing use of gas is limited to two sectors—manufacturing and electricity generation—with use in other sectors, primarily services and personal consumption, declining. The net result is that final demand for gas declines, but intermediate demand continues to increase, although at a declining rate. The projected price increases for gas reduce its use in service and final use sectors, where its place is taken by petroleum and electricity, but gas remains important as an input into manufacturing and to electricity generation because of its nonprice advantages—it is clean burning, easily handled and it has many uses as a material, rather than a fuel, input.

The composition of inputs into the electricity generation sector is shown in Table 13. The dominant feature here is the projected

TABLE 13

COMPOSITION OF INPUTS TO U.S. ELECTRICITY GENERATION, 1970-2000 (PERCENT OF TOTAL INPUT)

	1970	1975	1980	1985	2000
COAL	45.4	41.5	37.3	33.2	21.6
PETROLEUM	12.7	11.5	9.4	7.9	5.3
GAS	24.4	20.6	16.4	13.6	8.8
HYDRO, NUCLEAR, OTHER	17.5	26.4	36.9	45.2	64.3

increase in hydro and nuclear, virtually all the increase being in nuclear, generation. In 1970 hydro and nuclear provided 17.5 percent of total input but this proportion increases to 64 percent by 2000. Coal input, although growing rapidly in absolute terms, declines in relative importance as an input from supplying almost half of total inputs in 1970 to one-fifth in 2000. Similarly, although petroleum and gas inputs increase in absolute terms, this increase is relatively slow and they show a steady decline in proportion to total input.

The changing composition of inputs into electricity generation reflects relative prices and is undertaken by the electricity generation sector in response to market forces. But, these forces also reflect, in part at least, a basic characteristic of U. S. fuel supply possibilities. Coal, uranium, and hydro power are not only relatively abundant but also have a low economic opportunity cost, for, apart from electricity absorbed in uranium enrichment, they have comparatively few alternative uses, whereas petroleum and gas fuels are not only in relatively short supply but also have many alternative uses that render their opportunity cost, for use in electricity generation, very high. In these respects, and omitting nuclear safety from consideration, the forecast outcome of market forces, as they affect inputs to electricity generation, appears to be entirely consistent with the objectives of currently proposed national energy policies.

☐ **Alternative projections.** This section presents summary comparisons between our energy projections and alternative projections. This comparison is conducted for two purposes: first, to obtain an independent check that our forecasts are of reasonable orders of magnitude and, second, to pinpoint the differences between economically based forecasts and technologically based forecasts. The alternative forecasts considered are those prepared by the Committee on the U. S. Energy Outlook of the National Petroleum Council,[26] referred to as NPC, the U. S. Department of the Interior forecasts prepared by Dupree and West, denoted DW, and, for electricity, various forecasts reported by Chapman, Mount, and Tyrrell,[27] denoted CMT.

The fossil-fuel projections of our interindustry model, denoted HJ, of the NPC, and of Dupree and West are shown in Table 14.

TABLE 14

ALTERNATIVE FORECASTS OF U.S. ENERGY UTILIZATION*

	DW		HJ		NPC	
	Q BTU	GROWTH RATE	Q BTU	GROWTH RATE	Q BTU	GROWTH RATE
COAL						
1970	12.922		12.922		12.922	
1975	13.825	1.4	13.260	0.5	16.1	4.5
1985	21.470	4.5	16.538	2.2	22.7	3.5
2000	31.360	2.6	26.891	3.3		
PETROLEUM						
1970	29.614		29.614		29.614	
1975	35.090	3.5	33.696	2.6	39.8	6.1
1985	50.700	3.7	44.547	2.8	56.2	3.5
2000	71.380	2.3	58.148	1.8		
GAS						
1970	22.029		22.029		22.029	
1975	25.220	2.7	24.727	2.3	21.3	-0.7
1985	28.390	1.2	28.145	1.3	21.1	-0.1
2000	33.980	1.2	28.681	0.1		
TOTAL INPUT						
1970	67.444		67.444			
1975	80.265	3.5	77.099	2.7		
1985	116.630	3.8	105.326	3.2		
2000	191.900	3.4	161.241	2.9		

*DW IS DUPREE AND WEST, HJ IS OUR PROJECTION, NPC IS THE NATIONAL PETROLEUM COUNCIL. GROWTH RATES ARE AVERAGE ANNUAL PERCENTAGE RATES OF GROWTH. Q BTU IS QUADRILLION BTU.

For 1975 our forecasts and those of Dupree and West are similar, although for each fuel and for total input, ours are lower. The higher fuel prices incorporated into our projections reduce fuel use below the extrapolations of past trends used by Dupree and West. This same relationship continues for 1985 and 2000; i.e., our forecasts and those of Dupree and West are reasonably close, but the divergence increases over time, with our energy use always being below that in Dupree and West. This indicates first, that the orders

[26] See National Petroleum Council [36].

[27] See Chapman, Mount, and Tyrrell [1970]; a detailed report on the econometric model underlying these projections is given by Mount, Chapman, and Tyrrell [34].

of magnitude of our forecast are consistent with past energy trends used by Dupree and West, and second, that simple extrapolations of these trends are insufficient. Expected fuel price increases and changes in the nonenergy sections of the economy do exert a significant influence working to depress energy use below historical trends. The NPC forecasts for 1975 do not, at the present time, appear very realistic and the 1985 NPC forecasts are also open to question. The fact that our projections are well below those of NPC gives further weight to the need to take account of price and other factors that affect demand.

Consider now the various projections of electricity consumption, given in Table 15. The striking impression gained from these

TABLE 15

ALTERNATIVE FORECASTS OF U.S. ELECTRICITY CONSUMPTION (TRILLION KILOWATT HOURS)

PROJECTION*	1970	1975	1980	1985	1990	2000
CORNELL-NSF	1.57	2.15	2.92	3.96	5.38	10.25
CMT-HIGH	1.53	2.14	3.05	—	5.66	9.89
-MEDIUM	1.53	1.98	2.38	—	3.01	3.45
-LOW	1.53	1.88	2.07	—	2.11	2.01
DW	1.53	2.13	3.00	4.14	—	9.01
FPC	1.53	—	3.07	—	5.83	—
HJ	1.53	1.94	2.59	3.36	—	6.98
NPC	1.59	2.29	3.29	4.54	—	—

*CORNELL-NSF IS CORNELL-NATIONAL SCIENCE FOUNDATION WORKSHOP; CMT IS CHAPMAN, MOUNT, AND TYRRELL; DW IS DUPREE AND WEST; FPC IS FEDERAL POWER COMMISSION; HJ IS OUR PROJECTION; NPC IS NATIONAL PETROLEUM COUNCIL.

SOURCES: CORNELL-NSF, FPC, NPC FORECASTS ARE GIVEN IN CHAPMAN, MOUNT, AND TYRRELL (1972), TABLE 1, PAGE 3. THE CMT "LOW" FORECAST ASSUMES THAT THE REAL ELECTRICITY PRICE DOUBLES BY 2000, THE "MEDIUM" FORECAST CORRESPONDS TO THE FPC ESTIMATES OF A 19-PERCENT REAL PRICE INCREASE OVER 1970-1990. "HIGH" CORRESPONDS TO A 24-PERCENT REAL PRICE DECLINE OVER 1970-1980 AND A 12-PERCENT DECLINE OVER EACH OF THE 1980-1990 AND 1990-2000 INTERVALS.

forecasts is one of great variation—the projected 2000 consumption ranges from 2.0 to 10.3 trillion Kwh. Our forecasts are never the extreme entry but are always towards the lower end of the range of forecasts. To the extent that such a diversity of forecasts permits any conclusion, we can say that the above conclusions are reinforced. Our projections are in line with other forecasts but, by taking price and other economic factors into account, our projections are comparatively low.

All the alternative energy forecasts considered here, apart from Chapman, Mount, and Tyrrell's electricity forecasts, have been prepared without explicit assumptions about the growth of real or relative fuel prices, without explicit consideration of interfuel substitutions and without explicit account of the interactions between energy and the rest of the economy or of future developments in the rest of the economy. This lack of analysis of the underlying forces at work in determining energy demand is a serious conceptual drawback. The omission is particularly significant in view of the present need to use energy models to aid the formation of future U. S. energy policy. Our interindustry model, however, avoids these problems and can provide a more realistic basis for policy analysis.

6. Energy policy: a Btu tax

Introduction. This section describes the application of the interindustry energy model to the analysis of one specific energy policy—a Btu tax designed to secure energy independence. This application serves to illustrate the methodology of the model and provides an evaluation of a specific tax proposal currently under consideration. The actual tax considered is a uniform rate of tax levied on the energy content of all fuels used outside the energy generation sector; such a tax is proposed in the Energy Revenue and Development Act of 1973, at present under consideration in the Senate Finance Committee.[28]

The starting point for the analysis of this tax is the base case set of energy projections which have been presented above. These projections are based on the assumption of no major new developments in energy policy and so can be used as a reference point against which the changes induced by policy changes are measured. In particular, the fuel imports in these forecasts define the objective of the policy, which is to reduce these imports to nominal levels by a target date, which we take to be 1985.

The tax has the effect of creating a wedge between the price paid by the energy consumer and that received by the producer. Since the forecast supply prices already equal average costs, the tax is translated into an increase in the fuel selling price, leaving the supply price unchanged, apart from indirect impacts on the supply price due to any production cost increases that are caused by the tax. Therefore, the tax leads to the reduction of fuel imports solely by acting on the demand side of the energy equation. This approach is the opposite of much of the present debate about energy policy, which is supply oriented. It emerges that demand based policies can be extremely effective and have the power to produce energy independence even without accompanying supply expansion policies.

The Btu tax is inserted into the model by means of price markups that increase the sales prices of fuels above the output price received by the seller. These markups vary for the five fuels since the Btu content of each dollar of fuel output is different for each fuel (the costs of Btu's obtained from the various fuels are later given in Table 18). The tax is assumed to be levied only on energy as it emerges from the fuel sectors so that energy inputs into fuel production, including the generation of electricity, are not subject to the tax. Also, exports of fuels, mainly coal, are considered exempt from the tax.

After price markups have been inserted, the model is solved to obtain a new set of economic and energy projections. This involves solving for new output prices and input-output coefficients, new final demand components, and the associated industry output levels and interindustry transactions. This information is then used to determine the energy deficits—the excess of U. S. demand for each fuel over the domestic output, both being calculated in terms of the new set of prices—which must be made up by fuel imports. This procedure captures both the direct and indirect effects of energy conservation—the direct saving of energy as an input by the sub-

[28] This section is based on Hudson and Jorgenson [24], presented as testimony at hearings by the Senate Finance Committee, January 16, 1974.

stitution of other inputs for energy as well as the indirect saving of energy produced by substituting, in production and consumption, nonenergy intensive for the energy intensive goods and services.

☐ **Tax policy and conservation.** We proceed in two steps, first examining the effects of the Btu tax by assessing the impact of various rates of tax on the energy and nonenergy sectors of the economy in 1980 and second, considering the specific tax structure that would be required to achieve the objective of independence from energy imports by 1985. The definition of energy independence used is for zero imports of gas and for only nominal imports of petroleum, such as would be required for fueling of U. S. aircraft and ships abroad. An arbitrary limit of one quadrillion Btu of energy imports was adopted for 1985; this limit is less than 10 percent of actual 1973 fuel imports.

The impact on energy use of various rates of Btu tax is summarized in Table 16. These figures show that the Btu tax can induce

TABLE 16

IMPACT OF BTU TAXES ON TOTAL ENERGY INPUT, 1980

TAX RATE ($/MILLION BTU)	0	0.1	0.3	0.5
TOTAL ENERGY INPUT (QUADRILLION BTU)	90.484	88.890	86.004	83.440
CHANGE FROM BASE		-1.593	-4.479	-7.043
CHANGE/TAX RATE		-15.930	-14.930	-14.070
PERCENT CHANGE FROM BASE		-1.760	-4.950	-7.780
INTERMEDIATE USE (QUADRILLION BTU)	70.356	69.321	67.415	65.692
CHANGE FROM BASE		-1.035	-2.941	-4.664
CHANGE/TAX RATE		-10.350	-9.800	-9.330
PERCENT CHANGE FROM BASE		-1.470	-4.180	-6.630
FINAL USE (QUADRILLION BTU)	20.127	19.569	18.589	17.748
CHANGE FROM BASE		-0.558	-1.538	-2.379
CHANGE/TAX RATE		-5.580	-5.130	-4.760
PERCENT CHANGE FROM BASE		-2.770	-7.640	-11.820
ENERGY IMPORTS (QUADRILLION BTU)	15.886	14.311	11.435	8.871
IMPORTS IN TOTAL INPUT (%)	17.600	16.100	13.300	10.600
TAX REVENUE ($ BILLION)	0	6.026	17.400	28.003

significant reductions in energy use and that it has the potential for reducing demand sufficiently to secure energy independence. The highest rate shown, $0.5 per million Btu, leads to a decline of 7 quadrillion Btu, or 7.8 percent in energy use relative to the no tax projection. This reduction is made up of substantial cuts in both final and intermediate uses of energy. The greater part of the reduction comes from the decline in energy input to intermediate production, but the relative cutback in use is greater in final demand where energy input is reduced by 11.8 percent, compared to the 6.6 percent fall in intermediate use.

Our results indicate that although final users of energy may be more responsive to price increases than business users, the sheer volume of energy absorbed in production requires that, for maximum effect, energy conservation policies give at least as much weight to reducing intermediate as to reducing final demand. The response of both intermediate and final users to the tax varies with the tax level with, in both cases, the tax having a diminishing marginal impact on energy use. The decline in effectiveness is, however, gradual so that a reasonable first approximation is that each dollar of tax per million Btu reduces total energy input by about 15 quadrillion Btu. This corresponds to an elasticity of Btu use to Btu price of approximately —0.24.

The detailed adjustments by energy users to the imposition of the tax are shown in Table 17. These figures show, first, the differ-

TABLE 17

IMPACT OF BTU TAXES ON INPUT PATTERNS, 1980

TAX RATE ($/MILLION BTU)	0	0.1	0.3	0.5
ENERGY INPUT, INCLUDING USE OF ELECTRICITY (QUADRILLION BTU)				
MANUFACTURING	18.908	18.669	18.222	17.909
SERVICES	10.496	10.335	10.032	9.752
ELECTRICITY GENERATION	27.403	27.289	27.063	26.840
PERSONAL CONSUMPTION	20.474	20.025	19.224	18.527
PERCENTAGE CHANGE IN TOTAL INPUTS:				
ENERGY INPUT		-1.76	-4.95	-7.78
CAPITAL INPUT		0.11	0.29	0.49
LABOR INPUT		0.14	0.39	0.61
INPUT-OUTPUT COEFFICIENTS FOR TOTAL ENERGY INPUTS:				
AGRICULTURE	0.0245	0.0239	0.0226	0.0216
MANUFACTURING	0.0235	0.0233	0.0232	0.0231
TRANSPORT	0.0447	0.0440	0.0426	0.0413
SERVICES	0.0190	0.0189	0.0186	0.0183
ELECTRICITY GENERATION	0.1864	0.1859	0.1850	0.1844
INPUT-OUTPUT COEFFICIENTS FOR MANUFACTURING:				
COAL	0.0026	0.0026	0.0025	0.0025
PETROLEUM	0.0069	0.0068	0.0068	0.0067
ELECTRICITY	0.0085	0.0086	0.0087	0.0088
GAS	0.0055	0.0054	0.0052	0.0051
TOTAL ENERGY	0.0235	0.0233	0.0232	0.0231
CAPITAL SERVICES	0.1201	0.1200	0.1199	0.1198
LABOR SERVICES	0.2887	0.2889	0.2893	0.2897
MATERIALS INPUT	0.5164	0.5163	0.5162	0.5160

ent degrees of energy conservation in the four major energy consuming sectors—manufacturing, services, electricity generation, and personal consumption. Substantial economies are made in the energy input to each sector, but there is a wide difference in the proportionate response: energy used in personal consumption is reduced by 9.5 percent in response to the $0.5 tax rate, with services use being cut by 7.1 percent, manufacturing use by 5.8 percent, and input into electricity generation by 2.1 percent.

Energy use can be split into two broad categories—discretionary use and process use. Discretionary use includes inputs for comfort functions such as heating and cooling as well as personal services such as automobile travel. Process use covers fuel inputs for driving machinery, heating materials, turning generators, and so on. Discretionary uses are, typically, characterized by greater flexibility than process uses, firstly because the associated capital stock is generally less durable and easier to replace with more energy efficient capital, and secondly because discretionary uses are input to the generation of psychologically, rather than technically, specified performance, e.g., desired miles driven rather than energy inputs required per ton of alumina to be smelted, so the level of use within the existing capital stock can be more readily varied. Given this categorization, it is apparent that personal consumption includes a high ratio of discretionary to process energy use while this ratio for service activity is a little lower, for manufacturing lower still, and for electricity generation lowest of all. This ordering is precisely that observed in the sectoral response to the increase in energy prices.

More specifically, the input-output coefficients in Table 17 show the processes of adjustment to the tax induced increases in energy prices. Two types of adjustment can be seen. First, there is a substitution of the now relatively less expensive nonfuel inputs for the relatively more expensive fuel inputs. This is reflected in the total energy input coefficients which, for each sector, decline as the energy tax increases. The manufacturing input-output coefficients show the outcome of this substitution process: energy inputs decline, as do inputs of materials and inputs of capital, with inputs of labor services increasing to replace the three reduced inputs. Thus, in manufacturing, energy-capital-materials complementarity means that adjustments to economize on the expensive energy input lead to the reduction in the use of energy, capital, and materials and to the adoption of more labor-intensive production techniques.

The shift away from energy in manufacturing is relatively small in terms of changes in input-output coefficients, but it represents a substantial amount of energy. A similar process of substitution away from energy takes place in all sectors. The overall result, however, is a little different from that in manufacturing. The total reduction in energy use is made possible by an increase in inputs of both capital and labor. That is, on an economy-wide basis, energy-capital substitutability, such as the use of insulation or smaller cars to save energy, dominates the complementarity relation that characterizes manufacturing. The net effect is that the 7.8 percent fall in energy input is accommodated by a 0.6 percent increase in the demand for labor and a 0.5 percent increase in capital use and does not result in a comparable reduction in potential output.

The second set of adjustments occurs within energy input as interfuel substitution takes place. The tax has the effect of increasing different fuel prices by different extents, depending on the energy content of each fuel. Coal prices are increased to the greatest extent, with gas prices next, then petroleum, with electricity prices being increased the least. The input-output coefficients for fuel use in manufacturing, shown in Table 17, illustrate the resulting substitutions. Coal use declines only slightly for, despite the sharp price increase, coal remains the cheapest source of energy as well as being, for technical reasons, used in some production, such as steel, regardless of the price changes. The petroleum and gas coefficients both decline more noticeably. Electricity use, however, increases since electricity has become a relatively less expensive fuel and since its flexibility in use permits it to substitute for petroleum and gas.

The impact of the Btu tax on the price and consumption of each fuel is shown in Table 18. Coal prices increase by the largest pro-

TABLE 18

IMPACT OF BTU TAXES ON ENERGY CONSUMPTION AND PRICES, 1980

TAX RATE ($/MILLION BTU)	0	0.1	0.3	0.5
FUEL CONSUMPTION (QUADRILLION BTU)				
COAL	14.304	14.237	14.104	13.971
PETROLEUM	39.176	38.459	37.138	35.940
ELECTRICITY	8.854	8.817	8.744	8.672
GAS	26.888	26.126	24.791	23.650
CHANGE IN CONSUMPTION FROM BASE				
COAL		-0.47	-1.40	-2.33
PETROLEUM		-1.83	-5.20	-8.26
ELECTRICITY		-0.42	-1.24	-2.06
GAS		-2.83	-7.80	-12.04
AVERAGE FUEL PRICES				
COAL ($/TON)	16.3800	17.3100	19.1600	21.0200
REFINED PETROLEUM ($/BARREL, WHOLESALE)	11.8400	12.3800	13.4600	14.5500
ELECTRICITY ($/KWHR)	0.0256	0.0259	0.0266	0.0273
GAS ($/TH CU.FT.)	1.4800	1.5500	1.7000	1.8400
CHANGE IN PRICES FROM BASE (PERCENT)				
COAL		5.68	16.94	28.33
REFINED PETROLEUM		4.57	13.71	22.85
ELECTRICITY		1.32	3.96	6.60
GAS		4.87	14.62	24.36
PRICE OF ENERGY TO TAXED USERS ($/MILLION BTU)				
COAL	0.71	0.81	1.01	1.21
PETROLEUM	2.21	2.31	2.51	2.71
ELECTRICITY	7.46	7.56	7.76	7.96
GAS	1.42	1.52	1.72	1.92

portion, with the average increase of 28 percent, for the $0.5 tax rate, comprising a 70-percent increase in price to nonfuel purchasers and virtually no increase in price in fuel production or export uses. The gas price increases by 24 percent for the highest tax rate, with the wholesale price of refined petroleum products increasing by 23 percent and electricity rising by only 7 percent. The small price increase for electricity leads to a correspondingly small demand decline, only 2 percent. This small decline also implies that coal input into electricity, the main use of coal, declines by only a small proportion. The only other sizeable domestic use of coal is in the manufacturing sector, but this use is relatively insensitive to price as discussed above. The price of gas increases substantially and this, with the opportunities to economize in its use and to substitute other fuels for it, leads to a decline of 12 percent in its use. Similarly, petroleum prices rise substantially, causing a decline of 8 percent in the consumption of petroleum. The average price elasticities implied by these consumption responses are: —0.08 for coal, —0.36 for petroleum, —0.31 for electricity, and —0.49 for gas.

The effects of the Btu tax are not restricted to the energy

sectors. There are also effects on the input structure of other production, as have been outlined above, on the costs and prices of these other goods, on the demand for these goods in intermediate uses, and on the level and composition of real final demand. These effects are summarized in Table 19, which gives the tax induced

TABLE 19

IMPACT OF BTU TAXES ON NONENERGY PRICES AND QUANTITIES, 1980 PERCENTAGE CHANGE FROM BASE CASE

	PRICE			QUANTITY		
TAX RATE ($/MILLION BTU)	0.1	0.3	0.5	0.1	0.3	0.5
GROSS OUTPUT:						
AGRICULTURE	0.17	0.51	0.82	-0.18	-0.53	-0.85
MANUFACTURING	0.16	0.50	0.82	-0.10	-0.30	-0.49
TRANSPORT	0.21	0.64	1.05	-0.17	-0.52	-0.84
SERVICES	0.06	0.22	0.34	-0.08	-0.22	-0.36
FINAL DEMAND:						
CONSUMPTION	0.27	0.86	1.34	-0.12	-0.33	-0.53
INVESTMENT	0.16	0.48	0.74	-0.11	-0.31	-0.51
GOVERNMENT	0.12	0.38	0.56	-0.08	-0.22	-0.36
GNP	0.23	0.69	1.07	-0.09	-0.26	-0.42

changes in nonenergy prices and quantities demanded. All prices are increased as the effects of the higher fuel prices work through the production cost structure. The process of substitution towards the relatively less expensive inputs lessens the impact of the fuel price increases on average costs, but the net effect is still an upward movement in costs, and in prices in general.

The price increases by sector are in line with the importance of energy in the sector's inputs: transport prices increase the most, followed by agriculture and manufacturing, with service prices rising the least; this ranking is the same as the ranking of these sectors in terms of energy input-output coefficients. The price increases are not, however, very sizeable. Even for the $0.5 tax rate, the price increases range from 0.34 percent for services to 1.05 percent for transport. Similarly, the aggregate consumption price index and the GNP price deflator increase by only about 1 percent since the small nonenergy price rises dominate the larger fuel price increases in these price indices. The quantity changes induced by the energy tax are correspondingly small. The new prices, and the decline in real incomes, lead to a reduction and redirection of real final demand, but the resulting change in total real consumption and real GNP are very small, only of the order of 0.5 percent.

Two alternative bases for the Btu tax were examined. These were the cases in which the tax was levied only on final consumption of energy and in which the tax was levied only on intermediate inputs to the nonenergy sectors. Both taxes do reduce total energy input, the final use tax by 2.8 percent and the intermediate tax by 5.3 percent for a tax rate of $0.5 per million Btu. This compares with the 7.8 percent reduction for the general tax. Thus, although the energy conservation impact of a Btu tax diminishes as the base is restricted, even a tax on one of the narrow bases does lead to nonnegligible cutbacks in energy consumption. So, if political considerations were to preclude a final use tax, the intermediate Btu tax could still play an effective role in an energy policy package.

☐ **Import independence.** Our central conclusion is that a Btu tax can induce significant reductions in energy consumption, corresponding to even larger relative reductions in fuel imports, at the cost of comparatively minor changes in production patterns, prices, or demand. Our next step is to apply these results to formulate a specific Btu tax program that would result in import independence by 1985. The tax rate required to reduce energy consumption in 1985 to a level that requires no fuel imports is $1.35 per million Btu. The effects of this tax are summarized in Table 20. The revenue yield

TABLE 20

IMPACT OF AN ENERGY INDEPENDENCE BTU TAX, 1985

TAX RATE ($/MILLION BTU)	0	1.35
TAX REVENUE ($ BILLION)	0	76.942
TOTAL ENERGY INPUT (QUADRILLION BTU)	105.326	88.244
PERCENTAGE CHANGE FROM BASE CASE		-16.2
IMPORTS OF PETROLEUM (QUADRILLION BTU)	9.628	0.975
IMPORTS OF GAS (QUADRILLION BTU)	3.635	0
IMPORTS IN TOTAL INPUT (PERCENT)	12.6	1.1
FUEL CONSUMPTION (QUADRILLION BTU)		
COAL	16.538	15.730
PETROLEUM	44.547	36.255
ELECTRICITY	11.499	10.780
GAS	28.835	21.926
PERCENTAGE CHANGE IN FUEL CONSUMPTION FROM BASE CASE		
COAL		-4.89
PETROLEUM		-18.61
ELECTRICITY		-6.25
GAS		-23.96
PERCENTAGE CHANGE IN PRICES FROM BASE CASE		
COAL		49.15
PETROLEUM		42.20
ELECTRICITY		15.57
GAS		47.16
AGRICULTURE		1.78
MANUFACTURING		1.70
TRANSPORT		1.98
SERVICES		0.71
CONSUMPTION		2.41
GNP		2.04

of this tax is $76.9 billion. This is a substantial withdrawal from the private spending stream and, unless returned through other policy measures, would have a significant deflationary impact. This revenue would be more than ample to sustain a major research and development effort on energy supply as well as to cover other government spending or to permit reductions in other taxes.

The reduction in energy use is secured through the effect of the tax in increasing fuel prices. In fact, a rise of around 40 percent in fuel prices, compared to the base case, is required to obtain the necessary cutback in energy use. Average coal prices are raised by 49 percent by the tax, gas prices are increased by 47 percent, petroleum prices by 42 percent, and electricity prices by 15 percent. The induced reductions in fuel consumption range from 5 percent for coal and 6 percent for electricity, to 19 percent for petroleum, and 24 percent for gas. The total reduction in energy input is 17.1 quadrillion Btu, which is 16.2 percent of the base case level. This reduction corresponds to a greater reduction in energy input per dollar of Btu tax rate than emerged from the 1980 simulations. The reason for this is the lags involved in responses to higher energy prices, particularly the lags involved in the replacement of capital stock with more energy efficient capital.

These reductions in fuel consumption are greater than are strictly necessary for energy independence. Coal and electricity suffer no supply deficit and reduction in the use of these fuels does not directly contribute to the independence objective. Gas use is cut back substantially by the tax, by much more, in fact, than the 10 percent reduction that would be required to eliminate gas imports. Thus, although the Btu tax does, fortuitously, induce conservation primarily in the two fuels whose consumption must be reduced if the independence objective is to be achieved, it is comparatively inefficient in achieving this objective, for it is so broad in its effects that the tax rate required to achieve petroleum independence produces excessively large reductions in the use of other fuels. The application of a general instrument in pursuit of such a specific objective results in unnecessary economic cost. Since excess petroleum demand is the binding constraint in the independence objective, a specific petroleum tax would be a more efficient instrument.

The effects of the energy tax on the nonfuel sectors are not large. Prices increase by an average of only 2 percent, for the large fuel price increases are dominated, in the consumption and GNP price indices, by the smaller increases in the prices of nonfuel goods. The input substitution processes, described above, permit production to accommodate the large reductions in energy input and the increases in fuel prices without corresponding decreases in output or increases in total costs.

The specific rate timetable of a Btu tax that would achieve energy independence by 1985 is a variable. The rate structure is subject to constraints, the rate must be zero in 1974 and $1.35 per million Btu in 1985, but the path between these end points is somewhat arbitrary. One possible rate timetable, with its associated macro-economic effects, is presented in Table 21. But, if the Btu tax program were to be implemented, its rate structure should satisfy certain additional conditions. First the tax must be regarded as a permanent measure, for, in the absence of other policies, continued energy independence requires that energy demand continually be depressed by taxes, and this will probably mean steadily increasing rates of tax. Second, the rate level should be increased fairly rapidly to the $1.35 level for, in view of the delays in incorporating energy conservation measures into production and final consumption, a high tax rate over much of the 1974 to 1985 period would not only induce substantial conservation but would induce it early enough

TABLE 21

A BTU TAX FOR ENERGY INDEPENDENCE IN 1985

	1975	1980	1985
BTU TAX RATE ($/MILLION BTU)	0	0.5	1.35
ENERGY INPUT (QUADRILLION BTU)	77.1	83.4	88.2
ENERGY IMPORTS (QUADRILLION BTU)	15.9	8.9	1.0
REVENUE FROM TAX ($ BILLION)	0	28.0	76.9
CONSUMPTION OF ENERGY (QUADRILLION BTU)			
COAL	13.3	14.0	15.7
PETROLEUM	33.7	35.9	36.3
ELECTRICITY	6.6	8.7	10.8
GAS	24.7	23.7	21.9
PERCENTAGE CHANGE FROM ZERO TAX CASE			
REAL CONSUMPTION	0	-0.53	-1.22
CONSUMPTION PRICE	0	1.34	2.41
REAL GNP	0	-0.42	-0.99
GNP PRICE DEFLATOR	0	1.07	2.04

for its full effects to be felt by 1985. Third, it is desirable that the rate structure be known for several years into the future so as to reduce any economic incentive for delaying energy conservation in the hope that energy prices would fall in the future.

Our results show that a Btu tax can induce sufficient fuel conservation to secure energy independence by the mid-1980s. Further, it suggests that the cost of achieving this is not catastrophic. But it does not follow that this is the best or only means for securing independence. Arguments could be made for a policy mix that stimulated supply expansion as well as demand reduction. Also, arguments could be made for a mix of policies on the demand side, such as a Btu tax, regulations requiring building insulation, taxing of particularly heavy energy uses, and so on. And it might be argued that a general policy, such as a Btu tax, is inefficient, for it reduces demand for all fuels when the critical problem is evidenced only in petroleum. More fundamentally, the desirability and definition of energy independence could be debated *ad infinitum*. Our results beg the question of underlying objectives, but they do indicate that the Btu tax does provide a means of securing energy independence.

7. Conclusion

■ We conclude with a brief overview of our methodology, comparing our framework for the analysis of energy policy with alternative approaches. Our model replicates each of the components of the overall transactions flow—purchase of primary inputs, sales of goods and services between sectors, formation of product prices, and purchase of output by final users. These aspects of economic activity are brought into consistency by means of simulated market processes. All decision units react to the same set of prices, and prices and quantities adjust so that all markets are cleared and each production sector covers its costs. The heart of the model lies in a series of submodels of production behavior, one for each of the nine domestic producing sectors. This set of production relationships provides the basic information used to determine relative output prices and the corresponding set of input patterns.

The sectoral production models are based upon price possibility frontiers. Within each sector, the price frontier expresses output price in terms of the prices of inputs and the production efficiency of that sector. Each frontier contains first- and second-order terms so that both the average requirements for each input and the interrelationships among inputs are captured. Thus, although the specification is in terms of prices, it captures the same information concerning input requirements, complementarities, and substitutabilities as the more traditional production function.

The primary input prices are generated within our macro-econometric growth model. Productivities are exogenous to the producer submodels so the nine sectoral output prices can be solved from the nine price frontiers. The simultaneous determination of prices permits the derivation of a set of prices for all produced output which not only takes account of production constraints and interrelationships and primary input prices, but also integrates all these price determinants into a consistent framework so that all sectors are simultaneously charging the minimum price that covers all their costs, including a return to capital. In short, given primary input prices and underlying production information, the model finds the equilibrium price system for the economy.

This approach to prices and production also permits the input pattern for each sector that is best suited to the prevailing set of prices and costs to be chosen from the infinity of patterns that are possible. Thus, producers can react to the prevailing structure of relative prices by adjusting their input patterns and substituting, within the limits set by the complementarity and substitutability information contained in the price frontiers, relatively less expensive for relatively more expensive inputs. In this way, a critical feature of the economy, the response of input patterns to prices, is systematically included in our model. This feature of the model is very important from a practical point of view in accommodating the detailed operation of the price system. Conceptually, it frees input-output analysis from the assumption of fixed coefficients, and extends input-output analysis by making the coefficients variable and by making them fully endogenous and linking them to price behavior.

After the analysis of production relationships has been completed, the next step in implementing our model is to examine the components of final demand. These components are produced by three final demand submodels. The initial inputs into these submodels are the prices charged for the output of each of the domestic producing sectors, the prices of primary inputs that enter directly into final use, and the total current dollar expenditures in three of the final demand categories—personal consumption, private investment, and government purchases. These expenditure totals are produced by the use of our macro-econometric model of U. S. economic growth. Our growth model provides levels of relative prices of consumption, investment, capital, and labor within the framework employed to generate final demand in real terms. The model represents a major extension of macro-econometric model building to incorporate elements of both aggregate demand and aggregate supply.

EDWARD A. HUDSON AND
510 / DALE W. JORGENSON

We begin with the set of prices that is consistent with equilibrium demand and supply and with the levels of prices of primary inputs. We find the pattern of input-output coefficients associated with this price regime. We determine the level and composition of final demand that are associated with these prices. The input-output coefficients have been determined already, so the balance equations of input-output analysis are sufficient to find, in constant dollar terms, the total sectoral outputs and the pattern of industry purchases and sales necessary to sustain the final demand. This step imposes a final set of market clearing relationships on our simulated economy with the condition of equality between the real demand and supply of every commodity's being added to the condition of equality between value of demand and supply and of equality between receipts and expenditures for every sector.

The output from the interindustry model comprises prices for each of the supplying sectors, the matrix of constant dollar transactions for the entire structure, and, by combining these two elements, the matrix of current dollar transactions. The transactions matrices cover the whole economic structure—energy as well as nonenergy sectors. The energy information contained in these transactions matrices is already extensive, covering both volumes and prices. It can readily be extended, however, to produce the data forms traditionally used in energy analysis—energy flows in British thermal units, energy flows in physical units, and energy prices in terms of physical units. This further information for Btu's can be generated by inserting known values of the Btu content per constant dollar of each fuel, the volume in physical units can be obtained by using known physical units per constant dollar ratios, and fuel prices can be generated by applying the base period prices to the price indices simulated in the model.

We conclude this overview of our framework for energy policy analysis by comparing the conceptual properties of our model with those of the traditional energy balance approach to energy analysis. The fundamental advantage of our approach is that, rather than viewing energy in isolation, we view it as one of the many interacting parts that make up the economic system. This perspective permits the systematic analysis of all the factors that influence energy on both demand and supply sides and, equally important, it permits the explicit linkage of energy developments to those variables such as employment, incomes, and consumption, that are the ultimate ends to which energy use is only a means.

More specifically, our model incorporates the influence of fuel prices on the level and composition of energy use and, further, it incorporates the effects of the level and pattern of nonenergy activity on energy use as well as the reverse linkage of energy prices and supplies to nonenergy price input, output, and consumption patterns. These interrelationships are critical and it is essential, for both forecasting and policy purposes, to recognize them. Some examples of these linkages are the severe impact of the recent oil shortages on the output of and the incomes generated by the automobile and tourist industries, the implications for energy use of the secular trend of demand towards service activities, and the economies in fuel use induced by the recent increases in fuel prices.

One aspect of these linkages in which our model represents a particularly important advance over energy balance procedures is the variation of input patterns in response to relative prices. This is of great importance in energy analysis in view of the widespread ability to substitute between fuels, for example between coal and oil in electricity generation, between oil and gas in industrial heating, between gas and electricity in home cooking, and so on. This substitution is not merely a matter of switching fuels, for the associated capital stock must also be changed to permit the use of a different fuel. Our production models, by treating all inputs simultaneously, take account of the possibilities for both interfuel substitutions and substitutions of fuel for nonfuel input, but do this in such a way that the constraints implied by complementarities between inputs are recognized, so that a consistent analysis of the entire input picture is obtained.

Our investigation of the effects of a Btu tax serves to demonstrate the usefulness of our framework for energy policy analysis. The basic properties of the model are illustrated by the result that, in the 1980 simulations for example, energy input can be reduced by 8 percent at the cost of only a 1-percent increase in average prices and a 0.4-percent decrease in real income. In other words, the flexibility of the economy in adapting to changing resource availabilities, and the power of the price system in securing this adaptation, mean that substantial reductions in energy use can be achieved without major economic cost. The analytical property of our model that incorporates this flexibility is the endogenous formation of prices and the endogenous determination of the response of production patterns and final demand to these prices. Also the integration of the various components of the model by means of interindustry analysis secures an overall consistency in the simulation of the market process. These features, permitting price formation and the reaction of producers and consumers to price changes, combined in a simultaneous model of the entire economy, represent a major advance over traditional interindustry and energy balance analysis.

References

1. Almon, C., Buckler, M. R., Horwitz, L. M., and Reimbold, T. C. *1985: Interindustry Forecasts of the American Economy*. Lexington, Mass.: Heath, 1974.
2. Berndt, E. R., and Christensen, L. R. "The Specification of Technology in U. S. Manufacturing." Discussion Paper 73-17, University of British Columbia, November, 1973.
3. ———. "Testing for the Existence of a Consistent Aggregate Index of Labor Inputs." *The American Economic Review*, Vol. 64, No. 3 (June 1974), pp. 391–404.
4. ———. "The Translog Function and Substitution of Equipment, Structures, and Labor in U. S. Manufacturing, 1929–1968." *Journal of Econometrics*, Vol. 1, No. 1 (March 1973), pp. 81–114.
5. Berndt, E. R. and Jorgenson, D. W. "Production Structure," Chapter 3 in Jorgenson *et al.* [26].
6. Berndt, E. R. and Wood, D. W. "Technology, Prices, and the Derived Demand for Energy." Discussion Paper 74-09, University of British Columbia, May 1974.

7. Burmeister, E. and Dobell, A. R. *Mathematical Theories of Economic Growth,* New York: Macmillan, 1970.
8. Burrows, J. C. and Domencich, T. *An Analysis of the United States Oil Import Quota.* Lexington, Mass.: Heath-Lexington, 1970.
9. Carter, A. P. *Structural Change in the American Economy.* Cambridge: Harvard Univ. Press, 1970.
10. ——— and Brody, A. *Contributions to Input-Output Analysis.* Amsterdam: North-Holland, 1970.
11. Chapman, D., Mount, T. and Tyrrell, T. "Electricity Growth: Implications for Research and Development." Testimony before the Committee on Science for Astronautics, U. S. House of Representatives, June 16, 1972.
12. Christensen, L. R. and Jorgenson, D. W. "Intertemporal Optimization and the Explanation of Consumer Behavior." Paper presented at the Winter Meetings of the Econometric Society, December, 1968.
13. ———. "The Measurement of U. S. Real Capital Input, 1929–1967." *Review of Income and Wealth,* Series 15, No. 4 (December 1969), pp. 293–320.
14. ———. "Measuring Economic Performance in the Private Sector," in M. Moss, ed., *The Measurement of Social and Economic Performance,* Studies in Income and Wealth, No. 37, New York: Columbia Univ. Press, 1973, pp. 233–351.
15. ———. "U. S. Income, Saving, and Wealth, 1929–1969." *Review of Income and Wealth,* Series 19, No. 4 (December 1973), pp. 329–362.
16. ———. "U. S. Real Product and Real Factor Input, 1929–1967." *Review of Income and Wealth,* Series 16, No. 1 (March 1970), pp. 19–50.
17. Christensen, L. R., Jorgenson, D. W., and Lau, L. J. "Conjugate Duality and the Transcendental Logarithmic Utility Function." *Econometrica,* Vol. 39, No. 4 (July 1971), pp. 255–256.
18. ———. "Transcendental Logarithmic Production Frontiers." *Review of Economics and Statistics,* Vol. 55, No. 1 (February 1973), pp. 28–45.
19. ———. "Transcendental Logarithmic Utility Functions." *The American Economic Review,* forthcoming.
20. Dupree, W. and West, R. *United States Energy through 2000.* U. S. Department of the Interior, Washington, D. C., December, 1972.
21. Faucett, J. and Associates. *Data Development for the I-O Energy Model,* Final Report to the Energy Policy Project, Washington, D. C., May, 1973.
22. Goldberger, A. S. *Impact Multipliers and Dynamic Properties of the Klein-Goldberger Model.* Amsterdam: North-Holland, 1959.
23. Hickman, B., ed. *Econometric Models of Cyclical Behavior.* Studies in Income and Wealth, No. 36. New York: Columbia Univ. Press, 1972.
24. Hudson, E. A. and Jorgenson, D. W. "Tax Policy and Energy Conservation." Testimony before the Committee on Finance, U. S. Senate, January 16, 1974.
25. ———. "U. S. Economic Growth, 1973–2000," in Data Resources, Inc., *Long Term Projections of the U. S. Economy,* Lexington, Mass., April, 1974, Chapter 17, pp. 161–236.
26. Jorgenson, D. W., Berndt, E. R., Christensen, L. R. and Hudson, E. A. *U. S. Energy Resources and Economic Growth,* Final Report to the Energy Policy Project, Washington, D. C., September, 1973.
27. Klein, L. R. and Goldberger, A. S. *An Econometric Model of the United States, 1929–1952.* Amsterdam: North-Holland, 1955.
28. Lau, L. J. "Econometrics of Monotonicity, Convexity, and Quasiconvexity." Technical Report No. 123, Institute for Mathematical Studies in the Social Sciences, Stanford University, March, 1974.
29. Leontief, W. W. "Introduction to a Theory of the Internal Structure of Functional Relationships." *Econometrica,* Vol. 157, No. 4 (October 1947), pp. 361–373.
30. ———. *The Structure of the American Economy 1919–1939.* New York: Oxford Univ. Press, 1951.
31. ———. *Studies in the Structure of the American Economy.* New York: Oxford Univ. Press, 1953.

32. MALINVAUD, E. *Statistical Methods of Econometrics.* Amsterdam: North-Holland, 1970.
33. MORISHIMA, M. "A Dynamic Leontief System with Neo-Classical Production Functions." *Equilibrium Stability and Growth.* Oxford: Oxford Univ. Press, 1964, pp. 54–92.
34. MOUNT, T. D., CHAPMAN, L. D. AND TYRRELL, T. J. "Electricity Demand in the United States." Oak Ridge National Laboratory, June, 1973.
35. NATIONAL PETROLEUM COUNCIL. *U. S. Energy Outlook: An Initial Appraisal 1971–1985.* Washington, D. C., July, 1971.
36. ———. *U. S. Energy Outlook.* Washington, D. C., December, 1972.
37. SAMUELSON, P. A. "Nonsubstitution Theorems," in J. Stiglitz, ed., *The Collected Scientific Papers of Paul A. Samuelson,* Cambridge: M.I.T. Press, Vol. 1, pp. 513–536.
38. SUITS, D. B. "Forecasting and Analysis with an Econometric Model." *The American Economic Review,* Vol. 52, No. 1 (March 1962), pp. 104–132.
39. U. S. FEDERAL POWER COMMISSION. *The 1970 National Power Survey.* Washington, D. C.: December, 1971.
40. VACCARA, B. N. "Changes over Time in Input-Output Coefficients for the United States," in Carter and Brody, eds., [10], pp. 238–260.
41. WALRAS, L. *Elements of Pure Economics,* translated by W. Jaffé. Homewood, Ill.: Irwin, 1954.

[23]

RICHARD T. ELY LECTURE

The Economics of Resources or the Resources of Economics

By Robert M. Solow*

It is easy to choose a subject for a distinguished lecture like this, before a large and critical audience with a wide range of interests. You need a topic that is absolutely contemporary, but somehow perennial. It should survey a broad field, without being superficial or vague. It should probably bear some relation to economic policy, but of course it must have some serious analytical foundations. It is nice if the topic has an important literature in the past of our subject—a literature which you can summarize brilliantly in about eleven minutes—but it better be something in which economists are interested today, and it should appropriately be a subject you have worked on yourself. The lecture should have some technical interest, because you can't waffle for a whole hour to a room full of professionals, but it is hardly the occasion to use a blackboard.

I said that it is easy to choose a subject for the Ely Lecture. It has to be, because twelve people, counting me, have done it.

I am going to begin with a quotation that could have come from yesterday's newspaper, or the most recent issue of the *American Economic Review*.

> Contemplation of the world's disappearing supplies of minerals, forests, and other exhaustible assets has led to demands for regulation of their exploitation. The feeling that these products are now too cheap for the good of future generations, that they are being selfishly exploited at too rapid a rate, and that in consequence of their excessive cheapness they are being produced and consumed wastefully has given rise to the conservation movement.

The author of those sentences is not Dennis Meadows and associates, not Ralph Nader and associates, not the President of the Sierra Club; it is a very eminent economic theorist, a Distinguished Fellow of this Association, Harold Hotelling, who died at the age of seventy-eight, just a few days ago. Like all economic theorists, I am much in his debt, and I would be happy to have this lecture stand as a tribute to him. These sentences appeared at the beginning of his article "The Economics of Exhaustible Resources," not in the most recent *Review*, but in the *Journal of Political Economy* for April 1931. So I think I have found something that is both contemporary and perennial. The world has been exhausting its exhaustible resources since the first cave-man chipped a flint, and I imagine the process will go on for a long, long time.

Mr. Dooley noticed that "th' Supreme Coort follows the iliction returns." He would be glad to know that economic theorists read the newspapers. About a year ago, having seen several of those respectable committee reports on the advancing scarcity of materials in the United States and the world, and having, like everyone else, been suckered into reading the *Limits to Growth*, I decided I ought to find out what economic theory has to say

* Professor of economics, Massachusetts Institute of Technology.

about the problems connected with exhaustible resources. I read some of the literature, including Hotelling's classic article—the theoretical literature on exhaustible resources is, fortunately, not very large—and began doing some work of my own on the problem of optimal social management of a stock of a nonrenewable but essential resource. I will be mentioning some of the results later. About the time I finished a first draft of my own paper and was patting myself on the back for having been clever enough to realize that there was in fact something still to be said on this important, contemporary but somehow perennial topic—just about then it seemed that every time the mail came it contained another paper by another economic theorist on the economics of exhaustible resources.[1] It was a little like trotting down to the sea, minding your own business like any nice independent rat, and then looking around and suddenly discovering that you're a lemming. Anyhow, I now have a nice collection of papers on the theory of exhaustible resources; and most of them are still unpublished, which is just the advantage I need over the rest of you.

A pool of oil or vein of iron or deposit of copper in the ground is a capital asset to society and to its owner (in the kind of society in which such things have private owners) much like a printing press or a building or any other reproducible capital asset. The only difference is that the natural resource is not reproducible, so the size of the existing stock can never increase through time. It can only decrease (or, if none is mined for a while, stay the same). This is true even of recyclable materials; the laws of thermodynamics and life guarantee that we will never recover a whole pound of secondary copper from a pound of primary copper in use, or a whole pound of tertiary copper from a pound of secondary copper in use. There is leakage at every round; and a formula just like the ordinary multiplier formula tells us how much copper use can be built on the world's initial endowment of copper, in terms of the recycling or recovery ratio. There is always less ultimate copper use left than there was last year, less by the amount dissipated beyond recovery during the year. So copper remains an exhaustible resource, despite the possibility of partial recycling

A resource deposit draws its market value, ultimately, from the prospect of extraction and sale. In the meanwhile, its owner, like the owner of every capital asset, is asking: What have you done for me lately? The only way that a resource deposit in the ground and left in the ground can produce a current return for its owner is by appreciating in value. Asset markets can be in equilibrium only when all assets in a given risk class earn the same rate of return, partly as current dividend and partly as capital gain. The common rate of return is the interest rate for that risk class. Since resource deposits have the peculiar property that they yield no dividend so long as they stay in the ground, in equilibrium the value of a resource deposit must be growing at a rate equal to the rate of interest. Since the value of a deposit is also the present value of future sales from it, after deduction of extraction costs, resource owners must expect the net price of the ore to be increasing exponentially at a rate equal to the rate of interest. If the mining industry is competitive, net price stands for market price minus margi-

[1] *The Review of Economic Studies* will publish a group of them in the summer of 1974, including my own paper and others by Partha Dasgupta and Geoffrey Heal, Michael Weinstein and Richard Zeckhauser, and Joseph Stiglitz, from all of which I have learned a lot about this subject. I would especially like to thank Zeckhauser for conversation and correspondence, and for the kind of reading of the first draft of this Lecture that one only dares to hope to get because it is so close to Christmas. The final version reflects his comments.

nal extraction cost for a ton of ore. If the industry operates under constant costs, that is just market price net of unit extraction costs, or the profit margin. If the industry is more or less monopolistic, as is frequently the case in extractive industry, it is the marginal profit—marginal revenue less marginal cost—that has to be growing, and expected to grow, proportionally like the rate of interest.

This is the fundamental principle of the economics of exhaustible resources. It was the basis of Hotelling's classic article. I have deduced it as a condition of stock equilibrium in the asset market. Hotelling thought of it mainly as a condition of flow equilibrium in the market for ore: if net price is increasing like compound interest, owners of operating mines will be indifferent at the margin between extracting and holding at every instant of time. So one can imagine production just equal to demand at the current price, and the ore market clears. No other time profile for prices can elicit positive production in every period of time.

It is hard to overemphasize the importance of this tilt in the time profile for net price. If the net price were to rise too slowly, production would be pushed nearer in time and the resource would be exhausted quickly, precisely because no one would wish to hold resources in the ground and earn less than the going rate of return. If the net price were to rise too fast, resource deposits would be an excellent way to hold wealth, and owners would delay production while they enjoyed supernormal capital gains.

According to the fundamental principle, if we observe the market for an exhaustible resource near equilibrium, we should see the net price—or marginal profit—rising exponentially. That is not quite the same thing as seeing the market price to users of the resource rising exponentially. The price to consumers is the net price plus extraction costs, or the obvious analogy for monopoly. The market price can fall or stay constant while the net price is rising if extraction costs are falling through time, and if the net price or scarcity rent is not too large a proportion of the market price. That is presumably what has been happening in the market for most exhaustible resources in the past. (It is odd that there are not some econometric studies designed to find out just this. Maybe econometricians don't follow the iliction returns.) Eventually, as the extraction cost falls and the net price rises, the scarcity rent must come to dominate the movement of market price, so the market price will eventually rise, although that may take a very long time to happen. Whatever the pattern, the market price and the rate of extraction are connected by the demand curve for the resource. So, ultimately, when the market price rises, the current rate of production must fall along the demand curve. Sooner or later, the market price will get high enough to choke off the demand entirely. At that moment production falls to zero. If flows and stocks have been beautifully coordinated through the operations of futures markets or a planning board, the last ton produced will also be the last ton in the ground. The resource will be exhausted at the instant that it has priced itself out of the market. The Age of Oil or Zinc or Whatever It Is will have come to an end. (There is a limiting case, of course, in which demand goes asymptotically to zero as the price rises to infinity, and the resource is exhausted only asymptotically. But it is neither believable nor important.)

Now let us do an exercise with this apparatus. Suppose there are two sources of the same ore, one high-cost and the other low-cost. The cost difference may reflect geographical accessibility and transportation costs, or some geological or chemical difference that makes extraction cheap at one site and dear at the other. The important thing is that there are cost differences, though the final mineral product is identical from both sources.

It is easy to see that production from both sources cannot coexist in the market for any interval of time. For both sources to produce, net price for each of them must be growing like compound interest at the market rate. But they must market their ore at the same price, because the product is identical. That is arithmetically impossible, if their extraction costs differ.

So the story has to go like this. First one source operates and supplies the whole market. Its net price rises exponentially, and the market price moves correspondingly. At a certain moment, the first source is exhausted. At just that moment and not before, it must become economical for the second source to come into production. From then on, the world is in the single-source situation: the net price calculated with current extraction costs must rise exponentially until all production is choked off and the second source is exhausted. (If there are many sources, you can see how it will work.)

Which source will be used first? Your instinct tells you that the low-cost deposit will be the first one worked, and your instinct is right. You can see why, in terms of the fundamental principle. At the beginning, if the high-cost producer is serving the market, the market price must cover high extraction costs plus a scarcity rent that is growing exponentially. The low-cost producer would refrain from undercutting the price and entering the market only if his capital gains justify holding off and entering the market later. But just the reverse will be true. Any price high enough to keep the high-cost producer in business will tempt the low-cost producer to sell ore while the selling is good and invest the proceeds in any asset paying the market rate of interest. So it must be that the low-cost producer is the first to enter. Price rises and output falls. Eventually, at precisely the moment when the low-cost supply is exhausted, the price has reached a level at which it pays the high-cost producer to enter. From then on, *his* net price rises exponentially and production continues to fall. When cumulative production has exhausted the high-cost deposit, the market price must be such as to choke the demand off to zero—or else just high enough to tempt a still higher-cost source into production. And so it goes. Apart from market processes, it is actually socially rational to use the lower-cost deposits before the higher-cost ones.

You can take this story even further, as William Nordhaus has done in connection with the energy industry. Suppose that, somewhere in the background, there is a technology capable of producing or substituting for a mineral resource at relatively high cost, but on an effectively inexhaustible resource base. Nordhaus calls this a "backstop technology." (The nearest we now have to such a thing is the breeder reactor using U^{238} as fuel. World reserves of U^{238} are thought to be enough to provide energy for over a million years at current rates of consumption. If that is not a backstop technology, it is at least a catcher who will not allow a lot of passed balls. For a better approximation, we must wait for controlled nuclear fusion or direct use of solar energy. The sun will not last forever, but it will last at least as long as we do, more or less by definition.) Since there is no scarcity rent to grow exponentially, the backstop technology can operate as soon as the market price rises enough to cover its extraction costs (including, of course, profit on the capital equipment involved in production). And as soon as that happens, the market price of the ore or its substitute stops rising. The "backstop technology" provides a ceiling for the market price of the natural resource.

The story in the early stages is as I have told it. In the beginning, the successive grades of the resource are mined. The last and highest-cost source gives out just when the market price has risen to the point where the backstop technology becomes competitive. During the earlier phases, one imagines that resource companies keep a careful eye on the prospective costs associated with the backstop technology. Any laboratory success or failure that changes those prospective costs has instantaneous effects on the capital value of existing resource deposits, and on the most profitable rate of current production. In actual fact, those future costs have to be regarded as uncertain. A correct theory of market behavior and a correct theory of optimal social policy will have to take account of technological uncertainty (and perhaps also uncertainty about the true size of mineral reserves).

Here is a mildly concrete illustration of these principles. There is now a workable technology for liquefying coal—that is, for producing synthetic crude oil from coal.[2] Nordhaus puts the extraction-and-preparation cost at the equivalent of seven or eight 1970 dollars per barrel of crude oil, including amortization and interest at 10 percent on the plant; I have heard higher and lower figures quoted. If coal were available in unlimited amounts, that would be all. But, of course, coal is a scarce resource, though more abundant than drillable petroleum, so a scarcity rent has to be added to that figure, and the rent has to be increasing like the rate of interest during the period when coal is being used for this purpose.

In the meanwhile, the extraction and production cost for this technology is large compared with the scarcity rent on the coal input, so the market price at which the liquefied-coal-synthetic-crude activity would now be economic is rising more slowly than the rate of interest. It may even fall if there are cost-reducing technological improvements; and that is not unlikely, given that research on coal has not been splashed as liberally with funds as research on nuclear energy. In any case, political shenanigans and monopoly profits aside, scarcity rents on oil form a larger fraction of the market price of oil, precisely because it is a lower cost fuel. The price of a barrel of oil should therefore be rising faster than the implicit price at which synthetic crude from coal could compete. One day those curves will intersect, and that day the synthetic-crude technology will replace the drilled-petroleum technology.

Even before that day, the possibility of coal liquefaction provides a kind of ceiling for the price of oil. I say "kind of" to remind you that coal-mining and moving capacity and synthetic-crude plant cannot be created overnight. One might hope that the ceiling might also limit the consuming world's vulnerability to political shenanigans and monopoly profits. I suppose it does in some ultimate sense, but one must not slide over the difficulties: for example, who would want to make a large investment in coal liquefaction or coal gasification in the knowledge that the current price of oil contains a large monopoly element that could be cut, at least temporarily, if something like a price war should develop?

The fundamental principle of the economics of exhaustible resources is, as I have said, simultaneously a condition of flow equilibrium in the market for the ore and of asset equilibrium in the market for deposits. When it holds, it says quite a lot about the probable pattern of exploitation of a resource. But there are more than the usual reasons for wondering whether

[2] As best one can tell at the moment, shale oil is a more likely successor to oil and natural gas than either gasified or liquefied coal. The relevant costs are bound to be uncertain until more research and development has been done. I tell the story in terms of liquefied coal only because it is more picturesque that way.

the equilibrium conditions have any explanatory value. For instance, the flow market that has to be cleared is not just one market; it is the sequence of markets for resource products from now until the date of exhaustion. It is, in other words, a sequence of futures markets, perhaps a long sequence. If the futures markets actually existed, we could perhaps accept the notion that their equilibrium configuration is stable; that might not be true, but it is at least the sort of working hypothesis we frequently accept as a way of getting on with business. But there clearly is not a full set of futures markets; natural-resource markets work with a combination of myopic flow transactions and rather more farsighted asset transactions. It is legitimate to ask whether observed resource prices are to be interpreted as approximations to equilibrium prices, or whether the equilibrium is so unstable that momentary prices are not only a bad indicator of equilibrium relationships, but also a bad guide to resource allocation.

That turns out not to be an easy question to answer. Flow considerations and stock considerations work in opposite directions. The flow markets by themselves could easily be unstable; but the asset markets provide a corrective force. Let me try to explain why.

The flow equilibrium condition is that the net price grow like compound interest at the prevailing rate. Suppose net prices are expected by producers to be rising too slowly. Then resource deposits are a bad way to hold wealth. Mine owners will try to pull out; and if they think only in flow terms, the way to get out of the resource business is to increase current production and convert ore into money. If current production increases, for this or any other reason, the current price must move down along the demand curve. So initially pessimistic price expectations on the part of producers have led to more pressure on the current price. If expectations about future price changes are responsive to current events, the consequence can only be that pessimism is reinforced and deepened. The initial disequilibrium is worsened, not eliminated, by this chain of events. In other words, the market mechanism I have just described is unstable. Symmetrical reasoning leads to the conclusion that if prices are initially expected to be rising too fast, the withholding of supplies will lead to a speculative run-up of prices which is self-reinforcing. Depending on which way we start, initial disequilibrium is magnified, and production is tilted either toward excessive current dumping or toward speculative withholding of supply. (Still other assumptions are possible and may lead to qualitatively different results. For instance, one could imagine that expectations focus on the price level rather than its rate of change. There is much more work to be done on this question.)

Such things have happened in resource markets; but they do not seem always to be happening. I think that this story of instability in spot markets needs amendment; it is implausible because it leaves the asset market entirely out of account. The longer run prospect is not allowed to have any influence on current happenings. Suppose that producers do have some notion that the resource they own has a value anchored somewhere in the future, a value determined by technological and demand considerations, not by pure and simple speculation. Then if prices are now rising toward that rendezvous at too slow a rate, that is indeed evidence that owning resource deposits is bad business. But that will lead not to wholesale dumping of current production, but to capital losses on existing stocks. When existing stocks have been written down in value, the net price can rise toward its future rendezvous at more or less the right rate. As well as being destabilized by flow reactions, the market can be stabilized by capitalization reactions. In fact the two stories can be made to merge: the reduction in flow price coming from increased current production can be read as a signal and capitalized into losses on asset values, after which near-equilibrium is reestablished.

I think the correct conclusion to be drawn from this discussion is not that either of the stories is more likely to be true. It is more complex: that in tranquil conditions, resource markets are likely to track their equilibrium paths moderately well, or at least not likely to rush away from them. But resource markets may be rather vulnerable to surprises. They may respond to shocks about the volume of reserves, or about competition from new materials, or about the costs of competing technologies, or even about near-term political events, by drastic movements of current price and production. It may be quite a while before the transvaluation of values—I never thought I could quote Nietzsche in an economics paper—settles down under the control of sober future prospects. In between, it may be a cold winter.

So far, I have discussed the economic theory of exhaustible resources as a partial-equilibrium market theory. The interest rate that more or less controls the whole process was taken as given to the mining industry by the rest of the economy. So was the demand curve for the resource itself. And when the market price of the resource has ridden up the demand curve to the point where the quantity demanded falls to zero, the theory says that the resource in question will have been exhausted.

There is clearly a more cosmic aspect to the question than this; and I do not mean to suggest that it is unimportant, just because it is cosmic. In particular, there remains an important question about the social interest in the pace of exploitation of the world's endowment of exhaustible natural resources. This aspect has been brought to a head recently, as everyone knows, by the various Doomsday forecasts that combine a positive finding that the world is already close to irreversible collapse from shortage of natural resources and other causes with the normative judgment that civilization is much too young to die. I do not intend to discuss those forecasts and judgments now—this convention already has one session devoted to just that—but I do want to talk about the economic issues of principle involved.

First, there is a proposition that will be second nature to everyone in this room. What I have called the fundamental principle of the economics of exhaustible resources is, among other things, a condition of competitive equilibrium in the sequence of futures markets for deliveries of the natural resource. This sequence extends out to infinity, even if the competitive equilibrium calls for the resource to be exhausted in finite time. Beyond the time of exhaustion there is also equilibrium: supply equals demand equals zero at a price simultaneously so high that demand is choked off and so low that it is worth no one's while to lose interest by holding some of the resource that long. Like any other competitive equilibrium with the right background assumptions, this one has some optimality properties. In particular, as Hotelling pointed out, the competitive equilibrium maximizes the sum of the discounted consumer-plus-producer surpluses from the natural resource, *provided* that society wishes to discount future consumer surpluses at the same rate that mine owners choose to discount their own future profits.

Hotelling was not so naive as to leap from this conclusion to the belief that *laissez-faire* would be an adequate policy for the resource industries. He pointed to several ways in which the background assumptions might be expected to fail: the presence of externalities when several owners can exploit the same underground pool of gas or oil; the considerable uncertainty surrounding the process of exploration with the consequent likelihood of wasteful rushes to stake claims and exploit, and the creation of socially useless windfall profits; and, finally, the existence of large monopolistic or oligopolistic firms in the extractive industries.

There is an amusing sidelight here. It is not hard to show that, generally speaking, a monopolist will exhaust a mine more slowly than a competitive industry facing the same demand curve would do. (Hotelling did not explore this point in detail, though he clearly knew it. He did mention the possibility of an extreme case in which competition will exhaust a resource in finite time and a monopolist only asymptotically.) The amusing thing is that if a conservationist is someone who would like to see resources conserved *beyond* the pace that competition would adopt, then the monopolist is the conservationist's friend. No doubt they would both be surprised to know it.

Hotelling mentions, but rather pooh-poohs, the notion that market rates of interest might exceed the rate at which society would wish to discount future utilities or consumer surpluses. I think a modern economist would take that possibility more seriously. It is certainly a potentially important question, because the discount rate determines the whole tilt of the equilibrium production schedule. If it is true that the market rate of interest exceeds the social rate of time preference, then scarcity rents and market prices will rise faster than they "ought to" and production will have to fall correspondingly faster along the demand curve. Thus the resource will be exploited too fast and exhausted too soon.

The literature has several reasons for expecting that private discount rates might be systematically higher than the correct social rate of discount. They fall into two classes. The first class takes it more or less for granted that society ought to discount utility and consumption at the same rates as reflective individuals would discount their own future utility and consumption. This line of thought then goes on to suggest that there are reasons why this might not happen. One standard example is the fact that individuals can be expected to discount for the riskiness of the future, and some of the risks for which they will discount are not risks to society but merely the danger of transfers within the society. Since there is not a complete enough set of insurance markets to permit all these risks to be spread properly, market interest rates will be too high. Insecurity of tenure, as William Vickrey has pointed out, is a special form of uncertainty with particular relevance to natural resources.

A second standard example is the existence of various taxes on income from capital; since individuals care about the after-tax return on investment and society about the before-tax return, if investment is carried to the point where the after-tax yield is properly related to the rate of time preference, the before-tax profitability of investment will be too high. I have nothing to add to this discussion.

The other class of reasons for expecting that private discount rates are too high and will thus distort intertemporal decisions away from social optimality denies that private time preference is the right basis for intertemporal decisions. Frank Ramsey, for instance, argued that it was ethically indefensible for society to discount future utilities. Individuals might do so, either because they lack imagination (Böhm-Bawerk's "defective telescopic faculty") or because they are all too conscious that life is short. In social decision-making, however, there is no excuse for treating generations unequally, and the time-horizon is, or should be, very long. In solemn conclave assembled, so to speak, we ought to act as if the social rate of time preference were zero (though we would simultaneously discount future *consumption* if we expect the future to be richer than the present). I confess I find that reasoning persuasive, and it provides another reason for expecting that the market will exhaust resources too fast.

This point need not be divorced so completely from individual time preference. If the whole infinite sequence of futures markets for resource products could actually take place and find equilibrium, I might be inclined to accept the result (though I would like to know who decides the initial endowments within and between generations). But of course they cannot take place. There is no way to collect bids and offers from everyone who will ever live. In the markets that actually do take place, future generations are represented only by us, their eventual ancestors. Now generations overlap, so that I worry about my children, and they about theirs, and so on. But it does seem fundamentally implausible that there should be anything *ex post* right about the weight that is actually given to the welfare of those who will not live for another thousand years. We have actually done quite well at the hands of *our* ancestors. Given how poor they were and how rich we are, they might properly have saved less and consumed more. No doubt they never expected the rise in income per head that has made us so much richer than they ever dreamed was possible. But that only reinforces the point that the future may be too important to be left to the accident of mistaken expectations and the ups and downs of the Protestant ethic.

Several writers have studied directly the problem of defining and characterizing a socially-optimal path for the exploitation of a given pool of exhaustible resources. The idea is familiar enough: instead of worrying about market responses, one imagines an idealized planned economy, constrained only by its initial endowment, the size of the labor force, the available technology, and the laws of arithmetic. The planning board then has to find the best feasible development for the economy. To do so, it needs a precise criterion for comparing different paths, and that is where the social rate of time preference plays a role.

It turns out that the choice of a rate of time preference is even more critical in this situation than it is in the older literature on optimal capital accumulation without any exhaustible resources. In that theory, the criterion usually adopted is the maximization of a discounted sum of one-period social welfare indicators, depending on consumption per head, and summed over all time from now to the infinite future. The typical result, depending somewhat on the particular assumptions made, is that consumption per head rises through time to a constant plateau defined by the "modified Golden Rule." In that ultimate steady state, consumption per head is lower the higher is the social rate of discount; and, correspondingly, the path to the steady state is characterized by less saving and more interim consumption, the higher the social rate of discount. That is as it should be: the main beneficiaries of a high level of ultimate steady-state consumption are the inhabitants of the distant future, and so, if the planning board discounts the future very strongly, it will choose a path that favors the near future over the distant future.

When one adds exhaustible resources to the picture, the social rate of time preference can play a similar, but even more critical, role. As a paper by Geoffrey Heal and Partha Dasgupta and one of my own show, it is possible that the optimal path with a positive discount rate should lead to consumption per head going asymptotically to zero, whereas a zero discount rate leads to perpetually rising consumption per head. In other words, even when the technology and the resource base could permit a plateau level of consumption per head, or even a rising standard of living, positive social time preference might in effect lead society to prefer eventual extinction, given the drag exercised by exhaustible resources. Of course, it is part of the point that it is the planning board in the present that plans for future extinction: nobody has asked the about-to-become-defunct last generation whether *it* approved of weighting its satisfactions less than those of its ancestors.

Good theory is usually trying to tell you something, even if it is not the literal truth. In this context, it is not hard to interpret the general tenor of the theoretical indications. We know in general that even well-functioning competitive markets may fail to allocate resources properly over time. The reason, I have suggested, is because, in the nature of the case, the future brings no endowment of its own to whatever markets actually exist. The intergenerational distribution of income or welfare depends on the provision that each generation makes for its successors. The choice of a social discount rate is, in effect, a policy decision about that intergenerational distribution. What happens in the planning parable depends very much—perhaps dramatically—on that choice; and one's evaluation of what happens in the market parable depends very much on whether private choices are made with a discount rate much larger than the one a deliberate policy decision would select. The pure theory of exhaustible resources is trying to tell us that, if exhaustible resources really matter, then the balance between present and future is more delicate than we are accustomed to think; and then the choice of a discount rate can be pretty important and one ought not to be too casual about it.

In my own work on this question, I have sometimes used a rather special criterion that embodies sharp assumptions about intergenerational equity: I have imposed the requirement that consumption per head be constant through time, so that no generation is favored over any other, and asked for the largest steady consumption per head that can be maintained forever, given all the constraints including the finiteness of resources. This criterion, like any other, has its pluses and its minuses and I am not committed to it by any means. Like the standard criterion—the discounted sum of one-period utilities—this one will always pick out an *efficient* path, so one at least gets the efficiency conditions out of the analysis. The highest-constant-consumption criterion also has the advantage of highlighting the crucial importance of certain technological assumptions.

It is clear without any technical apparatus that the seriousness of the resource-exhaustion problem must depend in an important way on two aspects of the technology: first, the likelihood of technical progress, especially natural-resource-saving technical progress, and, second, the ease with which other factors of production, especially labor and reproducible capital, can be substituted for exhaustible resources in production.

My own practice, in working on this problem, has been to treat as the central case (though not the only case) the assumption of zero technological progress. This is not because I think resource-saving inventions are unlikely or that their capacity to save resources is fundamentally limited. Quite the contrary—if the future is anything like the past, there will be prolonged and substantial reductions in natural-resource requirements per unit of real output. It is true, as pessimists say, that it is just an assumption and one cannot be sure; but to assume the contrary is also an assumption, and a much less plausible one. I think there is virtue in analyzing the zero-technical-progress case because it is easy to see how technical progress can relieve and perhaps eliminate the drag on economic welfare exercised by natural-resource scarcity. The more important task for theory is to try to understand what happens or can happen in the opposite case.

As you would expect, the degree of substitutability is also a key factor. If it is very easy to substitute other factors for natural resources, then there is in principle no "problem." The world can, in effect, get along without natural resources, so exhaustion is just an event, not a catastrophe. Nordhaus's notion of a "backstop technology" is just a dramatic way of putting this case; at some finite cost, production can be freed of dependence on exhaustible resources altogether.

If, on the other hand, real output per unit of resources is effectively bounded—cannot exceed some upper limit of productivity which is in turn not too far from where we are now—then catastrophe is unavoidable. In-between there is a wide range of cases in which the problem is real, interesting, and not foreclosed. Fortunately, what little evidence there is suggests that there is quite a lot of substitutability between exhaustible resources and renewable or reproducible resources, though this is an empirical question that could absorb a lot more work than it has had so far.

Perhaps the most dramatic way to illustrate the importance of substitutability, and its connection with Doomsday, is in terms of the permanent sustainability of a constant level of consumption. In the simplest, most aggregative, model of a resource-using economy one can prove something like the following: if the elasticity of substitution between exhaustible resources and other inputs is unity or bigger, and if the elasticity of output with respect to reproducible capital exceeds the elasticity of output with respect to natural resources, then a constant population can maintain a positive constant level of consumption per head forever. This permanently maintainable standard of living is an increasing, concave, and unbounded function of the initial stock of capital. So the drag of a given resource pool can be overcome *to any extent* if only the initial stock of capital is large enough. On the other hand, if the elasticity of substitution between natural resources and other inputs is less than one, or if the elasticity of output with respect to resources exceeds the elasticity of output with respect to reproducible capital, then the largest constant level of consumption sustainable forever with constant population is—zero. We know much too little about which side of that boundary the world is on—technological progress aside—but at least the few entrails that have been read seem favorable.[3]

Perhaps I should mention that when I say "forever" in this connection, I mean "for a very long time." The mathematical reasoning does deal with infinite histories, but actually life in the solar system will only last for a finite time, though a very long finite time, much longer than this lecture, for instance. That is why I think it takes economics as well as the entropy law to answer our question.

I began this lecture by talking of the conditions for competitive equilibrium in the market for natural resources. Now I have been talking of centralized planning

[3] See pp. 60–70 in William D. Nordhaus and James Tobin.

optima. As you would expect, it turns out that under the standard assumptions, the Hotelling rule, the fundamental principle of natural-resource economics, is a necessary condition for efficiency and therefore for social optimality. So there is at least a prayer that a market-guided system might manage fairly well. But more than the Hotelling condition is needed.

I have already mentioned one of the extra requirements for the intertemporal optimality of market allocations: it is that the market discount future profits at the same rate as the society would wish to discount the welfare of future inhabitants of the planet. This condition is often given as an argument for public intervention in resource allocation because—as I have also mentioned—there are reasons to expect market interest rates to exceed the social rate of time preference, or at least what philosophers like us think it ought to be. If the analysis is right, then the market will tend to consume exhaustible resources too fast, and corrective public intervention should be aimed at slowing down and stretching out the exploitation of the resource pool. There are several ways that could be done, in principle, through conservation subsidies or a system of graduated severance taxes, falling through time.

Realistically speaking, however, when we say "public intervention" we mean rough and ready political action. An only moderately cynical observer will see a problem here: it is far from clear that the political process can be relied on to be more future-oriented than your average corporation. The conventional pay-out period for business is of the same order of magnitude as the time to the next election, and transferring a given individual from the industrial to the government bureaucracy does not transform him into a guardian of the far future's interests. I have no ready solution to this problem. At a minimum, it suggests that one ought to be as suspicious of uncritical centralization as of uncritical free-marketeering. Maybe the safest course is to favor specific policies—like graduated severance taxes—rather than blanket institutional solutions.

There is another, more subtle, extra requirement for the optimality of the competitive market solution to the natural-resource problem. Many patterns of exploitation of the exhaustible-resource pool obey Hotelling's fundamental principle myopically, from moment to moment, but are wrong from a very long-run point of view. Such mistaken paths may even stay very near the right path for a long time, but eventually they veer off and become bizarre in one way or another. If a market-guided system is to perform well over the long haul, it must be more than myopic. Someone—it could be the Department of the Interior, or the mining companies, or their major customers, or speculators—must always be taking the long view. They must somehow notice in advance that the resource economy is moving along a path that is bound to end in disequilibrium of some extreme kind. If they do notice it, and take defensive actions, they will help steer the economy from the wrong path toward the right one.[4] Usually the "wrong" path is one that leads to exhaustion at a date either too late or too soon; anyone who perceives this will be motivated to arbitrage between present and future in ways that will push the current price toward the "right" path.[5]

It is interesting that this need for some-

[4] This sort of process has been studied in a different context by Frank Hahn and by Karl Shell and Joseph Stiglitz.

[5] For example, suppose the current price is too low, in the sense that, if it rises according to the current principle, the demand path will be enough to exhaust the resource before the price has risen high enough to choke demand to zero. A clever speculator would see that there will be money to be made just after the date of exhaustion, because anyone with a bit of the resource to sell could make a discrete jump in the price and still find buyers. Such a speculator would wish to buy now

one to take the long view emerged also when the question at hand was the potential instability of the market for natural resources if it concentrates too heavily on spot or flow decisions, and not enough on future or stock decisions. In that context too, a reasonably accurate view of the long-term prospects turns out to be a useful, maybe indispensable, thing for the resource market to have.

This lecture has been—as Kenneth Burke once said about the novel—words, all words. Nevertheless, it has been a discourse on economic theory, not on current policy. If some of you have been daydreaming about oil and the coming winter, I assure you that I have been thinking about shadow prices and transversality conditions at infinity. If I turn briefly to policy at the end, it is not with concrete current problems in mind. After all, nothing I have been able to say takes account of the international oil cartel, the political and economic ambitions of the Middle Eastern potentates, the speeds of adjustment to surprises in the supply of oil, or the doings of our own friendly local oligopolists. The only remarks I feel entitled to make are about the long-run pursuit of a general policy toward exhaustible resources.

Many discussions of economic policy—macroeconomics aside—boil down to a tension between market allocation and public intervention. Marketeers keep thinking about the doughnut of allocative efficiency and informational economy and *dirigistes* are impressed with the size of the hole containing externalities, imperfections, and distributional issues. So it is with exhaustible resources. One is impressed with what a system of ideal markets, including futures markets, can accomplish in this complicated situation; and one can hardly miss seeing that our actual oligopolistic, politically involved, pollution-producing industry is not exactly what the textbook ordered. I have nothing new to add to all that. The unusual factor that the theory of exhaustible resources brings to the fore is the importance of the long view, and the value of reasonable information about reserves, technology, and demand in the fairly far future.

This being so, one is led to wonder whether public policy can contribute to stability and efficiency along those lines. One possibility is the encouragement of organized futures trading in natural resource products. To be useful, futures contracts would have to be much longer-term than is usual in the futures markets that now exist, mostly for agricultural products. I simply do not know enough to have an opinion about the feasibility of large scale futures trading, or about the ultimate contribution that such a reform would make to the stability and efficiency of the market for resource products. But in principle it would seem to be a good idea.

The same considerations suggest that the market for exhaustible resources might be one of the places in the economy where some sort of organized indicative planning could play a constructive role. This is not an endorsement of centralized decision-making, which is likely to have imperfections and externalities of its own. Indeed it might be enough to have the government engaged in a continuous program of information-gathering and dissemination covering trends in technology, reserves and demand. One could at least hope to have professional standards govern such an exercise. I take it that the underlying logic of indicative planning is that some comparison and coordination of the main participants in the market, including the

and hold for sale then. But that action would tend to raise the current price (and, by the fundamental principle, the whole price path) and reduce demand, so that the life of the resource would be prolonged. The speculation is thus corrective.

government, could eliminate major errors and resolve much uncertainty. In the case of exhaustible resources, it could have the additional purpose of generating a set of consistent expectations about the distant future. In this effort, the pooling of information and intentions from both sides of the market could be useful, with the effect of inducing behavior that will steer the economy away from ultimately inferior paths. It is also likely, as Adam Smith would have warned, that a certain amount of conspiracy against the public interest might occur in such sessions, so perhaps they ought to be recorded and the tapes turned over to Judge Sirica, who will know what to do with them.

REFERENCES

P. Dasgupta and G. Heal, "The Optimal Depletion of Exhaustible Resources," *Rev. Econ. Stud.*, forthcoming, 1974.

F. H. Hahn, "Equilibrium Dynamics with Heterogeneous Capital Goods," *Quart. J. Econ.*, Nov. 1966, *80*, 633–646.

H. Hotelling, "The Economics of Exhaustible Resources," *J. Polit. Econ.*, April 1931, *39*, 137–175.

W. D. Nordhaus, "The Allocation of Energy Resources," *Brookings Papers on Econ. Activ.*, forthcoming.

——— **and J. Tobin**, "Is Economic Growth Obsolete?" in National Bureau of Economic Research, *Economic Growth*, 50th Anniversary Colloq. V, New York 1972.

K. Shell and J. E. Stiglitz, "The Allocation of Investment in a Dynamic Economy," *Quart. J. Econ.*, Nov. 1967, *81*.

R. M. Solow, "Intergenerational Equity and Exhaustible Resources," *Rev. Econ. Stud.*, forthcoming, 1974.

J. E. Stiglitz, "Growth with Exhaustible Natural Resources," *Rev. Econ. Stud.*, forthcoming, 1974.

M. Weinstein and R. Zeckhauser, "Use Patterns for Depletable and Recyclable Resources," *Rev. Econ. Stud.*, forthcoming, 1974.

[24]

Intergenerational Equity and the Investing of Rents from Exhaustible Resources

By JOHN M. HARTWICK*

Invest all profits or rents from exhaustible resources in reproducible capital such as machines. This injunction seems to solve the ethical problem of the current generation shortchanging future generations by "overconsuming" the current product, partly ascribable to current use of exhaustible resources.[1] Under such a program, the current generation converts exhaustible resources into machines and "lives off" current flows from machines and labor. Under such a program one might assume that in some sense the total stock of productive capital was never depleted since ultimately the exhaustible resource stock will be transmuted into a stock of machines and, given that machines are assumed not to depreciate, no stock either of machines or of exhaustible resources is ever consumed. If in this sense the stock of productive capital is not being depleted, what can one say about the time path of current output and current consumption per head? For the case of per capita consumption remaining constant over time, one could say that no generation was better off than another. Intergenerational equity was being achieved.[2] For simplicity, we shall assume *ZPG* or a constant population so we need only ask what happens to the time path of aggregate consumption. Let me restate the problem in brief: if society invests all rents from exhaustible resources in reproducible capital goods, and invests only this amount, i.e., consumes the remainder of the product given population constant, will consumption and output rise, remain constant, or fall over time?

I shall formally set this problem out below and solve it for the case of a Cobb-Douglas technology. The Cobb-Douglas technology has the important property that each input (in particular, the flow of minerals from an exhaustible resource) is essential for producing a positive output of the single produced commodity. Thus the economy cannot exhaust any natural resource and continue to have positive consumption and output. Beckmann (1974, 1975), Solow, and Solow and Wan have used the Cobb-Douglas technology in their analyses of utilization of exhaustible resources in aggregate dynamic models.

Production in the model at period t will be assumed to require inputs of reproducible capital $k(t)$, flows of mineral from an exhaustible resource $y(t)$ and labor. The labor force is constant so we can set it at one unit. The $k(t)$, $y(t)$, commodity output $x(t)$, and consumption $c(t)$ are defined in per capita terms. The technology $f(k(t), y(t), 1)$ will be assumed to exhibit constant returns to scale so that $f(\cdot)$ is homoge-

*Associate professor, Queen's University.

[1] The idea for this paper arose after hearing a seminar by Anthony Scott on resource policy. He estimated the returns Canadians might receive in 1975 if they had invested all resource royalties in assets yielding the current rates of interest prevailing since 1911. The interest each year was to be consumed. He labeled such a strategy as a "Saudi Arabian" program!

[2] See Kenneth Arrow for a systematic exploration of savings rules and intergenerational equity within the context of a model of accumulation of reproducible capital in the absence of exhaustible resources. The current interest in intergenerational equity was aroused by remarks of John Rawls (sec. 44). One should of course consult Rawls, Arrow, and Robert Solow for an introduction to the diverse notions of intergenerational equity which have been proposed for consideration in the current investigations. Rawls was concerned with the problem of balancing the relative burden of savings on early generations with the burden on later generations. Capital was being *accumulated* over some part of a society's program of consumption and investment. With exhaustible resources, one must be concerned with *forestalling decumulation* of society's productive capital in order to achieve some notion of intergenerational equity.

neous of degree one. The value of $x(t) = f(\cdot)$ will be zero if any argument of $f(\cdot)$ is zero. That is, each input is essential. The marginal productivities $\partial f/\partial k$ and $\partial f/\partial y$ are assumed to be positive; $\partial^2 f/\partial k^2$ and $\partial^2 f/\partial y^2$ are assumed to be negative. Let $f_k \triangleq \partial f/\partial k$, $f_y \triangleq \partial f/\partial y$, $f_{kk} \triangleq \partial^2 f/\partial k^2$, $f_{ky} \triangleq \partial^2 f/\partial k \partial y$, and $f_{yy} \triangleq \partial^2 f/\partial y^2$. A D before a variable will indicate the time derivative of that variable (for example, $Dk \triangleq dk/dt$). At any instant of time, the product $x(t)$ is completely divided between current consumption $c(t)$, investment Dk and extraction costs $ay(t)$, where a is the cost measured in units of the single produced commodity of extracting one unit of the exhaustible resource. Thus, we have our accounting relation

$$x(t) = c(t) + Dk + ay(t)$$

Our savings or investment function is

$$Dk = (f_y - a)y(t) \tag{1}$$

Efficiency of exhaustible resource extraction requires that the rate of return from a unit of reproducible capital equal the rate of return from owning a unit of deposits of the exhaustible resource.[3] In price terms, this condition is characterized by the current capital gain on mineral deposits being equal to the interest rate or rate of return on reproducible capital. In our one-commodity world, this condition is satisfied by *the rate of change in the marginal product of the mineral being equal to the marginal product of reproducible capital.* This is sometimes referred to as the Hotelling Rule. It characterizes the efficient exploitation of an exhaustible resource. That is

$$\frac{d \log (f_y - a)}{dt} = f_k \tag{2}$$

or

$$f_{yy} Dy + f_{yk} Dk = f_k (f_y - a) \tag{2'}$$

Relations (1) and (2) define the dynamics of the economy. There are two differential equations in the variables $y(t)$ and $k(t)$. We require initial values $k(0)$ and $y(0)$ in order to define the time paths of $y(t)$ and $k(t)$. We shall assume that $k(0)$ and $y(0)$ are selected so that the initial stock of exhaustible resource S is precisely sufficient to sustain the economy over infinite time. We shall remark below that there exists a finite S which will yield the consumption path below. By definition, $dS/dt = -y(t)$, where the stock S is defined in per capita terms.

Aggregate output is rising, constant, or falling over an interval of time as $Dx \gtreqless 0$. Now from the definition of the production function, we get

$$Dx = f_k Dk + f_y Dy \tag{3}$$

For the case of the Cobb-Douglas technology, we have

$$\overset{*}{x} = k^{\alpha} y^{\beta} 1^{\gamma}$$

with $\alpha + \beta = 1$ and $f_k \triangleq \alpha x/k$ and $f_y \triangleq \beta x/y$. Also $f_{yy} \triangleq \beta x(\beta - 1)/y^2$ and $f_{yk} \triangleq \alpha\beta x/yk$.

For the case of the Cobb-Douglas technology, (2′) becomes

$$f_y Dy - xDy/y + f_k Dk = (y/\beta) f_k (f_y - a)$$

and substituting for Dk from (1), we get

$$\beta[f_y Dy + f_k(f_y - a)y] = f_y Dy + f_k(f_y - a)y \tag{4}$$

Since $0 < \beta < 1$, equation (4) can only be satisfied if $f_y Dy + f_k(f_y - a)y = 0$ but $f_y Dy + f_k(f_y - a)y$ is the right-hand side of (3). Thus we have established that x will be constant over time and since $c(t) = (1 - \beta)x(t)$, (recall $f_y y = \beta x$), we have the result that consumption per head will be constant over time. Given the finiteness of natural resource stock, it will be necessary over infinite time to have the current flow of resources extracted asymptotically approach zero as time tends to infinity. By Solow's definition of intergenerational equity—namely per capita consumption remaining constant over time—we have established that the savings investment rule (invest all net returns from exhaustible re-

[3]Since in a well-behaved problem (e.g., the case with a Cobb-Douglas production function) f_y always increases as $t \to \infty$, one has only to assure that the extraction costs are such that $(f_y - a) > 0$ at t_0.

sources in reproducible capital) implies intergenerational equity. A perusal of the mathematics of Solow's paper indicates that this result was implicit in his mathematics—to preserve $Dc = 0$, society should invest the current returns from the utilization of flows from the stock of exhaustible resources.

We have in fact obtained the rule for a model with nonzero extraction costs. Solow had no extraction costs in his formulation. He proved that the existence of a solution required that $\beta < \alpha$. We take this as a necessary condition for existence. It implies that the share of output ascribable to natural resources be less than that share ascribable to reproducible capital—a condition which empirical results indicate is unambiguously satisfied. To be precise, the only model in which the existence of a solution with c positive over infinite time and S finite has been established is the above Cobb-Douglas case with extraction costs set at zero. On reading the above note Solow pointed out that the rule "invest exhaustible resource rents and c will remain constant" is very general. To see this substitute from (1) and (2) in the relation $Dx = f_k Dk + f_y Dy$ for f_k and Dk to get

$$Dx = \left\{ \frac{df_y}{dt} \Big/ (f_y - a) \right\} (f_y - a) y + f_y Dy$$

$$= \frac{d(f_y y)}{dt} = \frac{d(Dk + ay)}{dt}$$

Given $x = c + Dk + ay$ we conclude that $Dc = 0$ regardless of whether $Dx = 0$. Thus we have established, for general technologies, the rule: "the investment of current exhaustible resource returns in reproducible capital implies per capita consumption constant." For the Cobb-Douglas case $Dk + ay = f_y y = \beta x$. Thus from above we have $Dx = \beta Dx$ and since $\beta \neq 1$, $Dx = 0$. If there is depreciation of reproducible capital at the rate δ per unit capital per unit time, then net capital accumulation is currently of the amount $dk/dt + \delta k(t)$ and given our savings rule, equation (1) becomes $Dk + \delta k = (f_y - a) y(t)$. Reworking the steps for solving for Dc above reveals that $Dc = -\delta f_k k$. Hence our savings investment rule will not provide for the maintaining of per capita consumption constant over time. The current decline in per capita consumption is simply the amount of the produced commodity required to offset the current amount of depreciation in the reproducible capital. Arrow's results would not turn on whether he has reproducible capital depreciate; he does not explicitly treat depreciation. In my 1976 paper the present model is extended to cover cases of many exhaustible resources. The intergenerational equity result has also been established in a Uzawa two-sector model with an exhaustible resource.

REFERENCES

K. Arrow, "Rawls' Principle of Just Saving," *Swedish J. Econ.*, Dec. 1973, *75*, 323–35.

M. J. Beckmann, "A Note on the Optimal Rate of Resource Exhaustion," *Rev. Econ. Stud.* Symposium, 1974, 121–22.

———, "The Limits to Growth in a Neoclassical World," *Amer. Econ. Rev.*, Sept. 1975, *65*, 695–99.

J. M. Hartwick, "Substitution Among Exhaustible Resources and Intergenerational Equity," *Rev. Econ. Stud.*, forthcoming.

John Rawls, *A Theory of Justice*, Cambridge, Mass. 1971.

R. M. Solow, "Intergenerational Equity and Exhaustible Resources" *Rev. Econ. Stud.* Symposium, 1974, 29–46.

——— **and F. Y. Wan**, "Extraction Costs in the Theory of Exhaustible Resources," *Bell J. Econ*, Autumn 1976, *7*, 359–70.

[25]

A Neoclassical Analysis of the Economics of Natural Resources

J. E. Stiglitz

I
Introduction

The oscillations in the general views on the prospects for the future, alternating between the despair of imminent and inevitable doom and the euphoria of an impending new millenium, have a remarkable regularity about them, perhaps matching that of the long business cycle. The nineteenth century concern about the implications of the second law of thermodynamics, as represented, for instance, in the work of Henry Adams, has its twentieth century counterpart in Georgescu-Roegen, and Malthus's worry about food scarcity has its twentieth century counterpart in Meadows and Forrester. There are those who would suggest that the nineteenth century preachers of inevitable progress have been replaced by the modern neoclassical economist who, if he does not believe that we are on the verge of the new millenium, finds it difficult to take seriously the prognostications of imminent doom.

What are we to make of these continuing controversies? How could what appears to be essentially the same controversy continue for so long (not always, of course, with the same intensity)? Those who forecast imminent disaster in the nineteenth century were clearly wrong. Why were they wrong? Were they wrong only in the time predicted for disaster but correct in their long-run analysis?

There are, I think, two reasons for the failure to resolve these disputes. The first is that the issues have not been posed in such a way as to admit an unambiguous answer. The second is that, when properly posed, it becomes clear that some of the more important issues cannot be answered on the basis of presently available evidence.

This paper attempts to pose what I consider the appropriate ques-

The author is indebted to V. Kerry Smith for his helpful comments on an earlier draft.

tions and to suggest what kind of evidence might resolve them. There are, of course, important policy issues in which one's position inevitably depends on the answers to these unresolved questions. What is one to do until these are obtained? Some preliminary thoughts on this question constitute the third objective of this paper.[1] However, there are a few distinctions which will help clarify the scope of my subject.

First, it is obvious that continued exponential growth is impossible, if only because eventually, at a strictly positive growth rate, the mass of people would exceed the mass of the earth. I am not concerned here with such very long-run problems. (Similarly, I am not concerned with long-run problems arising from the laws of thermodynamics.) I am concerned here with the more immediate future.

Second, it is important to distinguish between natural resources which can be treated as *private goods,* such as coal, oil, and iron, and those which are basically *public goods,*[2] such as air and water. Whether public intervention is required in the allocation of the former is at least a moot question about which we shall have much to say later; but in the case of the public goods, appropriate allocation inevitably will require some type of government action. There is a third category of natural resources—those which are normally publicly managed but which are really private goods, as for example, forests. A clear distinction has to be made between these resources and true public goods.

Although it is often clear whether the good is a public good, a private good, or a publicly provided private good, there are some ambiguous cases. The fish in a lake are often thought of as a public good although fish caught by one individual may reduce the amount available to others, and property rights for fishing in a lake could be given to an individual. The appropriate method of organizing the allocation of such goods is one of the more important issues in the economics of natural resources.

Third, I have not yet defined what I mean by natural resources. Presumably a natural resource is any commodity or factor which is provided by nature and not produced, or producible, by man. Although such a definition is not very precise, it will suffice for our purposes. (Most of what we think of as natural resources, such as oil, require human activity

[1] I was asked by RFF to present a survey of the neoclassical view of the problem of the scarcity of natural resources. I have chosen to provide an "interpretive review" rather than an exhaustive review of the literature; for the latter, see Peterson and Fisher [1].

[2] The distinction I have in mind is based on the traditional definition of public goods: those involving difficulties in appropriability and those in which the marginal cost of an extra person enjoying the benefits of the good is zero. Thus coal can be marketed just like any other produced good and if one person burns a unit of coal, it is not available for use by any other individual. It would be hard to charge for the use of air.

to convert them into a useful form and indeed to extract them from the ground. Moreover, the supply of many natural resources, say fish, can be affected by human activity.) It will, however, be useful to consider some polar cases. These are *exhaustible natural resources,* the supply of which is fixed and cannot be augmented; *renewable natural resources* such as fish, the supply of which can be increased (renewed) after utilization; *inexhaustible but nonaugmentable resources,* for example, land; and *recyclable resources.*

In fact, many cases do not fall into these neat categories; land—at least arable land—can be augmented as well as depleted. A resource which was perfectly recyclable would essentially be inexhaustible but nonaugmentable, while so long as some of the resource cannot be recovered for reuse, a recyclable resource is like an exhaustible natural resource.

We used to think of air as an inexhaustible resource. Clean air, we now recognize, is a resource which can be exhausted, but in most cases it should be viewed as a renewable natural resource: by taking appropriate action, a dirty atmosphere can be changed into a clean one if there is sufficient time.

I am primarily concerned here with the first category of natural resources, that is, exhaustible, nonrenewable, natural resources, since they are considered to pose the most serious problem. Much of what I shall have to say is, however, applicable with appropriate modifications to the other categories as well.

The paper is divided into eight sections. Section II identifies the major issues; sections III–VI are concerned with each of the major issues in turn; section VII is concerned with a brief discussion of two important recent policy issues, and our concluding remarks are contained in section VIII.

II
The Issues

There are several distinct issues which have become confused in many of the popular discussions of resource availability. This section is intended to help identify and distinguish them.

Viability. Is it possible to have sustained economic growth? This is concerned not with the question of whether it is likely that we face imminent doom if we do not take remedial action soon (the forecasting issue), but with whether doom is *inevitable.* Is there anything we can do? This is one of the questions to which neoclassical economics has provided a clear answer. There are conditions under which growth cannot be sustained and conditions under which it can. In section III we identify these conditions.

Forecasting. Forecasting is concerned with predicting what the world will be like in twenty-five, fifty, or a hundred years. One might be interested in a forecast to satisfy one's curiosity by gazing into a crystal ball. But information is valuable mainly to the extent that it leads to alteration in behavior, and usually the forecasters have some policy prescription in mind. Thus their forecasts have a qualification, "unless a certain prescribed course of action is taken." Forecasting usually entails an extrapolation of past events into the future; this extrapolation requires certain, usually quite strong assumptions, and a great deal of the disagreement concerning predictions is concerned with precisely those assumptions. We talk about these assumptions at greater length in section IV.

There are two distinctive issues on which the forecasts of the future are supposed to provide us with some information. (1) The more popular versions focus on some cataclysm; the economy, in any form similar to that which we presently know, is not viable. "We shall be unable to sustain current energy consumption rates." "The temperature of the earth will increase by x° and New York City will be under water." "We shall suffocate from . . . poisoning." (2) The more mundane versions are concerned with estimating the magnitude of the increase in natural resource prices and gauging the effect of this increase on our standard of living.

Efficiency. The question of whether we are wasting our resources is quite separate from the question of whether doom is imminent or inevitable. We might not be facing imminent doom, yet we could still be misusing our resources. Much of the popular literature seems to take the position that doom is not inevitable but because of wastage of resources, it is nevertheless likely to occur, perhaps soon, and it is in this sense that the three issues are closely linked.

There are a number of reasons that the economy might not allocate natural resources efficiently, some arising out of peculiarities of the private market, and some arising from particular government actions. These are discussed in section V.

Intertemporal Equity. The economy could be efficient, resources might be being used efficiently, but still, because of the rate at which we use our resources, our descendants will be poorer. There is a tradeoff between our consumption and that of our descendants, and the market solution may be inequitable. We shall have a few words to say on this in section VI.

III
The Viability Issue: Is Sustained Growth Possible in the Presence of Exhaustible Natural Resources?

Any analysis of the economics of natural resources should, I think, begin with the question of why the scarcity of natural resources is different from the scarcity of any other factor of production (or any commodity, for that matter). The layman's response to this question would probably be that if we run out of corn we can produce more corn. With natural resources there is simply nothing we can do. This, of course, is not a completely satisfactory answer.

For there to be a meaningful natural resource problem, several conditions need to be satisfied:

1. A resource must be in limited supply relative to current usage rates; thus, if there are several hundred years' worth of reserves[3] of some natural resource, there is, at most, a very long-run problem.

2. It must be nonrenewable and nonrecyclable. Natural resources (such as metals) do not disappear after they are used; rather, they become available for reuse. Only that part which is not renewable—or more precisely, that part, for which the cost of renewing to a marketable state will exceed the market prices—is exhausted in economic terms. The net utilization rate of natural resources may thus be considerably less than the gross.

3. It must be essential, that is, it must be required for production (or consumption).

4. There cannot exist substitutes for it; for example, capital cannot be substituted for it in production.

5. It must be impossible to improve the efficiency with which the resource is utilized beyond some point; for there to be an immediate problem, we must be near that point. Technical change which increases the productivity of a resource is referred to as "resource-augmenting technical progress," since its effect is equivalent to an increase in the stock of the resource.

[3] I have used the term "reserves" in the casual way that it is usually used; it is important to observe, however, that the natural economic definition of reserves is not that conventionally employed, for example, by geologists. Since the amount of a natural resource which can be extracted depends on the expenditure on extraction, reserves is an economic, not a geological concept. The appropriate definition entails an analysis of how much can be extracted, not at current market prices, but at market prices that will prevail as the supply becomes smaller; there is some presumption that this may be significantly larger than that implied by the conventional definitions.

6. It must be impossible to develop a substitute for the given resource.

Several of the recent articles on the economics of natural resources have attempted to make more precise the conditions under which it is possible to sustain a constant per capita consumption [2–4]. Most of these conditions are fairly intuitive.

The assumption that the natural resource is essential implies that the isoquants (between the natural resource and the other produced factors of production) never hit the axis of the produced factor. (See figures 2-1 to 2-7 where the other factor of production is taken to be capital. In figure 2-1 natural resources are essential; in figure 2-2 natural resources are not essential.) Note that what is crucial for the question of whether the natural resource is essential is not any property of the production (isoquant) function when there is a large input of natural resources. If the isoquant has relatively little curvature (i.e., is flat, as in figure 2-3) we might be tempted to extrapolate the curve and to infer that the natural resource is nonessential, but the curvature could easily be changed as the resource input becomes small, so that although at present it is easy to substitute, say capital, for resources, eventually it becomes quite difficult. Conversely, it may be quite difficult to substitute now (the isoquant can still be quite curved) but when the resources become sufficiently scarce, we may be able to switch to a completely different technology, for which the substitution is relatively easy, and the isoquant does hit the axis (figure 2-4).

The concept that economists usually use for measuring the ease of substitution is the "elasticity of substitution," which gives the percentage change in the input ratio (say, of natural resources to capital) which would be engendered by a percentage change in the relative price of capital to natural resources.[4] If a 1 percent change in relative prices does not give rise to a change in factor inputs (figure 2-5), the elasticity of substitution is zero; if we discontinue using the factor whose price has risen, we say that the elasticity is infinite. The central case where a 1 percent change in factor prices gives rise to a 1 percent change in factor inputs is called the Cobb–Douglas production function.[5] If the elasticity of substitution between natural resources and capital and/or labor is

[4] The production function $Q = \phi([aK^\rho + (1 - a)R^\rho]^{1/\rho}, L)$ where Q is output; K is the stock of capital; R is the flow of natural resources used in production; L is the labor supply has a constant elasticity of substitution between K and R: $\partial Q/\partial K = \phi[aK^\rho + (1 - a)R^\rho]^{1/\rho - 1}aK^{\rho-1}$; $\partial Q/\partial R = \phi[aK^\rho + (1 - a)R^\rho]^{1/\rho-1}(1 - a)R^{\rho-1}$

$$-1/\sigma \equiv \text{elasticity of substitution} = \frac{d \ln [(\partial Q/\partial K)/(\partial Q/\partial R)]}{d \ln K/R} = \frac{1}{(\rho - 1)}.$$

[5] The Cobb–Douglas production function can be written: $Q = K^{\alpha_1}R^{\alpha_2}L^{\alpha_3}$, $\alpha_1 + \alpha_2 + \alpha_3 = 1$. With a constant population, a sustained per capita consumption is possible if $\alpha_1 > \alpha_2$.

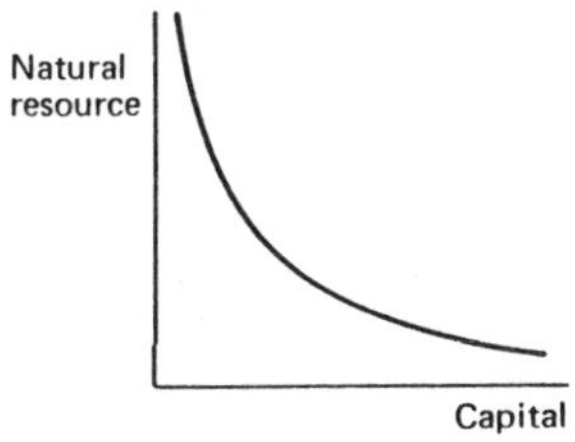

Figure 2-1. Natural resource essential.

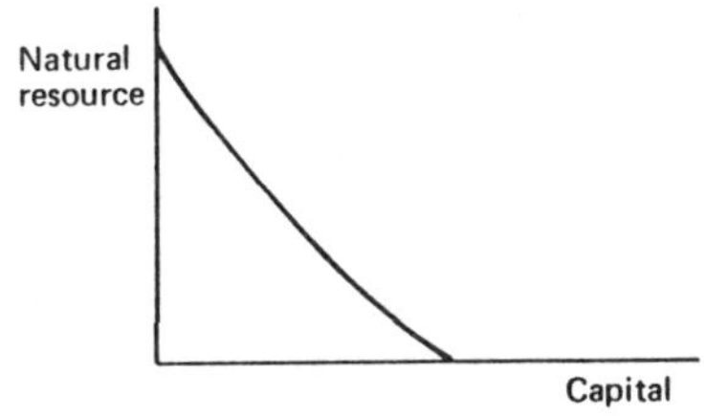

Figure 2-2. Natural resource not essential.

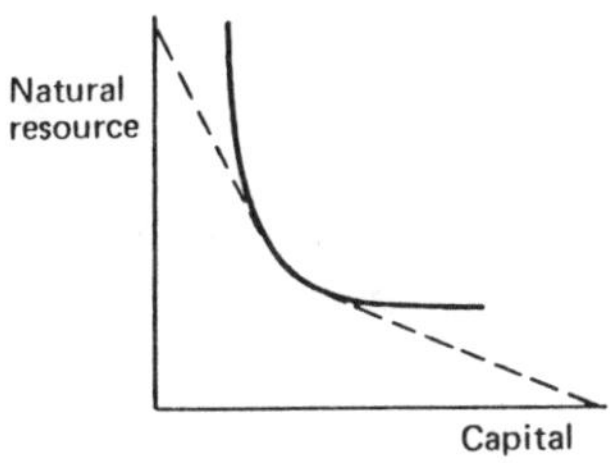

Figure 2-3. Local high elasticity: resource essential.

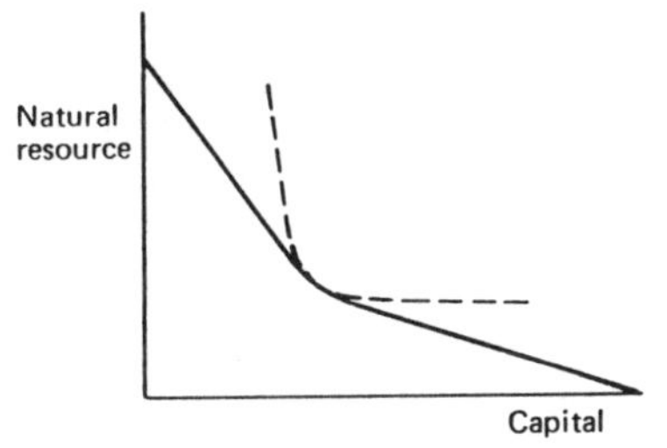

Figure 2-4. Local low elasticity: resource inessential.

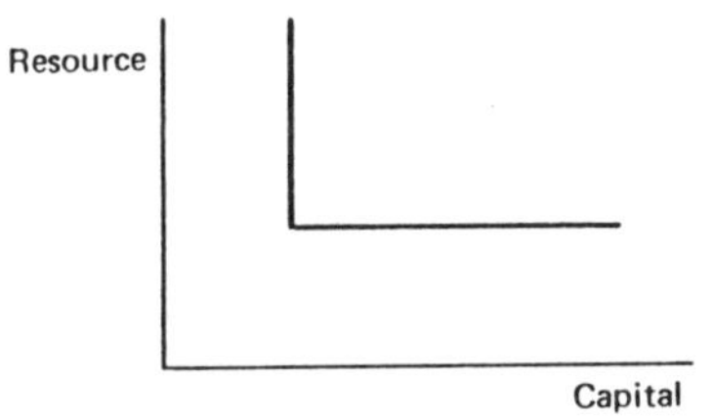

Figure 2-5. Zero elasticity of substitution.

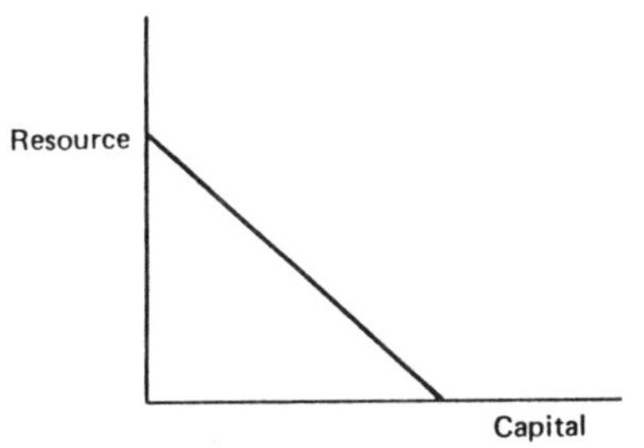

Figure 2-6. Infinite elasticity of substitution.

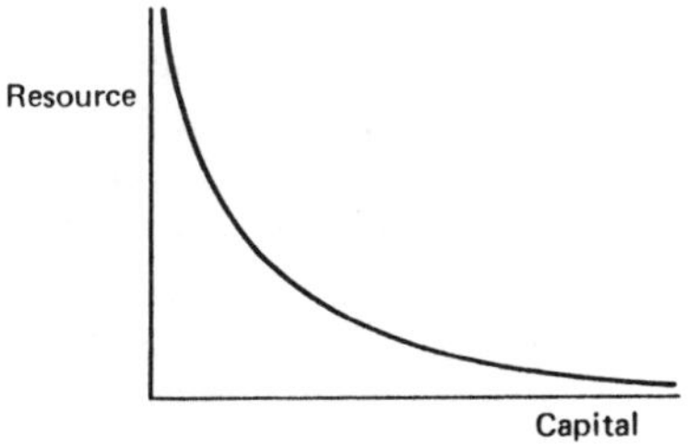

Figure 2-7. Unity elasticity of substitution.

constant, then a sustained constant per capita consumption is feasible if (1) the elasticity is greater than unity, or (2) if the elasticity is unity and if the share of capital exceeds that of natural resources.

If the elasticity of substitution is not constant, what is crucial is what happens to the elasticity asymptotically as resource input goes to zero. In these cases the produced input is sufficiently substitutable for the natural resource that the decrease in supply of the natural resource can be compensated for by an increased supply of capital. Of the two cases, the Cobb–Douglas case is clearly the most interesting for there natural resources are essential in the sense that some input of the natural resource is required for production (the isoquants never do hit the axes). But a small input of natural resource can be compensated for by a sufficiently large input of capital, and whether that is feasible for the economy depends simply on the relative shares of the two.

The conditions under which technical change makes sustained per capita consumption viable are also easily interpretable. The development of a produced substitute can be interpreted as changing the isoquant so that in fact it does hit the natural resource axis. The other condition is that there be a strictly positive rate of resource-augmenting technical progress.[6] If the input of the natural resource were to decline at the rate of the resource-augmenting technical progress—clearly a feasible policy —the effective input would be constant, and with a constant population a constant per capita level of consumption would thus be viable.

IV
The Forecasting Issue

The way we have posed the problem makes it easy to see what is required if we are to forecast the future: we need to know to what extent it will be possible to substitute produced goods for natural resources and we need to know the likelihood of technical changes which will lead either to the development of substitutes or to an increase in the efficiency with which we use what resources we have.

This problem can be approached in several different ways. The "engineering approach" examines the set of presently available techniques and attempts to assess the difficulties associated with developing new technologies. There seems to be an "informed consensus" that within the next 100 years technologies for the production of essentially a boundless supply of energy at "reasonable cost" (for example, solar energy)

[6] A production function with factor-augmenting technical progress may be written: $Q = F(K, R, L, t) \equiv F[\mu(t)K, \gamma(t)R, \lambda(t)L]$.

One unit of resource at t is a perfect substitute for $1/\gamma(t)$ units at time zero.

will become available. On other issues, for example, whether the atmosphere will be affected by certain technological developments, there seems to be less of a consensus. The economist, I suspect, has relatively little to say on these technological issues. The economist's approach is, in some sense, much cruder. He attempts to look at the past and to extrapolate from that into the future. It is not, however, obvious what the appropriate method of extrapolation is. The crucial question is what is to be taken as a constant. For instance, is the average *rate* of growth of the input of oil and the *stock* of presently discovered oil to be taken as given? If so, we are probably indeed in trouble. Or is the rate of increase of discovered resources to be taken as given, in which case prospects are not quite so bleak. And if the rate of growth is to be calculated, over what period should we make the calculation? How is one to choose among these alternative hypotheses and calculations?

Economists approach this problem by formulating a model which is sufficiently general to encompass the entire range of competing hypotheses as special cases and then subjecting the model to empirical testing. In principle then, we should be able to ascertain, at least by extrapolating from past experience, which hypothesis is more reasonable.

If this were as easily done as said, there would be far fewer disputes. Three problems are repeatedly encountered. (1) The data may not be able to distinguish among hypotheses; that is, two alternative hypotheses could be equally consistent with past observations, yet have very different implications for the future. (2) Most tests involve some kind of parameterization; thus, the actual test performed is not of the hypotheses in their most general form. (3) The structure of the economy may have changed, making "testing" of alternative hypotheses on the basis of past data inappropriate.

Let me illustrate these points with the issue at hand. We identified in our earlier discussion two parameters determining the future viability of the economy: the elasticity of substitution and the rate of technical progress augmenting natural resources. I focus my attention here on attempts to measure the former. The question, as we noted earlier, is whether the elasticity of substitution between natural resources and, say, capital, is greater or less than unity (again, recall that what is crucial is the limiting elasticity as the input of natural resources goes to zero). Note that no simulation exercise will ever resolve the question of the correct value of a parameter such as the elasticity of substitution. Moreover, if we know the value of the elasticity, analytical methods can completely answer the question of whether the economy is viable in the long run; we do not need to resort to simulation. What simulation can do—if we have a reasonable model with reasonable estimates of the parameters—is to give us some feel for how long the long run is (for example, if there is a "long-run natural resource problem," is it likely to make itself evident in 50 or 500 years?). Simulation may also enable us to identify the crucial parameters but, here as elsewhere, direct analytical methods are likely to be less ambiguous.

The easiest way of obtaining an estimate of the elasticity of substitution entails introducing one further assumption: that of competitivity, that the prices of (returns to) factors of production equal the value of the marginal productivity of the factor. With this additional assumption, if the elasticity of substitution were much less than unity, then we should expect the share of various natural resources in national income to have changed dramatically; those which have become relatively scarce should have a rapidly increasing relative share, those for whom unexpectedly large supplies have been found (and hence, whose scarcity had previously been overestimated) should have a falling share. In fact, at least in the aggregate, such dramatic shifts do not for the most part seem to have occurred if we take as our period of calculation the past 100 years; obviously if we look only at the past few years we could obtain quite different results. In the one case—land—where there has been a marked change in the share over the past several hundred years, it has been in the wrong direction. The share of land, for instance, in the past was as high as 50 percent; today it is much less in spite of a much larger ratio of population to land.

Resource optimists might argue that the share, say of land, has declined either because the elasticity of substitution is not low and/or because the effective supply of natural resources (land), taking account of technical change, has increased.

Resource pessimists might argue against these results in a number of different ways. (1) The competitive assumption on which the analysis was based might not be true, but for them to argue that the elasticity is low, they must argue further that the degree of monopoly in natural resources has changed in a significant way. (2) The future may not be like the past: it is at this point then that recourse has to be made to the "engineering approach" discussed earlier. (3) The particular parameterization implicit in the above calculation that the elasticity of substitution is constant is not correct: for example, they might argue that as resources become scarcer, the elasticity declines.

I am not aware of any convincing test of this hypothesis; my own guess is that the data are consistent with the hypothesis of a constant elasticity of substitution, but that if some parameterizations involving nonconstant elasticities were employed, nonzero point estimates of the rate of change of the elasticity would be observed. I am not sure whether the point estimates would side with the pessimists or optimists, but the value might well vary according to the parameterization adopted. My own suspicion is that the standard errors would be large enough to allow pessimists and optimists to continue arguing their case.

No matter how the test turned out, either side could argue against the result on the grounds that the analysis had been carried on at too high a level of aggregation. A real test, it might be argued, entails an analysis at a much more disaggregated level, taking into account different productivities of different firms within the industry, elasticities of substitution within each sector, and intersector elasticities of demand. Thus, for instance, changes in the distribution of firms within a sector (say, the difference between best practice and average practice) would result in the aggregate elasticity of substitution appearing not to be constant, although the "best practice" production function does have a constant elasticity.[7] Similarly, even if the production of neither commodity A nor commodity B allows the substitution of capital for natural resources, if commodity A uses less of the resource relative to capital than does commodity B, by substituting commodity A for commodity B, we can reduce our utilization of natural resources. Thus, estimating the full potential for substituting other factors for natural resources requires an understanding of all the possible patterns of substitution available in consumption and production.

Once we recognize possible methods of substituting produced for nonproduced goods, a further kind of technological adaptation becomes possible: tastes may change to reflect the changing environment. Although there is a widespread feeling that tastes are "endogenous," that they respond to changing circumstances, the extent to which changed patterns of consumption reflect a rational response to changing prices (the changed patterns of consumption reflect the changes in the best method of obtaining the "basic wants" but the underlying utility function

[7] For a more extended discussion of the relationship between micro and macro production function see K. Sato [5, 6].

Many of the attempts to estimate the elasticity of substitution are based on cross-sectional data, rather than the heuristic time series approach that we have employed above. There are a number of problems involved in using these cross-sectional estimates as a basis of predicting the long-run elasticity of substitution; this is not the place for an extended discussion of these econometric problems except to note one that has not received sufficient attention in the literature. Consider the traditional method of estimating constant elasticity functions, originally introduced by Arrow and coauthors [7], entailing regressing marginal product (wage) on average product. Assume that labor is not homogeneous and one cannot observe how many "efficiency units" are associated with any particular individual. Then, if some firms hired mostly efficient workers and paid them a high wage, and if some firms hired mostly inefficient workers and paid them a low wage, it would appear as if there were a positive elasticity of substitution, even though the elasticity might have been zero.

remains unchanged) rather than a change in basic attitudes and preferences remains a moot question.

Where does this leave us? The kinds of tools economists have used in the past are not likely to resolve this issue for those who are firmly committed to one side or the other in this debate. Yet it seems that what economic theory does remind us of is "that there is more than one way to skin a cat." If the resource pessimists are correct that we are going to be facing a serious resource problem in the immediate future, they must convincingly show that (1) within each sector the elasticity of substitution is low and the demand elasticities are also low, so that as resources become scarcer we do not, or cannot, substitute less resource-intensive commodities for more resource-intensive commodities; (2) the prospects of adapting tastes to the new set of economic circumstances are poor; and (3) the prospects are bleak for technical changes that would enable us better to use what resources we have.

The more mundane forecasting problem of estimating the likely impact of increasing scarcity of natural resources on our standard of living not only requires a knowledge of some of the same critical parameters, but also entails information concerning a few other parameters.

Elasticity of Demand. If the elasticity of demand for natural resources is high (either because there are readily available substitutes in production or because there are readily available substitutes in consumption which are less resource intensive), then increasing scarcity will not be a serious problem. Note that while for the question of viability we needed to know what happened to the elasticity as the resource input became arbitrarily small, here we only need to know the elasticity associated with levels of input not too different from those at present—a far easier question.

Costs of Extraction. There is, however, an argument that to assess the impact of increasing scarcity in the short run we need not know the demand elasticity; in competitive markets prices will be rising in such a way as to make net royalties (price minus cost of extraction) rise at the rate of interest.[8] The level of prices and the patterns of consumption are affected by the nature of the demand curve, but not changes in the price level.[9] Thus, to forecast real price movements all we need to do is fore-

[8] Consider a competitive firm. It must decide whether to extract a unit of oil today, receiving net $p_t - c_t$ where p is the price and c_t is the extraction cost; or to extract it tomorrow, receiving net $p_{t+1} - c_{t+1}$ whose present value is $(p_{t+1} - c_{t+1})/(1 + r_t)$ where r is the rate of interest. In equilibrium, it must be indifferent, that is, $(p_{t+1} - c_{t+1})/(1 + r_t) = p_t - c_t$ or in continuous time $(\dot{p} - \dot{c})/[p(t) - c(t)] = r(t)$.

[9] Of course, changes in the consumer surplus, that is, welfare, associated with given changes in the price will depend on the elasticity of demand, so that the impact of a given increase in prices will be greater the lower the demand elasticity.

cast real interest rates and movements in real costs of extraction. The former are not likely to vary much in the immediate future (if anything, the increased scarcity of natural resources will lower the real return on capital and the rate of increase in prices). Thus, predicting future price movements comes down to predicting movements in the cost of extraction. There are two offsetting effects here. On the one hand it is in general optimal to first extract natural resources from deposits with lower extraction costs; as these are used up, extraction costs increase. Now, if natural resources were not scarce so that the price were determined by the extraction costs (as for a conventionally produced commodity), then this would raise the price. But for a natural resource, the increased extraction cost actually leads to a lower *rate* of increase of prices.[10] The second effect, which in the past has dominated, is that technological change has lowered the cost of extraction.[11] The lowering of the cost of extraction enables royalties to rise, even with a constant consumer price. As a consequence, in the recent past, there has been no significant increase in the real costs of extraction for many commodities, and a decline for several. If this trend continues, then it is not likely that an increase in resource scarcity will significantly lower standards of living.

New Deposits. The third factor which is important in determining the likely movements of prices in the short run is discoveries of new deposits. To the extent that such discoveries are anticipated (in a probabilistic sense), they should have no effect on price movements; but to the extent that the actual discoveries are greater or less than anticipated, prices will rise or fall (or rise at a faster or lower rate than they otherwise would have). Since by definition it is only the "unanticipated" discoveries which affect the price path, there is no way that we can say anything about the direction of these effects (similarly with respect to the discovery of substitutes for natural resources).

Finally, it should be noted that exhaustible resources constitute a small fraction of national income so that even a significant rise in price will have only a small effect on standards of living. (This is not to say that there might not be costly adjustments associated with significant price rises.)

[10] $\dot{p}/p = (1 - c/p)r$ so an increase in c lowers $\dot{p}/p$. This argument assumes that c is not changing over time.

[11] One ought really to include in the costs of extraction the total costs associated with discovery, bringing the resource from below ground to above ground, refining, and delivering the resource to the user. The "consumer price" may be lowered because of a reduction in any of these costs. Note that much of the technological change has resulted from better methods of extracting the mineral of interest from the ore. Some technological change may be indistinguishable from "returns to scale": as the economy grows, it may be possible or economical to have larger mines and refineries.

In short, it is not likely that resource shortages will lead to a significant lowering of standards of living in the immediate future. The arguments are reinforced if one believes that there are incentives within the market for estimating resource scarcities (as we shall argue in the next section), so that prices reflect the judgments of those participating in the market concerning the scarcity of resources. To argue that we will face, in the near future, a significant lowering of the standard of living requires arguing that the market systematically underestimates future demands and overestimates available supplies. A fuller analysis than we have been able to present here would recognize more explicitly that there is uncertainty associated with forecasts of demand and supply; it would then calculate the consequences, say, of underestimating demand by one standard deviation and overestimating supply by one standard deviation. If demand elasticities are low, such errors may be associated with large welfare losses. The market, however, presumably at least partially takes these risk considerations into account. The extent to which it does this is discussed more fully in the next section.

V
The Efficiency Issue

There is, admittedly, something satisfying about knowing whether the economy, as we know it, will come to an end in fifty years or so, or not. But there are those who seem to feel that establishing the fact that the economy faces dire problems with respect to natural resources has obvious implications for the role of the government. On the contrary, I would argue that the existence of a natural resource problem has no immediate implications: it is neither a necessary nor a sufficient condition for governmental intervention in the markets for natural resources. The market could be doing as well as could be done, and the economy could still be facing a doomsday; and the market could be doing a quite bad job of resource allocation, and yet there might be no doomsday in store.

This section is concerned with the *efficiency* issue as opposed to the equity issue, which we treat in the next section. Efficiency has a very precise definition: Is it possible to increase the consumption (welfare) of any individual (consumption at any date) without decreasing it at any other? If it is, we are clearly wasting resources. Efficiency is not, of course, the only objective of economic policy; we are also concerned with the intertemporal distribution of income. A pattern of growth which left our grandchildren with few resources might be efficient, yet very undesirable.

The object of this section is to analyze the efficiency with which the market allocates resources; the role of the government in resource markets (as in other markets) is customarily justified in terms of some kind of market failure. We consider here the major possible sources of market failure.

In the discussion below we show that there are a number of reasons to expect that the market allocation of natural resources might not be efficient; some of the effects lead to excessively rapid rates of resource utilization, others lead to excessive conservation; in still other cases we are able to show only that the market may not be efficient but are, in general, unable to argue whether the market is too profligate or too conservation minded. The conclusion of the analysis is that there is no overriding case that we are using our resources too rapidly, but that there are a number of actions which the government could take which would probably lead to a more efficient utilization of our natural resources.

We first show that the problems associated with monopoly are probably less important than for other commodities, while the problems associated with imperfect risk and futures markets, and costly information, are perhaps more important. The most convincing case for government intervention is in the area of support for research and development.

The government has, of course, intervened extensively in the market for natural resources; these interventions have probably been the most important source of inefficiency in the allocation of resources, and although one perhaps should not infer from this that future interventions would be equally as unsuccessful, there is little reason to believe that the efficiency with which the government allocates resources would be drastically better in the future than it has been in the past (we do not present any argument for why there should be systematic inefficiencies associated with governmental regulation; these have been discussed elsewhere in the literature). We now turn to a detailed discussion of each of the major sources of market and nonmarket failure.

Monopoly. In the wake of the oil embargo there developed a widespread belief that the noncompetitive nature of the oil industry raised prices and led to resource inefficiencies. For conventional *commodities,* monopoly results in the raising of prices and a decrease in consumption. For natural resources the effect is less clear, for if fewer natural resources are consumed today, there will be more to consume in the future; thus, raising the price today means that the price will be lower at some date in the future.

The fundamental principle defining efficient intertemporal allocation of resources is that the present discounted value of the (net) marginal product[12] must be the same at all dates, that is, the value of the marginal product rises at the rate of interest. This is ensured in a competitive economy by having the price rise at the rate of interest. A monopolist will equate the present value of *marginal revenue* at all dates, so marginal revenues rise at this rate of interest.[13] Thus, if price is proportional to marginal revenue, that is, for a constant (over time as well as for different levels of consumption) elasticity demand function, the *monopoly and competitive allocation are identical.* If the elasticity increases over time, as one might expect with development of substitutes, or if there are positive extraction costs, then the monopolist adopts an excessively *conservationist* policy.[14]

Absence of Futures Markets. The presumption that the market allocates resources efficiently follows from the more general result that, under certain conditions, a competitive economy provides efficient (Pareto optimal) allocation of resources. One of the conditions is that there be a full set of markets, including markets for trading into the future and trading different risks. These are obviously absent and their absence may have important implications.

Before discussing these, there is one fallacious argument which does need to be disposed of: it is sometimes argued that because our children

[12] Net of extraction costs. For simplicity we assume here that extraction costs are negligible.

[13] These basic principles of the intertemporal allocation of resources were originally discussed by Hotelling. They are just the application to this problem of the fundamental principles of efficient resource allocation. A unit of, say, oil must have the same *value* in all its uses. Among its alternative uses are its consumption today versus its consumption next period. Postponing consumption by a unit lowers output this period by the marginal product of oil today MP_{t+1}; it raises it by the marginal product of oil next period MP_{t+1}, but a unit of output next period is worth less than a unit of output today, if the rate of interest is positive. We require: $MP_t = (MP_{t+1})/(1 + r)$ where r is the rate of interest.

[14] These results are discussed in greater length in Stiglitz [8]. They apply only for the case of an elasticity of demand exceeding unity (which it must be at the point at which the monopolist is operating since otherwise revenues could be raised by lowering output). Gilbert [9] has analyzed models in which the elasticity is less than unity, but there is a limit price, that is, a price above which a substitute is introduced or developed. Gilbert [9], Salant [10], Stiglitz [11], and Cremer and Weitzman [12] have analyzed the behavior of a single large producer facing a competitive fringe. They have employed different models, entailing slightly different solution concepts; Salant has analyzed a Nash equilibrium, while Gilbert and Stiglitz have used a Stackleberg equilibrium with the monopolist acting as the leader; the monopolist knows the response of the competitive fringe to his actions. Gilbert has shown that the behavior of the large producer depends critically on whether the competitive fringe is supply constrained. If it is, it can be shown that as the supply of the competitive fringe is increased, the supply of the large producer will be increased. Newbery [13] has shown how the argument of Stiglitz [11] for random taxation can be reinterpreted in this context as establishing that under certain conditions it is optimal for a monopolist facing a limit price to randomize his prices.

and grandchildren are not present today, the market will systematically underrate their importance; that it will be reluctant to undertake investment decisions lasting longer than the lifetime of the typical investor (for example, planting and harvesting hardwood trees). (One guise in which this argument sometimes appears is that the social rate of time preference ought to be lower than the market rate of interest.) This overlooks the fact that the investor can sell the asset (for example, the hardwood forest) to another investor (say, in the next generation) to obtain his return; the forest does not have to be felled for him to get a return on his investment.

The consequence of the absence of futures and risk markets is that investors have to *guess* about future prices and have to absorb the risks associated with the uncertainty. Thus, whether the market pursues an excessively conservationist or excessively profligate consumption policy depends on whether there are systematic errors in the forecasts of future prices (demands); but even if it turned out that in some past period there were systematic errors, that does not imply that there is a need for government intervention, or indeed that government intervention is desirable. There are strong incentives within the market for discovering systematic errors—incentives which are, in general, lacking within a bureaucratic framework. Because the rewards obtained within the private sector depend on the accuracy of their forecasts, speculators and firms will strive to obtain better forecasts; and when they fail they bear the brunt of the costs. This is not so with a government allocation.

Moreover, when individuals differ in their judgments—as they are likely to—the market provides a systematic way of aggregating their beliefs, although it may be objected that the beliefs of the wealthy carry more weight than those without financial resources. On the other hand, little is known about how a bureaucracy would aggregate their beliefs.

Although it does seem to me that these considerations strongly suggest that the basic responsibility for the intertemporal allocation of our resources remains with the market, the absence of futures markets does result in at least three distinct problems for which remedial action by the government might be desirable.

The first is the case of the individual who believes that the market is underestimating the magnitude of resource scarcity. This person purchases the resource (or claims on the resource), holding it until the date (which he believes will surely come) when the scarcity is recognized. At that date the price rises and he will have made a capital gain. In effect he transfers resources that would have been consumed today to some date in the future, capturing the difference in (present) prices as his return. But if the shortage is not recognized within his lifetime, then there will be no way in which he can capture the returns; thus the individual has no incentive to enter the market.

Similarly, if an individual believes that the market overestimates future demand (underestimates supply), he would engage in a long-term contract for future delivery; he believes that eventually the market will recognize that prices are too high, so that prices will fall and he will make as his profit the difference between the contracted price and the lower spot price then prevailing. There are two difficulties with this. First, the long-term contracts required may not (generally do not) exist. (There is an apparent asymmetry in the market associated with holding positive and negative quantities of an asset.) Second, in order to capture the returns, the error in market forecasts has to be recognized within the individual's lifetime.

In fact, in the case where the market is excessively conservationist, the error might never be reflected in market price. The price of the natural resources may be rising at the rate of interest so that over the short run there is an efficient allocation of natural resources, but the initial price may be too low or too high. In the former case it can be shown that in finite time with a Cobb–Douglas production function the price will become infinite, but if the initial price is too high, nothing will happen except that we will never use up all our resources; thus short-run perfect foresight (short-run efficiency) can be consistent with an excessively profligate consumption policy for a finite time, but it can be consistent with an excessively conservationist policy forever [15].[15]

These considerations do, I think, provide a convincing argument that the market allocation may not be efficient, but it is not clear whether there is a bias for or against conservation.

Second, the efficient intertemporal allocation of resources requires the acquisition of information concerning future demands and supplies. Although the fundamental theorem of welfare economics establishes the Pareto optimality of the market allocation of resources with a given information structure, there is no theorem establishing its optimality when information has to be acquired. On the contrary, it can be established that the market may not be Pareto optimal [9, 16]; there may be excessive expenditures on information (arising from unnecessary duplication, as when different individuals attempt to acquire the same piece of information, for example, about the value of some lease) or there may be too

[15] Indeed, the problems associated with excessively high rates of capital accumulation (savings) which were the center of so much of the discussion in the pure theory of growth in the 1960s (see, e.g. [14]) are resolved by the presence of natural resources: there can never be a problem of oversaving in capital since with very low rates of interest, lower than the rate of growth, resources will be exhausted in finite time; but there can be a problem of oversaving of resources [15].

little expenditure because information is partially a public good. In particular, if the market prices adjust to reflect information, then individuals who do not do the research themselves may, by observing the market prices, obtain the information indirectly (thus part of the total benefits of obtaining information accrue to individuals other than those who purchase it directly) [17].

Third, many natural resource markets seem to be characterized by excessive instability, and this instability may in part be a result of the absence of future markets. There are strong a priori reasons to believe that adjustment processes for natural resources may not work well; we showed earlier that equilibrium in the market requires the expected rate of change of (net) price to be equal to the rate of interest. If for some reason price rises slightly faster than the rate of interest (for example, because some individuals believe that there is an impending resource scarcity and hold back supplies), then that has two effects: those who base their expectations of future price movements on past movements will extrapolate the high rate of price increase into the future, and thus expect the price to rise faster than the rate of interest; they will then hold back their supplies as well, leading to further increases in price in a destabilizing way.[16] Second, the reduced supplies will lower the rate of interest if resources and capital are complements in production (as one would normally expect); thus the gap between the expected rate of price increase and the rate of interest is again increased. (For a fuller discussion, see Stiglitz [15].)

These considerations suggest that there may be a role for the government in indicative planning in helping to forecast future demands and supplies, but it should be emphasized that this role is far different from one of actually allocating the resources themselves.

Absence of Risk Markets. The absence of risk markets means that those who hold stocks of natural resources may not be able to insure themselves against the risks of a price fall because of the discovery of a substitute or of a new supply of the resource.

The effect of this is actually very hard to analyze. There is a simple intuitive argument: viewing the holding of a resource (not extracting it) as a risky investment, an increase in uncertainty makes it less attractive, thus increasing the rate of consumption.

But this misses the fact that the risk depends on the stock (the amount by which the price falls when a substitute is discovered depends

[16] Note that the instability arises from individuals having to forecast future prices; futures markets, were they to exist, might equilibrate the process. The expectations formation process may, under not unreasonable conditions, be very destabilizing.

on the stock available at that date) and that what is critical for the intertemporal allocation of the resource is how the risk varies over time.

In one simple model in which the probability of discovery of a substitute is constant per unit time and there is risk neutrality [with respect to income but not with respect to consumption of oil, i.e., the indirect utility function is of the form $V(p, Y) = V(p)Y$], for large stocks of the natural resource, uncertainty leads to greater conservationism, whereas for small stocks, it leads to a more profligate policy.[17] In another model where the stock of the resource is unknown and the only information that is acquired over time is that one had at least as much as one had consumed, it has been shown that uncertainty always leads to more conservation [18].

Uncertainty undoubtedly has an effect on intertemporal allocation, but for some purposes this is not as important as the question of what policy implications this has. For policy purposes it is inappropriate to compare an allocation of one market structure with an allocation which would emerge from another, except if the policy to be considered is a change in the market structure; the risks associated with uncertain future supplies and demands borne by private individuals in imperfect risk markets are real "costs"; and the fact that under some idealized world in which these risks are not borne by these individuals the intertemporal allocation of oil would be different is of interest, but of no direct policy import. An appropriate question to ask, for instance, is whether a Pareto optimal improvement could be made within our market structure by taxing or subsidizing the return to holding oil; or to put it another way, to ask whether the intertemporal allocation of natural resources arising from a stock market provides a "constrained" Pareto optimum. The answer, in general, is no, but the magnitude and direction of bias depend on a number of factors, for example, on differences in attitudes to risk in the economy [11]. A full analysis of this problem is presently under way. Until this is completed, it is difficult to ascertain how important this potential inefficiency of the market is.

Common Resource Problems. As we noted earlier, there are several distinct categories of natural resources; most of this paper is concerned with private exhaustible natural resources. Many of our most

[17] One might have thought that the effect of uncertainty is to raise the effective interest rate, thereby increasing the rate at which the price of a natural resource increases, hence leading to a lower price (more consumption) today. Although there is an effect going in this direction, there is another effect, which may be called the precautionary effect. Because one does not know when the substitute will arrive, one must hold back a supply of the resource as a precaution against the contingency that it does not arrive until some date far in the future.

important natural resources are, however, "public"—such as atmosphere—and for these the presumption that the market provides an efficient allocation of resources is reversed; only under the most unusual circumstances would one expect the market solution to be efficient. Individuals and firms will not, for instance, take into account the effect of their activities in lowering the quality of the air.

Most of these resources are "renewable": a stock of fish can be maintained, provided fish aren't consumed too rapidly; a "reasonable" quality of air may be maintained so long as the flow of pollutants is not too high. We are concerned here, however, only with three aspects of the problem: whether this will lead to an "exhaustion" of our natural resources and, even if it does not lead to exhaustion, whether it will lead to a significant lowering of the standard of living of succeeding generations, and whether there are any simple remedies for these problems.

There are some cases where the failure to charge for "common resources" has led to near exhaustion of a renewable natural resource: for example, whales, and there is evidence that if governments had not taken action, the air in certain cities would have become so polluted as to be almost unbreathable.

Most of the resources with which we are concerned here are renewable resources; hence there is a sustained level of per capita consumption which can be maintained with these resources with a constant population. Whether a sustained per capita level of consumption can be maintained with a growing population depends on exactly the same kinds of parameters we discussed in previous sections, for example, the substitutability between the given natural resource (now viewed as a constant flow) and other factors of production, the development of new substitutes, and the continuation of technical progress that augments resources.

There are, however, strong reasons to believe that with common resource problems the rate of extraction will be too rapid. Consider, for instance, a pool of oil which is available to anyone who wishes to take from it (or who purchases some land over the pool). Clearly, the intertemporal tradeoff of the individual is considerably different from when the resource is completely privately owned. For now the individual does not ask what will he get if he sells a unit today or tomorrow. He observes that if he does not extract it today, someone else will, and there will be no oil available for him to extract tomorrow. Thus market equilibrium will be associated with excessively rapid rates of extraction.

There are a variety of remedies for these problems. For many common resource problems a simple solution is to put the entire pool under a single management (for example, unitization of the oil pools, or giving a monopoly for fishing within a lake; obviously where there is more than one governmental jurisdiction involved, as for fishing in international waters, arriving at a solution will be more difficult). For other resources, such as "atmosphere," either governmental regulation or taxation is required. The relative merits of alternative methods are too broad to be discussed here.

Nonmarket Failures. The market failures discussed above appear to be the major ones. There are, in addition, a number of sources of nonmarket failure, of government interventions leading to inefficiencies both in the rate and pattern of extraction. These inefficiencies are perhaps even more important than those discussed above because they result in what appears to be shortages, when in fact they would not exist were it not for the government's actions; this is particularly true of the regulatory practices described below.

Taxation. There are several provisions of the tax code which may have an important effect on the allocation of natural resources.

Depletion Allowances. The impact of a depletion allowance at a constant rate may be less than is generally thought; if extraction costs were zero and the sector were competitive, it would have no effect. The reason for this is simple: the depletion allowance only changes the pattern of resource extraction to the extent that it changes the intertemporal tradeoffs between extracting today and extracting tomorrow. A depletion allowance increases receipts today but it increases receipts tomorrow by exactly the same percentage, and hence has no effect on intertemporal tradeoffs. If extraction costs were positive and constant, they would lead to faster extraction (than in the absence of the depletion allowance).[18] If the sector were monopolized, it would also lead to a faster rate of extraction if the monopoly was excessively conservation minded (as it would be with positive extraction costs and a declining elasticity of demand), offsetting the effects of the monopoly. If the monopoly were excessively profligate, the depletion allowance would lead to a still faster rate of consumption.

Although the effects of a constant depletion allowance may be smaller than is widely assumed, the effects of the gradual removal of the depletion allowance are larger than is widely assumed; if a depletion allowance is to be removed, it should be removed as quickly as possible. For again, the gradual removal of the depletion allowance means that it

[18] See Stiglitz [16]. Let the depletion rate be d. Assume extraction costs are constant, then equilibrium with the depletion allowance requires $[(1 + d)\dot{p}]/[(1 + d)p - c] = r$ or $\dot{p}/p = r(1 - [c/p(1 + d)])$ from which it is clear that the depletion allowance raises the rate of price increase and hence must lower the price initially.

is more profitable to extract now than later, and this encourages depletion.

Other provisions of the depletion allowance probably have serious implications for the structure of the industry; what implications this has for economic efficiency is not clear. For instance, any provision which assigns a depletion allowance for some categories of firms and not for others ought to result in most of the extraction being conducted by those who receive the depletion allowance. Limiting the amount of the depletion allowance as a percentage of profits, unless computed on a well-by-well or mine-by-mine basis, will also affect the structure of the industry.

Although the depletion allowance would have no effect on the pattern of extraction from known reserves with zero extraction costs, it does affect the incentive for discovering oil. The total stock of oil which is likely to be discovered is greater, and because of that, present prices are lower and consumption rates higher. Similarly, if there is a positive extraction cost, oil, which it would never pay to extract without a depletion allowance, will be extracted. But, apart from considerations of monopoly, the cost of discovery and extracting this oil will exceed its benefits, and so such a policy is undesirable.

If the removal of the depletion allowance has not been anticipated, the immediate effect of the removal will be to raise the price of oil, both because of the slower rate of extraction and the smaller stock of oil that will eventually be extracted. These effects may be disguised at present because the short-run effect of a gradual removal of the depletion allowance is to increase the supply of oil.

Immediate Write-Off of Drilling Expenses. With immediate write-off of capital expenditures, the corporation tax can be viewed as a pure profits tax, and is thus nondistortionary.[19] But since other sectors are not so treated, there is a relative distortion; investment in oil exploration is encouraged relative to other kinds of investment. There is no justification for this favorable treatment.

Inability to Write Off Immediately Expenditure on Lease (or Land) Acquisition. For expenditures other than drilling, the oil sector is treated like any other sector. But this symmetric treatment does induce a distortion in the rate of extraction relative to what it would be in the absence of taxation. Consider a firm which purchases land under which there is oil. After the oil has been extracted, the land will be worth less. The firm can take a capital loss on the reduction in land value, and thus reduce its tax liability. The present value of the tax write-off is increased by accelerating extraction. It would probably be desirable to allow imme-

[19] Without interest deductibility of debt. Immediate write-off with interest deductibility would appear to act like a capital subsidy.

diate write-off of these expenditures. In order to discourage excessive allocation of resources to this sector relative to others, other taxes would have to be imposed.

Special Treatment of Capital Gains. Since the return to holding oil is the increase in its price, if this increase in price is subjected to capital gains taxation, the return to holding oil will exceed the return to holding conventional assets, and there will be excessive conservation. This effect could be significant because the difference in rates is large. Since the sale of oil is treated as ordinary income, this distortion requires that firms that do the exploration hold on to the land until the date of extraction. They then sell their land; the purchaser extracts the oil, the income from the oil being perfectly offset by the reduction in the value of the land.

Of the various provisions discussed above, several have been introduced to offset the tax-induced biases of the general tax structure (or so proponents of the provision argue). The depletion allowance is justified as a simple alternative to depreciation, but the present tax code essentially allows triple depletion; immediate write-off of drilling expenses, depletion allowances, and write-off of the loss in value of leases or land after extraction is completed. Other proponents of these provisions admit that it constitutes favorable treatment but argue that the favorable treatment is required to offset the distortionary effects of the corporation tax structure, which unduly penalizes risky and capital-intensive sectors. Elsewhere it has been argued that the sector is not more risky than other sectors, that the corporation tax does not discourage (but rather encourages) risk taking, that it does not penalize capital-intensive industries, and that there are second-best arguments for taxing the oil industry at a higher rate than other industries because the share of rents in that sector is larger.

Leasing Policy. The terms on which the government leases oil probably have a significant effect, not so much on the total rate of extraction as on the pattern of extraction. Efficiency requires that oil be taken from tracts of land with low extraction costs before it is taken from tracts with high extraction costs. (Similarly, tracts with low exploration costs should be explored before tracts with high exploration costs.) One of the provisos in government leases is the so-called diligence clause, requiring some development of the tract within five years. It pays the firm to extract at the end of five years so long as price exceeds extraction costs; but in the absence of the clause it would pay the firm to postpone extraction on tracts with high extraction costs.

Moreover, in models of competitive bidding it can be shown that the return in excess of extraction costs is a function of the number of bidders; this, it can be shown, leads to too little expenditure on exploration in areas in which there are a large number of "informed" firms, that is, firms who have extensive information about the tract; and it induces firms to nominate for leasing and explore fringe areas where there are likely to be few bidders and high exploration costs. This is inefficient.

In addition to these inefficiencies, there is an inefficiency arising from excessive and, from a social point of view, misdirected prebidding exploration, the object of which is not to discover oil but to ascertain how much should be bid for the lease. This has a private but not a social return. (It leads, for instance, to inefficiencies in the location of wells; for example, locating wells at boundaries of tracts to find out about neighboring tracts.)

It is not easy to devise more efficient policies. Recent proposals for developing pools as a single unit avoid the obvious inefficiencies arising when there are several tracts within a pool owned by different firms (the usual common-pool problem, where each firm fails to take account of its effect on the extraction of others; this normally leads to excessively rapid rates of depletion). The royalty bidding experiments showed that firms would bid very high royalty rates, but the consequence was that the well would be abandoned if the extraction costs as a percentage of price plus the royalty rate exceeded unity. Since these would be tracts with high extraction costs from which extraction should be postponed, provided they could eventually be leased again, this is not as great a disadvantage as it might seem.

Regulatory Policies. The government has engaged in both long-term and short-term attempts at price regulation. In the case of natural gas, the government has attempted to keep the price low; in the case of oil, following the embargo, they have attempted to keep it from rising to its scarcity value. Both policies have serious effects. Because the price was not allowed to rise with the rate of interest, there was an incentive to extract as quickly as possible. Because the price was kept at a very low level, consumers had little incentive to conserve on natural resources.

The effects of price regulation in the oil embargo are not as apparent. The immediate effects were the usual inefficiencies associated with prices which are fixed below a market-clearing level, for example, lines at gas stations. It might be thought that because the supply was inelastic, there would be no effects on the total rate of consumption; although this is true in the short run, it is not in the long run. If firms and individuals come to believe that when there is an event such as an oil embargo prices will be kept below their market-clearing level, they will not have the incentive to take the efficient precautionary policies; firms will extract too much oil today (since they obtain a return on holding a precautionary stock that is less than its social value) and individuals will adopt technologies of consumption involving less flexibility than is socially optimal.[20]

VI
The Intertemporal Equity Issue

One of the major issues underlying the debate concerning natural resources is whether, with present policies, our descendants will be left in an impoverished state, without natural resources. What is of concern here is an equity issue—the welfare of our descendants versus the welfare of the present generation. There are two important observations to be made in this respect.

First, one should not view equity in a narrow sense of simply looking at the division of natural resources between present and future generations; the present generation may give future generations fewer natural resources (this is inevitable in the case of exhaustible natural resources), but it will give future generations a higher level of technology and more capital. One has to look at the relative welfare of the different generations and there is a strong presumption that future generations may be better off than the present generation. On grounds of equity it might be argued that we should consume even more now (including more natural resources).

Second, the appropriate instrument to use for obtaining more equitable distribution of welfare (if one believes that the present distribution is not equitable) are general instruments, for example, monetary policy directed at changing the market rate of interest. These do not lead to an inefficient allocation of resources. (Obviously, prior to using such general instruments, policies directed at correcting inefficiencies in the allocation of resources, as discussed in the previous section, have to be adopted.)

There has been an extensive recent literature concerned with the optimal pattern of consumption of natural resources. Although the models employed in this literature are obviously overly simple, they do give us a method of obtaining a rough check on whether the present allocation,

[20] Indeed, recent regulatory policies have punished those who have adopted flexible technologies: electricity generating plants which were convertible to coal were required to do so, forcing those who had adopted the flexible technology to pay more for their fuel than those who had not.

with all its market distortions, is different in a significant way from an optimal allocation. We solve for the optimal rate of extraction, using the utilitarian criterion, that is,

$$\max \int U(c)e^{-\delta t}e^{nt}\,dt$$

where $U(c)$ is the utility associated with a per capita consumption flow of c, n is the rate of growth of population, and δ the rate of discount, subject to the national income constraint,

$$C + \dot{K} = F(K, R, L, t) = K^{\alpha_1}R^{\alpha_2}L^{\alpha_3}\,e^{\lambda t}$$

(where F is a constant return to scale production function, which we have specialized as the Cobb–Douglas production function, with λ being the rate of technical progress; K is aggregate capital; L is labor; R is resource flow; and C is aggregate consumption), and to the resource constraint

$$\int R(t)\,dt = S$$

where S is the stock of the resource today.

The exact nature of the optimal consumption path (which, unlike the case without resources, may not be monotonic) depends on the elasticity of marginal utility, the rate of time preference, the rate of population growth, and the rate of technical change.[21] If the utility function is logarithmic $U = \ln c$, then it can be shown that the optimal rate of extraction is just equal to the pure rate of time discount minus the rate of population growth $(R/S) = \delta - n$. If $\delta - n$ is a number around 3 percent, this means that we should use up 3 percent of our remaining stocks each year, that is, at *current* rates of consumption, we have approximately thirty-three years' supply left, a number which is certainly smaller than our stock of energy-producing materials.

VII
Two Policy Issues

In the preceding two sections we discussed several policy issues in the context of whether a sustained level of per capita consumption growth is feasible and whether there are systematic misallocations of natural resources arising from market failure. In this section I wish to look at two further policy questions which have received widespread discussion.

[21] The general formula is that, asymptotically: $R/S = ([\delta(1 - \alpha_1) - \nu\lambda])/(1 - \alpha_1 - \alpha_2\nu) - n$ where ν is the elasticity of utility [4].

Operation Independence. A careful analysis of the costs associated with attaining energy independence (autarchy) would likely lead to a realization of the undesirability of that policy. Here, I wish to draw attention to two aspects of the debate.

The first question that ought to be posed is whether there is any reason to believe that the market has taken insufficient precautions for the contingencies of an embargo or a large price rise. The fact that individual economic agents have not "fully" insured against these contingencies is to be expected, given the high cost of such insurance. Our previous analysis suggested that in the absence of government regulatory policies (which do not allow the price to rise when scarcity values rise), there is no particular reason to believe that the market will make insufficient provision for these contingencies.[22]

Second, with respect to measures designed to obtain greater independence through increasing domestic supply, there is really a question of the intertemporal allocation of dependency; increased extraction today may reduce the costs of an embargo today, but would increase it in the future. Indeed, the argument of nondependency was one of the main ones used in support of the oil quotas in the 50s and 60s; this led to a higher level of consumption of our domestic supply, leaving us now with less domestic oil than we would have had then. (Thus, although we were better "protected" then we are less well protected now.)

R and D Policy. It has recently been suggested that some kind of price guarantee is required in order to elicit the appropriate levels of investment in R and D. The patent system is probably not an effective instrument in encouraging investment in R and D for energy (or natural resource) substitutes; the immediate effect of a discovery is to lower the price of the natural resource,[23] hence there is likely to be a long delay

[22] Obviously, if the market's assessments of the probabilities of an embargo or price rise are wrong, then it will take incorrect precautionary actions. If it underestimates these probabilities, then it will take insufficient precautionary actions. This can be corrected (assuming that the bureaucracy has reason to believe that it has better information about the probabilities, and that releasing this information to the market will not affect the market allocation, presumably because they refused to believe the bureaucrats) but tariff policy may not be an effective instrument. If individuals have point estimates of the price next period and production is not capacity constrained, then they always wish either to extract everything this period, everything next period, or are indifferent. Hence, with elastic foreign supplies, tariffs cannot control the intertemporal domestic supply allocation (although they do control domestic consumption). However, quotas will work.

[23] In a recent unpublished paper, Steinmuller [19] has reinterpreted the introduction of coal in Britain: it is not that the exhaustion of the forests led to the introduction of coal, but rather that less expensive methods of extracting coal made wood uneconomic as a fuel; although coal extraction was cheaper than planting and cutting a forest, it was not cheaper than cutting a forest; hence the immediate effect was to exhaust the forests and it was only later that it became economic to extract coal.

between invention and innovation. In addition, the lowering of the price is a social benefit which (apart from engaging in speculative activity) the inventor cannot capture. Although this does suggest that some kind of government price support program might be desirable, the price support should take the form of long-run contracts rather than a price support for the market as a whole.

VIII
Concluding Remarks

In this paper we have taken the view that natural resources are basically no different from other factors of production. There are presently extensive possibilities of substitution between resources and other factors (capital) and, with further research, there are likely to be further ways of substituting other factors for natural resources and making what resources we use go further. There is not a persuasive case that we face a problem from the exhaustion of our resources in the short or medium run, and although there are undoubtedly market failures leading to inefficiencies in resource allocation, there is no reason to believe they are worse here than elsewhere. Indeed, in the case of the effects of monopoly there is reason to believe that they are significantly less. In addition, a number of sources of inefficiency arising out of government intervention have been identified.

Although this is the conclusion that I, and I think many other economists would come to, I should emphasize that many of the points raised are moot. Although the economics of resource allocation in a static environment without uncertainty are well understood, there are many gaps in our understandings of dynamic economies with imperfect information, imperfect risk markets and imperfect futures markets, and these limitations are probably more important here than for conventional commodities. Further, at a number of points we have had to make empirical judgments (for example, about the value of the elasticity of substitution) about which there clearly can be disagreement. Perhaps more important than reaching agreement on the answer is reaching agreement on how we might resolve our disagreements. It is my hope that in this paper I have presented a framework within which this may be done.

References

1. F. M. Peterson and A. C. Fisher, "The Exploitation of Renewable and Nonrenewable Natural Resources," *Economic Journal* vol. 87 (December 1977) pp. 681–721.

2. P. Dasgupta and G. M. Heal, "The Optimal Depletion of Exhaustible Resources," *Review of Economic Studies*, Symposium on the Economics of Exhaustible Resources, 1974, pp. 3–28.
3. R. M. Solow, "The Economics of Resources or the Resources of Economics," *American Economic Review* Papers and Proceedings (May 1974) pp. 1–14.
4. J. E. Stiglitz, "Growth with Exhaustible Natural Resources: Efficient and Optimal Growth Paths," *Review of Economic Studies*, Symposium on the Economics of Exhaustible Resources, 1974, pp. 123–138.
5. K. Sato, "Micro and Macro Constant Elasticity of Substitution Production Functions in a Multi-firm Industry," *Journal of Economic Theory* vol. 1 (1969).
6. K. Sato, *Production Functions and Aggregation* (Amsterdam, North-Holland, 1975).
7. K. Arrow, H. Chenery, B. Minhas, and R. M. Solow, "Capital Labor Substitution and Economic Efficiency," *Review of Economics and Statistics* vol. 43 (1961) pp. 225–250.
8. J. E. Stiglitz, "Monopoly and the Rate of Extraction of Exhaustible Resources," *American Economic Review* vol. 66 (1976) pp. 655–661.
9. R. Gilbert, "Imperfect Competition and the Allocation of Oil," Working Paper No. 87, Department of Economics, University of California at Berkeley, 1976.
10. S. Salant, "Nash-Cournot Equilibrium for an Exhaustible Resource Like Oil," Federal Reserve Board, mimeo, 1975.
11. J. E. Stiglitz, "Utilitarianism and Horizontal Equity: The Case for Random Taxation," Institute for Mathematical Studies in the Social Sciences, Report, Stanford University, 1976.
12. J. Cremer and M. Weitzman, "OPEC and the Monopoly Price of World Oil," *European Economic Review* vol. 8, no. 2 (August 1976) pp. 155–164.
13. D. Newbery, "Feasible Price Stability May Not be Desirable," Institute for Mathematical Studies in the Social Sciences, Working Paper 68, Stanford University, April 1976.
14. P. Diamond, "National Debt in a Neoclassical Growth Model," *American Economic Review* vol. 55 (December 1965) pp. 1126–1150.
15. J. E. Stiglitz, "Growth with Exhaustible Resources: The Competitive Economy," *Review of Economic Studies*, Symposium on the Economics of Resources, 1974, pp. 139–152.
16. J. E. Stiglitz, "Information and Economic Analysis," unpublished monograph.
17. S. J. Grossman and J. E. Stiglitz, "Information and Competitive Price Systems," *American Economic Review* vol. 66, no. 2 (May 1976).
18. R. Gilbert, "Resource Depletion Under Uncertainty," Department of Economics, Stanford University, mimeo, 1975.
19. W. E. Steinmuller, "The Seventeenth Century Energy Crisis: a Critique of the Timber Famine Theory," unpublished paper, Department of Economics, Stanford University, July, 1976.

Bibliography

Barnett, H. J., and C. Morse. *Scarcity and Growth: The Economics of Resource Availability* (Baltimore, Johns Hopkins University Press for Resources for the Future, 1963).

Dasgupta, P. "Some Recent Theoretical Explorations in the Economics of Exhaustible Resources," in H. W. Gottinger, ed., *Systems Approaches and Environmental Problems* (Göttingen, Vandenhoek and Ruprecht, 1974) pp. 193–214.

——— and J. E. Stiglitz. "Uncertainty and the Rate of Extraction under Alternative Institutional Arrangements," Department of Economics, Stanford University, mimeo, 1975.

Herfindahl, O. C. "Depletion and Economic Theory," in Mason Gaffney, ed., *Extractive Resources and Taxation* (Madison, University of Wisconsin Press, 1964).

Hotelling, H. "The Economics of Exhaustible Resources," *Journal of Political Economy* vol. 39 (1931) pp. 137–175.

Nordhaus, W. "The Allocation of Energy Resources," *Brookings Papers on Economic Activity* no. 3 (Washington, D.C., Brookings Institution, 1973).

Pindyck, R. S. "Gains to Producers from the Cartelization of Exhaustible Resources," World Oil Project Working Paper, Massachusetts Institute of Technology, 1976.

Salant, S. W. "No End to the Age of Zinc: The Length of the Optimal Program when Depletion Affects Extraction Costs," International Finance Discussion Paper 73, Board of Governors, Federal Reserve System, 1976.

Solow, R. M. "Intergenerational Equity and Exhaustible Resources," *Review of Economic Studies*, Symposium on the Economics of Exhaustible Resources, 1974, pp. 29–45.

Stiglitz, J. E. "Tax Policy and the Oil Industry," report prepared for the Energy Policy Project sponsored by the Ford Foundation, partly published as chapter 3, "The Efficiency of Market Prices in Long Run Allocations in the Oil Industry," in G. Brannan, ed., "Studies in Energy Tax Policy" (Cambridge, Mass., Ballinger, 1975), pp. 55–100.

———, P. Dasgupta, R. Gilbert, G. Heal, and D. Newbery. "An Economic Analysis of the Conservation of Natural Resources," Federal Energy Administration Report (forthcoming).

Sweeney, J. "Economics of Depletable Resources: Market Forces and Intertemporal Bias," Department of Engineering Economics, Stanford University, mimeo, 1974.

[26]

Energy Vol. 6, pp. 187–195
© Pergamon Press Ltd 1981. Printed in Great Britain

0360-5442/81/0201-0187$02.00/0

ENERGY AND TECHNICAL PROGRESS

BRUCE HANNON and JOHN JOYCE

Energy Research Group, Office of Vice Chancellor for Research, University of Illinois at Champaign-Urbana, 130A Observatory, Urbana, IL 61801, U.S.A.

(*Received* 28 *January* 1980)

Abstract—When we use the Cobb–Douglas equation to investigate economic growth, we require data on the levels of the capital and labor inputs, and corresponding data on the level of output, at different points in time. We conclude that the inclusion of energy in the standard Cobb–Douglas production function does not explain the contribution of technological process (Hicks-neutral) to a rise in the rate of output unless, possibly, one is able to make the seemingly unreasonable assumption that labor should not appear in the production function.

An interesting future project will be a statistical analysis of the historic trend in GNP, a technological change and the consumption of available energy.

INTRODUCTION

Since 1925, when Cobb and Douglas introduced their aggregate production function,[1] economists have tried to explain the growth of the American economy by the growth of the factor inputs in such a production function. Historically, only the aggregate inputs of capital and labor were considered and the production function was written

$$Y = AK^{\alpha}L^{\beta}, \tag{1}$$

where Y is national income, A a constant of proportionality, K some measure of the capital input (usually the gross or net capital stock), L the labor input, and α and β the output elasticities, respectively, of capital and labor.

The aggregate production function, in Cobb–Douglas form [Eq. (1)]. along with consumption and investment functions etc., has long been used as one of the principal equations in macroeconomic models of the economy. In this paper, we will ignore demand. Equation (1) says that if the capital and labor inputs are given, output is uniquely determined.

When we use the Cobb–Douglas equation to investigate economic growth, we require data on the levels of the capital and labor inputs, and corresponding data on the level of output, at different points in time. In this way, we seek to explain changes in output by changes in inputs and to understand better what the exact relationship is. The problem is more complicated than it might appear at first glance. There is a two-fold data problem. On the one hand, we are aggregating many different types of capital inputs, labor inputs, and the diverse outputs of the economy. On the other hand, we must be aware of the fact that there are substantial qualitative changes in the form of these inputs and outputs as we go backward or forward in time. Both of these problems have been the subjects of books and articles.

The point here is not to discuss and resolve them thoroughly, which might not be possible in any case, but simply to mention them as examples of the data problems that plague all the work in this field.

The labor input is measured by person-hours of labor per year, and each hour of each worker is treated as equal to each hour of every other worker. This method does not take account of the qualitative changes in the makeup of the work force, which are many. The age and sex breakdowns vary over time and affect productivity, e.g. older workers are likely to be more productive than younger workers. We also neglect, by our method, the effect of "human capital" formation on productivity. "Human capital" is formed when workers forge present earnings to invest time in school or training that will enable them to obtain higher wages in the future. Over the period 1929–69, from which our data are taken, the average level of educational attainment was increasing, in general, in the U.S. We should suspect that this trend affects the relative value of an average input man-hour in different years.

The dollar value of the capital stock is usually used as a proxy for the capital input. Two series for the value of the aggregate capital stock are generally calculated by the statisticians: one consists of the purchase prices of all capital in service in year t, called the gross capital stock in year t; the other consists of the sum of the values of all the different pieces of capital in year t after allowance has been made for yearly depreciation and is called the net capital stock in year t. These values for total capital stock must be considered approximate at best. Constant dollars are used in order to try to put all the data on a uniform footing. The use of the aggregate capital stock as a proxy for the capital input is somewhat suspect, but defining what exactly does constitute the true capital input is a serious problem. If we want to be strict, production machinery can be considered an input only when it is actually in use. The walls of a building, on the other hand, provide a constant sheltering service to whatever is inside that building and should be evaluated differently. There are an infinite number of different type of capital goods that fall between the above two extremes; the problem of defining a truly satisfactory capital input number seems insoluble. If it were not impossible, the empirical data problem surely would be.

It is clear that some sweeping assumptions have to be made in the acceptance of some definition of the capital input. We must first distinguish between a change in the capital input arising from an addition to or retirement from the stock, and a change in the capital input owing to greater or lesser use of the existing stock. One assumption we will make is that the stock of capital is used evenly. We assume that each discrete unit comprising the total stock has a "normal" rate of utilization. This "normal" rate can be left undefined; it will be used only in a relative sense. If we assume the aggregate stock to be in a "normal" state of utilization, and then assume a 5% increase in the aggregate capital input, with no additions to the stock proper, we mean that each discrete unit of capital is being used at a rate 5% greater than "normal". Although this prescription is undoubtedly incorrect, it is at least a step toward clarifying what we mean by "capital input". We need not define a "normal" rate of usage since our method uses an index number for the data. We will assume that deviations from normal usage of the capital stock are reflected in labor force employment rates. While there is surely a correlation between the rate of employment of the labor force and that of the capital stock, there is no telling how great or small this correlation may be. Therefore, with tongue firmly in cheek, we multiplied our capital stock indices by the labor force employment rate to arrive at a capital input index. We should add that Solow applied a similar correction in his path-breaking work in this area.[2]

TECHNICAL PROGRESS

Past work has often assumed that the Cobb–Douglas aggregate production function exhibits constant returns to scale; i.e. it is homogenous of degree one. This means that the exponent, α and β, should sum to one. The advantage of this prescription, theoretically, is that the function is homothetic; the marginal rate of factor substitution remains unchanged when the input factor quantities are multiplied by any scalar. Theory tells us that the output elasticities of capital and labor [the exponents α and β in Eq. (1)] should be equal to the fractions of the national income earned by capital and labor, respectively. Over the last century, labor's share of national income has stood between 2/3 and 3/4, with capital holders receiving the remainder. In order to test how well our data from 1929 to 1966 fit the constant returns to scale and factor share assumptions, we regressed Eq. (1), in log-linear form (see the methodology section), without imposing any *a priori* constraints on α and β. Ordinary least-squares regression gave values of $\alpha = 0.234$ and

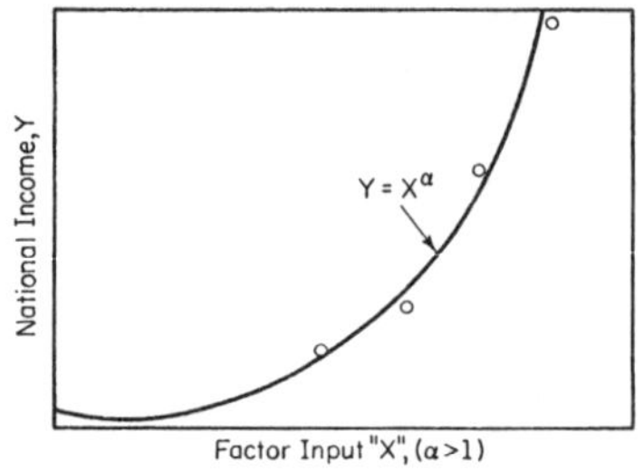

Fig. 1. Increasing returns to scale.

$\beta = 0.852$. These sum to 1.086, indicating increasing returns to scale, and puts the distributive shares at 0.785 for labor, 0.215 for capital.

Graphically, a production function with increasing returns to scale has an increasing slope. Figure 1 illustrates a one-factor production function with increasing returns to scale. Some hypothetical data points are included.

Equation (1) has no time dependence. Equal amounts of capital and labor in any two years should produce equal amounts of output. Over any span of years, the quality of capital goods and that of consumer goods and services, and of the labor force, changes considerably. Technical innovation and changing consumer preferences are constantly altering the forms of these goods and services so that they are not strictly comparable over time. As mentioned before, the qualities that describe the aggregate labor force are also changing over time so that not even the unit of labor input (i.e. the man-hour) will measure the same thing in diverse years. Taken together, these forces give rise to another almost insurmountable data problem.

This is one of the reasons for which theorists have suggested including some sort of time dependence in the aggregate production function. The usual method for achieving this is to add an exponential time term, of the form† e^{rt}, to the function. With this addition (and subscripts on the variables to indicate year 5), Eq. (1) becomes

$$Y_t = Ae^{rt}K_t^\alpha L_t^\beta. \tag{2}$$

The variable r is now interpreted as the rate of increase of technical progress, where technical progress is broadly defined to include anything that augments output without changing the quantity of capital or labor.

An exponential technical progress term has the effect of shifting the entire production function upward each year by a constant factor. Using the same data we used in regressing Eq. (1), where we found increasing returns to scale, we regressed Eq. (2), again in log-linear form. The results were $\alpha = 3.04$, $\beta = 0.692$, and $r = 0.0138$.

We now have $\alpha + \beta = 0.996$ (this was not statistically different from $\alpha + \beta = 1$) and a rate of increase of technical know-how of 1.38% per year.

Technical progress has always been difficult for economists to explain fully. It is not a physical quantity that can be measured explicitly, but it can be defined strictly as follows. Given two points in time t_0 and t_1, with t_1 later than t_0, there has been technical progress over the period $[t_0, t_1]$ if it is possible to produce a greater quantity of output at t_1 than was possible at t_0, with identical factor inputs at both times. Mathematically, we write a production function $y = F(\bar{x}, t)$ with $\bar{x}$ a vector of factor inputs, time coordinate 5 and physical output y. We have had technical progress from t_0 to t_1 if $y_1 = F(\bar{x}, t_1) > y_0 = F(\bar{x}, t_0)$. The chief discussion, in the past, has been over the question of the neutrality of technical progress. Was technical progress embodied in new capital,

†We also tried $Y_t = AtK_t^\alpha L_t^\beta$ and $Y_t = At^\gamma K_t^\alpha L_t^\beta$; these gave inferior statistical fits.

or in the changing makeup of the labor supply, or was it (Hicks) neutral, continuing to increase output without changing the actual capital stock or the labor force? Solow, in an early article on the subject,[2] found that technical progress was essentially neutral; more important, he attributed 7/8 of the growth rate in GNP to technical progress and only 1/8 to increases in the capital and labor inputs. This last result is extraordinary: economic growth cannot be well explained by increases in the capital stock and the labor force; we must instead resort to a mysterious concept that we call "technical progress"

In 1963, Nelson[3] investigated this problem, using data series running from 1929 to 1960. Table 1 is constructed from data taken from his paper.

Nelson tried to correct for variations in the level of business activity by appealing to the concepts of "potential GNP", and "potential man-hours". These items are defined and discussed by Okun in his 1962 paper.[4] Nelson did not regress his data to choose α and β, the capital and the labor output elasticities. Instead, he specified them as $\beta = 0.75$, which is about what we would expect from looking at the national income accounts, and $\beta = 0.5$, which he tries simply as a sensitivity test of the results. These assumptions, when combined with a constant returns to scale assumption, also specify α, and make the estimation of r a simple matter.

ENERGY

The central question which we tried to examine in this paper was the connection between energy consumption and technical progress. Fossil fuel energy has become totally indispensable to the modern industrial economy, as necessary for production as machinery or labor. In addition, the level of energy consumption has been increasing exponentially since the industrial revolution, and has totally altered the ways in which we live and work. Empirical work has indicated that, in the aggregate, energy may be a substitute for labor and capital in production. In the short run, aggregate energy and capital are complements, but in the long run they can substitute.[5]

Mineral reserves, including energy, are of course finite. What distinguishes fossil fuels from the other raw material resources is that they can be used but once. After a fuel is burned, it is gone forever, whereas metals can be reused over and over again, if enough energy is at hand to reforge them.

We do not think that previous investigators have handled the question of a connection between energy use and technical progress. Solow includes the value of mineral reserves in his capital stock, but this does not really measure the extent of the energy *input* to the economy. Nelson uses as his capital stock the value of all reproducible capital, which clearly excludes energy. The true fossil fuel input is that amount that is extracted and consumed each year. Both Nelson's and Solow's methods assume implicitly that fuels will always be available as an economic good. If this assumption is wrong, the theory should be altered accordingly. It may be that the earth's fossil fuel reserves will be able to supply

Table 1. Historical growth rates of output, labor force, capital, and technical progress; from Ref. 3.

YEAR	Potential GNP	Potential Man-Hours	Gross Capital Stock	Technical Progress r (β = .75)	Technical Progress r (β = .5)
1929 - 1960	3.1	0.7	2.0	2.1	1.7
1929 - 1947	2.5	0.5	1.0	1.9	1.7
1947 - 1960	4.0	0.8	3.6	2.5	1.8
1947 - 1954	4.4	0.7	4.0	2.9	2.1
1954 - 1960	3.5	0.8	3.1	2.1	1.6

us into the distant future. If this is in fact true, there is no quarrel with Solow's or Nelson's method.

Recently, there has been some conjecture that we really are in danger of running short of accessible fuels. It is only in this case that the previous work needs re-examination. The question of interest becomes the following: Does the rate of technical progress, r, change when fossil fuels are put into the production function? If there is a reduction in r when fossil fuels are included, we might conclude that a scarcity of fuels in the future might cause a reduction in the rate of technical progress.

In order to try to answer this question, we included an energy input explicitly in a Cobb–Douglas type production function. If increased energy consumption was behind some of what we call technical progress, the value of r, the technical progress term, should be less than its value in the non-energy-inclusive production function.†

METHODOLOGY

In order to run a regression, we must put our production function [Eq. (2)] into the proper form for econometric analysis. We include a random disturbance term, w_t, with mean one and finite variance, in our function, which becomes

$$Y_t = Ae^{rt}K_t^{\alpha}L_t^{\beta}w_t. \tag{3}$$

Taking logarithms gives

$$\ln Y_t = \ln A + r^t + \alpha \ln K_t + \beta \ln L_t + \ln w_t \tag{4}$$

We define $\ln w_t = u_t$ so that u_t has mean zero and finite variance, while

$$\ln Y_t = \ln A + r^t + \alpha \ln K_t + \beta \ln L_t + u_t \tag{5}$$

Equation (5) is in the proper form for least-squares regression. In several of our runs, we added another term such as $\gamma \ln E$ where E was energy input or $\gamma \ln$ [El] with [El] the electricity input. For all of our time series, we indexed the initial year's value to 100 before taking logarithms. The computer program that we used in all of our least-squares regressions gives estimates of the values of all the unknown coefficients, along with estimates of their variances. In addition, a t-statistic is calculated, which measures the probability that a given coefficient is different from zero. The r^2 statistic is a measure of how well the estimated model explains the observed data. All of the regressions gave values for "A" that were statistically indistinguishable from one. No more need be said here about "A".

THE ENERGY-INCLUSIVE MODELS

Two different methods for including energy in a production function were tried. The first method treats the energy flow into the economy as part of the total capital input. We created a time series with which we attempt to measure the absolute value of the yearly input of reproducible capital and added to it the value of energy consumed in each year. In constructing our time series for the input of physical capital alone (as we mentioned in the introduction), we avoided the problem of determining an absolute level of capital input. If we want to combine capital and energy into a single "value of capital/energy" input this is no longer possible. We decided to use the amount of depreciation in the physical capital stock as an approximation of the absolute level of the physical capital input. This seems reasonable; physical depreciation is in some sense the actual input from the capital stock, and should vary in rough proportion to the amount of use that the stock undergoes.

†Berndt and Wood[11] assumed a translog cost function and determined the factor cost shares the tendency for substitution or complimentarity between capital, labor, energy, and materials from 1947 to 1971.

In constructing this series, we combined Denison's data [6] for the net value of the reproducible capital stock with data on yearly investment as given in the appropriate columns in Ref. 7. The method will now be described. Defining K_t and K_{t+1} as the values of the net capital stock in the years t and $t + 1$ and I_t as the amount of investment in year t, the depreciation D_t is given by

$$D_t = K_t + I_t - K_{t+1}. \tag{6}$$

The ratio D_{t/K_t} was calculated for each of the years 1947–1964. The mean value of the ratio is 0.097 with a surprisingly small standard error (0.007). We therefore multiplied our "value of net capital stock" time series by 0.097 to form our capital input series.

Data on the value of all energy produced, i.e. the value of all fossil energy at the mine mouth or the well-head (plus value of net imports) are given in Ref. 8. Adding the value of energy and the value of the capital input gives a new time series that we can use to investigate the effect of energy on technical progress. Our new production function (again with random error term w_t) takes the form

$$Y_t = Ae^{rt}(K + E)_t^{\alpha} L_t^{\beta} w_t. \tag{7}$$

If increased energy consumption had been understating the input of capital to the economy, the estimated value of "r" in Eq. (7) should be lower than that found in Eq. (2).

The second method we used was the addition of an independent energy term to the two-factor capital–labor–energy production function. This is written as follows:

$$Y_t = Ae^{rt}K_t^{\alpha} L_t^{\beta} E_t^{\gamma} w_t. \tag{8}$$

The E_t time series that we used measure the quantity of energy consumption, in physical units. We also entertained the suspicion that increasing use of electricity might be more closely connected to technical progress than energy *per se*, so we tested the following capital–labor–electricity production function:

$$Y_t = Ae^{rt}K_t^{\alpha} L_t^{\beta} [El]_t^{\gamma} w_t. \tag{9}$$

The $[El]_t^{\gamma}$ time series is also in physical units and was indexed to 1929 = 100 (as all our data were), before logarithms were taken.

DATA

We required time series data on GNP, labor input, capital stock, total energy consumption (physical units and dollars), investment, and electricity consumption. In order to avoid the vagaries introduced by the war years and the depression, we used the years 1929, 1940, 1941, 1947–69. Table 2 displays our data sources.

The data were indexed to 1929 = 100 and logarithms taken to put them into suitable regression form. Regression data are displayed in Table 3.

Table 2. Data sources.

GNP	Allen[9]
LABOR	Allen[9]
CAPITAL STOCK	Denison[6]
GROSS ENERGY	Sonenblum[8]
VALUE OF ENERGY	Sonenblum[8]
GROSS ELECTRICITY	Edison Electric Institute[10]
INVESTMENT	Survey of Current Business[11]

Table 3. Regression input data; all data (except time) are logarithms of indices with 1929 = 100.

Year	Time	GNP	Labor	Capital	Energy	Energy + capital	Electricity
1929	1.0	4.605	4.605	4.605	4.605	4.605	4.605
1940	12.0	4.715	4.670	4.317	4.609	4.398	5.067
1941	13.0	4.864	4.786	4.407	4.708	4.503	5.230
1947	19.0	5.025	4.885	4.555	4.880	4.698	5.672
1948	20.0	5.069	4.908	4.633	4.905	4.826	5.772
1949	21.0	5.070	4.881	4.662	4.832	4.805	5.805
1950	22.0	5.162	4.919	4.728	4.908	4.881	5.926
1951	23.0	5.238	4.999	4.797	4.987	4.939	6.050
1952	24.0	5.268	5.021	4.842	4.979	4.971	6.125
1953	25.0	5.312	5.039	4.882	5.009	5.017	6.241
1954	26.0	5.298	5.006	4.890	4.973	5.012	6.306
1955	27.0	5.371	5.030	4.939	5.064	5.061	6.463
1956	28.0	5.390	5.053	4.986	5.113	5.105	6.562
1957	29.0	5.404	5.057	5.026	5.113	5.145	6.614
1958	30.0	5.392	5.028	5.030	5.113	5.081	6.633
1959	31.0	5.454	5.053	5.066	5.147	5.161	6.730
1960	32.0	5.479	5.070	5.094	5.180	5.183	6.816
1961	33.0	5.498	5.066	5.108	5.196	5.179	6.870
1962	34.0	5.634	5.094	5.146	5.241	5.233	6.944
1963	35.0	5.601	5.107	5.175	5.280	5.256	7.013
1964	36.0	5.654	5.129	5.216	5.319	5.285	7.084
1965	37.0	5.715	5.166	5.270	5.359	5.309	7.151
1966	38.0	5.779	5.215	5.335	5.415	5.388	7.237
1967	39.0	5.804	5.242	5.392	5.447	5.438	7.301
1968	40.0	5.850	5.270	5.444	5.506	5.481	7.384
1969	41.0	5.876	5.298	5.495	5.556	5.530	7.466

RESULTS

The results of our least-squares estimations are shown in Table 4. Standard errors are displayed beneath the estimates of the coefficients.

Before we draw any conclusions from these results, an important fact about least squares regression must be noted. When we apply least squares to a model, we have already assumed that the model has been specified correctly. Errors in estimation (biased estimators) result when a relevant variable has been excluded. The addition of extraneous variables will not bias the estimators, or their estimated variances, provided the set of data points is infinite. If we have a finite set of data points and have included extraneous variables in our model, the variances of the coefficients of the variables that are relevant will be increased and the estimators for the irrelevant variables should be zero.

Our results cannot tell us which one of the models "*a*" through "*i*" of Table 4 is the correct one, but if one of them happens to be correct, the table gives the unbiased least-squares estimators. If we can narrow down the range of possible models to include only "*b*", "*c*", and "*d*", we can see that, no matter which one is correct technical progress increases at a rate of 1.3% per year. We can say that, whether or not energy consumption was a determinant of output, the rate of technical progress has had this constant value over the period 1929–69.

If capital and labor inputs are taken as given, a scarcity of fossil-fuels inputs will not stop our production possibilities from increasing at 1.3% per year, if any of these three models is correct. If "*c*" or "*d*" is correct, a scarcity of fuel will limit the maximum production that we can attain but the production function as a whole will continue its steady upward shift.

Table 4. Results of regressions; standard errors are given in parentheses.

Equation form	Values of the Coefficient				r^2 (statistic)
	r	α	β	γ	
a) $Y = AK^{\alpha}L^{\beta}$	-	0.234 (0.098)	0.852 (0.096)	--	0.99495
b) $Y = Ae^{rt}K^{\alpha}L^{\beta}$	0.0138 (0.0004)	0.304 (0.045)	0.692 (0.041)	--	0.99958
c) $Y = Ae^{rt}(K+E)^{\alpha}L^{\beta}$	0.0133 (0.0012)	1.383 (0.072)	0.610 (0.068)	--	0.99943
d) $Y = Ae^{rt}K^{\alpha}L^{\beta}E^{\gamma}$	0.0130 (0.0010)	0.224 (0.070)	0.571 (0.093)	0.202 (0.139)	0.99962
e) $Y = Ae^{rt}K^{\alpha}L^{\beta}[El]^{\gamma}$	0.0131 (0.0013)	0.300 (0.045)	0.686 (0.042)	0.001 (0.002)	0.99959
f) $Y = Ae^{rt}K^{\alpha}E^{\gamma}$	0.0084 (0.0011)	0.031 (0.102)	--	0.976 (0.097)	0.99895
g) $Y = Ae^{rt}L^{\beta}E^{\gamma}$	0.0130 (0.0012)	--	0.440 (0.098)	0.550 (0.102)	0.99944
h) $Y = Ae^{rt}L^{\beta}[El]^{\gamma}$	0.0168 (0.0021)	--	0.948 (0.025)	0.002 (0.003)	0.99877
i) $Y = Ae^{re}K^{\alpha}[El]^{\gamma}$	-0.0002 (0.0037)	0.990 (0.057)	--	0.007 (0.006)	0.99464

As an extreme condition, we eliminated labor from the model and found that a capital-energy and especially a capital-electricity model (*f* and *i*) allowed an accurate estimate of production without a term for technical progress (this was not true for a labor-energy model). The standard error on some of the exponents was quite large, however. These results sharply point up the deficiencies of the production function approach to distributing shares of production. We have only personal judgment as to which variables should appear in the function.

We conclude that the inclusion of energy in the standard Cobb–Douglas production function does not explain the contribution of technological progress (Hicks-neutral) to a rise in the rate of output, unless, possibly, one is able to make the seemingly unreasonable assumption that labor should not appear in the production function.

The results would seem to imply that the economy exhibits constant or slightly decreasing returns to scale (mean returns to scale = 0.990, standard deviation = 0.0159). We should emphasize here that the data refer to the years 1929, 1940, 1941, and 1947–69. It is possible that extension of these data into the past or the years after 1969 could alter these results. Data on the post-oil embargo years would be interesting but it seems very unlikely that any change that showed up in the results would be significant, considering the weight of the evidence presented here.

An alternative view of the production process is possible which contains more physical meaning. We could consider labor and capital as intermediate products and energy (more broadly, availability) as the single contributing factor to production. Availability is the measure of an entity to do mechanical work. To do work, it is only necessary that this entity possess a greater degree of order than its environment. Thus, of course, coal, oil, gas, and uranium possess a calculable amount of availability and so does iron ore, geothermal energy, solar radiation, rain, wind, etc. In any time period, these physical factors combine under the guidance of man (technology) and nature to allow the production of labor and capital and subsequently the production of our desired goods and services. Availability occurs in two forms. A flow of sunlight and geothermal energy and a stock of fossil and nuclear fuels and a stock of non-energy minerals such as iron ore, bauxite, etc. There is also an enormous amount of order stored in the earth, water and air, but their availability is very low on the average.

Joyce estimated the availability in the non-fuel minerals mined in 1968 at about 1.0 quads. A very rough estimate of the photosynthesized solar energy harvested in this year is 5.0 quads, while fossil and nuclear fuels and hydropower consumed was about 70 quads. The dominant consumption of availability was clearly from the stock. An interesting future project will be a statistical analysis of the historic trend in GNP, technological change and the consumption of availability. If such a view of the macro economy has meaning then to the extent that availability flows can be substituted for stocks, the output of the economy can be preserved.

Acknowledgements—We wish to thank Andrew Postlewaite, Department of Economics, Princeton, and Robert Solow, Department of Economics, Massachusetts Institute of Technology, for their many useful comments.

REFERENCES

1. C. W. Cobb and P. H. Douglas, *Am. Econ. Rev.* **18,** 139 (1928).
2. R. M. Solow, *Rev. Econ. Statist.* **39,** 312 (1957).
3. R. R. Nelson, *Am. Econ. Rev.* **54,** 575 (1964).
4. A. Okun, "Potential GNP: Its Measurement and Significance", presented at *Ann. Meeting of the American Statistical Association*, Minneapolis, Minnesota, 7–10 Sept. 1962. Proceedings of the Business and Economics Statistics Section, 98 (1962).
5. J. M. Griffin and P. R. Gregory, *Am. Econ. Rev.* **66,** 845 (1976).
6. E. F. Dension, *Accounting for U.S. Economic Growth 1929–69*, Brookings Institute, 1775 Mass. Ave., N.W., Washington, DC 20000 (1974).
7. "Survey of Current Business", Office of Business Economics, U.S. Dept. of Commerce, Washington, DC (Aug. 1966).
8. S. Sonenblum, "Energy and Economic Growth", Report to the State of California, Energy Resources Commission, Sacramento, California (1978).
9. E. Allen, "U.S. Energy and Economic Growth", Institute for Energy Analysis, Oak Ridge, Tennessee (Sept. 1976).
10. Edison Electric Institute, "Historical Statistics, 1972", Washington, D.C. (1973).
11. E. R. Berndt and D. O. Wood, *Rev. Econ. Statist.* **LVII,** 259 (1975).

[27]

PROCEEDINGS OF THE WORLD BANK ANNUAL CONFERENCE ON DEVELOPMENT ECONOMICS 1990

The Environment and Emerging Development Issues

Partha Dasgupta and Karl-Göran Mäler

Development economics for the most part has failed to recognize renewable environmental resources as economic goods. Yet the lives of the poor in developing countries are intimately tied to the fate of their local common-property resources—soil, fuelwood, water, and so on. Ignoring the loss of common-property renewable resources (via degradation, centralization, or privatization) in calculations of net national product can mask the destruction of resources available to a country's poor behind the false appearance of a growing national economy. This paper describes the difference between the problem of managing local common-property resources in developing countries and the problem of protecting the global commons such as the atmosphere and oceans. It outlines the calculations that would include the local commons in the national income accounting of capital assets and argues that the interests of those most dependent on the increasingly scarce local commons will be best served by placing control in their hands, while the central government provides infrastructure and educational support. To control the pollution of the global commons, on the other hand, the paper sketches a plan in which most countries would have a high incentive to participate—a global allocation of limited, tradable permits to discharge defined pollutants.

Environmental resources are of minor importance to poor countries. . . . They play an insignificant role in the process of economic development. . . . Such resources are luxury goods, and they loom large in public consciousness only when incomes are high. . . . Environmental resources are only a rich country's preoccupations. . . . They are a mere diversion created by economists not sensitive to the true needs of the poor in poor countries. . . .

These sentences will seem at once strange and recognizable. They will seem

Partha Dasgupta is professor of economics and philosophy at Stanford University, professor of economics at the University of Cambridge, and a research adviser to the World Institute for Development Economics Research (WIDER), Helsinki. Karl-Göran Mäler is professor of economics at the Stockholm School of Economics and a research adviser to WIDER. This study forms part of a larger work currently under way at WIDER. The authors thank Lawrence Goulder, Frank Hahn, James Mirrlees, Mohan Munasinghe, Henry Peskin, Robert Repetto, Shekhar Shah, Robert Solow, Joseph Stiglitz, Martin Weale, and John Whalley for their comments.

strange because today we all like to believe that the ideas they express are not true. At the same time, they will be recognizable to anyone who has delved into the literature on development economics and looked for environmental issues in it. Even were we to search through the vast body of writings on development planning, not to mention the literature on investment criteria, we would be hard put to discover any sign of environmental resources. Academic development economics does not yet acknowledge their existence, in that one will not find them in any recognized survey article, text, or treatise on the subject.

To cite only a few instances, Drèze and Stern (1987) and Stern (1989) are surveys of cost-benefit analysis and development economics, respectively. The former, a ninety-page article, contains precisely one sentence on the subject of nonrenewable and renewable resources (and it is to tell readers where to go if they wish to learn about such matters); the latter, an eighty-eight-page article, also contains a single sentence (and this also tells readers where to go should they wish to learn about such matters). A third example is provided by the two-volume *Handbook of Development Economics* (Chenery and Srinivasan, eds., 1988), which simply has no discussion of environmental resources and their possible bearing on development processes. The disconnection of academic development economics from a central aspect of the lives of people about whom development economics revolves would appear to be pretty much complete. Environmental resources appear in this literature about as frequently as rain falls on the Sahara.[1]

Despite this neglect, something like a literature on the environment and economic development is now emerging.[2] In this paper we therefore try to place this small literature in perspective and develop what appears to us to be some of the themes with the greatest research potential—that is, those which also have implications for policy in poor countries. This will enable us also to apply economic theory to obtain new insights into the allocation and management of environmental resources. We want to do this by providing a framework of thought which will encourage the reader to view environmental resources as *economic goods*. This will no doubt seem a banal intention, but it is remarkable how much of current writing on the environment displays an innocence of this point of view. Our approach is based on the conjecture that until environmental

1. Yet, environmental economics as an autonomous subject has developed rapidly in recent years. Much work has been done both at the analytical and empirical levels on the valuation of environmental resources (see, for example, Johansson 1987, 1990) and on the identification of appropriate environmental policy instruments (see, for example, Tietenberg 1990). However (and this is the point we are making in the text), both these sets of questions have been discussed in the context of western industrial democracies. We will confirm later in this article that the context matters very much in the field of environmental economics.

2. The incorporation of environmental resources into social cost-benefit analysis for development planning was attempted in Dasgupta (1982). Earlier treatises on social cost-benefit analysis in poor countries, such as Little and Mirrlees (1969, 1974) and Dasgupta, Marglin, and Sen (1972), contain no discussion of these issues. Readers wishing to study environmental economics should consult Ulph (1989) for a review of a number of texts and treatises written over the past fifteen years or so.

resources become a commonplace furniture of economic thinking and modeling, they will continue to be neglected in the design of policy and in the implementation of policy. Periodic "affirmative actions" on the environment is not the right way of going about things.

Now there are several routes we can follow for the purpose of introducing our subject. Here, we pursue one which is perhaps less popular today than it used to be: it is to seek a unifying principle concerning the "technology of production" of environmental goods and services. In other words, we start by viewing the matter from the purely *ecological* side of things. Later, we view the subject from the *institutional* end. These are complementary avenues of inquiry. Both need to be covered, but it is pretty much a matter of indifference with which one we begin.

I. Environmental Resources and Their Physical Characteristics

Environmental problems are almost always associated with resources that are regenerative or renewable but that nonetheless are in danger of exhaustion from excessive use.[3] The earth's atmosphere is a paradigm of such resources. That is, under normal courses of events the atmosphere's composition regenerates itself. But the speed of regeneration depends upon the rate at which pollutants are deposited into it, and it depends upon the nature of the pollutants (smoke discharge is clearly different from the release of radioactive material). Now, in talking of a resource we need, first of all, a way of measuring it. For the case in hand we have to think of an atmospheric quality index. The net rate of regeneration of the stock is the rate at which this quality index changes over time. This will depend upon the nature and extent of the pollutants which are discharged, and it will also depend upon the current index of quality; that is, the current level of the stock. These are immensely complex, ill-understood matters. There is a great deal of synergism associated with the interaction of different types of pollutants in the atmospheric sink (see Ehrlich, Ehrlich, and Holdren 1977). But the analytical point we are making here remains valid.

Animal, bird, plant, and fish populations are also typical examples of renewable natural resources, and there are now a number of studies which address the reproductive behavior of different species under a wide variety of environmental conditions, including the presence of parasitic and symbiotic neighbors. Land is also such a commodity, for the quality of arable and grazing land can be maintained by careful use. Overuse, however, impoverishes the soil and eventually produces a wasteland. (The symbiotic relation between soil quality and vegetation cover is, of course, at the heart of the current anxiety over Sub-Saharan erosion.)

3. Except for some remarks in section II, we are ignoring nonrenewable resources here for reasons of space. For an account of what resource allocation theory looks like when we include exhaustible resources in the production process, see Dasgupta and Heal (1979).

Underground basins of water often have a similar characteristic. The required analysis, however, is a bit more problematic in that we are concerned both about its quality *and* quantity. Under normal circumstances an aquifer undergoes a self-cleansing process as pollutants are deposited into it. (Here, the symbiotic role of different forms of bacteria, as in the case of soil and the atmosphere, is important.) But the effectiveness of the process depends, as always, on the nature of pollutants (for example, the type of chemicals) and the rate at which they are discharged. Furthermore, many aquifers are recharged over the annual cycle. But if the rate of extraction exceeds the rate of recharge, the water table drops, thereby raising extraction costs. In fact, for coastal aquifers the issue is not only one of depositing pollutants; excessive extraction alone may cause the groundwater table to drop too far, and saltwater intrusion can then destroy the basin. (For instances of these occurrences, see Dasgupta 1982 and Center for Science and Environment 1982, 1985.)

These examples suggest that a number of issues in environmental economics are within the subject matter of capital theory (this is very much the spirit of the exposition in Clark 1976). But there are added complications, among them the fact that the impact on the rate of regeneration of environmental resources of a wide variety of investment decisions is not fully reversible, and in some cases quite irreversible. (As always, one should not be literal. A very slow rate of regeneration produces a strong flavor of irreversibility.) In this limited sense, issues concerning what is usually labeled "pollution" can be studied in the same general sort of way as those concerning animal, bird, plant, and fish populations, aquifers, forests, and soil quality. This provides us with a unified way of thinking about matters. It allows us to use insights we draw from a study of one class of environmental resources when studying another. It forces us to pay attention to the intertemporal structure of economic policies.

If this were all, life would be relatively simple. But it is not all. We argue presently that admitting environmental resources into economic modeling ushers in a number of additional, potent complications for development policy. They arise out of the fact that for poor people in poor countries, such resources are often *complementary* to other goods and services. So an erosion of the environmental resource base can make certain categories of people destitute even while an economy on average is seen to be growing.

There is thus an intellectual tension between aggregate concerns (for example, the greenhouse effect, or the appropriate mix of natural resources and manufactured capital in aggregate production) that sweep across regions, nations, and continents, and concerns (for example, the decline in firewood or water availability) that are specific to the needs and concerns of poor people from as small a group as a village community. This tension should be borne in mind. Environmental problems present themselves in the form of different images to different people. At this stage of our understanding it would be a wrong research strategy to try to put them all together into one large basket. In the following section we ignore the detailed, institutional features which surround the use of *local* environmental resources in favor of studying the intertemporal nature of environmental policy in terms of aggregate considerations. This allows us to discuss in a simple manner the link between measures of net national product (NNP) and rules for social cost-benefit analysis of investment activity. As a by-product, it enables us to talk briefly about a matter much discussed today—sustainable development—that a decade and a half ago was shown to be the implication of a wider, richer inquiry going under a different name—optimal development. We will pick up the institutional trail in section III.

II. Determinants and Constituents of Well-Being and Net National Product

It is useful to remember that the kinds of resources we are thinking of here are, on occasion, of direct use in consumption (as with fisheries), on occasion in production (as with plankton, which serves as food for fish species), and sometimes in both (as with drinking and irrigation water). Their stocks, as we noted earlier, are measured in different ways, depending on the resource: in mass units (such as biomass units for forests, cow dung, and crop residues); in quality indexes (such as water and air quality indexes); in volume units (such as acre-feet for aquifers); and so on. When we express concern about environmental matters we in effect point to a decline in their stock. But a decline in their stock *on its own* is not a reason for concern. This is seen most clearly in the context of exhaustible resources, such as fossil fuels. To not reduce their stocks is to not use them at all, and this is unlikely to be the right thing to do.

To be sure, this is not the case with renewable natural resources, since we could in principle limit their extraction and use to their natural regeneration rate and thus not allow their stocks to decline. But this too may well be the wrong thing to do, in that there is nothing sacrosanct about the stock *levels* we have inherited from the past. Whether or not policy should be directed at expanding environmental resource bases is something we should try to *deduce* from considerations of population change, intergenerational well-being, technological possibilities, environmental regeneration rates, and the existing resource base. The answer cannot be pulled from the air.

That it can is a common misconception. Here is a recent example of this: "We can summarize the necessary conditions for sustainable development as constancy of the natural capital stock; more strictly, the requirement for nonnegative changes in the stock of natural resources, such as soil and soil quality, ground and surface water and their quality, land biomass, water biomass, and the waste-assimilation capacity of the receiving environments" (Pearce, Barbier, and Markandya 1988, p. 6).

The passage involves a "category mistake," the mistake being to confuse the *determinants* of well-being (for example, the means of production) with the *constituents* of well-being (for example, health, welfare, and freedoms). But leaving that aside, the point is not that sustainable development, even as it is defined by these authors, is an undesirable goal. It is that thus defined it has negligible information content. (We are not told, for example, what stock levels we ought to aim at.) This is the price that has to be paid for talking in terms of grand strategies. The hard work comes when one is forced to do the ecology and the economics of the matter.

So then, information about stocks on their own is not a sufficient statistic for well-being. What we are after are present and future well-being and methods of determining how well-being is affected by policy. And it is not an accident that the index which, when properly computed, can be used toward this end is net national product.[4]

Defining Net National Product

Net national product is held in ill repute today. (In what follows, we will be concerned only with the real value of NNP.) It is often thought that it is even in principle incapable of reflecting aggregate intergenerational well-being. This belief is not correct. Subject to certain technical restrictions, for any conception of the social good, and for any set of technological, transactional, informational, and ecological constraints, there exists a set of shadow (or accounting) prices of goods and services. These prices, when used in the estimation of NNP, ensure that small projects relative to the size of the economy that increase this index are at once also those that increase the social good.[5] A great many of these shadow prices will be person- or household-specific. Put more generally, they will be agent-relative, to reflect the fact that needs and incomes differ along gender, age, caste, or ethnic lines, and that many of the exchanges which take place in poor agrarian and pastoral societies are not anonymous; that is, such exchanges take place in personalized, or named markets.[6]

These considerations bring out the intellectual tension we spoke of earlier. The point is that in practice it is impossible to take all this into account in one

4. The theory of optimal development in economies with environmental resources was developed over fifteen years ago, and it produced a series of sharp prescriptions, much sharper than the recent literature on sustainable development. Implied in the theory of optimal development were rules for social cost-benefit analysis of investment projects and for estimating net national product. See, for example, Mäler (1974), Solow (1974), Dasgupta and Heal (1979), Dasgupta (1982), and Lind (1982).

5. The technical restrictions amount to the requirement that both the set of feasible allocations and the social ordering reflecting the social good are convex. The assumption of convexity is dubious for pollution problems, as was illustrated in Starrett's classic article (see Starrett 1972). Nevertheless, in a wide range of circumstances it is possible to separate out the "nonconvex" sector, estimate net product for the "convex" sector, and present an estimate of the desired index as a combination of the net product of the convex sector and estimates of stocks and their changes in the nonconvex sectors. This is a simple inference from the work of Weitzman (1970) and Portes (1971).

6. We are borrowing the latter terminology from Hahn (1971). As an example of a named market, consider the Hindu *jajmani* system, which supports a set of mutual economic obligations between landowners (peasant masters) and low-caste families. (See, for example, Wiser 1936, Epstein 1967, and Bowes 1978.) We are not suggesting that such personalized markets are efficient, even when we take account of existing transaction and informational costs. Nor are we suggesting that they lead to equitable allocations. The outcomes are often the result of implicit bargaining among people with vastly different bargaining strengths. See below in the text. See also Dasgupta (1990b) for further discussion.

grand measure of real NNP. The thing to do is to improve on our current ways of doing things and capture these concerns in a piecemeal fashion. This is the approach we will take here.

Current estimates of NNP are based on biased sets of prices. They are biased in that the prices that are imputed to environmental resources in situ are usually zero, even though we know that their accounting prices are positive. Thus the social profits of projects which degrade the environment are lower than the profits that are actually imputed to them. This means, in turn, that wrong sets of projects are chosen. In particular, resource-intensive projects look better than they in fact are.

One can go further. The bias would be expected to extend itself to the prior stage of research and development. When environmental resources are underpriced, there is little incentive for agencies to develop technologies which economize on their use. The extent of the distortion created by this underpricing will vary from country to country. Quite obviously, poor countries cannot compete with rich ones in the field of basic research. They have to rely on the flow of new knowledge produced in advanced industrial economies. Despite this, poor countries need to have the capability for basic research. This is because the structure of their shadow prices is likely to be quite different, most especially for non-traded goods and services. Even when it is publicly available, basic knowledge is not necessarily *usable* by scientists and technicians unless they themselves have a feel for basic research. Basic knowledge should not simply be taken off the shelf. Ideas developed in foreign lands ought not merely be transplanted to the local economy. They need to be adapted to suit local ecological conditions.

This is where estimating shadow prices (and more generally, creating incentives which correspond to these shadow prices) can be of considerable help. Adaptation is itself a creative exercise. Unhappily, it is often bypassed. The foreign technology is simply purchased and installed with little modification. There is great loss in this.

This leads to the question of how we should value environmental resources in situ; or in other words, what their shadow prices are in situ. It also leads to the question of how we should measure NNP once we have estimated shadow prices. As it happens, this latter question has a precise answer. How we should define NNP is therefore not a matter of opinion; it is a matter of fact.

The theory of intertemporal planning tells us to choose current controls (for example, current consumptions and the mix of current investments) in such a way as to maximize the current value Hamiltonian of the planning problem. As is now well-known, the current value Hamiltonian is the sum of the flow of current well-being and the shadow value of all the net investments currently being undertaken.[7] (Thus, the planning exercise generates the entire set of intertemporal shadow prices.) It is possible to show that the current value Hamilto-

7. The current value Hamiltonian will in general also contain terms reflecting the social cost of breaking any additional constraint that happens to characterize the optimization model. But we are avoiding such technicalities in the text.

nian measures the return on the *wealth* of an economy, measured in terms of well-being (see the appendix). The theory also tells us that if the objective of planning (that is, the intergenerational index of well-being) is stationary, then along the optimal program the Hamiltonian remains constant through time. (In saying this we are suitably normalizing the relevant variables. The best economic treatment of all this is still Arrow and Kurz 1970.)

This provides us with the necessary connection between the current value Hamiltonian and NNP. NNP is merely a linearized version of the current value Hamiltonian, the linearization amounting to representing the current flow of well-being by the shadow value of all the determinants of current well-being. In the simplest of cases, where current well-being depends solely on current consumption, NNP is reduced to the sum of the social (or shadow) value of an economy's consumption and the social (or shadow) value of the changes in its stocks of real capital assets.

The Hamiltonian calculus in fact implies something more. It implies also that along an optimal program NNP at any date is the maximum consumption the economy can enjoy at that date, were it made a condition that consumption be never allowed to fall. In short, NNP at any date along an optimal program is the maximum sustainable consumption at that date. This was not seen immediately as an implication of the mathematical theory of programming, although it should have been transparent from the work of Arrow and Kurz (1970) and Solow (1974). So a proof was provided by Weitzman (1976) in the context of economies which are capable of sustaining a steady economic state, and it was extended to the context of economies with exhaustible resources in Dasgupta and Heal (1979, pp. 244–46; see also Solow 1986).

Now "changes in capital stocks" means changes in *all* capital stocks—manufactured capital as well as natural resources. As of now, we know of no country which tries to account for environmental resource changes in its national income accounting. As we noted earlier, the effect on estimated growth rates could be significant were countries to do this, since it is entirely possible that growth in per capita GNP is currently being achieved in a number of poor countries via a deterioration of the environmental resource base. Growth in NNP over the recent past may well have been negative in a number of countries which are judged to have enjoyed positive growth.[8]

To understand how this can be, consider as an extreme hypothetical case a

8. See the calculations in Repetto and others (1989). Suggestions that conventional measures of NNP should be augmented by measures of *net nature product* (see the interesting work of Agarwal and Narain 1989) are prompted by this concern and are perfectly consistent with what we are saying in the text. By net nature product, Agarwal and Narain mean a measure of net changes in the environmental resource base. The correct measure of NNP is, of course, the sum of conventional NNP and net nature product. It is this aggregate which we are elaborating upon in the text. For tactical reasons it may well be desirable to present valuations of changes in the environmental resource base separately from conventional NNP figures. But we are concerned in the text with analytical issues. For this reason we are discussing an overall measure of NNP.

country which lives ("optimally") simply by exporting its exhaustible resources and importing consumption goods with the export revenue. Assume, too, that resources are costless to extract and that other than extraction there is no domestic production. NNP in this country, when correctly measured, would be nil. No matter that its current consumption may be high, the social value of the rate of disinvestment would exactly match the shadow value of its consumption rate. Furthermore, being exhaustible, exports must eventually go to zero, and so consumption must eventually go to zero. Such an economy cannot manage a positive, sustainable consumption path. It is this fact which is reflected in NNP, when correctly measured. We would not, however, know any of this from conventional national income accounts. NNP in this economy, as conventionally measured, could be large, and it would even be seen to be rising if the rate of disinvestment were stepped up (see Dasgupta and Heal 1979, pp. 244–46).

An important practical question (much discussed in the recent literature; see El Serafy and Lutz 1989) concerns the appropriate treatment in NNP of expenditure for protecting the environment. The answer is that it all depends upon whether such expenditures are for the purposes of redressing a flow of environmental damage or for enhancing stocks of environmental resources. To take an example, expenditures for liming lakes to counter the flow of acid rains maintain environmental quality. Were this not to be included in final demand, increased liming (which increases well-being) would be recorded as a decline in NNP. Similarly, capital expenditures (for example, the construction of stack-gas scrubbers, sewage treatment plants, and solar energy equipment) should also not be deducted from estimates of NNP, for they go to augment the durable-capital base. On the other hand, those expenditures which go to enhance resource bases, such as forests, get reflected in the value of changes in resource stocks. In order to avoid double-counting, such expenditures therefore should be excluded from NNP computations.

The general moral is that before writing down an expression for NNP it is best to express in a formal model the characteristics of the activities being undertaken in the economy under study. What passes for intermediate expenditure and what does not will be reflected in the Hamiltonian, and one can then readily construct the formula for the country's NNP, and one can simultaneously obtain the system of shadow prices. The procedure is a shade pedantic. But it provides a safeguard against mistakes. In the appendix we provide a formal account of this, and for completeness we sketch the manner in which future uncertainty can be taken into account.

Difficulties associated with the estimation of shadow prices and real NNP are compounded by the fact that unlike computers and tractors, environmental resources often affect current well-being directly as stocks, and not merely as service flows (an exception is noise pollution). Fisheries and aquifers are useful not only for the harvest they provide (this is the flow) but also as stocks, because harvesting and extraction costs are low if stocks are large. Tropical forests are beneficial not only because they may supply timber (this is the flow of service);

they are beneficial also as stocks, because they prevent soil erosion and, in the case of large tropical forests, help maintain a varied genetic pool, and contribute substantially to the recycling of carbon dioxide.

To take another example, transportation costs (in particular caloric costs) for women and children would be less were the sources of firewood not far away and receding. Likewise, air and water quality have direct effects on well-being; it is, let us remember, the concentration of pollutants which is relevant here. This has an implication for the current value Hamiltonian, and therefore for real NNP (see the appendix).

Estimating Shadow Prices

The prior question, of how we should estimate shadow prices for environmental resources, is a complex one. But it is not uniformly complex. For commodities such as irrigation water, fisheries, and agricultural soil, there are now standard techniques of evaluation. These techniques rely on the fact that such resources are inputs in the production of tradable goods (see, for example, Brown and McGuire 1967 for irrigation water; Cooper 1975, Clark 1976, and Dasgupta 1982 for fisheries; and Repetto and others 1989 for soil fertility). For others, such as firewood, and drinking and cooking water, the matter is more complex. The fact remains, though, that even these are inputs in production; specifically, inputs in household production. And this provides us with a method for estimating their shadow prices. To obtain them we will need to have an estimate of household production functions. In some cases (as with fuelwood) the resource input is a substitute for a tradable input (for example, kerosene); in others (as with cooking water) it is a complement (sometimes a weak complement) to tradable inputs (for example, food grain). Such information is in principle obtainable. It allows the analyst to estimate shadow prices. (See Schecter and others 1989 and Mäler 1990c. We should note, however, that applied work in this area has shied away from analyzing production in poor households in poor countries. There is an urgent need for research in this field. See section IV below.)

The approach we have outlined above (the "production function approach") allows one to capture only the "use value" of a resource, and its shadow price may well exceed this. Why? The reason is that there are additional values embodied in a resource stock. One additional value, applicable to living resources, is their "intrinsic worth" as living resources. (We plainly do not think that the value of a blue whale is embodied entirely in its flesh and oil, or that the value of the game in Kenyan game parks is simply the present value of tourists' willingness to pay!) It is pretty much impossible to get a quantitative handle on intrinsic worth, and so the right thing to do is to take note of it, keep an eye on it, and call attention to it whenever the stock is threatened.

Another source of value is more amenable to quantification. It arises from a combination of two things common to environmental resources: uncertainty in their future use values and irreversibility in their use. (Genetic materials in tropical forests provide a prime example.) The twin presence of uncertainty and irreversibility implies that even were the aggregate well-being function neutral to risk, it would not do to estimate the shadow price of an environmental resource solely on the basis of the expected benefit from its future use. Irreversibility in its use implies that preservation of the stock has an additional value—the value of extending one's set of future options. Future options have an additional worth precisely because with the passage of time more information is expected to be forthcoming about the resource's use value. (This additional worth is often called an *option value*.) The shadow price of a resource is the sum of its use value and its option value (see Arrow and Fisher 1974; Henry 1974; Dasgupta 1982; Fisher and Hanemann 1986; and Mäler 1989).

As we noted earlier, environmental resources in situ are mostly regarded as free in current economic exercises even while their shadow prices are positive. As we have also noted above, this leads to biases in policy and to biases in the design and installation of new technology. Recall also that even at the prior level of research and development they lead to biases, for there is little incentive for agencies to develop technologies which economize on the use of environmental resources. Indeed, an entire debate on whether economic and environmental considerations are in contraposition is a misplaced one; they are when economic calculations are biased. They would be consonant with each other if environmental resources were put on par with the kinds of goods and services which occupy the attention of economists most of the time.

III. Markets and Their Failure

All this has been from what one might call the operations-research side of things. It is an essential viewpoint, but it is a limited viewpoint. Its complement is the institutional side, with all its attendant difficulties. Indeed, it is the institutional side which has most often been the starting gate for environmental economics. Particular emphasis is placed upon the fact that markets for these goods either do not exist or are prone to malfunctioning when they do exist (see Mäler 1974; Baumol and Oates 1975; and Dasgupta 1982). By "markets" we do not necessarily mean price-guided institutions. Rather, we mean institutions which make available to interested parties the opportunity to negotiate courses of actions. And by "malfunctioning markets" we mean circumstances in which such opportunities are not present (because, say, property rights are unspecified); in which they are exploited at best partially (because, say, the bargainers do not know each other well; see Farrell 1987); or in which they are somewhat one-sided. (This often-one-sidedness of opportunities means that we are thinking of distributional issues as well and not merely those bearing on efficiency.)

Nowhere is this generalized form of market failure more common than in those hidden interactions which are *unidirectional*—for example, deforestation in the uplands, which often inflicts damages on the lowlands. As always, it pays to concentrate first on the assignment of property rights before seeking remedies. The common law, if one is permitted to use this expression in a universal context, usually recognizes the rights of polluters, not those of pollutees. Translated into our present example, this means that the timber merchant who has obtained a concession in the upland forests is under no obligation to compensate farmers in the lowlands. If the farmers wish to reduce the risk of heightened floods, it is they who have to compensate the timber merchant for reducing the rate of deforestation. Stated this way, the matter does look morally bizarre, but it is how things are. Had property rights been the other way around—recognizing pollutees' rights—the boots would have been on the other set of feet, and it would have been the timber merchant who would have had to pay compensation to farmers for the right to inflict the damages which go with deforestation. When the cause of damages is hundreds of miles away, however, and when the victims are thousands of impoverished farmers, the issue of a bargained outcome usually does not arise.[9] Thus, judged even from the viewpoint of efficiency, a system of polluters' rights in such an example would be disastrous.[10] We would expect excessive deforestation. Stated in another way, the private cost of logging would in this case be lower than its social cost.

Now when the social costs of production of environmental goods are higher than their private costs, resource-based goods may be presumed to be underpriced in the market. Quite obviously, the less roundabout, or less distant, is the production of the final good from its resource base, the greater is this underpricing, in percentage terms. Put another way, the lower is the value added to the resource, the larger is the extent of this underpricing of the final product. We may then conclude that countries which export primary products do so by subsidizing them, possibly on a massive scale. Moreover, the subsidy is paid not by the general public via taxation but by some of the most disadvantaged members of society: the sharecropper, small landholder, or tenant farmer; the forest dweller; and so on. The subsidy is hidden from public scrutiny; that is why nobody talks of it. But it is there. It is real. We should be in a position to estimate it. As of now, we have no such estimates.

Matters are usually quite different for economic and ecological interactions which are *reciprocal,* such as the use by several parties of a piece of grazing land. Many of the production and exchange contracts we see in poor agrarian and pastoral societies over the use of such resources are explicit ones, and compli-

9. Even under such conditions, bargained outcomes do arise sometimes, since community leaders, nongovernment organizations, and a free press have been known to galvanize activity on behalf of the relatively powerless. In recent years this has happened on a number of occasions in India in different sets of contexts. The most publicized one has been the Chipko movement, which involved the threatened disenfranchisement of historical users of forest products. This was occasioned by the state claiming its rights over what was stated to be "public property" and then embarking on a logging program. The connection between environmental protection and civil rights is a close one. There is absolutely no question that political and civil liberties are instrumentally useful for environmental protection.

10. The classic on this subject (Coase 1960) had an argument proving the neutrality of the assignment of property rights on allocative efficiency. Coase's theorem requires stringent assumptions, including the little-noticed one that there are only two parties involved. With more than two parties matters are different, for Shapley and Shubik (1969) and Starrett (1973) have shown that an economy can fail to possess a core allocation if there are polluters' rights over private "bads," such as household garbage. In their examples a core allocation, however, does exist with pollutees' rights.

ance with them is enforced by means of elaborate rules, regulations, and fines. (In the current literature, see in particular Feder and Noronha 1987; Wade 1987, 1988; Hecht, Anderson, and May 1988; Agarwal and Narain 1989; Chopra, Kadekodi, and Murty 1989; and Ensminger, forthcoming.) However, many contracts are merely implicit, the obligations they entail having often been codified over the years in the form of social norms. In such cases, associated social sanctions sometimes are imposed on violators of such norms, occasionally on those who fail to impose sanctions on violators, even more rarely on those who fail to impose sanctions on those who fail to impose sanctions on violators, and so on.[11] Thus, the fact that a piece of environmental property is not in private (or government) hands does not mean at all that there is institutional failure. The anthropological literature is replete with studies of communities that have developed elaborate patterns of monitoring and control over the use of what are today called common property resources.

Economic analysis is thought by some to have implied that common property resources can only be managed through centralized coordination and control (by "centralized" we mean the government or some agency external to the community of users). Referring to the problem of the commons in the theoretical literature, Wade (1987, p. 220), in a much-cited article, writes that "the prevailing answer runs as follows: when people are in a situation where they could mutually benefit if all of them restrained their use of a common-pool resource, they will not do so unless an external agency enforces a suitable rule." And he proceeds to describe enforcement mechanisms in his sample of villages which do not rely on external agencies.

But the literature has not implied the necessity of centralized or external control. The theory of games has unraveled the variety of institutional mechanisms (ranging from taxes to quantity controls) which can in principle support desirable allocations of common property resources. The theory makes clear, and has made clear for quite some time, that enforcement of the agreed-upon allocation can be carried out by the users themselves. In many cases this may well be the most desirable option. As always, monitoring, enforcement, information, and transaction costs play a critical role in the relative efficacy of these mechanisms, and we will have something to say about this in what follows. (For a formal, mathematical account, and an informal discussion of the possibilities implied by the formal analysis, see Dasgupta and Heal 1979, chap. 3, sections

11. By a social norm we mean a generally accepted injunction to follow a specified behavioral strategy. Social norms are internalized by people up to a point, so that the infinite chain of meta-norms we have just mentioned in the text is not required for sustaining reciprocity and cooperation. See in particular Wiser (1936), Polanyi (1944, 1977), Goody (1973), Scott (1976), Chambers, Longhurst, and Pacey (1981), Cashdan (1989), and a fine analytical discussion by Elster (1989). We are not suggesting that social norms are efficient, nor that they are necessarily equitable. In fact, inefficiencies and inequities abound. (See, for example, Popkin 1979, Beteille 1983, Iliffe 1987, and Elster 1989.) We are merely asserting that they exist, and that they support outcomes which would not prevail in their absence. Aumann (1981), Fudenberg and Maskin (1986), and Abreu (1988) provide a formal basis for seeing how norms can be sustained even when people have not internalized them, and how they can be both inefficient and inequitable. The required analysis makes use of the theory of repeated games. See Dasgupta (1990, forthcoming) for an elementary exposition.

4–5.) The confirmation of theory by evidence on the fate of different categories of common property resources is a pleasing success of modern economic analysis.[12]

IV. Public Failure and the Erosion of Local Commons

There is a vast difference between global and local commons. The open seas are common property resources, as are usually village ponds. As economic analysis makes clear, what are problems for the former are by no means problems for the latter. However, it is the global commons, and popular writings on them (for example, the influential article by Hardin 1968), which have shaped popular images of all common property resources. This has been most unfortunate because, unlike global commons, the source of the problems associated with the management of local commons is often not the users but other agencies. The images invoked by "the tragedy of the commons" are mostly not the right ones when applied to local commons. The point is that local commons (village ponds and tanks; pastures and threshing grounds; watershed drainage and river beds; and sources of fuelwood, medicinal herbs, bamboo, palm products, resin, gum, and so on) are in no society open for use to all. They are open only to those having historical rights, through kinship ties, community membership, and so on. Those having historical rights of use tend, not surprisingly, to be very protective of these resources. Local commons are easy enough to monitor, and so their use is often regulated in great detail by the community, as we noted earlier, either through the practice and enforcement of norms or through deliberate allocation of use (see, for example, Wade 1987).

The extent of common property resources as a proportion of total assets in a community varies greatly across ecological zones. In India they appear to be most prominent in arid regions, mountain regions, and unirrigated areas. They are least prominent in humid regions and the river valleys (see Agarwal and Narain 1989; and Chopra, Kadekodi, and Murty 1989). An almost immediate empirical corollary of this is that income *inequalities* are less where common property resources are more prominent. Aggregate income, however, is a different matter altogether, and it is the arid and mountain regions and unirrigated areas which are the poorest.[13] This needs very much to be borne in mind when policy is devised.

In an important and interesting article, Jodha (1986) used data from eighty villages in twenty-one dry districts from seven states in India to estimate that among poor families the proportion of income based directly on common prop-

12. Game-theoretic analyses of common property resources have almost invariably concentrated on the case of large numbers of users, where each user contributes a tiny amount to environmental degradation, but where the total effect, by virtue of the large numbers involved, is substantial. Repetto (1988) confirms that it is this class of cases, and *not* instances of large investment projects, which provides most of the bases of environmental degradation in poor countries.

13. As might be expected, even within dry regions dependence on common property resources falls with rising wealth across households. The interrelationship between destitution and the erosion of the rural environmental resource base is developed in a wider analytical context in Dasgupta (1989).

erty resources is for the most part in the range of 15 to 25 percent. This is a nontrivial proportion. Moreover, these resources are very much complementary to the sources of income from private property resources, which are mainly labor; milch and draft animals; land for cultivation and crops, but often not the stubble in the postharvest period; common agricultural tools, such as ploughs, harrows, levelers, and hoes; fodder-cutting and ropemaking machines; and seeds. Common property resources also provide the rural poor with partial protection in times of unusual economic stress. For landless people they may be the only nonhuman asset at their disposal. As it also happens, a number of such resources—such as fuelwood and water for home use, medicinal herbs, resin, and gum—are the responsibility of women and children.

A similar picture emerges from Hecht, Anderson, and May (1988), which describes in rich detail the importance of the extraction of babassu (a Brazilian palm tree) products among the landless in the Brazilian state of Maranhao. The complementarity between this extraction activity and agricultural work is striking, most especially for women. These extractive products are, as it happens, a particularly important source of cash income in the period between agricultural food crop harvests.

It is not difficult to see why common property resources matter greatly to the poorest of the rural poor in a society, nor, therefore, is it difficult to understand the mechanisms through which such people may well get disenfranchised from the economy even while in the aggregate it enjoys economic growth (a formal account of the processes through which this can occur is developed in Dasgupta 1989). If one is steeped in social norms of behavior and understands community contractual obligations, one does not calculate every five minutes how one should behave. One follows the norms. This saves on costs all around, not only for the individual as an "actor," but also as a policeman and judge. It is also the natural thing for one to do if one has internalized the norms. But this is sustainable so long as the background environment remains pretty much constant. It will not be sustainable if the social environment changes suddenly. One might even be destroyed. It is this heightened vulnerability, often more real than perceived, which is the cause of some of the greatest tragedies in contemporary society. For they descend upon people who are, in the best of circumstances, acutely vulnerable.

The sources which trigger destitution by this general means vary. The erosion of common property resource bases can come about in the wake of shifting populations (accompanying the growth process itself), rising populations, technological progress, unreflective public policies, predatory governments, and thieving aristocracies. There is now an accumulation of evidence on this range of questions, and in what follows we present an outline of the findings in three sets of studies.

In his work on the drylands of India, Jodha (1986) noted a decline in the geographical area covering common property resources ranging from 26 percent to 63 percent over a twenty-year period. This was in part due to the privatization of land, well over half of which in his sample had been awarded to the rural nonpoor. He also noted a decline in the productivity of common property resources on account of population growth among the using community. In an earlier work, Jodha (1980) identified an increase in subsistence requirements of the farming community and a rise in the profitability of land exploitation from cropping and grazing as a central reason for increased desertification in the state of Rajasthan in India. Jodha argued that, ironically, it was government land reform programs in this area, unaccompanied by investment in improving the productive base, which had triggered the process (for a formalization of the dynamics of such a process, see Dasgupta 1982, chap. 6).

Ensminger (forthcoming), in a study of the privatization of common grazing lands among the Orma in northeastern Kenya, indicates that the transformation took place with the consent of the elders of the tribe, and she attributes this willingness to changing transaction costs brought about by cheaper transportation and widening markets. The elders were, quite naturally, from the stronger families, and it does not go unnoted by Ensminger that privatization has accentuated inequalities. However, she provides no data to tell whether the process has increased the prevalence of destitution among the economically weak.

In an earlier, much-neglected work on the Amazon basin, E. Feder (1977, 1979) described how massive private investment in the expansion of beef-cattle production in fragile ecological conditions has been supported by domestic governments in the form of tax concessions and provision of infrastructure, and by loans from international agencies such as the World Bank. The degradation of vast tracts of valuable environmental resources was, not surprisingly, accompanied by the disenfranchisement of large numbers of small farmers and agricultural laborers from the economy, and it made traditional forest dwellers destitute, at best, and simply eliminated them, at worst.[14]

The sources of the transformation of common property resources into private resources described in these three sets of studies are, of course, quite different. Consequently, the ways in which they affected those with historical rights have been quite different. But each is understandable and believable. Because they are confirmed by economic theory, the findings of these case studies are almost certainly not unrepresentative. They suggest that privatization of village commons and forest lands, although hallowed at the altar of efficiency, can have disastrous distributional consequences, disenfranchising entire classes of people from economic citizenship (for alternative demonstrations of this theorem, see Cohen and Weitzman 1975 and Dasgupta and Heal 1979, chap. 3). They also

14. See also Dasgupta (1982, chap. 2) and Hecht (1985). The data suggest that during the decades of the 1960s and 1970s protein intake by the rural poor *declined* even while the production of beef protein increased dramatically. Much of the beef was destined for exports, for use by fast-food chains. These matters, which are an instance of the intricate link between economic, social, and financial institutions, have been taken up anew by Repetto (1988), Mahar (1988), and Binswanger (1989). The latter in particular has shown how in Brazil the exemption from taxation of virtually all agricultural income (allied to the fact that logging is regarded as proof of land occupancy) has provided strong incentives for the acquisition of agricultural lands by the higher-income groups and a general incentive for the acquisition of forest lands for the purposes of deforesting them.

show that public ownership of such resources as forest lands is by no means necessarily a good basis for a resource allocation mechanism. Decisionmakers are in these cases usually far removed from the site (living as they do in imperial capitals); they have little knowledge of the ecology of such matters; their time horizons are often short; and they are in many instances overly influenced by interest groups far removed from the resource in question.

All this is not at all to suggest that rural development is to be avoided. It is to say that resource allocation mechanisms which do not take advantage of dispersed information; which are insensitive to hidden (and often not so hidden) economic and ecological interactions (that is, general equilibrium effects); which do not take the long view; and which do not give a sufficiently large weight to the claims of the poorest within rural populations (particularly the women and children in these populations) are going to prove environmentally disastrous. It appears, then, that during the process of economic development there is a close link between environmental preservation and the well-being of the poor, most especially the most vulnerable among the poor. Elaboration of this link has been one of the most compelling achievements at the interface of anthropology, economics, and nutrition science.

The links between environmental degradation and an accentuation of deprivation and hardship can take forms which are even today not always appreciated. The responsibilities for gathering fuelwood and fetching water for domestic use in most rural communities fall upon women and children. When allied to household chores and their farming obligations, the workload of women in South Asia in terms of time is often one-and-a-half to two times that of men (see, for example, Fernandes and Menon 1987; Kumar and Hotchkiss 1988; B. Agarwal 1989). This workload has over the years increased directly as a consequence of receding resources. It is very much worth reminding ourselves that we are speaking of a category of people of whom more than 50 percent suffer from iron deficiency, of whom only a little below 50 percent suffer from wastage, and who in some parts of the world work fifteen to sixteen hours a day during the busy agricultural season. Thus, communities in the drylands of the Indian subcontinent and in Sub-Saharan Africa today often live miles away from fuel and fodder sources and permanent water sources. Surveys in East Africa have shown, for example, that women and children spend up to five hours a day collecting water during the dry season (see Food and Agricultural Organization 1987). The consequence is that anything between 10 and 25 percent of daily daytime energy expenditure is required for collecting water.[15]

15. See Chen (1983) for a review of the link between improved water supply and health benefits among the rural poor. We should note that a similar problem is associated with fuelwood collection. In northern India, for example, it is thought that some 75 percent of firewood for domestic use comes from twigs and fallen branches. From data that are now available from the drylands of India on time allocation on the part of women in fuelwood collection, the energy costs in this activity would seem to be also in the range of 10 percent to 25 percent. It should be noted that estimates of the energy-cost of collection are essential ingredients in the calculation of the shadow prices of fuelwood and water.

All this cannot but be related to the fact of high fertility and low literacy in rural areas of most poor countries. Poverty and the thinness of markets make it essential for households to engage in a number of complementary production activities: cultivation, cattle grazing, fetching water, collecting fuelwood, cooking food, and producing simple marketable products. Each is time-consuming. (Labor productivity is low not only because capital is scarce but also because, as we have just noted, environmental resources are scarce, too.) If it is to survive, a household simply has to accomplish these tasks each day, and a small household cannot do them all. Each household needs many hands. Children are needed as workers by their parents, even when parents are in their prime. Children are not merely a consumption good (as in Nerlove, Razin, and Sadka 1987), nor are they only a means to old-age security (as in Cain 1983). They are also of current use to parents. But a high rate of fertility and population growth further damages the environmental resource base, which in turn can provide further (private) incentives for large families, which in turn further damages the resource base, and so on—until some countervailing factors (whether public policy or some form of Malthusian check) stop the spiraling process. But by the time this happens millions of lives have suffered (for further development of these issues, see Dasgupta, forthcoming, and Nerlove and Meyer 1990).

Information concerning the ecology of local commons is often dispersed, and is usually in the hands of the historical users. There are exceptions, of course, but as a general rule this makes it desirable that the local commons be protected as commons and that decisions regarding local commons be left in the hands of the users themselves. This is because the local commons will almost certainly remain the single source of essential complementary goods for poor people for a long while yet. To be sure, it is essential not only that governments provide infrastructural and credit and insurance facilities, but that they make new information concerning technology, ecology, and widening markets available to the users. But there is little case for centralized control. Quite the contrary, there is a case for facilitating the growth of local, community decisionmaking—in particular, decisionmaking by women, who are for the most part the main users of such resources. The large, often fragmented literature on local common property resources is beginning to offer us an unequivocal picture that during the process of economic development the protection and promotion of environmental resources would be best served were a constant public eye kept on the conditions of the poorest of the poor in society. Environmental economics and the economics of destitution are tied to each other in an intricate web. We should not have expected it otherwise.

V. International Failure and the Erosion of Global Commons

Global commons pose a different type of problem. The impossibility of establishing adequate property rights to the atmosphere, to watersheds, and to large bodies of water, such as the oceans, are a cause of inefficiencies in the allocation of resources. In the case of the atmosphere (for example, over the matter of global warming), there is not even the option of "voting with one's feet." Furthermore, future generations are not directly represented in today's forum for decisionmaking. Their interests are included only indirectly through transactions between different coexisting generations. Thus the inefficiencies and inequities involved are not merely static ones but are intergenerational ones as well (see Mäler 1990b for a more detailed discussion of these issues). From this it follows that the international community needs consciously to design systems which improve upon existing resource allocation mechanisms.

The most complicated international environmental problems are, like the local commons, characterized by reciprocal externalities; that is, most countries that contribute to environmental damages also suffer from them. Emissions of greenhouse gases are an instance of this. A central problem is that "reciprocal" countries do not contribute to the damages in equal amounts. Thus, for a cooperative outcome to be achievable, some financial transfers will be necessary, if only in an implicit manner. Several such systems suggest themselves—debt relief for the preservation of the Amazon being among the most frequently talked about.

This is not to say that agreements cannot be reached without side payments; it is only to say that they will tend to be less efficient. Barrett (1990) has argued, for example, that one should not expect all countries to sign the Montreal protocol on emissions of chlorofluorocarbons (CFCs). (The protocol involves no side payments.) If an equilibrium exists, it can only involve *some* countries signing the protocol. The reason is that were only a few countries to sign the protocol, national benefits from further reduction in CFC emission would be high. This would induce more countries to sign. However, were many countries to sign the protocol, national benefits from further reduction would be small, and it would then not be worth a country's while to sign the agreement.

Direct (side) payments among countries for solving environmental problems have not been so common. When made, side payments have tended to be nonpecuniary; for example, trade and military concessions (see Krutilla 1966; Kneese 1988). Very recently, an agreement has been reached on reducing the production and use of CFCs in developing countries. This has involved the creation of an international fund for technological transfers to these countries. It is a most promising development, and it needs to be studied carefully for the purposes of further development.

One broad category of allocation mechanisms well worth exploring in the international context involves making the global commons quasi-private. The basic idea, which originated in Dales (1968), is similar to the principle currently being experimented with in the United States. The idea, if extended to the international sphere, would have the community of nations set bounds on the total annual use of the global commons, such as the atmosphere, have it allocate an initial distribution of transferable national rights which add up to the aggregate bound, and allow the final allocation among different users to be determined by competitive markets.[16]

For example, consider greenhouse gases. Suppose the community of nations desires to reduce emissions to a prescribed global level. Units of the various gases would then be chosen so that all gases have the same (expected) effect on global climate. In other words, at the margin the emission of one unit of any one gas would have the same (expected) climatic effect as the emission of one unit of any other gas. The scheme would allow countries to exchange permits for one gas for permits for any other. Countries would receive an initial assignment of marketable permits. As is well known, this scheme has informational advantages over both taxes and quantity controls on individual emissions. (See Dasgupta, Hammond, and Maskin 1981 for a formal analysis of optimal incentive schemes for pollution control.) Furthermore, were the permits to refer to *net* emissions (that is, net of absorption of carbon dioxide by green plants), the scheme would provide an incentive for countries with fast-growing tropical rain forests to earn export revenue by encouraging forest growth and then selling permits to other countries. The scheme also has the advantage that the necessary side payments required to induce all (or most) countries to participate in the agreement can be made through the initial distribution of emission permits. Countries which do not expect severe damages from global warming would also wish to participate were they to be provided initially with a sufficient number of permits (or rights).

The sticking point will clearly be over reaching an agreement on the initial distribution of permits among nations. (How a national government allocates the nation's rights among agencies within the country is a different matter.) But the point here is that if the bound set on annual aggregate greenhouse emissions is approximately optimal, it is always possible to distribute the initial set of rights in such a way that all countries have an incentive to join the scheme. For this reason one cannot overemphasize the fact that there are large potential gains to be enjoyed from international cooperation; a scheme involving the issue of marketable permits in principle offers a way in which all nations can enjoy these gains. The argument that "national sovereignty" would be endangered is in fact no argument, for the point about global commons is precisely that they are beyond the realm of national sovereignty.

VI. Summary

In this article we have tried to present a perspective on what we take to be the central emerging issues at the interface of environmental and development con-

16. See Tietenberg (1980, 1990) for reviews of the experience that has been accumulated with such schemes in the United States. See also Dasgupta (1982) and Mäler (1990a) for mathematical formalizations of these schemes under varying environmental circumstances. The motivation behind these formalizations is that they enable us to calculate the efficiency gains realizable by such resource allocation mechanisms.

cerns. The fact that for such a long while environmental and development economics have had little to say to each other is a reflection only of these academic disciplines; it does not at all reflect the world as we should know it. Poor countries are for the most part agrarian and pastoral, and it is but natural that the bulk of society in these lands depends crucially on renewable natural resources. One of the sobering lessons of the international development experience has been that the magnitudes of poverty and destitution have proved singularly resistant to reduction in many parts of the globe. It is beyond our competence to try to explain this, but there is growing evidence that acute poverty and environmental degradation are closely linked in most poor countries. We have argued that the poor in resource-exporting countries are very likely subsidizing these exports. A reasonable rule of thumb for the "environmentalist" would therefore be to keep a constant eye on the poorest of the poor. Their activities (for example, migration patterns of communities or time-use patterns of poor women) are often a good signal of the state of the environment. By the same token, it is fatuous to talk and write about poverty and development unless we simultaneously study the fate of environmental resources under alternative resource allocation mechanisms. The separation of environmental and development economics has proved to be enormously costly in terms of lost hopes and wasted lives.

We have argued that local and global commons pose quite different problems and that environmental damages at the local level have often been inflicted upon such communities (possibly unwittingly) by outside agencies—very often by their own governments. There are countries where information about the environmental resource base is almost wholly absent. If environmental resources are to be brought into line with other capital assets, they must as a minimum enter national income accounting. In this paper we have presented an outline of how to go about doing this in a meaningful way. We would imagine that it is the global commons which will occupy the international stage in the immediate future, as evidence accumulates on the mechanisms underlying their degradation. We have presented the bare bones of a resource allocation mechanism in which countries receive marketable permits for the use of the global commons and in which all nations have an incentive to participate.

Appendix

Consumption and Accumulation with Environmental Effects: A Definition of NNP

In this part of the appendix we present what we hope is a canonical model involving consumption and accumulation with environmental effects. Our aim is to display the connection between shadow prices, rules for project evaluation, and national income accounting in a context which is simple but which has at the same time sufficient structure to allow us to obtain a number of the prescriptions we alluded to in the text, especially in section II. Keeping to essential matters here, we ignore the kinds of "second-best" constraints (including disequilibrium phenomena) which have been the center of attention in the literature on project evaluation, as in Dasgupta, Marglin, and Sen (1972) and Little and Mirrlees (1974).

We consider an economy with a multipurpose, manmade capital good whose stock is denoted by K_1. If L_1 is the labor effort combined with this, the flow of output is taken to be $Y = F(K_1, L_1)$, where $F(.)$ is an aggregate production function. (In what follows we assume that all functions satisfy conditions which ensure that the planning problem defined below is a concave program. For example, we assume that $F(.)$ is concave.) The economy enjoys in addition two sorts of environmental resource stocks: clean air, K_2, and forests, K_3. Clean air is valued directly, whereas forests have two derived values: they help keep the atmosphere clean, and they provide fuelwood, which, too, is valued directly. Finally, we take it that there is a flow of environmental amenities, Z, which directly affects aggregate well-being.

Forests enjoy a natural regeneration rate, but labor effort can increase it. Thus we denote by $H(L_2)$ the rate of regeneration of forests, where L_2 is labor input for this task, and where $H(.)$ is, for low values of L_2 at least, an increasing function. Let X denote the rate of consumption of fuelwood. Collecting this involves labor effort; let this be L_3. Presumably, the larger the forest stock the less is the effort required (in calorie requirements, say). We thus assume that $X = N(K_3, L_3)$, where $N(.)$ is an increasing, concave function of its two arguments.

Output Y is a basic consumption good, and this consumption is also valued directly. However, we take it that the production of Y involves pollution as a by-product. This reduces the quality of the atmosphere both as a stock and as a flow of amenities. We assume, however, that it is possible to take defensive measure against both these ill effects. First, society can invest in technologies that reduce the emission of pollutants, and we denote the stock of this defensive capital by K_4. If P denotes the emission of pollutants, we have $P = A(K_4, Y)$, where A is a convex function, decreasing in K_4 and increasing in Y. Second, society can mitigate damages to the flow of amenities by expending a portion of final output, at a rate R. We assume that the resulting flow of amenities has the functional form, $Z = J(R, P)$, where J is increasing in R and decreasing in P.

There are thus four things that can be done with output Y: it can be consumed (we denote the rate of consumption by C); it can be reinvested to increase the stock of K_1; it can be invested in the accumulation of K_4; and it can be used, at rate R, to counter the damages to the flow of environmental amenities. Let Q denote the expenditure on the accumulation of K_4.

Now, the environment as a stock tries to regenerate itself at a rate which is an increasing function of the stock of forests, $G(K_3)$. The net rate of regeneration is the difference between this and the emission of pollutants from production of Y.

We can therefore express the dynamics of the economy in terms of the following equations:

$$dK_1/dt = F(K_1, L_1) - C - Q - R \tag{1}$$

$$dK_2/dt = G(K_3) - A[K_4, F(K_1, L_1)] \tag{2}$$

$$dK_3/dt = H(L_2) - X \tag{3}$$

$$dK_4/dt = Q \tag{4}$$

$$X = N(K_3, L_3) \tag{5}$$

$$Z = J\{R, A[K_4, F(K_1, L_1)]\}. \tag{6}$$

The current flow of aggregate well-being, W, depends positively upon aggregate consumption, C; the output of fuelwood, X; the flow of environmental amenities, Z; and the quality of the atmospheric stock, K_2. However, it depends negatively upon total labor effort, $L = L_1 + L_2 + L_3$. (As noted above, labor effort could be measured in caloric terms.) We thus have $W(C, X, Z, K_2, L_1 + L_2 + L_3)$.

Stocks of the four types of assets are given at the initial date; the instantaneous control variables are C, Q, R, X, L_1, L_2, and L_3; and the objective is to maximize the (discounted) sum of the flow of aggregate well-being over the indefinite future. We take well-being to be the numeraire. Letting p, q, r, and s denote the (spot) shadow prices of the four capital goods, K_1, K_2, K_3, and K_4 respectively, and letting v be the imputed marginal value of the flow of environmental amenities, we can use equations 1 through 6 to express the current value Hamiltonian, V, of the optimization problem as

$$\begin{aligned} V = {} & W[C, N(K_3, L_3), Z, K_2, L_1 + L_2 + L_3] \\ & + p[F(K_1, L_1) - C - Q - R] + q\{[G(K_3) - A[K_4, F(K_1, L_1)]\} \\ & + r[H(L_2) - N(K_3, L_3)] + sQ + v[J\{R, A[K_4, F(K_1, L_1)]\} - Z]. \end{aligned} \tag{7}$$

Recall that the theory of optimum control instructs us to choose the control variables at each date so as to maximize expression 7. (Notice that we have used equations 5 and 6, and so the controls are now C, Z, Q, R, L_1, L_2, and L_3.) Writing by W_C the partial derivative of W with respect to C, and so forth, it is then immediate that along an optimal program the control variables and the shadow prices must satisfy the conditions:

$$\text{(i) } W_C = p; \quad \text{(ii) } W_X N_2 + W_L = rN_2; \quad \text{(iii) } W_Z = v;$$
$$\text{(iv) } W_L = [(q - vJ_2)A_2 - p]F_2; \quad \text{(v) } W_L = -r dH(L_2)/dL_2;$$
$$\text{(vi) } p = vJ_1; \quad \text{and (vii) } p = s.$$

F_2 stands for the partial derivative of F with respect to its second argument,

L_1; and as mentioned earlier, $L = L_1 + L_2 + L_3$. We have used this same notation for the derivatives of $N(.)$, $J(.)$, and $A(.)$. Interpreting these conditions is instructive, for they tell us what kinds of information we need to estimate shadow prices. (We do not write down the intertemporal "arbitrage conditions" which these shadow prices must also satisfy, since we do not use them.) But for our purposes, the point to note is that we can immediately derive the correct expression for NNP from equation 7. It is the linear support of the Hamiltonian along the optimal program. In order to keep the expression for NNP from becoming overly cumbersome, let us denote by O^* the *vector* of all the nonprice arguments in the Hamiltonian function along the *optimal* program at any given date. Thus, $O^* = (C^*, Z^*, Q^*, R^*, K_1^*, K_2^*, K_3^*, K_4^*, L_1^*, L_2^*, L_3^*)$. It follows from taking the Taylor expansion around O^* that the linear support of the Hamiltonian is

$$V(O^*) + W_C C^* + W_X X^* + W_Z Z^* + W_{K_2} K_2^* + W_L(L_1^* + L_2^* + L_3^*) + pdK_1^*/dt + qdK_2^*/dt + rdK_3^*/dt + sdK_4^*/dt \quad (8)$$

where Z^* is equal to $J\{R^*, A[K_4^*, F(K_1^*, L_1^*)]\}$.

We may divide the whole expression by W_C to express NNP in aggregate consumption numeraire. It should also be recalled that by assumption W_L is *negative*. Now, for evaluating a marginal project $V(O^*)$ is irrelevant, as it is a constant term. We may as well then drop it. In this case, NNP in the optimizing economy, measured in well-being numeraire, is the remaining term of equation 8, namely:

$$\text{NNP} = W_C C + W_X X + W_Z J\{R, A[K_4, F(K_1, L_1)]\} + W_{K_2} K_2 + W_L(L_1 + L_2 + L_3) + pdK_1/dt + qdK_2/dt + rdK_3/dt + sdK_4/dt. \quad (9)$$

However, for international comparisons $V(O^*)$ is of interest, since countries differ in the stocks of capital they hold. In this case NNP should be measured on the basis of equation 8 (see Weale 1990). It is possible to show that there is a permissible representation of the well-being function for which

$$V(O^*) = \delta(pK_1^* + qK_2^* + rK_3^* + sK_4^*) \quad (10)$$

where δ (> 0) is the rate at which the flow of aggregate well-being is discounted. This means that the Hamiltonian measures the sum of the social returns on all of society's capital assets. In short, it is a measure of society's wealth (see Solow 1986).

Notice that all resources and outputs are valued at the prices which sustain the optimum program.[17] To highlight the points we want to make here, we have

17. One could alternatively think of a sequence of policy reforms and use shadow prices defined at the existing structure of production. Given that the planning program is by hypothesis concave, a sequence of such moves would take the economy ultimately to the optimum. For a simplified exposition of the connection between these two modes of analysis (reforms and optimization), see Dasgupta (1982, chap. 5).

chosen to work with a most aggregate model. Ideally (income) distributional issues will find reflection in the aggregate well-being function. These considerations can readily be translated into the estimates of shadow prices (see Dasgupta, Marglin, and Sen 1972).

Let us suppose that we are involved in the choice of projects. A marginal project is a perturbation of the optimal program. It can be expressed as a ten-vector (dC, dX, dR, dL_1, dL_2, dL_3, dI_1, dI_2, dI_3, dI_4), where $I_i = dK_i/dt$, ($i = 1$, 2, 3, 4); and dC, and so on, are small changes in C, and so on. A marginal project has no effect on the fourth term on the right-hand side of equation 9. (For the purposes of social cost-benefit analysis we could, therefore, simply ignore this term when estimating NNP.) Along an optimal program, the social profitability of the last project is nil; that is, its contribution to NNP is nil. This follows from the fact that the controls are chosen so as to maximize expression 7. All this is well-known, and our purpose here is to obtain some additional insights. Scrutiny of equation 9 tells us the following:

- Were wages to equal the marginal ill-being of work effort, wages would not be part of NNP. Put in other words, the shadow wage bill ought to be deducted from gross output when we estimate NNP. However, if labor is supplied inelastically, it is a matter of indifference whether the wage bill in this optimizing economy is deducted from NNP. Conversely, were we to recognize a part of the wage bill as a return on the accumulation of human capital, that part *would* be included in NNP.
- Current defensive expenditure, R, against damages to the flow of environmental amenities should be included in the estimation of final demand (see the third term in equation 9). Moreover, investments in the stock of environmental defensive capital should also be included in NNP (see the final term of equation 9). Now, expenditures which go toward enhancing the environment find expression in the value that is imputed to changes in the environmental resource stock. We may conclude, therefore, that such expenditures should not be included in estimates of NNP. (Notice the absence of sQ in equation 9.)
- The value of *changes* in the environmental resource base (K_2 and K_3) should be included in NNP. However, anticipated capital gains (or losses) are not part of NNP.

Uncertainty

In this part of the appendix we present a sketch of the arguments involved in the definition of NNP when there are random events in the future that are expected to have an effect on the value of the then-existing capital assets. As an example, we could imagine the discovery and installation of cleaner production technologies which make existing abatement technologies less valuable. For simplicity of exposition, we will assume that such discoveries are uninfluenced by policy; for example, research and development policy, although such policy can easily be incorporated. The analysis builds on Dasgupta and Heal (1974) and Dasgupta and Stiglitz (1981). However, these earlier studies did not address the measurement of NNP, which is our present concern.

It is most striking to consider discrete events. Thus, imagine that at some random future date, T, an event occurs which is expected to affect the value of the then-existing stocks of capital. We consider the problem from the vantage point of the present, which we denote by $t = 0$. Let us assume that there is a (possibly subjective) probability density function, π_t, over the date of occurrence. (We are thus supposing for expositional ease that the event *will* occur at some future date.) From this we may define the cumulative function Φ_t.

We take it that the social good is reflected by the expected value of the sum of the discounted flow of future well-being. Were the event in question to occur at date T, the economy in question would enter a new production and ecological regime. In what follows we continue to rely on the notation which was developed in the first part of the appendix. As is proper, we now proceed to work backwards. Thus, let K_i^T(with $i = 1, 2, 3, 4$) denote the stocks of the four assets at date T. Following an optimal economic policy subsequent to the occurrence of the event would yield an expected flow of aggregate well-being. This flow we discount back to T. This capitalized value of the flow of well-being will clearly be a function of K_i^T. Let us denote this by $B(K_1^T, K_2^T, K_3^T, K_4^T)$. It is now possible to show that until the event occurs, the optimal policy is to pretend that the event will never occur, and to assume that the flow of aggregate well-being is given, not by $W(.)$, as in the first part of the appendix, but by $(1 - \Phi_t)W(.) + \pi_t B(.)$. We may therefore conclude from the analysis of the first part of the appendix that NNP at any date prior to the occurrence of the event is given by the equation

$$\begin{aligned} \text{NNP} = {} & (1 - \Phi)[W_C C + W_X X + W_Z Z + W_{K_2} K_2 \\ & + W_L(L_1 + L_2 + L_3) + pdK_1/dt + qdK_2/dt + rdK_3/dt \\ & + sdK_4/dt] + \pi[B_{K_1}K_2 + B_{K_2}K_2 + B_{K_3}K_3 + B_{K_4}K_4]. \end{aligned} \tag{11}$$

(As in the first part of the appendix, we are suppressing the time index for clarity.) Notice that if the event is not ever expected to occur, then $\pi_t = 0$ for all t, and consequently $(1 - \Phi_t) = 1$ for all t. In this case, equation 11 reduces to equation 9. Note also that no marginal economic activity can affect the terms within the second pair of square brackets at date t. It follows that we could as well drop these terms from estimates of NNP. Now, were we to do this, equation 11 would *look* the same as equation 9. But their *values* would not be the same. This is because the shadow prices (for example, p, q, r, and s) appearing in equation 11 assume quite different values from those in equation 9: future possibilities hypothesized in this model economy are quite different from those assumed in the model economy of the previous section of the appendix. Finally, we should note that the shadow prices appearing in equation 11 are Arrow-Debreu contingent commodity prices.

References

Abreu, Dilip. 1988. "On the Theory of Infinitely Repeated Games with Discounting." *Econometrica* 56, no. 2: 383–96.

Agarwal, Anil, and Sunita Narain. 1989. *Towards Green Villages* (New Delhi, India: Center for Science and Environment).

Agarwal, Bina. 1989. "Rural Women, Poverty and Natural Resources: Sustenance, Sustainability and Struggle for Change." *Economic and Political Weekly* 24, no. 43: WS46–66.

Arrow, Kenneth J., and Anthony Fisher. 1974. "Preservation, Uncertainty and Irreversibility." *Quarterly Journal of Economics* 88, no. 2: 312–19.

Arrow, Kenneth J., and Mordecai Kurz. 1970. *Public Investment, the Rate of Return and Optimal Fiscal Policy*. (Baltimore, Md.: Johns Hopkins University Press).

Aumann, Robert. 1981. "Survey of Repeated Games." In M. Shubik, ed., *Essays in Game Theory and Mathematical Economics*. Mannheim, Germany: Wissenschaftsverlag, Bibliographisches Institut.

Barrett, Scott. 1990. "The Problem of Global Environmental Protection." *Oxford Review of Economic Policy* 6, no. 1: 68–79.

Baumol, William, and Wallace Oates. 1975. *The Theory of Environmental Policy*. Englewood Cliffs, N.J.: Prentice Hall.

Beteille, André, ed. 1983. *Equality and Inequality: Theory and Practice*. New Delhi, India: Oxford University Press.

Binswanger, Hans. 1989. "Brazilian Policies that Encourage Deforestation in the Amazon." World Bank Environment Department Paper 16. Washington, D. C. Processed.

Bowes, Pratima. 1978. *Hindu Intellectual Tradition*. New Delhi, India: Allied Publishers.

Brown, Gardner, and C. Bart McGuire. 1967. "A Socially Optimum Pricing Policy for a Public Water Agency." *Water Resources Research* 3, no. 1: 33–44.

Cain, Meade. 1983. "Fertility as an Adjustment to Risk." *Population and Development Review* 9, no. 4: 688–702.

Cashdan, Elizabeth, ed. 1989. *Risk and Uncertainty in Tribal and Peasant Economies*. Boulder, Colo.: Westview Press.

Center for Science and Environment. 1982. *The State of India's Environment*. New Delhi, India: Center for Science and Environment.

———. 1985. *The State of India's Environment*. New Delhi, India: Center for Science and Environment.

Chambers, Robert, R. Longhurst, and A. Pacey, eds. 1981. *Seasonal Dimensions to Rural Poverty*. London: Francis Pinter.

Chen, Lincoln C. 1983. "Evaluating the Health Benefits of Improved Water Supply through Assessment of Nutritional Status in Developing Countries." In B. Underwood, ed., *Nutrition Intervention Strategies in National Development*. New York: Academic Press.

Chenery, Hollis, and T. N. Srinivasan, eds. 1988. *Handbook of Development Economics*, vols. 1 and 2. Amsterdam: North-Holland.

Chopra, Kanchan G., K. Kadekodi, and M. N. Murty. 1989. *Participatory Development: People and Common Property Resources*. New Delhi, India: Sage.

Clark, Colin W. 1976. *Mathematical Bioeconomics: The Optimal Management of Renewable Resources*. New York: Wiley.

Coase, Ronald. 1960. "The Problem of Social Cost." *Journal of Law and Economics* 3, no. 1: 1–44.

Cohen, John S., and M. L. Weitzman. 1975. "A Marxian View of Enclosures." *Journal of Development Economics* 1, no. 4: 287–336.

Cooper, Richard. 1975. "An Economist's View of the Oceans." *Journal of World Trade Law* 9: 347–77.

Dales, J. H. 1968. *Pollution, Property and Prices*. Toronto: University of Toronto Press.

Dasgupta, Partha. 1982. *The Control of Resources*. Oxford, England: Basil Blackwell; Cambridge, Mass.: Harvard University Press; New Delhi, India: Oxford University Press.

———. 1989. "The Allocation of Hunger." Walras-Bowley Lecture, Econometric Society.

———. 1990. "The Environment as a Commodity." *Oxford Review of Economic Policy* 6, no. 1: 51–67. (To be republished in 1991 in A. Stevenson and D. Vines, eds., *Information Strategy and Public Policy*. Oxford, England: Basil Blackwell.)

———. Forthcoming. *An Inquiry into Well-Being and Destitution*.

Dasgupta, Partha, Peter Hammond, and Eric Maskin. 1981. "On Imperfect Information and Optimal Pollution Control." *Review of Economic Studies* 47, no. 4: 857–60.

Dasgupta, Partha, and Geoffrey Heal. 1979. *Economic Theory and Exhaustible Resources*. Cambridge, England: Cambridge University Press.

Dasgupta, Partha, S. Marglin, and Amartya Sen. 1972. *Guidelines for Project Evaluation*. New York: United Nations.

Dasgupta, Partha, and Joseph E. Stiglitz. 1981. "Resource Depletion under Technological Uncertainty." *Econometrica* 49, no. 1 (January): 85–104.

Drèze, Jean, and Nicholas H. Stern. 1987. "The Theory of Cost-Benefit Analysis." In A. J. Auerbach and M. Feldstein, eds., *Handbook of Public Economics*, vol. 2. Amsterdam: North-Holland.

Elster, Jon. 1989. *The Cement of Society: A Study of Social Order*. Cambridge, England: Cambridge University Press.

Ehrlich, Paul, A. Ehrlich, and J. Holdren. 1977. *Ecoscience: Population, Resources and the Environment*. San Francisco: Freeman.

El Serafy, Salah, and Ernst Lutz. 1989: "Introduction." In Y. J. Ahmad, S. El Serafy, and E. Lutz, eds., *Environmental Accounting for Sustainable Development*. Washington, D. C.: World Bank.

Ensminger, Jean. Forthcoming. "Co-opting the Elders: The Political Economy of State Incorporation in Africa." *American Anthropologist*.

Epstein, T. Scarlett. 1967. "Productive Efficiency and Customary Systems of Rewards in Rural South India." In R. Firth, ed., *Themes in Economic Anthropology*. London: Tavistock Publications.

Farrell, Joseph. 1987. "Information and the Coase Theorem." *Journal of Economic Perspective* 1, no. 2: 113–30.

Feder, Ernest. 1977. "Agribusiness and the Elimination of Latin America's Rural Proletariat." *World Development* 5: 559–71.

———. 1979. "Agricultural Resources in Underdeveloped Countries: Competition between Man and Animal." *Economic and Political Weekly* 14, nos. 30–32: 1345–66.

Feder, Gershon, and Raymond Noronha. 1987. "Land Rights Systems and Agricultural Development in Sub-Saharan Africa." *World Bank Research Observer* 2, no. 2: 143–70.

Fernandes, W., and G. Menon. 1987. *Tribal Women and the Forest Economy*. New Delhi, India: Indian Social Institute.

Fisher, Anthony C., and W. Michael Hanemann. 1986. "Option Value and the Extinction of Species." *Advances in Applied Micro-economics* 4.

Food and Agricultural Organization. 1987. *Fifth World Food Survey*. Rome.

Fudenberg, Drew, and Eric Maskin. 1986. "The Folk Theorem in Repeated Games with Discounting and with Incomplete Information." *Econometrica* 54, no. 3: 533–54.

Goody, Jack, ed. 1973. *The Character of Kinship*. Cambridge, England: Cambridge University Press.

Hahn, Frank H. 1971. "Equilibrium with Transaction Costs." *Econometrica* 39, no. 3: 417–40.

Hardin, Garrett. 1968. "The Tragedy of the Commons." *Science* 162: 1243–48.

Hecht, Suzanna. 1985. "Environment, Development and Politics: Capital Accumulation and the Livestock Sector in Eastern Amazonia." *World Development* 13, no. 6: 663–84.

Hecht, Suzanna, A. B. Anderson, and P. May. 1988. "The Subsidy from Nature: Shifting Cultivation, Successional Palm Forests, and Rural Development." *Human Organization* 47, no. 1: 25–35.

Henry, Claude. 1974. "Investment Decisions under Uncertainty: The Irreversibility Effect." *American Economic Review* 64, no. 4: 1006–12.

Iliffe, John. 1987. *The African Poor*. Cambridge, England: Cambridge University Press.

Jodha, N. S. 1980. "The Process of Desertification and the Choice of Interventions." *Economic and Political Weekly* 15.

———. 1986. "Common Property Resources and Rural Poor in Dry Regions of India." *Economic and Political Weekly* (India) 21 (July): 1169–81.

Johansson, Per. 1987. *The Economic Theory and Measurement of Environmental Benefits*. Cambridge, England: Cambridge University Press.

———. 1990. "Valuing Environmental Damage." *Oxford Review of Economic Policy* 6, no. 1: 34–50.

Kneese, Alan V. 1988. "Environmental Stress and Political Conflicts: Salinity in the Colorado River." Washington, D. C. Resources for the Future. Processed.

Krutilla, John. 1966. "The International Columbia River Treaty: An Economic Evaluation." In A. V. Kneese and S. Smith, eds., *Water Research*. Baltimore, Md.: Johns Hopkins University Press.

Kumar, Shiv K., and D. Hotchkiss. 1988. "Consequences of Deforestation for Women's Time Allocation, Agricultural Production, and Nutrition in Hill Areas of Nepal." International Food Research Institute Research Report 69.

Lind, Robert, ed. 1982. *Discounting for Time and Risk in Energy Policy*. Baltimore, Md.: Johns Hopkins University Press.

Little, Ian M. D., and James A. Mirrlees. 1969. *Manual of Industrial Project Analysis.* Paris: Organisation for Economic Co-operation and Development.

———. 1974. *Project Appraisal and Planning for Developing Countries.* London: Heinemann.

Mahar, Dennis. 1988. Government Policies and Deforestation in Brazil's Amazon Region." World Bank Environment Department Working Paper 7. Washington, D. C. Processed.

Mäler, Karl-Göran. 1974. *Environmental Economics: A Theoretical Inquiry.* Baltimore, Md.: Johns Hopkins University Press.

———. 1989. "Environmental Resources, Risk and Bayesian Decision Rules." Stockholm, Sweden: Stockholm School of Economics. Processed.

———. 1990a. "The Acid Rain Game, 2." Stockholm, Sweden: Stockholm School of Economics. Processed.

———. 1990b. "International Environmental Problems." *Oxford Review of Economic Policy* 6, no. 1: 80–108.

———. 1990c. "The Valuation of Environmental Resources." Stockholm, Sweden: Stockholm School of Economics. Processed.

Nerlove, Mark, and Anke Meyer. 1990. "Endogenous Fertility and the Environment: A Parable of Firewood." Philadelphia: University of Pennsylvania. Processed.

Nerlove, Mark, Assaf Razin, and Efraim Sadka. 1987. *Household and Economy: Welfare Economics of Endogenous Fertility.* New York: Academic Press.

Pearce, D., E. Barbier, and A. Markandya. 1988. "Sustainable Development and Cost-Benefit Analysis." Paper presented at the Canadian Environment Assessment Workshop on Integrating Economic and Environment Assessment.

Polanyi, Karl. 1944. *The Great Transformation.* New York: Holt, Rinehart and Winston.

———. 1977. *The Livelihood of Man.* New York: Academic Press.

Popkin, S. L. 1979. *The Rational Peasants: The Political Economy of Rural Society in Vietnam.* Berkeley: University of California Press.

Portes, Richard. 1971. "Decentralized Planning Procedures and Centrally Planned Economies." *American Economic Review* 61 (*Papers and Proceedings*): 422–29.

Repetto, Robert. 1988. "Economic Policy Reform for Natural Resource Conservation." World Bank Environment Department Paper 4. Washington, D. C. Processed.

Repetto, Robert, and others. 1989. *Wasting Assets: Natural Resources in the National Income Account.* Washington, D. C.: World Resources Institute.

Shapley, Lloyd, and Martin Shubik. 1969. "On the Core of an Economic System under Externalities." *American Economic Review* 59, no. 4: 678–84.

Schecter, M., M. Kim, and L. Golan. 1989. "Valuing a Public Good: Direct and Indirect Valuation Approaches to the Measurement of the Benefits from Pollution Abatement." In H. Folmer and E. van Ierland, eds., *Valuation Methods and Policy Making in Environmental Economics.* Amsterdam: Elsevier.

Scott, J. 1976. *The Moral Economy of the Peasant.* New Haven, Conn.: Yale University Press.

Solow, Robert M. 1974. "Intergenerational Equity and Exhaustible Resources." *Review of Economic Studies* 41 (Symposium on the Economics of Exhaustible Resources): 29–45.

———. 1986. "On the Intergenerational Allocation of Exhaustible Resources." *Scandinavian Journal of Economics* 88, no. 2: 141–56.

Starrett, David A. 1972. "Fundamental Non-Convexities in the Theory of Externalities." *Journal of Economic Theory* 4, no. 1: 180–99.

———. 1973. "A Note on Externalities and the Core." *Econometrica* 41, no. 1: 179–84.

Stern, Nicholas H. 1989. "The Economics of Development: A Survey." *Economic Journal* 99, no. 3: 597–685.

Tietenberg, T. H. 1980. "Transferable Discharge Permits and the Control of Stationary Source Air Pollution: A Survey and Synthesis." *Land Economics* 56: 391–416.

———. 1990. "Economic Instruments for Environmental Regulation." *Oxford Review of Economic Policy* 6, no. 1: 17–33.

Ulph, Alister. 1989. "A Review of Books on Resource and Environmental Economics." *Bulletin of Economic Research* 42, no. 3: 219–28.

Wade, Robert. 1987. "The Management of Common Property Resources: Finding a Cooperative Solution." *World Bank Research Observer* 2, no. 2: 219–34.

———. 1988. *Village Republics: Economic Conditions for Collective Action in South India*. Cambridge, England: Cambridge University Press.

Weale, Martin. 1990. "Environmental Statistics and the National Accounts." Cambridge, England: University of Cambridge. Processed.

Weitzman, Martin L. 1970. "Optimal Growth with Scale-Economies in the Creation of Overhead Capital." *Review of Economic Studies* 37, no. 4: 555–70.

———. 1976. "Welfare Significance of National Product in a Dynamic Economy." *Quarterly Journal of Economics* 90, no. 1: 156–62.

Wiser, W. H. 1936. *The Hindu Jajmani System*. Lucknow, India: Lucknow Publishing House.

576-92

[90)

[28]

NATURAL RESOURCE MODELING
Volume 4, Number 1, Winter 1990

Q210

Q30

D24

O40

Q32

GROWTH WITH EXHAUSTIBLE RESOURCES AND A MATERIALS - BALANCE PRODUCTION FUNCTION

L.S. GROSS
Keene State College
Keene, New Hampshire 03431

E.C.H. VEENDORP
Clark University
Worcester, Massachusetts 01610

ABSTRACT. The materials-balance production function proposed in this paper is a modified Cobb-Douglas-type function which is as neoclassical in nature as materials-balance considerations would seem to permit. Incorporation of such a production function in a neoclassical growth model severely limits the growth potential of an economy with exhaustible resources.

KEY WORDS: Economic growth, exhaustible resources, materials-balance, production function, substitution, technical progress.

1. Introduction. During the 1970s, the implications of scarcity of natural resources for economic growth became a controversial issue. Although some economists, such as Georgescu-Roegen and Daly, avowed that the finitude of resources ultimately limits all economic activity, others argued that the resource constraint can be eliminated by technical progress or substitution of renewable/reproducible inputs for exhaustible resources. In this paper, we propose a production function which captures some basic physical constraints on the use of natural resource inputs, and analyze the implications of incorporating such a production function in a neoclassical growth model with exhaustible resources. Given the limitations we place on production technology, recycling, and technical progress, steady growth appears to be insupportable.

2. Materials-balance and economic growth. Economic models often ignore the role of flows of materials and energy and relationships between the economy and the natural environment. Growing concern with depletion of resources and environmental pollution in the 1960s and 1970s forced economists to model the role of natural resources.

Some of them, including Boulding [1966], Daly [1968], Kneese, Ayres, and D'Arge [1970], and Georgescu-Roegen [1971], stressed the existence of a set of constraints on economic activity imposed by the laws of thermodynamics.

The relevance of thermodynamic laws for economic models continues to be controversial. This is particularly true for the Second, or Entropy, Law (see, for instance, Burness et. al. [1980] and responses by Daly, Burness and Cummings, and Norgaard [1986]). The significance of the First Law (conservation of mass) is more easily recognized. It was introduced to economists in the late 1960s as the principle of materials-balance: "[a]part from (temporary) accumulation in the form of capital goods, the laws of conservation of matter and of energy guarantee that the sum total of all waste flows to the environment from the economy must be equal to the sum total of all resources originally extracted from the environment."[1]

One implication of this principle is that consumption does not cause goods to disappear; rather, it merely transforms outputs from goods to waste. Stressing the pervasiveness of externalities that this implies, Kneese, Ayers and D'Arge [1970] formally modeled an economy constrained by the materials-balance principle, using a static general equilibrium model with fixed technical coefficients. More recently, Perrings [1986] applied the conservation of mass condition in a similar framework to explore the behavior of the system over time and the stability of global equilibrium. A materials-balance constraint has also been used in optimal growth models focusing on the relationships between pollution and growth paths (see, for instance, Smith [1972], D'Arge and Kogiku [1973], Kamien and Schwartz [1982]).

A second implication of materials-balance is the physical impossibility of producing material goods without natural resources, and this is the one stressed here. Making this constraint explicit in the modeling of economic growth may yield results that differ substantially from those derived using a traditional description of the production process.

The sustainability of economic growth, given exhaustible resources, has been analyzed extensively since the 1970s, but, with few exceptions, the models have ignored such physical constraints. Both Solow [1974] and Stiglitz [1974] analyzed the "survival" issue using a neoclassical growth model with a Cobb-Douglas aggregate production function having an exhaustible resource as an input. In this type of model, resources are essential for production in the sense that, without resources, no production can occur, but the resource input may be arbitrarily small provided that capital and/or labor inputs are sufficiently large. Thus, it was shown that a constant level of consumption could be maintained under three sets of conditions: (1) a constant elasticity of substitution (CES) production function with the elasticity of substitution between capital and resources greater than one; (2) a Cobb-Douglas aggregate production function with the share of capital exceeding the share of resources; or (3) a positive rate of ongoing exponential resource augmenting technical progress.

Several authors have since extended the model and relaxed various assumptions. Using a general-form production function with two inputs, capital (K) and resources (R), Kemp et. al. [1984] derived the condition for sustainable consumption in the absence of technical progress. This is the existence of some level of output Q^* for which there is a finite area under the Q^* isoquant between $K = K_0$ (the initial capital stock) and $K = \infty$. In any case, the specification of the aggregate production function is crucial to the outcome of the model.

3. A materials-balance production function. In order for models of economic growth to capture and evaluate a resource constraint, the role of natural resources in economic activity must be expressed in a physically realistic manner. If a simple aggregate production function is to be used, it should capture the constraint imposed by materials-balance that some minimum resource input is required in order to produce a given physical output. No material good can be produced without a more than equal input of resources, if both are expressed in comparable units of mass or weight.[2]

A description of the production process which maintains the spirit of materials-balance should have the following (related) properties:

(A) It should capture the difference between the nature of capital, labor, and resource inputs. This distinction is stressed by Georgescu-Roegen [1970], who classifies labor and capital inputs as funds or agents that transform resource inputs into product output, and resource inputs as flows in production.

(B) It should limit the amount of final product that can be produced with a given amount of resources. While one might be willing to accept the possibility that output is unbounded when any of the transforming agents is kept fixed, or even the possibility of positive production with a zero level of any one of these inputs, this is not the case for the resource input.[3]

(C) It should limit the degree of substitutability between resources and other factors of production. Although considerable substitution possibilities may exist between labor and capital funds, substitution between these transforming agents and the resource flow is clearly limited. A change in the amounts of capital and labor used cannot reduce the minimum input of resources required to produce a given physical mass of output, but can only reduce the amount of resource waste generated in the production process.

To assure asymmetry between the roles played by resources and other inputs, we propose the production function

$$Q = F(K, L, X), \quad K, L, X \geq 0,$$

with $X = R - Q$, where Q is output, K capital, L labor, R resources, and X a non-negative variable representing the amount of excess resources or resources wasted in production.[4] The function $F(\ )$ could represent any (traditional) production function. Since a linearly homogeneous Cobb-Douglas production function is commonly used in the related literature, we focus on this version of the materials-balance production function (MBCD)[5]

$$Q = K^{\alpha_1} L^{\alpha_2} (R - Q)^{\alpha_3}, \quad \alpha_1 + \alpha_2 + \alpha_3 = 1.$$

A production function with waste as one of the inputs is less peculiar than it might seem. Excess resources facilitate production in the same way as capital and labor do. The "efficient" level of waste will respond to input prices in the same way as other factors of production, with higher resource prices inducing management to use larger amounts of the other inputs to reduce waste.[6] Clearly, there are a number of production processes (paper, for example) where the amount of waste relative to output is significant, while in others (for example, gold jewelry) the "waste input" is quite small. Since all production processes will involve some amount of friction, however, the amount of waste must exceed zero.

The materials-balance production function satisfies properties A and B by assigning an asymmetric role to resources and by ensuring that the amount of resources used will always exceed the amount of final product produced. Our formulation clearly gives considerable weight to the resource input, which is both "necessary" and "important" in the sense of Dasgupta and Mitra [1983]. A zero level of resources precludes positive production, and the share of the resource input in output,

$$RQ_R/Q = \alpha_3 R/(R - Q + \alpha_3 Q),$$

must exceed zero. But the most characteristic feature of the materials-balance production function is that the iso-product curves are bounded away from the capital and labor axes but not from the resource axis. This is illustrated in Figure 1, which gives the isoquants of the MBCD for $Q = 5$, $L = 5$ and $Q = 10, L = 10$, using $\alpha_i = 1/3$.

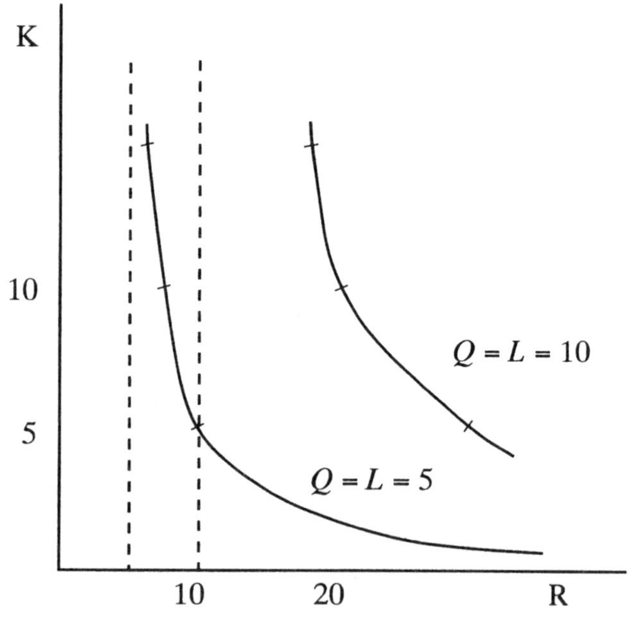

FIGURE 1. Isoquants for the MBCD Production Function ($\alpha_i = 1/3$).

To see how severely the MBCD production function limits substitution possibilities between resources and other inputs (property C), we calculate the elasticity of substitution for the two-factor MBCD production function

$$Q = K^{\alpha}(R-Q)^{\beta}, \quad \alpha+\beta=1.$$

Since the MBCD production function is linearly homogeneous with marginal products

$$\begin{aligned} Q_K &= \alpha Q(R-Q)/[K(R-Q+\beta Q)] \\ Q_R &= \beta Q/(R-Q+\beta Q) \end{aligned}$$

and second order partial derivatives

$$\begin{aligned} Q_{RR} &= \alpha\beta Q(Q-R)/(R-\alpha Q)^3 \\ Q_{KK} &= \alpha\beta(Q-R)QR^2/K^2(R-\alpha Q)^3 \\ Q_{RK} &= Q_{KR} = [\alpha\beta QR(R-Q)]/[K(R-Q+\beta Q)^3], \end{aligned}$$

we have

$$\sigma_{KR} = Q_K Q_R / Q_{KR} Q = 1 - \alpha(Q/R),$$

which is always positive but less than one. Specifically,

$$\lim_{K/R\to\infty} \sigma_{KR} = \beta, \quad \lim_{K/R\to 0} \sigma K_R = 1.$$

Capital is a fairly good substitute for resources when the capital-resource ratio is low (with little capital, resource waste is high) but substitution becomes more and more difficult as the capital-resource ratio increases. That substitutability may well differ "at the corners" is a conjecture that has been made by several authors (for example, Dasgupta and Heal [1979], who nonetheless use a CES production function in their analysis).

Although our MBCD production function stresses the asymmetry between resources and other inputs, it preserves the flavor of a neoclassical production function. Both marginal products are positive, although $Q_R < 1$, an obvious implication of property B. The second order partial derivatives satisfy the sign restrictions $Q_{KK} < 0$, $Q_{RR} < 0$, and $Q_{RK} = Q_{KR} > 0$. One can easily verify that the isoproduct curves are strictly convex.[7]

4. A growth model with materials-balance. Exhaustible resources and materials-balance will obviously limit the growth potential of an economy. Since the elasticity of substitution between capital and resources in the MBCD production function is always less that one, its use in Solow-Stiglitz type growth models without technical progress will generate the dismal result that a positive level of consumption is not sustainable.[8] In the following sections, we employ the general approach of Stiglitz (1974) to explore whether or to what extent the feasibility of some steady states can be salvaged if we allow for recycling or an expanded role for technical change.

a. *Capital or labor augmenting technical progress.* Imposing constant returns to scale but allowing for capital or labor augmenting technical change,[9] we use the production function

$$Q = K^{\alpha_1} L^{\alpha_2} X^{\alpha_3} e^{\lambda_t} \tag{1}$$

with $X = R - Q$ and $\alpha_1 + \alpha_2 + \alpha_3 = 1$. The growth rates of the relevant variables are therefore restricted by

$$g_Q = \dot{Q}/Q = \alpha_1 g_K + \alpha_2 g_L + \alpha_3 g_X + \lambda \tag{2}$$

with

$$g_X = [R/(R-Q)]g_R - [Q/(R-Q)]g_Q. \tag{3}$$

Total output is either consumed or invested. Assuming no depreciation, we can write

$$Q = C + \dot{K}$$

so that

$$g_K = sQ/K = (1-a)Q/K \tag{4}$$

where $s(a)$ is the propensity to save (consume). The labor force is assumed to grow exponentially:

$$g_L = \nu. \tag{5}$$

The intertemporal behavior of the economy is further constrained to adhere to the condition for efficient use of resources (the Hotelling rule) $Q_K = \dot{Q}_R/Q_R$. Using the expressions for the marginal products of capital and resources derived in the previous section in the efficiency condition, we obtain

$$g_R - g_Q = \alpha_1(Q/K)[(Q-R)/R]. \tag{6}$$

Writing $\theta = Q/K$ and $\rho = Q/R$ and using equations (3), (4), (5), and $g_\theta = g_Q - g_K = g_Q - s\theta$, we can replace (2) and (6) by the following differential equations in θ, C and ρ:

$$g_\theta = \frac{\alpha_2\nu + \lambda}{1-\alpha_3} - \frac{\alpha_2 s + \alpha_1\alpha_3}{1-\alpha_3}\theta \tag{7}$$

$$g_C = \frac{\alpha_2\nu + \lambda}{1-\alpha_3} + \frac{\alpha_1(s-\alpha_3)}{1-\alpha_3}\theta \tag{8}$$

$$g_\rho = \alpha_1\theta(1-\rho). \tag{9}$$

The output-capital ratio will approach the steady-state level

$$\hat{\theta} = (\alpha_2\nu + \lambda)/(\alpha_2 s + \alpha_1\alpha_3),$$

but the feasibility of this path is ruled out by (9) or

$$\dot{\rho} = \alpha_1\theta\rho(1-\rho),$$

which implies that ρ approaches unity and g_ρ approaches zero. Since g_Q approaches the (positive) level $s\hat{\theta}$ and since $g_R = g_Q - g_\rho$, g_R is bound to become and then stay positive, which is not sustainable if R is a finite exhaustible resource. Thus, positive exponential growth is not possible no matter how fast the labor force is growing and how high the rate of capital or labor augmenting technical progress is assumed to be.

b. *Recycling.* Recycling relaxes the resource constraint by enabling a fraction of the output being produced or of the excess resources used in production to be reused. Letting the first fraction be b_1 and the second b_2, net resource use is

$$R_N = R - b_1Q - b_2(R-Q) = (1-b_2)R - (b_1-b_2)Q$$

so that

$$\dot{R}_N = (1 - b_2)Rg_R - (b_1 - b_2)Qg_Q.$$

While it is conceptually possible to recycle all of the excess resources used in production as well as all of that portion of the final product that is consumed, complete recycling of the resources used for capital formation is certainly not possible.[10] Thus, we impose $b_1 < 1$ and $b_2 \leq 1$.

The same efficiency condition holds in a model which incorporates the above kind of recycling, so that we can use the results found in the previous section. With a constant rate of consumption, both g_R and g_Q approach the same level $(s\theta)$. Since $b_1 < 1$ and $R > Q$ implies $(1 - b_2)R > (b_1 - b_2)Q$, $\dot{R}_N$ will eventually turn positive. Recycling cannot prevent resource exhaustion although it does postpone the moment when it will take place.

c. *Output augmenting technical progress.* To expand the role of technical progress we generalize our materials-balance production function to

$$Q = K^{\alpha_1} L^{\alpha_2} (Re^{\mu t} - Q)^{\alpha_3} e^{\lambda t}, \tag{10}$$

which can be written as

$$Qe^{-\mu t} = K^{\alpha_1} L^{\alpha_2} (R - Qe^{-\mu t})^{\alpha_3} e^{(\lambda + \alpha_3 \mu - \mu)t}.$$

If $\lambda > \mu = 0$, technical progress is purely labor or capital augmenting, while, if $\mu > \lambda = 0$, it is purely resource augmenting. If $\lambda = (1-\alpha_3)\mu > 0$, technical progress is purely output augmenting since, in that case,

$$Qe^{-\mu t} = K^{\alpha_1} L^{\alpha_2} (R - Qe^{-\mu t})^{\alpha_3}.$$

Resource augmenting technical progress expands the "effective" stock of resources by increasing resource productivity, for example by enhancing the ability to extract and/or use lower quality resources or to recycle output and waste. However, given a fixed total physical quantity of depletable resources as well as upper limits to sustainable flows of renewable resources, these forms of technical progress are limited. New technologies may also increase the efficiency with which materials and energy are converted into goods and services, but the amount of output that can be obtained from given inputs of materials and energy is also limited. To use Baumol's [1986] example, no matter how fuel efficient automobiles are made, the amount of gasoline required to drive from New York to San Francisco cannot be reduced to zero.

It is not possible for technical progress to increase effective resource availability beyond such limits, so that materials-balance limits the physical mass of output. However, from the consumer's standpoint, the physical mass of output may be of little importance. Rather, the consumer purchases a certain bundle of services, and, given some relationship between output of mass and output of services, the potential for output augmenting technical progress exists. This type of technical progress increases the output of services from a given mass of output or enables the same bundle of services to be provided with a smaller output of mass. Nonetheless, the good which provides services must be embodied in some physical structure.[11]

Since purely resource augmenting technical progress seems to violate the spirit of materials-balance as well as the notion of exhaustible resources, we impose the restriction

$$\lambda \geq (1-\alpha_3)\mu \geq 0. \tag{11}$$

That is, the rate of technical progress is at most output augmenting, although even this ongoing exponential augmentation seems implausible.

Proceeding as in §4a, the new marginal products of capital and resources are

$$Q_K = \frac{\alpha_1 Q(Re^{\mu t} - Q)}{K(Re^{\mu t} - Q + \alpha_3 Q)}$$

$$Q_R = \frac{\alpha_3 Q e^{\mu t}}{Re^{\mu t} - Q + \alpha_3 Q},$$

while the efficiency condition becomes

$$\alpha_1 \left[\frac{Q}{K}\right]\left[\frac{Re^{\mu t} - Q}{Re^{\mu t}}\right] = g_Q - g_R - (1-\alpha_3)\mu\left[\frac{Q}{Re^{\mu t}}\right].$$

Using $\theta = Q/K$ and $\rho = Q/Re^{\mu t}$, this condition simplifies to

$$g_\rho = \alpha_1\theta(1-\rho) - (1-\rho+\alpha_3\rho)\mu. \tag{12}$$

The materials-balance production function (10) implies

$$g_Q = \alpha_1 g_K + \alpha_2 g_L + \alpha_3 g_X + \lambda$$

with

$$g_X = g_Q - [1/(1-\rho)]g_\rho.$$

Using $g_L = \nu$, $g_K = s\theta$, equations (7) and (9) are generalized to (12) and

$$(13) \qquad g_\theta = \frac{\alpha_2\nu + \lambda + \alpha_3\mu}{1-\alpha_3} - \frac{\alpha_1\alpha_3 + \alpha_2 s}{1-\alpha_3}\theta + \frac{\alpha_3^2\mu}{1-\alpha_3}\frac{\rho}{1-\rho}$$

Introducing the variable

$$z = \rho/(1-\rho) = Q/(Re^{\mu_t} - Q),$$

these differential equations can be simplified to

$$g_\theta = \frac{\alpha_2\nu + \lambda + \alpha_3\mu}{1-\alpha_3} - \frac{\alpha_1\alpha_3 + \alpha_2 s}{1-\alpha_3}\theta + \frac{\alpha_3^2\mu}{1-\alpha_3}z$$
$$g_z = -\mu + \alpha_1\theta - \alpha_3\mu_z.$$

Drawing a phase diagram for this system in θ, z space, we observe that both loci $g_\theta = 0$ and $g_z = 0$ are positively sloped with the first locus steeper than the second unless $s = 0$. For $s > 0$, the intersection occurs at

$$\theta^* = (\alpha_2\nu + \lambda)/\alpha_2 s$$
$$z^* = (\alpha_1\alpha_2\nu + \alpha_1\lambda - \alpha_2 s\mu)/\alpha_2\alpha_3 s\mu,$$

which is located in the first quadrant if

$$\mu < (\alpha_1\alpha_2\nu + \alpha_1\lambda)/\alpha_2 s$$

and in the fourth quadrant if the sign is reversed.

Figure 2 illustrates the first possibility. All trajectories will approach θ^*, z^* and will be potentially feasible if

$$g_R^* = g_\theta^* + s\theta^* - \mu - g_z^*/(z+1) = s\theta^* - \mu < 0.$$

Feasibility of trajectories approaching a positive level of z^* requires therefore that

$$(14) \qquad \frac{\alpha_2\nu + \lambda}{\alpha_2} < \mu < \frac{\alpha_1}{s}\left[\frac{\alpha_2\nu + \lambda}{\alpha_2}\right],$$

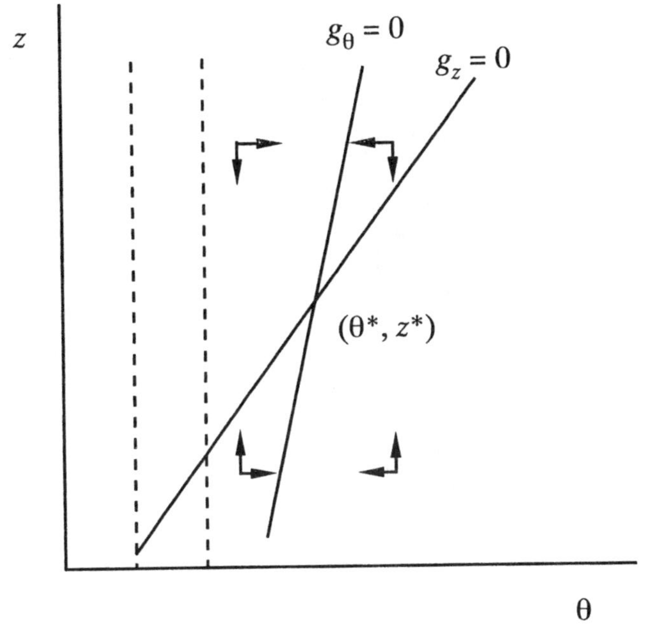

FIGURE 2. Phase Diagram for the MBCD Growth Model, Quadrant I Solution.

which in turn requires that

$$\alpha_1 > s$$

and

$$\alpha_2\nu + \lambda > 0.$$

However, condition (14) clearly violates the condition that technical progress is not resource augmenting, since (11) implies that

$$\mu \leq \lambda/(1 - \alpha_3) = \lambda/(\alpha_1 + \alpha_2) < (\alpha_2\nu + \lambda)/\alpha_2.$$

Figure 3 illustrates the second possibility that $g_\theta = 0$ and $g_z = 0$ intersect in the fourth quadrant. All time paths approach

$$\theta^0 = (\alpha_2\nu + \lambda + \alpha_3\mu)/(\alpha_1\alpha_3 + \alpha_2 s)$$
$$z^0 = 0.$$

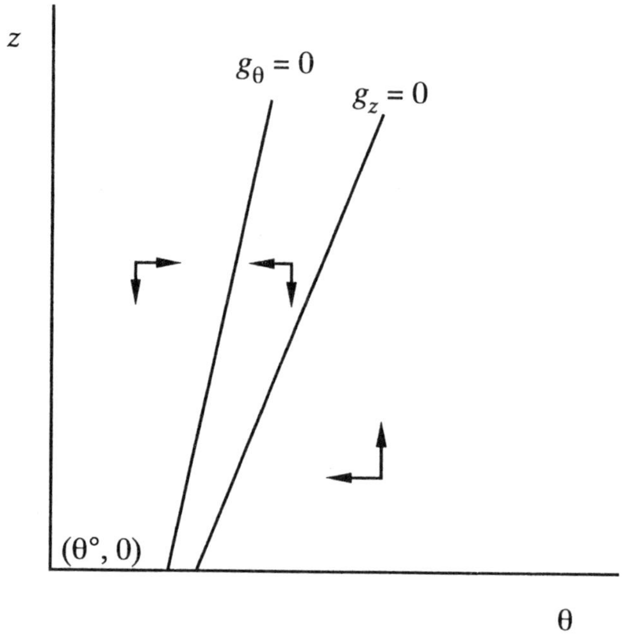

FIGURE 3. Phase Diagram for the MBCD Growth Model, Quadrant IV Solution.

Feasibility for this case requires that

$$g_R^0 = g_\theta^0 + s\theta^0 - g_\rho^0 - \mu = (s - \alpha_1)\theta^0 < 0$$

or

$$\alpha_1 > s.$$

Since this case applies when $\mu > \alpha_1(\alpha_2\nu + \lambda)/\alpha_2 s$, the feasibility condition $\alpha_1 > s$ implies that $\mu > (\alpha_2\nu + \lambda)/\alpha_2$, which again violates our restriction on the rate of technical progress.

Figure 4 depicts the special case that $g_\theta = 0$ and $g_z = 0$ are parallel lines. This occurs when all output is consumed ($s = 0$) in which case

$$g_\theta = \frac{\alpha_2\nu + \lambda + \alpha_3\mu}{1 - \alpha_3} - \frac{\alpha_1\alpha_3}{1 - \alpha_3}\theta + \frac{\alpha_3^2\mu}{1 - \alpha_3}z$$

$$g_z = -\mu + \alpha_1\theta - \alpha_3\mu z.$$

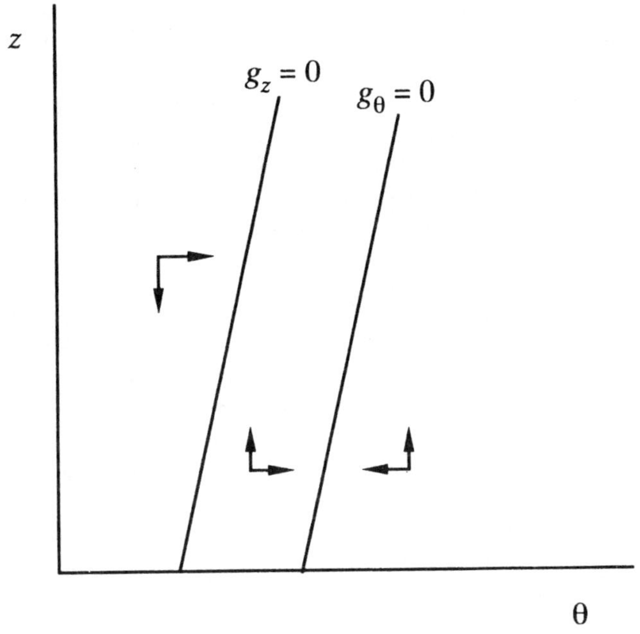

FIGURE 4. Phase Diagram for Growth with a Zero Savings Rate.

Inspection of Figure 4 shows that θ and z will eventually follow a path along which

$$g_\theta = g_z = \alpha_2 \nu + \lambda.$$

This path is feasible provided that

$$g_R = g_\theta - \mu - g_z/(z+1) < 0$$

or

$$\alpha_2 \nu + \lambda - \mu < 0.$$

Using the condition that the rate of technical progress be no more than the output augmenting rate, the condition for a feasible path requires that

$$\alpha_2 \nu < \alpha_3 \mu.$$

Thus, if people consume all they produce and the labor force grows relatively slowly, a constant rate of consumption is sustainable as long as there is sufficient output augmenting technical progress. However, if, by saving at a positive rate, current generations endow future ones with a larger capital stock, resource exhaustion is bound to occur. Limited substitutability between capital and resources is the basis for this counterintuitive result. A growing capital stock not only depletes resources in the production of the capital itself, but is used to produce even more output. Given the materials-balance constraint, this will speed up the depletion of the stock of resources.

Conclusion. Materials-balance severely limits the growth potential for an economy with exhaustible resources. Conventional wisdom has it that continued growth is feasible given sufficient substitution possibilities between resources and other inputs, and sufficiently high rates of technical progress. By obeying the law of conservation of mass, the materials-balance production function limits substitution to a greater degree than those typically used to explore the relationship between exhaustible resources and economic growth. In addition, materials-balance and the notion of a finite resource stock rule out technical progress of the purely resource augmenting type. Using our MBCD production function, steady growth is not feasible, even if one allows for the recycling of all output that is not invested and even if one allows for exponential technical progress of the labor, capital, or output augmenting variety.

It should be stressed that our model captures only one of the issues raised by the materials-balance principle, as we do not analyze the implications of the accumulation of non-recycled waste for continued growth of production and consumption. Hence, we have not addressed the dilemma posed by D'Arge and Kogiku [1973]: will we first run out of resources or will we first make the planet unfit for life?

ENDNOTES

1. Ayres [1978, p. 31]. Although use of the materials-balance principle predates this statement, it offers a particularly concise definition.

2. This materials-balance constraint does not pertain to the production of services. It is typically applied to the processing of materials, as in manufacturing and construction. We focus on the use of exhaustible resources for the production of tangible goods, where this constraint is clearly relevant.

3. A similar observation is made by Dasgupta and Heal [1979], who note that the laws of thermodynamics imply that production requires a certain minimum input of energy, limiting substitution between capital and energy. Such a minimum energy requirement could be met by using a CES production function with elasticity of substitution less than one. But, since the CES function treats all inputs symmetrically, the production of any particular level of output would then not only require a minimum input of energy (or resources), but also a minimum amount of capital. Anderson [1987] also suggests the use of a CES production function to capture thermodynamic constraints.

4. To allow for Q and R to be measured in different units, the same production function could be written as $Q = F(K, L, X')$, with $X' = R - \eta Q$ and $\eta > 0$.

5. The restriction $\Sigma\alpha_i = 1$ assures linear homogeneity: doubling, for instance, the levels of Q, R, and K clearly doubles the required amount of L.

6. Anderson [1987] also makes this observation, and distinguishes between economic and by-product waste.

7. The only other production function found in the growth literature which specifically addresses materials-balance and treats resources asymmetrically is the Bergstrom function used by Ayres and Miller [1980],

$$Q = R[1 - \exp(-F(K, L)/R)],$$

where $F(\ )$ could be any conventional production function. Checking the properties explored for the MBCD function in the two-factor version $Q = R[1 - \exp(-\gamma K/R)]$, we find that it limits substitution between K and R more severely than the MBCD function and that it is somewhat less neoclassical in nature.

8. We can also refer to the Kemp et. al. [1984] result concerning the shape of the isoquant. For the MBCD production function, any isoquant in capital-resource space approaches an asymptote at the minimum resource input required to produce a specific mass of output. Thus, the area under any isoquant between the initial capital stock and infinity is not finite, and positive consumption is not sustainable without technical progress.

9. Alternatively, this could be called Hicks-neutral, or even "waste augmenting," although the latter is difficult to interpret.

10. That complete recycling also violates the Second Law of thermodynamics is an argument made repeatedly by Georgescu-Roegen.

11. Much more could be said on the subject of expressing technical progress realistically. For example, Ayres and Miller [1980] argue that the neoclassical treatment of technical progress generally ignores the material and energy resources required for technical progress to occur and to be utilized or embodied in new capital. In many growth models, technical progress is implicitly assumed to depend only on time, and is thus invoked as a sort of *deus ex machina* which may rescue society from resource exhaustion. Whether modeled as exogenous or endogenous, however, the physical limits to technical progress noted in the text continue to apply.

REFERENCES

Curt L. Anderson [1987], *The production process: inputs and waste,* J. Environ. Econ. Manage. **14**, 1–12.

Robert U. Ayres [1978], *Resources, Environment, and Economics,* John Wiley and Sons, New York.

Robert U. Ayres and S.M. Miller [1980], *The role of technological change,* J. Environ. Econ. Manage. **7**, 353–371.

William J. Baumol [1986], *On the possibility of continuing expansion of finite resources,* Kyklos **39**, 167–179.

Kenneth E. Boulding [1966], "The economics of the coming spaceship earth," in *Environmental Quality in a Growing Economy* (H. Jarrett, Ed.), Johns Hopkins University Press, Baltimore.

H. Stuart Burness and R.G. Cummings [1986], *Thermodynamic and economic concepts as related to resource-use policies: reply,* Land Econ. **62**, 323–324.

H. Stuart Burness, Ronald Cummings, Glenn Norris and Inja Paik [1980], *Thermodynamic and economic concepts as related to resource-use policies,* Land Econ. **56**, 1–9.

Herman E. Daly [1986], *On economics as a life science,* J. Polit. Econ. **76**, 392–406.

——— [1986], *Thermodynamic and economic concepts as related to resource-use policies: comment,* Land Econ. **62**, 319–322.

Ralph C. D'Arge and K.C. Kogiku, *Economic growth and the environment,* Rev. Econ. Stud. **40**, 61–77.

P.S. Dasgupta and G.M. Heal [1979], *Economic Theory and Exhaustible Resources,* Cambridge University Press, Cambridge, Eng.

Swapan Dasgupta and Tapan Mitra [1983], *Intergenerational equity and efficient allocation of exhaustible resources,* Int. Econ. Rev. **24**, 133–153.

Nicholas Georgescu-Roegen [1970], *The economics of production,* Amer. Econ. Rev. **60.2**, 1–9.

——— [1971], *The Entropy Law and the Economic Process,* Harvard University Press, Cambridge, MA.

Morton I. Kamien and Nancy L. Schwartz, "The role of common property resources in optimal planning models with exhaustible resources," in *Explorations in Natural Resource Economics* (V. Kerry Smith and John V. Krutilla, Eds.), Johns Hopkins University Press, Baltimore.

M.C. Kemp, N.V. Long, and K. Shimomura [1984], "The problem of survival: a closed economy," in *Essays in the Economics of Exhaustible Resources* (M.C. Kemp and N.V. Long, Eds.), North-Holland, New York.

Allen V. Kneese, Robert U. Ayres and Ralph C. D'Arge [1970], *Economics and the Environment: a Materials Balance Approach,* Johns Hopkins University Press, Baltimore.

Tapan Mitra [1978], *Efficient growth with exhaustible resources in a neoclassical model,* J. Econ. Theory **17**, 114–129.

Richard B. Norgaard [1986], *Thermodynamic and economic concepts as related to resource-use policies: synthesis,* Land. Econ. **62**, 325–328.

Charles Perrings [1986], *Conservation of mass and instability in a dynamic economy-environment system,* J. Environ. Econ. Manage. **13**, 199–211.

Vernon L. Smith [1972], *Dynamics of waste accumulation: disposal versus recycling,* Quart. J. Econ. **86**, 600–616.

R.M. Solow [1974], *Intergenerational equity and exhaustible resources,* Rev. Econ. Stud. Supp. 29–45.

Joseph E. Stiglitz [1974], *Growth with exhaustible natural resources: efficient and optimal growth paths,* Rev. Econ. Stud. Supp., 123–137. [94]

[29]

Endogenous Technological Change

Paul M. Romer

University of Chicago

Growth in this model is driven by technological change that arises from intentional investment decisions made by profit-maximizing agents. The distinguishing feature of the technology as an input is that it is neither a conventional good nor a public good; it is a nonrival, partially excludable good. Because of the nonconvexity introduced by a nonrival good, price-taking competition cannot be supported. Instead, the equilibrium is one with monopolistic competition. The main conclusions are that the stock of human capital determines the rate of growth, that too little human capital is devoted to research in equilibrium, that integration into world markets will increase growth rates, and that having a large population is not sufficient to generate growth.

I. Introduction

Output per hour worked in the United States today is 10 times as valuable as output per hour worked 100 years ago (Maddison 1982). In the 1950s, economists attributed almost all the change in output per hour worked to technological change (Abramovitz 1956; Kendrick 1956; Solow 1957). Subsequent analysis raised our estimates of

Prepared for the conference "The Problem of Economic Development: Exploring Economic Development through Free Enterprise," held at the State University of New York at Buffalo, May 1988. I have benefited from the comments of many seminar and conference participants and two discussants (Rob Vishny, Buffalo, May 1988, and Dale Jorgenson, National Bureau of Economic Research Economic Fluctuations meeting, July 1988). Discussions with Gary Becker, Karl Shell, Robert Lucas, Gene Grossman, and Elhanan Helpman were especially helpful. Research assistance was provided by Danyang Xie. The original work was supported by National Science Foundation grant SES-8618325. It was revised while I was a visitor at the Center for Advanced Study in the Behavioral Sciences and supported by NSF grant BNS87-00864.

[*Journal of Political Economy*, 1990, vol. 98, no. 5, pt. 2]

the importance of increases in the effective labor force and the effective stock of capital in generating growth in output per worker (Jorgenson, Gollop, and Fraumeni 1987), but technological change has surely been important as well. The raw materials that we use have not changed, but as a result of trial and error, experimentation, refinement, and scientific investigation, the instructions that we follow for combining raw materials have become vastly more sophisticated. One hundred years ago, all we could do to get visual stimulation from iron oxide was to use it as a pigment. Now we put it on plastic tape and use it to make videocassette recordings.

The argument presented in this paper is based on three premises. The first is that technological change—improvement in the instructions for mixing together raw materials—lies at the heart of economic growth. As a result, the model presented here resembles the Solow (1956) model with technological change. Technological change provides the incentive for continued capital accumulation, and together, capital accumulation and technological change account for much of the increase in output per hour worked.

The second premise is that technological change arises in large part because of intentional actions taken by people who respond to market incentives. Thus the model is one of endogenous rather than exogenous technological change. This does not mean that everyone who contributes to technological change is motivated by market incentives. An academic scientist who is supported by government grants may be totally insulated from them. The premise here is that market incentives nonetheless play an essential role in the process whereby new knowledge is translated into goods with practical value. Our initial understanding of electromagnetism arose from research conducted in academic institutions, but magnetic tape and home videocassette recorders resulted from attempts by private firms to earn a profit.

The third and most fundamental premise is that instructions for working with raw materials are inherently different from other economic goods. Once the cost of creating a new set of instructions has been incurred, the instructions can be used over and over again at no additional cost. Developing new and better instructions is equivalent to incurring a fixed cost. This property is taken to be the defining characteristic of technology.

Most models of aggregate growth, even those with spillovers or external effects, rely on price-taking behavior. But once these three premises are granted, it follows directly that an equilibrium with price taking cannot be supported. Section II of the paper starts by showing why this is so. It also indicates which of the premises is dropped in growth models that do depend on price-taking behavior. The argument in this section is fundamental to the motivation for the particular model of monopolistic competition that follows, but it is more general than the model itself.

In the specific model outlined in Section III, a firm incurs fixed design or research and development costs when it creates a new good. It recovers those costs by selling the new good for a price that is higher than its constant cost of production. Since there is free entry into this activity, firms earn zero profit in a present value sense.

The conclusions of the model follow directly from this specification. On the basis of results from the static theory of trade with differentiated goods (see, e.g., Helpman and Krugman 1985), one should expect that fixed costs lead to gains from increases in the size of the market and therefore to gains from trade between different countries. Perhaps the most interesting feature of the equilibrium calculated for the model constructed here is that increases in the size of the market have effects not only on the level of income and welfare but also on the rate of growth. Larger markets induce more research and faster growth.

The analysis also suggests why population is not the right measure of market size and why the presence of a large domestic market in countries such as China or India is not a substitute for trade with the rest of the world. The growth rate is increasing in the stock of human capital, but it does not depend on the total size of the labor force or the population. In a limiting case that may be relevant for historical analysis and for the poorest countries today, if the stock of human capital is too low, growth may not take place at all.

These implications of the model are taken up briefly in the final sections of the paper. Section III describes the functional forms that are used to describe the preferences and the technology for the model. It defines an equilibrium that allows for both monopolistic competition and external effects arising from knowledge spillovers. Section IV offers a brief intuitive description of a balanced growth equilibrium for the model. Section V formally characterizes the equilibrium. Section VI describes the welfare properties of the equilibrium. Section VII discusses the connection implied by the model between trade, research, and growth. Algebraic details of the derivations are placed in the Appendix.

II. Rivalry, Excludability, and Nonconvexities

Economists studying public finance have identified two fundamental attributes of any economic good: the degree to which it is rivalrous and the degree to which it is excludable (Cornes and Sandler 1986). Rivalry is a purely technological attribute. A purely rival good has the property that its use by one firm or person precludes its use by another; a purely nonrival good has the property that its use by one firm or person in no way limits its use by another. Excludability is a function of both the technology and the legal system. A good is excludable if the owner can prevent others from using it. A good such as the code for a computer program can be made excludable by means of a legal system that prohibits copying or by means of encryption and copy protection schemes.

Conventional economic goods are both rivalrous and excludable. They are privately provided and can be traded in competitive markets. By definition, public goods are both nonrival and nonexcludable. Because they are nonexcludable, they cannot be privately provided or traded in markets. Public goods can be introduced into a model of price-taking behavior by assuming the existence of a government that can levy taxes. Basic scientific research is an example of a public good that could be provided in this way and that is relevant for modeling growth.

Rivalry and excludability are closely linked because most rival goods are excludable. (A parking space in a shopping center parking lot is an example of a good that is effectively nonexcludable because the cost of enforcing excludability is too high relative to the value of the good.) The interesting case for growth theory is the set of goods that are nonrival yet excludable. The third premise cited in the Introduction implies that technology is a nonrival input. The second premise implies that technological change takes place because of the actions of self-interested individuals, so improvements in the technology must confer benefits that are at least partially excludable. The first premise therefore implies that growth is driven fundamentally by the accumulation of a partially excludable, nonrival input.

To evaluate these claims, it helps to have a specific case in mind. The example of a nonrival input used in what follows is a design for a new good. The vast majority of designs result from the research and development activities of private, profit-maximizing firms. A design is, nonetheless, nonrival. Once the design is created, it can be used as often as desired, in as many productive activities as desired.

In this sense, a design differs in a crucial way from a piece of human capital such as the ability to add. The design is nonrival but the ability to add is not. The difference arises because the ability to add is inherently tied to a physical object (a human body) whereas the design is not.[1] The ability to add is rivalrous because the person who

[1] The original version of this paper used the terms "embodied" and "disembodied" to refer to the difference between an intangible such as the ability to add, which is tied to a specific person, and an intangible such as a design, which is not. This choice of terminology is not used in this revision because embodiment has another meaning in growth theory and because the notion of rivalry already exists in the public finance literature.

possesses this ability cannot be in more than one place at the same time; nor can this person solve many problems at once. As noted above, rivalry leads to a presumption that human capital is also excludable. Thus human capital can be privately provided and traded in competitive markets. In contrast, the design is nonrival because it is independent of any physical object. It can be copied and used in as many different activities as desired.

Like any scientific concept, nonrivalry is an idealization. It is sometimes observed that a design cannot be a nonrival good because it is itself tied to the physical piece of paper or the physical computer disk on which it is stored. What is unambiguously true about a design is that the cost of replicating it with a drafter, a photocopier, or a disk drive is trivial compared to the cost of creating the design in the first place. This is not true of the ability to add. Training the second person to add is as costly as training the first. For simplicity, the arguments here will treat designs as idealized goods that are not tied to any physical good and can be costlessly replicated, but nothing hinges on whether this is literally true or merely close to being true.

Nonrivalry has two important implications for the theory of growth. First, nonrival goods can be accumulated without bound on a per capita basis, whereas a piece of human capital such as the ability to add cannot. Each person has only a finite number of years that can be spent acquiring skills. When this person dies, the skills are lost, but any nonrival good that this person produces—a scientific law; a principle of mechanical, electrical, or chemical engineering; a mathematical result; software; a patent; a mechanical drawing; or a blueprint—lives on after the person is gone. Second, treating knowledge as a nonrival good makes it possible to talk sensibly about knowledge spillovers, that is, incomplete excludability. These two features of knowledge—unbounded growth and incomplete appropriability—are features that are generally recognized as being relevant for the theory of growth. What thinking about nonrivalry shows is that these features are inextricably linked to nonconvexities.

If a nonrival input has productive value, then output cannot be a constant-returns-to-scale function of all its inputs taken together. The standard replication argument used to justify homogeneity of degree one does not apply because it is not necessary to replicate nonrival inputs. Suppose that a firm can invest 10,000 hours of engineering time to produce a design for a 20-megabyte hard disk drive for computers. Suppose that it can produce a total of 2 trillion megabytes of storage per year (i.e., 100,000 units of the drive) if it builds a $10 million factory and hires 100 workers. If it merely replicates the rival inputs—the factory and the workers—it can double its output to 4 trillion megabytes of storage per year.

Suppose that the firm could have invested 20,000 hours of engineering time in the design work instead of 10,000 hours and, by doing so, could have designed a 30-megabyte hard disk drive that could be manufactured with the same factory and workers. When the firm doubles all its inputs, it uses a 20,000-hour design, two factories, and 200 workers and produces 6 trillion megabytes of storage per year, three times the original output.

More formally, if $F(A, X)$ represents a production process that depends on rival inputs X and nonrival inputs A, then by a replication argument, it follows that $F(A, \lambda X) = \lambda F(A, X)$. This replication argument assumes that X is an exhaustive list of the rival inputs. Because the focus here is on national economies, the argument neglects integer problems that may be relevant for a small market that gets stuck between n and $n + 1$ plants. The fact that it may not be possible to actually replicate all the inputs in the list X has no bearing on this argument about the properties of $F(\cdot)$.

If A is productive as well, it follows that F cannot be a concave production function because $F(\lambda A, \lambda X) > \lambda F(A, X)$. Because of the properties of homogeneous functions, it also follows that a firm with these kinds of production possibilities could not survive as a price taker. If disk drives sold for marginal cost, annual revenue for the firm would just equal interest payments on the capital and wage payments to workers. More generally, since $F(A, X) = X \cdot (\partial F/\partial X)(A, X)$, it follows that

$$F(A, X) < A \cdot \frac{\partial F}{\partial A}(A, X) + X \cdot \frac{\partial F}{\partial X}(A, X).$$

If all inputs were paid their value marginal product, the firm would suffer losses.

This point has been made many times before (Schumpeter 1942; Arrow 1962*b*; Shell 1966, 1967, 1973; Nordhaus 1969; Wilson 1975). Previous growth models have avoided this difficulty in various ways. Solow (1956) treats A as an exogenously provided public input (i.e., an input that is both nonexcludable and nonrival). Shell (1966, 1967) treats it as a public input that is provided by the government. In each case, the factor A receives no compensation, and every individual firm is assumed to be free to exploit the entire stock of A. These models are consistent with the first premise, that technological change drives growth, and the third, that the technology is a nonrival good, but they are inconsistent with the second premise. They both deny the role that private, maximizing behavior plays in generating technological change.

In an attempt to make the evolution of A responsive to market incentives, Arrow (1962*a*) assumed that an increase in K necessarily leads to an equiproportionate increase in knowledge through "learning by doing," but he still treats knowledge as a public good. Lucas (1988) assumed in effect that it is production of human capital rather than physical capital that generates this nonrival, nonexcludable good. Both of these papers make the production of a nonrival, nonexcludable good an unintentional side effect of the production of a conventional good.

The learning-by-doing formulation has the advantage that it makes the rate of accumulation of nonrival knowledge endogenous, but it is unsatisfactory because it takes the strict proportionality between knowledge and physical capital or knowledge and education as an unexplained and exogenously given feature of the technology. It preserves the public-good character of knowledge assumed by Solow and Shell but makes it a public good that is privately provided as a side effect. Like the other public-good formulations, it rules out the possibility that firms make intentional investments in research and development.

This formulation has the additional difficulty that it is not robust. The nonrival input produced through learning by doing must be completely nonexcludable. If it were even partially excludable, Dasgupta and Stiglitz (1988) show that decentralized equilibrium with many firms would not be sustainable.

In a partial equilibrium model of an industry in which firms face upward-sloping cost curves, Shell (1973) proposed a model with price taking in which expenditure on research was compensated out of quasi rents. Griliches (1979), again in an industry setting, made this formulation more explicit. He assumed that the production function takes the form $F(A_N, A_E, X)$, where A_E represents an excludable part of the benefits of research and development and A_N represents the nonexcludable part. Since A_E is excludable, it is accumulated intentionally. The nonexcludable part A_N is created as a side effect of producing A_E. He also assumed that the function $F(\cdot)$ is homogeneous of degree one in X and A_E taken together.

In an aggregate model of growth, I made the same kind of assumption (Romer 1986). To make the dynamic analysis in this paper simple, I reduced the dynamic model to one with a single-state-variable model by assuming that the excludable good A_E that the firm produces intentionally is used in fixed proportions with physical capital. As a result, the model ends up having dynamics similar to those of Arrow's learning-by-doing model, and the mathematical equations can be interpreted equally well in terms of learning by doing that is incidental to capital production.

The advantage of the interpretation that knowledge is compensated out of quasi rents is that it allows for intentional private investments in research and development. The difficulty is that it violates the logic of the replication argument. If the input A_E is truly the result of research and development, it is a nonrival good. In this case, the function $F(\cdot)$ must be homogeneous of degree one in X alone and the Dasgupta and Stiglitz (1988) argument applies. If a nonrival input is even partially excludable, nonconvexities are present and a decentralized equilibrium cannot be sustained. Stated in terms of cost curves, the replication argument implies that long-run cost curves are horizontal when all rival inputs are treated as variable. What appear to be quasi rents are merely competitive returns to rival factors that are in fixed supply. These quasi rents cannot be used to compensate both the innovation activity and the rival fixed factors. Even if these factors are in fixed supply at the aggregate level, they presumably have alternative uses and will not be supplied to an activity if they are not paid their marginal product.

An alternative approach to growth theory with price-taking competition is to dispute the first and third premises, that technological change drives growth and that knowledge about the technology is a nonrival input. Human capital models such as those presented by King and Rebelo (1987), Jones and Manuelli (1988), Rebelo (1988), and Becker, Murphy, and Tamura (this issue) treat all forms of intangible knowledge as being analogous to human capital skills that are rivalrous and excludable. There is no nonrival input like the technology and, hence, no nonconvexities or spillovers.

The only way to accept all three premises described in the Introduction is to return to the suggestion of Schumpeter (1942) and explicitly introduce market power. Shell (1973) described a model with a single monopolist who invests in technological change, but as he recognizes, it is difficult to give an aggregate interpretation to a model dominated by a single firm. In a recent paper (Romer 1987), I presented a model with market power but also with free entry and many firms. It builds on the model of monopolistic competition in consumption goods formulated by Dixit and Stiglitz (1977), applied in a dynamic setting by Judd (1985), used in a dynamic model of trade by Grossman and Helpman (1989*c*), and extended to differentiated inputs in production by Ethier (1982). The specification used here differs from that in my earlier paper primarily because it emphasizes the importance of human capital in the research process. The earlier model showed that scale is an important determinant of the rate of growth. The analysis here shows that the correct measure of scale is not population but human capital.

III. Description of the Model

The four basic inputs in this model are capital, labor, human capital, and an index of the level of the technology. Capital is measured in units of consumption goods. Labor services L are skills such as eye-hand coordination that are available from a healthy physical body. They are measured by counts of people. As used here, human capital H is a distinct measure of the cumulative effect of activities such as formal education and on-the-job training.

The concept of human capital as years of education or training that are person specific is close to the one used in labor market contexts (e.g., Heckman 1976; Rosen 1976). It corresponds to the practice in growth accounting applications that take account of changes in the quality of the labor force due to changes in observables such as the level of education and experience (see, e.g., Gollop and Jorgenson 1980). This concept of human capital is more limited than the notion used in theoretical models of growth based on unlimited human capital accumulation such as those presented by King and Rebelo (1987), Lucas (1988), and Becker et al. (this issue). These models implicitly combine a notion of knowledge that can outlive any individual with a labor market notion of human capital that does not.

The model used here separates the rival component of knowledge, H, from the nonrival, technological component, A. Because it has an existence that is separate from that of any individual, A can grow without bound. In the specific formulation used below, each new unit of knowledge corresponds to a design for a new good, so there is no conceptual problem measuring A. It is a count of the number of designs.

The formal model of the economy has three sectors. The research sector uses human capital and the existing stock of knowledge to produce new knowledge. Specifically, it produces designs for new producer durables. An intermediate-goods sector uses the designs from the research sector together with forgone output to produce the large number of producer durables that are available for use in final-goods production at any time. In practice, one might expect research on a new design and the production of the new good to take place within the same firm, and nothing in the analysis here rules this out; design work can take place either internally or in a separate firm that sells its patent to the firm that will produce the actual good. A final-goods sector uses labor, human capital, and the set of producer durables that are available to produce final output. Output can be either consumed or saved as new capital.

To keep the dynamic analysis simple and highlight the effects of interest, several simplifying assumptions are used. The first is that the population and the supply of labor are both constant. This rules out an analysis of fertility, labor force participation, or variation in hours worked per worker. The second is that the total stock of human capital in the population is fixed and that the fraction supplied to the market is also fixed. Thus the supply of the aggregate factors L and H

is fixed. The assumption on H is made largely for technical reasons. The dynamic analysis is greatly simplified by restricting attention to equilibria with constant growth rates. In a stationary population in which people have finite lives, the only feasible constant growth rate for total years of education or experience is zero. A more complicated dynamic analysis could consider the effects of the kinds of increases in H and L that have been observed historically, but even in this kind of analysis, H must ultimately approach an upper bound.

The other simplifying assumptions are extreme assumptions on factor intensities. One has already been made implicitly. Assuming that capital can be accumulated as forgone output is equivalent to assuming that capital goods are produced in a separate sector that has the same technology as the final-output sector. Forgoing consumption is then equivalent to shifting resources from the consumption sector into the capital sector. Also, the plausible assertion that research is relatively human capital– and knowledge-intensive is translated into an extreme specification in which only knowledge and human capital are used to produce new designs or knowledge. Labor and capital do not enter at all. These kinds of restrictions will reduce the analysis of the dynamics of this system to a system of equations that can be explicitly solved by doing algebra. Presumably, a relaxation of these assumptions that preserves the factor intensity orderings used here would not change the basic dynamics of the model.

Final output Y in this model is expressed as a function of physical labor L, human capital devoted to final output H_Y, and physical capital. The unusual feature of the production technology assumed here is that it disaggregates capital into an infinite number of distinct types of producer durables. For now, let these durables be indexed by an integer i. (Soon, the index i will be assumed to be a continuous variable instead of a discrete one to avoid integer constraints.) Only a finite number of these potential inputs, the ones that have already been invented and designed, are available for use at any time. Thus if $x = \{x_i\}_{i=1}^{\infty}$ is the list of inputs used by a firm that produces final output, there is some value A such that $x_i = 0$ for all $i \geq A$. Because A changes as new producer durables are invented, it is important to be able to describe final output as a stationary function of all conceivable input lists.

In this kind of environment, a simple functional form for output is the following extension of the Cobb-Douglas production function:

$$Y(H_Y, L, x) = H_Y^{\alpha} L^{\beta} \sum_{i=1}^{\infty} x_i^{1-\alpha-\beta}. \tag{1}$$

This production function differs from the usual production function only in its assumption about the degree to which different types of capital goods are substitutes for each other. In the conventional specification, total capital K is implicitly defined as being proportional to the sum of all the different types of capital. This definition implies that all capital goods are perfect substitutes. One additional dollar of capital in the form of a truck has the same effect on the marginal productivity of mainframe computers as an additional dollar's worth of computers. Equation (1) expresses output as an additively separable function of all the different types of capital goods so that one additional dollar of trucks has no effect on the marginal productivity of computers.

Among the various types of capital goods, one can imagine pairs of inputs that are close substitutes (trucks and trains), pairs that are complements (computers and communications networks), and many pairs that fall somewhere in between. The conventional formulation explores the case in which all durables are perfect substitutes. The model here considers the case in which all durables have additively separable effects on output. An investigation of complementarity as well as of mixtures of types of substitutability is left for future work.

Because the production function in equation (1) is homogeneous of degree one, output in the final-goods sector can be described in terms of the actions of a single, aggregate, price-taking firm. The sector that produces producer durables, however, cannot be described by a representative firm. There is a distinct firm i for each durable good i. A firm must purchase or produce a design for good i before commencing production. Once it owns the design, the firm can convert η units of final output into one durable unit of good i. As in the standard one-sector model, the formal specification here describes the sector that produces capital goods as a black box that takes final output in on one side and gives capital goods out of the other side. The correct interpretation of this formal description is that the forgone consumption is never manufactured. The resources that would have been used to produce the forgone output are used instead to manufacture capital goods. It is possible to exchange a constant number of consumption goods for each unit of capital goods if the production function used to manufacture capital goods has exactly the same functional form as the production function used to manufacture consumption goods.

Once a firm has produced a design for durable i, it can obtain an infinitely lived patent on that design. If the firm manufactures $x(i)$ units of the durable, it rents those durables to final-output firms for a rental rate $p(i)$. Since firm i will be the only seller of capital good i, it will face a downward-sloping demand curve for its good. Since the durables are assumed not to depreciate, the value of one unit of durable i is the present discounted value of the infinite stream of rental income that it generates.

There are many equivalent institutional arrangements that can support any given equilibrium. In what follows, the firm that owns the patent on the design for durable i is assumed to be the only one that manufactures it, but this is only a convenience. Whether the owner of the patent manufactures the good itself or licenses others to do so, it can extract the same monopoly profit. Design of new durables and manufacturing could take place within the same firm, but it is easier to describe the equilibrium if the research and development department is treated as a separate firm and designs are transferred for an explicit price. It is also easier to assume that the firm that buys a design and manufactures a differentiated producer durable rents its durables instead of selling them outright. In particular, this shows that there are market mechanisms that avoid the usual durable-goods-monopoly problem. Provided that the manufacturer of the durable could commit to levels of output, nothing would change if the durables were sold instead of rented. The analysis is further simplified by assuming that the durables do not depreciate. Adding depreciation would merely add a familiar term to the user cost of capital.

In parallel with the usual one-sector model and in conformity with national income accounting conventions, it is useful to define an accounting measure of total capital K as cumulative forgone output. Thus $K(t)$ evolves according to the rule

$$\dot{K}(t) = Y(t) - C(t), \quad (2)$$

where $C(t)$ denotes aggregate consumption at time t. Because it takes η units of forgone consumption to create one unit of any type of durable, this accounting measure K is related to the durable goods that are actually used in production by the rule $K = \eta \sum_{i=1}^{\infty} x_i = \eta \sum_{i=1}^{A} x_i$.

Thus H and L are fixed, and K grows by the amount of forgone consumption. It remains to specify the process for the accumulation of new designs, that is, for the growth of $A(t)$. As noted above, research output depends on the amount of human capital devoted to research. It also depends on the stock of knowledge available to a person doing research. If designs were treated as discrete indivisible objects that are not produced by a deterministic production process, the production technology for designs would have to take explicit account of both integer constraints and uncertainty. There is no doubt that both indivisibility and uncertainty are important at the micro level and over short periods of time. The simplifying assumption made here is that neither is crucial to a first-pass analysis of technological change at the aggregate level. Henceforth, the index i

for the different types of goods is treated as a continuous variable, and the sum in equation (1) is replaced by an integral:

$$Y(H_Y, L, x) = H_Y^{\alpha} L^{\beta} \int_0^{\infty} x(i)^{1-\alpha-\beta} \, di. \tag{1'}$$

(Here x must be interpreted as a function from an appropriately defined function space.) With this formal structure, the output of new designs produced by researcher j can be written as a continuous, deterministic function of the inputs applied. If the researcher possesses an amount of human capital H^j and has access to a portion A^j of the total stock of knowledge implicit in previous designs, the rate of production of new designs by researcher j will be $\delta H^j A^j$, where δ is a productivity parameter.

This formulation with a continuum of goods is close to that used by Judd (1985) in his discussion of patents. He also studies the process whereby goods are introduced as new patents are produced. The model here corresponds to Judd's discussion of the case in which patents are infinitely lived. The main differences are that Judd treats the differentiated goods as consumption goods rather than producer durables and uses a form of exogenous technological change to generate growth of productivity in the research sector. Here, growth in A by itself increases the productivity of human capital in the research sector.

Although other assumptions about secrecy and property rights could be considered, the equilibrium here is based on the assumption that anyone engaged in research has free access to the entire stock of knowledge. This is feasible because knowledge is a nonrival input. All researchers can take advantage of A at the same time. The output of researcher j is therefore $\delta H^j A$. If we sum across all people engaged in research, the aggregate stock of designs evolves according to

$$\dot{A} = \delta H_A A, \tag{3}$$

where H_A has the obvious interpretation of total human capital employed in research.

Equation (3) contains two substantive assumptions and two functional form assumptions. The first substantive assumption is that devoting more human capital to research leads to a higher rate of production of new designs. The second is that the larger the total stock of designs and knowledge is, the higher the productivity of an engineer working in the research sector will be. According to this specification, a college-educated engineer working today and one working 100 years ago have the same human capital, which is measured in terms of years of forgone participation in the labor market. The engineer working today is more productive because he or she can take advantage of all the additional knowledge accumulated as design problems were solved during the last 100 years.

The two functional form assumptions are that the output of designs is linear in each of H_A and A when the other is held constant. These assumptions, like the exclusion of the inputs L and $x(\cdot)$, are made largely for analytical convenience. Linearity in H_A is not important for the dynamic properties of the model, but weakening this assumption would require a more detailed specification of how income in the research sector is allocated to the participants.

Linearity in A is what makes unbounded growth possible, and in this sense, unbounded growth is more like an assumption than a result of the model. In what follows, it will become clear that the marginal product of human capital H_Y employed in the manufacturing sector grows in proportion to A. If A were replaced in equation (3) by some concave function of A—that is, if the marginal productivity of human capital in the research sector does not continue to grow in proportion to A—human capital employed in research would shift out of research and into manufacturing as A becomes larger. This will cause the rate of growth to slow down. Whether opportunities in research are actually petering out, or will eventually do so, is an empirical question that this kind of theory cannot resolve. The specification here, in which unbounded growth at a constant rate is feasible, was chosen because there is no evidence from recent history to support the belief that opportunities for research are diminishing. Moreover, linearity in A is convenient analytically, and assumptions about what will happen in the far future for values of A that are very large relative to the current level have very little effect on the question of interest: How do other variables in the model affect the rate of growth of A?

The crucial feature of the specification used here is that knowledge enters into production in two distinct ways. A new design enables the production of a new good that can be used to produce output. A new design also increases the total stock of knowledge and thereby increases the productivity of human capital in the research sector. The owner of a design has property rights over its use in the production of a new producer durable but not over its use in research. If an inventor has a patented design for widgets, no one can make or sell widgets without the agreement of the inventor. On the other hand, other inventors are free to spend time studying the patent application for the widget and learn knowledge that helps in the design of a wodget. The inventor of the widget has no ability to stop the inventor of a wodget from learning from the design of a widget. This means that the benefits from the first productive role for a design are completely excludable, whereas the benefits from the second are completely non-excludable. In an overall sense, this means that the nonrival design inputs are partially excludable.

In theory and in practice, there is always some ambiguity about what constitutes a design for a new and different good and what constitutes a copy of an existing design. In the model, this ambiguity is artificially resolved by the form of the production function Y. This functional form implies that new goods are never close substitutes for existing goods since all the producer durables enter into production in an additively separable fashion. (See Pascoa [1986] for a discussion of this general property of "no neighboring goods.") Although it greatly simplifies the analysis, this is not a realistic feature of the model. In particular, it rules out the possibility of obsolescence. Nonetheless, the general results here should be robust to more careful modeling of the nature of the interaction between different specialized producer durables. What matters for the results is that the knowledge is a nonrival good that is partially excludable and privately provided.

At the aggregate level, H_A and H_Y are related by the constraint $H_Y + H_A = H$. According to these equations, any person can devote human capital to either the final-output sector or the research sector. Implicitly this formulation neglects the fact that L and H are supplied jointly. To take the equations used here literally, one must imagine that there are some skilled persons who specialize in human capital accumulation and supply no labor.

To fix notation for prices, let spot prices at any point in time be measured in units of current output and let r denote the interest rate on loans denominated in goods. Let P_A denote the price of new designs, and let w_H denote the rental rate per unit of human capital. Because goods can be converted into capital one for one, the spot price for capital is one and its rate of return is r. Because of the assumption that anyone engaged in research can freely take advantage of the entire existing stock of designs in doing research to produce new designs, it follows from equation (3) that P_A and w_H are related by $w_H = P_A \delta A$.

Once a design has been produced, a large number of potential suppliers of the new good bid for the right to do so. Each of these firms takes the price P_A for designs, the price of one for capital goods, and the interest rate as given, but if it begins production it sets prices to maximize profits. Formally, it helps to let the rental price $p(i)$ for the ith durable lie in the range $\mathbb{R}_+ \cup \{\infty\}$ for its durable good. If no firm produces good i, its price can be understood to be $p(i) = \infty$.

Faced with a price list $\{p(i): i \in \mathbb{R}_+\}$ for all the producer durables, including infinite prices for the durables that have not been invented yet, the representative final-output firm chooses a profit-maximizing quantity $x(i)$ for each durable. Because it is a constant-returns-to-scale firm, its input demands are defined only after the scale of operation is pinned down. Let L and H_Y be the total amounts of labor and human capital that are used in the production of final-output goods. (The split of total H between H_Y and H_A remains to be determined.) Given values for H_Y and L, it is possible to derive the aggregate demand for the durables from a maximization problem that is conditional on them:

$$\max_x \int_0^\infty [H_Y^\alpha L^\beta x(i)^{1-\alpha-\beta} - p(i)x(i)]di.$$

Differentiating under the integral sign leads to an inverse demand function

$$p(i) = (1 - \alpha - \beta)H_Y^\alpha L^\beta x(i)^{-\alpha-\beta}. \tag{4}$$

(There is an important technical issue about what it means to the final-goods producer if prices change on a set of values of i that has measure zero. Because of the symmetry in this model, eq. [4] can readily be derived by a limiting argument. For a general discussion of this issue, see Pascoa [1986, 1990].)

The demand curve in equation (4) is what the producer of each specialized durable takes as given in choosing the profit-maximizing price to set. Faced with given values of H_Y, L, and r, a firm that has already incurred the fixed-cost investment in a design will choose a level of output x to maximize its revenue minus variable cost at every date:

$$\begin{aligned}\pi &= \max_x p(x)x - r\eta x \\ &= \max_x (1 - \alpha - \beta)H_Y^\alpha L^\beta x^{1-\alpha-\beta} - r\eta x.\end{aligned} \tag{5}$$

The flow of rental income is $p(x)$ times x. The cost is the interest cost on the ηx units of output needed to produce x durables. To keep the analysis simple, it is convenient to assume that the capital is putty-putty, so that the firm can solve this problem at every point in time, converting units of durables back into general capital, and avoid the interest cost if it decides to supply fewer units. At any date, the only sunk cost is the initial expenditure on the design. This assumption is harmless because the demand for durables is stationary in equilibrium, so no disinvestment ever takes place.

The monopoly pricing problem specified in equation (5) is that of a firm with constant marginal cost that faces a constant elasticity demand curve. The resulting monopoly price is a simple markup over marginal cost, where the markup is determined by the elasticity of demand, $\bar{p} = r\eta/(1 - \alpha - \beta)$. The flow of monopoly profit is $\pi = (\alpha + \beta)\bar{p}\bar{x}$, where $\bar{x}$ is the quantity on the demand curve (4) implied by the price $\bar{p}$.

Each producer of specialized durables must rent its output to a large number of final-goods producers that can operate at any scale. By assumption, it is not possible for the producer to monitor the use of its durables. As a result, price discrimination is not feasible. The best the firm can do is charge the simple monopoly price.

The decision to produce a new specialized input depends on a comparison of the discounted stream of net revenue and the cost P_A of the initial investment in a design. Because the market for designs is competitive, the price for designs will be bid up until it is equal to the present value of the net revenue that a monopolist can extract. At every date t, it must therefore be true that

$$\int_t^\infty e^{-\int_t^\tau r(s)ds}\pi(\tau)d\tau = P_A(t). \tag{6}$$

If P_A is constant (as it will be in the equilibrium described below), this condition can be put in a more intuitive form. Differentiating with respect to time t yields

$$\pi(t) - r(t)\int_t^\infty e^{-\int_t^\tau r(s)ds}\pi(\tau)d\tau = 0.$$

Substituting in the expression for P_A from equation (6) yields

$$\pi(t) = r(t)P_A. \tag{6'}$$

This equation says that at every point in time, the instantaneous excess of revenue over marginal cost must be just sufficient to cover the interest cost on the initial investment in a design. (This formulation of the intertemporal zero profit constraint is taken from Grossman and Helpman [1989*c*].)

The solution of the model for a balanced growth equilibrium given in Section V shows that the technology described above implies a negatively sloped linear relation between the rate of growth of output and the rate of return on investment. To close the model, it remains to specify preferences that imply a parallel relation between the rate of growth of consumption and the marginal rate of intertemporal substitution. This relation is easily derived for Ramsey consumers with discounted, constant elasticity preferences:

$$\int_0^\infty U(C)e^{-\rho t}dt, \quad \text{with } U(C) = \frac{C^{1-\sigma} - 1}{1 - \sigma} \quad \text{for } \sigma \in [0, \infty).$$

The implied intertemporal optimization condition for a consumer faced with a fixed interest rate r is that $\dot{C}/C = (r - \rho)/\sigma$. Preferences enter the solution of the model only through this relation between the consumption growth rate and the interest rate.

The consumers are endowed with fixed quantities of labor L and human capital H that are supplied inelastically. At time 0, consumers own the existing durable-goods-producing firms, and the net revenues of these firms are paid to consumers as dividends. Final-goods firms earn zero profits and own no assets, so they can be ignored in the specification of endowments.

An equilibrium for this model will be paths for prices and quantities such that (i) consumers make savings and consumption decisions taking interest rates as given; (ii) holders of human capital decide whether to work in the research sector or the manufacturing sector taking as given the stock of total knowledge A, the price of designs P_A, and the wage rate in the manufacturing sector w_A; (iii) final-goods producers choose labor, human capital, and a list of differentiated durables taking prices as given; (iv) each firm that owns a design and manufactures a producer durable maximizes profit taking as given the interest rate and the downward-sloping demand curve it faces, and setting prices to maximize profits; (v) firms contemplating entry into the business of producing a durable take prices for designs as given; and (vi) the supply of each good is equal to the demand.

IV. Discussion of the Model

A reasonable intuition for the behavior of this model can be inferred by considering the Solow (1956) model, in which the evolution of A is given exogenously, and the Uzawa (1965) model, in which the evolution of A is determined by the allocation of resources between a research sector and a final-goods sector. For a fixed amount of A, and therefore a fixed set of producer durables, the model is almost identical to the Solow model. Because of the symmetry in the model, all the durable goods that are available are supplied at the same level, henceforth denoted as $\bar{x}$. If they were not, it would be possible to increase profits in the producer durable sector by reducing the output of high-output firms and diverting the capital released in this way to low-output goods. Since A determines the range of durables that can be produced and since η units of capital are required per unit of durable goods, it is possible to solve for $\bar{x}$ from the equation $K = \eta A\bar{x}$. Then output Y can be written as

$$\begin{aligned} Y(H_A, L, x) &= H_Y^{\alpha}L^{\beta}\int_0^{\infty} x(i)^{1-\alpha-\beta}\,di \\ &= H_Y^{\alpha}L^{\beta}A\bar{x}^{1-\alpha-\beta} \\ &= H_Y^{\alpha}L^{\beta}A\left(\frac{K}{\eta A}\right)^{1-\alpha-\beta} \\ &= (H_YA)^{\alpha}(LA)^{\beta}(K)^{1-\alpha-\beta}\eta^{\alpha+\beta-1}. \end{aligned} \tag{7}$$

The last line of this equation shows that the model behaves just like the neoclassical model with labor and human capital augmenting technological change. In particular, it exhibits the usual diminishing returns to capital accumulation. Given the assumed form of preferences, a fixed level of A will lead to an equilibrium with a steady state in which the level of K is determined by the requirement that the marginal product of capital is equal to the discount rate. If A grew at an exogenously specified exponential rate, the economy would converge to a path on which K grows at the same exponential rate as A, just as it does in the Solow model. Along the transition path, the ratio of K to A would change, which implies that r and $\bar{x}$ would change as well. Along the balanced growth path, r, $\bar{x}$, and the ratio of K to A are all constant.

The nonconvexity evident in the expression for final output as a function of the primary inputs of the model (H, L, K, and A) is supported in a decentralized equilibrium that relies on monopolistic competition. In contrast, the nonconvexity present in equation (3) describing output of designs is supported through competition with external effects that arise from knowledge spillovers. In each case, the nonconvexity arises because the nonrival good A is an input in production. In the final-output sector, A matters indirectly because of its effects on the availability of the new $x(\cdot)$ goods. In the research sector, A enters directly.

Both spillovers and price setting seem essential to capturing the features of knowledge in a model of growth. There is little doubt that much of the value to society of any given innovation or discovery is not captured by the inventor, and any model that missed these spillovers would miss important elements of the growth process. Yet it is still the case that private, profit-maximizing agents make investments in the creation of new knowledge and that they earn a return on these investments by charging a price for the resulting goods that is greater than the marginal cost of producing the goods.

V. Solution of the Model for a Balanced Growth Equilibrium

The strategy for characterizing the model that is followed here is to solve for an equilibrium in which the variables A, K, and Y grow at constant exponential rates. This is generally referred to as a balanced growth equilibrium. The intuition from the Solow model suggests that such an equilibrium will exist if A grows at a constant exponential rate. The intuition from the Uzawa model suggests that it is possible for A to grow at an exponential rate because equation (3) for $\dot{A}$ is linear in A. It will grow at a constant rate if the amount of human capital H_A that is devoted to research stays constant. Verifying that a balanced growth equilibrium exists therefore reduces to the problem of showing that prices and wages are such that H_Y and H_A remain constant as Y, K, C, and A grow.

By focusing only on balanced growth paths, the analysis neglects the transient dynamics that arise when the economy starts from a ratio of K to A that differs from the ratio that is maintained along the balanced growth path. One should be able to study convergence to the balanced growth ratio of K to A using the tools used for studying the Solow and Uzawa models, but this analysis is not attempted here.

The first step in the solution of the model is to derive the relation between the growth rate of output and the rate of return on investment. Coupled with the relation between interest rates and growth rates implied by the preference side of the model, this will determine the rate of growth and interest rate. Given the intuition from the Solow model and the results derived in my earlier model of differentiated inputs (Romer 1987), it follows that along the balanced growth path, the ratio of K to A should be constant, which implies that $\bar{x}$ is constant as well. Because of the accumulation of both K and A, the wage paid for human capital in the final-output sector will grow in proportion to A, but by equation (3), the productivity of human capital in research also grows in proportion to A. Since the productivity of human capital grows at the same rate in both sectors, H_Y and H_A will remain constant if the price P_A for new designs is constant.

Figure 1 illustrates the behavior of the inputs in the model. All the producer durables that have been designed up to time t are used at the level $\bar{x}$ that indicates the height of the rectangle. The width is the measure of the number of designs or durables in use, $A(t)$. The area $A(t)\bar{x}$ is equal to total capital divided by η. Over time, $\bar{x}$ remains constant and A grows at a constant exponential rate.

It remains to check that this description of a balanced growth path is consistent with all the equilibrium conditions. As noted in the dis-

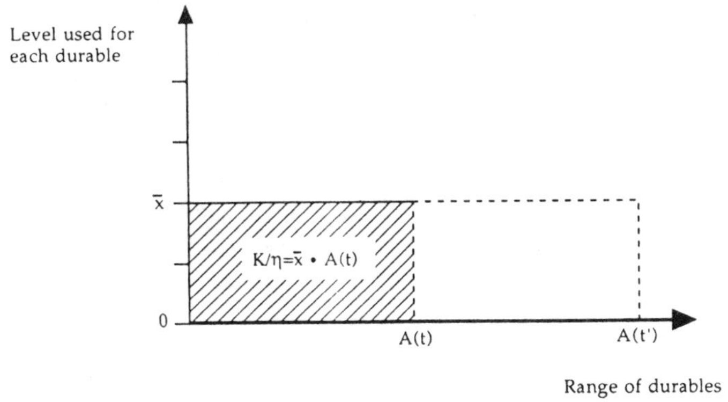

FIG. 1.—Producer durables used in production at time t and $t' > t$

cussion following the monopoly profit problem (5), the flow of profit that can be extracted by the seller of any particular durable input is $\pi = (\alpha + \beta)\bar{p}\bar{x}$. Since the present discounted value of this stream of profit must equal the price P_A of the design, it follows that

$$P_A = \frac{1}{r}\pi = \frac{\alpha + \beta}{r}\bar{p}\bar{x} = \frac{\alpha + \beta}{r}(1 - \alpha - \beta)H_Y^{\alpha}L^{\beta}\bar{x}^{1-\alpha-\beta}. \quad (8)$$

The last equality in this expression follows by using equation (4) to evaluate $\bar{p}$ in terms of $\bar{x}$.

The condition determining the allocation of human capital between the final-output and research sectors says that the wages paid to human capital in each sector must be the same. In the final-output sector, the wage for human capital is its marginal product. Since human capital receives all the income from the research sector, the wage there is $P_A\delta A$. To equalize returns to human capital in both sectors, $H_Y = H - H_A$ must be chosen so that

$$\begin{aligned} w_H = P_A\delta A &= \alpha H_Y^{\alpha-1}L^{\beta}\int_0^{\infty}\bar{x}^{1-\alpha-\beta}\,di \\ &= \alpha H_Y^{\alpha-1}L^{\beta}A\bar{x}^{1-\alpha-\beta}. \end{aligned} \quad (9)$$

Substituting P_A from equation (8) into equation (9) and simplifying yield

$$H_Y = \frac{1}{\delta}\frac{\alpha}{(1 - \alpha - \beta)(\alpha + \beta)}r. \quad (10)$$

For a fixed value of $H_A = H - H_Y$, the implied exponential growth rate for A is δH_A. From the monopoly pricing problem, we know that

$\bar{x}$ is constant if r is. The intuition from the Solow model suggests, and examination of the expression for final output

$$Y = H_Y^\alpha L^\beta \int_0^\infty \bar{x}^{1-\alpha-\beta}\, di = H_Y^\alpha L^\beta A \bar{x}^{1-\alpha-\beta}$$

from equation (7) shows, that output grows at the same rate as A if L, H_Y, and $\bar{x}$ are fixed. If $\bar{x}$ is fixed, then K must grow at the same rate as A, because total usage of capital is $A\bar{x}\eta$. Let g denote the growth rate of A, Y, and K. Since K/Y is a constant, the ratio

$$\frac{C}{Y} = 1 - \frac{\dot{K}}{Y} = 1 - \frac{\dot{K}}{K}\frac{K}{Y}$$

must also be constant. The common growth rate g for all these variables is therefore

$$g = \frac{\dot{C}}{C} = \frac{\dot{Y}}{Y} = \frac{\dot{K}}{K} = \frac{\dot{A}}{A} = \delta H_A.$$

Together with equation (10), the constraint $H_Y = H - H_A$ implies a relation between the growth rate g and the interest rate r:

$$g = \delta H_A = \delta H - \frac{\alpha}{(1-\alpha-\beta)(\alpha+\beta)} r, \tag{11}$$

which can be simplified as

$$g = \delta H_A = \delta H - \Lambda r, \tag{11'}$$

where Λ is a constant that depends on the technology parameters α and β,

$$\Lambda = \frac{\alpha}{(1-\alpha-\beta)(\alpha+\beta)}. \tag{12}$$

Implicitly, the allocation of H between the two sectors is constrained by the requirement that H_A must be nonnegative and that H_Y can be no larger than H. This implies that g is nonnegative. If this constraint is binding, equation (10) will hold as an inequality.

To close the model, it remains to impose the relation between the growth rate g and the interest rate r implied by the preference side of the model, $g = \dot{C}/C = (r - \rho)/\sigma$. Combined with equation (11), this gives an expression for g in terms of the fundamentals of the model,

$$g = \frac{\delta H - \Lambda\rho}{\sigma\Lambda + 1}, \tag{13}$$

where Λ is as defined in equation (12).

The expression for the growth rate suggests a minor technical restriction. For the integral in the consumer's preferences to be finite, the rate of growth of current utility $(1 - \sigma)g$ must be less than the discount rate ρ. Thus, for $\sigma \in [0, 1)$, $(1 - \sigma)\delta H/(\Lambda + 1)$ must be less than the discount rate ρ. If this does not hold, the integral can be infinite and some kind of overtaking criterion must be used to describe the behavior of the consumer.

VI. Interpretation and Welfare Properties of the Solution

Almost all the content of the model is contained in equation (11), which summarizes the effects of the technological side of the model, including the effects of imperfect competition in the market for producer durables. Part of the intuition behind equation (11) is easy to grasp. The opportunity cost of human capital is the wage income that can be earned instantaneously in the manufacturing sector. The return to investing human capital in research is a stream of net revenue that a design generates in the future. If the interest rate is larger, the present discounted value of the stream of net revenue will be lower. Less human capital will be allocated to research, and the rate of growth will be lower.

Other aspects of the intuition behind equation (11) are more subtle. The surprising feature of equation (11) is that neither L nor the parameter η is present. Equation (4) shows that an increase in L increases the demand faced by each monopolistic firm selling a durable good. From the monopoly problem (5), it is clear that a reduction in η reduces the costs of the monopolist and increases output $\bar{x}$. In either case, the stream of net revenue generated by a new design increases. Nonetheless, the amount of human capital devoted to research increases neither when L increases nor when η falls.

In a partial equilibrium analysis, one would always expect any change that increases the return to an activity to increase the allocation of resources to that activity. What a general equilibrium analysis emphasizes is that any intervention that increases the return to one activity can very well increase the return to some other activity that competes with the first activity for resources. An increase in L or a reduction in η (which increases $\bar{x}$) raises the return to human capital employed in manufacturing at the same time that it raises the return to human capital in research. For the functional forms used here, these two effects exactly cancel. The allocation of human capital between research and manufacturing does not change when L and η change.

This kind of exact cancellation is not a robust feature of the model.

If slightly different functional forms were used, the net effect of an increase in L or a decrease in η could be either to increase or to reduce the amount of human capital used in the research sector. For example, in an extension of this model (Romer 1990), I show how an increase in L could reduce research effort and the rate of growth. The correct inference from this model is that the effect of an increase in L or a decrease in η on the rate of growth is ambiguous, something that a priori theorizing cannot resolve.

This ambiguity contrasts with the strong and robust implication that reductions in the interest rate will speed up growth. Equivalently, from equation (12), it follows that any change in the preference parameters that acts to reduce the interest rate (an increase in patience captured by a decrease in the discount rate ρ, or an increase in the intertemporal rate of substitution, as captured by a decrease in σ) will increase research and growth. This implication follows directly from the assumption that the benefits of research come largely in the future and that the costs are incurred immediately.

From a policy point of view, the difference between a reduction in η and a reduction in the equilibrium interest rate is very important. A direct subsidy to investment in physical capital financed by a lump-sum tax is mathematically equivalent to a reduction in η. (An illustration of this easy result is given in Romer [1989].) On the basis of an intuition developed for the one-sector model, economists often identify the marginal product of physical capital with the market interest rate. If they were the same, anything that increases the capital stock and reduces the marginal product of physical capital would have the same growth-enhancing effect as a reduction in the interest rate. The intuition that subsidies to physical capital accumulation will also speed up growth in A was reinforced by the first generation of endogenous growth models. In Arrow's (1962*a*) model of learning by doing and in my first model (Romer 1986), the rate of growth of A was forced by assumption to be the same as the rate of growth of K. As a result, an intervention such as an investment tax credit that increased the accumulation of K necessarily increased the accumulation of A as well.

The model presented here shows that when the decision to invest in physical capital is uncoupled from the decision to invest in research, the effects of a subsidy to physical capital are quite different from the effects of a reduction in the market interest rate. If the fundamental policy problem is that we have too many lawyers and MBAs and not enough engineers, a subsidy to physical capital accumulation is a weak, and possibly counterproductive, policy response.

In the previous paper with differentiated producer durables (Romer 1987), I found that an increase in scale measured by L would increase the rate of growth. Here, an increase in L has no effect.

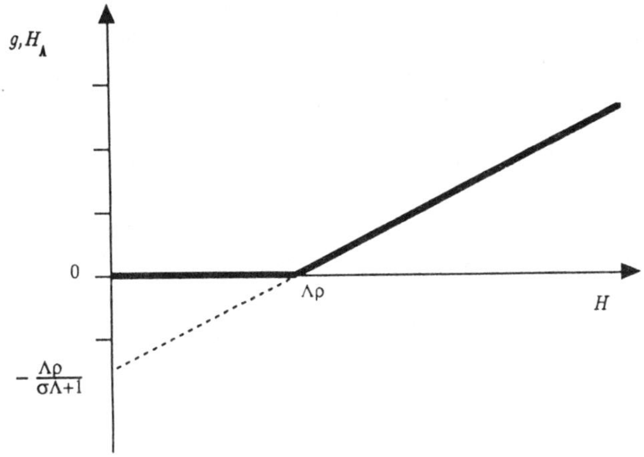

FIG. 2.—Growth rate and amount of human capital in research as a function of total human capital (for $\delta = 1$).

Instead, an increase in scale as measured by total human capital H has the effect of speeding up the rate of growth. This effect is illustrated in figure 2, which plots the rate of growth and the amount of human capital used in research as a function of total human capital. Both of these models have an underlying form of increasing returns in research. As a result, an increase in a scale variable induces an increase in the rate of growth. Human capital is the relevant scale variable in this model because it is the input that is used most intensively in research.

That the research sector in this model exhibits increasing returns is clear from equation (4). If the research technology exhibited constant returns to scale, a doubling of both the human capital and the stock of knowledge would leave the marginal product of human capital in research unchanged. Under the specification used here, a doubling of both leads to an increase in the marginal product of human capital in research. As a result, a permanent increase in the total stock of human capital in the population leads to an increase in the ratio of A to K and a more than proportional increase in the amount of human capital that is devoted to the research sector, as illustrated in figure 2. This implication is of interest from both a historical and a cross-sectional point of view. It is surely the case that the total level of human capital and the fraction of human capital devoted to research are higher now than they were at any time in the past. Moreover, the fraction of human capital devoted to research is apparently highest in the most developed countries of the world.

An extreme possibility illustrated in figure 2 is that if the total level of human capital is too small, stagnation may arise. If H is too low, the nonnegativity constraint on H_A is binding and growth does not take place. In this case, all the feasible growth rates for A are too small relative to the discount rate to justify the sacrifice in current output necessary for growth to take place. This result offers one possible way to explain the wide variation in growth rates observed among countries and the fact that in some countries growth in income per capita has been close to zero. This explanation is reminiscent of the explanation for the absence of growth in prehistoric time that is offered by some historians and anthropologists: civilization, and hence growth, could not begin until human capital could be spared from the production of goods for immediate consumption. This model cannot offer a complete explanation for these observations because it treats the stock of H (and of L) as given, but it does suggest directions for further work.

The effects of a subsidy to capital can be contrasted with a policy designed to encourage research. A subsidy to employment in the research sector that is financed through lump-sum taxes has the same effects on growth as an increase in the productivity parameter δ in equation (3). In the long run, the subsidy will cause an increase in the growth rate, a fall in P_A, and a reduction in $\bar{x}$ and in the ratio of K to A.

There are two reasons to expect that too little human capital is devoted to research. The most obvious reason is that research has positive external effects. An additional design raises the productivity of all future individuals who do research, but because this benefit is nonexcludable, it is not reflected at all in the market price for designs. The second and an equally important reason why too little human capital is devoted to research is that research produces an input that is purchased by a sector that engages in monopoly pricing. The markup of price over marginal cost forces a wedge between the marginal social product of an input used in this sector and its market compensation. Equation (7) shows that a new design (i.e., an additional unit of A) increases output at every date by an amount $H_Y^{\alpha}L^{\beta}\bar{x}^{1-\alpha-\beta}$; from the calculation in (5), the producer of a design captures only a fraction $1 - \alpha - \beta$ of this net benefit to society.

Both of these effects cause human capital to be undercompensated. The marginal product of capital in the manufacturing sector is equal to the wage w_H, but the marginal product in the research sector is higher than the wage because the price of the patent captures only part of the social value of the patent. From equation (9), (10), or (11), it follows that an increase in total human capital would lead to an increase in the amount of human capital employed in the research sector. As a result, in equilibrium, the marginal value of an additional unit of human capital is higher than the market wage. For simplicity, the model here has treated the stock of total human capital as being exogenously fixed, so undercompensating human capital has no welfare effects. In a more realistic model in which human capital is accumulated endogenously, the supply will be too low. As a result, a second-best policy for a government that cannot affect the allocation of human capital between different sectors would be to subsidize the production of human capital.

Within the confines of the model, the social optimum can be achieved by subsidizing the accumulation of A. Demonstrating this result rigorously starting from arbitrary initial conditions forces the analysis to depart from consideration of balanced growth paths. Any intervention designed to move an economy from one balanced growth path to another must consider the transition dynamics along the way, and an explicit analysis of these dynamics is beyond the scope of this paper. It is easy, however, to compare the growth rate along the balanced growth path that emerges from the equilibrium with the one that would emerge from the solution to a social planning problem. Because of the symmetry between the different producer durables in this economy, the optimal level of $x(i)$ is the same for all i between zero and A. This level is related to K and A by the constraint that $K = \eta x A$. Using this to express x in terms of K and A, we can therefore write the social planning problem for this economy as

$$\max \int_0^\infty \frac{C^{1-\sigma} - 1}{1 - \sigma} e^{-\rho t} dt,$$

subject to

$$\dot{K} = \eta^{\alpha+\beta-1} A^{\alpha+\beta} H_Y^{\alpha} L^{\beta} K^{1-\alpha-\beta} - C,$$

$$\dot{A} = \delta H_A A,$$

$$H_Y + H_A \leq H.$$

As is shown in the Appendix, the balanced growth solution to the first-order necessary conditions for this problem has a growth rate g^* that is given by

$$g^* = \frac{\delta H - \Theta\rho}{\Theta\sigma + (1 - \Theta)}, \tag{14}$$

where $\Theta = \alpha/(\alpha + \beta)$. The coefficient Λ from equation (12) is equal to this coefficient Θ times the markup from the monopoly sector, $1/(1 - \alpha - \beta)$. This accounts for part of the difference between the equilibrium and the socially optimal rate of growth. However, there is an additional effect because the term in the denominator in equation (14) also replaces the constant 1 from equation (13) with the expression $1 - \Theta$. This change reflects the effect of correcting for the external effects associated with the production of new ideas. Both of these changes in the expression for the rate of growth—the substitution of Θ for Λ and the substitution of $1 - \Theta$ for 1—cause the socially optimal allocation of human capital to research to be higher, and this causes the socially optimal rate of growth to be higher.

VII. Growth, Trade, and Research

The final observations about this model pertain to its implications for growth, trade, and research. These can be seen most simply by comparing the balanced growth equilibrium for two identical closed economies that operate in isolation with the balanced growth equilibrium that would obtain if the economies had always been fully integrated. In isolation, the common growth rate is given by g from equation (13), with H set equal to the amount of human capital in each country. In the second case, the growth rate is found by replacing H by total worldwide human capital, $2H$. Both the fraction H_A/H of total worldwide human capital devoted to research and the rate of growth increase.

This thought experiment suggests why a decision to engage in trade may be important even for a country that has a large population, such as China or India. If access to a large number of workers or consumers were all that mattered, having a large population would be a good substitute for trade with other nations. The model here suggests that what is important for growth is integration not into an economy with a large number of people but rather into one with a large amount of human capital. Many of the details of trade between different economies of this kind remain to be worked out,[2] but since growth seems to be correlated with the degree of integration into worldwide markets but not closely related to population size or density, the results from this model seem promising.

The most direct test of this implication of the model would come from a controlled experiment in which the level of research activity was monitored both before and after a country was opened to trade with the rest of the world. Sokoloff (1988) reports historical data on a natural experiment that comes close to this test. He finds cross-sectional variation that supports the model: counties in the United States in the early nineteenth century that had access to navigable waterways had higher rates of patenting than counties that did not.

[2] Since the first draft of this paper was written, Grossman and Helpman (1989*a*, 1989*b*) have explored detailed models of trade and growth with this kind of underlying technology.

More convincingly, he shows that over time, the introduction of water transportation (because of either the construction of a new canal or the dredging of a river) was followed by a sharp increase in the rate of patenting in counties adjacent to the waterway.

In subsequent work, Sokoloff and Khan (1989) examine the time-series variability in patenting for specific individuals. They find that there was a substantial group of people with broad general knowledge who moved in and out of research in response to aggregate disturbances (primarily business cycles and a trade embargo by the British) just as one would expect if human capital that could be used to do research had alternative uses in manufacturing and commercial trade. They conclude that at least in the nineteenth century, there was a fairly elastic short-run supply of human capital for use in research. Combined with the prior evidence that patenting activity does respond to changes in the size of the market, this offers some assurance that the basic mechanisms described in the model are relevant for historical experience.

VIII. Conclusions

The model presented here is essentially the one-sector neoclassical model with technological change, augmented to give an endogenous explanation of the source of the technological change. The most robust welfare conclusion from the model is that because research projects exchange current costs for a stream of benefits in the future, the rate of technological change is sensitive to the rate of interest. Although all the research is embodied in capital goods, a subsidy to physical capital accumulation may be a very poor substitute for direct subsidies that increase the incentive to undertake research. In the absence of feasible policies that can remove the divergence between the social and private returns to research, a second-best policy would be to subsidize the accumulation of total human capital.

The most interesting positive implication of the model is that an economy with a larger total stock of human capital will experience faster growth. This finding suggests that free international trade can act to speed up growth. It also suggests a way to understand what it is about developed economies in the twentieth century that permitted rates of growth of income per capita that are unprecedented in human history. The model also suggests that low levels of human capital may help explain why growth is not observed in underdeveloped economies that are closed and why a less developed economy with a very large population can still benefit from economic integration with the rest of the world.

Appendix

Calculation of the Balanced Growth Social Optimum

To derive the necessary conditions for the social optimization problem given in the text, construct the current-valued Hamiltonian:

$$\mathcal{H} = \frac{C^{1-\sigma}}{1-\sigma} + \lambda[\eta^{\alpha+\beta-1}A^{\alpha+\beta}(H - H_A)^{\alpha}L^{\beta}K^{1-\alpha-\beta} - C] + \mu\delta H_A A.$$

The necessary conditions follow by maximizing $\mathcal{H}$ with respect to the control variables C and H_A, and from the equations for the evolution of the multipliers λ and μ:

$$\dot{\lambda} = \rho\lambda - \frac{\partial\mathcal{H}}{\partial K}, \quad \dot{\mu} = \rho\mu - \frac{\partial\mathcal{H}}{\partial A}.$$

The first-order condition for maximizing $\mathcal{H}$ with respect to C gives the usual expression relating marginal utility and the multiplier λ:

$$C^{-\sigma} = \lambda. \tag{A1}$$

If the symbol Δ is used to represent the term $\eta^{\alpha+\beta-1}A^{\alpha+\beta}(H - H_A)^{\alpha}L^{\beta}K^{1-\alpha-\beta}$ from the Hamiltonian, the first-order condition for maximizing $\mathcal{H}$ with respect to H_A can be written as

$$\Delta = (H - H_A)\frac{\delta\mu}{\alpha\lambda}A. \tag{A2}$$

Then with equation (A2), the evolution equation for μ can be simplified to yield

$$\frac{\dot{\mu}}{\mu} = \rho - \delta\left(\frac{\alpha+\beta}{\alpha}H - \frac{\beta}{\alpha}H_A\right). \tag{A3}$$

For a balanced growth equilibrium, it must be the case that $\dot{\mu}/\mu = \dot{\lambda}/\lambda$ and that $\dot{C}/C = \dot{A}/A$. With equation (A1), these can be combined to yield $-\sigma(\dot{A}/A) = \dot{\mu}/\mu$. Then combining the evolution equation for A, $\dot{A}/A = \delta H_A$, with equation (A3) gives an equation in H, H_A, and the basic parameters of the model:

$$-\sigma\delta H_A = \rho - \delta\left(\frac{\alpha+\beta}{\alpha}H - \frac{\beta}{\alpha}H_A\right). \tag{A4}$$

Equation (14) in the text can be derived from this equation by solving for H_A and using the fact that the growth rate g is given by $g = \delta H_A$.

References

Abramovitz, Moses. "Resource and Output Trends in the United States since 1870." *A.E.R. Papers and Proc.* 46 (May 1956): 5–23.

Arrow, Kenneth J. "The Economic Implications of Learning by Doing." *Rev. Econ. Studies* 29 (June 1962): 155–73. (*a*)

———. "Economic Welfare and the Allocation of Resources for Invention." In *The Rate and Direction of Inventive Activity: Economic and Social Factors*, edited by Richard Nelson. Princeton, N.J.: Princeton Univ. Press (for NBER), 1962. (*b*)

Becker, Gary S.; Murphy, Kevin M.; and Tamura, Robert. "Human Capital, Fertility, and Economic Growth." *J.P.E.*, this issue.

Cornes, Richard, and Sandler, Todd. *The Theory of Externalities, Public Goods, and Club Goods.* Cambridge: Cambridge Univ. Press, 1986.

Dasgupta, Partha, and Stiglitz, Joseph E. "Learning-by-Doing, Market Structure and Industrial and Trade Policies." *Oxford Econ. Papers* 40 (June 1988): 246–68.

Dixit, Avinash K., and Stiglitz, Joseph E. "Monopolistic Competition and Optimum Product Diversity." *A.E.R.* 67 (June 1977): 297–308.

Ethier, Wilfred J. "National and International Returns to Scale in the Modern Theory of International Trade." *A.E.R.* 72 (June 1982): 389–405.

Gollop, Frank M., and Jorgenson, Dale W. "U.S. Productivity Growth by Industry." In *New Developments in Productivity Measurement and Analysis,* edited by John W. Kendrick and Beatrice N. Vaccara. Chicago: Univ. Chicago Press (for NBER), 1980.

Griliches, Zvi. "Issues in Assessing the Contribution of Research and Development to Productivity Growth." *Bell J. Econ.* 10 (Spring 1979): 92–116.

Grossman, Gene M., and Helpman, Elhanan. "Comparative Advantage and Long Run Growth." Working Paper no. 2809. Cambridge, Mass.: NBER, January 1989. (*a*).

———. "Growth and Welfare in a Small Open Economy." Working Paper no. 2970. Cambridge, Mass.: NBER, May 1989. (*b*)

———. "Product Development and International Trade." *J.P.E.* 97 (December 1989): 1261–83. (*c*)

Heckman, James J. "A Life-Cycle Model of Earnings, Learning, and Consumption." *J.P.E.* 84, no. 4, pt. 2 (August 1976): S11–S44.

Helpman, Elhanan, and Krugman, Paul R. *Market Structure and Foreign Trade: Increasing Returns, Imperfect Competition, and the International Economy.* Cambridge, Mass.: MIT Press, 1985.

Jones, Larry, and Manuelli, Rodolfo. "A Model of Optimal Economic Growth." Working paper. Stanford, Calif.: Stanford Univ., 1988.

Jorgenson, Dale W.; Gollop, Frank M.; and Fraumeni, Barbara M. *Productivity and U.S. Economic Growth.* Cambridge, Mass.: Harvard Univ. Press, 1987.

Judd, Kenneth L. "On the Performance of Patents." *Econometrica* 53 (May 1985): 567–85.

Kendrick, John W. "Productivity Trends: Capital and Labor." *Rev. Econ. and Statis.* 38 (August 1956): 248–57.

King, Robert G., and Rebelo, Sergio. "Business Cycles with Endogenous Growth." Working paper. Rochester, N.Y.: Univ. Rochester, 1987.

Lucas, Robert E., Jr. "On the Mechanics of Economic Development." *J. Monetary Econ.* 22 (July 1988): 3–42.

Maddison, Angus. *Phases of Capitalist Development.* Oxford: Oxford Univ. Press, 1982.

Nordhaus, William D. *Invention, Growth, and Welfare: A Theoretical Treatment of Technological Change.* Cambridge, Mass.: MIT Press, 1969.

Pascoa, Mario. "Monopolistic Competition and No-Neighboring-Goods." Working Paper no. 86-14. Philadelphia: Univ. Pennsylvania, Center Analytic Res. Econ. and Soc. Sci., 1986.

———. "Noncooperative Equilibrium and Chamberlinian Monopolistic Competition." *J. Econ. Theory* (1990), in press.

Rebelo, Sergio. "Long Run Policy Analysis and Long Run Growth." Working paper. Rochester, N.Y.: Univ. Rochester, 1988.

Romer, Paul M. "Increasing Returns and Long-Run Growth." *J.P.E.* 94 (October 1986): 1002–37.

———. "Growth Based on Increasing Returns Due to Specialization." *A.E.R. Papers and Proc.* 77 (May 1987): 56–62.

———. "Capital Accumulation in the Theory of Long Run Growth." In *Modern Business Cycle Theory,* edited by Robert J. Barro. Cambridge, Mass.: Harvard Univ. Press, 1989.

———. "Capital, Labor, and Productivity." *Brookings Papers Econ. Activity* (1990), in press.

Rosen, Sherwin. "A Theory of Life Earnings." *J.P.E.* 84, no. 4, pt. 2 (August 1976): S45–S67.

Schumpeter, Joseph A. *Capitalism, Socialism, and Democracy.* New York: Harper, 1942.

Shell, Karl. "Toward a Theory of Inventive Activity and Capital Accumulation." *A.E.R. Papers and Proc.* 56 (May 1966): 62–68.

———. "A Model of Inventive Activity and Capital Accumulation." In *Essays on the Theory of Optimal Economic Growth,* edited by Karl Shell. Cambridge, Mass.: MIT Press, 1967.

———. "Inventive Activity, Industrial Organisation and Economic Growth." In *Models of Economic Growth,* edited by James A. Mirrlees and Nicholas H. Stern. New York: Wiley, 1973.

Sokoloff, Kenneth L. "Inventive Activity in Early Industrial America: Evidence from Patent Records, 1790–1846." *J. Econ. Hist.* 48 (December 1988): 813–50.

Sokoloff, Kenneth L., and Khan, Zorina. "The Democratization of Invention during Early Industrialization: Evidence from the United States, 1790–1846." Working paper. Los Angeles: Univ. California, 1989.

Solow, Robert M. "A Contribution to the Theory of Economic Growth." *Q.J.E.* 70 (February 1956): 65–94.

———. "Technical Change and the Aggregate Production Function." *Rev. Econ. and Statis.* 39 (August 1957): 312–20.

Uzawa, Hirofumi. "Optimum Technical Change in an Aggregative Model of Economic Growth." *Internat. Econ. Rev.* 6 (January 1965): 18–31.

Wilson, Robert. "Informational Economies of Scale." *Bell J. Econ.* 6 (Spring 1975): 184–95.

625-49
[94]

[30]

Economic Modelling 1994 **11** (3) 283–307

Q20
013

Dynamic macro modelling and materials balance

Economic–environmental integration for sustainable development

C51

Jeroen C.J.M. van den Bergh and Peter Nijkamp

In this paper a model is presented with the aim of studying the long-term relationship between an economy and its natural enviroment. The approach is based on two main elements: reciprocal interactions between on the one hand population growth, investment, allocation of inputs and outputs, technology and productivity, and on the other hand declining environmental quality, pollution and resource exhaustion; and a more realistic and consistent representation of the interdependence between various environmental effects, from extraction to emission, by using a materials balance perspective on economic processes. In order to develop the model attention is devoted to production functions that are consistent with materials balance. The model is multisectoral, including economic processes of extraction, production of commodities and investment goods, waste abatement and recycling. Environmental processes of resource regeneration and waste assimilation are linked through the economic and environmental sphere, and the model allows for endogenous development of resource-efficient technology.

Keywords: Sustainable development; Economic growth; Dynamic modelling

Environmental concern has become a major issue in our modern society. In the past years, there has been an avalanche of literature in the field of environmental–economic analysis and modelling. Much of this was based on a straightforward extension of conventional welfare-theoretic models encapsulating externalities in resource utilization. After the publication of the Brundtland Report (WCED [21]) there has been an increasing need to develop models focusing explicitly on ecologically sustainable economic development (SD). Up till now comprehensive attempts at building such models in a dynamic setting have been exceptionally rare. Furthermore, the objective of sustainable development also requires that environmental economics is extended to macroeconomic analysis (see Daly [8]), for which adequate models have to be constructed.

This paper addresses the design and use of a dynamic macro model which aims to provide an analytical framework for the study of the long-term development of an economy in relation to its natural environment. Particular attention will be devoted to the conceptual structure and specification of such a model. Subsequently, calibration and results of scenario analyses are presented. The model will serve to assist decision makers in distinguishing between various policy strategies (or scenarios) that give rise to sustainable and unsustainable developments.

In order to provide a sufficient background for a discussion of the model, its presentation is preceded by a discussion of economic–environmental modelling

The authors are with the Faculty of Economics, Free University, De Boelelaan 1105, 1081 HV Amsterdam, The Netherlands.

This study was supported by the Foundation for Advancement of Economic Research (Ecozoek), which reports under the Dutch Organization for Scientific Research (NWO), project No. 450-230-007. We are grateful to an anonymous referee for comments which have led to a considerable improvement of this paper.

Final manuscript received 4 March 1994.

0264-9993/94/030283-25 © 1994 Butterworth-Heinemann Ltd

in the next section, which focuses on materials balance production functions which have received little attention in the literature. The fifth section includes a discussion of the specification and calibration of the model. The results of scenario analysis with the model are presented and conclusions are drawn in a final section.

Economy–environment interactions and materials balance production functions

The approach behind the model proposed in the next section uses the insight that a long-run economic–environmental analysis, which is needed for issues of sustainable development, should be based on two main elements. First, the long-run relationship between the economy and the natural environment is characterized by twoway interactions between on the one hand population growth, investment, technology and productivity, and on the other hand declining environmental quality and resource exhaustion. Second, a more realistic representation of the interdependence between various environmental effects, from extraction to emission, can be realized by adopting a materials balance perspective on economic processes. The result of including these two elements means that direct mutual impacts between the economy and the environment, as well as indirect economic–environmental influences are taken into account, by separating effects from and on production, consumption and welfare. Wilkinson [20] uses the idea of ecological disequilibrium to link economic change to environment–economy relationships. Inclusion of such environmentally influenced economic change in models has been performed by Faber and Proops [9] and Van den Bergh [6]. Related models stem from applied systems theory, notably in the fields of biophysical 'macroscopic mini-models' derived from energy language diagrams (Odum [16]), and global modelling (Meadows *et al* [15]). The main shortcomings of these models in comparison with the one proposed here are, first the lack of substitution between productive inputs or components of welfare, and secondly the lack of economic feedback mechanisms, as will be discussed later.

For an investigation of economic–ecological integration at both a theoretical and operational level of modelling we may – as indicated above – also include materials balance conditions to account in a consistent way for material flows that lead to various interlinked effects. Although the use of materials balance models was propagated more than two decades ago (see Ayres and Kneese [3]; Kneese *et al* [14]), they have seen unfortunately few applications. Furthermore, the combination of non-linear models and materials balance conditions is rare, in both theory and applications. The main reason is that materials balance or materials accounting can be done much more easily with linear production functions. Exceptions are Faber *et al* [10], Van den Bergh [4] and Ruth [17].

The concept of materials balance applies to all natural and economic processes. It means that materials in a physical system are not lost, and that material inputs in processes end up in either stock accumulation or material output flows. In addition a derived result has to be mentioned here, namely that the materials input is larger than the useful goods output, especially in view of spillage and auxiliary materials (like water and fertilizer in agriculture) (see Ayres and Kneese [3]).

In formalizing the materials balance principle in environmental economic models, the following steps are required:

(i) relevant variables should be in material units;
(ii) where necessary, transformations must be modelled between (variables in) material units and other units; and
(iii) materials balance conditions should be specified for economic variables in the economic system, for ecological/physical variables in the environmental system, or as a supplement to a description of economic–environmental interactions (which include both economic and environment variables).

Production functions can be formulated in various ways to satisfy the materials balance principle. Application of materials balance to the production process expresses that all material input must end up somewhere; in final or capital goods or in waste. The link between production theory and materials balance is rarely touched upon in the literature. Some theoretical and conceptual steps taken in this direction were set out by Anderson [1] and Smith and Weber [18]. Some properties for a Cobb-Douglas production function that satisfies materials balance are derived by Gross and Veendorp [12]. They show with a standard economic growth analysis that such a function sets a limit to growth for the case of an economy that obtains its material inputs from a non-renewable resource. In what follows we will devote more detailed attention to characteristics of non-linear production functions that are consistent with the law of materials balance. This will lead to several possible formulations, out of which two will turn out to be useful for the model in the next section, namely those given in Equations (4) and (6).

The production process may be envisioned as a transformation of resource inputs into goods and waste outputs by actors (funds/agents; see Georgescu-Roegen [11]). We may expect considerable potential for substitution between subcategories of actors –

labour and capital – since they serve a similar role in the production process, therefore, they are aggregated into the variable 'actors' (A).[1] An increase in the use of agents may reduce the amount of waste output. Substitution is possible within the category of resource inputs to production (R). Substitution between the categories of actors and resources is limited. This is not assumed *a priori*; it follows from the application of the materials balance condition to the production function.[2]

A materials balance can be expressed as follows: (i) the inequality $R>Q$ (R and Q denote the levels of material input and goods output from production, respectively) as a minimal consistency condition; or a more strict condition, such as $R>Q+x$ (x is a lower bound for waste residuals from production), based on knowledge of technical and physical constraints; (ii) the equality $R=Q+W$ (W is waste residuals from production) if material accounting is strived for (and possible); clearly, (ii) encompasses (i) since W is always positive.

The equations in (1) show a general relationship between the output of goods Q on the one hand, and all factors involved in its production on the other hand. These factors include A, R and W. The materials balance principle is explicitly stated in terms of an equality or inequality condition that relates the total resource input to the output of produced goods and waste. The parameter t in this (and the subsequent) production functions denotes the change resulting from technical progress. In the formal representation of (1) A, R and W are treated identically, ie their conceptual difference is not made explicit in $F(\cdot)$, but becomes clear only after the material constraint is added.

$$Q=F(A,R,W,t)$$
$$R=Q+W \tag{1}$$

All partial derivatives of $F(\cdot)$ are positive.[3] We have here a very aggregate description of the production process, namely only in material terms. Specific characteristics of the final product are not considered, so that a simple waste production function can be derived, as $W=R-F(A,R,W,t)$. Therefore, an *ex post* relationship can be established between the production functions for useful output and waste, ie after the application of the materials balance condition.

The second type of formulation of production subject to a materials balance shown in (2) starts with two separate (*ex ante*) independent production function for goods and for waste.

$$Q=F_1(A,t) \quad \text{and} \quad W=F_2(A,t)$$
$$R=Q+W \quad \text{or} \quad Q+W\leqslant R_c \tag{2}$$

The formulation in (2) with the equality constraint can be interpreted as follows: a given actor (and activity) level A determines the levels of useful and waste outputs; the sum of these gives the resource requirement. The interpretation based on the inequality constraint is that for a given amount of resources R_c, the inequality constraint should be satisfied, namely by choosing an actor level A such that the sum of the resulting values of Q and W is feasible.

As a special case of such a constrained process we may distinguish between actors allocated to production and to an activity (denoted by the function $F_3(\cdot)$) which reduces the amount of waste resulting from the (regular or main) production process (denoted by $F_1(\cdot)$ and $F_2(\cdot)$). It is assumed that the available levels of capital, labour and resources are given. This is formalized in (3).

$$Q=F_1(A_1)$$
$$W=F_2(A_1)-F_1(A_2)$$
$$A_1+A_2\leqslant A_c$$
$$Q+W\leqslant R_c \tag{3}$$

The conditions that apply to these functions are that all derivatives be positive, and that F_2 always has a higher value than F_3. In contrast to the formulation in (2), one may add the objective of maximizing Q or minimizing W to the formulation in (3).[4]

[1] In a dynamic analysis the distinction between these two types of actors is relevant in order to deal with distinct process of capital accumulation and labour/population dynamics, and for accounting of locked up resource material in capital stocks.

[2] Substitution between actors and resources is conceptually different from substitution within each category; the latter can be referred to as 'replacement'. The first is only substitution *ex post* ie after more efficiency in resource use is realized by an increased (intensity of) use of 'actors' as inputs in production.

[3] Notice that the negative effect of waste or pollution on production is not dealt with here. The fact that the partial derivative of F with respect to W is positive must be interpreted as follows: for a given combination of actors and resources, and in the absence of materials balance conditions, production can increase by working faster and leaving more waste per unit of actor.

[4] These two optimization problems have the same solutions in (A_1, A_2). In that case all available resource materials will be used, ie for optimal values of A_1 and A_2 we will have $F_1+F_2-F_3=R_c$. Then, choosing values for A_1 and A_2 to maximize F_1 is the same as choosing them to maximize the expression $(R_c-F_2+F_3)$. The latter can be rewritten as R_c + maximum($-W$), so that this formulation implies the same solution values for A_1 and A_2 (and consequently for Q and W) as minimizing W. We can see an analogy between the relationship between these physical optimization problems and the duality in microeconomics (eg profit maximization and cost minimization).

A third type of formulation of materials balance production functions uses a production function F that (automatically) satisfies the consistency restriction of a materials balance ie $F(A, R) \leqslant R$ for all values of $R > 0$. This can be accomplished in two ways. The first is shown in (4), and is characterized by taking the minimum of any general production function and some share of the resource input:

$$F(A, R, t) = \min\{F(A, R), a(t) \cdot R\} \tag{4}$$

The parameter $a(\cdot)$ describes technological efficiency in resource use; it falls between zero and one, and has a positive derivative. It can be defined as:

$$\begin{aligned} a(t) &= 1 - \frac{W}{R} \\ &= 1 - \frac{F_2(A, R, t)}{R} \end{aligned} \tag{5}$$

where we regard waste in terms of a production process as formalized in (2). From (5) it is clear that $a(\cdot)$ is to be interpreted as the efficiency of resource use in production at time t (ie with the technology available at t), which has a lower bound of zero.

The second way, shown in (6), uses a function that is based on a resource efficiency coefficient $r(\cdot)$ which can be regarded as a variable coefficient that relates useful output to resource input. This coefficient is increasing in all its arguments:

$$Q = r(A, R, t) \cdot R \qquad 0 \leqslant r(\cdot) < 1 \tag{6}$$

Resource efficiency in production is thus assumed to be improved either by increasing the intensity of the production activity factors relative to the resource input (indicated by an activity-resource ratio A/R), or by technological progress (indicated by t).

Since the materials balance condition implies a linear (in)equality, it complies easily with linear types of models well, such as fixed proportions and linear production functions (see Van den Bergh [4], Appendix 4). From the above distinction between the three types of representations of production processes satisfying materials balance, it is clear that we can choose between various specifications of materials balance production functions. In the next section we will see applications of such functions.

A dynamic macro model based on materials balance

In this section we will analyse the implications of production combined with materials balance for long-term development, by presenting a macro model with capital ('actor') accumulation, technical progress, and production with renewable and non-renewable resources and emission of waste residuals.

The model describes two phenomena: the material flows (and processes based upon these flows) in the economy, the natural environment and their interdependencies; and the feedback mechanisms from economy, environment and value systems to economic actions and policies. The latter occurs over a long-term horizon. A link therefore exists between the present model and economic growth models with ecological variables (see Kamien and Schwartz [13]; for a critical overview from the perspective of sustainable development see Van den Bergh and Nijkamp [5]).

The material flow is shown in Figure 1. It describes an economy with six sectors. The model has a supply orientation reflecting a long-term horizon. The model includes economic and materials balance equations, economic activity, and production functions. These production functions satisfy materials balance conditions, as discussed in the previous section. Economic activities include resource extraction, production, waste treatment (abatement), recycling, research and development of environmental technology, and environmental cleaning. Some of these activities are taken to be competitive for economic (financial) means, and can to some extent be stimulated by government policies.

In Figure 2 the economic–environmental relationships represented in the model are displayed. They include flows as well as impacts upon processes. An environmental quality indicator related to stocks of resources and pollution is used as a state variable for codetermination of the level of natural processes of regeneration and assimilation. Environmental quality is also assumed to have an impact upon the efficiency of final goods production (eg agriculture) and the quality of extracted renewable resources (eg the stock of fish). It is important to note that changes in non-economic variables, viz. technology and population, are also included.

The description of the economic activities module is based upon production and activity functions for six sectors: final goods production; investment goods production; waste treatment; recycling activity; renewable resource extraction; and non-renewable resource extraction. The functions satisfy, where necessary, materials balance conditions, and may include a mix of environmental and economic production factors, available resource materials (in inventory), environmental quality and the technological progress indicator. The capacities of each sector are influenced by feedback mechanisms related to investment allocation, that become operational when sustainable development

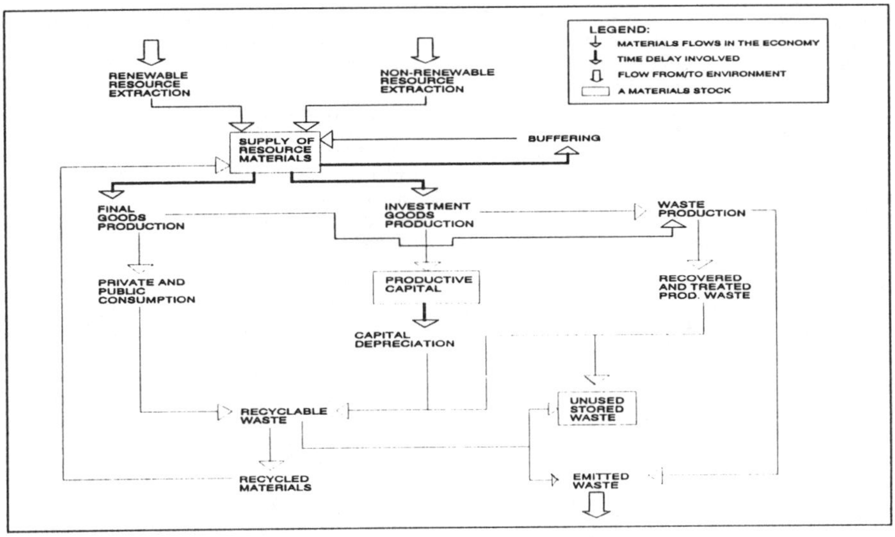

Figure 1. Material flow diagram of the economy.

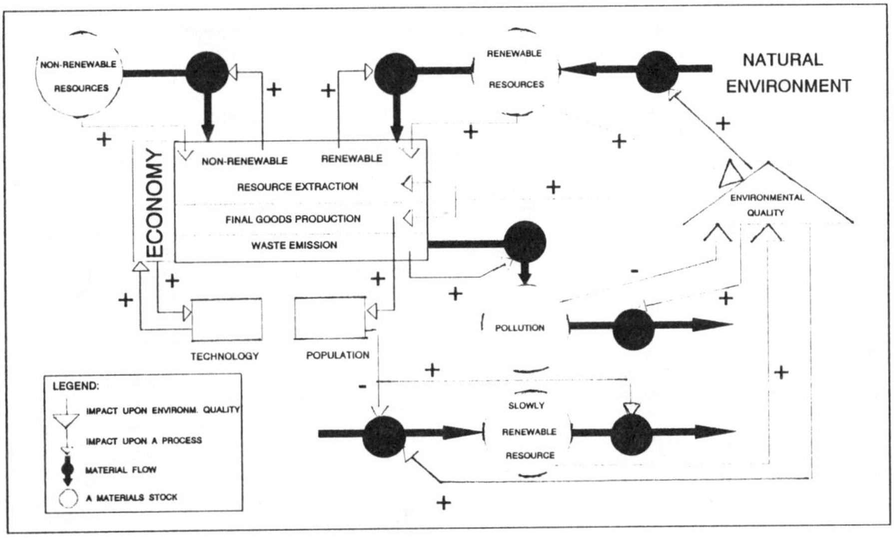

Figure 2. Material and non-material economic–environmental relationships.

conditions with respect to resource use and/or waste emission are violated (see the fourth section). The precise choices of investment allocation mechanism will be formalized in the context of specific development scenarios.

The first economic sector is an aggregate of all final goods producing sectors (including relevant types of agriculture).[5] Output (Q) is produced with capital (K_Q). It requires resource inputs $c_Q(T_{rd})$ per unit of output. A higher overall environmental quality (E) affects positively the productive efficiency. The availability of resources (R_{sup}) in inventory (ie supply of resource materials) imposes an upper limit on the output level of production. Therefore, we have the output function as in (7). It is in agreement with the third type of materials balance production function formulation as given in (4):

$$Q=\min\left\{F_Q(K_Q, E), \frac{R_{sup}}{c_Q(T_{rd})}\right\} \tag{7}$$

It is clear that output Q cannot exceed resource input and that some material loss in production is inevitable. This means that $c_Q(T_{rd})>1$. Technological change is assumed to generate more efficiency in terms of a lower ratio of resource input to Q, so that $dc_Q(x)/dx<0$.

A second activity is capital goods formation (I). This may be interpreted as the production of machines and new technology. The level of output is directly related to the amount of capital in this sector (K_I). An upper limit to the output level is set by taking the ratio of the supply (inventory) of resource materials (R_{sup}) less the amount required for the first (Q) production process to the per unit of output required input in this sector; thereby it is assumed that Q production is more important than I production for general welfare, and is therefore given priority for resource use; $c_I(T_{rd})$ is like $c_Q(T_{rd})$ made dependent on the state of technology T_{rd}. Again, (8) is of the general form as given in (4):

$$I=\min\left\{F_I(K_I), \frac{R_{sup}-c_Q(T_{rd})*Q}{c_I(T_{rd})}\right\} \tag{8}$$

Next, we consider two other economic sectors that are related to environmentally beneficial activities. Both sectors use capital. One process represented in (9) treats production waste in such a way that not all of it is directly carried to the natural environment. The amount of waste treated (R_{wa}) is determined by the level of waste arising directly from the two 'real' production processes ($W_{Q,I}$), the amount of capital used (K_{wa}), and the state of technology (T_{rd}). The latter two affect the effectiveness coefficient (f_{wa}) of abatement positively:

$$R_{wa}=f_{wa}(T_{rd}, K_{wa})*W_{QI} \tag{9}$$

A second process represented in (10) is concerned with recycling of materials. Its output R_{rec} is determined by the waste flow suitable for recycling (W_{rec}) and the state of technology, which positively affect the effectiveness coefficient (f_{rec}) of recycling:

$$R_{rec}=f_{rec}(T_{rd}, K_{rec})*W_{rec} \tag{10}$$

The functions f_{wa} and f_{rec} have values between 0 and 1, so that abated (recycled) waste does not exceed production waste (waste amenable for recycling): $R_{wa}<W_Q$ ($R_{rec}<W_{rec}$); all first partial derivatives are positive. Both (9) and (10) are special cases of the general materials balance production function formulation as given in (6).

Finally, two resource based economic activities remain to be described: renewable and non-renewable resource extraction (R_N and R_S, respectively), both using capital (K_N and K_S, respectively). The production functions, given in (11) and (12), are based upon respective stocks or resources; in the case of renewable exploitation, the environmental quality is also significant since it determines the part of the stock that is of sufficient quality (eg clean water, or healthy, non-toxic resources, normal size fish):

$$R_N=\min\{F_N(K_N, N, E), R_{N,perc}\} \tag{11}$$

$$R_S=\min\{F_S(K_S, S), R_{S,perc}\} \tag{12}$$

All production functions (F_Q, F_I, F_N, F_S) have positive partial derivatives. Extraction rates are limited from above by the perceived or acceptable levels of extraction ($R_{N,perc}$ and $R_{S,perc}$), which are based on a combination of ethical and ecological considerations. Although somewhat similar to the formulation in (4) the production functions in (11) and (12) should not be confused with materials balance production functions *per se*. $R_{S,perc}$ will in general be beneath a level of material availability as represented by actual physical stocks in the ground, since they represent perceived or acceptable levels from an ethical/value point of view. This will be discussed further below.

The economic dynamics incorporate changes in capital stocks, population, technology/knowledge, stored waste, and the inventory of resource materials. The changes in capital stocks depend on the allocation of investment ($I_i, i\in K, K=\{Q, I, wa, rec, N, S\}$) and the rates of depreciation ($D(K)$). The stock of capital

[5] For an explanation of symbols, the reader is referred to Appendix 1.

depreciates at a rate given by $D(K)$ which is strictly monotonically increasing:

$$\frac{dK_i}{dt}=I_i-D_i(K_i) \qquad i\in K \tag{13}$$

The equation to describe technological progress is based on sustainable development feedback (e_e), total investment (I), governmental support for R&D (O_{rd}), and the Kaldor-Verdoorn effect (dZ/dt); δ falls in between 0 and 1 and where the two remaining coefficients are positive and increasing in e_e:

$$\frac{dT_{rd}}{dt}=\alpha(e_e)*I+\varepsilon(e_e)*\left[O_{rd}+\delta*\frac{dZ}{dt}\right] \tag{14}$$

The population change is determined by the population level (Pop) and the level of material consumption (C) per capita (used as a welfare indicator):

$$\frac{dPop}{dt}=B\left(\frac{C}{Pop}\right)*Pop \tag{15}$$

A stock of stored waste that cannot be reused (S_{wa}) is filled from regained production waste (a part s_1) and waste left after recycling (a part s_2; $0<s_1,s_2<1$):

$$\frac{dS_{wa}}{dt}=s_1*R_{wa}+s_2*(W_{rec}-R_{rec}) \tag{16}$$

Two state variables are introduced to represent time delays. The inventory (buffer) of resource materials R_{sup} is introduced to allow for a time delay between the processes of recycling and resource extraction on the one hand, and the reuse of materials in production of final goods and investment goods on the other. This inventory can be regarded as the total supply of resources (then its initial condition has to be in accordance with this interpretation). This inventory originates from four sources: renewable and non-renewable resource extraction, recycling and buffering (inventory). The latter will occur when the amount of materials needed for production is smaller than that in the stockpile:

$$\frac{dR_{sup}}{dt}=R_N+R_S+R_{rec}-(c_Q(T_{rd})*Q+c_I(T_{rd})*I) \tag{17}$$

The choice for aggregation of renewable and non-renewable resources is to see how the use of one type develops over time relative to the use of the other. It is implicitly assumed then that the units of measurement of both R_N and R_S are such that they can be aggregated via simply adding the respective variables. This may require an additional assumption that extraction also involves some processing into a sufficient quality or concentration of required materials.

Finally, Z is an indicator for that part of total production output in which a change has a strong impact on materials saving progress in production technology (see the equation for T_{rd}):

$$\frac{dZ}{dt}=Q+I-Z \qquad Z(0)=Q(0)+I(0) \tag{18}$$

The initial conditions are $K_i(0)=K_{i0}$, $(i\in K)$, $T_{rd}(0)=0$, $Pop(0)=Pop_0$, $S_w(0)=0$, $R_{sup}(0)=R_{sup,0}$, $Z(0)=Q(0)+I(0)$, with all parameters non-negative.

Next, three economic balance conditions are required: Equation (19) reflects the fact that production output of the final goods sector equals consumption (C) plus social R&D outlays (O_{rd}); (20) reflects the idea that production of capital goods equals the sum of sectoral investments (I_i); and in (21) its is stated that total capital is the sum of sectoral capital stocks:

$$Q=C+O_{rd} \tag{19}$$

$$I=\sum_{i\in K} I_i \tag{20}$$

$$K=\sum_{i\in K} K_i \tag{21}$$

As far as the description of the ecological system is concerned, an aggregate ecological model should be consistent with a macroeconomic or regional system in terms of geographical coverage. Therefore, it should describe the essential features of a collection of various (possibly interacting) homogeneous ecological systems. Such a general model would have to be able to deal with the several functions and characteristics of ecological systems:

(i) regenerative capacity;
(ii) assimilation of pollution;
(iii) resource supply;
(iv) storage of waste materials;
(v) non-material services for consumption;
(vi) decreasing performance of all functions for increasing levels of pollution and/or decreasing resource stocks, and for an increasing level of other disturbances.

The equations in (22) show a model that describes three types of environmental processes: slow and regular regeneration of resource stocks (B and N, respectively), and assimilation of pollution (P). B can be regarded as semirenewable, ie falling in between a renewable and non-renewable resource (see Swallow

[19]); one may think of such categories as soil, land and water or even rain forests and similar slowly regenerating and sensitive natural systems. N may, for instance, denote biotic resources or ecosystems. The respective stocks and the overall environmental quality (E) have a non-negative relationship with the rates at which each process occurs. E is a function of N, P and B. This means that all environmental processes depend indirectly upon all three environmental stocks. A fourth environmental stock is that of non-renewable resources (S), which is assumed to have no natural environmental linkages with the other stocks or their natural processes. The exogenous impacts upon N and P are resource extraction (R_N) and waste emission (W_{em}) respectively. B is influenced by the pressure of the scale of the economy (indicated by economic capital K), the size of the population (Pop), and the intensity of extractive activities (indicated by the rates of renewable and non-renewable resource exploitation R_N and R_S). S decreases by extraction (R_S):

$$
\begin{aligned}
&E = H(N, B, P)\\
&\mathrm{d}N/\mathrm{d}t = G(N, E) - R_N\\
&\mathrm{d}P/\mathrm{d}t = -A(P, E) + W_{em}\\
&\mathrm{d}B/\mathrm{d}t = [b_1(E) - b_2(\mathrm{d}K/\mathrm{d}t) - b_3(\mathrm{d}Pop/\mathrm{d}t)\\
&\qquad - b_4(R_N) - b_5(R_S)] * B\\
&\mathrm{d}S/\mathrm{d}t = -R_S
\end{aligned}
\tag{22}
$$

The initial conditions are $N(0) = N_0$, $P(0) = P_0$, $B(0) = B_0$, $S(0) = S_0$. All functions and variables are non-negative. $H()$ is increasing in N and B and decreasing in P. The natural growth function satisfies $G(N_{min}, E) = G(N_{max}, E) = 0$, and for some (N, E), with $0 < N_{min} < N^* < N_{max}$, $E > 0$: $G(N^*, E) > 0$, where N_{min} is the minimum viable level of the regenerative resource and N_{max} denotes its maximum level (also referred to as carrying capacity). The natural assimilation function A is increasing in P. G and A are increasing in E. All $b_i (i = 1, 2, 3, 4, 5)$ and their first derivatives are positive.

Finally, the following *material balance conditions* apply to the flows within the economy and to the natural environment. Total waste output ($W_{Q,I}$) from 'real' production is equal to the inputs less the outputs:

$$W_{Q,I} = [c_Q(T_{rd}) - 1] * Q + [c_I(T_{rd}) - 1] * I \tag{23}$$

This 'dirty' waste is treated first (producing R_{wa} in Equation (9)) before it goes to one of three categories (see Figure 1), namely unused stored waste, waste amenable for recycling, or emitted waste, as reflected by Equations (16), (25) and (26), respectively.

The indicator of the demand for new resources is based on the amounts needed as inputs to production of Q and I less the recycled amounts from the present period:

$$R_{new} = c_Q(T_{rd}) * F_Q(K_Q, E) + c_I(T_{rd}) * F_I(K_I) - R_{rec} \tag{24}$$

Waste amenable for recycling purposes is the sum of private and public consumption, discarded capital and a part of treated production waste:

$$W_{rec} = Q + \sum_{i \in K} D_i(K_i) + (1 - s_1) * R_{wa} \tag{25}$$

The result W_{rec} goes into the recycling process (Equation (10)). As is made clear in Equations (16) and (26), the expression 'amenable for recycling purposes' indicates that only part of W_{rec} will be recycled (ie recycling is strictly less than 100%), and the remaining part ends up as waste that is being stored or emitted.

Emission of waste to natural media (W_{em}) thus equals the sum of non-treated production waste and the part of waste remaining after recycling that is not stored into S_{wa}:

$$W_{em} = (W_{Q,I} - R_{wa}) + (1 - s_2) * (W_{rec} - R_{rec}) \tag{26}$$

All material flows in the system have now been modelled. The feedback from environment and the value system, which was mentioned at several stages, still remains to be formalized. All variables necessary to perform this exercise are available now.

Economic feedback using sustainable development conditions

Sustainable development feedback can be regarded as a process of adjustment to meet or approximate sustainable development conditions (see van den Bergh [4] Chapter 2). Such conditions reflect goals associated with intergenerational equity and the maintenance of natural environmental qualities and quantities. We make a distinction between two types of conditions. First, constraints may be set on the level of welfare (for a whole generation or per capita), where the choice is to require that welfare exceeds a given subsistence level, and follows a non-decreasing time path. A second type is formed by constraints on stocks or flows (interchangeable) in and between economic and environmental systems. The notion of stock constancy may serve as an illustration of sustainable development conditions of the latter type.

The completed model allows for feedback from environmental quality, overall resource scarcity, unsustainable renewable resource exploitation, and unsustainable waste emission pollution to decisions regarding investment activities and technology. Conditions based on regeneration and assimilation on the one hand, and concern for future generations and natural environment on the other, are central in the feedback mechanism. They determine the acceptable or aspired to extraction and emission levels.

In the model this proceeds as follows. At the core are sustainable development (flow) conditions that apply to the flows of resources and waste of the type mentioned above. These are transformed, via behavioural or social control parameters, into variables that indicate the perceived levels. This involves an ethical choice by society or decision makers with regard to the degree of concern for the well-being of future generations. Market mechanisms (prices) and social controls (regulations and price corrections) are implicitly taken into account. They provide incentives that have an effect on the model outcome via the reaction of investment allocation and technological progress. For instance, a strong incentive may result from scarcity of resources through increasing prices. Low regeneration or assimilation rates, or increasing (decreasing) stocks of pollution (renewable resources), may induce social actions. These may take the form of controlling investment through land-use policies, subsidies on environmentally beneficial investment, or artificially higher prices for resource materials. They may also stimulate certain technological changes by subsidies or time paths for required levels of resource use or waste generation.

On the basis of social value considerations with respect to future generations and the relationship with nature, as well as ecological considerations with regard to regeneration and assimilation, we can devise the following three indicators: acceptable renewable ($R_{N,perc}$), non-renewable ($R_{S,perc}$), and total (R_{perc}) resource extraction:

$$R_{N,perc} = \max\{0, d^* p_N{}^* N + (1-d)^* G(N, E)\} \qquad (27)$$

The dummy variable d may reflect an objective of sustainable use ($d=0$) or a less strict objective ($d=1$), which requires an additional parameter for intergenerational concern, p_N (falling between 0 and 1); the latter shows high degrees of moral concern for future generations by way of taking low values. A second interpretation of this parameter is that it shows variable degrees of caution (prudence or risk aversion) in the face of much uncertainty:

$$R_{S,perc} = p_S{}^* S \qquad (28)$$

The parameter p_S is similar to p_N for non-renewable resources. Therefore d, p_N and p_S reflect value stances:

$$R_{perc} = R_{N,perc} + R_{S,perc} \qquad (29)$$

Thus, total perceived resource availability is simply the sum over the two sources. Notice that this is not simply a sum of two types of resource, but of perceived relevant amounts.

A variable R_{short} indicates the scarcity of resources by taking the ratio of R_{perc} and the required extraction of resources (R_{new}, see Equation (24)). A value higher than one means that there is no reason for an sudden strong economic reaction:

$$R_{short} = \min\{1, R_{perc}/R_{new}\} \qquad (30)$$

Other specifications may serve the same purpose, but (30) has the advantage of generating a dimensionless number.

Similar to the approach for resource feedbacks, one can also take a ratio of waste assimilation $A(P, E)$ and waste emission to obtain an indication of the degree of unsustainability of waste emission (too much waste emission, indicated by W_{too}):[6]

$$W_{too} = \min\{1, A(P, E)/W_{em}\} \qquad (31)$$

Since the variables R_{short} and W_{too} have values ranging from zero to one, the following functional structure ensures a range similar to that of E for the ecological effect variable e_e that appears in the model of endogenous technological progress in (14) and later in the investment (feedback) decision process (equations in (52)):

$$e_e = R_{short}{}^* W_{too}{}^* E \qquad (32)$$

On the basis of the indicator e_e we can include a feedback from the environmental state and the ethical stance to economic actions or decisions. This means that, in addition to feedback to production technology (Equation (14)), investment rules can be stated that incorporate the relevant indicator. This will be done in the context of specific scenarios later in the paper.

[6] Sustainable waste emission is used analogously to sustainable resource use in a strict sense (applied to renewable resources): the controlled flow of waste emission is such that the stock of pollution remains at the same level, or, in other words, such that the assimilative capacity is not exceeded. Alternatively, we may define it as avoiding a decrease in the stock of renewable resources, since it is linked to the assimilative capacity. In fact, sustainable management could be defined as a choice of combinations of resource extraction (use) and residuals emission, such that the resource size – and thus the assimilative capacity – is not reduced.

Specification and calibration

This section gives information on the calibrated model that was used for numerical simulations. Functional specifications, calibrated initial conditions and parameter values are discussed. The calibration was performed in order to satisfy conditions on initial values that reflect several aspects of supply–demand balance and environmental sustainability, so that a stationary or stable pattern is accomplished over a certain period of time. These conditions are presented in Appendix 2. The calibration also aimed at producing a base case model that represents a rather pessimistic opinion about the ultimate state of the economic system far in the future, namely one that collapses. This is done for the sake of comparative analysis of scenarios ie tracing the differential effects of certain additional developments or policies as compared to the base case. Such a choice certainly does not mean to imply that a base case will give rise to a collapsing system in the very long tun. Such a prediction cannot be provided with any appropriate degree of accuracy. Accordingly, the present model is not aiming to solve predictive issues.

The aggregation level of the model makes it difficult to base functional forms and parameter values on empirical data with straightforward interpretations. Estimated values for parameters are partly chosen in a realistic range. Some of these represent fractions, which are logically constrained. Others are based on, for instance, notions of maximum efficiency (such as in the case of recycling). The motivation for such choices are mentioned hereafter. Other variables have a long-run physical interpretation, so that their units of measurement are to be interpreted accordingly. Some values are set fixed over all scenarios, after calibrating to meet the above mentioned goals. Variations were allowed in other variables (controls) on the basis of their interpretation, policy relevance, strategic value, or because of the model sensitivity. It is hoped that the model base case and the various scenarios thus represent a realistic setting, which may be a country with an economy that is to a large extent based on natural resources and environmental conditions within its national boundaries.

The specifications and parameter choices that resulted are as follows. First, consistency requirements and functional characteristics are given. Then, the choices for specific functional structures and parameter values (in the basic scenario) are motivated. Finally, the realistic ranges are given for some important, uncertain, policy or behavioural parameters (the latter two types may change over time). All functions and variables are non-negative. The following set of consistency conditions applies to the functions in the model:

$$c_Q(\cdot), c_I(\cdot) > 1$$

$$f_{wa}(\cdot,\cdot), f_{wa}(\cdot,\cdot) \leq 1$$

$$s_1, s_2 \in [0, 1] \qquad (33)$$

The functions in the model are specified as follows:

$$F_Q(K_Q, E) = \min\left\{\frac{E + a_Q}{E_{crit} + a_Q}, 1\right\} * k_Q * K_Q$$

$$c_Q(T_{rd}) = \frac{T_{rd} + b_Q}{T_{rd} + c_Q} \qquad (34)$$

To establish physical capital coefficients (capital divided by total production), the Dutch data given in Table 1 are illustrative (based on *CBS Yearbook*, data 1988). The capital size was derived by assuming an average price of capital equal to 2 guilders per kilogram. In physical terms a value for k_Q can be compared with that for food in the table; we choose $k_Q = 5$. It is assumed that $c_Q(T_{rd}(0)) = 1.4$, so that with $T_{rd}(0)$ set at 100 (index) a possible choice for (b_Q, c_Q) is (600,400). Estimates are that as a result of environmental deterioration (pollution, soil erosion, water extraction) in agriculture, 5 to 10% damage is caused to crops in some developed countries and 10 to 20% in developing countries; damage caused to recreation, housing etc is not as simple to quantify. Based mainly on agricultural data we take for $E = 0.5$ $(E + a_Q)/(E_{crit} + a_Q) = 0.9$; then for $E_{crit} = 0.8$, $a_Q = 2.2$:

$$F_I(K_I) = k_I * K_I$$

$$c_I(T_{rd}) = \frac{T_{rd} + b_I}{T_{rd} + c_I} \qquad (35)$$

It is assumed that the capital sector is more capital

Table 1. Crude estimates of physical capital coefficients (thousand million).

	Production (kg) (1)	Capital (kg) (2)	Physical capital coefficient: (2)/(1)
Food, clothing	10	45	4.50
Oil	70	10	0.14
Chemicals	16	35	2.19
Construction	20	3	0.15
Metal	20	11	0.55
Total	136	104	1.31

intensive (physically) than the final goods sector (eg electronics, chemical, metal, construction materials), so that $1/k_I > 1/k_Q$. The efficiency of resource used is assumed to be somewhat higher in the capital sectors than in others: $c_Q(T_{rd}(0)) \leq 1.4$, so that a possible choice for (b_I, c_I) is (550,400):

$$f_{wa}(T_{rd}, K_{wa}) = a_{wa} * \frac{T_{rd}}{T_{rd} + b_{wa}} * \frac{K_{wa}}{K_{wa} + c_{wa}} \quad (36)$$

We choose $a_{wa} = 0.95$ (maximum abatement is 95% of waste subject to abatement); $K_{wa} = 5$. Initially 65% abatement; the contribution of the technology and capital terms in the multiplicative specification are assumed to be equal initially, so that $(0.95*100)/(b_{wa}+100) = 0.65^{0.5} = 5/(c_{wa}+5)$, or $b_{wa} = 18$ and $c_{wa} = 1.2$:

$$f_{rec}(T_{rd}, K_{rec}) = a_{rec} * \frac{T_{rd}}{T_{rd} + b_{rec}} * \frac{K_{rec}}{K_{rec} + c_{rec}} \quad (37)$$

We choose $a_{rec} = 0.8$ (maximum recycling is 80% of waste subject to recycling); $K_{rec} = 5$. Initially 15% recycling; also here, the contributions of the technology and capital terms are assumed to be equal initially, so that $(0.8*100)/(b_{rec}+100) = 0.15^{0.5} = 5/(c_{rec}+5)$, or $b_{rec} = 107$ and $c_{rec} = 8$:

$$F_N(K_N, N, E) = a_N * K_N^{\alpha_N} * (N * E)^{\beta_N} \quad (38)$$

We assume $\alpha_N = 0.8$, $\beta_N = 0.2$ and $a_N = 1.8$:

$$F_S(K_S, S) = a_S * K_S^{\alpha_S} * S^{\beta_S} \quad (39)$$

We assume $\alpha_S = 0.7$ (non-renewable resource extraction more capital intensive than renewable resource extraction), and $\beta_S = 0.3$ (stronger negative effect of decreasing stocks in comparison with renewables) and $a_S = 0.9$. In both extraction sectors we thus assume constant returns to scale:

$$D_i(K_i) = \delta_i * K_i \qquad i \in K \quad (40)$$

Because of new trends capital depreciation is assumed to take place at a faster pace in the final goods (Q) sector, where we assumed $\delta_Q = 0.04$ or an average lifetime of 20 years. In all other sectors we assume a rate of 0.025:

$$\alpha(e_e) = 1 - \alpha * e$$

$$\varepsilon(e_e) = 1 - \varepsilon * e \quad (41)$$

In (42), for $e = 1$ (maximum) the minimum levels of functions $\alpha(\,)$ and $\varepsilon(\,)$ are obtained, which we choose equal to 0.002 and 0.004 respectively. This leads to parameter values $\alpha = 0.098$ and $\varepsilon = 0.096$. We assume $\delta = 0.5$:

$$B\left(\frac{C}{Pop}\right) = \frac{a_B - \dfrac{C}{Pop}}{100} \quad (42)$$

In (43) it is assumed that population increases at a rate depending on the difference between C/Pop (estimated initially at around 5) and a level a_B. The initial growth is assumed 0% so that $a_B = 5$. The functional structure applies especially to medium developed contries:

$$e_e = R_{short} * W_{too} * E \quad (43)$$

This simple multiplicative form restricts e_e to the same range as each of the three parts, namely in between zero and one:

$$H(N, B, P) = \max\{0.2, \min\{1, N/N_{crit}\} * \min\{1, B/B_{crit}\} * \min\{1, P_{crit}/(P+1)\}\} \quad (44)$$

The following values are assumed: $E_{min} = 0.2$, $N_{crit} = N_*$ (where $G(N, E)$ has a maximum), based upon the initial levels of B and P, the expected $B_{crit} = 100$ and $P_{crit} = 450$. Initially the E level is approximately 0.5:

$$G(N, E) = \min\{E/E_{crit}, 1\} * r * N * \left(1 - \frac{N}{\min\{E/E_{crit}, 1\} * C_N}\right) \quad (45)$$

$N(t)$ is initially assumed at 10 000, a third of its maximum value $C_N = 30\,000$. r is 0.05, so that regeneration at a maximum environmental quality ($E = 1$ and $N = N_* = 15\,000$:

$$M(P, E(= \min\{E/E_{crit}, 1)\} * a_P * P^{b_P} \quad (46)$$

$a_P = 0.5$ and $b_P = 0.9$, in order to include 'decreasing returns to scale' in waste assimilation. Initially P is equal to 100 in the base scenario:

$$b_1(E) = a * E$$

$$b_2\left(\sum_{i \in K} \frac{dK_i}{dt}\right) = b * \sum_{i \in K} \frac{dK_i}{dt}$$

$$b_3\left(\frac{dPop}{dt}\right) = c * \frac{dPop}{dt}$$

$$b_4(R_N) = d * R_N$$

$$b_5(R_S) = e * R_S \tag{47}$$

$a = 0.005$, $b = 0.00005$, $c = d = 0.0001$, $e = 0.00004$, in order to obtain terms of similar size.

Finally, the critical damage level for natural regeneration, assimilation, and slow regeneration is equal to that for economic production (see above), namely 0.8.

The distribution of Q and I is as follows:

$$C = a_C * Q + b_C$$

$$O_{rd} = Q - C \tag{48}$$

The consumption propensity is set equal to 0.5, and the fixed consumption level is 30. Investment allocation equals all sectoral relative growth rates of capital:

$$I_i = \delta_i * K_i + \left(I - \sum_{j \in K} \delta_j * K_j\right) * \frac{K_i}{\sum_{j \in K} K_j} \tag{49}$$

The initial economic capital stock values are such that supply and demand for both new capital and for resources are in balance for the basic scenario (see Appendix 2): $K_Q(0) = 30$, $K_I(0) = 4$, $K_{wa}(0) = K_{rec}(0) = 5$, $K_N(0) = 15$, $K_S = 30$. The other stocks are at the beginning as follows: $X(0) = Y(0) = Z(0) = 0$ since the structure of their equations is such that they are set to zero each time; immediately after the beginning they will attain their 'right' values. $T_{rd}(0) = 100$, $S_w(0) = 0$, $Pop(0) = 100$. The initial values of N, P and B are chosen such that $E(0) = 0.5$: $N(0) = 10\,000$, $B(0) = 100$, $P(0) = 200$, $S(0) = 3000$.

In order to find the initial conditions of the economic stocks, the following set of equations has to be solved (see Appendix 2): first, the balance between resource materials supply and demand:

$$\frac{T_{rd} + 600}{T_{rd} + 400} * 5 * K_Q + \min\{(E + 2.2)/(E + E_{crit}), 1\} + \frac{T_{rd} + b_I}{T_{rd} + 400} * k_I * K_I = R_{rec} + R_N + R_S \tag{50}$$

with R_{rec}, R_N and R_S given as above; second, the balance between investment goods supply and demand for a stationary path:

$$0.04 * K_Q + \delta_K * (K_{wa} + K_{rec} + K_N + K_S) = (k_I - \delta_K) * K_I \tag{51}$$

For an initial choice of $T_{rd} = 100$, $N = 15\,000$, $P = 100$, $B = 100$ – so that $E(0) = 1$ (maximal environmental quality) the choice (K_Q, K_I, K_{wa}, K_{rec}, K_N, K_S) = (43, 3.6, 5, 5, 30, 30) satisfies these conditions. Initial values for R_{sup} and S may be chosen so that the initial conditions mentioned in Appendix 2 are satisfied.

In Table 2 we show the values of parameters in the basic scenario.

Scenario and sensitivity analysis

In this section the results of scenario and sensitivity analyses will be shown, based on the model formulated

Table 2. Parameter values in the simulation model.

Parameter	Variable	In base scenario	In equation
Critical environmental quality level: below which negative impact upon regeneration, assimilation and production	E_{crit}	0.8	Q, N, B, P, R_{new}
Capital productivity in investment sector	k_I	1	I, R_{new}
Parameters in technology effect on resource use in investment sector	b_I	550	$I, R_{new}, W_{Q,I}$
Scale factor abatement function	a_{wa}	0.95	R_{wa}
Scale factor recycling function	a_{rec}	0.8	R_{rec}
Rate of capital depreciation	δ_i	0.025	W_{rec}, $dK_i/dt, i \in K \setminus \{Q\}$
Sustainable development feedback to investment in R&D	α	0.098	dT_{rd}/dt
Population growth parameter	a_B	5	Pop
Fraction of treated waste going into unused stored waste	s_1	0.5	dS_w/dt, W_{rec}
Fraction of recycled waste going into unused stored waste	s_2	0.5	dS_w/dt, W_{em}
Sustainability dummy	d	0	$R_{N,perc}$
Fraction of renewables' stock regarded available	p_N	–	$R_{N,perc}$
Fraction of non-renewables' stock regarded available	p_S	1	$R_{S,perc}$
Critical renewables' level for environmental quality determination	N_{crit}	15 000	E
Critical slow renewables' level for environmental quality determination	B_{crit}	100	E
Critical pollution level for environmental quality determination	P_{crit}	450	E
Intrinsic growth of renewables	r	0.05	dN/dt, R_{short}
Carrying capacity of renewables	C_N	30 000	dN/dt, R_{short}
Assimilation function scale parameter	a_P	0.5	dP/dt, W_{too}
Population growth function parameter	a_B	0.005	dB/dt

Table 3. Scenarios for analysis; empty cells denote the same value as in the base case scenario.[a]

Scenarios	K_I	N	P	S	d	*inv*
0. Base case	3.6	15 000	100	3000	1	0
1. No ethical concern			599		0	
2. Low environmental quality		10 000	599		0	
3. Low non-renewables				1000	0	
4. Low quality and ethical concern		10 000				
5. Moderate growth	5				0	
6. Strong growth	7				0	
7. Extreme growth	10				0	
8. Sustainability and ethical concern						1
9. Extreme growth, sustainability and ethical concern	10					1
10. Extreme growth and sustainability	10				0	1

[a] K_I, N, P, S, d and *inv* stand for initial stocks of capital in the investment sector, initial stocks of natural resources, pollution and non-renewable resources, and two dummy variables for the presence or absence of ethical concern and the feedback to investment allocation, respectively.

above. First we will outline the scenarios and discuss an investment correction scheme that deals with the sustainable development feedback to capital investment in the economy for some scenarios. Next, results of simulations are presented and discussed. The scenario analyses serve two purposes. First, they will show the type of patterns that can be generated by the model. Second, certain types of (policy) strategies and exogenous scenarios can be illustrated by it. The sensitivity analysis is restricted here to changes in certain initial environmental stocks, as these may have a large impact upon the dynamic behaviour of the system. Table 3 lists 11 scenarios under which model simulations were performed. The results of these in terms of generated time patterns for certain indicators are shown in Figures 3 to 10. Table 3 indicates which control variables in scenarios 1–10 differ from the base case scenario. These variables were chosen in order to take the value system, environmental quality, non-renewable resources, and the pace of growth into account. The base case scenario 0 represents a stationary economy (no capital stock changes) with a balance of the demand for a supply of investment goods, maximum environmental quality, a balance of demand for a supply of resource materials, and maximum ethical concerns for future generations. Each of the other scenarios deviates from this base case scenario in at least one of these respects. Scenario 1, no ethical concern, differs from the base case by including a minimum of ethical concern, as reflected by the dummy d (see Equation (27)). In the scenarios 2, low environment quality, and 4, low quality and ethical concern, the environmental quality variable E is initially equal to 0.5 as a result of the chosen initial – lower than in the base case – stock values for renewable resources and pollution. The scenarios 5, 6 and 7, moderate growth, strong growth and extreme growth, represent moderate, strong and very strong *ex ante* growth schemes, based on a choice of the initial capacity for growth, given by the stock of capital in the investment sector. Scenarios 8 to 10, sustainability and ethical concern, extreme growth, sustainability and ethical concern, and extreme growth and sustainability, try to differentiate the effects of extreme growth, and ethical concern, given the active sustainability feedback to investment decisions. This means that investment is reallocated from final output to abatement and recycling when the indicator e_e (taking values between or equal to 0 and 1) is below its maximum. The correction terms that are applied respectively to investment in final goods production, waste abatement, and recycling are given in (52). The parameter values chosen are as follows: $part_Q = part_{wa} = 0.5$; the pattern of Q was similar when $part_{wa}$ was changed in 0 or 1. The dummy *inv* denotes whether or not the feedback to the investment allocation is working (1 and 0, respectively). In scenario sustainability and ethical concern the initial allocation and level of investment are based on the assumption of a stationary state. In scenario 9 strong growth is combined with endogenous investment allocation. Scenario 10 differs from 9 by suppressing strong ethical concern, so that reactions in both investment allocation and technological progress will be less than under scenario 9.

$$correction\ I_Q = cor_{I_Q} = -inv * part_Q * (1 - e_e) * I_Q$$

$$correction\ I_{wa} = part_{wa} * cor_{I_Q}$$

$$correction\ I_{rec} = part_{rec} * cor_{I_Q}$$

$$part_{wa} + part_{rec} = 1 \tag{52}$$

Since environmental quality is higher for scenarios 0, 1 and 3 than for scenarios 2 and 4, the initial production output is also higher (see Figure 3). For

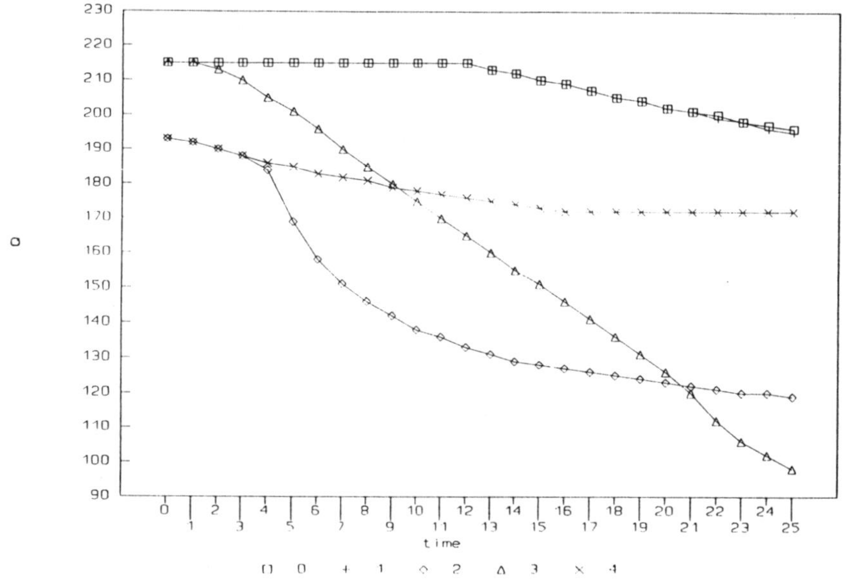

Figure 3. Development of final goods production over time for scenarios 0–4.

scenarios 2 and 3, environmental quality is the single cause of decreases in output. For the other three scenarios resource scarcity is an additional cause, as it leads to under-investment and decreasing stocks of capital over time (see Figure 4). However, the curve for scenarios 4 in Figure 3 shows that more concern for future generations and natural environment gives rise to a slowing down of the decline in output. Notice that the difference between scenarios 0 and 1 is negligible, due to the fact that a stronger ethical concern (as in the base case scenario 0) does not give rise to impacts on (tempered) growth as long as the system behaves in a sustainable pattern over the time period considered. However, when a non-stable pattern is followed, the differential outcomes of scenarios 2 and 4, which is due to the same difference in control settings, is not negligible.

The pace of technological progress is apparently influenced by three factors: investment, environmental effects and ethical concerns. With specific attention (or ethical concern) for future and environment, and with a non-optimal environmental quality, technological progress is seen to be highest in scenario 4 (see Figure 5). In scenario 4 two out of three factors are active. Scenario 2 shows also high technological progress, because the system described by it suffers from very low environmental quality (one factor). Scenarios 0 and 1 have an equal pace of technological progress as a result of equal capital investment (both initially stationary state paths). Scenario 3 shows the slowest technological progress, partly since the environmental effects are missing. To observe this completely, one should compare the environmental quality for each scenario in Figure 6. This may be interpreted as few non-renewable resources (eg fossil energy resources) giving rise to output decline and neutral or propitious environmental effects. Supply patterns of resource materials (Figure 7) differ widely because of the following factors: renewable resource stock, non-renewable resource stock, extraction and recycling capital, technological progress affecting extraction and recycling efficiency of capital. Concern for future generations allows the supply of materials for the present choice of initial natural resource stocks to keep increasing (scenarios 1 and 4). A low environmental quality causes materials supply to be initially lower for scenario 2 than for 3, but at a later stage this relationship inverts. This change results from a favourable turn in the development of the capital stock and technology.

In Figure 8 different growth patterns are shown. Here, we can compare long-term effects of variation

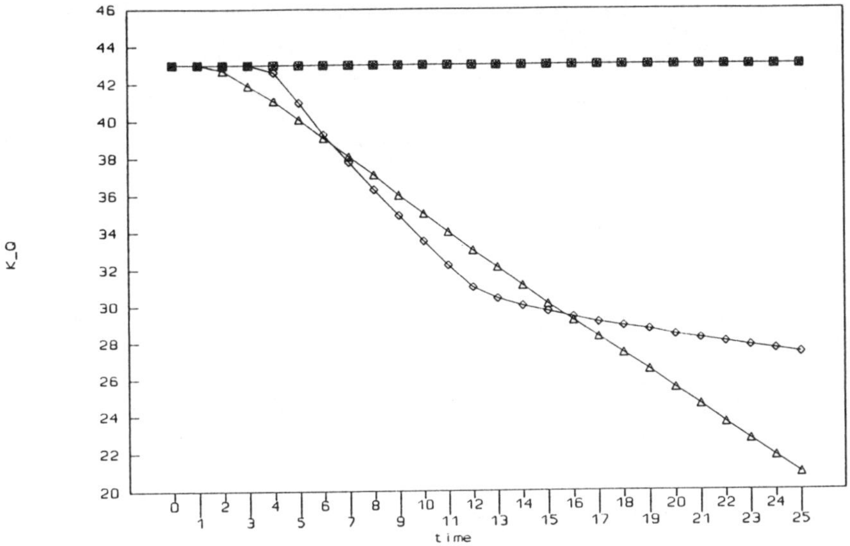

Figure 4. Development of capital in the final production sector over time for scenarios 0–4.

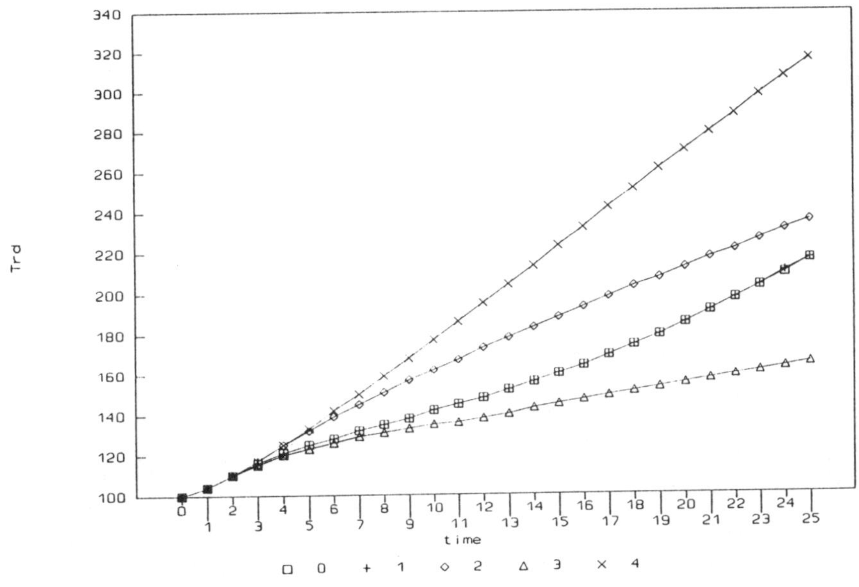

Figure 5. Technological progress for scenarios 0–4.

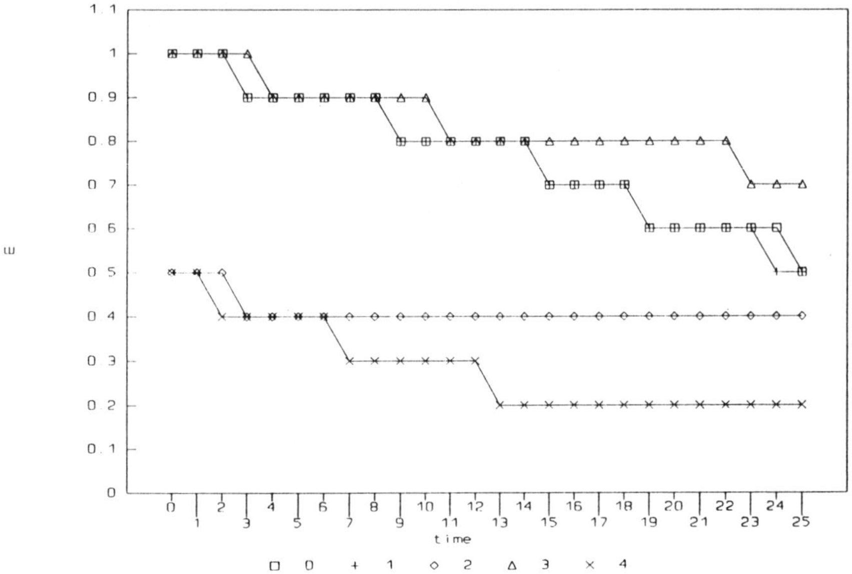

Figure 6. Environmental quality over time for scenarios 0–4.

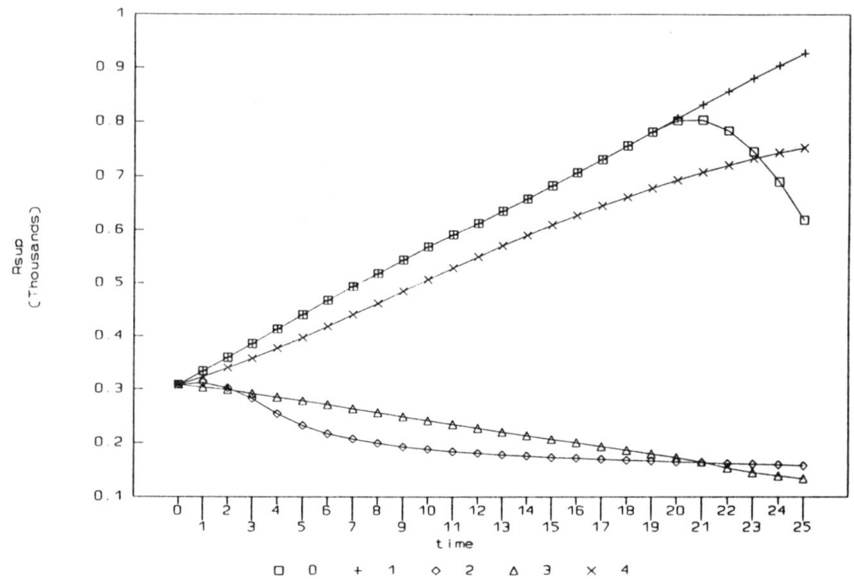

Figure 7. Total supply of natural resource materials over time for scenarios 0–4.

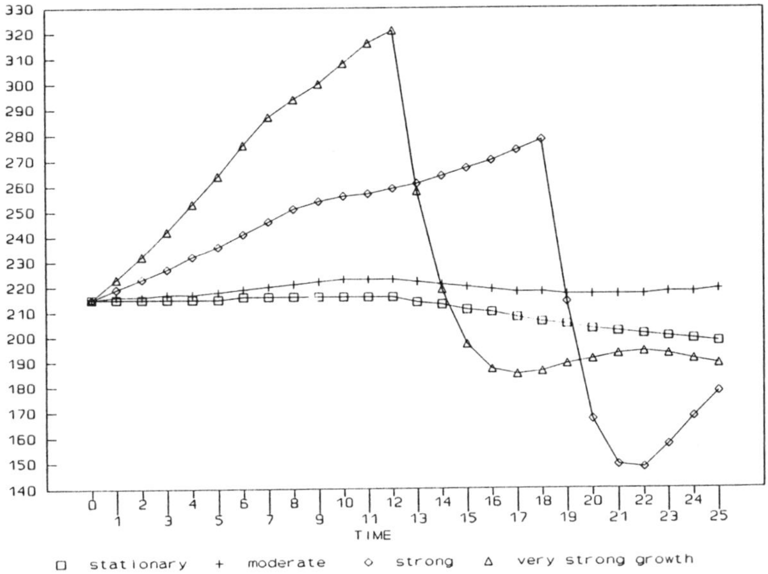

Figure 8. Final goods production over time for scenarios 0, 5, 6 and 7.

in the initial capacity for growth on output. Only moderate growth seems sustainable over the time period shown. Strong growth leads to a quickly decreasing environmental quality, and a collapse in output. The size of the fall in output levels is comparable for strong growth and extreme growth. As a consequence, it seems not to matter (relatively) how fast the growth, since the collapse will be relatively hard. However, the output level under strong growth is catching up with that under extreme growth at the end of the period shown (the environmental quality levels are also equal). Figures 9 and 10 show the results in terms of final goods output of various scenarios. We have taken a time horizon of 100 years since we wanted to investigate the very long-term effects of a completely different development of the sectoral structure. Here the investment allocation is endogenously influenced by the state of the environment and ethical concern, which we referred to in Section 4 as sustainable development feedback (according to Equations (52)). From Figure 9 it is clear that this reaction (called 'adjustment') has a positive effect on output in the long run, as compared without such a reaction. Collapses followed by growth are evaded ie patterns are smoothed out. Initially the output decreases faster under the investment allocation reaction, but from time 40 on output is higher. The collapse followed by recovery and a second, catastrophic collapse are explained by the fact that a collapse of the economy allows for decreasing environmental pressure and new economic opportunities for growth. However, the growth path is so steep, because of the relatively large supply of investment goods, that environmental pressure increases along with it. This causes the final collapse. In Figure 10 patterns can be compared between extreme growth (without adjustment of investment allocation) extreme growth, sustainability and ethical concern (with investment reaction and ethical concern) and extreme growth and sustainability (minimum ethical concern for future generations and/or natural environment). It is clear that without much ethical concern the reaction is too weak for output to stay on a growth pattern. When adjustment of investment allocation is based on the strong notion of ethical concern, the growth pattern can be maintained for a very long period of time. Finally, a sharp fall in output results, caused by both the negative impact on the environment of maintained growth and the investment reaction diverting investment from final goods production to waste abatement

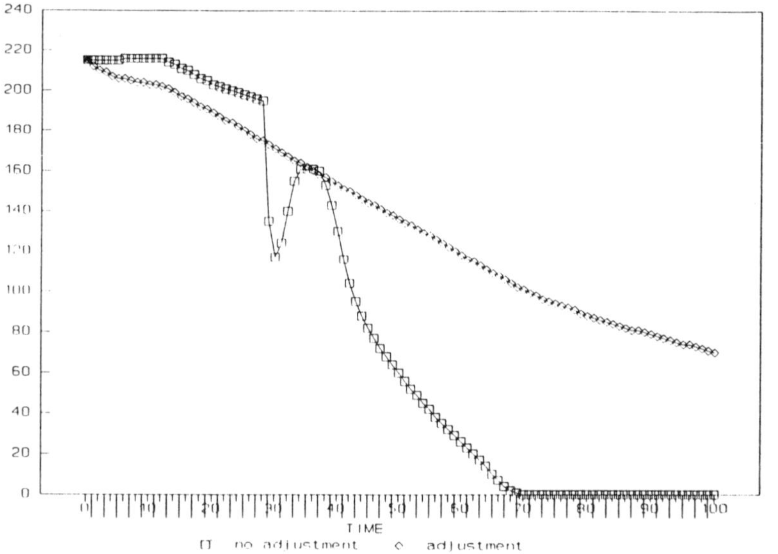

Figure 9. Development of final goods production over time for scenarios 0 (no adjustment) and 8 (adjustment).

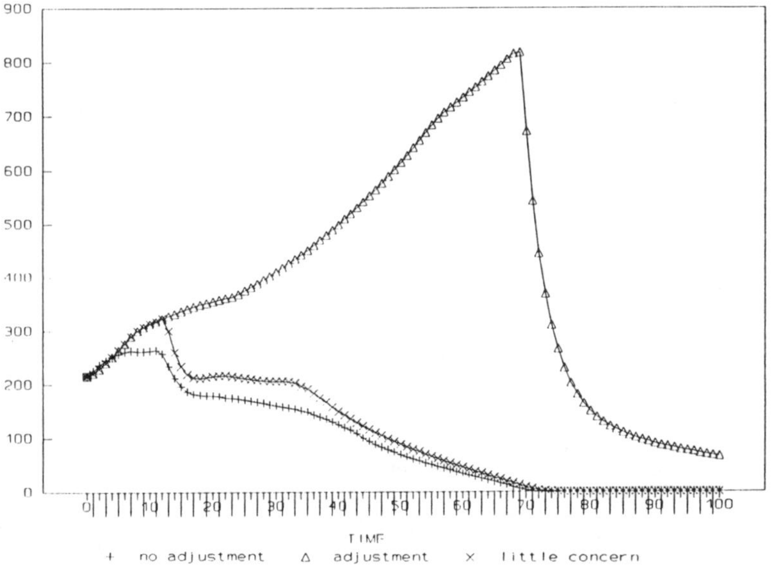

Figure 10. Final goods production over time for scenarios 7 (no adjustment) and 10 (little ethical concern).

and recycling sectors. The pattern finally seems to stabilize on a positive lower level. The conclusion may be that the development of sectoral structure has a large impact in the long run on the level of output of final goods.

Finally, sensitivity analysis was also performed to investigate the model reactions to changes in certain initial conditions. Especially the natural renewable resources N and pollution P are open to different interpretations, and therefore their levels were varied. The results are shown in Figures 11 to 14, for various combinations of renewable resource and pollution stocks, all under the base case scenario. The results show that patterns depend strongly on initial values of these variables, and that patterns of collapse are usually followed by recovery, though at a lower level than the initial one. This can be explained by the fact that a collapse of output may go along with a decrease of environmental pressure from both less resource use and less waste residuals being generated. This depends first of all on the magnitude of a fall in environmental quality and resource scarcity that causes an economic breakdown. Furthermore, whether recovery can occur, especially at a fast speed, depends also on certain characteristics of the economic system that determine its recovery capacity, such as technological progress, and capacities of abatement and recycling activities relative to production capacities. In Figure 11 one can find patterns that have a short recovery followed by a second collapse (or a phase-wise collapse). Here, the environmental breakdown causes an economic collapse, which allows environmental pressure to be weakened. The stationary path up to time 19, however, did not stimulate technological progress very much, so that the economic recovery capacity was weak, giving rise to the second collapse after time 22. A general conclusion is that the two-way economic–environmental interactions cause interesting feedback loops. One cannot simply make the general statement that a better initial environmental state will lead to better economic performance over a certain period of time. One has to realize that a better environment may give incentives for high growth and (*ex post*) undesirable or wrong directions of technological progress.

Conclusions

At the end of this study some retrospective remarks are in order. Economic modelling for the purpose of obtaining insight into the long-term behaviour of an

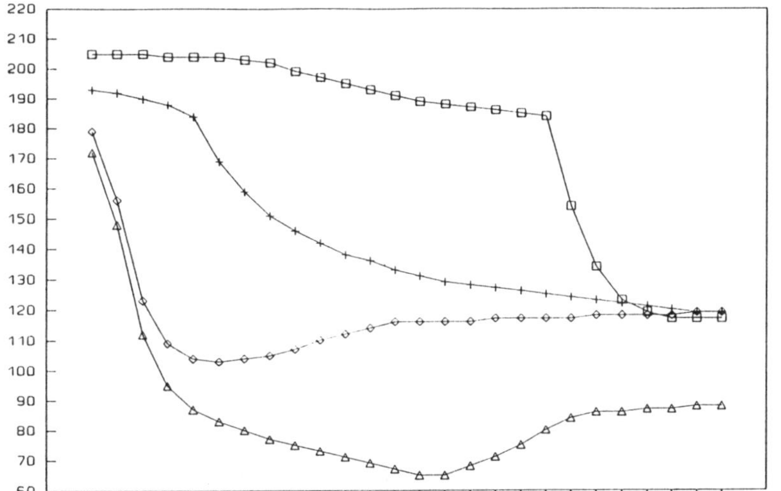

Figure 11. Development of final goods production over time for an initial renewable resource stock $N(0) = 10\,000$ and four different initial pollution stock levels.

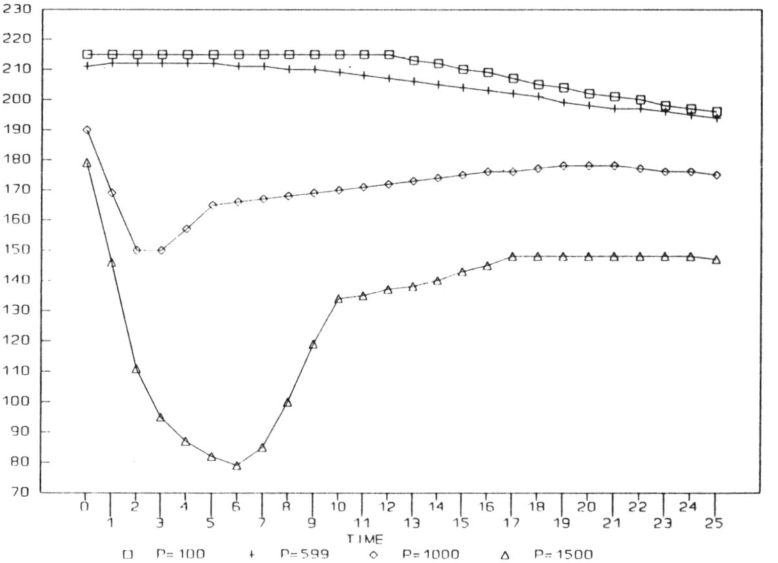

Figure 12. Development of final goods production over time for an initial renewable resource stock $N(0) = 15\,000$ and four different initial pollution stock levels.

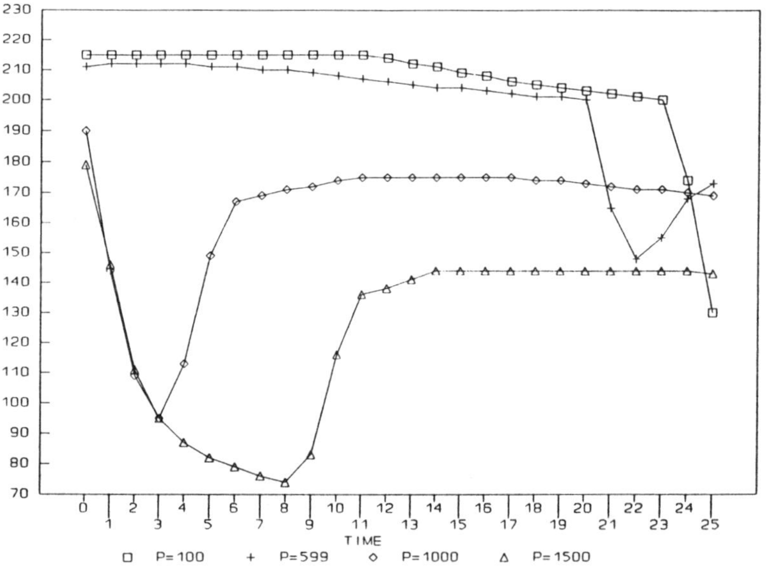

Figure 13. Development of final goods production over time for an initial renewable resource stock $N(0) = 20\,000$ and four different initial pollution stock levels.

Dynamic macro modelling and materials balance: J.C.J.M. van den Bergh and P. Nijkamp

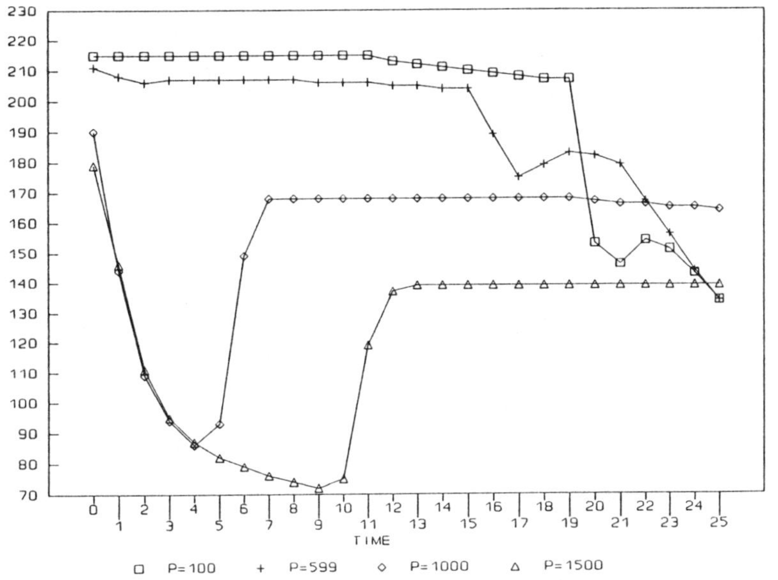

Figure 14. Development of final goods production over time for an initial renewable resource stock $N(0)=25\,000$ and four different initial pollution stock levels.

economic system faced by environmental and resource constraints requires an approach that combines dynamic models of growth and technology with environmental processes. Not only the adverse effects of economic sectors on environmental quality and resource availability can then be dealt with, but also the impact of environmental degradation and resource scarcity on economic performance. This may take the direction of physical effects, such as on health and productivity, and in addition, as was also pursued here, a focus on feedback mechanisms to economic decisions on technology and investment allocation. Such feedbacks were based on sustainable development conditions and the intensity of ethical concern about the environment and the future. More concern was interpreted here to imply a more cautious approach to designating resource availability and environmental quality as sufficient. This was one important characteristic of the model presented. The other important element is inclusion of the materials balance principle, via specification of production functions and materials' accounting between phases of extraction, production, consumption, emission and recycling.

The advantage of this approach is that it provides a consistent and general framework for dealing with complex dynamic interactions between an economy and its environment, especially taking account of the particularly relevant economic reaction mechanisms that are available and may become operative in such a context. The approach extends growth theory, materials balance accounting, and integrated modelling by combining complex interactions within both the economy and the environment. As a result, the model that was developed here is not analytically solvable. Therefore, it was calibrated to fulfill certain conditions in a base case scenario, linked to equilibrium and sustainability. Logical and realistic or plausible values were used for choosing parameter values. Subsequently, simulation experiments were carried out, focusing on initial economic and environmental stocks, feedback mechanisms, sustainability conditions and ethical concerns.

It is clear from the simulation and sensitivity analysis results that interesting and non-trivial insights can be obtained. The model can be regarded as helpful in our understanding (ie deduction) of how economy and environment interact in the long run, when material and physical relationships, and behavioural feedbacks are regarded. However, the important interpretations of the results should not be that within a known period of time the system will break down. More important are the types of patterns, economic reactions and intermediate dynamics of collapse and recovery. The results lead to an important observation: permanent reductions in environmental quality of environmental stocks seem to make it very difficult to sustain a positive trend or even a constant level of economic performance. We may even draw the conclusion that cautious behaviour does not necessarily lead to worse performance in the long run, but may certainly protect against adverse long-term developments. Furthermore, the model is very sensitive to changes in initial environmental and economic stocks, while patterns often tend to return to stable levels, even though possibly preceded by strong growth and collapse patterns. This is a clear non-linear characteristic of the model, which is tempered mostly by internal consistency provided by materials balance.

The inclusion of a materials flow and materials balance in dynamic models with non-linear processes has been shown to be able to provide a realistic and helpful tool for investigating long-term relationships between economy and natural environment. Application of this approach is limited by on the one hand data limitations, and on the other hand the problem of aggregation of physical information. One should either choose experimental models like the present one, or adopt a more specific disaggregated and often partial approach. The resulting model system provides a direction for further model oriented research for long-term environmental policy and scenario analysis.

References

1 C.L. Anderson, 'The production process: inputs and wastes', *Journal of Environmental Economics and Management*, Vol 14, 1987, pp. 1–12.
2 R.U. Ayres and A.V. Kneese, 'Production, consumption and externalities', *American Economic Review*, Vol 59, 1969, pp 282–297.
3 R.U. Ayres and A.V. Kneese, 'Externalities, economics and thermodynamics', in F. Archibugi and P. Nijkamp, eds, *Economy and Ecology: Towards Sustainable Development*, Kluwer, Dordrecht, 1989.
4 J.C.J.M. van den Bergh, *Dynamic Models for Sustainable Development*, PhD dissertation, Thesis Publishers/ Tinbergen Institute, Amsterdam, 1991.
5 J.C.J.M. van den Bergh and P. Nijkamp, 'Aggregate dynamic economic-ecological models for sustainable development', *Environment and Planning A*, Vol 23, 1991, pp 1409–1428.
6 J.C.J.M. van den Bergh, 'A framework for modelling economy-environment-development relationships based on dynamic carrying capacity and sustainable development feedback', *Environmental and Resource Economics*, Vol 3, 1993, pp 395–412.
7 CBS, *CBS Yearbook*, Central Statistical Office (CBS), SDU, The Hague, 1991.
8 H.E. Daly, 'Elements of environmental macroeconomics', in R. Costanza, *Ecological Economics: The Science and Management of Sustainability*, Columbia University Press, New York, 1991.
9 M. Faber and J.L.R. Proops, *Evolution, Time, Production and the Environment*, Springer-Verlag, Heidelberg, 1990.
10 M. Faber, H. Niemes and G. Stephan, *Entropy, Environment and Resources: An Essay in Physico-Economics*, Springer-Verlag, Heidelberg, 1987.
11 N. Georgescu-Roegen, *The Entropy Law and the Economic Process*, Harvard University Press, Cambridge MA, 1971.
12 L.S. Gross and E.C.H. Veendorp, 'Growth with exhaustible resources and a materials-balance production function', *Natural Resource Modeling*, Vol 4, 1990, pp 77–94.
13 M.I. Kamien and N.L. Schwartz, 'The role of common property resources in optimal planning models with exhaustible resources', in V.K. Smith and J.V. Krutilla, eds, *Explorations in Natural Resource Economics*, Johns Hopkins University Press, Baltimore, MD.
14 A.V. Kneese, R.U. Ayres and R.C. D'Arge, *Economics and the Environment: A Materials Balance Approach*, Johns Hopkins Press, Baltimore, MD, 1970.
15 D.H. Meadows, J. Richardson and G. Bruckmann, *Groping in the Dark: The First Decade of Global Modeling*, Wiley, New York, 1982.
16 H.T. Odum, 'Models for national, international and global systems policy', in L.C. Braat and W.F.J. van Lierop, eds, *Economic-Ecological Modelling*, North-Holland, Amsterdam, 1987.
17 M. Ruth, *Integrating Economics, Ecology and Thermodynamics*, Kluwer Adademic Publishers, Dordrecht, Netherlands, 1993.
18 J.B. Smith and S. Weber, 'Contemporaneous externalities: rational expectations, and equilibrium production functions in natural resource models', *Journal of Environmental Economics and Management*, Vol 17, 1989, pp 155–170.
19 S.K. Swallow, 'Depletion of the environmental basis for renewable resources: the economics of interdependent renewable and non-renewable resources', *Journal of Environmental Economics and Management*, Vol 19, 1990, pp 281–296.
20 R. Wilkinson, *Poverty and Progress: An Ecological Model of Economic Development*, Methuen, London, 1973.
21 WCED, *Our Common Future*, World Commission on Environment and Development, Oxford University Press, Oxford/New York, 1987.

Appendix 1

Model notation

The set $K = \{Q, I, wa, rec, N, S\}$ denotes the six sectors.

Stock variables

B	= a slowly renewable resource (soil, land, water)
K	= total economic capital
K_Q	= productive sector capital
K_I	= investment sector capital
K_{wa}	= waste abatement/treatment capital
K_{rec}	= recycling capital
K_N	= renewable resource extraction capital
K_S	= non-renewable resource extraction capital
N	= the stock of renewable resources
P	= stock of pollution in natural media or organisms
Pop	= human population level
R_{sup}	= inventory (supply) of natural resource materials
S	= the stock of non-renewable resources
S_{wa}	= the stock of useless, stored waste
T_{rd}	= progress indicator of environmental technology
Z	= artificial variable (total 'real' output delayed)

Flow variables

C	= consumption
E	= indicator for overall environmental quality
I	= total investment in replacement and new capital
$I_i (i \in K)$	= investment in sector i
O_{rd}	= social investment in research and development
Q	= output of final goods sector
R_{dem}	= total productive and consumptive demand for resources
R_N	= renewable resource extraction
$R_{N.perc}$	= subjective/perceived availability (rate) of renewable resources
R_{new}	= required extraction of resources
R_{perc}	= subjective/perceived availability (rate) of all resources
R_S	= non-renewable resource extraction
$R_{S.perc}$	= subjective/perceived availability (rate) of non-renewable resources
R_{short}	= indicator for insufficiency of perceived resource supply
R_{rec}	= recycled resource materials
R_{short}	= perceived shortage of resource supply to demand
R_{wa}	= abated/treated waste
e	= ecological effect indicator for technical progress
W_{em}	= emitted waste
$W_{Q.I}$	= gross waste from final and investment goods sectors
W_{rec}	= waste amenable for recycling
W_{too}	= indicator for unsustainability of waste emission

Fractions

A	= assimilation function
B	= population growth rate
$b_i (i = 1,\ldots,5)$	= regeneration and damage functions of slowly renewable recourses
c_Q	= ratio of resource input to material output in final goods sector
c_I	= ratio of resource input to material output in investment goods sector
$D_i (i \in K)$	= discarded capital
F_Q	= unrestricted production function final goods sector
F_I	= unrestricted production function investment goods sector
F_N	= unrestricted production function renewable resource extraction sector
F_S	= unrestricted production function non-renewable resource extraction sector
f_{wa}	= part of production waste that is abated/treated
f_{rec}	= part of waste amenable for recycling that is recycled
G	= regeneration function of renewable resource capacity
H	= environmental quality function
α	= general investment effect on technology
ε	= effect of social R&D and production growth on technology

Parameters

d	= dummy for basing resource availability on stock or sustainable flow
p_N	= fraction of stock of renewables regarded as available for use now
p_S	= fraction of stock of non-renewables regarded as available for use now
s_1	= fraction of treated waste ending up in unused stored waste
s_2	= fraction of recycled waste ending up in unused stored waste
E_{crit}	= below this value of environmental quality negative impacts on regeneration, assimilation, and production
a_Q	= environmental effect parameter in commodity production
b_Q, c_Q	= parameters of function $c_Q()$
b_1, c_1	= parameters of function $c_I()$
k_Q, k_I	= capital parameter in functions $F_Q()$ and $F_I()$

a_{wa}, b_{wa}, c_{wa} = parameters of the waste abatement/treatment process function
$a_{rec}, b_{rec}, c_{rec}$ = parameters of the recycling process function
a_N, α_N, β_N = parameters of function $F_N()$
a_S, α_S, β_S = parameters of function $F_S()$
$\delta_i(i \in k)$ = depreciation of capital
δ = production increase parameter in technology formation equation
α, ε = parameters in functions $\alpha()$ and $\varepsilon()$
$N_{crit}, B_{crit}, P_{crit}$ = parameters of environmental quality function $H()$
r, C_N = intrinsic growth and carrying capacity parameters in regeneration function $G()$
a_P, b_P = parameters in assimilation function $M()$
a_B = parameter in population growth function $B()$

Appendix 2
Requirements for initial values

Here we will state conditions for initial values of capital stocks. We have to solve a set of equations reflecting a balance of demand and supply initially on markets for investment goods and resource materials. Furthermore, we give conditions on economic stocks of capital as having initially sustainable resource extraction and sustainable waste emission. Finally, we list the conditions for stationary growth paths (proportional growth of all sectors).

The balance between the supply of and the demand for investment goods requires a certain sectoral distribution of capital such that:

$$F(K_I) = \sum_{i \in K} I_i \tag{53}$$

holds, and, if we require at least stationary development of all sectors, the following inequality is satisfied:

$$F(K_I) \geq \sum_{i \in K} D_i(K_i) \tag{54}$$

A second important balance is that between the demand and supply for resource materials. For a stationary path it is formulated as $R_{new} + R_{rec} = R_N + R_S$. It implies that $Q = F_Q(K_Q, E)$ and $I = F_I(K_I)$, and leads to

$$\begin{aligned} &c_Q(T_{rd})^* F_Q(K_Q, E) + c_I(T_{rd})^* F_I(K_I) \\ &\quad = R_{rec} + \min\{F_N(K_N, N, E), R_{N,perc}\} \\ &\quad + \min\{F_S(K_S, S), R_{S,perc}\} \end{aligned} \tag{55}$$

For a stationary path, where for all positive t and all $x \leq t$ $W_{rec}(t-x) = W_{rec}(t)$, $R_{rec}(t-x) = R_{rec}(t)$, $T_{rd}(t-x) = T_{rd}(t)$ etc we can solve it as follows. In the case where the perceived renewable and non-renewable resource amounts exceed the extraction capacities (based on $F_N()$ and $F_S()$), we can derive that a balance between supply and demand occurs for sectoral capital stock sizes satisfying, for given state of technology, environmental quality, and resource stocks, the equality in (56). We will use this to make our choice of initial values of the sectoral stock variables, so that initially a balanced economy is represented. Because of non-linear relationships between capital and activity levels, proportional sectoral capital growth may lead to an unbalanced development of activity levels, and a gap between supply and demand for resources:

$$\begin{aligned} &[c_Q(T_{rd}) - f_{rec}(T_{rd}, K_{rec}) \\ &\quad -(1-s_1)^* f_{wa}(T_{rd}, K_{wa})^* f_{rec}(T_{rd}, K_{rec}) \\ &\quad ^*(c_Q(T_{rd}) - 1)]^* F_Q(K_Q, E) \\ &\quad + [c_I(T_{rd}) - (1-s_1)^* f_{wa}(T_{rd}, K_{wa})^* f_{rec}(T_{rd}, K_{rec}) \\ &\quad ^*(c_I(T_{rd}) - 1)]^* F_I(K_I) - f_{rec}(T_{rd}, K_{rec})^* \sum_{i \in K} D_i(K_i) \\ &\quad - F_N(K_N, N, E) - F_S(K_S, S) = 0 \end{aligned} \tag{56}$$

When the extraction capacities exceed (ethical-ecological) acceptable resource extraction levels, the equation to be solved becomes (57) instead of (56):

$$\begin{aligned} &[c_Q(T_{rd}) - f_{rec}(T_{rd}, K_{rec}) \\ &\quad -(1-s_1)^* f_{wa}(T_{rd}, K_{wa})^* f_{rec}(T_{rd}, K_{rec}) \\ &\quad ^*(c_Q(T_{rd}) - 1)]^* F_Q(K_Q, E) \\ &\quad + [c_I(T_{rd}) - (1-s_1)^* f_{wa}(T_{rd}, K_{wa})^* f_{rec}(T_{rd}, K_{rec}) \\ &\quad ^*(c_I(T_{rd}) - 1)]^* F_I(K_I) - f_{rec}(T_{rd}, K_{rec})^* \sum_{i \in K} D_i(K_i) \\ &\quad - d^* p_N^* N - (1-d)^* G(N, E) - p_S^* S = 0 \end{aligned} \tag{57}$$

Trying to establish a dynamic path which satisfies (56) or (57) may be impossible; it is at least very hard to do. It requires that the equations have multiple solutions so that new values of K_i on a development path are in the solution space.

Sustainable use of renewable resources requires a stock of renewable resource extraction capital that satisfies $R_N \leq G(N, E)$ or:

$$F_N(K_N, N, E) \leq G(N, E) \tag{58}$$

This can be written as:

$$[c_Q(T_{rd})-f_{rec}(T_{rd},K_{rec}) \\ -(1-s_1)^*f_{wa}(T_{rd},K_{wa})^*f_{rec}(T_{rd},K_{rec}) \\ ^*(c_Q(T_{rd})-1)]^*F_Q(K_Q,E) \\ +[c_I(T_{rd})-(1-s_1)^*f_{wa}(T_{rd},K_{wa})^*f_{rec}(T_{rd},K_{rec}) \\ ^*(c_I(T_{rd})-1)]^*F_I(K_I)-f_{rec}(T_{rd},K_{rec})^*\sum_{i\in K}D_i(K_i) \\ -F_S(K_S,S)\leqslant G(N,E) \quad (59)$$

Sustainable waste emission, defined as $W_{em}\leqslant A(P,E)$, requires sectoral stocks of capital that satisfy inequality (60):

$$\{[1-f_{wa}(T_{rd},K_{wa})+(1-s_1)^*(1-s_2)^*f_{wa}(T_{rd},K_{wa}) \\ ^*(1-f_{rec}(T_{rd},K_{rec}))]^*(c_Q(T_{rd})-1) \\ +(1-s_2)^*(1-f_{rec}(T_{rd},K_{rec}))\}^*F_Q(K_Q,E) \\ +[1-f_{wa}(T_{rd},K_{wa})+(1-s_1)^*(1-s_2)^*f_{wa}(T_{rd},K_{wa}) \\ ^*(1-f_{rec}(T_{rd},K_{rec}))]^*(c_I(T_{rd})-1)^*F_I(K_I) \\ +(1-s_2)^*(1-f_{rec}(T_{rd},K_{rec}))^*\sum_{i\in K}D_i(K_i)\leqslant A(P,E) \quad (60)$$

Finally, in order that a stationary path be generated, at least the following conditions must be satisfied: $I_i=D_i(K_i)$ $(i\in K)$, $B(C/Pop)=0$, $R_N=G(N,E)$ and $W_{em}=M(P,E)$ and $b_1(E)=b_2(\mathrm{d}K/\mathrm{d}t)+b_3(\mathrm{d}Pop/\mathrm{d}t)+b_4(R_N)+b_5(R_S)$. The first of these can be solved by choosing, for a given total capital stock K, K_I such that $F_I(K_I)-D_I(K_I)+K_I=K$, subsequently allocating $K-K_I$ to the remaining sectors (such that other conditions are satisfied).

[31]

Environment and Development Economics 1 (1996): 85–101

Measuring sustainable development: progress on indicators*

DAVID PEARCE, KIRK HAMILTON and GILES ATKINSON
Centre for Social and Economic Research on the Global Environment (CSERGE), University College London, 136 Gower Street, London, WC1E 6BT, UK

ABSTRACT. The search for sustainability indicators should be guided by a theory of sustainable development (SD). In this paper we investigate two such theoretical frameworks and the indicators that they suggest. Indicators associated with weak sustainability are characterized by aggregative indicators such as green national income. We conclude, however, that a more promising offshoot of green accounting is measures of genuine savings (i.e. savings adjusted for loss of assets). To achieve SD, genuine savings rates must not be persistently negative. Strong sustainability indicators accord a more central role to the conservation of critical natural assets within the broader goal of prudently managing a nation's portfolio of assets over time. We discuss two approaches—carrying capacity and resilience—and conclude that, while measures of resilience are potentially attractive, more research is required regarding the resilience–SD link. However, an important conclusion that we can make is that, even in an economy operating under a strong sustainability regime, genuine savings are still key indicators of SD and are complementary to measures of changes in stocks of critical natural assets.

1. Aim of the paper

In this paper we briefly survey the operational meanings attached to the concepts of sustainable development and environmental sustainability. We then contrast two apparently consistent characterizations of sustainability. The first, 'strong sustainability' (SS), derives more of its foundations from ecological science than does the second, 'weak sustainability' (WS). Nonetheless, both SS and WS are informed by ecology as we show. The fundamental distinctions between SS and WS are: (a) that the former denies to a greater or lesser extent substitutability between natural assets and other assets—human and manufactured assets; and (b) that SS stresses 'discontinuity' and 'non-smoothness' in ecological systems and hence in the economic damages to which ecological impairment gives rise.[1] In effect, SS has at its starting-point ecological imperatives, and this dictates the

* An earlier version of this paper was presented at the Resource Policy Consortium Symposium, Washington, DC, 12–13 June 1995. CSERGE is a designated research centre of the UK Economic and Social Research Council (ESRC). This paper is part of a research programme on environmental sustainability funded by Directorate General XII of the European Commission. We are grateful for the comments of two anonymous referees. All errors remain the responsibility of the authors.

[1] The conventional environmental economics literature addresses this issue in terms of non-convexities. See Baumol and Oates (1988) and Burrows (1979).

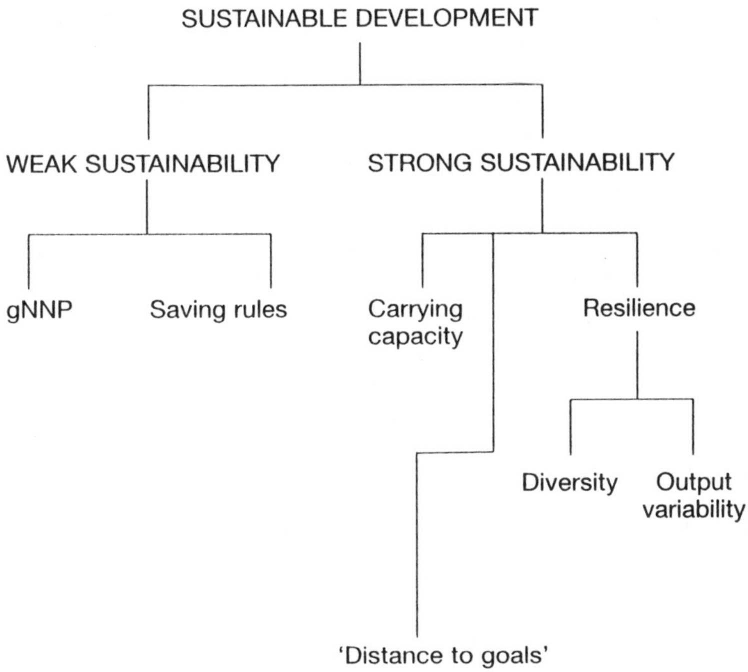

Figure 1. *Sustainability paradigms: candidate indicators.*

subsequent form of economic analysis. By contrast, WS begins with standard assumptions in economics, and this in turn shapes the form in which ecological and environmental concerns are evaluated.

The choice between SS and WS, we argue, has to do with issues of fact about the world in which we live. Since the facts themselves are uncertain, choosing between SS and WS will invariably depend on attitudes to behaviour under uncertainty. In terms of indicators of sustainability, different indicators are relevant to the two paradigms as illustrated in Figure 1. The SS paradigm dictates indicators that focus primarily on ecological assets, functions and processes. In turn, such indicators tend to stress 'limits' to the deterioration of ecological assets.[2] Indicators in this category include:

(a) Measures of resilience, most of which have yet to be developed. One suggestion involves indicators of biological diversity, since resilience is assumed to be a function of diversity (Common and Perrings, 1992; Arrow *et al.*, 1995). For agriculture-dependent systems, an indicator of yield variability may also be relevant (see below).

[2] In this respect, the SS paradigm has common elements with the earlier Limits to Growth literature; see Meadows *et al.* (1972).

(b) Measures of carrying capacity, e.g. incorporating supply/demand ratios for resources.
(c) 'Distance to goal' approaches in which deviations of ambient concentrations from sustainability 'targets' are aggregated to derive an overall performance indicator (Hammond *et al.*, 1995). The green national accounting counterpart of this is offered by Hueting *et al.* (1992) and is couched in terms of the costs of reaching these goals.

The WS paradigm, on the other hand, emphasizes the substitutability of manufactured and natural assets, and hence focuses on aggregate measures such as:

(a) 'green' national income
(b) genuine savings.

In the main, these indicators are intended to measure environmental performance at the national level, but it is as well to note in passing that the distinction between WS and SS also affects microeconomic indicators; in particular, it affects cost–benefit measures.[3]

2. Sustainability models

Economists have pursued a rigorous, though by no means complete, interpretation of sustainable development (SD). SD is defined as some indicator of human well-being which does not decline over time (Pearce *et al.* 1989; Pezzey, 1989). The indicator of well-being is not discussed here but could be some suitably modified measure of per capita real national income, making due allowance for existing unmeasured components such as environmental services (see, Hamilton, 1994; 1995a, b, c). At this point there is no significant divergence between WS and SS. Both paradigms broadly accept this definition of SD. The approaches are differentiated with regard to the conditions required to satisfy the achievement of SD. Taking the capital base as the means whereby future well-being will be sustained, this implies some form of conservation or maintenance of capital as a condition for SD. This 'constant capital' concept underlies the contributions of Solow (1986) and Mäler (1991).

The basic debate about sustainability models relates to the conditions set for the capital stock. In strong sustainability, SD is achieved by conserving certain components of the natural capital stock, as well as by setting a constraint of non-decline in the overall stock of capital. In weak sustainability the operative constraint is the overall stock of capital since all forms of capital are assumed to be substitutable. Although natural capital is a form of wealth whose services contribute to well-being it has no 'special role' as such in this scheme. In terms of indicators, then, SS will tend towards environmentally focused measures, supplemented by a weak sustainability measure, whilst WS will focus on assets generally, with no special highlighted measure for natural capital. Expressed this way the *information* required for SS and WS indicators is the same: both have to have measures of natural capital to be operative, and both have to measure other capital

[3] See Barbier *et al.* (1990); von Amsberg (1992).

assets as well. Non-substitutability is alleged to give rise to aggregation problems (see Faucheux *et al.*, 1994): natural capital has to be measured separately from human and manufactured and in non-comparable terms. But the 'constant capital' rule can be preserved even in this context since SS requires that each index be non-declining (assuming also that natural capital can be aggregated). The 'capital aggregation' problem, if it exists at all, is not therefore fatal for sustainability indicators.

For exhaustible resources the optimal growth path is not sustainable if resources are essential for production and the pure rate of time preference is positive. If the elasticity of substitution between produced assets and natural resources is less than 1, then the constant utility path is infeasible.[4] The degree of substitutability is therefore critical in determining whether WS is possible with exhaustible resources. If the elasticity of substitution is greater than or equal to 1, then 'savings rules', that investment in produced assets should offset the value of resource depletion, provide a test of sustainability.

Strong sustainability is not, strictly speaking, about elasticities of substitution that are equal to 0, since this would imply that capital and resources must be used together in fixed proportions. The essential idea underlying SS is that a given amount of a resource must be preserved intact, in order that it may continue to provide critical services. This could be conceived in two ways: (i) some amount of the resource is to be conserved, in which case there is only the residual that can be exploited; or (ii) the resource does not need to be kept intact, but if the stock size falls below a certain critical level then catastrophic consequences result.

For case (i) consider the tropical rainforest. If preserving some quantity of the rainforest is considered to be critical for the long-term well-being of humanity, the effect of this preservation is to reduce the quantity of forest that can be considered to be an economic resource. Other things being equal, therefore, the effect of the strong sustainability policy will be to reduce the quantity of harvest that can be carried out sustainably from the remaining stock. This will have the effect, under standard assumptions about the production function, of increasing the price of the resource; a different mix of natural resource and produced capital inputs into production will therefore ensue. However, the savings rule, as applied to the non-conserved stock, will still be the determinant of sustainability, because the savings rule is concerned with the *change* in the real value of stocks rather than with the *absolute size* of the stocks.

The key indicators for this living-resource economy operating under a SS regime will be twofold: are stocks of critical natural assets declining? and are genuine savings rates persistently negative? A positive answer to either of these questions would be an indication of unsustainability.

Case (ii) might be an upland forest, where exploitation beyond a certain point leads to catastrophic losses, as a result of soil erosion for instance. Here an optimal resource tax that is a function of both the size of the remaining stock and the level of harvest can be designed to ensure that the

[4] See Dasgupta and Heal (1979, ch. 10) or Hamilton (1995c).

long-run steady-state stock is greater than or equal to the critical stock size Again the savings rule determines sustainability.

The effect of these sorts of quantitative restrictions is to reduce the present value of consumption that would be attained if the restrictions did not exist. This is only an apparent decline in welfare, however, because of the significant welfare losses that would result if the SS regime is not followed. The key point is that savings rules provide essential indicators of sustainability whether SS or WS is the development paradigm.

3. The sustainability paradigm and its measurement

In the WS world there is nothing special about the environment as such. However, this does not permit overly rapid depletion of non-renewable resources or imply that environmental degradation does not matter. In the latter case these actions involve a loss in future well-being unless accompanied by adequate compensation in the form of investments in alternative forms of wealth. In the case of say air pollution, compensation must be equivalent to the social costs of emissions (see below). The SS paradigm requires arguments to the effect that environmental assets are special, i.e. do not have practical substitutes. Pearce *et al.* (1990) present candidate arguments:

(a) Irreversibility: manufactured capital can be created, destroyed and created again in ways which are not true of natural assets (especially living species). While the creation of a manufactured or new technology may be deemed to be an adequate compensation to future generations for an irreversible transformation of a natural asset, this assumes that we have perfect knowledge of future relative prices (Hamilton, 1995a). If future generations value nature much more highly, then such compensations will be either unfeasible or inadequate. Some human capital may also be characterized by irreversibility attributes, e.g. indigenous knowledge embodied in social custom.
(b) Uncertainty: there is extensive uncertainty about the workings of ecological systems and hence about the consequences of impairing the functionings of those systems.
(c) Scale: the existence of thresholds and discontinuities may well result in large-scale damage once thresholds are exceeded.

The combination of (a) to (c) would dictate a 'precautionary approach' to the degradation of natural capital (Dasgupta, 1982), and hence something akin to a SS rule.

If these differing notions of sustainability are to have operational consequences, then measures of progress towards sustainable development are required. As Hammond *et al.* (1995) note, indicators can ideally both quantify and simplify complex phenomena for decision-makers. Below we therefore explore the types of indicator that have been proposed for measuring weak and strong sustainability.

4. Indicators of weak sustainability

A good illustration of the theory of indicators based on weak sustainability is provided by the literature on 'green' national accounts (see, e.g., Mäler,

1991; Hartwick, 1993; Hamilton, 1994). This work arises from a concern that economic indicators such as gross national product (GNP) do not reflect the depletion and degradation of the environment and so may lead to incorrect development decisions, in much the same way that cost–benefit analyses that do not include the values people place on the environment may yield poor investment decisions. Some basic results from this theoretical work are set out below.

If we assume a simple economy in which a single composite good is produced by production function $F(K,L,R)$, and that this good can be either consumed or invested, then we wish to maximize the present value of utility for an aggregate utility function $U(C,E)$, where C represents consumption and E the level of services provided by the environment. K and L are (manufactured) capital and labour respectively, and R is the harvest of resource from a stock S that grows by an amount $g(S)$. Environmental services are negatively related to the level of a stock X of pollution in the environment, which dissipates naturally by an amount $d(X)$ and is augmented by pollution emissions related to the level of production, $e(F)$.

From this maximization problem we obtain the following expressions for the basic national accounting aggregates (see Hamilton, 1995b, for details):

$$\text{GNP} = F(K,L,R) = C + I$$
$$\text{NNP} = C + I - r(R - g) - p(e - d).$$

Here I is investment, r is the net resource rental rate and p is the marginal social cost of pollution emissions. If the notion of 'consumption' is extended to include the value of environmental services E, then a broader measure of economic welfare (MEW) is:

$$\text{MEW} = C + I - r(R - g) - p(e - d) + vE$$

where v is the price consumers would be willing to pay for a unit of environmental service. As expected, v can be expressed as the ratio of the marginal utilities of consumption and environmental services,

$$v = U_E / U_c.$$

While these green measures of net national product are important, in the sense that they are truer measures of income, their direct policy implications are limited. As argued by Hamilton (1994), the shifts in the *level* of income entailed by 'greener' measures of income do not, of themselves, suggest any policy consequences, while the policy signals from the *growth rates* of greener income, as compared with growth rates in traditional income measures, are likely to be equivocal. Pearce and Atkinson (1993) concentrate instead on adjusted savings measures, while Hamilton (1994) develops the concept of *genuine savings*. In the context of the green accounting aggregates just presented, we may define this as:

$$S_g = I - r(R - g) - p(e - d).$$

That is, genuine saving is a measure of net saving which accounts for the depletion of natural resources and the accumulation of pollutants. There are two important ways in which ecological processes enter this measure. First, growth of living resources is an addition to both NNP and genuine savings, so that management of natural resources that increases their total growth can be considered to be a form of wealth creation. Second, the natural dissipation of pollution stocks is also wealth-creating, in the sense that there is a decreasing liability in the national balance sheet in the form of pollution stocks.

The models of green national accounting presented above ignore geography for the sake of simplicity, but the question of transboundary pollution is clearly important in the case of air pollution. Without developing extensions to the formal model, the following line of argument is offered. First, regarding adjustments to income, an extension of the *polluter pays principle* to the domain of national accounting seems appropriate. This means that the estimates of the unit marginal social costs of pollution in a given country should include all costs, including those in other nations. These unit marginal costs should then be multiplied by the total level of emissions in that country. The argument for this treatment of transboundary pollution in the case of savings rules is, if anything, even stronger. Some portion of a given country's total savings should, at least notionally, be set aside in order to compensate the recipients of pollution emitted and transferred across international boundaries (Hamilton and Atkinson, 1995).[5]

Genuine saving measures have proven to be important empirically, as evidenced by Figure 2 for sub-Saharan Africa. The topmost curves on this figure represent standard national accounting identities: gross investment less the value of net foreign borrowing equals gross saving; and deducting the value of depreciation of produced assets from gross saving yields net saving. Genuine saving then is the result of deducting the value of resource depletion and the value of pollution emissions (limited to carbon dioxide in Figure 2) from net saving.

The welfare significance of genuine saving is that persistent negative genuine savings rates must lead to non-sustainability, in the sense that the level of welfare of the country will eventually decline (Pezzey, 1994; Hamilton and Atkinson, 1995). Figure 2, which aggregates many of the countries of sub-Saharan Africa, therefore says that not only was Africa's economic performance poor over the 1980s, as suggested by standard economic indicators, but the situation with regard to future well-being is even worse. Sub-Saharan Africa has been dissipating its wealth, both human-made and natural, at a significant rate over this decade.

What are the policy implications of low and even negative genuine savings? It could be argued that for countries with low income and, *ex hypothesi*, high discount rates, 'mining' natural resources is a perfectly rational response to economic circumstances. This is certainly plausible for countries where the preponderance of the economy consists of subsistence

[5] Opposite adjustments to income and saving have been made when resources are traded internationally (Asheim, 1986, 1994; Hartwick, 1994): i.e. upward adjustments for resource exporters (and corresponding downward adjustments for resource importers). This assertion rests upon the validity of a number of assumptions. In particular, if a country is a price-taker *vis-à-vis* international markets then the standard Hartwick–Solow rule should apply.

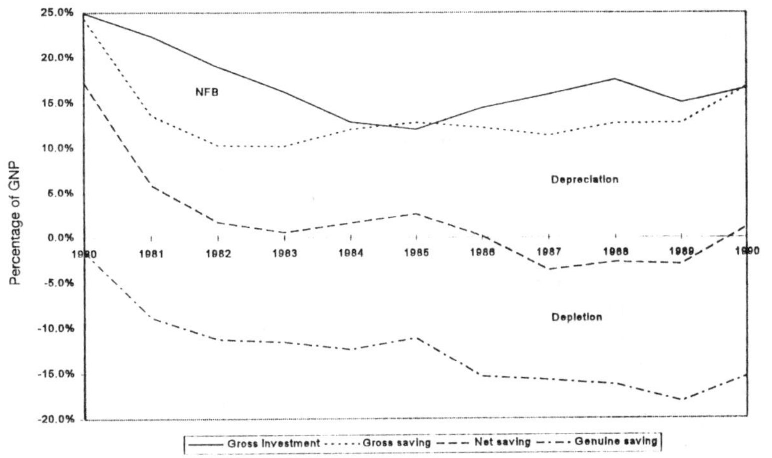

Figure 2. *Investment and saving in sub-Saharan Africa. NFB, net foreign borrowing.*

activities. Genuine saving calculations simply emphasize in quantitative form what is already known in qualitative form. The fact remains, however, that many policy-makers are unaware of resource 'mining' in this context. As it happens, the policy implications of genuine savings measures are richer than this.

With regard to natural resources policy, there is the question of whether royalties are being captured by governments, so that there is an incentive for overexploitation, and whether resource access and tenure regimes, particularly with regard to living resources, do not also encourage excessive harvest rates. There is the related issue of the investment of resource rents, whether in infrastructure or education, a key component of sustainable development policy. And there are the broader issues of macro- and microeconomic policy that play important roles in establishing savings levels.

Simply stating that resource rents should be invested and not consumed, or that investment must offset broader environmental degradation costs, is not sufficient as a policy prescription, however. Equally important is the *quality* of investment. While each unit of savings should be put to its most productive use in principle, in practice many investments, especially in developing countries, have been wasteful. And while public investments have traditionally been in physical infrastructure, it is important to recognize that investment in human capital is one of the surest ways to increase future well-being (see, e.g., World Bank, 1995, ch. 8).

5. Indicators of strong sustainability

More stringent requirements based on stronger conditions for achieving sustainability will result in a different set of indicators. Indicators of strong sustainability have been suggested by ecological economists. They represent a loosely assembled body of ideas, and in common with other dissenting schools the most notable unifying theme is a mistrust of neoclassical (environmental) economics, in particular the perceived inability of mainstream economists to integrate ecological imperatives into their analysis. However, some progress is being made in defining ecological economics (see, e.g., Perrings and Turner, 1995). Furthermore, Arrow *et al.* (1995) identify two core elements of research in ecological economics: carrying capacity and resilience.

(a) Carrying capacity

The notion of a carrying capacity is drawn from biology. It states that a given area can only support a given population of a particular species, and at this upper limit—the carrying capacity—population will have reached its maximum sustainable level. In order to apply this concept of a saturation point to human populations, we have to consider not only the level of population but also the level of economic activity. Furthermore, the composition of economic output can vary significantly, adding a further complication. Inevitably, in developing an indicator a number of brave assumptions have to be made. Some attention has been given to estimates of global carrying capacity, especially where these calculations are well below existing population levels or wildly in excess of predicted stationary levels (see, respectively, Ehrlich, 1992; Kahn and Simon, 1984).

More usually within ecological economics, carrying capacity is defined in terms of exceeding limits—to be set by ecological criteria. In moving to actual indicators of carrying capacity we require a more detailed specification of ecological limits. Usually, these consist of 'sustainability constraints' with respect to commercial or environmental resources: e.g., pollution should not exceed the assimilative capacity of the environment; harvest of renewable resource should not be greater than natural growth. (It is more difficult to think of an appropriate constraint governing the use of non-renewable resources on these terms where growth is by definition zero.) Underlying these constraints is the goal of maintaining the stock of *each* resource intact. If the definition of sustainability is non-declining human well-being, then what is argued is that human well-being is decreased unless *all* specified ecological constraints are observed.

Carrying capacity indicators tend to be extremely pessimistic regarding judgments about technology, i.e. the ability of people to expand the carrying capacity of the earth (Cohen, 1995). For example, Ehrlich (1992) speculates that, at industrialized levels of economic activity per capita, the level of global population that would not exceed carrying capacity thresholds is two billion, about one-third of the *prevailing* population. Although the genuine savings rule also assumes constant technology, the framework underlying the savings principle does at least allow for substitution between assets. In the carrying capacity schema each asset is to be maintained intact, and presumably the scale of economic activity or population is set by whichever constraint is reached first, as suggested by Rees and Wackenagel (1992). This is known as Liebig's law and implies that

$$CC = \min CC_i.$$

In effect, those who favour carrying capacity indicators are arguing that there is zero substitutability between assets. The difficulties in resolving this argument are well known. Empirically speaking, little is known about what might constitute *critical natural capital*, i.e. those environmental assets for which few or no substitutes exist (Pearce *et al.*, 1994). While lack of substitutability may be an appropriate characterization of some classes of environmental assets, it is too much to argue that this is true of all such assets. Neither is it the case that limiting factors will remain the same over time (Cohen, 1995).

A variant of this empirical approach to carrying capacity is to link population levels to very simple indicators of resource use. The commonly cited examples are land use for food production, fuelwood production and water use.

Table 1 provides an illustration of estimates of carrying capacity populations with respect to fuelwood availability in Africa. These estimates are based on the UN Food and Agriculture Organization's (FAO) data on fuelwood demand (production) and fuelwood supply (regeneration). Obviously, if the ratio of demand to supply is greater than unity, then current rates of harvest must be unsustainable. The carrying capacity estimate simply links this information to data on population levels in 1990 and projected levels in 2000. The estimates also take account of the ecological concern that persistent rates of overharvest will *reduce* the sustainable yield, and therefore population, over time. Hence, Nigeria's estimated sustainable population is 138 million in 1990. Yet by the end of the century this is predicted to fall to about 132 million. Owing to population growth over that period the estimate predicts that Nigeria's population may well be unsustainable by 2000, at least in terms of a fuelwood constraint. It is worth noting that Nigeria has oil resources but that pressure on fuelwood re-

Table 1. *Fuelwood availability and sustainable populations in Africa (millions)*

Country	*Sustainable population in 1990 (actual population)*		*Sustainable population in 2000 (predicted population)*	
Algeria	4	(25)	4	(31)
Burkina Faso	4	(9)	3	(12)
Chad	4	(6)	4	(7)
Ethiopia	37	(51)	36	(67)
Gambia	>1	(1)	>1	(1)
Kenya	14	(24)	14	(31)
Madagascar	16	(12)	15	(16)
Malawi	2	(9)	2	(11)
Mali	8	(9)	7	(12)
Morocco	9	(25)	9	(30)
Nigeria	138	(96)	132	(128)
Senegal	6	(7)	6	(10)
Sudan	21	(25)	21	(33)
Tanzania	16	(25)	15	(33)
Tunisia	>1	(8)	>1	(10)
Uganda	2	(16)	2	(22)
Zimbabwe	10	(10)	9	(12)

Source: adapted from Atkinson (1993); Population data: World Bank (1994).

sources persists owing to the lack of suitable refineries (Armitage and Schramm, 1989). All of the countries listed in Table 1 are either unsustainable in 1990 or are predicted to be unsustainable by the year 2000. Some of these are not surprising, being countries located in typically arid North Africa with few forest resources. Pressure on forest resources in other countries such as Malawi has been well documented elsewhere (e.g. Pearce and Warford, 1993).

Arrow *et al.* (1995) have argued that carrying capacities are not fixed or static in the way that ecological economics has in the main considered them. Thus, the maximum sustainable population (MSP) is given by

MSP = Annual resource yield/Minimum per capita requirement.

MSP can be raised by mining the stock where the resource is renewable, but this is clearly unsustainable. On the other hand, 'yields' can be increased by technological change in the case of renewable resources, and depletion profiles can be extended by technology in the case of non-renewable resources. Substitutes may also be found (e.g. kerosene or biomass charcoal for fuelwood). In addition, interactions between various carrying capacity constraints are seldom considered (Cohen, 1995). More importantly, carrying capacity calculations have limited relevance when trade is possible, since the scarce resource can be imported in exchange for another asset in which the exporting nation has a comparative advantage. Neither do these indicators address the measurement of progress in protecting critical natural assets. At best they might be thought of as simple poverty measures that introduce the notion of sustainable use of resources.

It is therefore difficult to pin down the concept to any single meaningful indicator. Instead, Arrow *et al.* (1995) propose that a more general indicator be employed, such as that set out by Vitousek *et al.* (1986). This describes pressures in terms of how much net primary productivity (NPP) human beings currently appropriate, as an indicator of human appropriation of ecosystems under a variety of scenarios. Note that this indicator is not couched in terms of a population carrying capacity, although implicit judgments concerning such a limit may need to be inferred if we are to suggest what levels of appropriation are to be considered excessive.

(b) Resilience

Many ecological economists have come to interpret sustainability as resilience along the lines introduced by Holling (1973, p. 12). Resilience determines the persistence of relationships within a system and is a measure of the ability of a system to absorb changes and still persist. In turn the *degree of resilience* of the system determines whether ecological productivity (e.g. ecosystem functioning) is largely unaffected, decreases either temporarily or permanently or, in the extreme, collapses altogether, as a result of stress or shock (Conway and Barbier, 1990).[6] A system that is unable to

[6] *Stresses* are small and predictable changes such as increasing erosion and salinity. Although small, these forces can have large cumulative effects. *Shocks* on the other hand are relatively large, temporary and unpredictable such as a new crop pest or a rare drought. For a discussion of these topics see Conway and Barbier (1990).

respond is, in some sense, unsustainable if the stresses and shocks are themselves not capable of control, or, for some reason, are unlikely to be controlled. In turn, capability for response to stress and shocks is usually, but not necessarily, thought to be correlated with diversity of capital, in the sense of either a wide portfolio of natural and produced assets or a wide portfolio of natural biological assets (Holling, 1973; Conway, 1985, 1992; Common and Perrings, 1992).

A measure of the degree of resilience could be interpreted as an indicator of the *degree of sustainability* of the system. However, it is less clear what this means for the sustainability of human development. In the green national accounting literature this link is provided by models that connect changes in human welfare to resource depletion and environmental degradation. These result in alternative measures to GDP (e.g. measures of economic welfare) and lead directly to a notion of genuine saving. Hence, green accounting tells us how resource and environmental issues are linked to sustainable human development. The task for the resilience school is to provide a similar link between changes in resilience and sustainability in this way. Although Common and Perrings (1992) have offered one such possible interpretation based on measures of biodiversity this preliminary work is not yet suggestive of a set of feasible indicators, although standard measures of species diversity and keystone species are clearly candidates as we discuss below.

It is clear that resilience is not something that can be observed directly, and so the search for indicators leads in the direction of measuring *inputs* that are thought to contribute to resilience, or the *outputs* that are believed to be affected by changes in resilience. Examples of the former are indicators of biodiversity. If resilience is positively related to biodiversity, as Common and Perrings (1992) and Arrow *et al.* (1995) tentatively suggest, then indices of diversity might be a useful input-based indicator of resilience in ecosystems. The ecology literature offers many such measures (Krebs, 1985), and a useful summary of information about species and habitat diversity is provided by WCMC (1992).

But problems remain. The first of these, as noted above, is that it is not at all obvious how given measures of diversity–resilience map into sustainability. A related problem is the absence of any clear baseline to assess the degree of sustainability using these indicators. The genuine savings approach outlined in section 4 has a natural measure of the degree of sustainability, since zero genuine saving defines the borderline between sustainability and non-sustainability. Zero or 'low' diversity would appear to qualify for low sustainability on a diversity measure, but there is no obvious scaling involved. Unless our criterion is to be non-declining diversity, when does sustainability become threatened on a diversity index?

Even if non-declining diversity (say, from existing levels) is to be our benchmark, there is a second problem of an empirical nature in that, where available, data are often constrained to a single point in time, whereas the relevant measure for sustainability purposes is the change in diversity, not the amount of diversity.

Finally, meaningful interpretations of the data if they did exist are often unclear. For example, the appropriate spatial scale of the index is not

Table 2. *Crop yield variability in India 1955–89*

Crop	*CY pre-green revolution*	*CY post-green revolution*	*Change in CY (%)*
Rice	5.1	7.0	+37
Wheat	7.2	5.8	−19
Coarse cereals	3.5	7.3	+108
Pulses	8.3	8.5	+2
Foodgrains	3.8	5.4	+42
Groundnut	5.5	12.1	+220
Cotton	8.8	8.9	+1
Sugarcane	5.2	4.4	−15

Source: Pearce and Prakesh (1993) adapted from Ninan and Chandrashekar (1991).

simple to determine: are such indicators meaningful on a national scale or relevant to particular ecosystems regardless of national boundaries? It is likely that national measures of diversity will tell us little about resilience or sustainability in general, although they may be useful in giving a preliminary answer to a different question, namely, where to direct conservation funds most effectively (see, e.g., Moran *et al.*, 1995).

If we restrict our focus to agricultural systems, then a mixture of input- and output-based indicators is suggested. An example of an indicator emerging from *output-based measures* of the loss of resilience, is the variability in crop yields. Upward trends in production may be associated with increasing variability of yields from year to year. In the limit, without any counteracting mechanism, these fluctuations may become so extreme that output collapses. A measure of this variability is the coefficient of variation. Thus, one indicator might be *changes in the coefficient of variation of crop productivity over time.*[7] For example, Hazell (1984) shows that in India, while the annual growth rate of cereal production was 2.7 per cent between 1952/3 and 1977/8, the coefficient of variation around this trend is 4.5 per cent per annum before 1965/6 and 6 per cent per annum thereafter. Table 2 shows estimates of crop yield variability in India for the period 1955–89. While other (e.g. climatic) factors need to be controlled for in an analysis, this period also coincides with the structural shift embodied in the green revolution and reveals that, in all but two cases, greater output has been achieved at the expense of greater variation in yields over time. In the case of coarse cereals and groundnuts this change in variation has been highly significant. Conway (1985, 1987, 1992) has provided a framework for this kind of analysis—agroecosystem analysis—whereby indicators of a number of desirable attributes of the system (e.g. productivity, resilience and equity) can be assessed. The extent to which farmers can either insure against loss of income in low productivity years or smooth

[7] This is actually a measure of stability (i.e. ability of a system to return to an equilibrium state after a small, temporary disturbance arising from normal fluctuations and cycles in the surrounding environment). In practice, stability and resilience are likely to be closely and positively linked, and hence the distinction may not be critical (Pearce and Prakesh, 1993).

their consumption by saving a portion of high productivity harvests also requires further investigation.

Indicators of diversity as discussed above are not irrelevant to this analysis. One of the principal explanations of the loss of resilience/sustainability in agricultural systems is *changes in biological diversity*. While there are a number of different definitions of biodiversity (Pearce and Moran, 1994), here we refer to *genetic diversity* (i.e. the genetic information contained in the genes of each plant, animal and micro-organism). The significance of genetic diversity is often highlighted with reference to global agriculture and food security. The majority of the world's human population rely on a small number of staple food species (often HYVs), which in turn rely on supply of genes from their wild relatives to supply new characteristics, for example to improve resistance to pests and diseases (Cooper *et al.*, 1992). Anderson and Hazell (1989) cite statistical evidence for variability induced by genetic uniformity of crops.

Of course, other factors are implicated in this process, not least an increased reliance on artificial fertilizers, pesticides and technology. Perrings (1994) suggests that the need to continually substitute produced assets for natural assets in the form of genetic diversity in agriculture can be seen as a cost of the loss of resilience. As such he suggests that the loss of sustainability of a system can be 'measured by the value of increasing quantities of herbicides, pesticides, fertilisers, irrigation and other inputs needed to maintain output at or above current levels ... [and] ... where output fails, the costs of relocation where soils or water resources have been irreversibly damaged' (Perrings, 1994, p. 39). Here, the weak sustainability aspects of substitutability—produced assets for natural assets—are linked to the strong sustainability consideration of ecological functions and the properties of ecosystems. The key question that future research must answer is to what extent the substitution option is as effective as in the past.

6. Conclusions

The search for sustainability indicators should be guided by a theory of sustainability. This paper suggests one definition of sustainability—non-declining per capita human well-being—and two interpretations of the conditions necessary to achieve this: strong and weak sustainability (SS and WS). Depending on the state of belief about these conditions, the choice of sustainability indicator is predetermined. Advocates of SS will be inclined towards ecological indicators in which non-substitution is assumed. Advocates of WS will stress substitution and hence aggregative indicators such as 'green' national income or 'genuine' savings.

In terms of progress, the WS paradigm is further advanced with numerous studies of green national income (Repetto *et al.*, 1989; Bartelmus *et al.*, 1993; van Tongeren *et al.*, 1993) and the first results for genuine savings (Hamilton and Atkinson, 1995; World Bank, 1995). One of the lessons from this type of green accounting is that natural processes, such as the growth of living resources and the dissipation of pollution stocks, are themselves wealth-creating. Savings rules, as an offshoot of green accounting, are likely to provide useful measures of sustainable development. This in turn will mean that sustainable development can be more than a vague commitment on the part of governments and instead become a measurable concept with wide-ranging policy implications.

Advocates of SS have indicated the direction in which SS indicators should go. Carrying capacity measures are familiar but appear most relevant to a limited number of countries, while global estimates of carrying capacity are far too speculative. Output variability has some potential, but a better understanding of the factors giving rise to it is required. Finally, indicators based on ecological–economic resilience need significant effort before they are suitable for indicating SS. However, measures of resilience has more appeal than the carrying capacity approach in that indicators of biodiversity are stressed as central to the measurement problem, and many would now agree that diversity is a critical natural asset (see Schulze and Mooney, 1994). As we have seen, however, practical indicators of biodiversity fall short of what we require to measure sustainability. For these reasons, we argue that resilience, whilst attractive from an ecological economics standpoint, has, at the moment, little to offer for the development of indicators.

References

von Amsberg, J. (1992), 'Sustainability and project evaluation: case studies', mimeo, University of British Columbia.

Anderson, J. and P. Hazell (1989), *Variability in Grain Yields: Implications for Agricultural Research and Policy in Developing Countries*, Baltimore: Johns Hopkins University Press.

Armitage, J. and G. Schramm (1989), 'Managing the supply and demand for fuelwood in Africa', in G. Schramm and J.J. Warford, eds., *Environmental Management and Economic Development*, Washington DC: World Bank.

Arrow, K., B. Bolin, R. Costanza, P. Dasgupta, C. Folke, C.S. Holling, B.-O. Jansson, S. Levin, K.-G. Mäler, C. Perrings and D. Pimentel (1995), 'Economic growth, carrying capacity and the environment', *Science* **268**: 520–21.

Asheim, G.B. (1986), Hartwick's rule in open economies', *Canadian Journal of Economics* **86**: 395–402.

Asheim, G.B. (1994), 'The concept of net national product in an open economy', paper presented to the International Symposium on 'Models of sustainable development. Exclusive or complementary approaches of sustainability?', Université Panthéon-Sorbonne, Paris.

Atkinson, G. (1993), 'Carrying capacity as a measure of sustainability', mimeo, CSERGE, University College London and University of East Anglia.

Barbier, E., A. Markandya and D.W. Pearce (1990), 'Environmental sustainability and cost–benefit analysis, *Environment and Planning A* **22**: 1259–1266.

Bartelmus, P., E. Lutz and S. Schweinwest (1993), 'Integrated environmental and economic accounting: a case study for Papua New Guinea', in E. Lutz, ed., *Toward Improved Accounting for the Environment*, Washington DC: World Bank.

Baumol, W. and W. Oates (1988), *The Theory of Environmental Policy*, 2nd edn, Cambridge University Press.

Burrows, P. (1979), *The Economic Theory of Pollution Control*, Oxford: Martin Robertson.

Cohen, J.E. (1995), 'Population growth and earth's human carrying capacity', *Science* **269**: 341–348.

Common, M. and C. Perrings (1992), 'Towards an ecological economics of sustainability', *Ecological Economics* **6**(1): 7–34.

Conway, G. (1985), 'Agroecosystem analysis', *Agricultural Administration* **20**, 31–55.

David Pearce et al.

Conway, G. (1987), 'The properties of agroecosystems', *Agricultural Administration* **24**, 95–117.

Conway, G. (1992), 'Sustainability in agricultural development: trade-offs with productivity, stability and equitability', mimeo, Imperial College London.

Conway, G. and E. Barbier (1990), *After the Green Revolution: Sustainable Agriculture for Development*, London: Earthscan.

Cooper, D., R. Vellvé and H. Hobbelink, eds., (1992), *Growing Diversity: Genetic Resources and Local Food Security*, London: Intermediate Technology Publications.

Dasgupta, P. (1982), *Control of Resources*, Oxford: Basil Blackwell.

Dasgupta, P. and G. Heal (1979), *Economic Theory and Exhaustible Resources*, Cambridge University Press.

Ehrlich, P. (1992), 'Ecological economics and the carrying capacity of earth', *2nd Meeting of the International Society for Ecological Economics, 3–6 August 1992*, Stockholm.

Faucheux, S., G. Froger and J. O'Connor (1994), 'The costs of achieving sustainability: the differences between "environmentally corrected national accounts" and "sustainable income" as information for sustainability policy', *Cahier de C3E*, Paris.

Hamilton, K. (1994), 'Green adjustments to GDP', *Resources Policy* **20**(3): 155–168.

Hamilton, K. (1995a), 'Measuring progress towards sustainable development', Issue paper for OECD Group on Economic and Environmental Policy Integration, Paris.

Hamilton, K. (1995b), 'National wealth and sustainable national income', mimeo, CSERGE, University College London and University of East Anglia.

Hamilton, K. (1995c), 'Sustainable development, the Hartwick rule and optimal growth', *Environmental and Resource Economics* **5**: 393–411.

Hamilton, K. and G. Atkinson (1995), 'Valuing air pollution in the national accounts', London Group on National Accounts and the Environment, Washington DC.

Hammond, A., A. Adriaanse, E. Rodenburg, D. Bryant and R. Woodward (1995), *Environmental Indicators*, Washington DC: World Resources Institute.

Hartwick, J. M. (1993), 'Notes on economic depreciation of natural resource stocks and national accounting', in A. Franz, and C. Stahmer, eds., *Approaches to Environmental Accounting*, Heidelberg: Physica-Verlag.

Hartwick, J. M. (1994), 'Sustainability and constant consumption paths in open economies with exhaustible resources', *AERE Conference, 5 June 1994*, Boulder, CO.

Hazell, P. (1984), 'Changing patterns of variability in world cereal production', in J. Anderson and P. Hazell, eds., *Variability in Grain Yields: Implications for Agricultural Research and Policy in Developing Countries*, Baltimore: Johns Hopkins University Press.

Holling, C.S. (1973), 'Resilience and Stability of Ecological Systems', *Annual Review of Ecology and Systematics*, **4**: 1–24.

Hueting, R., P. Bosch and B. de Boer (1992), 'Methodology for the calculation of sustainable national income', *Statistical Essays M44*, Central Bureau of Statistics, Voorburg.

Krebs, C.J. (1985), *Ecology*, 3rd edn, New York: HarperCollins.

Mäler, K.-G. (1991), 'National accounts and environmental resources', *Environmental and Resource Economics* **1**: 1–15.

Meadows, D.H., D.L. Meadows, R. Behrens and A. Randers (1972), *Limits to Growth*, New York: Universe Books.

Moran, D., D.W. Pearce and A. Wendelaar (1995), 'Investing in biodiversity conservation: a cost-effectiveness index for ranking priorities', mimeo, CSERGE, University College London and University of East Anglia.

Ninan, K. and H. Chandrashekar (1991), 'The green revolution, dryland agriculture

and sustainability insights from India', paper presented to the 21st International Conference of Agricultural Economists, Tokyo.
Pearce, D.W. and G. Atkinson (1993), 'Capital theory and the measurement of sustainable development: an indicator of weak sustainability', *Ecological Economics* **8**: 103–108.
Pearce, D.W. and D. Moran (1994), *The Economic Value of Biodiversity*, London: Earthscan.
Pearce, D.W. and T.R. Prakesh (1993), 'Sustainability as resilience', mimeo, CSERGE, University College London and University of East Anglia.
Pearce, D.W. and J.J. Warford (1993), *World Without End: Economics, Environment and Sustainable Development*, Oxford University Press.
Pearce, D.W., A. Markandya and E. Barbier (1989), *Blueprint for a Green Economy*, London: Earthscan.
Pearce, D.W., E. Barbier and A. Markandya (1990), *Sustainable Development: Economics and Environment in the Third World*, London: Edward Elgar, and London: Earthscan.
Pearce, D.W., G. Atkinson and W.R. Dubourg (1994), 'The economics of sustainable development', *Annual Review of Energy and Environment* **19**: 457–474.
Perrings, C. (1994), 'Ecological resilience in the sustainability of economic development', *International Symposium on Models of Sustainable Development*, Paris.
Perrings, C., R.K. Turner and C. Folke (1995), 'Ecological economics—paradigm or perspective?', *European Association of Resource and Environmental Economists Conference*, Umeå.
Pezzey, J. (1989), 'Economic analysis of sustainable growth and sustainable development', *Environment Department Working Paper No. 15*, World Bank, Washington DC.
Pezzey, J. (1994), 'The optimal sustainable depletion of non-renewable resources', mimeo, University College London.
Rees, W. and M. Wackenagel (1992), 'Appropriated carrying capacity: measuring the natural capital requirements of the human economy', *International Society of Ecological Economists Conference*, Stockholm.
Repetto, R., W. Magrath, M. Wells, C. Beer and F. Rossini (1989), *Wasting Assets: Natural Resources in the National Accounts*, Washington DC: World Resources Institute.
Schulze, E.-D. and H.A. Mooney (1994), *Biodiversity and Ecosystem Functions*, Berlin/Heidelberg: Springer-Verlag.
Simon, J.L. and H. Kahn (1984), *The Resourceful Earth*, Oxford: Basil Blackwell.
Solow, R.M. (1986), 'On the intergenerational allocation of exhaustible resources', *Scandinavian Journal of Economics*, **88**: 141–149.
van Tongeren, J., S. Schweinfest, E. Lutz, M. Gomez Luna and F. Guillen (1993), 'Integrated environmental and economic accounting: a case study for Mexico', in E. Lutz, ed., *Toward Improved Accounting for the Environment*, Washington DC: World Bank.
Vitousek, P., P. Ehrlich, A. Ehrlich and P. Matson (1986), 'Human appropriation of the products of photosynthesis', *Bioscience* **34**(6): 368–73.
World Bank (1994), *World Development Report 1994*, Washington DC: World Bank.
World Bank (1995), *Monitoring Environmental Progress: Environmentally Sustainable Development*, Washington DC: World Bank.
World Conservation Monitoring Centre (WCMC) (1992), *Global Biodiversity: Status of the Earth's Living Resources*, London: Chapman and Hall.

Name Index

Global Aspects of the Environment
Volume I

Environmental Analysis and Economic Policy

Series Editors: Kenneth Button
Professor of Public Policy
The Institute of Public Policy, George Mason University, USA

Peter Nijkamp
Professor in Regional, Urban and Environmental Economics
Free University, Amsterdam, The Netherlands

1. Global Aspects of the Environment (Volumes I and II)
   Robert U. Ayres, Kenneth Button and Peter Nijkamp
2. Environment, Land Use and Urban Policy
   David Banister, Kenneth Button and Peter Nijkamp
3. Environmental Valuation (Volumes I and II)
   Kenneth G. Willis, Kenneth Button and Peter Nijkamp
4. Environment and Transport
   Yoshitsugu Hayashi, Kenneth Button and Peter Nijkamp
5. Environmental Economics and Development
   J(Hans). B. Opschoor, Kenneth Button and Peter Nijkamp
6. Environmental Instruments and Institutions
   Thomas Tietenberg, Kenneth Button and Peter Nijkamp
7. Ecosystems and Nature
   R. Kerry Turner, Kenneth Button and Peter Nijkamp

Wherever possible, the articles in these volumes have been reproduced as originally published using facsimile reproduction, inclusive of footnotes and pagination to facilitate ease of reference.

For a list of all Edward Elgar published titles visit our site on the World Wide Web at http://www.e-elgar.co.uk

Global Aspects of the Environment Volume I

Edited by

Robert U. Ayres

Sandoz Professor of Environment and Management, Professor of Economics and Director of the Centre for the Management of Environmental Resources, European Business School, INSEAD, France

Kenneth Button

Professor of Public Policy, The Institute of Public Policy, George Mason University, USA

and

Peter Nijkamp

Professor in Regional, Urban and Environmental Economics, Free University, Amsterdam, The Netherlands

ENVIRONMENTAL ANALYSIS AND ECONOMIC POLICY

An Elgar Reference Collection
Cheltenham, UK • Northampton, MA, USA

Published by
Edward Elgar Publishing Ltd
Glensanda House
Montpellier Parade
Cheltenham
Glos GL50 1UA
UK

Edward Elgar Publishing, Inc.
136 West Street
Suite 202
Northampton
Massachusetts 01060
USA

A catalogue record for this book is available from the British Library.

Library of Congress Cataloguing in Publication Data

Global aspects of the environment / edited by Robert U. Ayres,
Kenneth Button and Peter Nijkamp.
(Environmental analysis and economic policy: 1)
Includes bibliographical references and index.
1. Environmental sciences. 2. Ecological engineering. 3. Human ecology. I. Ayres, Robert U. II. Button, Kenneth. III. Nijkamp, Peter. IV. Series. An Elgar reference collection.
GE105.G53 1999
333.7'2—dc21 99–29678
CIP

ISBN 1 85898 741 5 (2 volume set)

Printed and bound in Great Britain by Biddles Ltd, Guildford and King's Lynn

Contents

Acknowledgements

The editors and publishers wish to thank the authors and the following publishers who have kindly given permission for the use of copyright material.

Academic Press, Inc. for article: Herman E. Daly (1992), 'Is the Entropy Law Relevant to the Economics of Natural Resource Scarcity? – Yes, of course it is!', *Journal of Environmental Economics and Management*, **23** (1), July, 91–5.

American Association for the Advancement of Science for articles: Bruce Hannon (1975), 'Energy Conservation and the Consumer', *Science*, **189**, 11 July, 95–102; Cutler J. Cleveland, Robert Costanza, Charles A.S. Hall and Robert Kaufmann (1984), 'Energy and the U.S. Economy: A Biophysical Perspective', *Science*, **225**, 31 August, 890–97.

American Institute of Physics for article: Ann P. Kinzig and Robert H. Socolow (1994), 'Human Impacts on the Nitrogen Cycle', *Physics Today*, November, 24–31.

American Scientist for article: W.I. Vernadsky (1945), 'The Biosphere and the Noösphere', *American Scientist*, **33** (1), January, 1–12.

ASME International for article: R.U. Ayres and M. Narkus-Kramer (1976), 'An Assessment of Methodologies for Estimating National Energy Efficiency', *ASME Winter Meeting*, 1–20.

Blackwell Publishers Ltd for article: William C. Clark (1989), 'The Human Ecology of Global Change', *International Social Science Journal*, **121**, August, 315, 317–21, 323–45.

Blackwell Science Ltd for excerpt: M.J. Chadwick (1975), 'The Cycling of Materials in Disturbed Environments', in M.J. Chadwick and G.T. Goodman (eds), *The Ecology of Resource Degradation and Renewal*, 3–16.

Brookings Institution for article: William D. Nordhaus (1973), 'The Allocation of Energy Resources', *Brookings Papers on Economic Activity*, **3**, 529–76.

David Dyrssen and Daniel Jagner for excerpt: Lester Machta (1972), 'The Role of the Oceans and Biosphere in the Carbon Dioxide Cycle', in David Dyrssen and Daniel Jagner (eds), *The Changing Chemistry of the Oceans: Proceedings of the Twentieth Nobel Symposium held 16–20 August, 1971 at Aspenäsgården, Lerum and Chalmers University of Technology, Göteborg, Sweden*, 121–45.

Elsevier Science Ltd for articles: R. Stephen Berry, Peter Salamon and Geoffrey Heal (1978), 'On a Relation Between Economic and Thermodynamic Optima', *Resources and Energy*, **1** (2), October, 125–37; Paul P. Christensen (1989), 'Historical Roots for Ecological Economics – Biophysical versus Allocative Approaches', *Ecological Economics*, **1**, 17–36; Evan Mills, Deborah Wilson and Thomas B. Johansson (1991), 'Getting Started: No-Regrets Strategies for Reducing Greenhouse Gas Emissions', *Energy Policy*, **19**, July/August, 526–42; C. Bianciardi, A. Donati and S. Ulgiati (1993), 'On the Relationship between the Economic Process, the Carnot Cycle and the Entropy Law', *Ecological Economics*, **8**, 7–10; A. Azapagic and R. Clift (1995), 'Life Cycle Assessment and Linear Programming – Environmental Optimisation of Product System', *Computers and Chemical Engineering*, **19**, Supplement, S229–34.

Edward D. Goldberg and Werner Stumm for excerpt: J.M. Wood and E.D. Goldberg (1977), 'Impact of Metals on the Biosphere', in Werner Stumm (ed.), *Global Chemical Cycles and their Alterations by Man*, 137–53.

Rudolf B. Husar for his own article: Rudolf B. Husar and Janet M. Holloway (1983), 'Sulfur and Nitrogen over North America', in *Ecological Effects of Acid Deposition*, National Swedish Environment Protection Board, Report PM 1636, 95–115.

Institute of Electrical and Electronics Engineers, Inc. for article: Clark W. Bullard, III and Robert A. Herendeen (1975), 'Energy Impact of Consumption Decisions', *Proceedings of the IEEE*, **63** (3), March, 276–85.

International Society for Ecological Economics for excerpt: Sander de Bruyn, Jeroen van den Bergh and Hans Opschoor (1997), 'Structural Change, Growth, and Dematerialization: An Empirical Analysis', in Jeroen van den Bergh and Jan van der Straaten (eds), *Economy and Ecosystems in Change: Analytical and Historical Approaches*, Chapter 10, 201–28.

Kluwer Academic Publishers BV for excerpt: R.U. Ayres and A.V. Kneese (1989), 'Externalities: Economics & Thermodynamics', in F. Archibugi and P. Nijkamp (eds), *Economy and Ecology: Towards Sustainable Development*, Chapter 6, 89–118.

Macmillan Magazines Ltd for article: Robert Costanza, Ralph d'Arge, Rudolf de Groot, Stephen Farber, Monica Grasso, Bruce Hannon, Karin Limburg, Shahid Naeem, Robert V. O'Neill, Jose Paruelo, Robert G. Raskin, Paul Sutton and Marjan van den Belt (1997), 'The Value of the World's Ecosystem Services and Natural Capital', *Nature*, **387**, 15 May, 253–60.

National Academy Press for excerpt: Robert U. Ayres (1989), 'Industrial Metabolism', in Jesse Ausubel and Hedy E. Sladovich (eds), *Technology and Environment*, 23–49.

Regional Science Association International for article: Walter Isard (1968), 'Some Notes on the Linkage of the Ecologic and Economic Systems', *Papers of the Regional Science Association*, **XXII**, 85–96.

Southern Economic Association for article: Nicholas Georgescu-Roegen (1979), 'Energy Analysis and Economic Valuation', *Southern Economic Journal*, **45** (3), January, 1023–58.

Tellus for article: James E. Lovelock and Lynn Margulis (1974), 'Atmospheric Homeostasis by and for the Biosphere: The Gaia Hypothesis', *Tellus*, **XXVI** (1–2), 2–9.

Every effort has been made to trace all the copyright holders but if any have been inadvertently overlooked the publishers will be pleased to make the necessary arrangement at the first opportunity.

In addition the publishers wish to thank the Library of the London School of Economics and Political Science, the Marshall Library of Economics, Cambridge University and B & N Microfilm, London for their assistance in obtaining these articles.

Series Preface

Kenneth Button and Peter Nijkamp

The environment has been a dominant theme for research and public policy during the latter part of the twentieth century and there is no sign that this will change as we move to the new millennium. The global interest in environmental matters in part stems from the increased pressures that a mounting population and increased production puts on the planet's natural resource base. Additionally, as personal incomes rise and leisure time becomes more freely available in the developed world, concern with more immediate human needs give way to an interest in preservation and conservation for future generations and for others. Science has also contributed by both highlighting emerging environmental problems and discovering the importance of many components of the natural environment for medical and other uses.

Historically, the interest in the environment can be traced through several phases, each with its own particular emphasis. The interest is also far from new and in antiquity there were often laws governing such things as the removal of night soil and the use of noisy wagons into towns. Traditionally, also the natural environment was seen as a source of the necessities of life and was controlled through farming techniques to ensure that food was adequate to maintain the population. Land use was controlled and water resources were often regulated at the local and regional levels.

With the Industrial Revolution came population growth, urbanization and the accelerated exploitation of natural resources. Malthus advanced ideas of over population while others, such as Jevons, became concerned about the finite nature of such resources as coal. The new technologies contaminated water courses and the widespread use of carbon fuels led to atmospheric pollution. In cities, dirt, noise and urban decay led to new ideas of town planning and ideal cities.

Much of the interest in the latter part of the last century centred around public health. As knowledge of how disease was spread – urban authorities in particular sought to improve the local environment by such measures as sewage control and clean water supply to reduce the spread of germs and infection. This trend, much later, spread to policies embodied in various pieces of clean air legislation to reduce local atmospheric pollution that causes smog and other harmful effects.

Wealthy societies, and the better off within poorer societies, with the time and resources to expand, became concerned with the built environment and in shaping nature in ways they found aesthetically pleasing. Over the centuries this has led to landscaping of the countryside and the provision of parks and gardens in urban areas.

The twentieth century has seen many of these trends continue although some of the problems, such as depletion of natural resources, initially appeared less severe as new sources were found in Africa, South America and Asia. Difficulties in feeding and otherwise sustaining the burgeoning populations of these regions, however, resurrected Malthusian concerns over whether the planet could sustain itself. Increasing reliance on fossil fuels for energy and the geographical concentration of the main reserves of oil in politically unstable areas resurrected concern over natural resource reserves in the post Second World War period.

More fundamental to modern debates has been the global orientation of recent concerns. This has been accompanied by a much more popular interest in the state of the natural environment. Kenneth Boulding's development of the idea of 'spaceship earth' attracted much attention, as did the rather doom ladened prophesies of the Club of Rome, but this was subsequently overshadowed by the publication of the Brundtland Report, *Our Common Future*, with its emphasis on the need to maintain resources for future generations. With the report came the much touted, but universally ambiguous, concept of sustainable development.

Just as the academic interest in the environment has evolved so has the institutional setting in which the topic is debated and researched. In a very general way as our understanding of the environment has progressed and as technology has developed so there has been a movement out from micro-policy making in the hands of city councils, lords of the manor and the like to macro-policy making, involving global agencies such as the United Nations and World Bank. Indeed, one of the pressing issues of contemporary environmental debate is the appropriate jurisdiction for different levels of government.

For those interested in researching environmental topics, either from the standpoint of the beginner or of that of an expert wishing to expand an established basis of knowledge, there are practical problems. Environmental analysis extends across virtually all academic disciplines although many such as economics and engineering have specific sub-areas of interest specifically devoted to environmental issues. The multidisciplinary and transdisciplinary nature of environmental research makes it difficult to keep track of the key literature. To those new to the field there is the added problem of tracing back to find classic papers or research results. There are now a large number of specialist journals that deal with environmental topics but this is often adding to the problem of collecting information rather than reducing it. While these journals and periodicals provide explicitly environmental research findings, other general journals in economics, sociology, geography, engineering, medicine, biology and so on still continue to publish important findings.

The aim of this series of edited books is to provide both a set of reference volumes for those already in the field and to facilitate newcomers with accessible core material. The aim is not to offer readings on the purely scientific issues but rather to limit the papers to those concerned with social science aspects of the environment. This does not mean that we have excluded papers with a natural science orientation but rather the focus is on the softer debates and research. There are seven volumes in the series. The papers have all been previously published but do not constitute a set of traditional classics. While many of the articles have an established pedigree, equally there are less often referenced pieces. The aim of each volume is to paint a picture of how a topic area has developed, provide an indication as to some of the key contributions to this development but also look at some of the more recent literature that covers policy considerations.

The books cover a wide range of topic areas. At one extreme there are the macro-oriented volumes edited with Bob Ayres looking at the *Global Aspects of the Environment* while, in conjunction with Kerry Turner, there is a collection concerned with *Ecosystems and Nature*. Also at the aggregate level, edited with Hans Opschoor, there is a volume which focuses on *Environmental Economics and Development*. At the other extreme there are volumes concerned with micro-issues such as that along with Ken Willis that deals with *Environmental*

Valuation and with David Banister on *The Environment, Land Use and Urban Policy*. Transport's impact on the environment is multifaceted and can be severe hence there is a volume edited with Yoshi Hayashi, *The Environment and Transport* dealing exclusively with that subject. Finally, edited with Tom Tietenberg there is a collection that looks more generally at *Environmental Instruments and Institutions*.

We accept that there are many omissions from the collections and that others may have made different selections or have structured the material differently. Space limits what can be included and there is much more that could have been justified in a larger exercise (although this would add further to deforestation). We have also had to omit work by many eminent and important scholars for the same reason. As for structuring the material, a number of possible cuts are possible, the ones selected here just seem logical to us and would seem to fit with the needs of both researchers and, possibly, teachers.

Introduction

Robert U. Ayres, Kenneth Button and Peter Nijkamp

Getting an edited volume together, especially when it is part of a large series is never easy and involves considerable cajoling and manipulation. As one of us (Bob Ayres) put it, 'I do not quite know how it happened that I was persuaded to take on the task of editing this collection of papers. Everybody knows that editing is an unrewarding job. The publisher makes money, the authors are flattered and don't have to do any more work. (Admittedly, some of them are no longer around and only their ghosts or students are still capable of being flattered.) I really meant to refuse. In fact I did refuse a couple of times, for the usual very good reasons of being tired, overworked, etc. But, Peter Nijkamp is very persuasive and, obviously, I gave in. Flattery works almost every time.' Nevertheless, this collection has emerged on the tides of hope and despair.

Equally the work load is never easy and again Bob Ayres summed up the situation when he said. 'In retrospect, I guess I got a little bit intrigued with the idea of doing "my own" collection of important – or fairly important – papers. Well, "mostly" my own collection. (Button and Nijkamp have contributed some papers that I overlooked.) Also, there were a few miscellaneous constraints like not being too long (one author I would have liked to include never seems to write anything shorter than fifty pages), too mathematical, political, too abstract, too narrowly focused, etc.' Notwithstanding these limitations, we have witnessed the birth of this collection.

If the reader thinks that this collection is the result of years of surveys of expert opinion, painstaking bibliographical research and deep consideration of citation indexes, then he or she should be disabused of that ridiculous idea. All that can be said is that these papers are (mostly) ones that seem, as we approach the millennium, to have either made a significant contribution or, in some sense, serve as a recognizable milestone for one reason or another in the evolution of environmental economics, ecological economics or industrial ecology. We do not agree with all of them, by any means, but let them speak for themselves.

This volume of highlights of publications on global environmental issues differs from most other collections in two ways. First, it is deliberately interdisciplinary. Although Edward Elgar publishes economics books, the subjects covered here clearly overlap other fields, notably ecology, energy and materials flows. So some papers are not written by economists for economists. Instead, these are papers that economists should read, even if they have not yet done so. A second chance, so to speak. Second, a few of the papers were never published formally in the first place, for one reason or another, but seem, in retrospect, to have deserved greater attention than they were given at the time. A sort of retrospective archival service, if you like. These are from the so-called 'grey literature' – mainly reports written under contract to government agencies – but are worthy of revisiting.

Scope and Structure of the Volumes

A traditional task of the editor is to explain the 'theme' of the book and how each paper fits into that theme. As regards the theme, it seems to be covered perfectly well by the title. Frankly, if a paper was not thought worthwhile it would not be in this collection. At any rate there is no aspiration to write for the *Reader's Digest*, so the authors have been left to speak for themselves. However, since this collection is admittedly somewhat more idiosyncratic than most, it might be as well to explain 'where it is coming from', to use an obsolescent Americanism. Since Bob did the vast majority of the editing its structure can be seen as largely reflecting his own background (the general editors are too young to remember much of this!).

He was trained in theoretical physics, but had an unfortunate inability to specialize and become the world's expert in just one branch of the subject. This is not the way to get on in the academic world, and it led to frustration. So, he was persuaded to take a 'sabbatical' in 1962 at the Hudson Institute (then one year old) to study the potential environmental consequences of a hypothetical nuclear war.[1] Herman Kahn, founder and director of the institute was concerned about the possibility that, given an effective civil defense, nuclear war might seem to be sufficiently 'survivable' to tempt the USSR to make a preemptive first strike. In those days such scenarios were being taken seriously. The unresolved question was: would there be delayed but lethal environmental after-effects that would come back to haunt or even destroy the surviving power? Could such effects be described and documented?

He explored this depressing but fascinating question for four years, from 1962 through 1966. In the course of that period he encountered, for the first time the notion of an 'ecosystem' and the 'biosphere', the subject matter of the first two papers in this collection (Vernadsky Volume I, Chapter 1, and Hutchinson Volume I, Chapter 2). The conclusion of the study of the effects of nuclear war was not altogether unambiguous. But it was generally optimistic in the narrow sense that no single environmental mechanism could be identified that would certainly, or even probably, wipe out all human life on earth. He developed a healthy respect for the recuperative and self-regulating powers of nature, but also the beginnings of an appreciation of what 'nature' is, how it works in the very broad sense and what role it plays for humans.

The famous 1974 article by Lovelock and Margulis (Volume I, Chapter 5) – which did not exist at the time – is closely related to the issues he struggled with in 1963–66. It was one of the first clear statements of the 'Gaia hypothesis', namely that the earth can be considered to behave as though it is a single organism which defines and maintains the conditions for its survival. This proposition has since been both greatly admired and harshly attacked. Currently it has subdivided into 'strong' and 'weak' variants, of which the first has been largely discredited by mainstream evolutionary biologists, while the second weaker version – that the system as a whole self-regulates – is now regarded by many skeptics as true but not new, or even as a trivial consequence of co-evolution. Be this as it may, Lovelock's work has already been quite influential in biology. In fact the currently revived controversy over Darwinism probably derives some of its reflected heat from the Gaia controversy. The implications of bio-geo-chemical self-regulation for environmental and ecological economics are not yet so clear, but they seem certain to exist.

One of the topics not considered in the early study of the environmental consequences of nuclear war, but which should have been, was climate warming. (In fact, there was more concern at the time with the possibility of climate cooling, due to the massive quantities of dust and smoke that nuclear explosions would have injected into the stratosphere. This mechanism was the basis of the 'nuclear winter' controversy a few years later.) At any rate, Lester Machta's 1971 paper on the carbon dioxide cycle (Volume I, Chapter 4) was not the first on the subject, by several decades, but it was a milestone of sorts in terms of mobilizing official governmental interest in the subject. It is also both comprehensive and not excessively technical, although of course much more is known today than at the time.

Reverting for a moment to Bob Ayres history, in 1967 he left Hudson Institute and moved to Resources for the Future (RFF), as a visiting scholar, to finish a book on *Alternatives to the Internal Combustion Engine* (Robert U. Ayres and Richard McKenna 'Alternatives to the Internal Combustion Engine', Johns Hopkins University Press, Baltimore, MD, 1972). At RFF he encountered population economics, pollution, and environmental economics for the first time, although he did not immediately become aware of the related work that was going on in the neighborhood, so to speak. Input output (I–O) models were then first being applied to regional and environmental problems. One of the first applications of I–O to regional issues was Cumberland's 1966 paper (Volume II, Chapter 1; one of his many on the subject). Ronald Ridker and Bill Watson were using Clopper Almon's I–O models at RFF to assess the impact of population growth on America's future (also published as a book, Ridker, Ronald G. and William D. Watson *To Choose a Future*, Johns Hopkins University Press, Baltimore, MD, 1980). The best short paper one can now find from that era on population economics and 'optimum population' is that of Partha Dasgupta (Volume II, Chapter 17), though he was not aware of it then.

Herman Daly's 1968 article on economics as a life science (Volume II, Chapter 3) was one of the first to emphasize the links between economics and ecology. Garrett Hardin, an ecologist, wrote a landmark article for *Science* in 1968 in which he explained in simple and straightforward language – with real and horrifying examples – 'The Tragedy of the Commons' *Science*, 1968, **162**, 1243–8. In effect, he popularized a phenomenon previously discussed only in abstract terms by a few theoretical economists, namely the fact that 'public goods' can result in serious misallocation of resources. The 1968 paper by Walter Isard (Volume I, Chapter 3) was one of the first to address the connections between ecology and economics, also from a regional perspective. The 1992 article by Crocker and Tschirhart (Volume II, Chapter 6), though less well known, takes these ideas much further. These four articles, along with Daly's 1974 article on 'steady state economics' (Volume II, Chapter 20), taken together, can be regarded as key precursors of 'ecological economics', now a specialty with its own society and journal.

RFF itself was, in the late 1960s, the leading US center of economic studies of resource availability (about which more later) and, after 1966 (thanks to a Rockefeller Foundation grant), of environmental economics in general. The existence of market failures and 'externalities' was not a new discovery, of course, but their magnitude and pervasiveness – indeed, their primary importance for environmental problems – was a relatively new idea. One of the most often quoted (and reprinted) milestone articles was Kenneth Boulding's (1966, Volume II, Chapter 3) contribution to an RFF symposium on environmental quality and economic growth, where he coined the useful phrases 'cowboy economy' and 'spaceship economy'. Ayres's introduction to the subject was through Allen Kneese, who directed RFF's Quality of Environment program at the time. He has contributed in several fundamental ways to the development of environmental economics. Among Kneese's key contributions (which is mentioned here because it does not fit neatly anywhere else), is a little-known 1972 article with Ralph d'Arge (Volume II, Chapter 5) on the impact of international trade on environmental quality. Ralph d'Arge also made several important contributions, including being the perhaps first to have specifically addressed the problem of economic growth and environmental quality (1971, Volume II, Chapter 18). He continued this line of research in an important 1973 article co-authored with Kogiku (Volume II, Chapter 19) which will be mentioned again below.

RFF was, and still is, run by resource economists with a traditional academic perspective. In particular, RFF economists had already shown fairly conclusively, both in theory and in fact, that most extractive natural resources have become available as fast as they were demanded in the marketplace.[2] With only one significant exception, lumber, resource commodity prices in the US had declined monotonically, on a smoothed average basis, in real (inflation adjusted) terms since the mid nineteenth century when data began to be collected and recorded. The 'oil shock' of 1973–74 seemed, for a while, to mark the end of this long decline. But, as it turned out, the decline in real oil prices continued again after 1980. As far as resource prices are concerned, the end of the long decline is not yet clearly in sight.

Exposure to RFF and the theoretical basis for resource economics made Ayres immediately skeptical of the assertions of many neo-Malthusian 'doomsayers' of the late 1960s and early 1970s, from Paul Ehrlich and Barry Commoner to The Club of Rome[3] and its associated think-tanks. It was especially hard to ignore the *Limits to Growth* report produced for the Club of Rome by a group at Massettussets Institute of Technology using a new sort of dynamic simulation model pioneered by Jay Forrester.[4] Because of its explicit use of computer models, it attracted great attention from the media. Many millions of copies of this book were published within a couple of years in something like 20 languages. It became the focal point of a very intense controversy that still continues. But all the resource economists at RFF were sharply critical of the 'limits' thesis.

Responding, perhaps, to the worldwide publicity accompanying *Limits to Growth* it was no surprise that mainstream academic economists reacted with considerable scorn. In fact, many journal articles and books were written in the period 1974–79 to show that, subject to 'standard' assumptions of neoclassical economics, economic growth could continue more or less indefinitely without running into any limit attributable to scarcity of natural resources. A number of the papers reprinted in this collection were written during that period and in response to those concerns. Theoretical papers on the resource availability/sufficiency issue included in Volume II of this collection include the papers by d'Arge and Kokigu (Chapter 19), Daly (Chapter 20), Dasgupta and Heal (Chapter 21), Robert Solow (Chapter 23), Hartwick (Chapter 24) and Stiglitz (Chapter 25). The well-known 1974 paper by Stiglitz, Joseph 'Growth with Exhaustible Natural Resources: Efficient and Optimal Growth Paths', *Review of Economic Studies*, 1974, was not included on the basis that his 1979 paper covered enough of the same ground. Only the first two of these papers suggest that resource availability (including environmental resources) might constitute effective limits to economic growth.

During much of the 1968–79 period there was a related but more specialized debate in the US (and elsewhere) on 'energy futures'. Many massive reports were written and signed by prestigious committees. The main dispute (to reduce it to its bare bones) was between various representatives of the energy establishment, whose view was invariably that enormous sums needed to be invested in energy systems – from nuclear power plants to oil shale and coal gasification facilities – to secure the energy needed for future economic growth. Most of these studies were conducted outside the academic arena. However, several papers, for example Chapter 13 of Volume I by Nordhaus, and Hudson and Jorgenson (1974: Chapter 22, Volume II), combined economic projection models with energy supply models to address national energy policy in this period. Such hybrid models have been enormously elaborated since that time.

Some academics, notably Bullard and Herendeen (Volume I, Chapter 14), and Hannon (Volume I, Chapter 15), contributed to the debate in a different way, especially by focusing on the empirical links between energy consumption, industrial activity and consumption. One non-academic, Amory Lovins, challenged the conventional wisdom frontally. He boldly predicted a much lower rate of growth in energy consumption than any so-called expert. He was quite right. (Unfortunately no piece of his both short enough and 'academic' enough to reprint in this collection could be found, but he deserves to be included.) Chapter 23 in Volume I by Mills *et al.* on the subject of technological opportunities for energy savings is much in the Lovins tradition however.

The argument about the prospective roles of energy conservation and new supply began to attract physicists in the early 1970s. The American Physical Society (APS) sponsored a very influential summer study at Princeton in 1974 entitled 'Efficient Use of Energy: A Physics Perspective' (signed by nine authors). This book-length study succeeded in making a vital distinction between two definitions of efficiency, namely 'first law efficiency' and the 'second law efficiency'. The former is roughly the percentage of the heat content of a fuel that is not wasted; the latter is roughly a ratio between the minimum energy needed to accomplish some result by the most efficient means (the numerator) and the energy actually used to accomplish some function (the denominator). Both efficiencies are less than unity, by definition. The first law efficiency 'contains' the first law of thermodynamics in a sense, while the second law efficiency also takes into account the second law of thermodynamics (hence the names).

This distinction finally made it possible for someone other than an expert in the subject to make straightforward calculations of comparative efficiencies of such items of equipment as heating and ventilating systems, refrigerators and heat pumps. (Engineers were, and still are, taught a specialized and highly misleading jargon using terms like 'coefficient of performance' or COP, which permit comparisons between two units of the same type but give no clear idea of how to calculate the absolute performance of a heating or cooling system.)

Using this newly available 'textbook', with a colleague Marc Narkus-Kramer, Ayres (Volume I, Chapter 16) also contributed modestly to that energy conservation debate, by calculating the overall efficiency of the US economy: they estimated barely 4 per cent, give or take a little, which implies that there is a much larger potential for conservation than the 'experts' would admit. (The paper was delivered at an engineering society meeting in 1976 but never submitted to a journal.)

The entry of physicists into the energy debates in the mid 1970s also led to a burst of interest in what became known as 'net energy analysis'. Again, the underlying idea was to perform assessments of proposed new energy supply projects in terms of comparing expected future output with the energy inputs associated with initial capital investments and other startup costs. It is not proposed to summarize this episode, since in retrospect it must be regarded as something of an embarrassment. The idea itself was sound. If an energy supply project would consume more energy than it would produce, it should not be undertaken. However, several of the early studies – which will not be cited by name – were made by people with an axe to grind. The literature is replete with examples of misleading results arrived at by misspecification of the problem or by confusion of the costs of a prototype with the costs of a future fully developed system. To take just one example, investment in the development of photovoltaic technologies was undoubtedly inhibited (to an unknown degree) by results of net-energy studies carried out in the late 1970s allegedly 'proving' that the energy required to manufacture the silicon-based PV cells, and their associated infrastructure, would be greater than the energy generated by the cells during their operating lifetimes.

The two controversies – on limits to growth and net-energy analysis – served to cast a spotlight on Georgescu-Roegen's book *The Entropy Law and the Economic Process*, Harvard University Press, Cambridge MA, 1971. Georgescu-Roegen was seen to have provided some fundamental theoretical underpinning to the concerns about resource exhaustion that motivated *Limits to Growth*. He also was the strongest supporter of the idea that economic activity, especially growth, cannot be understood properly without explicit acknowledgment of the role of materials and energy (as well as entropy) in production (Volume I, Chapters 17 and 19). Later he, and others attacked the neoclassical growth theorists – especially 'Solow–Stiglitz' – for neglecting the fact that production involves physical transformations that must be consistent with the laws of thermodynamics. Christensen (Volume I, Chapter 10) and Daly (Volume I, Chapter 24), among others, have kept the controversy going. Yet, it appears that Georgescu-Roegen's arguments regarding the impossibility of utilizing solar energy for re-cycling materials on earth were flawed, as several recent authors have pointed out, namely Banciardi *et al.* (Volume I, Chapter 25).

Herman Daly (see Volume II, Chapter 20) has been the most effective defender and popularizer of Georgescu-Roegen's version of the 'limits to growth' thesis and simultaneously the strongest advocate of a stable 'no growth' economy – making him an opponent of the thesis of unlimited growth (subject to appropriate investment rules), associated – rightly or wrongly – with Solow, Stiglitz, Hartwick and Dasgupta and Heal (Volume II, Chapters 23, 25, 24 and 21 respectively), as mentioned above. Although many of his ideas were anticipated by Georgescu-Roegen and also by Ed Mishan (who especially emphasized the costs of growth), it was Daly who first explicitly introduced the idea of 'optimal growth' – taking social and environmental costs into account – to economics. This was a major conceptual step forward from the rather mechanistic 'limits to growth' thesis of the early 1970s, while also avoiding the 'no limits' optimism of some of the mainstream group.

Another outgrowth of the 'net-energy' fad in the late 1970s was a revival of interest in the old 'energy theory of value'. This notion has long been discredited, as far as economists are concerned. For a clear exposition of the reasons see Chapter 18 of Volume I by Berry, Salamon and Heal. However the fact remains that the empirical relationship between the size of the economy and the aggregate amount of energy consumed has always been close and predictable. This fact implies a causal link one way or the other. Such a relationship is what the energy industry relies on in its political arguments that economic growth requires energy supply-side growth.

The same set of data implies a strong correlation between energy and value. It equally implies a strong link between energy inputs and economic output (Hannon and Joyce, Volume II, Chapter 26). When the quantity of energy consumed by the economy is adjusted for quality, the correlation is even closer (Cleveland *et al.*, Volume I, Chapter 20). While it is *not* appropriate to conclude from this body of work that energy is after all a universal measure of value, it is highly appropriate to conclude that energy is a major factor of production and should be treated as such in future models of economic growth. The work of Rose and Chen (Volume II, Chapter 13) on structural decomposition is also very relevant.

Speaking of limits, Georgescu-Roegen and Daly have both assumed that the energy available to humans on the surface of earth for purposes of economic growth is the amount of solar energy that impinges on the earth's atmosphere. In 1979 and 1980 the US Department of Energy undertook a serious two-year assessment of the economics of solar satellites, capturing sunlight via photovoltaic cells, as an alternative energy source. The immediate conclusion, based partly on net energy analysis, was pessimistic. Since then, however, progress in solar cell technology, and other new concepts, appear to have overcome many of the earlier obstacles.

The downsizing of the space program has probably put an end to such grand dreams, at least for the next decade or so. Nevertheless, in principle, space technology together with photovoltaic technology offers the long-term potential of nearly unlimited, and relatively 'clean' energy that can be captured from the sun anywhere in the solar system and made available (by microwave beams) to virtually any location on the earth's surface. For this reason, among others, there is no fundamental resource limitation on human economic development. This paragraph is obviously a digression, but it is helpful to note that some of the differences between the views expressed by Georgescu-Roegen and Daly, *vis-à-vis* other views in this collection can be attributed to different degrees of technological optimism. Thus Chapter 22 of Volume I by Ayres and Kneese shares with Georgescu-Roegen and Daly a concern for the importance of material flows and the laws of thermodynamics, but is much more optimistic about technological potential.

From 1980 onward new ideas and concerns have come increasingly to the fore. On the one hand, there have been more and more studies of how specific materials, especially nutrients and toxins, 'cycle' through the economy as well as the biosphere. The simplest consequence of the first law of thermodynamics – 'mass balance' – is recognition of the fact that material inputs must eventually become material outputs and wastes (Ayres and Kneese, Volume II, Chapter 4). Applications of this generic principle at the material level have been relatively scarce, at least until recently, both in ecology and economics. An ecological perspective is provided in Chapter 6 of Volume I by Chadwick on materials cycling in disturbed environments; similarly, oceanographers Wood and Goldberg (Volume I, Chapter 7) traced the chemical cycles of metals in the oceans; geographers Husar and Holloway (Volume I, Chapter 8) examined the sources of sulfur and nitrogen in North America. Two physicists (Kinzig and Socolow, Volume I, Chapter 11) and ecologists have independently reviewed the evidence of change in the nitrogen cycle due to human activity, from different disciplinary perspectives. Taking a more global perspective is ecologist William Clark's article 'The Human Ecology of Global Change' (Volume I, Chapter 9). It is mentioned here largely because it is *sui generis* and it does not exactly fit anywhere else.

The need for process models to supplement or complement economic models has been increasingly recognized. Three discussions of this topic from the economics literature (among many possibilities) are included in the present collection (James, Volume II, Chapter 9; Anderson, Volume II, Chapter 10 and Azapagic and Clift, Volume I, Chapter 26).

As 'end-of-pipe' treatment of wastes has become more and more standardized in the industrialized world, it has become increasingly clear that this approach also has limits. Indeed, we are quickly approaching those limits. Yet the environment continues to degrade because for every incremental improvement in process (energy/materials) efficiency, economic growth causes total production and emissions to arise as much or more than the savings.

The source of the difficulty, in brief, is that materials extracted from the environment as inputs to the production process necessarily become wastes (Ayres and Kneese, Volume II, Chapter 4; Ayres, Volume I, Chapter 21). This phenomenon has led to a great deal of discussion about 'dematerialization' or 'eco-efficiency', meaning that in future a lot more service has to be delivered to the consumer with less materials attached. One approach is to extend useful product life. Another is to increase the use of secondary materials or 'waste mining' (Allen and Behmanesh, Volume II, Chapter 14). Problems associated with attempts to increase trade and recycling of secondary materials have not been neglected (Grace *et al.*, Volume II, Chapter 8; Fullerton and Kinnaman, Volume II, Chapter 15).

Environmental economics has also become increasingly empirical, as economists attempt to use cross-national statistics to verify hypotheses and project trends. Significant examples include Jänicke *et al.* (1988: Volume I, Chapter 11) and Simonis (Volume I, Chapter 12) on cross-national comparisons of structural changes related to energy consumption and heavy industry; Pearce *et al.* (1991: Volume II, Chapter 31) on measures of sustainability (based on comparing savings with depreciation of natural as well as man-made capital); de Bruyn and Opschoor (Volume II, Chapter 16) on the relationship between material consumption and income.

A good place to end is with the relationship between environmental constraints and economic growth. This topic is touched on more or less directly by Boulding in Chapter 2, by d'Arge in Chapter 18 and d'Arge and Kogiku in Chapter 19. It is the indirect subject of Chapter 20 by Daly on the steady-state economy (a topic Kneese and Ayres also wrote about in 1971 in an article not reprinted here). It is also the subject, at least indirectly, of Chapter 27 of Volume II by Dasgupta and Mäler. The subject has been revived recently, in the context of the 'new' theory of growth kicked off by Romer in 1986 (Romer, Paul M. 1986 'Increasing Returns and Long-Run Growth', *Journal of Political Economy*, **44** (5), October). (His 1990 article is reprinted here as Chapter 29 in Volume II). Two particularly apt contributions are those by Gross and Veendorp and van den Bergh and Nijkamp (Volume II, Chapters 29 and 30, respectively).

What have we learned and where do we go from here?

A very important question now arises: how can radical dematerialization – by sharply increasing resource productivity – be accomplished without sacrificing economic growth and prosperity? A macro-economic perspective can help to clarify the issues. At the aggregate level it is understood that investment (from savings) plays a comparatively small role, while 'technological progress' accounts for most growth (for example Robert M. Solow 1957, 'Technical Change and the Aggregate Production Function', *Review of Economics and Statistics*, **39**, August, 312–20). But technological progress is an undefined residual that encompasses many aspects of change, from urbanization to training and education. So-called 'growth accounting' studies have tried to identify and quantify these factors (for example Edward Denison: 1967, *Why Growth Rates Differ* Brookings Institution, Washington D.C.; 1974, *Accounting for US Economic Growth 1929–1969* Brookings; 1979, *Accounting for Slower Growth*, Brookings), but much serious theoretical and applied research remains to be done before we can really understand the economic aspects of global environmental change.

Notes

1. The Hudson Institute was a think-tank created in 1961 by Herman Kahn, Max Singer, William Brown and several other associates, originally from RAND Corp. Kahn was the author of a very controversial book entitled *On Thermonuclear War* (Kahn, Herman, 1960 Princeton University Press, Princeton N.J.), which explored the issue of nuclear deterrence and civil defense. He later wrote, or co-wrote, several other influential books on national security issues, including *Thinking about the Unthinkable*, 1962, Horizon Press, N.Y. and *On Escalation: Metaphors and Scenarios*, 1965, Praeger Press, N.Y. In the early years of the Institute the problem of nuclear deterrence and nuclear weapons control was its main research topic. Later the institute moved into other areas, notably 'future studies'.
2. The classic reference is the 1963 RFF study of resource scarcity by Harold Barnett and Chandler Morse, *Scarcity and Growth* (Johns Hopkins University Press, 1962).
3. The Club of Rome is a group of mainly European international notables, including several royals. It was co-founded by Alexander King, a British physicist who was Director of the Science and Technology Programme of the OECD during the 1960s. The group meets regularly in various locations to discuss world problems. It sponsors studies through its influence with such funding bodies as the Volkswagen Foundation.
4. Jay Forrester was an electrical engineer. During World War II he worked on servomechanisms and radar control systems. He became director of MIT's Digital Laboratory and was one of the designers of MIT's 'Whirlwind' computer and the SAGE air defense system (1946–56). In 1956 he was appointed Professor at MIT's newly created Sloan School of Business, where he began to apply some of the ideas of feedback control to business problems, and to explore and develop the capabilities of computers for dynamic simulation. A major early contribution inspired by him was a compiler that greatly simplified the creation of models based on linked sets of first-order differential equations. This led to the DYNAMO series of compilers. Forrester himself used such models to explore various systems. He emphasized particularly the capability of complex dynamic systems to display oscillatory, unstable and sometimes counter-intuitive behavior. This led to an influential book *Industrial Dynamics* (Forrester, 1961, *Industrial Dynamics*, MIT Press, Cambridge, MA) to explain the growth patterns of firms. In *Urban Dynamics* (Forrester, 1969, MIT Press, Cambridge, MA) he noted that cities with generous public housing programs and welfare schemes would reduce the available resources for job creation and simultaneously attract immigration of indigents from other places, thus making the local problems worse. In 1970 Forrester was invited to make a presentation of his ideas on complex dynamics to the Club of Rome. The project was funded by Volkswagen Foundation. The result was Forrester's *World Dynamics* (Forrester, Wright-Allen, 1971 Cambridge, Mass.) and Meadows *et al.*'s *Limits to Growth* (Meadows, Donella H., Dennis L. Meadows, Jorgen Randers, William W. Behreus III, 1972, *Limits to Growth: A Report for the Club of Rome's Project on the Predicament of Mankind*, Universe Books, New York), both of which modeled linkages between population, resource use and availability, pollution and productivity.

Part I
Ecological/Biological Perspective: Human Impacts on Biosphere, Gaia, etc.

[1]

AMERICAN SCIENTIST

A QUARTERLY PUBLICATION OF THE SOCIETY OF THE SIGMA XI DEVOTED TO THE ENCOURAGEMENT OF RESEARCH IN SCIENCE

VOL. 33 | JANUARY 1945 | NO. 1

THE BIOSPHERE AND THE NOÖSPHERE

By W. I. VERNADSKY†

Member of the Academy of Sciences of the U. S. S. R.

The following article is composed of an introductory abstract of a paper completed in 1938, and recently published in translation in the Transactions of the Connecticut Academy of Sciences (vol. 35, pp. 483–517), under the editorship of Professor G. E. Hutchinson, and a new essay, written in 1943 and translated from the Russian manuscript by Dr. George Vernadsky of Yale University. The two contributions together present the general intellectual outlook of one of the most remarkable scientific leaders of the present century.

The translation of the quotation under the frontispiece (from a letter to Professor A. Petrunkevitch) is as follows: I look forward with great optimism. I think that we undergo not only an historical, but a planetary change as well. We live in a transition to the noösphere. Cordial greetings, W. Vernadsky.

The table, reproduced in abstract below, that forms the main feature of the earlier paper, calls attention to many properties of living bodies that appear so elementary that they are in danger of neglect. It is an instructive experience to go through the table applying the criteria to the crystallizable viruses, the nature of which was not apparent when the earlier paper was written.—*Editor.*

THE BIOSPHERE

LIVING matter is the totality of all organisms present on the earth at any one time. It is usually such a totality that is important, though in dealing with the effect of man on the processes of this planet, a single individual may be of importance. The living matter of the earth may be regarded as the sum of the average living matter of all the taxonomically recognizable groups. Each of these groups is said to consist of *homogeneous* living matter.

Living matter exists only in the *biosphere* [1]. This includes the whole atmospheric troposphere, the oceans, and a thin layer in the continental regions, extending down three kilometers or more. Man tends to increase the size of the biosphere.

The biosphere is distinguished as the domain of life, but also, and more fundamentally, as the region where changes due to incoming radiation can occur.

Within the biosphere, matter is markedly heterogenous and may be distinguished as inert matter or living matter. The inert matter greatly predominates in mass or volume. There is a continual migration of

† *The sad news of Academician Vernadsky's death on January 6, 1945 has reached us in going to press—Editor.*

atoms from the inert matter to living matter and back again. All the objects of study in the biosphere are to be regarded as the *natural bodies* of the biosphere. They may be of varying complexity, inert, living, or bio-inert as in the case of soil or lake water. The study of all phenomena has a unity, leading to the production of a body of systematized knowledge, the *corpus scientiarum,* which tends to grow like a snowball; this *corpus* includes all systematized knowledge, and is contrasted to the results of philosophy, religion, and art where truth may be revealed intuitively; the systematized history of these activities belongs to the corpus.

Two concepts have been inadequately stressed in the past: (a) Pasteur was correct in regarding the preponderance of optically active compounds as the most characteristic general property of living matter and its products; this idea is of immense importance; (b) the functions of living organisms in the energetics of the biosphere have been seriously neglected. Biogeochemical energy may be expressed in the velocity with which the biosphere could be colonized by a given species. For certain bacteria, the limiting velocity of extension of a dividing chain of cells tending to embrace the whole circumference of the earth would tend to approach the velocity of sound.

Bearing in mind these introductory principles, the difference between living matter and inert matter of the biosphere is expressed in a table, here given in condensed form. The differences in this table are not merely differences with regard to energetics and chemical properties. They also involve a fundamental difference in the spacio-temporal manifestations of living and inert matter. It is suggested that the geometry appropriate to the bodies of living organisms may be different from that appropriate to inert bodies

I

A. Living natural bodies exist only in the biosphere and only as discrete bodies. They have never been observed to arise except from other living bodies. Their entry into the biosphere from cosmic space is hypothetical and has never been proved.

B. *Discrete inert forms are concentrated in the biosphere but are also found much deeper in the earth's crust. They are created in the biosphere, but also enter it from below in volcanic phenomena, and from cosmic space as meteorites and dust.*

II

A. Living natural bodies, in their cellular morphology, protoplasmic nature and reproductive capacity have a unity, which must be connected with their genetic connection with each other in the course of geological time.

B. *Inert natural bodies are extremely diverse and have no common structural or genetic connections.*

III

A. Chemical differences between right and left enantiomorphs characterize the state of physical space occupied by living organisms. Either left or right enantiomorphs predominate.

B. *The left and right enantiomorphs of the same chemical compounds have the same chemical properties in inert bodies. The numbers of dextral and sinistral crystals formed in an inert medium are the same.*

IV

A. New living natural bodies are born only from pre-existing ones. From time to time new generations arise differing from the preceding ones. The rise of the central nervous system has increased the geological rôle of living matter, notably since the end of the Pliocene.

B. *New inert bodies are created in the biosphere irrespective of the natural bodies that previously existed. In general the same kind of natural bodies are formed by inert processes, as were formed two billion years ago. New kinds of inert bodies appear only under the influence of living matter, notably man.*

V

A. There are no liquid or gaseous living bodies, though liquids and gases are present in the mesomorphous or solid living bodies. Spontaneous, largely self-regulated movement is characteristic of living bodies. This may be passive as in reproduction, but the effect of reproduction is for the biosphere to be colonized, by a process comparable to the expansion of a gas.

B. *Liquid and gaseous inert bodies take the form of the receptacles in which they are contained. Solid or mesomorphous inert bodies in general show no motion peculiar to the body as a whole.*

VI

A. There is a continual stream of atoms passing to and from living organisms from and into the biosphere. Within the organisms a vast and changing number of molecules are produced by processes not otherwise known in the biosphere.

B. *Inert natural bodies change only from outside causes, with the exception of radioactive materials.*

VII

A. The number of living natural bodies is quantitatively related to the size of the biosphere.

B. *The number of inert natural bodies is defined by the general properties of matter and energy, and is independent of the size of the planet.*

VIII

A. The mass of living matter has remained fairly constant, being determined by radiant solar energy and the biogeochemical energy of colonization, but apparently the mass increases towards a limit, and the process is not yet complete.

B. *The area of manifestation of inert natural bodies in the biosphere is limited by the size of the latter and increases only as the biosphere is expanded by the motion of living matter.*

IX

A. The minimum size of a living natural body is determined by respiration, and is of the order of 10^{-6} cms. The maximum size has never exceeded $n \cdot 10^4$. The range, 10^{10}, is not great.

B. *The minimum size of an inert natural body in the biosphere is determined by the degree of dispersion of matter and energy, i.e., the size of the ultimate particles of physics. The maximum size is determined by the size of the biosphere. The range is* 10^{40} *or more.*

X

A. The chemical composition of living bodies is a function of their own properties.

B. *The chemical composition of inert bodies is a function of the properties of the medium in which they are formed.*

XI

A. The number of kinds of chemical compounds in living bodies is connected with the kinds of individual organisms and probably reaches many millions.

B. *The number of different kinds of chemical compounds in inert bodies is limited to a few thousands.*

XII

A. The processes in living matter tend to increase the free energy of the biosphere.

B. *All inert processes, save radio-active disintegration, decrease the free energy of the biosphere.*

XIII

A. Living natural bodies are always mesomorphous, and except in latent conditions H and O as water predominate, with an extremely complicated mixture of other compounds. The chemical composition of any one kind of living matter, though not exhibiting stoichiometric relationships, is definitely determined and more constant than the isomorphous mixtures constituting natural minerals.

B. *The chemical composition of inert natural bodies may correspond to nearly pure chemical compounds with precise stoichiometric relations between the elements. In minerals solid solutions predominate.*

XIV

A. Isotopic ratios may be markedly changed by the processes in living matter.

B. *The isotopic ratios do not change markedly in the inert natural bodies of the biosphere, though outside the biosphere, deep in the crust, such changes may occur.*

XV

A. The vast majority of living natural bodies change their forms in the process of evolution. The rates at which these changes occur are, however, widely divergent.

B. *The majority of the inert bodies of the biosphere are stable, and so lack variety.*

XVI

A. The processes of living natural bodies are not reversible in time.

B. *All physiochemical processes in inert natural bodies are reversible in time.*

In everyday life one used to speak of man as an individual, living and moving freely about our planet, freely building up his history. Until recently the historians and the students of the humanities, and to a certain extent even the biologists, consciously failed to reckon with the natural laws of the biosphere, the only terrestrial envelope where life can exist. Basically man cannot be separated from it; it is only now that this indissolubility begins to appear clearly and in precise terms before us. He is geologically connected with its material and energetic structure. Actually no living organism exists on earth in a state of freedom [2]. All organisms are connected indissolubly and uninterruptedly, first of all through nutrition and respiration, with the circumambient material and energetic medium. Outside it they cannot exist in a natural condition.

In our century the biosphere has acquired an entirely new meaning; it is being revealed as a planetary phenomenon of cosmic character. In biogeochemistry we have to reckon with the fact that living organisms actually exist not on our planet alone, and not in the terrestrial biosphere only. It seems to me that so far this has been established beyond doubt only for all so-called "terrestrial planets," that is, for Venus, the Earth, and Mars [3].

The thought of life as a cosmic phenomenon was alive long ago, as evidenced by the archives of science, including Russian science. At the end of the seventeenth century, the Dutch scientist Christian Huygens (1629–1695) put forward that problem in his last work, "Cosmotheoros," which was published after his death. This book, upon the initiative of Peter the Great, was twice published in Russian in the first quarter of the eighteenth century, under the title, "The Book of Contemplation of the World" [4]. In it Huygens established the scientific generalization that "life is a cosmic phenomenon somehow sharply distinct from inert matter." I have recently called this generalization the "Huygens principle" [5]

Living matter, by weight, constitutes an insignificant part of our planet. Presumably, this is observed in the whole course of geological time; in other words, this relation is *geologically eternal* [6]. Living matter is concentrated in a thin but more or less continuous film on the surface of land in the troposphere, in the forests and fields, and permeates the whole ocean. Its quantity is calculated to be of the order of 0.25 per cent of the weight of the biosphere. On land it descends under the surface in non-continuous accumulations, probably down to an average depth of less than 3 kilometers.

THE NOÖSPHERE [7]

We are approaching the climax in the Second World War. In Europe war was resumed in 1939 after an intermission of twenty-one years; it has lasted five years in Western Europe, and is in its third year in our parts, in Eastern Europe. As for the Far East, the war was resumed there much earlier, in 1931, and is already in its twelfth year. A war of such power, duration and strength is a phenomenon unparalleled in the history of mankind and of the biosphere at large. Moreover, it was preceded by the First World War which, although of lesser power, has a causal connection with the present war.

In our country that First World War resulted in a new, historically unprecedented, form of statehood, not only in the realm of economics, but likewise in that of the aspirations of nationalities. From the point of view of the naturalist (and, I think, likewise from that of the historian) an historical phenomenon of such power may and should be examined as a part of a single great terrestrial *geological* process, and not merely as a *historical* process.

In my own scientific work the First World War was reflected in a most decisive way. It radically changed my *geological conception of the world*. It is in the atmosphere of that war that I have approached a conception of nature, at that time forgotten and thus new for myself and for others, a geochemical and biogeochemical conception embracing both inert and living nature from the same point of view [8]. I spent the years of the First World War in my uninterrupted scientific creative work, which I have so far continued steadily in the same direction.

Twenty-eight years ago, in 1915, a "Commission for the Study of the Productive Forces" of our country, the so-called KEPS, was formed at the Academy of Sciences. That commission, of which I was elected president, played a noticeable role in the critical period of the First World War. Entirely unexpectedly, in the midst of the war, it became clear to the Academy of Sciences that in Tsarist Russia there were no precise data concerning the now so-called strategic raw materials, and we had to collect and digest dispersed data rapidly to make up for the lacunae in our knowledge [9]. Unfortunately by the time of the beginning of the Second World War, only the most bureaucratic part of that commission, the so-called Council of the Productive Forces, was preserved, and it became necessary to restore its other parts in a hurry.

By approaching the study of geological phenomena from a geochemical and biogeochemical point of view, we may comprehend the whole of the circumambient nature in the same atomic aspect. Unconsciously such an approach coincides for me with what characterizes the science of the twentieth century and distinguishes it from that of past centuries. *The twentieth century is the century of scientific atomism.*

At that time, in 1917–1918, I happened to be, entirely by chance, in the Ukraine [10], and was unable to return to Petrograd until 1921. During all those years, wherever I resided, my thoughts were directed toward the geochemical and biogeochemical manifestations in the circumambient nature, the biosphere. While observing them, I simultaneously directed both my reading and my reflection toward this subject in an intensive and systematic way. I expounded the conclusions arrived at gradually, as they were formed, through lectures and reports delivered in whatever city I happened to stay, in Ialta, Poltava, Kiev, Simferopol, Novorossiisk, Rostov, and so on. Besides, in almost every city I stayed, I used to read everything available in regard to the problem in its broadest sense. I left aside as much as I could all philosophical aspirations and tried to rest only on firmly established scientific and empiric facts and generalizations, occasionally allowing myself to resort to working scientific hypotheses. Instead of the concept of "life," I introduced that of "living matter," which now seems to be firmly established in science. "Living matter" is the totality of living organisms. It is but a scientific empirical generalization of empirically indisputable facts known to all, observable easily and with precision. The concept of "life" always steps outside the boundaries of the concept of "living matter"; it enters the realm of philosophy, folklore, religion, and the arts. All that is left outside the notion of "living matter."

In the course of geological time living matter morphologically changes according to the laws of nature. The history of living matter expresses itself as a slow modification of the forms of living organisms which genetically are uninterruptedly connected among themselves from generation to generation. This idea had been rising in scientific research through the ages, until, in 1859, it received a solid foundation in the great achievements of Charles Darwin (1809–1882) and Wallace (1822–1913). It was cast in the doctrine of the evolution of species of plants and animals, including man. The evolutionary process is a characteristic only of living matter. There are no manifestations of it in the inert matter of our planet. In the cryptozoic era the same minerals and rocks were being formed which are being formed now [11]. The only exceptions are the bio-inert natural bodies connected in one way or another with living matter [12].

The change in the morphological structure of living matter observed in the process of evolution unavoidably leads to a change in its chemical composition [13].

While the quantity of living matter is negligible in relation to the inert and bio-inert mass of the biosphere, the biogenic rocks constitute a large part of its mass, and go far beyond the boundaries of the biosphere. Subject to the phenomena of metamorphism, they are converted, losing all traces of life, into the granitic envelope, and are no longer part of the biosphere. The granitic envelop of the earth is the area of former biospheres [14]. In Lamarck's book, "Hydrogéologie" (1802), containing many remarkable ideas, living matter, as I understand it, was revealed as the creator of the main rocks of our planet. Lamarck never accepted Lavoisier's (1743–1794) discovery. But that other great chemist, J. B. Dumas (1800–1884), Lamarck's younger contemporary, who did accept Lavoisier's discovery, and who intensively studied the chemistry of living matter, likewise adhered for a long time to the notion of the quantitative importance of living matter in the structure of the rocks of the biosphere.

The younger contemporaries of Darwin, J. D. Dana (1813–1895) and J. Le Conte (1823–1901), both great American geologists (and Dana a mineralogist and biologist as well) expounded, even prior to 1859, the empirical generalization that *the evolution of living matter is proceeding in a definite direction.* This phenomenon was called by Dana "cephalization," and by Le Conte the "psychozoic era." Dana, like Darwin, adopted this idea at the time of his journey around the world, which he started in 1838, two years after Darwin's return to London, and which lasted until 1842 [15].

Empiric notions of a definite direction of the evolutionary process, without, however, any attempt theoretically to ground them, go deeper into the eighteenth century. Buffon (1707–1788) spoke of the "realm of man," because of the geological importance of man. The idea of evolution was alien to him. It was likewise alien to Agassiz (1807–1873), who introduced the idea of the glacial period into science. Agassiz lived in a period of an impetuous blossoming of geology. He admitted that geologically the realm of man had come, but, because of his theological tenets, opposed the theory of evolution. Le Conte points out that Dana, formerly having a point of view close to that of Agassiz, in the last years of his life accepted the idea of evolution in its then usual Darwinian interpretation [16]. The difference between Le Conte's "psychozoic era" and Dana's "cephalization" thus disappeared. It is to be regretted that, especially in our country, this important empirical generalization still remains outside the horizon of our biologists.

The soundness of Dana's principle, which happens to be outside the horizon of our palaeontologists, may easily be verified by anyone willing to do so on the basis of any modern treatise on palaeontology. The principle not only embraces the whole animal kingdom, but likewise reveals itself clearly in individual types of animals. Dana pointed out that in the course of geological time, at least two billion years and probably much more, there occurs an irregular process of growth and perfection of the central nervous system, beginning with the crustacea (whose study Dana used to establish his principle), the molluscs (cephalopoda), and ending with man. It is this phenomenon that he called cephalization. The brain, which has once achieved a certain level in the process of evolution, is not subject to retrogression, but only can progress further.

Proceeding from the notion of the geological rôle of man, the geologist A. P. Pavlov (1854–1929) in the last years of his life used to speak of the *anthropogenic era* in which we now live. While he did not take into the account the possibility of the destruction of spiritual and material values we now witness in the barbaric invasion of the Germans and their allies, slightly more than ten years after his death, he rightly emphasized that man, under our very eyes, is becoming a mighty and ever-growing geological force. This geological force was formed quite imperceptibly over a long period of time. A change in man's position on our planet (his material position first of all) coincided with it. In the twentieth century, man, for the first time in the history of the earth, knew and embraced the whole biosphere, completed the geographic map of the planet Earth, and colonized its whole surface. *Mankind became a single totality in the life of the earth.* There is no spot on earth where man can not live if he so desires. Our people's sojourn on the floating ice of the North Pole in 1937–1938 has proved this clearly. At the same time, owing to the migthy techniques and successes of scientific thought, radio and television, man is able to speak instantly to anyone he wishes at any point on our planet. Transportation by air has reached a speed of several hundred kilometers per hour, and has not reached its maximum. All this is the result of "cephalization," the growth of man's brain and the work directed by his brain.

The economist, L. Brentano, illuminated the planetary significance of this phenomenon with the following striking computation: if a square meter was assigned to each man, and if all men were put close to one another, they would not occupy the area of even the small Lake of Constance between the borders of Bavaria and Switzerland. The remainder of the earth's surface would remain empty of man. Thus the whole of mankind put together represents an insignificant mass of the planet's matter. Its strength is derived not from its matter, but from its brain. If man understands this, and does not use his brain and his work for self-destruction, an immense future is open before him in the geological history of the biosphere.

The geological evolutionary process shows the biological unity and equality of all men, *Homo sapiens* and his ancestors, *Sinanthropus* and others; their progeny in the mixed white, red, yellow, and black races evolves ceaselessly in innumerable generations [17]. This is a *law of nature*. In a historical contest, as for instance in a war of such magnitude as the present one, he finally wins who follows that law. One cannot oppose with impunity the principle of the unity of all men as a law of nature. I use here the phrase "law of nature" as this term is used more and more in the physical and chemical sciences, in the sense of an empirical generalization established with precision.

The historical process is being radically changed under our very eyes. For the first time in the history of mankind the interests of the masses on the one hand, and the free thought of individuals on the other, determine the course of life of mankind and provide standards for men's ideas of justice. Mankind taken as a whole is becoming a mighty geological force. There arises the problem of the *reconstruction of the biosphere in the interests of freely thinking humanity as a single totality.* This new state of the biosphere, which we approach without our noticing it, is the *noösphere.*

In my lecture at the Sorbonne in Paris in 1922–23, I accepted *biogeochemical phenomena* as the basis of the biosphere. The contents of part of these lectures were published in my book, "Studies in Geochemistry," which appeared first in French, in 1924, and then in a Russian translation, in 1927 [18]. The French mathematician Le Roy, a Bergsonian philosopher, accepted the biogeochemical foundation of the biosphere as a starting point, and in his lectures at the Collège de France in Paris, introduced in 1927 the concept of the noösphere as the stage through which the biosphere is now passing geologically [19]. He emphasized that he arrived at such a notion in collaboration with his friend Teilhard de Chardin, a great geologist and palaeontologist, now working in China.

The noösphere is a new geological phenomenon on our planet. In it for the first time man becomes a *large-scale geological force.* He can and must rebuild the province of his life by his work and thought, rebuild it radically in comparison with the past. Wider and wider creative possibilities open before him. It may be that the generation of our grandchildren will approach their blossoming.

Here a new riddle has arisen before us. *Thought is not a form of energy.* How then can it change material processes? That question has not as yet been solved. As far as I know, it was first posed by an American scientist born in Lvov, the mathematician and biophysicist Alfred Lotka [20]. But he was unable to solve it. As Goethe (1740–1832), not only a great poet but a great scientist as well, once rightly remarked, in science we only can know *how* something occurred, but we cannot know *why* it occurred.

As for the coming of the noösphere, we see around us at every step the empirical results of that "incomprehensible" process. That mineralogical rarity, native iron, is now being produced by the billions of tons. Native aluminum, which never before existed on our planet, is now produced in any quantity. The same is true with regard to the countless number of artificial chemical combinations (biogenic "cultural" minerals) newly created on our planet. The number of such artificial minerals is constantly increasing. All of the *strategic raw materials* belong here. Chemically, the face of our planet, the biosphere, is being sharply changed by man, consciously, and even more so, unconsciously. The aerial envelope of the land as well as all its natural waters are changed both physically and chemically by man. In the twentieth century, as a result of the growth of human civilization, the seas and the parts of the oceans closest to shore become changed more and more markedly. Man now must take more and more measures to preserve for future generations the wealth of the seas which so far have belonged to nobody. Besides this, new species and races of animals and plants are being created by man. Fairy tale dreams appear possible in the future: man is striving to emerge beyond the boundaries of his planet into cosmic space. And he probably will do so.

At present we cannot afford not to realize that, in the great historical tragedy through which we live, we have elementally chosen the right path leading into the noösphere. I say elementally, as the whole history of mankind is proceeding in this direction. The historians and political leaders only begin to approach a comprehension of the phenomena of nature from this point of view. The approach of Winston Churchill (1932) to the problem, from the angle of a historian and political leader, is very interesting [21].

The noösphere is the last of many stages in the evolution of the biosphere in geological history. The course of this evolution only begins to become clear to us through a study of some of the aspects of the biosphere's geological past. Let me cite a few examples. Five hundred million years ago, in the Cambrian geological era, skeletal formations of animals, rich in calcium, appeared for the first time in the biosphere; those of plants appeared over two billion years ago. That calcium function of living matter, now powerfully developed, was one of the most important evolutionary factors in the geological change of the biosphere [22]. A no less important change in the biosphere occurred from seventy to one hundred and ten million years ago, at the time of the Cretaceous system, and especially during the Tertiary. It was in this epoch that our green forests, which we cherish so much, were formed for the first time. This is another great evolutionary stadium, analogous to the noösphere. It was probably in these forests that man appeared, around fifteen or twenty million years ago.

Now we live in the period of a new geological evolutionary change in the biosphere. We are entering the noösphere. This new elemental geological process is taking place at a stormy time, in the epoch of a destructive world war. But the important fact is that our democratic ideals are in tune with the elemental geological processes, with the laws of nature, and with the noösphere. Therefore we may face the future with confidence. It is in our hands. We will not let it go.

Borovoe, July 22/Moscow, December 15, 1943.

NOTES AND REFERENCES

1. The concept of "biosphere," that is, of the "area of life," was introduced into biology by Lamarck (1744–1829) in Paris in the early nineteenth century, and into geology by Suess (1831–1914) in Vienna at the close of the same century. On the biosphere, see W. Vernadsky, *Ocherki geokhimii*, 4th edition, Moscow-Leningrad. Index; *Biosfera* (The Biosphere) Leningrad, 1926; French edition, Paris, 1929.
2. That remarkable scientist, Caspar Wolf (1733–1794), a member of the St. Petersburg Academy of Sciences, who though not a Russian by birth, dedicated his whole life to Russia, clearly expressed the connection between organisms and their medium in a book published in German in St. Petersburg in 1789, the year of the great French Revolution, under the title, *Von d. eigenthüml. Kraft d. Vegetabl. sowohl auch d. animal. Substanz als Erläuterung zu zwei Preisschriften über d. Nutritions-kraft* (On the Specific and Active Force Proper to the Vegetal and Animal Substance).

Unlike the great majority of the biologists in his time, he tended toward the ideas of Newton rather than those of Descartes.

It it to be regretted that the manuscripts left after Wolf's death have been, as yet, neither studied nor published. In 1927 the Commission on the History of Knowledge at the Academy of Sciences decided to do this work, but it could not be accomplished because of the constant changes in the Academy's approach toward the study of the history of science. Now that work at the Academy has been reduced to a minimum, which is harmful to the cause.

3. See my article on *The Geological Envelopes of the Earth as a Planet,* Izvestiia of the Academy of Sciences, Geographical and Geophysical Series, 1942, p. 251. Cf. H. Spenser Jones, *Life on Other Worlds,* New York, 1940; R. Wildt in Proc. Amer. Philos. Soc. 81 (1939), p. 135. A Russian translation of Wildt's study, regrettably not in full (which is not indicated in the paper) appeared in the Astronomicheskii Zhurnal, vol. XVII (1940), no. 5, p. 81ff. By now, a new study by Wildt has appeared, *Geochemistry and the Atmosphere of Planets* (1941), but, to our regret, no copy of it has so far reached us.

   The Biogeochemical Laboratory of the Academy of Sciences in Moscow, now renamed the Laboratory of Geochemical Problems, in cooperation with the Institute of Microbiology of the same Academy (corresponding member, B. L. Isachenko) set the problem of cosmic life as a current scientific problem as early as 1940. This work was stopped by the war, but will be resumed at the first opportunity.
4. It would deserve a new edition in modern Russian, with commentaries.
5. See *Ocherki geokhimii,* pp. 9, 288, and my book *Problemy biogeokhimii* (The Problems of Biogeochemistry) III (in press).
6. *Problemy biogeokhimii,* III.
7. The word "noösphere" is composed from the Greek terms *noos,* mind, and *sphere,* the last used in the sense of an envelope of the earth. I treat the problem of the noösphere in more detail in the third part of my book, now being prepared for publication, on *The Chemical Structure of the Biosphere of the Earth as a Planet, and its Surroundings.*
8. It should be noted that in this connection I came upon the forgotten thoughts of that original Bavarian chemist, C. Schoenbein (1799–1868) and of his friend, the English physicist of genius, M. Faraday (1791–1867). As early as the beginning of the eighteen-forties, Schoenbein attempted to prove that a new division should be created in geology—geochemistry, as he called it. See W. Vernadsky, *Ocherki geokhimii* (Studies in Geochemistry), 4th edition, Moscow-Leningrad, 1934, pp. 14, 290.
9. On the significance of KEPS see A. E. Fersman. *Voina i strategicheskoe syrie* (The War and Strategic Raw Materials). Krasnoufimsk, 1941, p. 48.
10. See my article, *Out of my Recollections: The First Year of the Ukrainian Academy of Sciences,* to appear in the Jubilee volume of the Ukrainian Academy of Sciences, in commemoration of its twenty-fifth anniversary.
11. In accordance with modern American geologists as, for example, Charles Schuchert (Schuchert and Dunbar, *A Textbook of Geology,* II, New York, 1941, p. 88ff.), I call the Cryptozoic era that period which formerly had been called the Azoic, or the Archaeozoic, era. In the Cryptozoic era the morphological preservation of the remnants of organisms dwindles almost to nothing, but the existence of life is revealed in the organogenic rocks, the origins of which arouse no doubts.
12. On the bio-inert bodies see W. I. Vernadsky, *Problems of Biogeochemistry, II.* Trans. Conn. Acad. Arts Sci., vol. 35 (1944), pp. 493–494. Such are, for example, the soil, the ocean, the overwhelming majority of terrestrial waters, the troposphere, and so on.
13. This problem urgently needs experimental verification. It has been set forth by the Laboratory of Geochemical Problems in collaboration with the Palaeontological Institute of the Academy of Sciences, in the plan of our work for 1944.
14. See my basic work referred to in Note 7.
15. One should not fail to note here that the expedition during which Dana came to his conclusions concerning cephalization, the coral islands, and so on, actually and historically was closely connected with the exploration of the Pacific by Russian navigators, especially Krusenstern (1770–1846). See D. Gilman, *The Life of J. D. Dana,* New York, 1899. The chapter on the oceanic expedition in this book was written by Le Conte. Le Conte's book, *Evolution* (1888), has not been accessible to me. His autobiography was published in 1903: W. Armes, Editor, *The Autobi-*

ography of Joseph Le Conte. For his biography and bibliography see H. Fairchild in Bull. Geol. Soc. Amer. 26 (1915), p. 53. It was the report on the Russian travels, published in German in 1827, which spurred the American lawyer, John Reynolds, to insist on the organization of a similar American scientific oceanic expedition. Owing to Reynolds' persistence, the expedition eventually materialized, but not until eleven years afterwards, in 1838. This was the Wilkes expedition, which finally proved the existence of the Antarctic.

On Reynolds see the Index in *Centenary Celebration: Wilkes Exploring Expedition of the U. S. Navy, 1838–1842,* Proc. Amer. Philos. Soc., 82, No. 5 (1940). It is to be regretted that our expeditions in the Pacific, so active in the first half of the nineteenth century, were later discontinued for a long time (almost until the Revolution), following the death of both Emperor Alexander I (1777–1825) and Count N. P. Rumiantsov (1754–1826)—that remarkable leader of Russian culture who equipped the "Riurik" expedition (1815–1818) out of his private funds.

In the Soviet period K. M. Deriugin's (1878–1936) expedition should be mentioned; its precious and scientifically important materials have been so far only partly studied and remain unpublished. Such an attitude toward scentific work is inadmissible. The Zoological Museum of the Academy of Sciences must fulfill this scientific and civic duty.

16. D. Gilman, *o.c.*, p. 255.
17. I and my contemporaries have imperceptibly lived through a drastic change in the comprehension of the circumambient world. In the time of my youth it seemed both to me and to others that man had lived through a historical time only, within the span of a few thousand years, at best a few tens of thousand of years. Now we know that man has been consciously living through tens of millions of years. He consciously lived through the glacial period in both Eurasia and North America, through the formation of Eastern Himalaya, and so on. The division of historical and geological time is leveled out for us.
18. The last revised edition of my *Ocherki Geokhimii* (Problems of Geochemistry) appeared in 1934. In 1926 the Russian edition of *Biosfera* (The Biosphere) came out, and in 1929 its French edition. My *Biogeokhimicheskie Ocherki* (Biogeochemical Studies) was published in 1940. The publication of *Problemy biogeokhimii* (Problems of Biogeochemistry) was begun in 1940. (A condensed English translation of Part II appeared, under the editorship of G. E. Hutchinson, in Trans. Conn. Acad. Arts Sci., vol. 35, in 1944.) Part III is in press. *Ocherki geokhimii* was translated into German and Japanese.
19. Le Roy's lectures were at once published in French: *L'exigence idéaliste et le fait d'évolution,* Paris, 1927, p. 196.
20. A. Lotka, *Elements of Physical Biology,* Baltimore, 1925, p. 405 f.
21. W. S. Churchill, *Amid These Storms: Thoughts and Adventures,* New York, 1932, p. 274 f. I plan to return to this problem elsewhere.
22. I deal with the problem of the biogeochemical functions of organisms in the second part of my book, *The Chemical Structure of the Biosphere* (see Note 7).

[2]

ON LIVING IN THE BIOSPHERE

G. E. HUTCHINSON

In the article below, which is from a paper presented in the symposium on "The World's Natural Resources" (AAAS Centennial, September 13–17), Professor Hutchinson claims originality only for some of the computations and part of the interpretation—most of the fundamental data are easily accessible in the scientific literature. Professor Hutchinson is on the staff of the Department of Zoology at Yale and consultant in biogeochemistry, American Museum of Natural History.

IN discussing the subject of "The World's Natural Resources," I want first to make a number of general observations that will provide an intellectual framework into which our developing knowledge, both academic and practical, may be fitted. We live in a rather restricted zone of our planet, at the base of its gaseous envelope and on the surface of its solid phase, with temporary excursions upwards, downwards, or sideways onto or into the oceans. These regions in which we can live and which we can explore are characterized by their temperature, which does not depart far from that at which water is a liquid, and by their closeness to regions on which solar radiation is being delivered. This zone of life is spoken of as the biosphere. Within it, certain natural products can be utilized in both biological and cultural life. It is customary to consider these resources as either material or energetic, but the two categories are not easily separable; contemporary solar radiation is an energetic resource, coal and oil are to be regarded as material resources valuable for their high energy content, which we may call, epigrammatically, fossil solar radiation. There is a third very important though inseparable aspect, namely, the pattern of distribution. Most fossil sunlight, or chemical energy of carbonaceous matter, is diffused through sedimentary rocks in such a way as to be useless to us. Schrödinger says that we feed on negative entropy, and I am almost tempted to regard pattern as being as fundamental a gift of nature as sunlight or the chemical elements.

The first major function of the sunlight falling on the earth's surface is as the energy of circulation of the oceans and atmosphere. The second is to increase the mobility of water molecules, to become latent heat of evaporation, and so to keep the water cycle operating. The third major function is photosynthesis. Apart from atomic energy, and a little volcanic heat which presumably is actually of radioactive origin, all industrial energy is solar and due to one or the other of these three processes.

The material requirements of life are extremely varied. Between thirty and forty chemical elements appear to be normally involved. Industrially, some use appears to be found for nearly all the natural elements, and some of the new synthetic ones also. Looking at man from a strictly geochemical standpoint, his most striking character is that he demands so much—not merely thirty or forty elements for physiological activity, but nearly all the others for cultural activity. What we may call the anthropogeochemistry of cultural life is worth examining. We find man scurrying about the planet looking for places where certain substances are abundant; then removing them elsewhere, often producing local artificial concentrations far greater than are known in nature. Such concentrations, whether a cube of sodium in a bottle in the laboratory, or the George Washington Bridge, have usually been brought into being by chemical changes, most frequently reductions, of such a kind that the product is unstable under the conditions in which accumulation takes place. Most artifacts are made to be used, and during use the strains to which they are submitted distort them, and they become worn-out or broken. This results in a very great quantity of the materials that are laboriously collected being lost again in city dumps and automobile cemeteries. The final fate of an object may depend on many factors, but it is probable that in most cases a very large quantity of any noncombustible, useful material is fated to be carried, either in solution or as sediment, into the sea. Modern man, then, is a very effective agent of zoogenous erosion, but the erosion is highly specific, affecting most powerfully arable soils, forests, accessible mineral deposits, and other parts of the biosphere which provide the things that *Homo sapiens* as a mammal and as an educatable social organism needs or thinks he needs. The process is continuously increasing in intensity, as populations expand and as the most easily eroded loci have added their quotas to the air, the garbage can, the city dump, and the sea.

The most important general consideration to bear in mind in discussing the dynamics of the biosphere and its inhabitants is that some of the processes of significance are acyclical, and others, to a greater or less degree, cyclical. By an acyclical process will be meant one in which a permanent change in geochemical distribution is introduced into the system; usually a concentrated element tends to become dispersed. By a cyclical process will be meant one in which the changes involved introduce no permanent alteration in the large-scale geochemical pattern, concentration alternating with dispersion. The cyclical processes are not necessarily reversible in a thermodynamic sense: in fact, they are in general no more and no less reversible than the acyclical. Most of the cyclical processes operate because a continuous supply of solar energy is led into them, and sunlight will provide no problems for the conservationist for a very long time. It is important to realize that most of the acyclical processes are so slow that man appears as an active intruder into a passive pattern of distribution. They are safer to disturb because we know what the result of the disturbance will be. If we mine the copper in a given region sufficiently assiduously, we know that ultimately there will not be any more copper available there. Cyclical processes involve complex circular paths, regenerative circuits, feedback mechanisms, and the like. Small disturbances of such processes may merely result in small temporary changes, with a rapid return to the previous steady state. This has been beautifully demonstrated in the experiments of Einsele, who added single massive doses of phosphate to a lake, changing for a time, but only for a time, its entire chemistry and biology. This stability does not imply that if large disturbances strain the mechanism beyond certain critical limits very profound disruption will not follow; in fact, the very self-regulatory mechanisms that give the system stability against small disturbances are likely to accentuate the disruption when the critical limits are transcended. In disturbing cyclical processes we usually do not know what we are doing.

The most nearly perfect cyclical processes are those involving water and nitrogen. Some losses to the sediments of the deep oceanic basins must occur, but they are very small and are doubtless fully balanced or more than balanced by juvenile water and perhaps by molecular nitrogen and ammonia of volcanic origin.

The least cyclical processes are those in which material is removed from the continents and deposited in the permanent basins of the ocean. With one or two exceptions, the delivery to the deep-water sediments of the ocean is of little significance. Most of the mechanical and chemical sedimentation in the oceans takes place in relatively shallow water. The uplifting of shallow water sediments constitutes an important method of completing cycles. One, and perhaps two, exceedingly important elements are, however, sedimented less economically. Calcium, during the Paleozoic, was mainly precipitated in shallow water, but since the rise of the pelagic foraminifera in the Mesozoic, a great deal of calcium, along with an equivalent amount of carbon and oxygen, has been continually diverted to regions from which it is unlikely ever to be removed. At present the sedimentary rocks of the world are an adequate biological and commercial source of calcium, but, with progressive orogenic cycles, less and less of the element will be uplifted (Kuenen), and whatever organisms inherit the earth in that remote future will have to face the problem of the biosphere "going sour on them." For phosphorus, the case is less well established, but there is probably a slow loss in the form of sharks' teeth and the ear bones of whales (Conway), which are very resistant and which are known to be littered about on the floor of the abysses of the ocean.

It is desirable to consider two of the main geochemical cycles in order to gain an idea of the effect of man upon them. It must be admitted that we are ignorant of many matters of importance here. In the cycle of carbon we have a remarkable, possibly a unique, case in which man, the miner, increases the cyclicity of the geochemical process. It is generally admitted that our available store of carbon is ultimately of volcanic origin. A steady stream of carbon dioxide and lesser amounts of methane and carbon monoxide are entering the atmosphere from volcanic vents. Part, probably a major part, of this carbon dioxide is ultimately lost to the marine sediments as limestones; a very small part of it is then returned to the air, wherever lime kilns are in operation. Another part of the CO_2 entering the atmosphere is reduced in photosynthesis. A part of the organic matter so formed is fossilized, and a small part of this fossilized organic carbon is available as fuel, in the form of coal and oil. At the present time it appears that the combustion of coal and oil actually returns carbon to the atmosphere as CO_2 at a rate at least a hundred times greater than the rate of loss of all forms of carbon, oxidized and reduced, to the sediments (Goldschmidt). This particular process obviously cannot go on indefinitely. It concerns only the reduced carbon; to complete the cycle in the case of oxidized carbon, a great deal of energy would have to be supplied. It concerns only the reduced carbon which is aggregated. The poorest sources would be the poorest exploitable oil shales. Most of the reduced carbon is much more dispersed than this; in making an estimate of the total reduced carbon of the sediments, the commercially usable fuels constitute a negligible fraction that need not be considered.

Although the rate at which carbon dioxide is returned to the air by the human utilization of fossil fuels is so very much greater than is the primary production of carbon dioxide from volcanic sources, the rate is evidently a very small fraction—of the order of 1 percent—of the rate of photosynthetic fixation and subsequent respiratory liberation of CO_2 by the organisms of the earth. Since about 1890 a slight increase in CO_2 content in the air, at least at low altitudes over the land surfaces of the Northern Hemisphere, has been noted. This has been attributed to the accumulation of industrially produced CO_2, as the quantity of CO_2 that has appeared in the atmosphere is of the same order of magnitude as the total combustion of fuel (Callendar). In view of the small fraction of the total CO_2 production that industrial output represents, it seems very unlikely that merely adding an extra percent to the natural biological production should overload the cyclic process, so that it rejects quantitatively the additional load. It is known that the air at high altitudes still shows nineteenth-century values. The most reasonable explanation of the observed increase is that the photosynthetic machinery of the biosphere has been slightly impaired, probably by deforestation. It is clear that, in any intelligent long-term planning of the utilization of the biosphere, an extended study of atmospheric gases is desirable, even though for the moment it seems unlikely that the observed change is a particularly serious symptom.

The only other cycle that can be considered in any great detail is that of phosphorus. The chief event in the geochemical cycle of phosphorus is the leaching of the element from the rocks of the continents, and its transport by rivers to the sea. At the present time the rate of this transportaion is of the order of 20,000,000 tons of phosphorus per year for the entire earth. Part of this phosphorus, when it enters the sea, will ultimately be deposited in the sediments of the depths of the ocean. Such phosphorus will probably be largely lost to the geochemical cycle, as has just been indicated. The sedimentary rocks of the continents, therefore, will gradually lose phosphorus; there is some evidence that this has actually occurred (Conway). The main return path is by the uplifting of sediments formed in continental seas, which then undergo renewed chemical erosion. Of particular interest are methods by which concentrated phosphorus can be returned to the land surfaces. As far as is known, there are two such methods: The first is the formation of phosphatic nodules and other forms of phosphate rock in regions of upwelling in which water at a low *p*H, rich in phosphate, is brought up to the surface of the sea. The *p*H falls and an apatite-like phosphate is deposited. When the sea floor is later elevated, a commercial deposit may result. The second method is by the activity of sea birds, such as the guano birds of the Peruvian coast. There is little or no unequivocal evidence that guano deposits of great extent were formed prior to the late Pliocene or Pleistocene. Some of the well-known occurrences of rock phosphate, such as that of Quercy, have been explained in this way, but they are certainly not typical guano deposits. During the late Tertiary and Pleistocene, an extraordinary amount of phosphate was deposited on raised coral islands throughout the world, and bird colonies seem to provide the only reasonable agencies of deposition. The great deposits of Nauru, Ocean Island, Makatea, Angaur, the Daito Islands, Christmas Island south of Java, Curaçao, and some of the other West Indies all seem to have been formed in this way. This process is as characteristic of the time as is glaciation, though less grandiose. Its meaning is not clear, but it is probably connected with changes in vertical circulation of the ocean as glaciopluvial periods gave place to interpluvials. Today in certain regions massive amounts of guano are deposited, and it is probable that the oceanic birds of the world as a whole bring out from several tens to several hundreds of thousands of tons of phosphorus and deposit it on land. Only about 10,000 tons of the element are delivered in places where it is not washed away and where it can be carried by man to fertilize his fields.

The main processes that tend to reverse the phosphorus depletion of the continents are, therefore, the deposition of marine phosphorites on the continental shelves and subsequent elevation, and the formation of guano deposits. Both processes are evidently intermittent, and are quantitatively inadequate to arrest deflection of the element into the permanent ocean basins. Man contributes both to the loss and to the gain of phosphorus by land surfaces. He quarries phosphorite, makes superphosphate of it, and spreads it on his fields. Most of the phosphorus so laboriously acquired ultimately reaches the sea. At present the world's production of phosphate rock is about 10,000,000 tons per annum. This contains from 1,000,000 to 2,000,000 tons of elementary phosphorus. Human activity probably, therefore, accounts for from 5 to 10 percent of the loss of phosphorus from the land to the sea. Man also contributes to the processes bringing phosphorus from the sea to the land. This is done by fisheries. The total catch for the marine fisheries of the earth is of the order of $25–30.10^6$ tons of fish, which corresponds to about 60,000 tons of elementary phosphorus. The human, no less than the nonhuman, processes tending to complete the cycle seem miserably inadequate. It is quite certain that ultimately man, if he is to avoid famine, will have to go about completing the phosphorus cycle on a large scale. It will be a harder task than that of solving the nitrogen problem, which would have loomed large in any symposium on "The World's Natural Resources" fifty years ago, but possibly an easier problem than some of the others that must be solved if we are to survive and really become the glory of the earth.

The population of the world is increasing, its available resources are dwindling. Apart from the ordinary biological processes involved in producing population saturation already known to Malthus, the current disharmony is accentuated by the effect of medical science, which has decreased death rates without altering birth rates, and by modern wars, which one may suspect put greater drains on resources than on populations. Terrible as these conclusions may appear, they have to be faced. The results of the interaction between population pressure and decline of resource potential are further partly expressed in such wars, which are pathological expressions of attempts to cope with these and other problems and which now invariably aggravate the situation. It is evident that the fundamental causes of war lie in those psychological properties of populations which make them attempt solutions in a warlike manner, and *not* in the existence of the problems themselves. The two problems of war and of resources are, however, at present very closely interrelated; it is probably impossible to find a solution for one without progress in the solution of the other. Any kind of reasonable use of the world's resources involves better international relations than now exist. It is otherwise impossible to operate on a planetary basis, or to avoid the fearful material and spiritual expense of living in a world divided into two armed camps. The lack of international trust is the first difficulty in achieving rational utilization of the resources of the world; the second difficulty for the United States is what may be called the problem of the transition from the pioneering to the old, settled community. For a pioneer, life may be hard, but in good country there is "plenty more where these came from," whether lumber or buffalo tongues or copper. This attitude is incorporated into popular thought very deeply; in a crude form, it is now completely destructive, but it seems possible that attitudes might be developed that could utilize such a point of view. There is at least one thing of which we have plenty more than we use, whether we call it Yankee ingenuity, American know-how, or the human intellect. In some industrial fields there has been a notable series of triumphs of the kind that really will give plenty more of a number of things, and give them in a cyclic manner. The rise of the magnesium industry, and the utilization of the magnesium of sea water as a source of the metal is an interesting example. The production of plastics, though it is probably at the moment not as geochemically economical as it should be, is another. These, along with the development of silicate building materials not involving any particularly uncommon elements, all point the way to the kind of material culture that permits a reasonably high living standard, at least in certain directions, without devastation of the earth.

The future outlook for the world, particularly in food resources, has been put before the public in several recent books, notably those of Fairfield Osborn and William Vogt. Anyone with any technical knowledge understands that the dangers described in these books are real enough, more real and more dangerous perhaps than the threat of an atomic world war. The problem of getting action to forestall such dangers in a culture that has developed under conditions of potentially unlimited abundance just around the corner, is obviously extremely difficult. I do not think it is impossible. The first requirement is a faith that the job can be done. The very difficulty of the task, its apparent impossibility, may here and now prove a challenge that brings the desired response. The difficult we do at once, the impossible takes a little longer. I doubt that a direct appeal to fear will produce any results except a disbelief in the prophets of doom. Cassandra seems even more unpopular in modern America than in ancient Ilium. There would seem to be forces operating in society which tend to reverse the destructive processes, or which could be made to do so. One of the most immediate needs is to find out what they are and do everything possible to strengthen them. I will give one example. A number of industrial concerns—particularly in the chemical and pharmaceutical field, but also some engineering and publishing firms—have used in their advertising a legitimate pride in the learning, skillful research, and development that have gone into the manufacture of their product. It is quite certain that there are many cases in which one particular product, the result of considerable research, and of taking risks in development, actually reduces the drain on the natural resources of the country and the world. I should like to see a small systematic experiment, on the part of some such concern, in advertising in which it is pointed out that by buying this product one is letting the industrial skill behind the product operate for the benefit of one's children. I am fully aware that if this point of view were worked up skillfully enough by a few responsible, public-spirited corporations and put into a form that would pay the corporation, as well as the country and the world, there would be other less responsible concerns who would use the method when their product is actually not one sparing natural resources. This, however, seems to be a lesser evil than the total neglect of the commercial advertising field, which is one of the most potent in determining the values of the public and which at present is largely disruptive. I do not doubt that a professional cultural anthropologist could pick out a great many other fields that could be used to promote the idea of an expanding economy based on an abundance of human ingenuity rather than on an excess of raw materials.

There is one further point that I should like to develop. Though the pursuit of happiness is embedded deeply in the constitutional foundations of the United States, we do not know much about it. It is fairly certain that no metric exists that can be applied directly to happiness, but intuitively we may proceed a little way by arguing as if such a metric could exist. It is obvious that only a very few people, with a genius for sanctity, can be happy if half-frozen and starving. If the temperature be raised, and the food supply and other amenities increased, the possibility of happiness is obviously at first also increased, but beyond a certain increase in the environmental resources available no further increase in happiness would be expected and we might begin to look for a decline. The image of an overheated kitchen, filled with too much electrical equipment and catering to overfed people, will, if adequately evoked, have a nightmarish quality. In more formal language, if we could find a function of the environmental resources that expressed the relationship of happiness to those resources, it is reasonably certain that the function would not be monotonic. For every resource there seems likely to be an optimum level of consumption, but we do not know if this optimum level is, in any particular case, widely exceeded, so that gross overutilization is actually producing avoidable distress.

The problem is not by any means as simple as that of determining discrete optima. In any given society all the cultural values are probably interrelated to form a coherent system, so that the existence of one set of values may greatly modify the others. If, as seems possible, our attitude toward food leads a considerable section of our population to be definitely overweight, it is legitimate to inquire to what extent, by changes in the upbringing of our children, the psychological needs filled by food can be satisfied in other ways so that the psychologically optimum intake falls to a level nearer the psychological optimum for the individual and the moral optimum for the world. We might ask, for instance, how we can substitute the delights of ballet and Mozartian opera, which are geochemically very cheap, for part of those provided by hot dogs or apple pie and ice cream, which may in the long run prove too expensive to use except as a source of energy and essential nutrients. This example is chosen with a view to indicating that the kind of substitutions that might be considered need not be in the least puritanical. It may appear overintellectualized, but that at least is a guarantee that it is not inhuman. Indeed it raises the very interesting problem that those of us in the educational world have to face, namely, why we are raising a generation in the belief that the majority of constructive, complicated, difficult activities are boring duties, when the same generation shows us that in certain specific fields, mainly concerned with electronic amplifiers and with the explosive combustion of hydrocarbons, complicated activity can be very entertaining. What we have to do is to show by example that a very large number of diversified, complicated, and often extremely difficult constructive activities are capable of giving enormous pleasure. This is, in fact, the reason why it is essential that the teachers in our colleges and universities should be enthusiastic investigators in the fields of scholarship or practitioners and critics in their arts. It ought to be possible to show that it is as much fun to repair the biosphere and the human societies within it as it is to mend the radio or the family car.

[3]

REGIONAL SCIENCE ASSOCIATION: PAPERS, XXII, BUDAPEST CONFERENCE, 1968

SOME NOTES ON THE LINKAGE OF THE ECOLOGIC AND ECONOMIC SYSTEMS

by Walter Isard*

1. INTRODUCTION

This morning I want to describe in a *non-technical* manner some of our recent research, in particular some of the ways in which a new conceptual framework is being forged to link both the economic and ecologic systems. I also want to illustrate, again without supporting *technical* discussion, how we put into this framework some numbers that we judge to be valid for the quantification of the relationships involved. In this paper, I shall use very extensively materials already published[1] and those contained in a forthcoming book[2] which fully develops the technical materials underlying this research. While the empirical materials and basic concepts were developed by a team of research analysts, the shortcomings of the specific views to be presented in this paper are solely my own responsibility.

Before presenting the framework and empirical materials, let me first provide some perspective and background on our research. Let me quote from our forthcoming book on *Ecologic-Economic Analysis for Regional Development*. We state that "the combination of the two words, *ecologic* and *economic*, while an unusual one, is set forth as a true combination, in the sense of a synthesis of analyses of two systems within the world of actuality. Throughout this book, we stress the need for this synthesis. Having constantly in mind the mounting air, water, and sonic pollution problems of society, its solid waste problem, its need for open spaces, its diverse impingements on the environment, we assert that no longer can regional development and regional planning be treated in their traditionally narrow garb. Emphasis on the strict economics of such development and planning, with only passing consideration of physical environment and design, let alone social, political and other cultural factors, can no longer be tolerated. Whether we look at the problems of planning and development within the New England states, or metropolitan Budapest, or the environs of Lake Baikal, or in innumerable other regions of the world, we are confronted with the reality that control of the disturbance to the ecological and physical processes, and design of the environment are key elements of economic

* The author is associated with the Regional Science and Landscape Analysis Project, Harvard University and the Regional Science Research Institute, Philadelphia, U.S.A. He is indebted to the Environmental Sciences Services Administration (U.S. Department of Commerce), the National Science Foundation, the Federal Water Pollution Control Administration and Resources For the Future, Inc., for their financial support of this research.

[1] See Isard *et al.* [2].

[2] See Isard *et al.* [3].

development and planning work. Even conscious redesign of the environment may be."

The above view represents a new dimension which must be brought into our current-day analysis. It is around this new dimension that I wish to center my remarks this morning.

2. THE CONCEPTUAL FRAMEWORK

The conceptual framework which we have developed is in essence a framework for a linear system. It approaches the problem from a traditional input-output or, more strictly, a linear programming standpoint. In fact, my first step in developing this framework is to set down as the upper left major block of cells in Table 1 a traditional coefficient table, with columns representing sectors (industries and activities) and rows representing commodities associated with these sectors as outputs and resources. Thus we list by column some typical economic sectors (Agriculture, Textiles, . . . , Petroleum Refining, etc.), and by row some typical economic commodities (Wheat, Cloth, . . . , Crude Oil). Since those who are assembled here this morning are thoroughly aware of the nature of the input-output and linear programming techniques, and the virtues and limitations of a coefficient matrix for both description and projection, I need not spend any time on this first step.[3]

My second step involves the hypothesis that one can fruitfully view the ecologic system in input-output or activity analysis format. Thereby, the ecologic system is considered as a very large set of interdependent activities, involving as inputs and outputs many commodities, only a few of which correspond to those of the economic system.[4] Not only do these activities feed upon each other, that is provide each other with inputs, but also to the exogenous economic system they furnish final outputs or deliveries (in the sense of exports to the economic system). In turn these ecologic activities are dependent on the economic system (that is currently require or receive from it) various commodities (in the sense of imports) to be used as inputs. These imports from the economic system are at the same time exports or deliveries of the economic system to the world exogenous to it, while the exports from the ecologic system are at the same time imports into the economic system, again from the world exogenous to it. To the extent that one system's imports and exports are the other system's exports and imports respectively, to that extent the ecologic and economic systems can be effectively linked. And correspondingly the typical input-output or activity analysis coefficient table for the economic system can be many times enlarged to encompass a comprehensive accounting framework for a fused ecologic-economic system.

In effect my second step is to add three major blocks of cells to Table 1 to complement the major block in its upper left. The first of these major blocks, at the lower right of Table 1, relates to the ECOLOGIC SYSTEM and what might be designated its INTERPROCESS COEFFICIENTS (which parallel the INTERINDUSTRY

[3] The reader who is unfamiliar with these techniques may refer to Isard *et al.* [1; chaps. 8 and 10] and the references contained therein.

[4] For a standard discussion of ecologic systems, see Odum [4].

COEFFICIENTS of the ECONOMIC SYSTEM, in the upper left major block). The second new major block is at the lower left of Table 1. Here are listed in coefficient form the inputs and outputs of ecologic commodities (imports from and exports to the ecologic system), by economic sector. Finally, the third new major block is at the upper right of Table 1. Here are listed in coefficient form the inputs and outputs of economic commodities, by ecologic process.

3. ILLUSTRATIVE COEFFICIENTS

To point up the significance of the conceptual framework as briefly sketched in Table 1, as well as to cast light upon the problems of empirical implementation, we now present some representative types of coefficients.

Let us begin by considering the petroleum refining industry. In the upper part of Table 2 are listed its typical inputs and outputs of economic commodities in coefficient form. For example, the first coefficient, —.612006, is the dollar's worth of input of crude petroleum and natural gas per dollar output of the industry, and the

TABLE 1

		ECONOMIC ACTIVITIES						ECOLOGIC PROCESSES				
		Agriculture	Textile	...	Petroleum Refining	...	Sport Fishing	...	...	Plankton Production	Herring Production	Cod Production
ECONOMIC COMMODITIES	Wheat											
	Cloth											
	⋮			ECONOMIC SYSTEM: INTERSECTOR COEFFICIENTS				ECOLOGIC PROCESSES: THEIR INPUT AND OUTPUT COEFFICIENTS RE: ECONOMIC COMMODITIES				
	Crude Oil											
ECOLOGIC COMMODITIES	Water Intake											
	Alkalinity											
	⋮			ECONOMIC SECTORS: THEIR INPUT AND OUTPUT COEFFICIENTS RE: ECOLOGIC COMMODITIES				ECOLOGIC SYSTEM: INTERPROCESS COEFFICIENTS				
	Detritus											
	Plankton									+	−	
	Herring										+	−
	Cod						−					+

TABLE 2. Input-output Coefficients: RIS sector 2911
Petroleum Refining

Sector		Coefficient
RIS 1311	Crude Petroleum and Natural Gas	−.612006
1509	Construction, Maintenance, and Repair	−.001410
2652	Set-Up Paperboard Boxes	−.008054
2655	Fiber Cans, Tubes, Drums, and Similar Products	−.001557
2812	Alkalies and Chlorine	−.000935
2818	Industrial Organic Chemicals, n.e.c.	−.027180
2819	Industrial Inorganic Chemicals, n.e.c.	−.009190
2911	Petroleum Refining	+.954950
2992	Lubricating Oils and Greases	−.014954
3411	Metal Cans	−.005294
4811	Telephone Communications	−.001210
4890	Telegraph and Other Communications	−.000100
4911	Electric Utilities	−.007629
4920	Gas Companies and Systems	−.012336
4941	Water Supply	−.001082
4990	Sanitary and Other Systems	−.000142
6020	Interest	−.005088
6301	Insurance, Nonlife	−.001242
6510	Real Estate Services	−.007680
7301	Business Services excluding Advertising	−.017228
7310	Advertising	−.002312
7400	Research and Development	−.000390
7500	Automotive Repair	−.001040
9000	Local and State Taxes	−.008049
9100	Federal Income Tax	−.014755
9826	Office Supplies	−.000280
9842	Transportation Costs	−.040831
9888	Wages and Salaries	−.089378
9899	Residual	−.063589
WPC1001	Water Intake, Sanitary Use, 1,000 Gal./$ Output	−.000455
1002	" Production "	−.022827
1003	" Cooling "	−.114861
1011	Water Discharged, Sanitary Use "	+.000323
1012	" Production "	+.070219
1013	" Cooling "	+.050100
1031	Biochemical Oxygen Demand BOD, 5-Day, 1,000 Lb./$ Output "	+.000065
1032	Ultimate Oxygen Demand, UOD "	+.000076
1033	Chemical Oxygen Demand, COD "	+.000169
1041	Suspended Solids "	+.000084
1042	Settleable Solids "	+.000124
1047	Turbidity "	+.000365
1051	Alkalinity "	+.000051*
1052	Acidity "	+ n.a.
1061	Oils and Greases "	+.000012
10951	Phenols "	+.000003

* The pH factor for this industry is in the range of 4.5–10.7, with 7.0–8.0 most likely.

Source: Isard *et al.* [2; p. 94].

TABLE 3. Proposed Water Pollution Classification (WPC) Code

WPC Code	Water Related Item	Units
1000	Water Intake, Total*	1,000 Gal./$
1001	Sanitary Use	"
1002	Production Use	"
1003	Cooling	"
1004	Boiler Feed	"
1008	Irrigation	"
1009	Other, n.e.c.	"
1010	Water Discharge, Total	"
1011	Sanitary Use	"
1012	Production	"
1013	Cooling	"
1019	Other, n.e.c.	"
1020	Water Consumed	"
1031	Biochemical Oxygen Demand, BOD 5-Day	1,000 Lb./$
1032	Ultimate Oxygen Demand, UOD	"
1033	Chemical Oxygen Demand, COD	"
		"
1040	Solids, Total+	"
1041	Suspended Solids	"
1042	Settleable Solids	"
1047	Turbidity	"
1048	Color	"‡
1051	Alkalinity	"
1052	Acidity	"
1061	Oils and Greases	"
1062	Surfactants	"
1070	Pathogenic (Disease Causing) Organisms	—§
1080	Temperature	—§
1090	Other Pollutants	"
1095	Toxic Material#	"
1096	Radioactive Waste	—§
1099	Not Classified	"

* Cost of water intake is given by SIC code 4941—water supply.

+ 1040 = 1041 + 1042.

‡ In addition, color should be specified by kind by its wave length.

§ No one satisfactory measure was decided upon at this time, although thermal pollution may be specified most satisfactorily in terms of millions of BTU's per dollar output.

Phenols, which fall in this category, may be identified by a five-digit code such as WPC 10951.

Source: Isard *et al.* [2; p. 90].

next to last coefficient, —.089378, is wages and salaries per dollar output of the industry.

But, as is well known, the petroleum refining industry has inputs and outputs other than economic commodities. One set of these is concerned with water intake from, and pollutants discharged via water to the environment. To consider this set, we first need a workable classification of non-economic commodities. Such a classification with reference to water use and pollution has been developed and is presented in Table 3. For example, the fourth item in Table 3 is water intake for cooling (in 1000 gallons per dollar intake) while the last item but two is toxic materials. If we now list the set of noneconomic commodities of Table 3 along the left-hand tab in the lower half of Table 1, we are able to set down all the coefficients listed in the bottom half of Table 2 for petroleum refining. (For example we explicitly list in Table 1 water [intake] and alkalinity.) Thus we have taken a first step in linking the economic and ecologic systems empirically. Needless to say, petroleum refining involves other noneconomic commodities, for example air, and air pollutants; and other economic activities involve still other noneconomic commodities. Accordingly we need to extend the classification system of noneconomic commodities still further, and conduct further laborious work at obtaining the data for the development of coefficients.

Proceeding on, we may ask: from where do the inputs from the ecologic system to the economic system come? How are they produced? To begin to answer this question, let me shift away from the petroleum refining industry to one on which we have been conducting research recently, namely, the recreation industry. Let me also center my attention on the subject of recreational planning and development which has concerned us in the last year.

In the process of our research we have developed a rather fine classification system of recreational activities, and accordingly have developed the coefficients for a meaningful complex of these activities. One of these activities is sport fishing from boats with outboard motors. We therefore list this economic type activity at the head of one of the columns of left-hand part of Table 1 and set down its coefficients. One of these coefficients is of course fish; and in the context of the particular area of New England which we have studied, namely Plymouth Bay, a chief fish is cod. We have estimated that per 1,000 man days of sport fishing in outboards, 6,200 pounds of cod are required (in the sense of providing a satisfactory recreational outcome). Therefore we can set down that number in the cell corresponding to the cod row and sport fishing (outboard motor) column of Table 1.

Clearly, for cod to be an input in an activity of the economic system, it must be yielded as an output of a process of the ecologic system. We may designate this process the cod production activity. We have studied it, and we find that in the context of Plymouth Bay it requires, among others, the set of inputs listed in the last column of Table 4. That is the output of cod requires, as food inputs, herring, invertebrate carnivores and small fish. In turn, herring consumes (requires as inputs) plankton; and invertebrate carnivores and small fish consume herbivorous invertibrates which live on detritus. All these activities are part of a basic food chain in the ecologic system, and their input-output data are to be recorded in the lower right

TABLE 4. Food Chain Requirements for Cod Production

CLASSIFICATION NUMBER	COMMODITY	UNITS	MX 61–10 PLANKTON PRODUCTION	MX 60 ALL MARINE PLANT PRODUCTION	MX 6001 DETRITUS PRODUCTION	MX 718 PRODUCTION OF HERBIVOROUS INVERTEBRATES	MX 719 HERRING PRODUCTION	MX 722 SMALL FISH PRODUCTION	MX 721 CARNIVOROUS INVERTEBRATES (LARGE GASTROPODS) PROD.	MX 725 COD PRODUCTION
MX 61–10	PLANKTON	LBS.	+1				−10			
MX 60	ALL MARINE PLANTS	,,		+1	−1					
MX 6001	DETRITUS	,,			+1	−10				
MX 718	HERBIVOROUS INVERTEBRATES (MISCELLANEOUS)	,,				+1		−10	−10	
MX 719	HERRING	,,								-1.167
MX 722	SMALL FISH	,,					+1	+1		-1.167
MX 721	CARNIVOROUS INVERTEBRATES (LARGE GASTROPODS)	,,							+1	8.333
MX 725	COD	,,								+1

major block of Table 1 on ECOLOGIC SUBSYSTEM: INTERPROCESS COEFFICIENTS. Each of the production processes listed at the head of Table 4 is to be listed as a column in the right-hand part of Table 1 (which relates to ecologic processes) and every one of the commodities listed at the left-hand tab of Table 4 is listed among the ecologic commodities in the lower part of the left-hand tab of Table 1. A few of these are explicitly recorded in Table 1.

The food chain, however, represents only one basic set of linkages in the ecologic system. The proper study of the ecologic system—and its relation to the economic—must cover numerous other basic linkages. Biogeochemical cycles, such as the phosphorous cycle, and photosynthesis are obviously two sets of basic relationships which cannot be ignored. Consider, for example, photosynthesis. In Table 4, the −10 entry at the top of the herring production column indicates that 10 pounds

TABLE 5. Plankton and Related Production Activities

CLASSIFI-CATION	COMMODITY	UNITS	PMX 61–14 GROSS PHYTOPLANKTON PRODUCTION BY PHOTOSYNTHESIS	P1MX 61–15 UTILIZATION OF PHYTOPLANKTON MATTER BY RESPIRATION @ 2°	P2MX 61–15 UTILIZATION OF PHYTOPLANKTON MATTER BY RESPIRATION @ 7°	P3MX 61–15 UTILIZATION OF PHYTOPLANKTON MATTER BY RESPIRATION @ 12°	P4MX 61–15 UTILIZATION OF PHYTOPLANKTON MATTER BY RESPIRATION @ 17°C	PMX 61–11 PHYTOPLANKTON IN POUNDS (DUMMY UNIT CONVERSION PROCESS)	PMX 61–10 PLANKTON FORMATION	PMX 117 ZOOPLANKTON PRODUCTION
			PROCESS							
MX 11	SOLAR RADIATION	G. CAL PER CM² PER MIN.	−1.79							
MX 420	BELOW 0.55 MG. AT./M³ PHOSPHOROUS AS PHOSPHATE	MG. AT. PER M²	−0.06							
MX 42	ABOVE 0.55 MG. AT/M³	MG. AT. PER M²	−0.00							
MX 44	WATER TRANSPARENCY: $t = i/k$		−68.68						−0.86	
MX 61–11	PHYTOPLANKTON @ 3.98 g.c./M²	POUNDS						+1.00		−1.7
MX 61–12	PHYTOPLANKTON STOCK	GRAMS CARBON PER M²	±35.54							
MX 61–13	PHYTOPLANKTON	GRAMS CARBON PER M²		−.0201	−.0281	−.0400	−.0565			
MX 61–14	PHYTOPLANKTON-GROSS PRODUCTION BY PHOTOSYNTHESIS	GRAMS CARBON PER M² PER DAY	1.00					−277		
MX 61–15	LOSS OF PHYTOPLANKTON MATTER DUE TO RESPIRATION	GRAMS CARBON PER M² PER DAY		+1.00	+1.00	+1.00	+1.00			
MX 61–10	PLANKTON	POUNDS							+1.00	
MX 711	ZOOPLANKTON @ .68 g.c./M²	POUNDS							+0.14	+1.00

of plankton are required per pound of herring. Now, in the Plymouth Bay setting, one gram of plankton may be considered as a mixture consisting of 0.86 grams of phytoplankton, and 0.14 grams of zooplankton. The zooplankton largely grazes upon phytoplankton, and consumes phytoplankton as an input, as indicated in the last column of Table 5. The phytoplankton itself is produced via photosynthesis as indicated in the first column of Table 5. Note the inputs of light in the form of incident solar radiation at the ocean surface, measured in gram calories per square centimeter per minute, and the input of dissolved phosphate in sea water, etc. Lastly, in columns 2–5 of Table 5 we take into account the fact all the carbon fixed by phytoplankton photosynthesis is not available for consumption by zooplankton; some is used up by the phytoplankton's own respiration processes.

The above empirical materials illustrate the kinds of data which can be developed to fill in the cells in three of the major blocks of Table 1. We now turn to the question of data to fill in the cells of the fourth major block at the upper right of Table 1. Here, the seemingly difficult problem is resolved once the need for an appropriate classification of commodities is recognized.

Reconsider the commodity, cod. In Table 1 it appears as an ecologic (noneconomic) commodity in the lower half of the commodity classification system—a commodity which is yielded as an output (designated by a "+" sign) by an ecologic process, and used as an input (designated by a "—" sign) by an economic process. It could, however, be also classified directly as an economic commodity or resource, along side of land and labor; and the amount of cod which is provided by the natural environment, just as the amount of land and labor, represents the stock of a resource, which behaves as a constraint on the levels of economic activities which use these resources. Likewise the timber, or grass (for grazing purposes), or even the phosphate (basic to agricultural production) which natural production processes yield may be listed among the economic commodities in the upper half of the left-hand tab, and viewed as scarce economic resources whose use must not exceed current stocks. Given this viewpoint, then, the positive entries in the cells in the upper right block of Table 1 would represent outputs of ecologic processes which are provided as resources to the economic system; and negative entries would represent the use of these economic resources in natural production processes themselves, as the consumption of herring (an economic commodity) by cod, or the consumption of corn by wildlife, or the use of phosphorous in noneconomic processes in the phosphorous cycle.

However, this discussion should make clear the point that often a classification of commodities involves somewhat arbitrary decisions. Is land an economic commodity to be recorded in the upper half of the list of Table 1, or an ecologic commodity to be recorded in the lower half. Likewise with water, air, fish, livestock, etc. But while arbitrary elements do enter any classification system, for the purposes of our conceptual framework it makes no difference how we classify commodities, provided there is full accounting of all inputs and outputs of all commodities relevant for the problem being examined. Similarly, while a classification of processes as economic and ecologic may facilitate our comprehension of a problem, basically such classification is not essential for analysis, provided all processes relevant for a problem are identified and accurately described in an empirical manner.

4. REGIONAL CLASSIFICATION

There is one more general comment I wish to make. It is a well-known axiom in Regional Science that in any applied study the definition of relevant regions depends in large measure on the problems being attacked. Now, when we attack problems such as air pollution, water pollution, solid waste disposal, and sonic pollution (observe that these problems are of quite a different nature than those which have plagued regional analysts over the centuries), we clearly must introduce the physical environment properly into our conceptual framework. To do so requires in my mind a drastically different set of regions than those to which we have been accustomed. I envisage at least three new major types of regions, namely, LAND, AIR and MARINE. Each of these three major types of region is composed of sub-systems;

TABLE 6.

			LAND				AIR		MARINE	
			Economic		Ecologic		Economic	Ecologic	Economic	Ecologic
			Phila		Phila					
LAND	Economic	Phila								
	Ecologic	Phila								
AIR	Economic									
	Ecologic									
MARINE	Economic									
	Ecologic									

each has an economic sub-system, an ecologic sub-system, and other sub-systems. I have depicted in Table 6 these three major regions with their economic and ecologic systems. Note that this system of regional classification does not contain traditional regions such as the Philadelphia region or New England. However, such traditional regions can easily be incorporated. Under both the economic and ecologic sub-systems of the major region LAND, I have a sub-system region designated Philadelphia. A typical column could be the petroleum refining activity; and in that column would be listed all its input and outputs, some of which would be economic and noneconomic commodities that may be classified as land-type (such as crude oil, gasoline, and solid wastes); others would be commodities of marine-type (such as clean water from a river system, and water pollutants delivered to the marine region); and still others would be commodities of an air-type (such as unpolluted air itself, and particulates).

Similarly, I could list other subsystem regions (with their activities) relevant for the major environmental region LAND. I can also do so for the major environmental regions AIR and MARINE; but here the relevant subsystem regions may differ. The Philadelphia region need not be a meaningful subsystem region for the AIR and MARINE major regions.

In short, I have reached a point in my thinking where we must consider rather revolutionary changes in our conception of regions appropriate for planning purposes. In this matter the nature of our problems are compelling us to traverse quite unexpected ground in our effort to attack them effectively.

5. CONCLUSION

I now wish to conclude this set of notes with a few additional comments.

a) Our conception of regional planning has become greatly expanded. No longer can it be concerned with the economy and its interrelated sectors, nor with society and its interrelated behaving units. It must proceed further and encompass the interrelationships of economic and other social behaving units with the natural processes which are the essence of the ecosystem.

b) While the input-output and linear programming techniques are useful in depicting the interrelations between the economic and the ecologic systems, these techniques are only two of several that may prove to be efficient in depicting these interrelationships and in attacking problems which stem from such interrelationships. It is easily envisaged how nonlinear programming techniques, activity complex techniques, and comparative cost analysis may also be extended for use in ecologic analysis and in synthesized ecologic-economic analysis.

c) It is only by the development of such techniques that we can hope to reach viable or workable solutions to the pressing environmental problems which loom on the horizon.

REFERENCES

[1] Isard, W. *et al. Methods of Regional Analysis: an Introduction to Regional Science.* New York: The Technology Press of M.I.T. and John Wiley and Sons, 1960.

[2] ———, K. Bassett, C. Choguill, J. Furtado, R. Izumita, J. Kissin, E. Romanoff, R. Seyfarth and R. Tatlock. "On the Linkage of Socio-Economic and Ecologic Systems," *Papers, Regional Science Association*, Vol. 21 (1967), pp. 79–99.

[3] ———, C. Choguill, J. Kissin, R. Seyfarth, R. Tatlock, K. Bassett, J. Furtado and R. Izumita. *Ecologic and Economic Analysis for Regional Development*. Forthcoming, 1969.

[4] Odum, E. P. *Fundamentals of Ecology*. Philadelphia: Saunders Co., 1959.

[4]

The Role of the Oceans and Biosphere in the Carbon Dioxide Cycle

By Lester Machta

Part I. Prediction of Future Atmospheric Concentrations

Introduction

Carbon dioxide is known to play a role in the radiation budget of the atmosphere. The future growth and fluctuations of this gas, due either to natural or man-made causes, must be predicted and understood before forecasts of climate modification can be undertaken.

The three reservoirs through which carbon exchanges are: the atmosphere, the biosphere and the oceans. The prediction of future atmospheric carbon dioxide depends on: 1. The input of carbon dioxide from the combustion of fossil fuels; 2. The exchange between reservoirs; 3. The exchanges within parts of each reservoir; and 4. The stability of each reservoir. The present analysis considers a very simple set of reservoirs; assigns exchange rates between and within all reservoirs except between air and ocean; and employs certain estimates of the growth rate in the consumption of fossil fuels. It also makes the usual assumption, which can be challenged, that the characteristics of neither the oceans nor the biosphere alter with time. First, we shall examine the basis for calculating the exchange rates between air and sea surface; the one exchange left unspecified in the model.

Man Made $C^{14}O_2$

Nuclear weapon tests increased $C^{14}O_2$ over that present due to cosmic ray production. The U.S. Atomic Energy Commission and other U.S. organizations have monitored the troposphere and stratosphere for $C^{14}O_2$; the monitoring data have been analyzed and published (Telegadas, 1971). A sample analysis appears as Fig. 1, a vertical cross section from pole to pole. The heavy solid and dashed lines are schematic (or average) tropopause locations. The analysis consists of a series of isolines of concentration (thin solid and dashed lines) based on measurements at the position of the dots and crosses. The difference between dots and crosses simply reflects the organizations performing the $C^{14}O_2$ analyses in laboratories; only the dots have concentrations listed to their right in units of $10^5 C^{14}$ atoms/gram air in excess of the natural background

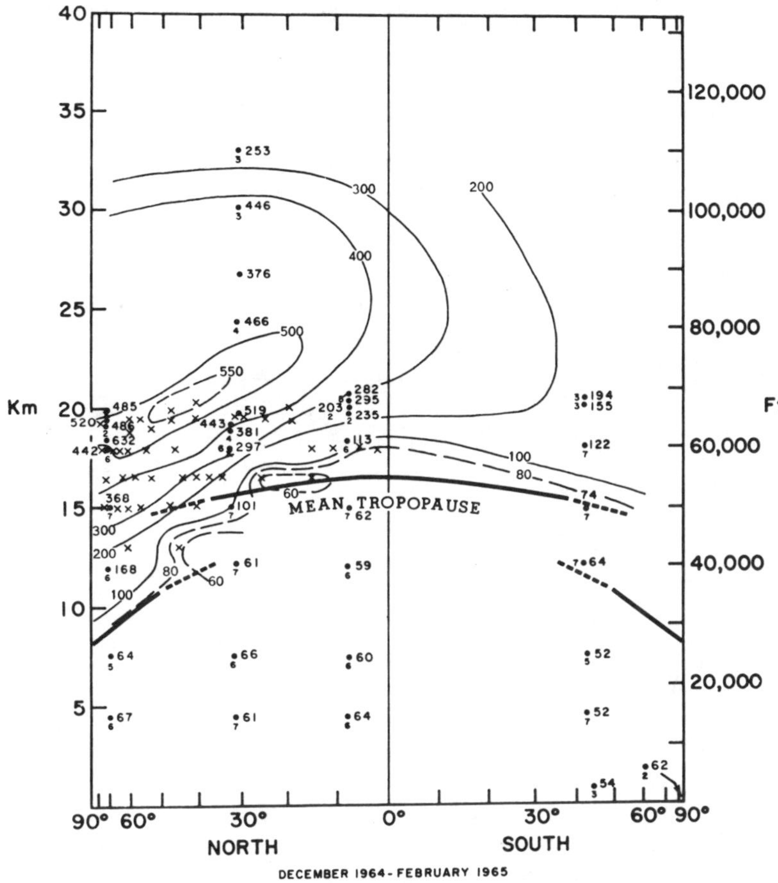

Fig. 1. A cross section from pole to pole showing concentrations of man made $C^{14}O_2$ (Telegadas, 1971). Heavy solid and dashed lines are schematic or average tropopause positions. Dots and crosses indicate locations near which samples have been collected. The difference between crosses and dots reflects the laboratory performing the radiocarbon analysis. Concentrations are expressed in 10^5 C atoms/gram air. The samples at altitudes above about 20 km were collected aboard balloons. All samples collected in the time interval December 1964 to February 1965 have been averaged. The thin lines (solid and dashed) are subjectively interpolated and extrapolated isolines of C^{14} concentration from the observed data.

from cosmic rays. The smaller numbers give the number of samples during the period represented by the figure (December 1964 through February 1965).

Fig. 1 contains slightly more information than is typical for the period 1955–1969 but is representative of the more important 1963–1969 interval. From a series of charts such as illustrated by Fig. 1, a time history of the tropospheric and entire atmospheric $C^{14}O_2$ is given in Fig. 2. The horizontal bars are the

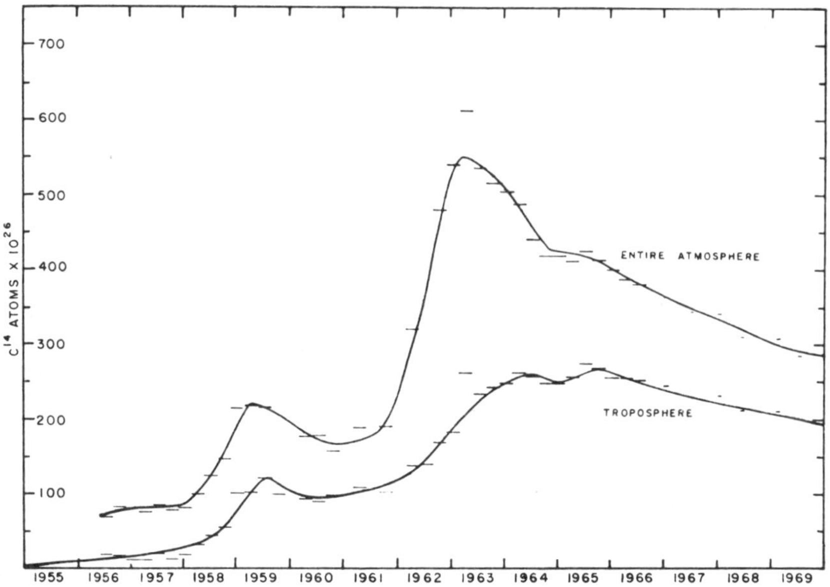

Fig. 2. The time history of C^{14} atoms produced by nuclear tests in the troposphere and in the whole atmosphere (up to about 30 km) as derived from a graphical intergration of the analysis in Fig. 1 assuming no change in the height of the troposphere and a non-varying concentration of atmospheric $C^{12}O_2$ (Telegadas, 1971).

actual numbers for given time intervals; the heavy lines, subjective best fits to the data.

Two periods of increasing C^{14} atoms can be recognized. The first ends in 1959 and the second in 1961 and 1962 although the troposphere continues to reflect increasing C^{14} atoms for several years after 1962. The decrease of the C^{14} in the entire atmosphere after say, 1962, implies its transfer to other reservoirs. Thus, it will be possible to select appropriate exchange rates to best fit the decrease in atmospheric $C^{14}O_2$.

Table 1 presents in the first four columns, the mean annual C^{14} content of the troposphere and entire atmosphere as taken from the smoothed curve in Fig. 2. The column labelled "corrected" reduces the atmospheric C^{14} content for the few large nuclear tests which took place in 1967–69. Note that only those periods have been chosen in the analysis which have no or only small injections of man-made C^{14}.

The lack of regularity in the observed annual changes in atmospheric C^{14} reflects the uncertainty and errors in the inventories of C^{14} due mainly to imperfect sampling by aircraft and balloons. No effort was made to reanalyze the cross sections such as shown in Fig. 1 to improve the smoothness in the

Table 1. *Man-Made $C^{14}O_2$*

Expressed in units of 10^{26} atoms of Carbon-14. (Telegadas, 1971)

		Selected years			
		Entire atmosphere		Observed yearly decrease	Predicted yearly decrease
Year	Troposphere	Observed	Corrected		
1955	5				
1956	15				
1957	24				
1958	50				
1959	110	220	220	50	29
1960	95	170	170		
1961	112				
1962	155				
1963	230	533	533	63	56
1964	255	470	470	55	58
1965	260	415	415	30	49
1966	250	385	385	39	41
1967	238	352	346	42	32
1968	220	318	304	30	24
1969	201	294	274		

trend of year to year changes. Rather, as will be seen later, a best fit minimizes the discrepancies between observed and predicted values over the entire period of selected years. The final column represents the predicted best fit which will be discussed later.

A Simple Model

Fig. 3 shows the elements of the simple model used below. It represents the ideas of many scientists (Dr George Woodwell, Brookhaven National Laboratory; Dr Jerome Olson, Oak Ridge National Laboratory, the present writer, Jonathan Machta, and most particularly, Dr C. David Keeling, Scripps Institution of Oceanography).

Two physical processes are employed. First order kinetics are used between troposphere and stratosphere, between troposphere and mixed layer of the ocean, and between mixed and deep layers of the oceans. This states that a fixed fraction, λ, of the content of one reservoir transfers to a neighboring reservoir in a unit of time. We assume that in the equilibrium state the mass of carbon dioxide (expressed in grams of carbon or atoms of C^{14}) times the exchange rate, in one reservoir equals the product of the same properties in the adjacent reservoir for which the exchange coefficient, λ, is applicable. The second process is the biospheric exchange which employs the concept of the net primary production, NPP. A mass of carbon equal to the NPP is transferred from troposphere to biosphere (or mixed layer to marine biosphere) in a year. This mass of carbon is tied up in the biosphere for a specified number of years,

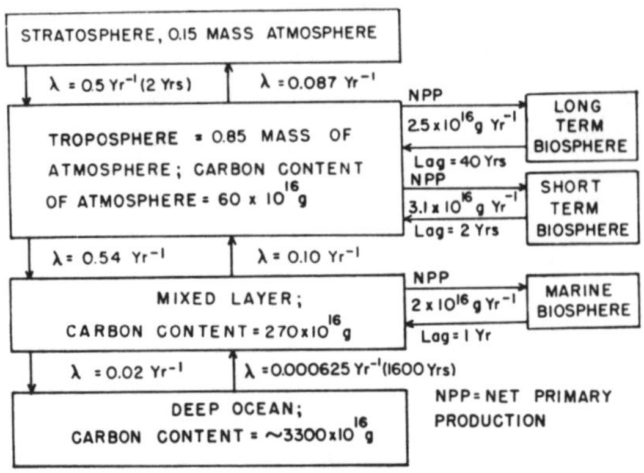

Fig. 3. The model of the three reservoirs of CO_2. The quantity, λ, denotes the fraction of the CO_2 in one reservoir being transferred to an adjacent reservoir according to the sense of the arrow. The transfer between troposphere and mixed layer is calculated from the history of the C^{14} data; all others are pre-assigned. The carbon content of the mixed layer and the exchange constant from mixed layer to deep ocean is derived from other assumed and computed parameters.

the lag, after which it returns completely to the reservoir from which it was initially taken, that is the troposphere or marine layer.

All but two of the exchange rates are assumed in advance. The exchange coefficients between troposphere and stratosphere are based on a stratospheric residence time of 2 years. The exchange coefficient from the deep ocean to the mixed layer derives from an age of 1 600 years of the deep ocean relative to the mixed layer. The exchange coefficient from the mixed layer to the deep ocean uses the assumption of equilibrium noted above. The carbon mass of the mixed layer is obtained when the two exchange rates between air and oceans are calculated from the man-made $C^{14}O_2$ data assuming that the oceans (both mixed and deep layers) contain about 60 times as much carbon as the atmosphere.

The net primary production and the lags represent the judgement of ecologists Woodwell and Olson. Their estimate of the total NPP is slightly larger than given by Leith (1970). It is recognized that the decay of forest organic material occurs over a range of years rather than the single value in Fig. 3 but the choice of these numbers will not be important for predicting the carbon dioxide concentrations in the year 2000. The NPP in the oceans follows Leith. The lag should be even smaller than one year, on the average, but this overestimate is also not significant to the present purposes. The mathematical formulation and the constants required for the calculations appear in the appendix. Two items deserve special attention, however, since they may not be obvious to the non-specialist.

The first item is the increased photosynthesis for certain parts of the land biosphere in the presence of higher carbon dioxide concentrations. Ecologists Woodwell and Olson estimate that very roughly half of the land biosphere is carbon dioxide limited. Further, they suggest (SCEP, 1970) that for each 10% increase in atmospheric carbon dioxide, the photosynthetic uptake of the carbon dioxide limited land biosphere will increase by about 5%. These biospheric estimates are provisional and very uncertain at this time.

The second factor is the dependence of the partial pressure of carbon dioxide on other carbon molecules in the sea. Thus the fractional change in the carbon dioxide partial pressure is ten times greater than the fractional change in the inorganic carbon content of the mixed layer. This buffering effect has the following consequences: Assume for the sake of explanation that the mixed layer has a carbon content equal to that of the atmosphere and that the mixed layer does not exchange with the deep ocean. Then if 11 units of carbon dioxide are added to the atmosphere, the equilibrium partition between air and mixed layer will not be 5.5 in air and 5.5 in ocean but rather 10 in air and only 1 in oceans. This 10 to 1 ratio may, according to Keeling, be as low as 6 to 1 or as high as 14 to 1.

Determining the Exchange Rates Between Troposphere and Mixed Layer

The model given in Fig. 3 and stated mathematically in the appendix permits prediction of the man-made $C^{14}O_2$ once the exchange between troposphere and mixed layer is specified. A trial and error procedure uses an assumed exchange rate and predicts the $C^{14}O_2$. Each year (1959–1960 and 1963–1969) the predicted change is subtracted from the observed change and the resultant difference is squared. The difference from 1959–60 is given a weight of one half because the $C^{14}O_2$ information in that period is inferior to that in the 1963–69 interval. All the squared differences are summed. The minimum value of the summation is selected.

The upper portion of Table 2 shows the results. The minimum squared error is achieved when 0.54/yr. transfers from troposphere to mixed layer and 0.10/yr in the reverse direction. Values near the best fit exchange rates give rise to errors almost as small as the minimum but, far removed, the squared errors are clearly larger.

The last column of Table 1 lists the year to year changes in man-made $C^{14}O_2$ given by the best fit calculations. Some of the discrepancies, such as the 1959–60 change are large but may not be entirely due to the errors produced by the model; that is, the observed changes also have an uncertainty.

The lower part of Table 2 describes some of the implications of the table. The exchange rate is the reciprocal of the mean residence time, so that 0.54/yr.

Table 2. *Squared error in predicted annual change of man-made* C^{14}, 10^{26} *atoms 1959–60 and 1963–69*

$\lambda_{OCEAN \to TROP}$, per year	$\lambda_{TROP \to OCEAN}$, per year				
	0.40	0.52	0.54	0.56	0.64
0.02	1 100	2 776	3 252	3 784	6 476
0.08	1 147	807	856	935	1 557
0.10	1 439	803	784	790	1 077
0.12	1 777	949	883	838	868
0.20	3 174	2 115	1 977	1 850	1 451

Minimum error occurs with:

$\lambda_{TROP \to OCEAN} = 0.54$ yr.$^{-1}$, mean residence time = 1.85 yr. for trophosphere or 2.15 yr for whole atmosphere.

$\lambda_{OCEAN \to TROP} = 0.10$ yr.$^{-1}$, mean residence time = 10 yr. for mixed layer.

$$\text{Mass of carbon in mixed layer} = \frac{\lambda_{TROP \to OCEAN}}{\lambda_{OCEAN \to TROP}} \times \text{Mass of carbon in troposphere } (50 \times 10^{16} \text{ g}) = 270 \times 10^{16} \text{ g.}$$

$$\text{Depth of mixed layer} = \frac{\lambda_{TROP \to OCEAN}}{\lambda_{OCEAN \to TROP}} \times \text{Depth of mixed layer with equivalent carbon mass of troposphere (50 m)} = 270 \text{ m}$$

is the same as a mean tropospheric residence time of 1.85 yrs. Since the literature usually deals with the whole atmosphere, one can convert 1.85 yrs to an equivalent mean residence time for the whole atmosphere of 2.15 yrs; since the troposphere contains 85 % of the mass of the whole atmosphere. This residence time is smaller than found by previous authors who usually quote values closer to 5 years. Because of the assumption that the same number of carbon atoms cross the air-sea interface in equilibrium conditions, the mass of carbon and the depth of the mixed layer can be estimated; the latter turns out to be 270 meters since 50 meters of the ocean has the same carbon mass as the troposphere.

Prediction of $C^{12}O_2$

The model now permits prediction of the changes in $C^{12}O_2$ due to the injection of fossil fuel $C^{12}O_2$ into the atmosphere.

Table 3 gives the latter information. Please note that the numbers in Table 3 should be 10 % greater than shown. The implications of this correction will be discussed further. However, the values in Table 3 have been used in the calculations shown in the subsequent tables and figures. The conversion of fossil fuel carbon into $C^{12}O_2$ is due to Keeling (1971*a*) who has modified a similar analysis (U. S. Government, 1965).

Table 3. *Annual release of fossil fuel Carbon Dioxide to the atmosphere*

Year	10^{16} gC	Year	10^{16} gC	Year	10^{16} gC	Year	10^{16} gC
1860	0.008	1900	0.048	1040	0.102	1980	0.512
61	0.008	01	0.050	41	0.106	81	0.530
62	0.008	02	0.052	42	0.110	82	0.548
63	0.008	03	0.054	43	0.114	83	0.566
64	0.010	04	0.056	44	0.120	84	0.586
65	0.010	05	0.056	45	0.124	85	0.608
66	0.010	06	0.058	46	0.128	86	0.628
67	0.012	07	0.060	47	0.134	87	0.650
68	0.012	08	0.062	48	0.140	88	0.674
69	0.012	09	0.064	49	0.144	89	0.696
70	0.012	10	0.072	50	0.148	90	0.722
71	0.012	11	0.072	51	0.158	91	0.746
72	0.014	12	0.074	52	0.160	92	0.772
73	0.014	13	0.076	53	0.162	93	0.800
74	0.016	14	0.076	54	0.168	94	0.828
75	0.016	15	0.078	55	0.182	95	0.856
76	0.016	16	0.078	56	0.196	96	0.886
77	0.018	17	0.080	57	0.202	97	0.918
78	0.018	18	0.080	58	0.214	98	0.950
79	0.020	19	0.082	59	0.224	99	0.982
80	0.020	20	0.086	60	0.238		
81	0.020	21	0.088	61	0.232		
82	0.022	22	0.088	62	0.244		
83	0.024	23	0.088	63	0.260		
84	0.024	24	0.088	64	0.278		
85	0.024	25	0.090	65	0.286		
86	0.026	26	0.090	66	0.300		
87	0.026	27	0.090	67	0.304		
88	0.028	28	0.092	68	0.320		
89	0.028	29	0.092	69	0.332		
90	0.028	30	0.084	70	0.346		
91	0.030	31	0.086	71	0.360		
92	0.030	32	0.088	72	0.376		
93	0.032	33	0.092	73	0.388		
94	0.034	34	0.094	74	0.404		
95	0.036	35	0.096	75	0.420		
96	0.038	36	0.098	76	0.436		
97	0.040	37	0.100	77	0.454		
98	0.040	38	0.104	78	0.473		
99	0.044	39	0.106	79	0.492		

(Fossil Fuel Production from U.N. Statistical Papers)

Conversion to atmospheric CO_2 as follows:

1860–1949, U.S. Gov't × 0.8
1950–1968, Keeling, 1971*a*
1969–1979, Annual growth rate of 4 %
1980–1999, Annual growth rate of $3^1/_2$ %

All annual releases of Carbon Dioxide listed in this table should be increased by 10 % because of computational error. In 1970, for example, the reported value of 0.346×10^{16} gC should be 0.381×10^{16} gC.

The changes in $C^{12}O_2$ predicted by the model should in principle be added to the natural background concentration prior to the industrial era, arbitrarily chosen to begin in 1860. However, the value of $C^{12}O_2$ concentration in the last century is known very imperfectly so that a prediction of future CO_2 concentra-

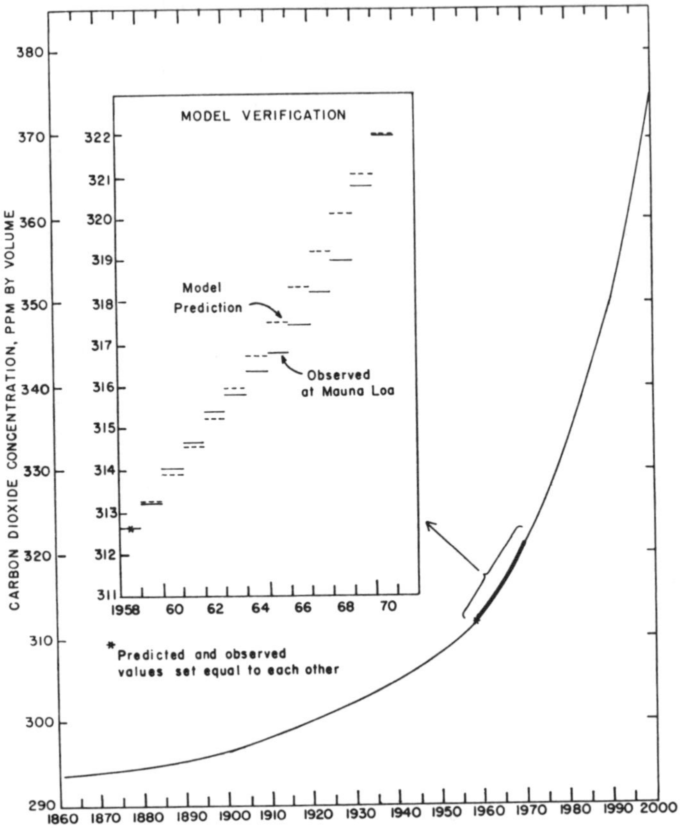

Fig. 4. The predicted time history of global atmospheric $C^{12}O_2$. Because of the 10% error in the amount of fossil fuel consumed correct prediction in the year 2 000 should be raised to 380 ppm and the value in 1860 reduced to about 290 ppm. Intermediate values should be corrected by a proportionate amount. The observed $C^{12}O_2$ concentrations at Mauna Loa are due to Pales and Keeling (1965) and Keeling (1971).

tions might better use a modern value from which to add or subtract the fossil fuel $C^{12}O_2$. The year 1958 has been chosen and the Mauna Loa value of about 313 ppm is taken as the world value (Pales and Keeling, 1965).

The results appear as the large graph in Fig. 4, the bold portion between 1958 and 1971 being amplified in the inset. Due to the 10% underestimate of the fossil fuel CO_2, the predicted value in the year 2000 should be raised by about 5 ppm to 380 rather than the plotted value of 375 ppm and the 1860 value decreased to about 290 ppm. The inset shows that over the 13 year period the average fit between the model prediction and the Mauna Loa record is almost perfect. But the year to year changes in the predicted values are, as might be expected, more regular than for the Mauna Loa changes. This irregularity in the Mauna Loa record will acquire significance later when the uncertainty in the prediction for the year 2000 is assessed.

The prediction for the year 2000 agrees with that made in the SCEP report even though the latter did not have the benefit of the $C^{14}O_2$ data. The value in 1860 of 290 ppm is also reasonable.

Further Consequences

Both the man-made $C^{14}O_2$ and the man-made fossil fuel $C^{12}O_2$ were added initially to the atmosphere but with time transferred to other reservoirs. Table 4 shows the partition of these gases among the three reservoirs in 1970 according to the model calculation. More than half of the $C^{12}O_2$ remains airborne; the majority of the rest going into the oceans and only 15 % tied up in the biosphere. In contrast, 60 % of the man-made $C^{14}O_2$ is found in the oceans and less than a third remains airborne; this despite the fact that the $C^{14}O_2$ has had less than 20 years to transfer out of the atmosphere while the $C^{12}O_2$ has been added to the atmosphere since 1860. The reason for the different behavior lies in the buffering action of the ocean for $C^{12}O_2$ while $C^{14}O_2$, being present in trace quantities, exerts no buffering effect. It is also possible to predict future $C^{14}O_2$ (assuming no further injections) as shown in Fig. 5. The scale is given as the C^{14}/C^{12} ratio taking into account the fact that $C^{12}O_2$ will also change with time as in Fig. 4. It is usual to measure the C^{14}/C^{12} ratio rather than the $C^{14}O_2$ concentration. The curve labeled "natural plus man-made" shows what may be found in say, future tree rings corrected for fractionation. The man-made C^{14}/C^{12} ratio appear as crosses and long dashes as a matter of interest and should be read on the right-hand scale.

Table 4. *Partition of man-made C^{14} and fossil fuel $C^{12}O_2$ in 1970 according to model calculations*

	$C^{14} \times 10^{26}$ atoms	%	$C^{12} \times 10^{16}$ g	%
Atmosphere	172	29	5.6	55
Stratosphere	82	14	4.8	47
Troposphere	90	15	0.8	8
Biosphere	66	11	1.5	15
Long term land	49	8	1.3	13
Short term land	13	2	0.2	2
Marine	4	1	0	0
Oceans	359	60	3.0	30
Mixed layer	345	58	2.5	25
Deep layer	14	2	0.5	5
Total	597	100	10.1	100

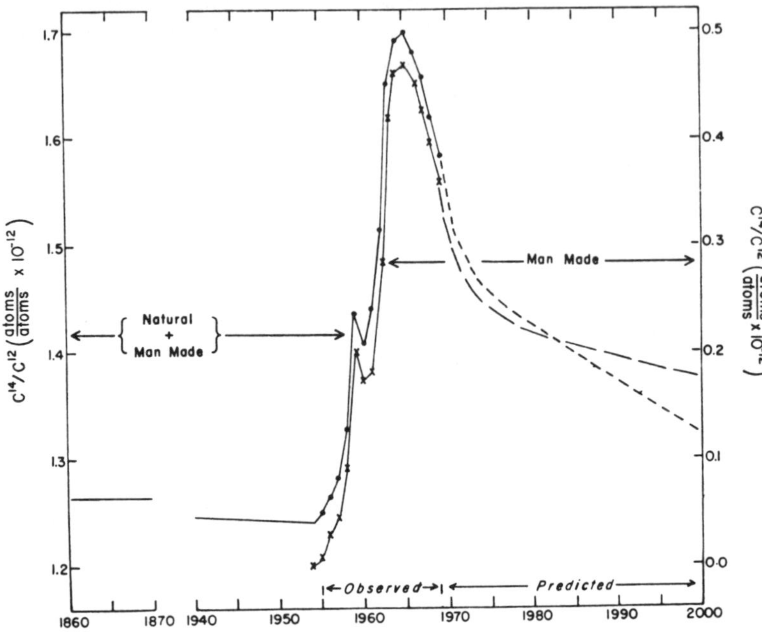

Fig. 5. Time history of the C^{14}/C^{12} ratio in tropospheric air according to the model calculations. The curve entitled "Natural + Man made", read on the left hand scale, should describe the actual measured ratios while the curve entitled "Man made" only reflects the ratio of man-made C^{14} atoms to man-made C^{12} atoms in the troposphere. The decrease from 1860 to 1950 is usually called the "Suess Effect". The ordinate scale shows that radiocarbon constitutes only a very small fraction ($\sim 10^{12}$) of the total carbon atoms in the atmosphere.

The portion of the graph between 1860 and 1954 reflects the dilution of cosmogenic $C^{14}O_2$ by man-made $C^{12}O_2$, the Suess effect. The curve beyond 1970 indicates that the consequences of adding $C^{14}O_2$ to the air by nuclear testing will not be compensated by increasing $C^{12}O_2$ up to the year 2000.

Deletion of the Biosphere

One can readily remove the biosphere from the model by setting the NPP equa to zero and ask for the best fit to a new set of air-ocean exchange parameters. The results of this exercise appear in Table 5 in which some of the features of the original model are compared with a two reservoir (air and ocean) system. Since the biosphere is unavailable, the transfer of carbon dioxide to the ocean must be faster (0.64/yr. vs. original 0.54/yr.) although the reverse coefficient (0.10/yr.) does not change. The squared error, the measure of how well the observed changes in $C^{14}O_2$ are reconstructed, becomes slightly but not significantly

Table 5. *Comparison of models with and without a biosphere*

	Biosphere + Oceans	Oceans only
$\lambda_{TROP \to OCEAN}$	0.54 y.$^{-1}$	0.64 y.$^{-1}$
Mean air residence time	2.15 y.	1.80 y.
$\lambda_{OCEAN \to TROP}$	0.10 y.$^{-1}$	0.10 y.$^{-1}$
Mean ocean residence time	10 y.	10 y.
Squared error	784	824
Depth of mixed layer	270 m	320 m
Man-made C^{14} atoms in mixed layer in 1969	342×10^{26}	410×10^{26}
Fixing 1958 CO_2 concentration as 313 ppm in atmosphere		
Concentration in 1860	295	293
Concentration in 1970	322	322
Concentration in 2000	375	381
$\left[\left(\frac{C^{14}}{C^{12}}\right)_{1950} \Big/ \left(\frac{C^{14}}{C^{11}}\right)_{1860} - 1\right] \times 100$	−1.8%	−1.1%

larger when the biosphere is deleted. The mixed layer of the ocean must also increase slightly to 320 meters depth. There are small changes in the predictions for the year 2000 which will be discussed below. Finally, the Suess effect drops from −1.8 % to −1.1 %. On the whole, however, there is nothing which derives from the model without the biosphere which looks more unreasonable (except that we know there is a biosphere) than the original formulation. Perhaps if the depth of the mixed layer is too large and the atmospheric mean residence time too short then these parameters become even less acceptable for a model without a biosphere.

Sensitivity Tests

Just as one may run the case with no biosphere, so have we run other cases changing model parameters. These appear in Tables 6*a* and 6*b*. Table 6*a* treats those items in which the ocean is affected; a repeat of the no biosphere case, a decrease in the age of the deep ocean and changing the buffering factor both up and down. Each variation provides detectable changes in the reported parameters but, focusing on the prediction in the year 2000, no number is more than 10 ppm below or above the original model prediction.

The first four columns in Table 6*b* (with the second column of Table 6*a*) show the role played by biospheric parameters including reducing the biosphere to zero and doubling the NPP. Only the change in growth limitation reduced to 0.05 rather than 0.25 produces a change more than 10 ppm in the year 2000 and that by only 11 ppm.

Table 6*a*. *Variation of model parameters*

Property	Model	No Biosphere	Deep ocean 400 y. not 1 600 y. old	Buffering factor 6 not 10	Buffering factor 14 not 10
$\lambda_{TROP \rightarrow OCEAN}$, y.$^{-1}$	0.54	0.64	0.56	0.54	0.54
$\lambda_{OCEAN \rightarrow TROP}$, y.$^{-1}$	0.10	0.10	0.12	0.10	0.10
Squared error	784	824	770	780	788
Average fraction fossil fuel CO_2 in troposphere	50 %	54 %	44 %	40 %	52 %
Assuming 313 ppm in 1958, the predicted concentration in					
1860	295	293	297	298	293
1970	322	322	321	321	323
2000	375	381	370	365	380

Thus it is concluded that changing parameters in the original model by even a very large amount still produces errors in the prediction for the year 2000 of about 10 ppm or less.

The final column adds a different dimension to the prediction; namely it questions the initial $C^{14}O_2$ data upon which the calculations are based. It is argued that the years soon after nuclear tests may have more irregular gradients of atmospheric $C^{14}O_2$ so that the inventories are less reliable. These years have

Table 6*b*. *Variation of model parameters*

Property	Model	Doubled net primary production	Growth limitation 0.05 not 0.25	Long term biosphere lag 20 y. not 40 y.	Omitting 1959–1960 1963–1964
$\lambda_{TROP \rightarrow OCEAN}$, y.$^{-1}$	0.54	0.40	0.50	0.54	0.32
$\lambda_{OCEAN \rightarrow TROP}$, y.$^{-1}$	0.01	0.08	0.08	0.10	<0.02
Squared error	784	749	788	786	Not comparable
Average fraction fossil fuel CO_2 in troposphere	50 %	43 %	52 %	50 %	34 %
Assuming 313 ppm in 1958, the predicted concentration in					
1860	295	297	293	294	300
1970	322	321	323	323	319
2000	375	369	386	377	357

been removed from the analysis in the least squares fit. The resulting transfer to the oceans is slower giving rise to what may be a more reasonable mean residence time of a CO_2 molecule in the atmosphere (over 3 years) but allowing only 34% of the fossil fuel $C^{12}O_2$ to remain airborne in contrast to the model's 55%. During the 1958–1968 period about 50% of the fossil fuel remained airborne if one accepts the Mauna Loa record as representative of the whole atmosphere. The observed airborne fraction thus agrees better with the original model using all the changes in $C^{14}O_2$ rather than the selected years in the last column of Table 6*b*.

Conclusion

A simple model of the exchange of CO_2 between air, ocean, and biosphere has, in part, been calibrated by new $C^{14}O_2$ data. The results using the 1958 value Mauna Loa as the base, predict concentrations of 380 ppm in the year 2000. Uncertainties in the parameters create errors generally below 10 ppm. But unevaluated uncertainties in future combustion of fossil fuels may produce changes larger than 10 ppm. Larger potential errors may follow from the imprecision in the world-wide inventories of the man-made $C^{14}O_2$.

Perhaps more serious, from a fundamental viewpoint, than the uncertainties and errors just noted, is the lack of full understanding of the transfer between air and ocean and the integrity of the biosphere. These comments are prompted by the unexplained changes in the growth rate of atmospheric $C^{12}O_2$ seen in Fig. 4. Relatively small changes in the sea surface temperature or in the net primary production (due to variable rainfall or sunshine in say, equatorial forests) may alter the future predictions. Thus, it is not known whether the 1960's, the period of the present analysis, is necessarily representative of longer time periods or of the future 30 years.

Part II. Seasonal Variations in Atmospheric Carbon Dioxide

Introduction

The most conspicuous feature of atmospheric carbon dioxide observations at clean-air locations, particularly in the northern hemisphere, is a seasonal variation. A pronounced decrease occurs during the summer five months with a somewhat slower recovery during the remaining seven months. There is little disagreement that the explanation for the decrease lies in photosynthetic uptake during the summer season. Nevertheless, the sea surface temperatures warm during the summer, and this should raise the air carbon dioxide in opposition to the sense of changes caused by photosynthesis.

Junge and Czeplak (1968) have made the first attempt to relate the seasonal variations of biospheric uptake (photosynthesis) and release to the air (decay and respiration) to atmospheric changes in carbon dioxide via a simple north-south mixing process for the northern hemisphere. Earlier Bolin and Keeling (1963) had derived source and sink information, in part, from the seasonal variation in observed atmospheric carbon dioxide.

In the present analysis, a more sophisticated diffusion model and more recent estimates by Leith of the seasonal variation in biospheric use of carbon dioxide than was available to previous investigators will aim to see if the land biosphere alone can account for the observed seasonal variation. The limited and imperfect Antarctic monitoring will be compared with that expected from the land biosphere because the ratio of land to oceans in the southern hemisphere is much smaller than in the northern hemisphere. If the oceans play a role in controlling the seasonal variation of atmospheric carbon dioxide, it should be most apparent in high southern latitudes such as in Antarctica.

Leith has provided an estimate of the seasonal variation in uptake and release of carbon dioxide by the land biosphere at about 50° N based on conditions over Europe. He also has mapped the world-wide distribution of net primary production. These two sets of data permit quantitative calculation of the seasonal source and sink functions near latitude 50° N. The growing season north of 50° N becomes shorter and, in winter with snow cover, even the decay largely ceases. Further, both the net primary production and geographical area decrease toward the pole. On the other hand, these latter properties increase toward low latitudes but the more uniform temperature and solar radiation throughout the year will result in more constant photosynthesis, decay and respiration throughout the year. While Junge and Czeplak used estimates of rainfall to judge the differences in summer over winter uptake and decay in low latitudes, this analysis has employed the amount of solar radiation. In practice, there is general agreement between the two approaches.

Table 7 lists the monthly source and sink strength in 20° bands for the land biosphere from 90° N to 50° S. Note that in the tropical rain forests between 10° N and 10° S, it is assumed that the decay and respiration approximately balance photosynthetic uptake of atmospheric carbon dioxide each month so that there is no entry in the table. While the values between 30° and 70° derive directly from Leith's data, all other numbers represent the author's judgment based on the principles stated in the previous paragraph and on Leith's map of net primary production.

Meteorological Diffusion Model

The meteorological diffusion model employs a flux-gradient process of transfer with the constant of proportionality called the eddy diffusion coefficient. Mixing

Table 7. *Carbon Dioxide release (+) to or uptake (−) from the atmosphere, 10^{14} gC month^{-1}*

	Biospheric sources and sinks					
	70°–90° N	50°–70° N	30°–50° N	10°–30° N	10°–30° S	30°–50° S₂
Jan.	0.0	1.00	1.56	1.12	−0.98	−1.64
Feb.	0.0	1.00	1.56	0.75	−0.61	−1.40
Mar.	0.08	1.20	1.84	0.0	0.0	0.60
Apr.	0.12	1.28	2.00	−0.75	0.61	0.80
May	0.12	0.44	0.68	−1.12	0.98	0.60
Jun.	−0.24	−3.16	−4.92	−1.49	1.22	0.40
Jul.	−0.20	−4.80	−7.48	−1.19	0.92	0.36
Aug.	−0.16	−4.16	−6.44	−0.75	0.61	0.36
Sep.	0.20	1.84	2.88	0.0	0.0	0.40
Oct.	0.08	2.40	3.72	0.75	−0.61	0.44
Nov.	0.0	1.76	2.76	1.19	−0.92	0.16
Dec.	0.0	1.20	1.84	1.49	−1.22	−1.08

is permitted in north-south and vertical directions. This is common practice in global transport problems since the west-east advection usually creates zonal homogeneity much sooner than mixing produces meridional or vertical uniformity.

The grid spacing in the north-south direction is 20° latitude, with the grid values at 80° (70–90°), 60° (50–70°) and so forth. In the vertical direction, the grid spacing is 2 kilometers centered at 0 km (0–1 km), 2 km (1–3 km), 4 km (3–5 km), ... 36 km (35–37 km), 38 km (37–39 km) and 40 km (39–40). The tropospheric horizontal eddy diffusion coefficients are about $1\text{–}5 \times 10^{10}$ cm^2/s, with slightly lower values at the 0–1 km interface. Winter values exceed those in summer, southern hemisphere values exceed those in the northern hemisphere and values near the equator are smaller than to the north or south. The tropospheric vertical eddy diffusion coefficients are in the range $4\text{–}8 \times 10^4$ cm^2/s. Values are smaller near the earth's surface than in the free air and, over the

Table 8. *Change in partial pressure of CO_2 at sea surface from end of winter to end of summer*

	% change due to			
	Biospheric uptake			
Depth of mixed layer	270 m	60 m	Warming of sea surface	Temperature change, °C
60° N	−6.5	−30	+17	+5
40° N	−2.2	−10	+20	+6
20° N	−0.4	−1.5	+ 7	+2
20° S	−0.5	−2	+13	+4
40° S	−2.2	−10	+13	+4
60° S	−5.0	−22	+ 6	+2

polar winter area they are especially small. The summertime values exceed those in winter and are highest in the temperate northern hemisphere (with its large heated land surface). After passing through a transition zone in the vicinity of the troposphere, the vertical eddy diffusion coefficient is less than 10^4 cm^2/s everywhere in the stratosphere. The depth of the troposphere varies with latitude and season according to classical climatology.

The numerical calculation is a time iteration process with a time step of one day. During each step, the appropriate amount of carbon dioxide is added to or taken from the bottom boxes in which the biosphere resides. A background value of about 320 ppm is given initially to every grid point. The program is run for five years but after the first year the seasonal variation is virtually identical for each of the other four years.

The Results

The results appear in Fig. 6 for both northern and southern hemisphere in the 0–1 km layer. The amplitude of the seasonal variation is significantly smaller, at all latitudes, in the southern than northern hemisphere. The amplitude of the seasonal variation has a maximum at 50–70° N. These amplitude changes with latitude and hemisphere are qualitatively in accord with observations, as far as is currently known. The eddy diffusion coefficients change abruptly each three months to accommodate their seasonal variability in the program. These abrupt changes in horizontal and vertical mixing partially account for the lack of smootheness in the calculated time history of the atmospheric carbon dioxide in Fig. 6.

Fig. 7 shows a few relevant quantitative comparisons between the month-to-month calculated and observed change in atmospheric carbon dioxide. The lower part of the figure compares predictions in the 70–90°S region with data at the South Pole. The amplitudes agree well while the phase is moderately good; the minimum is properly timed but the calculated maximum occurs two months earlier than observed.

The upper part of the figure compares Scandinavian aircraft data (Bolin& Bischof, 1970) with calculations at two different altitudes. At both altitudes, the phase is well represented but predicted amplitudes are about one-half of the predicted values. The reduction in amplitude with height is correctly modelled.

Discussion

It is argued that the relatively good fit between the model prediction of carbon dioxide from the biosphere and the observations at the South Pole suggests

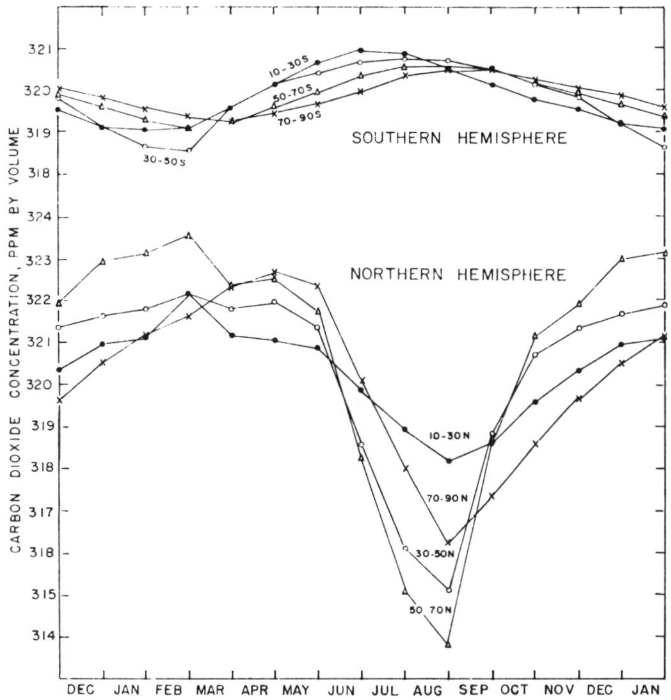

Fig. 6. Seasonal variation in ground level carbon dioxide concentration based on the uptake and release of carbon dioxide by the land biosphere and an atmospheric diffusion model. The absolute value of the concentration is arbitrary and has been fixed to reproduce conditions in the late 1960's.

negligibly small oceanic contributions to the seasonal variation despite the predominance of water in that hemisphere.

The poorer fit in the northern hemisphere cannot be used to argue that the seasonal variation is accounted for by the seasonal changes in biospheric uptake and release of carbon dioxide. Nevertheless, a significant fraction is so explained. Assuming the Scandinavian measurements to be characteristic of the 50–70° N band, two likely explanations for the imperfect fit are suggested. First, and more likely, the information on biospheric release and uptake of carbon dioxide may be in error. This information is admittedly uncertain.

The amplitude of the observed seasonal variation at Mauna Loa, 19° N (Pales and Keeling, 1965) is also greater by almost a factor of two over that calculated. Thus, if the discrepancy between the calculated and observed seasonal amplitudes is ascribed to a larger seasonal amplitude in the biospheric uptake and release of carbon dioxide, the correction to Table 6 must be both large and widespread.

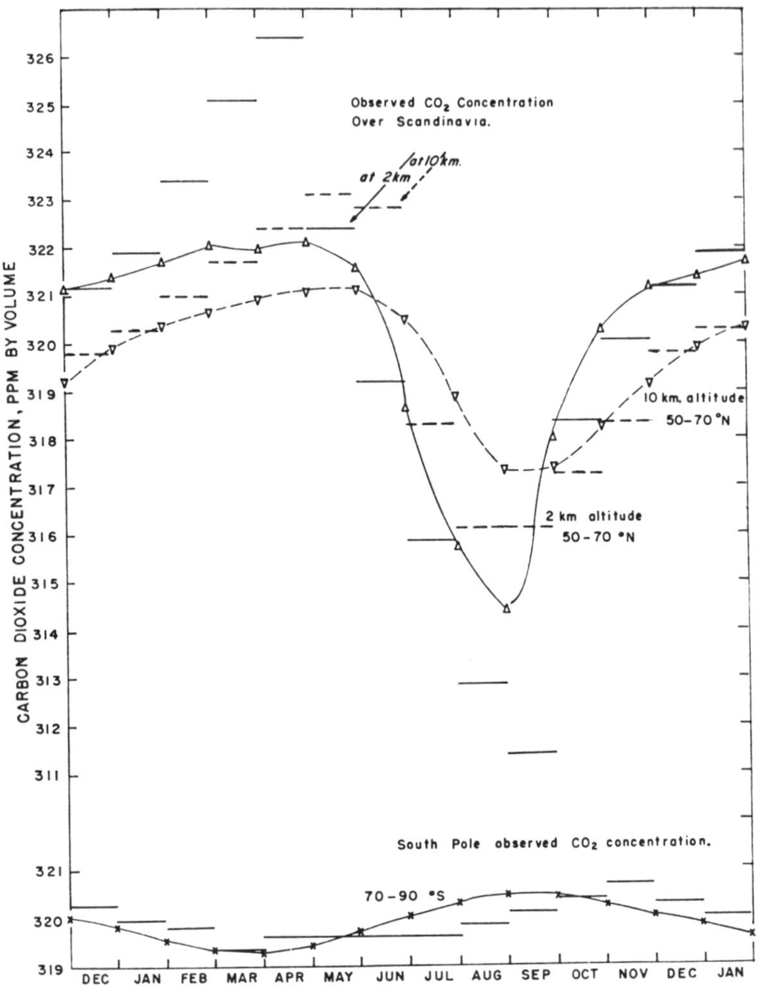

Fig. 7. Comparisons of the observed seasonal variations over Scandinavia (Bolin and Bischof, 1970) and at the South Pole (Brown and Keeling, 1965; Keeling, 1971) with those calculated by a diffusion model and land biospheric uptake and release of carbon dioxide.

Should the input data in Table 6 for the northern hemisphere be in error, it is probable that the same increase in seasonal amplitude ought to be applied equally to the southern hemisphere; the calculated amplitude of the 70–90° S band would also double. In that case, a compensating seasonal variation such as would occur from change in ocean temperature would be required to fit the calculated and observed data. The oceans would then play a role in reducing the southern hemisphere seasonal variation over that which would take place from the land biosphere alone.

A second possible explanation for the discrepancy in amplitudes in the northern hemisphere is the seasonal variation in the consumption of fossil fuels. It is estimated that this can only add, at most, one or two ppm to the calculated amplitudes since the total annual increase if confined to the northern hemisphere troposphere would amount to only about 3.5 to 4.0 ppm. The seasonal component will be only a fraction of the annual increase.

It is unlikely that changes in the treatment of atmospheric transport can account for the difference in the computed and observed seasonal variation of carbon dioxide.

The apparent absence or near absence of any contribution of the oceans to a seasonal variation in atmospheric carbon dioxide may have one of two likely explanations. First, the response time of carbon dioxide release to or from the ocean surface may be too slow to permit a seasonal response (Bolin, 1960).

The second reason is a compensating effect of the marine biosphere. The latter will extract more carbon from the upper layers of the water in the spring or summer when the solar radiation is more intense and the sea surface slightly warmer. Table 8 shows a comparison of the change in partial pressure of carbon dioxide in the mixed layer due to one estimate of photosynthetic uptake and decay using Leith's maps of annual net primary production in the sea and due to changes in sea surface temperature through the summer. The latter has been calculated on the assumption that a one degree Celsius warming increases the partial pressure of carbon dioxide in the sea by a little over 3 %. The fractional reduction of inorganic carbon due to its transfer to organic carbon by photosynthesis creates a tenfold greater fractional decrease in partial pressure of sea water carbon dioxide. To compute the fractional reduction of inorganic carbon requires knowledge of the amount of carbon initially in the reservoir. The table uses two examples, one for a mixed layer 270 m deep (as in Part I) and a second more realistic summertime depth of 60 m.

The comparison of fractional changes in partial pressure for biospheric seasonal variation is seen to be of the same order of magnitude but of opposite sign to that due to the warming of the surface waters. This may also contribute to the explanation of a lack of significant contribution to the atmospheric seasonal variation of carbon dioxide by the oceans.

Conclusion

Despite uncertainties in the true seasonal variation of biospheric uptake and release of carbon dioxide to the atmosphere, the present analysis confirms, in a semi-quantitative way, its dominant role in observed seasonal variation. More important, the analysis suggests the lesser role of the oceans even in the southern hemisphere with its large water area. In fact, the oceans, if important for this purpose, would produce the opposite changes in atmospheric carbon dioxide than have been observed because of its temperature variations. The reasons for the lesser role of the oceans may be its slow response time to temperature changes and the compensating effects of the marine biosphere.

The support of the Fallout Studies Branch, Division of Biology and Medicine, US Atomic Energy Commission is gratefully acknowledged. The assistance of many others including Mr Milton Smith, who programmed the several mathematical models, and Miss Y. Causey and Jonathan Machta who helped in the computations, is appreciated. Above all, I thank Dr. C. David Keeling of Scripps Institution of Oceanography without whose guidance much of the work could not be performed.

References

Bolin, B., Tellus, *XII*, 274 (1960).
Bolin, B. & Keeling, C. D., J. Geophys. Res., *68*, 3899 (1963).
Bolin, B. & Bischof, W., Tellus, *XXII*, 431 (1970).
Brown & Keeling, C. D., J. Geophys. Res., *70*, 6077 (1965).
Junge, C. E. & Czeplak, G., Tellus, *XX*, 422 (1968).
Keeling, C. D. Private Communication (1971).
Keeling, C. D. "Carbon Dioxide and other Carbonaceous Emissions from Fossil Fuels", To be published (1971 a).
Lieth, H., "Phenology in Productivity Studies in "Ecological Studies", 4 (ed. J. Jacobs *et al.*). Springer Verlag, New York, Heidelberg, Berlin, 1970.
SCEP "Man's Impact on the Global Environment", Report of the study of Critical Environmental Problems, MIT Press, Cambridge, Mass., 1970.
Telegadas, K., U. S. Atomic Energy Commission Report, HASL-243 (1971).
United Nations "World Energy Supply", Statistical Papers, Series J., United Nations, New York, many years up to 1968.
U. S. Government, "Restoring the Quality of Our Environment", Appendix Y4, p. 114, Washington, D. C. (1965).

Appendix

The transfer equations between troposphere, stratosphere, biosphere, mixed and deep layers of the oceans are applied to three kinds of carbon dioxide. The first is $C^{12}O_2$ due to the combustion of fossil fuels starting in 1860. The second is cosmogenic $C^{14}O_2$ which is in quasisteady state and the third is man-made $C^{14}O_2$ due to weapon's testing beginning in 1954.

The transfer between troposphere and stratosphere, between troposphere and the mixed layer of the oceans and between the mixed and deep layers of the oceans obeys first order kinetics; that is a fixed fraction of the content of each reservoir is transferred to its adjacent reservoir during a given time step. On the other hand, the transfer to and from the several biospheric reservoirs depends on the net primary production or the fraction of carbon converted to living matter during a growing season. For simplicity, the return of living carbon back to the reservoir with which it is in contact also occurs in a single year with the following lags: short-term land biosphere, 2 years; longterm land biosphere, 40 years; and marine biosphere, 1 year.

The predictive equations for the troposphere are:

$$T(t) = T(t-1) - \lambda_{T\to M} T(t-1) + b\lambda_{M\to T} M(t-1) - \lambda_{T\to S} T(t-1) + \lambda_{S\to T} S(t-1) - \\ - fP_L T(t-1)/T_0 + fP_L T(t-41)/T_0 - fP_S T(t-1)/T_0 + fP_S T(t-3)/T_0 \\ + I(t) \quad (1)$$

$$T^{14}(t) = T^{14}(t-1) - \alpha_2 \lambda_{T\to M} T^{14}(t-1) + \frac{1}{\alpha_2} \lambda_{M\to T} M^{14}(t-1) \\ + b\lambda_{M\to T} \frac{M(t-1)}{M_0 + M(t-1)} [M_0^{14} + M^{14}(t-1)] - \lambda_{T\to S} T^{14}(t-1) \\ + \lambda_{S\to T} S^{14}(t-1) - \alpha_1 P_L \frac{T^{14}(t-1)}{T_0 + T(t-1)} - \alpha_1 P_S \frac{T^{14}(t-1)}{T_0 + T(t-1)} \\ + \alpha_1 P_L \frac{T^{14}(t-41)}{T_0 + T(t-41)} + \alpha_1 P_S \frac{T^{14}(t-3)}{T_0 + T(t-3)} \quad (2)$$

No predictive equation for the transfer of man-made $C^{14}O_2$ from the troposphere. Rather its transfers from and to the troposphere are computed and stored.

$$TBR(t) = -\alpha_2 \lambda_{T\to M} TB(t-1) + \frac{1}{\alpha_2} \lambda_{M\to T} MB(t-1) + b\lambda_{M\to T} \frac{M(t-1)}{M_0 + M(t-1)} \\ \times MB(t-1) - \alpha_1 P_L \frac{TB(t-1)}{T_0 + T(t-1)} + \alpha_1 P_L \frac{TB(t-41)}{T_0 + T(t-41)} \\ - \alpha_1 P_S \frac{TB(t-1)}{T_0 + T(t-1)} + \alpha_1 P_S \frac{TB(t-3)}{T_0 + T(t-3} \quad (3)$$

For the mixed layer, the transfer equations are:

$$M(t) = M(t-1) + \lambda_{T\to M} T(t-1 - b\lambda_{M\to T} M(t-1) - \lambda_{M\to D} M(t-1) + \lambda_{D\to M} D(t-1) \quad (4)$$

$$M^{14}(t) = M^{14}(t-1) + \alpha_2 \lambda_{T\to M} T^{14}(t-1) - \frac{1}{\alpha_2} \lambda_{M\to T} M^{14}(t-1) \\ - b\lambda_{M\to T} \frac{M_0^{14} + M^{14}(t-1)}{M_0 + M(t-1)} \times M(t-1) - a_1 P_M \frac{M_0^{14} + M^{14}(t-1)}{M_0 + M(t-1)} \\ + \alpha_1 P_M \frac{M_0^{14} + M^{14}(t-2)}{M_0 + M(t-2)} - \lambda_{M\to D} M^{14}(t-1) + \lambda_{D\to M}(t-1) \quad (5)$$

$$MB(t) = MB(t-1) + \alpha_2 \lambda_{T\to M} TB(t-1) - \frac{1}{\alpha_2} \lambda_{M\to T} MB(t-1)$$

$$- b\lambda_{M\to T} \frac{MB(t-1)}{M_0 + M(t-1)} \cdot M(t-1) - \alpha_1 P_M \frac{MB(t-1)}{M_0 + M(t-1)}$$

$$+ \alpha_1 P_M \frac{MB(t-2)}{M_0 + M(t-2)} - \lambda_{M\to D} MB(t-1) + \lambda_{D\to M} DB(t-1) \quad (6)$$

The predictive equations for the deep ocean are:

$$D(t) = D(t-1) + \lambda_{M\to D} M(t-1) - \lambda_{D\to M} D(t-1) \quad (7)$$

$$D^{14}(t) = D^{14}(t-1) + \lambda_{M\to D} M^{14}(t-1) - \lambda_{D\to M} D^{14}(t-1) \quad (8)$$

$$DB(t) = DB(t-1) + \lambda_{M\to D} MB(t-1) - \lambda_{D\to M} DB(t-1) \quad (9)$$

The predictive equations for the stratosphere are:

$$S(t) = S(t-1) + \lambda_{T\to S} T(t-1) - \lambda_{S\to T} S(t-1) \quad (10)$$

$$S^{14}(t) = S^{14}(t-1) + \lambda_{T\to S} T^{14}(t-1) - \lambda_{S\to T} S^{14}(t-1) \quad (11)$$

The predictive equations for the short-term biosphere are:

$$B_S(t) = B_S(t-1) + fP_S \frac{T(t-1)}{T_0} - fP_S \frac{T(t-3)}{T_0} \quad (12)$$

$$B_S^{14}(t) = B_S^{14}(t-1) + \alpha_1 P_S \frac{T^{14}(t-1)}{T_0 + T(t-1)} - \alpha_1 P_S \frac{T^{14}(t-3)}{T_0 + T(t-3)} \quad (13)$$

$$BBS(t) = BBS(t-1) + \alpha_1 P_S \frac{TB(t-1)}{T_0 + T(t-1)} - \alpha_1 P_S \frac{TB(t-3)}{T_0 + T(t-3)} \quad (14)$$

The predictive equations for the long-term biosphere are:

$$B_L(t) = B_L(t-1) - fP_S \frac{T(t-1)}{T_0} - fP_S \frac{T(t-41)}{T_0} \quad (15)$$

$$B_L^{14}(t) = B_L^{14}(t-1) + \alpha_1 P_L \frac{T^{14}(t-1)}{T_0 + T(t-1)} - \alpha_1 P_L \frac{T^{14}(t-41)}{T_0 + T(t-41)} \quad (16)$$

$$BBL(t) = BBL(t-1) + \alpha_1 P_L \frac{TB(t-1)}{T_0 + T(t-1)} - \alpha_1 P_L \frac{TB(t-41)}{T_0 + T(t-41)} \quad (17)$$

The predictive equations for the marine biosphere are:

$$B_M^{14}(t) = B_M^{14}(t-1) + \alpha_1 P_M \frac{M_0^{14} + M^{14}(t-1)}{M_0 + M(t-1)} - \alpha_1 P_M \frac{M_0^{14} + M^{14}(t-2)}{M_0 + M(t-2)} \quad (18)$$

$$BBM(t) = BBM(t-1) + \alpha_1 P_M \frac{MB(t-1)}{M_0 + M(t-1)} - \alpha_1 P_M \frac{MB(t-2)}{M_0 + M(t-2)} \quad (19)$$

The predictive equations for the world atmosphere

$$(\Delta W)_P(t) = TBR(t) + [T^{14}(t) - T^{14}(t-1)] + [S^{14}(t) - S^{14}(t-1)] \quad (20)$$

The definition of symbols follows:

T =carbon content in 10^{16} g of the troposphere due to the carbon dioxide from the man-made combustion of fossil fuels

S =same—for the stratosphere

M =same—for the mixed layer of the oceans

D =same—for the deep layer of the oceans

B_s =same—for the short-term land biosphere

B_L =same—for the long-term land biosphere

B_M =same—for the marine biosphere

Each of the above symbols with a "14" superscript represents the cosmic ray C^{14} content of each reservoir in 10^{26} atoms of C^{14}.

TB =C^{14} content in 10^{26} atoms of the troposphere due to the $C^{14}O_2$ from the man-made production of C^{14}

MB =same for mixed layer of the oceans

DB =same for deep layer of the oceans

BBS =same for short-term land biosphere

BBL =same for long-term land biosphere

BBM =same for marine biosphere

TBR =the net amount of man-made C^{14} in 10^{26} atoms resulting from the transfer out of and into the troposphere in one unit of time end at time t.

t =time; $t-1=$ one unit of time before t.

T_0 =total quantity of carbon as $C^{12}O_2$ in the troposphere in 1860; taken as 50.68×10^{16} g based on an atmospheric concentration of CO_2 of 290 ppm by volume.

T_0^{14} =total quantity of C^{14} atoms as $C^{14}O_2$ in the troposphere in 1950; taken as 323.3×10^{26} atoms of C^{14} based on an atmosphere C^{14} concentration of 74×10^5 atoms g^{-1} (air).

M_0+D_0 =total carbon content of the oceans in 1860, taken as about 60 times the carbon content of the troposphere.

λ_{ij} =fractional transfer of the contents of reservoir i to reservoir j in one year. The reservoirs are: T=troposphere; S=stratosphere; M= mixed layer of the oceans and D=deep layer of the oceans.

M_0 is obtained from the assumption of a steady state transfer across the sea surface:

$$M_0 = T_0 \frac{\lambda_{T\to M}}{\lambda_{M\to}}$$

$M_0^{14} = C^{14}$ content of the mixed layer of the ocean in units of 10^{26} atoms due to cosmic ray production and is obtained from:

$$M_0^{14} = 0.95 \frac{\lambda_{T\to M}}{\lambda_{M\to T}} T_0^{14}$$

$$\lambda_{T\to S} = 0.087 \text{ yr.}^{-1}$$

$$\lambda_{S\to T} = 0.5 \text{ yr.}^{-1}$$

$$\lambda_{D\to M} = 0.000625 \text{ yr.}^{-1}$$

$$\lambda_{D\to M} = \frac{61\,T_0 - M_0}{M_0} \cdot \lambda_{D\to M}$$

P_S = net primary production into the short-term land biosphere; taken as 3.1×10^{16} gC/yr.

P_L = net primary production into the long-term land biosphere; taken as 2.5×10^{16} gC/yr.

P_M = net primary production into the marine biosphere; taken as 2.0×10^{16} gC/yr.

α_1 = fractionation factor for carbon transferring from air or water to the biosphere; $(C^{14}/C^{12})_B/(C^{14}/C^{12})_{T,M} = 0.964$ where the subscripts T, M refer to either the tropospheric air or the ocean water.

α_2 = fractionation factor for carbon transferring from air to water; $(C^{14}/C^{12})_M/(C^{14}/C^{12})_T = 0.984$

f = fraction of land biosphere whose growth is assumed to be CO_2 limited, taken as 0.5, times a factor representing the fractional growth of the CO_2 limited land biosphere from a given fractional increase in atmospheric CO_2, taken as 0.5. The value of f is therefore 0.25.

b = buffering factor which accounts for the change in partial pressure of CO_2 in the mixed layer resulting from changes in the carbon content of the mixed layer; taken as 10.

[5]

Atmospheric homeostasis by and for the biosphere: the gaia hypothesis

By JAMES E. LOVELOCK, *Bowerchalke, Nr. Salisbury, Wilts. England* and LYNN MARGULIS, *Department of Biology, Boston University, 2, Cummington Street, Boston, Mass. USA*

(Manuscript received May 8; revised version August 20, 1973)

ABSTRACT

During the time, 3.2×10^9 years, that life has been present on Earth, the physical and chemical conditions of most of the planetary surface have never varied from those most favourable for life. The geological record reads that liquid water was always present and that the pH was never far from neutral. During this same period, however, the Earth's radiation environment underwent large changes. As the sun moved along the course set by the main sequence of stars its output will have increased at least 30 % and possibly 100 %. It may also have fluctuated in brightness over periods of a few million years. At the same time hydrogen was escaping to space from the Earth and so causing progressive changes in the chemical environment. This in turn through atmospheric compositional changes could have affected the Earth's radiation balance. It may have been that these physical and chemical changes always by blind chance followed the path whose bounds are the conditions favouring the continued existence of life. This paper offers an alternative explanation that, early after life began it acquired control of the planetary environment and that this homeostasis by and for the biosphere has persisted ever since. Historic and contemporary evidence and arguments for this hypothesis will be presented.

It is widely believed that the abundance of the principal gases N_2 and O_2 is determined by equilibrium chemistry. One of the larger problems in the atmospheric sciences is that of reconciling this belief with the uncomfortable fact that these same gases are cycled by the Biosphere with a geometric mean residence time measured in thousands of years. The more thoroughly the inventory of an individual gas is audited the more certain it seems that inorganic equilibrium or steady state processes determine its atmospheric concentration but the same audit frequently further reveals the extent of its biological involvement. A lucid account of contemporary information on the problem of the cycle of gases is in the paper of Junge (1972).

This paper presents a new view of the atmosphere, one in which it is seen as a component part of the biosphere rather than as a mere environment for life. In this new context the incompatibilities of biological cycles and inorganic equilibria are seen as more apparent than real.

A starting point is a consideration of the profoundly anomalous composition of the Earth's atmosphere when it is compared with that of the expected atmosphere of a planet interpolated between Mars and Venus. Thus on Earth the simultaneous presence of O_2 and CH_4 at the present concentrations is a violation of the rules of equilibrium chemistry of no less than 30 orders of magnitude. Indeed so great is the disequilibrium among the gases of the Earth's atmosphere that it tends towards a combustible mixture, whereas the gases of Mars and Venus are close to chemical equilibrium and are more like combustion products.

The anomalous nature of the atmosphere has been known since Lewis & Randall (1923) first commented that at the pE and pH of the Earth the stable compound of nitrogen is the NO_3^- ion in the oceans; gaseous nitrogen should not be present. In spite of reminders by Hutchinson

(1954) and Sillén (1966) this anomaly has remained unnoticed in the debate on atmospheric cycles.

Given the chemical composition of a planetary atmosphere it is possible to infer the presence or absence of life, Hitchcock & Lovelock (1967). To do this the entire ensemble of reactive gases constituting the atmosphere needs be considered and when this is done information is made available which is otherwise inaccessible when each gas is considered separately in isolation. This approach applied to the present problem of the anomaly of the chemical distribution of the gases of the atmosphere, offers a strong suggestion that the Earth's atmosphere is more than merely anomalous; it appears to be a contrivance specifically constituted for a set of purposes.

This paper examines the hypothesis that the total ensemble of living organisms which constitute the biosphere can act as a single entity to regulate chemical composition, surface pH and possibly also climate. The notion of the biosphere as an active adaptive control system able to maintain the Earth in homeostasis we are calling the 'Gaia' hypothesis, Lovelock (1972). Hence forward the word Gaia will be used to describe the biosphere and all of those parts of the Earth with which it actively interacts to form the hypothetical new entity with properties that could not be predicted from the sum of its parts.

Theoretical basis

The fundamental problem underlying the formal recognition of an unfamiliar living association, such as Gaia, is that of recognising life itself. We are so preprogrammed to recognise life instinctively that the logical basis of the recognition is rarely questioned. Thus the discovery of a member of new and bizarre animal species raises the question, what is it, never the question, is it life? This instinctive recognition passes beyond organisms to systems provided that they are not too big to be seen. Thus a bees nest is recognised as a purposeful structure, quite different from a cluster of non-social insects. The intricate organisation of an ecosystem such as a tropical rain forest, however, was not recognised until the evidence of the interdependence of its parts and the economy of the cycling of essential elements was discovered. With the total ecosystem, Gaia, apart from the shadowy evidence of those primitive beliefs in her reality we are dependent upon physical rather than biological evidence. There is little doubt that living things are elaborate contrivances. Life as a phenomenon might therefore be considered in the context of those applied physical sciences which grew up to explain inventions and contrivances, namely thermodynamics, cybernetics and information theory.

The first cautious approach to a classification of life, reached general agreement as follows 'Life is one member of the class of phenomena which are open or continuous reaction systems able to decrease their entropy at the expense of free energy taken from the environment and subsequently rejected in a degraded form' (Bernal, 1951; Wigner, 1961).

This may also be expressed in the form of the equation of continuity for entropy (Denbigh, 1951).

$$\varrho \frac{dS}{dt} + \operatorname{div} S = \theta$$

Where θ is the rate of internal creation of entropy, ϱ the density and S the entropy; div S is the outflow of entropy and $\varrho(dS/dt)$ the rate of change of entropy in the enclosed region; θ must by second law be zero or positive; the possibility that div S can be large and positive makes possible a negative trend for $\varrho(dS/dt)$.

This classification is broad and includes also phenomena such as vortices and flames and many others. Life differs from such primitive processes of the abiological steady state in the singularity, persistence and size of the entropy reduction it sustains. Although limited, this phenomenological description of the class of process, which includes life, is helpful in our search for proof of the existence of Gaia in two ways. Firstly by serving to define the boundary of the internal region where entropy is reduced and secondly by suggesting that the recognition of a living entity can be based upon the extent of its physical and chemical disequilibrium from the background environment.

On the matter of boundaries, it is obvious that a man, as an example of a living entity, takes in free energy in the form of the chemical potential difference between food and oxygen and sustains a low internal entropy through excretion of waste chemicals and heat. To a man, the environment to which entropy is discarded includes the atmosphere and his boundary is therefore his skin. It might seem pointless therefore on Earth to seek the existence of a general living system, Gaia, in terms of entropy reductions within the atmosphere which clearly for some species is a sink for degraded products and energy but this neglects the fact that photosynthetic life uses premium radiation direct from the sun to sustain a high chemical potential gradient within the atmosphere on a planetary scale. For a tree, the boundary within which entropy is reduced is not its surface in contact with the atmosphere but rather the interface between the sun with the atmosphere as an extension of the tree. The tree produces not only food for consumers but also the equally important gas, oxygen which does not accumulate within the tree waiting to be eaten.

When the whole assembly of life is so seen it is clear that the true boundary is space. The outgoing entropy flux from the Earth indeed from Gaia 'if she exists', is long wavelength infra red radiation to space. This then, is the physical justification for delineating the boundary of life as the outer reaches of the atmosphere. There is also to a lesser extent an inner boundary represented by the interface with those inner parts of the Earth as yet unaffected by surface processes. We may now consider all that is encompassed by the bounds as putative life. Whether or not Gaia is real will depend upon the extent to which the entropy reduction within a compartment such as the atmosphere is recognisably different from the abiological steady state background.

On the matter of recognition a debt is owed to the fertile concept of information theory, Shannon & Weaver (1963). It has been demonstrated, for example by Evans (1969) that the classical properties, entropy and free energy, have exact information theoretic equivalents. Thus the information (I) of a system can be defined as

$$I = S_0 - S$$

where S_0 is the entropy of the components of the system at thermodynamic equilibrium and S the entropy of the system assembled. This relationship can be transferred directly from information theoretic to classical thermodynamic terms as follows:

$$I = (E + PV - TS - \Sigma N, N_i)/T$$

Table 1. *The partial pressures in millibars of CO_2, N_2 and O_2 on Mars, Earth and Venus and on two model abiological Earth's*

Model (D) after life had been deleted and Model (I) a geometric mean interpolation between Mars and Venus

Gas	Venus	Earth	Mars	Abiological earth models D	I
CO_2	90 000	0.3	5	0.3–1 000	300
N_2	1 000	780	0.05	1	30
O_2	0	210	0.1	1	0.3

where the right hand side of the equation expresses information in terms of temperature (T) pressure (P) internal energy (E), volume (V), entropy (S), and the chemical potential (N) of the molecules present; it follows that information is a measure of disequilibrium in the classic sense and recognisability in the information theoretic sense.

By examining the extent to which the atmosphere is in chemical and physical disequilibrium both within itself and with the surface of the Earth we have a measure of the extent to which it is recognisable as a separate identity against a neutral background equilibrium state. Whether or not it is seen to be a component part of Gaia will depend upon the size of the disequilibrium revealed.

Thermodynamic evidence

We can determine the degree of departure from a conceivable abiological steady state atmosphere of the Earth by two means; firstly, by comparing the Earth's atmosphere with Mars and Venus which may be taken as representative of lifeless planets and secondly by comparing the present atmosphere with the atmosphere to be expected if life were deleted from Earth.

Table 1 shows the atmospheric composition of Mars and Venus compared with that of a hypothetical abiological Earth interpolated between them. Also shown is the atmospheric composition to be expected of the Earth if life were deleted. This was derived as follows: nitrogen reacts under the influence of solar

Table 2. *Gases in the Earth's atmosphere with origins in the Biosphere*

The source strengths, residence times, concentrations, and concentrations expected if the atmosphere was a thermodynamic equilibrium

Gas	Atmospheric abundance	Anticipated equilibrium concentration	Departure from equilibrium	Residence time years	Output megatons YR^{-1}	Source
N_2	0.8	E-10	E 10	3E6	E 3	Denitrifying bacteria
CH_4	1.5E-6	E-35	E 29	7	2E3	Anaerobic fermenting bacteria
N_2O	3E-7	E-20	E 13	10	6E2	Denitrifying bacteria
NH_3	1E-8	E-35	E 27	E-2	1.5E3	Nearly all organisms
CH_3I	1E-12	E-35	E 23	E-3	30	Marine algae

UV, electrical discharges, ionisation radiation and combustion with both O_2 and CO_2 to give oxides of nitrogen. By such processes in time all nitrogen would be converted to the stable NO_3^- ion dissolved in the oceans. At the pE even of Mars the reverse reactions converting NO_3^- to N_2 do not proceed. In the course of the removal of N_2 most of the O_2 also would go from the atmosphere. The continuation of water and CO_2 photolysis would ensure the production of some oxygen, so that nitrogen removal could go to completion. In the end the atmosphere would apart from argon be dominated by CO_2 like those of Mars and Venus.

The model of a lifeless Earth coincides well with that of a planet interpolated between Mars and Venus, the slight chemical disequilibrium is due to dissociative effect of sunlight and the escape of hydrogen. The present Earth is clearly recognisable against the background of the lifeless model.

The terrestrial planets orbit in the outer reaches of the sun's hydrogen atmosphere and their interiors are reducing hence any concentration of oxygen in a planetary atmosphere is indicative of disequilibrium. Mars has 0.01 % of oxygen in its atmosphere and Venus less than this. Such small oxygen concentrations are quite consistent with the abiological steady state consequent upon the upper atmosphere photolysis of water vapour and CO_2 on Mars and perhaps to a lesser extent on Venus. It is well to consider that Venus has probably lost oceans of water by photolysis and other reactions, yet still has no O_2. The presence of nitrogen on Venus does not indicate disequilibrium since Venus is both hot and acid and bereft of water, these are all conditions favouring nitrogen rather than NO_3^- as the stable compound. That nitrogen has not yet been detected on Mars is in agreement with the expectation of equilibrium chemistry. A lifeless Earth would similarly be without atmospheric nitrogen. These arguments were used successfully to predict that Mars would lack N_2 before this fact was discovered (Lovelock & Giffen, 1969).

The anomalies of abundance are more marked with the lesser gases as is shown in Table 2 which lists their concentrations, residence times, sources and expected concentrations, were chemical equilibrium determining.

The simultaneous and large fluxes of N_2 CH_4, NH_3, N_2O and O_2 are all inconsistent with an abiological model. The degree of departure from equilibrium expectations is to be measured in 10's of orders of magnitude. In these circumstances there is almost a certainty that the atmosphere is a part of the biosphere, for nowhere, but in living systems is so intensive and constant a disequilibrium revealed. Conversely there is a neglible possibility that the atmosphere is a neutral background source of materials which life merely recycles.

The atmosphere as a contrivance

If we assume the Gaia hypothesis, and regard the atmosphere as a contrivance, then it is reasonable to ask what is the function of its various component gases. Outside the Gaia hypothesis such a question would rightly be condemned as circular and illogical but in its context such questions are no more unreasonable than asking, for example, what is the function of fibrinogen in blood. The function of oxygen and nitrogen in sustaining the cycle of raw material and energy is evident. Less obvious is the function for example of CH_4. This gas is made on a scale requiring the shunting of as much as 5 % of the photosynthetic energy of the entire biosphere. In the Gaia context it must surely have an important purpose of biology is normally efficient and parsimonious over waste. Such a possible important function for methane has been proposed (Lovelock & Lodge, 1972). It is seen as a kind of molecular hydrogen balloon whose purpose is to carry excess hydrogen to the upper atmosphere where it can escape and thus sustain the present oxygen tension of the Earth. The details of this notion are given in the reference quoted. Wofsy et al., (1972) suggest that such is the magnitude of the upward transfer of methane to the stratosphere where it is oxidised to water vapour that the stratosphere is a source of water vapour to the troposphere rather than a sink. It follows that much of the water vapour in the upper atmosphere which is photolysed to produce oxygen is a biological product resulting from the oxidation of biological methane.

As with CH_4 the biosphere uses a great amount of energy for NH_3 production. When the total production of acids by the oxidation of nitrogen and sulphur is taken into account the ammonia production by the biosphere is found to be just sufficient to sustain a rainfall pH near 8, the optimum for life. One penalty of oxidising environments is the tendency to low pH as the elements such as carbon, nitrogen and sulphur oxidise. The huge biological ammonia production conveniently answers this need. Can it be accidental and have always been so? Other atmospheric gases and vapours for example dimethyl sulphide, dimethyl selenide and methyl iodide may serve in the mass transfer of essential elements between the land and the sea.

A gas wholly in disequilibrium, a biological product on a grand scale, is nitrous oxide. It is made by soil microorganisms and some penetrates to the stratosphere where it reacts to give NO_x. By the Gaia hypothesis it must have an important atmospheric purpose, could this be concerned with the regulation of the position or density of the ozone layer? In the summary there is overwhelming evidence that the atmosphere apart from its content of noble gases is a biological product. It may also be a biological contrivance; not living but as essential a part of the biosphere as is the shell to a snail or the fur to a mink.

Systems analysis

One set of tests which can be applied to prove the existence of a control system are those which make changes in the variables thought to be controlled. It is possible to show for example that the core temperature of a man remains constant when the environmental temperature ranges between 0 and 45°C. With the Earth it is reasonable to suspect that at least the following variables may be controlled: (1) Atmospheric, oceanic and soil chemical composition. (2) Surface temperatures in the 'core' region i.e. between 45° N and 45° S latitudes. (3) Surface and ocean pH.

It is not usually possible to vary the global environment as a test but perturbations affecting these variables have accompanied the evolution of the solar system. Since life began, 3.5×10^9 years ago, the solar output has probably increased between 30 and 100 % (Sagan & Mullen, 1972). The most recent estimates in which the current low rate of neutrino emission from the sun is taken into account is 70 % (Dilke & Gough, 1972). It is even suggested that in addition to the slow exponential increase in solar output, fluctuations of 10 % may occur over short periods i.e. 10 M years. The other major environmental change is that of redox potential. At its formation the Earth was highly reducing with hydrogen as a predominant atmospheric gas. It is suggested that 3.5×10^9 years ago the Earth's pE was -5 at which gases such as CH_4, NH_3, H_2S etc. would be stable in their own chemical right. Hydrogen would be escaping to space but for some time there would be sufficient to regenerate the hydrides dissociated by solar UV. Such an atmosphere could according to Sagan & Mullen (1972) effectively retain heat. It would permit surface temperatures similar to those of present in spite of a much reduced solar output. This is due principally to the radiative properties of gases such as ammonia and the others mentioned above. These were probably the conditions under which life evolved.

Soon after the evolution of life and possibly because of it, Ycas (1972) the pace of hydrogen loss accelerated. By 3.2×10^9 years ago the pE had risen above -4 at which level NH_3 is no

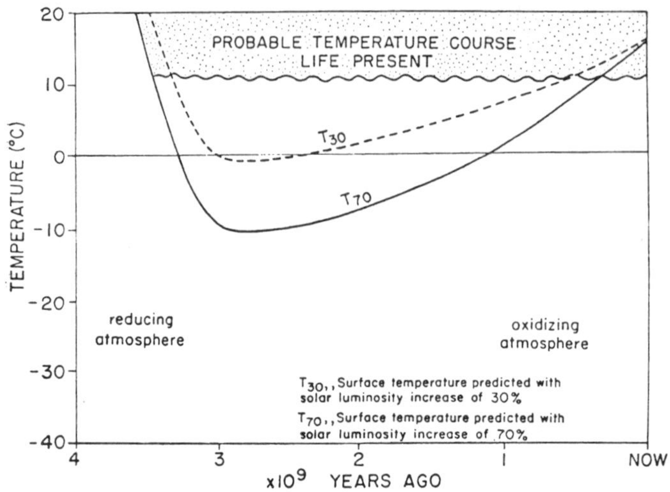

Fig. 1. The mean surface temperature of the Earth to 3.5×10^9 years B. P. Shaded area, temperature range deduced from the geological record and the persistence of life. Solid and dashed lines, the temperatures expected for a lifeless Earth assuming a 70 % (solid line) or a 30 % (dashed line) increase in solar output.

longer stable in its own right (Sillén, 1966). Had the laws of chemistry and physics operated soon after this time NH_3 and other polyatomic reducing gases would have vanished leaving an atmosphere of nitrogen and CO_2 at a concentration not substantially different from the present. If we assume that the solar output was in fact lower at this time such an atmosphere could not possibly have sustained present day temperatures the short term equilibrium temperature would have been -5 and -16°C. If the large albedo of ice and snow cover is additionally taken into account the equilibrium temperature could even have been lower.

Chemical changes of the evolving earth and the physical changes of its radiation environment are shown in Fig. 1. The geological record and the fact of the persistence of life provide strong evidence that at no time in the past 3.5×10^9 years did the oceans freeze or the mean temperature rise above 50°. Indeed the evidence suggests that apart from glacial episodes, the temperature has always been constant. Glaciations do not usually affect regions between latitudes 45° N and 45° S where 70 % of the surface resides.

Sagan & Mullen (1972) propose that the continuous biological synthesis of ammonia before oxygen appeared gave an atmospheric concentration of ammonia sufficient to sustain by its greenhouse effect equable and tolerable temperatures. This is supported by the fact that the current biological ammonia production, which is 2×10^9 tons per year, might in a neutral or reducing atmosphere have sustained the concentration of ammonia said to be needed to provide the desired radiative properties namely 10^{-5} by volume. However, if we assume that the solar output was indeed lower in remote times, the mere blind production of a gas such as NH_3 by the biosphere would not have been likely to suffice in the control of surface temperatures by its radiative properties. With the reduced solar output and the possibility of positive feedback on cooling, any failure of the ammonia harvest or the evolution of a vigorous NH_3 consumer, of which there are many today, would have set in train a course for catastrophic irreversible cooling. A mechanism which plausibly can oppose an unfavourable trend is not enough, there must also be a system able to sense the trend and actively control the means for its opposition. It must be observed that there are other 'greenhouse' gases than NH_3. CO_2 at high concentrations could have performed a similar function as could other polyatomic gases. To sustain a constant temperature for 3×10^9 years in the face of chemical and physical environmental changes strongly implies the presence of an active process for thermostasis. Life at its origin can be considered to have fed on the blanket of gas which kept it warm. Its continued survival required the early development of the capacity to recognise potentially adverse changes and of processes which could oppose such changes. It is not the purpose of this article to consider in detail the exact mechanism which could have been used to this end. The change of atmospheric composition is a powerful technique but other changes such as in the surface albedo the rate of evaporation of water, are at least conceivable. A more detailed account of speculations on possible mechanisms is given in a forthcoming paper (Margulis & Lovelock, 1973). Our purpose is to stress that the present knowledge of the early environment suggests strongly that a first task of life was to secure the environment against adverse physical and chemical change. Such security could only come from the active process of homeostasis in which unfavourable tendencies could be sensed and counter measures operated before irreversible damage had been done.

In addition to the possibility of remorseless environmental change set by the evolution of the sun, there were also major changes in the chemical environment. These were caused in part by the evolution of life itself but they also reflected upon the physical environment through the change in atmosphere composition. Prominent amongst these changes must have been the first appearance of free oxygen in the atmosphere which is thought to have taken place between 1 and 2×10^9 years ago. It is worth digressing a moment to consider what a catastrophic air pollutant oxygen must have been. Its presence in the air must have destroyed a vast range of species and driven underground into anaerobic muds others from which prison they have never returned. Not only was the appearance of oxygen itself a biological crisis but also the rapid reduction of pH which inevitably would accompany oxidation of nitrogen and sulphur compounds and the increased rate of removal of ammonia must have posed other serious problems. Firstly in the maintenance of pH itself and secondly in sustaining a sufficient concentration of ammonia as this gas was depleted by its new role of neutralising acidity. Again the details of the atmospheric changes of this time

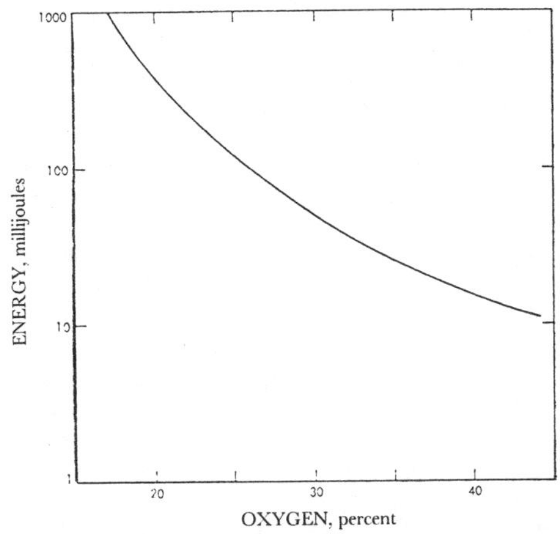

Fig. 2. Energy required to ignite propane in mixtures of oxygen and nitrogen.

are still not known although it is conceivable that the CO_2 concentration rose as the pH fell and to some extent offset the effects of the loss of NH_3. It is unlikely however that this alone would have been sufficient to sustain tolerable temperatures at this period when the solar output may have been at least 10 and possibly 40 % less than now. Nevertheless sufficiently constant temperatures were sustained and life persisted. Even today the regulation of oxygen concentration may represent an important task. Fig. 2 shows how drastically the probability of the ignition of inflammable material rises with increasing oxygen concentration. Above 25 % even trees in rain forests might be in hazard from fire. With these crises as with those earlier the blind adaptation to change seems unlikely to have been enough.

Conclusions

We have presented some of the evidence concerning the Gaia hypothesis. The hard core geochemists will still no doubt argue that the cycling of gases through the biosphere is a passive process and does not determine atmospheric composition. He may compare it with the cycling of water from the oceans to the land. The delivery of rain to the land surfaces involves a disequilibrium of gravitational and osmotic potential and is driven by solar energy. The vegetation merely borrows the rain and repays at the rate of borrowing. We think that it is all a matter of degree. Sunlight may distil water from the sea later to fall as rain on the land but sunlight does not spontaneously at the Earth's surface split oxygen from water and drive reactions leading to the synthesis of intricate compounds and structures. The disequilibrium characteristic of water cycling is minor when compared with that of cycling of the atmospheric gases.

To those who are convinced that the atmospheric gases are biological products but are reluctant to accept the notion of homeostasis we say: if life has merely a passive role in cycling the gases of the air then the concentrations will be set by equilibrium chemistry; in fact they most certainly are not. If life actively cycles the gases then we ask how could such a system be stable in the long run without homeostasis?

The purpose of this paper is to introduce the Gaia hypothesis at least for entertainment and for the induction of new questions about the Earth. Proof of Gaia's existence may never approach certainty but further evidence is more likely to come from the study of the contemporary Earth. Astronomical evidence is notoriously fickle and although geological evidence is rather more certain one learns less about a person from the study of his grandfather's bones than from talking to him face to face.

Acknowledgement

We acknowledge the unstinted help and advise of our colleagues S. R. Epton, P. B. Fellgett, J. P. Lodge and P. G. Simmonds. We thank Shell Research Ltd., for financial support to one of us (J. E. L.)

REFERENCES

Bernal, J. D. 1951. *The physical basis of life*. Routledge and Kegan Paul, London.

Denbigh, K. G. 1951. *The thermodynamics of the steady state*. Methuen and Co., London.

Evans, R. B. 1969. A proof that essergy is the only consistent measure of potential work. Ph.D. Thesis, Dartmouth College, University Microfilms, Ann Arbor, Michigan.

Hitchcock, D. R. & Lovelock, J. E. 1967. Life detection by atmospheric analysis. *Icarus 7*, 149–159.

Hutchinson, G. E. 1954. Biochemistry of the terrestrial atmosphere. In Chapter 8 of the *Solar system* (ed. Kuiper). University of Chicago Press, Chicago.

Junge, C. 1972. The cycle of atmospheric gases-natural and man made. *Quart. J. R. Met. Soc. 98*, 711–729.

Lewis, G. N. & Randall, M. 1923. *Thermodynamics and the free energy of chemical substrates*, Chapter 39. McGraw Hill, New York.

Lovelock, J. E. 1972. Gaia as seen through the atmosphere. *Atmospheric Environment 6*, 579–580.

Lovelock, J. E. & Giffen, C. E. 1969. *Planetary atmospheres: compositional and other changes associated with the presence of life, advanced space experiments* (Ed. O. L. Tiffany and E. Zaiteff) pp. 179–193. American Astronautical Soc., Washington, D.C.

Lovelock, J. E. & Lodge, J. P. 1972. Oxygen in the contemporary atmosphere. *Atmospheric Environment, 6*, 575–578.

Margulis, L. & Lovelock, J. E. 1973. Biological modulation of the Earth's atmosphere. Icarus, in press.

Sagan, C. & Mullen, G. 1972. Earth and Mars: evolution of atmospheres and surface temperatures. *Science 177*, 52–56.

Schrodinger, E. 1944. What is Life? Cambridge University Press.

Shannon, C. E. & Weaver, W. 1963. *The mathematical theory of communication*. University of Illinois Press.

Sillén, L. G. 1966. Regulation of O_2, N_2, and CO_2 in the atmosphere: thoughts of a laboratory chemist. *Tellus 18*, 198–206.

Tribus, M. & McIrvine, E. C. 1971. Energy and information. *Scientific American 224*, No. 3, 179–188.

Wigner, E. P. 1961. *The logic of personal knowledge*. Routledge and Kegan Paul, London.

Wofsy, S. C., McConnell, J. C. & McElroy, M. B. 1972. Atmospheric CH_4, CO and CO_2. *J. Geophys. Res. 77*, 4477–4493.

Ycas, M. 1972. Biological effects in the early atmosphere. *Nature 238*, 163–164.

[6]

The cycling of materials in disturbed environments

M. J. CHADWICK *Department of Biology, University of York, U.K.*

Introduction

Commoner (1972) gives some cogent advice to the enthusiastic yet relatively untutored 'ecologist' that current interest in environmental matters seems to have produced. He suggests that a persistent effort to answer the question 'Where does it go?' can yield surprisingly valuable information about an ecosystem. He stresses that nothing 'goes away' when it is discarded. Materials are merely converted from one molecular form to another and become transferred from place to place. Although the concept of materials cycling has been implicit in much early ecological work, and even explicit in agriculturally based work, it is only within the last twenty-five years that the universal applicability and widespread importance of this concept has been generally recognized. This recognition, in turn, has been instrumental in calling attention to the really serious nature of environmental perturbations which otherwise might have been regarded as merely rather unfortunate local incidents.

Basic biogeochemical cycles

Ecologists recognize that functional attributes of an ecosystem are as important as the structural attributes that occupied so much early ecological work. Ecosystem function stresses the interrelationship between organisms and the non-living components of the system. Many naturally occurring elements pass from the abiotic environment to living organisms. After molecular rearrangement they are eventually released to the environment again as the products of excretion and decaying organic matter. This back-and-forth movement or flux of elements has been referred to as a *biogeochemical cycle* (Hutchinson 1944). That the cycle is driven by a one-way flow of energy has been emphasized by Odum (1963) when relating the two major functional attributes of an ecosystem.

In any biogeochemical cycle some forms of the element are far less accessible to organisms than other more available forms that are relatively easily utilized by organisms. It is usual, therefore, to distinguish between the relatively unavailable or *storage pool* of an element and the exchangeable reservoir or *active pool*. The active pool usually forms the source of an element for organisms to utilize and incorporate into their biomass, which can then be distinguished as the *utilized pool*. All three pools may exist in a range of different forms and as well as there existing fluxes between one pool and another, fluxes also exist between different forms within the same pool. All fluxes can be characterized by rate constants that depend on a whole range of environmental factors. In the diagrams that follow (Figs 1 to 5) the pools (A—storage pool; B—active pool; C—utilized pool) are distinguished as being of relatively different capacities although the sizes do not represent exact quantities.

In some cycles adjustments are made between the various pools with relative ease. These are the *gaseous cycles* involving carbon, nitrogen or oxygen, which are considered to be *cyclic* or perfect because the natural negative feedback pathways (c and d in the diagrams) balance well the positive feed back pathways (a and b in the diagrams), when subjected to some rather unnatural influences. Other biogeochemical cycles, the *sedimentary cycles*, are characterized by relatively large amounts in the unavailable storage pool. These cycles tend to be less perfect and are regarded as *acyclic*. Hutchinson (1948) has pointed out that man, acting as 'a mighty geological agent' (Vernadsky 1945) has speeded up many of these cycles. Thus man dislodges many cycles from their relatively *natural form* (1) to a considerably *perturbed form* (2) characterized by a high degree of acyclism. If the resource is to be renewed, to enable continuing reutilization, it must be established in a *recycling form* (3).

It is the purpose of this paper to examine a number of these cycles in their relatively natural and perturbed forms and to indicate the effect of recycling. This will be done using extremely simplified 'figure-of-eight' cycle diagrams for the nitrogen cycle, phosphorus cycle, mercury cycle, a cycle involving water utilization and disposal following extraction from aquifers and nutrient cycles on colliery spoil tips.

Nitrogen cycle

The level of nitrogen availability, in a form utilizable by plants, is extremely important in controlling biomass production in an ecosystem. Nitrogen is

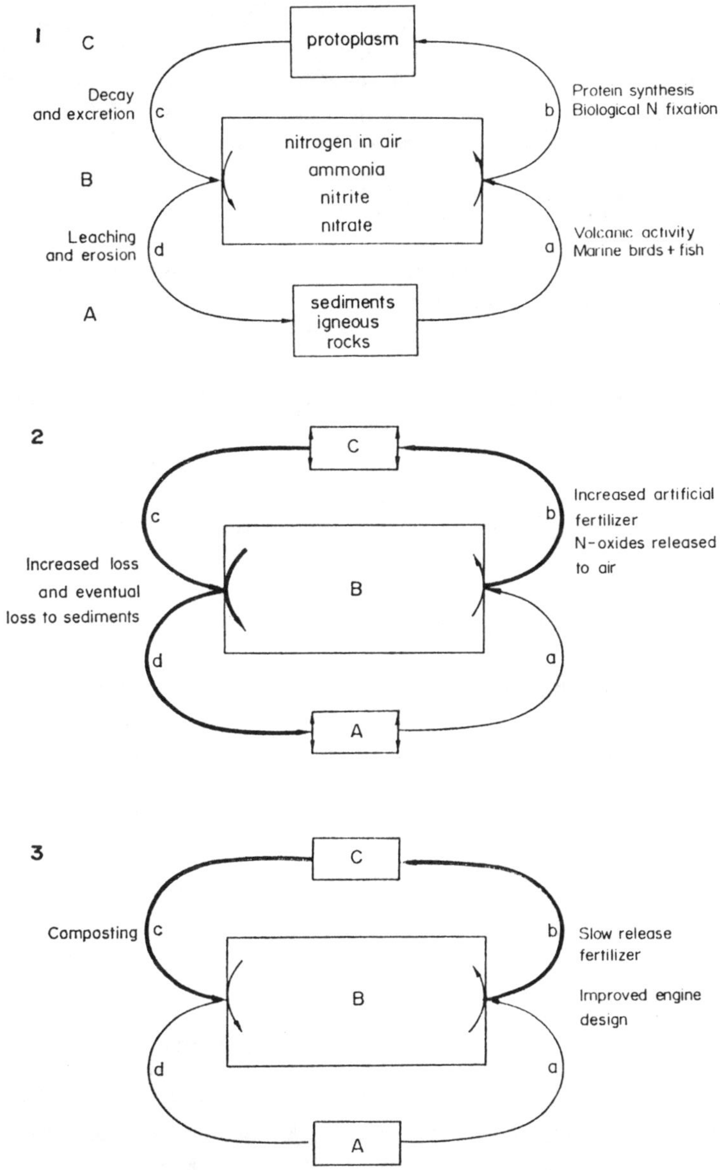

Figure 1. The nitrogen cycle. (1—'natural' cycle; 2—perturbed cycle; 3—recycling cycle; A—storage pool; B—active pool; C—utilized pool; a and b—positive feedback pathway; c and d—negative feedback pathway. Arrows in 2 indicate increasing or decreasing pool.)

required in protein synthesis and is supplied in a utilizable form by biological nitrogen fixation by microorganisms (Fig. 1 b). Under natural conditions this may occur at a rate of about 10 lb per acre per annum. Under agricultural conditions this rate can be greatly increased by the use of leguminous crops. Non-biological fixation in the atmosphere supplies only relatively small amounts. Excretion and the decay of organic matter releases fixed nitrogen (Fig. 1 c) and leaching and erosion causes losses from the active pool (B) to sediments that, with igneous rocks, make up the storage pool (A). Volcanic activity and the activities of marine birds and fish return nitrogen compounds to the active pool (Fig. 1 a).

Massive intrusion by man into the cycle is thought to be threatening the whole integrity of the nitrogen cycle (Commoner 1970). This is because nitrogen has hitherto been represented mainly in reduced forms (nitrogen gas or combined with hydrogen rather than oxygen in living things). The effects of man's intrusion on a massive scale is shown in Fig. 1–2. Technology introduces nitrogen into the environment almost entirely in oxidized forms. Power plants, automobiles and rapid increase in artificial nitrogenous fertilizer use increase the transfer represented by b, c and d in Fig. 1–2 but have little effect on a. Surface waters become polluted and as well as the disruption of biological systems nitrogen oxides in both water and air constitute potential human health hazards.

Figure 1–3 indicates that at least some improvement towards a cyclic condition can be achieved. Improved fertilizer practices (and improved engine design) can reduce the rate of loss from the active to the storage pool and the increased return of organic matter to the soil can give a constant rate of release into the active pool for subsequent uptake by plants. More balanced fluxes are established between positive and negative feedback controls leading to a more cyclic condition.

Phosphorus cycle

Phosphorus, like nitrogen, is an essential constituent of protoplasm and often limits biomass production. However the large reservoir of this element is in the storage pool, represented by phosphate rocks. Phosphate release to the active pool occurs only slowly (Fig. 2–1 a) whereas plant uptake is usually more rapid (Fig. 2–1 b). Fixation in the soil and loss to marine sediments (Fig. 2–1 d) may also be rapid. Man ahs intervened to make available in large quantities the phosphates necessary for rapid crop growth and other purposes. Phosphate deposits, that took millions of years to accumulate, are being mined extensively and much of the phosphate produced eventually finds its way into ecosystems, particularly aquatic ones, causing their altera-

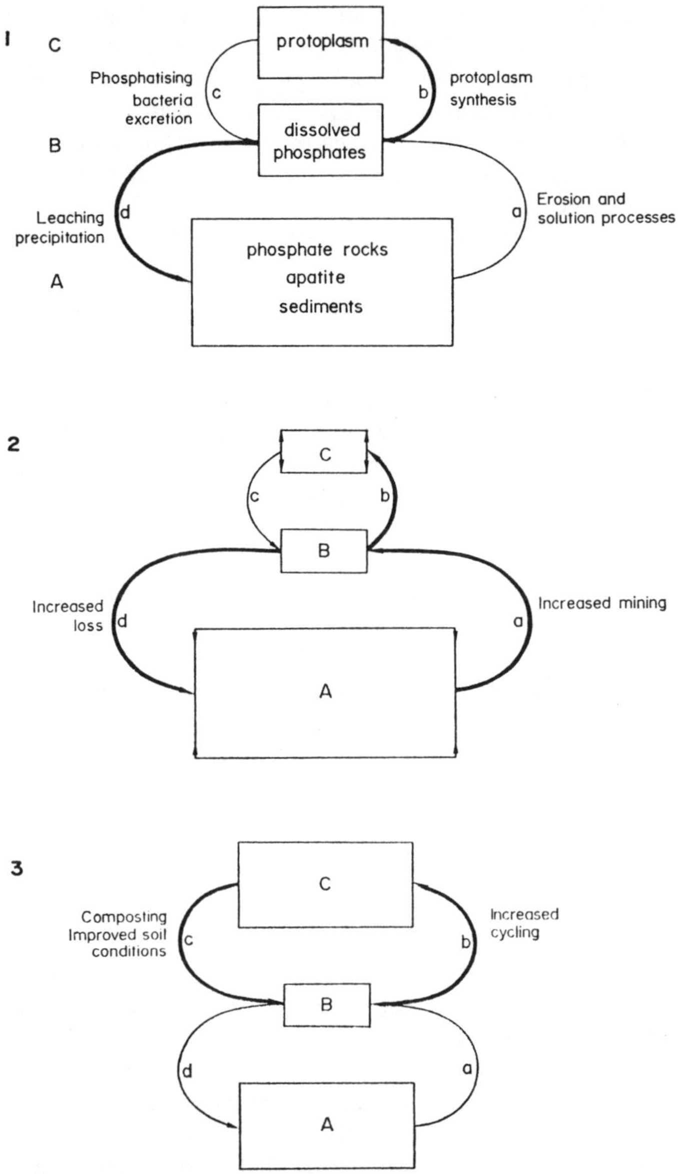

Figure 2. The phosphorus cycle. (See Fig. 1 for notation.)

tion and disruption (Hasler 1970). Eventually much of the phosphorus is lost again to the storage pool in a widely dispersed form, making it difficult to recover (Fig. 2–2). The depletion of phosphorus reserves, coupled with a largely acyclic characteristics suggest that eventually a greater degree of recycling must take place whereby composting procedures and methods of improving soil conditions to allow increased uptake of naturally released phosphate, are established (Fig. 2–3). It would be necessary to eliminate the major causes of eutrophication of acquatic systems.

Mercury cycle

Mercury is one of the less abundant elements in the Earth's crust. Under natural conditions small amounts passed from the storage pool to the active pool, mainly by weathering (Rankama & Sahama 1950). Dissolved mercury was taken up by plants and entered organisms to be returned again to the soil during decay processes (Fig. 3–1 b and c). Little information is available on natural levels and a close study of the cycling of this element has only been made since the industrial and agricultural use of mercurials has become common practice. However, mining of cinnabar and the extraction of the metal to meet industrial demand has undoubtedly caused increased rates of cycling in the biosphere. Increasing demand, coupled with the depletion of high-grade mercury mines suggests the development of a severe shortage of the element over the next thirty years (Lovering 1969). Thus the situation illustrated by the perturbed cycle (Fig. 3–2) will have to be transmuted to a recycling situation (Fig. 3–3) or alternative compounds developed.

Water utilization and renovation

The principles illustrated by general biogeochemical cycles considered on a world scale can be applied to more specific situations. In a degraded environment the object will be to move towards resource renewal by developing a recycling-type cycle out of a perturbed-type cycle. Both water utilization and disposal and vegetation establishment and maintenance on colliery spoil illustrate attempts at such management procedures.

In Britain, as elsewhere, groundwater is commonly extracted on a large scale for both public supply and industrial use. In Britain, chalk formations and the Bunter Sandstone form the principal aquifers and the rates of extraction exceed those of natural replenishment. The result is that large volumes of the aquifers are dewatered, giving rise to problems of continued extraction but also additional problems arise due to the return of the water,

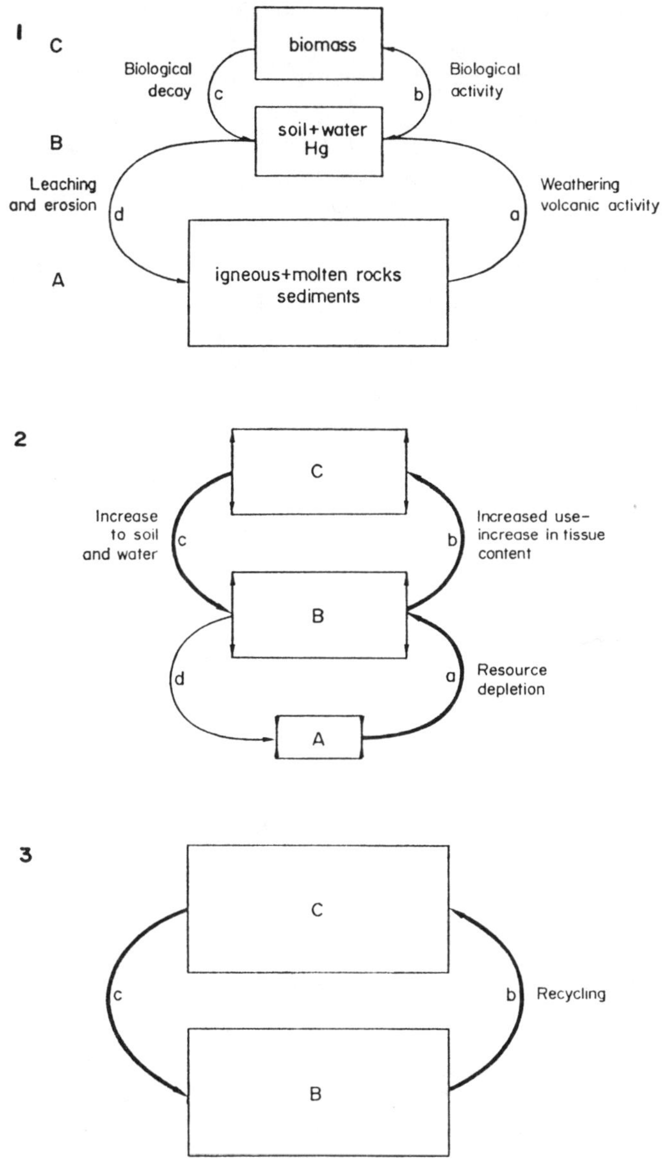

Figure 3. The mercury cycle. (See Fig. 1 for notation.)

as effluent, to surface water courses. Thus water, in an unwanted form, is discharged where it is not required and at the same time water sources, at the point they are required, are depleted. The natural and perturbed cycles are illustrated in Fig. 4–1 and 2.

It has seemed desirable to try and work out a system whereby application of the polluted water, following utilization, to a soil—vegetation system could result in renovation of the water applied, recharge of the aquifer and also possibly an increase in plant yield due to the irrigation and fertilization effect of the applied water. In a well-documented scheme in Pennsylvania (Kardos 1967; Parizek & Myers 1969; Sagmuller & Sopper 1967; Sopper 1968; Sopper & Sagmuller 1966) sewage effluent was applied to woodland and agricultural crops. These, together with the physicochemical and microbiological system represented by the soil, played a complementary role in water quality improvement. Crop yields were increased and the aquifers replenished, allowing continued water extraction. The polluting effect of the effluent on streams and rivers was considerably reduced. A similar pilot scheme in Britain, with a number of modifications, has been described by Chadwick *et al.* (1974). The modifications included the use of badly polluted river water from a tributary of a major river in place of sewage effluent and its application to natural heathland vegetation in place of tree and agricultural crops. This represents an attempt to move towards the type of recycling system shown in Fig. 4–3.

Table 1 gives an indication of the improvement in water quality obtained for three elements and also the excess of each element found in the irrigated vegetation over the amounts in similar unirrigated stands of vegetation.

Table 1. Amounts of three nutrients applied to lowland heath vegetation in polluted water, the excess of each nutrient in the irrigated vegetation (tops) and amounts in the soil water at 1·58 m depth (from Chadwick *et al.* 1974).

Nutrient	Amount applied	Amount in vegetation tops	Amount in soil water at 1·58 m
N	21·50	14·30	14·83
P	2·35	1·11	0·07
K	18·80	7·20	6·30
		(all in g. m^{-2})	

Improvement was also obtained in the alkalinity (reduction of 97%), total hardness (59%) calcium hardness (67%), conductivity (24%) sodium (5%) and anionic detergent (82%). This was using a very high rate of application (1,565 mm total, at a daily rate of 152 mm).

The attempt to renovate water by this method appears to be successful for at least some of the elements responsible for water eutrophication.

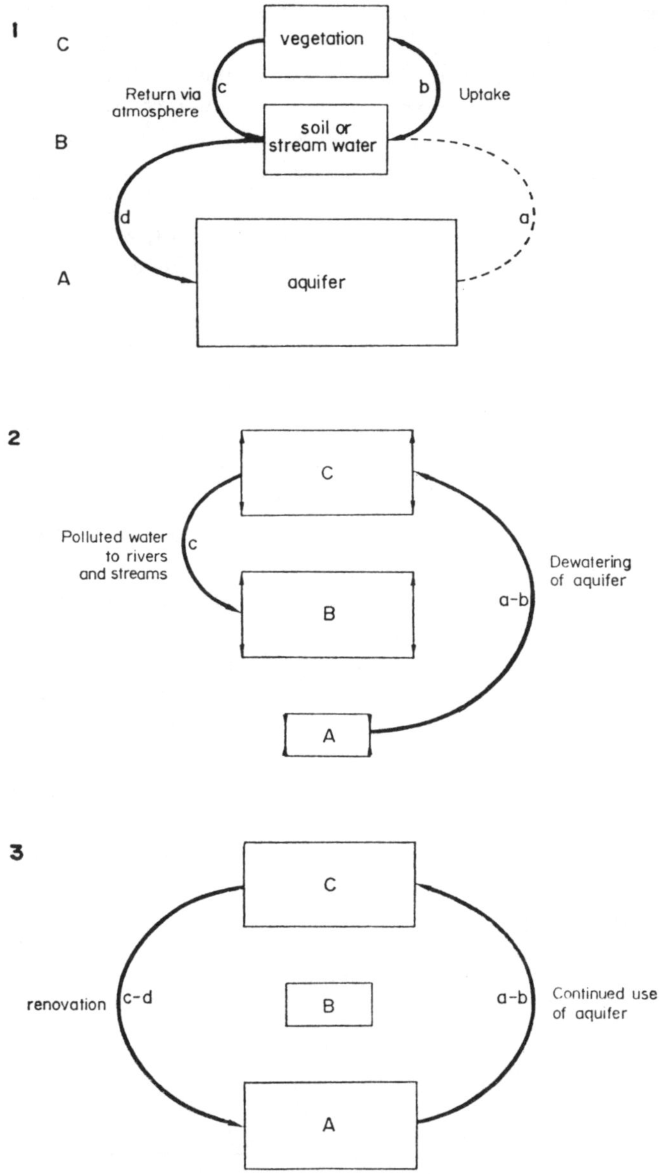

Figure 4. The cycle of water following groundwater extraction. (See Fig. 1 for notation.)

Although only partially successful in the case of nitrogen, the situation is probably better than indicated in Table 1. Most of the nitrogen applied in the ammonium form was apparently removed (96%) but much of it was probably converted to nitrate. The effect of irrigation was to increase soil pH from 3·4–4·0 to 4·7–5·4. Under higher pH conditions the ammonia oxidizing bacteria increased considerably, the nitrite oxidizers to a lesser extent and the denitrifiers hardly at all. Given faster growing plant species (conifers, grasses or arable crops) it is anticipated that nitrogen removal would have reached a higher level. Thus the indications are at least hopeful for the establishment of a recycling system.

Vegetation establishment and maintenance on colliery spoil

Studies aimed at elucidating nutrient cycles on colliery spoil heaps, in relation to reclamation work, have been carried out for some years at the University of York (Dennington, unpublished material). One of the main aims of the work has been to ascertain the input and output of nutrients from the system in an attempt to anticipate maintenance problems.

The sporadic vegetation that establishes naturally on undisturbed colliery spoil usually has available in the active pool only low levels of mineral nutrients, released only slowly from the storage pool by weathering processes (Fig. 5–1 a and b) and supplied in rainfall. Nutrients are also only slowly released from dead organic matter as the usually acid spoil conditions reduce bacterial activity so that a mat builds up on the colliery spoil surface (Fig. 5–1 c). In addition, topographical features of most tips lead to the loss of considerable amounts of nutrients in run-off water and by erosion of particulate matter (Fig. 5–1 d). Although this situation does not represent a 'natural' cycle it represents the situation before man attempts any management practices on the sites.

Table 2. Total annual nutrient input ($kg.ha^{-1}$) for three colliery spoil sites.

Site	Na	K	Ca	Mg	P	Total N	NO_3	NH_4	Organic N	Precipit cm yr^{-1}
Bullcroft	16·10	2·84	22·42	5·10	0·25	9·34	3·47	4·31	1·56	64·23
Mitchell	8·62	2·08	11·33	2·59	0·18	8·34	3·10	4·59	0·93	63·78
Maltby	10·67	3·72	16·44	5·35	0·67	8·71	3·01	4·13	1·57	63·29

Table 2 gives the total annual input of six nutrients in rainfall obtained by fortnightly collection and analysis at three colliery spoil sites in south Yorkshire. They represent levels capable of maintaining vegetation at a low level of productivity except that a proportion will be lost in run-off and a considerable proportion will be supplied during the winter months when

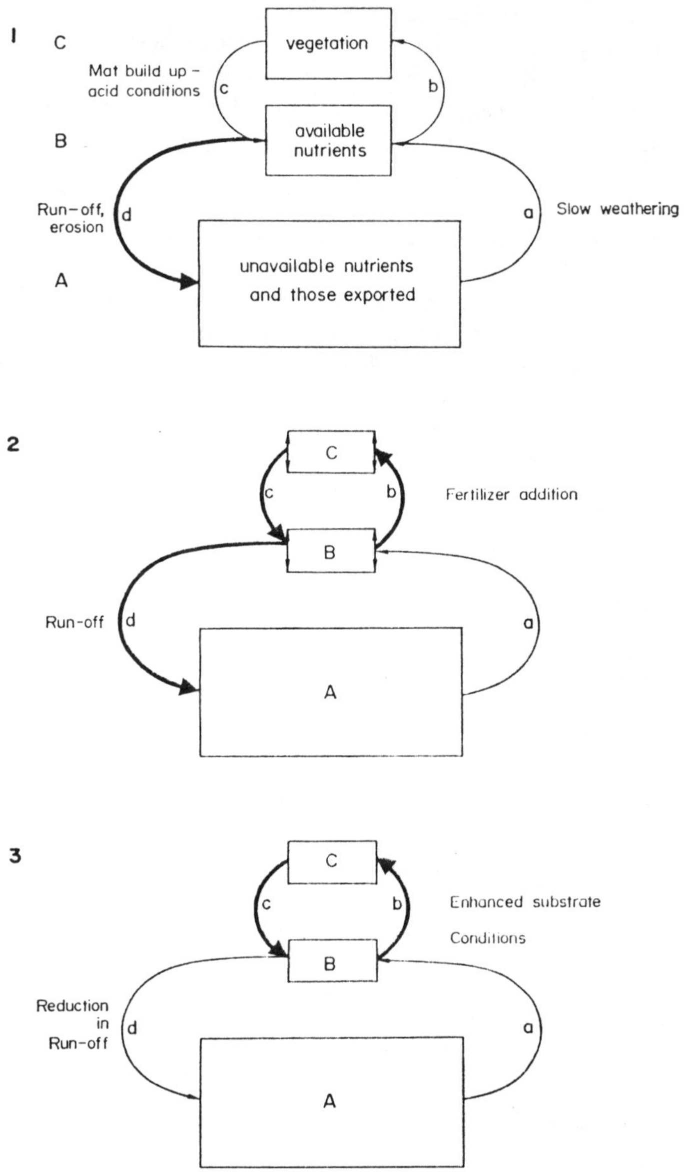

Figure 5. The nutrient cycle on colliery spoil heaps. (See Fig. 1 for notation.)

growth will be slow or have ceased entirely. For these reasons nutrients must be supplied artificially in order to establish vegetation successfully and often to maintain it. When this is done not only is the transfer from the active to the utilized pool speeded up (Fig. 5–2 b) but an improvement in substrate conditions gives increased bacterial activity and an increased rate of transfer from the utilized to the active pool (Fig. 5–2 c). Until a closed vegetational cover is established, however, it may also mean the rate of nutrient loss by run-off and erosion is still high. The effect on run-off of establishing grass species on colliery spoil is shown in Table 3. Although the effect obviously depends on both the intensity and amount of rainfall, it is not unusual to find that run-off from bare colliery spoil is three times as great as on grassed areas. It is thus possible to reduce run-off and establish something approaching a cyclic biogeochemical cycle shown in Fig. 5–3.

Table 3. Percentage run-off from grassed and bare colliery spoil sites.

Rainfall over period (cm)	Percentage run-off grassed	bare
0·203	3·35	9·85
0·584	3·42	3·42
1·727	3·94	10·42
1·854	3·56	10·00
1·941	4·74	79·27

A degraded resource can therefore become renewed but a comparison between the way in which certain nutrients are partitioned in the ecosystem (Table 4) suggests that in a relatively mature Douglas-fir ecosystem (Cole *et al.* 1969) the soil system may represent a much greater part of the active pool than for colliery spoil ecosystems that have undergone reclamation within the last ten years. Renewal to a stable state obviously requires long-term management procedures.

Table 4. Nutrient supply represented by annual precipitation supply and available soil nutrients compared with the nutrient content of the standing crop for acid colliery spoil and a second growth Douglas-fir ecosystem.

		K	Ca	P
Colliery Spoil	Precipitation	0·56	3·53	0·19
	Vegetation	1·00	1·00	1·00
	Soil	0·10	4·24	1·61
Forest	Precipitation	negligible	0·01	negligible
	Vegetation	1·00	1·00	1·00
	Soil	1·06	2·23	58·70

Conclusions

The cycling of materials is one of the two main functional attributes of any ecosystem. It is important that the cycle be understood for the widespread cycling of single nutrients. It is equally important to elucidate the cycles for specific resource management activities. The investigation of biogeochemical cycles has been responsible for demonstrating the adverse effects of massive intervention by man in some of the cycles. For many cycles, however, there are parts of the cycle that are imperfectly understood and quantitative relationships have not been established. Much information is required on residence times in the atmosphere and considerable study must be undertaken to ascertain the significant parameters that need to be determined for any cycle. For example, it has been suggested recently that particle-borne acid may be a better index of potential health hazards than the measurements of sulphur dioxide concentrations that are currently made (Brosset 1973).

Where management practices are aimed at the renewal of degraded environments it is of equal importance to determine the existing cycles and the effect of management practices on them. A positive attempt to rectify acyclic tendencies and establish balanced cycles is required rather than the following of simple rule of thumb procedures.

It is against this background that the contributions to this Symposium are made.

Summary

The importance of biogeochemical cycles in relation to disturbed environments is emphasized. Comparisons are made between the cycling of nutrients essential for plant growth and the cycling of more 'exotic' elements and compounds. The application of recycling procedures in resource renewal, in both aquatic and terrestrial situations, is stressed.

References

Brosset C. (1973) Air-borne acid. *Ambio* **2**, 2–9.

Chadwick M.J., Edworthy K.J., Rush D. & Williams P.J. (1974) Ecosystem irrigation as a means of groundwater recharge and water quality improvement. *J. appl. Ecol.* **11**, 231–47.

Cole D.W., Gessel S.P. & Dice S.F. (1969) Distribution and cycling of nitrogen, phosphorus, potassium and calcium in a second-growth Douglas-fir ecosystem. In *Symposium on Primary Productivity and Mineral Cycling in Natural Ecosystems*, University of Maine Press.

16 *M. J. Chadwick*

COMMONER B. (1970) Threats to the integrity of the nitrogen cycle: nitrogen compounds in soil, water, atmosphere and precipitation. In *Global Effects of Environmental Pollution* (Ed. by S.F. Singer), Reidel, Dordrecht, Holland.

COMMONER B. (1972) *The Closing Circle.* Cape, London.

HASLER A. D. (1970) Man-induced eutrophication of lakes. In *Global Effects of Environmental Pollution* (Ed. by S. F. Singer), Reidel, Dordrecht, Holland.

HUTCHINSON G.E. (1944) Nitrogen and the biogeochemistry of the atmosphere. *Amer. Scient.* 32, 178–95.

HUTCHINSON G.E. (1948) On living in the biosphere. *Scient. Monthly* 67, 393–8.

KARDOS L.T. (1967) Waste water renovation by the land—a living filter. In *Agriculture and the Quality of our Environment* (Ed. by N.C. Brady), A.A.A.S. Monograph 85.

LOVERING T.S. (1969) Mineral resources from the land. In *Resources and Man.* Freeman, San Francisco.

ODUM E.P. (1963) *Ecology.* Holt, Rinehart & Winston, New York.

PARIZEK R.R. & MYERS E.A. (1969) Recharge of groundwater from renovated sewage effluent by spray irrigation. In *Proc. 4th Amer. Water Res. Conf.*

RANKAMA K. & SAHAMA T.G. (1950) *Geochemistry.* Chicago.

SAGMULLER C.J. & SOPPER W.E. (1967) Effect of municipal sewage effluent irrigation on height growth of white spruce. *J.For.* 63, 822–3.

SOPPER W.E. (1968) Waste water renovation for reuse: key to optimum use of water resources. *Water Res.* 2, 471–80.

SOPPER W.E. & SAGMULLER C.J. (1966) Forest vegetation growth responses to irrigation with municipal sewage effluent. In *Proc. 1st Pan Amer. Soil Conserv. Cong.*

VERNADSKY W.I. (1945) The biosphere and the noosphere. *Amer. Sci.* 33, 1–12.

[7]

Global Chemical Cycles and Their Alterations by Man, ed. W. Stumm, pp. 137-153. Berlin: Dahlem Konferenzen 1977.

Impact of Metals on the Biosphere

J. M. Wood and E. D. Goldberg
Freshwater Biological Institute, College of Biological Sciences
University of Minnesota, Navarre, MN 55392, USA

INTRODUCTION

Before the evolution of organic matter in living systems, a number of inorganic equilibria existed which were influenced by the various physical constraints on the earth. Upon this matrix of inorganic compounds, evolution led to the synthesis of both coordination complexes and organometallic complexes. Initially, metal-ligand interactions occurred in a strictly anaerobic environment. Many primitive coordination complexes still exist; for example the ferredoxins represent electron transfer proteins designed on a system of solubilizing, by protein association, the very insoluble iron sulfide complex. Later, molecules of more adaptable design evolved which could function in a variety of catalytic processes (e.g. DNA-polymerase which has a zinc atom at its active site). A second example is the iron complex heme which initially may have evolved to facilitate the reduction of soluble inorganic ions such as sulfate and nitrate. The appearance of molecular oxygen in the biosphere caused the adaptation of heme complexes such that they could function in the reduction of oxygen to water. This reaction clearly evolved as a detoxification mechanism since "activated" molecular oxygen is exceedingly toxic to the anaerobic bacteria (4). The scene was set for the evolution of complexes which could deal with toxic elements so as to give organisms selective advantages. The environment is of critical importance to the survival of the

Manuscript received: 1976, September 20; revised: 1977, January 10.

cell. It is clear that the initial inorganic matrix in the earth'e crust forced the evolution of natural biogeochemical cycles for all elements including the toxic ones. Since we are dealing with natural biological cycles, we should not be surprised to find inorganic and organometallic equilibria for toxic elements in aqueous systems. Clearly, perturbations of these natural biological cycles occur as a result of the mining and mobilization of toxic elements by advanced industrial societies (2). In what follows we identify research areas which are fundamental to our understanding of biogeochemical cycles of metals and give examples of how these natural cycles have been affected by man's activities.

Abundance and Availability

There is a very delicate balance between the use of metals for important catalytic processes that occur in cells of higher organisms and the bioaccumulation of metals to a toxic level. This balance is not only affected by the availability and abundance of metals in the earth's crust, but also by microbial interconversions of metals to coordination and organometallic complexes in the environment. Figure 1 depicts the abundances of elements in crustal materials of the earth. There is a generally decreasing concentration with increasing atomic number.

The involvement of iron, for example, in biochemical processes, may be a consequence of (1) its high abundance, (2) the unique properties of its compounds (in the aqueous systems, the divalent forms are soluble and the trivalent forms are insoluble, existing as particulate phases or colloids) and its ability to participate in electron exchange reactions. On the other hand, titanium and aluminum are virtually unavailable for bioaccumulation because they form hydroxides that are

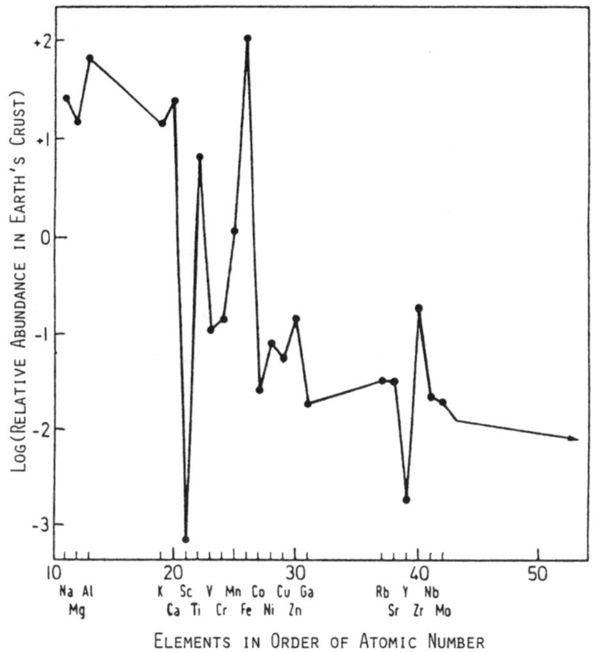

FIG. 1 - Abundance of metals in earth's crust.

very insoluble in water. Nickel and chromium are rarely found in significant concentrations inside cells; they are made biologically unavailable as a consequence of formation of stable octahedral complexes with silicates in the soil. This type of assessment shows that few metals are available, and even fewer are directly involved in biocatalysis. Besides sodium (Na^+) and potassium (K^+), which play very critical roles, only 11 other metals have been shown to be involved in important catalytic reactions in the cell. These metals either function at the active sites of enzymes or regulate enzyme activity. Typical reactions catalyzed by metalloenzymes include oxidation/reduction, hydration and dehydration, hydrolysis, or the reverse of the aldol condensation.

We find that in the metalloenzymes discovered to date by biochemists the following metals rank with respect to the number of biological complexes they form:

Fe > Mg, Mn > Ca > Zn > Cu > Co > Mo > Ni > V > Cd.

Metals of biological concern can be conveniently divided into three groups: (1) light metals, or those that are usually transported as mobile cations in aqueous solution, e.g., Na^+ and K^+; (2) transition metals, or those that are essential in electron transfer reactions e.g., Fe, Co, Cu, Mo, and may be toxic in high concentrations; and (3) heavy metals and metalloids, or those that may be required for metabolic activity at trace levels by the cell, but which as somewhat higher concentrations are toxic to the cell, e.g., Hg, Pt, Au, Pd, Sn, Be, As, Se, Te.

In order to understand the movement of toxic elements in geochemical cycles, it is necessary to examine several factors. Chemical form should be determined by distinguishing the state (solution, colloidal, and/or particulate pahses) and charge (3,5-8). The importance of speciation cannot be overemphasized; chemical behavior of any element in the environment depends on the nature of its components. In addition, the following studies are critical: the involvement of the metal species in oxidation-reduction reactions and the distribution of the species between solution, vapor and solid phases by adsorption-desorption, precipitation-dissolution, ion exchange processes, and volatilization. A geochemical approach that considers various environmental reservoirs and rates of exchange between them provides information on the overall fate of metals; the sources and geographical location of toxic metals are also important, especially in relation to the movement of toxic metals caused by man's activities.

Jernelov (1974) presents a comprehensive set of environmental characteristics and properties that, when determined, permit an evaluation of potential undesirable inputs of a metal: (1) production and emission in relation to natural flux; (2) residence times in various reservoirs; (3) bioaccumulation, both "passive" and "active"; (4) physical and chemical properties relating to dispersion (e.g., volatility, adsorption tendencies, dissociation-association reactions, formation of insoluble precipitates)(17,18,20-26,28-30,32,33); (5) toxicity to aquatic organisms; (6) toxicity to man and other mammals; (7) long-term biological effects on ecosystem metabolism; and (8) transformation reactions by organisms. Jernelov's monograph examines selected toxic elements in detail (16).

Table 1 classifies elements in terms of their relative toxicity and availability. Biologists should give careful consideration to those metals in category B. Category C deserves notice as well, because insoluble toxic elements can be rendered soluble by man-made chelating agents such as polyphosphates and nitrilotriacetic acid (NTA), both of which are added to our water supply via detergents. For example, the very toxic lanthanides will complex with polyphosphate or NTA.

Toxic metals that form organometallic compounds, especially the metalalkyls (e.g. methylmercury), deserve special concern. Most metalalkyls are poisonous to the central nervous systems of higher organisms, and these compounds do accumulate in cells (34,35). Metalalkyls that are stable in water and can be synthesized include the following toxic elements: Hg, Sn, As, Se, Te, Pd, Pt, Au, Tl, and Pb (13,16,36,40,42).

Mercury, lead, and cadmium are metals of great concern because of their high toxicity and wide dispersal in the environment. The case of mercury pollution has clearly demonstrated the profound importance of understanding biologically-mediated transformation reactions that yield metalorganic compounds with a high potential for bioaccumulation and toxicity. It has also illustrated the value of fundamental information on physical-chemical states of an element (vapor, aqueous, liquid, solids), its association-dissociation reactions (14,15), oxidation-reduction processes [HG(II), Hg(I), and Hg^0], adsorption of molecules and ions in solution, and distribution equilibria between phases (biota, sediments, minerals). Experience with mercury on a global level has clarified the importance of its physical-chemical properties in the cycling of the element through the air-water-land system. Under natural, or "preman" conditions, mercury vapor is a key species in the geochemical cycle of mercury through the atmospheric, oceanic, and soil reservoirs. As shown by Garrels et al (9), present-day mining and industrial emissions have resulted in mobilization and high flux rates in the mercury cycle. It is estimated that man's activities have increased the flow of mercury from the land to the oceans via rivers almost four-fold. On a planetary basis, the mercury cycle is now more clearly understood as far as the relative contributions of vaporization of elemental mercury from the land and oceans, man-caused emissions, rainout, river transport, and

TABLE 1 - Classification of Elements According to their Toxicity[1]

Noncritical	Very toxic and relatively accessible	Toxic but very insoluble or very rare
Na	Be	Ti
K	Co	Hf
Mg	Ni	Zr
Ca	Cu	W
H	Zn	Nb
O	Sn[4]	Ta
N	As[4]	Re
C	Se[4]	Ga
P	Te[4]	La[3]
Fe	Pd[4]	Os
S	Ag	Rh
Cl	Cd	Ir
Br	Pt[4]	Ru
Fe	Au[4]	Ba
Li	Hg[4]	
Rb	Tl[4]	
Sr	Pb[4]	
Al	Sb	
Si	Bi	
F[2]		

1. Elements omitted from this table should not be neglected in the environmental sense. For example, iodine and manganese are important elements, but they fit more than one category for the above classification.

2. Some may argue with this designation, but we do add fluoride to drinking water.

3. All the lanthanides are very insoluble and some are very rare.

4. Metals having metal alkyls which are stable in aqueous systems which have been reported to be biomethylated.

oceanic sedimentation are concerned (Fig.3). On a local, regional or ecosystem basis (rivers, lakes, coastal environments, soils), the potential rate-determining role of biomethylation of mercury is now recognized, although few predictive models are available; we have but limited information on actual methylation and degradation rates under specific environmental conditions.

Experience with the mercury cycle can be extrapolated to other metals, and these can immediately be separated into metals that are capable of forming stable metal alkyls and those that are not. Methyl-B_{12} is an environmentally important molecule because it can react with a number of metal ions or salts of metalloids (such as arsenic and selenium), to produce metalalkyls that bioaccumulate in the central nervous system of higher organisms. Column B in Table 1 contains toxic elements that react with methyl-B_{12}.

The Importance of Microorganisms to the Movement of Toxic Elements

Microorganisms are exceedingly versatile in the way they metabolize natural substances. If microorganisms do not degrade a particular compound, higher organisms are unlikely to be able to do so. Therefore, a study of microbial interconversions of toxic compounds can contribute substantially to environmental science. Toxic elements move through the environment under natural conditions by way of biogeochemical cycles (Fig. 2). Man's activities provide new sources of toxic elements by moving them out of the geocycles and making them available to the biological cycles. The availability of alternative metabolic pathways facilitates the movement of toxic compounds in the environment. When confronted with a toxic substance, microorganisms frequently adapt to detoxify it. However, the resulting product may be either more or less toxic to higher organisms (41,43,45,46).

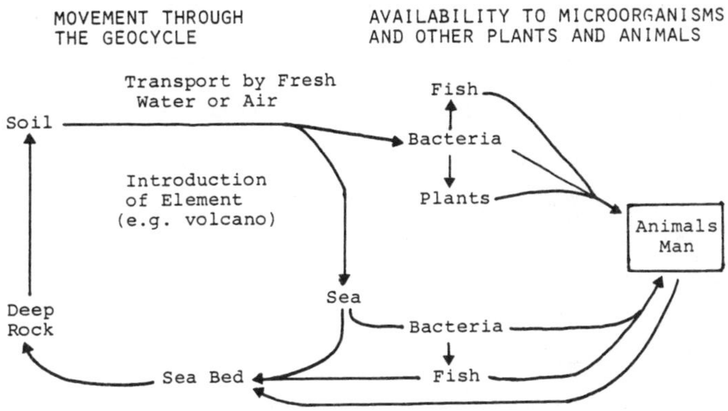

FIG. 2 - Movement of toxic elements through the geocycle.

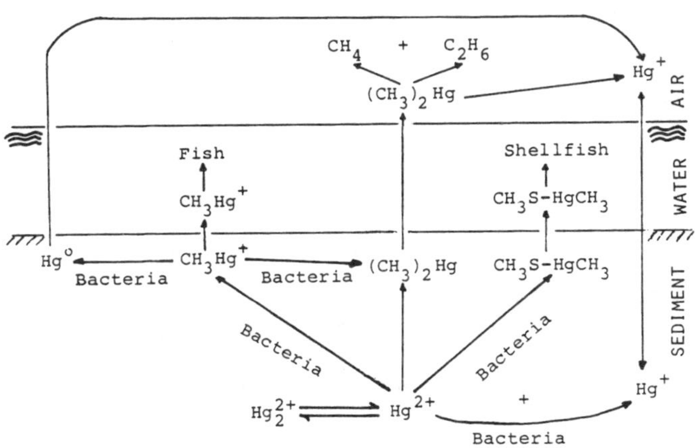

FIG. 3 - Biological cycle for mercury.

Monitoring the steady state concentration of an individual metal in a given ecosystem is of limited value because of the interrelationships between cycles for different metals. Analysis of metal ions with similar chemical, physical, and biological properties allows one to predict the movement of individual metal species in the total system. For example, inorganic cobalt is fixed by certain primitive species of bacteria and incorporated into the Vitamin-B_{12} molecule. Vitamin-B_{12} forms the active site of a number of important enzymes, for example, enzymes involved in the biosynthesis of methane gas and of the essential amino acid methionine. In addition, Vitamin-B_{12} coenzyme is the active site for a group of enzymes called the "mutases", and for the enzyme ribonucleotide reductase, which catalyzes a crucial step for DNA synthesis in some organisms. After synthesis by microorganisms, Vitamin-B_{12} enters food chains and finally is consumed by man. A special protein called "Intrinsic Factor" has evolved in higher organisms, which complexes with Vitamin-B_{12} (molecular weight of approximately 1,400) to form a much larger entity with a molecular weight of approximately 65,OOO (19). This complex is transported into the blood stream, where a second group of B_{12}-binding proteins, called the "transcobalamins are responsible for transport of B_{12} in circulatory systems and function in cellular uptake of the vitamin (19,44).

In addition to their role in the transport of cobalt, microorganisms can readily change the oxidation state of the metal, which in turn changes the chemical reactivity of the cobalt atom. For Vitamin-B_{12}, the cobalt atom can exist in three alternative oxidation states:

$$\mathrm{Co(III)} + e \longleftrightarrow \mathrm{CO(II)} + e \longleftrightarrow \mathrm{Co(I)}$$

The Co(I) variety of Vitamin-B_{12} is very important because it is a good reducing agent and it can be alkylated to produce stable alkyl-B_{12} derivatives. For example, methyl-B_{12} is synthesized in biological systems by the reaction of $Co(I)B_{12}$ and N^5 methyltetrahydrofolic acid:

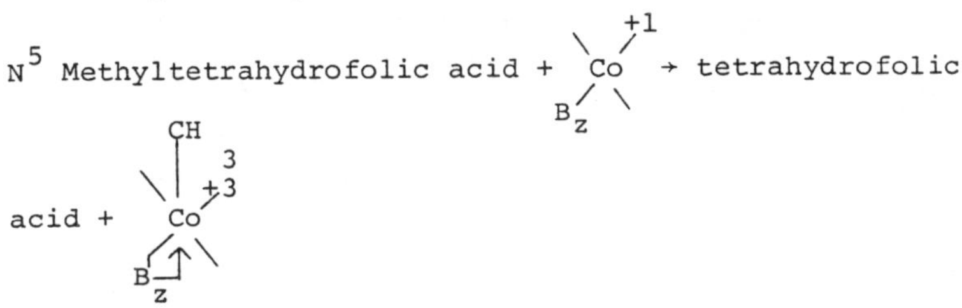

Methyl-B_{12} holds a key position in the movement of toxic elements in the biosphere because of its involvement in the biomethylation of a number of heavy metals and metalloids (37-39).

The Mobilization of Metals by Man

Technological societies are introducing heavy metals to the environment at rates that are comparable to those in the major sedimentary cycle(10). Releases from agricultural, industrial, and social activities can be directly to the atmosphere, to rivers and to the oceans by commercial and industrial sewer outfalls, or by dumping on land and sea. Two high temperature processes, the combustion of fossil fuels and the production of cement, are responsible for the release of metals to the atmosphere in amounts that are slightly less than those carried from the continents to the oceans by rivers (Table 2). Since the greater part of these industrial activities take place in the mid-latitudes of the northern hemisphere, changes in heavy metal compositions of natural waters and of air will be most evident at these latitudes. Specific uses of some metals can allow their entry to the environment in amounts that are similar to those in weathering cycles.

TABLE 2 - Fluxes of metals to the atmosphere from fossil fuel combustion and from cement production and to the oceans from the rivers. (Units of 10^9 grams/year.)

Element	Fossil Fuel Combustion	Cement Production	Weathering Mobilization
Zn	7	32	720
Pb	3.6	30	110
As	0.7	3.2	62
Se	0.45	0.71	7.2
Hg	1.6	0.1	2.5

For example, copper and cupric oxides are used in anti-fouling paints for the protection of ships. To prevent the attachment and subsequent growth of organisms, the copper must leach into the water coating the hull of the ship at a rapid enough rate to interfere with the organisms' activity and yet not so rapid as to exhaust the copper in a short time. Goldberg, (10) suggests that copper leaching from merchant ships and naval vessels is about one percent of the flux of copper from the rivers to the oceans. The influence of the leached copper may be evident in harbors, coastal areas or shipping lanes.

The impact of the dispersion of metals by man upon the oceans is most clearly seen in the coastal waters, the estuaries, lagoons and harbors. In the sediments of the Baltic Sea, there are enrichments of cadmium by seven-fold over the concentrations in materials deposited a century ago. Of lead, by four-fold; of zinc by three-fold and of copper by two-fold. These increases were attributed to the combustion of coal with the subsequent dispersal of fly-ash containing these heavy metals.

Similar enrichments of heavy metals in coastal U.S. deposits have been reported. Usually, the contributions of anthropogenic metals from winds, sewer outfalls, storm runoff, and river runoff are extremely difficult, if not impossible, to ascertain. But the fingerprints of man's technology are readily seen in the enhanced metal contents of his airs and waters.

Historical Records of Man's Mobilization of Metals

The history of man's mobilization of heavy metals may be recorded in sedimentary deposits -- marine, lacustrine, glacial and estuarine as well as in those of reservoirs and fresh waters associated with urban living. In addition exoskeletal materials of organisms may hold such records.

The records of anthropogenic lead fallout from the atmosphere are preserved in glaciers. Increasing entries of lead to the atmosphere over the past several thousand years resulted from the increasing uses of this metal by man. There are especially sharp increases beginning in the 1940s as a consequence to the combustion of lead alkyls in gasolines. Changes in lead fluxes to coastal environments are seen in their sediments. For example, of the 24 tons of lead burned daily in internal combustion engines on the Los Angeles area, 0.5 tons are calculated to have been deposited in the 12,300 square kilometers coastal zone.

Pollution records of radioactive strontium-90 fallout from bomb-testing has been found in the annual layers deposited by corals in Entiwetok Atoll. Similarly, levels of mercury in aqueous environments over the past century have been sought in museum specimens of fish and in their recently caught counterparts (10).

REFERENCES

[1] Baes, C.F., Jr. 1973. The Properties of Cadmium, pp. 29-46. In Cadmium, The Dissipated Element. Edited by W. Fulkerson and H.E. Goeller. ORNL.

[2] Bowen, H.J.M. 1975. Trace Elements in Biochemistry, Academic Press.

[3] Crerar and Barnes. 1974. Geochim et Cosmochim Acta 38,2: 279-300.

[4] Dagley, S. 1975. A Biochemical Approach to Some Problems of Environmental Pollution. Essays in Biochemistry, edited by P.N. Campbell. Biochemical Society.

[5] Eaton, A.D. 1974. Marine Geochemistry of Cadmium, Thesis, Harvard University.

[6] Friberg, L., et al. 1974. Cadmium in the Environment, 2nd ed. Cleveland: CRC Press.

[7] Gardiner, J. 1974a. The Chemistry of Cadmium in Natural Water - I. A study of Cd complex formation using the cadmium specific ion electrode. Water Research 8: 23-30.

[8] Gardiner, J. 1974b. The Chemistry of Cadmium in Natural Water - II. The adsorption of cadmium on river muds and naturally occurring solids. Water Research 8: 157-64.

[9] Garrels, R.M.; Mackenzie, F.T.; and Hunt, C. 1975. Chemical Cycles and the Global Environment. Los Altos: Wm. Kaufmann, Inc.

[10] Goldberg, E.D. 1976. The Health of the Oceans. The Unesco Press. Paris: 172 pp.

[11] Guy, R.D.; Chakrabarti, C.L.; and Schramm, L.L. 1975. The Application of a Simple Chemical Model of Natural Waters to Metal Fixation in Particulate Matter. Can. J. Chem. 53: 661.

[12] Hohl, H., and Stumm, W. 1975. Interactions of Pb_2 with Hydrous γ-Al_2O_3. 49th Nat. Colloid Sympos., Potsdam, N.Y.

[13] Jarvie, A.W.P.; Markall, R.N.; and Potter, H.R. 1975. Nature 255: 217-218.

[14] Jellinek, H.H.G., and Sangal, S.P. 1972. Complexation of Metal Ions with Natural Polyelectrolytes. Water Research 6: 305.

[15] Jenne, E.A. Mercury in Waters of the United States 1970-71. U.S. Department of the Interior Geological Survey.

[16] Jernelov, A. 1974. Heavy Metals, Metalloids, and Synthetic Organics. In, the Sea, V. 5, edited by E.D. Goldberg.

[17] Kubota, J., et. al. 1974. Lead, Cd, Zn, Cu, and Co in Streams and Lake Waters of Cayuga Lake Basin, New York. Environ. Sci. Technol. 8: 243.

[18] Leckie, J.O., and Nelson, M.B. 1975. Role of Natural Heterogeneous Sulfide Systems in Controlling the Concentration and Distribution of Heavy Metals. 2nd Int. Sympos. Environ. Biogeochem., Burlington, Ont., Canada. April.

[19] Lien, E.L.; Ellenbogen, L.; Law, P.Y.; and Wood, J.M. Studies on the Mechanism of Cobalamin-Binding to Hog Intrinsic Factor. In, Journal of Biological Chemistry 249: 890-894.

[20] Morel, F.M., and Morgan, J.J. 1972. A Numerical Method for Computing Equilibria in Aqueous Chemical Systems 6: 58.

[21] Morel, F.M.; McDuff, R.E.; and Morgan, J.J. 1973. Interactions and Chemostatis in Aquatic Chemical Systems: Role of pH, pE, Solubility, and Complexation. In, Trace Metals and Metal-Organic Interactions in Natural Waters, edited by P.C. Singer. Ann Arbor Science.

[22] Morel, F.M.; Westal, J.C.; O'Melia, C.R.; and Morgan, J.J. 1975. Fate of Trace Metals in Los Angeles Co. Wastewater Discharge. Environ. Sci. Technol. 9: 756.

[23] Morgan, J.J., and Sibley, T.H. 1975. Chemical Models for Metals in Coastal Environments. ASCE Specialty Conference on Ocean Engineering, Dover, Delaware.

[24] Ramamoorthy, S., and Manning, P.G. 1975. Equilibrium Studies of Solutions Containing Al^{3+}, Cab_2^+ and Cysteine, Orthophosphate and a Carboxylic Acid. J. Inorg. Nucl. Chem. 37: 363-67.

[25] Rohatgi, N., and Chen, K.Y. 1975. Transport of Trace Metals by Suspended Particulates upon Mixing with Seawater. Journal of the Water Pollution Control Federation 47: 2298.

[26] Schindler, P.W. 1975. The Regulation of Trace Metal Concentrations in Natural Water Systems: A Chemical Approach. Int. Assoc. Great Lakes Res. Sympos. Geneva Park, Ontario: October.

[27] Schmidt, U., and Huber, F. 1975. Nature 259: 157-158.

[28] Sibley, T.H., and Morgan, J.J. 1976. Equilibrium Speciation of Trace Metals in Freshwater: Seawater Mixtures. Caltech, Keck Laboratories.

[29] Sillen, L.G., and Martell, A.E. 1964. Stability Constants of Metal-Ion Complexes. London: The Chemical Society; + Supplement No. 1 (1971).

[30] Stumm, W., and Brauner, P.A. 1975. Chemical Speciation. In, Chemical Oceanography, edited by J.P. Riley and G. Skirrow. 2nd Ed., Academic Press.

[31] Sugiura et. al. JACS (in press) (attached).

[32] Vuceta, J. 1976. Adsorption of Pb(II) and Cu(II) on α-Quartz from Aqueous Solution. Thesis, Caltech.

[33] Walker, M.O., and Williams, D.R. 1974. Thermodynamic Considerations in Coordination. XVII. Formation Constants for Cd(II)-Amino Acid Complexes as Determined by Glass and Solid-State Cadmium-Electrode Potentiometry 1186-89. J.C.S.Dalton.

[34] Westoo, G. 1967. Methylmercury in Fish. Acta Chemica Scandanavica 21: 1790-1797.

[35] Westoo, G. 1967. Methylmercury in Food. Var Foda 19: 121-127.

[36] Wong, P.T.S.; Chan, Y.K.; and Luxon, P.L. 1975. Nature 253: 263-264.

[37] Wood, J.M. 1971. Environmental Pollution by Mercury, pp. 39-56. Advances in Environmental Science and Technology, Volume II, Edited by J.N. Pitts and R.L. Metcalf.

[38] Wood, J.M. 1972. Interconversions of Poisonous Elements in the Environment, pp. 87-95. SOCMA medal lecture. American Paint Journal, October.

[39] Wood, J.M. 1973. Toxic Elements in the Environment, pp. 1-8. Revue International d'Oceanographie, Medicale, Edited by M. Aubert. CERBOM 31: 32.

[40] Wood, J.M. 1974a. Biological Cycles for Elements in the Environment. Science 183: 1049-1054.

[41] Wood, J.M. 1974b. Metabolic Cycles for Toxic Elements in the Environment, pp. 105-112. A Study of Kinetics and Mechanism, Edited by P. Krenkel.

[42] Wood, J.M. 1975a. Biological Cycles for Elements in the Environment. Die Naturwissenschaften 8: 357-365.

[43] Wood, J.M. 1975b. Mechanisms for the Biosynthesis and Neurotoxicity of Methylmercury, pp. 377-389. Organotransition-Metal Chemistry, Edited by Y. Ishii and M. Tsutsui.

[44] Wood, J.M., and Brown, D.G. 1972. The Chemistry of Vitamin B_{12}-enzymes, pp. 47-105. Structure and Bonding, Springer Verlag II.

[45] Wood, J.M.; Kennedy, F.S.; and Rosen, C.G. 1968. The Synthesis of Methylmercury Compounds by Methanogenic Bacteria. Nature 220: 173-175.

[46] Wood, J.M.; Penley, M.W.; and DeSimone, R.E. 1972. Mercury Contamination of Man and His Environment. In, Technical Reports, No. 137: 49-65. International Atomic Energy Commission, Vienna.

96-116

[83]

[8]

In: *Ecological Effects of Acid Deposition.*
National Swedish Environment Protection Board – Report PM 1636 (1983).

φ25

Sulfur and nitrogen over North America

RUDOLF B. HUSAR and JANET M. HOLLOWAY

Washington University, Center for Air Pollution Impact and Trend Analysis (CAPITA), Box 1185, St. Louis, MO 63130, USA

ABSTRACT

Coal and oil combustion and smelting of metals results in sulfur oxide emission rate of about 15 Tg S/yr over North America (NAM), excluding Mexico. Historically, the estimated NAM emission rate fluctuated between 10-15 Tg S/yr since 1910. The current nitrogen oxides emission rate over NAM is about 6 Tg N/yr from manmade sources and has increased steadily since the turn of the century. For current emissions the potential proton donor capability of SO_x emissions (about $1x10^{12}$ eqv) accounts for two thirds of the total equivalents, but the fraction of nitrogen contribution has been and probably will continue to increase. Since the early part of the century the SO_x emission density has shifted from New York, New England toward West Virginia, Kentucky, and Tennessee. Wet deposition data from five monitoring networks over NAM yield an average sulfur wet deposition of 0.3 g S/m^2,yr (6.2 Tg S/yr). About 20-40% of this total NAM S deposition is attributed to natural background sources. For the eastern third of North America (ENA) the average excess sulfur deposition rate is about 0.6 Tg S/m^2,yr (3.7 Tg/yr) and the "background" is estimated 12-25% of the ENA total. Over ENA the wet deposited manmade sulfur accounts for only 25-30% of the ENA emitted sulfur. The remaining 70-75% is evidently dry deposited and exported to the Atlantic. Similar analysis of precipitation nitrate data gives 2.3 Tg N/yr for NAM, 20-40% of which is attributed to "natural" sources. About 20% of the manmade NO_x emissions are accounted for by wet deposition. The remainder is evidently dry deposited or exported from the continent. The ammonia wet deposition NAM is about 2 Tg N/yr with the highest deposition rates recorded in midwestern cornbelt regions. A tentative emission inventory using cattle and pig population and fertilizer consumption data suggests that these three sources could be responsible for much of the deposited ammonia. The annual average airborne sulfate concentration over the eastern U.S. is about 3 $\mu g\ S/m^3$. Visual range observations from an extensive synoptic network are used as a surrogate for sulfate to explore spatial-temporal trends over the past three decades. The strongest increase since the 1950's was recorded over the southeastern U.S. in the summertime and in the northwestern U.S. and adjacent Canada in the wintertime. In the Boston-Washington corridor there has been a decrease of the wintertime haziness. The haze trend data support the obvious but far reaching statement that every source region, even in the era of "long range transport", impacts on itself most.

INTRODUCTION

An examination of the trends over the past three decades reveals that there has been a remarkable shift in the nature and perception of air pollution over the industrialized regions of eastern North America:

1. Until the 1950's most of the concern was focused on primary emissions from smokestacks and tailpipes, including flyash, SO_2, CO, NO_x, hydrocarbons, etc. By the 1970's much of the attention was directed toward secondary pollutants, i.e. species that are formed within the atmosphere as chemical reaction products of primary emissions. These include O_3 arising from NO_x and hydrocarbons, sulfuric acid and ammonium sulfate, the reaction products of SO_2; and nitric acid, the product of NO_x and hydrocarbon emissions.

2. In the 1960's gaseous pollutants, e.g. SO_2, NO, NO_2, CO, were generally considered more noxious than the particles. By the late 1970's, the major effects of air pollution on health and welfare were attributed to particles, specifically fine particles which are carriers of sulfates, nitrates and organics as well as of hydrogen ion.

3. Since the secondary pollutants, such as O_3, H_2SO_4 and HNO_3, are not formed until about one day after the emission of their precursors, their region of impact is far downwind (order of 500 or more km) from the source at any given time. Hence, there has been a shift in interest from the near source impact (<50 km) of primary emissions to the long distance impact (500 km) of secondary pollutants.

4. In the 1960's much of the impetus for the control of air pollutants arose from their effects on human health. By the 1970's, in addition to the health concerns, the effects on welfare including acidification of precipitation and deterioration of visual air quality have gained more importance.

5. As a direct consequence of the shift to secondary pollutants and long range transport, the "airsheds" (i.e. the regions of impact) have shifted from the old Air Quality Control Regions (AQCR) having an order of 100 km size to larger multistate regions. This political readjustment to the needs imposed by long range transport has not yet been accomplished.

6. As in Europe, "airsheds" or regions of impact cover subcontinental areas and consequently long range transport of secondary pollutants constitutes an international problem requiring attention from the respective departments of state as well federal environmental agencies.

Currently, in eastern North America, the per capita emission rate of sulfur and nitrogen is about 300 g S and 100 g N per day. All of that sulfur and nitrogen returns to the earth surface, but how long does it reside in the atmosphere, what transformations will occur during atmospheric transport and what is the location and chemical/biological fate of the deposited material? Both sulfur and nitrogen are regarded by life scientists as "friendly" elements, as food rather than hazard, as essential to life as water, energy and carbon dioxide. However, excess exposure to sulfur and nitrogen compounds may be life threatening. How much did man's activity perturb the natural cycle of these elements and what is the tolerance range of the ecosystem?

Source-Receptor Relationship

The source identification of primary emissions is facilitated by the simple fact that their concentration is highest near the source. This allows the simple backward tracing, sometimes by visual observation of plumes as pointers to the stack of their origin. Unfortunately, for the quantification of the source identification of secondary pollutants SO_4 or NO_3, currently not a single direct method exists. Some of the main reasons are listed below:

a. Sulfates and nitrates are formed many hours or days after their emission as illustrated schematically in Fig. 1. Hence, their formation rate in the atmosphere can not be measured at the source. Furthermore, the secondary pollutant formation process depends on such factors as solar radiation and the concentration of other pollutants, e.g. hydrocarbons. Thus, the amount of secondary material from a given emission rate of precursor gases is not constant but depends on the environment.

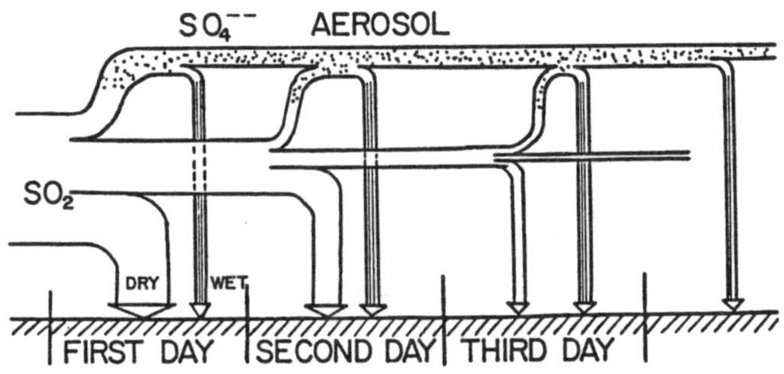

Fig. 1. During its residence time SO_2 is transformed to SO_4 and removed by dry and wet deposition. The sulfate itself is subject to dry and wet deposition. Typically, about one half of the SO_2 is transformed or removed during the first day of its atmospheric residence.

b. The residence time of sulfates in the atmosphere is from a few days to a week and the average transport distance at any given day exceeds 500 km as illustrated in Fig. 2. Within that distance the contributions of many sources are generally superimposed.

c. Long range transport of primary-secondary chemical complex results in the superposition and chemical interaction of different types of sources (e.g. enhanced reactions of a power plant plume imbedded in a reactive urban plume). In the case of secondary fine particle formation (Fig. 3), sulfates, nitrates and organics from different sources are frequently merged into a single internal mixture within each particle. At present only the bulk chemical composition is known for mixture of particles from different sources. Inherently, this precludes a simple separation of source contributions.

Fig. 2. Comparison with scales of transport for one day (500 km), two days (1000 km) and three days (1500 km) with the size of the North American continent.

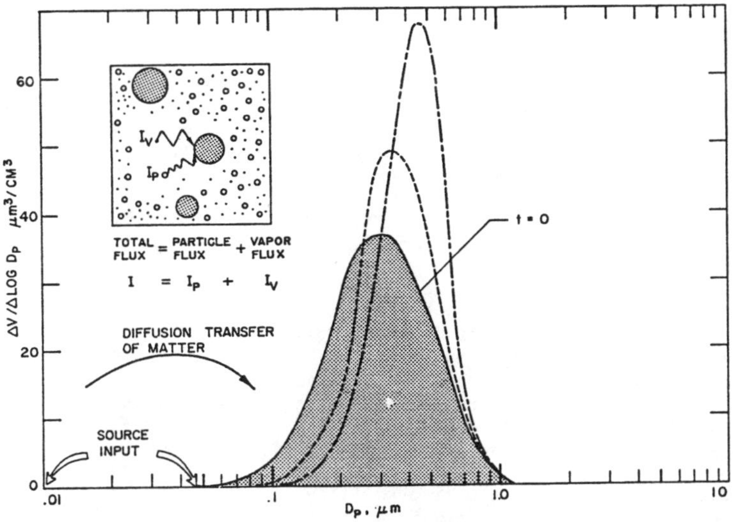

Fig. 3. Illustration of aerosol growth mechanism by coagulation (particle flux, I_P) and by condensation (vapor flux, I_V).

Because of the complexities of the regional scale source receptor quantification for secondary pollutants, a number of diverse techniques are currently utilized. These include receptor oriented methods such as chemical "finger printing" of fine particles for their general source type, back trajectory analysis to show the general direction of the source region from the receptor region, and the trend analysis of emissions and ambient concentrations. Other source oriented approaches include direct field observations of transport, transformation and removal. Finally, in the "diagnostic modeling" approach, pioneered in Scandinavia, simultaneous use is made of existing source inventories, wind fields and ambient concentration data as input into the diagnostic model with the task of deciphering what is happening between the source and the receptor. In what follows the emission, transport, concentration and deposition data are summarized for North America.

The above points and questions constitute the motivation and context for the present summary report in which the fine particles, SO_x, and NO_x patterns over North America are assessed.

MANMADE SO_x AND NO_x EMISSION PATTERN AND TRENDS

Coal and oil combustion, along with the smelting of metals, constitutes the bulk of the anthropogenic sulfur emissions to the atmosphere over North America. Currently, extensive emission inventories exist for manmade SO_x and NO_x but they are rather opaque with regard to the sources of information and procedures used in their construction. For purposes of historical emission estimates, it is more advantageous to use the fuel consumption and metal production data directly, along with estimates of emission factors, since historical fuel use and smelting data are more readily available.

In 1974, the sulfur content of the U.S. coal mined east of the Mississippi River weighed by the production of each mining region was about 2.4%, and the coal energy content was 26 megajoules/kg. Hence, the sulfur content is 0.87 g S/megajoule of coal. The emission rate of sulfur is then the (coal consumption rate) x (the average sulfur content) x (the fraction of S emitted to the atmosphere). Tentatively, it is estimated that in 1970 all of the coal sulfur is emitted (emission fraction=1.0), but that in 1850, this emission fraction was only 0.6; i.e. 40% of the sulfur being retained in the ash or lost otherwise. Tentatively, the S emision fraction was linearly interpolated between 1850 and 1970. The resulting U.S. sulfur emission estimates over the past century are shown in Fig. 4. Evidently, since 1910, the national sulfur emission from coal combustion ranged between 10 and 15 million tons S/yr.

The sulfur content of crude oil consumed in the U.S. is about 0.5% by weight, and the energy content is about 5.8 million Btu/barrel, or 42 megajoules/kg. Accordingly, the average sulfur content for crude oil is 0.12 kg/megajoule . The national sulfur emissions from crude oil products, using an emission factor of 0.12 g/megajoule are also shown in Fig. 4a. Currently, these estimates have an uncertainty factor of about ±30%.

Sulfur emissions trends from smelting of metals are even less certain. From emission inventories, compiled in the 1970's, an average emission factor of 1 million tons S/million tons of copper produced was obtained. Assuming this factor to be applicable historically, the S emissions from smelters were estimated from the historical copper production rate (Fig. 4a).

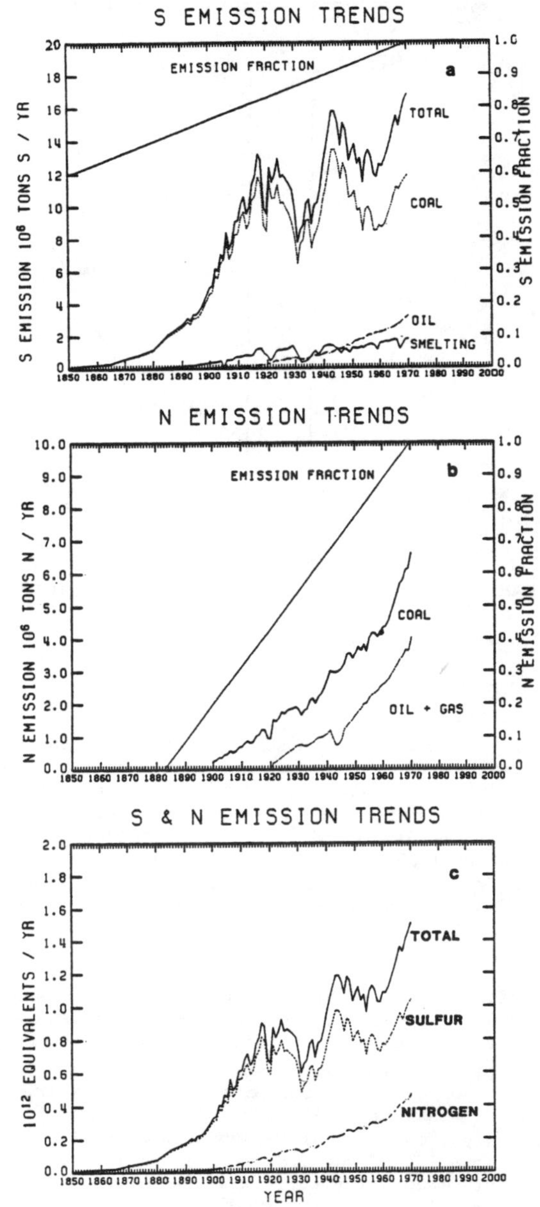

Fig. 4. Estimated historical U.S. emission trends for a. SO_x, b. NO_x, and c. sum of SO_x and NO_x equivalents.

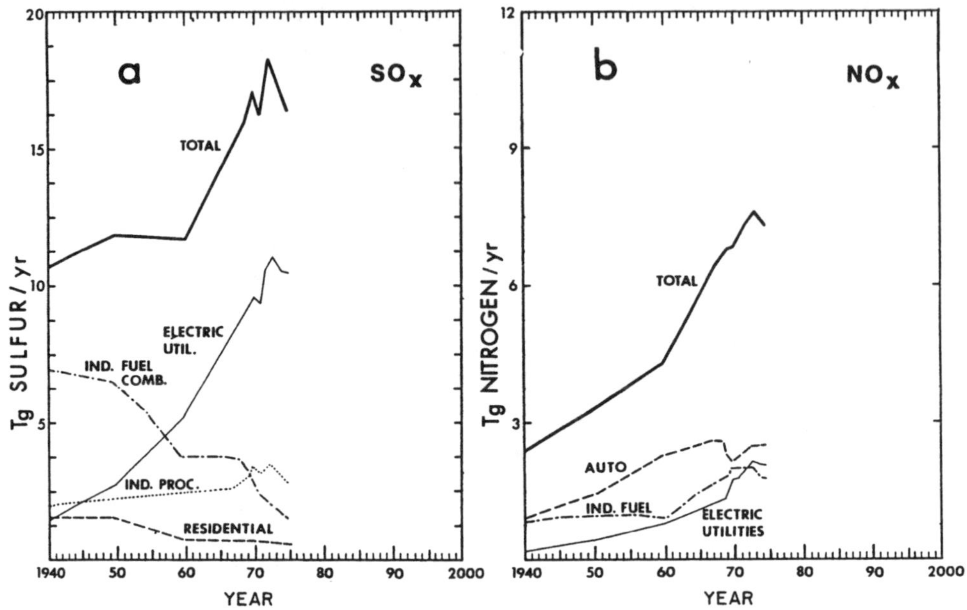

Fig. 5. Emission trends for SO_x and NO_x by source category. (U.S. EPA 1976)

The total sulfur emission trend estimate from coal, petroleum and smelting shows that the coal combustion constitutes the most significant sulfur source, exceeding 70% even in the time period 1950-1970. The historical trend estimates of sulfur emission therefore, closely reflect the coal consumption trends: a sharp rise from 1880 to 1910; roughly constant at 12 million tons until 1930; a reduction to 8 million tons in the 1930's; followed by a strong peak of about 15 million tons during the World War II. In the 1950's there was again a reduction to 12 million tons/yr following a sharp rise in the 1960's to 16 million tons/yr. More detailed emission inventories are available for the 1970's which show a roughly constant emission trend of 15 million tons/yr.

The sulfur emission trends are broken down according to energy sectors in Fig. 5a. From 1940 to 1960 the reductions of sulfur emissions from industrial, residential and transportation fuel consumption were more than balanced by increases of electric utility sulfur emissions. By 1973, electric utilities accounted for 60% of the sulfur emissions, mostly from coal (53%). The bulk of the national S emissions are located east of the Mississippi River (75%). Within this region, coal and oil combustion accounts for 92% of the sulfur emitted. Of the 25% of the U.S. emissions west of the Mississippi, 62% are due to mining and industrial activities (smelting, refining) and only 38% are due to fuel combustion. In the 1970's there has been a further shift to SO_x emissions from the non-electric to electric utility sector, increasing their share from 53% in 1970 to about 70% by 1980.

The spatial distribution of sulfur emissions have been shifting away from the industrialized northeast toward the southcentral states: Kentucky, Tennessee, North and South Carolina and Alabama, in accordance with the spatial shifts in the coal consumption. Assuming an average coal sulfur content of 2.4%, a set of coal consumption and sulfur emission density maps are given in Fig. 6 for 1889, 1919, 1957, and 1975. The key changes are the decline in the industrialized northeast and the increase in the eastern sunbelt region. A more detailed sulfur emission map for North America for all the sources is shown in Fig. 9a.

Nitrogen emissions to the atmosphere arise primarily from the "fixation" of atmospheric nitrogen during high temperature combustion. A fraction of the nitrogen emissions also results from the nitrogen contained in the fuels. Hence, the emission factors and emission rates strongly depend on the nature of combustion. Historical nitrogen emissions, therefoe, can only be estimated crudely. Since 1940, the NO_x emissions have been estimated by U.S. Environmental Protection Agency (EPA) Fig. 5b, and indicate about a threefold increase by the 1970's. The sharp rise in the 1960's is attributed to the increased emissions from industrial and electric utility sources. Coal combustion in high temperature boilers, such as utility boilers, produces approximately 5 kg N/ton of coal, or 0.2 g N/megajoule of coal. In industrial boilers, the corresponding nitrogen emission factor is 0.1, and for domestic and railroad consumption is 0.4 g N/megajoule.

For fuel oil consumption, the power plant emissions factor is about 0.1 g N/megajoule, and for residential and industrial uses it is taken to be 0.07. The nitrogen emission for gasoline combustion is taken to be 1.5 g N/vehicle mile, the average mileage 13 mpg, and gasoline energy content of 0.14 million Btu/gallon. The resulting nitrogen emission factor is 0.14 g N/megajoule. The total emission trend of nitrogen from fuel oil, gasoline and coal is depicted in Fig. 4b, showing a steady increase since the turn of the century.

For some purposes, such as the estimation of the total deposition of acidifying compounds, it is instructive to examine the combined emission trends of sulfur and nitrogen. For this purpose the relevant measure of acidifying potency has units of equivalents, S/16 and N/14. The sum of sulfur and nitrogen emission equivalents since 1850 is displayed in Fig. 4c. It is apparent that even currently about two thirds of the potential proton donor capability is from sulfur emissions, but the fraction of nitrogen contribution has been steadily increasing. It is probable that the nitrogen fraction will continue to increase.

The above emission trend estimates of nitrogen and sulfur compounds, although rather crude, convey two points:

1. The sum of sulfur and nitrogen emissions (in equivalents) has increased by only about 50% since the turn of the century, and NO_x contributed most of the increase.

2. There has been a substantial relocation of the emission pattern from the industrialized northeast (Pennsylvania, New York, New England) to the southcentral states (Kentucky, Tennessee, North and South Carolina, Alabama).

103

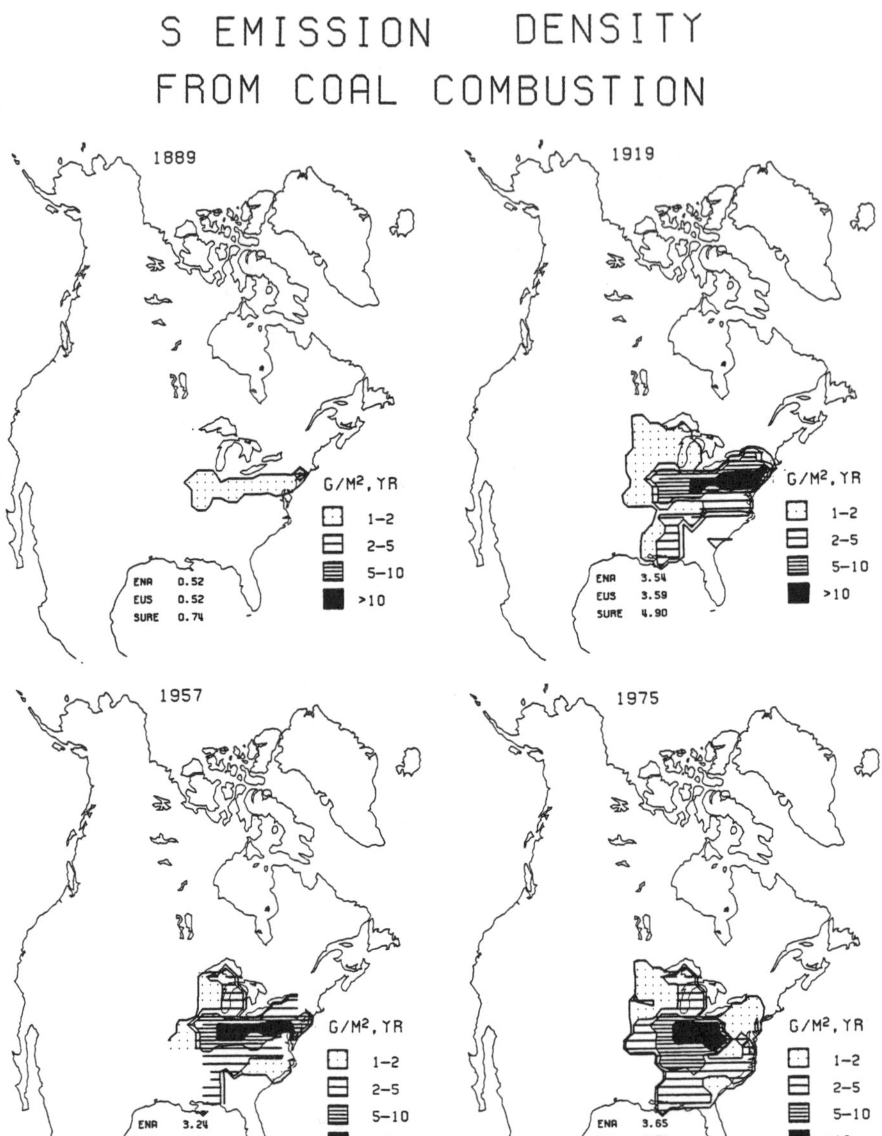

Fig. 6. Sulfur emission density estimated from historical state by state coal consumption data for 1889, 1919, 1957, and 1975.

WET DEPOSITION DATA FOR SULFATE, NITRATE AND AMMONIA

Motivated by increasing concern about "acid rain", there have been at least five rain chemistry networks operational over North America since 1978. The locations of the Multistate Atmospheric Power Production Pollution Study network's (MAP3S) 8 event sampling stations, the National Atmospheric Deposition Program's (NADP) 80 weekly sampling stations, the Electric Power Research Institute's (EPRI) 9 event stations with two analyzed collectors each, the Canadian Atmospheric Network for Sampling Air Pollutants' (CANSAP) 60 monthly sampling stations, and the Canadian Air and Precipitation Monitoring Network's (APN) 4 event sampling stations, are shown in Fig. 7a. The yearly average rainfall rate measured by the above stations for 1977-1980 is given in Fig. 7b. It should be noted that the NADP station density in the southwestern U.S. is sparse and data are available only for about one year; in the eastern U.S. and Canada the data base extends for two or more years. We are fully aware that the deposition maps shown below will change in the future for many reasons. We also recognize that the present report would be considerably stronger if error and uncertainty analysis would have been used more extensively. The bibliographical documentation is also embarassingly meager. Two significant digits are given for accounting purposes only.

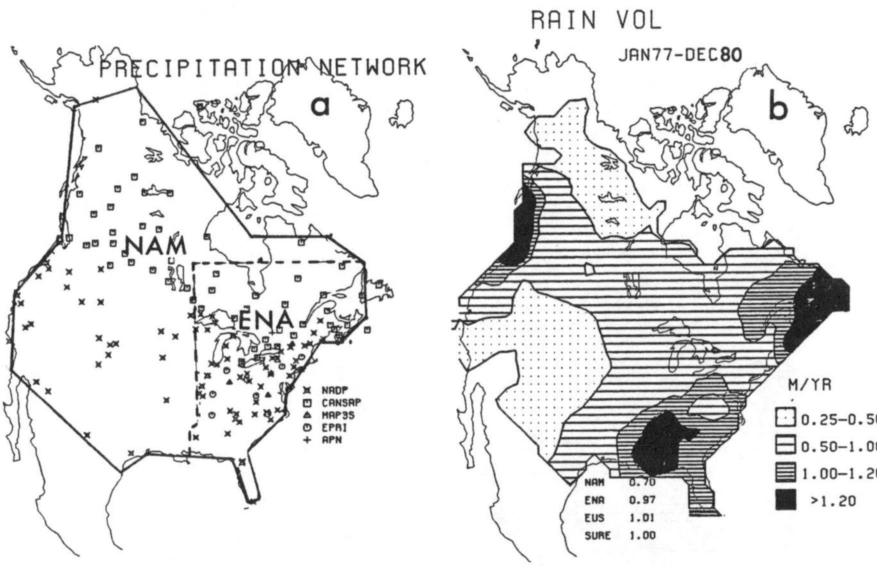

Fig. 7. a. Location of 161 precipitation chemistry sampling stations over North America (NAM) and eastern North America (ENA) operated by 5 networks. b. Yearly average rainfall maps obtained from the precipitation chemistry networks using all available data between January 1977 and December 1980.

Sulfate

The yearly average sulfur deposition and precipitation weighed concentration fields arising from the five network data set for the time period 1977-1980 is given in Fig. 8. Some of the sulfur in precipitation is due to sea salt and its contribution may be estimated either from Na or Cl data. Taking the sea salt sulfate to be 0.047xCl, the marine sulfate deposition over the continent was calculated to be only about 6% of the sulfur wet deposition over North America. For the sites more than 100 km from the coastline the marine sulfur deposition was insignificant. In the following discussion S refers to the excess beyond the sea salt sulfur.

For the North American continent (NAM), excluding Mexico, the average sulfur wet deposition is 0.34 g S/m^2,yr. The area considered is $18x10^{12}$ m^2 and the total wet S deposition is thus 6.2 Tg S/yr. The highest sulfate deposition rate and concentration in precipitation occurs in the region surrounding the eastern Great Lakes. The sulfate wet deposition there exceeds 1 g S/m^2,yr. For the region east of the Mississippi River and south of the James Bay (defining here eastern North America, ENA, area $5.8x10^{12}$ m2) the sulfate wet deposition is about 0.63 g S/m^2,yr. The lowest sulfate deposition rate of about 0.1 g/m^2,yr is measured in northwestern Canada and southwestern U.S. Both of these regions have less than 0.5 m/yr of rainfall.

The weighted average sulfate concentration in precipitation (Fig. 8) ranges between 15 μeqv/l in remote U.S. and Canadian regions to about 70 ueqv/l in the vicinity of the Great Lakes. Hence, while the average deposition rate varies ten fold over the continent, the average precipitation sulfate concentration increases only five fold from remote regions to industrial hot spots.

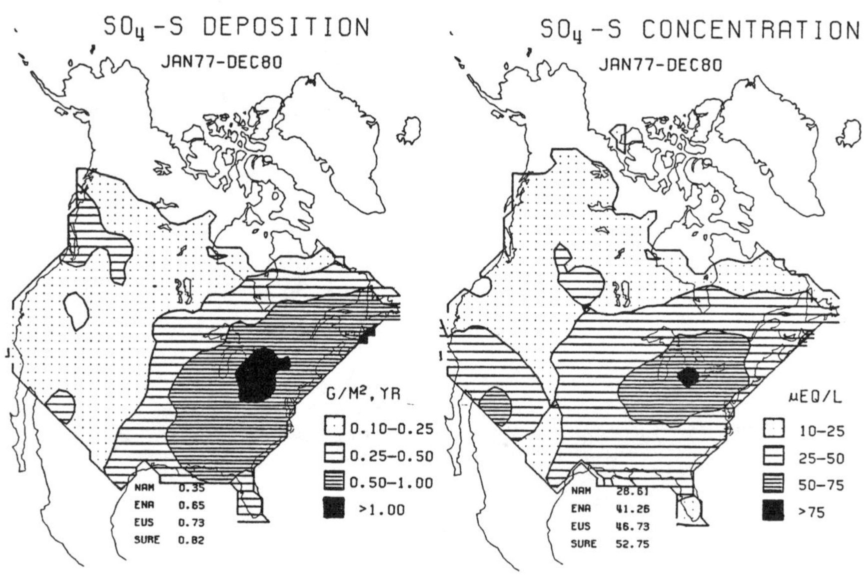

Fig. 8. Maps of sulfate wet deposition rate and precipitation weighed average concentration over North America.

There are few areas of North America for which it can be safely assumed that the sulfate deposition values represent the "natural background". Considering the wind pattern and the anthropogenic S emission fields over the continent, a possible area uninfluenced by man may be northwestern Canada, inland from the Pacific Coast. The measured sulfur wet deposition rate in that region is 0.1 g S/m^2,yr, compared to the North American average of 0.34 g S/m^2,yr. In estimating the natural global sulfur cycle from rain chemistry data, Granat et al.(1976) have chosen 0.15 g S/m^2,yr as the representative background deposition over nondesert land areas. Considering the severe uncertainties of the deposition estimates, 0.07 g S/m^2,yr would also seem reasonable as a North American average background. Scaled to the nondesert land area of the world (120×10^{12} m^2), the global S wet deposition from the above "background" sources would be 9-18 Tg S/yr, which is below the estimate of 18 Tg S/yr by Granat et al. (1976). An increasing data base of precipitation sulfur values obtained in the southern hemisphere, e.g. Stallard and Edmond (1981), from the Amazon region, and data from the polar icecaps dictates that continental average background deposition outside the 0.07 to 0.15 g S/m^2,yr range is unlikely. A documentation of this statement is best available in Granat et al. (1976), but will require continuous reevaluation as the global data base expands.

The lower (0.07 g S/m^2,yr; 1.3 Tg/yr) and the upper bound (0.15 g S/m^2,yr; 2.5 Tg/yr) of the estimated "natural background" sulfur deposition constitutes 20 to 40 % of the measured total sulfur deposition (0.34 g S/m^2,yr; 6.2 Tg/yr) over North America. For eastern North America the average excess sulfur deposition rate is about 0.63 g S/m^2,yr (3.7 Tg/yr) and the "background" (.5-1.0 Tg S/yr) deposition would account for 12 to 25 percent of the total ENA wet deposition. This is a substantial upward revision of previous estimates of "natural contribution" for North America (Galloway and Whelpdale 1980).

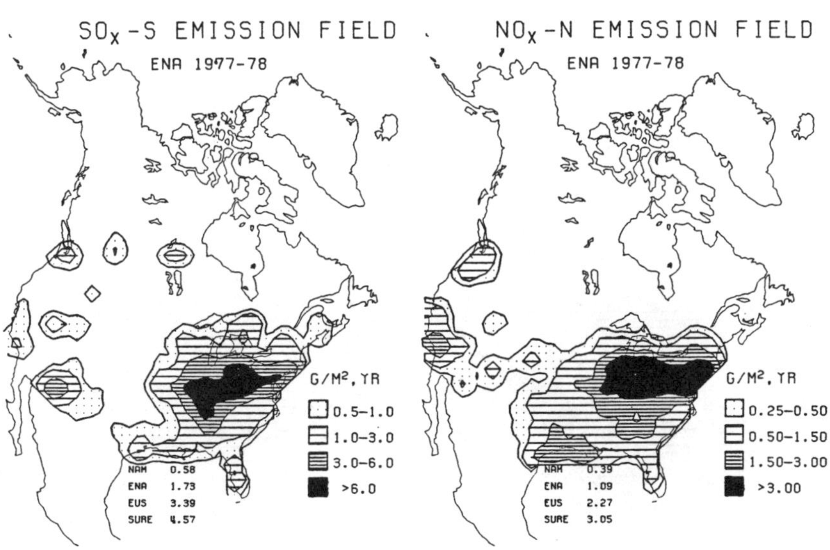

Fig. 9. Emission density of manmade sulfur and nitrogen oxides over North America.

It is also instructive to compare the measured wet deposition pattern and rates to the emission field of manmade sulfur over North America (Fig. 9a). The average emission density over eastern North America is 1.9 g S/m^2,yr while the average wet deposition over the same region is 0.63 g S/m^2,yr, i.e. about 30% of the known manmade emissions. If we further assume that the natural sources contribute on the average 0.07 g/m^2,yr, the measured wet deposition of sulfur amounts to only 25-30% of the manmade sulfur. The remaining 70-75% of the manmade sulfur is then either dry deposited as SO_2 or SO_4, or exported to the Atlantic by the prevailing winds.

Nitrate

The average deposition and concentration pattern of nitrate over NAM is given in Fig. 10. The average NAM wet deposition rate is 0.13 g N/m^2,yr corresponding to 2.4 Tg/yr. The highest deposition rates (0.4 g N/m^2,yr) are observed in the area surrounding eastern Great Lakes. The average over ENA is 0.23 g N/m^2,yr. Using again the deposition data for remote western Canada (0.03-0.06 g N/m^2,yr), approximately an order of magnitude increase may be observed from remote to high emission regions. Similarly as for the sulfate, the nitrate concentration in rain has only a five fold increase from remote (3-6 μeqv/l) to industrialized regions(25 μeqv/l).

If we take the "background" nitrate wet deposition rate to range between 0.03-0.06 g N/m^2,yr over the entire continent, the "background" wet removal will contribute 0.5-1.0 Tg N/yr. As for sulfate wet deposition, the "background" nitrate deposition would account for 20-40% of the measured nitrate deposition in precipitation. Over eastern North America, the corresponding background contribution would be 10-25% of the total measured wet deposition. Taking 0.03-0.06 g N/m^2,yr as representative over the

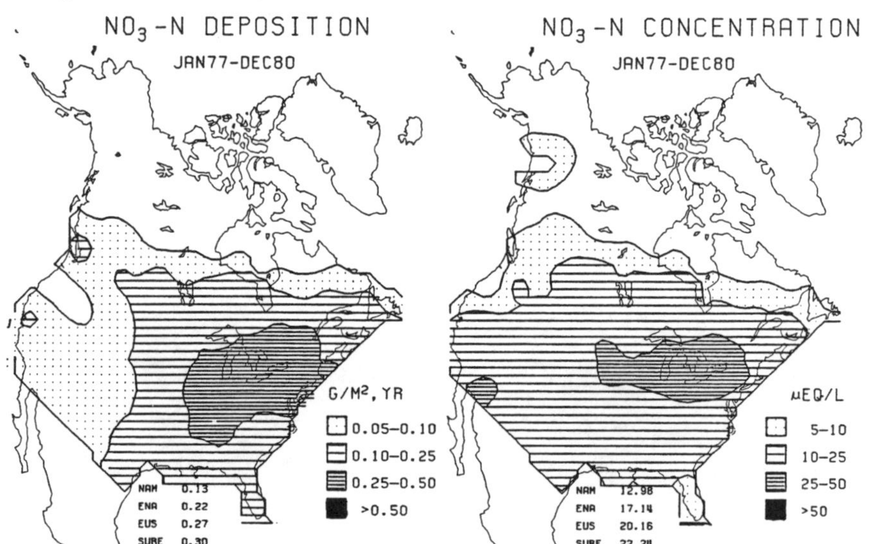

Fig. 10. Maps of nitrate wet deposition rate and precipitation weighed average concentration.

nondesert land areas (1.2×10^{14} m^2) of the world, the global nitrate wet deposition would range between 3 and 7 Tg N /yr. This is an order of magnitude less than the total (dry and wet removed, natural and manmade) nitrate deposition (32-83 Tg N/yr) estimated by Soderland and Svennson (1976). Using the average measured NAM nitrate wet deposition of 0.13 g N/m^2,yr and applying it to the global nondesert land area, yields 16 Tg N/yr which is still a factor of two to five less than the nitrate deposition (dry and wet) estimated by Soderland and Svennson (1976).

An emission map (Fig. 9b) for manmade NO_x over eastern North America shows that the highest emission density is in excess of 3 g N/m^2,yr in the Great Lakes region and it roughly coincides with the location of the highest deposition density (Fig. 10). The comparison with the deposition data also reveals that wet deposition accounts for only 20% of the known manmade NO_x emissions over eastern North America. The remaining 80% is thus either dry deposited or exported from the continent.

Ammonia

The yearly average ammonium ion deposition field is shown in Fig. 11. The North American average deposition is 0.11 g N/m^2,yr (2.0 Tg N/yr) and for eastern North America 0.16 g N/m^2,yr (1.0 Tg N/yr). The highest deposition rates, ranging 0.25-0.50 g N/m^2,yr, extend from the Great Lakes to the Rocky Mountains, an area generally known as the corn belt region. West from the Rockies and in northern Canada the deposition rate is below 0.1 g N/m^2,yr. Conspicuously, the industrialized northeastern part of the U.S. does not show high ammonia deposition values. Here again it is useful to compare the measured NAM ammonium wet deposition rates (2 Tg N/yr), scaled to the globe (13 Tg N/yr), to the global dry and wet ammonium/ammonia deposition rates (91-186 Tg N/yr), estimated by Söderland and Svennson (1976).

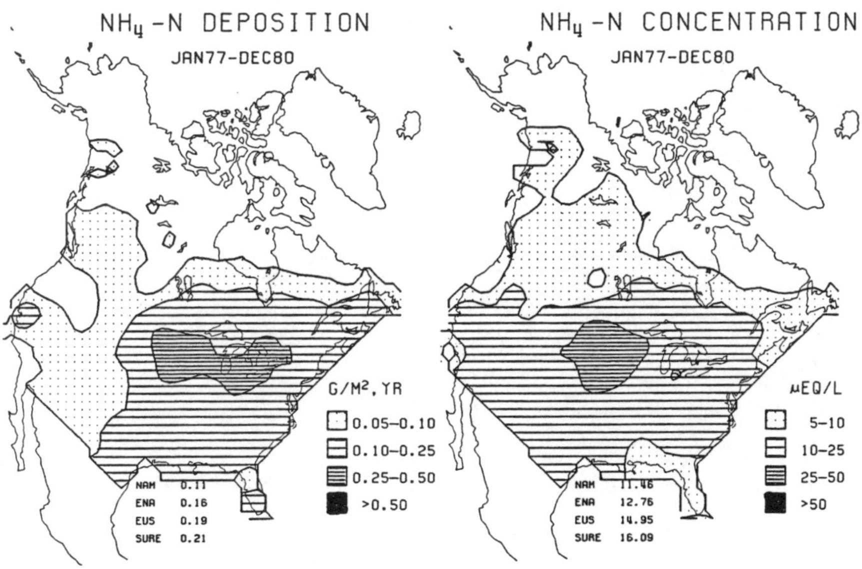

Fig. 11. Maps of ammonia wet deposition rate and precipitation weighed average concentration.

Possible sources of ammonia. Unlike for SO_x and NO_x, an emission inventory of manmade or "known" ammonia sources over North America currently does not exist. As an aid to interpret the measured ammonia wet deposition field we have constructed a tentative inventory for the U.S. The emission factors for domestic animals were taken from Bottger et al.(1978). For nitrogen fertilizer it was assumed that 10% N is volatilized to the atmosphere. Minor contributions from known industrial sources were also included. The resulting ammonia emission density map is shown in Fig. 12. The total U.S. ammonia emission from the above "known" sources is estimated at 3.4 Tg N/yr with the highest emission density in the corn belt region exceeding 1 g N/m^2,yr.

There is a rough coincidence of the area of high measured deposition rate and estimated emission density, both extending through the cornbelt region. The estimated U.S. ammonia emission rate from "known" U.S. sources is 3.4 Tg N/yr, which is comparable to the wet deposition integral of 2 Tg N/m^2,yr for NAM. However, considering the severe uncertainties of the "known" source estimates, little significance is attached to the emission values beyond suggesting that the inventory is within the right order of magnitude.

In spite of the uncertainties associated with the spatial-temporal coverage, sampling and analytical procedures, and the interpretation of the wet deposition data, it is most gratifying that such a continental scale data base currently exists for North America. Prudent use of such a data base will undoubtedly provide us with a much improved perspective on both scientific and political aspects of the "acid rain" problem.

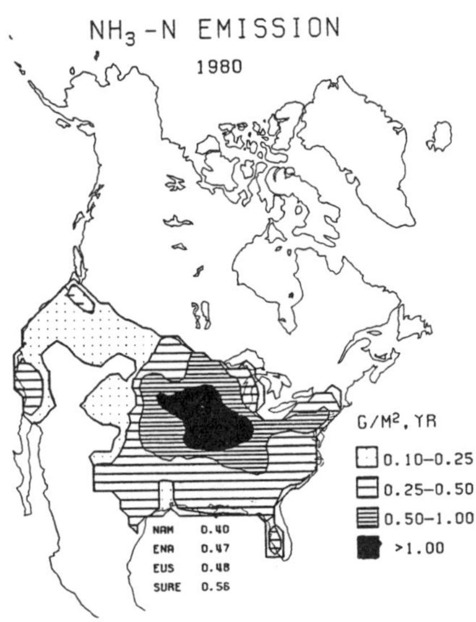

Fig. 12. Estimated emission density of ammonia emissions for the U.S.

SULFATE CONCENTRATIONS IN AIR

There are two aerosol sulfate concentrations data bases available for the U.S. The National Air Surveillance Network (NASN) has monitoring sites located primarily in urbanized areas, sampling for one day every sixth day, and the chemical analysis is performed on high volume filter samples. The second network was operated for the Electric Power Research Institute (EPRI) as part of the Sulfate Regional Experiment (SURE) covering the northeastern U.S. for six one month periods. The SURE sampling sites were located outside of urban-industrial centers providing daily or six hourly sulfate concentrations.

The annual average sulfate concentration for the eastern U.S. arising from the NASN network is shown in Fig. 13a. The SURE sulfate data averaged for August and October 1977 and January, April, July and October 1978 are also shown (Fig. 13b). It is apparent that the SURE data for 1978 indicate about 20-30 percent lower sulfate concentration than the NASN network. The discrepancy is probably due to the combination of the following factors: the NASN data are primarily from urban sites, the sampling and chemical analysis of NASN filters is likely influenced by a positive artifact from SO_2, and, finally the sampling periods did not overlap.

Both networks indicate that the highest airborne sulfate concentrations occur in the region between the Great Lakes and the Ohio River. However, the NASN network indicates a strong hotspot (15 $\mu gSO_4/m^3$) in the Ohio River valley, and about a factor of two less in Iowa and Minnesota. The SURE network shows a much smoother pattern having also about 6 $\mu gSO_4/m^3$ in the upper Midwest but less than 10 $\mu gSO_4/m^3$ in the Ohio River region.

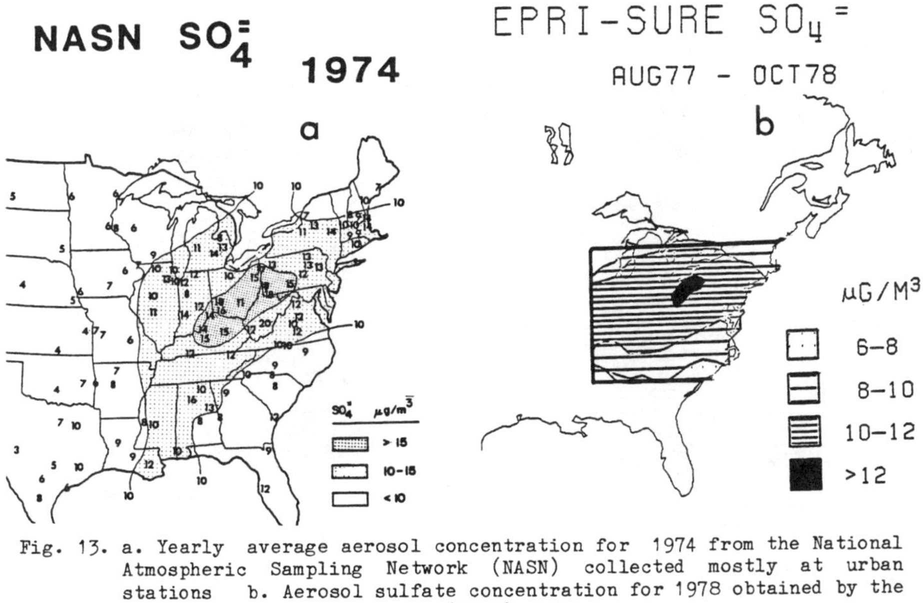

Fig. 13. a. Yearly average aerosol concentration for 1974 from the National Atmospheric Sampling Network (NASN) collected mostly at urban stations b. Aerosol sulfate concentration for 1978 obtained by the Sulfate Regional experiment (SURE) network.

VISIBILITY

Atmospheric optical data are much more abundant on a continental scale than chemical composition data, with the exception of the water vapor content. Hence, if a reasonably well defined relationship can be established between, say the visual range or turbidity and the fine particle content of the atmosphere, then the extensive meteorological observations by human observers may provide an attractive surrogate. Unfortunately, there are many problems inherent in the interpretation of the visibility data obtained by routine meteorological observation networks. These include the subjectivity of the human observer, the lack of suitable visual targets, and the numerous natural phenomena that perturb the visual environment (rain, fog, blowing dust, natural haze, etc.). The spatial and trend analysis of the visibility data, therefore needs to be conducted with utmost caution. The data presented below arise from the analysis of 147 U.S. and 177 Canadian stations from 1948 to 1980.

The yearly average extinction coefficient (3.9/visual range, km) from noon observations in the absence of precipitation and fog is shown in Fig. 14. At midlatitudes (40-60 degrees N) there are three hazy regions: surrounding the Great Lakes, the Mississippi Delta and southern California including the San Joaquin Valley. A conspicuously hazy region also exists north of the arctic circle, the cause of which is unknown and will be ignored here. The lowest mean extinction coefficient (< 0.1 1/km) occurs in the U.S. Rocky Mountain region. The long term availability of visibility data also permits the examination of secular trends of continental haze from about 1950 to 1980. The five year average map for 1950-1954 shows substantially lower haziness than the 1976-1980 period. The increase in the extinction

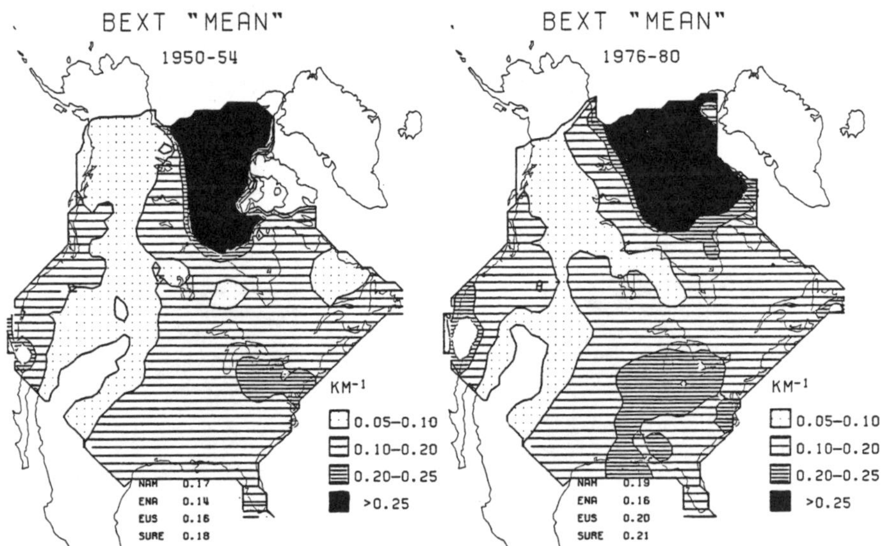

Fig. 14. Yearly average noon extinction coefficient over North America for 1950-54 and 1976-1980.

coefficients between 1950-1954 and 1976-1980 is also displayed in Fig 15. The important changes in the three decades occured in the southeastern U.S. (mostly in the summer months) as well as over the northwestern U.S. and adjacent southwestern Canada(mostly in the winter months).

A more detailed trend pattern for eastern U. S. haze is shown in Fig. 16. It is apparent that in some areas such as in New England, the haze decreased for the first quarter of the year. Other regions, such as surrounding the Smoky Mountains exhibit a strong increase, particularly during

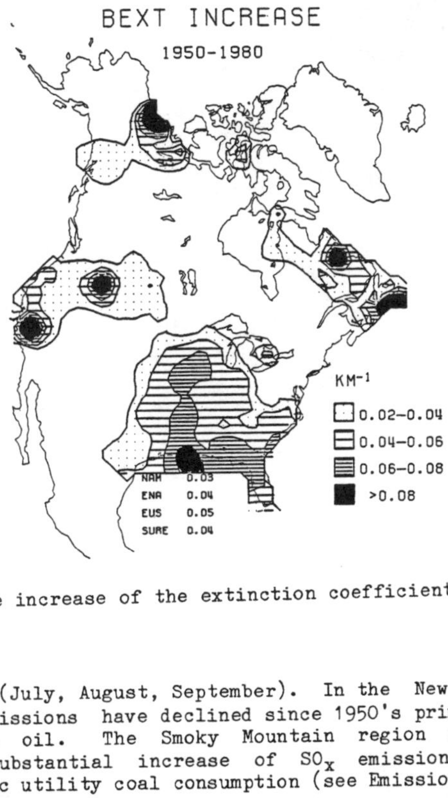

Fig. 15. Map of the increase of the extinction coefficient from 1950 to 1980.

the third quarter (July, August, September). In the New England, New York area the SO_x emissions have declined since 1950's primarily because of the shift from coal to oil. The Smoky Mountain region on the other hand, experienced a substantial increase of SO_x emissions primarily due to increasing electric utility coal consumption (see Emission section above).

Beyond demonstrating that substantial changes have occured in the optical environment, the visibility data demonstrate that within the eastern U.S. some subregions may exhibit a decline, others an increase of haziness. This implies that in spite of "long range transport" subregions covering several U.S. states will exhibit trends in fine particle concentration consistent with their own emission trends. In other words, the long term haze data base confirms the most revealing conclusion of the European OECD project that every source region impacts on itself more than on any other region.

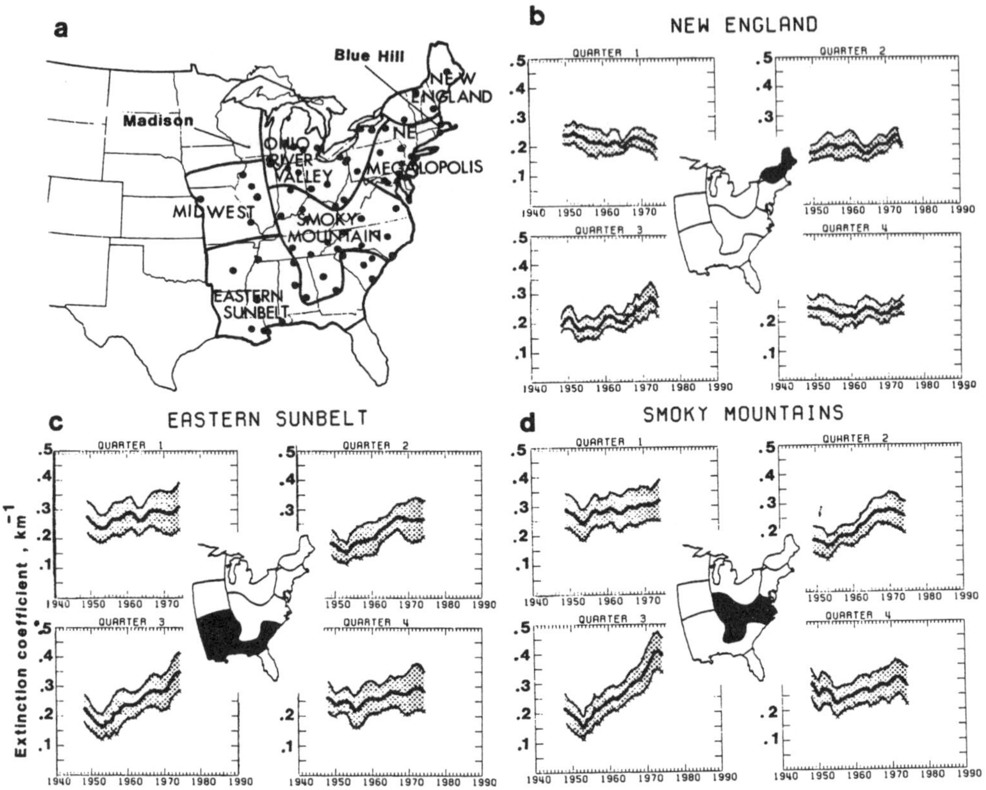

Fig. 16. a. Location of 70 eastern U.S. sites where detailed trend analysis was performed. The trend lines indicate mean and arithmetic standard deviation among the stations within each region. b.-d. Trends of extinction coefficient by yearly quarters for New England, southeastern sun belt and the Smoky Mountain region.

FUTURE TRENDS

There are at least five key factors that will determine the future trend of SO_x and NO_x emissions in the U.S.

1. Population growth and spatial shifts. In the 1960's there has been a remarkable redistribution of the population from the industrialized northeast to the "sunbelt" region of the U.S. The southward movement of the people caused an increase in the economic activity in the "sunbelt" (or vice versa?), which then resulted in an increased emission of air pollutants. Any prediction of population growth and redistribution trends for the remainder of this century is most unreliable.

2. Vigor of the U.S. economic activity. From the 1950's through 1973 all economic indices show a steady upward trend, which was paralleled (even if not causing) a steady growth in the national energy consumption. In 1973, following the oil embargo, there was an abrupt reduction in the U.S. energy consumption followed by a roughly constant energy demand until 1981. These drastic changes in the energy demand and resulting SO_x emisssions in the U.S. were not caused by slow exhaustion of the fuel resources, but rather by economic and political "instabilities" such as the great depression of the 1930's, World War II and the oil embargo of 1973. Like in the case of the mesoscale weather, events growing out of instabilities can not be predicted.

3. Energy efficiency of the economy. The amount of energy consumed divided by the Gross National Product (GNP) has been extensively used as a measure for the energy efficiency of different nations. For the U.S., there is evidence that since 1973 the national energy demand has been reduced by 15-20%. It is likely that the increasing price of fossil fuels will further reduce the energy demand/$GNP. However, while the national energy demand is a measurable quantity, the GNP is a dubious measure of the nation's economic well being.

4. The fuel mix. The U.S. energy consumption trend by fuel type shows that the nation has shifted from the coal based economy in the first half of the century to oil and natural gas since World War II. Since 1970, the energy supplied by coal remained constant at 20%. However, because of its high sulfur emission factor (about 1 g S/megajoule) compared to oil (0.1 g S/megajoule), coal consumption dominates the sulfur emission trends. Another remarkable change occured in the U.S.: within the past decade the contribution of the western low sulfur (<0.5% S; 0.25 g S/megajoule) coal has increased from less than 10% to over 30% of the national coal supply. Consequently, in spite of an increase in coal consumption from about 600 to 800 million tons there was no increase in SO_x emissions from coal.

5. Source emission controls for SO_x. Sulfur oxides emissions may also be reduced by flue gas desulfurization. At present we are not in position to assess the role of this technology in the emission trends for the remainder of this century.

Predicting the future trends of sulfur emissions in the U.S. will thus remain a highly subjective guess. It is our guess that these emissions will remain at roughly the same level (say, within 30%) as the current emissions of about 15 Tg S/yr for the remainder of this century.

The research described in this article has been funded in part by the United States Environmental Protection Agency through Cooperative Agreement 806606, Dr. William E. Wilson, Project Officer and grant R 806548, Dr. Lowell Smith, Project Officer, and benefited from the generous assistance of Dr. Janja Djukic Husar. It has not been subjected to Agency review and therefore does not necessarily reflect the views of the Agency and no official endorsement should be inferred. The opinions expressed are those of the authors and should not be construed as representing Agency policy.

REFERENCES

Bottger, A., Ehhalt, D.H. and Gravenhorst, G. 1978. Atmospherische Kreislaufe von Stickstoffen und Ammoniakum.- Berichte der Kernforschunsanlage Julich: 1558.

Galloway, J.N. and Whelpdale, D.M. 1980. An atmospheric sulfur budget for eastern North America.- Atmos. Environ. 14:409.

Granat, L., Rodhe, H. and Hallberg, R.O. 1976. The global sulphur cycle. In Nitrogen, Phosphorus and Sulphur-Global Cycles. Svensson, B.H. and Söderlund R., Eds.-SCOPE report 7. Ecol. Bull. (Stockholm) 22:89.

Söderlund, R. and Svensson, B.H. 1976. The global nitrogen cycle. In Nitrogen, Phosphorus and Sulphur-Global Cycles. Svensson, B.H. and Söderlund, R., Eds.-SCOPE report 7. Ecol. Bull. (Stockholm) 22:23.

Stallard, R.F. and Edmond, J.M. 1981. Geochemistry of the Amazon. 1. Precipitation chemistry and the marine contribution to the dissolved load at the time of peak discharge.- J. Geophys. Res. 86:9844.

U.S. Environmental Protection Agency. 1976. National emissions report (1973).

[9]

The human ecology of global change*

William C. Clark

Introduction

The need to view human activity as an integral component of the geosphere–biosphere system has been emphasized since the earliest writings on global change. The Russian mineralogist Vladimir Ivanovitch Vernadsky, recapitulating in 1945 themes he had first articulated in lectures and books 20 years earlier, described his concept of

> the biosphere, the only terrestrial envelope where life can exist. Basically, man cannot be separated from it. He is geologically connected with its material and energetic structure. (Vernadsky, 1945)

Echoing the Italian geologist Stoppani, Vernadsky argued that the most environmentally significant aspect of the human connection was not technology *per se*, but rather the sense of global knowledge and communication engendered by that technology.[1] He portrayed this 'noosphere' or 'realm of thought' as

> a new geological phenomenon on our planet. In it for the first time man becomes a large-scale geologic force. Chemically, the face of our planet, the biosphere, is being sharply changed by man, consciously, and even more so, unconsciously. (Vernadsky, 1945)

Vernadsky and his predecessors had neither the data nor the instrumentation to convert their insights into useful analytical tools for describing and understanding global interactions between environment and processes of human development.[2] Over the last 50 years, however, and especially in the quarter-century since the International Geophysical Year, the necessary measurements, models, and concepts have rapidly accumulated. We now know that human activities have fundamentally transformed the face of the earth, as well as the diversity and distribution of its biota. These activities have induced chemical fluxes at the continental and even global scale that are comparable to or greater than those occurring in nature. They have long since changed local climates and may now be altering the heat and water fluxes of the entire planet (Bolin and Cook, 1983; Holdgate et al., 1982; Nriagu and Pacyna, 1988; Turner et al., forthcoming). These global changes in the earth's environment have become the focus of the new International Geosphere–Biosphere Programme (ICSU, 1988).

William Clark is a professor at Harvard University's John F. Kennedy School of Government. His address: 79 Kennedy Street, Cambridge, MA 02138, USA. Trained as a biologist, his research interests are now focused on the policy and management issues associated with large-scale, long-term environmental change. He recently led the programme on Sustainable Development of the Biosphere at the International Institute for Applied Systems Analysis in Austria, and is a member of the US National Academy of Sciences' Committee on Global Change.

The reciprocal impacts of environmental change on human societies have also become significantly clearer over the last half-century. Simplistic theories of environmental determinism have been replaced by an increasingly sophisticated appreciation of the ways in which the physical environment shapes the challenges and opportunities that face communities, regions, and states (Chisholm, 1982; White, 1988b). Still needed, however, is a research paradigm that takes Vernadsky's insights seriously and that harnesses the full range of scholarship necessary to address the interactions among human, ecological, and physical systems involved in global change. Such a paradigm is beginning to emerge. As articulated by Robert Kates at ICSU's 1984 Ottawa symposium on Global Change:

> [t]he research opportunity over the next decade is to examine the boundaries of the sustainable development of the earth. Basic scientific research will provide much improved knowledge of human-induced change in the biosphere, the capacity of natural systems to absorb such change, and the ability of human societies to adjust or adapt their behaviour. The focus will be on the convergence of problems, methods, and theory. From such fundamental knowledge a scientific and truly human ecology can emerge and it should be part of an international geosphere–biosphere program. (Kates, 1984, p. 493)

Over the past several years, a growing number of scholars and institutions have been exploring the challenges and opportunities outlined by Kates.[3] This article seeks to review and summarize these explorations and their recommendations, with a view towards identifying the principal issues and questions that merit central attention in a research programme on the human ecology of global change. In attempting to cover this broad area, I will focus on the interactions of social and environmental systems. Related and equally important questions that arise within the social or environmental sciences, or primarily in connection with policy decisions, I leave for others to explore.

The article proceeds from a general discussion of relevant themes to certain specific conclusions regarding priority research needs for understanding the human ecology of global change. I begin with a broad overview of the principal elements involved in human interactions with the global environment. Next comes a discussion of major unresolved questions regarding the character, causes, and consequences of these interactions. Many such questions will be addressed through the normal course of research now planned or under way in a number of countries and disciplines. In addition, however, a number of especially complex and interdisciplinary research challenges are identified that might best be addressed through new international initiatives on the human ecology of global change. Three such challenges are highlighted in the final section.

Principal elements of the human system involved in global change

This section outlines the elements of the human system that require special attention in studies of global change. Three basic dimensions of the human role in global change are discussed: the interactions between human and environmental systems; the choices that individuals, governments and other organizations make in efforts to alter or manage those interactions; and the underlying elements of social organization, structure and culture that shape both interactions and choices.

Interactions

The most basic elements of the interactions between human and environmental systems are suggested in Figure 1. The figure reflects the central fact that both the human and the environmental systems of the earth change in response to their own internal dynamics, to external perturbations over which neither exerts appreciable control, and to their interactions with one another. Two forms of interaction are of central importance for the purposes of this review. The first concerns the sources of environmental change that result from demographic, economic, institutional, technological, agricultural, and behavioural changes in the human system. The second concerns the consequences for human well-being that result from climatic, chemical, and biotic changes in the environmental system. Also requiring special attention are the temporal and spatial scales at which important interactions occur between human and environmental systems.

Sources. In principle, human processes drive global change by altering the flows of energy and materials among components of the geosphere–biosphere system (Orians, forthcoming). In practice, the most important sources of alterations involve the following:

- the release of pollutants as varied in their effects as DDT, carbon dioxide, and nuclear radiation;
- the sequestering or redistribution of other

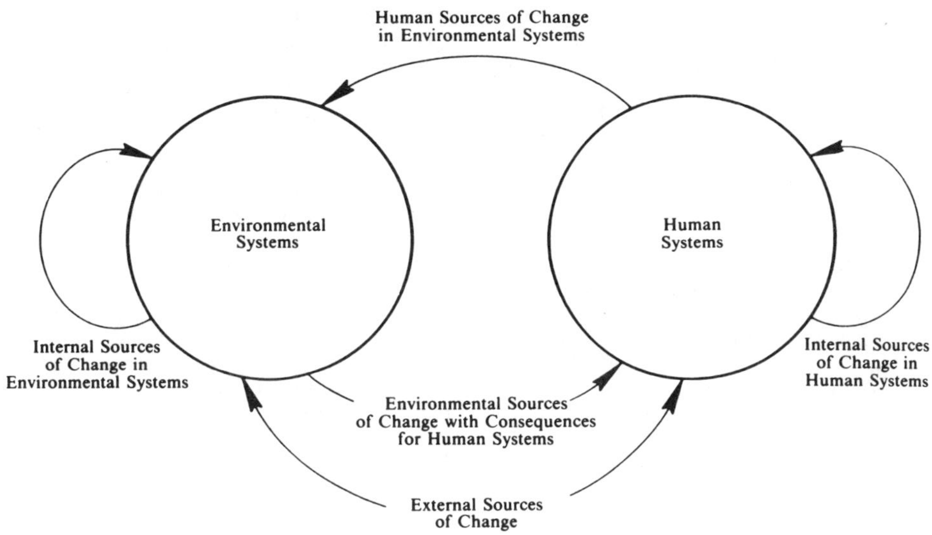

FIGURE 1. Interactions between human and environmental systems

materials and energy ranging from phosphorus to soil organic matter to running water;

the direct transformation of physical structures (e.g. terracing), surface properties (e.g. albedo), and habitats (e.g. wetland drainage);

the direct removal of species from the biotic system through harvest, the direct addition of species to the system through transport from other areas (invasion), or low- and high-tech versions of genetic engineering (domestication); and

interactions among the above.

The human activities that have contributed most importantly to these sources of global change include agricultural and industrial production, and energy consumption (Bolin and Cook, 1983; Clark and Munn, 1986; Turner et al., forthcoming). Within these broad categories, there is a need to identify which specific activities are most significantly implicated in existing global environmental changes, and which additional activities might become implicated in the future.

Consequences. Research on the human consequences of environmental change has established that the degree of threat experienced by a society is a function of four interactive variables: risk, exposure, vulnerability and response (Kasperson and Kasperson, 1988; Kates et al., 1985b; Kotlyakov et al., 1988; Whyte and Burton, 1980).

In terms of the framework presented here, risk is defined in terms of the actual or estimated changes in selected environmental variables. One of the clearest lessons of environmental impact studies is that assessments are ineffective if they seek to develop comprehensive lists of all environmental risks affected by human activity. More useful have been studies that focus on a modest number of 'valued environmental components' (Beanlands and Duinker, 1983). These are simply attributes of the environment that people choose to value. Which components are valued in any particular situation will depend on many of the considerations of scale, choice, and culture discussed below. Scientists, policy makers, and other interests must therefore negotiate the valued environmental components that merit priority attention in assessing the risks posed by global change.

Notions of social vulnerability are central to understanding the human consequences of global change (Brown, 1978; Timmerman, 1981). Individuals and societies can cope with a wide range of environmental changes, but at different costs and within different limits. Studies of human response to natural hazards, climate change, and nuclear war show that both costs and limits change with time and are mixed functions of the environment itself plus the underlying demographic, organizational, and developmental characteristics of the human system as discussed later in this review (Burton et al., 1978; Harwell and Hutchinson, 1985; Parry et al., 1988). Some tentative generalizations have begun to emerge. But a general understanding of social vulnerability to environmental variations remains a distant if urgent goal.

Differences in the exposure of various social groups and geographical regions to globally distributed environmental changes severely complicate both expert and lay assessments of environmental threats. Recent work on 'total exposure assessment' to air pollutants has shown how misleading broad average exposure estimates can be (e.g. Ott, 1985; K. Smith, 1988; Spengler and Soczek, 1984). Concepts relating to heterogeneities in exposure and empirical estimates of such exposures will be central elements in an understanding of the human ecology of global change (e.g. Vaupel and Yashin, 1986).

The human choices that constitute societies' responses to global change are reviewed later.

Scale. A consensus is emerging that studies of global change need to devote particular attention to interactions that become significant on temporal scales ranging from decades to centuries and spatial scales ranging from large regions to the globe as a whole (Clark, 1986; ICSU, 1988). These are much coarser scales than those at which most research on human systems has focused. On the other hand, many aspects of long-term global change have their primary sources and consequences at relatively fine scales (Holling, 1986; White, 1988a). Moreover, notions of sustainability are strongly dependent on the linkages among regions simultaneously exposed to global environmental change. Coupling observation and explanation across multiple scales therefore becomes a central requirement for understanding global change (Risser, 1986; Rosswall et al., 1988).

A long tradition of attention to space and time dimensions in geography, economics and history has produced a relatively sophisticated view of the difficulties involved.[4] This experience shows that much confusion and unproductive debate results when scholars working at different temporal or spatial scales contrast unlike situations without recognizing the problems and limits of transference (Chisholm, 1982). In order to minimize such problems in efforts to understand the human ecology of global change, a first requirement is the explicit identification of what scales are involved in each effort to document or explain specific interactions between environmental and social systems. Beyond this, it is important to know which human processes are likely to interact most strongly with the environment at the coarse (decadal and regional) scales that are central to global change. Initial studies suggest that special attention should be given to global, long-term studies of topics such as the life cycle dynamics of major industrial processes, fuel substitution in energy systems, urbanization, labour absorption in the agricultural sector, and the conditions limiting the extent of major crop zones (Clark, 1987). Finally, understanding is needed on the ways in which certain fine-scale phenomena of human systems (e.g. technical innovations) cascade 'up-scale' to yield consequences significant for global change.

Choice

A fundamental asymmetry in the interactions between human and environmental systems does not appear in Figure 1, but is nonetheless essential to an understanding of global change. For while the response of environmental systems to human activities is entirely reactive, the response of human systems to changes in the environment has both reactive and proactive elements. Human behaviour can respond not only to actual environmental changes that have already occurred, but also to people's perceptions and assessments of possible future changes that they hope to encourage or avoid. This reflexive or anticipatory potential of human systems raises the prospect of conscious environ-

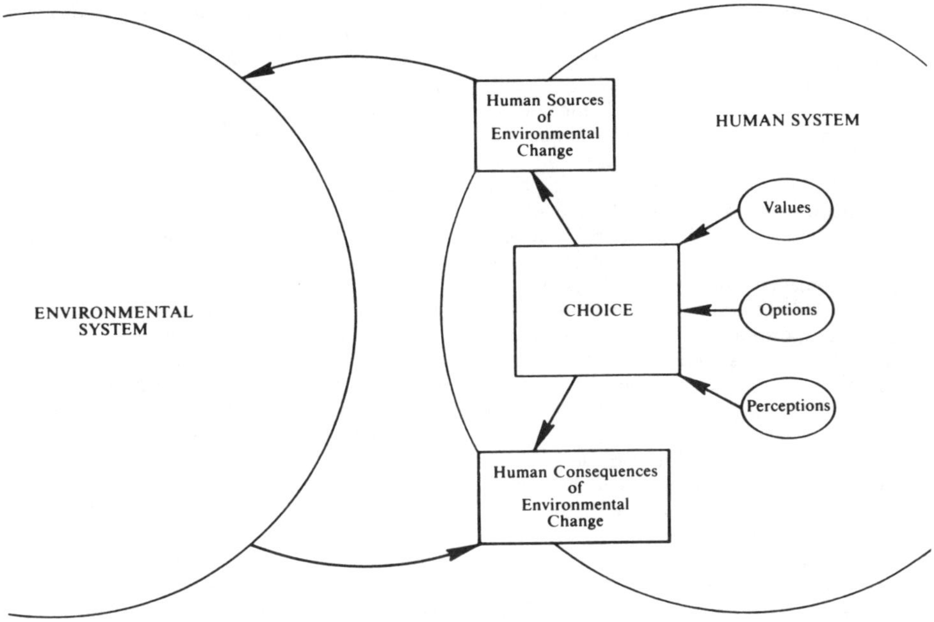

FIGURE 2. The role of choice behaviour in the interactions between human and environmental systems. This figure is based on a model of human choice and environmental perception proposed at the China–US Workshop on the Human Dimensions of Global Change (Tang and Jacobson, 1988), and causal structure of hazard management proposed by Hohenemser et al. (1985: Figure 11).

mental management. It also focuses attention on the central role that studies of human choice and behaviour must play in efforts to understand global change.

The central position of choice behaviour in the interactions between human and environmental systems is suggested in Figure 2, which draws heavily on research concerning the human ecology of natural and technological hazards. One important aspect of that research is its treatment of choice behaviour as a potential modifier of both the human sources and the human consequences of environmental change. Another is its recognition that virtually all human choices relating to environmental change entail a significant degree of risk-taking and uncertainty, and thus almost inevitably result in a significant degree of surprise (White, 1988a). Coupled with broader studies in the behavioural sciences, decision analysis and policy-making, the research on hazards suggests that three factors play significant roles in shaping human choices relating to environmental change: values, options, and perceptions.

Values. Values, in the present context, can be viewed as an indication of what people think they want from the interactions between human and environmental systems. Research has shown strong associations between positive valuations of the environment and concrete behaviours that sustain environmental systems (Darley and Gilbert, 1985). Conversely, a strong argument has been made that most human choices that degrade the global environment are governed by values that weight short-run benefits to people over long-run damages to the environment and the foundation for sustainable resource development that it provides (Bandura, 1986).

The important roles of human values in the interactions between people and the environment are rapidly changing, as reflected in the tremendous expansion and strengthening of the environmental movement over the last quarter-century (White, 1988a). At the local level, this shift can be seen in the explosive growth of environmental groups for self-help and neighbourhood action around the world. At the national level, political parties have begun to give environmental issues central places on their agendas. Expenditures for environmental protection have grown to constitute 1 or 2 per cent of the GNP in most industrialized countries (Holdgate et al., 1982). At the international level, the evolving position of environmental values is reflected in successive reports of special UN World Commissions. Whereas commissions formed earlier in this decade had dealt with security or development issues in isolation, the Brundtland Commission's 1987 report on *Our Common Future* emphasized the connections among environment, development, and security. It stressed the need to pursue all three objectives in concert (WCED, 1987).

Which values are most relevant to global change? Classic economic notions of efficiency are clearly no longer the dominant values guiding many of today's decisions that are seen to have an environmental component. Increasingly important are values reflecting alternative conceptualizations of development, a sense of stewardship for the planet, and obligations to future generations or neglected populations (White, 1988a). How such values are factored into environmental choices, how they enter policy processes, and how their character and reach vary over time or among cultures, are central to an understanding of global change.

Options. If values reflect what people want, options reflect what they can get. Clearly, not everything people value can be obtained; what can be obtained may often be reached via alternative paths. Moreover, the value and option dimensions of choice are related. Research on policy making has shown that people generally decide what they want only in the light of what they think they can get (March and Olsen, 1976; Wildavsky, 1979). Reciprocally, what they can get at any given time often reflects options developed in response to previously unobtainable desires.[5] Not surprisingly, one of the greatest contributions that formal analysis actually makes to practical policy is through enlarging the range of available options (Schelling, 1983).

Such options can be grouped under three interrelated categories: technological, organizational, and economic. Technological options concern the alternative ways in which basic processes of resource use, manufacturing, service provision, and waste control can be carried out. Examples range from selective harvest versus clear-cutting for forest management, to fossil versus nuclear sources of electrical energy. Organizational or institutional options range from basic modes of social structure, through alternative regulatory arrangements, to various legal structures. Examples include market versus control economies, pollution permits versus effluent taxes, and international versus regional treaties and conventions. Finally, a focus on economic options emphasizes the fundamental requirement that societies be able to afford choices undertaken in pursuit of their values. While maximizing efficiency may not always be an appropriate criterion for choice, even the most ardent conservationists are coming to realize that environmentally sustainable development must be economically sustainable as well (Barbier, 1987; Madden, 1987; Mellor, 1988; WCED, 1987).

Perceptions. In contexts as complex as global change, perceptions and assessments will inevitably differ from reality for reasons of fundamental ignorance or uncertainty – i.e. the actual causes and effects of change are not understood by anyone. Such fundamental inaccuracies increase as one moves along the chain of causation from outcomes cast in terms of altered material and energy flows in the physical environment, to ecosystem-level impacts of such alterations, to consequences for individuals and entire social systems (Schneider, 1983). These difficulties are only partially overcome by the wide array of formal assessment methods developed over the last two decades.[6]

Inadequate education and poor communication of expert assessments limit the accuracy of perceptions for choice. But experience suggests a wide range of other potentially limiting factors that are characteristic of the choosers themselves. These may include the proximity of an individual to environmental damage; exposure to mass media, education, variety of life experience, age, cultural and organizational context, and others. Not the least of these is bias: individuals and institutions have a significant ability to hear what they want to know (Douglas and Wildavsky, 1982; Tang and Jacobson, 1988). Explaining and predicting interactions between human and environmental systems require understanding of different peoples' perceptions of global change and their roles in it, the factors that cause variation or distortion in those perceptions, and the steps that can be taken to make formal assessments more accurate and useful.

Culture

Interactions with the global environment, as well as choices regarding environmental management, are ultimately grounded in a wealth of underlying social factors and historical contexts that might be called 'culture'. Patterns of global environmental change have often been described without reference to cultural factors. But because of the integral role of human systems in the dynamics of global change, some understanding of why societies function as they do will be required for explaining and predicting interactions between people and their environments.

A long-standing debate exists on the relative importance of various cultural factors as causes of environmental change and as determinants of their consequences (e.g. Garcia, 1981). That debate is sure to intensify with increased attention to the problems of global change. In fact, however, the practical difficulty is not to imagine ways in which cultural variation might influence global change. Rather, it is to bound the consideration of contributing cultural factors in a way that leads to efficient explanations, and thus keeps research from becoming synonymous with research on social systems in general.[7] It thus seems prudent to focus initial studies on the dimensions of culture that scholarship indicates are almost certainly central to the long-term, large-scale interactions of human and environmental systems. These include the structure, distribution, and growth of human populations, the modes of social, political and economic organization, and the resulting state of agricultural, industrial, and general economic development.

Population.[8] Population characteristics are clearly most fundamental, with direct implications for resource use, waste production, and social vulnerability (Repetto, 1987). Despite continuing declines in fertility rates throughout much of the world, the population of the earth is almost certain to double within the next century. Migration-induced changes in the distribution of population are even more dramatic than population growth *per se*. Fifty years ago, less than one quarter of the world's population lived in urban areas; fifty years from now, more than half will (UN, 1985a). Moreover, trends toward very large cities will almost certainly entail nonlinear implications for human interactions with the environment (Gu et al., forthcoming). Despite years of discussion on the interactions of population, resources, and environment, however, there is still only limited understanding of how the elements of long-term, large-scale human population dynamics (e.g. fertility, migration, age distribution, and life expectancy) affect either sources of environmental change or the implications of those changes for people.[9]

Organization. A second set of underlying human dimensions of global change can be grouped under the heading of social institutions or organizations. In the broadly defined sense employed here, 'organization' includes such forms of human interaction as the family, markets, corporations, command and regulatory aspects of government, voluntary associations, religious structures, laws, education, the media, and a variety of international arrangements. The mix and strength of such mechanisms clearly vary, but we know little about the environmental implications of the different organizational means of seeking human well-being. We do not know much more about what determines the efficacy of alternative organizational approaches to incorporating environmental considerations in the social calculus (Tang and Jacobson, 1988). Clearly, the nation state is one of the most important organizational structures involved, but international institutions and regimes are becoming increasingly important (Kay and

Jacobson, 1983; Majone, 1986; Keohane and Nye, 1987). And underlying structures determining local property rights and access to resources also have a substantial impact on what people actually do to their environments.

Perhaps the most significant organizational trend for studies of global change is the worldwide growth in the reach and power of human institutions that has been gathering momentum over the last several hundred years, and seems likely to accelerate into the future. Most organizational forms, over most of their history, have been preoccupied with more or less immediate goals of physical, economic, and spiritual security. It is only relatively recently that a few organizational structures have emerged with self-professed goals of environmental protection and environmentally sustainable development. Especially needed is a deeper understanding of those emerging organizations with global reach and those capable of forcing particularly rapid change.

Development. As important as population growth, structure, and distribution may be, they represent only the ultimate foundations on which rest the human activities that are the proximate sources and receivers of global change. Trends in resource use *per capita* must be considered along with trends in population density if we are to understand variations in total pressures on the environment exerted by different societies (Clark, 1988a; Goldemberg et al., 1987). Similarly, the character of a region's resource use, productive activities, and trade relations are all important determinants of its vulnerability to environmental change (Burton et al., 1978; Chisholm, 1982; Parry et al., 1988). In general, efforts to understand the underlying cultural dimensions of global environmental change must also look to long-term, large-scale changes in how people produce and consume goods and services. Industrial and agricultural processes have for centuries been the activities with the greatest consequences for environmental change. In the latter part of the present century, energy-related activities have also become important developmental dimensions of global change (Clark, 1986; Turner et al., forthcoming). For example, prior to 1950 the majority of carbon dioxide released to the atmosphere through human activities came from biomass burning. Since then, the majority has come from fossil fuel combustion (Bolin, 1986). Consumptive end use processes are also increasingly evident as major agents of global change (Ayres and Rod, 1986).

Two basic issues arise in efforts to understand the past and future role of major development trends in global change. The first concerns the concept of 'sustainability', defined by the Brundtland Commission as the capacity to provide for the needs of the present without diminishing the options of future generations (WCED, 1987).[10] Despite the importance and popularity of the sustainability theme, there is still little understanding of just what constitutes a sustainable, as opposed to an unsustainable, development path. We need not only more careful case histories of how environmental change has influenced development, but also a framework of concepts and causal hypotheses for use in synthesizing and generalizing the cases.[11]

A second and related developmental dimension concerns alternative paths or models of development. What feasible and desirable paths would reflect a more balanced view of physical, economic, and environmental well-being or security than has traditionally been the case? What international processes would best catalyse and legitimize the search for such alternatives? What series of development indicators would eliminate the worst imbalances and biases of current measures, and provide more meaningful and useful documentation of progress toward the sustainable development of the earth?[12]

Unresolved questions

Human sources of global change

An early target of research on global change should be documentation and understanding of the most important ways in which human processes drive or force changes in the environmental system. Experience with the study of the greenhouse problem and stratospheric ozone depletion shows that accurate histories of emissions resulting from human processes are necessary to differentiate among competing theories or models. The same experience shows that using superficial trend-extrapolation scenarios of future emissions resulting from human processes can seriously misdirect the attention of the research community. Finally, more informed social choices regarding environmental management require better understanding not only of how, but also why, the human forcing of global change varies over time, space, and culture.

Identifying human activities that drive global change. The first requirement is for a more systematic identification of which human-induced changes in energy and chemical flows, water use, habitat extent, or other variables constitute the most significant 'inputs' to climatic, biogeochemical, or biotic dynamics. For a few aspects like the greenhouse effect or stratospheric ozone depletion, this preliminary identification of 'input' linkages between the human and other components is relatively well in hand: research can confidently focus on a specific list of radiatively active trace gases and halocarbons. For most other aspects, however, much basic research on human forcing of interactions with the environment needs to be done.

Figure 3 gives a simple conceptual framework for the process of explicitly and systematically identifying the important linkages among components of the geosphere–biosphere system. The entries in the 'human component' cell of the figure are drawn from the NASA report *Earth System Science: A Closer View*, one of the most recent and systematic efforts to set priorities for linkage information.[13] Those are nonetheless preliminary and incomplete, reflecting only partial reviews of the knowledge about human activities that researchers working on the natural components of global change believe they need in order to test their models and make predictions. Moreover, to be maximally useful such priority lists must also specify the scale, resolution and accuracy required of the input data if they are to be useful. Needed as well is information on the threshold rates or quantities above which specified anthropogenic inputs, although not now of great concern, would become potentially important agents of global change.

Despite these demanding requirements, however, the kind of perspective suggested in Table 1 needs to be developed systematically as a useful first step in setting priority targets for studies on the energy flows, chemicals, habitats, water flows, physical properties and biological entities that constitute the most important links between human activities and rest of the geosphere–biosphere system.

Once a priority list of linkage variables has been defined, there remains the question of which specific human activities are capable of changing the fluxes of those variables. Even in relatively simple cases, this analysis is not trivial (e.g. Wuebbles and Edmonds, 1988). Establishing the required knowledge base will require close collaboration between scientists studying the relevant non-human components of global change and specialists in the workings of relevant technologies and land use practices (especially chemical, water, mechanical, and agricultural engineers). Table 2, drawn from one such collaborative effort, suggests the wide range of human processes that force changes in just a few of the chemical species of importance to global change.

Input–output relationships in human activities. How much alteration of the relevant material or energy flows is created per unit of human source activity? How do these 'intensive' relationships between human processes and material and energy flows vary across space, time, and culture?

Answering these questions will require basic quantitative process studies on the transformation of human activity 'inputs' such as coke production or rice paddy cultivation into 'outputs' such as methane flux to the atmosphere. Engineering expertise will again be essential to the task. Typical of the most complete work on individual inputs and outputs is the work on quantities of carbon dioxide produced per unit energy derived from various fossil fuels (Marland and Rotty, 1983). Even in the carbon dioxide case, however, comparable input/output coefficients for various types of land use conversion activities are still underdeveloped.

The relative simplicity of the I/O structure for the case of carbon dioxide emissions is potentially misleading as a guide to research requirements in this complex area. In the more general situation suggested in Table 2, a given pollutant flow will be altered through several human activities. Complex sequences of reaction, deposition, and remobilization may be involved within the human system en route to a final measured 'output' into the environment. At

TABLE 1. A framework for defining linkages among components of the geosphere–biosphere system.

Needed from \ Needed by	Human component	Biotic component	Biogeochemical component	Climatic component
Human component	(Internal social dynamics)	Land use transforms Yield management ?	Land use transforms øCO_2, NO_x, SO_x ?	Land use transforms ø Greenhouse gas ø Aerosols Albedo change Water distribution
Biotic component	Changes in major ecological zone boundaries Changes in productivity potential	Internal (biotic dynamics)	?	?
Biogeochemical component	'Air quality' 'Water quality' ø Toxics in soil	?	(Internal biogeochemical dynamics)	?
Climatic component	Regional temp. statistics Regional ppt statistics Regional ice statistics	?	?	(Internal climatic dynamics)

In the 'looking outward' or systems approach of this framework, attention is focused on what studies of each component need to know from other components in order to proceed with their work (Holling, 1978). The entries in the cells of the figure are illustrative only, reflecting the needs from and by human component studies as defined in NASA's ESSC (1988). Note that 'ø' indicates fluxes.

each stage, process understanding is usually imperfect, as is monitoring data for estimating fluxes and pools. An important methodological advance in dealing with such complexities in I/O assessments has therefore been 'materials/energy balance accounting' (Ayres, 1978 and 1989). This approach takes advantage of conservation principles to balance amounts of energy and materials drawn into the human system with amounts exported or stored at any given time. It focuses attention on the fact that all energy and materials used in human development have to go somewhere, even when monitoring data fail to find them. It has helped in the 'discovery' of unsuspected pollutant sources in seemingly innocuous or irrelevant human activities (Ayres and Rod, 1986; Ayres et al., 1988). More systematic use of balance accounting would almost certainly be useful in future efforts to answer questions concerning the I/O relations between human and other components of the geosphere–biosphere system.

An especially difficult but necessary aspect of I/O studies will be their historical dimension. Long-term changes in the completeness of combustion or depth of ploughing may be more important for certain environmental emissions than changes in the total amount of combustion or extent or arable land. Ignoring historical changes in the I/O relations can lead to serious errors in understanding the human role in global change. For example, the uncritical use of con-

TABLE 2. Human processes that force changes in selected chemical constituents of the atmospheric environment.

Emission source	Environmental component						
	NO_x	SO_x	HCl	Sea Salt	CH_4	CO	N_2O
Production/distribution of mineral fuels							
Coal					X		
Oil					X		
Natural Gas					X		
Combustion of mineral fuels							
Coal	X	X	X			X	X
Oil	X	X	X				X
Gasoline						X	
Other						X	
Natural Gas	X	X	X				X
Industrial processes							
Coke Production					X		
Pig Iron Production						X	
Smelting of							
Copper		X					
Lead		X					
Zinc		X					
Agricultural							
Biomass Combustion	X				X	X	X
Nitrogen Fertilizers	X						X
Rice Paddy Cultivation					X		
Ruminant Animal Grazing					X		
Miscellaneous							
Industrial uses, refrigerants, consumer products							
Landfills					X		
Oceans				X			
Wetlands					X		

These illustrative data are drawn from a study by Darmstadter et al. (1987, Vol. 2: B3).

temporary emission coefficients for assessment of historical carbon monoxide emissions in the US has almost certainly resulted in cumulative estimates that are significantly too low (Darmstadter et al., 1987, App. A, p. 26). Table 3 shows the kinds of results that careful collaborative research can produce.

The changing magnitude of human forcing. What are the total amounts or strengths of the relevant human source activities? How do these 'extensive' measures of the human ecology of global change vary across space, time, and culture? Ultimately, the need is for something approaching a theory of world development, cast in terms appropriate to produce relevant human forcing functions needed for the understanding of global environmental change. The underlying research questions were summarized in a recent Social Science Research Council report on changes in the global environment (SSRC, 1988).

Among the earliest efforts to address such questions at the global scale were the various world systems modelling efforts of the 1970s (e.g. CEQ, 1980; Meadows et al., 1972). The shortcomings of these efforts are well known and can be traced to weaknesses in data, methodology, and conceptual foundations (Brewer, 1986; Greenberger et al., 1983; Meadows and Robinson, 1985; OTA, 1982). A number of more modest efforts, focused on particular sec-

TABLE 3. Reconstruction of historical methane emission coefficients, reflecting changes in human economy and technology

	Methane emissions coefficients (metric tons CH_4/metric ton of fuel)					
	1800	1860	1890	1920	1950	1980
Anthracite	–	0.005	0.006	0.007	0.007	0.007
Appalachian bituminous (underground)	0.005	0.005	0.005	0.005	0.005	0.005
US Average bituminous	–	–	0.005	0.005	0.005	0.005
Coking (based on coal used for coking)	–	–	0.27	0.054	0.03	–
Gas (based on unaccounted potential production of associated gas)	–	0.30	0.25	0.22	0.20	–
Gas distribution (based on gas marketed)	–	–	0.03	0.02	0.01	0.01

This example is drawn from the study by Darmstadter et al. (1987, Vol. 2: A83) on the impact of world development on the atmosphere. The listed coefficients are expressed as tons of CH_4/ ton of fuel. Dashes reflect missing data.

tors of the human system, have since been carried out. Several in the fields of population, agriculture, forestry, and energy modelling, are relevant to studies of global change.[14] The best of these sectoral studies are often contradictory in ways that can only partially be resolved through subsequent reanalysis (Toth et al., 1988). No credible integrated and dynamic model yet exists of long-term global changes in human activities that force environmental change. Prospects for such a model, while still distant, are nonetheless improving and need to be pursued. As regards data, many of the most important contemporary human releases of materials and energy are monitored through national and international environmental networks. A careful review is needed, however, of the adequacy of such monitoring. On the methodological side, another recent SSRC study has illustrated how much the social and natural sciences can learn from each other as they struggle with their parallel problems of modelling and predicting global change (Land and Schneider, 1987). Finally, initial attempts to provide conceptual foundations for long-term global studies of human development can be seen in historians' debates on 'the modern world system' and investigations of 'long-wave' phenomena in economic life (Ausubel, 1989; Braudel, 1984; Vasko, 1987; Wallerstein, 1974). Although most of this work has ignored environmental and resource dimensions, some impressive recent efforts of geographers and historians are beginning to formulate a theory of global human ecology (Chisholm, 1982; Richards, 1986; Turner et al., forthcoming).

Human consequences of global change

Formal assessments of global change and its consequences aim to help make human perceptions more useful and effective guides to action. Research to improve assessments is therefore central to an improved understanding of the interactions between human and environmental systems. Summarized below are the principal questions regarding assessment that should be addressed in the early stages of a research programme on the human ecology of global change.

Determining the environmental dimensions of human vulnerability. A basic research task is to identify what kinds of change, and what rates of change, are those to which people in different cultural settings are most vulnerable. More precisely, what information concerning the character, timing, and location of possible changes in the earth's biogeochemical, climatic, hydrologic, and biotic processes is needed as inputs to studies attempting to understand and expand the boundaries of sustainable development?[15]

The conceptual framework required for addressing this question is thus the same as that introduced in Figure 3. In this case, however, knowledge of the human system's sensitivities rather than that of the environmental system must provide the point of departure. Without explicit guidance from scholars regarding the kind, scale, and resolution of information needed to assess important aspects of social vulnerability to environmental change, natural scientists cannot be expected to focus on the few aspects of change that are most important to people (Chen and Parry, 1987).

How might scholars of human response best advance their task? Useful approaches are suggested by work in the study of natural hazards, climate change and the environmental effects of nuclear war (Burton et al., 1978; Harwell and Hutchinson, 1985; Kates et al., 1985a). A particularly instructive approach has been applied by Parry to address the response of sensitive agricultural systems to climatic variation and change (Parry et al., 1988). This begins by focusing on agricultural regions on the margin of economic viability with respect to temperature or moisture conditions. Rather than asking how these systems would respond to climate change in general, it analyses the social, economic, and agronomic characteristics of each region to assess the limits of climate change beyond which significant disruption or displacement of agricultural activity would be expected to occur. The challenge is then thrown back to the climatologists to estimate whether, and when, changes of such magnitude might be expected due to natural variability or anthropogenic forcing.

Efforts are now needed to move beyond climatic considerations in asking what kinds of environmental change are most important for people. An excellent foundation for this task is provided by the published proceedings of the Dahlem Conference on *World Resources and Development* (McLaren and Skinner, 1987).

Assessing syndromes of environmental transformation. Multidimensional 'syndromes' of environmental transformation are a central feature of global change (Regier and Baskerville, 1986). For example, the sustainability of forest resource use in some regions is simultaneously threatened by changes in land allocation, climate, and atmospheric chemistry. The resulting syndrome of forest resource degradation is the environmental change that must ultimately be explained and the policy problem that must ultimately be addressed. Other syndromes such as those associated with development of the rich and diverse farming landscapes of Normandy, can be viewed as positive transformations. In general, however, assessments must move beyond problem-by-problem formulations to provide a synoptic or integrated account of the overall environmental changes and consequences that result from specific patterns or strategies of human development.

To date, however, most of the relatively few assessments that have addressed large-scale environmental changes have dealt with the relationships between single environmental components, e.g. acid deposition, and single development sectors, e.g. forestry. Even the most ambitious works – for example the programmes of the US National Research Council and the Organization for Economic Co-operation and Development on environmental impacts of energy production, or the Scientific Committee On Problems of the Environment's programme on climate impact assessment – have dealt only with the impacts of change in a single valued environmental component across a range of human activities, or the impacts of a single human activity across a range of valued environmental components (Brooks and Hollander, 1979; Kates et al., 1985a; Torrens, 1984). The assessment of multiple threats or, more generally, change syndromes has been largely ignored.

One notable exception is provided by the experiments in syndrome assessment conducted in recent years by Thomas Graedel of AT&T–Bell Laboratories and Paul Crutzen of the Max Planck Institute (Crutzen and Graedel, 1986; Darmstadter et al., 1987). A sample of their work, dealing with valued components of the atmospheric environment, is reproduced in Figure 3. The relative simplicity of the figure should not be allowed to disguise either the tremendous amount of research necessary to provide the findings it portrays, nor the incomplete nature of the results obtained.

Linking spatial scales in assessments of global change. A central question is how to link glo-

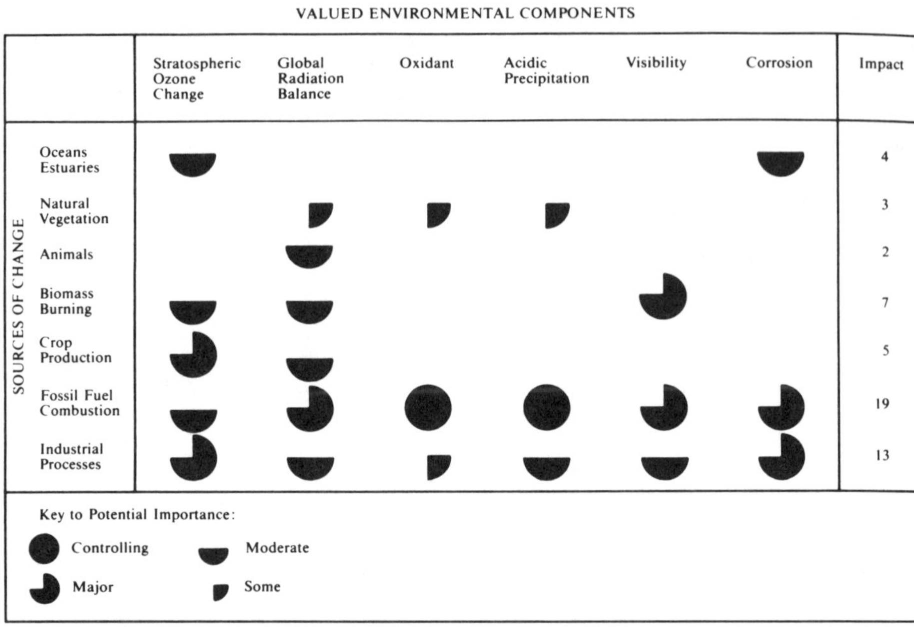

FIGURE 3. Assessing syndromes of environmental change. An example of the required integrated assessment perspective is suggested in this figure, developed to show the overall consequences for the atmospheric environment of changes in a wide range of human and natural processes (Crutzen and Graedel, 1986). Each cell of the matrix reflects one of the classical single-cause, single-consequence assessments. Although there is no mechanistic way that cell values can be quantitatively combined, each 'column total' would in principle represent the net effect of all sources of change on a single environmental component. Each 'row total' would indicate the net effect of a single source of change on a wide range of environmental components. (Significantly, these 'row totals' are equivalent to the integrated sectoral perspective on sustainable human activities and global change recently advanced by the U.N.'s Brundtland Commission on Environment and Development.) The matrix as a whole constitutes a qualitative synopsis of global (atmospheric) change.

bal change to local conditions at intermediate or regional scales. Much of what is interesting or worrisome in global change seems to consist of people in one set of places taking actions that have their major consequences for others elsewhere.[16] Much of sustainability relates to importing needs or exporting wastes from one place to another. Finally, the values, options, and perceptions central to human choices regarding global change differ widely from place to place on earth. The uniqueness of and interactions among places are thus central to the human meaning of global change and to the prospects for managing sustainable development. Yet they are poorly captured (if addressed at all) by current assessment practices. Only the barest outlines of an approach to improve this situation have been sketched (Chisholm, 1980; Clark, 1987). At least two related research tasks merit early attention.

Mapping vulnerabilities. Work in the assessment of climate impacts has demonstrated the utility of focusing on places and peoples that are especially vulnerable or sensitive to specific kinds of climatic change and variation (Parry, 1985). The search for such regions and social groups should now be extended to include consideration of other environmental components. For example, which human societies would be most seriously at risk under scenarios of continued depletion of stratospheric ozone?

Even more important, and difficult, is the task of identifying places and people that are likely to be particularly vulnerable to the 'syndromes' of multiple environmental changes. Boundaries of the geographic regions or social groupings most threatened by each individual component of environmental change are generally not entirely congruent. It will therefore be necessary to search for situations where especially worrisome changes in several valued environmental components overlap at particular locations. The concepts and tools of geographic information systems seem especially well suited to the required mapping task. A useful guide to what will be required is provided by efforts to map areas with a high risk of forest mortality under multiple stresses of acid deposition, tropospheric oxidants, drought, and other factors (UNECE, 1986). More broadly, efforts should be undertaken to examine previous episodes of rapid regional environmental change to identify the determinants of vulnerability (e.g. Glantz, 1988).

Global linkages in the assessment of social vulnerability. It is equally essential that the global linkages among regions not be ignored. As was recognized some time ago, for example, the social significance of a prolonged drought in the North American grain belt would depend strongly on what was happening to the weather in other major grain-growing regions of the world. Nevertheless, most contemporary assessments of global climate change still manage only to tally consequences for a selection of regions considered independently, as though they were on different planets.[17] Related shortcomings characterize the country-by-country approach to recent assessments of forest damage in Europe (Nilsson and Duinker, 1987).

Improvements of concepts and methods for assessing the human implications of linkages among regional consequences of global change are badly needed. They will require close collaboration between natural scientists concerned with global patterns of environmental change (e.g. the 'teleconnections' of climatology), and social scientists concerned with the economic, political, and other human processes around the world. Preliminary studies seeking to address the linkage issue suggest that global models of resource and commodity trade are likely to be a necessary component of such assessments (e.g. Binkley, 1988; Williams et al., 1988). Substantial use of explicit scenarios of environmental change and human development will also be required (Brewer, 1986).

Incorporating values in assessments of global change. The values that figure so prominently in the determination of human choice are not reflected adequately in most environmental assessments. The special problems raised by global environmental change – among them multigeneration time horizons, multination spatial scales, and the prospect of irreversibility – increase both the difficulties and the importance of doing better. Research in environmental economics has made some progress in addressing the value issues that arise in the treatment of externalities and the valuation of non-market goods. The field has nonetheless found it extremely difficult to provide a balanced treatment of values other than efficiency in its assessment calculus (Kneese and Schulze, 1985). Other disciplines, including those concerned with legal and ethical issues, have begun to address some of the key issues of equity and multigenerational tradeoffs in a global change context (e.g. Brown-Weiss, 1984). Progress has nonetheless been slow.

Valued environmental components. Experience in environmental impact assessment emphasizes the importance of focusing assessments on a relatively few 'valued environmental components' like summer soil moisture or tropical species diversity (Beanlands and Duinker, 1983). The list of components is always subject to revision, but efforts to exclude nothing simply devalue everything. Scientists have an important input to make to the selection of valued environmental components, but so do the policy makers who will be asked to use the results and the public who will be asked to pay the bills. Procedures need to be designed that will help different groups affected by global change to identify and articulate the components they value most. Special attention is needed on how to give fair consideration to the values of traditionally under-represented groups – the poor, the uneducated, the unborn. Also needed are fora and negotiating processes that can promote agreement among groups on the core valued environmental components to be addressed in mutually supported research. A variety of workshop methods have evolved over the last decade to facilitate such negotiations for small scale environmental problems. But these approaches must still be evaluated and adapted to the context of long-term, global interactions between human and environmental systems (Sonntag, 1986).

Attitudes towards risk and uncertainty. People's inevitably incomplete understanding of global change means that all assessments are bound to be riddled with uncertainties. This raises a number of important value-related questions that should be addressed. For example, studies of human attitudes toward risk have shown that different individuals or organizations confronting the same objective uncertainties will place significantly different values on actions to remove the uncertainty or avoid its possible consequences. Put simply, some people will be more risk-averse than others; these differences will have significant implications for the choices societies make when confronted with environmental threats (Fischhoff et al., 1981; Kleindorfer and Kunreuther, 1986). Moreover, we know that the form in which risks are communicated to people can make a tremendous difference in their resulting behaviour. Most research, however, has focused on perceptions of environmental risks very different from those posed by global change. We know very little, for example, about how people evaluate highly uncertain predictions of high-consequence, large-scale, relatively irreversible events.[18] Better understanding is needed of how such values can be measured, how they vary with space and time, and how they can be incorporated in assessments of global change. The implications of ways in which such assessments are communicated through technical and mass media also need study. Even more fundamentally, it will be important to follow up the leads of anthropological research that suggest how cultural factors shape the ways in which people and organizations value uncertain environmental risks (Douglas, 1966, 1986; Douglas and Wildavsky, 1982; Gerlach and Rayner, 1987).

Human management of global change

A major challenge of coming decades is to learn how long-term, broad-scale interactions between human activities and the earth's environment can be managed to increase the prospects for sustainable improvements in well-being. Management is not the same as prediction or even understanding, for management can be improved despite the enormous uncertainties and downright ignorance that will continue to make detailed predictions illusory. A central question is whether we are in fact improving our management of environmental change and, if so, which forms of social action are most effective in what situations (White, 1988a). Improvements in the management of global change can be defined in terms of their ability to increase social choice and decrease vulnerability in the face of uncertain futures of both environmental change and human objectives.[19]

Which management options should actually be adopted in the face of global change is a matter for resolution through the policy process. But increasing the range of management options, and characterizing their likely performance, should be a central focus for invention, imagination, and research applied to the human component of global change.

Broadly, efforts to manage global change involve changes in technologies, institutions, or behaviour of individuals. No serious effort has yet assessed the relative efficacy of efforts to employ these three mechanisms for managing environmental change. Such an examination is especially important for a global change programme because of the rapid increases in scale of the environmental transformations that must be managed. It is not clear that there is much precedent for dealing with the long-term, broad scale issues that are increasingly at the forefront of concern.[20]

Technology and the management of global change. Technical change is rapidly accelerating, mostly in pursuit of objectives that have little direct connection with environment. On the scale of years, technological innovations can be expected to make only a limited difference in the management of global change.[21] On a scale of decades, however, the consequences – intended and incidental – can be tremendous.

Work on life-cycle dynamics of major technical processes suggests that 20 to 40 years is sufficient for oil to displace coal as a world energy source, for the steel industry to shift from open hearth to electric arc processing, or for world turpentine production to shift from biotic to chemical feed stocks (Ausubel and Herman, 1988; Clark, 1987). Each of these developments, and many others like them, have had major implications for environmental change, though few were undertaken with environmental issues in mind. Turning to intentional use of technology to manage environmental change, options exist to eliminate atmospheric emissions of sulphur dioxide and carbon dioxide due to fossil fuel production (e.g. Haefele et al., 1986) and to reduce radically the danger posed by halocarbons to stratospheric ozone. But between these observations of technical feasibility and useful assessment of the management options that technologies might offer for dealing with global change are a number of basic research questions that have been only partially addressed by existing studies:

- What are the major technological trends that are likely to restructure the nature of interactions between environment and development over the next several decades (Ausubel and Sladovich, 1989)?
- What technological opportunities appear most promising in the light of what we understand about the quantities and qualities of the key flows of materials and energy involved in global change (NAE, 1988)?
- What are the necessary conditions to induce technical innovations that would relax the constraints posed by global change (Runge, 1986)?
- What are realistic penetration and diffusion times for such innovations in the world market (Marchetti, 1983; Nakicenovic, 1988; Pry, 1973)?
- How can both innovation and diffusion of technologies supportive of sustainable development be encouraged in an increasingly internationalized technology market (Guile and Brooks, 1987)?
- To what extent will it be most productive to pose the above questions not for individual technologies, but rather for groups of tightly interacting and therefore 'co-evolving' technologies?

Institutions and the management of global change. Humans organize their responses to global change through a wide range of institutions and other structures. The fundamental research challenge is to understand the relative effectiveness of alternative institutions as mechanisms for management of interactions between human and environmental systems. The institutions to be examined include not only those involved in classic market and national regulatory functions, but also an increasing array of voluntary and international organizations. All these structures, plus the bodies of policy, law, and practice they reflect, are evolving rapidly in terms of reach and power. Each country's management options are increasingly influenced and constrained by the institutions and policies of other countries. A dynamic, global perspective on the changing efficacy of alternative institutions for managing global change is therefore necessary, backed by appropriate comparative and historical research. A limited body of existing scholarship on national and international mechanisms for managing large-scale environmental change provides the foundations on which such research can build (e.g. Caldwell, 1984; Carroll, 1983; Feldman, 1988; Kay and Jacobson, 1983). Among the specific questions to emerge from recent reviews, the following stand out as meriting high-priority attention:

- What is the special place of the nation-state in managing the interactions between human and environmental systems? How are economy/environment interactions treated in national policy processes? What determine the prospects for inter-state co-operation on environmental problems? How do domestic interests reflect on such foreign policy decisions.[22]
- What influences 'the pace or response-time with which scientific information about threats of significant perturbations in the geosphere and biosphere is translated into government action' (Jacobson and Shanks, 1987, p. 29)? More generally, which factors place environmental issues on national and international agendas? Which keep them there through time? Which institutional mechanisms might

facilitate anticipatory as opposed to reactive agenda setting for issues of global change (Tang and Jacobson, 1988)?

What strategies for environmental protection – what mix of education, markets, regulatons, and laws – seem to work best for dealing with large-scale, long-term problems in specific cultural contexts (Tang and Jacobson, 1988)? What are the 'attitudinal and political preconditions for effective regulation or resource management programs' (Jacobson and Shanks, 1987, p. 29)?

To what extent can methods of environmental protection be transferred from one country to another? What facilitates the transfer? What are the limits and barriers to one country's application of strategies successfully employed in another (Tang and Jacobson, 1988)?

Despite the generally poor performance of both market and regulatory approaches to problems involving public goods and externalities, 'some good solutions to borderline cases, such as research and development, are available through a third option, the "not-for-profit" sector. This sector has played a very helpful role in such areas as public health and agricultural research associated with the "Green revolution"'. What are the conditions under which it might play an equally important role in managing global change (McLaren and Skinner, 1987, p. 544)?

To what extent could trade and investment policies play a role in the management of global change? Do policies that encourage the spread of multinational business enterprises: (1) accelerate the diffusion of advanced pollution control and resource-management techniques; or (2) redistribute environmental degradation to less developed countries (Tang and Jacobson, 1988)?

What are the 'prospects for, and the strength of obstacles to, effective international co-ordinated regulation and resource management programs addressed to problems of international and global environmental degradation' (Jacobson and Shanks, 1987, p. 29)?

What new international frameworks, such as the concept of a 'planetary trust' developed by Edith Brown-Weiss in the context of UN University studies on sustainable development, might provide useful tools for the management of global change (Brown-Weiss, 1984)?

Behaviour and the management of global change. Ultimately, it is certain patterns of human behaviour that lead to environmental degradation, and other patterns that result in sustainable development. Research needs to establish how relevant human behaviours are shaped, and how they can be altered as part of efforts to manage the long-term, large-scale interactions between people and their environments. A growing body of scholarship in economics, psychology, attitude change, communications networks, and social diffusion provides substantial foundations upon which the needed research can be built.[23] With some exceptions, however, this work has tended to focus on the determinants and control of individual behaviours. Global change problems, in contrast, highlight the importance of collection action and organizational behaviour in shaping long-term, large-scale interactions between people and the environment.[24] A fundamental challenge in studies of the human ecology of global change is to work towards a theory of human system behaviour that encompasses all these relevant levels of social organization. On the way to such a comprehensive theory, several specific questions merit early attention:

Why do some persons exhibit behaviours that contribute relatively strongly to the forcing of global environmental change, while others choose to behave in ways that contribute relatively little?[25] To what extent do cultural factors, including income, social context,[26] access to relevant information or immediate experience play a role in determining differences in relevant behaviours?

Can some groups, organizations (e.g. corporations) or societies be identified as having exhibited behaviours consistent with environmentally sustainable development over extended periods? If so, what special characteristics set these groups apart? What, if any, is the distinctive role of 'traditional' knowledge and practice in such cases?

Given that most global environmental changes are both long-delayed and common-property in nature, people have limited incentives to change immediate behaviour to reduce those changes. Neither markets nor state regulatory systems have proved particularly effective in dealing with the public good and externality

Cultural dimension of human ecology: the anthropocentric conception of the universe results in treating animals and nature with contempt. Doisneau/Rapho.

dimensions of this situation (McLaren and Skinner, 1987, p. 544). How can the needed collective behaviour changes best be motivated and sustained? What can be learned from the recent rise of apparently powerful grassroots environmental movements and other changes in the valuation of environmental concerns?

What is the role of technical information and expert assessment in shaping behaviours relevant to global change? How can such knowledge be framed and communicated in ways that maximize its impact on the public and on support for social action?[27] In particular, experience in smaller-scale environmental problems strongly suggests that technical information will be more effective in changing behaviours when it is conveyed along with assessments of alternative management actions, their likely consequences, and the prospects for fair distribution of costs and sacrifices (Brooks, 1986; White, 1988a). In what ways can such experience be used to improve the ability of technical information on global change to modify behaviours of individuals, organizations, and nations? What special challenges are posed by the international character of global change?

Finally, as in the case of research on institutional questions, there is a need to provide fora in which behavioural issues relevant to the management of global change can be systematically pursued in close connection with research characterizing the likely nature, extent, and timing of global environmental change *per se*.

Selected research challenges[28]

The objective of this section is to isolate a small number of crosscutting research challenges that, due to their importance, complexity, and interdisciplinary character, might best be addressed through new international initiatives on the human ecology of global change.

Human causes of global change: land use and industrial metabolism

One major challenge is to build a better understanding of the processes underlying global land use change. Use of the earth's land surface is a key focus of the interactions between human and environmental systems. Human activities are transforming land surfaces in ways that have profound implications for ecosystems, biogeochemical fluxes, and at least regional climates (Bolin and Cook, 1983; Turner et al., forthcoming). Conversely, changes in the global environments have major implications for human land use (Jäger, 1988; Parry et al., 1988). Explaining the large-scale, long-term environmental changes of the last several centuries and predicting such changes for the future each require a deep understanding of the human processes underlying global land use change. A useful beginning in this direction has been made by the SCOPE/ICSU land transformation project (Wolman and Fournier, 1987) and the Dahlem Conference on resources and development (McLaren and Skinner, 1987).

A related challenge is a better understanding of the 'metabolism' of productive and consumptive processes through which industrial societies force changes in the earth's environment. Such understanding does not now exist, except for isolated technologies and waste products.[29] The central goal of an industrial metabolism study would be to understand and document how processes of industrial production transform resource inputs into outputs that must be absorbed and processed by the environment. A materials and energy balance approach would be central to the study. Useful beginnings have been made in the work of Ayres and his colleagues (Ayres et al., 1988, Ayres and Tarr, forthcoming) and a recent study of the US National Academy of Engineering (Ausubel and Sladovich, 1989).

Usable knowledge of global change

A second major challenge is to make knowledge about global change more useful as a guide to human action. As noted, technical information, popular perceptions, and fundamental values interact closely in shaping human choices. Unfortunately, research in understanding the components of choice has rarely been as integrated as the components themselves. The goal of this study would be to break new ground in developing an integrated understanding of how to improve the utilization of knowledge in human choices relating global change. Toward this end, the study would focus on determining how the interplay of assessment methods, perceptions, and values might be modified in ways that make possible more informed human choices on problems that arise through the long-term, large-scale interactions between peoples and their environments.

Institutions for managing global change

A third major challenge is to understand better the ways in which institutions at all scales interact in shaping the human system's capacity for coping with global change.

The institutional challenges posed by global change are substantial and varied. They range from providing for the basic research and monitoring that generate the technical knowledge base, through facilitating consensus building on required responses, to implementing co-ordinated actions in a great variety of local, regional, and national contexts. In a few cases, notably the recent protocol on protection of the ozone layer, interactions among institutions have evolved over a period of decades to the point that what began as the concern of a few scientists is beginning to translate into co-ordinated and effective action at the international scale. In numerous other cases such as desertification, however, institutional mechanisms have been inadequate to translate concern into better management. Unfortunately, we have little understanding of why our institutions have coped better with some long-term, large-scale environmental problems than with others. We know even less about how present institutional structures might be redesigned better to cope with the increasing scale and urgency of global change (see, e.g. Feldman, 1988).

Individual studies have, of course, illuminated important components of the overall environmental management picture. Thus we have treatments of the roles played by markets, international law and treaties, international organizations in the public and private sectors, international mechanisms for the exchange of scientific information, behavioral 'regimes' based on informally shared norms, voluntary organizations, and so on (e.g. Caldwell, 1984; Carroll, 1983; Kay and Jacobson, 1983; Keohane and Nye, 1977). There also exist a number of comparative studies examining different styles of environmental decision-making at the national scale (e.g. Brickman et al., 1985; Enloe, 1975; Lundqvist, 1980; Vogel, 1986). Lacking, however, is a strategic or synoptic approach that puts these individual pieces into perspective, assesses their relevance and limitations with regard to the problem of managing sustainable development in the face of global change, and focuses attention on missing dimensions of present understanding.

A useful study of institutions for managing global change would almost certainly require some reconceptualization of the dynamics of international institutions, their role in an increasingly interconnected world, and their interactions with domestic politics and agendas (Keohane and Nye, 1987). Previous studies have shown that much can be learned from comparative historical analyses of institutional performance in the emergence of large scale environmental issues (Kay and Jacobson, 1983).

Notes

* An earlier version of this review was prepared as part of the US National Academy of Sciences' efforts to shape a preliminary plan for US participation in the International Geosphere–Biosphere Program (NRC, 1988). Comments on earlier drafts from the following have contributed significantly to its content and style: Jesse Ausubel, Richard Bishop, Harvey Brooks, Robert Chen, Jerome Clubb, Philip Converse, Chester Cooper, Joel Darmstadter, George Demko, John Firor, Roland Fuchs, Robin Gregory, Harold Guetzkow, Wayne Holtzman, Harold Jacobson, Robert Kagan, Roger Kasperson, Roberta Miller, James Mitchell, Sherry Oaks, Ted Parson, Steve Rayner, John Richards, William Riebsame, E. Fred Roots, Norman Rosenberg, Mark Rosenzweig, Thomas Schelling, Stephen Schneider, T. Paul Schultz, Eugene Skolnikoff, Peter Timmerman, Barbara Torrey, Amos Tversky, Edith Brown Weiss, Duzheng Ye and Dorothy Zinberg. I owe a special debt to Robert Kates and Gilbert White on whose constructive criticism and insightful comments I have drawn especially heavily.

1. Stoppani, writing in 1873, argued that man constituted a new geological force, and designated ours as the 'anthropozoic era'. (*Corso di Geologia*, Vol. ii, cap. xxxi, sect. 1327 Milan). For a summary of modern thinking on the noosphere concept, and its connections to Lovelock's notions of 'Gaia', see Glazovsky (1988) and Serafin (1988).

2. For example despite their seminal character and originality, both the work of George Perkins Marsh (1864) and, nearly a century later, the first major interdisciplinary review of *Man's Role in Changing the Face of the Earth* (Thomas, 1956), are ultimately anecdotal works that infer a global assessment on the basis of experience in a relatively small number of intensely studied areas.

3. See, in particular, the work of the International Institute for Applied Systems Analysis (Clark and Munn, 1986); the Brundtland Commission (WCED, 1987); the Dahlem Conference on Resources and World Development (McLaren and Skinner, 1987); the Royal Society of Canada (Braybrooke and Paquet, 1987); the joint programme of the International Federation of Institutes for Advanced Study, United Nations University, and International Social Science Council (IFIAS, 1987; IFIAS et al., 1988); the US National Science Foundation (Jacobson and Shanks, 1987); the US Social Science Research Council (Land and Schneider, 1987; Rockwell, 1988); the US National Academies of Science (NRC, 1988) and Engineering (Ausubel and Sladovich, 1989); the Swedish Council for Planning and Co-ordination of Research (Svedin and Heurling, 1988); the European Science Foundation (J. Smith, 1988); and an increasing number of bilateral studies (e.g. Kotlyakov et al., 1988; Tang and Jacobson, 1988).

4. Economists have devoted extensive attention to what Thomas Schelling (1978) has called the connections between 'micro motives and macro behaviour'. Geographers have made significant progress in understanding connections between indivdidual human actions and regional environmental consequences at the level of the landscape (Kotlyakov et al.,

1988). And historians are increasingly pursuing comparative, transnational approaches to the study of long-term global interactions between people and their environments (Richards, 1986).

5. A relevant but neglected body of work on the mechanisms of this process has accumulated under the rubric of 'induced innovation' studies. For overviews, see Binswanger and Ruttan (1978), Ruttan (1984), and Runge (1986).

6. Examples include technology assessment (Shrader-Frechette, 1985), environmental impact assessment (e.g. Bisset, 1987; Munn, 1979), risk assessment (e.g. Covello et al., 1985; Whyte and Burton, 1980), and efforts to provide more comprehensive social impact assessments (e.g. Kates et al., 1985; Leistritz and Ekstrom, 1986; McAllister, 1982).

7. An analogous case was successfully argued early in the evolution of the natural science components of the IGBP. As a result, the natural science component of IGBP does *not* involve most environmental science, but rather focuses on the small but important subset of science necessary to explain certain long-term, global, interactive environmental changes.

8. This section draws heavily on the report of the US–China workshop on the human dimensions of global change (Tang and Jacobson, 1988).

9. See, for example, the materials of the Population, Resources and Environment Program (PREP) of the American Association for the Advancement of Science (AAAS).

10. For a review of alternative definitions and their problems, see B.J. Brown et al. (1987).

11. Some of the issues, initiatives, and difficulties are set forth in Clark and Munn (1986), Jacobs and Monroe (1988), and Liverman et al. (1988).

12. The problem of developing more useful indicators of sustainable development is beginning to be addressed in a number of efforts. For a sample of current work, see Liverman et al. (1988), IIED/WRI (1987), and Daly (1988). The long and not altogether happy history of work on social indicators is relevant to this effort.

13. Earth System Science Committee (1988), Figure 2.4.2. Note that an earlier and more detailed effort, though restricted to a single component of the geosphere–biophere system, is available in the NRC (1984) report on Global Tropospheric Chemistry.

14. Representative of the best work is UN (1985b) on population, UN Food and Agriculture Organization (1981) for agriculture, Kallio et al. (1987) for forestry, and Edmonds and Reilly (1985) and Nordhaus and Yohe (1983) for energy. An overview of other global sectoral forecasts relevant to global change is given in Toth et al. (1988).

15. A strong case for such management-driven definitions of data needs is made for the case of studies on climate warming in Jäger (1988).

16. As John Firor (National Center for Atmospheric Research) put it: 'How do we deal with the possibility that burning fuel in Japan may flood Bangladesh, or cutting trees in Brazil may impact forest migrations in Kentucky?' (Letter to William Clark, 15 April, 1988).

17. Stewart and Glantz (1985) have pointed out this difficulty in the ostensibly global climate impact assessment by the National Defense University. The problem persists even in the most recent work, however, e.g. the Villach/Bellagio review of policies for dealing with climatic change (Jäger, 1988).

18. For some initial findings in a rigorous comparative perspective, see Kates et al. (1985).

19. The preceding paragraph is a paraphrase from the planning documents of the International Institute of Applied Systems Analysis's biosphere programme as published in Clark (1986).

20. Partial exceptions worth examining for what they have to say about contemporary concerns for global change are the limited test ban treaty for nuclear weapons, the law of the sea, and the international ozone protocol. For perspective, see Feldman (1988), Gerlach and Rayner (1988), McLaren and Skinner (1987) and Kay and Jacobson (1983).

21. The role played by the development of substitutes for certain Chlorofluorocarbons in the promotion of the Montreal Protocol on the Ozone Layer may be, but probably is not, something of an exception. See Office of Technology Assessment (1988).

22. These perspectives were articulated by Eugene Skolnikoff of MIT in a letter to William Clark, dated 22 September, 1988. More generally, see Keohane and Nye (1977 and 1987).

23. See, for example, Bandura (1986), Lindzey and Aronson (1985), Rogers and Kincaid (1981), Schelling (1978), Winett (1986), and Stokols and Altman (1987).

24. For a general review of the problems of collective action, see Olson (1971). For a recent anthropological perspective on organizations as the relevant units of behaviour, see Douglas (1986). A broader social view is sketched by Gerlach and Rayner (1988).

25. This question has been usefully addressed in the context

of behaviours that concern energy use by Cook and Berrenberg (1981).

26. The term 'social context' is used here in the sense employed by Douglas and Wildavsky (1982).

27. The wording of this question is from Albert Bandura, in a comment on an earlier version of this report (letter to Dr Dan Druckman of National Research Council staff, dated 9 May, 1988). Bandura cites the work of attitude change theory (McGuire, 1969), decision theory in risk perception (Kahneman et al., 1982) and social cognitive theory (Bandura 1986) as providing relevant guidelines.

28. These challenges are developed more fully in Clark (1988b).

29. The case of chlorofluorocarbon emissions and stratospheric ozone depletion offers a simple example of the need for some form of future scenario as part of a global change research effort. Much attention has been devoted to the creation of such scenarios, and the most sophisticated recent probabilistic assessment (Hammitt et al., 1987) constitutes a significant improvement over early trend extrapolations. Nonetheless, it could do no better than assume that chlorofluorocarbon use in the long run would increase proportionally with global economic growth, and that economic growth would follow the trajectories assumed in the best models of energy use and carbon dioxide emissions noted earlier.

References

AUSUBEL, J. H., 1989. Long Term Regularities in Technological Development: Implications for Environment. In J. H. Ausubel and H. E. Sladovich (eds.), *Environment and Technological Change*. Washington, D.C.: National Academy Press.

AUSUBEL, J. H.; HERMAN, R., 1988. *Cities and their Vital Systems: Infrastructure Past, Present and Future*. Washington, D.C.: National Academy Press.

AUSUBEL, J. H.; SLADOVICH, H.E. (eds.), 1989. *Environment and Technological Change*. Washington, D.C.: National Academy Press.

AYRES, R. U., 1978. *Resources, Environment and Economics: Applications of the Materials/Energy Balance Principle*. New York: Wiley.

AYRES, R. U., 1987. The Industry–Technology Life Cycle: An Integrating Meta-model? *Research Report* No. RR-87-3. Laxenburg: International Institute for Applied Systems Analysis.

AYRES, R. U., 1989. Industrial Metabolism. In J. H. Ausubel and H. E. Sladovich (eds.), *Environment and Technological Change*. Washington, D.C.: National Academy Press.

AYRES, R.U.; ROD, S.R., 1986. Reconstructing an Environmental History: Patterns of Pollution in the Hudson–Raritan Basin. *Environment*, Vol. 28, No. 4, pp. 14–20, 39–43.

AYRES, R. U. et al., 1988. An Historical Reconstruction of Major Pollutant Levels in the Hudson–Raritan Basin, 1880–1980. *NOAA Tech. Memorandum* NOS OMA 43. Rockville, MD: National Oceanic and Atmospheric Administration.

AYRES, R. U.; TARR, J., forthcoming. The Environmental History of the Hudson-Raritan Basin. In B. L. Turner and others (eds.), *The Earth Transformed by Human Action*. New York: Cambridge University Press.

BANDURA, A., 1986. *Social Foundations of Thought and Action: A Social Cognitive Theory*. Englewood Cliffs, N.J.: Prentice-Hall.

BARBIER, E. B., 1987. The Concept of Sustainable Economic Development. *Environmental Conservation*, Vol. 14, No. 2, pp. 101–10.

BEANLANDS, G. E.; DUINKER, P. N., 1983. *An Ecological Framework for Environmental Impact Assessment in Canada*. Hull, Quebec: Canadian Federal Environmental Assessment Review Office.

BINKLEY, C. S., 1988. A Case Study of the Effects of CO_2-induced Climatic Warming on Forest Sector Growth and the Forest Sector. Part B. Economic Effects on the World's Forest Sector. In M. L. Parry, T. R. Carter, and N. T. Konijn (eds.), *The Impact of Climatic Variations on Agriculture, Vol. 1. Assessments in Cool, Temperate and Cold Regions*. Dordrecht, Holland: Reidel, pp. 197–218.

BINSWANGER, H.; RUTTAN, V. W., 1978. *Induced Innovation: Technology, Institutions and Development*. Baltimore, MD: Johns Hopkins University Press.

BISSET, R., 1987. Methods for Environmental Impact Assessment: a Selective Survey with Case Studies. In A. K. Biswas and Q. Geping (eds.), *Environmental Impact Assessment for Developing Countries*. London: Tycooly.

BOLIN, B., 1986. How Much CO_2 Will Remain in the Atmosphere? In B. BOLIN et al. (eds.), *The Greenhouse Effect, Climate Change, and Ecosystems*. Chichester: Wiley, pp. 93–115.

BOLIN, B.; COOK, R.B. (eds.), 1983. *The Major Biogeochemical Cycles and their Interactions*. Chichester: Wiley.

BRAUDEL, F., 1984. *The Perspective of the World*. New York: Harper & Row.

BRAYBROOKE, D.; PAQUET, G., 1987. Human Dimensions of Global Change: The Challenge to the Humanities and the Social Sciences. *Trans. Royal Society of Canada, Ser. V. II*, pp. 271–91.

BREWER, G. D., 1986. Methods for Synthesis: Policy Exercises. In W. C. Clark and R. E. Munn (eds.), *Sustainable Development of the Biosphere*. Cambridge: Cambridge University Press, pp. 455–73.

BRICKMAN, R.; JASANOFF, S.; IGLEN, T., 1985. *Controlling Chemicals: The Politics of Regulation in Europe and the United States*. Ithaca, N.Y.: Cornell University Press.

BROOKS, H., 1986. The Typology of Surprises in Technology, Institutions and Development. In W. C. Clark and R. E. Munn (eds.), *Sustainable Development of the Biosphere*. Cambridge: Cambridge University Press, pp. 455–73.

BROOKS, H.; HOLLANDER, J. M., 1979. United States Energy Alternatives to 2010 and Beyond: The CONAES Study. *Annual Review of Energy*, Vol. 4, pp. 1–70.

BROWN, B. J.; HANSON, M. E.; LIVERMAN, D. M.; MERIDETH, R. W. JR., 1987. Global Sustainability: Towards Definition. *Environmental Management*, Vol. 11, No. 6, pp. 713–19.

BROWN, H., 1978. *The Human Future Revisited: the World Predicament and Possible Solutions*. New York: Norton.

BROWN-WEISS, E., 1984. The Planetary Trust: Conservation and Inter-generational Equity. *Ecology Law Quarterly*, Vol. 11, pp. 495–581.

BURTON, I.; KATES, R. W.; WHITE, G. F., 1978. *The Environment as Hazard*. New York: Oxford University Press.

CALDWELL, L. K., 1984. *International Environmental Policy: Emergence and Dimensions*. Durham, N.C.: Duke University Press.

CARROLL, J., 1983. Environmental Diplomacy. Ann Arbor, Mich.: University of Michigan Press.

CEQ (COUNCIL ON ENVIRONMENTAL QUALITY), 1980. *Global 2000: Report to the President*. Washington, D.C.: US Government Printing Office.

CHEN, R. S.; BOULDING, E.; SCHNEIDER, S. H. (eds.), 1983. *Social Science Research and Climatic Change*. Dordrecht, Netherlands: Reidel.

CHEN, R.; PARRY, M. L. (eds.)., 1987. *Climate Impacts and Public Policy*. Laxenburg: International Institute for Applied Systems Analysis.

CHISHOLM, M., 1980. The Wealth of Nations. *Transactions of the Institute of British Geographers, New Series*, Vol. 5, pp. 255–75.

CHISHOLM, M., 1982. *Modern World Development*. London: Hutchinson.

CLARK, W. C., 1986. Sustainable Development of the Biosphere: Themes for a Research Program. In W. C. Clark and R. E. Munn (eds.), *Sustainable Development of the Biosphere*. Cambridge: Cambridge University Press, pp. 474–90.

CLARK, W. C., 1987. Scale Relationships in the Interactions of Climate, Ecosystems, and Societies. In K. C. Land and S. H. Schneider (eds.), *Forecasting in the Social and Natural Sciences*. Dordrecht: Reidel, pp. 337–78.

CLARK, W.C., 1988a. Visions of the 21st Century: Conventional Wisdom and Other Surprises in the Global Interactions of Population, Technology and Environment. In *Perspectives 2000: Visions of Canada in the Year 2000*. Ottawa: Economic Council of Canada.

CLARK, W. C., 1988b. Human Dimensions of Global Change. In National Research Council, *Understanding Global Change*. Washington, D.C.: National Academy Press.

CLARK, W. C.; MUNN, R. E. (eds.), 1986. *Sustainable Development of the Biosphere: Human Activities and Global Change*. Cambridge: Cambridge University Press.

COOK, S.W.; BERRENBERG, J.L., 1981. Approaches to Encouraging Conservation Behavior: A Review and Conceptual Framework. *Journal of Social Issues*, Vol. 37, pp. 73–107.

COVELLO, V. T. et al. (eds.), 1985. *Environmental Impact Assessment, Technology Assessment, and Risk Analysis: Contributions from the Psychological and Decision Sciences*. NATO ASI Series G. Ecological Systems. Vol. 4. Berlin: Springer.

CRUTZEN, P. J.; GRAEDEL, T.E., 1986. The Role of Atmospheric Chemistry in Environment–Development Interactions. In W. C. Clark and R. E. Munn (eds.), *Sustainable Development of the Biosphere*. Cambridge: Cambridge University Press, pp. 213–50.

DALY, H. E., 1988. On Sustainable Development and National Accounts. In D. Collard, D. Pierce and D. Ulph (eds.), *Economics and Sustainable Environments: Essays in Honor of Richard Lecomber*. New York: Macmillan.

DARLEY, J. M.; GILBERT, D. T., 1985. Social Psychological Aspects of Environmental Psychology. In G. Lindzey and E. Aronson (eds.), *Handbook of Social Psychology, Vol. 2*. New York: Random House, pp. 949–91.

DARMSTADTER, J. et al., 1987. *Impacts of World Development on Selected Characteristics of the Atmosphere: an Integrative Approach. Vols. 1 and 2*. ORNL/Sub/86-2203/1. Springfield, Va.: NTIS.

DOUGLAS, M., 1966. *Purity and Danger: An Analysis of the Concepts of Pollution and Taboo*. London: Routledge and Kegan Paul.

DOUGLAS, M., 1986. *How Institutions Think*. Syracuse, N.Y.: Syracuse University Press.

DOUGLAS, M.; WILDAVSKY, A., 1982. *Risk and Culture*. Berkeley, Cal.: University of California Press.

EDMONDS, J.; REILLY, J., 1985. *Global Energy*. New York: Oxford University Press.

ENLOE, C., 1975. *The Politics of Pollution in a Comparative Perspective*. New York: McKay.

ESSC (EARTH SYSTEMS SCIENCE COMMITTEE), 1988. *Earth System Science: A Closer View*. Washington, D.C.: NASA.

FAO (FOOD AND AGRICULTURE ORGANIZATION), 1981. *Agriculture: Towards 2000*. FAO Economic and Social Development Series No. 23. Rome: FAO.

FELDMAN, D. L., 1988. *Managing Global Environmental Change Through International Cooperation: Lessons from Prior Resource Management Efforts*. ORNL/TM-10914. Oak Ridge, Tenn.: Oak Ridge National Laboratory.

FISCHHOFF, B. et al., 1981. *Acceptable Risk*. New York: Cambridge University Press.

GARCIA, R., 1981. Drought and Man: the 1972 Case History. In R. Garcia (ed.), *Nature Pleads Not Guilty, Vol. 1*. New York: Pergamon.

GERLACH, L. P.; RAYNER, S., 1987. Culture and the Common Management of Global Risks. Paper delivered at the Annual Meeting of American Sociological Association, Chicago.

GERLACH, L. P.; RAYNER, S., 1988. *Managing Global Climate Change: A View from the Social and Decision Sciences*. ORNL/6390. Oak Ridge, Tenn.: Oak Ridge National Laboratory.

GLANTZ, M. (ed.), 1988. *Societal Responses to Regional Climatic Change: Forecasting by Analogy*. Boulder, Colo.; Westview Press.

GLAZOVSKY, N. F., 1988. *Struktura noosphery i zadachi geographii* (Noosphere Structure and Tasks of Geography), Vol. 1, pp. 38–48.

GOLDEMBERG, J.; JOHANSSON, T. B.; REDDY, A. K. N.; WILLIAMS, R. H., 1987. *Energy for a Sustainable World*. Washington, D.C.: World Resources Institute.

GREENBERGER, M.; BREWER, G. D.; HOGAN, W.; RUSSELL, M., 1983. *Caught Unawares: The Energy Decade in Retrospect*. Cambridge, Mass.: Ballinger.

GU, B.; JAING, M.; MA R.; NI S.; TORREY, B. B., forthcoming. The Tectonics of Population: Effects of China's Urbanization on its Environment'. *Environment*.

GUILE, B. R.; BROOKS, H. (eds.), 1987. *Technology and Global Industry*. Washington, D.C.: National Academy Press.

HAEFELE, W.; BARNERT, H.; MESSNER, S.; STRUBEGGER, M.; ANDERER, J., 1986. Novel Integrated Energy Systems: The Case of Zero Emissions. In W. C. CLARK AND R. E. MUNN (eds.), *Sustainable Development of the Biosphere*. Cambridge: Cambridge University Press, pp. 171–93.

HAMMITT, J. K. et al., 1987. Future Emission Scenarios for Chemicals that may Deplete Stratospheric Ozone. *Nature*, Vol. 330, pp. 711–16.

HARWELL, M. A.; HUTCHINSON, T. C., 1985. *Environmental Consequences of Nuclear War. Vol. II. Ecological and Agricultural Effects*. (SCOPE Report No. 28). Chichester: Wiley.

HOHENEMSER, C.; KASPERSON, R.E.; KATES, R. W., 1985. Causal Structure. In R. W. Kates, C. Hohenemser and J. X. Kasperson (eds.), *Perilous Progress: Managing the Hazards of Technology*. Boulder, Colo.: Westview Press, pp. 24–42.

HOLDGATE, M. W.; KASSAS, M.; WHITE, G. F., 1982. *The World Environment 1972–1982*. Dun Laoghaire: Tycooly International.

HOLLING, C. S. (ed.), 1978. *Adaptive Environmental Assessment and Management*. Chichester: Wiley.

HOLLING, C. S., 1986. The Resilience of Terrestrial Ecosystems: Local Surprise and Global Change. In W. C. Clark and R. E. Munn (eds.), *Sustainable Development of the Biosphere*. Cambridge: Cambridge University Press, pp. 292–317.

ICSU, 1988. *The International Geosphere–Biosphere Programme: A Study of Global Change. A Plan for Action*. Report No. 4, Paris, International Council of Scientific Unions.

IFIAS (International Federation of Institutes for Advanced Study), 1987. *The Human Response to Global Change: Prospectus for an International Research Program*. Toronto: IFIAS.

IFIAS et al., 1988. *Human Dimensions of Global Change: A Research Programme on Human Interactions with the Earth*. Report on Programme Objectives from a Symposium on the Human Dimensions of Global Change, Tokyo, September. Toronto: IFIAS.

IIED/WRI (International Institute for Environment and Development/World Resources Institute), 1987. *World Resources 1987*. New York: Basic Books.

Jacobs, P.; Monroe, D. (eds.), 1988. *Conservation with Equity: Strategies for Sustainable Development*. Ottawa: Canada Department of the Environment.

Jacobson, H. K.; Shanks, C., 1987. *Report of the Workshop on an International Social Science Research Program on Global Change*. Institute for Social Research, University of Michigan, Ann Arbor, Mich.

Jäger, J., 1988. Developing Policies for Responding to Climate Change. Summary of the discussions and recommendations made at workshops held in Villach, Austria and Bellagio, Italy, World Meteorological Organization. *World Climate Program Publ. No. 1 WMO/TD-No. 225*. Geneva: WMO.

Kahneman, E.; Slovic, P.; Tversky, A. (eds.) 1982. *Judgment Under Uncertainty: Heuristics and Biases*. New York: Cambridge University Press.

Kallio, M.; Dykstra, D. P.; Binkley, C. S. (eds.), 1987. *The Global Forest Sector: An Analytical Perspective*. Chichester: Wiley.

Kasperson, R. E.; Kasperson, J. X., 1988. *Nuclear Risk Analysis in Comparative Perspective*. Winchester, Mass.: Allen & Unwin.

Kates, R. W., 1984. The Human Use of the Biosphere. In T. Malone and J. Roederer (eds.), *Global Change*. Cambridge: Cambridge University Press, pp. 491–3.

Kates, R. W.; Ausubel, J.; Berberian, M. (eds.), 1985a. *Climate Impact Assessment: Studies of the Interaction of Climate and Society*. Chichester: Wiley.

Kates, R. W.; Hohenemser, C.; Kasperson, J.X. (eds.), 1985b. *Perilous Progress: Managing the Hazards of Technology*. Boulder, Colo.: Westview Press.

Kay, D. A.; Jacobson, H. K., 1983. *Environmental Protection: The International Dimension*. Totowa, N.J.: Allanheld, Osmun.

Keohane, R. O.; Nye, J. S. Jr., 1977. *Power and Independence: World Politics in Transition*. Boston: Little, Brown.

Keohane, R. O.; Nye, J. S. Jr., 1987. Power and Independence Revisited. *International Organizations*, Vol. 41, No. 4, pp. 725–53.

Kleindorfer, P.; Kunreuther, H. (eds.), 1986. *Insuring and Managing Hazardous Risks; from Seveso to Bhopal and Beyond*. New York: Springer-Verlag.

Kneese, A. V.; Schulze, W. P., 1985. Ethics and Environmental Ethics. In A. V. Kneese and J. H. Sweeney (eds.), *Handbook of Natural Resource and Energy Economics, Vol. 1*. Amsterdam: North-Holland.

Kotlyakov, V. M.; Mather, J. R.; Sdasyuk, G. V.; White, G. F., 1988. Global Change: Geographical Approaches. *Proceedings, National Academy of Sciences*.

Land, K. C.; Schneider, S. H. (eds.), 1987. *Forecasting in the Social and Natural Sciences*. Dordrecht: Reidel.

Leistritz, F. L.; Ekstrom, B. L., 1986. *Social Impact Assessment and Man: An Annotated Bibliography*. New York: Garland.

Lindzey, G.; Aronson, E. (eds.), 1985. *Handbook of Social Psychology*. New York: Random House.

Liverman, D. M.; Hanson, M. E.; Brown, B. J.; Merideth, R. W. Jr., 1988. Global Sustainability: Towards Measurement. *Environmental Management*, Vol. 12, No. 2, pp. 133–43.

Lundqvist, L., 1980. *The Hare and Tortoise: Clean Air Policies in the United States and Sweden*. Ann Arbor: University of Michigan Press.

Madden, P., 1987. Can Sustainable Agriculture be Profitable? *Environment*, Vol. 29, No. 4, pp. 18ff.

Majone, G., 1986. International Institutions and the Environment. In W. C. Clark and R. E. Munn (eds.), *Sustainable Development of the Biosphere*. Cambridge: Cambridge University Press, pp. 351–8.

March, J. G.; Olsen, J. P., 1976. *Ambiguity and Choice in Organizations*. Bergen: Universitetsforlaget.

Marchetti, C., 1983. On the Role of Science in the Postindustrial Society. *Technological Forecasting and Social Change*, Vol. 24, pp. 197–206.

MARLAND, G.; ROTTY, R. M., 1983. Carbon Dioxide Emissions from Fossil Fuels: A Procedure for Estimation and Results for 1950–1981. *DOE/NBB-0036*. Springfield, Va.: NTIS.

MARSH, G. P., 1864. *Man and Nature: Or, Physical Geography as Modified by Human Action*. New York: Scribner (Reprinted 1965, Cambridge: Harvard University Press).

MCALLISTER, D. M., 1982. *Evaluation in Environmental Planning: Assessing Environmental, Social, Economic and Political Trade-Offs*. Cambridge, Mass.: MIT Press.

MCGUIRE, W.J., 1969. The Nature of Attitudes and Attitude Change. In G. Lindzey and E. Aronson (eds.), *The Handbook of Social Psychology, 2nd ed.* Reading, Mass.: Addison-Wesley, pp. 136–314.

MCLAREN, D. J.; SKINNER, B. J., 1987. *Resources and World Development*. Dahlem Workshop Report No. 6. Chichester: Wiley.

MEADOWS, D. H. et al., 1972. *The Limits to Growth*. Washington, D.C.: Potomac Associates.

MEADOWS, D. H.: ROBINSON, J. M., 1985. *The Electronic Oracle: Computer Models and Social Decisions*. Chichester: Wiley.

MELLOR, J., 1988. Sustainable Agriculture in Developing Countries. A special issue of *Environment*, Vol. 30, No. 9, pp. 6–40.

MILBRATH, L., forthcoming. *Envisioning a Sustainable Society*.

MUNN, R. E. (ed.), 1979. *Environmental Impact Assessment*. SCOPE: 5. Chichester: Wiley.

NAE (NATIONAL ACADEMY OF ENGINEERING), 1988. Background papers for a workshop on Technology and the Environment, Woods Hole, 24–26 August. Washington, D.C.: National Academy of Engineering.

NAKICENOVIC, N., 1988. Dynamics and Replacement of US Transport Infrastructures. In J. H. Ausubel and R. Herman (eds.), *Cities and their Vital Systems: Infrastructure Past, Present and Future*. Washington, D.C.: National Academy Press.

NILSSON, S.; DUINKER, P., 1987. The Extent of Forest Decline in Europe: A Synthesis of Survey Results. *Environment*, Vol. 29, No. 9, pp. 30–1.

NORDHAUS, W. D.; YOHE, G. W., 1983. Paths of Energy and Carbon Dioxide Emissions. In National Research Council, *Changing Climate*. Washington, D.C.: National Academy Press.

NRC (NATIONAL RESEARCH COUNCIL), 1981. Governing Board Committee on the Assessment of Risk. *The Handling of Risk Assessments in NRC Reports*. Washington, D.C.: National Academy Press.

NRC, 1983. *Changing Climate*. Washington, D.C.: National Academy Press.

NRC, 1984. *Global Tropospheric Chemistry*. Washington, D.C.: National Academy Press.

NRC, 1988. *Understanding Global Change*. Washington, D.C.: National Academy Press.

NRIAGU, J. O.; PACYNA, J. M., 1988. Quantitative Assessment of Worldwide Contamination of Air, Water and Soils by Trace Metals. *Nature*, Vol. 333, pp. 134–9.

OLSON, M., 1971. *Logic of Collective Action: Public Goods and the Theory of Groups*, (Rev. ed.). Cambridge, Mass.: Harvard University Press.

ORIANS, G., forthcoming. Sustainable Development. *Environment*.

OTA (US OFFICE OF TECHNOLOGY ASSESSMENT), 1982. *Global Models, World Futures, and Public Policy: A Critique*. OTA-R-165. Washington, D.C.: OTA.

OTA, 1988. *An Analysis of the Montreal Protocol on Substances that Deplete the Ozone Layer*. Washington, D.C.: OTA.

OTT, W. R., 1985. Total Human Exposure. *Environmental Science and Technology*, Vol. 19, pp. 880–6.

PARRY, M. L., 1985. The Impact of Climatic Variations on Agricultural Margins. In R. W. Kates, J. Ausubel, and M. Berberian (eds.), *Climate Impact Assessment: Studies of the Interaction of Climate and Society*. Chichester: Wiley, pp. 351–68.

PARRY, M. L.; CARTER, T. R.; KONIJN, N. T. (eds.), 1988. *The Impact of Climatic Variations on Agriculture, Vol. 1, Assessments in Cool, Temperate and Cold Regions. Vol. 2, Assessments in Arid and Semiarid Regions*. Dordrecht: Reidel.

PRY, R. H., 1973. Forecasting the Diffusion of Technology. *Technical Information Series Report No.* 73-CRD-220. Schenectady, N.Y.: Research and Development Center, General Electric.

REGIER, H. A.; BASKERVILLE, G. L., 1986. Sustainable Redevelopment of Regional Ecosystems Degraded by Exploitative Development. In W. C. Clark and R. E. Munn (eds.), *Sustainable Development of the Biosphere*. Cambridge: Cambridge University Press, pp. 75–101.

REPETTO, R., 1987. Population, Resources, Environment: An Uncertain Future. *Population Bulletin*, Vol. 42, No. 2, pp. 1–43.

RICHARDS, J. F., 1986. World Environmental History and Economic Development. In W. C. Clark and R. E. Munn (eds.),

Sustainable Development of the Biosphere. Cambridge: Cambridge University Press, pp. 53–71.

RISSER, P. G. (ed.), 1986. *Spatial and Temporal Variability of Biospheric and Geospheric Processes*. Proceedings of the St. Petersburg Workshop, International Council of Scientific Unions and National Research Council. Paris: ICSU Press.

Rockwell, R., 1988. Human Processes in Earth Transformation: A proposed council program on the social sciences and global environmental change. *Items 42* (1/2), pp. 16–18.

ROGERS, E. M.; KINCAID, D. L., 1981. *Communication Networks: Towards a New Paradigm for Research*. New York: Free Press.

ROSSWALL, T.; WOODMANSEE, R. G.; RISSER, P. G., 1988. *Scales and Global Change: Spatial and Temporal Variability in Biospheric and Geospheric Processes*. Chichester: Wiley.

RUNGE, C. F., 1986. Agricultural Development: A Commentary. In W. C. Clark and R. E. Munn (eds.), *Sustainable Development of the Biosphere*. Cambridge: Cambridge University Press, pp. 136–9.

RUTTAN, V. W., 1984. Social Science Knowledge and Institutional Change. *American Journal of Agricultural Economics*, Vol. 66, No. 5, pp. 549–59.

SERAFIN, R., 1988. Noosphere, Gaia, and the Science of the Biosphere. *Environmental Ethics*, Vol. 10, pp. 121–37.

SCHELLING, T., 1978. *Micromotives and Macrobehavior*. New York: Norton.

SCHELLING, T., 1983. Climatic Change: Implications for Welfare and Policy. In National Research Council, *Changing Climate*. Washington, D.C.: National Academy Press, pp. 449–82.

SCHNEIDER, S. H., 1983. CO_2 Climate and Society: A Brief Overview. In R. S. Chen, E. Boulding, and S. H. Schneider (eds.), *Social Science Research and Climatic Change*. Dordrecht: Reidel, pp. 9–15.

SHRADER-FRECHETTE, K. S., 1985. *Science Policy, Ethics, and Economic Policy: Some Problems of Technology Assessment and Environmental Impact Analysis*. Dordrecht: Reidel.

SMITH, J. H. (ed.), 1988. *Environment and Development*: Report on the ESF (European Science Foundation) workshop on the environment and development. Strasbourg, France: European Science Foundation.

SMITH, K. R., 1988. Air Pollution: Assessing Total Exposure in the United States. *Environment*, Vol. 30, No. 8, pp. 10–15.

SONNTAG, N. C., 1986. Commentary on Methods for Synthesis: Policy Exercises. In W. C. Clark and R. E. Munn (eds.), *Sustainable Development of the Biosphere*. Cambridge: Cambridge University Press, pp. 473–5.

SPENGLER, J. D.; SOCZEK, M., 1984. Evidence for Improved Ambient Air Quality and the Need for Personal Exposure Research. *Environmental Science and Technology*, Vol. 18, pp. 268a–280b.

SSRC, 1988. Background materials for a meeting on the role of the social sciences in studies of global change. New York: SSRC.

STEWART, T. R.; GLANTZ, M. H., 1985. Expert Judgment and Climate Forecasting: A Methodological Critique of 'Climate Change to the Year 2000'. *Climatic Change*, Vol. 7, pp. 159–83.

STOKOLS, D.; ALTMAN, I. (eds.), 1987. *Handbook of Environmental Psychology*. New York: Wiley.

SVEDIN, U.; HEURLING, B. (eds.), 1988. *Swedish Perspectives on Human Response to Global Change*. Stockholm: Swedish Council for Planning and Co-ordination of Research.

TANG, X.; JACOBSON, K., 1988. *Human Dimensions of Global Environmental Change: Proposals for Research*. Summary report of the Chinese–US workshop. Peking University, Beijing, 12–16 May. (Washington: International Programs Division, National Science Foundation).

THOMAS, W. L., Jr. (ed.), 1956. *Man's Role in Changing the Face of the Earth*. Chicago: University of Chicago Press.

TIMMERMAN, P., 1981. *Vulnerability, Resilience and the Collapse of Society*. Institute of Environmental Studies Monog. No. 1. Toronto: University of Toronto.

TORRENS, I. M., 1984. Environmental Effects of Energy Systems: The OECD COMPASS Project. *Environment International*, Vol. 10, pp. 419–29.

TOTH, F.; HIZSNYIK-TOTH, E.; CLARK, W. C. (eds.), 1988. *Scenarios of Socioeconomic Development for Studies of Global Environmental Change: A Critical Review*. Laxenburg: International Institute for Applied Systems Analysis.

TURNER, B. L.; CLARK, W. C.; KATES, R. W.; MATHEWS, J. T.; RICHARDS, J. R.; MEYER, W. (eds.), forthcoming. *The Earth as Transformed by Human Action*. Proceedings of an international symposium held at the Graduate School of Geography, Clark University, Worcester, MA. 25–30 October.

UN (United Nations), 1985a. *Estimates and Projections of Urban, Rural, and City Populations, 1950–2025: The 1982 Assessment*. New York: United Nations.

UN 1985b. *World Population Prospects, Estimates and Projections as Assessed in 1982*. UN Population Studies No. 865, ST/ESA/SER.A/86. New York: United Nations.

UNECE (United Nations Economic Commission for Europe), 1986. *Draft Report on the 1986 Forest Damage Survey in Europe*. Geneva: UNECE.

Unesco (United Nations Educational, Scientific and Cultural Organization), 1986. Man and the Biosphere Programme. Final report of the General Scientific Advisory Panel. *MAB Report. Series No. 59*. Paris: Unesco.

Vasko, T. (ed.), 1987. *The Long-Wave Debate*. Berlin: Springer-Verlag.

Vaupel, J.; Yashin, A., 1986. Cancer Rates Over Age, Time and Place: Insights from Stochastic Models of Heterogeneous Populations. *IIASA WP*-86-59. Laxenburg: International Institute for Applied Systems Analysis.

Vernadsky, V. I., 1945. The Biosphere and the Noosphere. *American Scientist*, Vol. 33, pp. 1–12.

Vogel, D., 1986. *National Styles of Regulation: Environmental Policy in Great Britain and the United States*. Ithaca, N.Y.: Cornell University Press.

Wallerstein, I., 1974. *The Modern World System*. New York: Academic Press.

WCED (UN World Commission on Environment and Development, the Brundtland Commission), 1987. *Our Common Future*. New York: Oxford University Press.

White, G. F., 1988a. *Greenhouse Gases, Nile Snails, and Human Choice*. Distinguished lecture series on behavioral science. Boulder: Institute of Behavioral Science, University of Colorado.

White, G. F., 1988b. Environmental .Anxiety. *Environment*, Vol. 30, No. 1, p. 1.

Whyte, A. V.; Burton, I. (eds.), 1980. *Environmental Risk Assessment*. (SCOPE report No. 15). New York: Wiley.

Wildavsky, A., 1979. *Speaking Truth to Power: The Art and Craft of Policy Analysis*. Boston: Little, Brown.

Williams, G. D. V.; Fautley, R. A.; Jones, K. H.; Stewart, R. B.; Wheaton, E. E., 1988. Estimating Effects of Climatic Change on Agriculture in Saskatchewan, Canada. In M. L. Parry, T. R. Carter, and N. T. Konijn (eds.), *The Impact of Climatic Variations on Agriculture. Vol. 1, Assessments in Cool, Temperate and Cold Regions*. Dordrecht: Reidel, pp. 219–379.

Winett, R. A., 1986. *Information and Behavior: Systems of Influence*. Hillsdale, N.J.: Erlbaum.

Wolman, M. G.; Fournier, F. G. A. (eds.), 1987. *Land Transformation in Agriculture*. Chichester: Wiley.

Wuebbles, D. J.; Edmonds, J. A., 1988. A Primer on Greenhouse Gases. *DOE/NBB-0083*. Washington, D.C.: US Department of Energy.

[10]

Ecological Economics, 1 (1989) 17–36
Elsevier Science Publishers B.V., Amsterdam – Printed in The Netherlands

HISTORICAL ROOTS FOR ECOLOGICAL ECONOMICS — BIOPHYSICAL VERSUS ALLOCATIVE APPROACHES

PAUL P. CHRISTENSEN
Department of Economics, Hofstra University, Hempstead, NY 11550 (U.S.A.)

ABSTRACT

Christensen, P.P., 1989. Historical roots for ecological economics — Biophysical versus allocative approaches. *Ecol. Econ.*, 1: 17–36.

A number of economists today are aware that the results of their science are more or less at variance with concrete fact and are alive to the necessity of perfecting it. They go wrong, rather, in their choice of means to that end. They try obstinately to get from their science alone the materials they know are needed for a closer approximation to fact, whereas they should resort to other sciences and go into them thoroughly—not just incidentally—for their bearing on the given economic problem. Until economic science is much farther advanced, "economic principles" are less important to the economists than the reciprocal bearings of the results of economics and the results of the other social sciences. Many economists are paying no attention to such interrelations, for mastery of them is a long and fatiguing task requiring an extensive knowledge of facts; whereas anyone with a little imagination, a pen, and a few reams of paper can relieve himself of a chat on "principles" (Pareto, 1935, para. 2022).

INTRODUCTION

Economics has become even more of an axiomatic and deductive science since Pareto's time (with economists asserting that the core theory can be deduced entirely from the principles of rational choice). Criticism of economics as an "armchair science" divorced from any "systematic understanding of a real economic system" has also increased (Leontief, 1982; Simon, 1986). Exploration of the "results of other social sciences" has not been absent, especially by institutional economists and students of economic psychology, but this work has not been enough by itself to constitute an alternative theory of economic activity. What has been missing in all of the various critiques is a reconstruction of the biophysical foundations of economic activity. It is not just the other social sciences which matter for

0921-8009/89/$03.50

18

economics but the facts and theories of physics, chemistry, biology, and ecology. Modern economic theories have neglected the implications of the basic physical principles governing material and energy use for an economic theory of production, for the operation of a production-based price system, for macro-economic (disequilibrium) dynamics, and for longer run growth processes and prospects.

Why has there been this neglect of basic bio-physical principles? According to Georgescu-Roegen, the mechanistic sins of modern economics (Marxian and neoclassical) can be traced to the Ricardian concept of land "which is expressly defined as a factor immune to any qualitative change (which) we could refer to simply as space" (1971, p. 2). But closer examination of the treatment of natural resources and the physical assumptions of production theory reveal an early attention to the physical side of economic activity in pre-classical, physiocratic, and early nineteenth century classical economics which is absent in modern theories.

Starting from an analogy between the nutrition of living bodies and the provisioning of the economy, early economic writers regarded production in terms of the transformation of materials and food taken from the land. Food and raw materials were *basic* commodities in the Sraffa (1960) sense of inputs entering into the production of all other goods directly or indirectly. Food and materials were used to produce machines and human labor which were employed in turn to extract more materials and food. This materials-processing, proto-energetic approach to production included application of the principle of mass conservation to manufacturing production. It also included an attempt by late classical writers to incorporate energy converters and sources of motive power into production theory.

Parodoxically, the formulation of the laws of thermodynamics in the 1840s and 1850s was not used to extend the classical production approach to include energy and energetic principles or to make a general application of the mass conservation principle. This was certainly due to the fact that the materials/energy foundation underlying classical production (and price) theory was both incomplete and only imperfectly understood. An additional difficulty was that the energy laws were stated exclusively in terms of energy conversion and dissipation in an isolated system. A unified physical understanding of the relation between matter and energy transformations in open systems could have come from physiology (or later from ecology) but economics from the 1860s was increasingly coming under the sway of analytical mechanics.

The adoption of the maximization model from analytical mechanics led to a shift from production dynamics to an analysis of exchange value. The result was a static optimization theory concerned primarily with efficiency and equilibrium which took its inputs (land, labor, and capital) and the tastes, drives, and preferences of consumers as given. Production was treated as an aspect of the allocation and pricing of the "primary" factors of production. Material and energy resources disappeared from theoretical view. If they were considered, it was assumed they could be treated like the other factors of production (i.e., land). Indeed, no essential physical distinction was made between any input. The result was that physical and technical assumptions (marginal productivity) were made about economic activity which were in conflict with the basic physical principles governing material and energy transformations (Georgescu-Roegen, 1971; Ayres, 1978, ch. 3).

Production is a set of sequential activities which feed on low entropy energy and materials. Physical and biological principles suggest the existence of a strong interdependence between material, energy and information flows and the machines and agents which extract, transform, and utilize these flows. Orthodox production theory with its assumption of the separable productivity of independently given primary factors is obviously ill-suited to such interdependence. This suggests the need for an alternative and more systems-based theory of production (and prices).

This paper argues that some of the tools for a reconstruction of a more ecologically based economics may be found in the older classical tradition. What follows is a consideration of (1) the development and physical assumptions of the classical model of production, (2) the neglect of material and energy resources and production interdependence in early neoclassical theory, (3) the relation of neoclassical price theory to the production model, (4) some theoretical implications of a synthesis of a biophysical model of material and energy flows with the classical production approach, (5) implications of a physically specified approach for macro economic dynamics, and (6) an extension of the classical model of production prices to include reproduction of environmental resources.

THE CLASSICAL PRODUCTION APPROACH

The classical model had its origins in a physiological analogy. Thomas Hobbes (1651) used a Harveian model of the circulation of "nutriments" by the blood as an analogy for the materials foundation of the commonwealth. Materials and foodstuffs extracted from the land and sea were carried via monetary exchanges through various channels of transformation and trade prior to use. Although this original physiological source was soon forgotten, the classical model developed by William Petty, Richard Cantillon (1755), the French Physiocrats, Adam Smith (1776), and their 19th century classical successors continued to embody this materials-based approach to production and prices.

20

The starting point of the classical approach was a theory of production. Production was a set of sequential activities: the extraction of materials and food preceeded the processing and fabrication of materials. This shaped a sectoral characterization of production inputs: labor, tools or machines, and food (the energy source for labor) were required in each sector. Fertile land or resource-bearing sites were specific to agriculture and mining. Raw materials were the basic input in manufacturing. The capital stock, following Smith (1776, book II, ch. 1) was divided into two broad categories: fixed capital (machines and structures) and the circulating capital (food, fodder, raw materials, working finance, etc.). Physical complementarity existed between fixed and circulating capital. There was no production without appropriate tools, materials, food and fodder.

A crucial distinction was made between land and industrial machinery. The Physiocrats, for example, regarded the land as *productive* because it yielded a *surplus* of output above the material input advanced at the beginning of production: one livre of seed planted yielded five livres of output. Artisan activities, by contrast, were *transformations* of raw materials. Industry buys raw materials from agriculture in order to work them up. Manufacturing gives raw materials a form, but it adds nothing to them materially.

A similar distinction was made by the classical writers. Malthus (1815, 1836) argues that only "the machinery of the land" could produce food and raw materials. This was something no industrial machine could do. The latter only transformed materials from one form to another (when supplied with the appropriate agencies to affect work). Ricardo (1817) similarly writes of the "original and indestructible powers of the soil". This is obviously something more than the mere space which Georgescu-Roegen (1971, p. 2) attributes to him. In an age before a distinction between energy and force had been made, powers referred to those agencies which had the capacity to transform matter (Locke, for example, writes of the power of the sun to melt wax and fire to make lead liquid). Powers were the capacity to bring about change.

Because fertile land was limited in supply, the classicals believed that agricultural output would be subject to diminishing returns. More labor and capital applied to a given plot of land or to a less fertile plot yielded a proportionately smaller return. Industrial machines, although they lacked independent productivity, could be replicated. Thus, given the availability of appropriate raw materials and energy sources, manufacturing output could be extended indefinitely. Manufacturing consequently was characterized by constant or increasing returns. In the absence of an appropriate increase in materials, more labor and capital would not supply more output. This was effectively a "law of zero returns". Manufacturing was subject to the conservation of matter. Double the labor and capital employed on the manufacture of cotton without increasing the quantity of raw materials and "the quantity of manufactured produce could not be sensibly increased" (Senior, 1836, p. 82).

While the classicals correctly recognized the inability of industrial machines to produce without appropriate materials and sources of power, they failed to make the same distinctions about production from the "machinery" of the land (from an engineering point of view, physical structures are passive machines). Of 18th and 19th century writers only the Italian, Pietro Verri, in his critique of the physiocrats, makes it clear that land does not "produce" or create its products any more than industrial machines.

> All the phenomena of the universe, whether they are produced by the hand of man or by the universal laws of physics, are not to be conceived of as an actual *creation* but only as a modification of existing materials.... This is equally the case...whether the earth, air, and water of the fields are transmutted into grain or when by the hand of man the secretions of an insect are transmutted into silk or some metal parts are organized into a repeating watch (Verri, 1773, pp. 21–22).

A specification of the material and energy flows through the land provides a physical basis for the empirically based "law of diminishing returns". Application of successive does of labor and capital to a given plot of land or to new fields of lower fertility can yield positive if diminishing returns because of the availability of extra material and energy resources available in the field: unutilized solar flow, CO_2 in the atmosphere, ground water, available nutrients in the soil, etc. Farming a plot more intensively draws more of these inputs into the production flow. The conservation of mass (as Verri recognized) applies to agricultural production in the same way it applies to manufacturing. For any given technique, there is no increase in output without an increase in material (and energy) input. Marginal products exist in agriculture because of the presence of slack or unutilized resources.

Under the influence of British technical writers such as Smeaton, Babbage and Ure, the post-Ricardian classical economists began to elucidate the importance of power in the new industrial technologies which had established Britain's industrial dominance. In reviews of Babbage and Ure, McCulloch attributed the industrial prosperity of the British nation to its exploitation of coal and coal-using technologies which overcame limited supplies of wood (McCulloch, 1835). The invention of steam engines permitted the draining of coal mines, thus removing a central bottleneck to energy production. Low cost coal was gradually substituted in technological innovations as a fuel in metallurgy and other industrial processes. The low cost production of metals in large quantities lowered the costs of producing machines (including steam engines) and other industrial structures. Low cost fuels permitted low cost and large volume manufacturing, enlarging domestic and international markets.

McCulloch confined himself to emphasizing the commercial importance of coal. The most sustained attempt to incorporate the new prime movers and inanimate sources of power into a theory of physical production was made by his contemporary, Senior (1836). Senior's starting point is a physical classification of production agents. These are labor and skills, natural agents, and abstinence (the source of capital). Capital is divided into fixed and circulating components in order to distinguish tools and instruments of production from the materials which will be embodied in product. Energy resources (food and coal for the laborers and machines) are included with the fixed capital since these are not embodied in output (Senior, 1836, p. 64). Implements, following Babbage and Ure, are further divided into two classes, those which produce power (steam engines, water wheels, etc.) and those which transmit and apply power.

Unfortunately Senior's successors, including Mill (1848), fail to sustain Senior's achievements in establishing a physical taxonomy of production inputs or his explicit linking of production returns in manufacturing to material use via the law of mass conservation. Mill, like Senior, classifies production inputs under three headings: labor, natural agents and capital. Nature provides materials transformed by labor and the (motive) powers which cooperate with and substitute for labor. But instead of maintaining the distinction between materials and motive powers, he subsumes the latter —the coals that drive the steam engine and the food that feeds the worker—under materials in order "to avoid a multiplication of classes (of) no scientific importance". When Senior objected to his inclusion of fuel with materials in an unsigned review (Senior, 1848) on the grounds that only the latter are embodied in materials, Mill replies that although his terminology is not in accord with the physical meaning of the word material such a distinction between materials and fuels is of "almost no importance to political economy" (Mill, 1871, pp. 34–35)! Mill's decision to forego any concern with the development of a scientific terminology appropriate to physical processes obscured the thrust of the materials-energy approach (and it may have influenced Marshall (1920) who explicitly chooses commercial rather than scientific terminology in his discussion of capital).

NEOCLASSICAL PRODUCTION THEORY

Neoclassical economics is distinguished from classical theory by the wholesale shift from a production and growth approach to a static analysis of an exchange economy. Individuals possess initial endowments of various resources (including their own labor) which they trade in markets to maximize their utility. Judgements about the economic value of various goods and services are determined by the subjective tastes and preferences of the individual decision makers (relative to the scarcity of the resources in question). The formation of these tastes, like the endowments of resources, is not examined. Preferences and endowments are simply given.

A full neoclassical theory of production was not formulated until the mid-1890s (by Wicksteed and Wicksell), two decades after the marginal utility approach to prices was established. Even then it was developed not as a theory of production per se but as a theory of the distribution of net output between the contributing factors of production. Each factor, it was assumed, makes a positive, if diminishing contribution to output (the utilization of all other factors is held constant). Conceptually this may be viewed as an attempt to generalize the Ricardian theory of differential rent (based on the differential productivity of land) to labor and capital. Capital is treated analogous to land. But agriculture provides a misleading analogy. Although land acreage was held constant as labor was increased, there was an "invisible" increase in the flows of materials and energy through the land into biomass. In manufacturing, plant and equipment could be held constant as labor was increased but an accounting had to be made of the additional materials and energy—the circulating capital—required to produce another increment of output. Marginal productivity theory presumes the independent productivity of *individual* inputs. This entirely ignores complementarity between inputs within techniques and suppresses materials and energy. If the classical's error was a failure to specify the material and energy resources flowing through the land, the neoclassicals extend this error to manufacturing and the capital stock.

The distorting influence of the marginalist program on the formulation of neoclassical production theory is evident from the beginning. Heterogeneity and complementarity are eliminated and replaced with some version of homogeneity and separability.

In his chapter on "the theory of capital", Jevons (1871) eliminates the distinction between fixed and circulating capital. Fixed capital he says can be considered as a long-lived version of the latter. He then reduces circulating capital to "nothing but" the subsistence of the workers: what they need for food and shelter to engage in long-lived projects (p. 226). Despite his earlier insistence on the importance of coal as the industrial source of power driving British economic growth (Jevons, 1864), he completely ignores machines, raw materials and industrial fuels in his discussion of capital. The marginalist program appears to have created a conceptual blindness in a scientist who well understood the physical significance of energy in production activity.

Early on in his great study, Menger (1871) presents a long discussion of the importance of raw materials and intermediate products in the production of goods of "higher" and "lower" orders (a classification based on distance from the consumer). He also clearly recognizes the existence of fixed proportions between inputs. But his imputation theory of prices requires a theory of substitution in order to assess the difference made by the presence or absence of an individual factor. He thus makes a universal assumption of variable proportions on the basis of the existence of variable proportions in agriculture. Chemical reactions, he says, may be characterized by fixed weights but in economics "the most ordinary experience" teaches that input quantities can be varied (p. 161). More land or more fertilizer can be employed to produce the same output. His mistake (the mistaken analogy with agriculture) is to confuse the possibility of substitution between various techniques of production (each with its individual material, energy, and machine requirements) with a general theory of "variable proportions" (the smooth substitution of one factor for another along an isoquant).

The fundamental step of arguing the independent contribution of individual inputs to output was first made by Walras (1874). Again agriculture is the source of the analogy: capital is put on the same conceptual footing as land. Just as "a field will grow us a crop year after year", he writes, "so machines, instruments, tools... engender incomes in the same way" (p. 213). Items of fixed capital are assumed to yield flows of "producer services" (producer goods) and "income" (consumer) goods. But how does a machine produce physical flows of output independent of material feedstocks and the energy required to do work?

Walras suppresses this question by vertically aggregating manufacturing and agricultural production. There are, he says, two kinds of production. The first kind (agriculture) is a combination of "nothing but" land, labor and capital services:

$$Ta,\ La,\ Ka \rightarrow RM$$

The second case (manufacturing) applies the services of land, labor, and capital to raw materials (p. 237):

$$RM + Tm,\ Lm,\ Km \rightarrow C$$

The second, however, is reduced to the "case of the direct combination of productive services alone" (p. 240):

$$T',\ L',\ K' \rightarrow C$$

where $T' = Ta + Tm$, $L' = La + Lm$, and $K' = Ka + Km$. Final output is produced directly from *primary* factors of production in one step by using larger amounts of the primary factors (the sum of those that would have been applied sequentially). The result of this reduced form equation was the elimination of raw materials and time from the representation of production.

The same neglect of raw materials is found in Marshall (1920) despite his professed sympathy with the classical model and his attempt to preserve the classical distinction between diminishing returns in agriculture and increasing returns in manufacturing. Marshall is well aware of the application of the conservation of matter in production. In his introductory remarks he writes that production "cannot create material things" but only change their form (1920, p. 63). This is repeated when he takes up agricultural production (p. 144). Agricultural crops are linked to the presence of chemical elements in the air, water, and soil (p. 146). But such discussion is entirely absent from his treatment of manufacturing. Nor is the idea of the conservation of materials mentioned here. Indeed, his discussion of this sector (conducted under the heading of capital as an agent of production) lacks any treatment of the physical side of production or even enumeration of the physical inputs of production. Manufacturing is treated solely in organizational terms.

Marshall's hesitancy in speaking about marginal products (and the strategic silence he maintains on raw materials in manufacturing) may stem from an awareness of the significance of materials in production and recognition of their incompatibility with marginalist equilibrium theory. That Marshall knows that marginal products do not exist independently of material flows is apparent in his definition of net product as the additional product obtained by the incremental application of a factor "after allowing for incidental expenses" (p. 432). But materials and fuel are hardly incidental to physical productivity. He affirms the importance of material flows to economic growth in his concluding chapter on production where he states that England's ability to secure the economies of specialized skills and machinery will depend on the future availability of food and other materials on easy terms (pp. 320–322); but these comments are not reflected in a theory of the production process.

The early neoclassicals probably would have disavowed any simple version of the marginal productivity hypothesis, but their neglect of raw materials, energy carriers, and complementarity in production sets the stage for such formulations in the 1890s. This occurred not in the context of a theory of production but as a theory of the distribution of output between the factors of production. "Each factor", Wicksteed (1894) writes, "receives a share of the product regulated by its marginal efficiency as a producer". The "(marginal) significance of each factor", he adds, "is determined by the effect upon the product of a small increment of that factor, all the others remaining constant" (Wicksteed, 1894, pp. 8–9). This, of course, simply assumes that an increase in any input, other inputs being constant, yields a positive increment of product (first partials are taken to be positive). This assumption of the existence of variable proportions in the short run is then

26

extended to provide a general theory of substitutability between inputs. These ideas are advanced without any discussion of relevant physical features of production activity.

NEOCLASSICAL PRICE THEORY

Having reduced all production to a standard model of production from primary "factors", a uniform model of prices for all markets could then be logically justified. Each market exhibits the same opposing forces of downward-sloping demand and upward-sloping supply curves with supply and demand operating entirely independent of each other. The assumption of upward-sloping supply was due to scarcity and diminishing returns. The downward slope of demand curves reflected diminishing marginal utility. It was also assumed that prices were flexible upwards and downwards, thereby establishing equilibrium between supply and demand in each market. Overall equilibrium in the economy as a whole was assured by price flexibility (to eliminate "false" trading at non-equilibrium prices an auctioneer was invoked which established equilibrium in all markets before any exchange took place). Futures markets were similarly assumed to extend this one-period multimarket model to provide equilibrium prices for all future periods.

The basic leitmotif of the exchange model is that market prices of each commodity are derived from the subjective preferences of individual consumers. This is an expression of the methodological individualism of the analytical approach. All explanations relate back to transactions between individual economic agents. Social wholes are built up from the actions of individual agents without feedback from the system or its subsystems to the individual. The essential premise is the given human preferences. There is a remarkable absence of any investigation of how a price system actually works or how preferences are actually formed.

Modern neo-Walrasian theory replaces arguments based on partial derivatives with a programming approach which assumes only that production sets are closed and convex. The starting point of analysis remains the same: the maximizing behavior of individual producers and consumers. Each individual consumer comes into the world with (1) a given initial endowment of resources (including know how and the capacity to labor), and (2) given tastes and preferences which guide consumption decisions. But this a priori conception of resource endowments presents a one-way logic running from given resources (and relative scarcities) to factor prices and from resource prices to factor proportions. It entirely suppresses the dynamic running from technology to resources. We simply do not know the size of existing resource pools and flows, rates of depletion, future discoveries, or the future technologies which will redefine available resources. The purpose of this theory is to demonstrate the existence and stability of an equilibrium set of prices given an initial allocation. It is not an analysis of resource use through time.

It is sometimes held that the transition to the convexity approach eliminates the need for marginal productivity assumptions. "Marginal concepts" Bliss (1975) writes, "are not primary but follow from the basic postulates of maximization". He then asserts that since equilibrium analysis is in no way dependent upon marginal concepts, it "can suffer no crisis on their account". But this ignores the link between maximization and marginalism and the fact that the neoclassical theory of price determination depends on marginal concepts. Bliss quotes, but does not resolve the dilemma posed by Schumpeter:

> ...mere recognition of the element of productivity does not help us much unless it is streamlined by the notion of marginal productivity, exactly as the element of utility will not produce any servicable theory of price unless streamlined by the notion of marginal utility (Schumpeter, 1954, p. 677).

Marginal utility and productivity cannot be dispensed with because they underly the adjustment mechanisms of the theory. Incremental adjustments to incremental changes maintain the story of equilibrated optimal outcomes. Maximization and incremental adjustment together obviate the investigation of physical and informational processes moving in historical time.

According to neoclassical theory, economic values not only are but should be derived from individual preferences. Since there is no way of discerning the preferences of unborn generations, this leaves us with the dictates of the market and the subjective judgements of the present generation as to the values which should be placed on resources and natural environments. While there is undeniably plenty of scope for trying to change those values, there is, from a theoretical standpoint, no way around the autonomy of the subjective judgments of admittedly present-minded individuals. There is nothing in the theory to suggest any intrinsic or internal requirements of environmental or social systems that should shape the process of resource valuation and use. The existence of market failure provides, of course, an argument for recourse to collective action but such action lies outside the bounds of the theory (and runs counter to its normative prescriptions).

A BIO-PHYSICAL APPROACH TO PRODUCTION

In the classical view production, not exchange, provides the starting point of economic theory. A bio-physical perspective extends the classical approach to include the low-entropy energy and materials extracted from environmental systems and eventually returned as waste. Production and exchange in economic systems becomes part of a larger totality of interdependent material, energy, and information exchanges.

28

Production inputs

From a physical point of view, the basic factors of production are material, energy, and information flows, and the physical and biological structures and agents which convert, transmit or apply materials, energy and information. As Kenneth Boulding (1986) points out, the inputs of traditional theory (land, labor, and capital) are each "hopelessly heterogeneous aggregates". Land is a heterogeneous aggregate of physical structures (storage and staging areas), materials (stocks and flows), and energetic potential. The capital stock is likewise a heterogeneous collection of structures, equipment, intermediate materials, fuels, etc. and requires a physical taxonomy to distinguish between fundamental physical classes. Labor in turn can be viewed as a self-directing (and programmable) cybernetic control mechanism and chemical energy converter. A suggestive physical classification of the inputs in a chemical processing plant has been recently made by Van Gool (1985).

A physical perspective also challenges the neoclassical conception of primary factors of production. In ecological terms, the primary net input is solar energy:

> Practically everything on the earth can be considered to be a direct or indirect product of past and present solar energy.... Fossil fuels and other natural resources represent millions of years of embodied sunlight. Environmental flows (such as winds, rain, and rivers) represent embodied sunlight of more recent origin (Costanza, 1980, p. 1219).

Since human economies import both materials and energy, the basic or primary inputs can be broadened to include anything not produced within the human system of production: the materials and energy and environmental services which have been produced by biological and geological activity past and present. What is regarded as a primary input is in part a question of the boundaries of the system in question. Also, just as natural resources, structures, and organisms can be regarded as embodied energy, they also represent embodied materials and embodied know-how (genetic and cultural). Indeed, from an economic perspective the latter may be the most important.

Input complementarity

A biophysical perspective also suggests a strong physical interdependence between the inputs employed within production techniques. Materials and energy and the equipment, structures, and agents that process these flows are used together. Machines and other capital equipment are designed with specific engineering characteristics appropriate to given fuels or to specific types of materials. Inputs are *complementary* within an activity. Production is not, therefore, a process that can usefully be described by the mechanistic method of " varying factors one at a time". Successive doses of one input, all others constant, is a methodological-mathematical construct that has little meaningful application to physical processes. Marginal products of individual capital goods simply do not exist.

Sectoral complementarity

Since inputs into one sector are produced by another sector, another aspect of complementarity and interdependence is the coevolution of economic techniques (technologies and organization) between activities and sectors of the economy. Coevolution is the reciprocal adaptive responses of two interacting species: the evolution of flower design and the hummingbird's beak, for example. Norgaard (1984) has applied the idea of coevolution between species to the coevolution of environment and society. The coevolutionary idea can also be directly applied to the evolution of techniques within the economy (an economic technique like a species is a "genetic" mechanism for the extraction and transformation of materials and energy).

Coevolution occurs when positive feedback initiates an ongoing reciprocal process of change between evolving systems or their components (see Norgaard, 1984; Levins and Lewontin, 1985). An example in economics is the emergence of a coal–iron–machine–engineering complex in the British industrial revolution. Another is the decision of Japanese strategic planners to build an industrial economy on the dynamic interdependence between steel, oil refining, petrochemicals, automobiles, industrial machinery and, more recently, electronics. The Taiwanese are considering building an automobile industry in order to develop the technological infrastructure and skills of a machine industry which would supply parts to autos. Marshall's (1920) concept of the internal synergies of an industrial district (economies external to individual firms) is relevant here.

Vertical disaggregation between extraction and manufacture

By presenting production as a one-step process of combining primary factors of production to obtain final products, neoclassical theory ignores both the time pattern of production and important differences in the conditions of resource extraction and resource processing. The extraction of energy and materials prior to processing, fabrication, and use imposes a temporal sequence on production. And differences in the quality, quantity, and timing of resource availability by geographic site impose constraints on primary sector production not present in manufacturing production where materials are purchased.

The high variability of resource availability by geographic site is a result of geological and biological production processes that have unevenly concentrated ores and soil nutrients. Bio-mass production is characterized not only by uneven nutrient availability but also by the condition of geographically dispersed and low intensity energy flow from the sun. Such production is geographically extensive compared to the highly concentrated energy of fossil fuels now employed in minerals extraction, processing and manufacture. High quality resource sites will have lower costs than low quality or low density sites. Extraction sector production, moreover, depletes existing reservoirs of materials which have to be replaced by ongoing climatological and biological cycles in the case of renewable resources, or by the discovery of new sites for exhaustible resources. The timing and extent of resource availability will have important implications not only for the structure of costs but also for price behavior and the dynamics of sector adjustment.

Processing and manufacturing technologies, as Malthus (1836) pointed out, are reproduced by the replication of existing techniques or by the discovery of newer, lower-cost techniques which can in turn be replicated (given necessary materials, energy, skills and so on). Additional production capacity will be obtained at constant or lower unit costs than previously installed capacity (for given input prices). Economies of scale in processing and manufacturing reinforce this cost structure (downward-sloping cost curves). In the presence of economies of scale and available resources, increases in demand lead to lower unit costs in production and set up a positive feedback loop between production and input costs.

Maximum power and efficiency

In energetically open systems, those species and economic technologies which get the most energy and use it the most effectively in competition with other systems will reproduce at a differentially higher rate (Lotka's Principle). Under conditions of limited (renewable) resources, equilibrium population levels may be established for non-evolutionary time scales. The shift from a largely renewable resource base to the exploitation of large stocks of coal and petroleum changes the possibilities for growth. Resources are used to produce machines which are used to extract more materials and energy, produce more machines, etc. This sets up a positive feedback (autocatalytic) loop. The new technology generates a larger surplus which is reinvested, the adoption of new technology is a function of the rate of investment, and the system grows at an exponential rate.

Obviously there are limits to such growth. Bacteria in a petrie dish can reproduce at an exponential rate as long as they have food and space. Most economists have dismissed concerns about resource scarcity by noting that resource prices have not risen faster than prices in general (and by assuming that should resource scarcity appear that technological change will create "new" resources). But this fundamentally misreads a historical record where growth and technological change have been based on the exploitation of large but not unlimited supplies of coal and oil. It is certainly possible that new virtually unlimited supplies of energy may be developed in the future. It is also not certain. In any case the major difficulty of the future may not be scarcity of resource "inputs" but the increasing problem presented by the inability of natural systems to absorb increasing levels of material and toxic wastes. The inconsistency of maximum power behavior with environmental limits indicates the need for control mechanisms to keep the economic subsystems in balance with environmental systems.

CLASSICAL/POST-KEYNESIAN PRODUCTION PRICES

The theoretical implications of a bio-physical approach to production extend beyond consideration of environment–economy interactions. They require a reformulation of the theory of interactions within the economy. This includes a sector-based model of prices and a short-run macro model of price and quantity dynamics which offers an explanation of the output, productivity, and inflationary effects of primary commodity price shocks unexplained by conventional theory.

A sectoral model of asymmetric price behavior was partially developed by classical theorists. Petty, Cantillon, the Physiocrats, and Smith formulated manufacturing prices in terms of the costs of raw materials and labor employed in production plus profit earned on the value of capital advanced. Longfield (1833) observed that manufactories used inventory and quantity adjustments to maintain equality between demand and supply. Prices were determined by costs of production. Agriculture and raw material prices were determined by demand and supply and were highly variable. Mill (1848) treated prices under two categories: demand and supply pricing when there is some constraint on supply; and cost of production prices when output can be increased without limit (but he does not make this distinction by sector and he reduces costs to direct and indirect labor).

Modern empirical work (Hall and Hitch, 1939; Wilson and Andrews, 1951; Wiles, 1963) has emphasised the ubiquity of a materials or prime-cost pricing model (also known as full-cost and mark-up pricing) in industrial sectors of the economy. Several Post-Keynesians, notably Kaldor (1979) and Sylos-Labini (1985) have combined a mark-up model for industrial sectors with a recognition that markets for agricultural products and industrial raw materials are sensitive to changes in short-run supply and demand (and to expectations). But the underlying physical conditions which shape this difference between "flex" and "fix" prices have not been considered.

The behavior of extraction sector and agricultural prices is shaped by the high variability in the quality and extent of resources by geographic site. When production cost conditions vary significantly by site, there is no firm cost-floor under prices (there would be, rather, a more steeply upward slanting floor). Under conditions of low demand, lowest cost producers may still be able to produce at a profit or even earn economic rents. But low prices dissuade new exploration and investment (setting the stage for a future escalation of prices as demand is restored). Increased demand and higher prices permit production from high-cost sites. Prices are sensitive to changes in demand and supply and exhibit significant instability. The instability of primary commodity prices has been widely noted (O.E.C.D., 1980).

Processing and manufacturing sectors with significant variable costs are characterized by cost-plus pricing behavior. Prices are set to cover the throughflow of materials, fuel, and the hire of labor plus a markup for overhead and profit on capital. These sectors have a floor under prices determined by costs (with economies of scale this is a downward-sloping floor). Costs (and prices) change when the costs of labor, raw materials, energy and other intermediate goods change; but prices are less sensitive to changes in demand. Faced with a fall in demand for its product, a firm cannot make any substantial cut in its prices until the prices of its inputs are cut. It lays off workers, reduces purchases of materials and fuel, etc. Quantity adjustments preceed price adjustments. A materials- and energy-based, cost-plus pricing model provides a theoretical rationale for the quantity adjustments of Keynesian theory.

Given cost-plus pricing in manufacturing sectors, the behavior of primary sector prices will have important effects on macro economic behavior. Low commodity prices and availability of resources set the conditions for a sustained economic expansion. Increases in demand lead to increases in output at stable prices. The existence of economies of scale lead to lower unit costs, increasing productivity and moderating price pressure. Low energy prices reduce the costs of extracting raw materials, transportation costs, and the costs of manufacturing including the costs of capital goods. They also contribute to rapid technical progress since mass-production technologies substitute cheaply produced machines and large energy supplies for labor.

Rising prices of primary commodities and energy are, likewise, passed on in manufacturing costs. An increase in energy prices, for example, increases the cost of production in extraction sectors; it increases the cost of manufacturing directly and indirectly via the higher cost of raw materials; and the higher cost of capital goods has a further effect on prices. This is a "price multiplier". There is also an important recessionary effect as higher commodity prices reduce the demand available for manufacturing goods leading to layoffs, loss of economies of scale, lower productivity growth, etc. Seven of the eight recessions since World War II were preceeded by a significant increase in the price of crude oil (Hamilton, 1983).

The connection between rising primary commodity prices, especially energy, and inflation is obvious in the historical record but has been difficult to explain in conventional models. According to the primary factor model of orthodox theory, an increase in energy prices has only a small inflationary effect. If energy costs are 5% of GNP, for example, then a doubling of energy prices would increase prices by 10% and only if there was *no* substitution away from energy. According to neoclassical theory, the energy price shocks of the 1970s explain only a small part of the price escalation of the mid and late 1970s. But as Robert Fri of Resources for the Future observes:

> Neither energy's share of GNP nor the slow rate at which energy consumption patterns change accounts for the sharp changes in production, productivity, unemployment, or inflation that occurred in 1974 and 1979. It appears that something amplifies the effect of oil price swings (Fri, 1987, p. 40).

ENVIRONMENTAL RESOURCES AND REPRODUCTION PRICES

A materials and energy approach to prices also indicates the incompleteness of the price system. Environmental resources are, in general, valued at the margin at costs of extraction (not reproduction). The effects of economic production on natural environments and the global system is likewise valued at costs of dumping or dispersal. Although orthodox theory recognizes the existence of such externalities in the divergence between private and social cost, any calculation of social costs is ordinarily made in terms of current market prices which themselves do not reflect the full cost of the use of resources or environmental systems.

The production prices of the classical model (Sraffa, 1960) provide one methodology for extending the range of price calculations. These are ostensibly *reproduction* prices. That is, they are formulated to ensure the replication of the components of the social-economic system through time. They have, therefore, an intrinsic determination shaped by the internal physical requirements of replacing commodities and services used up in production and consumption.

But this model of reproduction has been formulated in terms of produced commodities alone. The continued flow of necessary materials and sources of energy from natural systems has been taken as given. There is no calculation of the environmental costs of production of the resources taken from natural systems apart from their economic costs of extraction. Reproduction of the larger bio-social system through time suggests an extension of the concept of production prices from produced commodities to environmental resources and services. Most resources are not, of course, reproducible except on a geological time scale and their production requires the use of other resources (and non-reproducible energy). In a reproduction approach, exhaustible fuels, for example, would be priced either at the cost of renewable sources of power or in terms of the cost of exploiting alternative fuels whose supply is sufficiently large to avoid near-term exhaustion. Materials, likewise, would be priced in terms of the costs of recycling or in terms of the cost of substitutes. This would in effect reverse Marx's famous transformation of values into prices. Current prices would be recalculated to reflect reproduction " values" for critical environmental resources and services.

Prices alone, of course, are insufficient. The goal is the long-term viability of the environmental-economic system. Markets and prices are important social tools. They are one type of regulatory regime, but they have their limits. As Georgescu-Roegen suggested:

> The market mechanism has never been able to deal with bioeconomic ills. Whenever communities have been concerned with conserving resources—forests, fish, or game—or a healthy environment, they had to introduce legal quantitative restrictions (1981, p. 73).

The interdependence between economy and environmental resources (and potential long-term damage of industrial pollutants) suggests the need for quantity signals, physical limits and non-market regulatory systems. The physical underpinning of the classical model with its potential emphasis on the interconnectedness of social life and nature provides a rather different policy emphasis than the *laissez-faire, laissez-passer* dictum drawn by the Physiocrats and Adam Smith.

CONCLUSION

Classical economic theory provides a physical, materials-flow and proto-energetic approach to production and prices which had its origin in a provisioning metaphor drawn from physiology. Extension of this approach via the insights of modern physical and ecological science provides an alternative theoretical viewpoint to the equilibrium assumptions and marginalist adjustment mechanisms underlying neoclassical theory. It also provides an alternative conception of the internal physical and social reproduction requirements of a price (and regulatory) system which is absent from the marginalist approach and its emphasis on the subjective preferences of the present generation. Ecological and socio-economic systems are highly interdependent and a theory is needed which reflects that interdependence at the physical level. Obviously, the ideas of classical theory are only a beginning but they provide a possible starting point for developing a more ecologically (and socially) informed economic theory.

REFERENCES

Ayres, R.U., 1978. Resources, Environment, and Economics: Applications of the Materials/ Energy Balance Principle. Wiley, New York, N.Y.

Babbage, C., 1832. On the Economy of Machinery and Manufactures. Knight, London.

Bliss, C., 1975. Capital Theory and the Distribution of Income. North Holland, Amsterdam.

Boulding, K., 1986. What went wrong with economics? Am. Sci., 30: 5–12.

Cantillon, R., 1755. Essai sur la Nature du Commerce en General. A.M. Kelley, New York, N.Y.

Costanza, R., 1980. Embodied energy and economic valuation. Science, 210: 1219–1224.

Fri, R.W., 1987. New directions for oil policy. Environment, 29: 17–42.

Georgescu-Roegen, N., 1971. The Entropy Law and the Economic Process. Harvard Univ. Press, Cambridge, Mass.

Georgescu-Roegen, N., 1981. Energy, matter, and economic valuation: where do we stand? In: H.E. Daly and A. Umaña (Editors), Energy, Economics and the Environment. Westview Press, Boulder, Colo.

Hall, R.L. and Hitch, C.J., 1939. Price theory and business behavior. Oxford Economic Papers, pp. 212–245.

Hamilton, J.D., 1983. Oil and the macroeconomy since World War II. J. Polit. Econ., 91: 228–248.

Hobbes, T., 1651. Leviaton, or the Matter, Forme, and Power of a Common-Wealth. C.B. Macpherson (Editor), Penguin, 1968.

Jevons, W.S., 1864. The Coal Question; an Enquiry Concerning the Progress of the Nation, and the Probable Exhaustion of our Coal Mines. Macmillan, London.

Jevons, W.S., 1871. The Theory of Political Economy. Macmillan, London. (Reprinted by Penguin, 1970).

Kaldor, N., 1979. Further Essays in Economic Theory. Holmes and Meier, New York, N.Y.

Leontief, W.W., 1982. Academic economics. Science, 217: 104–107.

Levins, R. and Lewontin, R., 1985. The Dialectical Biologist. Harvard University Press, Cambridge, Mass.

Longfield, M., 1833. Lectures in political economy. In: R. Collison Black (Editor), Economic Writings of Mountiford Longfield. Kelley, New York, N.Y., 1971.

Malthus, T.R., 1815. An inquiry into the nature and progress of rent. In: The Pamphlets of Thomas Robert Malthus. Kelley, New York, N.Y., 1970.

Malthus, T.R., 1836. The Principles of Political Economy. Murray, London, 2nd ed.

Marshall, A., 1920. Principles of Economics. Macmillan, London, 8th ed.

McCulloch, J.R., 1835. Philosophy of manufactures. Edinburgh Rev., 61 (July).

Menger, C., 1871. Principles of Economics. Free Press, Glencoe, Ill., 1950 (Translation of: Grundsätze der Volkswirtschafts-lehre).

Mill, J.S., (1848) 1871. Principles of Political Economy. Longmans, Green, London, 7th ed.

Norgaard, R.B., 1984. Coevolutionary development potential. Land Econ., 60: 160–173.

O.E.C.D. (Organization for Economic Co-operation and Development), 1980. The Instability of Agricultural Commodity Markets. Paris.

36

Pareto, W., 1935. The Mind and Society. Harcourt, New York, N.Y.
Ricardo, D., 1817. On the Principles of Political Economy and Taxation. Murray, London.
Schumpeter, J.A., 1954. History of Economic Analysis. Oxford University Press, New York, N.Y.
Senior, N., 1836. An Outline of the Science of Political Economy. Kelley, New York, N.Y., 1965.
Senior, N., 1848. J.S. Mill on political economy. Edinburgh Rev., 178: 297–320.
Simon, H.A., 1986. The failure of armchair economics. Challenge, 29: 18–25.
Smith, A., 1776. An Inquiry into the Nature and Causes of the Wealth of Nations. E. Cannan (Editor), Modern Library, New York, N.Y., 1937.
Sraffa, P., 1960. The Production of Commodities by Means of Commodities: Prelude to a Critique of Economic Theory. Cambridge University Press, Cambridge.
Sylos-Labini, P., 1985. Weintraub on the price level and macro-economics. J. Post Keynesian Econ., 7: 559–574.
Van Gool, W., 1985. Towards a physical interpretation of production functions. In: W. van Gool and J. Bruggink (Editors), Energy and Time in Economic and Physical Sciences. Elsevier, Amsterdam, pp. 274–256.
Verri, P., 1773. Meditazioni sulla economia politica. In: Custodi (Editor), Italian Economists. Parte Moderna, XV.
Walras, L., 1874. Elements of Pure Economics. Allen and Unwin, London, 1954 (Translated by W. Jaffe).
Wicksteed, P.H., 1894. An Essay on the Co-ordination of the Laws of Distribution. Macmillan, London.
Wiles, P.J., 1963. Prices, Costs and Output. Blackwell, Oxford.
Wilson, T. and Andrews, P.W.S., 1951. Oxford Studies in the Price Mechanism. Clarendon Press, Oxford.

[11]

HUMAN IMPACTS ON THE NITROGEN CYCLE

Fertilizer production and other human activities have more than doubled the global rate of nitrogen fixation since preindustrial times. The resulting imbalance is contributing to ecosystem disruption, ozone depletion, greenhouse effects and other environmental problems.

Ann P. Kinzig and Robert H. Socolow

Humans are greatly perturbing the global nitrogen cycle. Perhaps the best evidence for this perturbation comes from air trapped in layers of quasipermanent ice in the Arctic and the Antarctic. Ice cores taken from these two polar regions give us a 2000-year record of the Earth's atmospheric composition. This record indicates a striking constancy in nitrous oxide concentrations, at approximately 285 parts per billion, for nearly 1500 years. (See figure 2.) Since about 1600 AD, however, nitrous oxide concentrations have been increasing, and the present-day atmospheric burden of this gas is greater than at any other time in the past two millennia. Furthermore, nitrous oxide concentrations continue to increase, currently at a rate of about 0.3% per year.[1] These variations indicate that many nitrogen flows are now larger than in preindustrial times, and other evidence suggests that human activity is responsible. (See figure 1.)

These increased flows of nitrogen should be of some concern to humankind. The anthropogenic disruptions of the nitrogen cycle have led to the eutrophication (or excessive fertilization) of coastal waters, characterized by increases in algal blooms and decreases in dissolved oxygen. Urban smog, potentially dangerous nitrate levels in drinking water, stratospheric ozone depletion, greenhouse effects and the leaching of nutrients from soil are further consequences of these increased nitrogen flows.

To understand how to alleviate these effects, we must first understand the nitrogen cycle itself. Nitrogen cycles operate on many spatial and temporal scales, from the fast, microscopic scale of amino acid production and degradation to the very slow, global tectonic processes whereby nitrogen is subducted below the Earth's crust and later returned to the biosphere by volcanoes. Our discussion will concentrate on global-scale biological nitrogen cycles (excluding tectonic cycling), with particular attention to nitrogen cycling within terrestrial ecosystems.

Ann Kinzig is a postdoctoral researcher in the department of ecology and evolutionary biology and the center for energy and environmental studies at Princeton University, in Princeton, New Jersey. Robert Socolow is the director of the center for energy and environmental studies and a professor of mechanical and aerospace engineering at Princeton.

The global biogeochemical cycle

In its simplest form, the global biogeochemical nitrogen cycle consists of three major reservoirs—the atmosphere, the oceans and terrestrial ecosystems—and the flows among them. (See figure 3.) Nearly all of the energy that drives the flows within the nitrogen cycle ultimately derives from the Sun; photosynthesis converts this radiant energy to chemical energy, which is then stored in living and dead organic matter.

Over 99.9% of the nitrogen in global biogeochemical reservoirs is stored in a form inaccessible to almost all living organisms: diatomic nitrogen either in the atmosphere or in solution in the ocean.[2] The strong triple bond of the N≡N molecule makes it nearly inert. To most of the biosphere, therefore, the nitrogen in the atmosphere is like the ocean to a thirsty person: amazingly abundant but not quite in the right chemical form. Only a few species of aquatic and terrestrial bacteria and blue-green algae can convert N_2 into ammonium (NH_4^+) through the process of biological nitrogen fixation. This biological pathway provides the dominant natural flow of fixed nitrogen on the planet. Lightning, however, provides a nonbiological pathway for nitrogen fixation, oxidizing N_2 to nitric oxide (NO), which is rained out as nitrate (NO_3^-) within days. Satellites record approximately 100 flashes of lightning per second (or 3 billion flashes per year); each flash on average oxidizes about 1 kilogram of N_2 to nitric oxide.[3] The product, a fixation rate of 3×10^{12} grams of nitrogen per year [3 Tg(N)/yr], is only a few percent of the rate of biological nitrogen fixation, though nitrogen was fixed exclusively by lightning and in the shock waves

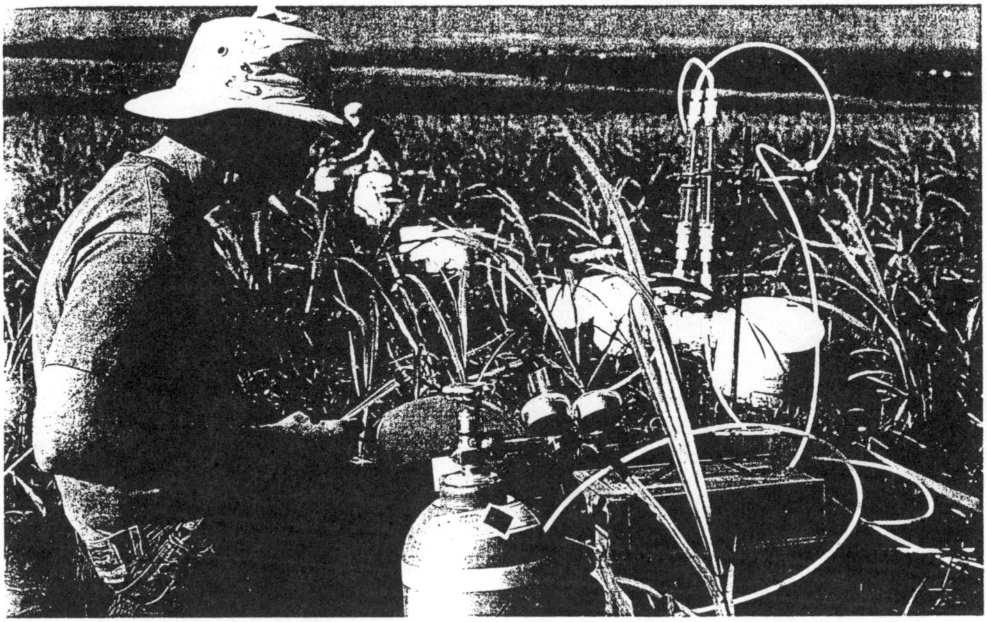

Measurements of emissions of trace gases (such as N_2O and NO) from fields of fertilized sugarcane in Maui, Hawaii, provide us with information on the global nitrogen cycle and how human activity is modifying it. See the box on page 29 for a description of this type of measurement. (Courtesy of Pamela Matson, University of California at Berkeley.) **Figure 1**

of meteors before the origin of life on this planet.[4] Humans have approximately doubled the global rate of nitrogen fixation, from the 130 Tg(N)/yr fixed in preindustrial times by terrestrial and aquatic bacteria and blue-green algae to the nearly 280 Tg(N)/yr fixed today.[4,5]

The nitrogen fixed in terrestrial and aquatic ecosystems must somehow be "unfixed," or else the atmosphere would eventually run out of nitrogen (though even at today's rate of fixation it would take 12 million years!). The N_2 molecule is reconstituted and returned to the atmosphere via denitrification. Except for a small amount of pyrodenitrification in forest, grassland and agricultural fires, denitrification is accomplished nearly entirely by another specialized group of bacteria.

The terrestrial cycle

"Opening" any of the three reservoirs presented in figure 3 reveals layer upon layer of internal cycles. We will describe here only the first layer of the terrestrial nitrogen cycle, which has four major reservoirs and four major flows. (See figure 4.) The four reservoirs are plants, soil microorganisms (bacteria and fungi), dead organic matter and inorganic nitrogen (ammonium and nitrate). Plants account for about 94% of the nitrogen in living organisms in terrestrial ecosystems, while microorganisms and animals contain only 4% and 2%, respectively, of the nitrogen in living organisms.[6] Soil invertebrates can play an important role in the nitrogen cycle through their redistribution and comminution (or pulverization) of detritus. Larger animals, however, do not usually play a significant role in nitrogen-cycling processes, except in some agricultural or heavily grazed systems.[7]

The four major processes that move nitrogen within the terrestrial nitrogen cycle are:
▷ death of plants and microorganisms
▷ mineralization, or liberation of nitrogen as ammonium, which occurs when microorganisms, in search of energy, attack amino acids and break the bond between a carbon atom and an amino group (NH_2)
▷ assimilation, or uptake of inorganic nitrogen by plants
▷ immobilization, or uptake of inorganic nitrogen by microorganisms.

Furthermore, ammonium is oxidized to nitrate (nitrification) within the reservoir of inorganic nitrogen.

In addition to the nitrogen flows within terrestrial ecosystems, there are flows of nitrogen into and out of ecosystems. The flows of nitrogen into an ecosystem include the anthropogenic and natural nitrogen fixation already mentioned. In addition, both wet and dry deposition can deliver nitrogen compounds from the atmosphere to terrestrial ecosystems. Nitrogen can move in surface runoff or in groundwater from one ecosystem to another. Nitrogen can be lost from a system through leaching, denitrification, ammonia volatilization and harvesting of plant material.

The analogy of the thirsty person at the seashore continues *within* the terrestrial nitrogen cycle, where the vast majority of nitrogen is unavailable to plants for a second reason. Plants require primarily inorganic forms of nitrogen—ammonium and nitrate—for their nutrition, but about 96% of the nitrogen in terrestrial ecosystems is bound up in dead organic matter. (See figure 3.) While

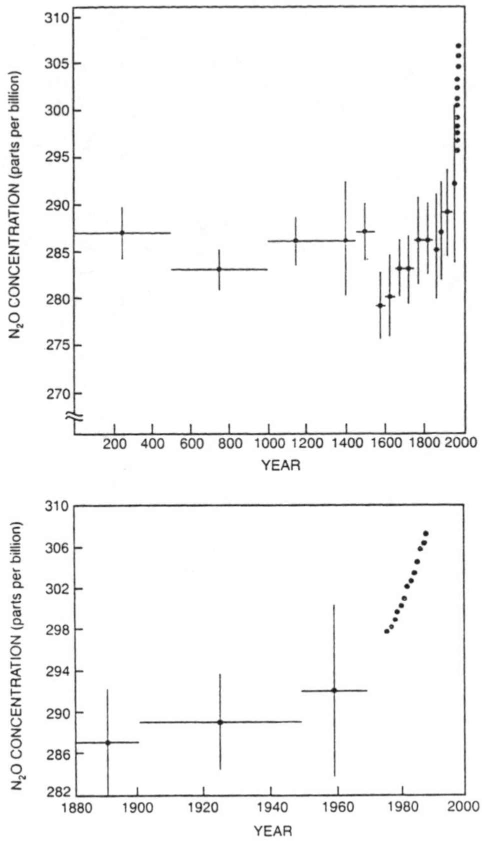

Atmospheric concentration of nitrous oxide (N_2O) over the past two millennia as measured directly from the atmosphere (red) and from bubbles trapped in layers of polar ice (blue). Horizontal bars indicate the period that each ice-core value is averaged over; vertical bars indicate 90% confidence limits for those averages. The dip near 1500 could be due to a reduction in microbial activity caused by the "Little Ice Age."[1] Human disruptions of the nitrogen cycle are probably responsible for the steeply increasing levels of N_2O in recent years. (Adapted from M. A. K. Khalil, R. A. Rasmussen, *J. Geophys. Res.* **97**, 14 651, 1992.) **Figure 2**

"labile" organic matter decomposes easily and quickly, releasing nitrogen for use by plants and microorganisms, other, "recalcitrant" dead organic matter can sequester nitrogen for hundreds or thousands of years.

Though frequently an insignificant reservoir in terms of quantity of nitrogen stored, soil microorganisms are crucial to the existence and control of the interreservoir flows of nitrogen, including the decomposition, mineralization, nitrogen fixation, nitrification and denitrification processes already discussed. Bacteria and blue-green algae are the most primitive organisms known, with a genesis that dates back to more than 3 billion years ago. They are also the most common living organisms on the planet. The biochemical versatility of populations of soil bacteria and fungi allows for the degradation of all natural organic compounds and many (but not all) man-made organic compounds as well.[8]

The relationship between plants and microorganisms in the soil system is simultaneously competitive and mutualistic. Plants and microorganisms compete for the inorganic nitrogen needed by both communities for biosynthesis and growth. On the other hand, microorganisms and plants are mutually dependent. The carbon that microorganisms need for biosynthesis, growth and energy is fixed by plants in photosynthesis and delivered to soil microorganisms through root exudates, litter fall and plant death. The nitrogen needed by plants for biosynthesis and growth is liberated by microbially mediated decomposition. Interesting questions arise as to how the evolutionary pressures exerted by this simultaneous competition and mutualism serve to structure ecosystems and to control the flows of nitrogen.[9]

Many terrestrial ecosystems have "closed" nitrogen cycles, in the sense that rates of nitrogen input or loss are small compared with internal rates of nitrogen cycling. Mature and undisturbed ecosystems, in particular, are associated with low losses of nitrogen to groundwater or to the atmosphere. These closed cycles can be attributed to a relative balance between nitrogen supply and plant and microbial nitrogen demand.[10] For example, the ability of plants and microorganisms to take up available nitrogen quickly can reduce the amount of nitrogen available for leaching or denitrification, "closing" the system to nitrogen losses. Managed or disrupted ecosystems, on the other hand, often have "open" nitrogen cycles, with inflows and outflows of nitrogen that are comparable to or even exceed internal cycling rates.[2] Fertilization and land clearing, and the consequent disruption of nitrogen production and uptake, are frequently associated with accelerated nitrogen loss due to denitrification and leaching.[11]

Perturbations due to human activities

The approximate doubling of the rate of nitrogen fixation is probably the most important perturbation of large-scale nitrogen cycling caused by human activity in the industrial age. The increased fixation is largely the result of production of nitrogen fertilizer, whose first step, making NH_3 from N_2, mimics at high temperatures and pressures (typically about 500 °C and 300 atmospheres) what bacteria and blue-green algae accomplish with enzymes at ambient temperature and atmospheric pressure. The rate of nitrogen fixation due to fertilizer production today is approximately 90 Tg(N)/yr.[5]

Because nitrogen availability limits crop growth for many combinations of soils and crops, it has been relatively easy to increase crop yields by supplementing natural sources of available nitrogen with synthetic sources, though one gets diminishing returns. An instructive model of a typical US corn field[12] shows a grain yield of 4 metric tons (4×10^6 grams) of corn per hectare in the absence of fertilizer, 7 t/ha when fertilizer is applied at 100 kg(N)/ha, and 8 t/ha when fertilizer is applied at 200 kg(N)/ha. Since corn is about 1.3% nitrogen by weight, the harvested grain contains 39 of the first 100 kilograms of nitrogen added as fertilizer, but only 13 of the second 100. (In the US, corn production is more commonly reported in bushels; a bushel is 30.3 liters, and there are about 25 kg of corn in a bushel.)

The large-scale use of fertilizer is a relatively recent phenomenon. Night soil, dung and recycled crop wastes have been the traditional means of returning nitrogen to the soil system. Figure 5 shows that the total global use of nitrogen fertilizer in 1960 was only about 10 million metric tons of nitrogen per year, about one-tenth of today's levels. Furthermore, nitrogen fertilizer use is likely to continue to increase so that a growing world population may be fed.[13]

The planting of leguminous crops (such as soybean and alfalfa) and the combustion of fossil fuels are the second and third most important human activities that increase nitrogen fixation rates. (See figure 6.) Leguminous crops host symbiotic nitrogen-fixing bacteria in their root nodules. The 250 million hectares of leguminous crops currently cultivated worldwide increase nitrogen fixation over preindustrial levels by about 40 Tg(N)/yr.[5]

Fossil fuel combustion fixes as NO nitrogen that had previously been sequestered either as atmospheric N_2 or as fuel-bound nitrogen deep underground. Globally the two processes are about equally important, and they fix nitrogen at a combined rate of about 20 Tg(N)/yr.[5] Atmospheric N_2 is the dominant source of nitrogen when liquid fuels are burned in gasoline and diesel engines and in aircraft gas turbines, while fuel-bound N is the dominant source of nitrogen in coal combustion at power plants. This difference arises in part because coal is an order of magnitude richer in nitrogen than are refined fuels such as gasoline, diesel and jet fuel (typically 1% versus 0.1% nitrogen by weight). Also, the formation of NO from atmospheric N_2 depends strongly on temperature, and the combustion temperatures in power plant boilers are generally lower than those in internal combustion or gas turbine engines, because there is less of a constraint to burn fuel rapidly.

Consequences of increased nitrogen fixation

The consequences of increased nitrogen fixation are manifold. They include disruptions of the nitrogen budgets of natural and managed ecosystems, leaching of nutrients from soils, effects on stratospheric ozone chemistry, alterations in the greenhouse effect and direct impacts on human health.

Nitrogen budgets of ecosystems. Nitrogen fixed by human activities disperses widely to unintended targets—sometimes up to hundreds of kilometers downwind or downstream. For example, nitrogen fixed during the combustion of fossil fuels and lofted into the atmosphere causes both wet and dry deposition of nitrate and nitric acid (HNO_3). Similarly, application of fertilizer based on urea ($CO(NH_2)_2$) or ammonium (NH_4^+) results in the volatilization of ammonia (NH_3), the conversion of ammonia to ammonium in the atmosphere and the deposition of ammonium in precipitation.[14]

Since many terrestrial ecosystems are nitrogen limited, nitrogen deposition may initially increase an ecosystem's net primary productivity. Inevitably, however, some species will assimilate the additional nitrogen more easily than others, causing differential growth of species and altering the species composition of the community. Altered species composition may reduce ecosystem biodiversity or otherwise change ecosystem function. Ecosystem productivity determines the rate at which the Earth's populations can harvest the goods supplied by ecosystems, such as food, fodder, fuel and fiber. In addition to supplying harvestable material resources, diverse and healthy ecosystems can reduce soil erosion from surface runoff, regulate local climate and help prevent outbreaks of pests and plant diseases. Many scientists recognize the urgency of characterizing the precise effects of anthropogenic disruptions on the flows of these (and many other) goods and services provided by ecosystems and collections of ecosystems. In particular, society needs a clearer view of the degree to which particular goods and services are unique to specific ecosystems or redundant across many different types of ecosystems.

A back-of-the-envelope calculation gives some indication of a relevant time scale for the terrestrial nitrogen deposition problem. Assuming that most of the terrestrially deposited inorganic nitrogen is either taken up by living organisms or stored in the most labile part (roughly 15%) of the dead organic matter pool, nitrogen stocks totaling approximately 20 000 Tg(N) are being affected by anthropogenic fixation. Figure 3 gives the net increase in nitrogen deposition on land (fixation minus runoff minus denitrification) as about 80 Tg(N)/yr. Dividing the

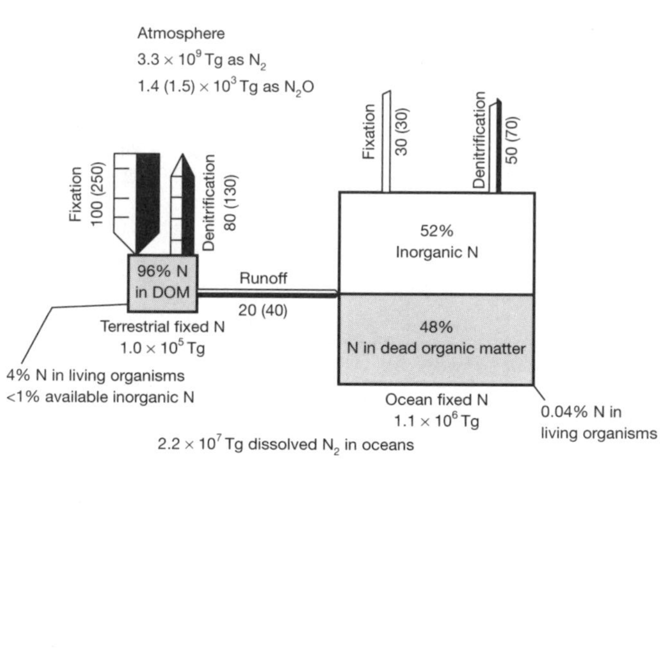

Global biogeochemical nitrogen cycle. Stocks of nitrogen are given in teragrams (Tg or 10^{12} grams); flows are given in Tg(N)/yr. Yellow portions of arrows indicate preindustrial flows; full arrows (and numbers in parentheses) indicate the current flows. Very little inorganic nitrogen (ammonium and nitrate) is available in terrestrial ecosystems because nitrogen transformations and uptake by plants and microbes are rapid. (Additional terrestrial stocks of inorganic nitrogen unavailable to living organisms, and not represented here, may be sequestered in the soil or held on the clay matrix.) Oceanic stocks of nitrogen include nitrogen stored in the deep oceans. Terrestrial stock estimates are taken from ref. 4, oceanic stock estimates from ref. 6, and flow estimates from ref. 5. Estimates of some global nitrogen stocks and flows can vary by as much as an order of magnitude.[23] **Figure 3**

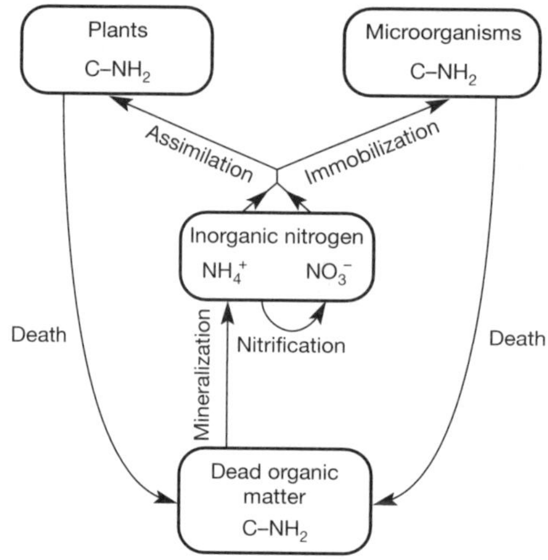

Terrestrial nitrogen cycle in simplified form, with the major chemical form indicated for each of the reservoirs. The global terrestrial flows via mineralization, total uptake by plants and microorganisms, and total death are each approximately 1200 Tg(N) per year. These flows occur *within* the terrestrial box of figure 3. **Figure 4**

stock by this flow yields 250 years as a rough estimate of the time required to double the stock of living and labile terrestrial nitrogen at the current rates of human perturbation of the nitrogen cycle. The analogous ratio for the greenhouse effect (stock of atmospheric carbon dioxide divided by the current annual increase in the stock) is similar—also roughly 200 years.

Nitrogen fixed or mobilized by human activities can also affect the nitrogen budgets of aquatic ecosystems. One result has been the eutrophication of coastal waters. (Productivity in freshwater ecosystems tends to be limited by the availability of phosphorus, while nitrogen frequently limits productivity in saltwater ecosystems.) The Baltic and North Seas, the western Mediterranean and the Atlantic seaboard (including the Narragansett and Chesapeake Bays) have all been affected by nitrogen-induced eutrophication.[15] The increased algal blooms and decreased oxygen supplies in eutrophic waters can have detrimental effects on fisheries and other aquatic populations.

Leaching of cations from soils. The surfaces of clay minerals and amorphous organic materials in the soil are negatively charged, and nutrient cations such as Ca^{2+}, Mg^{2+}, K^+ and Na^+ are held on this clay–organic matrix. These nutrient cations are gradually displaced from the soil matrix by hydrogen ions and thus become available for assimilation by plants. This cation-exchange capacity is a critical feature of the soil system: The retention of nutrients on the surface of soil particles prevents their being quickly leached away, beyond the reach of plant roots. The deposition of nitric acid on soil ecosystems, however, can increase the concentrations of hydrogen ions in soils (leading to acidic soils), which in turn can lead to mobilization of nutrients in excess of plant needs. The end result is that cation nutrients are leached away and lost from the system.

Acidification of soils can also mobilize aluminum and other metallic cations. Increased aluminum mobilization in terrestrial watersheds can elevate aquatic aluminum concentrations, killing fish and amphibians. Elevated aluminum concentrations in soil may be contributing to the *Waldsterben* (forest death) that has decimated some forests in central Europe. The mobilization of metals such as lead, cadmium, arsenic and mercury may cause similar toxicity problems.

Stratospheric ozone. A small but important subcycle of the global nitrogen cycle involves the flow of nitrous oxide (N_2O) from soils or oceans to the atmosphere. Nitrous oxide is produced as a by-product of nitrification and denitrification in soils and oceans. There are no known tropospheric chemical loss processes for N_2O; instead, N_2O migrates to the stratosphere, where it is primarily destroyed by photodissociation into O atoms and N_2. Approximately 10% of the N_2O in the stratosphere, however, is destroyed by activated atomic oxygen, yielding either nitric oxide (NO) or N_2 and O_2.[16]

The production of NO is important in stratospheric ozone (O_3) chemistry, since NO can catalytically destroy ozone via the following reactions:

$$NO + O_3 \rightarrow NO_2 + O_2$$
$$NO_2 + O \rightarrow NO + O_2$$
$$\text{Net:} \quad O + O_3 \rightarrow 2O_2$$

These cycles involving NO and NO_2 (collectively known as NO_x), and similar cycles involving chlorine, bromine and OH radicals (and leading to the same net reaction),

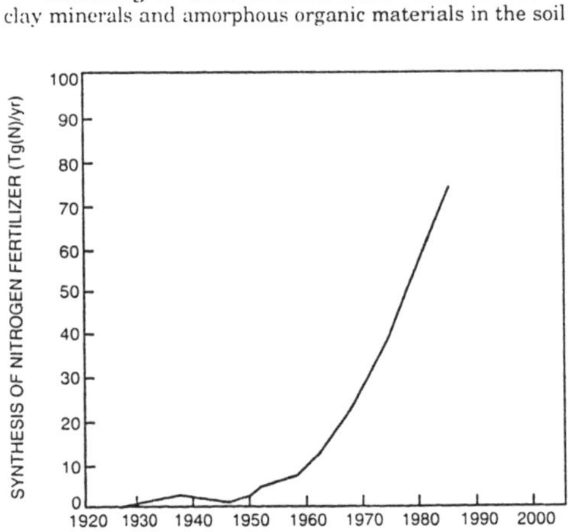

Global nitrogen fertilizer production from 1920 to 1985. The cumulative production over this 65-year period is 1000 Tg, with half of that occurring between 1976 and 1985. (Adapted from V. Smil, in *The Earth as Transformed by Human Action*, B. L. Turner II, W. C. Clark, R. W. Kates, J. F. Richards, J. T. Mathews, W. B. Meyer, eds., Cambridge U. P., Cambridge, England, 1991, p. 423.) **Figure 5**

Measuring N_2O Fluxes

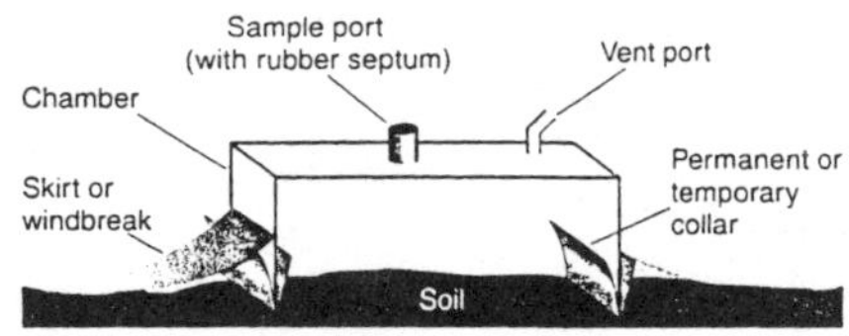

In the closed-chamber method for measuring fluxes of nitrous oxide (and other trace gases) a chamber is placed over the soil surface (as shown in figure 1). The efflux of gas from the soil is then proportional to the increase in gas concentration within the chamber over time—the slope $\Delta c/\Delta t$ of the line in a plot like that below. Gas samples are typically extracted from the sample port with a syringe for later analysis with, for example, a gas chromatograph. The gas in the chamber must be well mixed, either passively by diffusion (in small chambers) or actively with fans (in large chambers).

The data generally depart from linearity for reasons that experimenters identify and allow for. For example, as gas concentrations build up in the chamber, the flux from soil to air decreases due to the decreased concentration gradient. Also, surface winds may cause pressure fluctuations and turbulence around the chamber, resulting in the leakage of gas out of the bottom of the chamber. A windbreak can reduce this effect. The chamber itself can affect the fluxes by changing evaporation rates and surface temperatures. Consequently the chamber should remain on the surface for the shortest possible time, although experimental runs must be long enough to detect the changes in gas concentration. In a typical N_2O experiment samples are extracted every 5 minutes for 20 minutes.

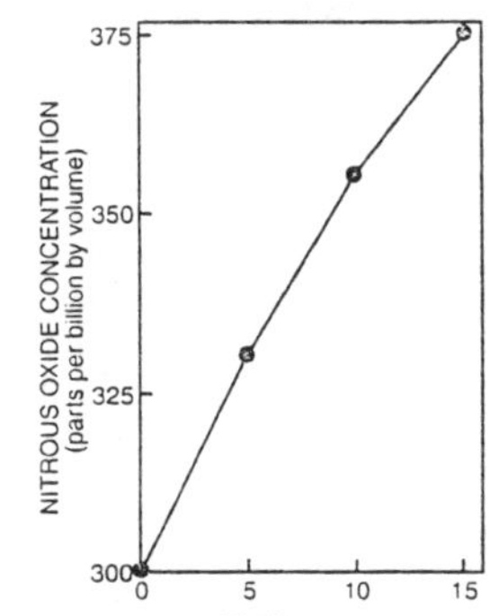

are among the principal reactions governing stratospheric ozone concentrations.[16,17] Ozone in the stratosphere absorbs incoming ultraviolet-B radiation (0.23–0.32 microns). DNA and proteins are sensitive to disruption at those wavelengths.

Attempts to balance the global nitrous oxide budget have not yet been successful, and our understanding of the cause of the imbalance is inadequate. Nevertheless it is believed that fertilizer application and changes in land use are contributing to the observed increase in atmospheric nitrous oxide concentrations through their effects on global nitrification and denitrification rates.[11] Current anthropogenically induced ozone losses, however, are attributable mostly to long-lived industrial chlorine- and bromine-containing compounds, especially the chlorofluorocarbons.

The impact of fertilizer use on stratospheric ozone via increased atmospheric N_2O concentrations should be compared with the similar impact of direct injection of NO into the stratosphere. The latter receives immense attention today in the aerospace community, which designs supersonic and hypersonic aircraft and their engines to meet anticipated regulatory constraints on NO emissions in the stratosphere and upper troposphere.

Greenhouse effects. Nitrous oxide is also of interest in greenhouse policy. The 150-year mean life of N_2O in the atmosphere is far longer than the mean life of any other oxide of nitrogen. The combination of its atmospheric concentration and its particular frequencies of absorption of infrared radiation make nitrous oxide an important contributor (along with water vapor, carbon dioxide, methane and ozone) to the natural greenhouse effect—the effect that makes life possible by raising the Earth's average surface temperature from about −18 °C to +15 °C. The surface temperature increases because gases in the atmosphere trap infrared radiation emitted by the Earth and reradiate it back down to the surface. Human activities have increased the atmospheric budgets of many greenhouse gases and have consequently increased this downward infrared radiation; increases in nitrous oxide concentrations have contributed 5% of this increase in downward infrared radiation.[1]

Anthropogenic nitrogen fixation has additional greenhouse impacts as well. In some ecosystems, ammonium deposition has been shown to reduce methane (CH_4) consumption by soils, perhaps due to the inability of certain enzymes to distinguish between NH_4^+ and CH_4.[18] This decrease in soil consumption may contribute to the observed increase in methane in the atmosphere, though the global significance of this mechanism is still unknown. In addition, the increased availability of nitrogen in ecosystems due to human activities may be increasing the storage of carbon in plants and soils via nitrogen fertilization. Because carbon and nitrogen compounds are both required for the synthesis of biological material and both are released by the decomposition of organic material, the carbon and nitrogen cycles are tightly linked. Increases in available inorganic nitrogen in ecosystems may stimulate plant and microbial growth and increase the storage of carbon and nitrogen in dead organic matter.[19] The magnitude of the increased carbon storage due to nitrogen fertilization will depend on the average C:N ratio of the reservoirs undergoing accretion. (See figure 7.)

In the absence of other perturbations or ecosystem responses, this effect of nitrogen fertilization would decrease the greenhouse effect by transferring carbon from the atmosphere to the biosphere. Anthropogenic nitrogen fertilization could be sequestering in terrestrial ecosystems some of the "missing" global carbon[5,20]—the as yet unidentified flow of carbon out of the atmosphere and into terrestrial or oceanic systems that seems to be required to balance the atmospheric CO_2 budget. A fully convincing global carbon budget, however, awaits additional research.

Direct impacts on human health. High concentrations of fixed nitrogen in both air and water have direct health impacts. The notorious photochemical smog that bedevils many of the world's cities today requires nitrogen oxides as precursors. Of all the instances of nitrogen fixation, probably none has received as much regulatory attention as nitrogen oxide emissions into urban air. Nitrate concentrations in water are also regulated, however, because while nitrates themselves are not particularly toxic, certain bacteria residing in vertebrate digestive tracts can convert the relatively benign nitrate (NO_3^-) into

Annual nitrogen fixation due to human activities alone (right) now approximately equals the estimated preindustrial fixation rate (left). **Figure 6**

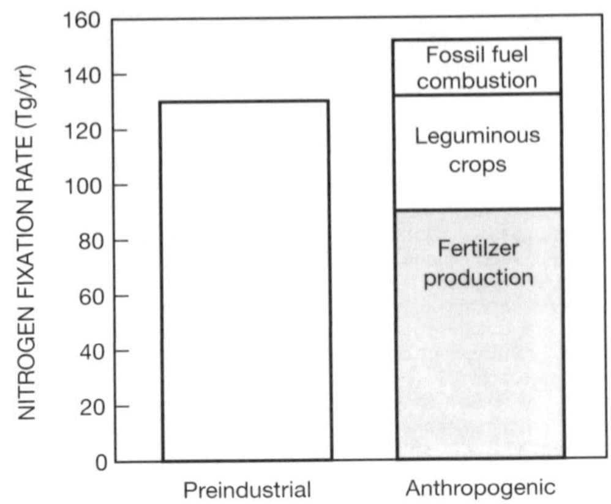

highly toxic nitrite (NO_2^-). Human infants and farm animals are particularly likely to house such bacteria.

Reducing anthropogenic nitrogen fixation

In the framework of global and regional environmental change, anthropogenic disruptions of the nitrogen cycle may be as significant as anthropogenic disruptions of the carbon cycle—even though the nitrogen cycle has received far less attention. In particular, fertilizer use and legume production should perhaps be of as much concern to humans as fossil fuel use and deforestation, the principal agents of disruption of the global carbon cycle.

Because strategies for reducing human impacts on the carbon cycle have been under development for several decades, looking to these strategies for guidance in addressing potentially harmful modifications to the nitrogen cycle seems a reasonable first step. In particular, there are likely to be many cost-effective strategies to achieve efficient use of nitrogen in food production, distribution and consumption that are analogs of the strategies for energy efficiency (effectively, efficient use of carbon) currently being developed and tested around the world.

Some examples of possible innovations in the production of food and fodder include:

▷ new crop combinations and improved soil management to increase the yield at a given level of fertilizer use

▷ improved mechanisms and strategies for retaining natural and synthetic nitrogen in the soil system for subsequent availability to plants

▷ bioengineered crops such as nitrogen-fixing corn.

Much can be achieved through wise management practices. For instance, many farmers use legumes in a crop rotation scheme, leaving the legumes in place and tilling them under, or returning the waste from cash-crop legumes. This increases nitrogen availability in the soil and also increases the soil's organic content, which can improve its water-holding capacity. Alternating a shallow-root crop with a more deeply rooted crop can reduce losses of nitrogen: The deeply rooted crop can assimilate nitrogen that has leached to lower layers of the soil system, and this nitrogen can then be returned to the surface soil in crop wastes.[21] As a complement to such technical innovations, a vigorous regime of regulations and market incentives (taxes, subsidies, tradable nitrogen-fixation permits and so on) would be appropriate.

Innovations in the distribution of food and fodder would include systems for recycling the nitrogen in wastes from feedlots and from sites of centralized food processing and consumption, as well as systems to reduce preconsumer spoilage (a particular problem in developing countries). Innovations in consumption (that is, innovations in end-use nitrogen efficiency) would include changes in the diets of people and domesticated animals. In particular, the diets of prosperous individuals could shift toward a more protein-efficient diet that substitutes grain for meat. (Production of 1 kilogram of beef or chicken in the US in recent years has required feed equivalent to about 14 or 2.5 kilograms of corn, respectively.[22])

Nitrogen management, of course, also involves crops grown for purposes other than food and fodder, including crops for fiber, paper, timber and energy. For each of these crops a sustainable nitrogen management strategy would include mechanisms for leaving in place or recovering and reapplying the nitrogen in crop residues.

Public policy for agriculture today does not reflect the priorities suggested in this article. Although minimizing soil erosion and the consequent transport of nitrogen has been a historical target of agricultural policy, fertilizer production and use—the most important source of disruption of the nitrogen cycle—awaits creative policymaking.

The challenges ahead

Throughout recorded history, our species has derived from the biosphere not only the goods and services necessary for physical survival but also aesthetic pleasure, creative inspiration and spiritual renewal. Yet our growing population and increasing reliance on resource-consumptive technologies have created the potential for unprecedented anthropogenic disruption of the biosphere. Protection of the biosphere therefore becomes a priority. But we cannot protect what we do not understand. Policies and practices designed to protect the biosphere against the grosser insults of human activity must rest on a deep understanding of environmental sciences across a wide range of spatial and temporal scales.

The nitrogen cycle is emerging as an important focus of environmental science. It has become clear that human actions are significantly disturbing the stocks and flows of nitrogen. It is not yet possible, however, to assess accurately the significance for human beings of these disturbances of nitrogen cycling, either in absolute terms or relative to the disturbance of the global carbon cycle. The disturbance of the carbon cycle is probably leading to global warming, and the disturbance of the nitrogen cycle is probably leading to global fertilization. At an earlier time of anthropocentricity and ecological innocence, one might have supposed that warmer is better and more fertile is better too. Ecology teaches otherwise. The Earth's ecosystems have evolved and adapted to past changes in environmental conditions, and will continue to evolve and adapt. But the changes being created by

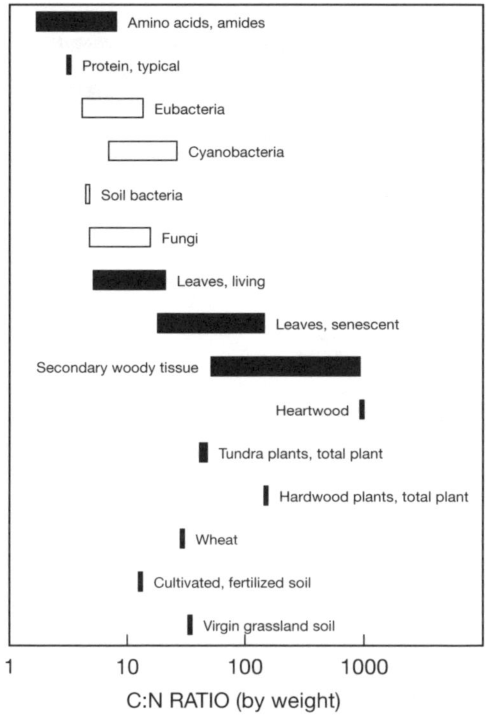

Ratio of carbon to nitrogen in various terrestrial materials. These ratios are often used to estimate global stocks and flows of nitrogen from the somewhat better known values for carbon. Horizontal bars indicate ranges of estimates; vertical bars indicate single-value estimates for which ranges were not available. Plant selection (breeding, species choice) can alter the C:N ratios for the vegetation in an ecosystem, affecting total carbon storage per unit of available nitrogen. C:N ratios serve as useful constraints and indicators in the development of coupled models of C and N cycles. (Data from refs. 7, 8, 19 and 24.) **Figure 7**

humans today may be too rapid for ecosystem adaptation to outpace ecosystem damage on time scales relevant to humans. As we come to understand anthropogenic disruptions of the biosphere more deeply, we may find the nitrogen cycle sharing center stage as an agent of long-term environmental change.

Nitrogen cycles are scarcely part of current public discourse. The global carbon cycle, by contrast, is better known to the public because of its association with the greenhouse effect and deforestation. Fossil fuel use has increasingly been the subject of innovations in technology and policy designed to reduce environmental impacts on the global carbon cycle. We may be well advised to seek analogous innovations for fertilizer use if we are to succeed in the grand task of conducting the human enterprise within environmental constraints.

* * *

The ideas in this article were developed in close collaboration with John Harte, John Holdren, F. Stuart Chapin and Mary Firestone (members of Kinzig's dissertation committee) and with Robert Ayres and William Schlesinger (coauthors of a recent book chapter with Socolow). We also acknowledge the specific assistance of Tom Bowman, William Chameides, Fred Dryer, Jonathan Dushoff, Harte, Aslam Khalil, Hiram Levy, Pamela Matson, Lori Perliski, Vern Ruttan and Schlesinger.

References

1. Intergovernmental Panel on Climate Change, World Meteorological Organization and the United Nations Environment Programme, *Climate Change: The IPCC Scientific Assessment*, J. T. Houghton, G. J. Jenkins, J. J. Ephraums, eds., Cambridge U. P., Cambridge, England (1990).
2. T. Rosswall, in *Some Perspectives of the Major Biogeochemical Cycles*, G. E. Likens, ed., Wiley, New York (1981), p. 25.
3. M. G. Lawrence, W. L. Chameides, P. S. Kasibhatla, H. Levy II, W. Moxim, "Lightning and Atmospheric Chemistry: The Rate of Atmospheric NO Production," to appear in *Handbook of Atmospheric Electrodynamics*, vol. 1, CRC Press, Boca Raton, Fla.
4. W. H. Schlesinger, *Biogeochemistry: An Analysis of Global Change*, Academic, San Diego (1991).
5. R. Ayres, W. Schlesinger, R. Socolow, in *Industrial Ecology and Global Change*, R. Socolow, C. Andrews, F. Berkhout, V. Thomas, eds., Cambridge U. P., Cambridge, England (1994), p. 121.
6. R. Söderlund, T. Rosswall, in *The Natural Environment and the Biogeochemical Cycles*, O. Hutzinger, ed., Springer-Verlag, New York (1982), p. 61.
7. J. Sprent, *The Ecology of the Nitrogen Cycle*, Cambridge U. P., Cambridge, England (1987).
8. E. A. Paul, F. E. Clark, *Soil Microbiology and Biochemistry*, Academic, San Diego (1989).
9. J. Harte, A. P. Kinzig, Am. Naturalist **141**, 829 (1993). A. P. Kinzig, "Mutualism and Competition Between Plants and Microorganisms: Implications for Nitrogen Allocation in Terrestrial Ecosystems," PhD dissertation, U. of Calif., Berkeley (1994).
10. P. M. Vitousek, W. A. Reiners, BioScience **25**, 361 (1975). P. M. Vitousek, L. R. Walker, in *Colonization, Succession, and Stability*, A. J. Gray, M. J. Crawley, P. J. Edwards, eds., Blackwell Scientific, Oxford (1986), p. 207.
11. J. D. Aber, J. M. Melillo, *Terrestrial Ecosystems*, Saunders, Philadelphia (1991). P. M. Vitousek, P. A. Matson, in *Biogeochemistry of Global Change*, R. S. Oremland, ed., Chapman and Hall, New York (1991), p. 193.
12. Natl. Res. Council, *Soil and Water Quality: An Agenda for Agriculture*, Natl. Acad. P., Washington, D. C. (1993).
13. K. M. Constant, W. F. Sheldrick, *World Nitrogen Survey*, World Bank, Washington, D. C. (1992).
14. W. H. Schlesinger, A. E. Hartley, Biogeochemistry **15**, 191 (1992).
15. H. W. Paerl, Can. J. Fisheries Aquatic Sci. **50**, 2254 (1993).
16. J. P. D. Abbatt, M. J. Molina, Annu. Rev. Energy Environment **18**, 1 (1993).
17. G. Brasseur, S. Solomon, *Aeronomy of the Middle Atmosphere: Chemistry and Physics of the Stratosphere and Mesosphere*, Reidel, Dordrecht, The Netherlands (1984).
18. P. A. Steudler, R. D. Bowden, J. M. Melillo, J. D. Aber, Nature **341**, 314 (1989).
19. E. B. Rastetter, M. G. Ryan, G. R. Shaver, J. M. Melillo, K. J. Nadelhoffer, J. E. Hobbie, J. D. Aber, Tree Physiol. **9**, 101 (1991).
20. D. W. Schindler, S. E. Bayley, Global Biogeochem. Cycles **7**, 717 (1993).
21. C. R. Carroll, J. H. Vandermeer, P. M. Rosset, *Agroecology*, McGraw–Hill, New York (1990).
22. US Dept. of Agriculture, *Agricultural Statistics 1989*, US Govt. Printing Office, Washington, D. C. (1989).
23. See, for instance, D. A. Jaffe, in *Global Biogeochemical Cycles*, S. S. Butcher, R. J. Charlson, G. H. Orians, G. V. Wolfe, eds., Academic, San Diego (1992), p. 263.
24. B. N. Richards, *The Microbiology of Terrestrial Ecosystems*, Longman Scientific and Technical, Harlow, England (1987). ■

[12]

The value of the world's ecosystem services and natural capital

Robert Costanza*†, Ralph d'Arge‡, Rudolf de Groot§, Stephen Farber||, Monica Grasso†, Bruce Hannon¶, Karin Limburg#°, Shahid Naeem**, Robert V. O'Neill††, Jose Paruelo‡‡, Robert G. Raskin§§, Paul Sutton|||| & Marjan van den Belt¶¶

**Center for Environmental and Estuarine Studies, Zoology Department, and †Institute for Ecological Economics, University of Maryland, Box 38, Solomons, Maryland 20688, USA.*

‡Economics Department (emeritus), University of Wyoming, Laramie, Wyoming 82070, USA.

§Center for Environment and Climate Studies, Wageningen Agricultrural University, PO Box 9101, 6700 HB Wageningengen, The Netherlands

||Graduate School of Public and International Affairs, University of Pittsburgh, Pittsburgh, Pennsylvania 15260, USA

¶Geography Department and NCSA, University of Illinois, Urbana, Illinois 61801, USA

#Institute of Ecosystem Studies, Millbrook, New York, USA

***Department of Ecology, Evolution and Behavior, University of Minnesota, St Paul, Minnesota 55108, USA*

††Environmental Sciences Division, Oak Ridge National Laboratory, Oak Ridge, Tennessee 37831, USA

‡‡Department of Ecology, Faculty of Agronomy, University of Buenos Aires, Av. San Martin 4453, 1417 Buenos Aires, Argentina

§§Jet Propulsion Laboratory, Pasedena, California 91109, USA

||||National Center for Geographic Information and Analysis, Department of Geography, University of California at Santa Barbara, Santa Barbara, California 93106, USA

¶¶Ecological Economics Research and Applications Inc., PO Box 1589, Solomons, Maryland 20688, USA

The services of ecological systems and the natural capital stocks that produce them are critical to the functioning of the Earth's life-support system. They contribute to human welfare, both directly and indirectly, and therefore represent part of the total economic value of the planet. We have estimated the current economic value of 17 cosystem services for 16 biomes, based on published studies and a few original calculations. For the entire biosphere, the value (most of which is outside the market) is estimated to be in the range of US$16–54 trillion ($10^{12}$) per year, with an average of US$33 trillion per year. Because of the nature of the uncertainties, this must be considered a minimum estimate. Global gross national product total is around US$18 trillion per year.

Because ecosystem services are not fully 'captured' in commercial markets or adequately quantified in terms comparable with economic services and manufactured capital, they are often given too little weight in policy decisions. This neglect may ultimately compromise the sustainability of humans in the biosphere. The economies of the Earth would grind to a halt without the services of ecological life-support systems, so in one sense their total value to the economy is infinite. However, it can be instructive to estimate the 'incremental' or 'marginal' value of ecosystem services (the estimated rate of change of value compared with changes in ecosystem services from their current levels). There have been many studies in the past few decades aimed at estimating the value of a wide variety of ecosystem services. We have gathered together this large (but scattered) amount of information and present it here in a form useful for ecologists, economists, policy makers and the general public. From this synthesis, we have estimated values for ecosystem services per unit area by biome, and then multiplied by the total area of each biome and summed over all services and biomes.

Although we acknowledge that there are many conceptual and empirical problems inherent in producing such an estimate, we think this exercise is essential in order to: (1) make the range of potential values of the services of ecosystems more apparent; (2) establish at least a first approximation of the relative magnitude of global ecosystem services; (3) set up a framework for their further analysis; (4) point out those areas most in need of additional research; and (5) stimulate additional research and debate. Most of the problems and uncertainties we encountered indicate that our estimate represents a minimum value, which would probably increase: (1) with additional effort in studying and valuing a broader range of ecosystem services; (2) with the incorporation of more realistic representations of ecosystem dynamics and interdependence; and (3) as ecosystem services become more stressed and 'scarce' in the future.

Ecosystem functions and ecosystem services

Ecosystem functions refer variously to the habitat, biological or system properties or processes of ecosystems. Ecosystem goods (such as food) and services (such as waste assimilation) represent the benefits human populations derive, directly or indirectly, from ecosystem functions. For simplicity, we will refer to ecosystem goods and services together as ecosystem services. A large number of functions and services can be identified[1–4]. Reference 5 provides a recent, detailed compendium on describing, measuring and valuing ecosystem services. For the purposes of this analysis we grouped ecosystem services into 17 major categories. These groups are listed in Table 1. We included only renewable ecosystem services, excluding non-renewable fuels and minerals and the atmosphere. Note that ecosystem services and functions do not necessarily show a one-to-one correspondence. In some cases a single ecosystem service is the product of two or more ecosystem functions whereas in other cases a single ecosystem function contributes to two or more ecosystem services. It is also important to emphasize the interdependent nature of many ecosystem functions. For example, some of the net primary production in an ecosystem ends up as food, the consumption of which generates respiratory products necessary for primary production. Even though these functions and services are interdependent, in many cases they can be added because they represent 'joint products' of the ecosystem, which support human

° Present address: Department of Systems Ecology, University of Stockholm, S-106 91 Stockholm, Sweden.

welfare. To the extent possible, we have attempted to distinguish joint and 'addable' products from products that would represent 'double counting' (because they represent different aspects of the same service) if they were added. It is also important to recognize that a minimum level of ecosystem 'infrastructure' is necessary in order to allow production of the range of services shown in Table 1. Several authors have stressed the importance of this 'infrastructure' of the ecosystem itself as a contributor to its total value[6,7]. This component of the value is not included in the current analysis.

Natural capital and ecosystem services

In general, capital is considered to be a stock of materials or information that exists at a point in time. Each form of capital stock generates, either autonomously or in conjunction with services from other capital stocks, a flow of services that may be used to transform materials, or the spatial configuration of materials, to enhance the welfare of humans. The human use of this flow of services may or may not leave the original capital stock intact. Capital stock takes different identifiable forms, most notably in physical forms including natural capital, such as trees, minerals, ecosystems, the atmosphere and so on; manufactured capital, such as machines and buildings; and the human capital of physical bodies. In addition, capital stocks can take intangible forms, especially as information such as that stored in computers and in individual human brains, as well as that stored in species and ecosystems.

Ecosystem services consist of flows of materials, energy, and information from natural capital stocks which combine with manufactured and human capital services to produce human welfare. Although it is possible to imagine generating human welfare without natural capital and ecosystem services in artificial 'space colonies', this possibility is too remote and unlikely to be of

Table 1 Ecosystem services and functions used in this study

Number	Ecosystem service*	Ecosystem functions	Examples
1	Gas regulation	Regulation of atmospheric chemical composition.	CO_2/O_2 balance, O_3 for UVB protection, and SO_x levels.
2	Climate regulation	Regulation of global temperature, precipitation, and other biologically mediated climatic processes at global or local levels.	Greenhouse gas regulation, DMS production affecting cloud formation.
3	Disturbance regulation	Capacitance, damping and integrity of ecosystem response to environmental fluctuations.	Storm protection, flood control, drought recovery and other aspects of habitat response to environmental variability mainly controlled by vegetation structure.
4	Water regulation	Regulation of hydrological flows.	Provisioning of water for agricultural (such as irrigation) or industrial (such as milling) processes or transportation.
5	Water supply	Storage and retention of water.	Provisioning of water by watersheds, reservoirs and aquifers.
6	Erosion control and sediment retention	Retention of soil within an ecosystem.	Prevention of loss of soil by wind, runoff, or other removal processes, storage of stilt in lakes and wetlands.
7	Soil formation	Soil formation processes.	Weathering of rock and the accumulation of organic material.
8	Nutrient cycling	Storage, internal cycling, processing and acquisition of nutrients.	Nitrogen fixation, N, P and other elemental or nutrient cycles.
9	Waste treatment	Recovery of mobile nutrients and removal or breakdown of excess or xenic nutrients and compounds.	Waste treatment, pollution control, detoxification.
10	Pollination	Movement of floral gametes.	Provisioning of pollinators for the reproduction of plant populations.
11	Biological control	Trophic-dynamic regulations of populations.	Keystone predator control of prey species, reduction of herbivory by top predators.
12	Refugia	Habitat for resident and transient populations.	Nurseries, habitat for migratory species, regional habitats for locally harvested species, or overwintering grounds.
13	Food production	That portion of gross primary production extractable as food.	Production of fish, game, crops, nuts, fruits by hunting, gathering, subsistence farming or fishing.
14	Raw materials	That portion of gross primary production extractable as raw materials.	The production of lumber, fuel or fodder.
15	Genetic resources	Sources of unique biological materials and products.	Medicine, products for materials science, genes for resistance to plant pathogens and crop pests, ornamental species (pets and horticultural varieties of plants).
16	Recreation	Providing opportunities for recreational activities.	Eco-tourism, sport fishing, and other outdoor recreational activities.
17	Cultural	Providing opportunities for non-commercial uses.	Aesthetic, artistic, educational, spiritual, and/or scientific values of ecosystems.

* We include ecosystem 'goods' along with ecosystem services.

much current interest. In fact, one additional way to think about the value of ecosystem services is to determine what it would cost to replicate them in a technologically produced, artificial biosphere. Experience with manned space missions and with Biosphere II in Arizona indicates that this is an exceedingly complex and expensive proposition. Biosphere I (the Earth) is a very efficient, least-cost provider of human life-support services.

Thus we can consider the general class of natural capital as essential to human welfare. Zero natural capital implies zero human welfare because it is not feasible to substitute, in total, purely 'non-natural' capital for natural capital. Manufactured and human capital require natural capital for their construction[7]. Therefore, it is not very meaningful to ask the total value of natural capital to human welfare, nor to ask the value of massive, particular forms of natural capital. It is trivial to ask what is the value of the atmosphere to humankind, or what is the value of rocks and soil infrastructure as support systems. Their value is infinite in total.

However, it is meaningful to ask how changes in the quantity or quality of various types of natural capital and ecosystem services may have an impact on human welfare. Such changes include both small changes at large scales and large changes at small scales. For example, changing the gaseous composition of the global atmosphere by a small amount may have large-scale climate change effects that will affect the viability and welfare of global human populations. Large changes at small scales include, for example, dramatically changing local forest composition. These changes may dramatically alter terrestrial and aquatic ecosystems, having an impact on the benefits and costs of local human activities. In general, changes in particular forms of natural capital and ecosystem services will alter the costs or benefits of maintaining human welfare.

Valuation of ecosystem services

The issue of valuation is inseparable from the choices and decisions we have to make about ecological systems[6,8]. Some argue that valuation of ecosystems is either impossible or unwise, that we cannot place a value on such 'intangibles' as human life, environmental aesthetics, or long-term ecological benefits. But, in fact, we do so every day. When we set construction standards for highways, bridges and the like, we value human life (acknowledged or not) because spending more money on construction would save lives. Another frequent argument is that we should protect ecosystems for purely moral or aesthetic reasons, and we do not need valuations of ecosystems for this purpose. But there are equally compelling moral arguments that may be in direct conflict with the moral argument to protect ecosystems; for example, the moral argument that no one should go hungry. Moral arguments translate the valuation and decision problem into a different set of dimensions and a different language of discourse[6]; one that, in our view, makes the problem of valuation and choice more difficult and less explicit. But moral and economic arguments are certainly not mutually exclusive. Both discussions can and should go on in parallel.

So, although ecosystem valuation is certainly difficult and fraught with uncertainties, one choice we do not have is whether or not to do it. Rather, the decisions we make as a society about ecosystems imply valuations (although not necessarily expressed in monetary terms). We can choose to make these valuations explicit or not; we can do them with an explicit acknowledgement of the huge uncertainties involved or not; but as long as we are forced to make choices, we are going through the process of valuation.

The exercise of valuing the services of natural capital 'at the margin' consists of determining the differences that relatively small changes in these services make to human welfare. Changes in quality or quantity of ecosystem services have value insofar as they either change the benefits associated with human activities or change the costs of those activities. These changes in benefits and costs either have an impact on human welfare through established markets or through non-market activities. For example, coral reefs provide habitats for fish. One aspect of their value is to increase and concentrate fish stocks. One effect of changes in coral reef quality or quantity would be discernible in commercial fisheries markets, or in recreational fisheries. But other aspects of the value of coral reefs, such as recreational diving and biodiversity conservation, do not show up completely in markets. Forests provide timber materials through well established markets, but the associated habitat values of forests are also felt through unmarketed recreational activities. The chains of effects from ecosystem services to human welfare can range from extremely simple to exceedingly complex. Forests provide timber, but also hold soils and moisture, and create microclimates, all of which contribute to human welfare in complex, and generally non-marketed ways.

Valuation methods

Various methods have been used to estimate both the market and non-market components of the value of ecosystem services[9-16]. In this analysis, we synthesized previous studies based on a wide variety of methods, noting the limitations and assumptions underlying each.

Many of the valuation techniques used in the studies covered in our synthesis are based, either directly or indirectly, on attempts to estimate the 'willingness-to-pay' of individuals for ecosystem services. For example, if ecological services provided a $50 increment to the timber productivity of a forest, then the beneficiaries of this service should be willing to pay up to $50 for it. In addition to timber production, if the forest offered non-marketed, aesthetic, existence, and conservation values of $70, those receiving this non-market benefit should be willing to pay up to $70 for it. The total

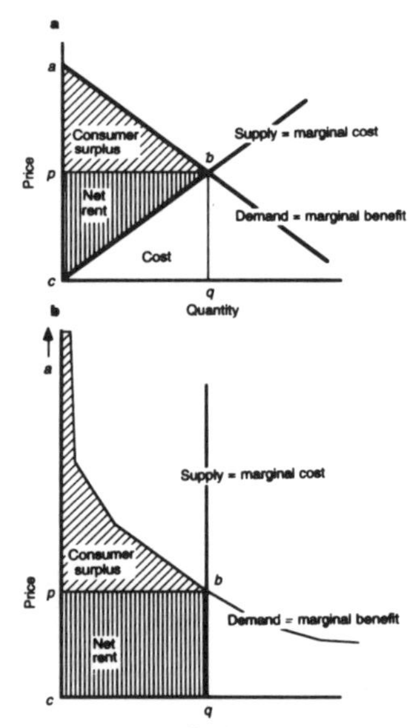

Figure 1 Supply and demand curves, showing the definitions of cost, net rent and consumer surplus for normal goods (**a**) and some essential ecosystem services (**b**). See text for further explanation.

Table 2 Summary of average global value of annual ecosystem services

Biome	Area (ha × 10^6)	1 Gas regulation	2 Climate regulation	3 Disturbance regulation	4 Water regulation	5 Water supply	6 Erosion control	7 Soil formation	8 Nutrient cycling	9 Waste treatment	10 Pollination	11 Biological control	12 Habitat/ refugia	13 Food production	14 Raw materials	15 Genetic resources	16 Recreation	17 Cultural	Total value per ha ($ ha^{-1} yr^{-1})	Total global flow value ($ yr^{-1} × 10^9)
		Ecosystem serivces (1994 US$ ha^{-1} yr^{-1})																		
Marine	36,302																		577	20,949
Open ocean	33,200	38							118			5		15	0			76	252	8,381
Coastal	3,102			88					3,677			38	8	93	4		82	62	4,052	12,568
Estuaries	180			567					21,100			78	131	521	25		381	29	22,832	4,110
Seagrass/ algae beds	200								19,002						2				19,004	3,801
Coral reefs	62			2,750						58		5	7	220	27		3,008	1	6,075	375
Shelf	2,660								1,431			39		68	2			70	1,610	4,283
Terrestrial	15,323																		804	12,319
Forest	4,855		141	2	2	3	96	10	361	87		2		43	138	16	66	2	969	4,706
Tropical	1,900		223	5	6	8	245	10	922	87				32	315	41	112	2	2,007	3,813
Temperate/boreal	2,955		88		0			10		87		4		50	25		36	2	302	894
Grass/rangelands	3,898	7	0		3		29	1		87	25	23		67		0	2		232	906
Wetlands	330	133		4,539	15	3,800				4,177			304	256	106		574	881	14,785	4,879
Tidal marsh/ mangroves	165			1,839						6,696			169	466	162		658		9,990	1,648
Swamps/ floodplains	165	265		7,240	30	7,600				1,659			439	47	49		491	1,761	19,580	3,231
Lakes/rivers	200				5,445	2,117				665				41			230		8,498	1,700
Desert	1,925																			
Tundra	743																			
Ice/rock	1,640																			
Cropland	1,400										14	24		54					92	128
Urban	332																			
Total	51,625	1,341	684	1,779	1,115	1,692	576	53	17,075	2,277	117	417	124	1,386	721	79	815	3,015		33,268

Numbers in the body of the table are in $ ha^{-1} yr^{-1}. Row and column totals are in $ yr^{-1} × 10^9, column totals are the sum of the products of the per ha services in the table and the area of each biome, not the sum of the per ha services themselves. Shaded cells indicate services that do not occur or are known to be negligible. Open cells indicate lack of available information.

value of ecological services would be \$120, but the contribution to the money economy of ecological services would be \$50, the amount that actually passes through markets. In this study we have tried to estimate the total value of ecological services, regardless of whether they are currently marketed.

Figure 1 shows some of these concepts diagrammatically. Figure 1a shows conventional supply (marginal cost) and demand (marginal benefit) curves for a typical marketed good or service. The value that would show up in gross national product (GNP) is the market price p times the quantity q, or the area $pbqc$. There are three other relevant areas represented on the diagram, however. The cost of production is the area under the supply curve, cbq. The 'producer surplus' or 'net rent' for a resource is the area between the market price and the supply curve, pbc. The 'consumer surplus' or the amount of welfare the consumer receives over and above the price paid in the market is the area between the demand curve and the market price, abp. The total economic value of the resource is the sum of the producer and consumer surplus (excluding the cost of production), or the area abc on the diagram. Note that total economic value can be greater or less than the price times quantity estimates used in GNP.

Figure 1a refers to a human-made, substitutable good. Many ecosystem services are only substitutable up to a point, and their demand curves probably look more like Fig. 1b. Here the demand approaches infinity as the quantity available approaches zero (or some minimum necessary level of services), and the consumer surplus (as well as the total economic value) approaches infinity. Demand curves for ecosystem services are very difficult, if not impossible, to estimate in practice. In addition, to the extent that ecosystem services cannot be increased or decreased by actions of the economic system, their supply curves are more nearly vertical, as shown in Fig. 1b.

In this study we estimated the value per unit area of each ecosystem service for each ecosystem type. To estimate this 'unit value' we used (in order of preference) either: (1) the sum of consumer and producer surplus; or (2) the net rent (or producer surplus); or (3) price times quantity as a proxy for the economic value of the service, assuming that the demand curve for ecosystem services looks more like Fig. 1b than Fig. 1a, and that therefore the area $pbqc$ is a conservative underestimate of the area abc. We then multiplied the unit values times the surface area of each ecosystem to arrive at global totals.

Ecosystem values, markets and GNP

As we have noted, the value of many types of natural capital and ecosystem services may not be easily traceable through well functioning markets, or may not show up in markets at all. For example, the aesthetic enhancement of a forest may alter recreational expenditures at that site, but this change in expenditure bears no necessary relation to the value of the enhancement. Recreationists may value the improvement at \$100, but transfer only \$20 in spending from other recreational areas to the improved site. Enhanced wetlands quality may improve waste treatment, saving on potential treatment costs. For example, tertiary treatment by wetlands may save \$100 in alternative treatment. Existing treatment may cost only \$30. The treatment cost savings does not show up in any market. There is very little relation between the value of services and observable current spending behaviour in many cases.

There is also no necessary relationship between the valuation of natural capital service flows, even on the margin, and aggregate spending, or GNP, in the economy. This is true even if all capital service flows had an impact on well functioning markets. A large part of the contributions to human welfare by ecosystem services are of a purely public goods nature. They accrue directly to humans without passing through the money economy at all. In many cases people are not even aware of them. Examples include clean air and water, soil formation, climate regulation, waste treatment, aesthetic values and good health, as mentioned above.

Global land use and land cover

In order to estimate the total value of ecosystem services, we needed estimates of the total global extent of the ecosystems themselves. We devised an aggregated classification scheme with 16 primary categories as shown in Table 2 to represent current global land use. The major division is between marine and terrestrial systems. Marine was further subdivided into open ocean and coastal, which itself includes estuaries, seagrass/algae beds, coral reefs, and shelf systems. Terrestrial systems were broken down into two types of forest (tropical and temperate/boreal), grasslands/rangelands, wetlands, lakes/rivers, desert, tundra, ice/rock, cropland, and urban. Primary

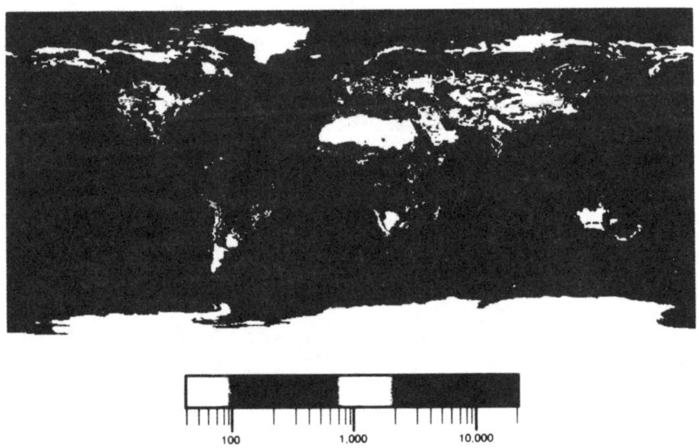

Figure 2 Global map of the value of ecosystem services. See Supplementary Information and Table 2 for details.

data were from ref. 17 as summarized in ref. 4 with additional information from a number of sources[18–22]. We also used data from ref. 23, as a cross-check on the terrestrial estimates and refs 24 and 25 as a check on the marine estimates. The 32 landcover types of ref. 17 were recategorized for Table 2 and Fig. 2. The major assumptions were: (1) chaparral and steppe were considered rangeland and combined with grasslands; and (2) a variety of tropical forest and woodland types were combined into 'tropical forests'.

Synthesis

We conducted a thorough literature review and synthesized the information, along with a few original calculations, during a one-week intensive workshop at the new National Center for Ecological Analysis and Synthesis (NCEAS) at the University of California at Santa Barbara. Supplementary Information lists the primary results for each ecosystem service and biome. Supplementary Information includes all the estimates we could identify from the literature (from over 100 studies), their valuation methods, location and stated value. We converted each estimate into 1994 US$ ha^{-1} yr^{-1} using the USA consumer price index and other conversion factors as needed. These are listed in the notes to the Supplementary Information. For some estimates we also converted the service estimate into US$ equivalents using the ratio of purchasing power GNP per capita for the country of origin to that of the USA. This was intended to adjust for income effects. Where possible the estimates are stated as a range, based on the high and low values found in the literature, and an average value, with annotated comments as to methods and assumptions. We also included in the Supplementary Information some estimates from the literature on 'total ecosystem value', mainly using energy analysis techniques[10]. We did not include these estimates in any of the totals or averages given below, but only for comparison with the totals from the other techniques. Interestingly, these different methods showed fairly close agreement in the final results.

Each biome and each ecosystem service had its special considerations. Detailed notes explaining each biome and each entry in Supplementary Information are given in notes following the table. More detailed descriptions of some of the ecosystems, their services, and general valuation issues can be found in ref. 5. Below we briefly discuss some general considerations that apply across the board.

Sources of error, limitations and caveats

Our attempt to estimate the total current economic value of ecosystem services is limited for a number of reasons, including:

(1) Although we have attempted to include as much as possible, our estimate leaves out many categories of services, which have not yet been adequately studied for many ecosystems. In addition, we could identify no valuation studies for some major biomes (desert, tundra, ice/rock, and cropland). As more and better information becomes available we expect the total estimated value to increase.

(2) Current prices, which form the basis (either directly or indirectly) of many of the valuation estimates, are distorted for a number of reasons, including the fact that they exclude the value of ecosystem services, household labour and the informal economy. In addition to this, there are differences between total value, consumer surplus, net rent (or producer surplus) and $p \times q$, all of which are used to estimate unit values (see Fig. 1).

(3) In many cases the values are based on the current willingness-to-pay of individuals for ecosystem services, even though these individuals may be ill-informed and their preferences may not adequately incorporate social fairness, ecological sustainability and other important goals[16]. In other words, if we actually lived in a world that was ecologically sustainable, socially fair and where everyone had perfect knowledge of their connection to ecosystem services, both market prices and surveys of willingness-to-pay would yield very different results than they currently do, and the value of ecosystem services would probably increase.

(4) In calculating the current value, we generally assumed that the demand and supply curves look something like Fig. 1a. In reality, supply curves for many ecosystem services are more nearly inelastic vertical lines, and the demand curves probably look more like Fig. 1b, approaching infinity as quantity goes to zero. Thus the consumer and producer surplus and thereby the total value of ecosystem services would also approach infinity.

(5) The valuation approach taken here assumes that there are no sharp thresholds, discontinuities or irreversibilities in the ecosystem response functions. This is almost certainly not the case. Therefore this valuation yields an underestimate of the total value.

(6) Extrapolation from point estimates to global totals introduces error. In general, we estimated unit area values for the ecosystem services (in $ ha^{-1} yr^{-1}) and then multiplied by the total area of each biome. This can only be considered a crude first approximation and can introduce errors depending on the type of ecosystem service and its spatial heterogeneity.

(7) To avoid double counting, a general equilibrium framework that could directly incorporate the interdependence between ecosystem functions and services would be preferred to the partial equilibrium framework used in this study (see below).

(8) Values for individual ecosystem functions should be based on sustainable use levels, taking account of both the carrying capacity for individual functions (such as food-production or waste recycling) and the combined effect of simultaneous use of more functions. Ecosystems should be able to provide all the functions listed in Table 1 simultaneously and indefinitely. This is certainly not the case for some current ecosystem services because of overuse at existing prices.

(9) We have not incorporated the 'infrastructure' value of ecosystems, as noted above, leading to an underestimation of the total value.

(10) Inter-country comparisons of valuation are affected by income differences. We attempted to address this in some cases using the relative purchasing power GNP per capita of the country relative to the USA, but this is a very crude way to make the correction.

(11) In general, we have used annual flow values and have avoided many of the difficult issues involved with discounting future flow values to arrive at a net present value of the capital stock. But a few estimates in the literature were stated as stock values, and it was necessary to assume a discount rate (we used 5%) in order to convert them into annual flows.

(12) Our estimate is based on a static 'snapshot' of what is, in fact, a complex, dynamic system. We have assumed a static and 'partial equilibrium' model in the sense that the value of each service is derived independently and added. This ignores the complex interdependencies between the services. The estimate could also change drastically as the system moved through critical non-linearities or thresholds. Although it is possible to build 'general equilibrium' models in which the value of all ecosystem services are derived simultaneously with all other values, and to build dynamic models that can incorporate non-linearities and thresholds, these models have rarely been attempted at the scale we are discussing. They represent the next logical step in deriving better estimates of the value of ecosystem services.

We have tried to expose these various sources of uncertainty wherever possible in Supplementary Information and its supporting notes, and state the range of relevant values. In spite of the limitations noted above, we believe it is very useful to synthesize existing valuation estimates, if only to determine a crude, initial magnitude. In general, because of the nature of the limitations noted, we expect our current estimate to represent a minimum value for ecosystem services.

Total global value of ecosystem services

Table 2 is a summary of the results of our synthesis. It lists each of the major biomes along with their current estimated global surface area, the average (on a per hectare basis) of the estimated values of the 17 ecosystem services we have identified from Supplementary Information, and the total value of ecosystem services by biome, by service type and for the entire biosphere.

We estimated that at the current margin, ecosystems provide at least US$33 trillion dollars worth of services annually. The majority of the value of services we could identify is currently outside the market system, in services such as gas regulation (US$1.3 trillion yr^{-1}), disturbance regulation (US$1.8 trillion yr^{-1}), waste treatment (US$2.3 trillion yr^{-1}) and nutrient cycling (US$17 trillion yr^{-1}). About 63% of the estimated value is contributed by marine systems (US$20.9 trillion yr^{-1}). Most of this comes from coastal systems (US$10.6 trillion yr^{-1}). About 38% of the estimated value comes from terrestrial systems, mainly from forests (US$4.7 trillion yr^{-1}) and wetlands (US$4.9 trillion yr^{-1}).

We estimated a range of values whenever possible for each entry in Supplementary Information. Table 2 reports only the average values. Had we used the low end of the range in Supplementary Information, the global total would have been around US$19 trillion. If we eliminate nutrient cycling, which is the largest single service, estimated at US$17 trillion, the total annual value would be around US$16 trillion. Had we used the high end for all estimates, along with estimating the value of desert, tundra and ice/rock as the average value of rangelands, the estimate would be around US$54 trillion. So the total range of annual values we estimated were from US$16–$54 trillion. This is not a huge range, but other sources of uncertainty listed above are much more critical. It is important to emphasize, however, that despite the many uncertainties included in this estimate, it is almost certainly an underestimate for several reasons, as listed above.

There have been very few previous attempts to estimate the total global value of ecosystem services with which to compare these results. We identified two, based on completely different methods and assumptions, both from each other and from the methods used in this study. They thus provide an interesting check.

One was an early attempt at a static general equilibrium input–output model of the globe, including both ecological and economic processes and commodities[26,27]. This model divided the globe in to 9 commodities or product groups and 9 processes, two of which were 'economic' (urban and agriculture) and 7 of which were 'ecological', including both terrestrial and marine systems. Data were from about 1970. Although this was a very aggregated breakdown and the data was of only moderate quality, the model produced a set of 'shadow prices' and 'shadow values' for all the flows between processes, as well as the net outputs from the system, which could be used to derive an estimate of the total value of ecosystem services. The input–output format is far superior to the partial equilibrium format we used in this study for differentiating gross from net flows and avoiding double counting. The results yielded a total value of the net output of the 7 global ecosystem processes equal to the equivalent of US$9.4 trillion in 1972. Converted to 1994 US$ this is about $34 trillion, surprisingly close to our current average estimate. This estimate broke down into US$11.9 trillion (or 35%) from terrestrial ecosystem processes and US$22.1 trillion (or 65%) from marine processes, also very close to our current estimate. World GNP in 1970 was about $14.3 trillion (in 1994 US$), indicating a ratio of total ecosystem services to GNP of about 2.4 to 1. The current estimate has a corresponding ratio of 1.8 to 1.

A more recent study[28] estimated a 'maximum sustainable surplus' value of ecosystem services by considering ecosystem services as one input to an aggregate global production function along with labour and manufactured capital. Their estimates ranged from US$3.4 to US$17.6 trillion yr^{-1}, depending on various assumptions. This approach assumed that the total value of ecosystem services is limited to that which has an impact on marketed value, either directly or indirectly, and thus cannot exceed the total world GNP of about US$18 trillion. But, as we have pointed out, only a fraction of ecosystem services affects private goods traded in existing markets, which would be included in measures such as GNP. This is a subset of the services we estimated, so we would expect this estimate to undervalue total ecosystem services.

The results of both of these studies indicate, however, that our current estimate is at least in approximately the same range. As we have noted, there are many limitations to both the current and these two previous studies. They are all only static snapshots of a biosphere that is a complex, dynamic system. The obvious next steps include building regional and global models of the linked ecological economic system aimed at a better understanding of both the complex dynamics of physical/biological processes and the value of these processes to human well-being[29,30]. But we do not have to wait for the results of these models to draw the following conclusions.

Discussion

What this study makes abundantly clear is that ecosystem services provide an important portion of the total contribution to human welfare on this planet. We must begin to give the natural capital stock that produces these services adequate weight in the decision-making process, otherwise current and continued future human welfare may drastically suffer. We estimate in this study that the annual value of these services is US$16–54 trillion, with an estimated average of US$33 trillion. The real value is almost certainly much larger, even at the current margin. US$33 trillion is 1.8 times the current global GNP. One way to look at this comparison is that if one were to try to replace the services of ecosystems at the current margin, one would need to increase global GNP by at least US$33 trillion, partly to cover services already captured in existing GNP and partly to cover services that are not currently captured in GNP. This impossible task would lead to no increase in welfare because we would only be replacing existing services, and it ignores the fact that many ecosystem services are literally irreplaceable.

If ecosystem services were actually paid for, in terms of their value contribution to the global economy, the global price system would be very different from what it is today. The price of commodities using ecosystem services directly or indirectly would be much greater. The structure of factor payments, including wages, interest rates and profits would change dramatically. World GNP would be very different in both magnitude and composition if it adequately incorporated the value of ecosystem services. One practical use of the estimates we have developed is to help modify systems of national accounting to better reflect the value of ecosystem services and natural capital. Initial attempts to do this paint a very different picture of our current level of economic welfare than conventional GNP, some indicating a levelling of welfare since about 1970 while GNP has continued to increase[31–33]. A second important use of these estimates is for project appraisal, where ecosystem services lost must be weighed against the benefits of a specific project[8]. Because ecosystem services are largely outside the market and uncertain, they are too often ignored or undervalued, leading to the error of constructing projects whose social costs far outweight their benefits.

As natural capital and ecosystem services become more stressed and more 'scarce' in the future, we can only expect their value to increase. If significant, irreversible thresholds are passed for irreplaceable ecosystem services, their value may quickly jump to infinity. Given the huge uncertainties involved, we may never have a very precise estimate of the value of ecosystem services. Nevertheless, even the crude initial estimate we have been able to assemble is a useful starting point (we stress again that it is only a starting point). It demonstrates the need for much additional research and it also indicates the specific areas that are most in need of additional study. It also highlights the relative importance of ecosystem services and the potential impact on our welfare of continuing to squander them. □

Received 23 January: accepted 2 April 1997.

1. de Groot, R. S. Environmental functions as a unifying concept for ecology and economics. *Environmentalist* **7,** 105–109 (1987).
2. Turner, R. K. *Economics, Growth and Sustainable Environments* (eds Collard, D. *et al.*) (Macmillan, London, 1988).
3. Turner, R. K. Economics of wetland management. *Ambio* **20,** 59–63 (1991).
4. de Groot, R. S. *Functions of Nature: Evaluation of Nature in Environmental Planning, Management, and Decision Making* (Wolters-Noordhoff, Groningen, 1992).
5. Daily, G. (ed.) *Nature's Services: Societal Dependence on Natural Ecosystems* (Island, Washington DC, 1997).
6. Turner, R. K. & Pearce, D. in *Economics and Ecology: New Frontiers and Sustainable Development* (ed. Barbier, E. D.) 177–194 (Chapman and Hall, London, 1993).
7. Costanza, R. & Daly, H. E. Natural capital and sustainable development. *Conserv. Biol.* **6,** 37–46 (1992).
8. Bingham, G. *et al.* Issues in ecosystem valuation: improving information for decision making. *Ecol. Econ.* **14,** 73–90 (1995).
9. Mitchell, R. C. & Carson, R. T. *Using Surveys to Value Public Goods: the Contingent Valuation Method* (Resources for the Future, Washington DC, 1989).
10. Costanza, R., Farber, S. C. & Maxwell, J. Valuation and management of wetlands ecosystems. *Ecol. Econ.* **1,** 335–361 (1989).
11. Dixon, J. A. & Sherman, P. B. *Economics of Protected Areas* (Island, Washington DC, 1990).
12. Barde, J.-P. & Pearce, D. W. *Valuing the Environment: Six Case Studies* (Earthscan, London, 1991).
13. Aylward, B. A. & Barbier, E. B. Valuing environmental functions in developing countries. *Biodiv. Cons.* **1,** 34 (1992).
14. Pearce, D. *Economic Values and the Natural World* (Earthscan, London, 1993).
15. Goulder, L. H. & Kennedy, D. in *Nature's Services: Societal Dependence on Natural Ecosystems* (ed. Daily, G.) 23–48 (Island, Washington DC, 1997).
16. Costanza, R. & Folke, C. in *Nature's Services: Societal Dependence on Natural Ecosystems* (ed. Daily, G.) 49–70 (Island, Washington DC, 1997).
17. Matthews, E. Global vegetation and land-use: new high-resolution data bases for climate studies. *J. Clim. Appl. Meteorol.* **22,** 474–487 (1983).
18. Deevey, E. S. Mineral cycles. *Sci. Am.* **223,** 148–158 (1970).
19. Ehrlich, R., Ehrlich, A. H. & Holdren, J. P. *Ecoscience: Population, Resources, Environment* (W.H. Freeman, San Francisco, 1977).
20. Ryther, J. H. Photosynthesis and fish production in the sea. *Science* **166,** 72–76 (1969).
21. United Nations Environmental Programme *First Assessment Report, Intergovernmental Panel on Climate Change* (United Nations, New York, 1990).
22. Whittaker, R. H. & Likens, G. E. in *Primary Production of the Biosphere* (eds Lieth, H. & Whittaker, R. H.) 305–328 (Springer, New York, 1975).
23. Bailey, R. G. *Ecosystem Geography* (Springer, New York, 1996).
24. Houde, E. D. & Rutherford, E. S. Recent trends in estuarine fisheries: predictions of fish production and yield. *Estuaries* **16,** 161–176 (1993).
25. Pauly, D. & Christensen, V. Primary production required to sustain global fisheries. *Nature* **374,** 255–257 (1995).
26. Costanza, R. & Neil, C. in *Energy and Ecological Modeling* (eds Mitsch, W. J., Bosserman, R. W. & Klopatek, J. M.) 745–755 (Elsevier, New York, 1981).
27. Costanza, R. & Hannon, B. M. in *Network Analysis of Marine Ecosystems: Methods and Applications* (eds Wulff, F., Field, J. G. & Mann, K. H.) 90–115 (Springer, Heidelberg, 1989).
28. Alexander, A., List, J., Margolis, M. & d'Arge, R. Alternative methods of valuing global ecosystem services. *Ecol. Econ.* (submitted).
29. Costanza, R., Wainger, L., Folke, C. & Mäler, K.-G. Modeling complex ecological economic systems: toward an evolutionary, dynamic understanding of people and nature. *BioScience* **43,** 545–555 (1993).
30. Bockstael, N. *et al.* Ecological economic modeling and valuation of ecosystems. *Ecol. Econ.* **14,** 143–159 (1995).
31. Daly, H. E. & Cobb, J. *For the Common Good: Redirecting the Economy Towards Community, the Environment, and a Sustainable Future* (Beacon, Boston, 1989).
32. Cobb, C. & Cobb, J. *The Green National Product: a Proposed Index of Sustainable Economic Welfare* (Univ. Press of America, New York, 1994).
33. Max-Neef, M. Economic growth and quality of life: a threshold hypothesis. *Ecol. Econ.* **15,** 115–118 (1995).

Acknowledgements. S. Carpenter was instrumental in encouraging the project. M. Grasso did the initial identification and collection of literature sources. We thank S. Carpenter, G. Daily, H. Daly, A. M. Freeman, N. Myers, C. Perrings, D. Pimentel, S. Pimm and S. Postel for helpful comments on earlier drafts. This project was sponsored by the National Center for Ecological Analysis and Synthesis (NCEAS), an NSF-funded Center at the University of California at Santa Barbara. The authors met during the week of June 17–21, 1996 to do the major parts of the synthesis activities. The idea for the study emerged at a meeting of the Pew Scholars in New Hampshire in October 1995.

Correspondence and requests for materials should be addressed to R.C. (e-mail: costza@cbl.cees.edu).

Supplementary Information is on www.nature.com. Paper copies are avaialble from Mary Sheehan at the London editorial office of *Nature*.

Part II
Physical Economics Perspective: Materials Energy, Entropy, Mass Balance

[13]

WILLIAM D. NORDHAUS

Yale University

The Allocation of Energy Resources

ABUNDANT ENERGY AT LOW COST is fundamental to a highly industrialized economy like the United States. The American way of life is hard to visualize without commuters, television, overheated houses, aluminum cans, and jet setters; yet it is equally difficult to conceive of substitutes for these hallmarks of American society if cheap energy were no longer available.

Given the dependence on energy, there has been perennial anxiety over the adequacy of the nation's resources for meeting its apparently insatiable appetite for energy. More recently, the concern for adequacy of energy has been embedded in a more general pessimism about the viability of economic growth on a finite world.[1] This new and pessimistic view about economic growth holds that growth is limited by a finite amount of essential, depletable natural resources. In the process of consuming finite resources, the world standard of living descends inexorably toward that of Neanderthal man.

Note: The research underlying this paper was supported by the National Science Foundation. I am grateful for helpful comments by Gary Haller, Tjalling C. Koopmans, Alan S. Manne, and members of the Brookings panel. In addition, Paul Krugman provided research assistance extraordinary. Remaining errors are, of course, my responsibility.

1. See Paul R. Ehrlich and Anne H. Ehrlich, *Population, Resources, Environment: Issues in Human Ecology* (W. H. Freeman, 1970); Jay W. Forrester, *World Dynamics* (Wright-Allen, 1971); Donella H. Meadows and others, *The Limits to Growth: A Report for the Club of Rome's Project on the Predicament of Mankind* (Universe Books, 1972); "A Blueprint for Survival," *The Ecologist*, Vol. 2 (January 1972), pp. 1–22.

If such a scenario is plausible, the crunch probably will be felt first in energy resources. For one thing, energy is an essential input in many processes, required by the laws of physics. Although more efficient use might save energy, it simply is not possible to heat houses, produce aluminum, run transportation systems, or generate electricity without it. With the exception of food, no other single commodity is so essential. Second, energy resources are nonrenewable. Aside from hydro, no significant use is now made of renewable sources (such as solar, geothermal, or gravitational energy, or wood) in the United States. Third, energy resources cannot be recycled. Once coal or petroleum is burned its energy dissipates beyond economical recapture. Finally, the enormous current and prospective energy consumption raises difficult environmental problems. No currently used fuel is completely clean, economical, and abundant.

Energy resources, then, are a likely test case for examining resource scarcity. In what follows, I will first explore the use of markets to allocate scarce resources over time, and then turn explicitly to an empirical estimate of the efficient allocation of energy resources.

The Role of Markets for Resources

In the United States, the prices of appropriable resources have for the most part been determined by market forces.[2] Why has public policy accepted a laissez-faire approach to resource pricing?

The intellectual basis for allowing market determination of prices lies in the theory of general economic equilibrium. This theory assumes that there are consumers with initial resources and given preferences, and producers operating with well-defined technical relations. The theory can embrace many time periods and uncertainty about the exact demand or supply conditions; but it assumes convex production and preference sets, and that markets exist for all goods, services, and contingencies. This means that there must be futures markets for, say, petroleum and coal in the year 2000; and there must be insurance markets for such contingencies as the failure of breeder processes to become economically viable. Also, all the costs and benefits of a particular process of production must be internalized to the

2. An appropriable resource is one for which all rewards or penalties from services or uses accrue to the owner.

decision maker. Under the above conditions a market system will have a general equilibrium of prices and quantities. There is nothing in such a market system that will ensure an equitable distribution of consumption over space or time. But the equilibrium will be efficient in the sense that there is no way of improving the lot of one consumer without worsening the lot of another. Expressed differently, the prices are appropriate indicators of social scarcity given the preferences and initial endowments of the society.

The application of the results of market equilibrium analysis to depletable natural resources is straightforward. In considering these I distinguish between *extraction costs*, the vector $z(t)$, or the marginal cost per unit of output excluding rents and royalties; and *royalties*, the vector $y(t)$,[3] which are a reflection of the presumed scarcity of a particular resource. The t refers to the time period.

Consider a world of certainty and a time horizon of T years.[4] There are $R(t)$ units of the resource remaining at any point of time, and extraction costs are zero up to the resource limit. If alternative assets yield a rate of return, $r(t)$, the equilibrium condition for some owners to hold and others to sell the resource is equality between the rate of capital gains on the resource and the interest rate:

$$\frac{\Delta y(t)}{y(t)} = r(t)\,, \tag{1}$$

where Δy is the change in y. Thus the resource price rises exponentially at the interest rate.

There is a family of solutions to equation (1), each having different levels of y. The unique solution depends on the terminal condition that all resources are used up at the end of the last period (T):

$$y(t),\ \text{such that } R(T) = 0\,. \tag{2}$$

There generally will be a unique set of prices satisfying (1) and (2).

3. I use "royalty" to denote the difference between price and marginal extraction cost, a concept similar to rents on land. Royalties have many other meanings in resource economics.

4. The terminal point can be a sticky issue. If there exists what I later call a "backstop technology" (roughly, a substitute process with infinite resource base), then T is the time at which transition to it is completed; if resources are finite and essential, and no backstop technology exists, T is the time of extinction. For an analysis of the second case, see Tjalling C. Koopmans, "Some Observations on 'Optimal' Economic Growth and Exhaustible Resources," Cowles Foundation Discussion Paper 356 (Cowles Foundation, March 1973; processed).

More interesting is the case where extraction costs are positive. Price, $p(t)$, is the sum of current extraction cost and royalty:

$$p(t) = z(t) + y(t)\,. \tag{3}$$

In efficient allocations, resources are extracted when their present value is maximized. The present value of the profit from selling a unit of the resource at time t when extraction cost is z_0 is $[p(t) - z_0]\exp(-rt)$, and this is maximized if t is chosen so that $\Delta p(t) = r[p(t) - z_0]$, or when

$$\frac{\Delta p}{p} = r\left(\frac{p - z_0}{p}\right) = \frac{ry}{p}\,. \tag{4}$$

If production of a resource with cost z_0 occurs for all t, then (4) must hold for all t. Moreover, for all periods when sales occur, $(p - z_0)\exp(-rt)$ is constant—this being a solution to (4). During periods when sales are occurring, $\Delta y/y = r$, so (1) is satisfied. Since $y = p - z_0$ is the royalty at time of extraction, the new condition for recovering a resource is that the expected rate of increase of the price of the resource be less than or equal to the interest rate times the share of royalties in the total resource price. This rate will always be less than the interest rate. If extraction costs are constant, royalties will again satisfy the exponential relation shown in (1); since the share of the royalty increases to unity, the resource price will accelerate toward a rate of increase of r.

In the analysis of programs developed below, the path of prices can be made considerably more explicit. Today's energy technology is highly dependent on resources that are very cheap to extract but relatively scarce when viewed over a very long time horizon. In this technology royalties to scarce low-cost resources may be relatively important in today's price. Over the next century or so, many low-cost energy resources will be largely depleted, leaving more abundant but also more expensive resources. Ultimately, if and when the transition is completed to an economy based on plentiful nuclear resources (either through breeder or fusion reactors), the economic importance of *scarcity* of resources will disappear, and capital and labor costs alone will determine prices. This ultimate technology—resting on a very abundant resource base—is the "backstop technology" and is crucial to the allocation of scarce energy resources.

An oversimplified example will show how the backstop technology enters. Consider two processes for generating electricity. One process uses one unit of petroleum per unit of output; petroleum resources are finite in supply (R recoverable units) and free to extract. The second process uses nuclear fuel, which is superabundant and free, and K dollars' worth of capital per unit of output. Assume that the rate of interest is r, and that demand is inelastic, with D units of electricity demanded per year. Clearly, the petroleum process will be used first, and the switch to the nuclear process (the backstop technology) will take place R/D years out.

Prices are easy to calculate along an efficient path. At the switch point $\hat{T}$, the price of electricity, p, is given by the cost of the backstop technology,

$$p(\hat{T}) = (r + \delta)K,$$

where δ is the depreciation rate and K is the capital requirement of the backstop technology. This implies that the price and therefore the royalty on petroleum at the switch point are also $p(\hat{T})$. But then the price and the royalty on petroleum along the efficient path from now to $\hat{T}$ are

$$(5)\quad y(t) = p(t) = p(\hat{T}) \exp[-r(\hat{T} - t)] = (r + \delta)K \exp[-r(\hat{T} - t)].$$

The royalty on the scarce resource is simply the switch price, $p(\hat{T})$, discounted back to the present.

There are three important elements in determining current royalty, $y(0)$: the cost of the backstop technology, the interest rate, and the switch date. The capital requirement of the backstop technology enters linearly. The interest rate enters positively as a linear function of the cost of the backstop technology and negatively as a discount factor applying to the switch date. For fixed $\hat{T}$, a higher interest rate lowers $y(0)$ if $\hat{T}(r + \delta) < 1$ and raises $y(0)$ if $\hat{T}(r + \delta) > 1$.

The switch date $\hat{T}$ enters in an exponential way in much the same way as the interest rate. Recall that $\hat{T} = R/D$. If the amount of resources doubles, or demand halves, the switch date is doubled. This lowers the royalty by a factor of $\exp(-rR/D)$. Such an effect is very powerful: if current royalty is one-tenth of the price of the backstop technology, then a change in supply or demand that doubles the switch date means that current royalty will fall to one-hundredth of the price of the backstop technology.

One further feature of efficient paths is worth mentioning. In realistic cases, there are extraction costs for the early technology, say $\bar{z}$. So the path of prices is given by

$$p(t) = \bar{z} + [(r + \delta)K - \bar{z}] \exp[-r(\hat{T} - t)].$$

The feature of this path is that the run-up of prices can be a big surprise. The royalty component may be small for a long time, then suddenly dominate. Thus, for a 10 percent interest rate, if royalty is 5 percent of price, prices in successive decades rise at rates of 8 percent, 19 percent, 41 percent, 76 percent, 112 percent, and up to a maximum of 159 percent. A high interest rate keeps royalties low initially, but when they rise they really take off. This acceleration can wreak havoc for producers who are locked into capital goods and have extrapolative expectations.

This simplified example illustrates the technique for estimating efficient energy prices in the next section. As equation (5) indicates, if the price of the backstop technology is low, if the switch date is far off, or if the interest rate is high, then the royalty on energy resources is relatively low. Conversely, if these conditions are reversed, the royalty on energy resources is high. The question explored in the next section is whether the current market-determined royalty on energy resources is close to that associated with an efficient path for the allocation of energy resources. Unfortunately, the calculation required to get the answer is extremely complex. Since there are many sources and grades of energy resources, many uses, and many demand categories, each with peculiar specifications, calculation of the optimal path and the switch points for different resources is cumbersome.

A further extension of the model considers the functioning of resource markets under uncertainty. The complete general equilibrium analysis discussed above requires not only a complete set of futures markets, but also a complete set of insurance markets or contingent commodities markets. The insurance markets would span all economically relevant events, such as whether and when breeder reactors become available; what the future course of population growth will be; what happens in Mideast politics; whether very large oil reserves in Alaska will be recoverable; whether environmental policy will be tough or lenient. In each case, a contingent commodity would be sold: for example, one barrel of crude in January 1984, if the trans-Alaska pipeline is not built. It can be shown that the price system is ex ante efficient as long as a complete set of futures markets exists.

An important difference between the model and the real world is that a full set of futures and insurance markets is not available. Although long-term contracts are often made—these being rough substitutes for futures markets—they are relatively rare; and I am unaware of any insurance markets for selling resources contingent on the state of the world.[5]

5. The sale, option, or leasehold arrangements currently employed for oil-, gas-, and ore-bearing lands are not good substitutes for futures markets (1) because they represent sale of rights to recover to producers (for example, coal companies) rather than

What are the possible consequences of the absence of a complete set of futures and insurance markets? Three problems might be serious.

The first complication concerns the appropriate discount rate to use in resource decisions. Recall from equation (1) that the equilibrium condition for the resource market is that prices rise at the same rate as the interest rate. In an uncertain world, this means that prices rise at the discount rate appropriate for the owners of the resource. It has often been argued that the discount rate used in the United States for private investment is generally too high. The sources of the positive differential between private and social discount rates are risk and taxes.

In the absence of perfect risk and insurance markets, the owners of resources will bear risks associated with price volatility, the incursion of competing resources into established markets, the advent of new technologies, and so forth. Many economists have argued that such risks are not always social risks because they can be widely spread over the population, or more precisely, because the effects of risk on output are very small relative to average income.[6] If this is the case, then the private discount rate will be above the appropriate social discount rate. A second force that points in the same direction is the existence of taxes on capital income. An investment in resources that has an annual rate of capital gain of r_b has an after-tax rate of return $r_a = r_b(1 - \tau)$ (τ is the tax on capital income). Again, if the pretax interest rate is the social discount rate, then the presence of capital taxes will make the equilibrium rate of capital gain too high. Tax rates vary greatly, of course, from virtually zero for petroleum extraction to more than 50 percent for capital gains on land or for royalties accruing to corporations; but the existence of general capital taxes causes a distortion in the required rate of return.

The distortion of the interest rate is a particularly serious problem in natural resources. As can be seen in equations (1) to (4), too high an interest rate casts a long shadow over the future. When royalties dominate the price, too high an interest rate tilts the entire price path in favor of the present, with the result that resources are consumed too quickly.

---

sales to ultimate consumers (say, utility companies); and (2) because they are spot markets or very short-run futures markets. In the cosmic framework of the ultimate exhaustion of fossil fuels or energy sources or phosphorus, these transactions cover a very short span.

6. Prominent in this discussion has been the work of Kenneth J. Arrow. See his *Essays in the Theory of Risk-Bearing* (Markham, 1971), especially Chap. 11.

The second possible complication involves myopic decisions. In the present context, "myopia" means that the planning horizons of economic agents are relatively short.

Myopic decisions are a possibility because of the absence of futures markets. Recall that expected capital gains play a crucial role in resource decisions. Take two polar strategies toward investment in resources—an in-and-out strategy (buying on the speculation of short-term gain) or a buy-and-hold strategy (buying with the intention of selling the ultimate commodity under the soil rather than the land). Up to now the buy-and-hold strategy was assumed to dominate. Say that investors have an in-and-out strategy, buying titles to resources with an eye to capital gains rather than to selling the resources directly, and, for simplicity, that all investors plan to sell after one period. This leads to the equilibrium condition outlined in equation (1) or (4) above: the market is in equilibrium when investors expect that the (one-period) capital gain on the asset is equal to the (risk-corrected) one-period interest rate. The most important point is that this condition has no unique solution; rather, the path depends on expectations. Indeed, a path with zero royalties will satisfy the myopic equilibrium—for a while.

The missing element in this system is the "global planner" (or speculator) who calculates the quantities demanded along a price path to see whether they are consistent with overall availabilities (the calculation implicit in equation 2). The reason why pricing of resources might be myopic is that very few planners have the ability, or perhaps even the desire, to check consistency for several decades.

The third complication that may arise in resource markets is price instability. Recall that the price includes a recovery cost and a royalty. At a given moment, the recovery cost is well determined by the current technology and factor costs, but the royalty is not; rather, it depends on future conditions of supply and demand. More precisely, the royalty calculation derives from knowledge about the paths of output and input prices over the indefinite future.

Given the structure of markets, the royalty may exhibit considerable instability. The instability results from the role price changes play in affecting both demand and expectations in spot markets. A supplier of a resource can observe only current and past resource prices. In line with earlier discussion, it seems reasonable to assume that, without information from futures markets, the quantity supplied will be positively related to the difference between the interest rate and the expected rate of price increase of the resource. Thus if prices are expected to rise more rapidly than the interest rate, suppliers will cut down on production in anticipation of future capital gains. The response of consumers, however, accelerates this process. As production is cut back, prices rise more rapidly, and as they do, producers expect even more rapid rises, leading to further production decreases. And so on. The same sort of instability can be seen for price reductions as well.

The result of this interaction of supply and demand response is that the royalty component of resource price may behave in an unstable manner. In the case described above, royalty determination in spot markets leads either to a dynamically unstable or to an inefficient path.

It can be argued on the basis of this discussion that the market mechanism now existing in the United States is an unreliable means of pricing and allocating exhaustible appropriable natural resources. The absence of futures and insurance markets rules out the theorems usually drawn from general equilibrium theory. The possibility of instability in resource markets is a further complication. The most serious potential obstacle to the market's (or anybody's) finding the correct price lies in determining the appropriate royalty, or scarcity rent, on exhaustible resources. The difference between exhaustible resources and other commodities is that the share of royalties is relatively small for these others. Unfortunately, an estimate of whether current usage is too fast or too slow cannot be made a priori; it can emerge only from a carefully constructed econometric and engineering model of the economy.

Efficient Allocation of Energy Resources

The upshot of the foregoing discussion is that markets, in their current form, may be unreliable ways to allocate exhaustible resources. Energy resources are perhaps the best example of this problem. They are essential, and perforce their consumption stretches over a very long period. They, and their products, have no futures markets. Because the availability of resources and future technologies are uncertain, so is the path of energy consumption. As if the basic economic problems were not sufficiently complicated, recently there has been considerable political interference, expressed through the operation of the petroleum market by exporting countries, the regulation of prices of natural gas and petroleum products by the U.S. government, and the setting of environmental standards by virtually everybody. It takes an act of faith to believe that "the market" can somehow see the proper allocation through this tangle of complexity, uncertainty, and politics.

What alternative exists to relying on an incomplete set of markets? In the face of this uncertainty, two basic approaches are open to finding the appropriate allocation.[7] The most appropriate strategy—called "indicative planning" by Meade—is to match up future supplies and demands in a simulated market: the government summons all the citizens to a meeting in Yankee Stadium, gives them a set of questionnaires (listing trial prices), and asks for their demands and supplies. The process continues until it yields a balanced set of supplies and demands.[8]

While this approach is intriguing, all the current and future citizens can hardly gather in Yankee Stadium. Meade considers a second technique—econometric forecasting:

> If one knows all the technological and behavioural relationships in the economy—that is to say, what outputs can be produced with what inputs and how citizens as entrepreneurs, workers, savers, consumers, etc., re-act to changes in prices, costs, incomes, interest rates, etc.—if one knows the starting point of the economy, that is to say the existing capital equipment and so on—and finally if one knows how the future exogenous variables will behave, and we are in fact assuming that there are no environmental uncertainties—then theoretically . . . one should be able to forecast the future course of all prices and quantities in all markets.[9]

What I propose to do is to find a middle ground between the two approaches. On the one hand, there are considerable data on the supply side of the energy market—enough to allow an intelligent guess as to how a profit-maximizing firm would behave when faced with a set of current and future prices for energy resources and products. On the other hand, there is considerable uncertainty about the demand relations. To determine the efficient allocation of energy resources then requires calculating how a complete set of spot and futures markets would allocate resources, given the best data available at the present time.

7. This line of thought was suggested by J. E. Meade in *The Theory of Indicative Planning* (Manchester University Press, 1970), and further spelled out in his *The Controlled Economy* (State University of New York Press, 1972).

8. See Meade, *Theory of Indicative Planning*, Chap. 4.

9. Ibid., p. 12.

The general procedure calls for calculating the allocation of different resources over time that minimizes the cost of meeting the demands, using the data on the time path of demand for various energy products, on the finite stocks of energy resources, and on the costs of alternative processes for transforming them into energy products.

MODELING DETAILS

Given the immense trade in energy products it is impossible to treat the allocation of energy resources as a problem for the United States alone. The purview is therefore the non-Communist world, which is broken into five regions: the United States, Western Europe, Japan, the Persian Gulf and North Africa, and the rest of the world (ROW). The problem is, furthermore, of a very long-run nature; in principle—as described below—it has an infinite time horizon. In practice, the analysis covers two hundred years, embedded in a longer-run model as described below.

At present there are four important energy resources: petroleum, coal, natural gas, and uranium-235. In the future oil shale and uranium-238 will probably join this list. On the demand side, the model specifies five demand categories: electricity, industrial heat, residential heat, and two transport categories.

Once this framework is determined, there remains little room for maneuver. The problem already strains reasonable computational budgets. In particular, the analysis cannot take elastic demand curves or exchange rate adjustments into account, although these are important problems.

DEMAND

The five demand categories for energy products have been broken down as follows:

1. Electricity.

2. Industrial nonelectric energy uses: process and space heating, rail, subway, and ship. There are virtually no constraints on how these demands are satisfied (except imposed environmental standards) and thus the cost of fuel is the major consideration.

3. Residential nonelectric uses: space and other heating. This use is

limited to fuels that are relatively easy and clean to process in small quantities.

4. Substitutable transportation: truck, bus, and 75 percent of automobile use. These uses are those for which fuels other than gasoline, particularly electricity, can be substituted relatively easily.

5. Nonsubstitutable transportation: air traffic and 25 percent of automotive. These are the long-distance uses that electricity cannot easily satisfy.

Table 1 shows an estimate of the energy consumption pattern by fuel and

Table 1. Per Capita Energy Consumption in the United States, by Type of Fuel and Demand, 1929 and 1968

Millions of Btu[a]

		Demand category					
			Heat		Transportation[b]		
Fuel	Year	Electricity	Industrial	Residential	Substitutable	Nonsubstitutable	Total demand
Petroleum	1929	0.5	12.7	17.8	12.1	4.0	47.2
	1968	7.8	22.2	27.6	45.9	19.7	123.3
Natural gas	1929	1.0	3.5	3.1	0.0	0.0	7.5
	1968	17.6	41.0	33.0	0.0	0.0	91.6
Nuclear	1929	c	c	c	c	c	c
	1968	0.7	0	0	0	0	0.7
Coal	1929	9.7	76.5	36.2	0	0	122.4
	1968	36.3	24.9	2.5	0	0	63.7
Hydro	1929	6.6	0	0	0	0	6.6
	1968	13.5	0	0	0	0	13.5
Total	1929	17.8	92.7	57.1	12.1	4.0	183.8
	1968	75.9	88.0	63.1	45.9	19.7	292.6

Sources: William H. Lyon and D. S. Colby, "Production, Consumption, and Use of Fuels and Electric Energy in the United States in 1929, 1939, and 1947," Report of Investigations 4805 (U.S. Bureau of Mines, 1951; processed); and Associated Universities, Inc., *Reference Energy Systems and Resource Data for Use in the Assessment of Energy Technologies* (AU, April 1972). Figures may not add to totals because of rounding.

a. The common measure of energy used in the present paper is the British thermal unit (Btu). Conversion factors are 5.8 million Btu per barrel of petroleum; 25.8 million Btu per ton of bituminous coal; 1,000 But per cubic foot of natural gas; and 3,413 Btu per kilowatt-hour.

b. Substitutable uses are those for which fuels other than gasoline can be substituted relatively easily: trucks and buses and 75 percent of automotive uses; nonsubstitutable uses are for aviation and 25 percent of automotive uses.

c. Technology unknown in 1929.

by demand category for 1929 and 1968.[10] A glance indicates the pervasive changes over the past forty years. They have taken two forms: (1) within specific categories of consumption there have been dramatic shifts in fuel composition (for example, the wholesale shift from coal to natural gas and petroleum, and the expanded use of both for industrial heat and electricity generation); and (2) the differential growth of demand categories (such as the shift from rail and water to automobile and jet transport, the rapid growth of electricity, and the decline in industrial heating uses).[11]

In face of the impossibility of detailing every conceivable kind of change, it seems reasonable in projecting future trends to focus on these two general kinds of structural change.

Unfortunately, it has not been possible to introduce the more general substitution by ultimate consumers of other products for energy products. Thus the demand for each specific category is a fixed path over time. To have it otherwise involves nonlinearities that are computationally very difficult. Most studies of the final demand categories indicate that price elasticities are quite low, lying mostly between zero and unity. On the other hand, studies that introduce interfuel substitution not surprisingly indicate considerably higher cross-elasticities of demand. The framework set out here is more pessimistic in one sense since it assumes no responsiveness of final demand to price. But it is undoubtedly more optimistic in its assumption that fuels are perfectly substitutable for meeting demand requirements. It would be desirable to test the sensitivity of the results to some price elasticity of final demand.

SUPPLY

The supply side is much more complicated, but fortunately the data are much better. Supply for each product involves three stages:

10. The common measure of energy used in this paper is the British thermal unit (Btu). Conversion factors are 5.8 million Btu per barrel of petroleum; 25.8 million Btu per ton of bituminous coal; 1,000 Btu per cubic foot of natural gas; and 3,413 Btu per kilowatt-hour.

11. Two nonenergy uses for energy inputs that have been omitted are needs for petrochemical feedstocks and for direct conversion of hydrocarbons into food. I have assumed that 10 percent of all natural and synthetic oil is reserved for nonenergy uses, which seems adequate for at least 150 years. After that period, inputs for these uses must come from even lower-grade resources, which are very abundant but relatively expensive.

Extraction is the crucial part of the model, for the resource availabilities are the basic constraining factor. Table 2 shows the distribution of each of the five major kinds of producible resources, by region. Given resource availabilities, the capital and labor requirements of resource extraction are taken into account.

Transportation costs in the model are based on estimated capital and current costs and distances involved.

The last step is *processing* the fuels to meet final demand. This is the most difficult question in that it involves some processes that are yet unproven. In Table 3, which is a schematic presentation of the different technologies, two things are especially interesting. First, it reveals a considerable range of interfuel competition, especially for electricity and heat. Second, it designates processes according to their state of technical development, starting with A, the current technology whose properties are relatively well known, and progressing to D, which is speculative and whose properties are little known.

Table 2. Recoverable Energy Resources, by Type of Fuel and Regions of the World, 1970[a]

Quadrillions (10^{15}) of Btu

Fuel	*United States*	*Western Europe*	*Persian Gulf and North Africa*	*Rest of the world*	*Total*
Fossil					
Petroleum					
Proven reserves	213	70	2,543	756	3,582
Unproven but recoverable	350	34	1,755	2,103	4,242
Coal	33,588	8,626	0	17,915	60,129
Shale oil	11,362	1,090	0	12,328	24,780
Natural gas	447	83	3,409	2,268	6,207
Total fossil	45,960	9,903	7,707	35,370	98,940
Nuclear					
U-235	...	...	...	...	1,504,100
U-238	...	...	...	...	206,970,000
Total nuclear	...	...	...	...	208,474,100
Total recoverable energy resources	...	...	...	...	208,573,040
Addendum: World energy consumption, 1965	...	...	...	...	154

Sources: Given in an appendix available upon request from author.

a. All fuels are calculated at their theoretical energy content. Nuclear fuels are not allocated by region. All quantities apply a conventional recovery rate to original resources in place.

Table 3. Technologies Used in Processing Energy Resources, by Type of Fuel and Demand

	Technology, by demand category[a]				
		Heat		*Transportation*[b]	
Fuel	*Electricity*	*Industrial*	*Residential*	*Substitutable*	*Nonsubstitutable*
Petroleum	Refine for use in standard oil-fired plant (A)	Refine for standard oil or gasoline power (A)			
Shale oil	Mine and retort shale (B) *and* Refine for use as petroleum (A)				
Coal	Use in standard coal-fired plant (A) *with:* Sulphur scrubbing (B) *or* Low-sulphur coal (A) *or* Gasified coal (B)		Gasification into either pipeline-quality high-Btu gas, or into low-Btu gas (B)	Production of synthetic crude (C) *and* Refine for use as gasoline or aviation fuel (A)	
Nuclear	Light-water reactor (A) *or* Breeder reactor (B)	Resistance heating or heat pump (A)		Medium-range electric automobile (C)	Hydrogen-fueled automobile and aircraft (D)
Natural gas	Use in standard gas-fired equipment (A)			Ruled out as uneconomical	

Sources: Developed by author. Underlying process data are given in an appendix available on request from author.

a. Technologies are designated according to their state of development: A = current widespread use; B = pilot plants operating currently; C = in development; D = speculative.

b. See Table 1, note *b*, for definitions.

Several potentially important technologies are not included. One is solar energy, indubitably an attractive resource but one whose current capital cost for, say, electricity generation is perhaps a hundred times that of conventional equipment.[12] Another is the production of alcohol from grain

12. Hoyt C. Hottel and Jack B. Howard, *New Energy Technology: Some Facts and Assessments* (MIT Press, 1971), p. 343.

crops, but this method would not become competitive until wheat, for example, fell to 50 cents a bushel (from the late 1973 price of $5). It seems a safe bet that no exotic new technologies will become dominant within the next decade.

Table 4 gives a rough idea of the resource costs *exclusive of royalties* of

Table 4. Cost of Intermediate Energy Products Exclusive of Royalties, by Source[a]

Energy source	*Cost in 1970 dollars (per million Btu)*	*Cost in 1970 dollars (per conventional unit)*[b]
CRUDE PETROLEUM[c]		
United States		
Category 1 Drilled reserves	0.05	0.29
Category 2 Undrilled reserves	0.42	2.41
Category 3 Undrilled reserves	0.52	3.02
Category 4 Undrilled reserves	0.81	4.70
Category 5 Undrilled reserves	1.15	6.17
Category 6 Undrilled reserves	3.35	19.45
Persian Gulf and North Africa		
Category 1 Drilled reserves	0.01	0.06
Category 2 Undrilled reserves	0.05	0.29
CRUDE OIL FROM SHALE		
United States		
25 gallons per ton of shale	0.96	5.58
10 gallons per ton of shale	2.00	11.59
COAL		
United States		
Liquefied	1.31	7.62
Gasified	1.19	1.19
Strip mined	0.29	6.07
Deep mined	0.47	9.69
ELECTRICITY GENERATION		
Petroleum	2.03	7.5 mills
Natural gas	1.62	5.5 mills
Coal	3.17	10.8 mills
Light water reactor	2.51	8.6 mills
Breeder reactor	3.12	10.6 mills
HYDROGEN (BY ELECTROLYSIS)	5.46	...

Sources: The underlying data for these estimates are in an appendix, available upon request from author.

a. All cost figures (except electricity) are at minehead or wellhead. Electricity costs are busbar. Costs include direct costs (capital and current costs), but exclude any shadow prices or royalties.

b. Per barrel for crude petroleum, crude oil from shale, and liquefied coal; per thousand cubic feet for gasified coal; per ton for strip-mined and deep-mined coal; per kilowatt-hour for electricity generation.

c. Petroleum was separated into six different cost categories for the United States and two categories for all other regions. Category 1, the lowest cost, is drilled reserves, while categories 2 through 6, the highest cost, are undrilled.

alternative processes for producing different fuels, all in 1970 prices. (With the exception of electricity, all prices are at the well or mine.) These are calculated at a 10 percent interest rate. The projected costs of foreign petroleum reserves are surprisingly low, on the order of 30 cents per barrel for undeveloped Mideast crude. On the other hand, the projected costs of synthetic fuels—such as shale oil, gasified coal, or liquefied coal—are all very high relative to both current prices and to the projected cost of natural fuels.

THE PROBLEM

The object of the problem is to determine the allocation of energy resources (over time, space, and different categories) that minimizes the discounted costs of meeting a set of final demands. In algebraic terms, the problem is to minimize

$$\sum_{(i,j,k,l,m)} c(i,j,k,l)x(i,j,k,l,m)(1+r)^{-m},$$

where

c = production cost
x = activity level
r = relevant interest rate

and, as subscripts,

i = country where resource is located
j = kind of resource
k = country demanding energy product
l = demand category
m = time period.

All activities are measured in terms of delivered thermal content of the final product. Thus $x(1,1,1,1,1)$ (which is explained in detail below) is U.S. electricity produced from U.S. petroleum during period 1 in delivered Btu.

The production cost of a given activity has three components:

$$c(i,j,k,l,m) = (1+r)^{-m}[ex(i,j) + tr(i,j,k) + pro(j,l)],$$

where

ex = extraction cost
tr = transport cost
pro = processing cost.

These in turn are

$$ex(i,j) = [r + \delta^e(i,j)]\, K^e(i,j) + L^e(i,j)$$
$$tr(i,j,k) = [r + \delta^t]K^t(i,j,k) + L^t(i,j,k)$$
$$pro(j,l) = \{[r + \delta^p]K^p(j,l) + L^p(j,l)\}/eff(j,l),$$

where

δ^i = depreciation rate, $i = ex, tr, pro$
K^i = investment requirement, $i = ex, tr, pro$
L^i = current inputs, $i = ex, tr, pro$
eff = thermal efficiency of process,

and the superscript e refers to the extraction sector, t to the transportation sector, and p to the processing sector.

The activities are

$x(i,j,k,l,m)$ = flow of resource j from area i to demand category l in country k and time period m,

where

$i = 1, \ldots, 4$ (the United States, Western Europe, Persian Gulf and North Africa, ROW)
$j = 1, \ldots, 17$ (six petroleum categories for the United States, two for other countries; two shale oil categories; stripping coal; deep coal; natural gas; four nuclear categories)
$k = 1, \ldots, 4$ (the United States, Western Europe, Japan, ROW)
$l = 1, \ldots, 5$ (electricity, industrial heat, residential heat, substitutable transport, nonsubstitutable transport)
$m = 1, 2, \ldots$ (1970, 1980, 1990, . . .).

The constraints are:

$$\text{Supply:} \quad \sum_{k,l,m} x(i,j,k,l,m)/eff(j,l) \leqq R(i,j)$$
$$\text{Demand:} \quad \sum_{i,j} x(i,j,k,l,m) \geqq D(k,l,m),$$

where $R(i,j)$ is resource availability of resource j in country i, and $D(k,l,m)$ is demand for product l in country k and period m.

THE REST OF THE ECONOMY

Although I am investigating only the energy sector, some thought must be given to the rest of the economy. This nonenergy sector is assumed to be unconstrained by resources. It produces nonenergy output from capital, labor, and energy according to a constant-returns-to-scale production function. To simplify matters I assume that the social saving is completely elastic with respect to the interest rate, so that the rate of return on capital is constant at rate *r*. In addition, whatever secular productivity increase occurs is assumed to be purely labor augmenting and proceeds at the same rate in both the energy and nonenergy sectors; so labor and other current inputs are always in efficiency units and the production function is unchanging. Thus, any increase in labor efficiency will raise output per worker accordingly.

The only other important simplification I make is to ignore the effect of energy prices on capital goods prices—which in turn affect energy prices. This is not serious, for the share of energy costs in the capital goods used in the energy sectors is quite small.[13] Only if energy prices rose by a factor of ten would this assumption need revision, but, as noted below, the projected rise in intermediate energy prices is much more modest than this.

In what follows, per capita income is assumed to grow as a result of labor-augmenting technological progress at 2 percent per annum for the United States, and at higher rates for other countries. As a result, the general price level in the United States will be falling 2 percent annually with respect to wage rates and per capita incomes. All calculations are presented in 1970 prices. Strictly speaking, this means that the price of capital goods or efficiency (productivity-adjusted) labor is considered to be a numeraire.

TERMINAL CONDITIONS AND DISCOUNT RATES

Because of the very long-run nature of the problem under consideration, particular attention must be paid to the terminal conditions. In principle, the planning period for essential exhaustible resources must cover the duration of man's habitation on the planet. It would be myopic, to say the least, to devise a rational plan for fifty years, only to find that consumption must be drastically reduced because the rest of the future had been ignored.

The concept that is relevant to this problem is the *backstop technology*, a set of processes that (1) is capable of meeting the demand requirements,

13. In 1963, the total direct and indirect energy-type inputs (energy mining, petroleum refining, and public utilities) were about 5 percent of the total for engines and turbines and for construction, mining, and oil field machinery.

and (2) has a virtually infinite resource base. The backstop technology may well be extremely expensive relative to current technology; nevertheless, if it exists, it assures that the planning problem at least has a feasible solution.

For example, consider a backstop technology for the automobile. With current output rates, available technology, and resources that are currently economical to recover, the resources for automobile transport will last perhaps seventy years. Resources for automobiles operating on electricity generated by breeder reactors will last approximately 100 million years. In some sense, the current stage of history is a transitory phase between dependence on cheap but scarce resources and dependence on more costly but abundant resources.

Thus the first question is whether the system is feasible over some indefinite period of time—say, a thousand years. If not, the problem of determining an efficient solution does not make any sense: in a programming framework, if the problem is infeasible prices are infinite.

Next, if the problem is feasible, the backstop technology is identified. Given the foregoing assumptions, the property of the model is that once transition to the backstop technology is reached, all prices will remain constant. An efficient program extending beyond $\hat{T}$, the time at which all energy is produced with the backstop technology, will have exactly the same solution in the transition phase, independent of the planning horizon. Once $\hat{T}$ is identified, the period beyond it can be ignored in the computations.

The existence of a backstop technology is relevant for the proper discount rate to use in calculations of the efficient program. In most situations this is 10 percent. It is supposed to be an index of the supply price for capital and of the opportunity cost of capital, not of the social rate of time preference.[14] There is sometimes confusion on this question, particularly in evaluating allocation of exhaustible resources. Recall from the discussion above that a high interest rate means low initial prices and high initial consumption. In a sense this pattern favors the present over the future,

14. The 10 percent figure used as the appropriate interest rate approximates the average pretax return on reproducible tangible capital, and as such is a reasonable estimate of the social productivity of investment. In 1968 the ratio of profit-type income (profits, interest, rents, and one-half of entrepreneurial income) to the replacement cost of private reproducible capital (all tangibles, including land, consumer durables, and institutional structures) was 10.6 percent.

but if the opportunity cost of investment is also high, investment in reproducible capital is a relatively more efficient way of increasing future consumption than holding sterile resources in the ground. Thus a high interest rate may encourage rapid depletion of petroleum and natural gas; but the resources saved by using these cheap resources can be put in the bank to grow at 10 percent annually, and then can be used to build coal liquefaction and gasification plants and breeder reactors in twenty or thirty years.

When no backstop technology exists, however, a high interest rate is definitely inappropriate. If no feasible solution exists—that is, if no backstop technology can be identified—the basic allocation has no solution. Strictly speaking, this leads to infinite prices for energy resources. It would then be very misleading to use the kind of analysis presented here since resource exhaustion implies extinction.

To summarize, the interest rate is an index of capital's productivity in an economy with an indefinitely feasible consumption plan. To use a lower rate to reflect the social rate of time preference is inappropriate unless there is evidence that the productivity of capital will be lower in the future.

The technique for calculating the efficient path grows naturally out of this discussion. In a program with four fifty-year periods, the backstop technology was reached in the fourth period. Thus, within the technological specification, an all electric-hydrogen basis for the linear programming model was reached in the fourth of these intervals, 2120–70. The expensive shale oil and the most expensive category of U.S. oil are saved for the period after 2070. The cheap shale oil and about 90 percent of the coal is saved for the period 2020–70. According to the efficient solution, during the next fifty years, 1970B2020, the economy will utilize a small fraction of the world coal resources, the low-cost U-235 to be used in conventional nuclear reactors, and all but the high-cost petroleum resources.

ENVIRONMENTAL CONSTRAINTS

The major shortcoming of the model outlined so far is the omission of environmental constraints. It has been argued, in fact, that environmental policies have played a major role in the current energy crisis by removing certain fuels (notably high-sulphur coal) from the resource base. Over the longer run, waste heat, carbon dioxide, and nuclear wastes may be constraints on overall energy consumption.

Stringent environmental constraints coupled with the inability to find technological solutions to them might very well mean drastically different results or even infeasibility of the basic problem. The approach followed in this paper is to specify a set of environmental standards that must be met by the various processes. By and large, these standards are at least as strict as existing laws. The following specific assumptions are made: (1) petroleum refineries can process crudes containing up to 2.5 percent sulphur, and produce clean gasoline and fuel oil with a sulphur range of 0.2 to 1.0 percent; (2) electricity generated with coal employs expensive equipment—either sulphur dioxide scrubbing or low-Btu gasification—which brings sulphur emissions down to current standards; (3) nuclear power production meets the current Atomic Energy Commission standards; and (4) the cost of all surface mining includes $5,000 per acre for reclamation. It should be emphasized that the estimated costs for meeting these standards are very high. For example, the reclamation for surface mining is probably ten times what is required by current law.

Perhaps either the costs are too low or the standards too lenient. But judging from the history of automobile emission standards, the problem of imposing standards is more a matter of time than of cost. The National Academy of Sciences estimates that the stratified charge engine can meet the 1975–76 standards at an annual average cost of about $70 per vehicle. This means a reduction in emissions of at least 90 percent compared with the 1970 level (and perhaps 95 percent for uncontrolled vehicles) for only 2 to 4 percent of the vehicle's total cost.[15] If these figures are at all representative of what can be done to improve environmental performance with sufficient money and time, then the provisions for environmental protection made here should be adequate. There are good economic reasons to expect that (with current technology) prices of energy resources will rise. On the other hand, there is no reason to rule out much cheaper long-run solutions to the brand new environmental constraints. After all, environmental resources have been free goods—and have been treated as such. The radical shift in relative prices, making environmental resources very costly goods, will promote technological change aimed at saving these resources, although this may take time.

15. "Report by the Committee on Motor Vehicle Emissions" (National Academy of Sciences, February 1973; processed), pp. 101, 116. For standards and uncontrolled performance, see Hottel and Howard, *New Energy Technology*, p. 297.

Results of the Basic Case

The basic case describes the efficient allocation of energy resources for two hundred years (five ten-year periods followed by two twenty-five-year periods and two fifty-year periods). It is calculated with an interest rate of 10 percent, and assumes that sufficient resources have been reserved to meet energy needs efficiently forever (see pp. 547 ff.).

PROCESSES

The first detail of the optimal solution is the set of least-cost technological processes. Table 5 shows the time path of U.S. processes over the planning horizon. The discounted cost incurred in using new technologies is low at the left and bottom (as in using nuclear fuel for electricity generation in the distant future), and high at the top right (as in using electric cars right away).

The first resources used are petroleum and natural gas reserves. These are already drilled and are almost costless, leading to the virtual exhaustion of domestic petroleum resources in the first decade. Proved reserves are the cheapest fuel simply because extraction is almost free and transport costs are low.

The set of processes for the next two decades relies exclusively on imported petroleum and imported liquefied natural gas, both low-cost resources. An efficient program for the period 1980 to 2000 does indeed involve heavy dependence on foreign supply, and the implications of this dependence for the U.S. balance of payments are discussed below.

The fourth and fifth decades are a transitional period. The first market that imported petroleum and gas lose is electricity generation and process heat. Shale oil and liquefied coal take over the bulk of the transport market at the end of the fifth decade.

The first part of the twenty-first century sees the world energy market progressively dominated by U.S. coal and shale oil. After 2020, virtually all energy processes outside of electricity generation are run on raw or processed coal and shale oil, and the U.S. resources are about half the known reserves in these categories.

The final stage is transition to the breeder technology, which starts about

Table 5. Time Path of Optimal U.S. Technological Processes in Solving the Energy Problem, by Demand Categories, 1970–2120 and Beyond

<table>
<tr><th rowspan="3">Period</th><th colspan="5">Fuel, by demand category</th></tr>
<tr><th rowspan="2">Electricity</th><th colspan="2">Heat</th><th colspan="2">Transportation[a]</th></tr>
<tr><th>Industrial</th><th>Residential</th><th>Substitutable</th><th>Nonsubstitutable</th></tr>
<tr><td>1970–80</td><td>Domestic natural gas</td><td>Domestic petroleum and natural gas</td><td>Imported oil</td><td colspan="2">Domestic and imported oil</td></tr>
<tr><td>1980–90</td><td rowspan="2">Imported liquefied natural gas</td><td colspan="4">Imported petroleum</td></tr>
<tr><td>1990–2000</td><td></td><td rowspan="2">Imported liquefied natural gas</td><td colspan="2" rowspan="2"></td></tr>
<tr><td>2000–10</td><td rowspan="2">Light-water reactor</td><td rowspan="5">Domestic deep coal</td></tr>
<tr><td>2010–20</td><td>Domestic high-cost natural gas</td><td colspan="2">High-cost domestic and imported petroleum</td></tr>
<tr><td>2020–45</td><td rowspan="2"></td><td>Domestic gasified deep coal and natural gas</td><td colspan="2">Domestic and imported low-cost shale oil and domestic liquefied coal</td></tr>
<tr><td>2045–70</td><td>Light-water reactor</td><td colspan="2">Domestic liquefied deep coal</td></tr>
<tr><td>2070–2120</td><td colspan="4" rowspan="2">Breeder reactor</td><td>Domestic liquefied deep coal and high-cost shale oil</td></tr>
<tr><td>2120 to indefinite future</td><td></td></tr>
</table>

Source: Developed by author. The underlying data are given in an appendix available upon request from author.

a. See Table 1, note *b*, for definitions.

2020. The use of shale oil and liquefied coal for transportation persists through 2120. But by 2120 all the fossil fuels have been exhausted and the economy is run completely on an electric-hydrogen technology with a resource base that is virtually infinite.

The pattern of interfuel substitution and the way in which processes unfold over time are sensitive to changes in parameters. The linearity of the objective function leads to extreme solutions. For these reasons the description in Table 5 should be regarded as suggestive rather than exact. It is somewhat surprising, for example, that nuclear generation of electricity is delayed until 2000. Partly this tardiness results because petroleum prices are much lower in the optimal solution than in the real world; partly because the actual level of prices reflects the substantial federal subsidy of the nuclear power industry; partly because of the rapid and unexpected run-up in prices of nuclear generating equipment reflected in the technological assumptions.[16]

But, while its details should not be taken literally, Table 5 spells out the inevitable transition from exhaustible fossil fuels to nuclear fuels; and this basic pattern is all but invariant to such things as modifications in cost.

PRICES

Perhaps the most important policy problem that is addressed by the results concerns the prices of scarce resources. The programming problem described on pages 545–46 above yields a set of shadow prices associated with the solution. The shadow prices, shown in Table 6, can be interpreted as the appropriate rent or royalty that a competitive market, operating with the same information applied here, would impute to scarce low-cost resources.

The striking thing about the basic results is that the royalties on almost all energy resources are very low. The highest in 1970 is 42 cents per barrel on U.S. drilled petroleum, corresponding roughly to costs for proved reserves. This figure is misleading since much of it represents simply quasi-rents on drilling equipment and past exploratory costs.

For Mideast oil, which must be transported to markets, the royalty is 18 cents per barrel, amounting to about one-half cent per gallon of gasoline

16. See Federal Power Commission, *The 1970 National Power Survey* (1971), Pt. 1, pp. 1-6-13 to 1-6-15, for a discussion of the recent rise in prices of nuclear generating equipment.

Table 6. Royalties (Shadow Prices) on Energy Resources, 1970, 1980, and 2000[a]

1970 dollars

Resource	1970	1980	2000
Petroleum (per barrel)			
Drilled			
United States	0.42	1.50[b]	17.93[b]
Persian Gulf and North Africa	0.18	0.46	1.94[b]
Undrilled			
United States	0.00	0.00	0.63
Persian Gulf and North Africa	0.17	0.44	1.77
Coal, United States (per ton)			
Eastern deep mined	0.07	0.18	1.23
Western strip mined	0.01	0.03	0.21
Shale oil, United States (per barrel)			
25 gallons per ton	0.02	0.06	0.37
10 gallons per ton	0.00	0.01	0.07
Natural gas (per thousand cubic feet)			
United States	0.16	0.32	0.59
Persian Gulf and North Africa	0.00	0.00	0.25
Nuclear fuel			
Inexpensive uranium (per million Btu)	0.01	0.02	0.17

Source: Derived from the program described in the text.

a. The royalties are the values of the dual variables on resources in the optimal solution. They exclude any future direct cost but include quasi-rents on past direct costs for drilled petroleum.

b. The resource is exhausted by this date and the royalties are therefore nominal.

(about 1 percent of the current retail price in the United States). For coal, the shadow price is even smaller—approximately 7 cents a ton for Eastern U.S. deep coal. In fact, the only relatively high shadow price is that on natural gas in the United States and Western Europe, which reflects the fact that it is a very cheap fuel—it needs no further refining and no expensive equipment to make it environmentally acceptable. As a result, it has a scarcity rent of 16 cents per thousand cubic feet for 1970 in the United States.

The fuel prices that emerge from the efficient solution are also of interest. They are the sum of shadow prices and the costs of extraction. Tables 7 and 8 show the time path for prices for the United States to 2010 and comparisons with actual prices since 1950. The miraculous outcome of this procedure is that the calculated prices seem to be much the same as the market prices (except for petroleum products and coal). This finding is particularly surprising given the many disparate sources for the technological data and the enormous aggregation needed to obtain resource categories.

Divergent trends in calculated prices appear among the different fuels. The path for calculated electricity prices shows a very gentle increase (1.1 percent annually over the next forty years) as full adaptation to a nuclear technology takes place. The calculated price of coal is almost constant, rising only 0.7 percent annually for forty years. The time path for petroleum prices is much steeper with calculated prices of crude oil and gasoline rising at around 4.6 and 3.5 percent annually. Natural gas also increases rapidly—3.9 percent annually—rising from a 1970 level of 21 cents per million Btu to a 2010 level of 97 cents. The major reason behind the projected rise of petroleum and natural gas prices is that, with the exhaustion of petroleum resources, the economy must turn to considerably

Table 7. Intermediate Prices for Energy Other Than Petroleum Produced in the United States, Actual 1950–70, and Projections to 2010

Period	*Electricity (before transmission; cents per kilowatt-hour)*	*Natural gas (cents per thousand cubic feet; average at wellhead)*	*Bituminous coal and lignite (f.o.b. at mine; average dollar value per ton)*
		Prices (1970 dollars)	
Actual			
1950	1.16	11.0	8.16
1960	0.98	18.3	6.14
1970	0.77[a]	17.1	6.26
Calculated			
1970	0.68	21.0	11.91
1980	0.76	37.6	12.07
1990	0.85	45.7	12.42
2000	1.03	64.1	13.34
2010	1.06	97.1	15.77
		Annual percentage change	
2010 from 1970 calculated	1.1	3.9	0.7
2010 from 1970 actual	0.8	4.4	2.3

Sources: Calculated values are derived from the program described in the text. The electricity price for 1970 is from U.S. Federal Power Commission, *The 1970 National Power Survey* (1971), Pt. 1, pp. I-1-3; earlier years assume a constant ratio of production to transmission and distribution costs and use the figures for the total from U.S. Federal Power Commission, *Typical Electric Bills: Typical Net Monthly Bills as of January 1, 1970, for Residential, Commercial, and Industrial Services (1970)*. Natural gas prices and coal are from U.S. Bureau of Mines, *Minerals Yearbook, 1951* (1954), and issues for *1961* (1962) and *1971* (1973). Prices include direct costs and royalties.

a. 1968.

Table 8. U.S. and European Prices of Petroleum Products, Actual 1950–73, and Projections to 2010

Period	*Crude oil*[a] *(per barrel)* United States	*Crude oil*[a] *(per barrel)* Western Europe	*Gasoline*[b] *(per gallon)* United States	*Gasoline*[b] *(per gallon)* Western Europe
	Prices (1970 dollars)			
Actual				
1950	4.34	n.a.	0.169	n.a.
1960	3.89	2.76	0.152	0.092
1970	3.23	2.38	0.126	0.059
Winter 1973–74	4.50–9.99[c]	12.27–22.80	0.166	0.43–0.67
Calculated				
1970	1.20		0.052	
1980	1.70		0.065	
1990	2.13		0.077	
2000	3.19		0.105	
2010	7.12		0.209	
	Annual percentage change			
2010 from 1970 calculated	4.6	4.6	3.5	3.5
2010 from winter 1973–74 actual	1.2 to −0.9	−1.3 to −3.0	0.6	−1.8 to −3.0

Sources: Calculated values are derived from the program described in the text. *United States:* Crude oil prices for 1950–70 are price of crude petroleum at wells (Oklahoma) reported in U.S. Bureau of Economic Analysis, *1971 Business Statistics*, Supplement to *Survey of Current Business*, p. 166. Winter 1973–74 crude oil figures are from the *New York Times*, December 25, 1973. Gasoline prices for 1950 and 1960 are from American Petroleum Institute, *Petroleum Facts and Figures*, 1971 edition, p. 458. Figures for 1970 and 1973–74 are derived from an adjustment of the most recent API gasoline price. *Western Europe:* 1960 and 1970 are derived from M. A. Adelman, *The World Petroleum Market* (Johns Hopkins University Press for Resources for the Future, 1972), pp. 365–66, 377. For crude oil, Adelman's realization less his calculated refinery margin is used. The figure for crude oil for 1973–74, from the *New York Times*, January 29, 1974.

a. For the United States, average annual price of mid-continent crude oil, except 1973–74, which is explained in note *c*. For Western Europe, the prices are monthly averages.

b. For the United States, average price for regular-grade gasoline at the refinery in Boston; for Western Europe, average of monthly Rotterdam prices of regular-grade gasoline.

c. $4.50 is the mid- to late December price for "old" oil produced in the United States. A comparable figure for domestically produced "new" oil is about $7.49. $9.99 is the posted price for Persian Gulf light crude oil effective January 1, 1974. As quoted in the *New York Times*, December 25, 1973, these prices in current dollars are, respectively, $5.25, $8.73, and $11.65.

n.a. Not available.

costlier processes—either shale oil or coal gasification and liquefaction. Thus in the efficient solution, refined oil can be delivered in the United States at $2.18 in 1970 while shale oil costs $5.58 a barrel and liquefied coal $7.62 (all exclusive of royalties). Before the technological transfer from natural oil to synthetic oil is made, the price of petroleum products must rise very significantly.

The solution for natural gas indicates that the actual 1970 price is below its efficient level. In fact, natural gas appears to be the only fuel that is underpriced relative to future availability. But natural gas prices have been controlled for two decades, and sporadic indications of shortages appeared in the early seventies. In the solution for the basic case, natural gas is underpriced by about 20 percent; in other solutions by much more. For example, in a solution of the model without free trade, calculated 1970 gas prices are 45 cents per thousand cubic feet—almost three times 1970 levels. Thus it appears that for natural gas, efficiency prices are substantially above their actual levels.

PETROLEUM PRICES

The major discrepancy between calculated and actual prices comes in crude petroleum and petroleum products. Because the winter 1973–74 prices in Table 8 are seriously distorted by the Mideast war and the current run-up of commodity prices, it is probably best to examine the price structure for 1970. For the United States, petroleum prices were far above the calculated long-run competitive supply price. The price of crude oil was \$3.23 a barrel, as against a calculated efficiency price of \$1.20—a markup of 169 percent over cost. For gasoline the price differential was of a similar magnitude—12.6 cents per gallon for the actual price against 5.2 cents for the calculated.

A good hint as to the source of the difference comes from the price for Western European petroleum products. Here the unregulated price—that is, without the import quotas, prorationing, and other impediments to market determination found in the United States—was quite a bit closer to the calculated long-run supply price. The price of crude in Western Europe was \$2.38 a barrel for 1970, about twice the calculated price.

What explains the discrepancy between the actual and calculated prices of petroleum in the United States—\$3.23 against \$1.20 for 1970? The first source of the discrepancy is the considerable interference with the free flow of petroleum into the United States. Until 1973 quotas were imposed on imports, so domestic prices were effectively determined by the domestic supply price. The next section reports a calculation that estimates the competitive supply price for domestic petroleum in a world without international trade in energy products at \$2.33 per barrel, about 90 cents below the actual price. The 90 cents is probably due to prorationing to suppliers.

Estimates of the benefit of efficient production are about $0.85 to $1.00 per barrel. In sum, it appears that the domestic price can be explained by import restrictions and prorationing.

While the discrepancy of the 1970 price is relatively easy to explain for the protected United States market, the discrepancy for the free market (Western European) price is more puzzling. As Table 8 shows, the 1970 Western European price was $2.38 a barrel versus $1.20 in the calculated path. Most of the differential of $1.18 can be reduced to payments to exporting countries. According to Adelman, the average payment per barrel in 1970 to the seven major exporting countries ranged from $0.81 to $1.09 per barrel, averaging $0.93.[17] In the efficient solution, royalties are $0.17 per barrel (see Table 6). Thus approximately $0.76 of the excessive $1.18 per barrel can be attributed to excessive royalties to producer countries. The remainder accrues in the form of additional profits—either excess transportation charges or rates of return to companies greater than 10 percent.

In the period since 1970, the royalty component has risen dramatically. At January 1974 posted prices, the royalties for most Mideast countries are about $7.00 per barrel—seven times the 1970 levels. These are almost twenty-five times the efficiency royalties shown in Table 6. As the discussion of the efficient path for royalties suggested, the difficulty with market allocation of resources indeed lies in proper determination of the royalty element! What lies behind the excessive royalties?

Three prominent sources of the very high royalties on petroleum are technology, monopoly, and instability. The first possibility is that participants in the energy market may be more pessimistic about the development of future technologies than I am. To test this possibility, I ran a few cases that rule out some of the technologies shown in Table 3. Within the basic model of free trade, anything but the most drastic pessimism did not seem to matter much. Thus I first assumed that all speculative technologies in Table 3 (D technologies) were one hundred times more expensive than assumed; then that all C and D technologies (thus including those now in development) were one hundred times more expensive. In both cases prices rose, but the price of refined petroleum rose by less than 10 percent even in the more pessimistic case.

Suppose that even the technologies currently in pilot plant or small-scale

17. Adelman, *World Petroleum Market*, p. 208.

operation (those designated B) will also be a hundred times more expensive than the current estimates. As a measure of the depth of this pessimism, this assumption implies that shale oil will cost at least $700 a barrel; that running a breeder reactor will cost almost a dollar a kilowatt-hour; and that coal cannot be economically used for electricity generation because the sulphur emission standards cannot be easily met. Even in this drastic case, prices on petroleum products do not reach current levels. Gasoline rises to 231 percent of the basic level—to 12 cents a gallon—and the price of crude petroleum rises about $2.50 over the basic solution. Natural gas, on the other hand, does have a much higher price—up to almost 80 cents per thousand cubic feet as against 21 cents in the basic solution. If technological pessimism lies behind the inflated level of petroleum prices, it must be a very deep pessimism indeed: it implies not only that engineers will be unable to solve the very difficult problems (like economical fusion, solar, or hydrogen power) but also that technologies that have actually operated in the past (like liquefaction of coal by the Germans in the Second World War) will be impracticable. This pessimism seems to plumb unreasonable depths, and so it is an unlikely explanation for the inflated level of petroleum prices.

A second possible reason is the presence of monopolistic restrictions. From the quantitative evidence presented here, most of the divergence of market price from the long-run competitive supply price appears attributable to government restrictions (oil import quotas and prorationing) and to excessive payments to producing countries. While oil companies obviously are not disinterested parties, little of the excessive price of crude petroleum seems to go directly to them. In 1970, only $0.25 of the excessive $2.03 is left unexplained by the government restrictions and country payments.

The other source of monopoly restriction is the control of price by oil-exporting countries. For the last few years, royalties to producing countries have been determined in bilateral negotiations between them and major oil companies in an arrangement that has led many observers—notably Adelman—to conclude that the inflated price was the outcome of monopolistic pricing by sellers.

Until recently, it was difficult to point to specific practices that were peculiarly monopolistic. While the pricing was "administered," until 1973 few significant quantitative restrictions were imposed and most countries were expanding production extremely rapidly. Moreover, many of the increases in posted price—especially those in 1973—merely brought the accounting price of crude oil up to the market price and transferred the windfall gains from company coffers to national coffers.

Thus monopolistic intent until 1973 (and behavior since October 1973) on the part of oil-exporting countries seems plausible. It is extremely difficult, on the other hand, to determine ex post whether observed prices result from monopolistic behavior or one of the other causes discussed here. Needless to say, the 1973 embargo *is* prima facie evidence of monopolistic restrictions on the part of some Arab producers; but this is quite different from earlier behavior.

A third possibility is that the discrepancy of the petroleum price arises because spot markets do not assess the royalty properly. I have argued above that resource markets might well exhibit incorrect—perhaps even unstable—pricing of scarce appropriable resources, because resource owners hold back on sales of petroleum resources in anticipation of a continuation of the very rapid rise in petroleum prices. If the basic calculation put forth here is correct, some of these owners will be unpleasantly surprised when they cannot realize the anticipated rate of return.

The wellhead price of Mideast crude for January 1974 is about $7.00 per barrel. If the market is misassessing royalties, some producers must think it worthwhile to curtail production and wait for higher prices. Given the extraction costs and a 10 percent discount rate, it would pay them to hold petroleum in the ground until 1980 if they thought that prices would rise at least to $13.50 a barrel; until 1990, if they thought the price would rise to $35 a barrel. If the estimates for the costs of synthetic fuels shown in Table 4 are close to accurate, it appears unlikely that any country will realize the implicit 1990 price for its petroleum exports.

Although the presence of an incorrectly assessed royalty is hard to prove, this possibility seems entirely consistent with the calculated and actual pattern of resource prices. If royalties were incorrectly assessed, how was the price of petroleum talked up so high? And how can it be talked back down to where it belongs?

The 1970 price of coal also seems a bit out of line, but this situation conceals important changes since 1970. The dramatic rise in the wholesale price of coal—from $7.64 per short ton in 1970 to $12.13 in October 1973[18]—is

18. The prices are for bituminous screenings for industrial use from *Survey of Current Business*, Vol. 53 (February and November 1973), p. S-35. The discrepancy between the 1970 figures here and in Table 7 is due to the fact that the series on average value used in Table 7 is not as up to date as the *Survey* figures for industrial screenings.

probably associated with sulphur restrictions and mine safety legislation. The data in Table 7 are based on estimates of capital and current costs under *current* standards, so the predicted price is not far off.

FINAL USE PRICES

Table 9 shows the time path of prices of energy products for the five final demand categories. The story is roughly the same as that told by Tables 7 and 8. Considering the first five time periods, the calculated rises for elec-

Table 9. Prices of U.S. Energy, by Demand Category, 1970, and Projections to 2120 and Beyond

	Demand category				
		Heat		*Transportation*[a]	
Period	*Electricity*	*Industrial*	*Residential*	*Substitutable*	*Nonsubstitutable*
	Prices (dollars per million Btu, delivered)				
Calculated					
1970	1.99	0.47	0.54	1.25	1.25
1980	2.27	0.59	0.68	1.58	1.58
1990	2.48	0.69	0.80	1.85	1.85
2000	3.03	0.77	1.10	2.53	2.53
2010	3.12	0.79	1.55	5.03	5.03
2020	3.12	0.84	2.53	6.29	6.29
2045	3.12	1.67	3.03	9.69	9.69
2070	3.12	3.12	3.12	39.35	46.77
2120 on[b]	3.12	3.12	3.12	39.35	46.77
Actual					
1970[c]	2.26	0.25	0.57	2.98	2.98
	Average annual percentage change				
2010 from 1970 calculated	1.1	1.3	2.7	3.5	3.5
2010 from 1970 actual	0.8	2.9	2.5	1.3	1.3
2120 from 1970 calculated	0.3	1.3	1.1	2.3	2.4
2120 from 1970 actual	0.2	1.7	1.1	1.7	1.9

Source: Derived from the program described in the text.

a. See Table 1, note *b*, for definitions.

b. The price structure for the period 2120 on represents the prices associated with the backstop technology discussed in the text.

c. The figures in Tables 7 and 8 are for electricity, coal, and natural gas used for the first three demand categories and for gasoline for the last two. The differences in levels between Tables 7 and 8, and 9 are accounted for by the thermal efficiencies of different end uses. Thus 1970 prices for transport would be 3.33 times the calculated gasoline price in Table 8, which reflects the 30 percent efficiency of automobiles.

tricity and industrial heating are rather gentle, slightly more than 1 percent annually. For residential heating and the transportation categories, the calculated rises are rather larger—2.7 percent and 3.5 percent annually. The average rise of energy prices for the five categories using 1970–80 weights is 2.4 percent annually for calculated prices and 1.3 percent annually using the actual 1970 prices as a base.

ALTERNATIVE INTEREST RATES

On the basis of the technological data and a 10 percent interest rate, current energy prices seem about right for electricity, coal, and natural gas. But the current prices of petroleum products—especially gasoline—seem far higher than is consistent with the long-run scarcity of energy resources.

Surprisingly, the shadow prices in the optimal solution were extremely insensitive to different specifications. Moderate changes in the assumptions about resource availabilities, growth rates for demand and for population, and capital or current costs always left the shadow prices surprisingly low.

One natural question concerns the importance of the interest rate in the outcome, for the 10 percent interest rate is perhaps debatable. Some might argue that the social rate of time preference should be used and might be lower than 10 percent (perhaps even zero). Others, accepting the return on investment as the appropriate criterion, might point out that the pretax return on corporate capital is closer to 20 percent.

It therefore seemed worthwhile to test the sensitivity of the optimal solution to rates of zero, 1, 5, 10, 20, and 30 percent. Royalties showed little sensitivity with higher interest rates because they were so small to begin with. At lower interest rates, however, they shifted significantly. The current royalty on undrilled U.S. petroleum rose from zero for 10 percent, to about 80 cents a barrel for 5 percent, then shot up to $12 a barrel for 1 percent, finally reaching $20 a barrel for zero percent.

The prices of final products formed a different pattern. As the discussion above (pp. 532–33) noted, the interest rate has an ambiguous effect on final demand prices, with a higher rate first lowering and then raising them over the period in which a resource is used. As it turns out, current prices of final energy products (except electricity) are *minimized* at a 10 percent interest rate; they rise roughly in proportion with the interest rate above that level. Below it, prices rise slightly as the royalty becomes larger, but only for transportation are the effects very large.

THE ROLE OF FOREIGN TRADE

This paper has described a pattern of resource utilization marked by close interdependence and vast trade flows among the regions of the world. In the optimal case depicted in Table 5, the United States relies heavily on foreign energy sources for much of the period from 1970 to 2020. In this allocation U.S. imports of petroleum are projected to rise from an average of 3.1 billion barrels per year in 1970–80 to 7.1 billion barrels in 1980–90, to fall sharply to 4.2 billion barrels in 1990–2000, to peak at 8.7 billion barrels in 2000–10, then to disappear after 2010. These figures compare with 1.2 billion barrels of crude and refined products imported in 1970, 1.7 billion barrels in 1972, and an import rate in mid-1973 of about 2.4 billion barrels. Western Europe and Japan show a similar dependency.

But all is flux, and the winter of dependency passes as quickly as it arrived. By 2020, the quantity of imported fuels in the calculated solutions drops to almost nothing, and American coal and shale oil come to dominate trade in fuels. U.S. coal exports are projected to start in the fourth decade (2000–10), then increase very rapidly to $44 billion annually in the next decade. American coal and shale oil exports dominate the final two periods before their exhaustion (2020–2070).

It is possible to calculate the balance of payments on energy account from the optimal program. To do this, I simply record the flows into and out of each region in the optimal solution. The results I present in Table 10 calculate the cost of the fuel at the port of export, thus excluding transport costs and any further processing (such as refining or electricity generation) that occurs at the point of consumption.

The results foreshadow a period of very large deficits on energy account, peaking in the 1990–2000 decade at almost $20 billion (or almost 1 percent of projected GNP, both in 1970 prices). The fourth decade embodies the transition to a coal technology, with the value of imports almost constant, the fifth a very large expansion of exports. In the final periods, covering the fifty years 2020 to 2070, the world's oceans will swarm with U.S. coal and shale oil in transit, with the value of exports exceeding $300 billion *annually* to 2045, and over $800 billion thereafter. Lest it appear that the United States would be turning into a new version of a banana republic, note that at their peak, energy exports total about 7 percent of projected GNP.

Again, the numbers are not to be taken too literally; yet they tell an important story. The United States may well become dependent for a time on

Table 10. Annual U.S. Trade Flows in Energy in the Optimal Solution, 1970–2070[a]

Billions of 1970 dollars at annual rates

			Energy payments surplus (+) or deficit (−)	
Period	*Exports*	*Imports*	*Amount*	*Percent of potential GNP*
1970–80	0	1.6	−1.6	−0.14
1980–90	0	8.8	−8.8	−0.57
1990–2000	0	19.2	−19.2	−0.94
2000–10	3.7	19.4	−15.6	−0.58
2010–20	44.2	49.9	−5.7	−0.16
2020–45	320.3	24.2	296.1	5.9
2045–70	836.0	0	836.0	7.3

Source: Derived from the program described in the text.
a. All values are calculated at port of export. Per capita potential GNP is projected to grow at 2 percent annually.

foreign petroleum and gas; but this period will not last forever. Sooner (because oil-exporting countries raise their prices above the competitively determined price) or later (as the oil is exhausted), the world must turn to the next fuel up the cost curve. By most reckonings this is likely to be coal or nuclear processes for stationary uses and liquefied coal or shale oil for mobile uses. The United States has these resources in abundance.

These optimistic remarks do not deny that *until* coal and shale oil become competitive, the United States dollar could have a rough time on foreign exchange markets; for that matter, Japan and Western Europe will have an even stormier time if balance on energy account is all that matters. But there are fairly clear limits to the U.S. dependence on foreign sources; and there is a clear point at which the rise in petroleum prices will force extensive substitution of domestic fuels, as the following section shows.

SELF-SUFFICIENCY?

A good question is whether the benefits of trade are worth the problems. The past months have demonstrated vividly that oil mixed with politics is a volatile brew. One way of calculating the gains from trade is to estimate the cost of self-sufficiency, or autarky—that is, of meeting all U.S. energy needs from domestic resources and all foreign energy needs from foreign sources.

This calculation calls for solving the basic problem without allowing any trade with the United States. The costs of meeting the U.S. energy requirements are compared in the basic case, with trade, and the autarkic case, without trade. The results are shown in Table 11. Not surprisingly, the costs of petroleum and natural gas rise dramatically in the absence of the large foreign supplies to meet demands for the next few decades. The royalty on petroleum rises from $0.42 a barrel in the basic case to $1.55 a barrel in the autarkic case. The change means that a barrel of crude oil at the refinery costs $1.20 in the basic case and $2.33 in the autarkic case. The royalty on natural gas rises similarly, from 16 cents to 40 cents per thousand cubic feet. The prices of different demand categories are also shown in Table 11.

The total cost of meeting the demands is shown at the bottom of Table 11. These figures indicate that free trade in energy products would be

Table 11. Energy Prices, Royalties, Costs, and Discounted Values, 1970–2120, for Basic and Autarkic Cases

1970 prices

Description	*Basic case—with foreign trade*	*Autarkic case—without foreign trade*	*Cost of autarkic in relation to basic case (percentage change)*
Price by demand category (dollars per million Btu)			
Electricity	1.99	2.58	30
Industrial heat	0.47	0.74	57
Residential heat	0.54	0.86	59
Substitutable transportation[a]	1.25	1.98	58
Nonsubstitutable transportation[a]	1.25	1.98	58
Resource royalties (cents per million Btu)			
Proven U.S. petroleum reserves	7.3	26.7	266
Deep coal	0.34	0.09	−74
Shale oil	0.37	0.24	−35
Natural gas	15.9	40.4	154
Cost of meeting demand requirements (billions of dollars)			
1970–80	252	371	47
1980–90	411	614	49
Discounted value of total program (billions of dollars)			
1970–2120	1,478	2,133	44

Source: Derived from the program described in the text.
a. See Table 1, note *b*, for definition.

moderately beneficial to the United States. Over the next twenty years, the total cost of meeting energy demand is $663 billion under free trade and $985 billion under self-sufficiency. Thus free trade costs one-third less than a regime of self-sufficiency. Free trade in energy products is worth $16 billion a year on the average. It should be stressed that this calculation of the cost of autarky assumes that foreign sources are competitively priced. The calculation obviously would be different in a real world of short-run disturbances or monopolistic pricing.

The autarkic case presents one significant modification of the basic case. If free trade does not prevail over the next few decades, the outlook for energy prices is clouded except for electricity. If foreign supplies dry up for any reason, the long-run supply price of crude petroleum at the refinery is projected to be twice as high ($2.33 a barrel against $1.20 a barrel), and the refinery price of gasoline 63 percent higher (8.5 cents per gallon against 5.2 cents per gallon), all for 1970 in 1970 prices. Still, the autarky prices are well below the current prices found for petroleum products (see Table 8). Thus, while prices might be higher under a regime of self-sufficiency, the current structure cannot be rationalized by the belief that trade will collapse for an indefinite period.

Implications for Energy Policy

To summarize the findings, this analysis has investigated the efficient allocation of energy resources over time by determining the cheapest way of meeting a growth path of final demands for energy products with a given stock of energy resources and a given set of processes for converting resources into products. After ensuring that the program was feasible for a very long time period, the procedure was to find the optimal path for consuming scarce resources and the prices associated with this path.

The basic case allows free trade in energy resources and assumes an interest rate of 10 percent. In this case, the scarcity rents or royalties on energy resources are quite modest, never more than 16 cents per million Btu (equivalent to about one dollar per barrel of petroleum). With 1970 as a basis of comparison, the final-demand prices associated with the optimal solution are not far from actual market prices, with one exception. The exception is petroleum prices, which were about 240 percent of the price calculated in the optimal program.

Subject to reservations discussed below, these results are quite suggestive for energy policy. *As a long-run policy* it would be unwise to jack up the prices of energy products in the interests of artificially preserving energy resources. Nor does a more drastic policy of permanent rationing of energy resources make sense. As long as investment yields around 10 percent, it seems best to use the cheap resources now and to put the real resources thereby saved to work on producing synthetic fuels later. Of course, any worthwhile effort aimed at reducing wasted energy should be applauded; but judging by their long-run scarcity prices, a dollar's worth of energy resources saved is no more deserving than a dollar's worth of idle labor or wasted capital saved.

For petroleum prices the lesson is, if anything, the opposite of the conventional story. Subject to the usual qualifications, the optimal solution indicates that the current price of crude oil is inflated considerably above its long-run competitive supply price. Before the 1973 Mideast war, crude oil was selling at about $4 per barrel and refined petroleum products were selling at about $6 a barrel; in current prices, the optimal solution indicates that the 1973 U.S. east coast price should be around $1.70 per barrel. I have not determined the precise source of this discrepancy, but it probably arises partly from excessive royalties to producing countries, partly from restrictions on imports into the United States, and partly from inefficient regulation. If this description is correct, then a policy that aims at further increases in the long-run price of gasoline would be pushing in the wrong direction.

Some dark spots mar this generally cheery landscape. The proviso about free trade is of considerable importance. The optimal program depicts a world economy with vast trade flows from energy-rich to energy-poor countries. In particular, in the last couple of decades of this century, the United States has a projected deficit on energy account of around $20 billion annually. This large a deficit must strain even an economy with roughly double the real GNP of the United States in 1970.

But if the calculations presented here are correct, free trade offers considerable gains. The optimal program costs the United States $16 billion less annually over the next twenty years than does a program without trade. In light of current shortages (especially in Western Europe and Japan), are the gains from trade for a highly exposed economy worth the risks? While self-sufficiency seems a laudable goal, many other policies are much cheaper. For example, if the gains from trade are $16 billion annually, half this amount will certainly finance an oil storage program that covers four years' imports at $6 a barrel. There are many ways to cover contingencies besides self-sufficiency. Dependence on foreign energy sources may impose costs in the sacrifice of minor political objectives or strain on foreign exchange, but it is hard to see how, short of outright war, these costs could outweigh the benefits of trade.

A second proviso is that the optimal solution depends to a certain extent on unproven technologies. The system simply cannot run very long without development of a breeder reactor, fusion technology, or some other process that rests on a virtually inexhaustible resource base. But time is not particularly pressing, and the economy can wait at least 100 years for this ultimate technology. The need for other sorts of technology is more pressing. In particular, some form of synthetic liquid fuel must be developed quite rapidly to replace petroleum when the latter is exhausted. Such processes are in development—shale oil and coal liquefaction being the most significant—but they have not yet proved their economic and environmental acceptability. On a pessimistic view about the viability of these technologies, the prices of petroleum products would be much higher than in the basic case.

The optimistic picture painted here contrasts vividly with the pervasive concerns about the "energy crisis." The results presented above, however, are relevant to the long-run availability of energy resources; they are not particularly helpful for managing short-run shortages such as the nation is currently experiencing. Recall that the model assumes very smooth functioning of markets, with perfect foresight about underlying forces such as demands, resources, prices, technologies, and government policies. In particular, in the optimal solution the capacity expansion is tailored precisely to the demand path, so that none of the important variables presents any surprises.

It seems likely, however, that the current energy crisis is the result of just such unforeseen shifts in underlying forces. The most obvious was the 1973 embargo on petroleum exports. World production in November 1973 was about 5 million barrels a day, or 10 percent, below the pre-embargo levels. A shortfall of this magnitude in such a crucial product obviously can induce serious short-run disruptions.

But these events have merely compounded problems brought about by earlier policies. For one thing, industrialized countries are locked into heavy dependence on imported petroleum by past decisions about capital equipment and insufficient attention to alternative sources. For another, the U.S. oil import quotas posed a quandary for oil companies in planning to meet future demands. As long as they were in effect, and as long as U.S. crude petroleum production was stagnant, refinery building in the United States made no sense. As a result, when the quotas were lifted in early 1973, refining capacity was insufficient to meet domestic demands. Moreover, the world boom of 1973 has spurred a very rapid expansion in petroleum demands that worldwide capacity could not meet at prevailing prices.

Petroleum did not stand alone as a scarce, expensive, raw material during 1973. In fact, before the cutbacks due to the embargo, most other commodity prices had risen even more. The wholesale cash prices of gasoline and fuel oil in the United States were up 10 and 20 percent, respectively, over a year earlier. while the Dow-Jones commodity index rose 94 percent.[19] It seems clear that the "energy crisis' is in part simply a reflection of the current "commodity crisis."

Another major disturbance was the 1967 and 1970 amendments to the Clean Air Act. Emission standards for automobiles and air quality standards suddenly changed the rules of the game. The standards for automobiles have resulted in a fuel penalty of 7 to 10 percent, increasing gasoline demand in an unforeseen way. The sulphur restrictions have been even more disruptive in that they simply removed high-sulphur coal and high-sulphur petroleum products from the eligible technology.

Direct interference with price has also been mentioned as a distorting influence. Wholesale producers' prices for natural gas have been regulated since 1954; and more recently, since August 1971, prices for refined petroleum products have been subject to a changing array of price controls. The price increases imposed by oil-exporting countries have been cited as an example of exercise of market power.

Nevertheless, while these disturbances may seriously impair the functioning of the industrialized economies over the next few years, a long-run policy based on the premise that energy resources are the nation's most precious resource would be a mistake. Many have proposed policies that treat energy as the only scarce resource—a kind of "Btu theory of value." If the results presented above are valid, they suggest that the current crisis should be viewed as the temporary effect of critical bottlenecks. They further suggest that stress should be laid on expansion of capacity in those

19. These comparisons are for September 28, 1972 and 1973; *Wall Street Journal*, October 1, 1973.

areas where resources are abundant—intensified drilling for oil and gas and heavier use of coal; and that greater attention should be paid to perfecting processes for producing clean synthetic fuels—particularly shale oil and liquefied coal.

If these conclusions are right, then the current "energy crisis" will blow over eventually. Real enough problems remain. Until supplies are expanded, the United States may experience very serious shortages or very high prices. In any case rising prices are likely over the long haul, especially for transportation; adaptation to new, potentially difficult, technologies will present a problem; and several lean years on foreign exchange markets loom ahead. But we should not be haunted by the specter of the affluent society grinding to a halt for lack of energy resources.

Comments and Discussion

Hendrik Houthakker: Bill Nordhaus is to be congratulated for blazing a new trail of analysis in this paper. He has taken a sophisticated approach to some of the policy issues surrounding the use of energy resources. Although his paper does not deal with current problems, it has important implications for energy policy for the longer run.

I would like to take issue with Nordhaus' contention that free markets cannot ensure an efficient pattern of resource allocation. This position is based on Debreu's view that a fully competitive economy must include conditional futures markets that cover all possible outcomes. But conditional futures markets do not exist for any commodity; the futures markets that are available operate on very different principles. Conditional futures markets involve a sharing of the general risks associated with future output. It is far-fetched to assume that conditional contracts could be devised to cover all contingencies, and partial coverage would result in frequent defaulting on contracts. Debreu's theory is not a realistic guide to the problems involved in allocating output and consumption through time. Furthermore, as Arrow has shown, security and asset markets can serve to cover the gaps in the risk coverage of existing futures markets.

Nordhaus appears to underestimate the importance of trade in existing assets. If price rises are anticipated for certain resources, the value of these assets will appreciate. Purchases of oil in the ground, long-term contracts, and similar transactions help to ensure that prices will not diverge markedly from the short-run efficient price. Admittedly, these asset markets function imperfectly but they serve to reduce large deviations between market prices and efficient prices and thus to eliminate the instability of the spot market emphasized by Nordhaus.

Many of the current energy problems stem from interference with the operation of the free market. The market mechanism has been severely restricted by federal regulation of electric power and natural gas. Rather than attribute energy problems to the failure of the market mechanism, I would put the blame on federal policies that have impeded the market.

A serious practical problem is Nordhaus' assumption that all demand elasticities are zero. He allows for substitution within fuel demand groups but not between energy products and other objects of consumption. He mentions that most studies of the final demand categories indicate that price elasticities range between zero and unity; but even within this range different values for the price elasticity of demand can have a substantial effect on the overall balance between supply and demand and on the calculation of the trend of supply. Allowing for some demand elasticity would not seem to make the computational problem insuperably difficult.

The origin of the reserve estimates underlying Nordhaus' calculations is important to the analysis. They are based on geologists' estimates of the amount of reserves that will be recoverable at a given price, and are derived from knowledge of the quantity of a resource that has been recovered from similar formations in the past. The assumed price may be the current price or some other price, and is rarely defined. These estimates are rough at best, and they should not be taken literally. However, Nordhaus' conclusions seem reasonable and I do not think that he's to be faulted for his use of the data.

I agree with Nordhaus' timetable, according to which nuclear energy technology is not relevant until the turn of the century. The nation has developed nuclear technology prematurely in an effort to promote science, and because regulatory policies on power are biased in favor of capital-intensive processes. I doubt that many nuclear power plants would have been built in a completely free market.

The events of late 1973 have demonstrated that efficiency, in the limited sense used here, is not the only dimension of the U.S. policy problem. Low-cost foreign supplies of energy can become suddenly expensive if there are monopoly elements in the market. And they can be interrupted, causing severe dislocations in the economy. So the actual calculations that Nordhaus makes are only indirectly relevant to the formation of policy. However, some of these political factors could be built into a more elaborate model of the kind he has pioneered.

Robert Solow: This is a fascinating paper. Details aside, it is clearly right in concept. The most striking result is the finding that the efficiency price of oil is low compared with the current market price. This does not mean that delivered oil is plentiful today; tightness in refining capacity and the recent cut-off of Arab oil have curtailed supplies. The immediate reduction in the supply of oil has to be met on its own terms and that problem is really beyond the scope of this paper. The current situation should be used to spur research on intermediate technologies, which have been bypassed in the rush to nuclear technology.

Nordhaus' discussion of the instability of the market mechanism overstates the issue somewhat. Monopoly aside, he covers the point formally by saying that he is illustrating a situation in which instability might arise with only a spot market for oil or without a global long-term planner. The rationale underlying this argument runs as follows: In order for a competitive market to be in flow equilibrium, the price of oil must be rising at the current rate of interest, since that is the means by which owners realize a return on their reserves. If the price of oil is rising more slowly than the rate of interest, then oil in the ground is a poor investment since it brings a lower rate of return than various kinds of reproducible capital. If there is only a spot market for oil, owners will produce and sell in an effort to get rid of their oil as quickly as possible. But since the demand curve for oil is negatively sloped, the increase in the quantity supplied will further depress its price. Because expectations about the rate of change of oil prices are generated by an adaptive expectations mechanism, reductions in the price will lead to even more pessimistic expectations about future prices. Thus, the adjustment process that takes place through the spot markets would result in a worsening of the initial disequilibrium.

But this kind of market disequilibrium is not restricted to exhaustible resources. It would also apply to the cheesiest kind of growth stock, like a security that does not pay dividends and never will, and whose only value in a portfolio is its current rate of appreciation. The real difference between the growth stock and oil is that oil has definite uses and its owners have in mind some longer-term rendezvous with a future price; to use an old terminology, the price of oil would be anchored by its value in use, as distinguished from its value in exchange.

As a result, if the price of oil now is expected to rise too slowly for equilibrium, petroleum owners will take a short-term capital loss on the value of their reserves, after which the price will rise at a rate approximately equal to the interest rate. While this version may be a bit overdrawn, it is no more so than the pure spot market story that leads Nordhaus to worry about extreme instabilities in price.

The 10 percent discount rate underlying Nordhaus' linear programming calculations is used as an estimate of the rate of return on reproducible capital. The basic issue here is similar to the one involved in public expenditure decisions. If resources for marginal public investment come from consumption, then the consumption rate of discount would be appropriate; alternatively, if resources are diverted from investment, then that discount rate should be applied. A weighted average of discount rates for consumption and investment would be appropriate when resources for public investment have been diverted from both activities. I might point out that even with a utility discount rate of zero, the consumption rate of discount would be positive if per capita incomes are expected to be higher in the future, because of the diminishing marginal utility of income. I do not know if 10 percent is exactly the right rate of discount; but I would not use a very different number.

Nordhaus has excluded from his model the really exotic energy technologies, such as fusion or solar power, that cannot be priced or dated accurately. The omission of such potential power sources gives a conservative bias to the paper which strengthens its conclusions. On the other hand, he also omits the use of oil as feedstocks to petrochemical industries, in which it is likely to have a very high value for a long time.

Environmental constraints could embody a significant cost and their exclusion from the model is unfortunate. I am particularly concerned here about the disposal of plutonium wastes; little is known about these costs but they could significantly affect the feasibility of breeder technology.

Nordhaus tested the validity of his results by increasing the current estimates of costs for intermediate technologies by a factor of 100; and he found that his conclusions were affected very little. Another kind of test would be to delay the introduction of intermediate technologies by a significant period of time, say, 100 years. Postponing a technology by 100 years would be equivalent to increasing its cost by a factor of 100 squared, or 10,000, and that might have a significant effect on the conclusions. If a new technology is totally unavailable until a distant date, the shadow price of oil might rise above the rendezvous price in the interim and dip down to its prescribed level when the alternative technology becomes available.

230-32 NA

General Discussion

Franco Modigliani seconded Solow's statement that the most striking result of the paper was the sizable difference between the actual and optimal petroleum prices. He noted two aspects of the model that would make it tend to overestimate optimal petroleum prices. Both the price-inelastic demand function for oil and Nordhaus' cost assumptions for substitute energy technologies—which some observers would regard as high—would bias upward the estimated petroleum price paths. He also cited three factors that explain why actual petroleum prices are so high: the very high future price of oil as a petrochemical feedstock; the fact that with a differential tax on capital gains, price might have to rise less rapidly to keep producers in equilibrium between present and future sales; and the monopolistic element in oil sales. Several other discussants suspected that monopoly pricing was important, citing as evidence the existence of the Texas Railroad Commission and the small number of international oil companies.

Nordhaus remarked that the influence of monopoly on petroleum prices would vary depending on the definition of the monopolist group. In his model, if the rest of the world is assumed to constitute the monopoly, then the predicted price rises to midway between the optimal competitive price and the actual price. However, if all producers of crude petroleum were to constitute the monopoly, then the price of petroleum would rise to about $5.50 per barrel, which is the price of shale oil.

William Poole observed that, despite opinions to the contrary, energy use in the United States did not appear profligate; per capita energy consumption did not even double over the period 1929–68. He also questioned the reliability of predictive models such as Nordhaus' in light of the uncertainty surrounding technological change. Today's energy technology was not envisioned fifty years ago; and the Nordhaus model would have predicted badly—it would have been far too pessimistic—if applied in 1929. Poole noted that pricing patterns in securities markets behave like a random walk with drift. Similar behavior by prices for oil-producing land would signify that prices were being determined by a process fundamentally different from the one modeled by Nordhaus.

In response to Poole, Nordhaus stated that the most a predictive model could offer was a "careful look around the corner." He noted that the history of technology forecasting was spotty and that many of the breakthroughs in the postwar period had not been foreseen. He added, however, that the scope of change might have been narrowed in the postwar period as nuclear processes came to dominate experimental technology.

James Duesenberry noted that oil producers will use a short time horizon in their planning efforts since technological change makes longer-term planning ineffective. The presence of important monopoly elements would encourage the industry to get high prices for oil products in the current period since the future path of prices for these products would be uncertain. A system of rational pricing would differ from the current situation in two important aspects. First, oil products would be cheaper. Second, the pattern of substitution among energy products would be different; for example, nuclear power would not have been introduced until it was economically viable. Arnold Packer added that uncertainty had a different impact on ideal resource allocation from the point of view of society than from that of the resource holder. If the emergence of a competitive resource is uncertain, because the stock of resources is not adequately known or because the feasibility of a backstop terminology is in doubt, society and the resource owner are motivated to opposite courses of action. The owner is moved to sell his resource now to avoid the risk of competition, while society is moved to husband the resource rather than risk shortages in the event that the competitor does not materialize.

Charles Holt expressed concern about the environmental implications of continued economic growth. Energy products impinge on the environment in a variety of ways and Holt questioned whether the model had attempted to account for these differential impacts.

Nordhaus answered that he had incorporated into his model the costs of meeting existing legal restrictions on environmental disruption—for example, the costs of reclamation, as much as $5,000 an acre, associated with strip mining. He noted that one kind of absolute environmental constraint that man might face would be a limit to the earth's tolerance for energy derived from nonhydro or nonsolar sources. It would be easy to incorporate this kind of a constraint into a linear programming model; this constraint would result in an initial reduction in prices for energy resources and a rise in final demand prices, since energy resources would have to be used up more slowly. Concerning petrochemical feedstocks, he noted that 10 percent of hydrocarbon fuels has been set aside for nonenergy uses; this should be adequate for at least 150 years. After that time, there are immense quantities of low-grade hydrocarbons, which were omitted from the resource estimates. Finally, he agreed that the problem of nuclear waste disposal was an unsolved issue. The problem, however, was not cost, but the "Faustian bargain" with future generations of bequeathing a low-probability risk of disaster. Unfortunately, we cannot even judge this probability until well after the bargain is made.

[14]

Energy Impact of Consumption Decisions

CLARK W. BULLARD, III, AND ROBERT A. HERENDEEN

Abstract—The energy cost of goods and services is computed, and applications are discussed. The method utilizes the data base of input-output economics, but entails additional analysis. Applications range over consumption options for individuals, business, industry, and government; from the total energy cost of bus versus auto travel to the national import-export balance.

I. Introduction

When ANYONE consumes anything, he consumes energy. Since energy reserves appear to be finite, and since every type of energy implies some kind of environmental impact, it is of interest to know the "energy cost" of the whole spectrum of goods and services produced by the U.S. economy. By energy cost we mean the total amount required to support all activities necessary for delivery of the product—from mining to final fabrication. In this paper, we describe a method for computing energy costs, based mainly on the data from input–output analysis but with additional analysis. In the Appendix, we discuss a similar method for computing "labor costs," the total man-hours required to produce a product.

One motivation for determining total energy cost lies in the magnitude of "direct" and "indirect" energy requirements. For example, only about 30 percent of America's energy is used in the private sector, that is, in homes and private transportation. An additional 40 percent is supplied to commerce and industry to be demanded by consumers through the purchase of nonenergy products. The remainder is utilized for governmental activities and exports.

Indirect energy requirements are thus extremely important. Consider an example from manufacturing: less than 10 percent of the energy needed to make an automobile is consumed directly by the manufacturer. Over 90 percent is required to produce the steel, glass, rubber, and other necessary inputs. We seek a method that accounts for these inputs explicitly.

To determine energy cost, we must model the physical flow of materials and energy through a system: the U.S. economic system. Data collection for this large complex system is extremely costly and time consuming; therefore, limited. This lack of data is the binding constraint, since in modeling physical systems the state of the art is quite advanced. In this paper, we have chosen not to start with a general theoretical development and simplify to accommodate data constraints. Instead, we begin with this constraint and basic physical principles and develop a model of increasing complexity. Since this approach maximizes use of existing data and provides immediate answers to a subset of policy questions, we believe it provides a useful framework for convincing government of the marginal utility of additional data required for more sophisticated models. The interested reader may refer to a set of papers outlining general theoretical approaches to problems such as this [1]–[4], and will note several conceptual similarities shared with the method developed in this paper. In particular, Krenz [4] has used an approach somewhat similar to ours. However, he assumed that energy is sold to all customers at the same price, which we have found leads to errors of at least a factor of two.

Our methods of analysis are described in Section II. In Section III, we discuss attempts, empirical and theoretical, to verify the accuracy of the approach. Section IV contains applications which cover such diverse subjects as the energy cost of an electric mixer and the energy impact of federal public works projects.

II. Computing Energy Cost

To determine the energy cost of a specific product, a detailed study of its manufacture, "vertical analysis," is often useful. Performing vertical analyses for a whole spectrum of products becomes too large a project to handle, so we sought a more all-encompassing approach which would probably sacrifice some accuracy. In Section II-B, after giving a brief description of vertical analysis, we present connections between the two techniques and discuss possible complications.

A. Vertical Analysis

Vertical analysis is best described by example. Suppose you want to know the energy cost of producing a car. First you would obtain the energy bill of the car manufacturer, which you would allocate on a per-car basis. In addition, you would need to know the nonenergy inputs to the car manufacturer, which you could also allocate on a per-car basis. Next, you would obtain the energy bill of the manufacturer of each of these inputs (steel, glass, tires, etc.), allocating those on a per unit basis (ton, gallon, etc.). It would also be necessary to determine their manufacturing inputs (iron ore, sand, etc.). There are many stages through which the original product must be traced. Theoretically this process could go on for an infinite number of steps. Practically speaking, however, the series can be truncated after several steps, but with some loss of accuracy. The error would be difficult to evaluate.

Vertical analysis can be used to treat very specific products. The actual physical quantity used to allocate energy can be selected as appropriate to the particular step in manufacture. These two advantages must be weighed against the problem of truncation along with the fact that vertical analyses are tedious and time consuming.

B. Energy Intensity of Goods and Services

Rather than laboriously performing a vertical analysis for each individual product, we will use a method for simultaneously treating all goods and services in the economy. It will be seen that the data base requirea is no different (with certain assumptions) from that required for vertical analysis, and the method is more tractable.

Consider an economic system composed of N sectors, each producing a unique output. We assume there is an energy

Manuscript received June 12, 1974; revised October 8, 1974. This work was supported by the National Science Foundation.

The authors are with the Center of Advanced Computation, University of Illinois, Urbana, Ill. 61801.

Reprinted from *Proc. IEEE*, vol. 63, pp. 484–493, Mar. 1975.

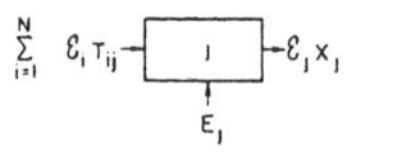

Fig. 1. The simplest energy balance diagram.

intensity ϵ_j, measured in Btu per unit output, associated with the output of each sector j.

Our data base is an $N \times N$ matrix of interindustry transactions T, a vector of total outputs X, and a vector of energy extraction (mining) data E. An element T_{ij} of the transactions matrix represents the amount of product i sold to sector j during a given time period (typically one year). The total output X_j of a sector is the sum of its sales to other sectors plus its sales for final consumption, Y_j:

$$X_j = \sum_{k=1}^{N} T_{jk} + Y_j. \tag{2-1}$$

With these data, we can now construct an energy balance diagram for each sector. Shown in Fig. 1, it simply states that the energy embodied in a sector's output is equal to that embodied in its inputs, plus energy extracted from the earth by that sector.

Based on Fig. 1, we can write a set of N energy balance equations

$$\sum_{i=1}^{N} \epsilon_i T_{ij} + E_j = \epsilon_j X_j. \tag{2-2}$$

Converting to matrix notation and solving, we have

$$\epsilon = E(\hat{X} - T)^{-1} \tag{2-3}$$

where $\hat{X}$ denotes a diagonal matrix with the elements of the vector X along the diagonal. Hereafter, the caret symbol over a matrix will indicate such diagonalization of a vector.

1) Relation to Input-Output Theory: It is sometimes convenient to define a new matrix A:

$$A_{ij} = T_{ij}/X_j. \tag{2-4}$$

Equation (2.3) now takes the form

$$\epsilon = E\hat{X}^{-1}(I - A)^{-1}. \tag{2-5}$$

The elements of a column of A are said to specify the "technology" of sector j. Each coefficient A_{ij} represents the amount of i purchased by sector j per unit output of j produced. In economics, a sector's production function is sometimes specified in terms of these fixed technological coefficients A_{ij}, and called an input-output (I/O) production function. Under the rather restrictive assumption that the matrix A is constant, independent of scale and time (the fundamentals of static I/O theory), an element $(I - A)^{-1}_{ij}$ represents the amount of i required directly and indirectly per unit of j delivered for final consumption [1].

The similarities with I/O theory are apparent, but there are several key differences. First, the units of A_{ij} in most applications of I/O theory are expressed in the units: dollars worth of i per dollar's worth of j. In our work we defined a set of N energy intensities, ϵ (one for each sector) and in Fig. 1 allocated these energies proportional to transactions T and outputs X. Thus we are free to define energy intensities in terms of physical units (e.g., Btu per ton of steel, cubic yards of concrete, etc.) or any other units expected to be the best allocators of embodied energy.

The second major difference is the severity of assumptions made. In I/O problems, one is usually interested in obtaining N^2 results, the elements of $(I - A)^{-1}$. It is, therefore, necessary to assume the entire matrix A to be constant, independent of scale and time. Such a restrictive assumption is merely sufficient, not necessary, to obtain the vector ϵ from (2-3) or (2-5). In our problem, we need only assume that vector ϵ obtained for the base year is independent of scale and time. That is, we require that a product has the same total energy intensity from one year to the next; not that all aspects of every production process remain identical.

Actually, the first calculation of energy intensities for 360 sectors of the U.S. economy was based on I/O theory [5], utilizing the published dollar-based inverse matrix [6]. Subsequent enhancements have removed difficulties associated with using redundant (physical and monetary) energy allocation data; Btu units have replaced dollars in the energy sector rows of T. We proceed with a review of these and other enhancements, continuing from Fig. 1 and (2-5).

2) Extension to Several Types of Energy: Up to this point, we have spoken in terms of total energy intensity: the amount of energy that must be extracted from the earth per unit of product j delivered for final consumption. If we define "coal intensity" in an exactly analogous way, Fig. 1 and the energy balance equations still apply—coal inputs to each sector equal coal outputs. Thus, the solution generalizes to

$$\epsilon = E\hat{X}^{-1}(I - A)^{-1} \tag{2-6}$$

where an element ϵ_{kj} of ϵ now represents the "energy of type k intensity of a unit of j." Similarly, the extraction data E_{ki} represent the (primary) energy of type k extracted from the earth by sector i. This is nonzero only for $i = k$, and its magnitude is equal to X_i, the total output (Btu) of sector i. Therefore, the product

$$(E\hat{X}^{-1})_{ij} = \begin{cases} 1, & i = j = \text{primary energy sector} \\ 0, & \text{otherwise.} \end{cases} \tag{2-7}$$

The multiplication in (2-6), therefore, extracts the energy sector rows from the inverse matrix, and identifies them as energy intensities ϵ.

In reality, some energy sectors process various types of energy into another form, such as electricity. Such "secondary" energy sectors do not extract input energy from the earth but receive it embodied in their purchases from the primary energy sectors. To determine the "electricity intensity" of goods and services, we simply extract the electric sector row of $(I - A)^{-1}$. Mathematically, this is equivalent to imagining an artificial extraction of electricity directly from the earth by the electric sector.

Obviously, primary and secondary energy intensities cannot be summed directly to obtain total primary energy intensities; coal used to make electricity would be double counted. The model contains 5 energy sectors: coal, crude oil and gas wells, refined petroleum, electricity, and utility gas. The first two are primary sectors; but electricity is mixed because it includes hydropower and nuclear, which are considered primary energy resources. We calculate the total primary energy intensities ϵ from the matrix ϵ using the formula

$$\epsilon_j = \epsilon_{1j} + \epsilon_{2j} + \frac{f}{\eta}\epsilon_{4j} \tag{2-8}$$

PROCEEDINGS OF THE IEEE, MARCH 1975

where f is the fraction of electricity produced from hydro and nuclear sources, and η is the efficiency of converting fossil fuels to electricity. This follows the established convention of defining the primary energy content of nuclear and hydro electricity as their fossil fuel equivalents.

3) Relation to Vertical Analysis: To understand the relation between the energy intensities obtained by vertical analysis and those found through matrix inversion, consider the series expansion:

$$(I - A)^{-1} = I + A + A^2 + A^3 + \cdots . \tag{2-9}$$

For this expansion to be valid the spectral radius (the maximum of the absolute values of the eigenvalues) of A must be strictly less than unity [7]. This condition is satisfied in our model. (Our A does *not* possess a column norm $\leqslant 1$ with at least one column sum < 1; this condition is sufficient, but not necessary.)

Terms on the right-hand side of (2-9) can be seen to represent successive terms in a vertical analysis. As a simple example, suppose Sector 1 is coal mining and Sector 2 is auto manufacturing. Then the second element of the first row of the inverse matrix will be the coal intensity of autos. Now consider each term on the right hand side contributing to this total. The first term is zero; the second term is the coal consumed directly by the auto sector per car produced; the third term represents coal used directly by the auto sector's suppliers to produce nonenergy inputs (steel, glass, tires, etc.); the fourth term adds the coal used to produce each of their inputs, etc.

In practice we often desire more detail than this affords—for example, differentiating a specific auto manufacturing line from the larger aggregated sector, "motor vehicles and parts." A "hybrid" analysis is used, replacing the first few stages of the expansion with more detailed problem–specific data, and using our average coefficients for the remainder. Thus from the identity

$$(I - A)^{-1} = I + A + A^2 + \cdots + A^{n-1} + (I - A)^{-1} A^n \tag{2-10}$$

we calculate

$$\epsilon' = E \hat{X}^{-1} [I + A' + (A')^2 + \cdots + (A')^{n-1} + (I - A)^{-1} (A')^n] \tag{2-11}$$

where the primes denote problem–specific data which may be expressed in physical units, or in dollars. Accuracy will depend on the choice of n.

C. Treatment of Imports and Exports

When we export goods, we export energy embodied within them. Imported goods on the other hand can be viewed as implied energy imports, displacing the need to consume energy domestically. We wish to construct our model in such a way that we obtain "domestic" energy intensities. This will permit us to apply the results to a variety of foreign policy issues and "the energy balance of trade".

Data from [6] include transactions and outputs of imported goods, as well as those domestically produced. We correct the data by subtracting imports from total outputs so each sector's energy use is allocated only to domestic outputs. Solving the equation implied by Fig. 2 gives

$$\epsilon = E(\hat{X} - \hat{Q} - T)^{-1} \tag{2-12}$$

where $\hat{Q}$ represents that portion of $\hat{X}$ composed of imported

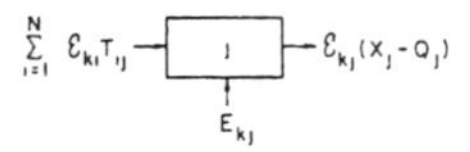

Fig. 2. Energy balance for a domestic sector.

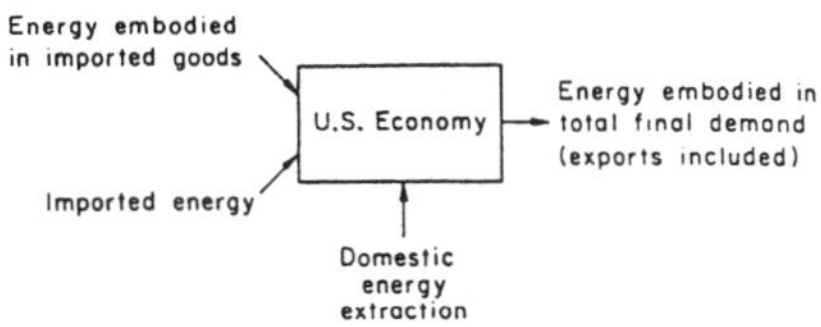

Fig. 3. Energy flow through the U.S. economy.

goods. The energy intensities ϵ now reflect the true energy intensity of products made exclusively from domestically produced inputs.

Some practical problems arise here because there are two classes of imported goods. Competitive imports are defined as those goods for which these are domestic counterparts, and are counted as part of the total output of the corresponding domestic sector. For example, in [6] the total output of the steel manufacturing sector X_{steel} includes imported as well as domestic steel. Fortunately, the vector Q is also recorded so the two types of outputs are distinguishable and (2-12) can be solved.

The other category of imports, noncompetitive, is defined as that for which no domestic counterpart exists. Examples include diamonds, bananas, and coffee. In [6], these products are not identified by type. Since they comprise only 20 percent of all imports, and only 1 percent of GNP, we ignore them in our analysis.

Exports are treated as part of final consumption, along with personal and government expenditures. The energy intensities of exported goods are likewise given by (2-12).

We can now construct a complete picture of energy flow through the U.S. economy. We assign to competitive imports the energy content of their domestic counterparts; this is equivalent to assuming that if we did not import them, we must produce them here at that energy cost. Fig. 3 shows the energy balance under this assumption.

The U.S. economy receives energy in three ways:

1) primary energy (coal, crude, hydro, nuclear) extracted domestically;
2) imported energy (almost exclusively petroleum) with the associated embodied energy penalty due to losses in extracting, refining, etc. carried out abroad;
3) energy embodied in other competitive imports.

Taking the product of energy intensities with total final demands, we obtain the total energy required to sustain that level of consumption,

$$\epsilon Y = e \tag{2-13}$$

where e is a vector of total requirements for each fuel type. This total can be decomposed into its components, domestic extraction and energy embodied in imports (including energy imports):

$$\sum_{j=1}^{N} \epsilon_{kj} Y_j = e_k = \sum_{j=1}^{N} E_{kj} + \sum_{j=1}^{N} \epsilon_{kj} Q_j \tag{2-14}$$

where the subscript k indicates fuel type and Q_j are competitive imports. Remember that noncompetitive imports have been neglected in this analysis.

D. Secondary Products

All interindustry data described thus far are collected from individual establishments and aggregated to N sectors. We have spoken of sector outputs as homogeneous quantities. In this section we shall discuss the realities of the situation, since outputs are not homogeneous.

If an establishment produces primarily automobiles (its primary product), it is placed in the automobile sector. It may also produce secondary products, items that are the primary outputs of other sectors. Therefore, a unit of output from sector i is not necessarily equivalent to a unit of (primary) product i; the output X_i may represent a heterogeneous mixture of products. This raises serious problems for applications, since we wish to know the energy intensity of goods and services, not sector outputs.

1) Activity-Based Analysis: The ideal solution to this problem would be to define a set of activities, one associated with each product. To define these activities, consider an establishment in the automobile sector whose output consists of 90-percent autos and 10-percent airplanes. We say that this single establishment engaged in two activities: a primary activity (making cars) and a secondary activity (making airplanes). If the establishment's accounting practices do not distinguish between purchases of inputs (say electricity) for making cars and making planes, it is not possible to construct a precise "activity-based" transactions matrix.

2) Makeup of the Transaction Matrix: To gain a thorough understanding of the secondary product problem and other aspects of this large and valuable data base, we present a brief outline of the method for assembling the data. Then we return to activity-based analysis.

Actually the transactions matrix of [6] is obtained by summing four $N \times N$ matrices:

$$T = DA + TF + MDT + IM \tag{2-15}$$

DA is the direct allocations matrix. Its elements DA_{ij} represent direct sales of primary product i to the purchasing sector j. The sum of elements of the airplane sector row of DA, plus total final demand for airplanes equals the total sales of airplanes, regardless of whether produced by the airplane sector, auto sector, or any other.

TF is the domestic transfers matrix. Its elements TF_{ij} represent the value of j, a secondary product, produced by sector i. Thus the value in producers' prices of airplanes produced in the auto sector would appear in the auto row and the airplane column of TF. It is important to note that if one looks only at the T matrix (15), the transferred output from the auto sector to the airplane sector is indistinguishable from the Chevys that Lockheed bought as fleet vehicles (the latter is a direct allocation).

MDT is the matrix of margins on domestic transactions. It has no direct bearing on secondary products, but is explained here to aid in understanding Section II-F. It contains nonzero elements only in the trade, transportation, and insurance rows. An entry represents the sum of domestic costs of distribution (transport, trade, insurance) associated with all of the inputs purchased by the consuming industry. It is important to note that the purchase of an airplane ticket for a GM executive appears as an entry in DA, while the sum of all air freight

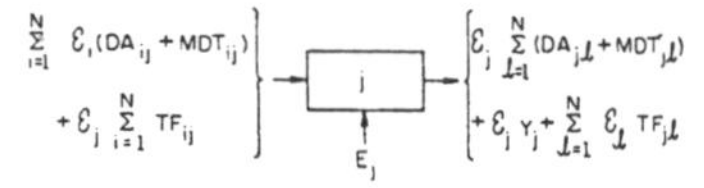

Fig. 4. Energy balance with secondary products.

charges on GM's purchased inputs appears in MDT. The two types of purchases are indistinguishable in T, appearing only as a transaction from the air transport sector to the auto sector.

IM is the transferred (competitive) imports matrix. It contains nonzero entries in one row, indicating the foreign port value of competitive imports. Note that this matrix would be deleted from (2-15) in accordance with the methods outlined in Section II-C. But in the source data, when IM is used to calculate T, competitive imports are effectively "transferred" to the corresponding domestic sector in a manner identical to the treatment of domestic transfers. There is one more feature of IM; it also has nonzero entries in the trade, transportation, and insurance rows. These fictitious purchases effectively bring the purchaser's price of imported goods to domestic port value. Summing the columns of IM yields the Q of Section II-C.

3) Secondary Product Adjustment: Utilizing the preceding data, several approaches have been proposed for determining the energy content of primary products. The first was used by Herendeen [5] to allocate a sector's energy consumption over only its products (primary and secondary), excluding that portion of its output produced as secondary products by other sectors. The advantage of this method was that it permitted use of the already published $(I - A)^{-1}$ matrix, avoiding costly inversion of an altered matrix. Its main disadvantage was that it did not attempt to allow for secondary products requiring more or fewer energy inputs.

Another method sometimes employed in econometric analyses [8] is based on the assumption that each product is produced with precisely the same technology (input proportions), regardless of the sector in which it is produced. This approach is more complete, since it corrects for transfers out of a sector as well as transfers into a sector. Its main disadvantage is that the assumption is evidently physically unrealistic because it results in numerous negative entries in the total requirements matrix $(I - A)^{-1}$. Such values are physically impossible and demonstrate that the assumption is too strong.

Remembering that our earlier calculation of energy intensities of sector outputs was possible under an assumption much less restrictive than that normally employed in I/O theory, we are able to make a similar assumption here: that the energy intensity of each product is independent of the sector in which it is produced. This does not require identical production technologies, only that the energy embodied directly and indirectly in the final product be the same. While this assumption is strong, it is not nearly as restrictive as needed in I/O analyses and it is the weakest possible within existing data constraints. Fig. 4 shows the energy balance for a typical sector (a group of establishments) under this assumption.

Solving the energy equation corresponding to Fig. 4, we obtain the energy intensity of products, rather than sector outputs.

$$\epsilon = E[\hat{Y} + \hat{D} - DA + \hat{M} - MDT - \hat{TF} + TF']^{-1} \tag{2-16}$$

where $\hat{Y}$ is a diagonal matrix whose elements are domestic final demands, and the elements of diagonal matrices $\hat{D}$, $\hat{M}$, and $\hat{TF}$ are the row sums of DA, the row sums of MDT, and the column sums of TF, respectively.[1] The superscript t denotes transpose.

In the absence of specific data, we cannot verify our assumption that a product's energy intensity is independent of its sector of origin. A sufficient indication that our assumption is in error would be the occurrence of negative values in ϵ, a physical impossibility. In terms of Fig. 4, this can occur if the energy ascribed to secondary products by our assumption exceeds energy embodied in inputs. In preliminary results with a 92-sector model, this happened for only two sectors. For both of these sectors, over 90 percent of outputs are secondary products, and it is not yet clear whether the source of the difficulty lies in our additional assumption, or slight inaccuracies in those sectors' direct energy input data. Work is currently underway to resolve this question, and alternative assumptions and methods of solution are being evaluated [9]. Accordingly, the numerical results presented here are derived from establishment-based data, pending refinement of the preceding methods for secondary product adjustments.

E. Capital Purchases

Since I/O tables are assembled only once every four or five years, and since their usefulness depends on the stability over time of the technological coefficients A, the transactions matrix is constructed exclusively from current account data. Thus interindustry transactions exclude flows of capital goods, partly because they could cause erratic annual fluctuations in the A matrix.

Therefore, capital purchases are normally listed as a component of final demand, not as interindustry purchases. In Fig. 1 the purchase of (say) machinery for manufacturing does not appear on the input side. Intuitively, we should expect it to be there; otherwise, the energy to produce it does not contribute to the final product. To correct this problem, we would like to allocate capital purchases, away from final demand, to interindustry purchases. This would require construction of a complete capital flow table, and separation of growth and replacement ("depreciation") components. In Section III-A, we discuss an example calculation.

F. Producers' versus Purchasers' Prices

The energy intensities obtained here are in terms of producers' prices—the price at the end of the assembly line. The actual price paid by a consumer includes margins ("markups") for transportation, insurance, and wholesale and retail trade. Each margin has its associated energy intensity. (For the energy sectors, whose output we measure in Btu, not dollars, we can state it this way: there is an additional energy penalty from marketing.)

In the normal final demand vector, transportation and trade margins are lumped with direct sales (i.e., final demand "buys" some retail trade services). If we want the total energy intensity of a product sold to final demand, we must use independent data to allocate margins on these sales to the appropriate products. Energy intensities in terms of purchaser's prices have been calculated for over 360 types of goods and services [10].

III. Applications: Checks on Accuracy

Questions relating to the accuracy of our approach fall into these following classes.

a) Conceptual problems persisting from the use of economic accounting data. Examples: the aggregation and capital problems.

b) Using dollars to allocate embodied energy, or actual errors in dollar data.

c) Technological change over time.

While we have not performed a comprehensive error analysis to answer all these questions, we will discuss these: For class a), we have incorporated a capital flow matrix, described in Section III-A. For classes b) and c), two approaches have been used. First, we have compared our results for energy cost of certain products with results of independent vertical analyses (Section III-B). Second, work is in progress to determine the sensitivity of energy cost to errors or time variation in the direct coefficients matrix A (Section III-C).

For empirical checks, we shall sometimes refer to our published results of calculations of energy intensity [5]. This earlier work used a method qualitatively similar to that offered in Section II, but differing slightly quantitatively. We shall indicate which method is used in each comparison. Example values for energy intensities are listed in Table I.

A. Capital Correction

The philosophy of energy cost implies that capital purchases, at least those for replacement, should be part of the interindustry transactions. The energy to produce them would no longer be viewed as being delivered to final demand, but instead would be part of the products they help produce. From the standard I/O data base we know the sum of all capital sales by selling sector (e.g., all airplanes sold as capital stock). We do not know who bought the planes. However, the Bureau of Economic Analysis of the U.S. Department of Commerce has independently produced a table of interindustry capital flows compatible with its 82-sector table [11]. We have expanded this to 92 sectors, and calculated the effect of including the capital flows.

With capital flows, Fig. 2 is modified as shown in Fig. 5. The effect is to increase or leave unchanged, (but not to decrease) the ϵ_{kj}. C_{ij} is the replacement capital flow, which is some fraction of the C_{ij} given in the capital flow table.

In our calculation [20], we assumed that half of each capital transaction that year was for replacement of obsolete or depreciated equipment. Because capital flow data are less reliable than current account transactions, and the uncertainty over the "typicalness" of 1963 capital flows, we consider this only an exercise in ascertaining an idea of the distribution of errors.

In Table II we list the distribution of changes in primary energy intensity resulting from including the capital flows. The dollar value of the replacement flows was about 4.5 percent of the GNP; this is roughly the expected average increase. The average appears close to this value. Only six of the 92

[1] The solution (2-16) requires that the $N \times N$ matrix be well-behaved. This requires that a unique primary output be associated with each sector. Actually this is not the case, since [6] contains several "dummy industries" (e.g., office supplies) in which no production occurs; all outputs are transferred in from the primary producing sector (dummy industries result from data acquisition constraints). The result is more equations than unknowns; the energy intensities of dummy sector outputs are linear combinations of other energy intensities. Eliminating such sectors, and adding their outputs to those of the appropriate primary sectors, we obtain a tractable system of simultaneous equations.

TABLE I
EXAMPLE OF ENERGY INTENSITIES (10^3 BTU/$); [5], 360-SECTOR MODEL

Primary Aluminum	380
Fertilizers	180
Airlines	152
Glass	101
Motor Vehicles	70
Cheese	62
Apparel	46
Hospitals	38
Computing Machinery	27
Banking	19
U. S. Average	82

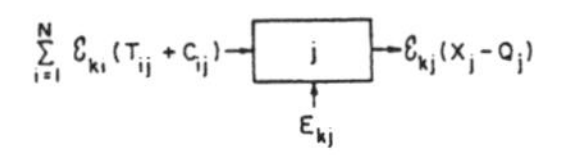

Fig. 5. Energy balance with capital flows.

TABLE II
EFFECT ON ENERGY INTENSITIES OF INCLUDING CAPITAL FLOWS: [20], 92-SECTOR MODEL

Percent Change of ε	Number of Sectors
0 - 3	18
3 - 6	49
6 - 12	19
12 - 18	5
18 - 24	1
> 24	0
	92

sectors show changes exceeding 12 percent. These are agricultural or service sectors.

B. Comparison with Results of Vertical Analysis

In Table III we compare energy costs computed from vertical analysis with our results. Comparisons are listed as the ratio (other study)/(our study). Performing this comparison is complicated by the following factors.

a) Matching the economic sector exactly with the specific product is usually not possible; e.g., metal containers includes paint cans and buckets in addition to beverage cans (this is the aggregation problem).

b) Primary/secondary product ratios are often low, thus casting doubt on the accuracy of establishment-based energy coefficients.

c) Price data (implied or actual) are often unavailable. Prices are needed to bridge the gap between our dollar-based approach and the usual physical basis of vertical analysis.

In theory, the preceding ratio should be $\leqslant 1$, since our approach is analogous to an infinite vertical analysis. (Neither method included a capital correction.) The results for 13 independent verifications yield an average value of 0.92, a spread of from 0.52 to 1.23, and a standard deviation of 0.23.

TABLE III
COMPARISON OF ENERGY COST COMPUTED BY OUR TECHNIQUE AND BY INDEPENDENT VERTICAL ANALYSIS: [12], 360-SECTOR MODEL

	Subject	Ratio(Other Result/Our Result)
1.	Auto (manuf.)	.81
2.	Finished steel	{ .89 1.00
3.	Primary copper	.52
4.	Primary aluminum	1.11
5.	Rolled & drawn aluminum	{ 1.21 1.23
6.	Aluminum castings	.95
7.	Bread	1.09
8.	Glass containers	.67
9.	Raw paper	.63
10.	Newspapers	.72
11.	Metal containers	1.13

Mean = 0.92, standard deviation = $\sqrt{\dfrac{\sum_{i=1}^{N}(x_i-\bar{x})^2}{N-1}} = 0.23$

C. Sensitivity of Results to Errors in A

Equation (2-6) represents a nonlinear operation on the matrix A. A is constructed from empirical data and is subject to error. We ask what errors are thereby implied in ϵ.

Attempts to obtain a closed form solution lead only to a rather weak inequality demonstrating that we cannot be sure that there is less than a factor of 5 (the condition number of the 360-order A matrix) multiplier on errors in A [13]. We expect that errors in A are at least ±10 percent, probably higher; this gives a least upper bound of ±50 percent for errors in ϵ, which we consider unacceptably high.

Work is now in progress [14] using numerical methods to improve these early estimates; these involve actual inversions of perturbed matrices under two assumptions: first, that the errors in A act in concert to give the worst possible error in $(I - A)^{-1}$, and second, that the errors in A vary stochastically. In addition, the work includes sensitivity analyses to identify important input data. All of these methods can be used to determine the effects of errors in published data [6] and to evaluate effects of expected technological change. Since we are dealing with N^2 technological coefficients (where N may be as large as 360), a systematic updating process must await development of these techniques.

IV. APPLICATIONS: SELECTED RESULTS

Almost all of the applications presented here are based on conceptually isolating some group—citizens, an industrial sector, the federal government—as a consumer, and analyzing the energy impact of changes in that group's consumption.

We first illustrate the idea of total energy cost for two family items, the private auto and an electric mixer (to indicate the range of applicability!). We then discuss results of this type of analysis applied to 1) consumer's choice of urban transportation (bus or car), and overall family spending pattern, and 2) federal government choice of alternative public works projects

TABLE IV
ENERGY COST OF THE AVERAGE PRIVATE CAR, 1963; [5]

Category	% of Total Energy
Fuel	
Content	57.2
Refining, etc.	11.9
Retail	1.3
Oil	
Content	0.5
Retail	0.2
Car	
Manufacture	9.9
Retail	3.4
Repairs, parts, maintenance	3.3
Parking	8.8
Tires	
Manufacture	0.8
Retail	0.2
Insurance	2.7
Highway construction (taxes)	4.8
	100.0

(highways versus mass transit). We also use energy costing techniques to analyze an industry's total energy dependence through its inputs of goods as well as energy and apply this notion to the nation's total import-export energy balance. Finally, we calculate the price impact (on all products) resulting from a tax on energy.

A. Total Energy Cost of an Automobile and an Electric Mixer

Results for the average automobile are listed in Table IV. Every auto-associated expenditure is converted to energy terms. This requires allocating each expenditure to the appropriate sector and multiplying by the corresponding energy intensity. The net effect is that the actual fuel in the tank accounts for only about 57 percent of the total. An additional 13 percent goes to the "energy cost of energy" (refining losses, etc.), which leaves 30 percent attributable to manufacture of the car, parts, maintenance, road construction, etc. Thus a multiplier of about 1.7 must be applied to the direct gasoline energy to obtain total impact. This same approach must be applied to other transportation modes before a true energy-per-service-mile comparison can be made.

Table V shows the total energy cost of an electric mixer. We account for the energy of manufacture and sale, and the expected service lifetime. Assuming that the mixer (125 W) is used 13 min per week, we find that the operational energy must be multiplied by 1.58 to yield total energy impact. This factor varies for different appliances. For the big users of energy such as refrigerators, stoves, etc., the multiplier is less than 1.1.

B. Energy Impact of Urban Bus and Auto Transportation

This question has been treated by Hannon and Puleo [15]. National average figures for all cars and buses were not adequate, so the specific urban transportation systems had to be analyzed individually by the hybrid technique. Expenditures and fuel consumption data for 38 urban bus systems were obtained and converted to energy terms as described; likewise for the car, with attention to the relatively poor gasoline economy in city traffic. Additionally, average load factors were used, leading to the results in Table VI.

TABLE V
ENERGY COST OF AN ELECTRIC MIXER

$$E_{tot} = E_{op} + E_{mfg}/\tau$$

where

E_{tot} = total primary energy for year

E_{op} = total primary energy to operate per year

E_{mfg} = total primary energy to manufacture and sell

τ = lifetime in years

(Maintenance and disposal energies are assumed negligible)

E_{op} is obtained from industry data (125 watts, 10 KW hr./yr. [21]). It is converted to primary terms by dividing by the energy efficiency of electricity generation [5]. E_{mfg} is obtained by multiplying purchase price ($20 [22]) by the energy intensity for appliances, converted to purchasers' price [10].

Thus:

E_{op} = 10 KW hrs. x 3412 Btu/KW hrs. x 1/.256 = 1.33×10^5 Btu/yr.

E_{mfg} = 1.08×10^6 Btu

We assume τ = 14 years; hence $E_{mfg}/\tau = 7.71 \times 10^4$ Btu/yr.

$E_{tot} = 1.33 \times 10^5 + 7.71 \times 10^4$

$= 1.33 \times 10^5 (1 + 0.58)$ Btu/yr.

TABLE VI
ENERGY IMPACT OF URBAN CAR VERSUS URBAN BUS [15]

Energy Intensity (Btu/pass. mile)	Car	Bus
Direct fuel	5500	2600
Total energy	8900	5300
Trip length (av.)	8.3 mi.	3.8 mi.
Passengers (av.)	1.9 (incl. driver)	12

We mention a problem associated with this comparison: the two options differ in their total expense to the consumer. Therefore, the total energy comparison "should" include the energy impact of spending of the saved money. There are two tacks. One is to try to predict consumer spending, a task we have foregone. The other is to compute the effective energy intensity of the saved money (Btu/dollars), and present the consumer with a list of spending alternatives rated according to their energy intensities. These are listed in [15].

C. Total Energy Impact of a Family's Expenditures

Herendeen [16] has computed the energy impact of all expenditures by American families as a function of income. Personal consumption expenditure data from the Bureau of Labor Statistics were converted to energy terms. Fig. 6(a) shows some of the expenditures, illustrating that direct energy consumption such as gasoline seems to level off with increasing income, while purchases of education and transportation other than the car (normally not considered energy purchases) do not. These have been converted to energy terms in Fig. 6(b). The energy impact of energy purchases indeed does level, but that of nonenergy purchases does not. For the lowest income group, energy purchases account for $\frac{2}{3}$ of the total, while for the highest income group, the fraction has dropped to $\frac{1}{3}$. This suggests that estimates of the effects on consumers of energy shortages will be quite misleading if based on direct use only, especially for the more affluent.

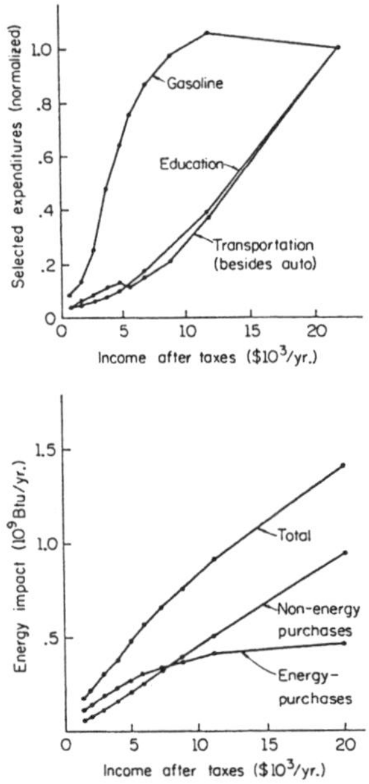

Fig. 6. Family energy impact versus income.

These results are averaged over geographical location, age, race, education and size of family. A question which arises here is: are there different spending patterns, resulting in significantly different energy impacts within the same income class? (Remember that even savings or investments have an energy impact—they help build shopping centers.) So far this question is unanswered.

D. Energy and Labor Impacts of Government Spending

As we show in the Appendix, we can derive "labor intensity" analogously to energy intensity. Knowledge of the two allows one to search for ways to conserve energy while not causing a decrease in employment. Bezdek and Hannon [17] have analyzed several federal government "public works" programs this way. The problem is to break down the program into expenditures compatible with the model's sectors, and then apply energy and labor intensities. In Table VII, we list their results for the highway trust fund and several programs which have been suggested as alternatives.

These are several programs which reduce energy impact, yet do not decrease (and sometimes increase) employment. Even if money is given back to individual consumers through tax relief, and then is spent by them in an "average way," energy requirements decrease.

E. Industrial Energy Dependence

With the concept of energy intensity, an industry's dependence on energy is seen to involve material and service inputs as well as actual energy. An industry may find that the effect of an energy shortage is to interrupt its supply of a critical energy intensive material more drastically than its intrinsic supply of energy. Or, if energy conservation is a goal, greater results may be obtained from a policy of conserving energy intensive materials (e.g., aluminum) than by turning down thermostats.

TABLE VII
ENERGY AND EMPLOYMENT IMPACT OF SEVERAL FEDERAL SPENDING OPTIONS (DATA NORMALIZED TO UNITY FOR HIGHWAY CONSTRUCTION, [17])

Program	Energy per $(1963)	Jobs per $(1963)
Highway Construction	1.00	1.00
Rail & Mass Transit	0.38	1.03
Waste Treatment Construction	0.58	1.01
National Health Insurance	0.36	1.65
Tax Relief	0.77	1.07

The method we have described in Section II is adaptable to this sort of energy analysis of inputs. Referring to Fig. 2, we write:

$$g_{kij} \equiv \frac{\epsilon_{ki} T_{ij}}{\epsilon_{kj}(X_j - Q_j)} \times 100$$

g_{kij} is defined thus: of the total amount of energy of type k required to produce 1 unit of product j, a percentage g_{kij} entered through j's purchase of i

$$\sum_{i=1}^{N} g_{kij} + \frac{E_{kj} \times 100}{\epsilon_{kj}(X_j - Q_j)} = 100.$$

The g_{kij} are calculated from base year data. It is attractive to treat them as constants, independent of X_j. Since we have already required that the ϵ_{kj} be constants, this requires that $T_{ij}/(X_j - Q_j)$ be constant. This is the constant technology assumption of I/O theory. The assumption was not necessary to compute energy intensities ϵ_{kj}, but it is necessary to attach significance to the g_{kij}. Table VIII presents an energy input analysis of two industries. The input sectors have been aggregated for simplicity, but the receiving sectors are at the 360-sector level of detail, For both receiving sectors (motor vehicles and parts, and seeds and feed grains) we see that energy inputs from energy sectors are sometimes smaller than those from certain materials inputs (e.g., primary metals for cars) [18]. Such an accounting suggests access points for energy conservation, but the actual conservation potential depends on the nature of the industry's dependence on that commodity.

In Table VIII we see that a large energy input appears to come from "self-sales." This is a consequence of aggregation. It may represent an actual "self-use," but more likely it results from the sales between dissimilar firms (preliminary and final assembly plants, for example) grouped in the same sector. This large self-energy is not that troublesome, however. First, it is somewhat arbitrary; the actual value of self-sales does not affect the value of ϵ. Second, the relative values of energy inputs from all other sectors (except self-sales) are not affected.

F. National Import-Export Energy Balance

Once we have obtained the energy intensities for all sectors we can quantify the import-export energy balance. Recalling Fig. 3, we note that there are 3 ways for energy to enter the

PROCEEDINGS OF THE IEEE, MARCH 1975

TABLE VIII
ENERGY INPUT ANALYSIS OF TWO SECTORS: [18], 360-SECTOR MODEL

Feed Grains

Input Sector	ε_{ij} (%)
Refined petroleum	51
Electric utilities	3
Agriculture	11
Construction	1
Chemicals, paints	20
Heavy machinery	1
Finance, insurance	5
Other	8
	100

Motor Vehicles

Input Sector	ε_{ij} (%)
Coal mining	3
Electric utilities	4
Gas utilities	2
Chemicals, paints	5
Stone, clay, glass	2
Primary metals	28
Fabricated metal products	9
Heavy machinery	4
Motor vehicles	34
Other	9
	100

economy through imports:

1) Btu content of imported energy. This is just $\Sigma_l Q_l$, where l runs over the energy sectors.
2) implied primary energy penalty on this imported energy; this is $\Sigma_l (\epsilon_l - 1) Q_l$;
3) embodied primary energy in imported goods; this is $\Sigma_{i \neq l} \epsilon_i Q_i$.

Analogous terms apply to exports. All quantities are known, leading to Fig. 7. Imports and exports are expressed as a percentage of the total U.S. energy budget, which we define consistent with the approach here, as the sum of domestic and all imported energy, actual and embodied.

We see that while the U.S. was a net importer of actual energy, it was a net exporter of energy embodied in goods. The combined effect is that the U.S. was still a net energy importer.

This suggests an interesting way to look at foreign trade policies. If energy embodied in exports exceeds that of imports, domestic energy supplies are being depleted by trade. As our economy becomes more open, the energy flux associated with trade will grow. A framework which explicitly recognizes this flow will be necessary for evaluating policies directed toward energy self-sufficiency.

G. An Energy Conservation Tax

It has been suggested that a tax be levied on energy use, to encourage conservation and to raise the revenue needed to solve energy-related problems [19]. Under the assumption that all taxes are passed through 100 percent in the form of consumer price increases, it was found that an energy tax

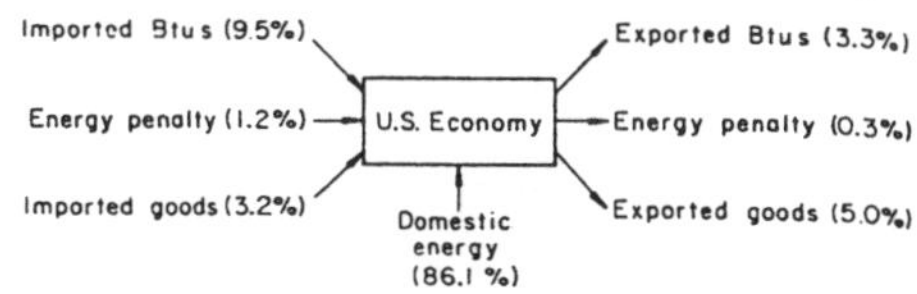

Fig. 7. U.S. import–export energy balance, 1963.

TABLE IX
POTENTIAL PRICE IMPACTS OF ENERGY TAX SCHEMES (CENTS PER DOLLAR DELIVERED FOR FINAL CONSUMPTION, 1963 PRODUCER'S PRICES), [19]

	Btu Tax	Ad Valorem Tax
Electricity	11.5	23.7
Plastics	4.5	2.5
Air Transport	3.1	2.3
Metal Cans	2.8	1.4
Motor Vehicles & Parts	1.4	1.1
Hospitals	.8	1.0
Doctors, Dentists	.3	.4

(20¢/million Btu) could raise very large amounts of revenue (about $15 billion) and would result in only slight price increases for most goods and services. However, tax impacts would be distributed in such a way that prices of some energy intensive products would rise substantially compared to their less energy intensive competitors.

It was determined that a "per Btu" tax would be substantially less regressive than an *ad valorem* tax, which would reflect inequities in utility rate structures. Revenues would be adequate to provide per capita income tax credits in an amount equivalent to a subsistence energy use level, thus transferring the burden of the tax to heavy energy users.

The method was based on results of [5], and the "per Btu" tax was taken to be directly proportional to energy content of goods and services. In an analogous manner, monetary data on energy use from [6] were used to calculate the "energy value content" of goods and services for determining the impact of the *ad valorem* tax, assuming that each sector passes on its increased dollar costs. During the computation, those elements of the matrix product corresponding to flows of secondary (already taxed) energy were taxed only on their "energy value added" (e.g., energy burned in refineries) to avoid double counting. Estimated price increases on some goods and services under the two taxing schemes are presented in Table IX. The 20¢/million Btu tax was compared to an 18 percent *ad valorem* tax because it could raise equal revenue.

APPENDIX
LABOR INTENSITY

In Section IV, we mentioned labor intensity of goods and services. The concept is analogous to energy cost ϵ_j; energy enters the economic system as it is extracted from the earth, while labor enters directly from the domestic labor force. The "labor balance" of a sector is shown in Fig. 8. Here λ_j is the labor intensity (total man-years, direct and indirect), per unit of X_j. L_j is the man-years used directly in industry j. Embodied labor is thus allocated by transactions of X_j as before. There is a question whether dollars or Btu's are better allocators of embodied labor for the energy sectors (perhaps

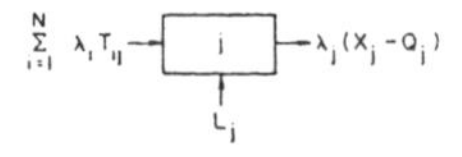

Fig. 8. Labor balance.

the high price of residential electricity reflects more labor intensive sales, promotions, and service). In the example quoted [17], dollars were used for all T_{ij} although it is not necessary.

Solving the equation implied by Fig. 8 gives:

$$\lambda = L(\hat{X} - \hat{Q} - T)^{-1}. \quad \text{(A-1)}$$

Equation (A-1) resembles (2-12). The main difference is that while the matrix E has only a few nonzero entries (primary energy out of the earth), the vector L has a nonzero entries for almost every sector; every industry receives output from the labor force, which is a part of final demand and not one of the N industrial sectors.

Secondary product and capital corrections can be included in the labor intensity calculation; the methods are analogous to those for energy.

Acknowledgment

The authors wish to express appreciation to N. Abbott for data management and computer programming.

References

[1] W. Leontief, *The Structure of American Economy, 1919-1939.* New York: Oxford Univ. Press, 1941.

[2] R. Ayres and A. Kneese, "Production, consumption, and externalities," *Amer. Econ. Rev.*, June, 1969.

[3] R. Tummala and L. Connor, "Mass-energy based economic models," *IEEE Trans. Syst., Man, Cybern.*, vol. SMC-3, pp. 548-555, Nov. 1973.

[4] J. Krenz, "Energy per dollar value of consumer goods and services," *IEEE Trans. Syst., Man, Cybern.*, vol. SMC-4, pp. 386-388, July 1974.

[5] R. Herendeen, "An energy input-output matrix for the United States, 1963: User's guide," Cent. Advanced Computation, Univ. Illinois, Urbana, Doc. 69, Mar. 1973.

[6] "Input-output structure of the U.S. economy: 1963, vol. 1-3," Bur. Econ. Analysis, U.S. Dep. Commerce, 1969. (Available from U.S. Gov. Printing Office.) BEA has published several amplifying articles: "Input-output structure of the U.S. economy: 1963," *Survey of Current Business*, Nov. 1969; "Personal consumption expenditures in the 1963 input-output study," *Survey of Current Business*, Jan. 1971. In addition, they have published a report, available on request: "Definitions and conventions of the 1963 input-output study," Apr. 1972. The complete tables are also available on tape from BEA.

[7] E. Isaacson and H. Keller, *Analysis of Numerical Systems.* New York: Wiley, 1966, sec. 1.1.

[8] L. Koenig and P. Ritz, "Secondary product adjustment with redistribution," Nat. Planning Assoc., Washington, D.C., NRAC Tech. Rep. 67, 1967.

[9] C. Almon, "How to purify an input-output matrix," Interindustry Forecasting Project, Univ. Maryland, College Park, Res. Memo 2, 1967.

[10] R. Herendeen, unpublished computations.

[11] "Interindustry transactions in new structures and equipment, 1963," *Survey of Current Business*, Aug. 1971. See also a more detailed but unpublished version (367 producing, 83 consuming sectors) available from the Bureau of Economic Analysis, U.S. Dep. Commerce, Washington, D.C.

[12] K. Kirkpatrick, "Independent verification of I/O energy results," Cent. Advanced Computation, Univ. Illinois, Urbana, Tech. Memo., 26, July 1974.

[13] B. Noble, *Applied Linear Algebra.* Englewood Cliffs, N.J.: Prentice Hall, 1969, Sec. 13.4.

[14] A. Sebald, "Sensitivity analysis of linear static matrix models," Cent. Advanced Computation, Univ. Illinois, Urbana Doc. 22, Nov. 1974.

[15] B. Hannon and F. Puleo, "Transferring from urban cars to buses: The energy and employment impacts," presented at the Int. Symp. Effects of the Energy Shortage on Transportation Balance, Pennsylvania State University, University Park, May 1974. (Also CAC Doc. 98.)

[16] R. Herendeen, "Affluence and energy demand" presented at the 98th ASME Winter Annu. Meet., Detroit, Mich., Nov. 1973 (Paper 73-WA/ENER-8). Also in *Mechanical Engineering*, vol. 96, no. 10, Oct. 1974. (Also CAC Doc. 102.)

[17] R. Bezdek and B. Hannon, "Energy, manpower, and the highway trust fund," *Science*, vol. 185, p. 669, Aug. 1974.

[18] C. Bullard and R. Herendeen, "Energy use in the commercial and industrial sectors of the U.S. economy, 1963," Cent. Advanced Computation, Univ. Illinois, Urbana, Doc. 105, Nov. 1973.

[19] C. Bullard, "Energy conservation through taxation," Cent. Advanced Computation, Univ. Illinois, Urbana, Doc. 95, Feb. 1974.

[20] K. Kirkpatrick, "Effect of including capital flows on energy coefficients, 1963," Cent. Advanced Computation, Univ. Illinois, Urbana, Tech. Memo. 32, Aug. 1974.

[21] Electric Energy Assoc., *Annual Energy Requirements of Electric Household Appliances*, EEA 201-73, 1973.

[22] *Merchandising Week*, vol. 104, no. 9, Feb. 28, 1972.

[15]

11 July 1975, Volume 189, Number 4197

SCIENCE

Energy Conservation and the Consumer

A tax placed on energy and adjusted to wage levels would ease a change to a more labor intensive economy.

Bruce Hannon

In this article I seek to explain what is known about the connections among consumer activities, the direct and indirect demands of these activities for energy and employment, and the consumer's options for energy conservation. When the consumer's activities demand employment, they require the consumer's own services, but when they demand energy they require the stored energy resources of the earth. This fact underpins the three dilemmas of energy conservation that I will discuss in detail. The consumer's control over the energy he uses to heat, cool, and light his residence or fuel his auto will not be discussed; the methods for conservation in these areas are well understood.

Individual control over energy use is viewed as an alternative to governmental assumption of that control. Energy conservation is necessary because of (i) the environmental effects of unbridled consumption; (ii) the long lead time and massive capital allocations necessary for future increases in energy supply; (iii) the instabilities associated with a large dependence on foreign energy supplies; (iv) the need for an enduring national goal that unifies the nation and does not require massive new economic growth; and (v) because energy is a fundamental ingredient in any economic system. This last point indicates the necessity to focus on the conservation of energy rather than steel, for example. A proper plan to conserve energy would conserve steel. In general, capital can be recycled but, of course, energy can not. Labor can not only be recycled, it can be multiplied.

During the industrial revolution in the United States there has been an unprecedented thrust to substitute the energy from fossil fuels for human labor to accomplish material transformations (toolmaking) and to achieve mechanization. From the worker's and consumer's view, this substitution has generally meant less drudgery in the work place, greater buying power, and more leisure. To the industrialist, the substitution of tools and energy for labor has provided a means to control and predict production costs and, consequently, profits.

Material and energy costs have been low, relative to labor, as well as predictable for almost 50 years. Machines and fuels are incapable of striking for higher wages. Economic growth seems to have absorbed those workers displaced by mechanization. Thus technology, material and energy use, and employment and economic growth appear to be intimately linked. But now the finiteness of resources comes into view as this nation's striving for a better life spreads internationally. We are beginning to understand the links among heavy consumption, environmental damage, and degradation of the quality of life. We are beginning to see the need for one more adaptive act of humankind, the need for a group morality.

It should be within the reach of each individual consumer to perceive that he must forgo certain forms of energy consumption now in order to ensure their availability to his offspring or to his own generation in the future. As the human body eventually reaches a condition of zero growth, then so must that population as a whole. Such an achievement, however, would require our casting out such mathematically impossible maxims as "the greatest good for the greatest number" and substituting, perhaps, "life-styles of elegant frugality." Then we would act to preserve the knowledge of how to live healthfully, limit our numbers, and minimize our per capita consumption of resources, particularly energy. We would recognize that, because people's wants are infinite and because resources are finite, only relative wants are important; and that economic growth increases the disparity in relative wants, eases our severe interpersonal and interindustry competition, and increases the uncertainty of future events.

Economic growth and development will probably continue until the assured available energy is almost entirely demanded for maintenance of the society. Stabilization of economic growth now is often argued as being a means of reducing the inevitable future trauma of living on an energy budget. There would obviously be problems, however, with the egalitarian national life-style that would be required to achieve stabilization now. We would have to recognize that the rising world population will result in the rising collision of individual and national freedoms as well as in the accelerated depletion of stored energy; and that the concept of individual and corporate ownership, apparently an adaptation to resource shortages, will probably give way to public or societal ownership, as the shortages worsen. It is unclear whether a totally democratic society could exist with severe shortages. If it could not, smaller and smaller fractions of the population would participate in resource allocation decisions. How to control the decision-makers would then become a problem. As long as the population at large could perceive an equitable distribution of the dwindling resources then,

The author is director of the energy research group at the Center for Advanced Computation, University of Illinois at Urbana-Champaign, Urbana 61801

presumably, the society could remain stable. Plato's concept of leaders or "guardians" deprived of all physical wealth and steeped in leadership seems an appropriate model for controllers of a stable society with limited resources.

Before we plan to conserve energy or to increase our energy supplies, we must understand the total energy and employment cost of every good and service. Only then will we be able to determine the energy conserved by consuming one good or service instead of another, or the energy cost of substituting a new technology, a new manufacturing process, for example, for another. The changes that may have to be made in order to reduce our energy consumption can be thought of as efficiency improvements in our current life-style. As I shall show, many such changes may be resisted and, unless they are fully understood, they may not provide significant net energy savings.

To calculate the energy cost of one unit of an item, one must determine the direct inputs of goods and services required to produce that item. Because each of these inputs has its own set of inputs, and so on, one obtains such a multitude of small inputs that leaving off the next round does not significantly change the total energy requirements of the original item. I have used the economic input-output data published by the U.S. Department of Commerce (*1*), and the data of Herendeen (*2*), to calculate the direct and indirect energy and employment demands of a number of personal consumption activities and to analyze trade-offs between human and mechanical energy.

Substituting Energy for Labor: The First Dilemma

If conserving energy or any material resource is judged to be the main path to our survival, the impact of such a path on employment must be known. Since we, as laborers, decide our future, and since we have repeatedly increased our wages and controlled the cost of the resource base, the potential redistribution of labor resulting from a conservation program is of primary importance.

Let us observe the problem from the point of view of an industrial decision-maker. He compiles his current capital, materials, labor, and energy costs and projects them on the basis of his expectations of the future. Trend analyses would have shown that his raw materials and energy costs are either stable, unless his product is derived from agriculture, or declining, as in the case of electricity, his principal energy substitute for labor. In terms of dollars, tool costs would have been rising primarily because of their labor content, but the most rapidly rising, and the most unpredictable, costs would have been those for manual labor. An important factor in the industrialist's decision-making process would be his anticipated relative unit costs of energy and of labor. The ratio of these two costs are formed into an index (wage divided by price of electricity) and plotted for the years 1926 to 1973 in Fig. 1. In that period the ratio grew by a factor of 10. It grew by 230 percent between 1950 and 1970, during which time electrical machinery costs increased by about 54 percent (current dollars). The curve in Fig. 1 shows a leveling to a declining trend during recession periods, possibly indicating an attempt of the economic system to equilibrate resource and labor costs.

Total costs of electricity relative to total product costs are small. In 1963, one average dollar of industrial sales caused only about 2.5 cents of electricity sales (*3*), and about 65 cents went for wages, directly and indirectly (*4*). As shown in Fig. 1, the rapidly rising cost of capital since 1950 has probably deterred a more rapid switch to a more electric, less labor intensive production system. Since 1950, the marginal productivities of labor and capital have been rising relative to that of electricity, while the ratio of the marginal productivities of labor and capital has been relatively neutral. This capital cost, however, contains the cost of building and heating machinery which cannot be considered a substitute for labor. The generally higher cost of capital should therefore aid in the return to a more labor, less electric intensive system. The point is twofold: the cost of electricity has failed to represent its importance in the marketplace, and the situation has grown worse with time.

If we take energy, labor, and capital to be factors required for production, then, as the cost of energy is increased relative to the cost of labor, the economic system will adjust, in theory, such that the marginal productivity (a change in gross national product resulting from a unit change in factor use) of energy will increase relative to that of labor. If the wage levels and labor productivity are constant, this change means lower energy use. In other words, Fig. 1 indicates that the productivity of labor has, historically, increased more rapidly than, and at the expense of, electrical energy productivity. To reverse this process, the price of electricity must rise relative to wages. This same phenomenon appears to be true for overall energy and labor use (*5*, *6*).

The overall effects of economic growth on energy and labor demand were suggested by an investigation (*7*) in which it was demonstrated that if, in 1963, the economic sectors had been allowed to grow at the proportional expense of all the remaining sectors, most industries would have become more energy demanding and less labor demanding. In a special case study (*8*, *9*) it was shown that the system that produces beverages would be less energy demanding and more labor demanding if the beverages were delivered in refillable rather than throwaway containers. Most important was the type of labor that would be involved in the change. Jobs would be lost in the highly organized, high-wage

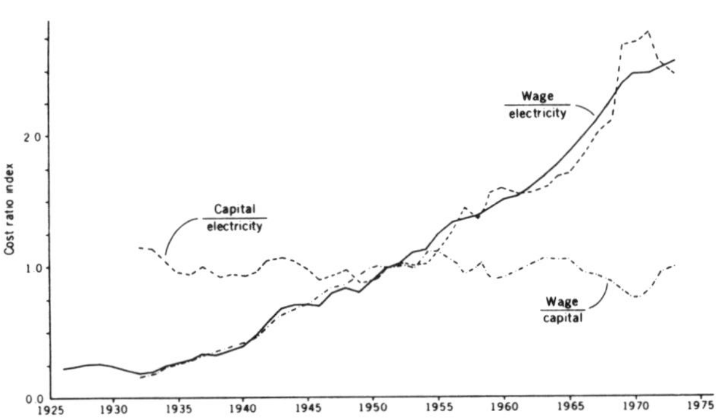

Fig. 1. Industrial labor, electricity, and capital cost ratios in the United States, 1926 to 1973, in current dollars. Based on manufacturing worker's hourly wages, the industrial price of a kilowatt-hour of electricity, and the yields on AAA corporate bonds (1951 = 1.0). Data from the Department of Commerce (*1*), Edison Electric Institute (*21*), and Moody's *Industrial Manual* (*22*).

paying, can-maker's plants and would be gained in the low-wage paying, relatively nonorganized, retail sector. The lack of acceptance of this plan was the direct result of opposition from organized labor and intensive lobbying by can- and bottle-making interests.

Another example of increasing energy and decreasing labor utilization is found in a comparison of the energy and employment costs of protein production (*3*). The production of a unit of milk protein requires about 2.7 times as much energy as a unit of cheese protein, and yet requires only about one-fifteenth as many jobs. This result derives from the nature of the different processes: milk contains mostly water and it must be homogenized, pasteurized, and trucked, each of which operations is an energy but not labor intensive process.

Herein lies the first dilemma. Energy conserving policies would increase overall employment in general by decreasing the number of high-wage jobs and increasing the number of low-wage jobs, but the people holding the former belong to the most highly organized unions. Thus organized labor, after struggling to achieve reasonable wages and personal dignity in the marketplace, and having slowly relinquished much dignity because of the inevitable relationship between high wages and assembly lines, for example, would now have to be asked to forgo their higher wages in order to avoid the general unemployment induced by declining energy supplies. Such a situation would be tenable only if prices and profits were thoroughly controlled. It also follows that technological developments have, historically, resulted in jobs being denied to marginally employed workers.

Income and Energy Use: The Second Dilemma

The relation between personal income and energy use can best be introduced by asking three questions:

1) What fraction of total energy use does the U.S. consumer control (directly and indirectly) through his purchase of goods and services?

2) How does this direct and indirect fraction vary with income?

3) How might consumers proceed to change their purchasing habits so as to reduce their energy dependence?

The first question is answered in Table 1, in which the data indicate the importance of personal consumption expenditures. Such expenditures demanded two-thirds [one-third of this amount for the automobile (*10*)] of all U.S. energy and 62 percent of all employment in 1963. Government activities accounted for about 13 percent of all energy consumption and employed about 21 percent of the total labor force. Capital formation demanded about 10 percent of all energy and 12 percent of all employment. Exports accounted for most of the remaining energy (coal) and labor consumed. The fractions of the various energy types allocable to personal consumption is surprisingly constant, except for coal. Most coal was consumed in steelmaking and electrical generation and about 16 percent was exported (directly and indirectly) in 1963. In the near future it will be possible to compare the data of Table 1 with similar data from 1967.

Figure 2 provides the answer to the second question. Family income after taxes is plotted against the direct, indirect, and total energy demand associated with the spending of that income (*10*). For example, the poor generally traveled by bus (when they traveled) and the rich usually traveled by plane. Both are indirect energy uses. Middle income people traveled by auto, which represents a direct energy use. The rate of direct energy use slowed as income increased, but the rate of indirect energy use did not. Together they yielded a total energy demand that was nearly linear with income. Here we find the second dilemma: spending money in any way demanding energy, and the extra dollar spent required almost the same energy when spent by a poor family as it did when spent by a rich one. It appears that the only way to save energy is to reduce income, unless the variation of energy consumption about the "total" line is very large. However, as I shall show in the next section, saving energy means saving dollars and that leads to additional energy use.

From the supporting data for Fig. 2 (*10*), it is evident that the average family used about 55 percent of its total energy directly. It is also evident that two-thirds of the lowest income group's total energy use was direct. Just the opposite is true of the upper income group: two-thirds of its total energy use was indirect. A further examination of these data shows that modification of the behavior of the upper middle income class would be the most fertile area for energy conservation. Herendeen is now investigating the energy-income redistribution effects in view of the slight nonlinearity of the total energy line in Fig. 2.

The third question is more difficult to answer. Consumer spending patterns would appear to be a function of income level. That is, as family income increases, buying patterns change. For example,

Table 1. Percentages of energy and labor (direct and indirect) for final demand activities, during 1963. The data are from Hannon and Abbott (*15*). The definitions of the activities are from (*1*).

Activity	Energy type: Coal	Crude oil	Refined petroleum	Electricity	Gas	Total primary energy	Labor*
Personal consumption expenditures	53.2	71.3	70.8	69.2	72.2	66.6	62.2
Gross private fixed capital formation	14.9	8.6	7.3	10.5	10.5	10.3	12.1
Net inventory change	1.2	1.3	1.5	0.9	1.0	1.3	0.9
Exports†	15.9	6.6	7.7	4.9	5.0	9.0	3.6
Federal government purchases, defense	6.6	5.6	6.1	6.5	4.7	5.9	8.4
Federal government purchases, other	1.0	0.9	0.9	0.4	0.9	0.9	2.5
State and local government purchases, education	2.4	1.8	1.6	3.0	2.0	2.0	4.6
State and local government purchases, health, welfare, and sanitation	1.0	0.8	0.7	1.1	0.9	0.8	1.5
State and local government purchases, safety	0.2	0.3	0.3	0.2	0.2	0.2	0.9
State and local government purchases, other	3.6	2.9	3.1	3.3	2.5	3.1	3.3
Total final demand	100.0	100.0	100.0	100.0	100.0	100.0	100.0

*Man-year basis: includes household and government workers but not military personnel. †Dollar value of imports not used.

meat becomes a larger part of the diet; travel is more frequent and by faster modes; and more investments are made. As income rises, more goods and services are consumed, resulting in rising indirect energy demands.

As an example of the utility of Fig. 2, let us consider a person living in New York. It was reported in 1973 (*11*) that the average New Yorker is much less energy demanding than the average U.S. citizen, even though his personal income is 25 percent higher than average. This, it was assumed, was because New Yorkers live in apartments and use mass transit heavily. But the data of Fig. 2 suggest that if a person has a 25 percent higher than average income, then his energy use is also nearly 25 percent higher than average. It is apparently difficult and expensive in both dollars and energy to service people living in a city. Or, perhaps living in a city induces a more frenetic, consequently more costly, pattern of behavior.

Figure 3 with the enlargement of its central portion shows the relative (direct and indirect) energy and labor cost per 1963 dollar of personal consumption on each of the 83 different consumption categories of the U.S. Department of Commerce. In general, the services, except transportation, seem to be distributed along a roughly horizontal (constant energy intensity) line with high-wage services on the left and low-wage services to the right (for example, physicians on the left, hospitals on the right). Goods and transportation services are distributed along a roughly vertical line which joins the service line at about its lower-third point. The energy purchases are, logically, highest on this line.

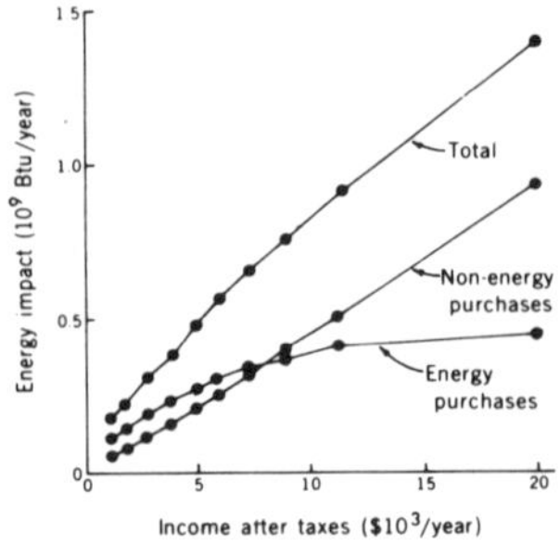

Fig. 2. Direct, indirect, and total family energy impact plotted against family income, 1960 to 1963. From Herendeen (*10*).

The possible lack of independence among some of the personal consumption items is demonstrated by the averaged point "owning and operating an auto" in the upper-center portion of Fig. 3, and by comparing this point with the points "gasoline and oil," "new and used cars," "auto repair," "auto insurance," and "tires." Knowledge of the individual auto expenditures could guide the consumer on the energy and employment effects of his marginal dollar expenditures, but the overall effect of owning an auto is given by the "average" point.

Figure 3 gives the total energy and employment effects produced by the spending of a consumer dollar on the 83 different consumption categories. The figure does not show the secondary effects. Suppose, for example, that you purchase a new car. The energy and employment effects shown in the figure represent only those effects which occur in the web of industries which supply the auto industry (including the fabricating, transporting, and retailing). These effects do not include the energy and employment demands created by the personal consumption of the workers who receive your new car money from the manufacturer, nor the energy and labor demanded by the workers of the producers of these goods as they too spend their wages. As money passes through each new set of hands, some is saved and taxed and eventually the effect dies out. If one assumes that for every dollar spent on a new car, two more are spent via the multiplier effect on average personal consumption, then each new car dollar spent, which represents one dollar of growth in the gross national product, demands 60,000 British thermal units (new car) and 172,000 Btu (two average 1963 personal consumption expenditure dollars) for a total of 232,000 Btu (1 Btu is equivalent to 1055 joules). The expansion from 60,000 to 232,000 Btu demonstrates the effects of the growth dollar on energy demand. It has a similar effect on employment demand.

Table 2. Selected results on the total dollar, energy, and labor impacts of consumer options in transportation during 1971. Data are expressed in terms of the requirements to move a million passengers 1 mile (1.6 km) (NA, not applicable).

Transportation mode	Load factor	Thousands of dollars (1971)	Energy*		Jobs
			10^6 Btu	Direct (%)	
		Intercity transportation			
Car	2.9 people	55	5900	51	3.7
Plane	53% full	58	9800	73	3.8
Bus	47% full	39	2700	51	3.1
Train	37% full	44	4000	58	7.2
Electric commuter†	31% full	128	9900	11	8.5
		Urban transportation			
Car‡	1.9 people	69	8900	58	4.2
Bus‡	12.0 people	105	5300	57	8.3
Motorcycle	1.1 people	57	4200	49	1.55
Bicycle	1.0 people§	26§	1300	59§	1.7‖
Walking	1.0 people	NA	710	51	0.7

*Vehicle transportation fuel only. †The "PATH" commuter system, New York-New Jersey, 1971 [see Penner (*16*)]. Other intercity data from Sebald and Herendeen (*17*). ‡Data from Hannon and Puleo (*18*). §From Hirst (*19*). ‖Center for Advanced Computation estimate based on Hirst (*19*).

Table 3. Intercity transportation. The energy that would be saved by shifting from each transportation mode to another. Plus or minus signs preceding numbers indicate, respectively, an increase or a decrease in energy use; plus or minus signs after the numbers indicate, respectively, an increase or decrease in dollar cost. Calculated from Table 2.

Shifting from	Shifting to			
	Car	Plane	Bus	Train
Car		+355,000+	200,000	633,000
Plane	355,000		-374,000	414,000
Bus	+200,000+	+374,000+		+260,000+
Train	+633,000+	+414,000+	-260,000	

Respending Dollars Saved: The Third Dilemma

Let us now look more deeply into the energy and labor needed to supply an entire service. Herendeen (*2*) has shown that all transportation consumed (directly and indirectly) about 42 percent of the U.S. energy in 1963, with the personal auto accounting for half of this demand. Such knowledge prompted us at this laboratory to investigate energy use and transportation options. Table 2 shows the dollar, energy, and employment costs of ten alternative passenger modes. The intercity modes were developed for an approximate 300-mile (482.7-kilometer) intercity distance in order to allow the five modes to be considered competitive. All costs are very sensitive to load factors. In terms of dollars and energy, the plane is easily the most expensive, and the train is the most employment intensive. The intercity bus costs the least in dollars and in energy.

The urban passenger has a variety of modes to choose from, as shown in Table 2. These modes are increasingly unpopular but decreasingly energy expensive as one moves down the list. Energy consumed in walking was that used to supply the food consumed by the average person for the energy used in excess of that used by the body in the resting position.

It has been pointed out by Bullard (*12*) that the important factor in energy conservation is the rate at which energy is saved on the transfer from one activity to another. Tables 3 and 4 show rates of energy savings (British thermal units saved per dollar saved) for shifts from each transportation mode to the other. The rates vary from about 17,000 to 633,000 Btu per dollar. In all cases except in the urban transportation shifts from car to bus, dollars are saved if energy is saved, and vice versa. For example, the traveler who switched from urban bus to bicycle would save energy (and dollars) at the rate of 51,000 Btu per dollar. If he were not careful to spend his dollar savings on an item of personal consumption which had an energy intensity greater than 51,000 Btu per dollar then his shift to the bicycle would have been in vain.

It is easy to notice that in every instance, a change in transportation mode that would conserve energy would also save dollars (except in the case of changing from urban car to bus transportation).

Table 4. Urban transportation. The energy that would be saved by shifting from each transportation mode to another. Plus or minus signs preceding numbers indicate, respectively, an increase or a decrease in energy use; plus or minus signs after the numbers indicate, respectively, an increase or decrease in dollar cost. Calculated from Table 2.

Shifting from	Shifting to				
	Car	Bus	Motorcycle	Bicycle	Electric commuter
Car		100,000+	392,000-	177,000	+ 17,000+
Bus	+100,000-		0	51,000	+200,000+
Motorcycle	+392,000+	0		61,000	+ 80,000+
Bicycle	+177,000+	+ 51,000+	+ 61,000+		+ 84,000+
Electric commuter	- 17,000-	-200,000-	80,000-	84,000	

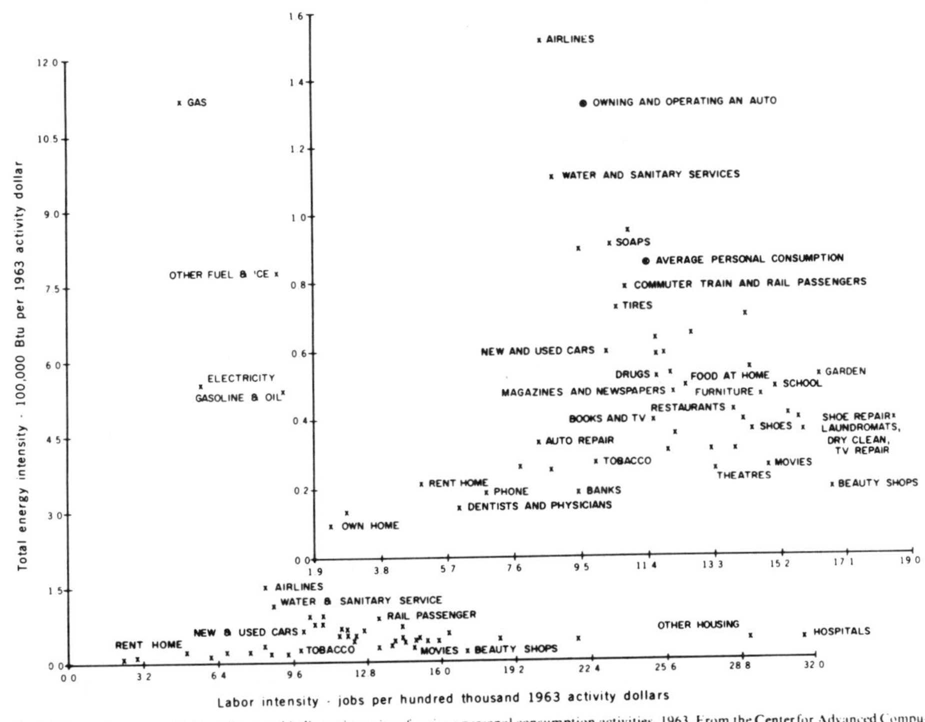

Fig. 3. The total energy and labor (direct and indirect) intensity of various personal consumption activities, 1963. From the Center for Advanced Computation's energy-employment model, August 1974. The inset is an enlargement of the central portion of the data.

This provides us with the third dilemma: that is, what does the consumer do with the dollars he saves after he has shifted to a cheaper mode of transportation? He can spend it or save it. In either case, energy will be required to provide for this freed expenditure. The second dilemma indicates that the consumer will adjust to increasing wealth through a change in purchasing habits, and his total energy use will rise proportionately to his income. However, armed with certain information, the consumer can purposefully direct his income so that he does save energy on balance. It should be clear, however, that he can never save more energy by redirecting certain portions of his income than he can by becoming that much poorer.

Table 5 is a rank order listing by energy intensity of the top 20 (in terms of dollars) personal consumption activities. Data on the activities of owning an auto and owning a home are listed so that they can be compared with the energy intensity of the average personal consumption dollar. Labor intensities are also given so that the interested reader may determine the impact of his own respending on the labor situation.

In the above example, the bus passenger turned bicycle rider (see Table 4) could spend his dollar savings on anything below and including "private investment" in the activity list of Table 5, and save some energy. The net energy is found by subtracting the number in Table 5 from the transfer savings of Table 4. If, for example, his savings were spent on "restaurants," the net energy savings would be 51,000 minus 32,400, which equals 18,600 Btu per dollar saved. Also from Tables 2 and 5, one can calculate the net impact on labor at four jobs gained for every million dollars saved on the bus to bicycle transfer.

Most of the savings rates are so large in Tables 3 and 4 that they could be overcome only if the consumer spent his savings directly on electricity or gasoline and oil. However, the consumer must be aware that the purchase of an extra car, for example, means greater purchases of gasoline and oil as well.

An interesting case arises in the shift from urban car to bus. Here the consumer saves 100,000 Btu per extra dollar spent on bus transportation. This extra dollar must be produced by reducing expenditure on other things, since urban bus transportation is more expensive than car (Table 2). Suppose the extra dollars came from a reduction in average personal consumption spending; then from Table 5, the total energy savings rate would be 170,000 Btu per dollar.

Table 6 shows the same phenomena as Tables 2 to 4. But in this case, home appliance alternatives are compared. The annual total dollar, energy, and employment costs were compiled for manufacturing, operating, and maintaining each appliance. From these data the energy savings rate can be calculated and compared to the data of Table 5 to determine the net energy savings.

Significant energy savings rates are obtainable if fewer appliances are used and if gas is used instead of electricity in stoves, dryers, and water heaters. This savings rate is reduced, if not reversed, by the energy demands of the alternative consumption activity.

Table 5 shows that the federal tax dollar and the private investment dollar are relatively l[illegible] in energy intensity. This fact indicates [illegible]at shifts from average personal consumption to federal spending or investment would reduce energy demands. However, many of these activities, for example, building highways or power plants, foster accelerated energy use and lower employment demands than personal consumption. Heavy construction programs can therefore be viewed as energy-wasteful disemployment programs. It is tempting for the energy conservation policy-maker to devise methods to divert the personal consumption dollars into federal spending or private capital formation. If such a procedure were followed for the purpose of conservation, it would be truly deceptive not to focus these dollars on processes which lead to a lowering of the average national energy intensity [see (*13*) for example].

In general, if all consumers were to make such carefully precalculated consumption shifts as outlined above, national income would be preserved, the energy intensity of personal consumption would be lowered, and employment would be maintained if not increased—and it would certainly be redistributed. However, the amount of net energy savings might be small because of the respending effect, and because of the fact that people will not want or need a sufficient quantity of the lower energy goods and services, and because of the likelihood of too few people actually being aware of their options. In any event, there is a limit to the savings that can be realized by such life-style changes which preserve national income.

Conclusion: Resolving the Three Dilemmas

The problem of energy conservation by the consumer has been examined in terms of three dilemmas: First, when wages increase relative to costs, then energy use increases through the process of mechanization. Second, energy use and income are linearly connected such that the spending of an average additional dollar of income demands nearly the same amount of energy, regardless of one's income level. Thus, doubling one's income doubles one's energy use so that the phenomenon of "rising expectations," so ingrained in the

Table 5. The energy and labor intensity of the 20 activities in which personal consumption expenditures (PCE) are highest in terms of dollars. Ranked in order of decreasing energy intensity during 1971. From Hannon and Puleo (*18*).

PCE sector description	Energy intensity (Btu per dollar)	Labor intensity (jobs per $1000)	PCE sector description	Energy intensity (Btu per dollar)	Labor intensity (jobs per $1000)
Electricity	502,500	0.0436	(Federal taxes)	(36,300)	(0.082)
Gasoline and oil	480,700	0.0729	Women's and children's clothing	38,100	0.1000
(Housing)	(144,000)	(NA)	Restaurants	32,400	0.0875
(Auto ownership)	(111,500)	(.081)	Men's and boys' clothing	31,400	0.0984
Cleaning preparations	78,100	0.0733	Religious and welfare activity	27,800	0.0863
(Average PCE)	(70,000)	(0.080)	Private hospitals	26,100	0.1718
Kitchen and household appliances	58,700	0.0551	Automobile repair and maintenance	23,500	0.0483
New and used cars	55,600	0.0775	Financial interests except insurance	21,500	0.0784
Other durable house furniture	54,600	0.0894	Tobacco products	19,800	0.0585
(Private investment)	(45,600)	(0.066)	Telephone and telegraph	19,000	0.0585
Food purchases	41,100	0.0852	Rented home	18,300	0.0350
Furniture	36,700	0.0917	Physicians	10,700	0.0325
			Own home	8,300	0.0167

American worker, is limited by the amount of available energy. Third, saving energy usually means saving money, the respending of which reduces, if not eliminates, the energy first thought saved. Not respending the savings means a reduction in national income (gross national product) and money saved too quickly means a reduction in employment. But this last effect is somewhat abated by the general increase in labor intensity that is associated with the lower average energy intensity. This effect could also be thwarted by each worker agreeing to work fewer hours. Both solutions, that is, not respending one's savings and working fewer hours, would eventually result in decreased per capita wealth. It should also be pointed out that convenience demands energy, for example, high speed autos and airplanes, and one-way packaging. Giving up convenience is the equivalent of the consumer's accepting a lower implicit wage, and saving energy.

There probably are no popularly acceptable solutions to these three dilemmas. If wages, as well as proprietor's incomes, were reduced relative to the cost of energy, or if the population decreased, or both, net energy consumption would be reduced without creating widespread unemployment. However, strong labor unions attest to the irreversibility of wage rates.

Perhaps an initial energy resource tax, geared proportionately to the wage level, could be placed on fuels, particularly on those used to make electricity. Such a tax would thrust our most important resource into a position in the marketplace competitive with its chief substitute, manpower. The tax revenue could be used to reward those who reduced their energy use or those who were severely disadvantaged by the tax. In effect, this money could be used to cover the initial costs of converting to a life-style with lower energy demands. With energy costing more, the severance tax revenue would generally be spent in a manner compatible with reducing our energy requirements. Also, in the long run, the tax would imply lower real wages because of its demanding a lower labor productivity through the substitution of labor for energy.

It is appropriate that an already energy intensive nation take steps to increase its productivity per unit of energy expended. Raising the cost of energy relative to the cost of its substitute appears to be the best way for us to achieve such an increase in that it induces through the market mechanism the change to a more labor intensive technology. The tax approach might also give us time and revenue to redirect technology into improving productivity from available energy and into easing the transition to lower labor productivity. While this relative cost change would probably be produced by an economic recession, through a reduction in the average real wage, the uncontrolled wage reduction that might result under such conditions could also cause much unemployment unless the technology of production could very rapidly become more labor intensive. The ratio of the energy tax to wages should be increased at a rate which maximally increases productivity from energy without creating an unacceptable level of unemployment.

The impact on prices of the two basic types of energy taxes (one based on the unit of energy, the other on the dollar value of energy) were estimated by Bullard (*6*). The tax based on the British thermal unit is preferred by those favoring energy conservation because it does not discriminate against small energy users who normally pay a higher unit price for energy than do large users.

Another method that would conserve energy, and that is more fundamental than taxation, would be energy rationing through the use of coupons. Energy storage would be guaranteed to the consumer by his being given coupons that could be used for different types of energy, such as gasoline or electricity, at different exchange rates. Thus the consumer would have the option of conserving energy through the process of ownership of a stored versatile fuel. This concept implies the existence of an energy interest rate that depends on consumer decisions to forgo present energy consumption in order to ensure a future supply. The energy coupons would be purchased from the federal government which consequently would need to "own" the energy resources. Such an ownership would allow the society to jointly set a national energy interest rate which would dictate how rapidly the proved energy reserves of the society could be consumed.

In the long run, we must adopt energy as a standard of value (*7*) and perhaps even afford it legal rights (*14*). For energy, like no other entity, cannot be recycled or multiplied. It flows from the sun and from the energy inventories of the earth. Shall we live off the flux or the storage? How fast and under what circumstances shall we draw it from storage? We would answer these fundamental questions correctly if we set energy as the basis of value and structured our economic and legal system accordingly.

As a final point, we must consider the major problem related to the veracity of the energy crisis. Is the crisis contrived or real? The conservative approach is to believe that it is real and act accordingly. If the crisis is real, we shall be viewed by future generations as terribly wise; if the crisis is false, we shall have had an instructive relief from the dancing mirage of our manifest destiny.

Table 6. The dollar, energy, and labor impacts of consumer activities in the home during 1971 (including manufacturing, operating, and maintaining the home). Herendeen and Sebald (*20*).

Type or number of appliances	Requirement per year: Dollars (1971)	10^6 Btu	Hundredths of a job	Energy savings rate* (Btu per dollar)
Kitchen appliance sets†				
Spartan (4 appliances)	152.50	64.9	1.09	+304,000
Moderate (11 appliances)	252.50	95.3	1.85	+254,000
Plush (20 appliances)	337.90	117.0	2.47	
Clothes dryer				
Gas	27.6	8.1	0.23	+500,000
Electric	39.4	14.0	0.27	383,000
Outdoor	3.4	0.2	0.03	
Stove				
Gas	33.5	13.4	0.26	+198,000
Electric	50.7	16.8	0.36	
Refrigerator (12 cubic feet‡)				
Conventional	45.0	11.1	0.35	+608,000
Frost-free	55.7	17.6	0.40	
Water heating				
Gas	46.0	37.7	0.22	+325,000
Electric	104.0	56.6	0.50	
Electronic media				
Black and white television	21.7	5.4	0.18	+ 79,000
Color television	67.2	9.0	0.63	120,000
Radio	4.1	1.2	0.03	

*When the activity is compared with the next one down the list; plus and minus signs mean, respectively, increases or decreases in energy use and dollar cost. †Including a stove, clothes dryer, and water heater all run on gas. ‡Twelve cubic feet are equivalent to 0.33 cubic meter.

References and Notes

1. U.S. Department of Commerce, *Input-Output Structure of the U.S. Economy, 1963* (Government Printing Office, Washington, D.C., 1969).
2. R. A. Herendeen, *Use of Input-Output Analysis to Determine the Energy Cost of Goods and Services* (Center for Advanced Computation, Document No. 69, Univ. of Illinois, Urbana, 1973).
3. B. Hannon, *Technol. Rev. (M.I.T.)* **76** (No. 4), p. xx (February 1974).
4. A. J. Walderhaug, "The composition of value added in 1963," *Survey of Current Business* (Government Printing Office, Washington, D.C., 1973), pp. 34–44.
5. U. S. Department of Commerce, *Handbook of Labor Statistics* (Government Printing Office, Washington, D.C., 1972), p. 220.
6. C. W. Bullard, *Energy Conservation Through Taxation* (Center for Advanced Computation, Document No. 95, Univ. of Illinois, Urbana, 1973).
7. B. Hannon, *Ann. Am. Acad. Polit. Soc. Sci.* **410**, 139 (November 1973).
8. ———, *Environment (St. Louis)* **14** (No. 2), 14 (March 1972).
9. H. Folk, *Two Papers on the Employment Effects of Mandatory Deposits on Beverage Containers* (Center for Advanced Computation, Document No. 73, Univ. of Illinois, Urbana, 1972).
10. R. A. Herendeen, *Am. Soc. Mech. Eng. Tech. Digest* 73-WA/ENER-8 (1973).
11. Resources for the Future, *Patterns of Energy Consumption in the Greater New York City Area* (Washington, D.C., July 1973).
12. C. W. Bullard, Energy Research Group, Center for Advanced Computation, Univ. of Illinois, Urbana.
13. R. Bezdek and B. Hannon, *Science* **185**, 669 (1974).
14. C. D. Stone, *Should Trees Have Standing?* (Kaufmann, Los Altos, Calif., 1974), p. 12. In this classic piece, Stone argues for legal rights for natural objects in the same sense that corporations have them (neither are "persons"). It is not too difficult therefore to wish to extend legal rights to such nonphysical entities as energy, the main purpose being to provide legal controls over the rate of its consumption and depletion and even over the purposes for which it is used.
15. B. Hannon and N. Abbott, *Dollar, Energy and Employment Impacts of Personal Consumption Activities* (Center for Advanced Computation, Document No. 128, Univ. of Illinois, Urbana, 1974).
16. P. Penner, *Dollar, Energy and Labor Cost: Electric Commuter, Intercity Truck, Intercity Bus, Motorcycle, Walking* [Technical Memos. 24, 19B, 31, 30, 29, respectively (Center for Advanced Computation, Univ. of Illinois, Urbana, 1974)].
17. A. Sebald and R. Herendeen, *The Dollar, Energy and Employment Impacts of Air, Rail and Automobile Passenger Transportation* (Center for Advanced Computation, Document No. 96, Univ. of Illinois, Urbana, 1974).
18. B. Hannon and F. Puleo, *Transferring from Urban Cars to Buses: The Energy and Employment Impacts* (Center for Advanced Computation, Document No. 98, Univ. of Illinois, Urbana, 1974).
19. E. Hirst, *Energy Use for Bicycling* [Report ORNL-NSF-EP-65 (Oak Ridge National Laboratory, Oak Ridge, Tenn., 1973)].
20. R. Herendeen and A. Sebald, *The Dollar, Energy and Employment Impacts of Certain Consumer Options* (Center for Advanced Computation, Document No. 97, Univ. of Illinois, Urbana, 1974).
21. Edison Electric Institute, *Historical Statistics of the Electric Utility Industry Through 1970* (Edison Electric Institute, New York, 1970), table 45S and supplements.
22. *Moody's Industrial Manual 1974* (Moody's Investor Service, New York, 1974), vol. 1, pp. 247, 248.

23. I thank Hugh Folk, Clark Bullard, Robert Herendeen, and Lee Schipper for their helpful suggestions. Supported by the RANN Program of the National Science Foundation and the Ford Foundation's Energy Policy Project.

[16]

An Assessment of Methodologies for Estimating National Energy Efficiency

Introduction

The ordinary citizen typically sees each physical object as practically inseparable from its service-content. However, the philosophical economist perceives the abstract service as independent of any particular product or means of delivery. To reduce the comparison to familiar terms: most of us tend to see a "dinner" primarily in terms of the combination of specific foods provided: meat, potatoes, bread, salad, etc. The philosophical economist, on the other hand, regards the dinner as a collection of separable experiences related to eating and drinking, to be assessed in terms of appearance, aroma, flavor, texture, plus functional utility (i.e., nutritive value).

To the unsophisticated citizen, therefore, it is difficult to divorce the idea of standard of living from the particular bill of goods and services which society produces (and consumes) at the moment. Achieving a higher standard-of-living through economic growth seems to suggest producing more of the same types of products, with occasional improvements in quality of performance. It is always assumed, implicitly, that the opportunities for improvement are specific and limited and will, sooner of later, be exhausted.

The world as seen by the economics profession is fundamentally different. Standard of living is not associated with a particular "bill of goods" at all. It is perfectly reasonable to project continuation of historical trends toward improved efficiency and performance, even beyond the known limits of existing technologies, by admitting the possibility of fundamentally new and different alternative approaches. The possibilities of substitution are limited only by fundamental physical constraints.

With regard to energy and other nonrenewable resources, the most common view – among citizens as well as the resources industries – is that fossil and nuclear energy resources on the earth's surface, constitute a limited "stockpile" which would, be implication, be used up eventually. This view is essentially Malthusian, and is in philosophical agreement with the Club of Rome's "Limits to Growth" hypothesis, except possibly with regard to the assumed size of the stockpile and the implications for public policy. The energy industry currently accepts that remaining petroleum and natural gas reserves are sufficient only for a few (i.e., three to eight) more decades; but it is established doctrine in the industry that coal and nuclear power (with breeder reactors) are effectively "unlimited." The pattern of intermediate and final demand for energy (per unit GNP) – the "bill of goods" mentioned earlier – is also typically regarded as relatively static, except for changes which can be specifically foreseen.

The economics profession, however, tends to be anti-Malthusian. Resource economists normally argue that shortages will be temporary because higher prices inevitably stimulate the development of alternative sources of supply, or alternative means of satisfying intermediate or final demands. Since goods are only produced to provide non-material "final services," there is no fixed relationship between the extent to which final demand is satisfied and the quantities of physical materials or energy required along the way. The range of substitution possibilities is broad. The "energy substitution ladder," illustrated in Table 1, illustrates this wealth of possibilities, ranging from the alternate crude fuels to alternate urbanization patterns designed to reduce the separation between residence and work place (or other potential destinations).

2

Table 1 – Energy Substitution Possibilities (Transportation-Related)

Level	Base case #1	Alternatives						
		#2	#3	#4	#5	#6	#7	#8
Type of Crude (Cra)	Domestic light	Athabasca Tar Sand	Same as # 1, 2	Same as # 1, 2	Bituminous coal	Same as # 1, 2, 5	Any	Any
Extraction Location	On shore	Same as #1	*	*	Anywhere e.g. offshore	*	*	*
Extraction Process	Pump (and steam recovery)	Steam + conveyor + retort process	*	*	Strip mining of underground	*	*	*
Refining Process	Thermal cracking and distillation	Same as #1	Thermal + catalytic cracking + reforming	*	Powdering + H-refining (to remove S)	*	*	
Finished Fuel	Gasoline	*	Same as #1	Fuel oil	Carbon (coal) slurry	Same as #4, 5	*	
Conversion of Chemical to Intermediate Energy	Spark ignition + combustion in cylinder of ICE	*	*	Compression ignition + combustion in cylinder of ICE	Combustion in combustion chamber of ECE	*	*	
Conversion of Intermediate Energy Form to (Mechanical Work)	Expansion of gases (combustion products) against piston & crank	*	*	Same as #1	Expansion of working fluid against piston and crank	Expansion of working fluid against turbine blade	*	
Conversion of Mechanical Work to Intermediate Final Product	Engine drives automobile	*	*	*	Same as #1	Same as #1	Engine drives bus/taxi	
Conversion of Intermediate Final Product to Final Service	Automobile carries driver/passenger	*	*	*	*	*	Bus carries passenger	
Satisfaction of Intermediate Need	Trip to work to earn income	*	*	*	*	*	Same as #1	Trip to work eliminated by co-location of workplace and residence
Final Need Satisfied	All needs requiring expenditure of money	*	*	*	*	*	*	Same as #1

The relevance of the preceding discussion to the main topic of this paper can now be stated: most of the analysis carried out to date of future opportunities for energy conservation have been flawed by an excessively narrow perspective. We hope to introduce some corrective broadening notions.

Structure of energy flows in the economy

The output of the economy is measured in terms of goods and services. However, as we have take pains to emphasize in the introduction section, the "consumption" of goods is tantamount to a process of conversion to final services of one kind or another. Thus, food produces gustatory and nutritional services, automobiles produce personal transportation services, etc. Apart from food and fuels, the material goods that consumers purchase are not – in most cases – physically depleted. They merely become waste products for disposal.

Energy is not a final service, as such. Rather, it is an intermediate commodity that is consumed in order to produce final services or – more commonly – in order to produce service-generating material goods. Of course "final" goods, in turn, are merely the end-

3

products of a chain of transformations that can be traced back to raw material extracted from the earth's surface.[1]

It is comparatively easy to assess the energy required directly to operate an automobile or to heat a house. These quantities correspond to consumer purchases of gasoline and household heating oil or gas (or electricity), respectively. However, it is not enough to compute direct energy consumption by a car alone. Clearly, the energy used in manufacturing the vehicle, making the iron, steel, aluminum, glass and rubber, refining the petroleum, and extracting the raw materials must be taken into account. Moreover, the car is useless without highways, so allowance must also be made for the energy consumed by earth-moving machinery and the energy embodied in asphalt and concrete.

Similarly, heating oil can only be burned in a furnace of some sort, and the heat cannot be distributed without pipes, radiators, air-ducts and/or blowers. Moreover, the warm air must be contained in a house, with structural walls, roof, doors, windows and (presumably) insulation.

Substantial quantities of energy are consumed in the manufacture of these things, and not a little is actually "embodied" in materials such as fiberglass insulation, window glass, ceramic tiles, bricks and concrete.

Unfortunately, the way in which energy statistics are normally compiled does not match well with the conceptual scheme we have elaborated in the foregoing. Virtually all energy data are organized in terms of four or five standard categories, viz.

- Electricity generation
- Industry (including mining)
- Transportation (including passenger cars)
- Commerce and government
- Households (including farms but excluding passenger cars).

Households are sometimes combined with the "commercial/government" category. Except for electric power generation, all these categories are considered to be independent "end-users." Only electricity itself is commonly regarded as an intermediate product and reallocated among the other categories. (Steel, concrete, and plastics, for instance, are thus implicitly regarded as "final products.")

The energy flows normally accounted for are displayed in Figure 1. However, it is evident upon inspection that indirect flows, reflecting energy "embodied" in materials, or energy consumed in the production of intermediate services for industry, are omitted. They are included, for contrast, in Figure 2. The second figure also separates household consumption from commercial and government activities – which are intermediate – and includes private automobiles with the former. Thus, all consumption activities normally classified by economists as "final" are now lumped together for convenience.

---

1 Not forgetting the contributions of human labor and technology.

4

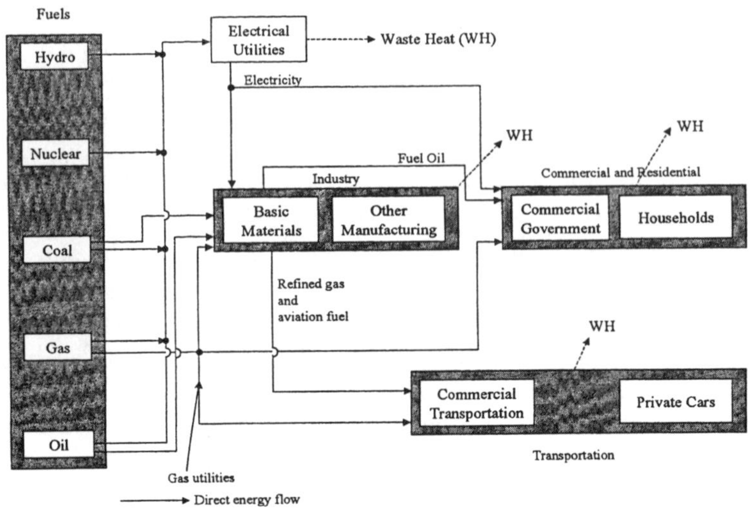

Figure 1 – Energy flow, conventional view

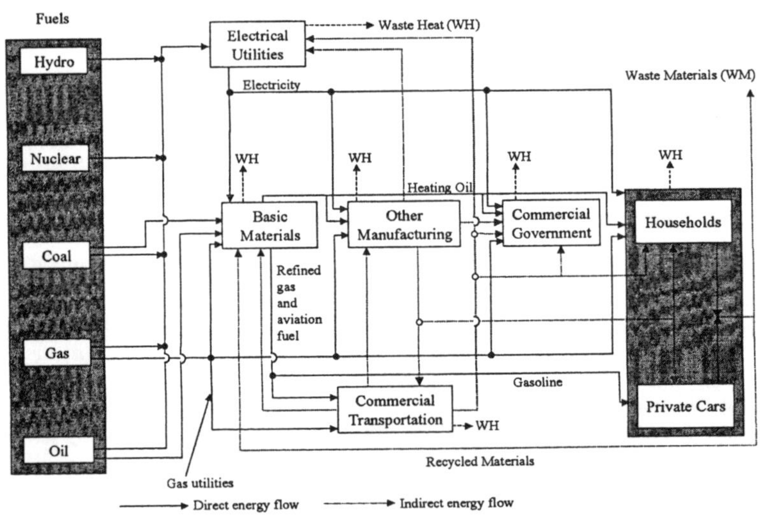

Figure 2 – Energy flow, economist's view

In the discussion and analysis which follows, the differences between Figures 1 and 2 will be emphasized, inasmuch as the kind of thinking that underlies Figure 1 has influenced energy policy in subtle, but critical ways. We wish to stress the alternative conceptualization exemplified by Figure 2, and its policy implications.

5

Energy efficiency considerations

Efficiency – in the sense of optimal allocation of resources – is a notion that is central to the discipline of economics. However, it is one thing to discuss economic efficiency in abstract terms. It is quite another thing to measure the efficiency of energy use as such, and offer specific recommendations as to how it can be increased.

As we have noted already, there are several distinct stages of transformation from raw materials extraction through materials and fuel processing, fabrication of products, household use, and final disposal. The applicable concept of energy efficiency varies from stage to stage. Where both inputs and outputs of a process can be measured meaningfully in energy units, the problem of selecting a measure of efficiency is relatively straightforward (though not quite trivial).

However, where energy is consumed by a materials-processing industry and some of it is "embodied" in the output material, awkward difficulties begin to intrude. Where the transformation process primarily involves non-energy materials – as in fabrication or construction – these difficulties are substantially compounded. And where the process involves generation of a final service that cannot be measured either in energy or material terms, a very careful and essentially unconventional type of analysis is indicated.

Thermodynamic Efficiency Measures

We will begin by discussing the concept of energy efficiency as it applies to energy-conversion and energy/materials transformations. This topic has received a good deal of attention in the past several years (3-5)[2], and it appears sufficient to summarize the major points without detailed justification. A list of possible measures of efficiency is given in Table 2, which follows. In reading the table, it is important to note that we have normally equated "useful work" with either mechanical energy (e.g., rotary motion) or electric energy. Of course, in the case of an electric motor or generator, the input energy is in one of these two forms and output is in the other, and inter-conversion does involve some losses.

2 Numbers in parenthesis designate References at end of paper.

6

Table 2 – Figures-of-Merit for Various Thermodynamic Processes

	Measure	Example(s)	Asymptotic Limit
(1)	$\frac{\text{Work Out}}{\text{Work In}}$	Electric Motor or Generator Torque Converter	1.0
(2)	$\frac{\text{Work Out}}{\text{Enthalpy In}}$	Heat Engine	$1 - T_A / T_H$
(3)	$\frac{\text{Heat Transferred}}{\text{Enthalpy In}}$	Burner or Boiler	1.0
(4)	$\frac{\text{Heat Added to Reservoir}}{\text{Work In}}$	Heat Pump	$[1 - T_A / T_R]^{-1}$
(5)	$\frac{\text{Heat Removed from Reservoir}}{\text{Work In}}$	Refrigerator	$[T_A / T_C - 1]^{-1}$
(6)	$\frac{\text{Available Work Out}}{\text{Available Work In}}$	Chemical or Electrochemical Process	1.0

Where: T_A = Ambient Temperature
T_R = Receptor (warm) Reservoir Temperature
T_H = Hot Reservoir Temperature
T_C = Cold Reservoir Temperature

It may be noted, further, that all of the thermal energy in a reservoir at infinite temperature can be converted into work, but when the reservoir temperature is finite, some of the heat energy is unavailable in principle. Entropy is the measure of this unavailability. Thus, the temperature at which heat energy is produced or used is a critical quantity in determining the efficiency with which it can be used.

The most widely used figures-of-merit are numbered 1 to 3 in Table 2. The general form which encompasses all three cases is as follows:

$$\eta = \frac{\text{energy transfer (of a desired kind)}}{\text{energy input}}$$

This ratio has been called "first law" energy efficiency (5) since it measures the total energy transfer from the initial state to a final state. Cases 4 and 5 are called "coefficients of performance" since they have values greater than unity. These measures are specialized to heat transfer processes and are not true efficiencies.

A more fundamental measure which is applicable to any process is number 6, which takes into account the change in availability of energy as a result of the process. We introduce the symbol

$$\xi = \frac{\text{available work of final outputs}}{\text{available work of inputs}} = \frac{B_f}{B_o}$$

Thermodynamically, the available work of an active system is the maximum amount of useful work, B, the active system could do on another (passive) system as both systems move to an equilibrium final state, with respect to the atmosphere

$$B_o - B_f = E_o - E_f + P_o(V_o - V_f) + T_o(S_o - S_f)$$

7

Where E, P, V, T, and S refer to internal energy, pressure, volume, temperature and entropy, respectively. The useful work, W, done on the passive systems by the active systems is defined by the first law of thermodynamics as the change in enthalpy, minus the heat that is lost to the atmosphere, and the (non-useful) work done on the atmosphere. It can be shown that the useful work done by the active system is maximized when there is no entropy change in either the active system or the atmosphere. This is the case if, and only if, the process occurs very slowly and reversibly, whence

$$W_{max} = B_o - B_f \bullet$$

The available work "in" the initial state is the work required to produce this state reversibly from a system in equilibrium with the ambient atmosphere. Similarly, the available work, B_f, "in" the final state is, therefore, equal to the minimum available work, B_{min}, needed to produce the final state starting from a system in equilibrium with an ambient atmosphere. These quantities are independent of any specific process. From the foregoing remarks, it can be seen that our definition coincides with that used in Reference (5).

$$\frac{B_f}{B_o} = \frac{B_{min}}{B_o} \bullet$$

A list of calculated typical first and second law efficiencies is given in Table 3. The industrial process efficiencies given are substantially higher than the entropy conversion efficiencies at the top of the table, because much of the available work in the input material remains available in the finished material. This is particularly noteworthy in the case of petrochemical-based plastics, where nominal second-law efficiencies as high as 80 percent may be achieved. Obviously, the key to "systems" efficiency in the use of materials is recycling. If a material once produced is retained perpetually in inventory (as is, for instance, gold), then its available energy is preserved. However, when a material is dissipated or degraded by use to the point of being unrecoverable, then its available energy content is lost. This is an aspect of efficiency of materials utilisation, which we discuss next.

8

Table 3 – Calculated Efficiencies, Various Processes

Energy Conversion Process	First Law η	Second Law ε	Source
Burner to Steam-Electric Generator	0.33	0.33	Ref 5
Burner to Process Heat, 1000° F	0.6	0.65	*
Burner to Process Steam, 250° F	0.6	0.225	*
Electric Heater to Hot Water, 120° F	0.75	0.045	*
Burner to Hot Water, 120° F	0.50	0.029	*
Burner to Space Heat, 70° F (room)	0.60	0.028	*
Burner to Space Heat, 110° F (at the register)	0.60	0.074	*
Burner to Space Heat, 160° F (at the furnace plenum)	0.75	0.145	*
Otto Cycle Engine, constant speed	0.32	0.32	*
Engine Driven Heat Pump to Stream (250° F)	1.18 (cop)	0.299	*
Engine Driven Heat Pump to Space Heat (110° F)	2.45 (cop)	0.202	*
Industrial Electric Drive	0.90	0.90	*
Automobile Engine to Rear Wheels, 60 mph	0.14	0.14	*
Automobile, Federal Urban Driving Cycle	0.09	0.09	*
Materials Transformation Process			
Coke Oven	na	0.88	Ref 4
Pig Iron, Blast Furnace	na	0.77	*
Steel, Open Hearth Furnace	na	0.66	*
Steel, Electric Arc Furnace	na	0.80	*
Steel, Basic Oxygen Furnace	na	0.93	*
Iron and Steel Finishing	na	0.415	*
Aluminum, Bayer-Hall Process	na	0.45	IR&T (A)
Aluminum, Alcoa Process	na	0.59	IR&T (B)
Chlorine Manufacturing	na	0.40	IR&T (B)
Ammonia Manufacturing	na	0.50	IR&T (B)
Pulp and Paper	na	0.29	IR&T (A)
Cement	na	0.25	IR&T (A)
Glass Manufacturing	na	0.22	IR&T (B)

IR&T (A) = adopted from Reference 4
IR&T (B) = adopted from Reference 6

One possible point of confusion in Table 3 should be cleared up at this point. Most physicists and engineers are accustomed to thinking that energy conversion processes are limited by the so-called "Carnot" efficiency, which, of course, reflects the second law of thermodynamics. But if B_{min} is limited by the Carnot efficiency, surely present electric generating plants are much closer to their ultimate limits than the 0.33 efficiency shown? The answer is that B_{min} is not Carnot-limited. The reason for this apparent contradiction is the fact that the Carnot limit applies only to a limited class of processes, viz., cyclic (reversible) heat engines operating between a high temperature reservoir and a low temperature reservoir. It does not apply to fuel cells, for instance.

Materials Utilization Efficiency

As previously emphasized, finished materials are not consumed as such; they are either embodied in durables or consumed (i.e. used up) in the operation, or maintenance of a machine or appliance. Structural materials belong to the first category; fuels, lubricants, detergents, bleaches, solvents, and so forth belong to the second category.

9

The efficiency with which a material is used can only be defined exactly, with respect to a specific application or task in terms of the quantity of material required in principle vis-a-vis the quantity that is actually used.

For example, one might consider the use of tin for plating sheet steel used in cans. A minimum quantity of tin can be specified based on the thickness required to form a rust-proof protective coating. Any extra use is not required by the function and represents inefficiency in the design of manufacturing process.

But if we select narrowly defined existing uses as the norm, we would implicitly exclude the possibilities for saving tin (or other materials) arising from technological substitution. In actual fact, the use of tinplate has been steadily declining as "tin" cans are replaced by aluminum cans or glass, plastic or paper containers. How can we define "efficiency" so as to take such possibilities into account? Clearly, at this level of product detail, there is no way to answer such a question realistically. To deal with the question at all, it is necessary to aggregate products into groups, such as containers, appliances, apparel, furniture, motor vehicles, or houses.

Materials incorporated in "durable" goods should be treated conceptually on a different basis from materials that are incorporated in "consumables". Opportunities for materials savings in the first category will arise primarily from improvements in performance and extended lifetime. Performance improvements will arise from advances in metallurgy, rheology, and solid-state physics. Newly developed "high-strength low alloy" steel offer a case in point: they permit unambiuous reductions in the size and weight of structural members.

The situation is complicated, however, by substitutions between different materials, e.g., aluminum for copper or stainless steel, plastics for cast iron or zinc castings, cast ceramics in place of superalloys, etc.

To estimate the efficiency of materials use in existing applications, one would have to define the "ultimate" material for each application and determine its embodied available energy. Tables 4(a) and 4(b) display a number of distinct materials and applications in terms of the properties of materials required (7).

10

Table 4(a) – Uses of Durable Materials, Based on Properties

Major Properties / Durable Materials	1	2	3	4	5	6	7	8	9	10	11
	Incompressability inertia	Rigidity Weather-Resistance	Rigidity Impenetra-bility	Rigidity Strength Toughness Temperature Resistance	Flexibility Inelasticity Toughness Impenetra-bility	Flexibility Elasticity Toughness Impenetra-bility	Flexibility Inelasticity Toughness Tensile Strength	Flexibility Inelasticity Toughness Absorbency	Thermal/ Electrical Conductivity Moderate Flexibility Temperature Resistance	Thermal/ Electrical Non-conductivity (Dielectric capacity	Flexibility Toughness Impenetra-bility Adhesion
Use (category) *Material (Type)*	Ballast/Filler Aggregate *Crushed stone* *slag* *gravel* Hydraulic fluids *High-male silicones oil*	Load-bearing Beams, Rods, Spars, Spokes *Concrete* *steel* *wood (for furniture)* Load-bearing Walls *Concrete* *brick* *masonry* *wood* Pavement *asphalt* *Concrete* *Stone (cut)* *Tile* *(????)* + Toughness & Strength Frames for Vehicles Machines Bridges *steel* *coal iron* *reinforced concrete* *aluminium alloy*	Structural Skin (exterior wall, roof) *wood* *plywood* *aluminium* *asbestos* *shingle* *plastic* Interior Wall Floor, Ceiling *plaster* *plasterboard* *plywood* *plastic* Constant Dimension Containers: Dishes, Tubs, Tanks *cast iron* *lead, copper* *galv. iron* *PVC* *ceramic* *concrete* + Transparency Windows *glass*	Engine Parts Block Pistons, etc. *cast iron* *cast aluminium* *cast ceramic* Axles Bearings Gears, Cams Turbine Wheels Valves *steel forging* *alum. forging* *ceramic* *superalloy* Burner Parts Manifold *steel* *stainless steel* *superalloy* *ceramics* + Hardness Cutting Tools *carbine alloy sinter*	Wall, Floor Covering *paper* *plastic* *linoleum* *carpet, felt* *woven fabric* Furniture Covering *plastic* *leather* Tubes, Hoses *plastic (PVC)*	Seals Gaskets Bearings *rubber* Shoe Soles Heels Cushions *foam rubber* *solid rubber* *plastic* *leather* Tires *rubber*	Load Bearing Cables Wires *steel* Rope String Cord, Net *cotton* *jute, help* *nylon, rayon* Power Transmission Belts *rubberised fabric* *chain* + Weather Resistance Wire Mesh (screen, fence) *aluminium* *plastic-coated steel*	Materials for Printing Books *paper* *plastic "mat"* Materials for Clothing Furnishings, etc. *natural fibre* *cotton* *wool* *synthetics* *paper* *leather*	Motor Generator Parts *copper* *aluminium* Electrical Wire Conductors *copper* *aluminium* *silver* + Liquids Secondary Battery Electrolytes *acid, alkali* *molten salt* *organic conductors* + Impenetra-bility Heat Exchanger Parts *copper* *aluminium* *steel*	Non-structural Insulating Material *feathers* *down* *asbestos* *fibreglass* *plastic foam* *PBC* *silicones (fluid)* + Rigidity Load Bearing Insulators *porcelain (ceramic)* *glass*	Binders *tar* *asphalt* *plastic (epoxy)* *cement* *sulfur* Coatings *plastic* *heavy oils* + Non-conductivity Electrical Insulation for Wire

Table 4(b) – Uses of Nondurable Materials

Critical Property Combination / Non Durable Materials	Solids			Non Solids						
	Rigidity Impenetra-bility	Flexibility & Inelasticity Toughness Impenetra-bility	Flexibility & Inelasticity Absorbency	Surface Modification	Solvent Power	Thermo-Dynamic Properties	Chemical Reactivity	Biological Activity-Antognistic	Biological Activity-Supportive Synergistic	Fuels and Additives
Use (category) *Material (type)*	Constant Dimension Containers Cans Drums Boxes *steel* *tirplate* *aluminium* *plastic foam* *wood boxes* *paperboard* + Trans-parency Bottles *glass* *plastic*	Variable dimension containers Bags Tubes Sacks *paper* *aluminium foil* *tin (tubes)* *woven fabric* *plastic film* + Trans-parency Clear Film *plastic*	Absorbent Tissue/Padding *paper* *cotton lint* + Toughness Materials to Print on *groundwood (newsprint)* *paper*	Wettings agents Detergents Cleaners Flocculants + Low Viscosity Lubricants	Solvents *alcohols* *hydro-carbons* *halogenated HC* + Low Viscosity Spreaders	Antifreeze *ethylene glycol* *propylene glycol* Snow-melting Agents *NaCl* $Ca(Cl_2)$ Freezing Agents *ether* Aerosol Propellants F_{11} F_{12} CO_2	Bleach $CaMo10_3$ Acid $H2SO_4$ *HCl* HNO_3 *HF* Alkali *NaOH* *KOH* $Ca(CH)_2$ $NaHCO_3$ Water Softener + Conductivity Primary Cell Electrolytes	Antibiotics Bactericides & Fungicides Herbicides Rodenticides Insecticides Food Preserv-atives Other Pharma-ceuticals	Fertilizers Vitamins and Amino Acids Sweeteners and Flavors Growth Hormones Foods	Gas Liquid Solid

We know of no comprehensive attempt to assess the future potential for improvement in materials performance or the implications for embodied energy. Such a study would obviously be a very difficult and challenging one. Its relevance to the overall national energy use picture may be assessed roughly from the energy allocation table, presenter later (Table 6). Of the 23 quadrillion Btu's of energy used by industry in 1970 (counting purchased electricity at its heating value, not its original fuel value), 14.3 quads were consumed in the refining of petroleum and the production of iron and steel, aluminum, cement, glass, paper, and basic organic an inorganic chemicals. The remaining 8.7 quads, not explicitly accounted for, also include increments for minor metals such as copper, lead and zinc, ceramics and clay products, building materials, paints, synthetic rubber and plastics, synthetic fibers, pharmaceuticals, fertilizers, and food processing operations.

Altogether probably no less than 18 quads were used in 1970 for processing materials (including food) from the raw state to "finished" form; while no more than 5 quads were used for all other industrial activities such as mining, cutting and machining of metals, weaving and knitting of textiles, printing, coating and coloring, fabrication, assembly, packaging and construction.

Of this total, most of the iron and steel, all of the cement and a substantial fraction of other metals and glass go into durables, while most of the paper and almost all of the chemicals are used for consumables (including finished fuels). It is interesting to note that plastics, synthetic rubber and synthetic fibers together account for only about half of the petrochemical feedstocks, and a very large fraction of these synthetic materials are used for packaging or other very temporary purposes. Only a small fraction of these materials – perhaps 20 percent – go into textiles, building materials or motor vehicle tires which have a normal service life of more than a few months to a year. The remainder of the petrochemical feedstocks end up as commercial solvents, surfactants (detergents), gasoline additives, antifreeze, dyes, pesticides and pharmaceuticals. These products are essentially discarded as they are used.

From and energy standpoint, there are three ways to reduce the energy embodied in durable materials. The first, which we frankly cannot evaluate quantitatively – but which is unquestionably of major importance – is to reduce or eliminate the amount of the material that is required. This can be done by improved product design, upgraded performance characteristics, or substitutions of information-intensive technologies for materials/energy-intensive technologies.[3]

The other two ways of increasing the efficiency of embodied energy are, respectively, to increase product lifetimes and to increase the recycling rates of their contained materials. These two methods are not quite independent, of course; the lifetime of a product is much less relevant if the materials incorporated in it are completely recycled (as, for example, in the case of automotive lead-acid storage batteries). The recycling may be equivalent to re-use.

The most striking opportunity for increasing the efficiency of energy use by extending product lifetimes and/or recycling rates would be to eliminate the use of energy-intensive throwaway containers – especially glass or plastic bottles and aluminum cans – in favor of returnable and re-usable containers. Studies (8, 9) have suggested that the "returnable" system must be presumed to be less – probably much less – than 33 percent efficient in utilizing embodied energy.

---

3 The oft-discussed substitution of advanced telecommunications technology for some types of transportation is an illustration of the latter.

12

The energy required to recycle metals, glass or paper is generally much less than the energy required to produce these materials initially. This is particularly true for aluminum, where the energy consumed per unit of secondary material is only 5 percent of the energy consumed per unit of primary product. Hence, the extent to which aluminum could be recycled, but is not, is a measure of inefficiency of energy use. The recycling efficiencies of major durable materials as of 1972 are shown in Table 6. On an energy-weighted base, no more than 4 percent of consumable materials (mainly paper) are recycled. For durables, the figure is probably closer to 20 percent.

The efficiency of recycling of embodied materials varies greatly from sector to sector. Thus, metals embodied in heavy machinery of transport vehicles is almost completely recycled; this is also true of scrap metal generated in the manufacturing process. Nonmetallic materials are not as efficiently utilized however. On the other hand, solid wastes generated in the commercial and household sectors are not recycled efficiently, except for certain categories of paper and office machines. Liquid and airborne wastes are almost entirely discarded in all cases.

Obviously, some of the losses indicated in Table 5 arise from essentially dissipative uses, such as throwaway containers, packaging materials, publications, and chemical coatings or additives. Almost all of the nonrecycled lead, for instance, is lost as tetraethyl lead (TEL), an octane additive to gasoline. Much zinc is not recovered because it is used as a plating material (galvanized iron or steel sheet) or for small "hardware" items. Much copper wire is lost in discarded electrical appliances, junk auto bodies, etc. In the long run, however, we are inclined to believe that virtually all intrinsically dissipative uses can – and must – be eliminated.

Table 5 – Materials Utilization Efficiency

Material	Fraction of Industrial Energy Inputs	Durable Fraction (a)	Consumption Fraction	Recycled Fraction (b)
Ferrous Metals	0.18	0.94	0.06	0.26
Aluminum	0.01	0.89	0.11	0.48
Cement	0.08	1.00	- 0	0
Glass	0.014	0.30	0.70	0.05
Paper	0.097	- 0	1.00	0.19
Chemicals	0.22	0.20 (est)	0.80 (est)	- 0.05
Asphalt	0.05	1.00	- 0	- 0
Other	0.345	0.40 (est)	0.60 (est)	- 0.05
Energy Weighted Average Fraction		0.50	0.50	

(a) Based on 1972 materials allocation to consumption categories, taken from IR&T-433-R, R. U. Ayres, M. Markus-Kramer, A. Watson, "The Analysis of Resource Recovery and Waste Reduction Using SEAS", May 1976.

(b) Based on Reference 10 and author's estimates

Energy Utilisation in Production of Services: Utility Efficiency

We now come to the major area of ambiguity in the assessment of energy efficiency. In a previous Section, we examined energy conversion and energy/materials transformation which can be analyzed meaningfully in thermodynamic terms. But it must not be forgotten that none of the outputs of the processes considered in that section are consumed or valued as such. This point has been made in the foregoing, with respect to materials. This caveat is particularly important to understand where the output of a process is a "pseudo-service" such as space-heat or torque transmitted to the rear wheels of a car.

In order to assess the full potential for energy conservation, we must examine the efficiency with which these "pseudo-services" generate actual final services to humans. To do this, we must recognize that the ultimate object of space-heating is not environmental conditioning but physical comfort.

Similarly, the ultimate object of using an automobile is not to travel – except in the case of teenager's "joy riding," – but generally to accomplish some definite errand or to arrive at a desired destination. A trip is rarely an end in itself, and passenger miles (or vehicle miles) are simply not a realistic measure of the service output of automobile transportation, although often used for this purpose.

More to the point, the purpose of the trip is to move the passengers and their belongings, not the car. The second-law efficiency for supplying energy to the rear wheels of a car (Table 3) does not reflect inefficiencies arising from the fact that the vehicle "shell" is larger than it really needs to be to accommodate the average payload, nor the fact that it is much bigger and heavier than it has to be to serve its primary function of providing a comfortable moving environment for the passengers. In particular, much of the weight of the vehicle is attributed to the fact that it must carry a large and bulky engine (and fuel tank), which, in turn, must be large because of the weight and bulk of the car. Weight is also considered as a contributor to good riding quality, but that function can equally well be provided by improving the smoothness of the road. (Or, equivalent inertia could be provided without weight by means of a gyroscopic flywheel).

How small and light could a basic vehicle be? If motive power and energy were not carried on board, little more than a shell or "pod" would be needed: a magnesium and plastic "bubble-car" might be sufficient, perhaps weighing as little as 250 lb or so. The details of an external drive mechanism need not be specified here; it is sufficient to point out that a number of such systems already exist (though most of them are crude) ranging from cable cars to trolleys. Our point, here, is that any excess weight that is carried for purposes extraneous to providing a comfortable environment for the passengers is probably a measure of inefficiency. Assuming an average of 1.5 passengers (weighing 150 lb each) plus luggage, the typical payload of a car would be around 250 lb. If the minimum weight of a passenger-containing shell is about 250 lb, the payload efficiency of a subcompact car (2,250 lb curb weight) would be around $\frac{500}{2500} \sim 0.20$. In a more extreme case, the payload efficiency of a 3,750 lb curb weight car would be $\frac{500}{4000} \sim 0.125$. Thus, the overall efficiency of a car (or any other vehicle) in terms of transporting "payload" would be much less than the thermodynamic efficiency of the power plant/drive train considered alone.

In the case of freight, payload efficiency tends to be somewhat higher, of course. Trucks, trains, ships, and aircraft may even carry maximum payloads exceeding their own weight, though typical load factors are much less than unity.

The same kind of argument is clearly applicable to other services such as environmental conditioning in structures, washing, cooking, illumination, communication, etc. An "ultimate" space heating system, for instance, would not necessarily, heat all the air in a house, or even a room. Rather, it would "sense" the momentary location of a person and direct a gentle, purified, and slightly cooled of warmed stream of air (perhaps supplemented by an infrared beam) to the occupied part of the room. The extent to which unoccupied space – not to mention inert structural elements – are unnecessarily heated or cooled represents an inefficiency not reflected in the purely thermodynamic calculation given in Table 3.

Washing and rinsing – especially with water – offers another example of "payload" inefficiency. Heat is primarily used to increase the "dissolving power" of the cleaning agents (principally water). It is possible – sometimes even necessary – to use unheated fluids for cleaning. It is also quite likely that other means such as ultrasonic sound, can be substituted more widely in place of sensible heat for "dirt loosening" purposes. We know of no thermodynamic analysis of this subject, but it seems very likely that such an analysis would show that washing – both in homes and in industrial or commercial establishments – utilizes heat energy very inefficiently.

The same principle also applies to cooking. Why waste energy unnecessarily heating the air and raising the temperature of pots and pans or cooking water that is later discarded? The "ultimate" cooking system will heat only the food, and to the minimum necessary degree. (Radar ovens are now exploiting this possibility; but the generation of microwaves is very inefficient at present – certainly < 10 percent.) At most, it is likely that no more than 20 percent of the heat energy used for cooking is really necessary – probably much less.

Illumination is yet another example. In the first place, unoccupied rooms need not be illuminated at all. Sensors could automatically turn off lights in the absence of people. Secondly, the most intense illumination is needed only for reading and office work. Low levels of lighting would be sufficient for most other activities. Focused or guided beams of bright light would be much more efficient than uniform high level illumination of all parts of a room. Flexible transparent plastic "light pipes" offer substantial possibilities for more efficient distribution of light for working. Third, most sources of illumination are still quite inefficient in terms of the fraction of their output that lies in the visible part of the spectrum. Incandescent bulbs are only about ~ 5 percent efficient in this sense, while fluorescent lights are about twice as good. The "ultimate" light generating system will probably be based on solid-state light emitting diodes (LEDs), which will convert nearly 100 percent of the input electric energy into visible light. Considering all of the foregoing, it seems plausible to us that better room and work lighting could be achieved by an ultimate system requiring less than 5 percent of the energy now expended for this purpose.

Television and radio offer similar opportunities for savings. It is patently obvious that most of the power used by a TV set is wasted and dissipated as heat; very little is ultimately converted into light or sound. We have no figures, but overall conversion efficiencies are certainly much less than 10 percent. (Even this reflects rapid improvements in the last several decades).

Table 6, which follows, represents our rough initial guesstimates of the efficiencies with which various final services are currently produced, as compared to conceivable "ultimate" systems. A more detailed analysis of the efficiency of these and other services would be well worth undertaking.

15

Table 6 – Allocation of Energy among Service-Generating Functions

Service Function	Fraction of U.S. Energy (Reference 12)	"Second Law" Thermal Efficiency ξ (Reference 5)	Utility of Payload (b) Efficiency ??	Composite Efficiency ξ x ??	Total available Energy in x 10^{15} STU (1971)	Available Energy Loss x 10^{15} STU (1971)
Transportation						
Truck	0.049	0.11	- 0.5	- 0.055	3.39	3.21
Bus	0.00196	0.2	- 0.3	- 0.060	0.136	0.127
Rail (diesel-electric)	0.0098	0.3	- 0.5	- 0.150	0.68	0.578
Air	0.0196	0.3	- 0.1	- 0.030	1.36	1.319
Military, Pipeline, etc.	0.0392	0.3	- 0.2	- 0.060	2.7	2.556
Automobile	0.127	0.11	- 0.20	- 0.022	8.8	8.6
Total or Average	0.247	0.163			17.14	16.438
Commercial (a)						
Space Heat	0.0605	0.06	- 0.2	- 0.012	4.16	4.11
Water Heat (Laundry)	0.0096	0.03	- 0.5	- 0.015	0.66	0.654
Cooking	0.0017	< 0.2 (b)	- 0.3	< - 0.060	0.12	0.113
Refrigeration	0.0096	0.04	- 0.5	- 0.02	0.663	0.6507
Air Conditioning	0.0157	0.05	- 0.2	- 0.025	1.086	1.059
Other (including Light)	0.0149	< 0.05	< 0.1	< 0.005	1.025	1.021
Total or Average	0.112	0.055			7.79	7.414
Residential						
Space Heat	0.086	0.06	- 0.3	- 0.018	5.952	5.84
Water Heat (Wash)	0.022	0.03	- 0.2	- 0.006	1.569	1.56
Cooking	0.0086	< 0.2 (b)	- 0.3	< 0.06	0.595	0.559
Clothes Drying	0.0023	< 0.1 (b)	- 0.2	< 0.02	0.162	0.159
Refrigeration	0.0086	0.04	- 0.5	- 0.02	0.595	0.583
Air Conditioning	0.0054	0.05	- 0.3	- 0.015	0.379	0.372
Other (including Light)	0.016	< 0.05	- 0.1	< 0.005	1.136	1.13
Total or Average	0.1506	0.0615			10.45	10.2

(a) Asphalt paving is omitted from this "end-use" category and included among basic materials.
(b) Author's estimate.

Available energy balance for an exemplary industry (Steel)

It is interesting to display a detailed available energy-balance for one multi-process industry in order to see how the various processes are linked. It is intuitively clear that the overall efficiency for a sector involving a network of processes – some of which occur in sequence – must be lower than the efficiency for any single stage. This fact is quantitatively demonstrated in the base of the steel industry.

Overall energy-use for 1969 was based on Bureau of Mines data [Reference (11)]. The information also is displayed in tabular form in Table 7.

It is significant that about 79 percent of the available energy inputs to the iron and steel industry are wasted. The remainder is "embodied" in the finished products. This is not a particularly high efficiency when it is recognized that a fairly substantial part of the input energy is in the form of scrap iron and utility electricity – neither of which is adjusted for original fuel inputs. If it were not for recycling, the apparent losses would be substantially greater.

16

Table 7 – Available Energy Balance, Steel Industry

Process	Coal	Coke	Fuel Oil	Utility Gas	BP Gas	Total (a) Electric	Utility Electric	Oxygen	Pig Iron	Scrap Iron	Input Steel	Output Steel	Waste Heat	Out/In
Coke Oven	2.21	-	-	-	-	0.024	-	-	-	-	-	-	0.25	0.88
Sinter/Pelletize	-	-	-	0.05	0.38	0.015	-	-	-	-	-	-	0.15	-0
Onsite Electric	0.05	-	0.12	0.14	0.29	-	-	-	-	-	-	-	0.40	0.33
Blast Furnace	-	1.38	-	0.04	-	0.048	0.018	-	-	-	-	-	0.34	0.77
Oxygen Plant	-	-	-	-	-	0.013	0.004	-	-	-	-	-	0.33b 0.01	0.28
BCF	-	-	-	-	-	-		0.004	0.34	0.10	-	-	0.03	0.93
Open Hearth	-	-	0.065	0.084	0.042	0.025	-	-	0.28	0.15	-	-	0.23	0.66
Electric Furnace	-	-	-	-	-	0.033	0.045	-	0.006	0.11	-	-	0.23	0.80
Soak, Soil, Weet, Treet	-	-	-	0.35	0.19	0.135	0.065	-	-	-	0.97	0.79	0.02	0.48
Other (Cast Iron, forging)	0.10	0.07	0.084	0.06	0.04	0.03	0.022	-	0.05			0.12[c]	0.30	0.30
Total (1971)	2.36	1.45	0.269	0.73	0.64	0.32	0.164			0.36			2.99	

a Onsite electric and steam plus utility electric

b Blast furnace gas not accounted for

c Assuming 15% of steel goes through a forging process.

Available energy balance for basic materials industries

Apart from available energy losses associated with petroleum refining (about 8 percent of the inputs) and gas separation (2 percent), the energy consumed by the "industry" sector is mainly used for the processing of basic materials. About a third of the available input energy is "embodied" in the finished outputs, the remainder being process losses.

The flows of energy from raw materials into finished fuels and finished materials are tabulated in Table 8 for 1971.[4] Overall, the industrial sector utilized 25.25 quadrillion Btu's of available energy in 1971, of which 3 quads were lost in petroleum refining and natural gas plants. This is properly attributable to the production of finished fuels rather than materials. However, 22.25 quadrillion Btu's were consumed in basic materials manufacturing, of which roughly one third was "embodied" in the output materials.

The term "embodied energy" is potentially misleading. As regards the first stage of ore reduction, the chemical energy required to separate the metal oxide molecules into constituent elements is recoverable (in principle) by re-oxidation. In fact, iron fillings, aluminum or magnesium powder can be "burned" in air.[5]

4 Table 4 refers to a different year than Table 3 because the best available sectoral database was for 1971. We used 1969 data for the steel industry, however. The data for transport commercial and residential consumption was extrapolated from a 1969 base. Hence, it differs slightly from Bureau of Mines data of 1971. However, the table illustrates a "trial" energy balance.

5 Thermit is a practical example on the combustibility of iron. Magnesium flash powder is another case in point.

17

Table 8 – Available Energy Allocation by Sector, 1971

Sector	Coal	Natural gas	Crude Oil	Resid. Oil	Distil. Oil	Gasoline Diesel & Aviation	Petrochem. Feed	Nuclear Heat	Hydraul. Power	Total Utility Elec.	Total Input Energy	Waste Heat Out	Total Output	Eff. Out/In
Fossil Fuel Electric Plants	7.74	3.87	-	2.38	-	-	-	-	-	-	13.99	9.72	} 4.69	-
Nuclear Electric	-	-	-	-	-	-	-	0.65	-	-	0.65	0.25		-
Hydro. Electric	-	-	-	-	-	-	-	-	1.14	-	1.14	0.23	0.915	-
Total Utility Elec.	7.74	3.87	-	2.38	-	-	-	0.65	1.14	-	15.78	10.20	5.59	0,354
Gas Separation	-	21.38	-	-	-	-	-	-	-	-	21.98	0.40	21.58	0,58
Petroleum Refin.	-	-	32.31	-	-	-	-	-	-	-	32.31	2.60	29.71	0,924
Ferrous Metals (a)	2.36	0.73	-	[0.27]		-	-	-	-	0.32	4.04 b	3.20	0.84	0.21
Aluminum	-	-	-	-	-	-	-	-	-	0.20	0.20	0.115	0.085	0,44
Cement	0.63	0.69	-	[0.40]		-	-	-	-	0.05	1.77	1.33	0.44	0.25
Glass		0.29	-	-	-	-	-	-	-	0.025	0.315	0.245	0.07	0,22
Pulp and Paper (b)	0.37	0.39	-	[0.24]		-	-	-	-	0.085	2.17	1.54	0.63	0.29
Chemicals	0.66	1.50	-	[0.23]		-	2.30	-	-	0.26	4.95	2.80	2.15	0.43
Asphalt	-	-	1.11	-	-	-	-	-	-	-	1.11	0	1.11	1,0
Other ??	1.87	3.82	-	[1.05]		-	-	-	-	1.39	7.69	5.39	2.30	0.3
Total Industry (c)	5.26	7.42	32.31	[2.19]		0	2.30	0	0	2.33	22.25	14.62	7.625	0.33
Transport	-	0.75	-	-	-	14.37	-	-	-	0.02	17.14	16.44	0.70	0,04
Commercial	0.304	2.27	-	-	3.97	-	-	-	-	1.45	7.79	7.61	0.18	0,04
Household	-	5.27	-	-	3.29	-	-	-	-	1.79	10.45	10.20	0.35	0,04
Total (d)	13.30	21.58	32.31	[11.93]		16.37	2.30	0.65	1.14	5.59	69.38	68.26		0.016

(a) Inputs to Ferrous Metals include recycled scrap iron with available energy ¯0.36 x 10^{15} STU
(b) Inputs include pulp and waste paper material with available energy of ¯1.09 x 10^{15} STU
(c) Inputs include available energy in scrap iron, paper pulp and waste paper
(d) Sum of Energy Inputs column must be adjusted to eliminate double counting of electrical inputs to sectors. To reconcile with "raw" fuel inputs, and add the gas/oil refinery losses (3.00).

However, it is not so clear how available energy used for the elimination of impurities or in heat treatment to modify grain structure could ever be recovered. Yet this energy is "embodied", in the sense that a minimal expenditure is required to achieve the desired state.

The obvious way of recovering "embodied" energy is via recycling. This can be done at any level of processing. Thus, truck tires can be recycled without reduction to constituent elements. New rubber is simply overlaid on the old tire casing. Old engine blocks can be re-bored and rebuilt. High chrome stainless steel and superalloys are also typically recycled as alloys,[6] rather than being refined to pure metals.

Generally, scrap reprocessing requires much less available energy than reduction of virgin ore. The ratio is nearly 20 to one for aluminum and four to one for iron. The most efficient possible economy from an energy standpoint would recycle all materials for which the net energy required for secondary recovery is less than that expended on primary production. Since recycling is clearly infeasible where materials are used dissipatively, as in protective coatings, coloring agents, cleaning agents, bleaches, fuel additives, lubricants, pesticides, anti-freeze, water softeners (and many other examples), the most efficient possible economy would also avoid such uses. The usual practical problems of recycling – inhomogeneous feedstock and undesirable impurities – can be controlled, in principle, by a combination of rationalized product design, improved secondary material recovery and sorting technology, and improved sensing and monitoring technology.

National energy perspective

The energy balance of the United States economy as a whole can now be estimated in terms of the following flows:

6 By blending.

18

Raw Fuels → Loss

↓ ↓

Finished Fuels →Finished Materials

↓ ↓ ↓

Durable Goods Consumable Goods

↓ ↓

Final Services

In terms of this representation, there are the three distinct "pathways" by which final services are generated, viz.

RM → FM → CG → FS
RM → FM → DG → FS
RM → FF → DG → FS

Each pathway can be weighted in terms of the total amount of energy involved, and each step can be characterized in terms of an efficiency. For the transformations between raw materials and finished fuels or finished materials, a thermodynamic definition of efficiency is entirely appropriate and adequate.

However, where services are "generated" by consumption goods or, by the consumption of energy in durable goods (such as cars or appliances), the situation is more complicated, as we have already explained. The efficiency of utilization of energy by a machine or an appliance has a thermodynamic aspect. But one must also consider the efficiency of utilization of the materials from which the device is constructed. In practice, we can roughly equate this with the fraction of each material that is recycles. Moreover, one must also consider the efficiency with which the machine or appliance delivers its output services to the ultimate human recipient – as distinct form providing those services "to" other machines, structures of unappreciative elements of the environment. We have called this utility efficiency or payload efficiency.

It would be a major undertaking to estimate utility efficiencies for all energy-intensive services on a consistent basis. Methodological questions need elucidation, and many ambiguities need to be clarified. However, based on the crud estimates we have given in Table 6, it would appear that this area offers significant long-term opportunities for energy conservation. An exploration of the magnitude of the potential would seem to have high priority in view of widespread public misconceptions about the overall efficiency of energy use in the United States economy.

Table 9 – Efficiency of the U.S. Economy as a Whole

Conversion Step	Fraction of U.S. Energy In (b)	Total Available Energy In x10^{15} BTU (1971) (c)	Efficiency Definition	Estimated Efficiency of Step (c)	Fraction of U.S. Energy Lost (c)	Available Energy In x10^{15} BTU (1971) (c)
RM → FF (a)	1.00	69.38	Available Energy, Finished Fuels (Out) / Available energy, raw Fuels (In) (a)	0.01	0.19	13.20
FF → FM	0.30	22.25	Available Energy, Finished Materials (Out) / Available energy, All Sources (In) (b)	0.33	0.21	14.62
FM →CG	0.055	3.81	Fraction of Finished Materials Allocated That is Recycled to Durable Goods (DG)	0.20	0.043	3.04
FF→CG→FS	0.51	35.38	Minimum Finished Fuels Input to Durables Required to Yield Final Services / Actual Fuels Input to Durables	0.04	0.49	34.24
FM→CG→FS	0.055	3.87	Fraction of Finished Materials Allocated to Consumable Goods (CG) That is Recycled	0.04	0.054	3.76
TOTAL	1.00	69.38		0.016	0.984	68.26

(a) Including Electricity
(b) Not including available energy in recycled iron and steel, chemical feedstock and wood pulp. Assuming no available energy in metal ores.
(c) Including terms omitted above, for convenience.

Table 9 presents a summary of our results in terms of the basic conversion steps presented at the beginning of this section. The estimated efficiencies for the first two steps – RM → FF, RM → FM – are strictly thermodynamic (second law). Recycling inefficiencies and "payload" inefficiencies enter the picture for later steps and account for the much lower values toward the bottom of Table 9. Table 10 shows how our sectoral efficiency estimates compare with others that have been extensively quoted. The most surprising result, of course, is the national total. Based on our analysis, the efficiency of the United States economy is in the range between 1 and 3 percent. While estimates of utility or payload efficiency are obviously very rough and crude, we have tried to make them fairly conservative. In each case, we are satisfied that a detailed analysis would probably lead to a lower estimate of the efficiency than that shown in Table 6.

Table 10 – Published Efficiency Estimates (1970)

	Reference 13	Reference 14	This Paper
Extraction	Not Considered	0.46	Not Considered
Electricity generation	0.35	0.30-0.35	0.32
Industry	0.76		0.33
Basic Materials		0.50	
Other Industry		0.45	
Agriculture		0.30	
Transport	0.27	0.20-0.25	0.04
Residential & Commercial	0.80	0.45	0.04
Household (ex auto)			

Two important implications for energy policy can be drawn from this analysis. First, the potential for energy conservation is much greater than has been generally realized. On benefit/cost grounds, conservation should receive greater emphasis than it has to date in ERDA's programming. Second, the greatest opportunities for savings are to be found among the energy-consuming service-producing machines and appliances in the transportation, commercial and residential sectors. While major savings can be identified in the industrial area, on the whole, industry is a much more efficient energy consumer than the service sectors.

20

References

1 Fisher, I., Nature of Capital and Income, New York, 1906.

2 Knight, F., Risk, Uncertainty and Profit, Boston and New York, 1921.

3 Korach, E.G., ed., Technology of Efficient Energy Utilization, Report of a NATO Science Committee Conference, Oct. 8-12, 1973, Pargamon Press, New York, 1975.

4 Gyftopoulos, E.P., Lazaridis, L.J., and Widmer, T.F., Potential Fuel Effectiveness in Industry, a Report to the Energy Policy Project, Ford Foundation, Ballinger Press, Cambridge, 1974.

5 Ross, M., and Socolow, R., et al., "Efficient Use of Energy: A Physics Perspective," Report of the Summer Study on Technical Aspects of Efficient Energy Utilization, American Physical Society, Jan. 1975.

6 Hall, E.H., et al., "Evaluation of the Theoretical Potential for Energy Conservation in Seven Basic Industries," Battelle Columbus Laboratories, Ohio, July 1975.

7 Ayres, R.U., "Technology and Materials Substitution," presented at the 1976 Joint Annual Meeting of the American Physical Society, Feb. 2-5, New York, 1976.

8 Hannon, S.B., "System Energy and Recycling: A Study of the Beverage Industry," Center for Advanced Computation (CAC), University of Illinois, 1972.

9 Herendeen, R., and Sebald, A., "The Dollar, Energy and Employment Impact of Certain Consumer Options," CAC, 1973.

10 McHale, J., Hale, M.C., "Human Requirements, Supply Levels and Outer Bounds," Aspen Institute for Humanistic Studies, 1975.

11 Minerals Yearbook, Bureau of Mines, 1971.

12 "Patterns of Energy Consumption in the United States," Stanford Research Institute, 1970.

13 Joint Committee of Atomic Energy, "Understanding the National Energy Dilemma," Washington, D.C., 1973.

14 "Study on Measures Taken, of Which Might be Taken, to Achieve Increased Economy and Efficiency in the Extraction, Conversion, Transport and Use of Energy in the ECE Region," UN Economic Commission for Europe, E/ECE/883, Feb. 19,1975, p. 54.

This case was written by R.U. Ayres and M. Narkus-Kramer, International Research and Technology Corp., Arlington, Va. It is intended to be used as a basis for class discussion rather than to illustrate either effective or ineffective handling of an administrative situation. It is not a reflection of any particular political thinking.

[17]

Process Analysis and the Neoclassical Theory of Production

To abuse a term is to use it without any attempt at explaining its meaning. In this sense, "process" has been abused in all sciences, but in none as much as in social sciences. Most curiously, in economics the greatest abuse has taken place where one would least expect it to happen, namely, in production theory. Neoclassical economists as well as the standard economists of latter days have never paused to describe the process of production in some operational manner so that you and I may know what they meant by the term. In comparison with our classical forefathers—who went to great pains to describe and analyze some processes of production, as Adam Smith, for example, did in his famous illustration of the pin factory—modern economists have found intellectual comfort in pure symbolism, so that they have gradually stopped considering even the traditional classification of the production factors.

Glaring evidence of the modern economist's craving for hollow symbolism is the fact that to this day Philip H. Wicksteed's presentation of the concept of production function constitutes the standard approach to the topic. "*The product being a function of the factors of production we have* $P=f(a, b, c, \cdots)$," Wicksteed [17, p. 4] said, and economists, generally, still define this fundamental concept in the same cavalier fashion.[1] If we have changed anything, we have replaced "product" and "factors of production" by the vapid terms "output" and "inputs," a substitution which only increases the reader's illusion that he is offered a cogent analytical definition. Now everyone can rest satisfied with the simple etymological translation: "input" is what we put in, and "output" is what is put out.

To be sure, symbolism has been the soul of science ever since man began to organize his knowledge about actuality. Yet symbolism, if not supported by an operational interpretation of each symbol (or at least of each primary symbol), silently but unfailingly leads the student away from the most arduous and most important task of any special science, that of bringing the human mind in closer contact with actuality. The neglect of clarifying even partially the concept of production function is all the more puzzling in view of the "practical" nature of the economic science as attested, in particular, by the immense number of works which only compute one gigantic "concrete" production function after another.[2] In any case, the omission is not a matter of purely academic interest only. On the contrary, as I have argued in a series of essays [6, 8, 9, 10, Ch. ix], it falsifies our understanding of the production process, a fact responsible for several important blank spots in neoclassical theory of production. One such blank spot concerns the fundamental difference between productive processes in agriculture (or other strongly seasonal activities) and productive processes in manufacturing. An-

[1] This summary presentation of the concept of production function does not only characterize most textbooks—some widely used, e.g., Stonier and Hague [16, p. 219], Leftwich [11, p. 109], Samuelson [15, pp. 515ff]—but it also appears in the writings of some consecrated pundits of our profession, e.g., Frisch [3, p. 41] and Samuelson [14, p. 57].

[2] It is not only because of this neglect that the relevance of these production functions must be questioned. The other reason pertains to the current econometric practices which also involve a chasm between the nature of statistical observations and the stochastic axioms of multivariate analysis. For this last point see Georgescu-Roegen [7].

other covers the difference between the economy of the productive processes in primary activities, mining and agriculture, and the productive processes in secondary activities. Still another concerns one of the most important factors of economic development, namely, the economy of capital utilization. In the present paper, I propose to review and expand these points and to present some additional results concerning the analysis of cost of production and of factor allocation.

Contradictions, Omissions, and Surmises

If our theory of production is to be an adequate logical representation of actuality, it is absolutely necessary that before making even one step further we insist on knowing what corresponds in actuality to every symbol—including F—in the popular formula

$$(1) \qquad Q = F(X, Y, \cdots, Z),$$

by which standard economics describes any "static" production process. The symbol F must also be included because "function" by itself is an ambiguous term even in mathematics. However, the point is that in the standard theory not even the other symbols are connected in some definite operational manner with data observable in actuality. A few careful authors do go beyond Wicksteed's hollow definition. But they tell two different stories. Some conceive the production function (1) as a relation between the *quantity* of product Q and the *quantities* of factors $X, Y, \cdots, Z$. Others conceive the production function as a relation between the output *per unit of time* q and the inputs *per unit of time* $x, y, \cdots, z$, briefly, as a relation between *rates* of flow:[3]

$$(2) \qquad q = f(x, y, \cdots, z).$$

A clear symptom of the standard economist's lack of respect for epistemological problems is the fact that no such economist seems to have thought of the possibility that the two viewpoints may not be equivalent. Even authorities use both definitions interchangeably.[4] Having been exposed by a long rote to this equivalence, we have become so firmly convinced of its validity that some of my fellow econometricians, on hearing for the first time its denunciation, immediately protested. As Patinkin admitted subsequently, the novelty so astounded him that he could but be certain that the argument misinterpreted the property of homogeneous functions.[5]

The argument, however, is extremely elementary. By definition we have $Q = tq$, $X = tx$, $Y = ty, \cdots, Z = tz$, for *any* interval of time t. If the two definitions are equivalent, (1) and (2) yield straightforwardly

$$(3) \qquad tf(x, y, \cdots, z) \equiv F(tx, ty, \cdots, tz).$$

And from this identity we deduce, first, that

$$(4) \qquad f(x, y, \cdots, z) \equiv F(x, y, \cdots, z),$$

and, second, that this common function is homogeneous of the first degree. The presupposed equivalence between (1) and (2) implies, therefore, that absolutely every production process is indifferent to scale, a position which, I hope, is no longer defended by anybody.[6]

An ad hoc attempt to save the day against the foregoing argument led to the suggestion that the quantities are accumulated variable flows:

$$(5) \qquad Q(t) = \int_0^t q(t)dt, \qquad X(t) = \int_0^t x(t)dt, \cdots, Z(t) = \int_0^t z(t)dt.$$

This idea makes matters worse. Instead of (3), we have

$$(6) \qquad f[x(t), y(t), \cdots, z(t)] = \sum x(t)\frac{\partial F}{\partial X},$$

for any $x(t), y(t), \cdots, z(t)$, which requires that the production function be a simple linear expression of the inputs [8, p. 43]

$$(7) \qquad f(x, y, \cdots, z) = ax + by + \cdots + cz.$$

The standard economist's lack of interest in arriving at a clear idea of the process of production is reflected also in his silence on the measurability of the "output" and "inputs" involved in (1) or (2). True, from the confusing arguments concerning the measurability of utility there emerged the idea that cardinal measurability is a bogey. A carpenter, we are told,

[3] See the sample list references given in [9], notes 4 and 5.

[4] Ragnar Frisch [3, p. 43] uses the two definitions on one and the same page.

[5] See "Discussion" in [6, p. 528].

[6] In fairness to Wicksteed, one should note that the modern incongruity cannot be laid at his door; he explicitly (although without any explanation) assumed that all production functions are homogeneous of the first degree [17, p. 33].

may very well lay down his measuring rod once, four times, eight times, · · · and count "one," "two," "three," · · · . Perhaps this idea prompted recent theorists of production to be content with any form of measurability for the factors and the product. As far as pure symbolism is concerned, such a position can raise no objection. But the situation changes if the production function is to reflect a part of actuality. In actuality, practically everything we buy and sell is cardinally measurable, for otherwise we would not be able to speak of uniform prices, i.e., of prices per unit. Even land and labor, which exist in a broad spectrum of qualities, are sold on the basis of a cardinal measure—acres or hours.[7]

Actually, of all students, economists should be the last to accept the doctrine that only ordinal measure counts. For if this doctrine were true, our most powerful tools of analysis would go overboard. If commodities were not cardinally measurable by one method or another, then the principle of decreasing marginal substitution, of increasing or decreasing returns, for instance, would lose the basis on which they are defined. One could, in that case, cause the isoquants to have practically any shape one pleases [5, p. 234; 8, pp. 38ff].

The fact that ordinal scales allow us a tremendous freedom of manipulation has led some to surmise that in each case the scales of the factors can be chosen so that the production function be homogeneous of the first degree (an idea that aims at denying the existence of optimum plants). Joan Robinson [13, pp. 109, 332ff] was first to argue that such a feat could be achieved if factors are measured in "efficiency" units. While she retracted her error at the first opportunity, others persisted in it.[8] That the surmise is not true as a general proposition may be shown by a simple example. Let $q=xy+y$ be the production function in case the product and the factors are measured in usual units. Let us assume that with the "efficiency" scales $x=h(u)$, $y=k(v)$, the production function $q(u, v)=h(u)k(v)+k(v)$ is homogeneous of the first degree. The supposition entails $h(0)=0$, $k(0)=0$. If we make $u=0$ in the defining identity

$$h(\lambda u)k(\lambda v)+k(\lambda u) = \lambda[h(u)k(v)+k(v)], \tag{8}$$

we obtain $k(\lambda v)=\lambda k(v)$. The introduction of this relation in (8) yields $h(\lambda u)=h(u)$, which is absurd.[9]

Finally, there is another thesis that aims at denying the existence of optimum scale of any process. The thesis is that if all pertinent factors are taken into account all natural laws (and hence any process of production as well) are expressed by homogeneous functions of the first degree. As Samuelson [14, p. 84] pointed out, the thesis is operationally idle. It can be proved that it is, in addition, downright incongruous [10, p. 107]. For if in any relation $y=f(x_1, x_2, \cdots, x_n)$, assumed to be the complete expression of a natural law, m factors are left out, the observations of those retained will generally fill a subspace of $n-m$ dimensions. Consequently, the observations of the $n-m$ retained factors could not possibly reveal to us the existence of any law.

Process: An Analytical Tangle

There is, however, some reason why the meaning of "process," a term so frequently used, is hardly clarified in the special scientific literature. Process implies Change, and Change is the most baffling concept in philosophy. Perhaps Change is only illusion—as many philosophers maintain. Perhaps everything is permanently there, just as the paintings of a museum which gradually emerge into our apprehension as we walk from one room to another. Be this as it may, to explain Change is the highest aim of any special science, even though we usually proclaim that science can study only what does not change. This proclamation seems intended to conceal the true difficulty of science in general, which is that our Understanding cannot conceive an action without an agent. The verb "becomes," like all verbs, requires a subject (in grammar as well as in our comprehension). And the rub is that "Jim became tired" is a very unsatisfactory thought. How can we conceive Jim as the *same* person before and after the event of his becoming tired? The fact that there is something to which we refer as "Jim" both before and after that event does not suffice in the least to establish sameness. Witness the

[7] On the issue of cardinality, see Georgescu-Roegen [5].

[8] E.g., McLeod and Hahn [12, pp. 132ff]. The stronger and hence the more inept position that factors can be measured in efficient units even when they differ qualitatively is basic in many econometric works on technological change.

[9] In [8, p. 39 n] I stated without proof that also the assertion by Dorfman [2, p. 82] that one can define the inputs and outputs "in such a way that all production functions are linear" is not true. This refutation is a little more involved because of the additional degree of freedom supplied by the scale of the product.

40 Energy and Economic Myths

fact that there is an entity to which we refer as "the President of the United States" both before and after the Inauguration. Sameness raises formidable problems of epistemology. Yet, curiously, most of them seem hardly touched by philosophers. For good measure, think of a cosmic event perceived by one observer as a flash, and as a slight wave of warmth by another who travels at a greatly different relative speed. Unsuspected though the fact may be, the issue of sameness is, as we shall see presently, a most crucial one in arriving at an analytical representation of a production process.

One may circumvent all these difficulties by subscribing to the basic tenet of Dialectics, which is that Being is Becoming. However, science cannot be erected on this foundation. For science must distinguish between "object" and "event," that is, between Being and Becoming. It is this distinction that draws the line between Dialectics and Analysis. Analysis offers science the great advantages of precision and easier description of actuality. But these advantages have a price: the endless paradoxes and logical contradictions that emerge from almost every analytical framework to remind us of the dialectical nature of actuality as well as of our thought. Actuality—we must stress the point—is seamless. Hence, violence is done to it as Analysis slices it into discretely distinct pieces in order to facilitate our Understanding.[10]

The upshot is that any analytical science can study only partial processes, i.e., only slices of actuality. Every *analytical* process, therefore, can be but a partial process. To determine such a process, we need an *analytical boundary*. This boundary must include the *duration* of the process and the "geographical" *frontier*, which separates the process from the rest of actuality at all times. The conclusion is clear: no analytical boundary, no analytical process.

Several points now deserve pointed emphasis. Since actuality is seamless we can, in principle, draw a boundary wherever we please. However, a student is always guided by some purpose, proper to the domain of his inquiry. Second, the analytical boundary is a void, for otherwise we would need boundaries of boundaries of boundaries . . . , and would be engulfed into an infinite regress. Third, what an analytical process does can be described only by listing everything that crosses its frontier and at what time. If we are interested in studying also what happens inside an analytical process, we must divide it into other processes by drawing new boundaries.[11] Lastly, in drafting the list of the elements that cross the frontier, one must bear in mind that, a partial process being an artificial slice of actuality, such a process is inexistent both before the origin of its duration, $t=0$, and after the end of that duration, $t=T$.

Within this framework "input" and "output" acquire an operational as well as analytical definite meaning. However, Analysis must make now another heroic step, which is to assume that the input and output elements, C_k, exist in a finite number of discretely distinct and measurable qualities. On this basis, the analytical coordinates of a process form a point in a functional space,

$$(9)\quad [\overset{T}{\underset{0}{I_1}}(t), \overset{T}{\underset{0}{I_2}}(t), \cdots, \overset{T}{\underset{0}{I_n}}(t); \overset{T}{\underset{0}{O_1}}(t), \cdots, \overset{T}{\underset{0}{O_n}}(t)],$$

where $I_k(t)$ is the cumulative input and $O_k(t)$ the cumulative output of C_k up to time t (inclusive). The same element may appear both as input and as output. Think of a process consisting only of locomotion (which involves no qualitative change) or one in which electrical motors are used to produce electrical motors.

The analytical picture (9) may suit many a special science, but not economics. In economics, commodities play the same fundamental role as that played by cells in biology, by molecules in chemistry, by elementary particles in physics. Commodity fetishism, notwithstanding its loud denunciation by Karl Marx, is the indispensable foundation of economics. Witness the fact that if we reexamine the entire literature on economic production, Marx's included, in the light of the argument of this section, we see that every economist has drawn economic boundaries only in relation to some commodity. An engineer, for example, may draw a boundary between the furnace with melted glass and the rolling machines of a plate glass factory, but not so an economist in the past or at present. For melted glass is not and has never been a commodity. The conclusion is that if any ele-

[10] For a detailed discussion of these issues, see Georgescu-Roegen [10].

[11] A point that illustrates the ways in which Analysis may baffle us: if we push this process to the limit, the whole happening slips through our analytical mesh.

ment *that is not a commodity* is an input or an output of a production process represented by (9), a thorny problem confronts the economist. For, in that case, how can he set up that indispensable tool in any science, namely, the balance between inputs and outputs, which in economics means a value balance? And the fact is that numerous elements of any production process are not commodities proper—tired workers, worn-out tools, and waste are normal outputs, while free goods are normal inputs.

The solution of the impasse lies in the analytical fiction of the static process or, as Marx more properly described it, the process that reproduces itself. Obviously, for a process to be reproducible, some of its elements must remain intact despite the change caused by the process. This is the inevitable consequence of the epistemological position specific to Analysis—that there is both Being and Becoming. It is the common conception that in any production process a number of agents act upon some materials so as to transform them into products. In a static process, the distinction necessarily leads to the division of the analytical coordinates into two categories—the *flow* and the *fund* (not stock!) elements. A flow is an element that either is only "consumed" or only "produced" by the process; it is either only an input or only an output. A fund is both an input and an output; more precisely, it is a factor whose economic efficiency is maintained by the very process in which it participates. The distinction, obviously, breaks down in the case of locomotion. However, locomotion as such presents no interest whatsoever for the economist. Even when locomotion is apparently the case—as it is in an oil pipeline—the economist has all the reasons in the world for treating the oil at the source and the oil at the destination as two different commodities.

The distinction may, however, raise other, more delicate issues which must be handled with what Alfred Marshall so rightly called "delicacy and sensitiveness of touch." For we must not forget the fact that Analysis has to cut some slits in a Whole that has no joints or seams. Ricardian land provides the clearest illustration of the concept of fund, but a machine that is continuously maintained and repaired also fits the definition. Accordingly, a machine coming out of a process is a fund even though it may have no part whatsoever in common with the "same" machine that went into the process at some time in the past. Still more important is another point, which hardly needs any elaboration: the clover seed is a flow in producing clover fodder, but a fund in producing clover seed.

A host of economists have assailed the notion of reproducible (static) process on the ground that it is a far remove from reality. But this objection is as unfounded as that against the fiction of uniform motion. What do most factories do other than reproduce today the process of yesterday? The main difficulty of the notion of reproducible process is of an analytical nature. First, in order to maintain the efficiency of a machine intact, we need other machines. Machines and tools to maintain other machines and tools lead to a regress which may stop only if the reproducible process includes practically every production process in the world. But even this difficulty may be circumvented by introducing the necessary services among inputs. One cannot, however, dispose as easily the fact that workers, other than the self-sufficient farmers, are not maintained within the production process. This analytical blemish, however, does not deprive the notion of reproducible process of its usefulness.[12] Yet the notion carries with it some danger. Its analytical transparency has led and it still leads some economists to ignore the entropic nature of the economic process (which continuously and irrevocably degrades man's environmental dowry) and, as a result, to maintain that the solution to mankind's economic problem lies in the stationary state. The truth, sad though it is, is that this problem is the heart of the evolution of the human species and will end only with the end of that species.

The Elementary Process: A Fundamental Concept

We are now in the position to introduce the basic analytical element of production theory. By an elementary process we shall understand the reproducible process by which one unit (natural or appropriately chosen) of the particular output called product is produced from certain other specified elements. The process by which a table is made by a cabinetmaker from dressed lumber, prepared coating materials, ready-made hardware, is a good illustration in

[12] On the issues mentioned in the last paragraphs, the reader may consult Chapter ix in Georgescu-Roegen [10].

point. But the elementary process is the basis of any production, whether of an automobile, of a pound of steel, or of a transatlantic ship.[13] Given the distinction we can make between flow and fund coordinates, we shall measure the former in ordinary cumulative amounts (flows) and the latter in cumulative services. This idea may seem simple enough, but one point, often neglected, should retain our attention. There is a dimensional difference between flows and services as well as between rates of flow and rates of services. Curiously enough, the rate of service is independent of the time dimension. The rate of service of two workers is just two workers; only their service may be one hundred or eight man-hours depending on its duration. A further point is that "the flow of services" is a license which throws an opaque blanket over an important problem of value. For only the woolen fabric flows into the coat, not so the services of the needle. If you find a needle or a part of it in your new coat, it is only because of a regrettable accident.

In a generic fashion, let us denote the services of Ricardian land by $L(t)$, those of capital proper by $K(t)$, and those of labor by $H(t)$. Similarly, we may denote by $Q(t)$, $R(t)$, $I(t)$, $M(t)$, and $W(t)$ the flows of products, of natural resources, of manufactured articles, of maintenance supplies, and of waste. The complete description of any elementary process is

$$[\overset{T}{\underset{0}{Q}}(t), \overset{T}{\underset{0}{R}}(t), \overset{T}{\underset{0}{I}}(t), \overset{T}{\underset{0}{M}}(t), \overset{T}{\underset{0}{W}}(t); \quad \overset{T}{\underset{0}{L}}(t), \overset{T}{\underset{0}{K}}(t), \overset{T}{\underset{0}{H}}(t)], \tag{10}$$

which can also be represented by a series of graphs, one for each element [6, p. 515; 9, p. 3].

This analytical setup brings to light two points of exceptional economic importance. The first concerns the long-standing denial of the value of Ricardian land and natural resources. But there can be no doubt that Ricardian land must be included in a complete description of a productive process. In agriculture, in particular, Ricardian land plays a role wholly analogous to that of a fisherman's net; only, instead of fish, it catches solar energy and its by-products [6, p. 508]. As to the natural resources, we may recall that Karl Marx, of all the economists, recognized that no one can fish from a lake where there are no fish. The omission of these two factors in standard analytical economics as well as in most applied works may also be imputed to the neglect of standard economists for any epistemological clarification of the production function. The consequence has been that the economic process is now viewed as a simple circular affair, as a mechanical analogue, which, like a pendulum, just beats time, but makes no history. The actual economic process, on the contrary, is making its own history, through the continuous tapping of natural resources and the search for a more efficient use of these scarce economic factors.

The second point is that in absolutely any elementary process numerous factors are inevitably idle during parts of the duration. This *technical* idleness is the worst form of economic waste and a great hamper to economic progress. Unfortunately, natural as well as economic factors prevent its complete elimination. The natural factors pertain to season. Some activities, especially farming, have a temporal rhythm over which, for all practical purposes, man has hardly any control. One can begin at any time of the day, of the week, or of the year, an elementary process by which a table or an automobile is produced. But a corn grower must sow corn in the fields within a specific, very short period of the year if he wants to have a crop. The difference is of paramount importance. In the case of manufactured or mined products, we can arrange the elementary processes *in line* in such a manner that each fund shifts to another process as soon as it has finished its task in the previous one. This is the way any factory operates, like an assembly line even though one is not in direct view.[14] There are inexorable physical factors, therefore, that oblige the farmers all over the world (with only a few, but highly instructive exceptions[15]) to grow practically all their products by elementary process *in parallel* (or almost so). In this arrangement, processes begin and end at the same time, with the result that the idleness coefficient of most factors is reduced very little, if at all. Think how long a plow remains idle on a farm compared with the continuous use of a furnace in steel works, for example.

Patinkin, in his argument against the im-

[13] The complications inherent to joint products must be left out at this stage.

[14] The point may be clarified by a diagram as in Georgescu-Roegen [6, p. 517].

[15] The exceptions consist of a few spots around the equator. For their object-lessons, see Georgescu-Roegen [6, pp. 524ff; 10, p. 253].

portance and the relevance of this difference, maintains that in the agriculture of the world viewed as a single process funds are continuously employed. The point is obviously inept: the tractors in the Northern Hemisphere *are* idle during the time those in the Southern Hemisphere are active and both *are* often idle over long stretches of time.[16] On the other hand, the economic importance of the difference between the factory and the farming system is splendidly dramatized by the "chicken war" in Europe, a direct consequence of the transformation of the American chicken farms into "chicken factories." It is not excluded that with time we may find the means for growing pigs or even cattle in a factory system. But in spite of many headline claims progress in that direction does not seem at this time too encouraging. In any case, the old Marxist dream of "factories in the open air" is not going to become a general fact for the farmer, not until man conquers the cosmic power necessary for rearranging the position of the globe on the ecliptic.

For the economic historian, too, the analytical concept developed in this section is highly instructive in that it casts a great deal of light on the intimate connection between the factory system and the intensity of demand, on the one hand, and on the assumed correlation between that system and modern technology. Certainly, if only one table is demanded during the period T, production can be carried on only by elementary processes *in series*, with one process beginning when the previous one ends. Nowadays, this is the case for the production of large ships, of space rockets, for example. At one time, this was the general situation in every artisan shop. If elementary processes must be arranged in series, there is hardly any possibility for reducing the technical idleness. Nor is there any incentive or occasion for some division of labor and even of machinery. A cabinetmaker of old times would not have acted economically if he had divided his tasks between him and another worker or if he had used different saws for different cuts. On the other hand, there is absolutely no reason why a factory system should be incompatible with the technology of the ancient times. Indeed, nothing militates against the idea that in the busy shops work proceeded according to that system even in antiquity. In fact, all these remarks justify the view that with the gradual increase in demand the factory system came to be practiced first in some shops of the eighteenth century and that this evolution spurred the division of labor and induced the fever that caused the Industrial Revolution. The relation of cause to effect may after all be the reverse of that of the accepted doctrine in economic history. And from all one can judge, Adam Smith seems to have had in mind this causal order in his argument relating the division of labor to demand.

The Production Functions

We may next assume—a quite strong assumption, if one stops to think about it—that, for a given product, the representations by (10) of all possible elementary processes form some "surface" in the corresponding functional space. The equation of this surface expressed in terms of the product coordinate is the production function of the elementary processes in point,[17]

$$\overset{T}{\underset{0}{Q}}(t) = \mathfrak{F}[\overset{T}{\underset{0}{R}}(t), \overset{T}{\underset{0}{I}}(t), \overset{T}{\underset{0}{M}}(t), \overset{T}{\underset{0}{W}}(t); \overset{T}{\underset{0}{L}}(t), \overset{T}{\underset{0}{K}}(t), \overset{T}{\underset{0}{H}}(t)]. \tag{11}$$

This "function" is what mathematicians call *functional*. It relates functions to functions, not numbers to numbers as (1) or (2) do.

We may now consider complex processes, beginning with those in which elementary processes are arranged in parallel. One category is illustrated by the growing of corn from seed on several farms in the same geographic area. If the elementary process represents what one farm does, then the corresponding production function is immediately derived from (11):

$$n\overset{T}{\underset{0}{Q}}(t) = \mathfrak{F}[n\overset{T}{\underset{0}{R}}(t), n\overset{T}{\underset{0}{I}}(t), \cdots, n\overset{T}{\underset{0}{H}}(t)]. \tag{12}$$

This is a situation in which the familiar yet misleading tenet "doubling the inputs doubles the output" applies. But there are other cases of processes in parallel to which the tenet does not apply. If (10) is a process by which one bread loaf is baked in a bakery, the baking of a batch of loaves does not require the amplification of all coordinates. The *same* mixing ma-

[16] See "Discussion" in [6, pp. 528ff].

[17] Two remarks are in order here. First, it is not necessary that the production surface should cover the entire factor space. Second, normally $Q(T) = 1$ and $Q(t) = 0$ for $t \neq T$.

chine, oven, and building can take care of the multiple task.

Difficulties of the kind just mentioned render very tedious the derivation of the factory production function from (11). But this can be achieved in a direct fashion by considering the general representation (10) and observing that in a factory system all coordinates, flows and services, are proportional to the time the factory remains open. Hence, we have for *any* T

$$(13)\quad \overset{T}{\underset{0}{(qt)}} = \mathcal{G}[\overset{T}{\underset{0}{(rt)}}, \overset{T}{\underset{0}{(it)}}, \overset{T}{\underset{0}{(mt)}}, \overset{T}{\underset{0}{(wt)}}; \overset{T}{\underset{0}{(Pt)}}, \overset{T}{\underset{0}{(Ht)}}],$$

where P stands for all funds other than labor. As is easily seen, (13) is a *degenerate* functional which can be reduced in two ways to a point function. The first form is

$$(14)\qquad q = \Theta(r, i, m, w; P, H),$$

where every symbol represents either a *rate* of flow or a *rate* of service. The second form is

$$(15)\qquad Q = \psi(R, I, M, W; \mathcal{P}, \mathcal{H}; t)$$

where every symbol—except t—is either the *amount* of some flow or the *amount* of some service over the period t:

$$(16)\qquad \begin{aligned} R = rt, \quad I = it, \quad M = mt, \quad W = wt; \\ \mathcal{P} = Pt, \quad \mathcal{H} = Ht. \end{aligned}$$

These two forms uncover the root of the confusion created by the two different interpretations (1) and (2) and exposed in the second section, above. It is seen that if the production function refers to a factory process (or to a process consisting of identical factories in parallel), all variables must be *rates of flows or of services*, as in (14). If one insists on describing the factory process in terms of *quantities of flows and services*, the corresponding production must include also the *duration* over which these quantities are accumulated, as (15) shows. Furthermore, we also find that there is some homogeneity of the first degree here, namely, of ψ. It is this form that lends another justification to the tenet mentioned earlier, "doubling the factors doubles the output," provided we specify that the doubling includes the time of production t as well. We also obtain a relation analogous to (3). Only, its correct form is

$$(17)\qquad \begin{aligned} t\Theta(r, i, m, w; P, H) \\ = \psi(R, I, M, W; \mathcal{P}, \mathcal{H}; t), \end{aligned}$$

which yields

$$(18)\qquad \Theta = [\psi]_{t=1}.$$

The existence of two distinct forms by which a factory process may be represented reflects a normal duality, even though we may not be fully conscious of it. Form (14) describes only what the process is capable of doing and what it needs for being carried on. It is the same way in which we describe an electric bulb by "100 watts, 110 volts." By contrast, (15) tells us what the factory process has done, is doing, or will do during any period t of full activity. The point that in order to describe what a factory actually does we have to refer either to ψ, which includes the time of production t, or to multiply Θ by this factor vindicates Marx's insistence on the importance of the time of production and of the working day for the theory of value. By comparison, we can see how far removed from the basic economic facts standard theory of production is because of the superficial treatment of such a fundamental concept as the production process. In the standard theory of production there is no room for the time of production or for the working day. The result is that planners everywhere seem to ignore completely one of the most effective and readily available levers of economic development—which is the minimizing of plant idleness. Witness the fact that even in the least developed countries plants lie idle the greatest part of the time either because of an anachronic regimen of a 40-hour week or (and) because they are used only by one shift.

The Components of the Factory Production Function

If we try now to project in greater detail the production function (14) against actuality, we immediately see that any such production function is limitational, that is, the nonwasteful combinations of factors satisfy not one but several relations.[18] It is unquestionable that what a factory can do is what its physical plant is capable of doing. Indeed, a qualified engineer can determine the value of q by merely study-

[18] For an introduction to the problem of limitationality and its impact on marginal pricing, see Georgescu-Roegen [4].

ing the blueprints of the factory. This means that the production capacity of a plant q^* is determined only by L and K,

$$q^* = G^*(L, K), \tag{19}$$

and that the same is true of the labor power H^* necessary for utilizing that capacity,

$$H^* = H^*(L, K). \tag{20}$$

A factory, however, may operate below its capacity, in which case what it can do depends also on the labor power $H \leq H^*$ actually employed. A more general relation must therefore replace (19):

$$q = G(L, K, H) \leq q^*. \tag{21}$$

To express what a factory does, we need an additional coordinate, namely, the proportion of the day during which the factory works or, in other words, the working time of the factory. This coordinate is, of course, closely associated with the worker's working day. If we denote it by τ, $0 \leq \tau \leq 1$, and if we assume that H does not vary during τ—which is a reasonable assumption—the amount of *daily* production is given by[19]

$$Q = \tau G(L, K, H). \tag{22}$$

The complete analytical picture of a factory process involves several other categories of factors. The element that comes first to mind represents the inventories, nonspeculative, to be sure. Lacking some detailed inquiries into the matter, we can only surmise that the level of the technical inventories is determined by the size of the other funds considered so far:

$$S = S(L, K, H). \tag{23}$$

The next element is a fund of a peculiar nature. It consists of what we generally call "goods in process." The term is certainly misleading, for half-tanned hides and melted glass, for instance, can hardly constitute goods in the proper sense of the term. The surprising point is that in spite of all our heroic assumptions, it is just not possible to arrive at a picture of a static factory process that involves only commodities proper (besides free goods and waste). But the most peculiar thing about this fund is that in effect it depicts a Becoming, namely, the Becoming represented by all transformation stages through which the product is created from certain material inputs. For this reason we may refer to it as the *process-fund* of a factory. With the same caveat as for the inventory fund, we may assume that the process-fund is determined by

$$\Gamma = \Gamma(L, K, H). \tag{24}$$

There remain the flow coordinates. It stands to reason that for the maintenance flow we have

$$m = m(K, H), \qquad m = w_1, \tag{25}$$

where w_1 is the flow rate of waste correlated to the ingredients and parts of which m consists. The last equality merely expresses the well-known principle of conservation of matter-energy: no process whatsoever can yield an excess or a deficit in these terms. Next, there are the input flows of natural resources and manufactured goods that are transformed by the agents into the output flows of product and correlated waste. Since this transformation, too, must obey the principle of conservation of matter-energy, all these last flows must be related by a homogeneous function of the first order, $qt = g(rt, it, w_2t)$, which yields

$$q = g(r, i, w_2). \tag{26}$$

Finally, we must take into account the fact that what differentiates one factory process from another is the size of the waste flow, i.e., the amounts of the input flows that go to waste. But the size of the waste flow is determined by the structure of the physical plant as well as by the size and the kind of manpower employed. This means that

$$w_2 = w_2(L, K, H), \tag{27}$$

so that (26) must be replaced by

$$q = g[r, i, w_2(L, K, H)]. \tag{28}$$

It is thus seen that the factory production function (14) actually consists of seven main components—relations (21), (23)–(25), and (27)–(28)—and one auxiliary relation, (20). The purely economic coordinate, τ, determines then the daily production Q, the daily maintenance flow $M = \tau m$, and the proper daily flows $R = \tau r$, $I = \tau i$.

Nothing need be added to convince ourselves that this analytical picture of a factory process is a far cry from the crude, colorless formula to which standard economics reduces the descrip-

[19] If we wish to get closer to the real facts in applied endeavors, we must replace τ by a function of the worker's working day and of the number of shifts.

tion of the same process. Some aspects, however, deserve to be emphasized. First, in contrast with the standard formula, the new picture provides a clear analytical basis to some elements that are ignored by that formula. The elements are the working time τ, the production capacity q^*, and the employment H^* for this capacity. As a result, we can now supply specific definitions for several notions that are used but only loosely defined by standard economics. The ratio q/q^* measures the capacity utilization at *a point in time*, while τ measures the capacity utilization *over time*. The overall capacity utilization is measured by the product $\tau q/q^* = Q/Q^*$, where Q^* is the maximum daily production (obtainable for $\tau = 1$). We can also define in a specific way the frequently used notion of labor-capital ratio. If this coordinate is to reflect the organic composition of capital, it must necessarily refer not to the labor power actually employed (which often may be nill) but to the labor power called for by the existing capital. Hence, the labor-capital ratio must be expressed by H^*/K.

In all probability, the fact that τ, q^*, H^* are not parts of the standard analytical framework (which means the standard production function) is responsible for the absence of the corresponding data from current statistical information. The closest element revealed is the number of man-hours supplied during the year or, what comes to almost the same thing, the total wage bill for each industry. What we generally know, therefore, is only $\sum \tau_i H_i$ over the year. Such data have clearly no connection with H^*. They do not even provide some indirect index of the average employment during the year. Certainly, 1600 man-hours may be the result of 400 men each working four hours as well as 200 men each working eight hours.[20] The results of those applied works in which the capital intensity is measured by the ratio between the total wage bill and the value of fixed capital must, therefore, be viewed with great reservations.

The new analytical representation of a factory process enables us to see clearly through the arguments repeatedly advanced in support of the thesis that there are no optimum units, that doubling the inputs doubles the output. As already mentioned, the thesis is confirmed if the doubling is achieved by doubling the time during which the factory works—a point clearly expressed by (22). The homogeneity of the first degree of the function (26) also bears out the thesis provided that we formulate it correctly: if all inputs *and the waste output* are doubled, then the product output must necessarily double. However, the issue of the existence of optimum units does not hinge upon these two almost tautological aspects. Instead, it hinges upon whether doubling the funds, L, K, H in (21) doubles the potential flow rate q. That numerous defenders of the position that scale does not matter in production have ignored the last point may again be due to the fact that the standard production function makes no distinction between flow and fund elements.

Analysis of Cost and Allocation in a Factory System

Much of the standard picture found in every textbook which aims at showing how cost is derived from the production function and how the optimum allocation of liquid capital is achieved on the same basis must now be changed. But before proceeding, I should stress the fact that, in contrast with the complete silence of the standard theory on this restriction, the analysis presented here applies only to the factory. This case, we should note, is the simplest of all. Other types of processes present difficulties which must be left for another occasion.

The total cost T of a daily output Q always consists of two parts: τT_V, the cost that varies proportionately with τ, and T_F, the cost that is independent of τ. (The distinction is analogous but not identical to the standard distinction between variable and fixed cost.) We thus have the identities

$$T = \tau T_V + T_F,$$

$$(29) \quad T_V = \sum H p^H + \sum r p^r + \sum i p^i + \sum m p^m,$$

$$T_H = \sum L p^L + \sum K p^K + \sum S p^S + \Gamma p^\Gamma,$$

where the p are prices.

If the management can vary only τ, then the cost of producing a daily output Q, Q varying from $Q = 0$ to $Q = q \times 1$, is $T = T_F + (Q/q) T_V$. In this situation, the optimum τ is determined by

[20] This is the elementary reason why t must appear explicitly in (14). The point bears also upon the fact that the static Leontief system is generally described in terms of total flows and the total "flow of labor services," with the fault that one can never know the actual scale of operation. For this last point, see Georgescu-Roegen [10, pp.

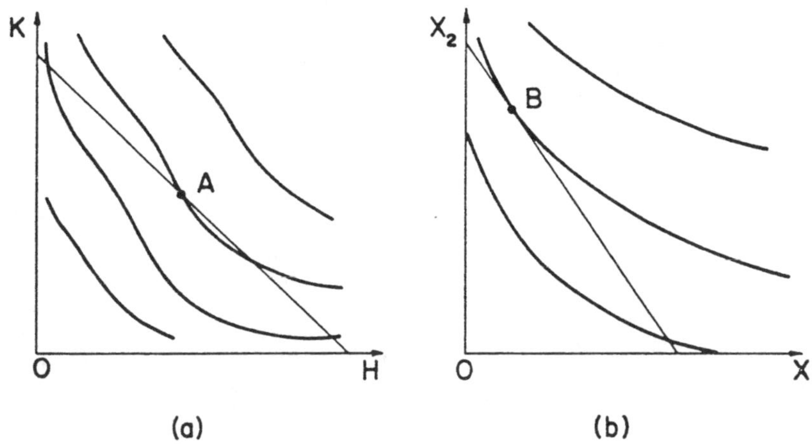

Figure 1

the maximization of profit $R-T$, where $R(Q)$ is the daily total receipts for the offer Q. If $R=pQ$, then the optimum is either $\tau=0$ or $\tau=1$ according to whether $p<$ or $>T_V/q$ and $0\leq\tau\leq1$ if $p=T_V/q$. If R has the usual parabolic shape, then optimum τ is given by $dR/dQ=T_V/q$.

If, on the other hand, τ is constant and all other elements may vary, then the total cost function must be determined in relation to the usual optimality criterion—the greatest daily output Q for every given total cost T. However, the assumption that τ does not vary is inept, even in the situations in which the worker's working day is limited by law. And since the analysis developed from this assumption is as complicated as the general case, to avoid duplication we may pass directly to the general case in which both τ and all production factors are free to vary—a problem that the standard theory of production could not, per force, even entertain. The main relevant results reached in this direction can be safely brought home by a simplified, albeit unrealistic, structure that leaves out the Ricardian land, the inventories, and the process-fund as well as the maintenance flow. If we also use a uniform notation for all remaining input flows, the production function is reduced to only two components,

$$(30)\quad Q=\tau G(H,K),\quad Q=\tau g[x,w_2(H,K)].$$

In the still more simplified structure in which there are only one kind of labor, one kind of capital, and two kinds of input flows, these relations may be represented graphically by two families of isoquants. These isoquants remain the same as τ varies; only the corresponding product values increase in the same proportion with τ. Figure 1a shows the isoquants of G, Figure 1b those of g.[21] We must, however, make a special note of the fact that, in general, the last map changes as the fund structure (H, K) changes. The graphical representation will enable us to pinpoint some of the results to which we now turn.

The optimalization problem in this case may be stated as follows: to find the maximum of $Q=\tau g$ for a given

$$(31)\quad T=\tau\left(\sum_a p_a^H H_a+\sum_c p_c^x x_c\right)+\sum_b p_b^K K_b,$$

and the additional constraint

$$(32)\qquad G(H,K)=g[x,w_2(H,K)].$$

In (31), p_a^H, p_b^K are the prices of the services of various kinds of labor and capital *per day*, and p_c^x are the prices *per customary unit of measurement* of the various input factors. The necessary condition for the maximum τg is $d(\tau g)=0$, to which we must add the constraints $dT=0$, $dG=dg$. Because of $d(\tau g)=0$ and of (32), the last relation may be replaced by the simpler one, $d(\tau G)=0$. We thus must have:

[21] The reason why the isoquants in Figure 1a are not necessarily convex toward the origin—as the standard theory assumes them to be—is revealed below.

$$dT = \tau\left(\sum_a p_a{}^H dH_a + \sum_c p_c{}^x dx_c\right) + \sum_b p_b{}^K dK_b + T_\gamma dr,$$

$$(33)\quad d(\tau G) = \tau\left(\sum_a G_a{}^H dH_a + \sum_b G_b{}^K dK_b\right) + G d\tau$$

$$d(\tau g) = \tau\left(\sum_a g_a{}^H dH_a + \sum_b g_b{}^K dK_b + \sum_c g_c{}^x dx_c\right) + g d\tau,$$

where the partial derivatives are denoted according to the pattern $G_a{}^H = \partial G/\partial H_a$.

To avoid the complications associated with a corner solution (which are irrelevant for the immediate argument), we shall assume that these linear equations may be simultaneously satisfied for differential elements that otherwise vary freely. This means that the corresponding linear forms are not independent. Hence:

$$(34)\quad \begin{aligned} &p_a{}^H - \lambda G_a{}^H - \mu g_a{}^H = 0,\\ &p_b{}^K - \tau\lambda G_b{}^K - \tau\mu g_b{}^K = 0,\\ &p_c{}^x - \mu g_c{}^x = 0,\\ &T_V - \lambda G - \mu g = 0. \end{aligned}$$

If τ is given, then the last relation must be eliminated.

In either case, from (34) we obtain

$$(35)\quad p_c{}^x / g_c{}^x = \mu,$$

$$(36)\quad \frac{p_a{}^H - \mu g_a{}^H}{G_a{}^H} = \frac{p_b{}^K - \tau\mu g_b{}^K}{\tau G_b{}^K} = \lambda,$$

for all a, b, and c.

Relations (35) show that for the optimal factor combination the marginal rates of substitution of the flow factors are equal to the price ratios. In Figure 1b, the corresponding budget line is, therefore, tangent to the isoquant at the optimal combination $B(x_1, x_2)$. This result does not differ from the standard theory of factor allocation. For comparison, we may recall that, if (1) is the production function and if $T = \sum p_i X_i$ is the total cost, instead of (33) we have

$$(37)\quad dT = \sum p_i dX_i = 0,\ dF = \sum F_i dX_i = 0,$$

where $F_i = \partial F/\partial X_i$. As above, we obtain

$$(38)\quad p_i - \mu F_i = 0,$$

and further

$$(39)\quad \frac{p_i}{F_i} = \mu = \frac{dT}{dQ} = \frac{T}{F^*},$$

where

$$(40)\quad F^* = \sum X_i F_i \cdot$$

But (36) offers the surprise that the ratios of the fund prices are no longer equal to the corresponding marginal rates of substitution derived from either relation (30). The result is that in Figure 1a the budget line of the fund factors, $\tau p^H H + p^K K = T_0$, is not necessarily tangent to the isoquant of the optimal combination $A(H, K)$. From (36), it is seen that the tangency obtains if and only if

$$(41)\quad \frac{g_a{}^H}{G_a{}^H} = \frac{g_b{}^K}{G_b{}^K}$$

for all a and b. In view of the definition of g by (30), the last relations become

$$(42)\quad \left(\frac{\partial w_2}{\partial H_a}\right) \Big/ G_a{}^H = \left(\frac{\partial w_2}{\partial K_b}\right) \Big/ G_b{}^K.$$

Recalling a main theorem of the Jacobian, we easily recognize that (42) represents the necessary and sufficient conditions for w_2 to be a function of G, in effect, of q. Hence, the second equation (30) may be solved for q and replaced by

$$(43)\quad q = h(x).$$

If this is the case, the isoquant map of the flow factors no longer shifts with the fund structure. The isoquants of the two families are now paired, $G = q^0$ is paired with $g = q^0$. Without some knowledge of the actual structures of factory production functions, it is hard to judge whether this case is realistic in any way. A reasonable guess would be that it is not. If waste depends only on the size of the product flow, it should be the same regardless of the extent to which capital is substituted for labor. Be this as it may, even if (43) is true, (36) shows that the optimal factory depends not only on prices but *also on* τ (if τ is fixed). And in either case, the marginal rates of substitution between funds are not equal to the price ratios unless $\tau = 1$.

The point that marginal productivity makes

no sense in the case of limitational factors is elementary. We must therefore be careful not to equate any of the partial derivatives contained in the foregoing calculations with the marginal productivity of the corresponding factor. An increase in the input flow rate x, for example, will not cause any increase in the product flow unless the fund factors are appropriately adjusted to handle that increase. However, the partial derivatives still enter into the basic formulas. These formulas may be derived either by blind calculus, as we have done above, or directly by determining the increments that match each other in the optimum manner [4]. For example, any increase dx must be accompanied by an adjustment of at least one fund factor, which means that the corresponding increments must satisfy the relations

$$(44) \quad g_a{}^H dH_a + g_c{}^x dx_c = dq, \quad G_a{}^H dH_a = dq.$$

Since the corresponding additional cost is $dT = \tau(p_c{}^x dx_c + p_a{}^H dH_a)$, (44) yields

$$(45) \qquad \frac{dT}{dQ} = \frac{p_c{}^x}{g_c{}^x}\left(1 - \frac{g_a{}^H}{G_a{}^H}\right) + \frac{p_a{}^H}{G_a{}^H} \cdot$$

If the adjustment is made by a change in K, the additional cost is $dT = \tau p_c{}^x dx_c + p_b{}^K dK_b$, and instead of (45) we have

$$(46) \qquad \frac{dT}{dQ} = \frac{p_c{}^x}{g_c{}^x}\left(1 - \frac{g_b{}^K}{G_b{}^K}\right) + \frac{p_b{}^K}{\tau G_b{}^K} \cdot$$

Now, since dT/dQ must be the same for any possible combination of increments, (45) and (46) yield (35) and (36). In fact, (45) and (46) are the correlative relations of $dT/dQ = p_i/F_i$ in (39). Finally, because of (34) we have straightforwardly $dT = \lambda d(\tau G) + \mu d(\tau g)$ or

$$(47) \qquad \frac{dT}{dQ} = \lambda + \mu,$$

a relation that mirrors $dT/dQ = \lambda$ of (39).

In order that the solutions supplied by (34) should correspond to a maximum of $Q = \tau g$, we must have $d^2(\tau g) < 0$ subject to the first two constraints of (33). The simple structure of Figure 1 will suffice to pinpoint the relevant results of this new problem. According to a classic theorem of quadratic forms, the condition that $d^2(\tau g) < 0$ subject to $dT = 0$, $d(\tau G) = 0$, is equivalent to a condition pertaining to the signs of the progressive sequence of the principal minors of the matrix[22]

$$(48) \quad \left[\begin{array}{ccccc|c|c} 0 & 0 & 0 & 0 & \tau G^H & \tau G^K & G \\ 0 & 0 & \tau p_1 & \tau p_2 & \tau p^H & p^K & T_V \\ 0 & \tau p_1 & \tau g^{11} & \tau g^{12} & \tau g^{1H} & \tau g^{1K} & g^1 \\ 0 & \tau p_2 & \tau g^{21} & \tau g^{22} & \tau g^{2H} & \tau g^{2K} & g^2 \\ \tau G^H & \tau p^H & \tau g^{H1} & \tau g^{H2} & \tau g^{HH} & \tau g^{HK} & g^H \\ \hline \tau G^K & p^K & \tau g^{K1} & \tau g^{K2} & \tau g^{KH} & \tau g^{KK} & g^K \\ \hline G & T_V & g^1 & g^2 & g^H & g^K & 0 \end{array}\right]$$

These signs must be alternatively negative and positive beginning with the first minor marked.[23] We may recall that in the standard case the same condition establishes the convexity toward the origin of the isoquants. In the present case, the same convenient general geometry does not obtain. Only the isoquants of Figure 1b must have this property, a result that follows from the sign of the first minor after p_1 and p_2 are replaced by μg^1, μg^2 on the basis of (34). The existence of a maximum output does not set any simple restriction on the shape of the isoquants $G = \text{const.}$ [4].

The introduction of the working day uncovers some unsuspected problems. To begin with, we may distinguish now two planning cost curves.[24] The first, $T(Q; \tau)$ represents the minimum total cost for any given Q when τ is fixed. This function is determined by (30), (31) and the system consisting only of the first three relations of (34). The second planning cost curve $T(Q)$ represents the minimum cost for every Q when τ is free to vary. In this case, the complete system (34) determines also the optimal value of τ for every Q. Naturally, $T(Q) \leq T(Q; \tau)$, the equality prevailing only for those values of Q for which the optimal value of τ coincides with the fixed τ.[25]

In case τ is not fixed, by (30) the last relation

---

[22] The new notations in the matrix are self-explanatory; the double superscript denotes a partial derivative of the second order. For the theorem, see, for example, Samuelson [14, p. 378].

[23] If τ is fixed, the same conditions prevail for the reduced matrix obtained by eliminating the last row and the last column from (48).

[24] The term "planning" describes better than "long-run" the situation in which absolutely no factor is fixed.

[25] The dichotomy goes further. There are also two types of *plant* cost curves—one in which τ is free to vary, $T(Q; K)$, and one in which τ is fixed, $T(Q; K, \tau)$. They are derived from (34) by making all $dK = 0$.

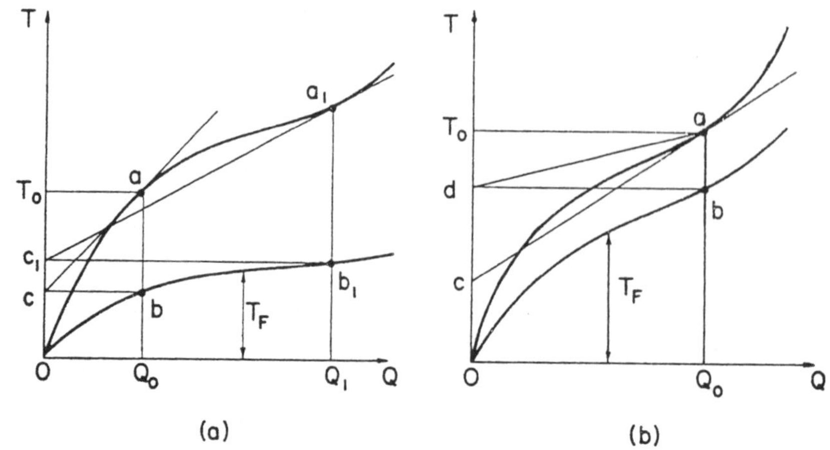

Figure 2

of (34) becomes $T_V = (\lambda + \mu)q$ which further yields

$$(49) \qquad T = (\lambda + \mu)Q + T_F.$$

Let now $T(Q)$ and $T_F(Q)$ be represented graphically in Figure 2a and let ac be the tangent at a to the total cost curve. By (47), the slope of this tangent is $\lambda + \mu$; hence $T_0c = (\lambda + \mu)Q_0 = ab$ is the variable, and $Oc = Q_0b$ is the fixed cost $T_F(Q_0)$. If all the variables of g are now kept constant and only τ is allowed to vary, the new total cost is represented by a segment of the straight line ca. All is in order if, as in the situation shown, the marginal cost is decreasing—i.e., if $d^2T/dQ^2 < 0$. For the resulting cost outside $Q = Q_0$ is greater than the minimum minimorum cost, $T(Q)$. However, if the marginal cost is decreasing—as for the scale Q_1—by varying only τ we obtain a total cost represented by a segment of the tangent a_1c_1 smaller than the minimum minimorum cost. We thus reach the conclusion that marginal cost dT/dQ must *always* be decreasing. The conclusion seems to make sense also from a different angle. From (49) we get

$$(50) \qquad \frac{dT}{dQ} = \mu + \lambda + Q\frac{d(\mu + \lambda)}{dQ} + \frac{dT_F}{dQ},$$

and if we take into account (47), this relation yields

$$(51) \qquad Q\frac{d^2T}{dQ^2} + \frac{dT_F}{dQ} = 0.$$

Since it stands to reason that $dT_F/dQ > 0$, we must have $d^2T/dQ^2 < 0$.

However, this conclusion is not necessarily true. A point which is all too often ignored is that calculus conditions for extrema under constraint fool us whenever the extremum is a corner solution. In case of τ, only $\tau = 1$ may be a corner solution. If this is the case, the last equation of (34) is no longer valid. Instead, it is replaced by the corner inequality $T_V < (\lambda + \mu)q$. Also, (49), (51) as well as the geometry of Figure 2a must be discarded. However, (47) still follows from the system (33) modified so as to take into account that $\tau = 1$.[26] Therefore, the intercept of the tangent ac still gives $T_0c = (\lambda + \mu)Q$. Only, the variable cost ab is now smaller than T_0c and hence the fixed cost Q_0b is greater than Oc, as shown in Figure 2b. The cost function obtainable by varying only τ is in this case represented by the segment da.[27] Since this cost is greater than the minimum minimorum, all is again in order. The point is that the planning marginal cost does not have to be always decreasing.

Let us also note that $T_V < (\lambda + \mu)q$ may obtain even for a scale for which $d^2T/dQ^2 < 0$, such as Q_0 in Figure 2a. But in that case, the optimal arrangement necessarily implies $\tau = 1$. For if

[26] In the modified system $\tau = 1$ and $d\tau = 0$.

[27] The corresponding optimal value being $\tau = 1$, the daily production cannot exceed Q_0.

$\tau < 1$, by increasing τ we could again obtain a cost for Q lower than $T(Q)$.

Concluding Remarks

Although the results presented in this paper represent only a first attempt at providing the concept of production process with an adequate analytical representation and at examining the bearings of such a representation on the theory of production, they teach a few general yet vital object lessons. The first such object lesson is that the analytical description of a concept which necessarily involves Change cannot possibly be reduced to the shallow, facile formula of the standard production function. The nature of Change would not have tormented past and present philosophical minds if the problem were that simple. Another object lesson is that time—not in the sense of a dimensional measure as in "sixty miles per hour," but in the full sense of duration—is one of the most important coordinates of any process, hence of any production process as well. Even in the case of a factory—a process which appears to start producing the instant the factory opens—the working time is a valuable economic coordinate which raises some intriguing questions of the sort examined in the last part of the preceding section. The analysis of the process of production brought to light also the fact that the economics of production must take into account not only the working time but also the time of idleness, an issue that distinguishes in a palpable way the economy of farming, especially, from that of manufacturing.

We have also seen that a fitting representation of the simplest production process (that of a factory), even if stripped of many essential but cumbersome components, does not lend itself to an analysis as simple as the seducing geometry of the familiar isoquant map and of the standard cost curves picture. Relations such as (45) and (46), although referring to a highly simplified structure, defeat any attempt at a simple, transparent translation in terms of specific factor productivities, to which the standard theory has accustomed us. The results reached for the total cost function, $T(Q)$, are somewhat simpler and more definite. However, the actual shape of this function is still an open question. The basis from which $T(Q)$ is derived being at least as complicated as the systems (35) and (36) are, it is hard to see how the structure of that function could be directly related to the shape properties of the functions that describe the factory process. In fact, because of the computational facilities offered by the Cobb-Douglas type function, no effort worthy of mention seems to have been directed toward determining the shape properties of the basic function (21) on which most standard works often rely.

The results presented in this paper, it must be emphasized, do not provide us with a set of engineering instructions to be applied in each particular circumstance. But they help us become aware of a series of new problems, some of a paramount importance for policy. It is vain to hope that the production theory may reach the stage when its general analysis will yield practical recipes. There is one crucial fact which we all recognize under examination but otherwise ignore completely. The passage from one fund structure to another in the production of the "same" product does not mean (outside some truly exceptional instances) that some capital funds are increased and some labor funds are decreased (or vice versa). Such a substitution involves above all a qualitative change. True, nothing stands in the way of introducing as many different qualities as we may wish in our production functions *on paper*.[28] Such a feat may indeed clarify some issues, but it does not provide us with an operational basis for cost analysis. In actuality cost analysis is not derived from some gigantic production functions involving all possible qualities of factors. Instead, the decisions are made by comparing the costs of a few possible processes. It would seem, therefore, that the cost analysis of the type propagated by Jacob Viner and amended by E. H. Chamberlin [1, App. B], although not as high in our esteem as that based on production functions, may after all be the more appropriate for the task. As far as factor allocation is concerned, there is only one practical counsel: every marginal dollar must bring the same physical return, which is all that we can really extract from (45) and (46). From all we know, only cost is a fact; the production functions are analytical fictions in a broader sense of the term than the formulas of the natural sciences are. The latter are calculating devices, the former are analytical similes which only help our Understanding to deal with a complex actuality pervaded by qualitative change. All the more necessary it is that these similes should be as faithful as Analysis can allow them to be.

[28] Cf. Georgescu-Roegen [5].

52 Energy and Economic Myths

References

[1] Chamberlin, Edward H., *The Theory of Monopolistic Competition*, 6th ed., Cambridge, Harvard University Press, 1948.

[2] Dorfman, Robert, *Application of Linear Programming to the Theory of the Firm*, Berkeley, University of California Press, 1951.

[3] Frisch, Ragnar, *Theory of Production*, Chicago, Rand McNally, 1965.

[4] Georgescu-Roegen, Nicholas, "Fixed Coefficients of Production and Marginal Productivity Theory," *Rev. Econ. Stud.* 3:40–49, Oct. 1935. Reprinted with a 1964 Postscript in Nicholas Georgescu-Roegen, *Analytical Economics: Issues and Problems*, Cambridge, Harvard University Press, 1966, pp. 279–299.

[5] ———, Chapter 11 in this volume.

[6] ———, Chapter 5 in this volume.

[7] ———, Chapter 10 in this volume

[8] ———, "Chamberlin's New Economics and the Unit of Production," Ch. 2 in *Monopolistic Competition Theory: Studies in Impact*, ed., R. E. Kuenne, New York, John Wiley & Sons, Inc., 1967, pp. 31–62.

[9] ———, Chapter 4 in this volume.

[10] ———, *The Entropy Law and the Economic Process*, Cambridge, Harvard University Press, 1971.

[11] Leftwich, Richard H., *The Price System and Resource Allocation.*, rev. ed., New York, Holt, Rinehart and Winston, 1960.

[12] McLeod, A. N., and F. H. Hahn, "Proportionality, Divisibility, and Economics of Scale: Two Comments, I and II," *Quart. J. Econ.* 63:128–137, Feb. 1949.

[13] Robinson, Joan, *Economics of Imperfect Competition*, London, The Macmillan Company, 1933.

[14] Samuelson, Paul A., *Foundations of Economic Analysis*, Cambridge, Harvard University Press, 1948.

[15] ———, *Economics*, 8th ed., New York, McGraw-Hill Book Company, 1970.

[16] Stonier, Alfred W., and Douglas C. Hague, *A Textbook of Economic Theory*, 2nd ed., New York, Longmans, Green and Co., 1958.

[17] Wicksteed, Philip, *An Essay on the Co-ordination of the Laws of Distribution*, (London, 1894), Reprints of Scarce Tracts, No. 12, London School of Economics and Political Science, London, 1932.

THE ECONOMICS OF PRODUCTION

RICHARD T. ELY LECTURE

Nicholas Georgescu-Roegen

For the last twenty years or so I have singled myself out among my fellow econometricians for arguing with all the means at my disposal that not every element of the economic process can be related to a number and, consequently, that this process cannot be represented in its entirety by an arithmomorphic model. At the same time, I have insisted that in our haste to mathematize economics we have often been carried away by mathematical formalism to the point of disregarding a basic requirement of science; namely, to have as clear an idea as possible about what corresponds in actuality to every piece of our symbolism. Curiously, in the home of quantity, in the natural sciences, this position does not constitute a singularity. On the contrary, essentially the same words of caution have come from many a high authority in physics—such as Max Planck or Percy William Bridgman, for example.[1] But even some engineers have raised their voices against blind symbolism. The recent remarks by a well-known British engineer are worth quoting at length:

> Contrary to common belief it is sometimes easier to talk in mathematics than to talk in English; this is the reason why many scientific papers contain more mathematics than is either necessary or desirable. Contrary to common belief it is also often less precise to do so. For mathematical symbols have a tendency to conceal the physical meaning that they are intended to represent; they sometimes serve as a substitute for the arduous task of deciding what is and what is not relevant; . . . . It is true that mathematics cannot lie. But it can mislead.
>
> However, the dangers of over-indulgence in formula spinning are avoided if mathematics is treated, wherever possible, as a language into which *thoughts may only be translated after they have first been* [*clearly*] *expressed in the language of words*. The use of mathematics in this way is indeed disciplinary, helpful, and sometimes indispensable.[2]

The topic of this lecture—the economics of production—presents, I believe, sufficient interest by itself. But in choosing it, I have been guided also by the fact that it may serve as a substantial illustration of the harm caused by the blind symbolism that generally characterizes a hasty mathematization.

I

What has come to be known as "the production function" is quite an old item in the economist's analytical paraphernalia. As we may recall, it was introduced in 1894 by Wicksteed with one simple remark: "*the product being a function of the factors of production we have* $P=f(a, b, c, \ldots)$."[3] This paradigm of imprecision apparently sufficed to make us accept Wicksteed's simple symbolism as an adequate analytical representation of any production process and use it indiscriminately in every kind of situation. And as the usage of the vapid terms "input" and "output" became widespread, popular manuals came to treat the subject in an even more cavalier manner than Wicksteed's. A typical presentation is that the production function expresses symbolically the fact that "the output of the firm depends on its inputs."

But even consummate economists have accepted Wicksteed's formula without any ado. They only felt that the meaning of the variables involved ought to be explained. The greater number of such authors adopt the position that the formula shows the quantities of inputs (or of factors) necessary to produce a certain quantity of output (or of product). Accordingly, all symbols in a production function,

$$Q = F(X, Y, Z, \cdots), \tag{1}$$

stand for quantities.[4] Others conceive the same function as a relation between the inputs per unit

[1] Max Planck, *The New Science* (New York, 1959), pp. 43, 158–59; P. W. Bridgman, *The Logic of Modern Physics* (New York, 1949), p. 50.

[2] Reginald O. Kapp, *Towards a Unified Cosmology* (New York, 1960), p. 111. My italics.

[3] Philip H. Wicksteed, *The Co-ordination of the Laws of Distribution* (London, 1894), p. 4.

[4] For a small yet representative sample, see A. L. Bowley, *The Mathematical Groundwork of Economics* (Oxford, 1924), pp. 28–29; J. R. Hicks, *The Theory of Wages* (London, 1932), p. 237; E. Schneider, *Theorie der Produktion* (Vienna, 1934), p. 1; A. C. Pigou, *The Economics of Stationary States* (London, 1935), p. 142; P. A. Samuelson, *Foundations of Economic Analysis* (Cambridge, Mass., 1948), pp. 57–58; K. E. Boulding, *Economic Analysis* (3rd ed., New York, 1955), p. 585; Sune Carlson, *A Study on the Pure Theory of Production* (New York, 1956), p. 12; Ragnar Frisch, *Theory of Production* (Chicago, 1965), p. 41.

of time and the output per unit of time; i.e., as a relation

$$(2) \qquad q = f(x, y, z, \cdots),$$

in which all symbols stand for rates of flow.[5]

Curiously, no one seems to have been intrigued by the existence of these entirely distinct interpretations. Instead, many economists—including some analytical authorities—have used both definitions indifferently, sometimse even on the same page.[6] The undeniable inference is that the economic profession considers relations (1) and (2) as two completely equivalent ways of representing any production process whatsoever. Yet behind this belief there lies an analytical imbroglio which is easily brought to light.

We need only recall that the production function is a tool associated with a static process or, to use a more explicit expression, with a steady-going process. For such a process, the following relations

$$(3) \qquad Q = tq, \; X = tx, \; Y = ty, \cdots$$

hold for any time interval t and for the quantities of product and of factors corresponding to that interval. With the aid of these relations and (2), relation (1) becomes

$$(4) \qquad tf(x, y, z, \cdots) = F(tx, ty, tz, \cdots).$$

And since this relation must be true for any t, it follows, first, that f and F are one and the same function,

$$(5) \qquad f(x, y, z, \cdots) \equiv F(x, y, z, \cdots),$$

and, second, that this function is homogeneous of the first degree. Therefore, the tacit presumption that the forms (1) and (2) are equivalent implies that the returns to scale must be constant in absolutely every production process.

Nothing, I believe, need be added to convince ourselves that this imbroglio is the direct consequence of our acceptance of Wicksteed's symbolism without first probing its validity as an analytical mirror of actuality. This conclusion raises a new and troublesome issue. Does either of the forms, (1) or (2), constitute an adequate representation of a process of production and, if so, what kind of process may be represented by it? For a start, let us try to examine it in its broad lines.

[5] G. Stigler, *The Theory of Competitive Price* (New York, 1942), p. 109; T. C. Koopmans, "Analysis of Production as an Efficient Combination of Activities," in *Activity Analysis of Production and Allocation*, ed. T. C. Koopmans (New York, 1951), p. 35.

[6] E.g., Frisch, *op. cit.*, p. 43.

II

Before anything else, we should note that for no other branch of economics is the concept of process as essential as for the economics of production. But, widely used though the word "process" is in sciences and philosophy, the literature seems to offer no specific definition of it. Now and then, the concept is merely associated with change. However, change is a notoriously intricate notion which has kept philosophers divided into two opposing camps: one maintaining that all is only being; the other, that all is only becoming. Obviously, science can follow neither of these teachings. Nor can it, unfortunately, embrace Hegel's dialectical synthesis that being is becoming. Analytical science must distinguish between object and event. Consequently, it must embrace the so-called "vulgar" philosophy—according to which there are both being and becoming—and cling to it to the very end. The upshot is that science must find a way to represent a process analytically.

It is obvious that, for this purpose, we must retain one point of dialectics; namely, that change and, hence, process cannot be conceived otherwise than as a relation between some entity and its complement in the absolute whole. In viewing a living tree as a process we oppose that tree to everything else—to "its other," in Hegel's terminology. Only for the absolute whole—the universe in its eternity—has change no meaning: such a totality has no complement. The notion of partial process, therefore, implies some slits cut into the absolute whole. But as a long series of thinkers, from the ancient Anaxagoras to the modern Niels Bohr, have taught us, this whole is seamless.[7] However, in this case as in all others, analysis must proceed by some heroic simplifications and totally ignore their ultimate consequences.

The first heroic step is to divide actuality into two parts—one representing the partial process in point; the other, its environment (so to speak)—separated by a boundary consisting of an arithmomorphic void. For if the boundary would not be such a void, we would get three parts instead of two and, as is easily seen, we would be drawn into a dialectical infinite regress. So, all that exists in actuality at any moment must belong either to a process or to its environment. The basic element of the analytical picture of a process is, therefore, the boundary. No analytical boundary, no analytical process.

[7] See Fragment 8 in J. Burnet, *Early Greek Philosophy* (4th ed., London, 1930), p. 47; Niels Bohr, *Atomic Physics and Human Knowledge* (New York, 1958), p. 10.

Now, precisely because actuality is a seamless whole we can slice it wherever we may please. And, Plato to the contrary,[8] actuality has no joints to guide a carver. As economists we know only too well the unsettled issue of where the natural boundary of the economic process lies. Only our particular purpose in each case can guide us in drawing the boundary of a process. So, every scientist slices actuality in the way that suits best his own objective—an operation that cannot be performed without some intimate knowledge of the corresponding phenomenal domain.

An analytical boundary, as conceived here, must consist of two components. Like a frontier, one component separates the process at any time from the rest of actuality, although we must not think that this frontier is necessarily geographical in nature or rigidly determined. Witness the process of thought itself or that of an acorn growing into an oak. The second component is the duration of the process, determined by the time moments at which the process we have in mind begins and ends. Naturally, these moments must be at a finite distance; otherwise, we would not know all that has gone into the process or all that the process does. Nor must we allow them to coincide. For, to recall Whitehead's admonition,[9] a durationless process, an event at an instant of time as a primary fact of nature, is nonsense.[10] We can then choose the time scale so that the process begins at $t=0$ and ends at $t=T$, with $T>0$. T is the duration of the process, but for a production process we may prefer, instead, Marx's term: the time of production.

The next point is truly crucial: in saying that a given analytical process begins at $t=0$ and ends at $t=T$ we must take the underscored words in their strictest sense. Before $t=0$ and after $t=T$ the analytical process is out of existence. That is, in conceiving such a process we must totally abstract from it what happens outside the duration we have assigned to the process. The mental operation is clear: an analytical process must be viewed as a hyphen between one *tabula rasa* and another.

Our next problem is to arrive at an analytical description of the happening, associated with a given process. Because of the principle, "No analytical boundary, no analytical process," analysis must renounce any hope of including in this description the happenings inside an analytical process. Indeed, in order to describe analytically what happens inside a process, we must divide it by a new boundary into two new processes to which the same rule will apply. And so on ad infinitum. The analytical description of a process, therefore, reduces to recording everything that crosses the boundary in either direction. In connection with this picture we can endow the terms input and output with quite precise meanings.

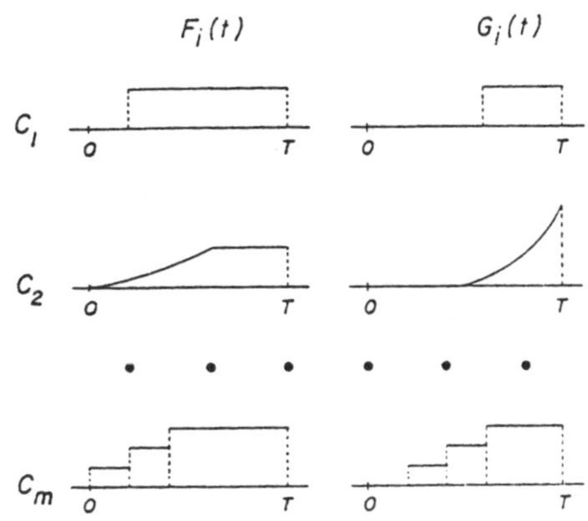

FIGURE 1

Analysis now needs to take another heroic step —to assume that the number of elements involved is finite and that every element is cardinally measurable (which implies that every element is a homogeneous entity). If $C_1, C_2, \ldots, C_m$ denote the distinct elements, the analytical description is complete if for every C_i we have determined two nondecreasing functions $F_i(t)$ and $G_i(t)$ over the closed interval $[0, T]$, the first function showing the cumulative input, the second the cumulative output up to time t (inclusive). Any analytical process—whether in economics or any other domain—may therefore be represented graphically by a series of curves, as in Figure 1.

In a plastic image, the coordinates of an analytical process may be likened to continuously reported data of import and export, with one important detail. Since in describing a process analytically we must begin and end with a *tabula rasa*, this hall (in which we are now gathered) must be listed both as input and as output in the process consisting of the delivery of this lecture. In the analytical approach we are not interested in how this hall came into being or in its use before or after this lecture. However, we must recognize

[8] Plato, *Phaedrus*, p. 265.

[9] Alfred North Whitehead, *An Enquiry Concerning the Principles of Natural Knowledge* (2nd ed., Cambridge, England, 1925), p. 2; also his *The Concept of Nature* (Cambridge, England, 1930), p. 57.

[10] All this does not mean that, in the next stage of our inquiry, we cannot arrive at the excluded cases by a passage to the limit.

64 Energy and Economic Myths

that, as the result of every use, the hall suffers some wear and tear, imperceptible though this may be. Similarly, in any production process the same person must be listed as a rested worker among inputs and as a tired worker among outputs. A tool, too, may go in new and come out used. But even though we recognize the rested and the tired worker as being the same man, we must treat the former as a different element from the latter. Each element of an analytical process—as we have decided—must be completely homogeneous, a condition that does not always cover sameness.

These cases, of the worker and of the tool, raise a troublesome problem for the economist. The reason is that our material of study is the commodity. We slice the economic domain into units of production and units of consumption because at the boundaries thus drawn we can catch every commodity. Drawing a boundary in a glass plant between the melted glass furnace and the rolling machines would serve none of the economist's purposes: at this time, melted glass is not a commodity.[11] Briefly, the economist cannot afford to abandon his commodity fetishism any more than the chemist, for example, can renounce his fetishism of the molecule.

The difficulty, which at bottom is related to qualitative change, is that even though we cannot avoid including "tired worker" and "used tool" in the list of outputs of any production process, neither category fits the usual notion of commodity. Our entire analytical edifice would collapse if we were to accept the alternative position that the aim of economic production is to produce not only the usual products but also tired workers and used tools.

A new heroic step is needed to eliminate this difficulty. It consists of the familiar, old fiction of a process in which capital is maintained constant. The fiction does raise some analytical issues, for if all tools and all workers are to be maintained at a constant level of efficiency, any production process will have to include most of the enterprises and households in the world. Factually, however, the fiction is not more, not less reasonable than that of frictionless movement in mechanics. A simple glance at the activity inside a plant or a household suffices to convince us that efforts are constantly directed not toward keeping durable goods physically self-identical (which is quite impossible), but toward maintaining them in good working condition. And this is all that counts in production. The only factor we need neglect is the daily wear and tear of the worker. This is not too much to demand, since the worker is daily restored in the household.

The elements may now be divided into two relevant categories. In the first category we shall place those elements that appear as input and as output and are related by reason of sameness or of equality of quantity. A piece of Ricardian land, a motor, the amount of clover used as seed in growing clover seed (not clover fodder!), or a worker, illustrate this category. To elements such as these I propose to refer as funds so as to emphasize their economic invariableness. The other elements, which appear only as input or only as output, constitute the category of flows. Obviously, since the fund elements are maintained, the process may be activated again provided that the necessary inflows are still forthcoming. Labeled variously as a static or as a stationary process, or, still, as a diagram of simple reproduction, this fiction constitutes the fundamental element in the analysis of production from the classical to the hypermodern school. Reproducible, however, seems to describe the process better. The analytical picture we have thus reached is worth stressing: in a reproducible process, the fund elements are the immutable agents that transform some input flows into output flows. No picture of a process—whether static or dynamic—is complete if it does not include both categories of elements.[12] And the essential difference between these categories calls for a different representation of the fund coordinates. A flow coordinate will continue to be represented, according to the case, by the cumulative input or the cumulative output; i.e., by a quantity of some substance. Because in case of a fund the input and the output are economically the same substance, the coordinate of a fund may be represented by the difference $F_i(t) - G_i(t)$. But to maintain a convenient symmetry with the flow coordinates, we may use instead the cumulative amount of that intangible entity usually called the service of the fund.

III

In the case of a production process, the elements may be classified into some fruitful categories. The inflows that are transformed by the agents may come either from nature or from other

[11] That technological innovations may change this situation is evidenced by ready-mix cement and brown-and-serve rolls, for example, which only recently have become commodities.

[12] A point on which I insisted long ago: cf. my article "Aggregate Linear Production Function and Its Applications to von Neumann's Model" in *Activity Analysis of Production and Allocation*, ed. T. C. Koopmans (New York, 1951), pp. 100–01.

production processes; we shall denote them generically by (R) and (I). There also are inflows, (M), earmarked for maintenance. The output flows consist of products, (Q), and waste, (W). Finally, the funds include Ricardian land (L), capital equipment (K), and—to use Marx's very appropriate term—labor power, (H). With these notations, the analytical representation of a reproducible process is

$$(6)\quad [R_0^T(t),\ I_0^T(t),\ M_0^T(t),\ Q_0^T(t),\ W_0^T(t);\ L_0^T(t),\ K_0^T(t),\ H_0^T(t)];$$

that is, a set of functions, which defines a point in an abstract (functional) space.

This is a far cry from the notion inherited from Wicksteed, according to which a process is represented by a point in the ordinary (Euclidean) space. The superiority of (6) over the point representation needs no elaborate argument. In (6) we have a complete set of instructions on how to set up the corresponding process. The form also reminds us continuously that a process has a duration, a time of production. Nothing is missing from it.

The difference yields an entirely new form for the production function. Since by a production function we must still understand the set of all processes that transform the same input flows into an outflow of the same product, from (6) it follows that the production function must be a relation among functions, instead of numbers. This relation, which may be written after the old pattern as

$$(7)\quad Q_0^T(t) = \mathfrak{F}[R_0^T(t),\ I_0^T(t),\ M_0^T(t),\ W_0^T(t);\ L_0^T(t),\ K_0^T(t),\ H_0^T(t)],$$

is what the mathematicians call a functional.

The results just reached call for numerous observations. Here, I can take up only a few and touch upon them briefly.[13]

First, the reason why I have excluded no element from the categories listed in (6), is that the scholar must never prejudge. Even an economist must first arrive at a complete description of a process and only then decide which elements may be left out because they are economically irrelevant. Nature does not indeed have a cashier's window where we may pay her for the elements (R); yet it would be utterly inept to ignore in the economics of production the fact that natural resources are neither inexhaustible nor uniformly distributed over the globe. The type of economic model now in vogue, which assumes that growth normally proceeds at a constant rate, simply blots out the most numerous and most critical cases—such as Somaliland or our own Appalachia, for instance. One may feel even more tempted to leave out the waste category, on the ground that waste by definition has no value. But again, as we have come to recognize recently on an increasing scale, the existence of waste is not an innocuous aspect of the economic process.

Second, we should not fail to observe that, since a fund enters a reproducible process and comes out without any impairment of its economic efficiency, service is the only way by which it can participate in the production of the product. While it is true that the cloth—an inflow element—effectively passes into the coat, the same cannot apply to the needle—a fund element. And if one finds the needle in the coat just bought, it certainly is a regrettable accident. The point is that the problem of how the contribution of a fund affects the value of the product is not as directly simple as in the case of a flow factor.

Third, both the value of a fund's service and that of its maintenance flow must, in principle, be imputed to the value of the product. Contrary to Marx's teachings—which have gradually infiltrated the thinking of many a standard economist—there is no economic double counting in this. No worker, no lecturer, can discharge his duties by sending to the shop or to the classroom only that "definite quantity of muscle, nerve, brain, etc., [which] is wasted" during work—as Marx claimed.[14] When one works, one must be present with his entire fund of mental and physical capabilities. The same is true for all other funds: the bridge must be there in its full material existence before we can cross the river. If it were true that we could cross a river on the maintenance flow of a bridge or drive the maintenance flow of an automobile on the maintenance flow of a highway, there would be little financial difficulty in saturating the whole world with all the river crossings and automotive facilities. Economic development could be brought about everywhere with practically no waiting.

IV

As with almost everything else, among the various processes we may envisage in production there is one process that fits the epithet "elementary." It is the process by which every unit of the product—a single piece of furniture or a molecule of gasoline—is produced. The process is

[13] For greater details see Chapter IX in my forthcoming volume, ***Entropy and the Economic Process*** (Harvard Univ. Press).

[14] Karl Marx, *Capital* (Chicago, 1932), Vol. I, p. 190.

directly observable in the shop of a cabinetmaker, but it can be easily determined even in a large-scale enterprise. Whatever the product, one thing is certain about the elementary process. In relation to it, most of the funds are idle over large periods of time. The plow is needed only a few days during the whole production time of growing a corn plant; the same is true for the saw or the plane in the production of a table. There is no exception to this rule. And, I contend, one of the most important aspects of the economics of production is how to minimize these periods of fund idleness, whether we are thinking of man, capital equipment, or land.[15]

Now, if only one table is demanded during the time of production, T, then obviously we need operate only one elementary process after another in succession. But if n tables, with $n>1$, are demanded during T, then two alternatives are open to us. We may start n processes at the same time and repeat the operation when they are ended. This is the arrangement in parallel. The second arrangement is the arrangement in line, in which equal batches of processes are begun one after another at intervals equal to an aliquot part of T.

It is obvious that the production function of a system in which the elementary processes are arranged in parallel is obtained from (7) by multiplying every coordinate by n:

$$(8)\quad [n\overset{T}{\underset{0}{Q}}(t)]=\mathfrak{F}\{[n\overset{T}{\underset{0}{R}}(t)], \cdots, [n\overset{T}{\underset{0}{W}}(t)]; [n\overset{T}{\underset{0}{L}}(t)], \cdots\}.$$

The point that deserves to be stressed is that the arrangement in parallel offers little or no economic gain. Most kinds of fund factors are now needed in an amount n times as great as in the elementary process. In addition, the idleness of each such fund factor is *ipso facto* amplified by n. The only exceptions are the fund factors that—like a large bread oven, for instance—may accommodate several elementary processes simultaneously. But even though the capacity of such a fund factor would be more fully utilized, its idleness period would remain the same.

The situation completely changes for the arrangement in line. If we assume away any incommensurabilities among the time periods involved in the schedule of the elementary process—an inevitable assumption in practice—and if n is sufficiently large, then a number of processes can be arranged in line so that no fund is idle at any time.[16] The situation is vividly exemplified by an assembly line, in which every tool and every worker shift without interruption from one elementary process to the next. The arrangement in line, however, describes any factory. In a factory, therefore, the economy of time reaches its maximum. This conclusion opens an avenue of utmost importance. To explore it, we may begin by determining the analytical representation of a factory process.

In a first approach we may consider the entire physical plant as one monolithic fund, P. Over an arbitrary interval $[0, t]$, during which the factory process is in operation, the coordinate of this fund is the function Pt. Similarly, the coordinate of labor power is Ht. And if for the convenience of diction we assume that all flow elements are continuous, their coordinates, too, are represented by linear homogeneous functions. Thus (7) becomes:

$$(9)\quad \overset{t}{\underset{0}{(qt)}} = \mathcal{G}[\overset{t}{\underset{0}{(rt)}}, \overset{t}{\underset{0}{(it)}}, \overset{t}{\underset{0}{(mt)}}, \overset{t}{\underset{0}{(wt)}}; \overset{t}{\underset{0}{(Pt)}}, \overset{t}{\underset{0}{(Ht)}}].$$

Let us note that this is a very special functional: first, every function involved in it depends upon a single parameter and, second, the value of t is entirely arbitrary. For these reasons, the functional degenerates into a point function.

There are two degenerate forms. The first is

$$(10)\quad q = \Theta(r, i, m, w; P, H).$$

This formula reminds us of one of the current interpretations mentioned in Section I; namely, that of relation (2). We should note, however, that θ involves two dimensionally different categories of variables. The lower case symbols represent flow rates of some substances. The upper case symbols stand for the rates of service per unit of time. Strangely, however, these last rates do not involve the time element at all: P stands for the plant, and H, for the total labor power—briefly, for quantities of some substances. The second degenerate form is

$$(11)\quad Q = \Psi(R, I, M, W; \mathcal{P}, \mathcal{H}; t).$$

Here, the symbols in roman capital letters are again quantities of some substances; those in script letters are services, and t is the period with which these quantities and services are associated. The form (11), in turn, reminds us of relation (1); i.e., of the quantity interpretation of Wicksteed's formula. The most important difference is that Ψ includes time as an explicit variable.

[15] A period of idleness is characterized by the constancy of the corresponding fund coordinate.

[16] The number of elementary processes that should be started each time is the smallest common multiple of the numbers of such processes that can be accommodated at the same time by each unit of the various funds. Batches should be started at intervals of T/d, d being the greatest common divisor of T and of the intervals during which the various funds are needed in an elementary process.

This difference bears upon the earlier argument that Wicksteed's production function is homogeneous of the first degree. Actually, Ψ is such a function—as easily follows from the identities

$$(12)\quad Q = qt, \cdots, W = wt, \mathcal{P} = Pt, \mathcal{H} = Ht,$$

analogous to (3). There is then an intimate relation between (10) and (11); namely,

$$(13)\quad t\Theta(r, i, m, w; P, H) = \Psi(R, I, M, W; \mathcal{P}, \mathcal{H}; t).$$

Hence,

$$(14)\quad \Theta = [\Psi]_{t=1}.$$

The imbroglio created by (5) is thus resolved. Of course, this does not mean that the factory process operates with constant returns to scale. The homogeneity of Ψ corresponds to the tautology that if we double the time during which a factory works, then the quantity of every flow element and the service of every fund will also double. The issue of returns to scale pertains, instead, to what happens if the fund elements are doubled. The point may be made still clearer.

A superficial inspection of any operating plant suffices to reveal that P consists of some Ricardian land, R, some capital equipment, K, some technical inventories, S, and a special fund, Γ, usually called "goods in process." The last term is definitely a misnomer: half-tanned hides or partly wired radio sets, for example, are not goods. Process-fund seems a more exact term because Γ is in effect a becoming frozen in its various phases. If a photograph of Γ would be projected part by part, as if it were a movie, we would witness the actual change of some input flows into product and waste flows. In spite of this varied composition of any plant, what a given plant can do depends on its buleprint alone, which in turn involves only L and K. And since what a plant can do is shown by the flow rate of its product, we have a first relation

$$(15)\quad q^* = G^*(L, K).$$

A second relation expresses the fact that, given the plant, we require a certain labor power, H^*, if we want to obtain the flow rate q^*. Hence,

$$(16)\quad H^* = H^*(L, K).$$

Should we man the plant with less labor power than H^*, the product flow rate would also become smaller than q^*. To account for this rather common situation, we need to put

$$(17)\quad q = G(L, K, H) \leq q^*.$$

But the fact that this relation looks very familiar should not mislead us: as (17) is defined here, if $q<q^*$, q does not necessarily decrease (and ordinarily does not) when L and K are decreased while H is kept constant. Actually, the ratio q/q^* measures the percentage of capacity utilized if H is the labor power employed.[17]

The next relations are self-explanatory:

$$(18)\quad S = S(L, K, H), \qquad \Gamma = \Gamma(L, K, H).$$

There remain the relations binding the other flow elements. The case of the maintenance flow, m, is simple: its size must depend on the amount of equipment to be maintained and the labor fund employed. In addition, by virtue of the conservation law of matter and energy, m must be equal to w_1, the flow rate of wear-and-tear waste—burned or discarded lubricating oil, broken saw bands, and the like. Hence,

$$(19)\quad m = m(K, H), \qquad m = w_1.$$

The same conservation law applies to all other flows. For example, the wood contained in a piece of furniture together with the scrap and the sawdust must exactly account for the wood introduced into the production of that furniture. In the case of a factory system, this relationship yields

$$(20)\quad qt = g(rt, it, w_2t),$$

where w_2 is the flow rate of waste arising from transformation alone. Since (20) must be true for any t, the function g must be homogeneous of the first degree. To this function we may indeed apply the old-time tautology that "doubling the inputs doubles the output." The basic error in some arguments about the returns to scale is to apply this tautology to (17) instead of (20). If L, K, and H are doubled, q does not necessarily double even if we double at the same time all flow inputs. The new factory may be more efficient or more wasteful in using the input materials, which leads us to put

$$(21)\quad w_2 = w_2(L, K, H).$$

Relation (10) is thus decomposed into seven basic relations, listed from (17) to (21), which together constitute the general representation of a factory process.[18]

We should now note that the picture at which we have arrived is analogous to the inscription "60 watts" on an electric bulb. That is, it tells us what the factory can do, not what it has done, is doing, or, above all, will do. Like the inscription

[17] As an ordinal measure of the utilized capacity we may use H/H^*. On this point, see note 21, below.

[18] Obviously, this analytical description will have to be completed with additional relations if the particular factory process happens to involve other limitationalities.

on the bulb, relation (17), for example, is true regardless of whether the factory works or is idle. To show what the factory does, we need an additional coordinate, which, under its various aspects, has deeply preoccupied Marx, but which, perhaps for easily understood reasons, is not found in the analytical tool box of the neoclassical economist. The coordinate is the time, δ, during which the factory works daily. The amount of the daily production, Q, follows immediately from (17):

$$Q = \delta G(L, K, H), \tag{22}$$

a relation which vindicates Marx's dear tenet that labor time measures value even though it has no value itself.[19]

V

So much for grounded-in-actuality symbolism. Let me devote my closing remarks to some of the object lessons of this symbolism.

I have stressed the fact that in any elementary process every agent is idle over some definite periods that depend not on our choice or whim but on the state of the arts. I have also argued that we can nonetheless eliminate this kind of idleness completely and that there is only one way to achieve this: to arrange the elementary processes in a factory system. Because of this extraordinary property, the factory system deserves to be placed side by side with money as the two most fateful economic innovations for mankind. I say "economic" and not "technical" because the economy of time achieved by the factory system is independent of technology. Nothing prevents us from using the most primitive technique of cloth weaving in a factory system.

To be sure, there is a second kind of idleness, which depends entirely on our decision: it is the idleness of the factory itself if δ is shorter than a full day. In view of these two kinds of idleness, the economics of production reduces to two commandments: first, produce by the factory system and, second, let the factory operate around the clock.

The first commandment calls for two observations. Even though we can draw the blueprint of a factory for any elementary process whatsoever, not every such factory is economically advantageous. For example, we do not build transoceanic "Queens" by processes in line. The reason is that we can build a "Queen" more quickly than it is demanded in relation to time. The much extolled progress of the industrial revolution may not after all be due only to technological innovations. For these innovations as well as the increased specialization of labor could not have come about unless an increased demand had already induced most craft shops to introduce the system in line. There can be little doubt about it: the factory system was born in an artisan's workshop, not in a factory.

The second observation is that to operate an arrangement of elementary processes in line it is absolutely necessary that we have the freedom to start a process at any time of the day, of the week, and of the year. Unfortunately, we do not always have this freedom. Seasonal variations—which result from the position of our planet relative to our main source of free energy, the sun—prevent us from adapting the factory system to a series of important productive activities. The most important instance is husbandry. For the overwhelming majority of localities, there is a very short and definite period of the year during which a corn plant, for instance, can be grown in the open space from seed. This is why farmers have to work their fields in parallel; that is, in a system of production that yields practically no economy of time. The global analytical representation of that system is (8), not (9). The upshot is that the open-air factories, about which socialist writers in particular have been continuously raving, will remain a utopian dream as long as we are unable to alter the orbit of the earth.

The association between agriculture and the idleness of all agents involved is by now a commonplace. Still, not much is known or even suspected about the importance of the related loss. Two simple illustrations may bring out this point. For the first, let us consider one of the exceptional localities—such as the Island of Bali—where, because of an almost constant climate throughout the year, rice could be grown in an open-air factory. In this case, every day the same number of hands would move over the fields with the same funds of plows, buffaloes, sickles, and flails to plow, sow, harvest, and thresh. Every day the villagers would eat the rice sown that very day, as it were, and they would no longer have to bear the burden of the debts specific to agriculture. But most important of all, we would also discover that, without diminishing the old production at all, there would remain a substantial number of superfluous workers as well as a substantial stock of superfluous equipment—a palpable measure of the overcapitalization of farming in comparison with manufacturing. The second illustration pertains to the current technique by which chickens are raised in the United States. In fact, in this country there are no longer any chicken farms—even though the term continues to be used. Instead, there are chicken factories, with elemen-

[19] Marx, *Capital*, Vol. I, pp. 45, 588.

tary processes arranged in line. The "chicken war" of yesteryear would not have come about if the difference between the old and the new techniques had not been so great as to exceed the shipping cost over the Atlantic plus the wage differential between this country and Europe.

But if not every production activity can be turned into a factory, we should at least try to render the idleness of the agents as small as possible in each particular case. In other words, to bring even a whole economy as near as possible to the functioning of a factory system should be the guiding thought of any planner at any level. In the activity of the countryside, the cottage industry propounded by the agrarians was one answer to this idea. In Romania (so I was told) tractors and drivers shuttle between the plain regions—where two crops are grown each year—and the hilly regions—where only one crop can be grown because of a shorter vegetation period. The necessary funds of tractors and drivers are thus substantially reduced at the cost of some gasoline, oil, and spare parts flows. Less costly solutions would be obtained by mixing several crops within the same locality, the crops being chosen so as to minimize idleness (and hence capital cost). Formally, the problem boils down to splicing graphic patterns with a minimizing condition—a problem of a special type of combinatorial analysis which, I am sure, will prove highly rewarding.

The second commandment is particularly relevant for the underdeveloped economies. In a rich country, it makes perfect sense to operate every factory with one shift, even if the shift be of six or four hours only. In a rich country, there also is no need for night shifts, except whenever technology imposes around-the-clock production. Briefly, in a rich country leisure is a commodity which people may prefer to higher income. Things are different in almost every underdeveloped country where—as every government pronouncement urges—the order of the day is not only development but rapid development. In such countries, the regimen of the eight-hour working day and the reluctance to use night shifts are anachronistic factors that work against the avowed aims.[20] There may be many reasons why planners as well as our planning theory have overlooked the simplest and the most direct lever of economic development; namely, the length of the working day. But one possible reason is that this element of the problem has been left out of the neoclassical representation of a production process. The same omission—we should note—vitiates also the familiar comparisons of the capital-output and capital-labor ratios computed from current statistical data. Since the theoretical apparatus ignores the working time, δ, the most sophisticated statistical agencies, too, have felt no need to include it in their usual collections. Thus, we are unable to obtain valid statistical estimates of K/q^* and K/H^*, the basic technical and theoretical elements.[21]

Another omission of the neoclassical representation is that, as a rule, only the funds (variously defined) are included in the production function. The fact that after a factory is built, production cannot go on unless the input flow factors are forthcoming, has thus been pushed away from the focus of attention. None other than an authority such as A. C. Pigou preached that "in a stationary state factors of production are stocks, unchanging in amount, out of which emerges a continuous flow, also unchanging in amount, of real income."[22] The omission of the input flow factors is not unrelated to the present race of all underdeveloped countries to build one factory after another without a thorough examination of the availability of the necessary flow inputs. I am confident that if the prospective economic plan of every country were realized by miracle overnight, we would discover that we have long since been planning for a world with an immense excess capacity of industrial production.

The thoughts I shared with you here may seem simple. Perhaps they are simple, once we have untangled the imbroglio hatched by blind symbolism. The economics of production, its elementary nature notwithstanding, is not a domain where one runs no risk of committing some respectable errors. In fact, the history of every science, including that of economics, teaches us that the elementary is the hotbed of the errors that count most.

[20] I may hasten to admit that (22) is only a first approximation formula: a factory working with one shift of ten hours will not produce 25 percent more than with a shift of eight hours. To take better account of facts, we should replace δ by a function of the number of shifts and the number of working hours of each shift. But this amendment does not affect in the least the validity of the statements just made.

[21] The difficulty is especially serious in the case of comparisons between two different industries. Even if we know that each industry has always used its full capacity, i.e., has worked with the corresponding H^*, the values of capital-labor ratios derived from the usual statistical data are neither comparable nor strictly relevant—unless we also know that both industries employed the same number of shifts. In fact, the Census of Manufactures provides no information on the number of shifts and on the percentage of utilized capacity.

[22] Pigou, *op. cit.*, p. 19.

[18]

Resources and Energy 1 (1978) 125–137. © North-Holland Publishing Company

ON A RELATION BETWEEN ECONOMIC AND THERMODYNAMIC OPTIMA*

R. Stephen BERRY and Peter SALAMON

Department of Chemistry, University of Chicago, Chicago, IL 60637, U.S.A.

Geoffrey HEAL

Department of Economics, University of Sussex, England

We present necessary and sufficient conditions that a least-cost production system operates within ε of its thermodynamic limit. The theorem can be extended, with the knowledge of the production function, to fix the economically-efficient level of thermodynamic efficiency or effectiveness. The analysis is also extended to cases in which several factors of production have different thermodynamic limits. The theorems, while weak, appear to represent the first precise relationship between economic and thermodynamic extrema.

1. Introduction

Our subject, the establishment of a first and very limited bridge between economics and thermodynamics, is unconventional in its scope. In consequence, it merits careful introduction for practitioners in both of the relevant disciplines. Creating such a bridge has been an elusive but tantalizing goal for a long time [Klein (1971) and Soddy (1922)]. We refer here to the physical content of thermodynamics, not to its mathematics, especially the calculus of variations for functions of several variables, which has been well integrated into economics by Samuelson (1947), or to its relational structure, the analysis of equilibrium for systems with several variables and constraints, which has also been used occasionally in economics [Samuelson (1947), Georgescu-Roegen (1971), Boulding (1964)]. The motivation to make such connections grew with the need for a better understanding of the role of energy in the functioning of the economy, and of how the market allocates energy resources.

Scholars from economics and from the natural sciences have studied energy problems, of course. Both groups have brought their own criteria of effective use of energy and related resources. The economist usually defines an efficient allocation as a Pareto optimum: that allocation of resources such

*This work was supported by a Grant from the National Science Foundation. Also the authors would like to thank Professor Thomas Veach Long, II and Dr. Bjarne Andresen for helpful comments.

that no other allocation would make any one individual better off without making at least one other worse off. Efficiency in this sense is a property of the entire economic system, and therefore refers to relationships among all the variables of the system, including labor as well as physical variables. Bounds in thermodynamic analyses are the natural limits of heat and work determined by the First and Second laws of thermodynamics and by the constraints that define the process under study. These bounds are given by the change in a generalized potential. These potentials exist for all reversible and all quasistatic processes [see Callen (1960), Salamon et al. (1977) and Gibbs (1948)],[1] so that if the range of possible changes in a process or set of processes is well-defined, thermodynamics can tell us the natural lower bound on the work required to make the system operate, or the maximum work that can be extracted, or the maximum heat that can be exchanged. Such an extremum is a condition that applies to physical resources, and thus to only a subset of all the economic variables.

As with many other economic questions, one recognizes that at least two kinds of extrema could be attained. We naturally ask whether there are any relations between them. If there are, we can then ask whether these relationships give us any useful tools for analyzing the allocation of resources. Rather than try to establish a complete structure, we have set ourselves the much more modest task of working forward in small steps, bringing the subject to the scientific community as it evolves. We have been troubled by misunderstandings that have grown out of ardent but unfruitful attempts to meld the two approaches. The physical scientist is tantalized by the notion that an efficient economic system 'ought', in some sense, to use physical resources in the most efficient way possible – meaning 'efficient' according to one of the thermodynamic criteria of merit. The economist, on the other hand, sees the variables governed by thermodynamics as a subset of all the variables of the complex economic system, and often argues that there is no reason that the existence of a thermodynamic extremum should have anything to do with the much more complex economic optimum. The former position has led some people to propose an 'energy theory of value' [see, for example, Hannon (1973), Odum (1973), and Gilliland (1975)]. This simplistic view, predictably, has drawn bitter attacks from economists who call up the criticisms of the antique but analogous labor theory of value [Huettner (1976), and Harcourt and Massaro (1964)]. We hope that our analysis here will begin to clarify some of the confusion in this debate.

We ask the indulgence of both the economists and the natural scientists who may read this. We are quite aware that many things we say are well known in one field or the other. But from our own experience, we believe

[1]A quasistatic process is one in which the system is, within limits of detectability, at equilibrium at every point between its initial and final states. The theorem for existence of potentials is actually much stronger: they exist for any path containing at least a countable set of points at which the system is, within limits of detectability, everywhere locally in equilibrium.

that many of these things are not well known to people in the other field, and their role in logic impels us to repeat them here. Moreover the following discussion deals only with static, equilibrium relationships between thermodynamics and economics. There may well be richer possibilities for systems out of equilibrium, and we hope to explore these, particularly by exploiting general principles of non-equilibrium thermodynamics [Salamon et al. (1977)]. For the present, we restrict our discussion to the more modest confines of equilibrium.

2. Background of the problem

The economic system we describe could be presented in terms of production or of consumption. In the former case, we want to minimize a cost function $\mathscr{C}$ associated with production of a good whose output is Q, or equivalently, maximize Q, if $\mathscr{C}$ is regarded as fixed by a budget. In the latter case, we would maximize the consumer's utility $U(Q)$ associated with consumption of an amount Q of the good. For convenience, we shall use the language of the producer's example; the conclusions hold equally for both, when the conditions are met.

We suppose that the output Q is described by some function of input variables, some of which, say $\{x\}$, including the fuels and non-fuel material inputs, enter into a thermodynamic description of the process, and some of which, say $\{y\}$, such as labor (in the sense of time devoted to monitoring controls), lie outside the domain of thermodynamics. We suppose that the production process allows for some substitution between 'thermodynamic' and 'non-thermodynamic' variables as well as among the physical quantities. The allowed possibilities of substitution must, of course, be well specified in order for the notion of a best process to have meaning. These possibilities may, at one extreme, be the limits of technology on line; at the other extreme, they may be only the laws of conservation and of thermodynamics. We shall discuss both the technologically-restricted case and the case of no special technological constraints.

We suppose that at least one of the input variables to the production system is a fuel or other input acting as a source of work in its precise physical sense, the integral of a generalized force through a generalized displacement. It is common but imprecise to think of this input as identical with energy. The reason is of course that energy may be extracted from a fuel either as work (the quantity we presume to be the desired form of energy in the present discussion) or as heat. Even if heat, e.g. as steam, is a desired joint product with electric power, we can safely assume here that work is the most valuable immediate product from the use of a fuel or other energy source – i.e., work sets the opportunity cost of the fuel. (This is not strictly true; some rate structures make a domestic heat set the opportunity cost, an almost incredible anomaly from a thermodynamic viewpoint.) We assume that the production process can be carried out reversibly or quasistatically.[2] The reversible limit is the thermodynamic limit of efficiency for a process unconstrained with respect to time; we assume that for processes constrained to operate in real time the thermodynamic extremum corresponding to a minimum in work required, or maximum in work produced, will always correspond to a quasistatic process. In other words, we suppose that quasistatic processes are always at least as good at turning stored potential into work as nonquasistatic processes. Given the statement about reversible processes and the assumption about real-time processes, we can call upon this theorem: for every reversible process, or for every quasistatic process, there exists a potential $\mathscr{P}$ such that the maximum work obtainable from the system (or the minimum work required to drive the system) from state A to state B (and in the interval from time t_A until time t_B, for the quasistatic process) is equal to $\mathscr{P}(B)-\mathscr{P}(A)=\Delta\mathscr{P}_{AB}$ [Salamon et al. (1977)]. Hence there is a natural bound to the capacity to make any system, however flexible, carry out a process.

It is important to recognize that $\Delta\mathscr{P}_{AB}$ is not necessarily a bound on the direct fuel energy required to drive the system. Generalized potentials usually include terms such as the product TS of the absolute temperature and entropy of the system. A process may be driven by the decrease in energy, by the increase in entropy or by some combination of both occurring in the driving part of the entire system.

Thermodynamic extrema usually represent unattainable limits. It is useful, therefore, to introduce a term to describe the merit of a system that operates by consuming an amount of work ε greater than the ideal limit. The ratio of actual work derived from a system to the upper bound of what can be derived (or the ratio of the minimum work required for a task to the actual work consumed for doing the same task) is called the *effectiveness*[3] of a process. As a convenient abbreviation, we term a process ε-effective if it derives a fraction $(1-\varepsilon)$ of the maximum work W_{max}, from its fuels and any other sources of work, that is, if its effectiveness is equal to or greater than $1-\varepsilon$.

3. Statement of the problem

The question which the following section will answer is this: Given that a production process operates at an economic optimum, what are the necessary and sufficient conditions that the process also operates at a thermodynamic optimum? We can state this more precisely in terms of a cost function

[2] *Quasistatic* means that temperatures and pressures remain well-defined locally at all times for the materials being processed.

[3] Sometimes called the 'second-law efficiency' also. The concept was introduced by Gibbs (1948).

$\mathscr{C}$, starting with the conditions

$$\left(\frac{\partial \mathscr{C}}{\partial x_i}\right)_Q = \left(\frac{\partial \mathscr{C}}{\partial y_k}\right)_Q = 0, \qquad \left(\frac{\partial^2 \mathscr{C}}{\partial x_i^2}\right)_Q \geqq 0, \qquad \left(\frac{\partial^2 \mathscr{C}}{\partial y_i^2}\right)_Q \geqq 0,$$

for all x_i, y_k, what other conditions are (1) necessary and (2) sufficient to insure that the work $W(x)$ extracted from the thermodynamic inputs $\{x\}$ satisfies

$$W_{\max} \equiv \Delta\mathscr{P} \leqq W(x)/(1-\varepsilon)?$$

That is, what added conditions make economic optimality imply that a process is ε-effective in a thermodynamic sense? (If the question is stated in terms of a production function, the inequalities are in the opposite direction, of course.)

Note that we are *not*, at this juncture, trying to argue whether conditions satisfying this question are desirable goals for resource allocation. Rather, we take the view that we should not ask whether or when thermodynamics is generally useful for normative decisions until we have accertained the logical relationship between thermodynamics and economics. Only then can we analyze the implications of the answer.

4. The representation of the technology

To represent the productive process, we suppose that we can write Q and $\mathscr{C}$ as C^1 (or R^1) functions

$$Q = Q(\{x\}, \{y\}),$$
$$\mathscr{C} = \mathscr{C}(\{x\}, \{y\}).$$

From the set $\{x\}$ we select out the subset E that provide the work to carry out the process and serve to define the boundary potential $\mathscr{P}$. The term 'work' is used in its broad thermodynamic sense, and therefore includes chemical and electrical work. This set E might be a single fuel, a bundle of resources in fixed proportions, or a set of partially or completely substitutable resources. Each choice corresponds to a different set of technological constraints and to a different potential $\mathscr{P}$. If the set E is a bundle in fixed proportions, then we can use the amount of any one of them as a surrogate for the entire set, and immediately infer the total cost $\mathscr{C}(e')$ of using any amount e' of the set, from a knowledge of the amount e_i' of the ith member of the set. Thus the case of 'energy' resources in fixed proportions need not be distinguished from the single fuel case. We shall deal with the case of a substitutable set separately from the single-fuel case. To make our questions non-trivial, there must be some possibility of substitution for at least one factor in E.

4.1. Single-fuel case

We are concerned here with the set I of all combinations of a single fuel e and the other variables (which we may call y now) that yield an output of at least Q. The boundary of this set comprises all combinations of e and y that yield the amount Q,

$$\partial I = \{e, y \,|\, Q(e, y) = Q\}.$$

This set is conventionally called an isoquant in economics. We shall make the usual assumption that I is a convex set, meaning that if the amount of at least one input is held constant and the amounts of all others are increased by a proportion α, then Q may be constant or rise by a proportion $\leqq \alpha$.

The cost-minimizing or profit-maximizing system will find the lowest total cost, which we can express as a linear function of the amounts e and y_i, and their corresponding prices P_e and P_i. Thus, we wish to

$$\text{minimize } \mathscr{C}(e, y) = P_e e + \sum P_i y_i,$$

$$\text{subject to } Q(e, y) = Q_0.$$

The convexity conditions ensure that at any point where the first-order conditions are met, the second-order conditions are also satisfied. Note that we do not need to analyze all of the implication of economic optimality, but only those of cost minimization.

We consider the simple case of a single variable in the set y, with y completely independent of thermodynamic considerations. Fig. 1 is a standard representation of production with inputs y and e, and isoquants Q_1, Q_2, Q_3.

The one piece of thermodynamic information put into this figure is the existence of the *asymptotes* $\mathscr{P}_1$, $\mathscr{P}_2$, $\mathscr{P}_3$ for the corresponding isoquants. The form of the isoquants is conventional, apart from this characteristic which reflects the existence of a lower bound to the work required to produce any given output. We can safely assume that the family of asymptotes constitutes a functional linear in Q, over most or all of the range of Q of any interest to us here, though the density of asymptotes could, in the general case, increase or decrease, corresponding to positive or negative returns to scale, in the thermodynamic limit.

Clearly all points on Q_1 to the left of the vertical line denoted by $e = e_1^* = \mathscr{P}_1(1+\varepsilon)$ are ε-effective. Hence whenever the economic system operates to

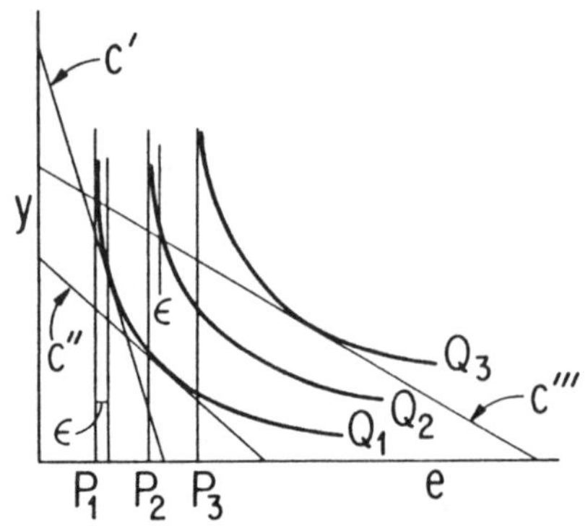

Fig. 1. Production isoquants Q_1, Q_2, Q_3 and lines of constant expenditure $\mathscr{C}'$, $\mathscr{C}''$ and $\mathscr{C}'''$, for production of the good Q from the input factors y and e. The vertical lines P_1, P_2 and P_3 are the thermodynamically-determined limits of the corresponding isoquants, with respect to use of the energy resource e. The vertical lines corresponding to levels of consumption of e that are ε greater than the limits P_1 and P_2 are also shown.

produce the amount Q_1 by using e_1^* of e and the corresponding amount of y as defined by $Q_1 = Q_1(e, y)$, the production system is ε-effective. This much is tautology and makes no use of the condition of cost minimization.

The condition of cost minimization for fixed output or output maximization for fixed costs is represented as follows. We take one input as the price numeraire or normalizer and measure all costs with respect to the cost of that factor. If cost is a linear function of all amounts, the surfaces of constant cost are hyperplanes in general or lines in the 2-factor case of fig. 1, as shown. (Linearity here is assumed for convenience; it is not necessary to the argument.) The condition of cost minimization is met when the output curve Q and the cost line or plane are tangent. The prices are the slopes of the cost lines, measured with respect to the numeraire – in fig. 1, the price of e in terms of y is $(\partial\mathscr{C}/\partial e)_y/(\partial\mathscr{C}/\partial y)_e = -\mathrm{d}y/\mathrm{d}e$.

The condition we seek is now immediately clear – when the price P_e (relative to the price P_y) is at least as great as the slope of the isoquant at a value of e situated at $e^* = \mathscr{P}(1+\varepsilon)$, the system operates with ε-effectiveness if it is optimized:

$$\left(\frac{\partial y}{\partial e}\right)_Q \leqq P_e^*(Q, e^*) \quad \text{and} \quad \left(\frac{\partial \mathscr{C}}{\partial e}\right)_Q = \left(\frac{\partial \mathscr{C}}{\partial y}\right)_Q = 0$$

imply

$$e \leqq \mathscr{P}(1+\varepsilon) \qquad \text{and} \quad y(Q, e) \geqq y(Q, e^*).$$

Whatever the price of e may be, the cost-minimized system must be ε-effective, if and only if the price is at least as great as $P^*(Q, e^*)$.

If there is more than one variable in the set of non-thermodynamic factors y, the argument can be extended quite simply. The asymptotes $\mathscr{P}$ become planes or hyperplanes of constant e, perpendicular to the e-axis, the isoquants become concave hypersurfaces and the lines of constant cost become planes or hyperplanes that cut all the axes. The points of minimum cost for a chosen level of production are the points at which the constant-cost planes are tangent to the isoquants. The condition of ε-effectiveness is met if the point of tangency of the isoquant surface and isocost plane lies at a point where $e \leqq e^* = (1+\varepsilon)\mathscr{P}$.

Fig. 1 puts quite graphically the difference between the viewpoints of the believer in a pure thermodynamic theory of value[4] (in relatively sophisticated form, not in its most naive way) and that of the conventional economist. Consider the constant-cost line $\mathscr{C}'''$, which we assume now is the locus of the highest-cost combination the buyer can afford. The standard economic arguments put the optimum mix of e and y at the point of tangency of $\mathscr{C}'''$ and Q_3, the highest isoquant with which it shares a common point. One sort of thermodynamic theory of value would decide on its ε and then have the system operate at the isoquant Q_2 that crosses at the e-value just ε greater than its asymptote. No higher isoquant can be reached within ε of its asymptote along $\mathscr{C}'''$, and Q_3 is the highest isoquant for which ε-effectiveness can be maintained. Clearly, the two approaches coincide if the price condition $(\mathrm{d}y/\mathrm{d}e)_e = (\mathrm{d}y/\mathrm{d}e)_Q$ is met at $(1+\varepsilon)\mathscr{P}$ or e^*.

A second kind of thermodynamic theory of value (suggested by an anonymous referee, to whom we owe thanks) formally more sophisticated than the first maximizes $(1-\varepsilon)$ subject to requirements that total costs $\mathscr{C}$ be less than or equal to a maximum $\mathscr{C}'''$, and that output be greater than or equal to a minimum Q_2. This model also has a simple solution, as the referee pointed out, if we assume that $Q(e, y)$ shows constant returns to scale. With this assumption, each straight line through the origin corresponds to a line of constant ε. The intersection of the isoquant $Q_2(e, y)$ with $\mathscr{C}'''$ consists of two points, in general, corresponding to two constant-ε rays and thus to two values of ε. The point to the left, with lower ε and higher y/e, is the optimum by this criterion.

Clearly, neither of these thermodynamic theories of value offers a solution compatible with usual criteria for choice. The first provides the maximum product subject to a fixed budget and an exogenous lower bound on thermodynamic effectiveness. The second provides the minimum acceptable product in order to maximize the thermodynamic function $1-\varepsilon$. If one wished to ascribe special value to saving energy, it appears from our analysis that it would be better to try to evaluate externalized costs and add them to

[4]See, for example, Hannon (1973), Odum (1973), and Gilliland (1975).

the direct cost of energy, thereby tilting the budget lines more steeply, than to introduce *ad hoc* criteria such as those in our two model thermodynamic theories of value. Energy theories of value have been criticized previously [Huettner (1976)], and, as we said, are subject to most of the same vulnerabilities as labor theories of value [see, for example, Harcourt and Massaro (1964), for a discussion of the labor theory].

The growing number of empirical studies of production systems justifies adding a simple corollary to the foregoing argument. Suppose our assumption about the shape of the isoquants is narrowed, by making the optimistic supposition that the isoquant contours Q_j are actually known, e.g. from engineering studies. Under this condition, the tangent point of the chosen budget line $C^{(i)}$ with the highest attainable isoquant Q_{max} is a known point for a real system. Hence the distance along the e-direction between this operating point and the corresponding thermodynamic asymptote $\mathcal{P}_{max}$ is completely determined. This distance is the amount of e expended beyond the thermodynamic lower bound. In other words, if the isoquants are known, specification of the budget line is sufficient to specify the economically optimal amount of thermodynamic waste!

This inference is *post hoc* not *ante hoc*, and therefore not very strong. The point of operation is selected by cost minimization or output maximization, which means that the prices and technology determine the optimal level of thermodynamic effectiveness, and not the reverse – at least within the confines of this little corollary.

The proposition just proved consists logically of four conditions:

(A) convexity of the production function,
(B) cost minimization,
(C) thermodynamic optimality, and
(D) the inequality condition on the slope of the production function.

The theorem says, in effect, that A and B imply C if and only if D. It might be tempting to turn the theorem around, to ask what conditions imply B. The well-known condition for this is the equality analogous to the inequality of D. But this equality, together with convexity, is enough to imply B, economic cost minimization, regardless of the thermodynamic condition C. Hence when conditions of economic cost minimization are stated in terms of prices and convex production functions, thermodynamic optimality adds no new insights.

4.2. *Partially or totally substitutable resources*

To amplify the picture, we turn to the case of substitutable thermodynamic factors. For example we may take e to be a fuel, as before, h as the amount of a high-grade ore and l as the amount of a low-grade ore. The low-grade

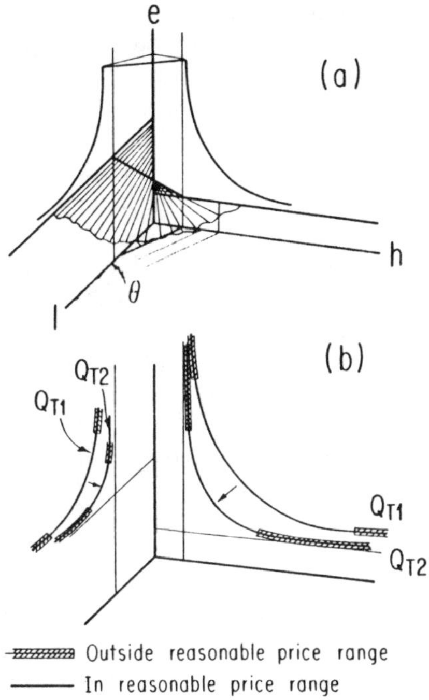

Fig. 2. The production space for a product using an energy resource *e*, and two kinds of ore, high-grade, *h*, and low-grade, *l*, all of which are substitutable factors. The thermodynamic limit is a surface, the ruled surface indicated in (a). In (b) are the isoquants Q_{T1} and Q_{T2} corresponding to two different levels of technology yielding the same amount of product. As technology improves, the isoquants are squeezed into the square corner defined by the thermodynamic and mass-conservation limits. If the producer were to achieve the physical extremum, he would find only a single point of operation for any fixed mix of *h* and *l* and a line of possible operating points, depicted in (a), if *h* and *l* remained substitutable. Note that if there are 'reasonable' upper bounds to the prices of the factors *h*, *l* and *e*, then the range of substitutability narrows as technology improves.

ore requires some energy e' per tonne of product for its processing; the high-grade ore requires $e''<e'$ per tonne of product. In effect, the high-grade fuel supplies thermodynamic potential that replaces some of e'. We could also imagine treating fuel mixes in this context, or any other set of physical factors among which at least a limited substitutability is possible. The fuel-and-two-ore example will illustrate the essential relationships. We assume labor and capital are fixed or occur in proportions fixed to the three inputs in this model, in order to keep the picture relatively simple.

The isoquants in this space have natural thermodynamic bounds of the two kinds. One is determined by mass balance and defines a set of planes parallel to the e-axis, whose angle θ (fig. 2a) is determined by $\tan\theta=$(tonnes

h/tonne of iron)/(tonnes l/tonne of iron). The other kind of asymptote is the natural limit on the requisite work, as provided by the fuel or fuel mix e, to carry out the process within the constraints of technology selected for the model. Because the minimum energy input now depends on the mix of ores, these asymptotes form a family of *ruled surfaces* whose intersections with the mass-conservation asymptotes are straight lines. The isoquants are presumably convex, roughly hyperbolic curves in any plane containing the e-axis.

The isocost surfaces in this system are planes in the positive octant. The condition that minimum cost assure thermodynamic 'optimization' in the sense of ε-effectiveness is, as previously, that the point of contact of the isocost plane and its tangent isoquant lie within e^* of the ruled-surface asymptote. A little care is required here, because one must specify whether the l/h mix is fixed, in which case the distance e^* must be taken parallel to the e-axis, or the l/h mix can be freely chosen, which means e^* can be taken as the e-component of the normal from the tangent point to the ruled-surface asymptote.

Just as a process will be ε-effective if the price of fuel is high enough, the process can be made comparably effective with respect to its use of mass. We call the mass effectiveness the ratio of actual amount of product per unit of mass used to the maximum amount of product that could be made within the constraints of the technology, with the same mass input. The process will be defined as μ-effective if the point of operation has a productivity of mass that is within μ of the ideal limit,

$$(\partial Q/\partial M)_{\text{actual}} \geqq (1-\mu)(\partial Q/\partial M)_{\text{ideal}}.$$

Let $M^* = (\partial M/\partial Q)_{\text{ideal}}(1-\mu)^{-1}$, the absolute mass used, per unit output, analogous to e^*. As in all the foregoing arguments, if the price of ore is high enough to put the contact point of the isocost and isoquant surfaces within $M^* - M_{\text{ideal}}$ of the mass asymptote, the process will be μ-effective.

We get a very interesting insight into the trade-offs in technological advance by examining the behaviour of a production system with physical substitutabilities and asymptotes. Fig. 2b suggests that it is very unlikely that a real system would be both ε-effective and μ-effective at the same time. If it were, the thermodynamic and mass conservation conditions would be acting simultaneously to constrain the system to operate within a very narrow range of substitutability, approaching the fixed-coefficients corner case as e^* and M^* are made smaller.

This narrowing is even more important when we conside what happens to the isoquants in the planes containing the e-axis, when technology is improved to make better (i.e., more nearly asymptotic) use of physical resources. In the region of the asymptotes, we cannot expect technological change to be Harrod-neutral, but Hicks-neutrality may or may not obtain. Whether the technological progress we envision is embodied or not is irrelevant to our model. However the kind of technological change that concerns us here is a sort that never decreases the average productivity of a physical resource, so that we can assume that the isoquants never move away from any asymptote when technology changes. For convenience, we shall refer to such technological change as resource-thrifty.

With the sketch of fig. 2b representing any plane containing the *e*-axis, we see immediately how resource-thrifty technological change acts: it *narrows* the range of substitutability from the wide arc of the 'old' isoquant Q_{T1} to the narrower arc on the isoquant Q_{T2}, associated with the new technology T2. This is the natural consequence of constraining the system to operate ever nearer to a point at which physical considerations determine all factor relationships, the unique mass-and-thermodynamic extremum.

It seems paradoxical that technological change could narrow the range of substitutabilities, rather than raise them, and of course the real change of substitutability never narrows because one could always go back to the old technology. However the practical consequence of resource-conserving technological change has been to move the isoquants as fig. 2b shows, and to be either labor-neutral or labor-augmenting, at least in the industries we have examined, coal mining and cement manufacture [Hebenstreit et al. (1976)]. In both these industries, the productivity of both labor and energy increased for long periods, as technology was improved. The paradox is only apparent, of course. Many processes do undergo technological change that augment substitutability. The narrowing range associated with resource-conserving change carries with it the implication that the change does not introduce possibilities of using heretofore-excluded inputs, or of new labor–resource substitutions that would allow for reducing labor. In this sense, the implication of narrowing substitutabilities suppose that the new technology is at least labor-neutral (and neutral with respect to any other non-thermodynamic variables), as well as that the isoquants move toward asymptotes.

References

Boulding, K., 1964, Meaning of the 20th century (Harper and Row, New York).

Callen, H., 1960, Thermodynamics (Wiley, New York).

Georgescu-Roegen, N., 1971, The entropy law and the economic process (Harvard University Press, Cambridge, MA).

Gibbs, J.W., 1948, Collected Works, vol. 1 (Yale University Press, New Haven, CT).

Gilliland, M.W., 1975, Science 189, 1051.

Hannon, B., 1973, Annals of the American Academy for Political and Social Sciences 410, 139.

Harcourt, G.C. and V.G. Massaro, Economic Journal LXXIV, 715.

Hebenstreit, S., T.V. Long, II, and R.S. Berry, 1976, in: Energy and environment, Proceedings of the Fourth National Conference, Oct. (Dayton, OH); extended report in preparation.

Huettner, D.A., 1976, Science 192, 101.
Klein, M., 1971, Ehrenfest, vol. 1 (North-Holland, Amsterdam).
Odum, H.T., 1973, Ambio 2, 220.
Salamon, P., B. Andresen and R.S. Berry, 1977, Thermodynamics in finite time, II; Potentials for finite-time process, Physical Review A15, 2094.
Samuelson, P., 1947, Foundations of economic analysis (Harvard, Cambridge, MA).
Soddy, F., 1922, Cartesian economics (Hendersons, London).

[19]

Energy Analysis and Economic Valuation*

NICHOLAS GEORGESCU-ROEGEN
Regional Research Institute, West Virginia University
and Faculté des Sciences Economiques, Université Louis Pasteur, Strasbourg

I. Introduction

Ever since the oil embargo of 1973–74 everyone speaks of the *energy* crisis. And whether we are "optimists" or "pessimists", by now most of us recognize that what will happen in the end to this crisis is not an idle question. Only economists still refuse to see the indissoluble relationship between the scarcity of natural resources and the economic process as a whole. As an excuse we hear, for example, that the limitations of natural resources cannot lead to any interesting conclusions [72, 43], as if scarcity were not the very element around which the economic system turns and spins. But the "official" position is systematically defended by the decision-makers of the most influential economic associations. No invited paper at the Tokyo World Congress of the International Economic Organization dealt with the limitations of natural resources. As explained by an official communication, the Program Committee was "very selective." Most curiously, the general theme of the Congress was "Economic Growth and Resources" and the year was 1977! All the more we should admire Mogens Boserup's candid verdict at the closing session.

> There was a remarkable degree of consensus in turning down, or rather ignoring, all the 'doomsday' attitudes and opinions on natural resources. And even apart from that particular issue, there was an almost complete absence of sharp confrontation of opinions.
>
> As we know, a gathering of economists which fails to produce disagreement on essential issues is a rare occurrence—and even a scandal, some would say . . . . A question [thus],immediately comes to my mind: Why do economists agree so largely on the issue of natural resources, not only at an I.E.A. meeting in Tokyo, but in the profession as a whole? [7].

*The author is Professor Emeritus, Vanderbilt University. He is deeply grateful to Jean-Paul Fitoussi, Egon Matzner, and William H. Miernyk for their sympathetic interest in the endeavors summarized in this paper. He wants also to thank the Earhart Foundation for a timely research fellowship.

A contrasting situation prevails in other circles, where new energy experts in ever increasing numbers now expatiate to their hearts' delight on the issue of accessible sources of energy, more often than not contributing to the confusion of both students and policy makers.

Some writers, however, did not have to wait for the warning spelled out by the oil embargo. Long before it, they re-examined with new insights the old problem of the relation between the supply of accessible natural resources, the size of population, and the well-being of the people. But what is highly significant is that, with practically no exception, these writers also argued that energy is the only necessary support of the economic process. This position is not, therefore, a product of the oil embargo; it represents the result of a genuine intellectual endeavor.

It is natural to refer to the belief that energy alone counts as the "energetic" dogma, even though this meaning is slightly different from those used by physics in the past. At first, "energetics"—a term coined by William Macquorn Rankine—denoted the science we now call "thermodynamics". Subsequently, "energetics" was used to denote a school of thought defended by a few illustrious scientists (such as Wilhelm Ostwald and George Helm in Germany and Pierre Duhem in France) influenced in part by the scientific epistemology of Ernst Mach [45]. Against the idea set forth by Ludwig Boltzmann that the laws of energy are the direct consequences of the laws of Newtonian mechanics applied to the motion of *material* particles, the energetic school maintained that, on the contrary, matter must in the ultimate analysis reduce to the only "substance", energy. The current view that only energy matters for mankind's specific mode of existence is not completely identical with that school's position, but at bottom the two are sufficiently similar to justify using the term "energetics" for its label.

The fierce controversy surrounding the old energetic school represented a purely academic issue at the time. The same is not true for the present energetic dogma. Being the currently dominating dogma, it determines the principles that guide policy makers concerned with the energy shortage and technology assessment. Some of us may think that the validity of these principles also constitute only a purely academic issue. For mankind as a whole, however, it is a truly vital matter.

In this paper, I propose, first, to explain in broad terms the technical reasons why the energetic dogma is wrong, why matter matters, too.[1] Second, I shall use the analytical representation of a multi-process for discussing the general problem of energy analysis. On this basis, I shall show where the

1. I may explain that from my first analysis of the entropic nature of the economic process I have maintained that both energy *and* matter irrevocably degrade from available into unavailable forms [29, 75, 93–6; 30, 142, 277–80]. At the time, believing that my thesis was common knowledge at least among natural scientists, I did not go into technical details. But the accentuation of the energetic position after the oil embargo convinced me that my belief was wrong; I then began offering specific arguments in support of my thesis [31–38].

recent claims that energy analysis is the rational basis for economic valuation go wrong. Third, on the basis of the impossibility of reducing matter to energy, I shall show that economic choice is a purely economic matter, not a physico-chemical one. Finally, I shall apply some of these results to the problem of what constitutes a *viable* technology, as distinct from *feasible* recipes. As an application, I shall investigate the viability of a technology based on direct use of solar radiation through the recipes known *at present*.[2] Surprising though it might seem to the current energetic temper, the conclusion of this last probing is that such a technology is not *viable*, that any direct use of solar radiation by any present recipe is a "parasite" of the current technology based on other sources. This result calls for a thorough reorientation of our present approach to technology assessment.

II. The Energetic Dogma

Different writers justify the modern dogma of energetics in different, sometimes in substantially different, ways. For example, one of the earliest propounders, Fred Cottrell [22] argued that all that mankind needs is to obtain *net energy*. The definition of this concept presents no apparent difficulties. If we use, for example, the energy equivalent of one ton of crude oil to extract ten tons of shale oil, a simple arithmetical substraction tells us that the resulting net energy is the equivalent of nine tons of oil. Twenty years later, one of the ablest American ecologists, H. T. Odum [53], revived Cottrell's idea and raised it to the rank of the only criterion of efficiency: the greater the net energy obtained by a process, the more efficient is that process.[3] Because of its simplicity, the principle has since been widely accepted and also defended with even greater force. Indeed, what could be more senseless than to use one ton of oil to obtain less than one ton of oil and nothing else?

However, this very idea prompts one to ask "why not relate efficiency to *net matter*?" Indeed, we use copper in the process of producing copper; hence, we thereby gain some net matter. Whatever we may do, we are faced with new snags. Copper mining also implies a *negative* net energy, whereas any power plant implies numerous *negative* net matters.

Even though Odum is not as explicit on this point as one would wish, we must grant him that net energy must be computed by subtracting from the gross output not only the energy directly used in the process but also the amount of energy necessary to produce or repair all *material* parts worn out by the process (Sec. VI). With this explanation, however, the slip of the energetic bias is showing badly. For without this bias, one may very well propose, by

2. Even though another official communication affirms that technology assessment is not an accepted preoccupation for the established profession, I owe no apology whatsoever for dealing with that problem in this paper.

3. A different approach focuses instead on gross energy (Sec. VI).

symmetry, to reduce everything to net matter, defined as the excess of matter output over the matter necessary to produce the matter and the energy used up. Such a proposal could point out that matter is used in a power plant to produce energy just as in another process energy is used to produce matter. The truth, as we shall see in time, is that neither net energy by itself nor net matter by itself can constitute a general principle of technology assessment (Sec. VIII).

Another mode of justifying energetics in recent times may be traced back to the now classic volume *The Next Hundred Years*, where we read that "All we need do is to add sufficient energy to the system and we can obtain whatever materials we desire" [11, 90, 95, 114]. The energetic gospel in this plain form has been spread by numberless other authors. But Harrison Brown and his associates have also justified it by the axiom that recycling of matter can, in principle, be complete [11, 90–92]. Interesting also is that they immediately added that "fundamentally there is no lower limit to the grade of an ore which can be processed," which obviously is a necessary implication of that axiom.[4]

The few physicists and chemists who have touched this issue also seem to support energetics. For example, Alvin Weinberg, quoted in [32], described energy as "an ultimate raw material" because "energy is convertible into most of the other requirements of life"—which can only consist of matter. Take also Glenn Seaborg, who argued that science will ultimately eliminate all technical inefficiencies so that with an abundant amount of energy we shall be able "to recycle almost any waste, . . . to extract, transport and return to nature when necessary all materials in an acceptable form, in an acceptable amount, and in an acceptable place so that the natural environment will remain natural and will support the continued growth and evolution of all forms of life" [65].

Taken at its face value, this is quite strong energetic position; it comes near to claiming that the whole planet could be maintained intact forever. But an even more stringent expression of the modern energetic dogma is due to Kenneth Boulding [4]: "There is, fortunately, no law of increasing material entropy."[5] This declaration exposes without any shadows the root of that dogma; matter does not count, only energy counts for mankind's economic struggle.

The foregoing variants seem to cover the entire spectrum of the energetic arguments. However, they all lead to one and the same analytical picture of the economic process in its relation to the environment. It is the picture

4. Energetics also permeates Brown's position in his earlier volume [10]. But, curiously, both volumes cite facts that go directly against the energetic dogma (see below, Sec. V).

5. Years after taking this position and yet arguing that a close relation exists between the Entropy Law and the economic process, Boulding [5], turned to arguing that entropy is a "negative potential" and hence is not a concept fit for explaining evolutionary development. But still more recently he came to realize that materials also are vital environmental elements [6].

represented by the multi-process matrix of flows and funds shown in Table I [30].[6] To avoid irrelevant issues, let us divide the economic process only into those consolidated processes and aggregated categories that are relevant for the present argument:

P_1: produces "controlled" energy, *CE*, from energy in situ, *ES*;
P_2: produces "capital" goods, *K*;
P_3: produces "consumer" goods, *C*;
P_4: recycles completely the material wastes, *W*, of all processes into recycled matter, *RM*;
P_5: maintains the population, *H*.

The special features of the foregoing picture must be well marked. First, neither a growing nor a declining economy can provide an acid test for the energetic dogma. Material growth cannot feed on an environmental flow of energy alone,[7] whereas a declining economy may very well need no flow of environmental matter. The test case must therefore be a stationary process, or, with Marx's more felicitous terminology, a reproducible one.

But regardless of the actual system, one point is beyond any doubt ever since Lord Kelvin observed more than one hundred years ago that energy is not lost but only becomes unavailable to us [74, 189; 78, 236–39]. All processes, therefore, necessarily produce dissipated (unavailable) energy, *DE*, which returns to the environment. In the energetic model, however, no matter leaves the economic process; all matter is completely recycled within that process. Thus, no matter has to be brought into the economic process from the environment. The only flows between the economic process and the environment are energy flows, namely, the input flow e_1 and the output flow $d = \Sigma d_i$.

Second, the representation of Table I reflects one elementary aspect of reality that needs unparsimonious emphasis in view of the "flow complex" that seems to dominate modern economic thought [29, 88; 30, 219]. Like all actual processes, the economic process has a material scaffold represented by its fund elements: capital equipment, K_i; people, H_i and H; and Ricardian land, L_i. We can never handle energy without a material lever, a material receptor, or a material transmitter. We ourselves are material structures without which no biological life can exist. In including the material funds (actually, the agents) in the analytical picture, I have assumed—a perfectly fair assump-

6. This analytical representation is both simple and safe. It does away with the bewildering flow diagrams used by ecologists and energy analysts (in which funds are ignored), and it avoids the analytical pitfall of "internal flow" that awaits the users of the input-output table [30, 253–62]. Also, the flow-fund model should not be confused with the flow-stock model introduced by Sir John Hicks in the theory of market disequilibriums. The function of the stocks is to accumulate flows and to decumulate themselves into flows. Funds participate in a process but without changing.

7. Obvious though this point may seem, it calls for some technical justifications, which will be provided later (Sec. III).

tion, I think—that the energetic position does not go so far as to claim that actual processes require no material structures of the kind we recognize side by side with energy at the macro-level.

Third, the output flow of capital, x_{22}, is destined to maintain the capital funds K_i in a reproducible condition; their wear-and-tear is thus compensated for by the maintenance flows x_{2i}. Similarly, the flows x_{i5} maintain the population H "intact". These are the elementary conditions for the (P_i)'s to be reproducible. And since in the case under consideration all flows must be expressed in physical units (calories or moles, for instance), the following equalities must always prevail as an aggregated translation of the conservation laws at the macro-level:

$$\begin{aligned} &d_1 = e_1 - x_{11}, && d_i = x_{1i} && (i = 2, 3, 4, 5), \\ &w_1 = x_{21}, && w_2 = x_{42} - x_{22}, && w_3 = x_{23} + x_{43} - x_{33}, \\ &w_4 = x_{44} - x_{24}, && w_5 = x_{25} + x_{35}. \end{aligned} \qquad (1)$$

Fourth, every recipe (P_i) is assumed to be *feasible*; that is, it can produce its product provided it is supported by the specified funds and is fed the specified inputs. But the feasibility of every recipe (P_i) does not necessarily

Table I. The Economic Process in Relation to the Environment According to the Energetic Dogma

Product	(P_1)	(P_2)	(P_3)	(P_4)	(P_5)
			Flow Coordinates		
CE	x_{11}	$-x_{12}$	$-x_{13}$	$-x_{14}$	$-x_{15}$
K	$-x_{21}$	x_{22}	$-x_{23}$	$-x_{24}$	$-x_{25}$
C	*	*	x_{33}	*	$-x_{35}$
RM	*	$-x_{42}$	$-x_{43}$	x_{44}	*
ES	$-e_1$	*	*	*	*
W	w_1	w_2	w_3	$-w_4$	w_5
DE	d_1	d_2	d_3	d_4	d_5
			Fund Coordinates		
Capital equipment	K_1	K_2	K_3	K_4	K_5
People	H_1	H_2	H_3	H_4	H_5
Ricardian land	L_1	L_2	L_3	L_4	L_5

imply the *viability* of the *technology* represented by all the processes together (an important point to be retained for further reference). The necessary and sufficient conditions for the viability of the technology of our reproducible economic system is given by the inequalities $x_{i5} \geqslant x^0_{i5}$, x^0_{i5} being a minimum determined by the standard of living, and the well-known relations

$$\sum' x_{1i} = x_{11}, \quad \sum' x_{2i} = x_{22}, \quad x_{35} = x_{33}, \quad \sum' x_{4i} = x_{44},$$
$$\sum' w_i = w_4, \tag{2}$$

where the prime accent shows that the variable subscript cannot be equal to the fixed one.[8]

III. Perpetual Motion of the Third Kind

Let us now recall that in the thermodynamic terminology established by Ilya Prigogine a system that can exchange only energy with its surroundings is *closed*. The economic process represented by Table I is therefore a closed system. In addition, this closed system is reproducible—i.e., it is a steady state, to use the thermodynamic terminology. According to the energetic dogma, it can provide internal *mechanical* work at a constant rate as long as a constant flow e_1 of environmental energy is forthcoming. Because of the theoretical importance of such a system for the energetic thesis and for other issues as well, I have proposed to refer to it as *perpetual motion of the third kind*.[9] And since my position is that this perpetual motion is impossible, by analogy with the negation of the other two perpetual motions by the first and second thermodynamic laws we may regard this impossibility as a fourth law of thermodynamics [31; 34–38].[10]

A technical point that must be dealt with before anything else in connection with this proposition concerns the famous Einsteinian equivalence between mass and energy, $E = mc^2$. For as even a Nobel laureate, Hannes Alfvén propounds, "matter, then, can be seen as a form of energy" [1]. The assertion only bespeaks the familiar bias in favor of energy. It is not because matter is not just mass, but some positive amount of mass and some positive

8. The case in which all w_i's are null, which does away with any need for recycling, corresponds to Boulding's tenet.

9. To my knowledge, only Zemansky [78, 193] uses the same term for a closely related system—namely, for a system in which work is not dissipated against friction, viscosity, etc. My definition, I believe, is more relevant analytically.

10. This law should not be regarded (as has proved possible quite often) as a corollary of the Entropy Law. According to this law, an *isolated* system—i.e., a system that can exchange neither matter nor energy with its surroundings—tends toward the Heat Death—or toward Chaos if one includes matter under the rule of that law [29, 75; 30, 142; 32, 8]. The confusion may stem from the loose practice of using "closed" instead of "isolated", e.g., [46].

Here it may be well to recall that a system which can exchange both matter and energy is *open*. The fourth logical category, a system that may exchange only matter, is factually impossible since any transport of matter involves energy as well.

amount of energy structured in the definite patterns of the chemical elements and their compounds, that Einstein's equivalence does not bear out that view. I wish to submit that there is an intrinsic asymmetry between matter and energy, which is due to the asymmetry between mass and energy. Indeed, as long as we do not recognize the irreducible asymmetry between the two terms of the Einstein formula—which with properly chosen units may be written in the perfectly symmetrical form $E = m$—we are just as justified to speak of a "crisis of matter" as of a "crisis of energy".

Energy is converted into mass (and vice versa) in numerous nuclear reactions which begin and end with some mass. In the familiar relations

Proton + Antineutrino ⇄ *Neutron* + *Positron*

Neutron + Neutrino ⇄ *Proton* + *Electron*

where the underlined terms have a positive mass. Pairs of anti-particles with positive mass may be entirely converted into pure energy, as it happens when an electron and a positron by colliding turned into a photon. To be sure, the reverse reaction also is *possible*. Photons may be converted into pairs of electrons and positrons, as it happened on a large scale immediately after the Big Bang while the temperature of the universe was still greater than 6×10^9 degrees Kelvin, which is the threshold of that reaction. But since such pairs are extremely unstable (like all pairs of matter and antimatter), they almost instantaneously disintegrate into pure energy. This is why positrons exist only in some intense astronomical phenomena or in high-energy laboratories. For photons to split into neutrons and antineutrons the temperature must be higher than 10^{13} degrees Kelvin, which is one hundred times higher than that believed to prevail at one hundredth of a second after the Big Bang. This last temperature—about 10^{11} degrees Kelvin—is greater than the hottest stars [75]. Nothing can be surmized about what happened before that particular moment, but what we know is that *material* particles—protons and neutrons were not produced then and cannot be produced now from pure energy.[11] And since without nucleons there can be no atoms, hence no matter, the present explanations of the origin of the chemical elements cannot do without assuming that the combined number of the protons and neutrons has always been the same as in the initial hot soup [73; 75]. To express the fact that nowhere in the universe matter can be created out of pure energy, the Einstein relation should be written $E + mc^2 = E_0 + m_0c^2$, $m_0 > 0$ whenever m includes some nucleon mass, that is, some matter. (The subscript identifies initial amounts.)

11. Only what is proven to work at the time has a legitimate place in the arguments of the sort developed here. Speculations have no proof value and may even be dangerous. For example, were the servants of science to persistently preach that we will certainly learn how to block out gravitation by some cavorite—the material discovered by Mr. Cavor in one of H. G. Wells's fantasies—people would be misled in building houses without stairs and without elevators.

Heavier elements are currently fused from lighter ones but only in stars, where the temperature reaches astronomical levels, between 10^7 and 10^{10} degrees Kelvin.[12] But at these temperatures matter exists only in a desaggregated state as plasma. Should a closed system become so hot, it could no longer produce any mechanical work (let alone harbor life).

Nuclear reactions are certainly going on at all times on our planet, too. Radioactive elements continuously decay. However, these phenomena as well as all nuclear reactions set up by us usually convert mass into energy, *not vice versa*. Mass is converted into energy even when we light up a match. This loss of mass is fantastically small because of the large value of c. But the difference between the weight of spent nuclear cores and their initial weight can be shown on ordinary scales. And at the Sun's dimension, 4,200,000 tons of mass are "lost" every second. Certainly, we can also convert energy into additional mass, but only in very special cases (usually, in laboratory installations) and in relatively small amounts and not into *nucleon* mass.

At the temperatures at which mechanical work can be performed, the overwhelming chemical elements are stable. In a closed system the amount of each element remains constant. "Materials are not destroyed",—Slesser, an energy specialist, tells us, "the iron molecule in iron ore is still an iron molecule when it is turned into steel or when it ultimately ends up as rust" [70]. Seaborg also argues that "we have here on earth the same amount of matter we have had since the dawn of history"—in truth, since the earth became a stable planet [65]. Only, both Slesser and Seaborg intended their observations as a support of energetics, in the same way in which Brooks and Andrews claimed that "the literal notion of running out of mineral supplies is ridiculous. The entire planet is composed of minerals" [9].

The simplest way to expose the fallacy of the Brooks-Andrews argument is to note that, by the same token, we could maintain that we cannot possibly run out of energy because the entire planet is full of energy. Indeed, the thermal energy contained in the ocean waters would alone suffice to maintain an undreamed-of industrial activity for billions of years to come. The rub is that all (or practically all) this fantastic amount of energy is unavailable for conversion into mechanical work by an engine of *finite* dimensions, which must by necessity operate in *cycles*. Indeed, as Planck suggested [57; 58], the thermal energy of a bath of constant temperature could be converted into mechanical work by an infinitesimally slow moving piston-and-cylinder of an *infinite* length (on the same familiar principles that govern the isothermal expansion of the gas in a Carnot cycle).[13]

12. Despite the immensity of these ovens the material universe still consists mainly of hydrogen, according to some estimations, 92.06 percent, with helium representing 7.82 percent [2].

13. Peculiar thermodynamic ideas such as this bring to mind Dirk ter Haar's statement that the concept of entropy "is not easily understood even by physicists," [30] may seem a severe judgment typical of an authority in a field. However, nowadays most physicists have only a

Certainly, the whole planet is made of matter. But the argument based on this point ignores the fact that, just like the thermal energy of the Earth, not all terrestrial matter is in available form. Matter also continuously degrades into an unavailable form.

Two factors may explain why this last fact is widely ignored. The first is the peculiar attraction our minds feel for all strains of mechanical models. The most probable reason is that we act on the material surroundings mainly by pushing and pulling. The mechanistic dogma had already lost its Laplacean grip on physics, when, in his Baltimore lectures (1894), Lord Kelvin confessed that he could not understand a phenomenon unless he was able to represent it by a mechanical model. It is the attraction of the mechanistic *Weltbild* that induces us to believe that matter cannot be definitively irrecoverable. Indeed, in mechanics, matter can only change its place, not its quality; hence any system may move back and forth without suffering any change.[14]

The second factor is that, curious though it may seem, the foundation of thermodynamics is energetic—as Rankine perceived it—for thermodynamics is concerned only with what happens to energy. To be sure, matter enters into the picture, but only as a support of chemical reactions (because they always involve energy transformations) and in problems of pure (nonchemical) mixtures (because unmixing necessitates work). Both these aspects have been introduced into thermodynamical theory by J. Willard Gibbs, who is thus regarded as the founder of "chemical energetics" [66].

Take the ultra familiar apparatus consisting of a piston-and-cylinder filled with some gas that is generally used to describe and justify the basic laws of thermodynamics and to prove Carnot's fundamental proposition that maximum efficiency is obtained only by a perfectly *reversible* engine. To circumvent the undeniable fact that because of friction no motion can be reversible, thermodynamic theory assumes that any motion is reversible if its speed is infinitesimally slow.[15] Such a speed does eliminate friction from the

superficial knowledge of thermodynamics (if any), so that some may go wrong even on the notion of heat (see *Journal of Economic Literature*, December 1972, p. 1268, and note 14, below).

14. A celebrated symptom of our mechanistic propensity was Ludwig Boltzmann's famous endeavor to explain irreversible phenomena by blending the perfect determinism of the reversible laws of Newtonian mechanics with probability . For the economist this hybrid construction presents a crucial interest. By maintaining that the regeneration of unavailable energy is only highly *improbable*, not *impossible*, the propounders of the probabilistic theory foster the belief in the possibility of cheating at the entropy game (just as we may cheat at any game of chance) or, as P. W. Bridgman ridiculed the thought some 50 years ago, of "bootlegging entropy" [29; 30; 32]. It is regrettable therefore that not all physicists know that Boltzmann's construction met with irrefutable criticism from some of the greatest physicists, for which see [25; 30]. Far sadder it is they do not know that through his pathbreaking contributions Ilya Prigogine showed that Boltzmann's "'mechanical theory' of the evolution of matter [is based on] intuitive arguments [and] the program was never realized, despite frequent affirmations to the contrary" [60–62]. If one has not gone beyond Boltzmann, one is apt to proclaim (as Auer does) that the Entropy Law sets no obstacle to endless economic growth [3].

15. It may be well to explain that in thermodynamics a motion is reversible only if the motion and *everything else in its surroundings* can be returned to the original situations [57; 78].

picture, but it introduces an even more essential obstacle. With an infinitesimally slow speed a piston would take an infinite time to move over any finite distance. Time and again, infinity steps in to set things out of bounds for us mortals.[16] Because a reversible engine thus exists on paper only, no actual engine can operate at maximum efficiency.

In the end, thermodynamics had to recognize the existence of friction as well as of a few other germane factors, which account not only for irreversibility in nature, but also for the fact that available energy cannot be completely converted into *useful* work. Part of this energy is always converted into irrecuperable heat.

Friction thus appears as a ghost, so to speak, in the backstage of the thermodynamic setup, a ghost that robs us of available energy. Indeed, thermodynamics did not move further to recognize and analyze the elementary fact that friction robs us of *available matter* as well. It did not tackle even the specific laws of the energy degraded by friction. This task was left to the engineers, but even they have established only tables for the forces of friction of the most frequently used materials. With the help of these tables, we can determine the work against friction W_f, to be used in the (still incomplete) transformation formula:

$$Q_a = W_u + W_f, \tag{3}$$

where Q_a is available energy and W_u is useful work (the internal energy of the system being assumed to remain constant).[17]

About the dissipation of matter by friction we still know practically nothing. One possible explanation of this conspicuous lacuna is the difficulty of explaining the phenomenon of friction. Mechanical laws applied to material particles cannot explain it. Its laws have a purely empirical basis and on closer and closer examination are usually found to be "'falser' and 'falser'" [27]. And a consummate student of the problem concludes that the subject of friction "remains a highly controversial one, and there are very few statements that can be made in this field which will find no opposition" [63].

But friction is not the "imperfection" only of matter in bulk that robs us of both energy *and* matter. There are neither perfectly rigid nor perfectly elastic materials; there are no perfect insulators and no perfect conductors; nor are there materials for which the force of friction is infinite (the opposite case of frictionless materials). And it is because of these numerous imperfec-

16. Recall Planck's infinite piston mentioned earlier.

17. The traditional formula is $Q = W$ because in thermodynamics work is defined only in the absence of friction [24]. Formula (3) was recently proposed by Silver [67]. But he, too, stopped short of mentioning the material effect of friction. Moreover, to cover all energy wastes, the formula should read $Q_a = W_u + W_f + Q_l$, where Q_l is the leakage of thermal energy—i.e., the amount of thermal energy that always descends a temperature gradient *without performing any work*.

tions (the list is not exhaustive) that the robber, the only robber, of energy and matter is matter itself.

All over the material world there is rubbing by friction, cracking and splitting by changes in temperature or evaporation, there is clogging of pipes and membranes, there is metal fatigue and spontaneous combustion. Matter is thus continuously displaced, altered, and scattered to the four corners of the world. It thus becomes less and less available for our own purposes.

The energetic dogma claims that this dissipation can be completely reversed provided there is enough available energy. But the operation must necessarily involve some material instruments. Because there are no perdurable material structures these instruments will necessarily wear out. They will have to be replaced by others produced by some other instruments, which will also wear out and will have to be replaced, and so on, in an unending regress. This regress is a sufficient ground for denying the possibility of complete recycling, just as the same kind of regress is the reason often invoked in thermodynamics against the possibility of completely erasing the changes caused in the *energy* structure by a natural process [24, 24].

Finally, another possible (and important) thought in support of complete recycling must be entertained. It relates to John von Neumann's proof that a universal Turing machine may be so designed that if left in a floating medium together with a large number of each one of its elementary parts it will reproduce itself [52]. Could not then the fund elements of the process represented by Table I constitute such a "machine"? This machine does "float" in a medium that, although closed, contains all the necessary elements for its reproduction, namely, the waste outputs of processes (P_i)'s. Yet the idea that the economic process could be a universal Turing machine must be rejected. Neumann's proof is a very ingenious paper-and-pencil exercise based on a requirement that makes the project impossible in practice. The troublesome requirement for a universal Turing machine is that its capacity for instruction must be limitless [30, 86–93]. And even if we softened this condition to "practically infinite," the machine would still require a practically infinite sequence of time intervals to reproduce itself, since each "move" of the machine requires a duration. Therefore, this last justification of complete recycling runs against the same familiar obstacle as all other ideal schemes: the unreachable infinity.

But the energetic thesis can be examined from still other analytical viewpoints.

IV. The Dissipation of Matter and Planck's Law

I have spoken of some asymmetry between energy and mass and between energy and matter. There is still another asymetry. Both mass and energy are homogeneous "substances". Energy is the same whether it is associated with a photon or with the force of the wind. This is certainly why the idea of reducing everything to energy has such a strong appeal. But there also is no qualitative difference whatever between the mass, say, of a proton and that of any other elementary particle. However, this last fact is of no value for the relationship between the economic process and the environment. In the economic process it is not mass as such that counts. What counts is matter in bulk (and, of course, energy). And the rub is that unlike mass and energy, matter is a highly heterogeneous category. Every chemical element has at least one property that characterizes it completely and hence renders it indispensable for some technical recipes.

We must therefore expect that, in contrast with the general theory of energy (thermodynamics), the study of the transformations of matter in bulk should be hard going, as we have just seen for the case of friction alone. It is a rather simple problem to understand how energy degrades as heat dissipates from the hotter to the colder bodies of a system, because in this way it becomes less and less available to be converted into mechanical work. Remember the famous Carnot principle: we must have a difference of temperature between the boiler and the condenser in order to obtain mechanical work by a cyclical engine. Without entering into the technical maze, one may see why the increase in the dissipation of energy is measured by the thermodynamic formula for entropy:

$$\Delta S = (\Delta Q/T)_{reversible} \tag{4}$$

where ΔQ is the heat transferred by conduction at the absolute temperature T.[18]

For the entropy of the simple mixture of two *distinct* ideal gases of the same pressure P and temperature T, we have Gibbs's famous formula:

$$\begin{aligned} S = & -R\left[m_1 \ln(m_1/m) + m_2 \ln(m_2/m)\right] + (c_1 m_1 + c_2 m_2)\ln T \\ & + (a_1 m_1 + a_2 m_2) - Rm \ln P, \end{aligned} \tag{5}$$

where $m_1 + m_2 = m$ are the corresponding moles, R is the gas constant, and a_i, c_i are physical characteristics of the gases. Mixing two gases of the same pressure and temperature thus increases the entropy by[19]

$$\Delta S = -R\left[m_1 \ln(m_1/m) + m_2 \ln(m_2/m)\right], \tag{6}$$

18. Even though "hotness" is one of the most common sensations, the flow of heat (as strictly conceived in thermodynamics) has no direct physical significance and there are no *direct* operations for measuring it [8; 24].

19. Because the same technical detail will come up later in Sec. VI, it may be well to add here that the enthalpy, H, does not change [48; 57; 58]. Enthalpy, which corresponds to the intuitive notion of heat content, is the amount of thermal energy necessary to bring the substance from absolute zero to its temperature state while pressure is kept constant. In practice, enthalpy is the calorific value of a fuel—i.e., the maximum amount of thermal energy that can be obtained by burning it.

a very familiar formula that nowadays is used on both solid and flimsy grounds.

Ordinarily, this formula is mentioned in connection with a paradox pointed out by Gibbs himself [66]. If the two gases are identical, (6) is still positive, although the mixture then causes no entropy change whatsoever. But another aspect of (6), stemming from an observation by Max Planck [58, 104], bears on the present argument. Planck suggested that in the case of (6) "it would be more to the point to speak of a dissipation of matter than of a dissipation of energy".[20]

According to either (6) or (5) the greatest dissipation for a given m occurs when $m_1 = m_2$. But let us consider the case in which $m_1 = 1$ and $m_2 = 10^{100}$. Then, ΔS is completely negligible. Yet, on the basis of the intuitive conception of dissipation, it stands to reason that, *if gas 1 is the valuable one*, from the particular human viewpoint the dissipation is far greater in the latter than in the former case. It is in the latter case that we can properly say that one mole of gas 1 has become unavailable to us. Indeed, to reassemble those molecules would be a task as formidable as reassembling the molecules of a small amount of ink spread over the Atlantic Ocean!

Undoubtedly, we can reassemble the scattered pearls of a necklace that broke in a room, in a theater, even somewhere in Manhattan provided we are prepared to expend enough time, enough energy, and wear out numberless objects in the process. The extrapolation of this macroscopic recipe to the microscopic level of molecules or even of small bits of matter is clearly unwarrented. From all we know, can one believe in the possibility of reassembling not all, but practically all the rubber molecules of our worn-out tires, all the lead dissipated through the exhaust pipes, or all the copper dissipated from our pennies through use? In a reasonably finite time as well? The object lesson is that at the microscopic level the same recipe that succeeds for reassembling the pearls of a broken necklace would require, among other hard-to-conceive facilities, an *infinite* time. The operation belongs therefore to the same category as the irreversible engine and Planck's scheme for using the heat of the oceans. Planck even concludes that, because $\Delta S > 0$, "diffusion like friction and heat conduction is an irreversible process" [57, 113; 58; 78].

The preceding analysis shows, however, that (6) measures in reverse the intensity of diffusion conceived so as to make sense in our actual manipulation of matter. According to that formula, mixing one carload of needles with one carload of hay creates a greater diffusion than losing one important needle in a carload of hay that corresponds to the task of complete recycling or of mining any rock. Although we can easily extract a metal from a rich ore, the

20. Understandingly, Planck did not say "dissipation of mass."

task is harder and harder as the metal content decreases, and for a content of, say, 10^{-100} it becomes impossible.

There remains however the issue of the thermodynamic interpretation of (6). For this, we may recall that in thermodynamic theory $T\Delta S$ represents the work necessary to restore the corresponding system to its original position. In the present case, this means to separate again the two gases completely. A theoretical blueprint for performing this separation was imagined by Jakobus Henrikus van't Hoff (the first Nobel Prize winner in chemistry). The device—the van't Hoff box—consists of a perfectly isolated cylinder with two pistons working in opposite directions; each piston consists of a semi-permeable membrane, one permeable only to gas 1, the other only to gas 2. The mixture is placed between the two pistons, which at first are far apart. As the pistons are pushed infinitesimally slowly against each other, the gases are separated since each gas gets behind its own semi-permeable membrane. It is easily shown that the work necessary for this pushing is indeed equal to the product of (6) by T [58; 78]. This result seems to provide a strong support to the purest energetic dogma. If we have enough energy (equal at least to $T\Delta S$), we can extract the entire amount of any gas from any mixture. Several snags, however, beset the actual implementations of this idea.

First, just as in actuality there are no completely frictionless, no perfectly elastic, and no other perfect materials, there are no perfect semi-permeable membranes. The separation therefore cannot be complete.[21] Second, with use all membranes become clogged [57; 58]. They wear out, just as does any other part of a mechanism; in the end they must be replaced, thus starting the endless regress mentioned earlier.[22]

Van't Hoff box constitutes at least an ideal procedure for unmixing gases, but no similar device exists for other mixtures. In practice, each mixture separation is achieved by some particular procedure, by chemical reactions, centrifugal or magnetic forces, etc. To be sure, the absence of a general blueprint does not prove that some ideal blueprint may not exist for each particular mixture. But several arguments plead against this thought.

Let us remember Maxwell's demon, the demon presupposed to separate the fast-moving from the slow-moving molecules of a gas. It is now generally believed that this miraculous demon has been "exorcised," so that, like any other living creature, it must consume a greater amount of available energy than it creates by the separation of "hot" from "cold" molecules. Now, to separate a mixture of, say, nitrogen and oxygen, we need a demon far more miraculous than Maxwell's. Indeed, Maxwell's demon does not have to bring back absolutely every molecule to its initial container. Moreover, it may safely

21. Another imperfection is responsible for the fact that in actuality some mixed gases always remain between the membranes no matter how much the pistons are pressed against each other.

22. Highly pertinent to the topic of this paper is the recent discovery (by R. K. Knoll and S. M. Johnson of NASA, Cleveland) that solar collectors also become progressively clogged.

leave some "hot" or "cold" molecules in the wrong container; it is only the average speed that counts. Our new demon, on the contrary, *must not leave even one single molecule mixed with those of the other kind*. To exorcise it, we must not only supply it with enough energy but also endow it with a material existence. And since matter keeps dissipating, the problem now boils down to whether our demon can recycle itself completely while recycling the gas mixture, also completely. The system of Table I depicts, in fact, such a demon. No wonder that the miraculous features of our demon are tacitly implied in many ideas concerning the unlimited renewability of material resources.

To Brown [10], it seems, belongs the idea of mining the whole crust by utilizing the energy of the radioactive elements thus obtained. Such a miraculous technology would come to an end only after all rocks had been crushed. The sad truth is that no blueprint has been put forward to describe how this technology may work. Instead, the idea is regarded as a real flight of fancy by practically all geologists, who generally support Peter Flawn's dictum that even "the average rock will never be mined" [28; 18; 68]. As Preston Cloud, who is known to have opposed with all his authority all minerological flights of fancy, tried to enlighten the layman, "it is the uncommon features of a rock that make it a candidate for mining!" This is why almost certainly your backyard is not a potential mine.

Brooks and Andrews denounced Flawn's dictum by invoking the fact that titanium is also mined (as a by-product) from presumably sub-average grades. A few other writers also proclaimed that "the world will not run out of geologic resources," assuring us that only the energy limit may prevent the mining of any ordinary rocks [21]. But most have a hard time remaining consistent when faced with such facts, for example, as the existence of a geochemical barrier cutting off the mining of copper at 16 times the average abundance]20, 129].

Let us also observe with necessity that even if the average rock could be mined, this would not entail that the entire mineral content of the crust is accessible. The distribution of grades has a peculiar bimodal shape and is thus very skew [69]. Hence, the overwhelming majority of rocks are much poorer than the average grade. Whether or not a mineral may be mined from a given rock depends on the specific theoretical cutoff for that mineral and for that type of rock, not on its average crustal abundance. Matter, to recall, is heterogeneous, so that for each substance and rock there is some specific mining procedure(s). This is why it seems well nigh impossible to establish a general formula for deciding whether or not a substance is available from a given mixture. In this lies the theoretical difficulty (hinted at earlier) of defining the unavailable form of matter by a general analytical formula.

Curves describing how the amount of rock crushed and the amount of energy used vary with respect to the rock's metal content are readily available in the literature. They all are asymptotic to the vertical axis [56]. Those for energy should actually be asymptotic to a line parallel to that axis, so as to reveal the existence of a theoretical cutoff, however small this may be.

As a last argument we may mention a statement by Planck, which although very important, seems to have passed unnoticed. At the end of an elaborate discussion of all types of mixtures, Planck arrived at a formula which entitled him to conclude that "neither a gas nor a liquid nor a solid body can ever be completely freed from the last traces of foreign contaminating substances" [57, 125; 58, 238]. Exceptions occur, but only at absolute zero [58, 239; 48, 46].

Now, according to the third law of thermodynamics, enunciated by W. Nernst, absolute zero cannot be attained in actuality. Therefore, Nernst's and Planck's negations are mutually bound. They also form a twin package. The former says that matter *in bulk* cannot be "purified" of thermal energy, the latter that it cannot be purified of any contaminant.

And what deserves strong emphasis is that, in contrast with the Entropy Law, the truth of these laws is not subject to any instrumental constraints. They are true in an absolute sense. To explain, *in theory* we can restore a partial system to exactly its original position, if we have a certain amount of energy and some specific apparatus at our disposal. There is no way, however, to circumvent the impossibilities affirmed by either Nernst's law or Planck's.

It is Planck's law for matter than provides us with a very important analytical argument against the energetic dogma, which necessarily implies that *recycling may be complete* and that *any rock can be mined*.

V. Matter Matters, Too

Complete recycling being impossible, even in a steady state the "transactions" between the economic process and the environment must necessarily consist of some available matter as well in order to compensate for the matter dissipated continuously and irrevocably. As Harrison Brown observed, if all the iron produced in the United States between 1870 and 1950 (about 2 billion tons) had still been in use in 1950, there would have been then 13.5 tons per capita, almost double the actual figure [10]. The same difference would be even more impressive were it established on the entire past production. We all know where the difference went. "Oxidation by the air, corrosion by liquids, and other general wear [certainly because of friction and metal fatigue] take heavy tolls" [10]. As is always the case, some iron is also "lost" during production processes. Although it is not possible to estimate the amount of iron "that has been lost never to be recovered", Brown [11] estimates that about 10 percent of the iron used for steel production is lost irrecoverably and that over 100 years the remaining iron becomes unavailable matter. Just to maintain the iron equipment existing in 1954 (without providing for any economic growth), a flow of ore convertible into about 0.3 tons of iron per person had to be mined each year.

Of course, this maintenance flow varies from substance to substance according not only to the changing technology but also to the size of the stock existing in the process. In the case of gold this flow is naturally small for several reasons—its chemical resilience, its particular use, and the relatively small amount of it in existence. Yet it is hardly exact to assert that "most gold ever mined is still available" [68]. The millions and millions of past bracelets, necklaces, coins, etc. have not all been kept at a Fort Knox.

The fact that the flow of dissipated matter increases with the size of the material stock must be recognized as a very important link in the argument of this paper. For it explains, on the one hand, why, even if some cutoffs may be so low that it may be almost impossible to ascertain them by a laboratory experiment, their effect at the scale of the entire planet and in the long run cannot be ignored. On the other hand, it explains why we are apt to believe in the perpetuity of the so-called natural cycles of oxygen, carbon, nitrogen, etc., that adorn all ecological manuals.[23]

As far as the economic process itself is concerned, we must not ignore the substantial dissipation of matter caused not by purely natural phenomena but by some activities of living creatures, of mankind's, above all. It is the dissipation of some vital elements by man's consumption of food and timber in places far away from the farm and the forest that produced those items. The practice—a consequence of the high and growing level of urbanization everywhere in the world—also wastes available energy. Most curiously, we are aware of this waste, but not of the waste of available matter. The difference is responsible for the fallacy that forests can provide "an endless supply of wood" because the flow of sunshine is practically endless [51]. Yet forests cannot be everlasting anymore than topsoil can retain its quality forever without outside intervention [30, 302].[24]

The conclusion of the foregoing analysis is that for the environmental transactions we must keep two separate books—one for matter, one for energy—for at the macro-level no practical procedure exists for converting energy into matter or matter of whatever form into energy. The relation of matter and energy is not like that between, say, dollars and yen, nor even between land and working implements in agricultural production (Sec. VII).

A new multi-process matrix must therefore replace that of Table I. In this

23. It is because of the immensity of every one of these stocks that the amount continuously leaving the cycle is not conspicuous over short periods. One of the telling facts is that practically all the carbon deposited as calcium carbonate on the bottom of the oceans will not re-enter the so-called carbon cycle. And this is only a minor troublesome factor of the global circulation of environmental carbon [76].

24. Very likely, it will be the more palpable waste—of the energy consumed by transportation—that, in case of a serious difficulty in obtaining environmental low entropy, will bring about a substantial amount of deurbanization [31].

new matrix (Table II), an additional process, (P_0), transforms matter *in situ*, *MS*, into controlled matter, *CM*. All other processes have the same roles as before and are identified by the same notations. But there are several important changes.

First, new flows, s_i, represent the dissipated matter, *DM*, produced by every process and passed into the environment. Second, the recycling process, (P_4), no longer recycles all material waste, as assumed in the energetic model. Since dissipated matter is irrecoverably lost, (P_4) can recycle only matter that is still available but has no longer a shape useful to us: broken bottles, broken pipes, run-down batteries, worn-out motors, etc. Because recyclable materials belong in garbage cans or in junk yards, for brevity, we may refer to them as "garbojunk", *GJ*.[25] Third, another inherent aspect of the economic process is represented by a flow of items that also are returned to the environment and are labeled here "refuse", *R*. This flow consists in large part of available matter and available energy, but in a form that is not potentially useful to us at present. The crushed rock of an open-pit copper mine, most urban waste, and nuclear garbage, for example, belong to this category.[26]

As in the case of Table I, the relations:

$$\Sigma' x_{0i} = x_{00}, \quad \Sigma' x_{1i} = x_{11}, \quad \Sigma' x_{2i} = x_{22}, \quad x_{35} = x_{33}, \quad \Sigma' x_{4i} = x_{44}, \quad \Sigma' w_i = w_4 \tag{7}$$

represent the viability of the steady state. However, because *R* may include both energy and matter, we can no longer write the relations for the conservation of these items separately, as in (1).

VI. Energy Analysis and Economics

My position has been (and still is) that the Entropy Law is the taproot of economic scarcity. In a world in which that law did not operate, the same energy could be used over and over again at any velocity of circulation one pleased and material objects would never wear out. But life could certainly not exist either.[27] In our world, everything that has some usefulness (desirability) for us consists of low entropy. It is for these reasons that the economic process is entropic in all its material fibers [29; 30; 32]. But I have also

25. The possibility of recycling garbojunk marks a second asymmetry between energy and matter. Recycling of matter is possible because some material objects are "durable" in the sense that, in contrast with energy, they are not *instantaneously* degraded through their participation in producing mechanical or other types of work. The point bears on the old economic controversy over repeated use versus instantaneous consumption.

26. In order to save space, the fund coordinates are not represented explicitly in Table II.

27. We often connect the Entropy Law with disorder, yet without it there would be no order in actual phenomena. The truth is that the Entropy Law is the law of orderly succession. For—a piquant observation—without the Entropy Law you would not dare take a bath, one half of the water might become *by itself* so hot as to scald your neck and the other so cold as to frostbite your toes.

Table II. The Actual Relationship Between the Economic Process and the Environment

Product	(P_0)	(P_1)	(P_2)	(P_3)	(P_4)	(P_5)
			Flow Coordinates			
CM	x_{00}	*	$-x_{02}$	$-x_{03}$	$-x_{04}$	*
CE	$-x_{10}$	x_{11}	$-x_{12}$	$-x_{13}$	$-x_{14}$	$-x_{15}$
K	$-x_{20}$	$-x_{21}$	x_{22}	$-x_{23}$	$-x_{24}$	$-x_{25}$
C	*	*	*	x_{33}	*	$-x_{35}$
RM	*	*	$-x_{42}$	$-x_{43}$	x_{44}	*
ES	*	$-e_1$	*	*	*	*
MS	$-M_0$	*	*	*	*	*
GJ	w_0	w_1	w_2	w_3	$-w_4$	w_5
DE	d_0	d_1	d_2	d_3	d_4	d_5
DM	s_0	s_1	s_2	s_3	s_4	s_5
R	r_0	r_1	r_2	r_3	r_4	r_5

maintained (without always being read correctly) that although low entropy is a necessary condition for usefulness, it is not also a sufficient one (just as usefulness is a necessary but not sufficient condition for economic value.) Poisonous mushrooms, for example, consist of low entropy [29, 94; 30, 282].

The entropic nature of the economic process notwithstanding, it would be a great mistake to think that it may be represented by a vast system of thermodynamic equations—as has been proposed, for example, by Lichnerowicz [50]. The entropic process moves through an intricate web of anthropomorphic categories, of utility and labor, above all. Its true product is not a physical flow of dissipated matter and energy, but the enjoyment of life—account being also taken of the drudgery of labor [30, ch. x]. And, in spite of some voices to the contrary, pleasure is not related by a definite quantitative law to the low entropy consumed. Nor is labor's "disutility" related by such a law to the expended low entropy. William Petty was right in teaching that nature is the mother and labor the father of wealth—only he ought to have said "of our existence." Therefore, even if we accepted the energetic view, in which the whole economic process is maintained only by the flow of environmental energy e_1, economic value could still not be reduced to energy.[28]

Nevertheless, the possibility of reducing prices to energy units, Btu's instead of dollars, has been in the focus of attention ever since the oil embargo prompted people to think of energy and to attempt an analysis of its circulation in man's activities. Gilliland [42] even argued that energy analysis is the natural way for doing away with the difficulty of adding apples and oranges, which is what economists have to do. Slesser [70] and R. S. Berry (quoted by Slesser) propound the strongest form of this dogma: to measure the cost of things in money "which is after all nothing more than a highly sophisticated value judgment" does not offer a firm basis for economic valuation. Actually, if economists were to look at scarcity in a more complete way, their "estimates would come closer and closer to the estimates [of] the thermodynamicists". This simply means to do away with economics and replace it by thermodynamics.[29] This position is so generally embraced that net energy analysis now constitutes the official criterion for technology assessment and energy policy in the United States [26].

All this is highly surprising in view of the fact that a great confusion still prevails even among the experienced practitioners of energy analysis. There is, first, the controversy concerning the difference between Odum's net energy analysis and the gross energy analysis based on some principles set up by a 1974 meeting of IFIAS, the International Federation of Institutes of Advanced Study (note 2, *supra*). Slesser, a representative of the latter school, claims that one has "yet to see a rigorous definition of net energy" [71]. It is true that even in his latest writings Odum is not sufficiently clear on many important points, such as whether the energy of labor should be taken into account. Moreover, he often confuses the reader by such requirements as that even money must be included in the general flow [54]. However, the situation is not much more enviable in the other camp, as is evidenced by a series of highly critical letters to the editor and small notes which failed to clear the

28. I have myself insisted that prices are parochial elements of the economic system and also that the market mechanism by itself it incapable of preventing environmental catastrophes. But I did not deny the necessity of the market mechanism for the allocation of resources and the distribution of income among the members of the *same generation* [32; 34].

29. Hayek [43, 51], as Keith Wilde recently reminded me, protested long time ago against "the various forms of social 'energetics' [as those propounded by] Ernest Solvay, Wilhelm Ostwald and Frederick Soddy." However, none of these authors held the same position as the modern energetists. What they claimed is that regardless of how economic value is established, the economic process cannot violate any of the natural laws, including the laws of thermodynamics. In fact, Ostwald was the first author to observe that from the dawn of mankind's existence technological progress has always consisted in increasing the power of our biological organs. And contrary to Hayek's accusation, Ostwald [55, 164] explicitly argued that "we would err if we measured value only in proportion to the amount of free energy." Only Soddy plunged into economic matters proper and devoted several works to a solution of the instabilities caused by credit creation. An interesting recent paper by Daly [23] aims at rehabilitating Soddy.

air.[30] One critic charged that because "the energy analysts disagree in their basic ground rules" energy analysis can be used "to prove whatever case you choose" [47]. Even one of the prominent energy analysts, P. F. Chapman, recognized that "there are almost as many methods [of analysis] as there are workers in the field" and energy analysis may follow four different aims and adopt three different methods [13]. Results, as he showed, may even be contradictory. There has been some controversy even over how to measure energy (as we shall presently see).

The issue of whether energy analysis provides an equivalent basis to the price system was recently entertained by David Huettner [46]. But, here again, the letters to the editor that his article provoked suggest that he, too, failed to make his case.[31] In deriving his price equations Huettner followed the fallacious practice in standard economics of ignoring the essential difference between flows—the material elements that are changed by a production process—and funds—the agents that perform the change [31; 32]. As a result, his price equations had exactly the same form as his equations for the presumed energy equivalents. Because of this identity the real difference between economic valuation and energy calculations could not be brought to light so as to settle the question raised by Price, "why energy analysis gives a different answer than the economic analysis" [59]. The flow-fund model used in my tables will settle this issue without much ado.

Let us consider first the simplest case, that of Table I, and begin with the net energy analysis. We must first decide what net energy is in that structure. I think that in the spirit of the Cottrell-Odum conception we may safely assume that net energy aims at determining how much controlled energy, *CM*, is accessible in all forms to the ultimate consumers. However, few additional observations now appear necessary. First, for the concept of net energy we do not have to distinguish between the various kinds of environmental energy, *ES*. In other words, it does not matter whether the net energy is derived from fossil fuels or from the wind, for example. Second, the dissipated heat must not be counted at all in computing net energy. That is, we must not add d_1 to the net output of energy x_{11}, or deduct d_i from the gross input x_{1i}. Finally, we must also not count in any way the energy spent by people in performing work or in consuming goods. To do so would be to mix economics with energy analysis. The analysis would then be falsified by a double counting resulting in an extensive cancellation, for in a steady state—which is still our testing ground—the total input flow of any environmental element is exactly equal to the corresponding total output flow.

Four possible definitions of net energy seem to deserve attention: a) x_{11}; b) the difference between x_{11} and the energy equivalent of x_{21}; c) x_{15}; and d) x_{15} plus the energy equivalent of x_{25} and x_{35}. The first suggestion can be easily dis-

30. *New Scientist*, 9, 16, and 23 January 1975.
31. *Science*, 15 April 1977, 259–62, especially M. Slesser. See also *Science*, 2 April 1976, 8–12.

carded. If x_{11} is, say, electricity produced from fossil fuels *in situ*, that electricity is not net energy even with respect to the process (P_1). For part of that electricity has been used in a roundabout process to produce, among other things, the material flow x_{21} necessary for compensating the wear and tear of K_1 during the mining of e_1 and its use in a thermoplant.

Suggestion (b) leads to

$$\textit{Net energy} = x_{11} - (x_{21})_e, \tag{8}$$

where $(x)_e$ denotes the energy equivalent of x. But this definition raises the troublesome question: what is the energy equivalent of a steel beam?

Here, again, various ideas have been at work. One simple method consists of taking into account only the energy *directly* used by a process because such data are readily available from official statistics. This method leads to

$$(x_{21})_e = x_{21}(x_{12}/x_{22}), \tag{9}$$

and

$$\textit{Net energy} = (x_{11}x_{22} - x_{12}x_{21})/x_{22}. \tag{10}$$

Obviously, this represents a substantial underestimation, for the production of x_{22} needs x_{42} units of *RM* as well. We must, therefore, find out $(x_{42})_e$ as well. We are thus involved in an algorithm that will not end until an energy equivalent is established for every type of product.[32] The result is a system of equations which recalls the Leontief system [16; 77].[33] Let us then denote by a_i the energy equivalent per unit of flow product of (P_i). From Table I we obtain:

$$\begin{aligned} \textit{Net energy} &= x_{11} - a_2x_{21}, \\ &-x_{12} + a_2x_{22} - a_4x_{42} = 0, \\ &-x_{13} - a_2x_{23} - a_4x_{43} + a_3x_{33} = 0, \\ &-x_{14} - a_2x_{24} + a_4x_{44} = 0, \end{aligned} \tag{11}$$

which by (2) yields

$$\textit{Net energy} = x_{15} + a_2x_{25} + a_3x_{35}. \tag{12}$$

This relation shows that suggestion (b) is equivalent to suggestion (d) and also that (c) would not do. It also shows that, for example, the average cost of K in units of controlled energy (net energy) is a_2.

32. It may be well to emphasize that the term "energy equivalent" does not imply a physical equivalence in the sense that a pound of copper, for example, may be converted into its energy equivalent or vice versa.

33. This is an almost insuperable task for the analyst even if a detailed input-output table in real terms is available. But this difficulty is wholly irrelevant to the issue under consideration now. Chapman, however, insists that in this method one should in *principle* consider a *subsystem*, not the whole national economy [13]. Moreover, in the illustrative application of this method to the energy sector, it is not clear how the energy equivalence of non-energy items was computed [16]. From repeated remarks on the matter, we may infer that for these items the energy equivalent of the dollar was used, a procedure that is totally incompatible with energy analysis.

Net energy defined by (11) can also be expressed as a function of the flow coordinates alone:

$$\textit{Net energy} = \begin{vmatrix} x_1 & -x_{21} & 0 \\ -x_{12} & x_{22} & -x_{42} \\ -x_{14} & -x_{24} & x_{44} \end{vmatrix} \div \begin{vmatrix} x_{22} & -x_{42} \\ -x_{24} & x_{44} \end{vmatrix} \tag{13}$$

This formula proves the extremely curious result that in the case of an energetic system, the net energy does not depend on the flows of the consumer goods industry.

Energy analysts are far from agreeing on whether or not we should attribute an energy equivalent to W. The difficulty is identical to that of attributing a cost to joint products in economics. But if we introduce the w_i's in (11) and denote by a prime the new energy equivalents and by z_w' the energy equivalent of w, by (2) we obtain immediately

$$a_i' - a_w' = a_i, \tag{14}$$

which means that the new equivalents are not completely determined. This result, which in some arguments may be quite troublesome, should have been expected.

We may turn now to the gross energy analysis. The stated aim of this analysis is to determine the amount of energy *in situ* "needed directly and indirectly to deliver a good or a service to the final customer" [49; 71]. However, "a quagmire of confusion" seems to prevail about the precise rules of how to achieve it [49]. Gross energy analysis has in view mainly the fossil fuels. For this reason, the recommended unit of energy is the calorific value, which represents the energy potentially available by burning a given amount of such fuels [16; 59].[34] A difficulty arises because there is no accepted calorific value for nuclear fuels. Nor can one speak of such a value for the energy input of a hydroelectric plant. The principles also provide that solar energy should not be counted as an input since it is "a free good" [19], a position which has its snags (and which does not affect the net energy approach). One author argues that labor and even profits "are also energy inputs" [77]. However, most of the gross energy analysts stick to the rules (also adopted above) according to which neither labor nor waste enter into the calculation [71]. There also are issues that concern both methods and about which no practitioner is explicit. One such issue is what precisely should be included in the capital cost x_{2i}.

Let now X denote the transposed matrix of the first four rows and four col-

34. In a more sophisticated form it is explained that we should consider only free energy, which determines the maximum mechanical work obtainable under normal pressure and temperature. Specifically, it is the Gibbs free energy, $G = H - TS$, that is meant [59; 70]. However, the enthalpy, H, is used instead (note 19, above), because in burning fuel under normal conditions ΔG does not differ substantially from ΔH [24, 73].

umns of Table I; let e denote the column vector $(e_1, 0, 0, 0)$; and let $b = (b_1, b_2, b_3, b_4)$ be the column vector of the energy equivalents in units of ES. We have

$$Xb = e\,. \tag{15}$$

From this and (1) we obtain

$$e_1 = b_1 x_{15} + b_2 x_{25} + b_3 x_{35}\,, \tag{16}$$

which shows how much energy *in situ*, b_i, is needed for a unit of every good consumed by the households.[35]

The comparison of (12) and (15) yields

$$Net\ energy = e_1 / b_1\,, \tag{17}$$

or, equivalently,

$$b_i = b_1 a_i\,, \qquad (i = 2, 3, 4)\,. \tag{18}$$

These results show, first, why the main issue in gross energy analysis turned around the proper unit for energy. If e_1 consists of fossil fuels, then Champan [13] is vindicated for insisting that one kWh of electricity should be counted as about four kWh's of calorific power, which simply means that $b_1 = 4$.

But the same relations prompt one to wonder why all the fight about which approach is the correct one since the two sets of energy equivalents are related by the simple relations (18). The truth is that, whereas a can be deduced from b, the converse is not true. This does not mean, however, that gross energy analysis is the better approach. According to the rule mentioned earlier, in the case of solar energy $e_1 = 0$, hence $b = 0$[36] and *Net Energy* $= \infty$. Briefly, within its own scope gross energy cannot discriminate between two technologies based on solar energy alone (also Sec. VIII). On the other hand, net energy analysis completely ignores the efficiency of the technologies by which resources *in situ* are transformed into controlled energy. As far as net energy analysis is concerned, it does not matter whether two tons or one million of tons of oil *in situ* are depleted for abtaining one ton of oil net.

Still reasoning on the basis of Table I, let us turn now to economic valuation, briefly, to what the normal prices would be in such an economic world. The point that can hardly be overemphasized in this connection is that in any economic system both the quantities represented by the flow elements *and* the services provided by the agents have value. Let $p = (p_1, p_2, p_3, p_4)$ be the column vector of the prices of the flow elements, and let P_K, P_H, P_L be the prices of services during their specific periods. The economic equations are

$$Xp = B\,, \tag{19}$$

35. It is easily seen that only b_3 is affected by the flow coordinates of (P_3). Also, the introduction of an energy equivalent b_w for W will lead to a system identical to (15) in which b is replaced by b', with $b' = b_1$, $b_i' = b_i - b_w$, $i \neq 1$.

36. This is so because if $|X| = 0$, then (15) has no solution.

where B is the column vector (B_1, B_2, B_3, B_4) and

$$B_i = P_K K_i + P_H H_i + P_L L_i. \quad (20)$$

We have further

$$p_1 x_{15} + p_2 x_{25} + p_3 x_{35} = \Sigma B_i. \quad (21)$$

which is the equation of the national budget.

Only in the highly unrealistic case, in which only the services of labor are included in the budgets B_i, can (19) determine all relative prices.[37] In actuality, the indeterminateness is removed by additional factors of a purely economic nature, such as tastes and income distribution.[38]

It is now perfectly clear that in *absolutely* no situation is it possible for the energy equivalents to represent economic valuations. Although the matrix of the price system (19) is the same as that of the systems of energy equivalents, (11) and (15), the former system cannot be equivalent to either of the latter. Actually, reducing economic value to energy is a more extreme position than the purest theory of labor value. To put in a homely way, according to the energetic portion one ounce of black caviar (which is mostly protein) should have the same price as one ounce of spaghetti (mostly carbohydrates) if it would take the same amount of either gross or net energy to produce them. Such an equivalence will never work.

VII. Global Analysis and Economic Choice

The analysis of the realistic case, in which not only energy but also matter counts, proceeds in the same way as for a generalized Leontief system in which there are several but distinct prime factors of production (say, uniform labor and uniform land). This means that the relations pertaining to one factor can be established by assuming that the other factors are in infinite supply [29, ch. 10].

Let us then abstract first from the MS. Let Y denote the transposed matrix of the first five rows and five columns of Table II; let f denote the column vector of the new gross energy equivalents $(f_0, f_1, f_2, f_3, f_4)$ and e the column vector $(0, e_1, 0, 0, 0)$. As before, we have

$$Yf = e, \quad (22)$$

which yields

$$e_1 = f_1 x_{15} + f_2 x_{25} + f_3 x_{35}, \quad (23)$$

37. In no actual economy is $P_K = 0$. Moreover, where institutionally $P_L = 0$, differential rent still gives rise to income transfers (illicit, naturally).

38. Should W be attributed a market price, p_w, the new prices, p', will satisfy analogous relations to those of note 35, above.

with

$$f_i = e_1 \Delta_{i1} / \Delta , \tag{24}$$

where Δ is the determinant of Y and Δ_{i1} the minor of the element of subscripts $(i, 1)$.

For gross matter equivalents we shall abstract from the *ES*. If g denotes the column vector of these equivalents and m the column vector $(M_0, 0, 0, 0, 0)$, then

$$Yg = m \tag{25}$$

from which we obtain

$$M_0 = g_1 x_{15} + g_2 x_{25} + g_3 x_{35} , \tag{26}$$

with

$$g_i = M_0 \Delta_{i0} / \Delta . \tag{27}$$

The corresponding formulae for the net energy follow straightforwardly:

$$\textit{Net energy} = e_1 / f_1 , \qquad \textit{Net matter} = M_0 / g_0 . \tag{28}$$

The conclusion is that to deliver a marginal unit of, say, C to the ultimate consumer we must consume f_2 units of energy *in situ* as well as g_2 units of matter *in situ*.

The upshot of the foregoing considerations is that, whatever the source of the energy used (solar or terrestrial), we must not ignore the depletion of the terrestrial deposits of available matter caused by any productive process. For all practical purposes, the Earth is a closed thermodynamic system, notwithstanding the meteorite fall and the material particles that occasionally escape the gravitation pull. In the very long run, therefore, some material elements will become more critical than energy for an industrial system of the prevailing sort [30; 32]. An increasing number of natural scientists have recently become convinced of this point and now even insist that a number of important elements are already approaching the pressing limit of scarcity [17; 69]. But sad to say, instead of following the old commandment and "beat our swords into plowshares," we keep on forging the plowshares of future generations into our present dreadful "swords".

Let us remember that energy and matter in bulk are not convertible into each other (Sec. VI), i.e., there is no relation $F(M, e) = \text{const}$. We do not have, therefore, a grid of isoquants which could enable us to reduce the economic choice pertaining to natural resources to physico-chemical calculations. Take the case of two technologies, $T_1(M_0^1, e_1^1)$ and $T_2(M_0^2, e_1^2)$, producing the same result and such that $M_0^1 > M_0^2$, $e_1^1 < e_1^2$. If they use terrestial resources, no proposition of physics or chemistry can tell us which technology is *economically* preferable. The nature of the issue is purely economic for it involves a multitude of factors of varying historical uncertainties and imponderabilities.

Since matter also matters, it would be misleading to reduce economic choice to energy alone. Actually, in some cases it is only matter that counts. Assume that the above technologies use "free" solar energy. The choice must now take into account net energy, *NE*, instead of gross energy. And how to choose between $T_1(M_0^1, NE_1)$ and $T_2(M_0^2, NE^2)$ when $M_0^1 > M_0^2$ and $NE^1 < NE^2$ is again an economic, not a purely technical problem. However, if $NE^1 = NE^2$, matter is decisive, T_2 is preferable, regardless of how much gross energy it uses.

One factor (besides those mentioned earlier) that may explain why matter has been ignored by the modern analysis of entropic transformations is the bonanza of fossil fuels that began about two hundred years ago and is still going on. This bonanza has an immense double advantage. It takes relatively little matter to extract fossil fuels from where they lie in the bowels of the earth *and* even less to convert them into industrial thermal energy. The same is not true for nuclear energy, which requires large installations for refining, enriching and conversion. The difficulties that now stand in the way of the direct use of solar energy by the *presently known* recipes also stem from the immense requirements of matter. From all we can judge now, the necessary amount of matter for a technology varies according to the intensity of the energy used. It is high for weak-intensity energy (as is the solar radiation at ground level) because such energy must be concentrated into a much higher intensity if it is to support the intensive industrial processes as those now supported by fossil fuels. And it is high for high-intensity energy because such energy must be contained (besides being "sifted" first).

VIII. Global Analysis and Technology Assessment

Nowadays we hear over and over again that nothing should stop us from using a technology based on solar energy because solar energy "is free, after all" [19]. However, absolutely every environmental energy *is* free, in the sense that nature does not have a check-out counter at which we should pay for resources *in situ*. Money royalties are instituted by man, not by nature. Perhaps by saying that solar energy is free we simply mean that it is "extremely abundant". And it is indeed abundant: its annual flow reaching the upper atmosphere is some twelve thousand times greater than the world's current consumption of energy from all sources! Unfortunately, abundance of energy *in situ* alone is not necessarily an advantage. This is the case of solar energy, which also has the great shortcoming of being extremely weak when it reaches us.

The direct use of solar energy constitutes now such a highly hopeful topic,[39] that it should be instructive to assess such a technology on the basis

39. E.g., *Congressional Record: Senate*, 31 July 1975 and 10 December 1975; Sylvia Potter, "We

of the ideas developed in this paper. Let us begin by recalling the necessary distinction between "recipe" and "technology" as well as the fact that *feasible* recipes do not necessarily constitute a *viable* technology (Sec. II). Legions of successful experiments represent feasible recipes that *at present* cannot be part of a viable technology. A striking example is the feasible recipe for quarring the Moon, which could not possibly replace now the mining of the Earth. Further, let us confine our argument to a technology based on those recipes for harnessing solar energy directly *that are known to work at present*.[40] We may safely include under the term "collectors" any devices used by these recipes.

For simplicity, we may divide the whole system into only three individual processes: (P_1) produces collected solar energy, SE, with the aid of collectors, CL, and some capital equipment, K; (P_2) produces collectors with the aid of SE and K; (P_3) produces K from mineral deposits by using SE (Table III).[41] Obviously,

$$x_{21} = x_{22}, \tag{29}$$

since collectors have no use outside (P_1).

It is legitimate to assume that all recipes (P_i) are feasible (P_1) is certainly feasible. Collectors also are currently produced, although by other sources of energy—mainly, by fossil fuels energy, FE. The same is true of K. But since energy is a homogeneous "stuff," fossil fuel energy could very well be replaced by collected solar radiation. The only possible snag may concern intensity. Unfortunately, the intensity of energy (expressed by dQ/dt) is another aspect—besides matter—ignored by thermodynamics, which cannot therefore cast any light on the issue.[42] However, we should not overlook the fact that proven recipes can raise the temperature of collected solar radiation to quite impressive levels. The solar oven Odeillo (in the Pyrenees) produces a temperature near to 4,000° centigrade, and has a power of about 65 kW. If we wanted a greater power, we may, conceivably, build as many Odeillos as needed. A solar power plant, however, requires an immense and elaborate installation. The plant projected by ERDA at Barston (California) involves not less than 1,700 polished mirrors, each of 400 square feet—a total of some 18 acres—moved by a very complicated mechanism that turns them so as to follow the sun while focusing on the boiler exactly and continuously. Yet its power is a modest 10 MW.

Stand on Threshold of Solar Energy Era," *News and Observer* (Raleigh, N.C.), 16 September 1975; "European Common Market Heats Up on Solar Energy," *International Herald Tribune*, 25 July 1977.

40. Accordingly, we do not consider the suggestion for collecting solar radiation in outerspace and transmitting the collected energy to the ground level. The recipe is still unproven.

41. Let us note that (P_1) must include a fund of CL besides the fund of K.

42. According to the standard formula $Q = W$, one should be able to send a rocket to the moon by spreading the necessary Q over time so thinly that it would amount to lighting up one match after another. The omission explains why many writers are puzzled by the paradox that solar energy, immensely abundant though it is, seems so very hard to be used directly for our *present industrial* needs.

For the technology depicted by Table III to be viable, we must have

$$y_1 = x_{11} - x_{12} - x_{12} > 0, \qquad y_3 = -x_{31} - x_{32} + x_{33} > 0, \tag{30}$$

where y_1 and y_2 represent the flows necessary for the *maintenance* of the corresponding fund factors (people and fixed capital).

The underlined term should be clearly understood. A viable technology is like a viable species, that is, once sprung from a previous technology all it needs to do is to maintain itself. To wit, the first bronze hammers were hammered by stone hammers, but during the succeeding age all bronze hammers were hammered by bronze hammers. Undoubtedly, to hammer a bronze hammer by stone hammers was a more demanding task than to hammer a stone hammer. But the threshold from an old to a new technology can be surmounted only by an additional cost *at the old prices*. If we ignore this point, we cannot see the weakness of the argument that solar technology is

Table III. Technology Based on Solar Energy

Product	(P_1)	(P_2)	(P_3)	Net flows
SE	x_{11}	$-x_{12}$	$-x_{13}$	y_1
C	$-x_{21}$	x_{22}	*	*
K	$-x_{31}$	$-x_{32}$	x_{33}	y_3

viable and the only thing necessary to bring it about is to find out how to produce collectors profitably. As the example of the Bronze Age shows, the viability of a technology requires only that its material scaffold be self-supporting, regardless.

Let us now consider the problem of prices. If (30) is fulfilled, there exists a price system that makes the system work. That is, if X denotes now the transposed matrix of Table III and p is the column vector (p_1, p_2, p_3), the system

$$Xp = B \tag{31}$$

has a positive solution, B being defined as in (20). Therefore, if no such solution exists for $B > 0$, the technology is not viable. But, curiously, the converse is not true: (30) may have a positive solution without the technology being viable.[43] On the other hand, the brute fact that we are not already living in a technology based on solar radiation does not prove that such a

43. See *Mathematical Note* at the end.

technology is not viable. It may very well be less efficient than the fossil fuel technology in current monetary terms or in terms of human effort. The issue is quite involved. However, other absences prevail over this inconclusive standpoint.

The most accepted explanation of why solar energy has not yet replaced other sources of power is that the necessary collectors cost too much. Save for this remediable situation, a solar technology capable of sustaining the modern industrial activity is in fact viable. The difficulty is "a cost problem, not a material problem" [64, 17]. But if the only obstacle were the money deficits incurred by solar recipes, one question would cry for an answer. Over the past five years, at least, hundreds of millions of dollars have been spent for the development of more efficient recipes. ERDA, in particular, has interspersed this country with numberless model homes and experimental windmills. Yet no breakthrough has taken place to increase the confidence in a viable solar technology. None of the richly endowed research outfits have achieved even a workable pilot of a combined (P_1) and (P_2) working in mutual support, let alone a pilot of full-fledged solar technology to prove its viability *independent of prices*. For when it comes to prove that a technical idea is workable, cost does not matter much. Otherwise, we would not have succeeded in proving that we can put a man on the Moon.

The obvious upshot is that at present it is not possible to produce collectors only by the solar energy they collect. Any use of a presently feasible recipe based on solar collectors, therefore, is a parasite of the current technology. And, like any parasite, it could not survive its host [38; 39; 40; 41].

This means that instead of (30), we have[44]

$$x_{11} < x_{12}, \qquad x_{11} < x_{13}, \tag{31}$$

even if we weaken the other condition to

$$-x_{31} - x_{32} + x_{33} = 0. \tag{32}$$

For a global analysis of the way solar energy is now harnessed, we must consider the flow matrix of Table IV, in which (29) and (31) are assumed to hold. The energy necessary to produce collectors by (P_2) now comes from a non-solar (fossil fuels) power plant $(P_4{}^0)$, which also provides the energy for the production of capital equipment by a new process, $(P_3{}^0)$. The argument will be even more instructive if we assume that the only net flow is x_{11}.

From the fact that $y_{33} > x_{33} = x_{31} + x_{22}$, it reasonably follows that $y_{43} > x_{13}$. Hence, $y_{44} = x_{12} + y_{43} > x_{12} + x_{13}$, and by (31)

$$y_{44} > 2x_{11}. \tag{33}$$

44. The inequality $x_{11} < x_{13}$ is a foregone result of the other inequality, given just the enormous amounts of heat needed to produce metals from ores.

Table IV. The Present Mixed Technology

Product	(P_1)	(P_2)	(P_3^0)	(P_4^0)	Net flows
SE	x_{11}	*	*	*	x_{11}
CL	$-x_{21}$	x_{22}	*	*	*
K	$-x_{31}$	$-x_{32}$	y_{33}	$-y_{34}$	*
FE	*	$-x_{12}$	$-y_{43}$	y_{44}	*

This proves that not only is (P_1) a parasite of fossil fuels, but that *globally the recipe consumes twice as much of the other energy than its net output.*[45]

A further result will clarify the real meaning of non-profitable cost. Since the mixed technology represented by Table IV effectively produces a net output, a system of positive prices exists. If we consolidate the budget equations of (P_1) and (P_2), we obtain

$$B_1 + B_2 + p_3(x_{31} + x_{32}) = p_1x_1 - p_4x_2. \tag{34}$$

Because of the first inequality (31), it follows that $p_1 > p_4$. In other words, in the mixed technology, the price of a BTU obtained from solar energy is higher than that obtained from fossil fuels.[46] It is thus seen that the fact that producing one BTU from solar energy is at present nonprofitable is not due to price; instead, it reflects a hidden waste of the solar recipe.

A successful pilot combination of (P_1) and (P_2) that needs only matter from the outside would constitute a significant achievement, but still not completely conclusive. This combination, which is represented by Table V, still requires process like (P_3^0) and (P_4^0). Instead of (31), we have

$$x_{11} > x_{12}, \qquad x_{11} < x_{13}. \tag{35}$$

Since even in this case, $y_{33} > x_{33}$, we obtain, as before,

$$y_{44} > x_{13} > x_{11} - x_{12}, \tag{36}$$

with the same result concerning the global energy deficit [41]. In this case, however, it does not seem necessary for this deficit to lead to $p_1 > p_4$.[47]

45. Even if we believed that because of the superior intensity of *FE*, y_{43} may not be necessarily greater than x_{13}, (33) would be replaced by a weaker, but still relevant inequality.

46. The fact that people neverhtheless buy home solar installations should not surprise us. People heat their homes with electricity even though this mode consumes more energy than using coal, for example. Also, people buy electric gadgets without necessarily deriving an amount of energy from them comparable with that which was used in the production of those gadgets.

47. Perhaps this result may not stand further probing. For it appears highly curious indeed for a solar BTU depending indirectly on fossil fuels to cost less than an input BTU.

Table V. Collectors Produced by Solar Energy

Product	(P_1)	(P_2)	(P_3^0)	(P_4^0)	Net flows
SE	x_{11}	$-x_{12}$	*	*	$x_{11} - x_{12}$
CL	$-x_{21}$	x_{22}	*	*	*
K	$-x_{31}$	$-x_{32}$	y_{33}	$-y_{34}$	*
FE	*	*	$-y_{43}$	y_{44}	*

The statement by Denis Hayes, that "we can use solar energy now [because] the technology is here" [44], probably reflects the overenthusiasm of a keen student of the problems of energy scarcity. The truth is that only the feasible recipes are here, just as numberless other such recipes are. But a viable technology is not.

The discovery of far more efficient recipes may change the picture radically. However, harnessing solar radiation is not a problem that has just recently confronted us, as was the case of the peaceful use of nuclear energy some forty years ago. At that time, one could have easily gone wrong about the possibility of using the newly discovered energy; none other than Lord Rutherford did. But solar collectors have been used on a substantial scale for almost one hundred years now, and during all this time there has been practically no true breakthrough [12]. Undoubtedly, the Sun is the only steady and completely healthy source of energy whether for a new "Wood Age" or some other kind of solar age. At present, however, it seems very unlikely however that it will enable mankind to fly in jets, to live in skyscrapers, and to travel in automobiles at one hundred kilometers per hours, for example.

To try persistently to find more efficient recipes is not only legitimate, it is imperative. But to claim that solar technology is here before it actually is, or to preach that "come what may we shall find a way" will only conceal from the public awareness the seriousness of the acute problem of natural resources and thus render any move toward an adequate policy to deal with that problem far more difficult than it actually is.

Mathematical Note

Let X be the matrix of Table III and let the prime accent denote inversion. By (30), the system

$$Xs' = w \tag{37}$$

has a positive solution $s = (1, 1, 1)$ for $w = y \geqslant 0$.[48] By Theorem 5 [29, 324–25], it has a solution $s > 0$ for any $w > 0$. Hence $|X| = 0$,[49] and by Theorem 4 [29, 323], the system

$$X'p' = B', \tag{38}$$

where $B = (B_1, B_2, B_3) > 0$, has a solution $p > 0$. This proves that for any viable technology and for any fund prices, there exists a set of positive prices for the flow elements.

Let us now assume that (38) has a positive solution. From Theorem 4, again, it follows that

$$X\lambda' = z'$$

has a solution $\lambda > 0$ for any $z > 0$. Hence, $\Omega \in \Gamma$, where Ω is the non-negative quadrant and Γ is the convey cone determined by (P_1), (P_2), and (P_3). Unless $X = I$, there must be some w such that $w \in \Gamma$ but $w \in \Omega$. Nothing warrants that this is not the case for y. For a simple example:

$$X = \begin{bmatrix} 4 & -2 & -3 \\ -1 & +1 & 0 \\ -1 & -2 & 5 \end{bmatrix}$$

48. The vector notation $a \geqslant$ b excludes the of case $a = b$.

49. Direct computation yields $|X| = x_{22}y_1(x_{31} + x_{32}) + x_{22}y_3(x_{11} - x_{12}) > 0$. Incidentally, the number of conditions of Theorem 7 [29, 326, 336] can be further reduced by one unit; if the minor of third order—in the precent case, $|X|$—is positive, so must the minor of second order be.

References

1. Alfvén, Hannes. *Atom, Man and the Universe*. San Francisco: H. F. Freeman, 1969.

2. Allen, C. W. *Astrophysical Quantitics*, 3rd ed. London: Athlone, 1973.

3. Auer, Peter L. "Does Entropy Production Limit Economic Growth?" in *Prospects for Growth: Changing Expectations for the Future*, edited by K. D. Wilson. New York: Praeger, 1977, pp. 314–34.

4. Boulding, Kenneth E. "The Economics of the Coming Spaceship Earth," in *Environmental Quality in a Growing Economy*, edited by H. Jarrett. Baltimore: Johns Hopkins University Press, 1966, pp. 3–14.

5. ———. "The Great Laws of Change," in *Evolution, Welfare and Time in Economics*, edited by A. M. Tang, F. M. Westfield and J. S. Worley. Lexington, Mass.: D. C. Heath, 1976, pp. 3–14.

6. ———, "Energy Policy: A Piece of Cake." *Technology Review*, December 1977, 8.

7. Boserup, Mogens, "Chairman's Report on Specialized Session III." Plenary Session, Fifth World Congress of the International Economic Association, Tokyo, 29 August–3 September 1977. (*Proceedings* forthcoming.)

8. Bridgman, P. W. *The Logic of Modern Physics*. New York: Macmillan, 1927.

9. Brooks, David P. and P. W. Andrews, "Mineral Resources, Economic Growth, and World Population." *Science*, 5 July 1974, 13–19.

10. Brown, Harrison. *The Challenge of Man's Future*. New York: Viking Press, 1954.

11. ———, James Bonner, and John Weir. *The Next Hundred Years*. New York: Viking Press, 1957.

12. Butti, Ken and John Perlin, "Solar Water Heaters in California, 1891–1930." *CoEvolution Quarterly*, Fall 1977, 4–13.

13. Chapman, P. F., "Energy Costs: A Review of Methods." *Energy Policy*, June 1974, 91–103.

14. ______, "The Energy Costs of Materials." *Energy Policy*, March 1975, 47–57.

15. ______, "The Ins and Outs of Nuclear Energy." *New Scientist*, December 1977, 866–69.

16. ______, G. Leach and M. Slesser, "The Energy Cost of Fossil Fuels." *Energy Policy*, September 1974, 231–43.

17. Chynoweth, A. G., "Materials Conservation: A Technologist's Viewpoint." *Challenge*, January-February 1976, 34–42.

18. Cloud, Preston. "Realities of Mineral Distribution," in *Man and His Physical Environment*, 2nd ed., edited by G. D. McKenzie and R. O. Utgard. Minneapolis: Burgess, 1974, pp. 185–98.

19. Committee for Economic Development (CED). *Key Elements to a National Energy Strategy*. New York, 1977.

20. Committee on Mineral Resources and the Environment (COMRATE). *Mineral Resources and the Environment*. Washington, D.C., 1975.

21. Cook, Peter, "Mineral Resources, Economic Growth, and World Population," *Science*, 5 July 1974, 13–19.

22. Cottrell, Fred. *Energy and Society*. New York: McGraw Hill, 1953.

23. Daly, Herman. "The Economic Thought of Frederick Soddy." Unpublished essay, 1978.

24. Denbigh, Kenneth. *Principles of Chemical Equilibrium*, 3rd ed. Cambridge, England: University Press, 1971.

25. Ehrenfest, Paul and Tatiana. *The Conceptual Foundations of the Statistical Approach in Mechanics*. Ithaca: Cornell University Press, 1959.

26. Energy Research and Development Agency (ERDA). *A National Plan for Energy Research, Development, and Demonstration: Creating Energy Sources for the Future*. Washington, D.C., 1975.

27. Feyman, R. P., R. B. Leighton and M. Sands. *The Feyman Lectures on Physics*, vol. I. Readings, Mass.: Addison-Wesley, 1966.

28. Flawn, Peter T. *Mineral Resources: Geology, Engineering, Economics, Politics*. Chicago: Rand McNally, 1966.

29. Georgescu-Roegen, Nicholas. *Analytical Economics: Issues and Problems*. Cambridge, Mass.: Harvard University Press, 1966.

30. ______. *The Entropy Law and the Economic Process*. Cambridge, Mass.: Harvard University Press, 1971.

31. ______, "A Different Economic Perspective." Paper read at the Boston Meeting of the American Association for the Advancement of Science, 21 February 1976.

32. ______. *Energy and Economic Myths: Institutional and Analytical Economic Essays*. New York: Pergamon Press, 1976.

33. ______, "Is Perpetual Motion of the Third Kind Possible?" Paper read at the Colloquium of the *Ecole Nationale Supérieure de Transportation*, Paris, 19 November 1976.

34. ______. "Economics and Mankind's Ecological Problem," in *U.S. Economic Growth from 1976 to 1986: Prospects, Problems, and Patterns*, Joint Economic Committee, Congress of the United States, Washington, D.C., vol. 7, pp. 62–91.

35. ______. "Bioeconomics: A New Look at the Nature of the Economic Activity," in *The Political Economy of Food and Energy*, edited by Louis Junker. Ann Arbor: University of Michigan, 1977, pp. 105–34.

36. ______, "The Steady State and Ecological Salvation: A Thermodynamic Analysis," *BioScience*, April 1977, 266–70.

37. ______. "Matter Matters, Too," in *Prospects for Growth: Changing Expectations for the Future*, edited by K. D. Wilson. New York: Praeger, 1977, pp. 293–313.

38. ______. "The Role of Matter in the Substitution of Energies" (Third International Colloquium on Petroleum Economics, Québec, 3-5 November 1977) ni *Resources énergétiques et coopération internationale*. Québec: Presses de l'Université Laval, 1978 (forthcoming).

39. ______, "Myths about Energy and Matter." *Growth and Change*, January 1979, pp. 16–23.

40. ______. "Matter: A Resource Ignored by Thermodynamics," in *Proceedings of the World Conference on Future Sources of Organic Materials* (Toronto, 10–13 July 1978), Toronto: Pergammon Press, 1979 (forthcoming).

41. ______, "Technology Assessment: The Case of the Direct Use of Solar Energy." *Atlantic Economic Journal*, December 1978, pp. 15–21.

42. Gilliland, Martha W., "Energy Analysis and Public Policy." *Science*, 26 September 1975, 1051–56, and 2 April 1976, 8–12.

43. Hayek, F. A. *The Counter-Revolution in Science*. Glencoe, Ill.: The Free Press, 1952.

44. Hayes, Denis, "We Can Use Solar Energy Now." *Washington Post*, 26 February 1978, D1–D4.

45. Hiebert, Erwin H. "The Energetics and the New Thermodynamics," in *Perspectives in the History of Science and Technology*, edited by Duanne H. D. Roller. Norman: University of Oklahoma Press, 1971, pp. 67–86.

46. Huettner, David A., "Net Energy Analysis: An Economic Assessment." *Science*, 9 April 1976, 101–104.

47. Kenward, Michael, "The Analyst's Precedent." *New Scientist*, 9 January 1975, 51.

48. Kirkwood, John G. and Irwin Oppenheim. *Chemical Thermodynamics*. New York: McGraw-Hill, 1961.

49. Leach, Gerald, "Energy Analysis." *New Scientist*, 16 January 1975, 160.

50. Lichnerowicz, Marc, "Economie et thermodynamique: Un modèle d'échange économique." *Economies et Sociétés*, October 1971, 1641–86.

51. Nash, Robert, "The Future of Wilderness: A Problem Statement." *Bulletin of the American Academy of Arts and Sciences*, May 1978, 18–24.

52. Neumann, John von. "The General and Logical Theory of Automata," in *Cerebral Mechanisms in Behavior: The Hixon Symposium*, edited by L. A. Jeffress. New York: Wiley, 1951, pp. 1–31.

53. Odum T. H., "Energy, Ecology, and Economics." *Ambio*, 1973, No. 6, 220–27.

54. ———, "Energy Analysis." *Science*, 15 April 1977, 260.

55. Ostwald, Wilhelm. *Die Energie*. Leipzig: J. A. Barth, 1908.

56. Page, Norman G. and S. C. Creasey, "Ore Grade, Metal Production, and Energy." *Journal of Research, U.S. Geological Survey*, January-February 1975, 9–13.

57. Planck, Max. *Theory of Heat*. London: Macmillan, 1932.

58. ———. *Treatise on Thermodynamics*, 7th ed. New York: Dover, 1945.

59. Price, John H. *Dynamic Energy Analysis and Nuclear Power*. London: Friends of the Earth, 1974.

60. Prigogine, Ilya. "Time, Structure and Entropy," in *Time in Science and Philosophy*, edited by Jiří Zeman. Amsterdam: Elsevier, 1971, pp. 89–100.

61. ———, "Irreversibility as a Symmetry-breaking Process." *Nature*, 9 November 1973, 67–71.

62. ———, C. George, F. Henin and L. Rosenfield, "A Unified Formulation of Dynamics and Thermodynamics." *Chemica Scripta*, 1973, No. 1, 5–32.

63. Rabinowicz, Ernest. *Friction and Wear of Materials*. New York: Wiley, 1965.

64. Rose, David J. "Materials Requirements for Emerging Energy Technology," in [20], Appendix to Section I, D.15–D.23.

65. Seaborg, Glenn T. "The Erehwon Machine: Possibilities for Reconciling Goals by Way of New Technology," in *Energy, Economic Growth and the Environment*, edited by Sam H. Schurr. Baltimore: Johns Hopkins University Press, 1972, pp. 125–38.

66. Seeger, Raymond J. *Men of Physics: J. Willard Gibbs*. New York: Pergamon Press, 1974.

67. Silver, R. S. *Introduction to Thermodynamics*. Cambridge, England: University Press, 1971.

68. Skinner, Brian J. *Earth Resources*, Englewood Cliffs, N.J.: Prentice-Hall, 1969.

69. ———, "A Second Iron Age Ahead?" *American Scientist*, May-June 1976, 258–69.

70. Slesser, Malcom, "Accounting for Energy." *Nature*, 20 March 1975, 170–72.

71. ———, "Energy Analysis." *Science*, 15 April 1977, 259–60.

72. Solow, Robert M., "Is the End of the World at Hand?" *Challenge*, March-April 1973, 39–50.

73. Tayler, R. J. *The Origin of the Chemical Elements*. London: Wykeham, 1972.

74. Thomson, Sir William (Lord Kelvin). *Mathematical and Physical Papers*, vol. I. Cambridge, England: University Press, 1881.

75. Weinberg, Steven. *The First Three Minutes*. New York: Basic Books, 1977.

76. Woodwell, G. M., R. H. Whittacker, W. A. Reiners, G. E. Likens, C. C. Delwiche and D. B. Botkin, "The Biota and the World Carbon Budget." *Science*, 13 January 1978, 141–46.

77. Wright, David J., "Goods and Services: An Input-Output Analysis." *Energy Policy*, December 1974, 307–14.

78. Zemansky, Mark W. *Heat and Thermodynamics*, 5th ed. New York: McGraw-Hill, 1968.

[20]

Energy and the U.S. Economy: A Biophysical Perspective

Cutler J. Cleveland, Robert Costanza
Charles A. S. Hall, Robert Kaufmann

Summary. A series of hypotheses is presented about the relation of national energy use to national economic activity (both time series and cross-sectional) which offer a different perspective from standard economics for the assessment of historical and current economic events. The analysis incorporates nearly 100 years of time series data and 3 years of cross-sectional data on 87 sectors of the United States economy. Gross national product, labor productivity, and price levels are all correlated closely with various aspects of energy use, and these correlations are improved when corrections are made for energy quality. A large portion of the apparent increase in U.S. energy efficiency has been due to our ability to expand the relative use of high-quality fuels such as petroleum and electricity, and also to relative shifts in fuel use between sectors of the economy. The concept of energy return on investment is introduced as a major driving force in our economy, and data are provided which show a marked decline in energy return on investment for all our principal fuels in recent decades. Future economic growth will depend largely on the net energy yield of alternative fuel sources, and some standard economic models may need to be modified to account for the biophysical constraints on human economic activity.

Stable consumer prices, full employment, and increasing per capita wealth have been economic and political goals in the United States since at least the 1930's. Aggregate economic growth has been the principal means for realizing these goals. On average, these goals were met from the mid-1940's to the early 1970's, when the U.S. economy grew at an average annual rate of 4 percent, recessions were relatively short and mild, and inflation rates rarely exceeded 4 percent. Since 1973, however, the United States and other Western nations have experienced irregular and even negative economic growth rates together with high unemployment, unprecedented inflation and budget deficits, and declining productivity rates.

These events seem to defy explanation by or even to contradict some of the most fundamental economic models that guided the prosperity of the preceding 40 years. A number of analysts have commented on the difficulties these models now encounter. Drucker (*1*) stated that "both as economic theory and as economic policy Keynesian economics is in disarray." Leontief (*2*) described many economic models as unable "to advance, in any perceptible way, a systematic understanding of the structure and the operations of a real economic system." Instead, they are based on "sets of more or less plausible but entirely arbitrary assumptions" leading to "precisely stated but irrelevant theoretical conclusions." Bailey and others (*3*) chronicled the failure, mutual conflicts, and frustrations of a number of economic models.

Glassman (*4*), responding to Leontief, suggested that greater diversity in economic theory is needed to supplement the conditioned expectations of formal theory. We agree, and present a different theoretical perspective for analyzing economic production based on relatively simple models that begin with the importance of natural resources, and fuel energy in particular. Our intent is not to replace standard economic models, nor do our models offer solutions for all the economic problems described above. Rather, our perspective, which in part has been presented by others (*5*), shows how some economic problems can be understood more clearly by explicitly accounting for the physical constraints imposed on economic production.

We examine the historical record of the last 90 years to test the hypotheses generated by our model. Empirical testing of economic theories is a difficult but essential procedure which is too frequently ignored. Simultaneous changes in variables make controlled observations difficult if not impossible. The empirical analyses of time series and cross-sectional data presented below cannot be used to prove hypotheses unequivocally, nor do they assure that the parameters will not change in new ways in the future. Empirical assessments, however, can be used to identify hypotheses that are consistent with reality and to reject hypotheses that are not.

Statement of Hypotheses

We approach macroeconomics from a thermodynamic perspective that emphasizes the production of goods, rather than the neoclassical perspective that emphasizes the exchange of goods according to subjective human preferences. Production is the economic process that upgrades the organizational state of matter into lower entropy goods and services. Those commodities are allocated according to human wants, needs, and ability to pay. Upgrading matter during the production process involves a unidirectional, one-time throughput of low entropy fuel that is eventually lost (for economic purposes) as waste heat. Production is explicitly a work process during which materials are concentrated, refined, and otherwise transformed. Like any work process, production uses and depends on the availability of free energy. The laws of energy and matter control the availability, rate, and efficiency of energy and matter use in the economy and therefore are essential to a comprehensive and accurate analysis of economic production.

Changes in natural resource quality affect the ease and cost of fuel and matter throughput in human economies because lower quality resources nearly always require more work directly and indirectly to upgrade them into goods and services. Technological change can counter changes in natural resource quality to varying degrees, but historically, many technical advances that have lowered unit labor costs have been realized by

Cutler Cleveland is a graduate research assistant in the Department of Geography, University of Illinois, Urbana 61801. Robert Costanza is an associate professor in the Coastal Ecology Laboratory, Center for Wetland Resources, Louisiana State University, Baton Rouge 70803. Charles Hall is an assistant professor in the Section of Ecology and Systematics, Cornell University, Ithaca, New York 14853. Robert Kaufmann is a research scientist in the Complex Systems Research Center, University of New Hampshire, Durham 03824.

increasing the quantity of fuel used directly and indirectly to perform a specific task. The degree to which technological change can offset declining resource quality as some basic natural resources are depleted (for example, fuel and metal ores) and/or mismanaged (some biotic resources) is an empirical question and cannot be easily predicted. Nevertheless, such resource changes have important implications for what is and is not possible in the economy. Economic theory and policy must incorporate the physical properties of resources if economic predictions are to be accurate and economic policies effective.

In the section below we present a series of specific hypotheses derived from our biophysical perspective, accompanied by an alternative example of a more traditional hypothesis. In our empirical assessments of how changes in fuel quantity and quality have affected the U.S. economy, we examine (i) various relations between fuel use, economic output, productivity, and inflation over the past 90 years and (ii) cross-sectional relations between direct and indirect fuel use and economic value for 1963, 1967, and 1972. We use real gross national product (GNP) as a measure of aggregate output in the time series analysis, while acknowledging some of the inadequacies of GNP as a measure of output and social welfare. As a measure of fuel use we sum the quantities of fossil, nuclear, and hydropower fuels used in the economy and analyze the effects of changing fuel mix by adjusting caloric heat measures for fuel quality. Except as indicated, nuclear and hydropower fuels are converted to heat equivalents based on the prevailing heat rate at fossil steam electric plants.

1) A strong link between fuel use and economic output exists and will continue to exist, both temporally and cross sectionally. The correlation is strengthened when adjustments are made for fuel quality and the sector in which fuel is combusted. Alternative hypotheses are that such a link never existed, or that it can be and has been substantially decoupled, especially as the price of fuel increases (*6*).

2) A large component of increased labor productivity over the past 70 years resulted from increasing the ability of human labor to do physical work by empowering workers with increasing quantities of fuel, both directly and as embodied in our industrial capital equipment and technologies. One alternative hypothesis views productivity as an exogenous technical driving force that has increased the productivity of capital and labor (*7*).

3) Changes in the general price level have been correlated with changes in the money supply relative to the physical supply of energy. This suggests that the rising real physical cost of obtaining energy and other resources from the environment is one important factor in inflation. Various economic models emphasize either monetary or fiscal measures for explaining inflation (*8*).

4) Energy costs of locating, extracting, and refining fuel and other resources from the environment have increased and will continue to increase despite technical improvements in the extractive sector. This reduces the supply of nonenergy goods producible from a given quantity of energy. One alternative hypothesis is that resource-augmenting technical change and/or the development of inexhaustible fuel supply systems will mitigate any foreseeable natural resource scarcity (*9*).

Our hypotheses are not necessarily inimical to standard economics. Rather, we believe such an approach provides a physical basis for some macroeconomic

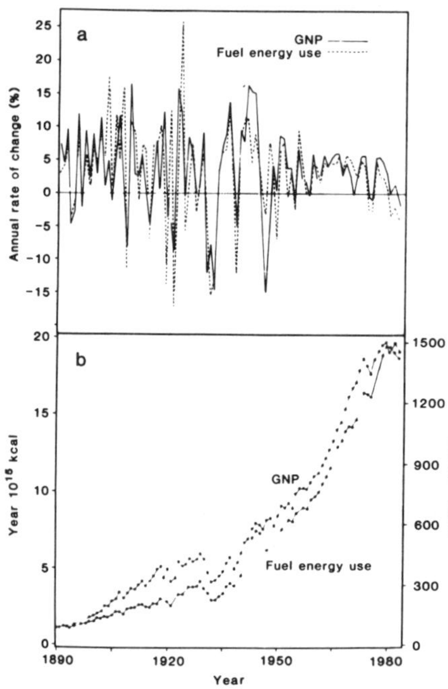

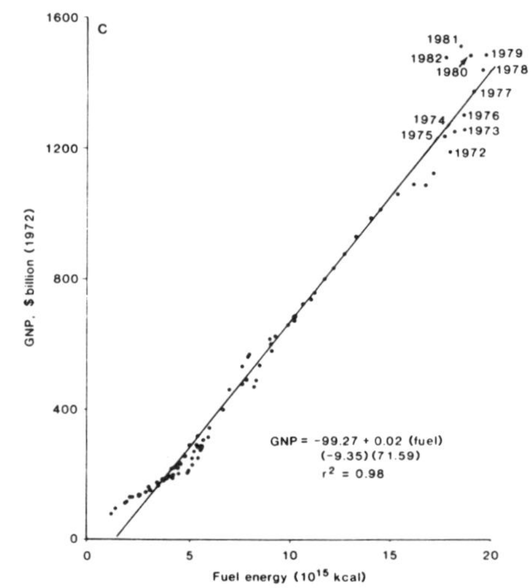

Fig. 1. (a) Annual rate of change in fuel use and real GNP in the United States from 1890 to 1982 (*55, 56*). Fuel use includes fossil fuels, nuclear, and hydropower. (b) Fuel use and real GNP per year. (c) Results of linear regression model between fuel use and real GNP in the United States from 1890 to 1982. The numbers in parentheses are *t*-statistics. Hydro and nuclear power converted to thermal units based on prevailing heat rates at fossil steam electric plants (*55, 56*).

phenomena. Such an analysis leads neither to an unrealistic cornucopian view of our future material condition, nor to one of "gloom and doom." We believe a physical analysis of economic production provides realistic assessments of the problems we face and some of the needed characteristics of any plausible solution.

Energy and Economic Production

The economic process is frequently depicted in basic economic texts as a closed system in which the flow of output is "circular, self-feeding, and self-renewing" (*10*). This model is seriously incomplete. In reality, the human economy is an open system embedded in a global environment that depends on a continuous throughput of solar energy. The global system produces the environmental services, foodstuffs, fossil and atomic fuels derived from solar and radiation energies, and various other resources that are essential inputs to the human economy. The human economy uses fossil and other fuels to support and empower labor and to produce capital. Fuel, capital, and labor are then combined to upgrade natural resources to useful goods and services. Economic production can therefore be viewed as the process of upgrading matter into highly ordered (thermodynamically improbable) structures, both physical structures and information. Where one speaks of "adding value" at successive stages of production, one may also speak of "adding order" to matter through the use of free energy (*11*).

Fuel quality as well as quantity limits economic production because fuels differ in the amount of economic work they can do per unit heat equivalent (kilocalorie). Petroleum, for example, can perform a more versatile array of tasks and do many of them more efficiently than coal (*12*). Per kilocalorie, petroleum is estimated to be 1.3 to 2.45 times as valuable as coal (*13*). Similarly, electricity can be converted to mechanical and heat energy at the point of application and can be controlled precisely, reducing the heat equivalents required to perform many tasks (*14*). One measure of the quality of electrical energy is the opportunity cost of transforming fossil fuels to electricity (3 to 4 kilocalories of fossil fuel per kilocalorie of electricity in 1983).

Another important quality of fuels is the amount of energy required to locate, extract, and refine them to a socially useful state. This aspect of fuel quality is measured by a fuel's energy return on investment (EROI), which is the ratio of gross fuel extracted to economic energy required directly and indirectly to deliver the fuel to society in a useful form. As the EROI for fuel declines, the energy opportunity costs of securing additional amounts increase, and increasing amounts of already extracted energy must be diverted from the production of nonenergy goods to extract a given quantity of new fuel. Net energy is a more relevant measure of fuel supply than gross energy because it represents the energy available to produce final-demand goods and services. At an absolute minimum, the aggregate EROI for fuels must be greater than 1 for an economic system to function, and probably much greater for it to grow. *Ceteris paribus*, economies with access to higher quality natural resources, particularly fuels with higher EROI, can do more economic work than those with lower EROI fuel resources.

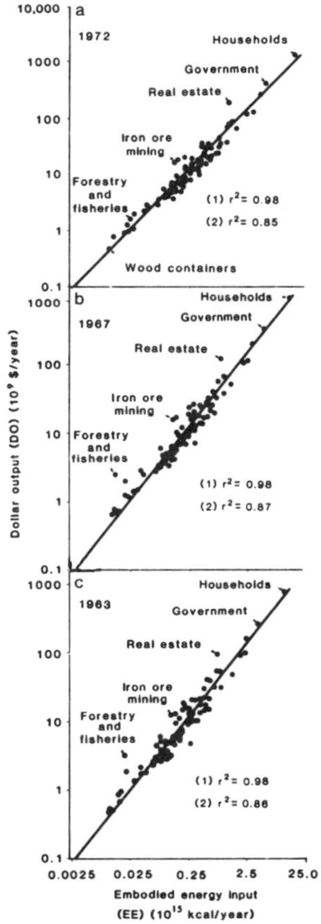

Fig. 2. Cross-sectional analysis of embodied energy inputs and dollar output for the U.S. economy in 1963, 1967, and 1972. Energy input includes direct fuel measured at the point of combustion and embodied in intermediate goods, labor, and government services purchased by each sector. For all three years r^2 = 0.98 if all points shown are included in the regression (option 1). Because households and government are relatively large sectors, including them in the regression may yield misleadingly high correlations. Excluding households and government from the regression yields r^2 values of about 0.86 (option 2). Excluding households and government from energy intensity calculations yields r^2 values of about 0.55 (*18, 24*).

Energy costs of capital and labor. Fuels, nonfuel minerals, capital, and labor are all necessary to produce economic output. Most standard models of production consider fuel and other natural resources to be qualitatively no different from other factors of production. As a result, many believe that natural resource inputs to production are "small potatoes compared to labor, or even to capital," and that "reproducible capital is a near perfect substitute for land and other exhaustible resources" (*15*). This view is inaccurate because free energy is required to upgrade and maintain all organized structures, including capital and laborers, against the ravages of entropy. It ignores the physical interdependence of capital, labor, and natural resources.

All goods and services (both economic and environmental) have quantifiable direct and indirect energy costs of production, termed their embodied energy. The embodied energy of a good or service can be calculated with input-output techniques developed by Herendeen and Bullard and by Hannon *et al.* (*16*), which were based on the pioneering economic work of Leontief (*17*). Early attempts to quantify the embodied energy of goods ignored the energy costs of labor, capital, and government services. These factors do have substantial energy costs. Labor consumes energy directly in the form of fuel and food, and indirectly as fuel energy embodied in shelter, clothing, education, and social services, and other commodities. These energy costs can be incorporated directly in calculations of embodied energy (*18*), or can be thought of as an energy opportunity cost (*19*) for labor, which is the amount of fuel that would have to be diverted from other uses to substitute for labor at the margin.

Standard production functions do not account for the important physical interdependence between energy and all other factors: the availability of all factors created by humans depends on the existence of free energy in the natural environment. Capital and labor are combined to extract energy from the environment, but they cannot create in a physical sense the free energy and matter from which they are derived. Thus, elasticities of substitution between natural resources and capital and labor calculated at the level of the firm or industry do not necessarily reflect true substitution possibilities over the economy as a whole. Including the direct and indirect energy costs of producing capital and labor reduces the degree to which capital and labor can be substituted for fuel in production (*20*).

Fuel use and economic output. Fuel use and economic output in the United States have been highly correlated for at least the past 90 years (*21*). This relation is shown in various ways in Figs. 1 and 2. The high coefficient of determination of Fig. 1c is consistent with the hypothesis that, at least in the past, economic output and fuel use have been tightly linked. While a causal relation from fuel use to GNP or vice versa cannot be verified, a strong contemporaneous link between the two variables is supported (*22*).

The results of statistical analyses of long time series of economic variables often are dominated by trends rather than correlations between annual variations in the variables. The validity of the statistical correlation of Fig. 1 was analyzed further with a Box-Jenkins transfer function analysis, a procedure that can remove the nonstationary components of the two time series. When this analysis is applied to the first differenced time series of real GNP and fuel use in the United States from 1900 to 1980, the results support a significant interrelation between the annual rates of change of GNP and fuel use (*23*).

Cross-sectional analysis of direct plus indirect fuel use and economic output in the United States reinforces the results of the time series analysis. Regression analysis of embodied fossil fuel, hydro, and nuclear energy use and dollar value of output across 87 sectors of the U.S. economy indicates a strong correlation between the two variables if the energy costs of labor and government services are included (*18*, *24*) (Fig. 2). This relation holds true for all the years for which the requisite national input-output tables are available.

Fuel efficiency. Despite problems inherent in measuring fuel quality, fuel use, and GNP as a measure of welfare (*25*), the fuel use/real GNP (E/GNP) ratio remains a popular and not inappropriate measure of fuel efficiency (*26*). The U.S. E/GNP ratio has declined 42 percent since 1929, about half of this since 1973 (Fig. 3a). This decline has been interpreted by some as meaning that factor substitution and conservation measures have decreased substantially the quantity of fuel used per unit of economic output and that similar improvements are possible in the future (*6*). We believe this interpretation of changes in the E/GNP ratio overestimates past improvements and potential future gains in actual energy efficiency.

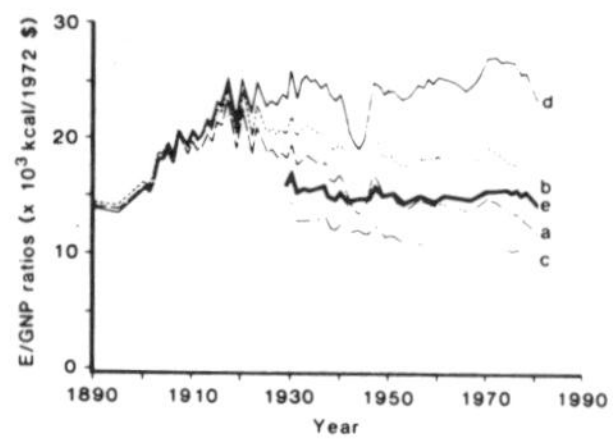

Fig. 3. Fuel/real GNP ratio for the United States. (a) No fuel or GNP modifications. (b) Fuel quality factors of 1.3 for petroleum, 1.0 for coal, and 4.0 for primary electricity and no GNP modifications for changes in household fuel use. (c) Fuel quality factors in (b) and GNP modified for effects of changes in household fuel use (*27*). (d) Fuel quality factors of 1.74 for petroleum, 0.92 for coal, 16.8 for electricity, and no GNP modifications. (e) Fuel quality factors as in (d) and GNP modified for changes in household fuel u e (*27*).

Figure 3 shows the sensitivity of the E/GNP ratio to corrections made for fuel quality and to GNP modifications to account for relative shifts in fuel use between sectors of the economy. Between 1929 and 1981, a period when the ratio was declining, three factors which are not normally considered as contributing to improved fuel efficiency account for 96 percent of the annual variation in the E/GNP ratio (*27*). The factors are (i) the proportion of total fuel use accounted for by petroleum, (ii) the proportion of total fuel use accounted for by primary electricity (hydro and nuclear), and (iii) the proportion of direct fuel use in final demand (that is, gasoline or electricity used by households) versus intermediate demand sectors (oil or electricity used in manufacturing).

An empirical examination of the relation between these factors indicates that 69 percent of the variation in the E/GNP ratio since 1929 can be attributed to changes in the type of fuel used. As the percentage of high-quality fuels such as petroleum and primary electricity increased, more economic work was done (more GNP produced) per heat equivalent burned, and the E/GNP ratio declined. Correcting fuel use data for changing fuel quality produces a smaller overall decline in the E/GNP ratio (Fig. 3, lines a, b, and d) and even a slight increase in the ratio if the quality factors derived from our regression analysis (*27*) are used (Fig. 3, line d). Thus, much of the decline in the E/GNP ratio has been due to our ability to expand the use of higher quality fuels.

A relative shift in direct fuel use from final demand sectors to intermediate sectors, or vice versa, also changes the E/GNP ratio. For example, a dollar's worth of fuel purchased by households represented 145,000 kilocalories in 1972, whereas a dollar's worth of *nonfuel* good or service purchased by households represented only 5,600 to 11,800 kilocalories (*28*). Thus, the E/GNP ratio is sensitive to the partitioning of fuel between direct household use and fuel use to produce goods consumed by households. Such relative shifts in the point of fuel combustion account for 27 percent of the variation in the ratio since 1929; they were most important during World War II, when petroleum use was rationed, and since 1973, when the high price of fuel has discouraged its direct use by households. Eighty-eight percent of the decline in the E/GNP ratio since 1973 can be explained by the declining proportion of GNP spent on fuel by households. Lines c and e in Fig. 3 show the effects on the ratio of including corrections for fuel quality and for relative shifts in fuel use between households and intermediate sectors. When these effects are accounted for, the corrected E/GNP ratio shows little or no long-term trend since 1929 (Fig. 3, line e).

We do not argue that there have been no energy efficiency improvements during this period. Even with corrections, the E/GNP ratio does show a modest decline since 1973, indicating that higher fuel prices have led to real efficiency improvements, as other analysts have suggested (*29*). The fuel quality and GNP modifications are an attempt to include important attributes of fuel use and social welfare not accounted for in uncorrected fuel use and GNP statistics. The E/GNP ratio is sensitive to such modifications. Our regression analysis suggests that technological change which has led to a decline in the E/GNP ratio has often relied on intensified use of higher quality fuels. Our analysis does not support the hypothesis that the shift toward a service-oriented economy, such as the United States has undergone since World War II, is a significant factor in the decline in the E/GNP ratio.

Labor Productivity and Technical Change

In many economic models, technological advance is presented as an exogenous driving force powered by advances in human knowledge that increase labor and capital productivity. Denison (*7*) states that advances in "human knowledge of how to produce things at low cost" are the most important causes of the increase in per capita national income observed between 1948 and 1973. In this and other analyses, technological change is not measured directly, but rather is assigned the residual of increases in per capita income after all "tangible" factors have been accounted for. Because energy and natural resources are not considered tangible factors by most analysts, a large residual remains. Griliches and Jorgenson (*30*) stated that relabeling changes in factor productivity as 'technical progress" or "advance in knowledge leaves the problem of explaining growth in total output unresolved."

From an energy perspective, productivity gains are facilitated by technical advances that enable laborers to empower their efforts with greater quantities of high-quality fuel embodied in and used by capital structures. As Cottrell (*5*) observed, "productivity increases with the per capita increase in available energy." Various empirical analyses support this view, and cross-sectional and temporal changes in labor productivity are correlated with the quantity of fuel used to empower a worker's efforts. Boretsky (*31*) noted that higher labor productivity rates in the United States than in Western Europe nations were associated with the substantially greater quantities of fuel used per employee in the United States. We found that in the U.S. manufacturing sector, output per worker-hour is closely related to the quantity of fuel used per worker-hour (*32*) (Fig. 4). A similar relation exists in the U.S. agricultural industry.

Such relations can be merged with standard economic models to explain historical changes in labor productivity. From 1900 to 1973 the real price of fuel declined relative to the wage rate, and fuel was substituted for labor in many processes. Labor productivity increased during this period. Since 1973 the price of fuel has risen relative to the wage rate, and labor has been substituted for fuel, thereby reducing productivity. While other factors certainly affected the decline in labor productivity in the 1970's, a biophysical analysis supports those analyses which indicate higher fuel prices as a significant contributor (*33*).

Energy and Inflation

The high rates of inflation that have recently plagued most industrialized nations can be explained by uniting the importance of fuel use and money supply. If an expanding money supply stimulates demand beyond the level that can be satisfied by existing fuel supplies, price levels must rise (*34*). A historical analysis indicates that changes in the ratio of money supply to fuel use are significantly correlated with changes in the consumer price index since 1890 (*35*) (Fig. 5). Manipulation of the monetary, and even fiscal, policy as a means of stimulating economic growth may now be less effective due to the increasing real physical cost of obtaining new quantities of fuel from the environment.

Natural Resource Quality from an Energy Perspective

The issue of natural resource scarcity has received considerable attention in recent years (*36*). Many suggest that the negative economic effects of depleting high-quality mineral deposits can be mitigated indefinitely through technical innovation and/or the use of more energy and capital structures to mine vast quantities of low-quality ore (*9*). Evidence for this hypothesis is that capital and labor inputs per unit output in the extractive sectors have either declined or remained stable throughout most of this century (*37*), a trend attributed to technical advance in those industries.

When analyzed from a physical perspective, the trend in the scarcity of some important natural resources is less reassuring. Technical improvements in the extractive sectors have made available previously uneconomic deposits only at the expense of more energy-intensive forms of capital and labor inputs (*38*). Physical output per kilocalorie of direct fuel input in the U.S. metal mining industries has declined 60 percent since 1939 (Fig. 6a), although a few exceptions to the trend are known (*39*). The energy cost per ton of metal at the mine

Table 1. Estimates of energy return on investment (EROI) ratios for some existing and proposed fuel supply technologies (*54*). Numbers in parentheses for electricity generation include a quality factor based on a heat rate of 2646 kcal/kWh.

Process	EROI
Nonrenewable	
Oil and gas (domestic wellhead)	
1940's	Discoveries >100.0*
1970's	Production 23.0, discoveries 8.0
Coal (mine mouth)	
1950's	80.0
1970's	30.0
Oil shale	0.7 to 13.3
Coal liquefaction	0.5 to 8.2
Geopressured gas	1.0 to 5.0
Renewable	
Ethanol (sugarcane)	0.8 to 1.7
Ethanol (corn)	1.3
Ethanol (corn residues)	0.7 to 1.8
Methanol (wood)	2.6
Solar space heat (fossil backup)	
Flat-plate collector	1.9
Concentrating collector	1.6
Electricity production†	
Coal	
U.S. average	9.0 (27.0)
Western surface coal	
No scrubbers	6.0 (18.0)
Scrubbers	2.5 (7.5)
Hydropower	11.2 (33.6)
Nuclear (light-water reactor)	4.0 (12.0)
Solar	
Power satellite	2.0 (6.0)
Power tower	4.2 (12.6)
Photovoltaics	1.7 (5.1) to 10.0 (30.0)
Geothermal	
Liquid dominated	4.0 (12.0)
Hot dry rock	1.9 (5.7) to 13.0 (39.0)

*Based on discovery rates reported by Hubbert (*44*) and the assumption that energy use in drilling was less than 1 barrel per foot [Hall and Cleveland (*44*)]. †Does not include energy in fuel.

mouth for industrially important metals such as copper, aluminum, and iron has risen sharply as their average grade declined. For all U.S. mining industries (including fossil fuels), output per unit input of direct fuel has declined 30 percent since 1939. This and other analyses (*40*) substantiate Brobst and Pratt's (*41*) statement that the cost and physical availability of fuel may well be the most important factors determining the limits to the economic exploitation of many nonrenewable resources.

Claims such as "it is simply not true . . . that average rock will never be mined" (*42*) to meet society's needs must be evaluated in light of the energy and environmental costs associated with mining and processing vast amounts of elements at or near their crustal abundance. Such costs had little negative economic impact prior to the 1970's, when domestic oil production was still increasing and real fuel prices were stable or declining. Energy costs of mineral extraction are especially important now because the energy costs of extracting fuel itself have increased substantially.

U.S. oil discoveries peaked in about 1930 and oil production in 1970 (*43*). For natural gas these dates were 1950 and 1973, respectively. As we have increasingly exhausted the possibilities of finding new large petroleum deposits, the rate at which we find new oil per unit of drilling effort in the lower 48 states has declined precipitously (*44*). The large increase in drilling effort since 1973 has not reversed this decline. As a result, the running average EROI for oil and gas at the wellhead has declined precipitously (Fig. 6b) (*45*). In Louisiana, a region that has accounted for 17 percent of all domestic oil and gas discovered and produced to date, the EROI for natural gas extraction declined from 100:1 in 1970 to 12:1 in 1981 (*46*). There has been a similar decline in the ratio of the energy in the petroleum we obtain from foreign sources compared to the energy required to make the goods and services we exchange for that petroleum (Fig. 6b) (*45*). Foreign suppliers acquired the leverage to raise oil prices dramatically in 1973 because domestic production could not keep pace with domestic demand, a gap that began in the late 1940's and was accentuated following the 1970 peak in domestic oil production. The bituminous coal industry shows a similar but less dramatic trend over the past 15 years. The EROI for coal at the mine mouth has decreased from about 80:1 in the mid-1960's to about 30:1 in 1977 (Fig. 6b) (*45*).

Declining resource quality and higher fuel prices impede economic growth by diverting increasing amounts of capital and labor to the extractive and resource processing sectors. Throughout most of this century, the real dollar value of the mining sector share of real GNP was relatively small and constant, averaging 3 to 4 percent (Fig. 7). This led some to conclude that natural resources were a small and unimportant factor of production (*47*). By 1982, however, more than 10 percent of real GNP was needed to extract mineral resources from the environment. Most of this increase was for fossil fuel purchases, which in 1981 were 4.5 times greater in inflation-corrected dollars than in 1972, despite the fact that total fossil fuel use was about the same in both years. Clearly, the cost of minerals is not "irrelevant" to our standard of living, as some suggest (*9*).

Alternative fuel sources. Because of the importance of net fuel supply, continued economic growth hinges in large part on our ability to develop new fuel

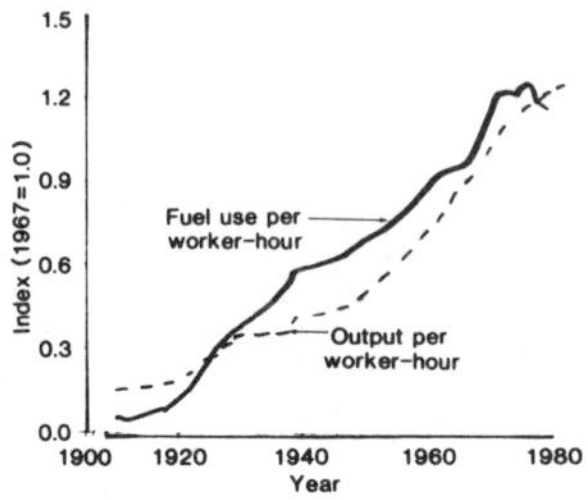

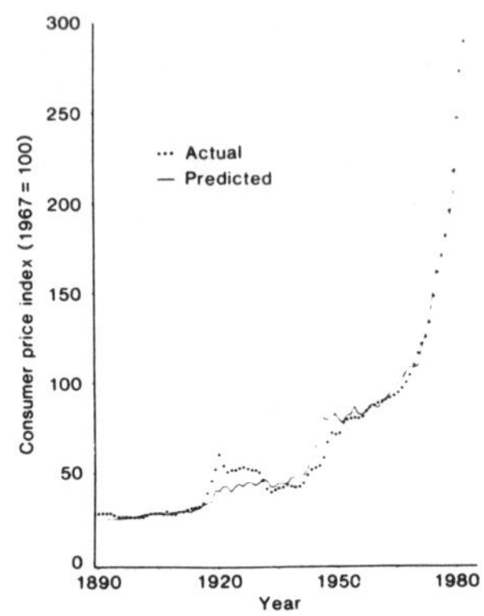

Fig. 4 (left). Productivity per worker-hour (as measured by real value added per worker-hour) and fuel use per worker-hour in the U.S. manufacturing sectors from 1909 to 1980. Fuel use includes fossil fuels and electricity (*32*). Fig. 5 (right). Results of regression model that predicts the level of the consumer price index (CPI) (1967 = 100) as a function of the ratio of the money supply (M2) to fuel use. Fuel use includes fossil fuels, nuclear, and hydropower. The curve represents values for the CPI as predicted by the ratio of M2 to fuel use. The points are actual CPI values (*35*).

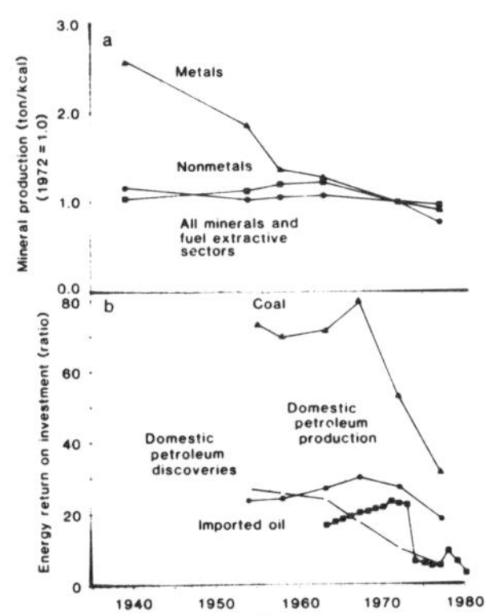

Fig. 6 (left). (a) Ratio of output per unit fuel (tons per kilocalorie) used by extractive sectors of the U.S. economy (Standard Industrial Code sectors 10 to 14). Index of physical output is from (*57*). Index of fuel use includes fossil fuels and electricity. (b) Net energy return (EROI) over time for U.S. fossil fuel sources. For domestic coal and petroleum (oil plus gas), energy use includes both direct use of fuels and electricity and energy embodied in purchased machinery installed and other supplies used as reported by the Census of Mineral Industries. EROI for imported petroleum is the ratio of energy in petroleum imports to energy embodied in goods exchanged for that petroleum (*45*). All the EROI ratios presented are conservative in that they exclude the energy costs of refining, transportation, labor, government, and environmental services. Fig. 7 (right). Real dollar value of fuel and nonfuel minerals as a percentage of real GNP in the United States from 1900 to 1981 (*58*).

sources with EROI's comparable to those we use today. As the values in Table 1 indicate, most alternative fuel sources have a positive but small EROI relative to fossil fuels. For proposed "inexhaustible" sources such as large-scale photovoltaics and fusion and breeder reactors, it is not yet clear whether the EROI will be large, marginal, or less than break-even. Current estimates of the EROI for new technologies are probably overly optimistic because the record shows we have routinely underestimated actual capital costs of new energy process plants by more than 100 percent (*48*). A recent survey of 40 nuclear power plants in various stages of construction in the United States indicates that they all will eventually cost an average of seven times their first cost estimates (*49*).

Conclusions

In one of the first detailed empirical analyses of fuel use in the United States, Tryon (*50*) stated in 1927 that

> Anything as important in industrial life as power deserves more attention than it has yet received from economists. . . . A theory of production that will really explain how wealth is produced must analyze the contribution of this element energy.

Toward this theory of production, our analysis emphasizes the economic importance of changes in the quality and availability of fuel and other natural resources faced by the United States. Declining energy return on investment for fuels and increasing energy costs for nonfuel resources have a negative impact on economic growth, productivity, inflation, and technological change. Some economic models that guided economic growth during the preceding period of high-quality resource abundance have become increasingly less powerful because they do not account for the importance of changes in natural resource quality and availability.

Some resource analysts have suggested that there was no physical reason for domestic oil demand to outstrip domestic production by almost 100 percent in the 1970's (*51*). The degree of depletion of domestic petroleum in the 1970's, however, was predicted accurately with physically based models over a quarter-century ago by Davis, Hubbert, and others (*44*). The "energy crisis" and ensuing economic problems cannot, therefore, be blamed solely on misguided regulatory policies, the monopoly power of the Organization of Petroleum Exporting Countries, a conspiracy among the multinational oil companies, or lack of proper economic incentive for the petroleum exploration and development industry. While these factors may have exacerbated the situation, underlying the energy crisis and the ensuing economic malaise was the declining physical availability of high-EROI petroleum, and a reliance on economic and political models that did not account for it. Market incentives in response to a quadrupling of real oil prices and a 280 percent increase in drilling effort between 1972 and 1981 made no significant reversal in declining oil and gas production and discovery rates in the United States.

If we are to sustain current levels of economic growth and productivity as minimum long-run goals, alternative fuel technologies with EROI ratios comparable to that of petroleum today must be developed, or there must be unprecedented improvements in the efficiency with which we use fuel to produce economic output. Many discount the decreasing availability of high-quality fossil fuel deposits, stating that such depletion is merely a signal of our impending transition to a society based on a "boundless supply of energy at reasonably low cost" (*52*) such as breeder or fusion reactors or direct solar power. But past experience with capital-intensive ventures such as fission and synfuels suggests that it would be unwise to assume a priori that fusion or any other proposed fuel source will necessarily have a large EROI. Although we should research aggressively all potential fuel technologies, particularly in regard to their potential EROI, we should also plan for the contingency that new high-EROI sources might not be found.

Based on our analysis, the economic recovery from the 1980–1982 recession was due in part to declining OPEC oil prices, which themselves were due to decreased world oil demand brought about by the worldwide recession. In effect, the EROI for imported oil rose recently because importing nations had to divert less of their output to trade for oil. Rising economic activity and fuel use, however, will again confront the economy with the physical limits of declining domestic fuel quality, and thereby increase the chances of a tight oil market in the near future (*53*). Our ability to cope with any economic contingencies will depend on the ability of our economic models to account for the biophysical constraints on human economic activity, and on the ability of our citizenry to accept and adapt to the realities of physical constraints imposed on our economic possibilities.

References and Notes

1. P. F. Drucker, in *The Crisis in Economic Theory*, D. Bell and I. Kristol, Eds. (Basic Books, New York, 1982).
2. W. Leontief, *Science* **217**, 104 (1982).
3. M. N. Bailey, *ibid.* **216**, 859 (1982); W. H. Miernyk, *The Illusions of Conventional Economics* (West Virginia Univ. Press, Morgantown, 1982).
4. R. B. Glassman, *Science* **218**, 108 (1982).
5. F. Soddy, *Wealth, Virtual Wealth, and Debt* (Dutton, New York, 1933); F. Cottrell, *Energy and Society* (McGraw-Hill, New York, 1955); H. T. Odum and E. C. Odum, *Energy Basis for Man and Nature* (McGraw-Hill, New York, 1976); N. Georgescu-Roegen, *The Entropy Law and the Economic Problem* (Harvard Univ. Press, Cambridge, Mass., 1971).
6. R. Stobaugh and D. Yergin, Eds., *Energy Future: Report of the Energy Project at the Harvard Business School* (Random House, New York, 1979); D. A. Huettner, *Science* **216**, 1141 (1982); M. H. Ross and R. H. Williams, in *Hearings Before the U.S. Senate Subcommittee on Energy of the Joint Economic Committee* (Series No. 91-952, Government Printing Office, Washington, D.C., 1977).
7. E. F. Denison, *Surv. Curr. Bus.* **59**, 1 (1979).
8. M. Friedman, "Monetary correction: A proposal for escalation clauses to reduce the costs of ending inflation" (Occasional Paper 41, Institute of Economic Affairs, London, 1974).
9. J. L. Simon, *The Ultimate Resource* (Princeton Univ. Press, Princeton, N.J., 1981).
10. R. L. Heilbroner and L. C. Thurow, *The Economic Problem* (Prentice-Hall, Englewood Cliffs, N.J., 1981).
11. P. C. Roberts, *Energy Policy* **10**, 171 (1982).
12. J. Dunkerley, in *International Energy Strategies*, J. Dunkerley, Ed. (Oelgeschlager, Gunn & Hain, Cambridge, Mass., 1980); M. Slesser, *Energy in the Economy* (St. Martin's, New York, 1978); F. Cottrell, in (*5*).
13. F. G. Adams and P. Miowic, *J. Ind. Econ.* **17**, 42 (1968); J. F. Alexander, D. P. Swaney, R. J. Roynstad, R. K. Hutchinson, in *Coal Burning Issues*, A. Green, Ed. (Univ. of Florida Press, Gainesville, 1980); M. Slesser, in (*12*).
14. W. D. Devine, *An Historical Perspective on Electricity and Energy Use* (Institute for Energy Analysis, Oak Ridge Associated Universities, Oak Ridge, Tenn., 1982).
15. R. M. Solow, *Am. Econ.* **2**, 5 (1978); W. Nordhaus and J. Tobin, in *The Measurement of Economic and Social Performance*, M. Moss, Ed. (Columbia Univ. Press, New York, 1973).
16. R. Herendeen and C. Bullard, *Energy Policy* **3**, 268 (1975); B. Hannon, R. A. Herendeen, T. Blazeck, "Energy and labor intensities for 1972" (Document 307, Energy Research Group, University of Illinois, Urbana, 1981).
17. W. W. Leontief, *The Structure of the American Economy, 1919, 1929: An Empirical Analysis of Equilibrium Analysis* (Harvard Univ. Press, Cambridge, Mass., 1941).
18. R. Costanza, *Science* **210**, 1219 (1980).
19. W. J. Baumol and E. N. Wolff, *J. Polit. Econ.* **89**, 891 (1981).
20. For example, D. Pimentel *et al.* [*Science* **182**, 443 (1973)] showed that between 1959 and 1970, the direct on-farm energy used by U.S. farmers to grow corn decreased 105 kilocalories per kilogram, and labor energy costs decreased 3 kilocalories per kilogram. The energy used off-farm to produce the capital that replaced the direct energy use increased 145 kilocalories per kilogram. Thus, the total energy used directly and indirectly to grow a kilogram of corn increased by 37 kilocalories during that period.
21. Time series of the link between fuel use and GNP have been analyzed extensively. See, for example, S. H. Schurr and B. Netschert, *Energy in the American Economy, 1850–1975* (Johns Hopkins Univ. Press, Baltimore, Md., 1960); H. J. Barnett, *U.S. Bur. Mines Inf. Circ. 7582* (1966); S. Brown and A. Lugo, *Management and Status of Marine Fisheries* (Council on Environmental Quality, Washington, D.C., 1981).
22. A. T. Akarca and T. V. Long, in *Energy Modelling II* (Institute of Gas Technology, Chicago, 1980).
23. Box-Jenkins transfer function analysis (software from BMDP2T/82) was applied to the fuel use and real GNP time series. The analysis was done on the first differenced series of the two variables. There were 80 data points (1900–1980). The estimated model is $DGNP_t = 9.31 + 0.43\ DGNP_{t-1} + 0.93\ DFUEL_t$, where DGNP and DFUEL are the first differences of real GNP and fuel use for year *i*, and 3.67, 0.10, and 0.11 are the respective estimated standard errors for the

parameters. A residual analysis showed that the remaining part is a random process. The mean of the residual is 0.0004, the standard error of the mean is 1.9860, and the *t* value of the mean for a test against zero is 0.0002. We thank K.-G. Maler for bringing this point to our attention.

24. R. Costanza and R. A. Herendeen, *Resources and Energy*, in preparation. The hypothesis that adding arbitrary household and government sectors to the input-output model would always yield a high correlation was tested theoretically and empirically and shown to be false.
25. See R. Turvey and A. R. Nobay [*Econ. J.* **75**, 787 (1965)] for a discussion of the problems of measuring fuel use. E. R. Berndt and D. O. Wood [*Am. Econ. Rev.* **69**, 342 (1979)] concluded that the E/GNP ratio yields a consistent and logical economic interpretation only under highly restrictive conditions. See H. E. Daly [*Steady-State Economics* (Freeman, San Francisco, 1977)] for a discussion of problems in GNP accounting.
26. H. H. Landsberg, *Science* **218**, 973 (1982); D. A. Huettner, in (*6*); *New York Times*, 18 May 1980, p. D1.
27. The regression equation is E/GNP = 60.41 − 52.33 (%PETRO) − 344.27 (%ELEC) + 10.41 (%PCE), $r^2 = 0.95$, $P < 0.0001$. The quantity %PCE is the percentage of personal consumption expenditures spent on direct fuel. This was algebraically manipulated to yield:

$$E + 52.33\,(\%PETRO)(GNP) + 344.27\,(\%ELEC)(GNP) = 60.41\,(GNP) + 10.41\,(\%PCE)(GNP) \quad (1)$$

The left side of the equation was set equal to a revised energy input E_R:

$$E_R = E + 52.33\,(\%PETRO)(GNP) + 344.27\,(\%ELEC)(GNP) \quad (2)$$

Then E_R was regressed on uncorrected fuel use data to calculate implied quality factors without GNP as a variable:

$$E_R = \alpha + \beta_1\,(\text{oil} + \text{gas}) + \beta_2\,(\text{coal}) + \beta_3\,(\text{hydro and nuclear power}) \quad (3)$$

The equation was fit yielding $\beta_1 = 1.74$, $\beta_2 = 0.92$, $\beta_3 = 16.8$, which are the implied quality factors. The α was not significant, $r^2 = 0.999$, $P < 0.0001$. This indicates that the economic value (or quality) of electricity relative to fossil fuels may be several times its conversion value. Equations 1, 2, and 3 were combined to calculate a revised GNP, GNP_R.

$$GNP_R = GNP\,[1 + 0.172(\%PCE)]$$

28. B. Hannon, *Energy Syst. Policy J.* **6**, 249 (1982).
29. E. Hirst, R. Marlay, D. Greene, R. Barnes, *Annu. Rev. Energy* **8**, 193 (1983).
30. A. Griliches and D. W. Jorgenson, *Rev. Econ. Stud.* **34**, 244 (1967).
31. M. Boretsky, *Am. Sci.* **63**, 70 (January 1975).
32. For this relation $r^2 = 0.99$, $P < 0.005$. Data are from U.S. Bureau of Census, Census of Manufacturers. Worker-hours for pre-1947 production were calculated by dividing production worker payroll by average hourly wage in manufacturing.
33. N. D. Uri and S. A. Hassanein, *Energy Econ.* **4**, 98 (1982).
34. H. T. Odum and E. C. Odum, in (*5*).
35. Both the (M2) money supply ($r^2 = 0.93$, $P < 0.001$) and fuel use ($r^2 = 0.89$, $P < 0.001$) alone are highly correlated with the CPI over the time period analyzed. The highest statistical correlation, however, is between the ratio of the two variables and the CPI ($r^2 = 0.98$, $P < 0.001$). If economic output is correlated with fuel use (E), the price level can be described as the ratio of money supply (M) to fuel use: $P = (M/E)(V/n)$. We assume that the ratio of the velocity of money (V) to fuel efficiency (n) has been relatively constant and unimportant.
36. G. M. Brown and B. C. Field, Jr., *J. Polit. Econ.* **86**, 229 (1978); V. K. Smith, *Land Econ.* **56**, 257 (1980); S. Devarajan and A. Fisher, *J. Polit. Econ.* **90**, 1279 (1982).
37. H. Barnett and C. Morse, *Scarcity and Growth: The Economics of Natural Resource Availability* (Johns Hopkins Univ. Press, Baltimore, Md., 1963).
38. N. Georgescu-Roegen, *South. Econ. J.* **41**, 3 (1975); T. S. Lovering, in *Resources and Man* (Freeman, San Francisco, 1969).
39. P. J. Kakela, *Science* **202**, 1151 (1978).
40. E. Cook, *ibid.* **191**, 677 (1976); T. S. Lovering, in (*38*); N. J. Page and S. C. Creasy, *J. Res. U.S. Geol. Surv.* **3**, 10 (1975).
41. D. A. Brobst and W. E. Pratt, *U.S. Geol. Surv. Circ. 862* (1973).
42. D. B. Brooks and P. W. Andrews, *Science* **185**, 13 (1974).
43. R. Nehring, *The Discovery of Significant Oil and Gas Fields in the United States* (Rand Corp., Santa Monica, Calif., 1981).
44. W. Davis, *Oil Gas J.* **56**, 105 (1958); M. K. Hubbert, in *Hearings Before the U.S. Senate Committee on Interior Insular Affairs* (Series No. 93-40, Government Printing Office, Washington, D.C., 1974); H. W. Menard and G. Sharman, *Science* **190**, 337 (1975); C. A. S. Hall and C. J. Cleveland, *ibid.* **211**, 576 (1981).
45. Oil and gas: C. A. S. Hall, C. J. Cleveland, R. Kaufmann, *Biophysical Economics* (Wiley, New York, in press) [see also (*44*)]; imported petroleum: R. Kaufmann and C. A. S. Hall, in *Energy and Ecological Modelling*, W. J. Mitsch, R. W. Bosserman, J. M. Klopatek, Eds. (Elsevier, New York, 1981), pp. 697–702; coal: C. Hall, C. J. Cleveland, M. Berger, *ibid.*, pp. 715–724. The decline in EROI for coal is due to (i) physical factors such as smaller seam thicknesses and a 1 percent per year decline in the average heat content of bituminous coal from 1955 to the mid-1970's; (ii) a relative increase in surface mining, which is more energy intensive than underground mining, and (iii) the effects on output of the Coal Mine Health and Safety Act of 1969.
46. R. Costanza and C. Cleveland, "Ultimate recovery of hydrocarbons in Louisiana: A net energy approach" (report to the Center for Energy Studies, Louisiana State University, Baton Rouge, 1983).
47. R. M. Solow, in (*15*); W. W. Hogan and A. S. Manne, in *Modeling Energy-Economy Interactions: Five Approaches*, C. J. Hitch, Ed. (Resources for the Future, Washington, D.C., 1977).
48. E. W. Merrow, S. W. Chapel, C. Worthing, *Review of Costs Estimates in New Technologies: Implications for Energy Process Plants* (Rand Corp., Santa Monica, Calif., 1979).
49. M. L. Wald, *New York Times*, 26 February 1984, p. 1.
50. F. G. Tryon, *J. Am. Stat. Assoc.* **22**, 271 (1927).
51. B. Commoner, *Poverty of Power* (Bantam, New York, 1976).
52. J. E. Stiglitz, in *Scarcity and Growth Reconsidered*, V. K. Smith, Ed. (Johns Hopkins Univ. Press, Baltimore, Md., 1979).
53. International Energy Agency, *World Energy Outlook* (OECD, Paris, 1982); D. Yergin, Baton Rouge *Morning Advocate*, 3 March 1983, p. 9B. Petroleum imports in the first quarter of 1984 were 35 percent greater than a year ago (U.S. Department of Energy, *Monthly Energy Review*).
54. Oil shale: C. G. Lind and W. J. Mitsch, in *Energy and Ecological Modelling*, W. J. Mitsch, R. W. Bosserman, J. M. Klopatek, Eds. (Elsevier, New York, 1981). Coal liquefaction: General Accounting Office, "DOE funds new energy technologies without estimating potential net energy yields" (report to Congress, July 1982). Geopressured gas: C. J. Cleveland and R. Costanza, *Energy* **9**, 35 (1984). Ethanol from sugarcane: C. S. Hopkinson, Jr., and J. W. Day, Jr., *Science* **207**, 302 (1980). Ethanol from corn: R. S. Chambers, R. A. Herendeen, J. J. Joyce, P. S. Penner, *ibid.* **206**, 789 (1979). Ethanol from corn residues: D. Pimentel *et al.*, *ibid.* **212**, 1110 (1981). Ethanol from wood: B. Hannon and H. Perez-Blanco, "Ethanol and methanol as industrial feedstocks," report to Argonne National Laboratories, Argonne, Ill., contract ANL 31-109-38-5154 (1979). Residential solar: B. Hannon, in *Energy, Economics, and the Environment*, H. E. Daly and A. F. Umana, Eds. (Westview, Boulder, Colo., 1981). Electricity Production U.S. average: D. Pilati, *Energy* **2**, 1 (January 1977); western surface coal: P. Penner, J. Kurish, B. Hannon, "Energy and labor cost of coal electric fuel cycles" (Document 273, Energy Research Group, University of Illinois, Urbana, 1979); hydropower: M. W. Gilliland, J. M. Klopatek, S. G. Hildebrand, *Net Energy of Seven Small-Scale Hydroelectric Power Plants* (Oak Ridge National Laboratory, Oak Ridge, Tenn., 1981); power satellite: R. A. Herendeen, T. Kary, J. Rebitzer, *Science* **205**, 451 (1979); power tower: A. J. Frabetti, "A study to develop estimates of merit for selected fuel technologies" (Development Sciences, Inc., East Sandwich, Mass., 1975); photovoltaics (low estimate): D. Grimmer, "Solar energy breeders" (Report LA-UR-78-2973, Los Alamos Scientific Laboratory, Los Alamos, N.M., 1978), quoted in Hannon, *op. cit.*; geothermal: R. Herendeen and R. Plant, "Energy analysis of geothermal electric system" (Document 272, Energy Research Group, University of Illinois, Urbana, 1979). See C. A. S. Hall *et al.* (*45*) and R. Kaufmann and C. A. S. Hall (*45*) for nuclear and other solar estimates.
55. Economic data are from the following sources: GNP, implicit price deflators, and consumer price index are from the *Economic Report of the President* (1982) and the U.S. Department of Commerce, *Historical Statistics of the United States, Colonial Times to 1970*; investment and money supply data are from the U.S. Department of Commerce, Bureau of Economic Analysis, *Survey of Current Business* (various issues), *Business Statistics: The Biennial Supplement to the Survey of Current Business* (1980), and *Long Term Economic Growth, 1860 to 1970* (1973); data on value added are from the U.S. Department of Commerce, Bureau of Census, *Annual Survey of Manufacturers and Census of Manufacturers* (various years); data on labor productivity are from the U.S. Department of Labor, Bureau of Labor Statistics, *Monthly Labor Review*.
56. Fuel and mineral use data are from the following sources: 1850–1947, S. H. Schurr and B. Netschert, in (*21*); 1947–1982, Energy Information Administration, U.S. Department of Energy, *Annual Report to Congress* (Government Printing Office, Washington, D.C., various years), U.S. Department of Commerce, Bureau of Census, *Historical Statistics of the United States* (Government Printing Office, Washington, D.C., 1976), and U.S. Department of Energy, *Monthly Energy Review* (various months). Data on output by extractive sectors are from the U.S. Bureau of Mines, *Minerals Yearbook* (various years), and from *Historical Statistics of the United States* (see above).
57. Index of physical output of extractive sectors (Standard Industrial Code sectors 10 to 14) from U.S. Bureau of Mines. Index of fuel use calculated from fuel use data reported by the U.S. Bureau of Census, *Census of Mineral Industries, 1954, 1958, 1963, 1967, 1972, 1977*.
58. Value of mineral consumption from U.S. Bureau of Mines, *Minerals Yearbook*, and U.S. Department of Energy, *Annual Report to Congress* (1981).
59. This article is in part adapted from C. A. S. Hall, C. J. Cleveland, R. Kaufmann, *Biophysical Economics* (Wiley, New York, in press). We thank J. Conrad, H. Daly, J. Day, J. Fruci, M. Gilliland, B. Hannon, R. Herendeen, M. Lavine, K. G. Maler, D. Pimentel, D. Robson, R. E. Turner, T. Viatoris and two anonymous reviewers for their critical review of earlier drafts. This work was funded in part by the Department of Geography, University of Illinois, Urbana; the Center for Energy Studies, Louisiana State University, Baton Rouge; the National Science Foundation (grant PRA-8003854); a grant from Carrying Capacity to the Complex Systems Research Center, University of New Hampshire, Durham; and the New York State College of Agriculture and Life Sciences, Cornell University, Ithaca.

[21]

Technology and Environment. 1989.
Pp. 23–49. Washington, DC:
National Academy Press.

Industrial Metabolism

ROBERT U. AYRES

We may think of both the biosphere and the industrial economy as systems for the transformation of materials. The biosphere as it now exists is very nearly a perfect system for recycling materials. This was not the case when life on earth began. The industrial system of today resembles the earliest stage of biological evolution, when the most primitive living organisms obtained their energy from a stock of organic molecules accumulated during prebiotic times. It is increasingly urgent for us to learn from the biosphere and modify our industrial metabolism, the energy- and value-yielding process essential to economic development. Modifications are needed both to increase reliance on regenerative (or sustainable) processes and to increase efficiency both in production and in the use of by-products.

In this chapter, mass flows for key industrial materials of environmental significance, and the waste emissions associated with them, are reviewed along with the environmental impact of the waste residuals; economic and technical forces driving the evolution of industrial processes; long-range tendencies in the development of the industrial metabolism; and some applications of "materials-balance" principles to provide more reliable estimates of outputs of waste residuals. The Appendix contains theoretical explorations of the biosphere and the industrial economy as materials-transformation systems and lessons that might be learned from their comparison.

Before presenting the main analytic framework, it is useful to begin with some positive examples of how industrial metabolism can shift in the direction of increased efficiency in materials flows and waste streams. It has been justly remarked that the history of the chemical industry is one of finding new uses for what were formerly waste products (Multhauf, 1967). One of the most interesting early examples of such an innovation was the Leblanc process (the predecessor of the Solvay process) for manufacturing sodium carbonate circa 1800. As a raw material, it made use of sodium sulfate, an unwanted by-product of the eighteenth-century process for manufacturing ammonium chloride (sal ammoniac). Sal ammoniac was used for cleaning metals and as a convenient source of ammonia, but sodium sulfate had no use. The Leblanc process reduced sodium sulfate to sodium sulfide by heating it in a furnace with charcoal. This, in turn, was heated with calcium carbonate (chalk), which induced a reaction producing the desirable sodium carbonate and a new waste product, calcium sulfide (Multhauf, 1967).

Meanwhile, the market for ammonium chloride failed, so sodium sulfate had to be produced in the same complex by reacting sulfuric acid with common salt. This yielded hydrogen chloride (hydrochloric acid) as another by-product. In this case, the hydrochloric acid quickly found a practical use in the manufacture of chloride of lime as a bleach for the rapidly growing textile industry. The other waste product, calcium sulfide, was not successfully used until the 1880s, as a source of sulfur for the manufacture of sulfuric acid.

One of the most important waste products of the early nineteenth century was coal tar, which was generated in large amounts by gasworks making "town gas" for illumination. It was a systematic search for useful by-products, initiated by German chemists, that resulted in the creation of the modern organic chemical industry. The synthetic aniline dyes introduced after 1860 were all essentially derived from chemicals obtained from coal tar.[1] Phenolic resins, aspirin, and the sulfa drugs are later examples of derivatives.

Until recently, even natural gas was regarded as a by-product of petroleum production. Although it was used almost from the beginning as a fuel for refinery processes and illumination in local areas, much of it was wasted by "flaring" (indeed, this is still true in remote areas of the world). It had no significant chemical uses until World War II, when natural gas became the feedstock for producing ethylene, and thence butadiene and styrene, the major ingredients of synthetic rubber.

Chlorine is a final example. It is manufactured jointly with sodium or sodium hydroxide (lye) by electrolysis of salt or brine. When this process was introduced in the 1890s, it was the sodium hydroxide that was wanted for a variety of purposes, including petroleum refining and soap manufacturing. At the time, chlorine was a low-value by-product, which was luckily available for use in municipal water treatment. The development of a number of valuable chlorine-based organic chemicals (e.g., the most

TABLE 1 Mass of Active Materials Extracted Commercially in the United States, 1960-1975 (million tons)

Material	Year 1960	1965	1970	1975
Food and feed crops (excluding hay)	267	295	314	403
Meat, fish, and dairy products	82	85	85	84
Cotton, wool, hides, and tobacco	5	5	4	4
Timber (15% moisture basis)	256	267	271	249
Fuels (coal, lignite, oil, gas)	990	1,176	1,458	1,392
Ores (Fe, Cu, Pb, Zn)	400	435	528	460
Nonmetallics	200[a]	240[a]	266	255
Total	2,200[a]	2,500[a]	2,926	2,847

[a]Estimated value.

NOTE: Vegetable material harvested directly by animals has been omitted for lack of data, along with some obviously minor agricultural and horticultural products. Figures for metal ores exclude mine tailings and gangue removed to uncover ore bodies. Inert construction materials such as stone, sand, and gravel have also been omitted. Inert materials account for enormous tonnages, but undergo no chemical or physical change except to the extent that they are incorporated in concrete or paved surfaces. The table also excludes soil and subsoil shifted during construction projects or lost by erosion.

SOURCE: U.S. Bureau of the Census (1960-1975).

common industrial solvents and refrigerants, pesticides, herbicides, and plastics such as polyvinyl chloride) actually reversed the position. By the 1950s and 1960s, chlorine was the primary product and sodium hydroxide was the less valuable by-product.

MASS FLOWS AND WASTE EMISSIONS

Our economic system depends on the extraction of large quantities of matter from the environment. Extraction is followed by processing and conversion into various forms, culminating in products for "consumption." The U.S. economy extracts more than 10 tons of "active" mass per person (excluding atmospheric oxygen and fresh water) from U.S. territory each year. Of the active mass processed each year, roughly 75 percent is mineral and "nonrenewable," and 25 percent is, in principle, from renewable (i.e., biological) sources as shown in Table 1. Of the latter, much is ultimately discarded as waste, although much of it could (in principle) be used for energy recovery.

It is difficult to estimate the fraction of the total mass of processed active materials that is annually embodied in long-lived products and capital goods (durables). None of the food or fuel is physically embodied in durable goods. Most timber is burned as fuel or made into pulp and paper products. At least 80 percent of the mass of "ores" consists of unwanted impurities (more than 99 percent in the case of copper). Of the final products made from metals, a large fraction is converted into "consumption goods," such as bottles, cans, and chemical products, and "throwaway" products such as batteries and light bulbs. Only in the case of nonmetallic minerals (if inert materials are ignored, as before) is as much as 50 percent of the mass embodied in durable goods (mainly portland cement used for concrete and clays used for bricks and ceramics). The annual accumulation of active materials embodied in durables, after some allowance for discard and demolition, is probably not more than 150 million tons, or 6 percent of the total. The other 94 percent is converted into waste residuals as fast as it is extracted.

Combustion and carbothermic reduction processes are the major sources of atmospheric pollutants today but by no means the only important ones; nor is the atmosphere the only vulnerable part of the environment. From a broader environmental perspective, the production and dispersal of thousands of synthetic chemicals—many new to nature, and some highly toxic, carcinogenic, or mutagenic—and the mobilization of large tonnages of toxic heavy metals are of equally great concern. The complexity of the problem is too great to permit any kind of short summary.

However, two points are worthy of emphasis. First, as noted above, most materials "pass through" the economic system rather quickly. That is to say, the transformation from raw material to waste residual takes only a few months to a few years in most cases. Long-lived structures are very much the exception, and the more biologically "potent" materials are least likely to be embodied in a long-lived form.

The second point is that many uses of materials are inherently dissipative (Ayres, 1978). That is, the materials are degraded, dispersed, and lost in the course of a single normal use. In addition to food and fuels (and additives such as preservatives), other materials that fall into this category include packaging materials, lubricants, solvents, flocculants, antifreezes, detergents, soaps, bleaches and cleaning agents, dyes, paints and pigments, most paper, cosmetics, pharmaceuticals, fertilizers, pesticides, herbicides, and germicides. Many of the current consumptive uses of toxic heavy metals such as arsenic, cadmium, chromium, copper, lead, mercury, silver, and zinc are dissipative in the above strict sense. Other uses are dissipative in practice because of the difficulty of recycling such items as batteries and electronic devices.

Tables 2 and 3, which summarize estimates of emission coefficients for various heavy metals by use category and annual emissions attributable to dissipative uses, make clear the heterogeneity of the materials flows. In some cases the dissipation is slow and almost invisible. For instance, paints (often containing lead, zinc, or chromium) gradually crack, "weather," and turn to powder, which is washed or blown away. Tires, which contain zinc (and cadmium) salts, are gradually worn away during use, leaving a residue on the roads and highways. Similarly, shoe leather, containing up to 2 percent chromium salts, is gradually worn away to powder during use. Incinerator ash contains fairly high concentrations of heavy metals from a variety of miscellaneous sources, ranging from used motor oil to plastics and pigments.

On reflection, many dissipative uses (food and fuel again excepted) are generally seen to be nonessential in the sense that technologies are theoretically available, or imaginable, that could eliminate the need for them. To take one example only: hydroponic agriculture in enclosed, atmospherically controlled greenhouses, with genetically engineered pest controls, would eliminate all losses of fertilizers and pesticides to watercourses by way of surface runoff.

ENVIRONMENTAL IMPACT OF WASTE RESIDUALS

Materials do not disappear after they are used up in the economic sense. They become waste residuals that can cause harm and must be disposed of. In fact, it is not difficult to show that the tonnages of waste residuals are actually greater than the tonnages of crops, timber, fuels, and minerals recorded by economic statistics. Although usually unpriced and unmeasured, both air and water are major inputs to industrial processes and they contribute mass to the residuals—especially combustion products. Residuals tend to disappear from the market domain, where everything has a price, but not from the real world in which the economic system is embedded.

Many services provided by the environment derive ultimately from "common property," including the air, the oceans, the genetic pool of the biosphere, and the sun itself. Distortions in the market (i.e., prices) are unavoidably associated with the use of common property resources. Clearly, environmental resources such as air and water have been unpriced or (at best) significantly underpriced in the past. For this reason, such resources have generally been overused.[2]

As noted, the total mass of waste residuals produced each year by industrial metabolism far exceeds the mass of active inputs derived from economic activities. This is because nearly half of the inputs other than air and water are fossil fuels (hydrocarbons), which combine with atmospheric oxygen and form carbon dioxide and water vapor. The carbon fraction of hydrocarbons ranges from 75 percent in methane to about 90 percent

TABLE 2 Consumption-Related Emissions Factors: Heavy Metals

Metal	Metallic Use[a]	Painting and Coating[b]	Paint and Pigments[c]	Electron Tubes and Batteries[d]	Other Electrical Equipment[e]
Silver	0.001	0.02	0.5	0.01	0.01
Arsenic	0.001	0	0.5	0.01	N.A.
Cadmium	0.001	0.15	0.5	0.02	N.A.
Chromium	0.001	0.02	0.5	N.A.	N.A.
Copper	0.005	0	1	N.A.	0.10
Mercury	0.050	0.05	0.8	0.20	N.A.
Lead	0.005	0	0.5	0.01	N.A.
Zinc	0.001	0.02	0.5	0.01	N.A.

N.A. = Not applicable.

[a] Alloys or amalgams (in the case of mercury) not used in plating, electrical equipment, catalysts, or dental work. Losses can be assumed to be due primarily to corrosion, except for mercury, which volatilizes.

[b] Protective surfaces deposited by dip coating (e.g., galvanizing), electroplating vacuum deposition, or chemical bath (e.g., chromic acid). The processes in question generally resulted in significant waterborne wastes until the 1970s. Cadmium-plating processes were particularly inefficient until recently. Losses in use are mainly due to wear and abrasion (e.g., silverplate) or to flaking (decorative chrome trim). In the case of mercury-tin "silver" for mirrors, the loss is largely due to volatilization.

[c] Paints and pigments are lost primarily by weathering (e.g., for metal-protecting paints), wear, or disposal of painted dyes or pigmented objects, such as magazines. Copper- and mercury-based paints volatilize slowly over time. A factor of 0.5 is assumed arbitrarily for all other paints and pigments.

[d] Includes all metals and chemicals (e.g., phosphorus) in tubes and primary and secondary batteries, but excludes copper wire. Losses in manufacturing may be significant. Mercury in mercury vapor lamps can escape to the air when tubes are broken. In all other cases it is assumed that discarded equipment goes mainly to landfills. Minor amounts are volatilized in fires or incinerators or lost by corrosion. Lead-acid batteries are recycled.

[e] Includes solders, contacts, semiconductors, and other special materials (but not copper wire) used in electrical equipment control devices, instruments, etc. Losses to the environment are primarily through discard of obsolete equipment to landfills. Mercury used in instruments may be lost through breakage and volatilization or spillage.

in anthracite. Petroleum is intermediate. Every ton of fuel carbon is converted into 3.67 tons of carbon dioxide emitted to the atmosphere.[3] It is thought that only about half of this amount remains in the atmosphere, but the carbon dioxide level of the earth's atmosphere has risen over the past century from about 290 parts per million (ppm) to around 344 ppm at present. Although the magnitude of the climatic warming (the "greenhouse

Metal	Chemical Uses Not Embodied[f]	Embodied[g]	Agri-cultural Uses[h]	Nonagri-cultural Uses[i]	Medical, Dental[j]	Miscella-neous[k]
Silver	1	0.40	N.A.	N.A.	0.5	0.15
Arsenic	N.A.	0.05	0.5	0.8	0.8	0.15
Cadmium	1	0.15	N.A.	N.A.	N.A.	0.15
Chromium	1	0.05	N.A.	1	0.8	0.15
Copper	1	0.05	0.05	1	N.A.	0.15
Mercury	1	N.A.	0.80	0.9	0.2	0.50
Lead	1	0.75	0.05	0.1	N.A.	0.15
Zinc	1	0.15	0.05	0.1	0.8	0.15

[f] Uses not embodied in final products include catalysts, solvents, reagents, bleaches, etc. In some cases a chemical is embodied but there are some losses in processing. Losses in chemical manufacturing are included here. Major examples include copper and mercury catalysts (especially in chlorine manufacturing); copper, zinc, and chromium as mordants for dyes; mercury losses in felt manufacturing; chromium losses in tanning; lead in desulfurization of gasoline; and zinc in rayon spinning. In some cases, annual consumption is actually loss replacement and virtually all of the material is dissipated. Detonators such as mercury fulminate and lead azide (and explosives) are included in this category.

[g] Uses embodied in final products other than paints or batteries include fuel additives (e.g., tetraethyl lead), anticorrosion agents (e.g., zinc dithiophosphate), initiators and plasticizers for plastics (e.g., zinc oxide), wood preservatives, and chromium salts embodied in leather. Losses to the environment occur when the embodying product is used, for example, gasoline containing tetraethyl lead is burned and largely (75%) dispersed into the atmosphere. However, copper, chromium, and arsenic are used as wood preservatives and dispersed only if the wood is later burned or incinerated. In the case of silver (photographic film), it is assumed that 60% is later recovered.

[h] Includes agricultural pesticides, herbicides, and fungicides. Uses are dissipative, but heavy metals are largely immobilized by soil. Arsenic and mercury are exceptions because of their volatility.

[i] Biocides used in industrial, commercial, or residential applications. Loss rates are higher in some cases.

[j] Includes primarily pharmaceuticals (including cosmetics), germicides, etc., as well as dental filling material. Most uses are dissipated to the environment through wastewater. Silver and mercury dental fillings are likely to be buried with cadavers.

[k] Miscellaneous emissions not counted elsewhere.

SOURCE: Ayres et al. (1988).

effect") due to the rising level of carbon dioxide is still quantitatively uncertain, the qualitative impact is likely to be adverse.

Carbon monoxide is, of course, toxic to humans and has been implicated in health problems among urban populations. It is less well known that carbon monoxide plays an active, and not necessarily benign, role in a number of atmospheric chemical reactions before it is ultimately oxidized to carbon dioxide. Inefficient combustion processes convert up to 10 percent

TABLE 3 Emissions from Consumptive Uses: Heavy Metals, United States, 1980 (metric tons)

Metal	Metallic Uses (Except Coatings and Electrical)	Protective Coverings		Electrical	
		Plating and Coating	Paints and Pigments	Batteries and Equipment	Other Electrical Uses, Instruments, etc.
Silver	0.83	[a]	0.0	0.15	0.67
Arsenic	0.04	0.0	0.0	1.97	0.0
Cadmium[b]	0.04	136.1	116	7.81	0.0
Chromium	151.80	155.7	6,490	0.0	0.0
Copper	11,074.00	0.0	0.0	0.0	0.0
Mercury[c]	0.0	0.0	0.0	195.91	17.52
Lead	1,249.00	0.0	48,500	8,510.00	0.0
Zinc[c]	514.00	8,778.0	77,750	63.00	0.0

[a] Included in first column.
[b] 1979.
[c] 1977.

SOURCE: Ayres et al. (1988).

of fuel carbon into carbon monoxide; carbothermic reduction of iron ore and other metals generates even higher percentages. However, the average percentage over all processes is much smaller. Actual emissions of carbon monoxide to the atmosphere in the United States were about 110 million tons in 1970 (mostly from automobiles and trucks), with a carbon content of 47 million tons. Emission controls reduced the net output of carbon monoxide in 1980 to about 85 million tons (U.S. Environmental Protection Agency, 1986).

The discovery of chlorofluorocarbons (CFCs) in the stratosphere has raised an even more frightening prospect: ozone depletion (see Glas, this volume). In apparent confirmation of this phenomenon, an "ozone hole" has recently appeared in the stratosphere over Antarctica. This "hole" has reappeared each spring for several years and seems to be growing (Clark, 1987; Miller and Mintzer, 1986). Chlorofluorocarbons are chemically inert gases, discovered in 1928 and produced since the 1930s mainly as refrigerants and solvents and for "blowing" plastic foams (see Friedlander, this volume). The major use is for refrigeration and air conditioning. In most uses, CFCs are not released deliberately, but losses and leakage are inevitable.

The problems revealed so far may be only the beginning. If (perhaps it is better to say "when") the ozone level in the stratosphere is depleted, the effect will be to let more of the sun's ultraviolet radiation through to

Chemical		Biocidal Poison Uses				
Industrial Catalysts, Reagents, Explosives, etc.	Consumer Uses, Additives, Extenders, Photography, etc.	Agricultural Pesticides, Herbicides, Fungicides	Nonagricultural Pesticides (Except Medical)	Medical, Dental, Pharmaceutical	Miscellaneous	Total
7.6	48	0	0	1.51	0.45	60
492.0	0	2,950	5,901	19.70	0	9,364
0	29	0	0	0	0.98	290
1,297.0	3,890	0	1,038	0	2,141.00	11,659
4,222.0	0	1,560	0	0	0	15,452
412.1	0	16	236	8.36	6.96	893
0	0	0	0	0	1,329.00	56,900
2,508.0	18,622	188	251	1,003.00	0	109,670

the earth's surface. One likely impact on humans is a sharp increase in the incidence of skin cancer, especially among whites. The ecological impact on vulnerable plant or animal species cannot be estimated, at present, but could be severe.

Methane, oxides of nitrogen (NO_x), and sulfur oxides (SO_x) are other residuals that have been seriously implicated in climatic or ecological effects. All three are generated by fossil fuel combustion, as well as other industrial processes. Like carbon monoxide, carbon dioxide, and the CFCs, they can be considered as metabolic products of industry. Methane is released in natural gas pipelines, petroleum drilling, coal mining, and several kinds of intensive agriculture (especially rice cultivation or cattle and sheep farming).

Nitrogen oxides are also coproducts of combustion. In effect, at high temperatures, some of the atmospheric nitrogen is literally "burned." Nitrogen oxides are implicated in many atmospheric chemical reactions (including those that create smog) and eventually oxidize to nitric acid, which contributes to acid rain. In the stratosphere, where nitrogen oxides are decomposed by ultraviolet radiation, atomic nitrogen also contributes to the destruction of ozone. Finally, oxides of nitrogen are "greenhouse gases" that contribute to climatic warming (Miller and Mintzer, 1986).

Sulfur oxides are generated by the combustion of sulfur-containing fuels—especially bituminous coal—and by the smelting of sulfide ores. Most copper, lead, zinc, and nickel ores are of this kind. In principle, sulfur can be recovered for use from all these activities, and the recovery rate is rising. However, the costs of recovery, especially from coal-burning electric power plants, are still far higher than the market value of the potential products (e.g., dilute sulfuric acid). Hence, for the present, calcium sulfites and sulfates—as well as SO_x—constitute a waste residual that must be disposed of. This is also true of fly ash.[4]

These examples show that although today's industrialized economic system may be in rough equilibrium in terms of supply and demand relationships, it is far from equilibrium in thermodynamic terms. Enormous quantities of fossil fuels and high-quality minerals are extracted each year to drive the economic engine. The economic system is stable somewhat in the way a bicycle and its rider are stable: if forward motion stops, the system will collapse. Forward motion in the economic system is technological progress.

EVOLUTION OF INDUSTRIAL PROCESSES

It is generally accepted by economists that the mechanism that normally drives the evolution of industrial processes is technological innovation. The primary incentive for taking risks appears to be economic. A new process that saves one link in the chain between raw materials and finished materials or final goods can usually be justified through savings in materials and energy inputs or capital requirements, if not both. Moreover, process technology is inherently easier to protect from piracy than product technology. As noted, most chemical products are intermediates used in the production of other chemicals. Thus, final products result from chains, or sequences, of processes with an overall energy conversion efficiency that is the arithmetic product of the conversion efficiencies at each stage.[5] If the typical chain has three steps, each of which has a conversion efficiency of 0.7, the overall conversion efficiency of the chain is about 0.34. A four-step chain would have an overall efficiency of around 0.24. That is, the available energy embodied in the final product might be somewhere between 25 and 35 percent of the available energy of the original feedstocks. Because primary feedstocks are essentially indistinguishable from fuels, efficiency improvements translate directly into cost savings.

Clearly, a powerful long-term strategy for improving overall effectiveness in production is the development of new processes to shorten these process chains, bypassing as many intermediates as possible. In other words, one would like to be able to produce final products such as polyethylene or synthetic rubber directly from first-tier intermediates or even from primary feedstocks such as ethane and propane.[6] Biological organisms differ from industrial organizations in that they are able to build complex molecules directly from elementary building blocks with relatively few intermediates. Thus, biotechnology offers a long-term prospect of radically higher production efficiencies and correspondingly lower costs (U.S. Office of Technology Assessment, 1982).

Another long-term strategy for increasing effectiveness is better use of by-products and wastes. When a process can be justified on the basis of the market for its primary product, by-product sales can be highly profitable. This positive motivation to seek new uses is compounded by the fact that, because of some of the environmental problems already noted, waste generation is being increasingly discouraged by environmental regulation.

Moreover, waste disposal is increasingly expensive, and the cost is becoming more uncertain. Firms that buried toxic chemical wastes many years ago in landfills (methods that were regarded as acceptable at the time) are sometimes finding themselves saddled retroactively with heavy costs of digging up the same wastes and disposing of them again in a safer manner. The chemical waste dump used by Hooker Chemical Company at Love Canal, near Niagara Falls, is one example. Meanwhile, suitable sites for disposal of hazardous solid or liquid wastes are becoming scarce.[7] Materials that are recovered and reused internally or embodied in marketable products that can be readily and effectively recycled are less likely to cause this kind of problem.[8]

The incentives for technological innovation in the area of waste reduction and disposal are not always operating, especially where there are massive market failures. In practice, the "unpaid" environmental damage costs have been deferred and many of them will have to be paid by later generations. But as these costs become larger and more visible, there will be growing political pressure to force the producers and users of fossil fuels (and other materials such as heavy metals) to pay the costs of abating the resulting environmental damages.[9] Notwithstanding resistance by energy users, it seems inevitable that in the long run these costs will be added to the prices of fuels and materials. This, in turn, will create significant economic opportunities for innovations in the area of "low-waste" technologies.

EVOLUTIONARY ECONOMICS AND GAIA

One view of evolution—and probably the most common view among scientists—can be characterized as "the myopic drunkard's walk." The drunkard's walk is not exactly random, but it tends to follow the path of least resistance in the short run. If a mutation or an innovation offers short-term advantages, it will be adopted. In this view, there is no long-range force or tendency to approach a distant goal. There is no Aristotelian "final cause." Most scientists tend to regard more teleological theories, including Lovelock's Gaia hypothesis (Lovelock, 1988), with skepticism, because of their aroma of mysticism.

In the case of biological evolution, indeed, it was difficult for a long time to suggest a likely mechanism whereby short-term advantages to the individual could in some cases be overridden in favor of longer-term benefits to the community or the species. Even when such a mechanism is suggested (e.g., the kinship hypothesis), there may be difficulties in explaining how it could have become programmed into the genetic code. Yet, such behavior is a biological fact. The only real argument is whether a "strict" myopic test of the immediate utility of every mutation would have directed biological evolution along pathways significantly different from those that have actually occurred.

In the case of economic evolution, there is no difficulty in identifying possible mechanisms for selecting goal-oriented pathways. The decision makers are humans with the ability to look and plan ahead. The question here is, How far ahead do industrialists plan? The answer surely varies with individuals and circumstances, but it can be decades and is seldom less than a few years. Thus, long-term goal orientation in economic life is not particularly rare.

To relate this general point to the present context, one might postulate a long-run evolutionary imperative favoring industrial metabolic technologies that result in reduced extraction of virgin materials, reduced loss of waste materials, and increased recycling of useful materials. For convenience, one might refer to this overall trend—if it exists—as an imperative to reduce materials intensiveness, or dematerialization (see Herman et al., this volume). In general terms, the same long-term tendency can be observed in the biosphere, as noted in the Appendix.

It must be pointed out that short-term economic incentives do not necessarily point in the direction indicated. For example, market forces appear to favor product differentiation and specialization, but these trends increase the costs of repair and recycling. In poor countries, such as India, there is virtually no such thing as "junk." Any manufactured product, no matter how old or obsolete, is likely to be repaired or rebuilt and retained in service as long as physically possible. When it can no longer be repaired, it will be disassembled and useful parts will be separated for further use; the remainder will be hand-separated by material (stainless steel, iron, aluminum, copper, rubber, plastic, paper) and recycled in "backyard" operations. By contrast, in advanced countries, manufactured products are becoming more and more complex and correspondingly difficult to repair. This is particularly true of electronic devices such as printed circuit boards and cathode-ray tubes.

Moreover, as products are designed to be more reliable, so that repairs are no longer "normal" and expected, disassembly is becoming more difficult and in many cases is actively discouraged. In fact, for warranty reasons, critical subassemblies are often sealed and must be either returned to the factory or discarded. Finally, the complexity of products is often reflected by the increasing complexity of materials, which makes recycling inherently more difficult. To take one simple example, worn-out woolen suits and dresses were once routinely collected and recycled, mainly into coats and blankets. The once-profitable wool-recycling industry has virtually disappeared, because most new clothes are "blends" of natural and artificial fibers that cannot be reprocessed economically.

The point of the last two paragraphs is that short-term economic incentives and trends are often inconsistent with the postulated long-term imperative. One must, therefore, also postulate other counteracting incentives and mechanisms. Without going into detail, such incentives and mechanisms (if they exist) must grow out of social and even political responses to perceived environmental problems. It is political action, ultimately, that creates the incentive structure—fiscal, monetary, and tax policy—and the regulatory environment within which economic incentives drive entrepreneurial activity. Having said all this, I believe that the "dematerialization imperative" is alive and well.

The question I now raise is the following: Does this (hypothetical) evolutionary imperative toward reduced material intensiveness have any additional specific implications for future industrial processes? Can it be used as a basis for forecasting? I think the answer is a qualified yes.

To take one example, a future industrial metabolism deriving its energy ultimately from nuclear fission, nuclear fusion, or the sun itself, rather than from fossil fuels, would necessitate a completely different chemical energy carrier system. The most likely bulk energy carrier appears to be hydrogen. Unless biotechnology offers an attractive alternative, it appears likely that hydrogen would be obtained from electrolytic or thermal decomposition of water. Of the two processes, thermal decomposition is more direct and consequently, if the scale is large enough, more efficient.[10] Intermediates that have been suggested for thermal decomposition processes include a variety of compounds of iron, calcium, strontium, mercury, copper, chromium, manganese, and vanadium with bromine or chlorine. Many, if not most, of these compounds are considered hazardous. In consideration of the extremely large scale on which any such process is likely to have to be carried out, it is obviously of the utmost importance that process intermediates be recycled with extremely high efficiency (of the order of 99.9 percent). This is a severe challenge for chemical engineering and one that will probably be a major concern in future decades.

As another example, if we postulate much higher prices for carbon-based fuels (to reflect the environmental damage associated with their use), the problem of mobile power takes on a new dimension. Electric propulsion based on nuclear electricity (from "inherently safe" high-temperature gas-cooled reactors) is one possibility favored by many. But practical implementation on a large scale requires a technological breakthrough of unlikely magnitude in the area of compact electrical energy storage systems. Hydrogen has also been suggested as a possible substitute for liquid fuel (most likely for jet aircraft), but liquid fuels are far more convenient for most purposes.[11]

The evolution of industrial metabolism can also be addressed from the perspective of eliminating dissipative uses of toxic heavy metals (arsenic, cadmium, chromium, copper, lead, mercury, nickel, silver, and zinc) and of halogenated hydrocarbons. All of these materials have been implicated in environmental problems ranging from "Minimata disease" (mercury poisoning) to erosion of the ozone layer in the stratosphere. In the long run, it appears likely that all of them will have to be replaced or used only in applications permitting an extremely high degree of recycling. It is probably safe to say that the industrial metabolism of the next century will recycle many of the waste products that are produced in the largest quantities today, notably sulfur, fly ash, and lignin wastes from the paper industry. Lignin wastes may yet turn out to be a useful growth medium for single-cell organisms, providing high-protein supplements for food products. By the same token, inherently dissipative uses of biologically active materials will have to decrease as the mistakes of the past are rectified. In particular, the underpricing of environmental and exhaustible resources must be reduced or even (temporarily) reversed.

APPLICATIONS OF MATERIALS-BALANCE PRINCIPLES

The materials-balance principle, a straightforward application of the first law of thermodynamics (widely used in the design of chemical engineering systems, for example), is a potentially valuable and underutilized tool for using economic data in environmental analysis. Frequently, a combination of input data (obtainable from economic statistics), together with technical process data available from engineering analysis, gives a more reliable estimate of waste residual outputs than direct measurements alone could be expected to do. This principle is particularly true when the pollutant of concern is produced in relatively small quantities and is emitted together with large amounts of combustion products or process wastewater.

A good example of the mass-balance principle has been taken from a study of environmental problems in the aluminum industry. One of the major environmental problems associated with aluminum smelting in the past was the emission of gaseous fluorides from the smelter. The source of fluorine is the electrolytes (molten cryolite and aluminum fluoride) used as a solvent for alumina in the electrolytic cell. An unavoidable side reaction in the cell breaks down these electrolytes and releases some of the fluorine at the anode. Exact "recipes" for the production of aluminum are known only by the aluminum companies. However, a materials-balance analysis for the year 1973 suggested that for each 100 kilograms of aluminum produced, 2.1 kilograms of cryolite and 3 kilograms of aluminum fluoride were consumed as inputs (Ayres et al., 1978). Based on these estimates and straightforward chemistry, the aluminum industry would have accounted for 40 percent of the known production of hydrofluoric acid in that year, which is consistent with both official (U.S. Bureau of the Census, 1960–1975) and unofficial estimates (Ayres et al., 1978). In the absence of fluorine recovery facilities, all of the fluorine consumed by the industry must have been emitted eventually to the environment. In other words, the total amount of cryolite and aluminum fluoride consumed by the aluminum industry was (and is) exclusively to replace fluorine losses.

It is interesting to note that the fluoride emissions calculated by using materials-balance principles were about twice as high as the Environmental Protection Agency's published estimates at the time. The latter were based on direct (but unreliable and difficult to verify) measurements taken at a few smelter sites. Assuming the production and use statistics for hydrofluoric acid were correct, one would have to believe that the indirect estimate based on materials-balance considerations was probably more reliable than the estimate based on partial and questionable direct measurements.

Another application of materials-balance methodology is in the reconstruction of historical emissions data. This is a problem of some importance to basic environmental science, because the cumulative impact of air or water pollution over long periods can be evaluated only in relation to a baseline of some sort. In this context it becomes important to know more about emissions in the past, when no measurements were made. To be sure, sediments and ice cores offer some help, but not enough. The picture can be clarified considerably, however, with the help of synthetic models using production and consumption data (which are often imperfect, but better known than emissions) together with engineering analysis of processes. Sometimes process information is not even needed, as when emissions are linked directly to inputs.

For example, fairly good historical estimates of SO_x emissions, required to analyze the long-term impact of acid rain, among other things, can be reconstructed easily from historical statistics on coal consumption and copper, lead, and zinc smelting (e.g., Gschwandtner et al., 1983). This is because the sulfur content of coal and metal ores can be assumed to be the same in the past as it is today, and until recently all of that sulfur was emitted straight to the atmosphere. Reconstruction of NO_x emissions data is slightly more complicated, but the approach is basically similar (Gschwandtner et al., 1983). More complex reconstructions of historical emissions data have been undertaken recently, e.g., for the Hudson-Raritan estuary (Ayres et al., 1988). Two examples of material process-product flows, taken from the above study, are presented in Figures 1 (cadmium) and 2 (chromium).

Another way of using the materials-balance approach is in the analysis of materials "cycles." The water cycle, the carbon cycle, and the nitrogen

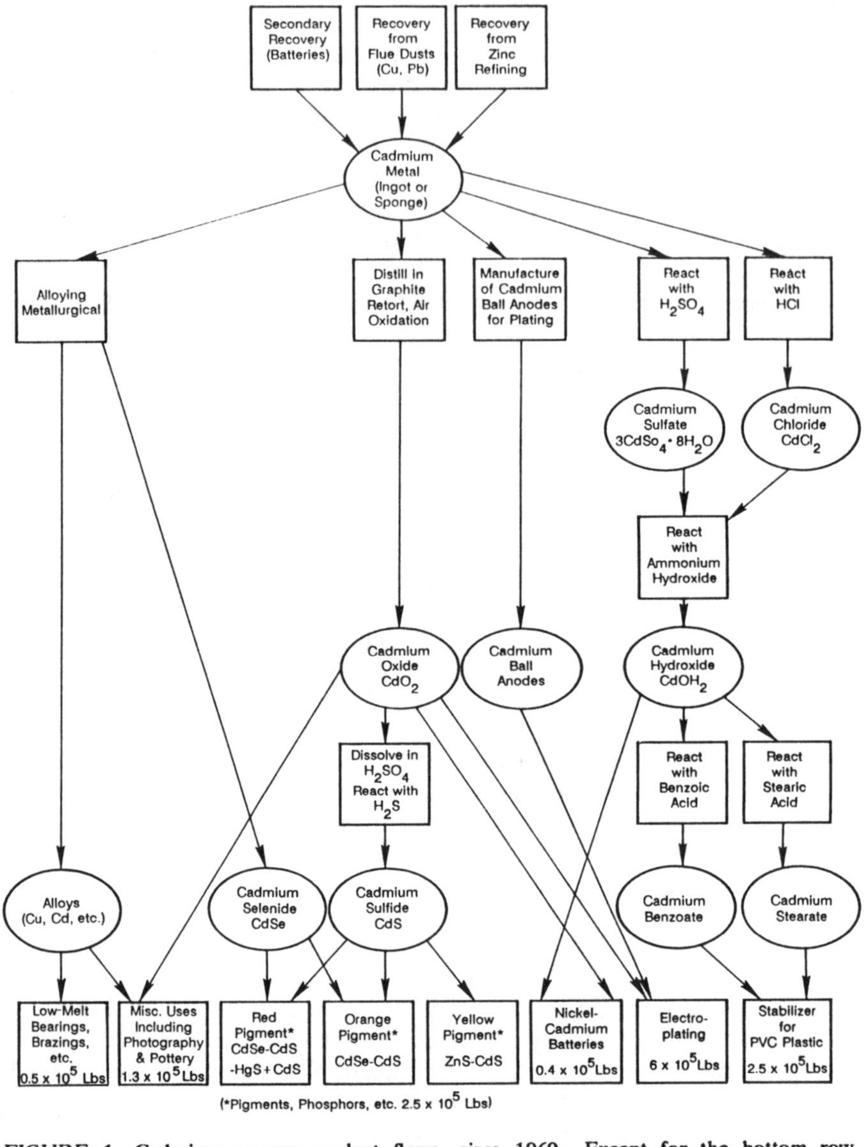

FIGURE 1 Cadmium process-product flows, circa 1969. Except for the bottom row, rectangles indicate processes; ellipses, products or reactions. SOURCE: Ayres et al. (1988).

cycle are familiar examples. The concept is also applicable, of course, to flows that are not truly cyclic, as in the case of arsenic (Figure 3). This concept has been widely used by geochemists, hydrologists, ecologists, and environmental scientists to organize and systematize their work. Such a presentation helps specify geographical scales of analysis. It also facilitates such comparisons as the relative importance of natural and anthropogenic sources. Finally, and potentially most important, it provides a starting point for detailed analysis of the effect of anthropogenic emissions on natural processes.

CONCLUSION

It appears that we have methods to describe our industrial metabolism better, both qualitatively and quantitatively. Initial analyses reveal several important points, for example, that many materials uses are inherently dissipative and thus pose difficulties for recycling. Analysis also shows that although residuals do not disappear from the real world of human health and environmental quality, they do tend to disappear from the market domain. Thus, many environmental resources are underpriced and overused. It is also clear that where the production and use of by-products are concerned, many industrial processes involve multiple steps, resulting in a low level of system efficiency, especially in comparison with biological systems. The sum of the argument here suggests that we should not only postulate, but indeed endorse, a long-run imperative favoring an industrial metabolism that results in reduced extraction of virgin materials, reduced loss of waste materials, and increased recycling of useful ones.

APPENDIX

The Biosphere as a Materials-Transformation System

Three salient characteristics mark the difference between the naturally evolved biosphere and its human-designed industrial counterpart, the "synthesphere." The first is that the metabolic processes of biological organisms are derived (by photosynthesis) from a renewable source: sunlight. The second characteristic is that the metabolism of living organisms (cells) is executed by multistep regenerative chemical reactions in an aqueous medium at ambient temperatures and pressures.[12] Most process intermediates are regenerated within the cell. Reaction rates are controlled entirely by catalysts (enzymes). The energy transport function is performed in all living organisms by phosphate bonds, usually in the molecule adenosine triphosphate (ATP).[13]

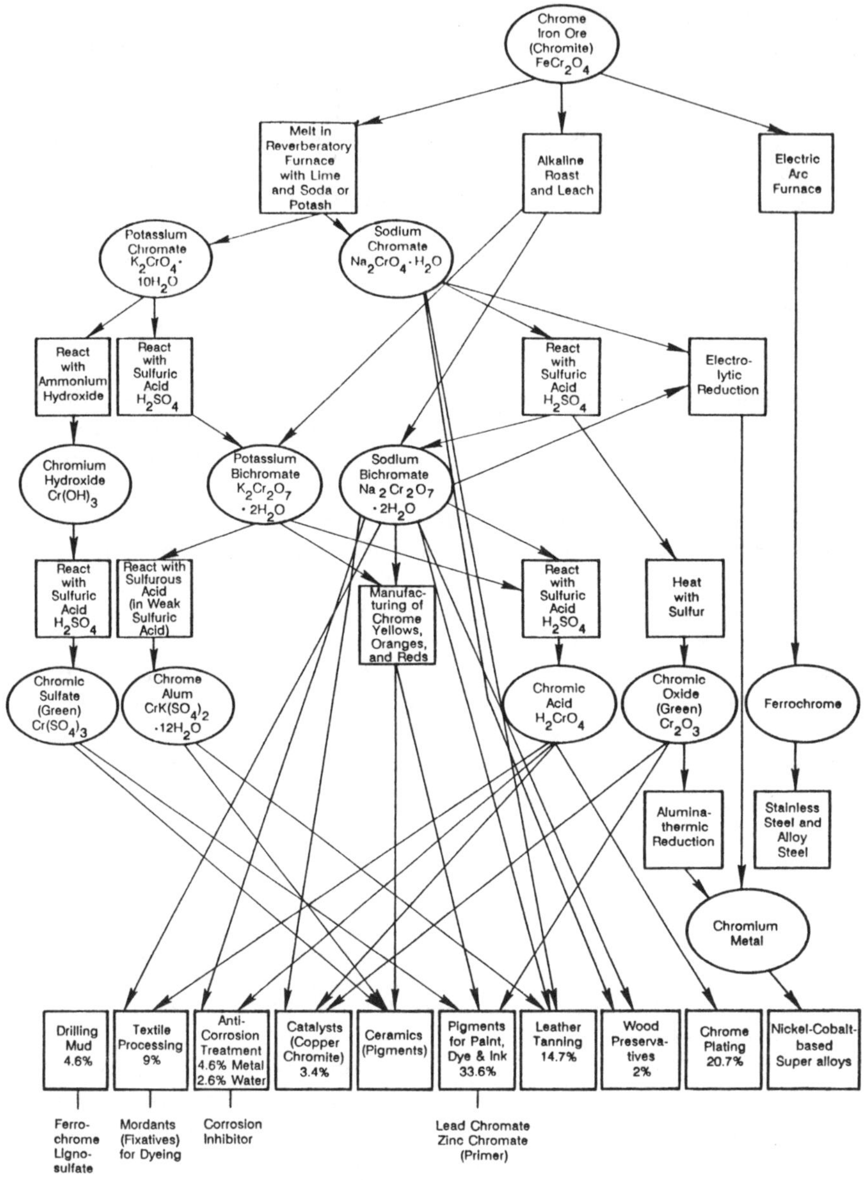

FIGURE 2 Chromium process-product flows, circa 1968. Except for the bottom row, rectangles indicate processes; ellipses, products or reactions. SOURCE: Ayres et al. (1988).

The third salient characteristic differentiating the biosphere from the industrial synthesphere is that, although individual organisms do generate process wastes—primarily oxygen in the case of plants and carbon dioxide and urea in the case of animals—the biosphere as a whole is extremely efficient at recycling the elements essential to life. Specialized organisms have evolved to capture nutrients in wastes (including dead organisms) and recycle them. A significant exception to this rule in the present geological epoch is the deposition of skeletal remains of zooplankton as sediments on the deep ocean floor. These remains are largely, but not entirely, calciferous.[14] Over geological time periods, some of this sedimentary material is likely to be recycled as chalk, limestone, or phosphate rock.

The biosphere as it now exists is a nearly perfect materials-recycling system, but this was not the case when life on earth began. The first and most critical evolutionary "invention," from which all else follows, was the process for replicating complex organic molecules. In effect, the information describing the entire living structure is stored as sequences of nucleic acids in the genetic substance known as deoxyribonucleic acid (DNA). The mechanism for storing, coding, transferring, and decoding that information apparently evolved some 4 billion years ago, before species differentiation. Both the code and the mechanism are common to all known living organisms.

The first cellular organisms, which appeared about 3.5 billion years ago, were prokaryotes (i.e., cells without nuclei). They obtained the energy needed to sustain the reproduction cycle from the anaerobic fermentation of organic molecules previously created by natural geophysical processes in an atmosphere containing no free oxygen. If the primitive atmosphere had contained oxygen, organic molecules could not have survived long enough to achieve the degree of complexity needed to construct self-reproducing systems.[15]

In cellular fermentation a molecule of glucose is split into two molecules of pyruvate. Energy is captured in the form of high-energy phosphate bonds in ADP (adenosine diphosphate) and ATP. Fermentation of a glucose molecule has a net yield of available energy in the form of two molecules of high-energy ATP, converted from the low-energy ADP form (Hinkle and McCarty, 1978). Further reactions in the cell convert the pyruvate to ethyl alcohol, lactic acid, and carbon dioxide, all of which are excreted as wastes. No oxygen is required. The fermentation-based forms of life could not have been the foundation for a sustainable ecosystem, however, because they were using up a finite stockpile of exploitable organic molecules.

The next great evolutionary innovation, about 3 billion years ago, was anaerobic photosynthesis. The first photosynthesizers were prokaryotic photobacteria. These organisms began to synthesize glucose from atmospheric carbon dioxide and sunlight, thus replacing the depleted organic "soup" of

42

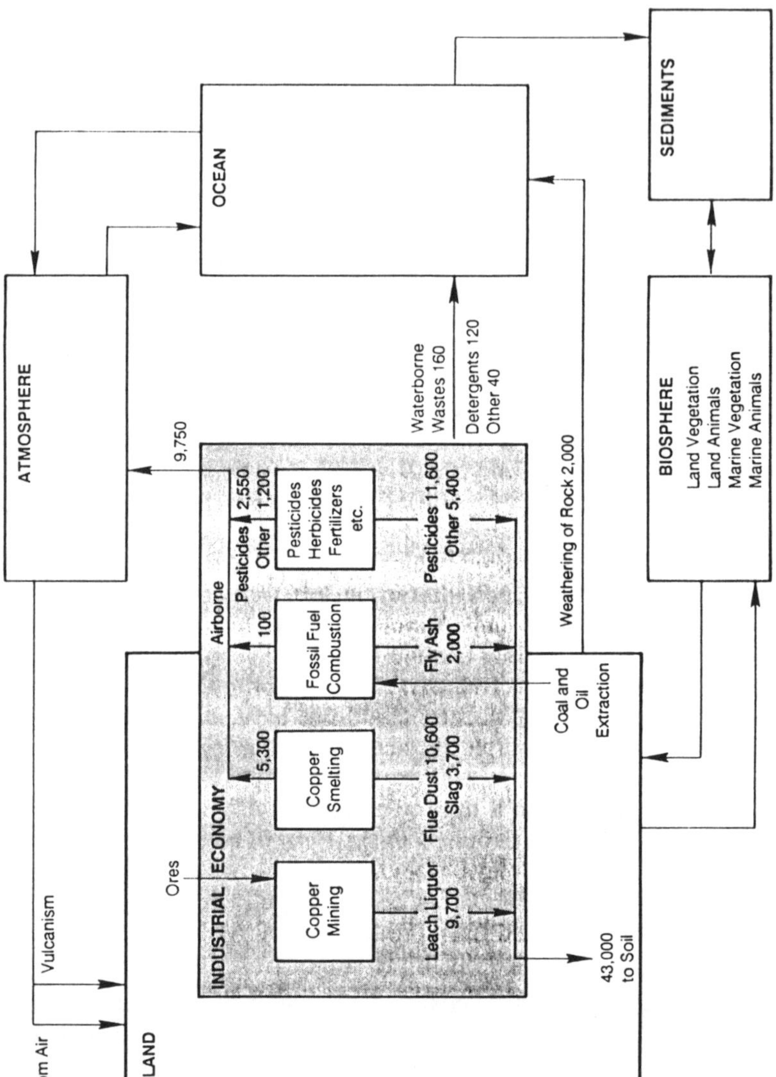

FIGURE 3 Simplified representation of arsenic pathways in the United States (metric tons), circa 1975. SOURCE: Ayres et al. (1988).

the primitive oceans. They also produced oxygen as a waste product. At first, the free oxygen was quickly removed by stromatolites, organisms that combined oxygen with iron dissolved in the oceans and precipitated as iron-rich reefs. This seems to have been the origin of hematite deposits being exploited as iron ores today. However, when the dissolved iron was used up, about 1.8 billion years ago, the oxygen level in the atmosphere began to rise, thereby increasing the rate of dissolution of all macromolecules. Thus, the ecosystem was still unsustainable, because it could neither tolerate nor recycle its own toxic wastes. The oxygen toleration problem was partially solved by a new class of aerobic photosynthesizers, the cyanobacteria (or blue-green algae). These appeared a little more than 2 billion years ago.

A third great evolutionary innovation—the substitution of respiration (using oxygen) for fermentation to obtain energy from organic molecules—solved the recycling problem. The respiration process begins the same way as fermentation, with the splitting of glucose into pyruvate (called glycolysis). However, in respiration, glycolysis is followed by a longer sequence of reactions, known collectively as the citric acid cycle. In effect, the pyruvates are oxidized enzymatically to carbon dioxide, with the formation of many more high-energy ATP molecules. In fact, each glucose molecule, when fully oxidized, yields 36 molecules of ATP, whereas the glycolysis stage alone yields only 2. The yield of available energy for further metabolic processes is, therefore, 18 times that of the fermentation process.

Because the respiration process is far more efficient than its precursor, aerobic respirators required much less organic material to sustain them. Thus, anaerobic organisms could not effectively compete with aerobic organisms in the presence of oxygen. (They still fill an environmental niche in sediments and deep oceans lacking free oxygen.)

With the "invention" of the citric acid cycle, the biosphere became sustainable within more self-defined system boundaries. Evolutionary developments since then—of which the most important were the development of the eukaryotes (cells with true nuclei) and the advent of sexual reproduction—made the system more diverse and more efficient. It is interesting to note that, despite the radical changes in energy metabolism which occurred, the basic scheme of macromolecular reproduction seems to have remained essentially unchanged for 4 billion years.

Industrial Transformation Processes

In contrast to modern biological processes, industrial processes are almost exclusively energized by the combustion of fossil fuels, which (by definition) are not regenerated within the system. In this sense, the industrial system of today resembles the earliest stage of biological evolution, when the most primitive living organisms obtained their energy from a stock of organic molecules accumulated during prebiotic times. Instead of regenerative cycles powered by solar (or nuclear) energy, industrial processes are linear sequences of discrete, irreversible transformations (Figure 4). The sequence begins with extraction of raw materials, followed by physical separation and elimination of impurities, and subsequent reduction or recombination into convenient "first-tier" intermediates. This category includes primary metals and other elements in pure form, cellulose, sodium carbonate, ammonia, methane, ethane, propane, butane, benzene, xylene, methanol, ethanol, acetylene, ethylene, propylene, and some others. These materials, in turn, are subsequently recombined into desired chemical and physical forms.

Almost all of the processes for reducing metals from ores or producing first-tier intermediates are endothermic, that is, driven by externally supplied heat.[16] Processes often use catalysts, and rates and directions are fine tuned by controlled variation of temperatures, pressures, and flow rates or dwell times. There are five major categories of endothermic processes: (1) dehydration; (2) calcination; (3) "reducing" processes for splitting metal (or other) oxides into their constituents; (4) dehydrogenation processes, of which the simplest is the splitting of the water molecule;[17] and (5) processes for combining or synthesizing molecules that do not combine spontaneously at ambient temperatures or pressures. Energy is obtained initially from combustion (oxidation) and delivered either by process steam or by direct contact with the oxidation products. Comparatively few industrial processes are electrolytic: the production of aluminum, sodium, and chlorine and the refining of blister copper are the primary examples.

The most familiar example of dehydration is the production of plaster of paris from the mineral gypsum. Brick and ceramic manufacturing are also based to some extent on this process. The major example of calcination (and the origin of the name) is the production of calcium oxide (quicklime) from calcium carbonate (limestone) by driving off carbon dioxide. This is a major element in the manufacture of portland cement. Both dehydration and calcination are accomplished by the simple application of heat. The so-called carbothermic reduction process by which iron ore is converted to pig iron is a typical example of the third category.[18] In this reaction, coke is partially oxidized to carbon monoxide, which in turn reacts with the ore—at appropriate temperatures—to reduce the iron oxides, while converting the carbon monoxide to carbon dioxide.[19] The temperature determines which way the reaction goes. If the temperature rises too high (above 1300° C), any pure iron that is present will reoxidize.

Ammonia production exemplifies both the fourth and the fifth categories. Dehydrogenation (water splitting) is an example of a process by which synthesis gas for ammonia is produced. In a first stage of the reaction, steam reacts with natural gas, in the presence of a catalyst, to produce

INDUSTRIAL METABOLISM **45**

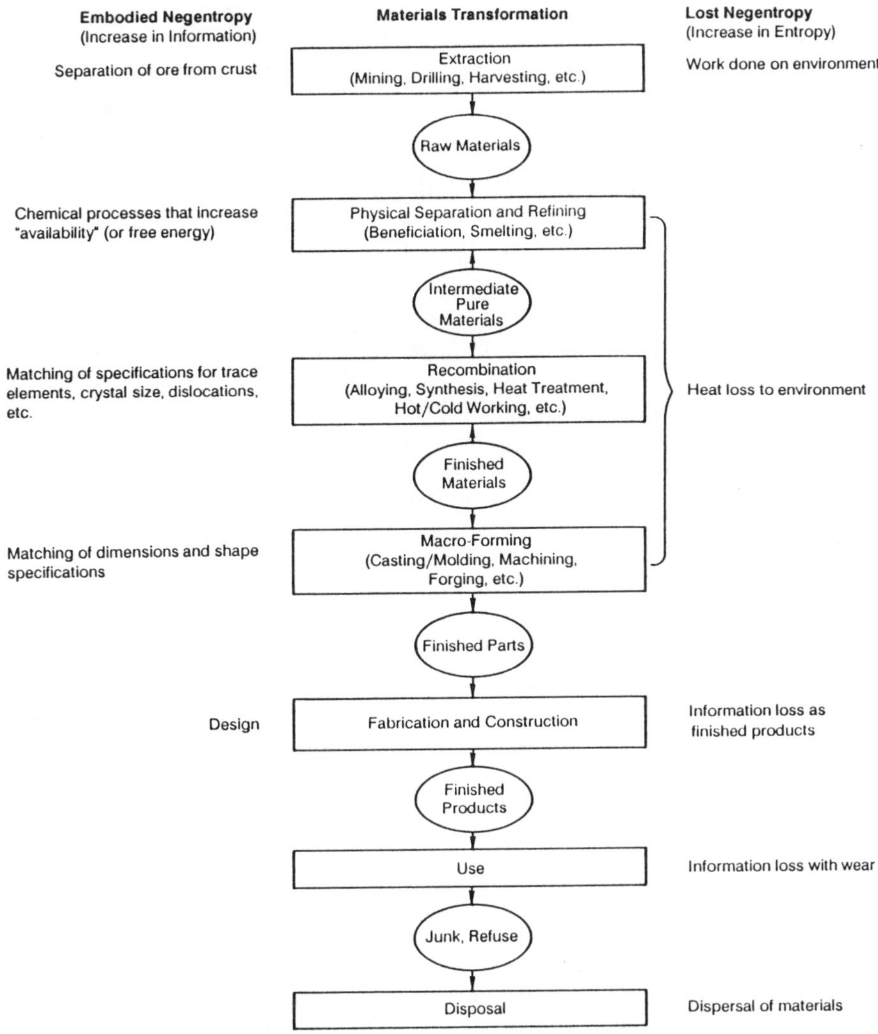

FIGURE 4 Representation of the economic system as a multistage system for the extraction, physical separation, recombination, formation, and consumption of materials. At the end of the sequence, materials are returned to the environment in a degraded form as waste. The process of creating a finished product (center column) concentrates information, or negentropy, in the product at the expense of increased entropy in the environment.

carbon monoxide and hydrogen. In a second stage, known as the water-gas shift reaction, carbon monoxide reacts with added steam to yield more hydrogen and carbon dioxide, which must be removed by a scrubber such as potassium carbonate. The hydrogen is then mixed with nitrogen gas (from the air) in a 3:1 ratio.[20] At very high pressures and temperatures, these ingredients combine endothermically—again in the presence of a catalyst—to form ammonia gas, the basis of virtually all nitrogen-containing compounds used by our industrial civilization.

Most of the reactions by which ammonia, chlorine, lime, sulfur, methanol, ethanol, acetylene, ethylene, propylene, or other first-tier intermediates are converted to other "downstream" compounds are exothermic and—in effect—self-energizing. For example, many second-tier intermediates are produced by controlled oxidation (e.g., sulfuric acid from sulfur, nitric acid from ammonia, acetaldehyde from ethanol or propane, acetic acid from acetaldehyde or butane, acetic anhydride from acetaldehyde, ethylene oxide from ethylene, propylene oxide from propylene, and so on). Most hydrogenation, chlorination, and hydrochlorination reactions are also exothermic, as are most reactions between strong acids and metals or hydroxides.

In effect, the first-tier intermediates are the energy carriers for subsequent reactions. They play a role somewhat analogous to that of ATP in biochemical systems. However, whereas the ATP is cyclically regenerated within the same cell, the first-tier intermediates are not regenerated but are physically embodied in downstream products. This is another fundamental difference between industrial metabolism in its present form and its biological analogue.

NOTES

1. The three major German firms—Bayer, Hoechst, and BASF—were merged in the 1920s into a single giant under the name I.G. Farbenindustrie. Farben is the German word for "color." The name reflected their common origin as synthetic dye (color) manufacturers.
2. An important corollary is that the underpricing of environmental resources corresponds to an underpricing of those exhaustible mineral resources whose subsequent disposal as waste residuals causes harm to the environment. This is because of the lack of any link between the market price paid (for coal, oil, or whatever) and the subsequent cost of waste disposal or—more important—of uncompensated environmental or health damages such as bronchitis, asthma, emphysema, cancer, soil acidification, the greenhouse effect, and so on. Here the distinction between renewable and nonrenewable resources is critical: although renewable resources can obviously create pollution problems, such as sewage, they are almost invariably localized in nature and can be abated at moderate cost. This is emphatically not the case for combustion products of fossil fuels or dispersion of toxic heavy metals.
3. The total world output of carbon dioxide from fuels and cement manufacturing has been estimated to be 5.1 billion metric tons for 1982, of which 26.7 percent was attributable to North America (Marland and Rotty, 1984).

4. Fly ash is primarily a by-product of coal combustion. At present it is being recovered fairly efficiently, from stack gases of large utility boilers and industrial furnaces, by means of electrostatic precipitators. However, the ash itself has become a large-scale nuisance because there exists no use or market for it. The amounts are large: over 50 million tons are generated annually in the United States alone. Several possible remedies exist. Fly ash is a potential "ore" for several metals, especially iron, aluminum, and silicon. These metals could probably be recovered commercially if, for example, bauxite became unavailable (Ayres, 1982). Alternatively, fly ash could be used as a substitute, or more likely as a supplement, for portland cement in the manufacture of concrete and concretelike products. (Its major disadvantage in this application is that concrete made with fly ash does not harden and set as rapidly as the commercial variety. This has obvious economic costs, but so does the disposal of fly ash in landfills.) Another use of fly ash—already demonstrated in France—is as a means for the permanent disposal of toxic liquid wastes in the form of a hard, impermeable rocklike material suitable for long-term storage.
5. The "chain" analogy is an oversimplification, of course, because many processes yield more than one useful product (the chlor-alkali industry is an obvious example) and many products also require two or more inputs. Thus, the structure of the system as a whole is more like a "tree."
6. For example, the first process to manufacture acetylene proceeded by way of calcium carbide production (from coke and limestone) but was displaced by direct dehydrogenation of hydrocarbon feedstocks. The first production of acrylonitrile involved a reaction between ethylene oxide (itself the third step in a chain beginning with ethane) and hydrogen cyanide (made from ammonia). This was replaced by an acetylene cyanation process (acetylene being made from methane) and finally by a propylene ammoxidation process (propylene reacting directly with ammonia). Similarly, the first process in the manufacture of acetaldehyde started with acetylene (from calcium carbide) or ethanol (from ethylene). A newer process made acetaldehyde directly from ethylene. Thus, in the first example the calcium carbide stage was bypassed. In the second example the oxidation of ethylene was avoided, and in the third example the conversion of ethylene to alcohol was avoided.
7. The problem has been dramatized by several recent episodes in which cities have attempted to dispose of solid wastes by using private contractors who, in turn, thought to transport them to countries where disposal rules are nonexistent or weakly enforced. For example, the city of Philadelphia contracted with Joseph Paolino & Sons to dispose of incinerator ash. Paolino, in turn, contracted with Amalgamated Shipping, a Bahamian concern, to transport the ash to the Bahamas. However, the Bahamian government barred the dumping, and the ship carrying the ash was subsequently turned away from ports in the Dominican Republic, Haiti (after it had dumped 2,000 tons of ash), Honduras, Costa Rica, Guinea-Bissau, and the Cape Verde Islands. The ship apparently succeeded in dumping its load of ash somewhere in the Indian Ocean, after more than two years (*New York Times*, November 10, 1988).
8. Admittedly, it can still happen. Asbestos and polychlorinated biphenyls are two examples of materials that were once thought to be safe but have subsequently come to be regarded as hazardous and for which the major producers or users have had to spend billions of dollars to collect and dispose of them safely. Nevertheless, there would seem to be an economic opportunity for a "high-tech" resource-recovery firm to go into business reconverting fly ash and incinerator ash into its most valuable components, light metals such as aluminum, iron, potash, and titanium (Ayres, 1982).

A mineralized glassy residue of sodium-silica and heavy metals would, of course, remain for disposal (though it might also find uses as a construction material).

9. This is a straightforward implication of the widely accepted, but sporadically enforced, "polluter must pay" principle.
10. A number of possible processes for large-scale thermal decomposition of water have been suggested and studied in some detail, e.g., by the European Atomic Energy Commission (EURATOM). See, for example, Marchetti (1973).
11. Ammonia, hydrazine, and other compounds have also been suggested, but there is no clear leader at this time. A new study of the economic and technical feasibility of non-carbon-based liquid fuels would be helpful in clarifying the choices.
12. It is true that some organisms have evolved to function in the deep oceans under conditions of high pressure and salinity; others have evolved to function in surface waters, saline or fresh; still others have evolved to function in environments with very little water. Nevertheless, the internal environment of every cell is aqueous and the pressure inside each cell is essentially the same as the external pressure of the environment in which the organism lives.
13. For a concise summary of the biochemistry of energy transport in cells, see Schopf (1978).
14. The so-called manganese nodules, which are accretions of iron, manganese, copper, cobalt, and other transition elements, are evidently the result of some combination (as yet imperfectly understood) of biological, chemical, and geological processes (see, for example, Morgenstein, 1973).
15. Even in the absence of oxygen, it would seem that dissolution of macromolecules should proceed faster than synthesis (Wald, 1954). The exact evolutionary mechanism leading to self-reproduction is still obscure. Fixation of nitrogen must have been accomplished during this early period, because the early atmosphere was nearly transparent to ultraviolet radiation and any free ammonia in the atmosphere would have been quickly destroyed.
16. The obvious exceptions are elements occurring naturally (such as sulfur) and hydrocarbons that can be obtained by physical separation from natural gas (methane, ethane, propane, butane) or coal tar (benzene, xylene, toluene). Cellulose occurs naturally in some very pure forms (e.g., cotton), but it is usually obtained from wood pulp by a chemical digestion process.
17. Other major examples of dehydrogenation include the splitting of methane, ethane, propane, and butane to produce acetylene, ethylene, propylene, butylene, and butadiene.
18. Pig iron is a solution of iron carbide in iron, with a typical carbon content of approximately 6 percent. The conversion of pig iron to pure (wrought) iron or steel requires removing this carbon and then adding any desired alloying elements.
19. Multistage reactions such as the reduction of iron ore and the synthesis of ammonia, involving an intermediate (carbon monoxide) that is produced by the reaction and later consumed, are quite common in industry. Less common are reactions involving an intermediate that is not produced within the process but is recycled. One of the first examples of such a process was the Solvay (ammonia-soda) process for manufacturing synthetic sodium carbonate from sodium chloride and calcium carbonate. In this process, ammonium hydroxide reacts with calcium carbonate to yield ammonium carbonate and calcium hydroxide. Ammonium carbonate is converted to ammonium bicarbonate. When this reacts with sodium chloride, sodium carbonate and ammonium chloride are produced. Finally, calcium hydroxide and ammonium chloride are reacted to recover ammonia (gas) for recycling and calcium chloride. The latter is a low-value by-product.

20. In principle, the nitrogen gas and steam could be produced together by partial oxidation of natural gas (or any hydrocarbon) in air to yield a mixture of nitrogen, steam, and carbon monoxide. The steam in the hot combustion products could then be reacted with additional natural gas to generate hydrogen and more carbon monoxide to be used as feedstock for the shift reaction. However, the presence of nitrogen before it is wanted complicates the engineering unreasonably.

REFERENCES

Ayres, R. U. 1978. Resources, Environment and Economics: Applications of the Materials/Energy Balance Principle. New York: John Wiley & Sons.

Ayres, R. U. 1982. Coalplex: An integrated energy/resources system concept. United Nations Environment Program Seminar on Environmental Aspects of Technology Assessment, United Nations, Geneva, November–December 1980.

Ayres, R. U., J. Cummings-Saxton, and E. Weinstein. 1978. Assessment of methodologies for indirect impact assessment. Final Report (IRT-468-R/a) prepared for U.S. Environmental Protection Agency. Washington, D.C.: International Research and Technology Corporation.

Ayres, R. U., L. W. Ayres, J. A. Tarr, and R. C. Widgery. 1988. An historical reconstruction of major pollutant levels in the Hudson-Raritan Basin: 1880–1980. National Oceanic and Atmospheric Administration Technical Memorandum NOS OMA 43, 3 vols. Rockville, Md.

Clark, R., ed. 1987. The Ozone Layer [Series: UNEP/GEMS Environment Library], Vol. 2. United Nations Environment Program, Nairobi.

Gschwandtner, G., K. C. Gschwandtner, and K. Eldridge. 1983. Historic Emissions of Sulfur and Nitrogen Oxides in the U.S. 1900–1980. Report to U.S. Environmental Protection Agency. Durham, N.C.: Pacific Environmental Services, Inc.

Hinkle, P. C., and R. E. McCarty. 1978. How cells make ATP: The prevailing theory is the "chemiosomotic" one. Scientific American 238(3):104–123.

Lovelock, J. 1988. The Ages of Gaia: A Biography of Our Living Earth. New York: Norton.

Marchetti, C. 1973. Hydrogen and energy. Chemical Economy and Engineering Review 5(1):7–25.

Marland, G., and R. M. Rotty. 1984. Carbon dioxide emissions from fossil fuels: 1950–1982. Tellus 36B(4):232–261.

Miller, A. S., and I. M. Mintzer. 1986. The Sky is the Limit: Strategies for Protecting the Ozone Layer, Research Report (3). Washington, D.C.: World Resources Institute.

Morgenstein, M., ed. 1973. Papers on the Origin and Distribution of Manganese Nodules in the Pacific and Propects for Exploration. Symposium, Hawaii Institute of Geophysics, Honolulu.

Multhauf, R. P. 1967. Industrial chemistry in the nineteenth century. Pp. 468–488 in Technology in Western Civilization: The Emergence of Modern Industrial Society—Earliest Times to 1900, M. Kranzberg and C. W. Pursell, Jr., eds. New York: Oxford University Press.

New York Times. November 10, 1988. After two years at sea, ship dumps U.S. ash. Vol. 138, 10(N), c4(L).

Schopf, J. W. 1978. The evolution of the earliest cells. Scientific American 239(3):110–138.

U.S. Bureau of the Census. 1960–1975. Statistical Abstract of the United States. Washington, D.C.: U.S. Government Printing Office.

U.S. Environmental Protection Agency. 1986. National Air Pollution Emission Estimates, 1940–1984 (EPA-450/4-85-014). Research Triangle Park, N.C.: Office of Air Quality Planning and Standards, U.S. Environmental Protection Agency.

U.S. Office of Technology Assessment. 1982. Genetic Technology: A New Frontier. Boulder, Colo.: Westview Press.

Wald, G. 1954. The origin of life. Scientific American 191(August):45–53.

[22]

EXTERNALITIES: ECONOMICS & THERMODYNAMICS

R.U. Ayres
A.V. Kneese

Preface

The term "thermodynamics" in the title, and throughout the paper, is used somewhat loosely. Our purpose in this paper is to explore connections between physical constraints (e.g., the "laws" of thermodynamics) and economic phenomena. When we say that a certain physical law implies, or does not imply, some economic relationship or consequence, we mean it in the special sense that when the physical constraint is superimposed on the various relevant economic assumptions or axioms, some conclusion follows. As any thermodynamic purist would point out, the laws of thermodynamics, taken by themselves, have no particular economic implications. We may also offend some purists by not offering formal definitions of numerous terms, such as "the economic system". This is because the exact definition would not in any case affect the argument.

Introduction

A major theme of this paper is the pervasiveness of "common property" problems and market resource misallocations associated with them. This is a consequence of the fact that the real economic system is an open system that must extract, process and discard great amounts of physical materials. All three processes involve significant externalities - market failures- particularly the final disposal of waste residuals. Economic theory has long recognized in a limited way the fact that real markets tend to be imperfect, notwithstanding the great emphasis placed on theorems derivable from the assumptions required for perfect markets. In other words, real markets tend, in practice, to violate some or all of the conditions needed to achieve a Pareto-optimum. A question of some interest to us is whether such an optimum still exists in the absence of complete and perfect "futures" markets for all resources, including environmental resources, and, if so, how it progressively deviates over time from the perfect market case. We argue that, if the dynamic optimum does exist, the current price structure and allocation of resources is quite far from it.

The standard Walrasian (neo-classical) model of an economic system consists of a set of production activities with cost functions, a set of independent resources, commodities and services, and a final demand function. It is assumed for convenience that all resources, commodities and services can be produced entirely from linear combinations of others in the system (i.e., the system is closed), with an appropriate expenditure of labor. The wages of labor constitute the income available to satisfy demand. In the original version, no joint products or co-products are permitted and (by assumption) ther are no "free goods" or wastes. In the 1870s Walras conjectured that such an economy would have an equilibrium set of prices and outputs, toward which the economy would always tend.

The existence of a static equilibrium was finally proved by Arrow and Debreu in 1954 [Arrow & Debreu 54]. This achievement has steered a generation of economists into the analysis of highly abstract mathematical models. Indeed, the rather tight restrictions of the original proof have been significantly loosened. The general equilibrium has even been extended to the dynamic case, with exhaustible resources (e.g. [Solow 74]), subject as before to the assumption of perfect future markets for resources. A great deal of theoretical superstructure has been added on to this basic model in recent years, e.g. [Dasgupta & Heal 79]. To borrow a quote attributed to Milton Friedman, "we curtsey to Marshall but we walk with Walras" [Hoos 83 p. 33].

But, however elegant the Walrasian model may be, it does not represent the real world at all well. It has been properly criticized as "a system devoid of human beings" [Seligman 62, p. 385] and a spuriously deterministic one, at that. Milton Friedman once described it as "a form of analysis without much substance" [ibid], and this criticism remains valid. Moreover, a theoretical objection can be raised: The Walrasian model is essentially closed. It posits abstract resources being converted endlessly into abstract goods and services "consumed" by abstract labor and finally back into resources. This is a kind of perpetual motion machine.

Of course, the word "consumption", so much used in economics, is a slippery and potentially misleading one. Materials and material products (unlike pure services) are not really consumed. The only thing consumed is their "utility". When material objects are worn out or obsolete and useless they are discarded. Fuels and a host of other products that are "used up", are really only dispersed or chemically transformed (e.g., into combustion products). All of the matter extracted and processed each year by the economic system is discarded sooner or later, (but in a thermodynamically degraded form). Only a small percentage of the total mass of materials processed is embodied in long-lived products like books, machines, or structures that last for more than a decade.

As we noted at the outset, the real economy is an open system which utilizes large quantities of material and energy resources provided by nature. These can be extracted, but not "produced" (as intermediate and final goods and services are), by linear combinations of labor, capital goods and other intermediate products. As Georgescu-Roegen has correctly pointed out, the real economy is a dissipative system, not a self-perpetuating one [Georgescu-Roegen 71].

The Walrasian paradigm is fatally flawed not so much by its closure, however, as by its assumption of perfect information availability to all transactors. In fact, a major theme of this paper is the necessity of tehcnological change in open systems, as the highest quality resources are exhausted and replaced by others requiring different technologies. (In an economy with fixed technologies, the mass of raw materials processed and the energy required for processing, per unit mass of final products, would necessarily increase over time even if population and per capita income did not).

However, the direction of technological change cannot be completely foreseen, though some trends are predictable. Even if resource stocks were completely known, there exists a fundamental and irreducible uncertainty, not only with respect to the nature of the next major innovation (say, in energy), but with respect to the secondary and indirect effects of it. Clearly, an industrial system based on solar cells or nuclear power would require very different materials and processes than an industrial system based on petroleum or coalburning steam engines. This uncertainty negates one of the necessary conditions for the existence of a dynamic general equilibrium. "Futures" markets involving technology are inherently imperfect and imperfectible. This means that the kind of economic growth that actually occurs in the real world tends to be discontinuous and otherwise inconsistent with the kind of growth permitted within conventional general equilibrium models of the economy.

The use of energy and matter by the economic system is, in some respects, similar to the use of energy and matter by an ecosystem. As with communities of living organisms, the fraction of each year's mass-energy intake that is actually embodied in the growth increment as new biomass is typically very small compared to the fraction that is used for metabolic activities. The term "industrial metabolism" is a convenient rubric for the economic analog of biological metabolism. We shall return to it later.

Economic Implications of Matter Conservation

The first law of thermodynamics states that the total amount of energy is conserved in all processes, hence in all transformations of materials. Since Einstein postulated the equivalence of matter and energy ($E=mc^2$) it follows that what is conserved is "mass-energy". Matter can, in principle, be created from energy or destroyed to produce energy, but as a practical matter, matter-energy conversion is extremely difficult to achieve under ordinary conditions. Nuclear bombs and nuclear power plants do make use of this equivalence, to produce very large amounts of energy from comparatively tiny amounts of matter. The reverse process - matter creation - is so rare that it is almost unobservable. (It was the dominant process in the universe, however, during the first picosecond after the "Big Bang"). But the essential point is that, with the exceptions noted, mass and energy are conserved independently in most processes on the earth's surface.

As noted above, the real economic system is dependent on the extraction of large quantities of matter from the environment, processing and conversion into various forms, culminating in final products for "consumption". Table 1 (below) shows the mass of "active" raw materials extracted for processing within the boundaries of U.S. for the years 1960, 1965, 1970 and 1975. Vegetable material harvested directly by animals has been omitted for lack of data, along with some obviously minor agricultural and horticultural products. Figures for metal ores exclude mine tailings and gangue removed to uncover ore bodies. Inert construction materials such as stone, sand and gravel have also been omitted. Inert materials account for enormous tonnages, but undergo no chemical or physical change except the extent they are incorporated in concrete or paved surfaces. We also exclude soil and subsoil shifted during construction projects or lost by erosion.

Evidently, more than 10 tons of active mass per person is extracted from its own territory by the U.S. economy, each year. The amount processed is somewhat larger, since the U.S. is now a net importer of many minerals and petroleum. Of course, quite a lot of the grain harvested in the U.S. is exported, which balances the international accounts to some degree. Of the active mass processed each year, roughly 75% is mineral and "non-renewable" while 25% is, in principle, from renewable sources. This point is important when we later come to address the issue of resource prices.

Table 1: Mass of Active Materials Extracted Commercially, U.S. (1960-75) (millions of tons)

Material	1960	1965	1970	1975
Food and feed crops (ex.hay)	267	295	314	403
Meat, fish & dairy products	82	85	85	84
Cotton, wool, hides, tobacco	5	5	4	4
Timber (15% moisture basis)	256	267	271	249
Fuels (coal, lignite, oil, gas)	990	1176	1458	1392
Ores (Fe, Al, Cu, Pb, Zn)	400	435	528	460
Non-metallics	200(e)	240(e)	266	255
Total	2200(e)	2500(e)	2926	2847

It is difficult to estimate the fraction of the total mass of processed active materials that is annually embodied in long-lived products and capital goods (durables). None of the food or fuel is physically embodied in durable goods. Most timber is burned as fuel or made into pulp and paper products. At least 80% of the mass of "ores" is unwanted impurities (more than 99% in the case of copper). Of the final products made from metals, a large fraction is converted into "consumption goods", such as bottles, cans, chemical products, and "throw-away" products such as batteries, light-bulbs, and so on. Only in the case of non-metallic minerals (ignoring inert materials, as before) is as much as 50% of the mass be embodied in durable goods (mainly portland cement used for concrete and clays used for bricks and ceramics). The annual accumulation of active materials embodied in durables, after some allowance for discard and demolition, is probably not above 150 million tons, or 6% of the total. The other 94% is converted into waste residuals as fast as it is extracted. In addition to raw materials, large amounts of energy, in a high quality ("available") form are also needed to drive the system. This energy is also degraded and finally discarded as low temperature heat. Entropy increases.

Figure 1 is a schematic representation of the economy as a materials-process-product transformation system. What it illustrates is that materials processing can be regarded as a series of changes in the energy and entropy states of "useful" materials. But the materials themselves (being conserved) do not disappear after they are "used up" in the economic sense. They become waste residuals that can cause harm and must be disposed of. In fact, it is not difficult to show that the tonnages of waste residuals are actually greater than the tonnages of crops, timber, fuels and minerals recorded by

Figure 1.

Representation of the Economic System As Materials- Process- Product Flow

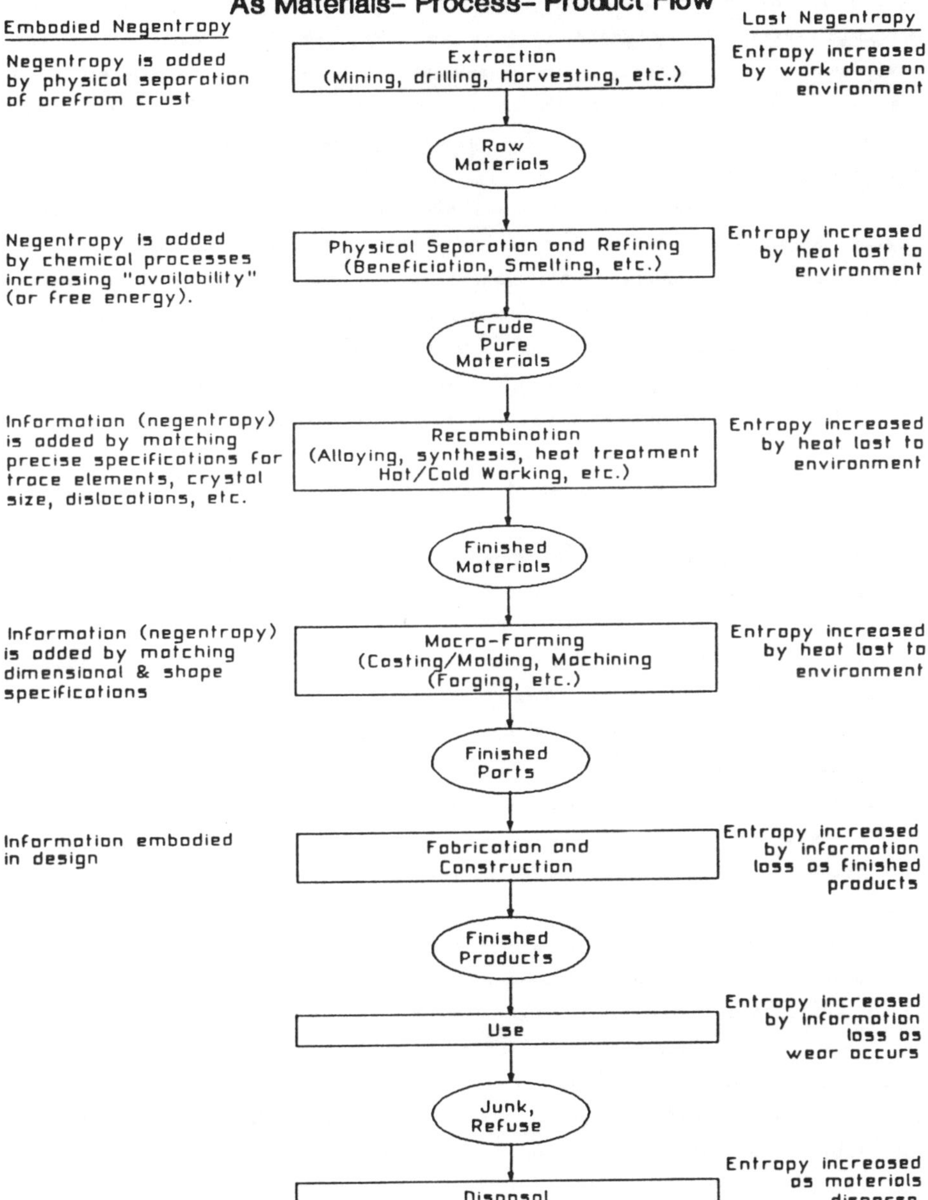

economic statistics. While usually unpriced and unmeasured, both air and water are major inputs to industrial processes and they contribute mass to the residuals - especially combustion products. Residuals tend to disappear from the "market" domain, but not from the external world in which the economic system is embedded (Figure 2). Many services provided by the environment are inherently from "common property" - including the air, the oceans, the genetic pool of the biosphere, and the sun itself. Distortions in the market (i.e., prices) are unavoidably associated with the use of common property resources.

One economic implication of the conservation of mass (part of the first law of thermodynamics), is that in an economic system depending heavily on the extraction of non-renewable resources and the use of common-property resources such as air or water, externalities cannot be rare or unusual. On the contrary, as we noted two decades ago [Ayres & Kneese 69], they are pervasive. Furthermore, increased population and higher levels of industrial output based on a fixed technology obviously would require the extraction of increasing amounts of raw materials. This, in turn, would lead to the increasing relative importance of external effects requiring regulatory solutions.

Among the latter, the most promising in the long run are to prohibit dumping of wastes except in carefully monitored locations, to assess the true costs of disposal on users of materials or material products, and to enforce a higher level of materials recycling activity than an unfettered market economy would adopt. The latter makes sense, because recycling simultaneously reduces both the need for extraction - which involves a considerable degree of environmental disturbance - and the need for disposal of waste residuals into environmental reservoirs that are becoming increasingly scarce. Clearly, environmental resources have been significantly underpriced in the past. For this reason, they have been over-used.

A corollary we did not see clearly enough twenty years ago is that the underpricing of environmental resources corresponds to an underpricing of those exhaustible mineral resources whose subsequent disposal as waste residuals causes harm to the environment. (Here the distinction between renewable and non-renewable resources is critical: while renewable resources can obviously create pollution problems, such as sewage, they are almost invariably localized in nature and can be abated at moderate cost. This is emphatically not the case for combustion products of fossil fuels or dispersion of toxic heavy metals, etc.) We argue, in fact, that exhaustible

96

Figure 2.

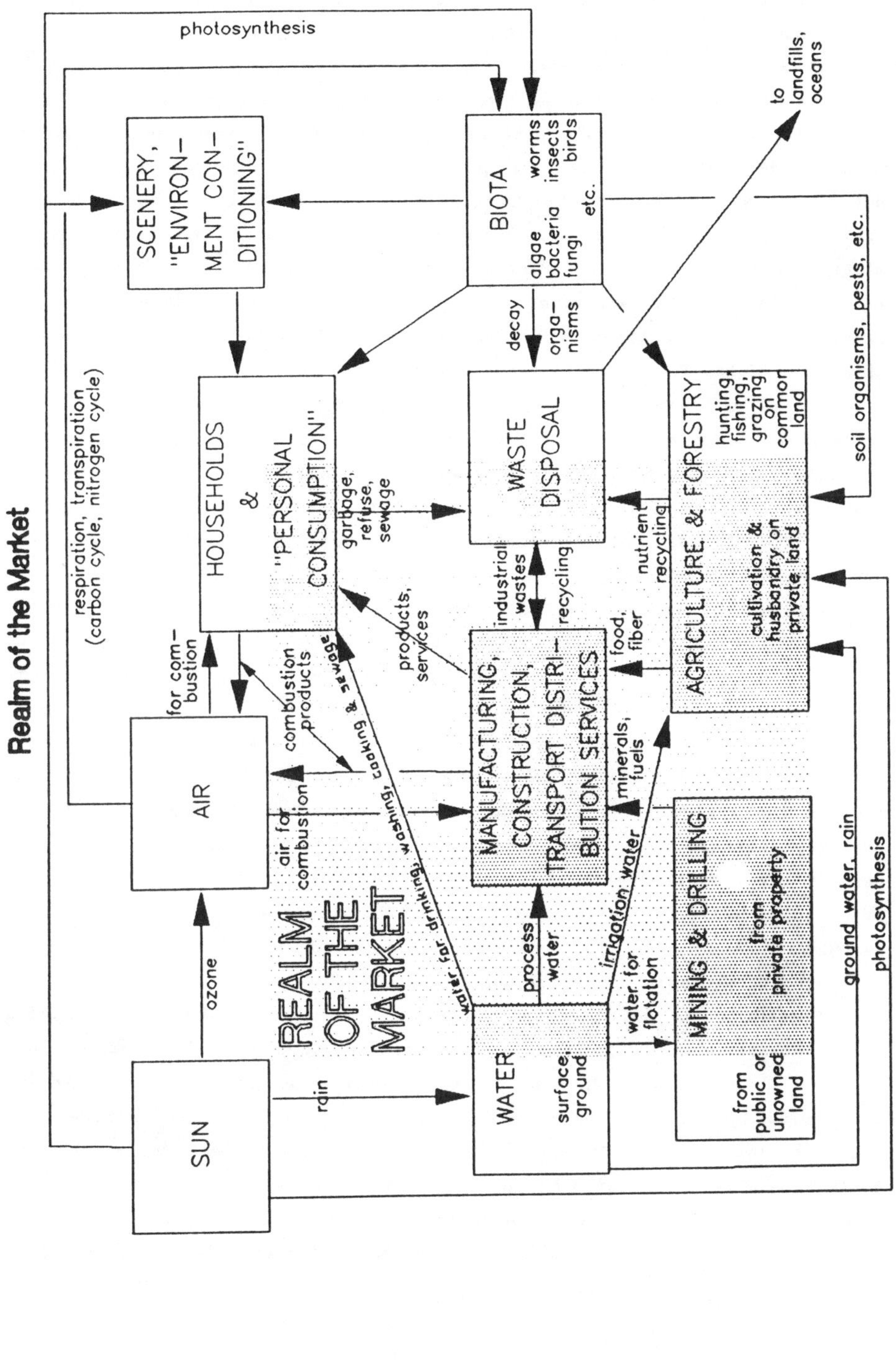
Realm of the Market
photosynthesis
respiration, transpiration
(carbon cycle, nitrogen cycle)
SCENERY, "ENVIRON- MENT CON- DITIONING"
BIOTA
algae
bacteria
fungi
worms
insects
birds
etc.
to landfills, oceans
soil organisms, pests, etc.
decay orga- nisms
HOUSEHOLDS & "PERSONAL CONSUMPTION"
WASTE DISPOSAL
AGRICULTURE & FORESTRY
hunting, fishing, grazing on common land
cultivation & husbandry on private land
garbage, refuse, sewage
industrial wastes
recycling
nutrient recycling
products, services
for com- bustion
combustion products
water for drinking, washing, cooking & sewage
food, fiber
MANUFACTURING, CONSTRUCTION, TRANSPORT DISTRI- BUTION SERVICES
minerals, fuels
AIR
air for combustion
REALM OF THE MARKET
MINING & DRILLING
from private property
from public or unowned land
process water
irrigation water
water for flotation
ground water, rain
photosynthesis
ozone
rain
SUN
WATER
surface, ground

resources have also been underpriced by a large factor, due to the lack of any link between the market price paid and the subsequent cost of disposal or of uncompensated environmental damage. The true dynamic equilibrium price of any exhaustible resource would have to correspond to a rate of use such that the discounted marginal environmental degradation (cost) resulting from the ultimate disposal of the material is balanced by a discounted stream of marginal benefits. The resource prices that would correspond to a true Walrasian dynamic equilibrium cannot be estimated quantitatively by any "partial equilibrium" approach. However, as an indication of magnitudes, it might be helpful to consider what the cost of energy would be if no incrases in atmospheric carbon-dioxide, or environmental acidity, were permitted. (These, among other constraints, would seem to be implied by the criterion of environmental "sustainability" currently in vogue).

One of the unresolved problems here is the choice of discount rate. While most economists appear to prefer a "market rate of interest", there is a school of opinion (with which we feel considerable sympathy) to the effect that a much lower -possibly zero rate- of discount is the appropriate one. This follows if (on ethical grounds) the interests of future generations are given equal weight with those of the present generation. Taking this position, which has been called "conservationist", no permanent degradation of the environment can be accepted because, without discounting, it results in an infinite cost. From an economic perspective, this is roughly what the notion of a "sustainable environment" implies.

Another important question is the following: Given that the economic system currently depends on processing large amounts of matter derived from outside the system (i.e. from the environment), does "thermodynamics" have anything further to say? In particular, is there any inherent relationship between the amount of matter that must be extracted and the aggregate value of economic services that can be produced? To simplify (perhaps excessively) is there any long-run lower limit to the amount of matter that must be processed to generate a unit of GNP?

The answer is no, both in practice and in theory. The trend toward a service-based economy in the U.S. is well-documented and to the point. Figure 3 shows the rise in employment in information-related activities and the corresponding long-term decline in agriculture, mining and manufacturing [Porat 77]. In general, jobs in the service sectors now contribute more to GNP as jobs in the sectors that process materials.

At the micro-level, too, there ar many examples of dramatic increases in the amount of economic value that can be delivered by a given amount of

98

Figure 3.

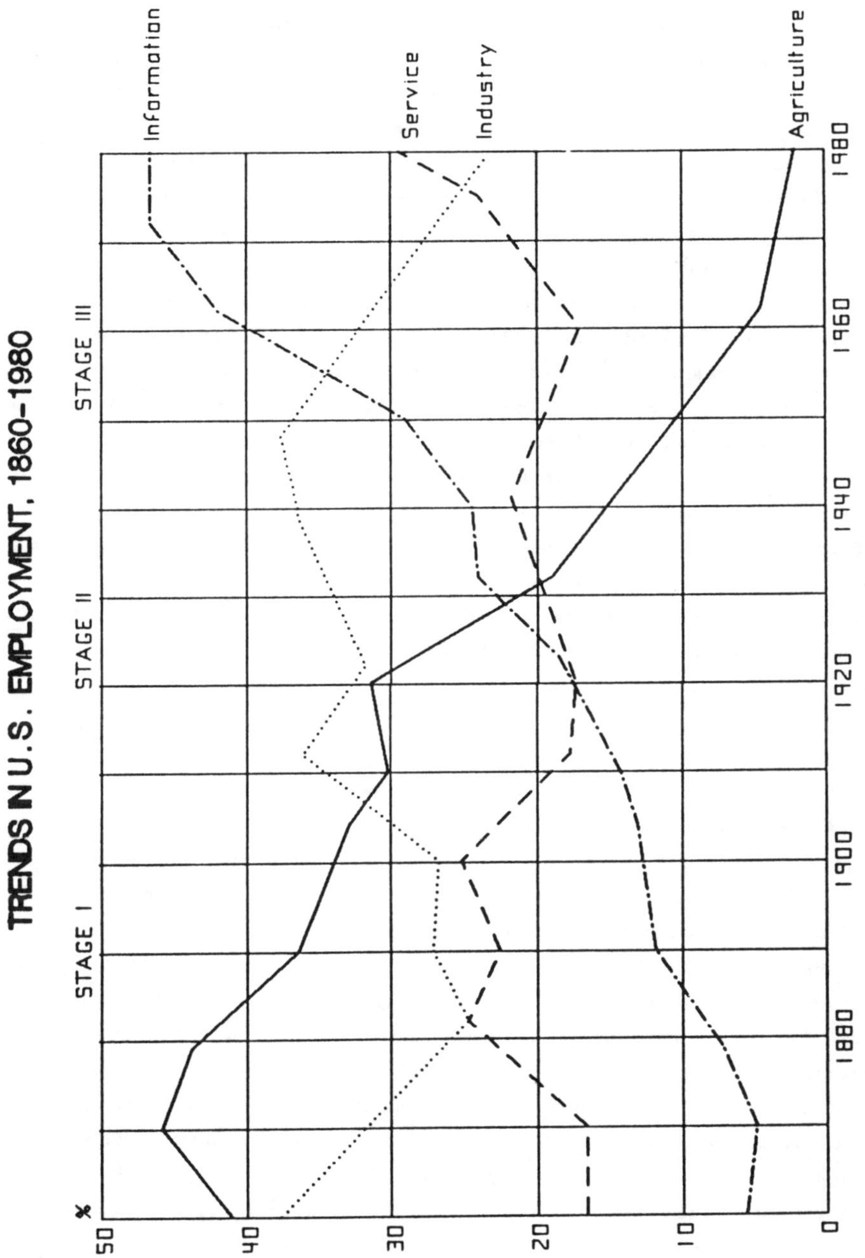

Source: [*Porat 77*]

matter. At the beginning of the industrial revolution, several tons of stone, wood and iron were required to deliver one shaft horsepower via a Newcomen steam engine. Today's aircraft jet engines require less than a pound of materials per (equivalent) shaft horsepower. The shrinkage of computers is even more remarkable. Vannevar Bush's differential analyzer (1935) weighed 100 tons. Though it was an analog computer, its essential functions could easily be performed by a sophisticated pocket calculator today. Yet several models were built nevertheless, to calculate ballistic trajectories, electronic orbits, and the like [Shurkin 85, p. 79]. Eckert and Mauchley's ENIAC (1946), the first purely electronic machine, weighed 30 tons and consumed 174 kilowatts, but it was hundreds of times faster than its predecessors [ibid. p. 139]. One of today's lap-top computers, thousands of times more powerful than ENIAC, weighs only 10 to 20 pounds. Yet the trend to miniaturization apparently has much farther to go, as functions now carried out by thumb-nail sized silicon "chips" may eventually be performed by individual custom-grown organic molecules.

The theoretical reason why physical considerations do not imply a lower limit to the Extracted Matter per unit of GNP (EM/GNP ratio) is that as long as energy is available there is no fixed upper limit - short of 100% - to the fraction of materials that can be recycled. In other words, energy is the only limiting factor. This point will be taken up again later in connection with the "second law", where it is more relevant.

Implications of Conservation of Energy

We have noted above that the "first law" of thermodynamics is better known as the law of conservation of energy. Given that the economic system is driven by a flux of thermodynamically available energy that must be provided from outside the system, and (as will be seen later) cannot be recycled, the corresponding question arises: Is there any inherent relationship between the amount of available (or free) energy that must be supplied and the economic value (GNP) that can be generated? Again, the answer is no. Again, we can argue the point on both empirical and theoretical grounds.

At the macro-level, the record is somewhat confusing at first glance. What it seems to show is an increase in energy-use per unit of GNP during the industrialization stages of development, followed by a decline as countries enter the "post-industrial" phase, with increasing contributions from the service sectors and, particularly, the information-intensive activities. In the case of the U.S. the period of increasing Energy/GNP ratio ended in the 1920s, and there has been a long-term decline since then, as shown in Figure 4. From 1929 to 1955 the decline was 24%. [Schurr et al 60, p. 177]. However, this decline was interrupted by a period of relatively constant ratio lasting from about 1950 until 1967. (An increase of 7% from 1968 through 1970 was probably due to the Viet Nam War). Real energy prices in the U.S. actually declined fby 10% from 1953 to 1973 [Dunkerley 80 p. 56]. After 1973 energy prices rose sharply and the long-term decline in E/GNP continued. There was a further 21% drop from 1976 through 1984.

The experience of other countries has been qualitatively similar, but peak E/GNP levels have been reached more recently. France and West Germany both peaked in 1957 [ibid], while Sweden, the Netherlands and Japan all reached their peaks in the mid-70s. The OECD as a whole experienced a drop in E/GNP ratio of 19% from 1973 to 1984. For developing countries, the ratio is still growing. Nevertheless, the record strongly suggests that E/GNP ratio's will decline once the industrialization process is well advanced.

Yet studies of the last three decades (for which reasonably good data are available for many countries) have also lent unjustified credence to the notion that the link between energy and GNP might be quite inelastic. A number of studies in the early 1970s suggested that energy supplies could not be curtailed without drastic reductions in standards of living. It was also suggested repeatedly by reputable analysts that the maximum potential for "conservation" was only of the order of 20%. For example, in 1977 the OECD estimated maximum potential conservation possibilities as follows [OECD 77]:

Energy sector	1%
Industry	4%
Transportation	7%
Residental/commercial	4-5%
Total	16-19%

This "maximum" saving was exceeded well within a decade.

The widespread assumption by economists that the E/GNP ratio is inelastic is parallelled by a pervasive misunderstanding of the distinction between energy use, and efficient energy use, to which engineers have also contributed. This misunderstanding amounts to a myth, viz. that roughly 50% of energy inputs are "used" by our economy (presumable efficiently)' while the other 50% is "lost". This summarizes the message conveyed by a very widely ciruclated pamphlet entitled "Understanding the National Energy

101

Figure 4.

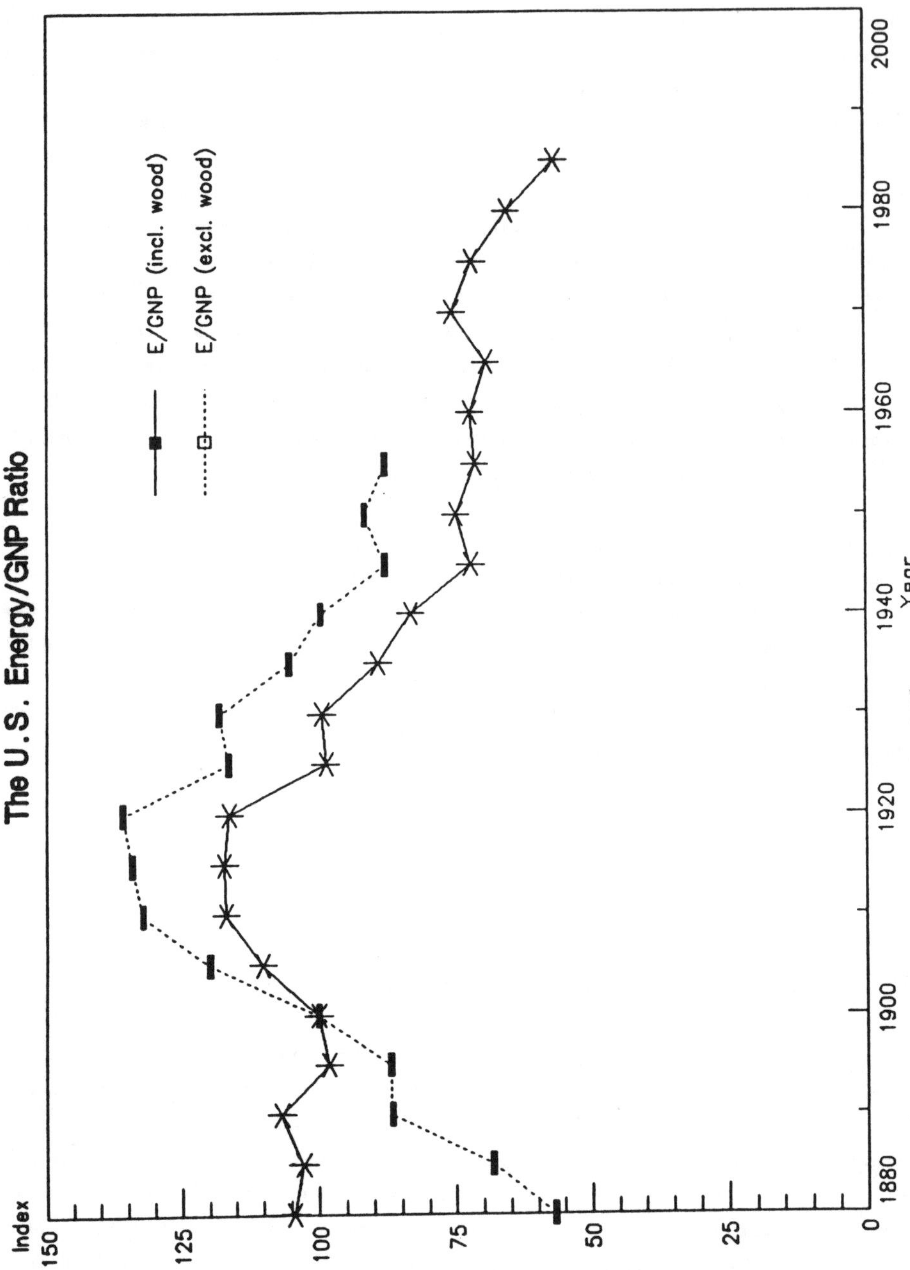

Sources: [*Schurr & Netschert* 60, *EIA* 86]

102

Dilemma", prepared with the help of Livermore National Laboratory and the staff of the Joint Committee on Atomic Energy [Bridges 73], and circulated by the JCAE. An updated version, entitled "Energy: A National Issue" was even published by the Center for Strategic and International Studies of Georgetown University [Murray 76].

According to the latter version, aggregate energy "loss" rates for various sectors were as follows:

Electricity Generation & Distribution	65%	(implied efficiency 35%)
Industry	30%	(implied efficiency 70%)
Transportation	75%	(implied efficiency 25%)
Residential/Commercial	30%	(implied efficiency 70%)
Total	51%	

If it were true, as these reports (and many others) suggested, that only 51% of all energy inputs to the economy were wasted, it must follow that the economic "engine" is already using energy with an average overall efficiency of 49%. If this were so, it also follows that very little more energy could be saved by conservation.

The above conclusion is wrong and, not to mince words, absurd. In the first place, it is based on a technically incorrect concept of thermodynamic efficiency. This was pointed out almost immediately in the report of a summer study supported by NSF and sponsored by the American Physical Society [Ross et al 75]. This report was later reprinted as an appendix to some congressional hearings, but had little impact. Yet, based on carefully documented calculations, the authors showed that the true thermodynamic efficiency of burning oil or gas in a furnace to supply warm air at a temperature of about 70° F to a room (the major use of energy in the residential/commercial sector), far from being 70%, is no more than 3%! Allowing further for the fact that many buildings are underinsulated (i.e. they leak heat), and many rooms are overheated, or are heated even when there are no occupants, the real efficiency of the current technology for space heating is much closer to 1% than 70%. Consequently opportunities for conservation are almost unlimited, a fact that is beginning to be cautiously recognized by scientists, at least [e.g. Rosenfeld & Hafemeister 88].

Though we cannot present additional details here, it must be said that the same basic conclusion can be reached for all of the other economic sectors. On an overall basis, even allowing for somewhat higher real efficiencies in the basic industrial sector, the overall efficiency with which the U.S. used energy in 1975 was probably less than 5% and perhaps considerably less [Ayres & Narkus-Kramer 76][Ayres 88]. The above figures do not reflect the possibility of fundamentally new technologies or new kinds of services requiring very little energy to deliver. Opportunities for conservation are correspondingly great. In terms of the E/GNP ratio, it seems likely that a fourfold reduction (or more) could be achieved even without inventing "radical" new technologies or extraordinary costs. Fuller documentation of this argument wil have to be presented in another paper, however [Ayres, 1988].

Implications of the Second Law: No Perpetual Motion

The second law of thermodynamics states, in essence, that "entropy" (a certain property of systems, which we do not propose to define rigorously here) increases in any irreversible process. A closed system is one that exchanges neither matter or energy with its surroundings. Indeed, except in exceptional and practically unrealizable situations, all real processes are irreversible, hence total entropy increases in virtually every real process. When entropy increases the energy in the system becomes less available to do "useful work". In engineering practice, entropic calculations are primarily useful for purposes of analyzing the efficiency of alternative energy-conversion systems and chemical reaction paths. In this domain, there are important economic tradeoffs between reaction rates, yields and efficiency, for example [Tribus & El-Sayed 81], [Kenney 84].

Most non-physicists are more familiar with simplified versions and (apparent) implications of this law. (One example of the latter is the so-called "heat-death" (Wärmetod) of the universe, which was popularized at the end of the 19th century in the context of debates about whether biological evolution is "contrary" to the laws of thermodynamics). A simple implication of the law is that all energy conversion systems and matter conversion systems require more available (or "free") energy inputs than they yield as outputs. In other words, all biological and industrial processes are less than 100% efficient. In still simpler terms, it is impossible to construct a perpetual motion machine - even an abstract one like the Walrasian economic system.

It follows from the second law, incidentally, that no material recycling process can be 100% efficient. A seemingly natural implication of this true statement is that stocks of matter must therefore be dissipated gradually, just as available energy is. Indeed, Georgescu-Roegen has reached this conclusion. Unfortunately he has elevated it into a "fourth law" of thermodynamics [Georgescu-Roegen 77,79]. If Roegen's "4th law" were valid, it would follow that even a continuous flux of solar energy would not be enough to ensure the possibility of a "sustainable" (stationary-state) economy. The earth would eventually "run down" like a mechanical clock, as its stocks of high-quality material resources (such as metal ores) were used up. Given the rate at which "grade deflation" for ores like copper seems to be occurring, the exhaustion-point would seem to be nearer a few hundreds of years from now than a few billions of years. Indeed, the sensational Club of Rome Report "Limits to Growth" assumed that reserves would be exhausted even sooner [Meadows et al 72]. The simplistic world-model presented in that book, and the whole notion of world-modelling, has been largely discredited since then.

Faced with a neo-Malthusian ("fixed resource base") argument, economists have always stressed the potential for technological substitution [Goeller & Weinberg 76], [Tilton 83]. For example, if copper becomes scarce, we switch to aluminum conductors; if tungsten becomes scarce, we can use molybdenum for tool steel; if cobalt and chromium are unavailable for "superalloys", we will develop ceramic turbine blades, and so on. Many such substitutions have already taken place and more will occur in the future. But this does not really contradict Georgescu-Roegen's "4th law" of matter dissipation. It merely delays the day when some "critical" material becomes unavailable and cannot be replaced by another. (For example, there is no technological substitute for phosphorus as a plant nutrient). In the long run, substitution is no substitute for at least some recycling.

Fortunately for the prospects of mankind beyond the next few centuries, there is a subtle but critical flaw in Georgescu's gloomy argument about the impossibility of recycling. It is perfectly true that no industrial process - including recycling - will ever be 100% efficient. But the earth is not a closed system as long as the sun shines, so energy availability is not a problem. Nor is it necessary to achieve thermodynamic perfection to separate useful materials from the terrestrial environment in sufficient quantities to operate an industrial economy. All that is needed is (1) enough solar energy and (2) a gravitational field strong enough to prevent matter from escaping into interstellar space. Given that the earth is a closed system with respect to matter (but not with respect to energy), and the fact that matter is conserved in a closed system, the most unfavorable limiting case is that one might have to extract materials from a "soup" with the average composition of the earth's crust.

In fact, a moment's reflection is enough to see that things can never get even that bad, by a considerable margin. Interatomic forces differentiate materials by chemical properties, such as reactivity, solubility, freezing/melting point and vaporization temperature. At the temperature of the Earth's crust some elements are gaseous, some compounds are liquid and some are solid. Some solids dissolve in water, others don't. The ocean will always have a much higher concentration of soluble salts (e.g. sodium, potassium and magnesium chlorides, bromides, etc.) than the crustal average while the atmosphere will always have a much higher concentration of other elements (such as oxygen, nitrogen and the rare gases). Natural photosynthetic processes also reconcentrate some elements, notably carbon, while hydrogeochemical processes reconcentrate others, including phosphorus, sulfur, calcium, manganese, copper and cobalt. In short, there is no danger of actually running out of materials, even the scarcest ones. A stationary-state economy will, however, have to utilize very low grade sources, in some cases, to replace whatever is "lost" through imperfect recycling. Some of the technical implications of this will be discussed later.

To summarize the last few paragraphs, the second law of thermodynamics does not imply (as argued by Georgescu-Roegen) that a sustainable, stationary-state economy is impossible on the earth, - at least as long as the sun continues to provide a flux of available energy (essergy). To be sure, the long-run prospects for economic growth (and human survival) are thermodynamically constrained, but no absolute limits are apparent from here. Thermodynamic considerations can also be integrated formally with traditional economic concepts to yield new insights into optimality conditions for resource and environmental management. In particular, the problem of optimizing resource extraction policy given a declining grade of resources can be treated more realistically (e.g. Faber et al 87).

Entropy, Self-Organization and Walras

The economy is like a living organism to the extent that it exhibits the attributes of self-organization and growth. For any real physical system to possess such attributes, two fundamental conditions are necessary. First, the system must be open and must receive an influx of available energy (e.g. sunlight) from outside. Second, the system must be non-linear and far from thermodynamic equilibrium. (Incidentally, it must be noted that self-organization cannot be attributed to the second law, although it can be described in thermodynamic language. It is a consequence, if anything, of the non-linearity of nature. It appears quite likely that non-linearity, in turn, is the root-source of irreversibility).

The words "far from equilibrium" require some explanation. In brief, thermodynamic equilibrium is that state of maximum uniformity and "randomness" (or entropy) toward which all spontaneous processes seem to tend, at least when they are fairly close. Near equilibrium, the equations that describe the relationships of state variables can always be linearized. However, far from equilibrium, the state equations of real systems are always non-linear [Nicolis & Prigogine 77]. One can define "far from equilibrium", then, as being far enough for the linear approximation to break down. Or, to put it another way, "far from equilibrium" means in the non-linear region of the equation of state. (In this region, the notion of entropy is not even necessarily well-defined, although it can sometimes be used).

An important attribute of non-linear equations must now be noted: they often have multiple solutions. In fact, as one moves away from the linear region, where there is a unique solution, the solution may "bifurcate" and new possible solutions can exist that do not evolve continuously into the limiting case. The physical configurations of matter that correspond to such alternate solutions of the equation of state typically exhibit a kind of order and persistence which has been called "self-organization" by Prigogine and others. Examples of self-organization can be found in most fields of science. The vortices that form in turbulent fluids, from sun-spots to hurricanes, are familiar examples.

Every living organism is far from thermodynamic equilibrium in the relevant sense, and is maintained by a flow of available energy from the outside. The same is true of the biosphere. Life on earth is energized by the sun. Entropy presumably continues to increase in the universe as a whole, but if one measures the entropy changes within a living organism, the reverse is true. Entropy within the (open) system boundaries actually decreases. This is reflected in the accumulation of energy-rich material (e.g. carbohydrates, proteins and fats) and, more subtly, in the organization and structure associated with life processes. This local decrease does not violate the second law of thermodynamics at all. It is just that the system in question is not a closed one. That it is also "far from equilibrium" is clear from the rapidity with which living organisms break down after death.

The human socio-economic system is evidently a self-organizing system. (The economic part of the system in question can be defined as the sum total of all human activities concerned with producing goods and services, irrespective of whether there exists a free market for them). It is clearly not a closed system, since materials are constantly cycling through it. It requires a flux of available (free) energy. Before the industrial revolution, most work expended in economic activity was derived from human muscles or those of animals. Ultimately, the source was the sun, whose energy was captured by green plants and thence cycled through the food-chain. Today, more than 90% of the energy utilized by industry is derived from fossilized hydrocarbons representing accumulations from millions of years of photosynthesis. Clearly, we are expending this resource at a fairly rapid rate, and if it continues the stockpiles we are using will be exhausted in a century or so.

Similarly, as noted already, we are using up and irreversibly dissipating the best quality ores of iron, copper, lead, nickel, chrome and a number of other metals. To adapt to this decrease in resource quality and availability, a number of major changes in our self-organizing economic system must occur. The "industrial metabolism" of the future will differ from that of today, as the present differs markedly from the past. More energy will be needed for extraciton and refining, deeper mines and wells will be dug, new separation processes will be introduced, more recycling will be done, substitutions will take place. Only the technological details are in doubt.

Incidentally, the present arrangement for disposing of waste materials is equally unsustainable, because it depends largely on dilution rather than recycling, and the dilution factor or absorption capacity permitted by land, water and air resources is, or soon will be, inadequate. As Kenneth Boulding pointed out very cogently some years ago, the days of the "cowboy economy", where human activity could be regarded as an insignificant disturbance to a natural environment of near-infinite capacity, must be replaced by a "space-ship" mentality, in which the earth is regarded as a finite - even small - place [Boulding 66]. This limitation on environmental dilution/absorption capacity has already reached the critical stage with regard to certain wastes, such as trash, toxic chemicals, chlorocarbons and carbon dioxide. Again, the economic system will continue to be self-organized, but the organization of our industrial metabolism must change.

What is not possible for our economic system, notwithstanding its self-organizing character, is to achieve a sustainable stationary state based on a continuation of the present technology of production and industrial structure. This is not an argument for or against particular political or institutional arrangements (e.g. capitalism vs. socialism). It is a plain statement that the input-output table describing the industrial metabolism of 2080 must look very different from the current one because different resources, requiring different technologies, will necessarily be used.

Incidentally, it is perhaps worth pointing out that, even if we could describe the technology of the (hypothetical) stationary state of the distant future - with its total dependence on solar energy in one form or another, its highly developed recycling processes for materials of all kinds, and its sophisticated methods of extracting metals and chemicals from extremely low grade sources (such as ocean water), - it would be economically irrational to adopt those technologies today. The costs would be excessive in relation to our current productive ability, and probably far beyond our present ability to pay. The optimum policy with respect to any exhaustible resource, as Herfindahl pointed out [Herfindahl 67], is to exploit the highest quality resources first and reinvest the profits to build up the capital and knowledge stock to permit lower grade resources to be used later. In effect, high quality resources in the ground are a form of capital. This capital can, of course, be invested wisely or squandered in excessive consumption. That is our choice.

In short, the most important economic implication of the second law of thermodynamics - as shorthand for the fact that the economic system is open, self-organizing and inherently dissipative, depending on external sources of free energy, raw materials and environmental resources - is the necessity for continued technological change and adaptation. This implies the impossibility of achieving a Walrasian equilibrium based on fixed technologies. At present we are depending on exhaustible stocks of all these resources. In the very long run the exhaustible high quality resources will run out (although the constituent atoms will not) and only the sun's energy will remain available to us. By that time a completely new and different industrial metabolism will necessarily have replaced the current one.

Although this paper is too short to go into details, we cannot resist pointing out again that the prevailing neo-classical economic model is, at bottom, predicated on perfect knowledge of all resource stocks and all technologies. Growth is permitted by the standard model, but technological "surprises" are not. Once the exhaustible resources are gone, the only kind of economic growth that is consistent with that model is growth of all sectors in exact proportion to each other. In biology, this kind of growth is seen in populations of bacteria or coral reefs, but not of complex organisms, where growth is inevitably accompanied by morphological change ("ontogeny recapitulates phylogeny"). The economic analog of a coral reef might be an economy of small, independent, self-contained farms. Needless to say, it does not exist today, if it ever did.

Implications for Industrial Metabolism, Past, Present and Future

As noted above, the total mass of residuals produced each year is likely to exceed the mass of active inputs. This is because nearly half of the inputs are fossil fuesl (hydrocarbons), which combine with atmospheric oxygen and form carbon dioxide and water vapor. The carbon fraction of hydrocarbons ranges from 75% in methane to about 90% in coal. Petroleum is intermediate. The average for all fossil fuels cannot be less than 82%, which translates into around 1200 tons of carbon in 1970. Imported oil would have increased this figure somewhat. In inefficient combustion processes up to 10% of this carbon is converted into carbon monoxide, while carbothermic reduction of iron ore and other metals generates even more, but most carbon monoxide is converted to carbon dioxide at the point of generation. Gross emissions of CO to the atmosphere in the U.S. were about 110 million tons in 1970 (mostly from automobiles and trucks), with a carbon-content of 47 million tons, or about 3.5% of the total. Emission controls reduced this to about 85 million tons net in 1980 [EPA 86].

Carbon monoxide is, of course, quite toxic to humans and has been implicated in health problems among urban populations. It is less well known that carbon monoxide plays an active, and not necessarily benign, role in a number of atmospheric chemical reacitons. Eventually, of course, all carbon monoxide oxidized to carbon dioxide. Thus, 1 billion tons of fuel carbon is converted ino 3.67 billion tons of carbon dioxide emitted to the atmosphere. (Note that the tonnage of carbon-dioxide emitted annually exceeds the total tonnage of active material inputs!) Not all of this remains in the atmosphere, but enough does to raise very serious concerns about the long term climatic impact of continued fossil fuel combustion. The problem is the "greenhouse effect". It appears that the carbon dioxide level of the earth's atmosphere has risen over the last century from about 290 ppm to around 340 ppm at present. While the baseline and the climatic impacts of the rising level of carbon dioxide are both still uncertain, it is already clear that "industrial metabolism" is influencing the natural environment in ways that appear increasingly adverse.

The discovery of chlorofluorocarbon gases in the stratosphere, here they can be ionized by the ultra-violet (U.V.) radiation of the sun, together with the ability of the chlorine ions to scavenge stratospheric ozone by a catalytic reaction, has raised an even more frightening prospect: ozone depletion. Apparent confirmation of this phenomenon has recently appeared in the form of an "ozone hole" in the stratosphere over Antarctica. Chlorofluorocarbons are industrial gases, discovered in 1928 and produced since the 1930s, whose inertness was once thought to be a great virtue. They are used commercially as refrigerants, solvents and for "blowing" plastic foams. In most uses they are not released deliberately, but losses and leakage are inevitable. Precisely because they do not break down in the presence of oxygen, they have now dispersed throughout the atmosphere. The problems revealed so far may only be the beginning. If (perhaps it might be better to say "when") the ozone level in the stratosphere is depleted, the effect will be to let more of the sun's U.V. radiation through to the earth's surface. The ecological impact on vulnerable species is incalculable, at present, but could be extremely severe.

Methane, oxides of nitrogen and sulfur oxides are other residuals that have been seriously implicated in climatic and/or ecological effects. All three are generated by fossil fuel combustion, as well as other industrial processes. Like carbon monoxide, carbon dioxide and the chlorofluorocarbons, they can be considered as metabolic products of industry. Methane is lost in natural gas pipelines, petroleum drilling, coal mining and several kinds of intensive agriculture (especially rice cultivation and sheep farming). Sulfur oxides are generated by combustion of sulfur-containing fuels- especially bituminous coal - and by the smelting of sulfide ores. Most copper, lead, zinc and nickel ores are of this kind. In principle, sulfur can be recovered for use from all these activities, and the trend is rising (Figure 4). But the costs of recovery, especially from coal-burning electric power plants, are far higher than the market value of the products (e.g. dilute sulfuric acid) and, for the present, constitute a waste residual that must be disposed of. This will only change if (or when) the economic value of sulfur rises very sharply above its present low level. This, in turn, depends on the discovery of new uses.

Nitrogen oxides are also co-products of combustion. In effect, at high temperatures, some of the atmospheric nitrogen is literally "burned". Again, in principle recovery for use (e.g. as nitric acid) is possible. But in comparison with the commercially dominant process of nitrogen fixation based on natural gas (the Haber-Bosch process), there is no exonomic justification for such recovery. It is only because dispersion of oxides of nitrogen is implicated in smog and acid rain that recovery technology is even being explored. Only a spectacular rise in the economic value of nitric acid would change this picture. Only new technology could lead to such a change.

Fly-ash is yet another by-product of combustion. At present it is being recovered fairly efficiently from stack-gases of large utility boilers and industrial furnaces, by means of electrostatic precipitators. But the ash itself has become a large-scale nuisance because there exists no use or market for it. The amounts are large: over 50 million tons per annum in the U.S. alone. At least two possible remedies exist, however. Fly-ash is a potential "ore" for several metals, especially iron, aluminum and silicon. These could probably be recovered commercially if, for example, bauxite became unavailable [Ayres 82b]. Alternatively, fly-ash could be used as a substitute, or more likely as a supplement, for portland cement in the manufacture of concrete and concrete-like products. Its major disadvantage in this application is that concrete made with fly-ash does not harden and set as rapidly as the commercial variety. This has obvious economic costs, but so does the disposal of fly-ash into land-fills. A final possibility worth mentioning is that fly-ash could be used as a medium for the disposal of toxic liquid wastes, yielding a hard, impermeable substance suitable for long term storage. This technology is already being developed in France.

Speculations along the above lines are not as "far out" as today's industrialists might believe. Several of the most valuable products of the chemical industry today began as unwanted by-products. Coal tar was one of the first. It became a valuable source of chemicals such as aniline dyes in the 19th century. Coke oven gas became not only a valuable fuel, but also the first commercial source of ammonia. Slag from blast furnaces was once a nuisance. Today it has a number of major uses and old slag heaps are even being "mined" in some locations. Natural gas was once an unwanted by-product of petroleum drilling - but not for long. Chlorine is a less well-known example, being originally a by-product of the manufacture of caustic soda (lye) by electrolysis of brine. Today chlorine is the primary product. Much of the petro-chemical industry is built on by-products of the natural gas industry: ethane, propane and butane. Helium is also taken from natural gas. Hydrochloric acid and sulfuric acid are increasingly being obtained as by-products, and many other examples could be given.

It is probably safe to say that the industrial metabolism of the next century will recycle a number of the major tonnage waste products of today, notably sulfur, fly-ash and lignin wastes from the paper industry. The latter may yet turn out to be a useful growth medium for single-cell organisms providing high-protein supplements for food products.

Combustion and carbo-thermal reduction processes are the major sources of atmospheric pollutants today, but by no means the only important ones. Nor is the atmosphere the only vulnerable part of the environment. From a broader environmental perspective, the production and dispersal of thousands of synthetic chemicals - many new to nature, and some highly toxic, carcinogenic or mutagenic - and the mobilizaiton of large tonnages of toxic heavy metals may be of equally great concern. The complexity of the problem is too great to permit any kind of short summary. However, two points are worthy of emphasis. First, as noted above, most material "pass through" the economic system rather quickly. That is to say, the transformation from raw material to waste residual takes only a few months to a few years in most cases. Long-lived structures are very much the exception, and the more biologically "potent" materials are least likely to be embodied in a long-lived form.

The second point, emphasized strongly some years ago [Ayres 78] but still worth repeating, is that many materials uses are inherently dissipative. That is, the materials are degraded, dispersed and lost in the course of a single normal usage. In addition to food and fuels (and additives such as preservatives), this applies to packaging materials, lubricants, solvents, flocculants, anti-freezes, detergents, soaps, bleaches and cleaning agents, dyes, paints and pigments, most paper, cosmetics, pharmaceuticals, fertilizers, pesticides, herbicides and germicides, and so on.

On reflection, most such uses (food and fuel again excepted) are generally seen to be non-essential in the sense that technologies are theoretically available, or imaginable, that could eliminate the need for them. To take one example only: hydroponic agriculture in enclosed, atmospherically controlled greenhouses, with genetically engineered "anti-pests", would eliminate all losses of fertilizers and pesticides to water-courses by way of surface runoff. It is obvious, however, that the techno-structure of a society employing such technologies would be very different from the present one. The developmental path "form here to there" involves fundamental structural changes not at all compatible with simple expansionary growth along an equilibrium "turnpike".

We conclude with a final point of a more methodological nature. It is that the "materials-balance" principle, a straightforward application of the first law of thermodynamics (widely used in the design of chemical engineering systems, for instance), is a potentially valuable and under-utilized tool for using economic data in environmental analysis. Not infrequently a combination of input data (obtainable from economic statistics), together with technical process data available from engineering analysis, gives a more reliable estimate of waste residual outputs than direct measurements could be expected to do.

One good example of this arises in the aluminum industry. One of the major environmental problems associated with aluminum smelting in the past was the emission of gaseous fluorine from the smelter. The source of the fluorine is the electrolytes (molten cryolite and aluminum fluoride) used as a solvent for alumina in the electrolytic cell. An unavoidable side-reaction in the cell breaks down these electrolytes and releases some of the fluorine at the anode. Exact "recipes" for the production of aluminum are only known by the aluminum companies, but a materials-balance analysis for the year 1973 suggests that for each 100 kg of aluminum produced, 2.1 kg of cryolite and 3 kg of aluminum fluoride were consumed as inputs [Ayres et al 78]. Based on these numbers and known chemistry the aluminum industry would have accounted for 40% of the known production of hydrofluoric acid in that year - consistent with both official (Dept. of Commerce) and unofficial estimates [ibid]. Note that, in the absence of recovery facilities all of this fluorine must have been emitted to the air, since consumption of cyolite and aluminum fluoride is exclusively to replace fluorine losses. It is interesting to observe, also, that the imputed emissions were about twice as high as EPA's published estimates based on direct (but unreliable and difficult to verify) measurements. Assuming the production and use statistics for hydrofluoric acid were correct, one would have to believe that the indirect estimate is more reliable than the direct one.

Another application of the materials-balance methodology is in the reconstruction of historical emissions data. This is a problem of some importance to basic enivironmental science, since the cumulative impact of air or water pollution over long periods can only be evaluated in relation to a baseline of some sort. In this context it becomes important to know more about emissions in the past, when no measurements were made. To be sure, sediments and ice-cores offer some help, but not enough. The picture can be clarified considerably, however, with the help of synthetic models utilizing production and consumption data (which is often imperfect, but better known than emissions) together with engineering analysis of processes. Sometimes process information is not even needed.

For example, fairly good historical estimates of sulfur oxide emissions (SOX) - needed to analyze the long-term impact of acid rain, among other things - can be reconstructed quite easily from historical statistics on coal consumption and copper/lead/zinc smelting [e.g. Gschwandtner et al 83]. This is because the sulfur content of coal and metal ores can be assumed to be the same in the past as it is today, and until recently all of that sulfur was emitted to the atmosphere. Reconstruction in the case of NOX is slightly more complicated, but basically similar [ibid]. More complex reconstructions of historical emissions have been undertaken recently, e.g. for the Hudson-Raritan estuary [Ayres et al 85].

Conclusions

Despite its length, this paper can be reduced to six relatively simple propositions.

1) Conservation of mass (a consequence of conservation of mass-energy) implies that materials extracted from the earth's crust must eventually return thence as waste residuals. The quantities involved are currently very large, and may well be sufficient to trigger irreversible changes in the earth's atmosphere.
2) The non-existence of any mechanism for pricing the environmental absorptive capacity (a common-property resource) needed to dispose of the waste residuals resulting from the use of fossil fuels and toxic heavy metals, has led to a massive underpricing of those resources, and a correspondingly excessive rate of use. The magnitude of the distortion involved also undermines the plausibility of existing equilibrium-type models.
3) There is no lower limit to the materials required to generate final goods and services (GNP), and the potential for materials conservation has been greatly underestimated.
4) Similarly, there is no lower limit to the energy required to generate final goods and services (GNP) and the potential for energy conservation has been massively underestimated in the past. In fact, the imperative for further economic growth in no way puts any long-term requirements on the amount of energy that will be needed. Even in the short term, energy conservation can easily accommodate any rate of growth that is likely to occur.
5) The fact that our economy is dissipative, and the fact that recycling cannot be 100% efficient (as implied by the second law of thermodynamics) do not imply the impossibility of reaching a sustainable stationary-state in the very long run - at least, as long as the sun shines.
6) The fact that our economy is self-organizing, and that it is currently not in equilibrium with the environment implies that our technology- as reflected by "industrial metabolism" - must and will change fairly radically. Such changes are not accommodated by existing economic growth models. Hence new approaches to economic growth modelling are needed.

116

References

Arrow, K.J. and Debreu, G. (1954), Existence of an Equilibrium for a Competitive Economy, Econometrica 22(3).

Ayres, R.U. (1978), Resources, Environment & Economics, John Wiley & Sons, N.Y.

Ayres, R.U. (1982), Coalplex: An Integrated Energy/Resource System Concept, UNEP Seminar on Environmental Aspects of Technology Assessment, United Nations, Geneva, Nov/Dec.

Ayres, R.U., Energy Inefficiency in the U.S. Economy, unpublished manuscript.

Ayres, R.U., Cummings-Saxton, J. and Weinstein, E. (1987), Assessments of Methodologies for Indirect Impact Assessment, Research Report (IRT-468-R/a), International Research & Technology Corporation, Washington DC. (NOTE: prepared for EPA).

Ayres, R.U. et al. (1985), An Historical Reconstruction of Major Pollutant Levels in the Hudson-Raritan Basin (3 Vols), Report by Variflex Corporation for the National Oceanographic & Atmospherican Administration (NOAA), Rockville, MD. Published as NOAA Technical Memorandum, 1988.

Ayres, R.U. and Kneese, A.V. (1969), Production, Consumption & Externalities, American Economic Review, June.

Ayres, R.U. and Narkus-Kramer, M. (1976), An Assessment of Methodologies for Estimating National Energy Efficiency, Winter Meeting, ASME, November.

Boulding, K.E. (1966), Environmental Quality in a Growing Economy, in: Essays from the Sixth RFF Forum, Johns Hopkins University Press, Baltimore, MD.

Bridges, J. (1973), Understanding the National Energy Dilemma (1973), Joint Committee on Atomic Energy, U.S. Congress, Washington DC.

Dasgupta, P. and Heal, G. (1979), Economic Theory and Exhaustible Resources, Cambridge U. Press, Cambridge UK.

Dunkerley, J. (1980), Trends in Energy Use in Industrial Societies, Research Report (R-19), Resources for the Future, Inc. Washington DC.

EPA (1986), National Air Pollution Emission Estimates, 1940-1984, Research Report (EPA-450/4-85-014), Office of Air Quality Planning and Standards, US Environmental Protection Agency, Research Triangle Park, NC, January.

Faber, Malte, Niemes, H. and Stephan, G. (1987), Entropy, Environment & Resources, Springer-Verlag, Berlin.

Georgescu-Roegen, N. (1971), The Entropy Law and the Economic Process, Harvard University Press, Cambridge MA.

Georgescu-Roegen, N. (1977), The Steady State & Ecological Salvation: A Thermodynamic Analysis, Bioscience 27(4), April.

Georgescu-Roegen, N. (1979), Myths About Energy & Matter, Growth & Change

117

10(1).

Goeller, H. and Weinberg, A. (1976), The Age of Substitutability, Science 191, February.

Gschwandtner, G., Gschwandtner, K.C. and Eldridge, K. (1983), Historic Emissions of Sulfur and Nitrogen Oxides in the U.S. 1900-1980, Research Report (Contract No. 68-02-3511 # 3), Pacific Env. Services Inc., Durham NC, October 5.

Herfindahl, O. (1967), Depletion and Economic Theory, in: M. Gaffney (ed.), Extractive Resources and Taxation, University of Wisconsin Press, Madison WI.

Hoos, I. (1983), Systems Analysis in Public Policy: A Critique, University of California Press. Revised edition.

Kenney, W.F. (1984), Energy Conservation in the Process Industries (Series: Energy, Science & Engineering: Resources, Technology & Management), Academic Press Inc., New York.

Meadows, D. et al. (1972), The Limits to Growth: A Report for the Club of Rome's Project on the Predicament of Mankind, Universe Books.

Murray, F.X. (1976), Energy: A National Issue, Center for Strategic & International Studies, Georgetown University, Washington DC. (NOTE: Update of Bridges 73).

Nicolis, G. and Prigogine, I. (1977), Self-Organization in Non-Equilibrium Systems, Wiley-Interscience, New York.

Organisation for Economic Cooperation and Development (1978), World Energy Outlook, OECD, Paris.

Porat, M. (1977), The Information Economy: Definition and Measurement, Research Report (Special Publication 77-12(i)), Office of Telecommunications Policy, Washington DC, May.

Rosenfeld, A. and Hafemeister, D. (1987), Energy Efficient Buildings, Scientific American.

Ross, M., Socolow, R. et al. (1985), Effective Use of Energy: A Physics Perspective, Research Report, American Physical Society, January.

Schurr, S.H. and Netschert, B.C. (1960), Energy in the American Economy, 1850-1975, Johns Hopkins University Press, Baltimore MD.

Seligman, B.B. (1962), Main Currents in Modern Economics, Free Press, Glencoe IL.

Shurkin, J. (1985), Engines of the Mind, Washington Square Press, New York. Revised edition. (NOTE: Pocket Books).

Solow, R.M. (1974), The Economics of Resources or the Resources of Economics, American Economic Review, 64.

Tilton, J.E. (ed.)(1983), Material Substitution: Lessons from Tin-Using Industries, Resources for the Future Inc., Washington DC.

Tribus, M. and El-Sayed, Y. (1981), The Strategic Use of Thermoeconomic Analysis for Process Improvement, AIchE, August 16-19. [118]

408-24 (US, Sweden, India)

[91] 813

Q25

Q28

[23]

Getting started

No-regrets strategies for reducing greenhouse gas emissions

Evan Mills, Deborah Wilson and Thomas B. Johansson

An integrated approach for choosing among energy supply- and demand-side measures shows that, compared to business-as-usual demand patterns, global greenhouse-gas emissions can be reduced well below current levels with net economic benefits *to society. Given these findings, a 'wait-and-see' stance towards new initiatives in energy and environmental policy is not economically justifiable. Achieving significant emissions reductions, however, will require commitments to policies aimed at enabling energy markets to function more efficiently and supporting legislation where market forces do not suffice.*

Keywords: Global warming; Greenhouse gases; Economics

Any strategy to reduce greenhouse gas emissions must involve the global energy system, because fossil-fuel-based energy consumption currently causes around 50% of global greenhouse forcing.[1] To stabilize atmospheric concentrations of the major greenhouse gases at today's levels will require immediate reductions of over 60% of current anthropogenic carbon dioxide (CO_2), nitrous oxide (N_2O) and chlorofluorocarbon (CFC) emissions and 15% to 20% reductions for methane (CH_4).[2] Can such reductions be accomplished, and would they be compatible with developing countries' ambitions and continued economic growth in the industrialized countries? If so, what would they cost?

These questions became especially relevant when the World Energy Conference (WEC) published projections in 1989 suggesting that global primary energy demand would grow by 76% by the year 2000 as economic growth proceeded, and that the associated CO_2 emissions would grow by 69% despite a three-and-a-half-fold increase in nuclear power capacity.[3]

Recent econometric studies suggest that significantly reducing CO_2 emissions will be very expensive.[4] This conclusion tempts policymakers to postpone any response to the climate change problem, and to wait instead for undisputed scientific proof that climate change is occurring. However, while the marginal costs of major reductions may be high, measures with no net cost compared to the 'business-as-usual' alternative are available for *beginning* to reduce emissions.

An energy strategy aimed at reducing greenhouse gas emissions at the least cost can combine different approaches: fuel switching, improved energy conversion efficiencies and improved energy end-use efficiencies. Many applicable technologies, especially end-use technologies, are presently available, but average efficiency levels do not come close to those of the best products on the market. Ongoing research and development work suggests that many more opportunities to reduce greenhouse gas emissions will become available. In this article, we describe a variety of existing and emerging energy supply and end-use technologies, their costs and their associated emissions savings, and case studies of their possible application in the USA, Sweden and the State of Karnataka, India. Finally, we present examples of policies to consider for realizing such opportunities.

Specific opportunities

Before making macro-level analyses covering entire sectors and geographical areas, it is necessary to review and compile data on specific measures for reducing greenhouse gas emissions. The net emissions associated with those measures, compared to a chosen base case, must then be estimated. These estimates can be combined with the incremental costs of choosing each measure *v* those for the

Evan Mills, Deborah Wilson and Thomas B. Johansson are with the Department of Environmental and Energy System Studies, University of Lund, Gerdagatan 13, S-223 62 Lund, Sweden.

0301-4215/91/060526-17 © 1991 Butterworth-Heinemann Ltd

base-case measures, in order to find the *net* cost (or benefit) of reducing emissions.

This section highlights some examples of the cost-effectiveness of measures to reduce greenhouse gas emissions, using economic evaluations reflecting a societal perspective. We compare fuels and technologies based on their costs without taxes and we use a societal discount rate of 6% (real) for calculating annual costs of supply and end-use efficiency investments. Our method of making these calculations is developed in Appendix 1 and a collection of examples is provided in Appendix 2.

To provide a more complete assessment of greenhouse gas emissions, we have considered more than just combustion-related CO_2. As a first step, we have included fuel-cycle emissions associated with the mining, processing, transporting and burning of fossil fuels. In some cases more than half of the emissions occur *before* the point of fuel combustion. We have also included greenhouse gases other than CO_2. Including a broader range of emissions (expressed in carbon equivalents) can lead to different rankings of measures. We found, for example, that if only CO_2 is counted, compressed natural gas (CNG) automobiles appear 'better' than gasoline-fueled cars. However, total greenhouse gas emissions are in fact greater for CNG automobiles after including the CO_2 releases from fuel production and related methane and N_2O emissions. The same reversal of rankings occurs for vehicles burning coal- and natural-gas-based methanol.

We find that all of the currently available end-use efficiency measures described in Appendix 2 (applicable to lighting, passenger cars, appliances, space heating, motors, etc) reduce emissions while providing a net economic *benefit* (ie savings) in the $100 to $500/tonne range.[5] Switching to biomass-based fuels for heating can also reduce emissions and provide a net economic benefit. Examples of biomass fuels include forest residues, agricultural wastes, by-products from the paper and pulp and sugar industries, and materials grown on plantations specifically for energy purposes. Burning renewably-grown biomass feedstocks results in zero net carbon emissions when the carbon emitted from burning the fuel is just equal to the carbon reabsorbed from the atmosphere as replacement biomass is grown.

For a given fuel, advanced commercially available electric power plants can also achieve emissions reductions at no net cost. Switching from coal to high-efficiency natural-gas-fired power plants avoids emissions and produces a net benefit of around $100/tonne of carbon equivalent. Switching to renewables at today's prices tends to result in net costs, but is expected to provide net benefits of around $100/tonne within 10 to 20 years.[6] Avoiding emissions by building new nuclear power plants would cost about $40/tonne of avoided carbon-equivalent emissions under today's US conditions *v* a benefit of $11/tonne for the industry's target costs.[7]

Efficiency opportunities

There are many cases in which a variety of technologies exist that serve the same function (ie, provide the same *service*) but require very different amounts of energy to do so. Efficient light bulbs, refrigerators, cars, motors, pumps, fans, and compressors are just a few examples. There is a large global potential for reducing energy requirements by introducing the most efficient technologies as replacements for the current equipment stock in the natural turnover cycle and as the stock grows. Many studies, four of which are described in this article, have been carried out that estimate the energy efficiency gains and emissions reductions achievable by applying efficient technologies. Physical and technical limits for these technologies are still far from being realized in most cases, thus continued research and development can continue to provide large returns.

Lighting. Lighting typically comprises 10 to 20% of electricity demand in developing and industrialized countries. Today, indoor lighting is often provided by incandescent lamps. As an example of providing lighting services with reduced energy input, a 15-watt compact fluorescent lamp can be used to replace a 75-watt incandescent lamp. The compact fluorescent lamp provides the same energy service (lumen output), but requires five-times less electricity to do so.

Efforts to promote energy efficiency often focus on lighting. In Europe, 33 utility programmes that used financial incentives to promote the use of compact fluorescent lamps achieved an average cost of conserved energy of ¢2.1/kWh, including all direct and indirect costs for lamps, administration, advertising, postage, etc.[8]

Significant amounts of energy can also be saved in conventional fluorescent lighting systems. The Swedish utility Vattenfall recently completed a lighting retrofit demonstration at their headquarters. By modifying their existing fluorescent systems to use efficient lamps, ballasts, daylighting controls, occupancy sensors and other measures Vattenfall achieved a cost-effective 71% reduction in the electricity used for lighting. The quality of office lighting was also improved markedly.

Passenger cars. The global fleet of gasoline-fueled vehicles emits larger quantities of greenhouse gases than any other category of end-use devices. In the USA the production, distribution and combustion of transportation fuels account for approximately 27% of all fossil-fuel-related CO_2 emissions.[9] Efficiency improvements effectively avoided more than 500 million tonnes (mt) of carbon-equivalent emissions between 1974 and 1985 (inclusive).[10] Emissions from new gasoline-fueled cars can be cut by over three-fourths through higher efficiencies, and can be virtually eliminated with vehicles that utilize non-fossil fuel sources.

Significant efficiency improvements have already been achieved in passenger cars: the average nominal (laboratory test) fuel efficiency for new passenger cars sold in the USA improved by 50% (to 8.4 litres/100km) between 1973 and 1986.[11] Of these efficiency gains 90% can be attributed to fuel-efficiency improvements as opposed to a shift in vehicle mix.[12]

A wide range of technologies are available today for further increases in the fuel-efficiency of new automobiles,[13] some of which can bring about improvements with negative to small positive costs. From an efficiency standpoint, one of the most advanced passenger-car prototypes is the Toyota AXV. This four- to five-passenger car has a fuel efficiency of only 2.4 1/100km.[14] A similar example is the Light Component Project prototype (LCP-2000) produced by Volvo. This four-passenger automobile passes the world's most stringent crash and emissions tests (ie, California's), achieves a fuel efficiency of 3.6 litres/100 km, and was designed to cost about the same to manufacture as an average subcompact car at a production level of 20 000 cars/year.[15] The Elsbett engine used in this car can be operated with biomass-derived fuels as well as gasoline.[16]

Many of the technological advances that can reduce fuel consumption in gasoline-fueled cars will be applicable to methanol cars as well. Because of the better combustion properties of methanol, methanol vehicles have an intrinsic potential for achieving fuel efficiencies that are up to 25% better than those for gasoline and compressed natural gas (CNG) vehicles.[17]

Efficient electric vehicle technology is in a rapid stage of development. General Motors, for example, has announced its intention to market its protoytype electric car: the Impact. This two-passenger car is expected to consume a mere 0.07 kWh/km (a gasoline equivalent of 1.7 1/100 km): only 14% of the primary energy consumption of the 1987 US stock-average gasoline-fueled vehicle.[18] As in the case of methanol, the amount of emissions from electric vehicles depends on the feedstock used, in this case to produce the electricity.

Supply opportunities

Modernization of power generation. The average conversion efficiency from primary fuels to electricity in new power plants has slowly improved in recent years. Vigorous R&D efforts underway during the 1980s resulted in maximum efficiencies of about 45% (on a higher heating value basis) for commercially available combined-cycle power plants fueled with natural gas.

The choice of fuel and conversion technologies has an important effect on the amount of greenhouse gas emissions from electric power production (Figure 1). Replacing average conventional coal-fired steam plants with pressurized fluidized-bed combustion plants (PFBC, also coal-based), reduces emissions by almost 20%. A *further* 48% reduction can be achieved by switching to natural gas and the best available combined-cycle power plants.

Power plant conversion efficiencies could be increased beyond currently-available levels with continued development of aeroderivative gas turbines. Through a combination of measures, such turbine designs currently reach 54% in natural-gas-fired systems.[19] The capital costs of these power plants ($500/kW) are estimated to be lower than those of new central-station, coal-steam plants with flue-gas desulphurization, nuclear or hydroelectric power plants.

The simultaneous production of electricity and useful heat (cogeneration) offers still higher emissions-reduction possibilities. The total system efficiencies of some cogeneration applications are well above 80%. Total system emissions can be minimized by combining the best available technologies to meet a given demand for electricity and heat. Figure 1 compares cogeneration systems with central-station plants that provide no useful heat. Cogeneration based on natural gas offers an additional 38% emissions reduction beyond the best available combined-cycle plants fired with natural gas. The plants with the lowest emissions also have the lowest costs.

The emergence of highly efficient gas turbines has provided new opportunities for utilizing gasified solid fuels such as biomass and coal. Advanced gas turbines using gasified coal have been demonstrated in coal-integrated-gasifier/gas turbines (CIG/GT). Advanced gas turbines fired with gasified wood (BIG/GT) have not been demonstrated, but detailed

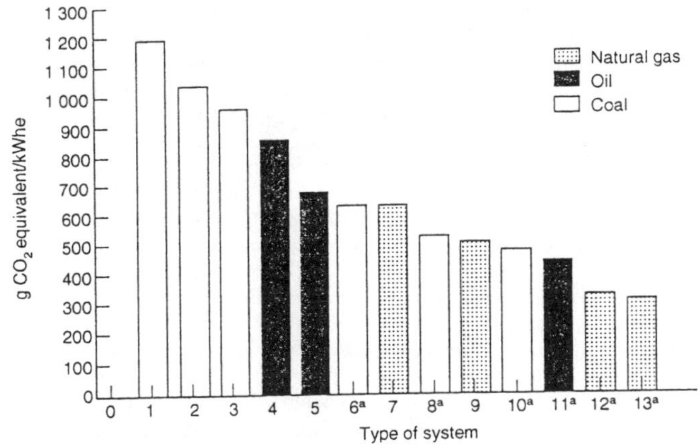

Figure 1. Greenhouse gas emissions from commercially-available power plants.

Notes: [a] = cogeneration. Greenhouse-gas emissions vary substantially among commercially-available technologies for producing heat and power. Central-station power plants are compared with cogeneration plants providing both useful heat and power. The energy requirement for electricity production using cogeneration technologies is taken as the total energy supplied minus that which would have been required to produce the heat independently (assuming a boiler efficiency corresponding to a lower-heating-value of 90%). All power plant efficiencies (Eff) are based on higher heating values. For the cogeneration systems, the power–to–heat ratios are given as E/H. The greenhouse gases are expressed as an equivalent amount of carbon dioxide (CO_2eq/kWhe). Methane and methane-related fuel-cycle emissions from coal, oil, and natural gas consumption are taken into account (see Appendix 1). 1. Average conventional steam turbine (coal, Eff 34%); 2. Best available steam turbine (coal, Eff 39%); 3. Pressurized fluidized bed combustion (PFBC) (coal, Eff 42%); 4. Average conventional steam turbine (oil, Eff 38%); 5. Best available combined-cycle gas turbine (oil, Eff 48%); 6. Cogeneration: average conventional steam turbine (coal, Eff 78%, E/H = 0.50); 7. Average combined-cycle gas turbine (natural gas, Eff 36%); 8. Cogeneration: best available steam turbine (coal, Eff 83%, E/H = 0.60); 9. Best available combined-cycle gas turbine (natural gas, Eff 45%); 10. Cogeneration: pressurized fluidized bed combustion (coal, Eff 86%, E/H = 0.65); 11. Cogeneration: best available steam turbine (oil, Eff 81%, E/H = 0.60); 12. Cogeneration: steam-injected gas turbine (natural gas, Eff 75%, E/H = 0.80) 13. Cogeneration: best available combined-cycle gas turbine (natural gas, Eff 77%, E/H = 1.0)

design studies indicate that it may be possible to commercialize them more quickly than CIG/GT schemes. This is because biomass is inherently cleaner and easier to gasify than coal and because the scale of demonstration plants is also appropriate for commercial applications.[20]

Modernization of sugar cane and alcohol industries. The sugar cane industries in developing countries offer a particularly promising area for the application of BIG/GT technologies. Sugar cane production yields two kinds of biomass fuel suitable for gasification: bagasse and barbojo. Bagasse is the residue from crushing the cane, and is thus available during the milling season; barbojo is the tops and leaves of the cane plant that could be harvested and stored for use after the milling season. Ogden *et al* calculate that by the year 2027 the 80 sugar cane producing developing countries could thus generate 70% more than their *total* 1987 electricity production from all sources. Moreover, the costs would be competitive with conventional sources of electricity based on fossil fuels.

The ethanol simultaneously produced from cane would be equivalent to about 9% of total oil-use in all developing countries in 1987. Odgen *et al* point out that were the electricity produced instead from coal, and were the alcohol and methane used instead of gasoline in the transport sector, the additional CO_2 emissions would equal nearly half of the total 1986 emissions from fossil fuels in developing countries.[21]

Solar electricity. Currently, about 1 500 MW of wind

capacity and 300 MW of solar thermal capacity is installed in the USA, mostly in California. The electricity costs from these systems are in the ¢5 to ¢10/kWh range, and ongoing development work is expected to lower costs to the point where they will be cost-competitive for wholesale power generation.[22] Perhaps the most promising solar technology is the solid-state photovoltaic (PV) cell, which converts sunlight directly to electricity. The costs are presently ¢30 to ¢35/kWh, and projected by five US National Laboratories to drop to ¢4 to ¢5/kWh by 2020 to 2030.[23] About 40 MW of photovoltaic capacity is currently installed worldwide, with the greatest market in remote applications where the high costs are acceptable. Projections based on development work in the PV industry suggest that photovoltaics will become competitive with fossil-fuel power plants within a decade:

> The development of large-scale, computer-integrated manufacturing lines should decrease the manufacturing costs of amorphous silicon solar cells to less than $1/$W_p$ by the early 1990s, leading to the development of both residential and central station utility applications in the mid-to-late 1990s.[24]

Global, national and regional examples

What are the likely future economic and emissions impacts of individual measures such as those we have described, if combined and applied at the global, national or regional levels? Evaluations aimed at answering this question must begin by assuming a certain level of economic growth. The economic growth rates (and embedded assumptions regarding growth in material output and demographic factors) can then be used to project the future demand for energy services. Once a demand level for energy services is determined, specific technologies for providing them can be evaluated and compared. Before proceeding to scenarios, we make a few points about projecting the future demand for energy services.

Energy, development and economic growth

History has shown that the link between economic growth and growth in energy demand has been broken, and there are good reasons to believe that there need not be a recoupling. In contrast to the trends for total energy consumption, electricity-use has tended to grow faster than GDP. However, growth in electricity/GDP in many countries including the USA, the UK, Japan, and West Germany slowed to zero in the mid-1970s and declined thereafter.[25]

The perhaps counter-intuitive decoupling of energy demand and GDP growth occurs as a result of three on-going types of change: (1) structural changes in consumption patterns, towards fewer energy- and materials-intensive products and activities, that take place as development in a country proceeds; (2) shifts towards products that are less energy- and materials-intensive; and, (3) improvements in energy conversion and end-use efficiencies.[26] The combined effect of these changes has been to reduce national energy intensity, even where there are no formal policies to promote the changes.

Significant errors can be introduced into energy demand projections if these fundamental trends are ignored. For example, even if developing countries choose to follow the conventionally projected development path, structural and technological trends will enable them to be much less energy-intensive than their northern neighbours. With an effective set of national and international policies, the trend toward reduced energy intensity can be accelerated in both industrialized and developing countries.

A global view

Goldemberg *et al* found that by using efficient end-use technologies all of the energy services associated with a significant increase in the standard of living in both developing and industrialized countries can be attained by the year 2020 with only a 10% increase in global primary energy-use.[27] This is much lower than the increases in global energy-use projected by the WEC and others. The scenario results in a 10% decline in fossil-fuel CO_2 emissions compared to 1980, using a conventional mix of energy sources, but relatively less reliance on coal. To achieve greater reductions, end-use efficiency measures need to be combined with increased use of non-fossil energy sources.

Industrialized countries

An example from the USA. The USA reduced its energy intensity by approximately 25% between 1973 and 1986. Today's supply and end-use technologies offer the prospect for significant additional efficiency gains.[28]

Possibilities for avoiding carbon emissions in the near term are shown as a function of cost for the USA in Figure 2. Each step in the figure corresponds to an end-use efficiency measure or a supply strategy that saves energy and reduces emissions at a given net benefit or cost compared to the current US Department of Energy (DOE) forecast for the year

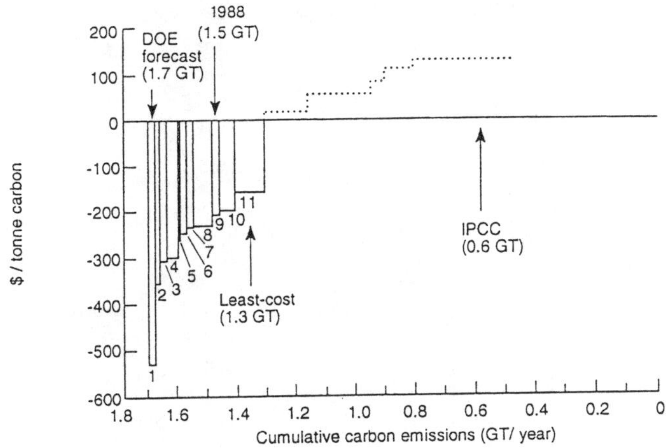

Figure 2. Net cost of avoiding emissions: USA in 2000.

Notes: The x-axis shows total national carbon emissions reductions achievable through the adoption of the 11 cost-ranked measures listed below.[52] The upper limit (1.7 GT) represents the current US DOE forecast for the year 2000. The IPCC label indicates the level of reductions necessary to stabilize atmospheric concentrations of greenhouse gases, according to the Intergovernmental Panel on Climate Change. The y-axis indicates the net cost of implementing each measure. Negative costs reflect a net economic benefit compared to the DOE forecast. The average net economic benefit for steps 1–11 is $231/tonne. 1. Raise the Federal gasoline tax by ¢12/litre within five years and spend part of the revenue on mass transit and energy-efficiency programmes; 2. Use white surfaces and plant urban trees to reduce air conditioning loads associated with the summer 'heat island' effect in cities; 3. Increase the efficiency of electricity supply through development, demonstration, and promotion of advanced generating technologies; 4. Raise car and light truck fuel-efficiency standards, expand the gas-guzzler tax, and establish gas-sipper rebates: new cars average 5.2l/100km and new light trucks average 6.7l/100km by 2000; 5. Reduce Federal energy-use through life-cycle cost-based purchasing; 6. Strengthen existing Federal appliance efficiency standards; 7. Promote the adoption of building standards and retrofit programmes to reduce energy-use in residential and commercial buildings; 8. Reduce industrial energy-use through R&D programmes, promotion of cogeneration, and further data collection; 9. Adopt new Federal efficiency standards on lamps and plumbing fixtures; 10. Adopt acid rain legislation that encourages energy efficiency as a means for lowering emissions and reducing emissions control costs; 11. Reform Federal utility regulation to foster investment in end-use efficiency and cogeneration.

2000. The forecast assumes 2.5% real economic growth.

In the DOE forecast, US carbon emissions would increase by 200 million tonnes/year by the year 2000 (13% of 1988 emissions). In contrast, 11 policy strategies for the year 2000 would lead to an emissions reduction of 170 mt (11%) from 1988 levels – similar to the global opportunity just described – and a cost *reduction* of $85 billion/year compared to the DOE forecast. The emissions reductions are accompanied by an average economic benefit (savings) of $231/tonne.[29]

These results reflect only 11 specific policies that pertain mostly to improved end-use efficiency, and are achievable over a period of 10 years. Renewable supply-side strategies are not included. Thus, the avoided carbon emissions represent only part of the long-run potential. For example, the use of biomass in electric power and heat production has not been included. This option is illustrated in the following example of an integrated supply–demand approach.

An integrated example from Sweden. The situation in Sweden presents an interesting case of balancing climate change issues with other energy policy goals. Sweden uses more nuclear-generated electricity per capita than any other country. In 1980 a public referendum called for a full phase-out of nuclear power. The referendum led to a parliamentary decision to phase out the country's 12 nuclear power plants by 2010. Sweden has also committed to holding CO_2 emissions at or below 1988 levels and

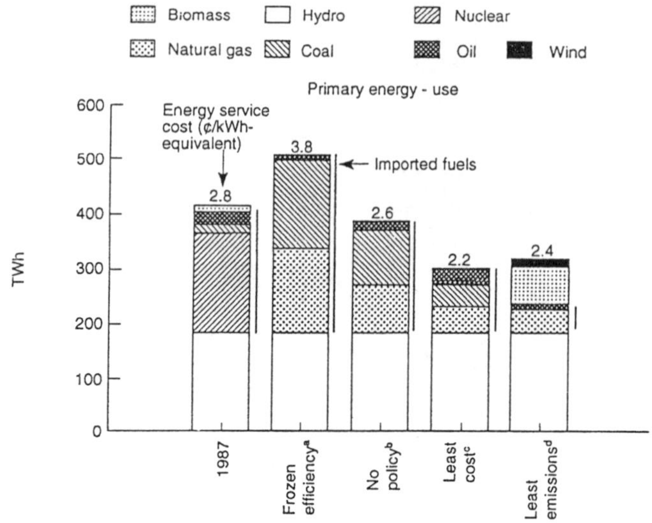

Figure 3. Swedish primary energy, energy service costs (efficiency investments and purchased energy) and energy import dependence: 1987 and scenarios for 2010.

Notes: All services derived from electricity, district heat, and 15 TWh of industrial process heat are incorporated. Electricity from hydroelectric, nuclear, and wind power is converted to its fuel equivalent using a 36% thermal efficiency. Electricity demand in the least-cost and least-emissions scenarios is 25% lower than 1987 levels. In contrast, electricity-use increases by 9% in the no-policy scenario. All scenarios include a complete phase-out of nuclear power by 2010, as called for by the Swedish government. [a] Frozen-efficiency baseline. End-use efficiencies do not improve beyond 1987 (base-year) levels. Electricity demand is met with existing non-nuclear power plants and by new efficient power plants half fueled by coal and half by natural gas. CHP is used extensively, both for industry and municipal district heating. [b] No-policy scenario. In this scenario, no new policies are implemented to increase the efficiency of electricity-use or to increase the use of renewable supply sources. The scenario includes only those efficiency improvements that are expected to result from a cost-driven average 50% increase in real electricity prices. The energy supply mix for electricity production is the same as in the frozen-efficiency baseline. [c] Least-cost scenario. This scenario goes beyond the no-policy scenario to show the impact on electricity demand if the most efficient technologies (for appliances, motors, lighting, etc) available today or near commercialization were introduced at the natural rate of capital turnover up to the year 2010. Also included is some fuel switching in heating systems that can currently use electricity in combination with other fuels. Only those measures costing less than the electricity they save are employed. The energy supply mix from the preceding scenario is retained. [d] Least-emissions scenario. This scenario begins with the end-use measures included in the least-cost scenario and introduces gasified biomass fuel for electricity production (replacing fossil fuels). A small amount of wind-generated electricity is also included. Natural gas is used after available biomass resources are allocated.

abstaining from constructing hydroelectric plants on the country's four remaining wild rivers. Hence, Sweden offers an acid test of energy planning in the face of potentially conflicting national policy goals.

Possibilities for meeting these goals were identified in a detailed assessment recently completed by Vattenfall (the Swedish State Power Board) and the University of Lund.[30] The boundary conditions encompass the energy services provided in the base-year (1987) by electricity, cogeneration, district heat, and some industrial process heat: 63% of total primary energy supply in Sweden.[31] Four electricity demand scenarios were developed, each incorporating the structural and demographic expansion associated with an anticipated 54% increase in real GNP to the year 2010, as assumed by the Ministry of Finance.

Figure 3 shows the types of fuels used to supply

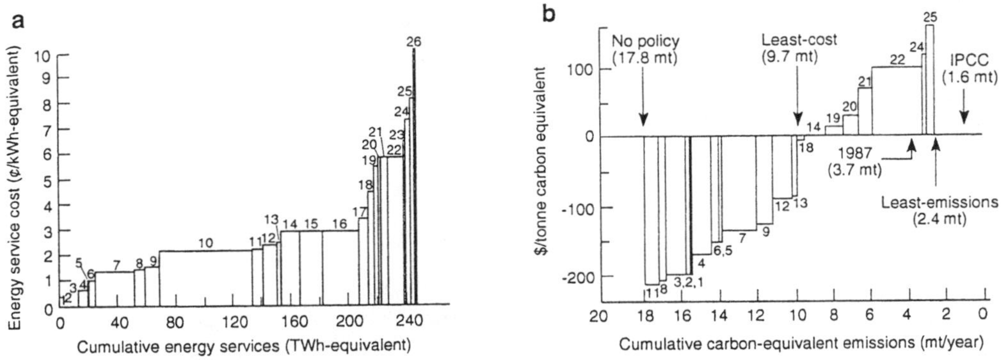

Figure 4. An integrated resource supply curve for Sweden in 2010 (Figure 4a) and the corresponding net costs of avoiding greenhouse gas emissions (Figure 4b).

Notes: Each step represents a particular supply or demand-side measure (see key) used to meet the total demand for energy *services*, ie the frozen-efficiency scenario (denoted in TWh-equivalents) projected for the year 2010. Figure 4a reflects the least-emissions scenario described in Figure 3. The width of each step in Figure 4a shows the energy provided or conserved by the measure. The height of each step is the measure's levelized cost/kWh-equivalent (using a 6% real discount rate). The net costs of avoiding emissions shown in Figure 4b reflect technology changes in the power and heat sector between 1987 and 2010 (eg existing hydroelectric plants are not shown). The horizontal axis shows greenhouse gas emissions (as carbon-equivalents) for the various scenario levels given in Figure 3. The vertical axis indicates the net cost of achieving the emissions reductions for each measure (using the method described in Appendix 1). Negative costs reflect a net economic benefit compared to the no-policy scenario.[53] Rankings shift in some cases between Figures 4a and 4b because the type and amount of avoided emissions varies depending on the measure. The IPCC label indicates the level of reductions necessary to stabilize atmospheric concentrations of greenhouse gases, according to the Intergovenmental Panel on Climate Change. *Key:* 1. Residential space heating (including effect of existing standards); 2. Conversion of large heat pumps; 3. Efficient electronic office equipment; 4. Efficient appliances (excluding lighting); 5. Efficient lighting – residential, incandescents; 6. Efficient lighting – commercial and industrial, incandescents; 7. Efficient motors, pumps, fans, compressors, etc; 8. Fuel switching to gas from electricity; 9. Efficient lighting – commercial and industrial fluorescents; 10. Existing hydroelectric plants; 11. Fuel switching to oil from electricity; 12. Miscellaneous efficiency improvements; 13. Efficient commercial food preparation; 14. Supplemental district heating (biomass); 15. Industrial process heat from cogeneration; 16. District heating from cogeneration; 17. Existing industrial cogeneration – electricity; 18. Heat pump – existing electric boiler; 19. Industrial cogeneration with biomass; 20. Cogeneration, biomass – (*v* existing cogeneration); 21. Cogeneration, biomass – (*v* new central station); 22. Cogeneration, biomass – (*v* new cogeneration); 23. Existing oil condensing plants; 24. Heat pump – existing direct electric; 25. Wind turbines; 26. Existing gas turbines.

the electricity and heat required in the four scenarios, the portion of this energy that is imported, and the resulting costs per unit of energy services provided (for purchased energy plus efficiency investments). Energy import dependence drops from today's value of 54% to 38% in the least-cost scenario and to 11% in the least-emissions scenario.

The means of providing energy services in the least-emissions scenario for 2010 are depicted in Figure 4a. This 'integrated resource supply curve' includes 26 steps representing demand-side efficiency improvements and supply options, including existing supply systems such as hydroelectric power, that are projected to be used in 2010. The steps are ranked by increasing cost of delivered energy services (¢/kWh-equivalent). The integrated resource supply curve improves upon the familiar conservation supply curve by explicitly incorporating the size and costs of various supply options, rather than assuming a generalized cut-off point beyond which energy-efficiency is not cost-effective. This is important given that supply-side options vary in cost and in their potential contribution to the energy system.

The cumulative greenhouse gas emissions avoided by each change from the existing system (20 steps) are shown in Figure 4b, where the measures are re-ranked in order of increasing cost of avoided emissions. The no-policy scenario results in annual emissions of 17.8 mt carbon-equivalent in the year 2010, a sixfold increase from 1987 levels. These emissions are reduced by 44% in the least-cost scenario but are 165% higher than base-year emissions. This situation is amended in the least-emissions scenario by bringing biomass (and a small amount of wind-generated electricity) into the supply mix. In this scenario, emissions decline to only 2.4 mt: 35% below actual 1987 levels.

The projected (no-policy scenario) emissions are avoided at an average net economic benefit to society of \$41/tonne. This net benefit, however, is \$102/tonne carbon-equivalent less than the benefit derived from the least-cost case. The \$102/tonne difference between the two cases can be viewed as the net economic cost (value) to Sweden of achiev-

Table 1. Alternative commercial energy-use scenarios for developing countries.[50]

	1985	2025[a] Goldemberg *et al*		RCWP[b]	RCW[b]
Final energy (EJ)					
Fuel	55.0	149.4	(2.5)	150.0 (2.5)	170.1 (2.9)
Electricity	5.6	26.5	(4.0)	35.1 (4.7)	39.9 (5.0)
Total	60.6	175.9	(2.7)	185.1 (2.8)	210.0 (3.2)
Primary energy (EJ)					
Fossil fuel	71.0	124.5	(1.4)	132.6 (1.6)	242.8 (3.1)
Hydro	6.5	20.5	(3.7)	33.2 (4.3)	28.9 (3.9)
Nuclear	0.7	1.4	(1.9)	13.8 (7.7)	14.3 (7.8)
Biomass	0.0	76.5		76.5	4.8
Solar	0.0	6.7		17.0	5.2
Total	78.2	229.6	(2.7)	273.1 (3.2)	296.0 (3.4)
Carbon emissions (GT)	1.5	2.2	(1.0)	2.9 (1.7)	5.4 (3.3)

[a] Average growth rates (in %/year) for the period 1985–2025 are given in parentheses.
[b] These are the projections made by the US Environmental Protection Agency (EPA) for its rapidly changing world (RCW) scenario and its rapidly changing world with policy (RCWP) scenario.[51]

ing reductions beyond the point where the individual measures can be implemented with no net cost to society.[32] The carbon taxes introduced in Sweden in 1991 are $150/tonne, suggesting that the $102/tonne premium is well within the bounds seen as reasonable by the Swedish government.

Developing countries

Future greenhouse gas emissions in developing countries are of special concern because of the rapidly growing demand for energy services there. Energy, however, need not constrain development of the material standard of living in these countries.[33] Encouraging technological 'leapfrogging' to reduce the cost of energy services (ie, the adoption of technologies that are more efficient than the average in industrialized countries) must become a major objective in development policy.

Two scenarios for developing countries (for supply and demand respectively) indicate that fossil-fuel carbon emissions can be constrained to 2.2 GT/year in 2025.[34] In the supply scenario, natural gas and oil each contribute one-third of the commercial fuel supply, and most of the balance is met with biomass (Table 1). In comparison, the US Environmental Protection Agency (EPA) has estimated that emissions in developing countries will grow to 5.4 GT/year by 2025.[35] The EPA estimate does not include new policy initiatives aimed at reducing emissions.

An integrated example from Karnataka, India. The official long range plan for power projects (LRPPP) for the electricity sector in Karnataka projects a more than threefold expansion of electricity generation and consumption between 1986 and 2000. In meeting this projection, the plan calls for annual expenditures of $3.3 billion/year and expansion of coal-fired electricity generation capacity, resulting in a 0.83 million tonne (127%) increase in annual carbon emissions from current levels.

An alternative scenario developed for Karnataka by the Department of Management Studies at the Indian Institute of Science would provide a much higher level of electricity services than the LRPPP plan.[36] Electricity demand in this scenario, however, only doubles between 1986 and 2000, as a result of investments in efficiency-improvement measures and the introduction of solar water heating and LPG stoves. The additional demand is met with cogeneration in sugar factories, mini/micro hydroelectric systems, and decentralized rural power generation based initially on biomass and later on photovoltaics. Annual expenditures of $0.6 billion are called for, and annual carbon emissions increase by 0.004 million tonnes. The net economic benefit of this scenario compared to LRPPP plan is over $3 000/tonne of avoided carbon emissions. The scenario illustrates that, even in developing countries, increasing electricity service levels and, thereby, living standards need not lead to emissions increases.

Policies

Although some of the technologies incorporated in the scenarios described above would be adopted under current market conditions, others would not. The opportunities lost when *inefficient* new vehicles, equipment and buildings are added to the stock are of primary importance.

Existing markets can not be relied upon to lead to

the sufficient use of technologies that cost-effectively reduce greenhouse gas emissions because: (1) biases towards conventional supply systems lead to under-investment in energy efficiency; (2) consumers lack sufficient information, financing and access to the full spectrum of available equipment; and, (3) energy prices normally do not reflect marginal costs or the costs of externalities, such as climate change.

Despite the fact that the global end-use scenario described earlier is consistent with plausible values of income and price elasticities, and with energy prices not much higher than those at present,[37] market failures are steering the energy sector towards a future with much higher energy demand and greenhouse gas emissions. New policies are required to remedy this situation.

The kinds of policies described below are within the sphere of government-level policymaking. We indicate the general nature of such policies and offer a few examples of their application. Specific policies and their implementation must be chosen to fit the cultural, political and market conditions of each nation.[38]

- **Redefining the mission of energy suppliers and creating new markets for energy efficiency or emissions offsets can help build markets for energy services rather than energy *per se*.**[39] Some US utilities have more than a decade of experience with operating informational and financial-incentive programmes to promote energy efficiency.[40] As noted earlier, a number of European electric utilities have used financial incentives to promote energy-efficient lighting. In a new entrepreneurial initiative, the Swedish utility Vattenfall is spending $150 million (1 billion kroner) to help finance the kinds of end-use technologies described earlier in the least-emissions scenario. Some electric utilities in the USA have gone a step further by implemented innovative auction systems in which vendors of energy-efficiency services are able to compete with power plant suppliers.

  Creating markets for tradeable emissions offsets is another promising and economically efficient way to stimulate investments in energy efficiency. This is exemplified by recent amendments to the US Clean Air Act that set limits on total SO_2 emissions by individual utilities. Utilities emitting less than their SO_2 allowance – eg by increasing energy efficiency – can sell the excess emissions rights to other utilities.
- **Governments and other large buyers of energy-using equipment can use innovative procurement systems to promote the design and commercialization of more efficient products.** The Swedish National Energy Administration, for example, has cooperated with the largest building-management companies in Sweden to submit efficiency-oriented procurement proposals to appliance manufacturers. As a first step, this group has invited the manufacturers to produce new refrigerator-freezer designs that feature improved energy efficiency. The result has been positive: new units will be commercialized in 1991 that achieve 55% lower electricity-use than the most efficient models now on the market, while eliminating CFCs from the insulation.[41]
- **Emissions taxes can be used to signal the need to lower emissions, to fund implementation of measures that achieve emission reductions and to create funds for research, development and demonstration (RD&D).** To generate US$30 billion/year would only require a $1/bbl-equivalent charge on fossil fuels.[42] Such a tax would be important, but, for institutional and other reasons, the charge would not in itself create a significant impact on the market.[43]
- **Energy performance standards should be introduced when market failures inhibit the attainment of optimal end-use efficiency goals.** Automobile efficiency opportunities offer an instructive example. The total cost of owning and operating an automobile is essentially constant over a wide range of fuel economies,[44] resulting in the lack of an incentive for improved energy efficiency. As a remedy for this type of problem, standards have been used in many countries. For example, the USA has fuel-efficiency standards for automobiles, buildings and household appliances, and standards are pending at the State or national level for motors, lighting and other technologies.[45] Energy-performance standards are a logical extension of safety and other standards now in place to protect both consumers and the environment. Society benefits from energy-performance standards that maintain energy service levels while yielding a reasonable return on the investments associated with increased energy efficiency.
- **RD&D priorities can be changed to reflect the most promising strategies for combating climate change.** The current funding priorities of International Energy Agency member governments are indicated in Figure 5. Nuclear ener-

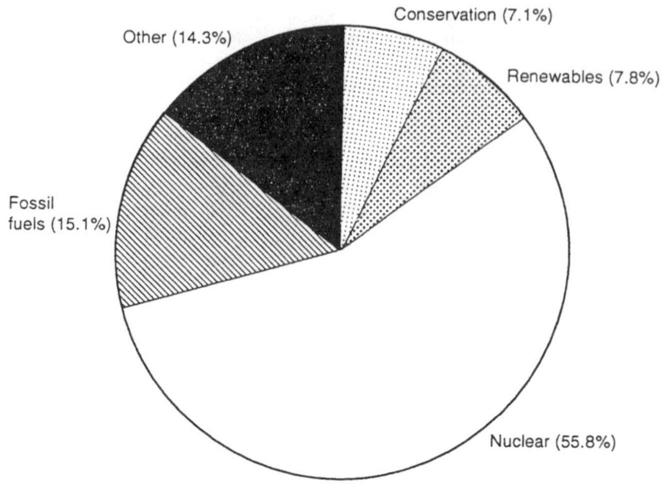

Figure 5. RD&D budgets (US$6.88 billion) in International Energy Agency governments, 1988.[54]

gy is receiving over half of the funding, despite its limited potential to reduce greenhouse gas emissions or the cost of energy services,[46] and despite the fundamental security problem of creating weapons-proliferation-proof nuclear energy systems.[47] As of 1988, end-use efficiency received only 7% of the countries' RD&D budgets and renewable energy sources received 8%. The share of total RD&D allocated to energy efficiency and renewable energy declined between 1977 and 1988.[48] In a strategy to abate global climate change, renewable sources and energy efficiency should become the major focus for government-sponsored R&D because of their long-term potential to reduce emissions, their security-enhancing benefits, their present costs and the small size of the existing industrial base.[49]

Observations and conclusions

The examples we have shown for the USA and Sweden suggest that industrialized countries can significantly reduce their energy sector greenhouse gas emissions within 10 to 20 years by implementing measures with no net economic costs to society. Our example from a developing country (the state of Karnataka in India) suggests that emissions can be constrained to current levels at a net economic benefit, even with a doubling of electricity supply and even greater growth in the level of energy services delivered.

A commitment to implementing policies can be justified on its economic merits alone. A wait-and-see policy is economically inefficient, and also forgoes non-economic benefits to society, such as enhanced national/international security and the reduced environmental impacts beyond those associated with climate change.

Our results show much lower costs for reducing greenhouse gas emissions than those arrived at using econometric models. Our methods differ from econometric methods in several important respects. We focus on emissions reductions that can be achieved now while ensuring a reliable supply of the energy *services* required for desired development and economic growth. Taking an end-use perspective enables us to incorporate new factors that cannot be accounted for by models based on observations of the past. It also enables us to identify existing market failures and to analyse policy options aimed at making markets work better.

Despite our conclusion that something, indeed quite a bit, can be done, the strategies that we have described are not sufficient to achieve the approximately 60% emissions reductions required to stabilize atmospheric concentrations of greenhouse gases in the long term. Ongoing and accelerated technological development, reforestation and structural and behavioural changes not analysed in this article offer prospects of further emissions reductions. It is important to begin *now* with measures that are economically justified, rather than waiting until a detailed strategy for meeting long-run emissions-reductions targets can be developed.

Support for this work was provided by the Swedish National Energy Administration and the Stockholm Environment Institute. An earlier version of this article was published in the Proceedings of the Second World Climate Conference, World Meteorological Organization, June 1991. For constructive review comments, the authors thank Dean Abrahamson, Randall Bowie, Lars Brinck, Tomas Ekwall, Jose Goldemberg, Leif Gustavsson, Paul Hofseth, Jill Jäger, Bill Keepin, Jon Koomey, Mattias Lundberg, Hans Nilsson, Lars Nilsson, Amulya Reddy, Henning Rodhe, Bo Wiman, and three anonymous reviewers.

[1]Intergovernmental Panel on Climate Change (IPCC), *Policymakers Summary of the Formulation of Response Strategies*, report prepared for IPCC by Working Group III.
[2]IPCC *Policymakers Summary of the Scientific Assessment of Climate Change*, report prepared for IPCC by Working Group I, 1990.
[3]World Energy Conference, *World Energy Horizons: 2000–2020*, Editions Technip, Paris, 1989.
[4]A.S. Manne and R.G. Richels, 'CO_2 emission limits: an economic cost analysis for the USA', *The Energy Journal*, Vol 11, No 2, pp 51–74. For a critique of Manne and Richels article, see R.H. Williams, 'Low cost strategies for coping with CO_2 emission limits', *The Energy Journal* Vol 11, No 4, pp 35–59.
[5]Monetary values are in US dollars throughout this article.
[6]Solar Energy Institute, *The Potential of Renewable Energy: An Interlaboratory White Paper*, Solar Energy Research Institute Report, SERI/TP-260-3674, 1990.
[7]Based on a total busbar cost of ¢5.68/kWhe for current light-water technology (6% real discount rate, 30-year amortization): plant size (1 100 MW), capital costs ($3 060/kW), and efficiency (33.4%). The capital cost drops to $1 670/kW for the target light-water reactor. These estimates are from the Electric Power Research Institute (EPRI). The nuclear fuel-cycle cost $0.84/GJ (¢1.11/kWhe: EPRI's projection for 1990 to 2000) and the operations and maintenance costs (¢0.91/kWhe) are the 1985 US averages for nuclear plants. This busbar-cost derivation is shown in R.H. Williams and E.D. Larson, 'Expanding roles for gas turbines in power generation', in T.B. Johansson, B. Bodlund and R.H. Williams, eds, *Electricity: Efficient End-Use and New Generation Technologies, and Their Planning Implications*, Lund University Press, Sweden, 1989, p 525.
[8]The indirect costs of implementing measures vary according to many factors. Performance standards, for example, can have a negligible cost per unit of energy saved while costs can be higher for conservation programmes, especially during initial 'learning' periods. European lighting efficiency programmes have shown implementation costs of only ¢0.3/kWh, ie approximately 1/20 the cost of the energy they save. See E. Mills, 'Evaluation of European lighting programmes: utilities finance energy-efficiency, *Energy Policy*, Vol 19, No 3, April 1991, pp 266–278.
[9]M. DeLuchi, R.A. Johnston and D. Sperling, 'Transportation and the greenhouse effect', *Transport Research Received*, No 1175, pp 33–44.
[10]Calculated using data from US Department of Energy, *Annual Energy Review 1988*, Report DOE/EIA-0384(88), Energy Information Administration, 1989.
[11]S.C. Davis, D.B. Shonka, G.J. Anderson-Batiste and P.S. Hu, *Transportation Energy Data Book: Edition 10*, Oak Ridge National Laboratory, Report ORNL-6565, Oak Ridge, TN, USA, 1989. Unless stated otherwise, single-figure fuel economies are weighted averages of highway (45%) and city (55%) mileage, as per the USEPA test procedure. Note: miles per gallon (mpg) = (235)/(litres per 100 km).
[12]This analysis applies to the 1976 to 1988 period. See M. Ross 'Energy and transportation in the United States', *Annual Review of Energy*, No 14, pp 131–171.
[13]D.L. Bleviss, *The New Oil Crisis and Fuel Economy Technologies: Preparing The Light Transportation Industry for the 1990s*, Quorum Books, 1988, p 268.
[14]Toyota, press release, 23 October 1985.
[15]Bleviss, *op cit*, Ref 13.
[16]R.W. Mellde, I.M. Maasing and T.B. Johansson, 'Advanced automobile engines for fuel economy low emissions, and multifuel capability', *Annual review of Energy*, No 14, pp 425–444.
[17]See S. Unnasch, C.B. Moyer, D.D. Lowell and M.D. Jackson, *Comparing The Impacts of Different Transportation Fuels on the Greenhouse Effect*, California Energy Commission, Consultant Report P500–89-001, 1989. The environmental implications of moving toward methanol vehicles are highly dependent on the feedstock used for methanol production. The prime attractiveness of methanol vehicles lies in the opportunity to use biomass feedstocks. However, creating a methanol vehicle fleet would also create a market for methanol produced from coal feedstocks. This eventuality should be avoided because producing, transporting, and burning coal-based methanol leads to higher greenhouse gas emissions than gasoline, compared on a per-vehicle-kilometre basis; see C. Difiglio *et al*, 'Cost effectiveness of future fuel economy improvements', *The Energy Journal*, Vol 11, No 1, pp 65–86. It should be noted that methanol vehicles are potential emitters of large quantities of formaldehyde: a highly toxic gas that can also contribute to the production of tropospheric ozone. Formaldehyde emissions can be kept very low, however, through the use of methanol engines with oxidation catalysts, see D. Sperling and M.A. DeLuchi 'Transportation energy futures', *Annual Review of Energy*, No 14, pp 375–424.
[18]This is the fuel economy in urban driving conditions. Due to the regenerative breaking system in the Impact, urban and highway mileage are about equal. Personal communication, Mr Sloane, Public Relations, General Motors Detroit, 2 August 1990. The gasoline equivalent was computed using an electrical conversion efficiency of 45% and assuming 3.96 MJ/kWh which includes 10% transmission and distribution losses.
[19]California Energy Commission, *Chemically Recuperated Gas Turbine*, Report number P500–90–001 (draft), 1990.
[20]E.D. Larson, P. Svenningson and I. Bjerle, 'Biomass gasification for gas turbine power generation', in Johansson *et al*, *op cit*, Ref 7.
[21]This discussion is based on J.M. Ogden, R.H. Williams and M.E. Fulmer, *Cogeneration Applications of Biomass Gasifier/Gas Turbine Technologies in the Cane Sugar and Alcohol Industries*, Center for Energy and Environmental Studies, Princeton University, NJ, USA, 1990. The electricity production potential is 2 780 TWh_e/year. The potential contribution of barbojo depends on how much of the resource can be covered cost-effectively. By utilizing only bagasse fuel in this estimate, 900 TWh could be produced, or 55% of 1987 electricity demand. (Based on note c to Table 13 of Ogden *et al*.) This scenario uses biomass-integrated gasifier technology with intercooled steam-injected gas turbines (BIG/ISTIG). The scenario is calculated assuming growth in sugar cane production of 3%/year (the historical rate of annual growth since 1960) over the 40-year period. Half of the growth is assumed to be committed to sugar production (equivalent to the World Bank projection of the sugar demand growth rate to 1995) and half to the alcohol production. This much more efficient use of potential cane resources would considerably improve the economics of the cane industry, and in effect turn it into an electricity-production industry with sugar and/or alcohol as marketable byproducts.
[22]Solar Energy Research Institute, *op cit*, Ref 6.
[23]*Ibid*.
[24]D. Carlsson, 'Low-cost power from thin-film photovoltaics', in Johansson *et al*, *op cit*, Ref 7, p 595.
[25]Personal communication, Lars Nilsson, Department of Environmental and Energy Systems Studies, University of Lund, July 1990, based on R. Summers and A. Heston, 'A new set of international comparisons of real product and price levels estimates for 130 countries, 1950–1985', *Review of Income and Wealth*, No 1-25; and United Nations, *Energy Statistics Yearbook*, New York, USA, 1988.
[26]R.H. Williams and E.D. Larson, 'Materials, affluence, and industrial energy use', *Annual Review of Energy*, No 12, pp 99–144.
[27]J. Goldemberg, T.B. Johansson, A.K.M. Reddy and R.H.

No-regrets strategies for reducing greenhouse gas emissions

Williams, *Energy for a Sustainable World*, Wiley-Eastern, New Delhi, 1988. Also summarized under the same title by the World Resources Institute, Washington, DC, 1987. The standard of living in developing countries is assumed to increase to that of the WE/JANZ regions, and by 50 to 100% in the industrialized countries.

[28]E. Mills, J.P. Harris and A.H. Rosenfeld, *Le Gisement d'Economies d'Energie aux Etats-Unis: Tendances, Perspectives et Propositions*, Energy Internationale, 1988–89, pp 169–193 (in French). Also as, 'Developing demand-side energy resources in the United States: trends and policies', Lawrence Berkeley Laboratory, Report No 24920, CA, USA.

[29]See *Notes* to Figure 2 for details.

[30]B. Bodlund, E. Mills, T. Karlsson and T.B. Johansson, 'The challenge of choices: technology options for the Swedish electricity sector', in Johansson *et al*, *op cit*, Ref 7.

[31]Using OECD conventions for counting nuclear and hydroelectric power.

[32]The marginal cost curve of avoided emissions is very steep. The marginal cost, in this case the cost of wind energy replacing fossil fuels, is $135/tonne. This reflects today's wind energy costs (¢7/kWh). At projected future costs of ¢3.3/kWh (see Table 1), the marginal cost becomes a net benefit of $41/tonne.

[33]J. Goldemberg, T.B. Johansson, A.K.N. Reddy and R.H. Williams, 'Basic needs and much more with one kilowatt per capita', *Ambio*, No 14, pp 190–200.

[34]J. Goldemberg, T.B. Johansson, A.K.N. Reddy and R.H. Williams, 'Energy for a sustainable world: an update, with emphasis on developing countries', presented at the *Bellagio Seminar on Energy Efficiency for a Sustainable World*; Goldemberg *et al*, *op cit*, Ref 33.

[35]US Environmental Protection Agency, *Policy Options for Stabilizing Global Climate*, D.A. Lashof and D.A. Tirpak, eds, Office of Policy, Planning, and Evaluation, Washington, DC, USA, (draft).

[36]A.K.N. Reddy, G.D. Sumithra, P. Balachandra and A. d'Sa, *Energy Conservation in India: A Development-Focused End-Use-Oriented Energy Scenario for Karnataka, Part 2 – Electricity*, Department of Management Studies, Indian Institute of Science, Bangalore, India, presented at the Bellagio Seminar on Energy Efficiency for a Sustainable World, Figures 24 and 32.

[37]Goldemberg *et al*, *op cit*, Ref 27, pp 479–481.

[38]See *Ibid*; H. Geller, 'Implementing electricity conservation programs: progress towards least-cost energy services among US utilities', in Johansson *et al*, *op cit*, Ref 7, pp 741–764; R.H. Williams, 'Innovative approaches to marketing electric efficiency', in Johansson *et al*, *op cit*, Ref 7, pp 741–764.

[39]D. Moskovitz, *Profits & Progress Through Least-Cost Planning*, National Association of Regulatory Utility Commissioners (NARUC), Washington, DC, USA, 1989; and O. de la Moriniere, 'Energy service companies: the French experience', in Johansson *et al*, *op cit*, Ref 7, pp 811–830.

[40]Geller, *op cit*, Ref 7.

[41]As the incentive to participate, manufacturers submitting proposals were paid approximately $16 000. As an incentive to exceed the target efficiency level, approximately $80/refrigerator will be paid to the manufacturer for each reduction of 15% in electricity-use beyond the target value given. The procurement guidelines also invited manufacturers to create an 'energy label' to compare the energy-use of their design to that of models already on the market. Personal communication, Hans Nilsson, National Energy Administration, Sweden, 10 August 1990.

[42]This is the annual tax proposed to support three climate change initiatives: (1) a CFC phase-out; (2) reforestation of 12 million hectares/year; and, (3) fossil-fuel energy conservation in the developing world. See J. Goldemberg, 'Policy responses to global warming', in J. Legget, ed, *Global Warming*, Oxford University Press and Greenpeace International, London, p 177.

[43]For a more detailed discussion of carbon taxes, see M. Grubb, *The Greenhouse Effect: Negotiating Targets*, The Royal Institute of International Affairs, London, UK, 1989.

[44]F. von Hippel and B.G. Levi, 'Automotive fuel efficiency: the opportunity and weaknesses of existing market incentives', *Resources and Conservation*, No 10, 1983, pp 103–124.

[45]For a description of the methodology used to determine appliance standard levels in the USA, see USDOE, *Technical Support Document: Energy Conservation Standards for Consumer Products-Refrigerators and Furnaces*, Report Number DOE/CE-0277, Washington, DC, USA, 1989. For proposed lighting and motor standards, see, for example, *An Act Requiring Minimum Energy Efficiency Standards for Lighting Fixtures, Lightbulbs, Floor Lamps, Table Lamps and Electric Motors*, House Bill number 5239, (1990), The Commwealth of Massachusetts; and *An Act Reducing The Greenhouse Effect by Promoting Clean and Efficient Energy Sources*, House Bill number 5277 (1990), The Commonwealth of Massachusetts, USA.

[46]B. Keepin and G. Kats, 'Greenhouse warming: comparative analysis of nuclear and efficiency abatement strategies', *Energy Policy*, Vol 16, No 6, December 1989, pp 537–642.

[47]R.H. Williams and H.A. Feiveson, *Diversion-Resistance Criteria for the Future Nuclear Power*, Princeton Center for Energy and Environmental Studies, Princeton, NJ, USA, 1989.

[48]International Energy Agency, *Energy Policies and Programmes of IEA Countries*, IEA, Paris, 1989.

[49]For examples of cost-effective government R&D in the buildings sector, see H. Geller, J.P. Harris, M.D. Levine and A.H. Rosenfeld, 'The role of Federal research and development in advancing energy efficiency: a $50 billion contribution to the U.S. economy', *Annual Review Energy*, No 12, pp 537–395; M.A. Brown, L.G Berry and R.K. Goel, 'Commercializing Government-sponsored innovations: twelve successful buildings case studies', Oak Ridge National Laboratory, Report No ORNL/CON-275, Oak Ridge, TN, USA, 1989.

[50]See Goldemberg, *et al*, *op cit*, Ref 33. In the low-carbon scenario, electricity demand grows by 4%/year, and would be met with the following generation mix: BIG/GT (32%), hydropower (31%), natural gas (17%%), PV/wind/solar thermal (10%), coal (6%), and nuclear (2%).

[51]United States Environmental Protection Agency, *op cit*, Ref 35.

[52]Steps 1 and 3–11 are adapted from H. Geller, *National Energy Efficiency Platform: Description and Potential Impacts*, American Council for an Energy-Efficient Economy, Washington, DC, USA, 1989. Step 2 involves measures to reduce ambient temperatures, ie, the 'heat-island effect' in cities, personal communication, Arthur Rosenfeld, Lawrence Berkeley Laboratory, 1989.

[53]The least-emissions case is the 'high-efficiency/environmental-dispatch scenario' and the no-policy case is the 'reference scenario/economic dispatch scenario' as described in Bodlund *et al*, *op cit*, Ref 30.

[54]International Energy Agency, *op cit*, Ref 48.

Appendix 1

Methodology for calculating the cost of avoiding greenhouse gas emissions

Quantifying and comparing greenhouse gases other than CO_2

In comparing the emissions from various fuels and technologies, we include the known effects of CO_2, CH_4 and tropospheric ozone (O_3) from methane oxidation and, in the case of automobiles, (N_2O).[55] We incorporate fuel-cycle emissions associated with mining, processing, transporting and burning the fuels. The greenhouse properties of each gas are expressed in terms of carbon-equivalents and added together. We use natural gas leakage rates of 1% of production for new distribution systems. Emissions from own-use (on-site) energy and those related to the energy embodied in materials are not included. Doing so would, in many cases, improve the attractiveness of efficiency and renewables in comparison to traditional energy systems.

Since the atmospheric residence times of greenhouse gases differ, it is necessary to choose a time period under which comparisons of the gases' relative greenhouse forcing will be made. In this article, we use a 20-year period in order to incorporate the effect of the gases on the rate of climate change and to give a fair basis for evaluation of measures and policies aimed at buying time in a near-term perspective (ie, 0 to 20 years). A long-term (100 years or more) perspective would discount the role of methane emissions in our comparisons.

The resulting carbon-equivalent emissions factors are (grams/MJ): Coal 30.6; oil 24.6; natural gas 17.2; and gasoline 25.7.[56] When shifting from a 20-year to a lifetime perspective (ie, integrating from zero to infinity), carbon-equivalent emissions decline by 19.6% for coal, 16.7% for oil, and 19% for natural gas.

$$\frac{\textit{Cost of a given measure} - \textit{Cost of base-case measure}}{\textit{Emissions from base-case measure} - \textit{Emissions from the given measure}} = \frac{\textit{Net cost (\$)}}{\textit{Net emissions (tonnes)}}$$

Figure 1.

Estimating the net cost (benefit) of avoiding greenhouse gas emissions

To compare various technical and fuel-choice measures for reducing emissions, to the corresponding base case, we use an indicator called the cost of avoided carbon-equivalent (CAC_{eq})[57] which is shown in Figure 1.

The following is an illustration based on conserving coal-based electricity with adjustible-speed motor drives (7th item in Appendix 2, Table 1). The levelized cost of conserved electricity is \$0.0111/kWh) *v* an electricity supply cost of \$0.0436/kWh (busbar cost). The electricity production results in emissions of 318 grams of carbon-equivalent/kWh_e. Choosing the efficiency strategy would result in a cost of avoided carbon-equivalent with a net economic benefit:

$$\frac{0.0111\ \$/\text{kWh} - 0.0436\ \$/\text{kWh}}{318\ \text{grams Ceq/kWh} - 0\ \text{grams Ceq/kWh}}$$

$$= -\$102/\text{tonne carbon-equivalent}$$

Emissions associated with the measure are zero for efficiency strategies, but are positive for substitution to fuels with net emissions of greenhouse gases.

[55]Deborah Wilson, 'Quantifying and comparing fuel-cycle greenhouse gas emissions from coal, oil and natural gas consumption', *Energy Policy*, Vol 18, No 6, July/August 1990, pp 550–562.
[56]Motor fuel emissions rates are derived from DeLuchi *et al*, *op cit*, Ref 9, based on Wilson, *Ibid*.
[57]For further detail on this method (as applied to carbon emissions), see F. Krause and J. Koomey, 'Unit costs of carbon savings from urban trees, rural trees, and electricity conservation: a utility cost perspective', *Proceedings of the Workshop on Saving Energy and Reducing Atmospheric Pollution by Controlling Summer Heat Islands*, Lawrence Berkeley Laboratory, Report No 27872, CA, USA, 1989.

Appendix 2

Examples of the cost of avoiding greenhouse gas emissions

In order to evaluate and rank the economic efficiency of technologies to reduce greenhouse gas emissions we have constructed a list of examples of specific measures and their costs (Tables 1, 2, and 3). These measures are grouped into three categories: electricity, heating and transportation and include efficient end-use technologies, conversion technologies and fuel substitution options. In each category, a base case is chosen (eg, power production with coal or natural gas) and its costs and emissions are compared with those of alternatives (eg, efficient lighting equipment). The Tables distinguish between available and emerging technologies. Data are not yet available to enable the consistent and comprehensive inclusion of indirect infrastructure-related costs.

The results incorporate CO_2 and other important greenhouse gases. The economic results are expressed in terms of the cost of avoided carbon-equivalent: CAC_{eq} (ie, the cost of a strategy to reduce emissions minus the cost of a base-case strategy, divided by the amount of emissions reduced (\$/tonne)). A negative CAC_{eq} corresponds to a net economic benefit, ie the cost of the emissions-reducing strategy is lower than the cost of the base-case strategy. Appendix 1 shows a sample calculation and describes our method of treating emissions in more detail.

The costs of avoided fuels are listed under the headings for each Table. When comparing the results, care must be taken to focus both on the *cost* of avoided emissions and the *amount* of emissions avoided. A small quantity of avoided emissions and a relatively large cost differential will lead to a large CAC_{eq}, as in the case of coal gasification for electric power production. The percentage-reduction values shown in the Tables put each measure in perspective.

No-regrets strategies for reducing greenhouse gas emissions

Table 1. Examples of avoided emissions and their costs: electricity.

Electricity[a]	Measure Resource cost	Avoided emissions		Cost of avoided carbon-equivalent (CACeq)
(Cost of avoided resource (coal): $0.44/kWhe)	($/kWh)	(g Carbon-eq/kWh)	(%)	($/tonne)
End-use efficiency[b]				
Available technologies				
Lighting (incandescent → compact fluorescent)	−0.011	318	100	−171
Lighting (efficient fluorescent tube)	−0.007	318	100	−159
Lighting (lamps, ballasts, reflectors)	0.013	318	100	−96
Refrigerator/freezer, no CFCs	0.018	318	100	−79
Freezer, automatic defrost, no CFCs	0.022	318	100	−67
Heat pump water heaters	0.034	318	100	−30
Variable-speed motor drive	0.011	318	100	−102
US field data, multifamily, leaking retrofits	0.038	318	100	−19
Retrofits in 450 US commercial buildings	0.026	318	100	−54
No-cost or behavioural measures	0	318	100	−137
Electricity production[c] (busbar costs)				
Available technologies				
Biomass steam-electric (woodfuel)	0.041	318	100	−9
STIG (gasified coal)	0.041	9	3	−313
STIG (natural gas)	0.027	163	51	−103
Wind (1988)	0.054	318	100	33
Solar thermal electric (1988)	0.114	318	100	221
Solar photovoltaics (1988)	0.231	318	100	588
Nuclear	0.057	318	100	41
Emerging technologies				
ISTIG (gasified coal)	0.034	57	18	−176
ISTIG (natural gas)	0.024	187	59	−106
Chemically recuperated gas turbine	0.029	204	64	−73
Solar thermal electric				
(2000)	0.043	318	100	−1
(2010)	0.036	318	100	−24
(2020)	0.031	318	100	−40
Solar photovoltaics				
(2000)	0.072	318	100	89
(2010)	0.050	318	100	22
(2020)	0.036	318	100	−24
Wind				
(2000)	0.033	318	100	−33
(2010)	0.027	318	100	−51
Nuclear – industry target for USA	0.040	318	100	−11
Fuel choice (STIG technology in all cases)				
Avoided resource costs (gasified coal): $0.071/kWh)				
Gasified coal → natural gas (1990)	0.027	155	50	−91
Gasified coal → biomass (sugar) (~2000)	0.033	309	100	−25

Notes: [a] Unless noted, the annualized costs of efficiency and supply measures are calculated with a 6% real discount rate and no taxes. Costs for electric power plants are amortized over a 30-year period. Efficiencies based on higher heating value (HHV) are used throughout this analysis. The reference (avoided) technology is a coal (2 × 500 MW) steam-electric plant, 34.6% efficiency, $1370/kW capital cost, and operations and maintenance (O&M) costs of ¢0.89/kWh. For all power plant costs: 70% capacity factor, coal price $1.79/GJ, natural gas price $2.10/GJ. [b] Three lighting examples: The first two measures result in a net benefit because reduced labour costs exceed incremental capital costs[58] (7% real discount rate); Refrigerator and freezer without CFCs: assumes a blend of hydrochlorofluorocarbons (HCFC-22, HFC-152a, and HCFC-124) with 97% lower ozone-depleting potential than the CFC-12 refrigerant currently used. For foam insulation, today based on CFC-11 as a blowing agent, alternatives are based on HCFC-141b and HCFC-123 which in fact have higher insulating values than current foams (7% real discount rate);[59] Variable-speed motor drive: ABB Corporate Research (personal communication, Lars Gertmar, 1988). Capital cost $104/kW (750 kW motor). Assumes installation costs equal to 8% of capital cost; 15-year lifetime; 4 000 hours/year; 35% reduction in electricity requirement; Heat-pump water heater: adapted from Brown *et al*, assuming 2 000 kWh/year reduction in electricity required, 10-year life;[60] US field data, multi-family space heating: includes measured data from 42 actual projects in the USA.[61] [c] STIG (Steam-injected gas turbine) for gasified coal: 2 × 50 MW steam-injected gas turbine (STIG), air-blown gasifier, hot-gas clean-up, 35.6% efficiency, $1 300/kW capital cost and ¢0.71/kWh O&M costs; STIG for natural gas: 4 × 51 MW, 40% efficiency, $410/kW capital cost and ¢0.29/kWh O&M costs; Chemically recuperated gas turbines: assumes 54% efficiency, $500/kW capital cost, 70% capacity factor.[62] O&M costs assumed same as for gas-fired STIG systems; Biomass-steam $1 500/kW capital cost, $0.005/kWh O&M, $1.5/GJ fuel cost, wind, solar, thermal, and solar photovoltaics.[63] These prices are based on the RD&D intensification scenarios developed by five US National Laboratories; Nuclear: see, *op cit*, Ref 7; ISTIG (intercooled STIG) coal: 110 MW, 42.1% efficiency, $1 030/kW capital cost and ¢0.60/kWh O&M costs; gas: 114 MW, 47% efficiency, $400/kW capital cost and ¢0.29/kWh O&M costs; Fuel choice: the coal base-case is gasified coal *v* natural gas:[64] Gasified coal *v* biomass: 111 MW BIG/ISTIG system.[65]

Table 2. Examples of avoided emissions and their costs: heating and fuel choice.

Heating and fuel choice	**Measure resource cost**	**Avoided emissions**		**Cost of avoided carbon-equivalent (CACeq)**
(Cost of avoided resource (oil): $5.56/GJ)	**($/GJ)**	**(g Carbon-eq/MJ)**	(%)	**($/tonne)**
End-use efficiency[a]				
Available technologies				
Low-emissivity window glass	1.75	25	100	−155
Flame retention head burners	0.93	25	100	−188
US field data, multi-family, heating retrofits	3.00	25	100	−104
No-cost or behavioural measures	0	25	100	−226
Fuel choice[b]				
Methanol (from sugar industry)(~2000)	11.53	31	100	−95
Biomass (1990)				
High price (Sweden)	3.00	25	100	−104
Low price (Brazil)	1.50	25	100	−165
Biomass in district heating (including capital costs + O&M)				
1990	9.94	35	100	63
2010	8.46	35	100	−4

Notes: [a] Low-emissivity window glass: transmittance reduction of 0.23 W/m^2-K in a climate with 3 316 HDD(C);[66] Flame retention head burners: assumes 18% fuel-requirement reduction, $300 incremental cost, 10-year life;[67] US field data, multi-family space heating: includes measured results for 111 actual retrofit projects around the USA.[68] [b] For comparison, the reference feedstock is coal; Methanol: near-term technology assumed to be the pressurized, steam/oxygen-blown, fluidized bed biomass gasifier that has been developed by the Institute of Gas Technology, in plant sizes of 101.5 million gallons/year production capacity. Costs are significantly lower ($8.56/GJ) for a 555.6 million-gallon-per-year plant.[69] Biomass in district heating: calculations are for Swedish conditions, assuming 4 000 hours/year operation, 20-year economic life, $370/kW capital cost, 88% thermal efficiency, $5.22/GJ fuel price (coal) in 1990 and $3.92/GJ in 2010.

Table 3. Examples of avoided emissions and their costs: transport.

Passenger cars[a]	**Measure resource cost**	**Avoided emissions**			**Cost of avoided carbon-equivalent (CACeq)**
End-use efficiency[b]	**($/GJ)**	**(kg/GJ)**	(%)		**($/tonne)**
(Cost of avoided resource: $12.63/GJ (high); $5.17/GJ (low)					
Available technologies					
US CAFE standards (16.8 to 8.7 l/100km)	0.86	26	100	−458	−168
Average (8.7 to 6.2 l/100km)	5.17	26	100	−290	0
No-cost or behavioural measures	0	26	100	−492	−201
Fuel Choice[c]	($/100 km)	(kg/100km)	(%)		
(Cost of avoided resource: $1.2/100km (low), $2.9/100km (high))					
Gasoline → compressed natural gas	–		Higher emissions than from gasoline		
Gasoline → methanol from natural gas	–		Higher emissions than from gasoline		
Gasoline → methanol from coal	–		Higher emissions than from gasoline		
Gasoline → methanol from biomass (present)	3.10	7.5	100	22	253
Gasoline → methanol from biomass (near-term)	1.70	7.5	100	−164	67
Electric cars (operating-cost comparison)					
Gasoline → electric car (natural gas)	0.19	6.4	86	−427	−158
Gasoline → electric car (biomass)	0.29	7.5	100	−352	−122

Notes: [a] Reference-technology emissions for passenger cars: gasoline (7 500 gCeq/100 km, at current US new-car average fuel economy of 8.4 litres/100 km); [b]Gasoline prices: Gasoline prices (excluding taxes) vary amongst industrialized countries. We show cost calculations for the two extreme ends of the range. The 'high' gasoline price is from Japan and the 'low' price from the USA. Fleet improvement: This reflects measures already used in the automobile industry, applied nationally (USA) without changing the size distribution of the car stock.[70] The manufacturing costs of these measures should be comparable in the other major automobile-producing countries. The potential application of the measures to new cars sold outside of the USA depends on the extent to which they are already in use;[71] [c] Fuel choice: Because the energy content of methanol and the energy efficiency of methanol cars differ from those of gasoline and gasoline-fueled cars, we show costs in terms of $/100km. For methanol cars, we use a vehicle efficiency that is 15% higher than that of gasoline-fueled cars. The assumed gasoline fuel efficiency is 8.4 litres/100km, the US average for new cars. The biomass feedstock price for methanol production is $2.33/GJ (both cases). The coal price is $1.48/GJ. Through an extensive literature review, Sperling and DeLuchi concluded that the vehicle price and maintenance costs for single-fuel methanol cars will become comparable to those for similar gasoline cars (ie, with same size, range, weight, vehicle life and power) in the near-term.[72] The only cost differential associated with owning and operating a methanol car, therefore, will be that of the fuel. Although methanol fuels (from natural gas or biomass) cost

No-regrets strategies for reducing greenhouse gas emissions

more per unit of energy than gasoline today,[73] part of this price differential will be balanced by the inherently better efficiency of methanol cars. Fuel-switching from gasoline to methanol manufactured from natural gas would increase the carbon-equivalent emissions per vehicle kilometre driven. The difference, however, is small and must be weighed against the potential role of natural gas as a transition fuel while biomass-based methanol supplies are made available; Electric cars: The calculations for electric cars are based on GM's prototype Impact discussed in the text, assuming the electricity is made with the currently available power plants (the first and third entries under the 'Electricity Production' heading in Table 1). The comparison is based strictly on operating costs.

[58]M.A. Piette, F. Krause and R. Verderber, *Technology Assessment: Energy-Efficient Commercial Lighting*, Lawrence Berkeley Laboratory, Report No 27032, CA, USA, 1989.
[59]See United States Environmental Protection Agency, *op cit*, Ref 51.
[60]Brown *et al*, *op cit*, Ref 49.
[61]C.A. Goldman, K.M. Greely and J.P. Harris, *Retrofit Experience in the U.S. Multifamily Buildings: Energy Saving, Costs, and Economics*, Lawrence Berkeley Laboratory, Report No 25248 1/2, CA, USA, 1988, p 24.
[62]California Energy Commission, *op cit*, Ref 19.
[63]Solar Energy Research Institute, *op cit*, Ref 6. The costs shown in this report include taxes (ie a fixed charge rate is used). We recalculated the costs using a 6% real discount rate and no taxes.
[64]Williams and Larson, *op cit*, Ref 7.
[65]Ogden *et al*, *op cit*, Ref 21.
[66]Geller *et al*, *op cit*, Ref 49.
[67]Brown *et al*, *op cit*, Ref 49.
[68]Goldman *et al*, *op cit*, Ref 61.
[69]R.H. Williams, *Will Constraining Fossil Fuel Carbon Dioxide Emissions Really Cost So Much?*, Center for Energy and Environmental Studies, Priceton University, NJ, USA, 1990.
[70]Personal communication, David L. Greene, Oak Ridge National Laboratory, 14 May 1990.
[71]D. Sperling and M.A. DeLuchi, 'Is methanol the transportation fuel of the future?', *Energy*, Vol 14, No 8, pp 469–482.
[73]Williams, *op cit*, Ref 69.

425-29

[92]

[24]

JOURNAL OF ENVIRONMENTAL ECONOMICS AND MANAGEMENT 23, 91–95 (1992)

Q30

Q20

A12

Is the Entropy Law Relevant to the Economics of Natural Resource Scarcity?— Yes, of Course It Is![1]

HERMAN E. DALY

Environment Department, World Bank, Washington, D.C. 20433

Received August 30, 1991

Several points in Jeffrey T. Young's paper [1], which gives a negative answer to the question in the title, require comment. The first part of the article does not elicit any quarrel from me in so far as it describes the view that I hold, namely that entropy is fundamental to economics. This view stems mainly from the work of Georgescu-Roegen. The only thing that I object to in this first part is the statement that "the entropy law is being recommended to us as the basis of a new energy and/or entropy theory of value." This is a totally incorrect statement taken from Burness *et al.* [2]. I took pains to correct their error in a comment in the same journal [3], and I would have appreciated it if Young had paid attention to that correction and avoided repetition of the error. For the rest, however, Young's model is, as he claims, "faithful to their conception of the economic process." It is significant that there is nothing in Young's model of the "entropy view" that implies or requires an energy or entropy theory of value, which makes the extraneous introduction of this error difficult to understand.

Young's fundamental criticism is that entropy is, "an anthropomorphic concept intimately associated with what is useful and, therefore, defined by current technology." The fact that entropy has an anthropomorphic aspect relating to usefulness in no way implies that it is defined by current technology. Also, Young thinks that the material interpretation is critical to the relevance of entropy to economics and that his criticism applies only to entropy as material dispersion. But in fact his arguments extend to energy as well. The obvious anthropomorphism of usefulness and also the relation to technology both apply to entropy in strict energy terms as well as in its looser material interpretation. The problem for Young is that recognition of the general nature of his criticism would involve him in proving far too much—namely that the entropy law is false in its classical formulation for energy! Yet he never explains why his arguments do not apply to energy as well as to matter.

Physicists routinely apply entropy to matter, and, although this extension may involve some difficulties, it is far more than a mere analogy. Our inability to reduce different forms of matter to a common denominator in the way we can do for energy prevents us from determining whether we eventually will use up more matter in the recycling effort than the amount recycled. But it remains clear that complete materials recycling would require ruinous amounts of energy and time.

[1] The views expressed in this article are those of the author and should in no way be attributed to the World Bank.

0095-0696/92 $5.00

To demonstrate his objection Young asks us to consider a world with two material resources: a known resource *a* and a newly recognized resource *b*. As *a* is depleted from concentrated mineral deposits and used, it eventually becomes dispersed in the environment. The material stock has become less orderly, more entropic. But suppose now the discovery that another mineral, *b*, is also useful, thanks to a new technology. Then if we add ordered deposits of *b* to the picture it might more than make up for the increased disorder or randomness in *a*, and thus entropy would have declined rather than increased as a result of new knowledge, which is anti-entropic according to Young. Why exactly the same observations would not apply to old and new energy sources is not explained. After all, energy does not disappear any more than matter does—it too is dispersed, and new discoveries of concentrated energy might more than offset energy degradation caused by economic activity in a given year.

This argument is strange in several other ways. New knowledge naturally will change any system in which it is discovered, regardless of whether that system incorporates the entropy law or the "canonical neoclassical assumptions." If we discover a novel resource, *b*, or even if we just discover more deposits of the same resource, *a*, the result is the same—namely, we must redescribe the state of the system, taking account of the new knowledge. That new description, based on new knowledge, would record a stock of low-entropy materials greater (and likewise in the case of energy) than that in the previous inventory. This does not mean that the economic process is not entropic or even that knowledge is anti-entropic—it only means that our description of the initial stock of low-entropy materials was incomplete in the light of new knowledge. Perhaps the upward bookeeping revision of inventory of low-entropy materials might be greater in a given year than the physical increase in entropy from resource extraction and use. That hardly reverses the entropic direction of economic activity. It may be claimed that new knowledge will expand available matter (and energy) faster than economic activity will convert it into unavailable matter (and energy). This could happen, but so could the reverse. New knowledge may reduce available matter–energy. For example, the greenhouse effect represents new knowledge that lowers the effective availability of fossil fuels because the capacity to absorb the dispersed CO_2 is less than previously thought. New knowledge may reveal new limits. The hole in the ozone layer is new knowledge. To suppose, as is usually done, that new knowledge will always expand the resource base and never contract it is to overspecify the content of new knowledge, which must always be something of a surprise—and not necessarily a pleasant one.

Georgescu-Roegen [4] claims to have discovered a fourth law of thermodynamics to the effect that matter will dissipate in a closed system (i.e., a system closed with respect to matter but with a plentiful throughput of energy), or, alternatively, that complete recycling is impossible. I am prepared to believe in common-sense evidence that for all practical purposes complete recycling is impossible, but I am not competent to assess the claim that it is physically impossible, so I leave that to the physicists. But for present purposes it does not matter whether Georgescu-Roegen is right or wrong about his fourth law.

What matters is a non sequitur in Young's fundamental argument: "Is *b* available matter when there are no known uses for it? If so, then how can we know that dissipated *a* is unavailable? The absence of a technology for using dissipated *a* would not mean it is unavailable matter. The point is that available matter is dependent on the existence of appropriate technologies. It is not a purely physical concept." If new knowledge can lead to a new use for low-entropy material resource, *b*, then why can it not find a technology for using dissipated old resource *a*? But a new use for a new source of low entropy tells us nothing at all about the possibility of finding a new use for the dissipated high-entropy remains of old resources. In both cases Young thinks we are doing the same thing—simply converting nonavailable into available matter through technology. To call both dissipated *a* and concentrated *b* (before discovery) by the same name, "unavailable matter," and then reason that if technology can turn previously "unavailable" *b* into a useful resource then surely it can do the same for newly unavailable *a* is a gross non sequitur. The reasons for the unavailability of *a* and *b* are entirely different—*a* is unavailable because it is physically dispersed, and *b* is unavailable because, although concentrated, we as yet have no knowledge of how to use it.

Does the fact that we discovered uses for aluminum imply that we can invent a technology to recycle all the particles of rubber scraped from tires on curbs and interstate highways? The difference is that there is a technology for using aluminum that is economic, but the known technologies for recycling rubber particles on highways are not economic. The main reason for that fact is that aluminum deposits are concentrated, and scraped rubber particles are highly dispersed. One recycling technology for rubber particles requires many people on their hands and knees using magnifying glasses and tweezers. That is not likely to be economic. Whether it is inevitable that more matter will be dissipated in the form of worn-out tweezers and skinned knees than will be recycled in the form of gathered rubber particles is a nice question that I cannot answer. However, it is clear to me that tweezer-based recycling of specks of rubber (or vacuum cleaning or sand blasting technology) will be ruinously expensive in terms of energy, labor, and time regardless of the exact balance of materials dissipation. Since disordered matter requires more energy for processing, and since that extra energy will at some point make the recycling of dispersed matter uneconomic, we need no rigorous law of material entropy with a physical (as opposed to an economic) limit on recycling matter. This economic limit stems from the physical fact that enormous amounts of energy, as well as of other materials, are required to recycle highly dispersed matter.

Even in the case of energy it is not that it is physically impossible to recycle degraded energy, but rather that it always costs more energy to carry out the recycle than the amount recycled. So recycling energy is always *uneconomic* (regardless of the price of energy!). This economic aspect of the entropy law (not only admitted but *emphasized* by Georgescu-Roegen as the main reason that economists need to pay attention to entropy) does *not* mean that entropy is subject to technology. Technology remains subject to entropy. Perpetual motion technologies remain undiscovered. If it were really so easy for technology to turn available matter (dispersed *a*) back into available matter (concentrated *a*), then why is someone not filtering sea water for molecules of gold and petroleum? And if recycling dispersed matter were as easy as finding new resources, then why have even neoclassical economists put the emphasis on substitution of new resources rather than on recycling of old resources, i.e., on finding *b* rather than recycling *a*?

It is possible to invent technologies that are able economically to exploit lower concentrations of minerals than were exploited previously. The tailings of old silver mines are used as ore for new processes. But the new process too ends up with tailings that it cannot reuse. New technology can more efficiently extract usefulness from the entropic degradation of the matter–energy throughout, but it cannot thereby become anti-entropic. Tailings that can be mined again are like water that has flowed part of the way to the sea, but still has a way to go. Mining tailings are not like having water run uphill.

Young states that "the model of entropic decay is not relevant for modeling open systems." This seems quite wrong to me. An organism is an open system that maintains itself in a kind of steady state that is far from thermodynamic equilibrium (death and decay). It does this by counteracting its tendency to entropic decay by "sucking low entropy from its environment," as Schroedinger [5] put it. The basic physical understanding of life is that it is an open system that imports low entropy from the environment and exports high entropy back to the environment, thereby maintaining its highly ordered structure in a quasi-steady state. It is one thing to recognize that open systems can *resist* entropic decay and something else to claim that entropic decay is *not relevant* to open systems.

Economies, like organisms, consist of ordered structures subject to entropic decay, but capable of building up their internal order at the expense of imposing greater disorder on their environment. Economies resist entropic decay by importing low-entropy matter–energy from the environment and exporting high-entropy matter–energy back to the environment. This throughput or metabolic flow is necessarily entropic in both cases. At the input end it takes low-entropy matter–energy from finite environment sources, and at the output end it returns high-entropy wastes to finite environmental sinks. The sources become depleted, and the sinks fill up and become polluted. The entropy law is supremely relevant because it says that sinks cannot serve as sources. Absolute scarcity arises from the *combination* of the first and second laws of thermodynamics, not from either alone. If sources and sinks were infinite (or could be created and destroyed) then it would not matter that the flow between them was entropic and irreversible; if sources and sinks were finite but no entropy law existed, then we could turn sinks into sources, recycle everything, and burn the same gallon of gas over and over. In the light of such basic facts it is difficult to understand how anyone could claim that the entropy law is not relevant to resource scarcity. Even if one considers the sun to be an infinite source and space to be an infinite sink it is only the total stock of solar energy that is quasi-infinite—its flow rate of arrival to earth is strictly limited. Nor would it be intelligent to increase it by "mining" the sun even if we could, because increasing the rate of insolation would be ecologically very disruptive.

We should not appeal to thermodynamics for temporally parochial measures of exchange value or relative scarcity, i.e., prices that measure scarcity relative to wants in the service of optimally allocating a given resource flow among alternative human uses in the present generation. But the question of absolute scarcity, of the optimal sustainable scale of the economic subsystem as a part of the overall ecosystem, cannot be understood without some help from thermodynamics [6].

With regard to the charge of "scientism" leveled at those who recognize and insist on the relevance of the entropy law to resource scarcity, I would like to plead innocent and to reverse the charge. It is not scientism to respect the effect of basic physical laws like gravity, entropy, and conservation of mass on economic activity. It is scientism to believe that applied science through technology can reverse these basic physical laws. The entropy law has implications that are unfriendly to the regnant ideology of growth as the economic sumum bonum. There is a strong tendency to resist such implications. But there is hope that they will not be resisted forever. The law of conservation of mass is now generally respected among economists, even though it has implications unfriendly to the growth ideology. Nevertheless, conservation of mass is still implicitly denied by devotees of the unlimited substitution of capital for resources in Cobb–Douglas production functions. Young, however shows signs of independence from the growth ideology by forthrightly pointing out that recognition of the first law of thermodynamics, "may constrain our optimism with regard to the sustainability of economic growth." If he continues to think about it, I would bet that he comes to see that the entropy law greatly increases the force of this constraint.

REFERENCES

1. J. T. Young, Is the entropy law relevant to the economics of natural resource scarcity?, *J. Environ. Econom. Management* **21**, 167–179 (1991).
2. S. Burness, R. Cummings, G. Morris, and I. Paik, Thermodynamic and economic concepts as related to resource-use policies, *Land Econom.* **56**, 1–9 (Feb. 1980).
3. H. E. Daly, Thermodynamic and economic concepts as related to resource-use policies: Comment, *Land Econom.* **63**(3), 319–322 (1986).
4. N. Georgescu-Roegen, Energy, matter, and economic valuation: Where do we stand?, *in* "Energy, Economics, and the Environment," (H. E. Daly and A. F. Umaña, Eds.) Westview Press, Boulder, CO (1981).
5. E. Schroedinger, "What is Life?," Macmillan Co., New York (1945).
6. H. E. Daly, Towards an environmental macroeconomics, *Land Econ.* **67**(2), 255–259 (1991).

Ecological Economics, 8 (1993) 7–10
Elsevier Science Publishers B.V., Amsterdam

On the relationship between the economic process, the Carnot cycle and the entropy law

C. Bianciardi [a], A. Donati [b] and S. Ulgiati [c]

[a] *OIKOS, International Foundation for Ecological Economics, Villa di Basciano, Siena, Italy*
[b] *Department of Chemistry, University of Siena, Pian dei Mantellini 44, 53100 Siena, Italy*
[c] *Department of Chemistry, University of Sassari, via Vienna 2, 07100 Sassari, Italy*

(Received 1 February 1992; accepted 10 September 1992)

ABSTRACT

Bianciardi, C., Donati, A. and Ulgiati, S., 1993. On the relationship between the economic process, the Carnot cycle and the entropy law. *Ecol. Econ.*, 8: 7–10.

The Carnot cycle, i.e., a reversible (ideal, not real) cycle carried out by an (ideal) heat-engine working with an (ideal) thermodynamic fluid, and the general concept of the thermodynamic cycle are sometimes used for comparisons with economic and biological (real) processes, supposing the latter resemble the former (Kummel, 1989; Mayumi, 1991). These comparisons provide a simple, useful way to account for very complex phenomena, but we should be aware that they are only a simplified model and not always applicable. Misunderstanding of this limitation (Khalil, 1990) may cause deep theoretical contradictions when a real process is considered.

THE CARNOT CYCLE AND THE ENTROPY FUNCTION

Let us recall the fundamental characteristics of the Carnot reversible cycle. It is accomplished by two (respectively, isothermal and adiabatic) expansions and two (again, isothermal and adiabatic) compressions, starting from a state A of the fluid and returning to it. The reversible ideal cycle can be viewed as a succession of equilibrium states and can be represented in a Cartesian system of axes p–V, called the Clapeyron plane. The first isothermal expansion is performed by adsorbing the quantity of heat Q_1 from a thermal source at the temperature T_1. The second expansion (adiabatic) is performed at the expense of the internal (incoherent) energy

Correspondence to: C. Bianciardi, OIKOS, International Foundation for Ecological Economics, Villa di Basciano, Siena, Italy.

Elsevier Science Publishers B.V.

of the fluid and hence by decreasing its temperature until T_2, lower than T_1, is reached. The expansions produce a coherent positive work on the environment. Then a coherent work is applied to the fluid by the environment so that an isothermal compression is performed, followed by an adiabatic one, reestablishing the initial conditions A. During the isothermal compression an incoherent energy $Q_2 < Q_1$ at the temperature $T_2 < T_1$ is delivered to the environment, which has the function of a cooling sink.

The net coherent work provided by the engine is

$$W = Q_1 - Q_2 \quad (1)$$

The efficiency η of an engine is, in general, defined as the ratio of delivered work W to the energy provided

$$\eta = (Q_1 - Q_2)/Q_1 \quad (2)$$

and it can be shown that for a Carnot cycle, and for every reversible cycle, the following expression holds:

$$\eta_{\text{rev}} = [(Q_1 - Q_2)/Q_1]_{\text{rev}} = (T_1 - T_2)/T_1 \quad (2')$$

Every reversible (ideal) engine cyclically working between T_1 and T_2 has the same efficiency that depends only upon these temperatures. It is easily shown that for a (real) irreversible engine working in a cycle between the same temperatures T_1 and T_2, the efficiency η_{irr} is

$$\eta_{\text{irr}} = [(Q_1 - Q_2)/Q_1]_{\text{irr}} < (T_1 - T_2)/T_1 = \eta_{\text{rev}} \quad (3)$$

The definition of the Entropy Function is possible by considering a reversible Carnot cycle. From the above expression of efficiency we can infer that

$$Q_1/T_1 = Q_2/T_2 \quad (4)$$

or, in a more general form,

$$\Sigma Q_i / T_i = 0 \quad i = 1, \ldots, n \quad (4.a)$$

where Q_i is the positive (received by the system) or negative (delivered) heat and T_i is the temperature of the source or sink.

For an irreversible engine the following expression holds:

$$\Sigma Q_i / T_i < 0 \quad i = 1, \ldots, n \quad (4.b)$$

A further generalization is possible, if $n \to \infty$. The summation of n discrete quantities is substituted by the integration of the infinitesimal quantities $\mathrm{d}Q / T$, giving

$$\int \mathrm{d}Q / T \leqq 0 \quad (5)$$

where the equality holds for reversible cycles.

If we refer to a reversible cycle beginning and ending at the state A, and if the state B belongs to this cycle, it can be easily demostrated that the integration

$$\int_{A}^{B} dQ/T \tag{6}$$

does not depend on the particular reversible transformation from A to B in the p–V plane. If the following definition is given for the Entropy Function

$$S_B - S_A = \int_{A}^{B} dQ/T \tag{7}$$

for every reversible transformation from A to B, it follows that the Entropy Function is defined only by its change between two states.

It can be shown that for every (real) irreversible process taking place in adiabatic isolated systems

$$S_B - S_A > 0 \text{ or } S_B > S_A \tag{8}$$

After this last result, we can recall the Entropy Law: the entropy of an isolated system always increases for every real process which occurs in it; no increase should be expected when an (ideal) reversible process is considered.

THE CARNOT CYCLE AND THE ECONOMIC PROCESS

In opposition to the well-known Georgescu-Roegen's paradigm (shortly expressed as: the economic process is subjected to the Entropy Law; G.R., 1971), Khalil (1990) has proposed a risky thesis that can be summarized as follows:

(1) the economic process is performed by a "purposeful" agency to generate coherent work;

(2) the Carnot cycle expression [(Total Energy Input – Wasted Energy)/Total Energy Input] has the same form as the economic process efficiency [(Input – Waste)/Input], "... thus, production can be conceived after the Carnot cycle" (Khalil, 1990);

(3) if the economic process resembles to the "purposeful" Carnot cycle, it cannot be subjected to the Entropy Law ("unpurposeful").

Lozada (1991) proposed a defense of G.R.'s paradigm, observing that a Carnot cycle is an ideal cycle, whereas economy is a real process. He also observed that the Entropy Law applies to both "purposeful" and "unpurposeful" processes, without any difference.

If the economic process does not resemble an (ideal) reversible Carnot cycle, it remains to be seen whether it can resemble a (real) irreversible thermodynamic cycle. If so, the Khalil efficiency for an economic process would be comparable to the efficiency of a thermodynamic cycle, as Khalil claims. This is evidently not so! In fact, the economic process is not even comparable to an irreversible thermodynamic cycle. A thermodynamic cycle, whether reversible and purposeful or not, is such that the system returns to the starting conditions (whence the name "cycle"), producing or absorbing free energy.

The economic process, to the contrary, is better regarded as a thermodynamic process starting from an initial state A and finishing in a final state B. Production is an activity (which may even be purposeful) such that the general resources (natural and/or technological) available in state A are used to "generate effects, products and services" (Khalil, 1990) as a final result in state B. It is clear that state A occurs before state B and that the economic process is an irreversible process which goes from A to B. Hence the efficiency of an economic process referred to by Khalil is only formally equivalent to thermodynamic cycle efficiency. The latter represents the efficiency per cycle of a heat engine, while the former represents the balance of a process which is not a cycle, i.e. the percentage of the resource in state A which is transformed into free effects in state B by an economic process.

On the other hand, the economic process consists of many real and irreversible thermodynamic transformations that always occur in economic production: any economic process can be viewed as production of goods and services and as the respective financial exchanges and investments of capital. Economic production is actually characterized by physical and/or bio-chemical processes which may or may not be cycles, depending on whether or not heat engines are used. In any case, like any irreversible process, all these processes involve heat exchanges inside the whole system and are, therefore, subjected to the Second Law and the Law of Entropy.

From this point of view, each economic process can be regarded as an irreversible transformation that increases the entropy of the isolated system in which it occurs.

REFERENCES

Georgescu-Roegen, N., 1971. The Entropy Law and the Economic Process. Harvard University Press, Cambridge, MA, 457 pp.

Khalil, E.L., 1990. Entropy Law and exhaustion of natural resources: is Nicholas Georgescu-Roegen's paradigm defensible? Ecol. Econ., 2: 163–178.

Kummel, R., 1989. Energy as a factor of production and entropy as a pollution indicator in macroeconomic modelling. Ecol. Econ., 1: 161–180.

Lozada, G.A., 1991. A defense of Nicholas Georgescu-Roegen's paradigm. Ecol. Econ., 3: 157–160.

Mayumi, K., 1991. Georgescu-Roegen's Fourth Law of Thermodynamics and the Flow-Fund model. Lecture held at the International Foundation for Ecological Economics OIKOS, Siena, Italy, on November 25th, 1991.

4 34-39
[95]

[26]

Pergamon

Computers chem. Engng Vol. 19, Suppl., pp. S229–S234, 1995

0098-1354/95 $9.50 + 0.00

LIFE CYCLE ASSESSMENT AND LINEAR PROGRAMMING -ENVIRONMENTAL OPTIMISATION OF PRODUCT SYSTEM

A. AZAPAGIC and R. CLIFT

Centre for Environmental Strategy, University of Surrey, Guildford, Surrey, GU2 5XH

φ70 c6\

ABSTRACT

Application of Linear Programming (LP) Modelling to Life Cycle Assessment (LCA) is a novel approach to analysing and managing the environmental performance of a complete product system. Not only can it solve the problem of allocation of environmental burdens in the Inventory stage of the LCA, but it also enables the environmental optimum of the system to be investigated. Once the environmental burdens have been allocated amongst different products on the basis of marginal changes in activities, they are aggregated to quantify environmental impacts and then reduced to a single environmental impact function which is treated as a new objective function of the LP model. The LP solution then gives the environmental optimum of the system and identifies places where improvements can be made. If multiobjective LP is applied, then the environmental performance of the system can be combined with other requirements, such as societal preferences or economics, thus giving the best compromise for the system performance. The idea behind this approach is illustrated by a simplified analysis of a system to produce thermoplastics.

KEYWORDS

Life Cycle Assessment; Environmental impact; Allocation; Linear Programming; System Analysis; Marginal values.

INTRODUCTION

Life Cycle Assessment (LCA), also known as "cradle to grave" analysis, represents a process to evaluate the resource consumption and environmental burdens associated with a product, process or activity, as well as their impacts on the environment (Fava et al., 1993). The process also encompasses identification and evaluation of opportunities for improving the environmental performance of a product or activity. The main stages of the LCA, as recognised by SETAC (Consoli et al., 1993), are:

1. *Goal definition and scoping*, where the boundaries of the system are defined;
2. *Inventory* of materials and energy use and of emissions and solid waste, known collectively as "burdens";
3. *Impact assessment*, comprising:
   -*Classification*, in which the burdens are aggregated into a smaller number of environmental impacts;
   -*Characterisation*, where the potential environmental impacts are quantified;
   -*Valuation*, by which the environmental impacts are reduced to a single measure of environmental performance;
4. *Improvement assessment* of the activities needed to reduce environmental impacts.

The methodology is still being developed and several unresolved problems remain to be addressed. One of these is allocation of environmental burdens in a system with more than one product. It has already been shown that representing the whole productive system by a Linear Programming (LP) Model can solve this problem through its marginal or shadow values (Azapagic and Clift, 1994; Azapagic and Clift, 1995a). This approach is appropriate because LCA is based on linear relationships between activities and environmental burdens, so that a linear system model should be used. Moreover, modelling a system through LP accommodates changes in the state of the system and so shows changes in the environmental burdens which result from marginal changes in activities. This property of LP models can also be used for allocation of environmental impacts in the Impact Assessment stage of the LCA. Furthermore, if the burdens are aggregated to a single environmental impact function, the LP solution can give the environmental optimum of the system as part of the Improvement stage. In addition, LP can incorporate economic and social aspects of the system analysis, thus offering answers for local and global system improvements. The interactions of different stages in LCA and LP modelling are shown schematically in Fig. 1.

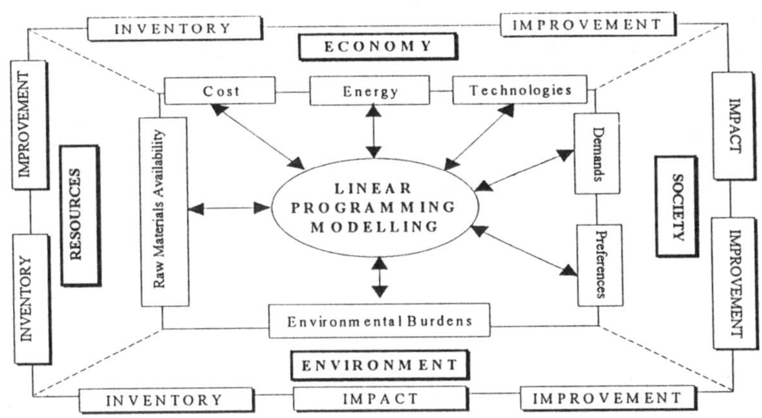

Fig. 1. Interactions between LCA and LP modelling

This paper outlines how LCA can be coupled with LP modelling to facilitate analysing and managing the environmental performance of the product system. The emphasis is on the approach rather than on a computational technique, and will be illustrated by a simplified example related to a system to produce thermoplastics.

LIFE CYCLE ASSESSMENT AND LINEAR PROGRAMMING

The LP model of a productive system in the context of LCA can be defined by n mass balance equations:

$$\sum a_{j,i}\, x_i \leq P_j - R_j,\quad i = 1,2,\ldots n \tag{1}$$

subject to demand D_j for products P_j, supply S_j of raw materials R_j and capacity C_i of process i:

$$P_j \leq D_j;\; R_j \leq S_j;\; x_i \leq C_i \tag{2}$$

where raw materials represent primary resources. The system is optimised on the objective function:

$$F = \sum f_i\, x_i \tag{3}$$

which is usually related to economic performance. Once the economic optimum has been found, environmental considerations are introduced through additional objective functions which include environmental burdens. In the Inventory and Impact assessment stages, these functions are used to allocate environmental burdens and impacts to different products in the system. At this stage, the system is not optimised on these functions. In the Improvement stage, however, optimisation on each of the environmental objective functions is performed to identify possible improvements in the system. Solutions of the economic and environmental optimisation usually conflict with each other. Multiobjective optimisation (Cohon, 1978) can resolve these conflicts and determine the best compromise.

Inventory - Allocation

Allocation in LCA refers to the problem at the Inventory stage of associating environmental burdens (such as resource depletion, emissions and solid waste) to each of the outputs from the system. Since allocation by LP modelling has been addressed elsewhere (Azapagic and Clift, 1994; Azapagic and Clift, 1995a), it will be mentioned here only briefly to introduce further possible applications of LP in LCA. For a linear system, the total environmental burdens, B_j, change in proportion to the output of the system:

$$B_j = \sum_i b_{j,i} x_i \qquad (4)$$

The coefficient $b_{j,i}$ quantifies the allocation of B_j to output x_i through the marginal or shadow values calculated as a part of the LP solution, and defined by the equation:

$$b_{j,i} = \partial B_j / \partial x_i \qquad (5)$$

Impact Assessment

Once identified, the environmental burdens are aggregated in the Classification/Characterisation stage to different impacts, such as green-house effect, ozone layer depletion, acidification, etc. The two major approaches used in LCA are known as "medium oriented" (Schmidt, 1991) and "problem oriented" (Heijungs, 1992). The details of these approaches need not be discussed here, but both aggregate the burdens to a smaller number of environmental impacts, forming the "environmental profile" of the system:

$$E_k = \sum_j e_{k,j} B_j \qquad (6)$$

which is another linear function of the individual burdens. In this way the LP structure of the model is preserved. The aggregated environmental impacts can then be allocated between outputs using marginal values of E_k. In the Valuation stage, these impacts are further aggregated to a single function by using a variety of techniques (Fava, 1993), discussion of which is beyond the scope of this paper. For our purposes, quantitative multi-criteria analysis, which assigns weights to impacts according to their importance, is used to give a single linear environmental impact function, the "Environmental Index":

$$EI = \sum_k w_k E_k \qquad (7)$$

Again, the shadow values can be used to identify products and activities in the system which make the greatest contribution to the overall environmental impact. This represents the starting point for the Improvement assessment.

Improvement assessment

The system can now be optimised on either of the objective functions defined by (4), (6) and (7), depending on the objective of the LCA study. However, system improvement cannot be carried out on the basis of environmental LCA only; technical, financial and social factors must also be considered. If multiobjective linear optimisation is applied, with the vector objective function defined as follows:

$$\underline{F} = [F_1, F_2, \ldots, F_m] \quad (8)$$

acceptable solutions, representing the compromise between conflicting objectives, can be found. This extension of the approach has been developed in a more recent paper (Azapagic and Clift, 1995b).

ENVIRONMENTAL OPTIMISATION OF PRODUCT SYSTEM -AN EXAMPLE-

The system under consideration is based around the commodities polypropylene (PP), expanded polystyrene (EPS), high density polyethylene (HDPE), low density polyethylene (LDPE) and high intensity polystyrene (HIPS). Operations and activities from the extraction of raw materials from the earth up to production of thermoplastic products are all included ("cradle-to-gate" approach: Fig. 2). For allocation purposes, the system is considered not to be constrained, i.e. raw material supply is unlimited. The objective function is taken to be profit, subject to constrained production of polymers. After the optimum solution has been found, each environmental burden, defined by eq. (4), is treated as an objective function, but the system is not optimised on the burdens. The total environmental burdens are estimated for the identified economic optimum and then allocated to each of the outputs through marginal values, eq. (5). The procedure is repeated for environmental impacts, eq. (6), and the single environmental impact function, eq. (7). To illustrate this process, Table 1 shows the values of allocated environmental burdens for SO_2, acidification (as aggregated impact of NOx and SO_2) and overall environmental impact, including green-house effect, acidification and nutrification (Heijungs, 1992). To create a single environmental impact function, eq. (7), arbitrary weighting factors have been used for illustration: 0.2, 0.4 and 0.4 respectively.

Table 1. Environmental burdens and impacts allocated through marginal values (MV) (objective function: Profit)

Commodity	Quantity (t)	MV for SO2 (kg/t)	MV for Acidification (kg/t)	MV for EI (kg/t)
PP	≤ 416000	11.0	21.3	469
EPS	≤ 33000	78.0	120.1	2155
HIPS	≤ 85000	28.0	55.3	1225
HDPE	≤ 32000	10.8	20.8	457
LDPE	≤ 550000	13.3	24.3	511
Raw material (Oil)	≤ 1600000	0.0	0.0	0

The environmental burdens and impacts are calculated for each part of the system, although only values for thermoplastic products are shown. As shown in Table 1, EPS has the highest marginal values for all environmental burdens, so change in its production will affect the environmental performance of the system the most: an increase of 1 t of its production, with other products kept constant, would increase the total environmental impact by 2.155 t.

The model discussed so far assumed that raw material supply (e.g. oil) is unlimited. In practice this is rarely the case, so it is interesting to see what happens when the system is constrained, i.e. oil availability is limited. Optimisation is again performed on profit, but the allocated environmental burdens are now different (Table 2): some are negative, which means that increasing their amount decreases the total amount of the burden in question (e.g. PP and HDPE). This is possible because these changes would happen at the expense of some other products in the system, which are much more "burden-intensive" (e.g. EPS and HIPS). If the system is then optimised on environmental impact function, EI, the optimum solution changes so that only commodities with negative burdens are produced (PP and HDPE). (Note that the sign of the marginal values in this case is opposite to the former one, because minimisation was performed.) For commodities whose production is slack (e.g. LDPE), the marginal burdens are zero, which means that their output can increase with no effect on the objective function.

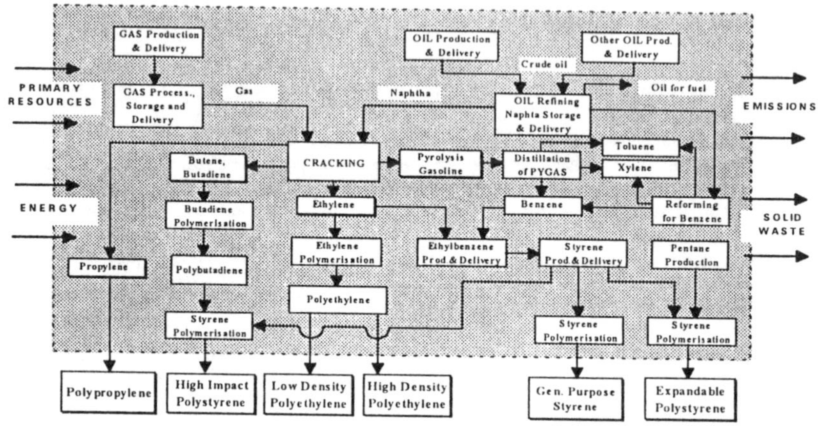

Fig. 2. Flow diagram of thermoplastic materials system

Table 2. Linear Programming as a tool in Improvement Assessment

Commodity	Constrained Quantity (t)	Estimated* Quantity (t)	MV for EI* (kg/t)	Estimated** Quantity (t)	MV for EI** (kg/t)
PP	≤ 416000	416000	**-41**	416000	**41**
EPS	≤ 33000	33000	**1470**	0	**0**
HIPS	≤ 85000	85000	**582**	0	**0**
HDPE	≤ 32000	32000	**-52**	32000	**52**
LDPE	≤ 550000	157812	**0**	309152	**0**
Raw material (Oil)	≤ 1600000	1000000	**386**	1000000	**-386**

*Objective function: PROFIT; **Objective function: EI

The solutions obtained in the two separate optimisations are different. If multiobjective LP is then performed, the best compromise between environmental performance and other requirements, such as economics or societal preferences, can be explored using the "Pareto surface" of non-inferior solutions. This further development of the system analysis has been explored in more detail in a more recent paper (Azapagic and Clift, 1995b).

CONCLUSIONS

Linear programming can be successfully combined with LCA to provide a powerful tool for analysing and managing environmental performance of a system while still satisfying other requirements, such as economic performance and societal preferences. It can be used in all stages of the LCA: for the allocation of environmental burdens in the Inventory stage, for evaluation of overall environmental impact in the Valuation stage, and as a management tool in the process of system improvement. Linear programming can offer more than other similar linear techniques because it can account for different states of the system and thus describe complex relationships between its different subsystems. Multiobjective linear programming can help in resolving conflicting demands on system performance.

NOMENCLATURE

a	Input/Output coefficients representing the amount of material used/produced in process
B	Burden (environmental burden or resource depletion)
b	Value of burden B allocated to output x
C	Capacity
D	Demand
E	Function of aggregated environmental burdens
e	Coefficient according to either "critical volume" or " problem oriented approach"
EI	Overall environmental impact or environmental index function
$\underline{F}$	Multiobjective (vector) linear function
F	Objective function
f	Coefficient in objective function
P	Products
R	Raw materials
S	Supply of raw materials
w	Weights in EI function
x	Output of activity (operation level of process)

Subscripts

i	number of activities/processes
j	number of commodities or burdens
k	number of aggregated environmental impacts
m	number of objective functions in multiobjective linear programming

REFERENCES

Consoli, F. et al., eds., (1993). *Guidelines for life cycle assessment: A 'Code of practice'*, SETAC, Brussels and Pensacola.

Azapagic, A. and R. Clift (1994). Allocation of environmental burdens by whole-system modelling - The use of linear programming. In: *Allocation in LCA* (Huppes, G. and F. Schneider, eds.), pp. 54-60. SETAC, Brussels.

Azapagic, A. and R. Clift (1995a). Whole system modelling and life cycle assessment. In: *The 1995 IChemE Research Event*, Vol. 1, pp. 429-431. IChemE, Rugby.

Azapagic, A. and R. Clift (1995b). The Application of Multiobjective Linear Programming to Life Cycle Assessment. Fifth SETAC-Europe Congress, Copenhagen, 25-28 June (submitted).

Cohon, J.L., (1978). *Multiobjective programming and planning*, Academic press, New York.

Fava, J., F. Consoli, R. Denison, K. Dickson and T. Mohin, eds., (1993). *A conceptual framework for life-cycle impact assessment*. SETAC, Brussels.

Heijungs, R., ed., (1992). *Environmental life cycle assessment of products: Guide*. MultiCopy, Leiden.

Schmidt, A., (1991). Evaluation (impact assessment). Paper presented at SETAC-Europe Workshop on Environmental Life Cycle Analysis, Leiden.

[27]

STRUCTURAL CHANGE, GROWTH, AND DEMATERIALIZATION: AN EMPIRICAL ANALYSIS

Sander de Bruyn
Jeroen van den Bergh
Department of Spatial Economics
Free University, De Boelelaan 1105
1081 HV Amsterdam, The Netherlands

Hans Opschoor
Institute of Social Studies, The Hague, The Netherlands

INTRODUCTION

Sustainable development in an expanding economy requires that the detrimental effect of growth on the environment should continuously be compensated. Initially, remedial environmental technologies can be successful. When such opportunities have been exhausted, more fundamental technological and structural changes are required.

From this point of view, a preventive environmental policy may be aimed at stimulating structural change in the productive side of the economy, leading to changes in the consumption of materials and energy. Structural change implies a reduction in the resource input of production, so that materials throughput will be reduced and less emissions and waste may be generated (Simonis 1989). Such a strategy may play a decisive role in achieving sustainability because the application of end-of-pipe technology may not be sufficient to reduce emissions such as of carbon dioxide. Economic growth will eventually boost these emissions up to the point where irreversible environmental changes may occur. Environmental protection, especially in the long run, will therefore necessarily have to be conceived in terms of structural policy. This has been called the "ecological/economic imperative" (Jänicke, Monch, and Binder 1993).

Many studies note that in developed economies the consumption of several materials and energy has declined, both in relative (per unit of GDP) and in absolute terms. This phenomenon has been labeled as "dematerialization" (Schmidt-Bleek 1993), "ecological structural change" (Simonis 1989) or "de-linking" (Jänicke et al. 1989; Opschoor 1990). According to Simonis (1989) this "ecological structural change" is the result of a combination of process integrated investments that have led to a more efficient use of energy and materials, and a shift in the structure of final demand towards less polluting products.

This chapter explores the question whether the de-linking of materials and energy consumption from economic growth has actually occurred in developed economies. First, a review is given of contributions that have concluded that de-linking may have taken place in developed economies. Special attention will be given to the methodologies that have been implied in these studies and the implication of these methodologies for the interpretation of the observed developments. Next, new analyses will be given that suggest a development of re-linking (i.e., increases in materials and energy consumption in several developed economies since the mid-1980s). This suggestion is further investigated using time-series analysis on energy, steel, and cement consumption for Japan, the Netherlands, and the United States which identifies economic growth as an important determinant of material and energy consumption. More explanations on the observed developments are provided using a decomposition method that is being applied to energy consumption in the Netherlands. The analysis presented in this chapter will provide some insight into the driving forces behind apparent de-linking and re-linking tendencies in the data.

A REVIEW OF EMPIRICAL INVESTIGATIONS

There is a vast body of literature in resource economics on empirical studies of the consumption of materials and energy (Barnett and Morse 1973; Malenbaum 1978; Bossanyi 1979; Larson, Ross, and Williams1986; Tilton 1990). Many of these investigations indicate structural reductions in the demand for materials and energy since the first oil crisis. Often the concept of the intensity of use is applied, defined as the ratio between the physical inputs (materials) and monetary outputs (GDP). Malenbaum (1978) presents the hypothesis that the relationship between the intensity of use and income over time can be described as an inverted U. This has been called the "intensity of use hypothesis." More precisely, this states that the intensity of use increases in early stages of economic development, then reaches a turning point and finally declines. The last stage is often referred to as de-coupling or de-linking materials and energy consumption from economic growth.

Tilton (1986) elaborates these reasons for de-linking:

1. Technological change, such as materials' saving production techniques and the substitution of metals by lighter chemical products, and
2. Structural changes in the economy, such as the shift from an economy oriented towards heavy industry to an economy primarily relying on services.

The inverted-U curve and falling consumption of energy and materials relative to income in developed economies have been confirmed by empirical observations on metal use (Larson, Ross, and Williams 1986; Tilton 1990; Valdes 1990) and energy use (Bossanyi 1979; Chesshire 1986; Goldemberg 1992). Other empirical studies indicate that Third World countries (Radetzki 1990; Suslick and Harris 1990) and COMECON-economies (Jänicke et al. 1989; Dobozi 1990) did not experience such a de-linking. As described by the intensity-of-use hypothesis, this implies that they have not yet reached the turning points.

Simonis (1989), Jänicke et al. (1989), Jänicke, Monch, and Binder (1993), von Weizsacker and Schmidt-Bleek (1994) and others have interpreted these findings from an environmental perspective. They pointed out that by reducing the resource input of production less emissions and wastes will occur that have a negative impact on the natural environment. This can be related to Daly's (1991) emphasis on the reduction of throughput, defined as the entropic physical flow of matter-energy from natural sources, through the human economy and back to nature's sinks (Daly 1991).

Some empirical evidence for a reduction in throughput was found in Jänicke et al. (1989), who compared the changes in a weighted sum of four indicators: the per capita consumption[1] of steel; the per capita consumption of energy; the production of cement; and the tons freight-transport between 1970 and 1985 for 31 industrialized countries. These four indicators were regarded as an approximation of the volume of throughput of a country.[2] For 1970 and 1985, each indicator has been calculated as the percentual deviation from the sample mean in these two years. The resulting (dimensionless) indicators are then aggregated into an Index of Environmental Deterioration (IED), giving equal weighting to all indicators (see Appendix 1 on the calculation of the IED).

In a cross-section analysis, Jänicke et al. (1989) regress the IED on the level of income and then compare 1970 to 1985. The results of their analysis are given in Table 10.1 (t-statistics were not provided).

The table shows that per capita income in 1970 gave a much better explanation for the environmental index than in 1985: the slope is less steep in 1985 than in 1970 and the measure of strength has decreased substantially.[3] The explanation for the decreasing slope is that the volume of (approximated) throughput increased in the poorer countries (e.g., in Bulgaria, Greece, Portugal, and Turkey), whereas it decreased in several richer countries. The fact that the measure of strength has decreased can be explained by the divergent developments of middle-income socialist countries (Czechoslovakia, DDR, Hungary, and Poland), which showed a high growth in materials and energy consumption, while the growth in materials and energy consumption of middle-income countries of the OECD (Ireland, Italy, New Zealand, and Spain) declined. These

Table 10.1. Relationships Between IED and GDP for 31 Countries

Regression Equation: $IED_t = \alpha + \beta GDP_t + e_t$

Years	1970	1985
Slope β	0.17	0.046
Intercept α	–1.2362	–0.3951
Measure of strength R^2	0.57	0.09

Note: Per capita GDP is measured in 1,000 1980 US$. Table taken from Jänicke et al. (1989).

developments are depicted in Figure 10.1 where arrows indicate linearized changes between 1970 and 1985 for various countries.

The analysis by Jänicke et al. has been interpreted as evidence of de-linking materials and energy consumption from economic growth through structural change in developed economies (Simonis 1989; RIVM 1991; Von Weizsacker and Schmidt-Bleek 1994). Several economies (e.g., Belgium, Denmark, France, Sweden, United Kingdom, and West Germany) showed "absolute structural improvements" (Jänicke et al. 1989) (i.e., their approximated throughput declined in absolute levels despite the growth in GDP).

The cross-section analysis by Jänicke et al. (1989) is in fact equivalent to the intensity of use hypothesis, as can be seen by Figure 10.2. This figure gives hypothetical intensity of use curves for four countries (C1,...,C4) with a shift-

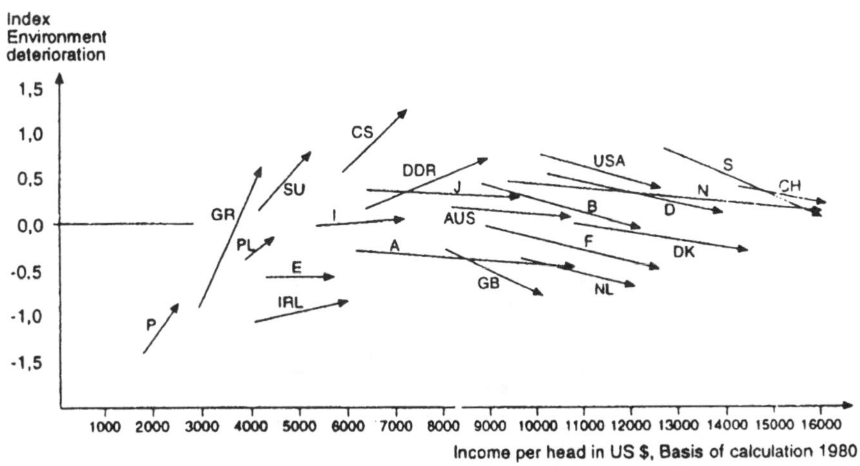

Source: von Weizsacker and Schmidt-Bleek (1994), after Jänicke et al. (1989).

Figure 10.1. De-linking economic growth from approximated throughput for some countries

ing inward for poorer countries resulting from the diffusion of technology as mentioned by Malenbaum (1978). Because of the diffusion of technological knowledge, poorer countries nowadays will have lower intensities than richer countries in the past for comparable income levels. The bold line in each curve illustrates the developments during the period of measurement (time 1,2,3) for each country. If these countries are included in a cross-section regression analysis for each period of time, the slope will decrease while the intercept will increase for later periods of time.

Concluding, the cross-section analysis of Jänicke et al. (1989) may confirm inverted-U shaped curves for the intensity of use of individual countries. Interpreted in this way, the Jänicke results can indeed be regarded as suggesting a process of de-linking materials consumption from economic growth for the more developed countries over the respective period.

Other investigations have disclosed similar tendencies for a number of pollutants. Inverted-U curves have been found in panel data analysis of the relationship between per capita income and air pollutants such as sulfur dioxide, nitrate, nitrite, and Suspended Particulate Matter (Shafik and Bandyopadhyay 1992; Selden and Song 1994).[4]

INTERPRETATION OF THE RESULTS OF PREVIOUS STUDIES

The results of Jänicke et al. (1989) and others have given an impetus for the suggestion that it would be possible to mitigate environmental problems in a

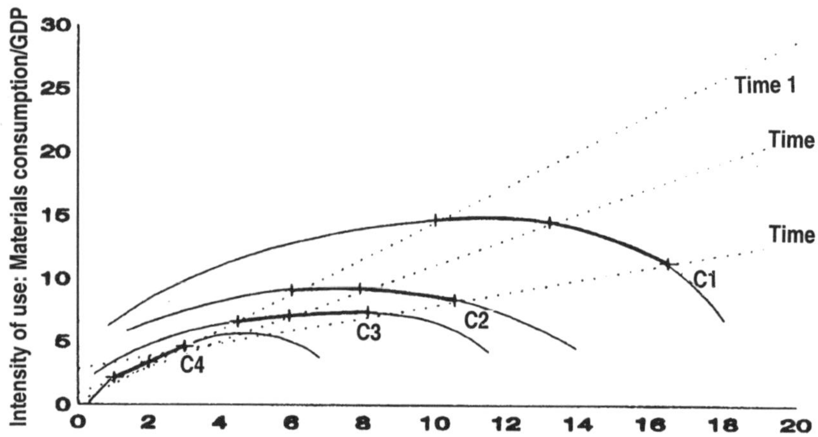

Note: The assumption here is that income rises over time.

Figure 10.2. An illustration of the outcomes of cross-section analysis in three periods, assuming inverted U-shaped intensity of use curves

growing economy through structural change. The perceived dematerialization is therefore an important issue in economic-environmental forecasting. The Dutch Environmental Planning Institute (RIVM) assumes that materials consumption per unit income in the Netherlands will decline by 20% in the year 2015 (RIVM 1991). But before interpreting the results from metal and energy demand studies in terms of dematerialization and structural change, we need to consider the following points.

First, only in closed economies does resource input in production equal emission and wastes in the long run.[5] International trade causes the inputs and outputs to diverge quite substantially. The perceived dematerialization in developed economies may therefore be entirely due to trade and relocation effects, and particularly to a relocation of energy and material-intensive industries to poor countries (Opschoor 1990). By importing the final products from these countries, production patterns are dematerialized, but consumption patterns are not. This phenomenon may be called import-substitution or exporting environmental pressure (Opschoor and Reijnders 1991). By importing products from dirty industries located abroad, developed countries deteriorate the environment in less developed countries. There is some evidence that this kind of import-substitution may be relevant. Al-Rawahi and Rieber (1991) show that the perceived decrease of the copper intensity of use in the U.S. economy could almost entirely be contributed to the increased import of embodied copper in (semi-)finished products. Total consumption of copper, both embodied and as raw material, grew almost at the same rate as income. For steel, however, Tilton (1990) reaches an opposite conclusion: the intensity of use in developed economies was lower when corrected for trade in steel embodied in finished products. The few case studies that are available, however, are too fragmentary to draw general conclusions about the effect of import-substitution on the intensity of use hypothesis. Moreover, the data on materials embodied in final products are being constructed from (monetary) trade statistics. This approach may not be very reliable (Tilton 1990). The materials and energy embodied in trade flows is a topic that deserves more attention in future research.

A second remark is that all dematerialization studies focus only on energy and a few types of metals. Wadell and Labys (1988) have proposed that dematerialization can generally better be described as "transmaterialization" since there is some evidence that lower quality bulk materials are being replaced by technologically superior lighter materials. Steel and cement are typically low quality materials, and plastics, platinum, titanium, and advanced ceramics are typical high quality materials. The demand for these is strongly increasing nowadays, whereby substitution is an important driving force. Therefore we need to incorporate more types of materials in dematerialization studies. Here we encounter a problem when aggregating several materials. Substitution of steel for plastics, for example in automobiles, will show up as dematerialization of the total product. Moll (1993) finds strong dematerialization tendencies for the United States if several materials are aggregated on weight. However, if the aggregation is performed on volume (m^3), no such de-linking can be found (Moll 1993). The question of which aggregation scheme is favorable from the environmental point of view is also a topic that deserves more attention.

A third remark is that none of the studies distinguishes between primary and secondary metals consumption. This distinction is not important if the purpose of the study is to predict or describe consumption. However, when we wish to equate materials and energy consumption with environmental pressure, the distinction between recycled and virgin materials is essential.

The above remarks indicate that a reduction in the consumption of energy and materials cannot be simply translated into environmentally beneficial dematerialization. It is clear that before anything definite can be said about dematerialization and de-linking, more data and alternative methods should be investigated. However, the problem is that few environmental data are available over a long time span (Jänicke et al. 1989). In the following sections we will therefore mainly consider alternative methods for expressing changes in the throughput of an economy, and search for explanations on the observed and estimated patterns.

IN-DEPTH CROSS-SECTION ANALYSIS

In this section we examine the study by Jänicke et al. (1989) in more detail, try to improve the method, and repeat the analysis using more data. The cross-section analysis by Jänicke et al. (1989) only contains samples for two years. No insight is obtained in the development of the approximated throughput during years in between. Hence the analysis is too general to draw strong conclusions about the stability of the de-linking process. It may be possible that the approximated throughput declined from 1970 to 1980 as a result of the two oil crises, but increased again after 1980. This would indicate that de-linking is not a continuous and stable process that autonomously leads towards dematerialization in the future. Also an observation can be made on the representation of the research results. Figure 10.1 showed the IED for each country for both 1970 and 1985 as the deviation from the sample mean of the per capita consumption of the selected indicators in 1970 and 1985, respectively. The arrows indicate developments in terms of this indicator for the two years. Since the indicators for both years are only measured in relative terms, Jänicke et al. cannot compare the developments in absolute levels between 1970 and 1985.[6] In other words, declining vectors in the figure do not automatically imply declining absolute levels of approximated throughput. Figure 10.1 only shows that the consumption of materials and energy in the countries with a lower per capita GDP rose relatively faster than in countries with a higher per capita GDP. Interpreted in this way, Figure 10.1 indicates convergence of levels of approximated throughput between countries over time and not necessarily de-linking. This fact, although recognized by Jänicke et al. seems to be overlooked by other studies referring to the work of Jänicke et al. (e.g., Simonis, 1989; Opschoor 1990; RIVM 1991; von Weizsacker and Schmidt-Bleek 1994).

In order to provide a more complete insight into the process of structural change and de-linking, the analysis of Jänicke et al. is here repeated, taking the above remarks into account. Data have been collected for steel, energy, and cement consumption and transport for a sample of nineteen countries[7] including all the years between 1966 and 1990 (see Appendix 2). This data collection diverges from the one by Jänicke et al. since fewer countries but more years have been included. Secondly, the indicator of cement production is replaced by cement consumption to diminish the sensitivity of the IED import substitution.

The IED for each country has been calculated as the deviation from the sample mean per capita consumption of the selected indicators. Contrary to the analysis of Jänicke et al., only one sample mean has been calculated as the average consumption over all countries and over all the years. Calculated in this way, absolute increases in throughput will result in an increase of the IED.

In order to obtain an overall and yet comprehensive view of the relationship between IED and GDP, the period under study was divided into four subperiods of seven years: 1966–1972, 1972–1978, 1978–1984, and 1984–1990. In the first period, economic growth was high and not disturbed by external shocks. The second period was characterized by the aftermath of first oil crisis in 1973, which influenced several economic parameters. The third period was marked by the second oil crisis and the subsequent negative growth rates in the beginning of the 1980s for most economies. The last period is characterized by a recovery of economic growth in the Western economies and by a disintegration of economic structures in formerly socialist economies. The IED and GDP of each country have been calculated as their average values in each period of seven years. This way yearly and incidental shocks in the economies or data are leveled out.

Jänicke et al. conclude that in 1985 the relationship between the IED and GDP, as measured by the coefficient of determination (R^2), was weakened and that the slope was less steep compared to 1970. The regression of IED on income for four periods with an improved IED calculation gives a more subtle representation of the development of this relationship (see Table 10.2). This table shows that the relationship between the IED and GDP was not stable over time. The second and third period give both a lower slope, a higher intercept, and a lower coefficient of determination (R^2) compared to the first period. This is similar to the conclusion of Jänicke et al. (see Table 10.1). However, this development of de-linking comes to an end in the fourth period. The slope rises, the intercept declines, and the coefficient of determination increases compared to the third period. Interpreted in the framework by Jänicke et al. this would mean that the de-linking phase for developed economies has come to an end in the last period under study.

Table 10.2. Descriptive Statistics of Cross Section Analysis for Four Periods

Regression Equation: $IED_t = \alpha_t + \beta_t GDP_t + \varepsilon_t$

Years	1966–1972	1972–1978	1978–1984	1984–1990
Slope β	0.0765	0.0552	0.0409	0.0435
(t- value)	(5.830)	(3.926)	(3.381)	(4.711)
Intercept α	-0.5892	-0.4188	-0.3938	-0.4624
(t- value)	(-6.725)	(-3.800)	(-3.734)	(-5.061)
R^2	0.666	0.476	0.402	0.566

Note: Income measured in 1,000 US$ per capita in 1985 prices.

Although, apparently, structural change has played a role in the second and third period, it has ceased to do so for developed economies in the 1980s. It seems that these economies are entering a period of re-linking rather than de-linking. This is clarified more explicitly in Figure 10.3, which presents the

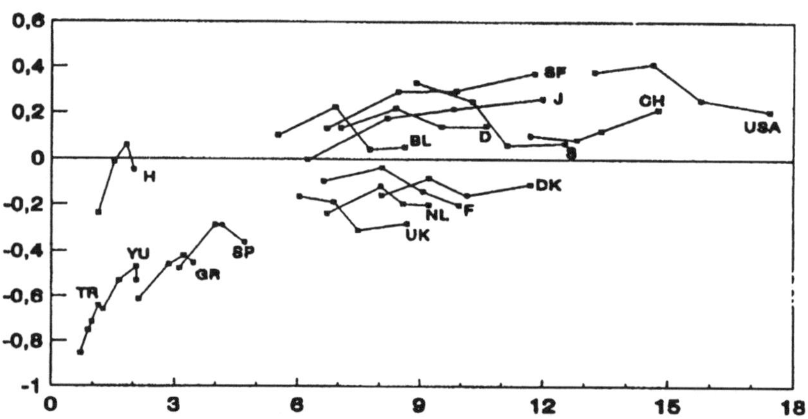

Note: Countries' symbols refer to international number plates, except for Spain (SP) and Belgium-Luxembourg (BL).

Figure 10.3. Developments of selected countries from 1966–1990 in levels of per capita GDP and IED

developments for several individual countries from period to period. Each dot indicates one period and the lines connect the dots for each country. As can be seen from this figure, three groups of countries can be distinguished. First, Turkey is an example of a group of countries that has not yet experienced any structural change. The IED has been rising faster than GDP in each period. Secondly, Greece, Hungary, Poland, Spain, and (former) Yugoslavia are countries that have experienced absolute improvements in the last period (1984–1990) compared to the third period (1978–1984).[8] After rapid increases in their levels of IED in the second and third period these countries now find themselves in a period of de-linking.

The group of richer countries shows more internal differentiation in the developments of the IED. However, most countries showed absolute improvements in the third period, between 1978–1984. This development of de-linking continued in the last period only for France and the United States. It is remarkable that all the other countries showed a rise in their levels of the approximated throughput in that period. For Finland and Japan this has been the case in each period, but the growth in their levels of IED has been declining.

Having noticed that the absolute improvements as measured and reported by Jänicke et al. (1989) are an exception (induced by the years of measurement) rather than a general rule, the question arises whether the developed countries are still de-linking in relative terms: the levels of approximated throughput may increase, but yet at a structurally slower rate than the increase in GDP. This would result in a continuously lower intensity of use. This is not the case for several countries as shown in Table 10.3.

This table shows for selected countries the developments of the intensities of use of steel, cement, energy, and transport during the 1980s. Overall, it appears that the process of structural change has been rather weak. In the Netherlands, United Kingdom, and Belgium–Luxembourg, the intensities of steel and cement have been increasing, especially after 1986. This implies that the consumption of these materials has grown at a higher rate than the economy. Only France and the United States experienced decreases over all periods. This apparent re-linking phase after a period of de-linking, forecasted by Pezzey (1989) and Opschoor (1990), and confirmed by some empirical investigations (c.f. Jänicke, Monch, and Binder 1992; Vial 1992), will be elaborated in the next section where we estimate several models for the consumption of energy, steel, and cement.

TIME SERIES ANALYSIS AND DE-LINKING

The analysis in the previous section suggests that in developed economies, for some materials, a period of de-linking may be followed by a period of re-

Table 10.3. Developments of the Intensity of Use in Selected Countries for Steel, Cement, Energy, and Transport (1966 = 100)

		Steel	Cement	Energy	Transport	IED
Belgium	1981	51.20	56.38	83.29	85.89	65.68
	1986	56.25	44.54	81.37	91.16	60.94
	1989	55.97	51.95	77.06	107.44	62.29
France	1981	62.34	63.25	98.28	81.31	68.92
	1986	50.88	48.30	99.50	68.26	58.21
	1989	50.98	47.58	95.32	69.87	57.85
Netherlands	1981	74.27	69.44	108.86	99.20	80.71
	1986	71.01	56.28	105.76	98.09	74.00
	1989	75.15	61.98	100.18	102.98	75.42
United Kingdom	1981	47.03	56.22	79.13	94.06	61.95
	1986	39.82	54.04	71.90	86.13	55.89
	1989	43.83	62.60	66.40	94.81	58.45
United States	1981	58.37	79.43	92.68	95.11	74.17
	1986	49.93	86.87	82.94	85.93	66.34
	1989	48.01	79.94	81.82	87.08	64.21

linking. This would not only invalidate the conclusions by Jänicke et al. (1989), but would also challenge the intensity of use hypothesis, which predicted continuously lower income elasticities of demand once economies have reached the turning points. The question is whether a de-linking as proposed by the intensity of use hypothesis is virtually nonexistent or whether high economic growth in the second half of the 1980s simply has resulted in an increase in demand that counteracted the materials and energy savings due to technological and structural changes.

To test the influence of economic growth on the demand for energy and materials, we construct three models that all use the following identity.

$$I_t = i_t \cdot Y_t \qquad (10.1)$$

where I_t is the demand for materials and energy, i_t is the intensity of use, and Y_t the level of GDP in year (t). The question here is thus whether the variation in the demand for I_t is mainly driven by variation in the level of GDP (the growth effect) or by the variation in the level of the intensity of use (the intensity of use effect). To answer this question we transform (10.1) into logarithms and take the first difference in order to obtain the growth rates of all variables.

$$\Delta \ln I = \ln I_t - \ln I_{t-1} = (\ln (i_t) - \ln (i_{t-1})) + (\ln Y_t - \ln Y_{t-1}) \qquad (10.2)$$

By adding an error term to (10.2), we construct three growth models that can be estimated.[9] These models differ in the way the change in the intensity of use is explained.

Model 1: The demand for materials and energy is solely explained by economic growth (i.e., the intensity-of-use effect is absent).

$$\Delta \ln I = \beta_1 (\Delta \ln Y) + u \qquad (10.3)$$

Model 2: The demand for energy and materials follows an inverted-U shaped curve (i.e., the intensity-of-use effect is explained by economic growth).

$$\Delta \ln I = \beta_1 (\Delta \ln Y) + \beta_2 \{(\ln Y_t)^2 - (\ln Y_{t-1})^2\} + u \qquad (10.4)$$

This model is the first difference of the translog models that have been used to estimate material demand according to the inverted-U shaped intensity of use curve (Suslick and Harris 1990).[10] It assumes that the intensity-of-use effect is dependent on economic growth. Under conditions of zero economic growth, the change in materials consumption will also be absent. This means that structural and technological changes responsible for the de-linking phase are a function of economic growth itself.

Model 3: The demand for materials and energy is a function of economic growth and the level of income.

$$\Delta \ln I = \beta_1 (\Delta \ln Y) + \beta_2 (\ln Y_t) + u \qquad (10.5)$$

According to this model, the intensity-of-use effect is explained by the level of income, and will result in more material and energy savings in rich economies than in poor economies.

In all models, it is expected that β_1 has a positive sign (representing the growth effect). Furthermore, in models 2 and 3, β_2 is expected to have a negative sign (representing the intensity-of-use effect).

These models have been estimated for the per capita consumption of steel, cement, and energy for each of the nineteen countries mentioned in the previous section. Income has been calculated as per capita GDP in US$1,000 constant (1985). The price changes of steel, cement, and energy for the U.S. market have been added as an additional explanatory variable (using parameter β_3).[11]

Table 10.4 gives the results for Japan, the Netherlands, and the United States. The first model offers a poor explanation in general. It is unlikely that the growth in the demand for materials can be explained by prices and growth in income only. The second model's explanation of the growth in materials and energy consumption is not much better than the first model. The parameter β_2 is nowhere significant at the 5% level, indicating that the intensity of use effect is either absent or poorly specified in this model. Furthermore the signs of steel and cement consumption in the Netherlands and steel consumption in the United States are the opposite of our expectations. The third model gives the best fit: all parameters have the expected signs and most of them are significant. The first parameter β_1 estimates the growth effect of the economy on materials consumption. For the countries selected in this table, this growth effect is significant at the 5% level for all materials and energy.

According to the estimators, an economic growth of 1% results in an increase in the steel consumption of 2.4% in Japan, 3.4% in the Netherlands, and even 4.8% in the United States. This growth effect is, however, counteracted by the intensity of use effect (parameter β_2). The intensity of use effect is appropriately specified as being dependent on the levels of income, being significant on the 5% level for all cases except for steel consumption in Japan and energy consumption in the Netherlands and the United States, which fall inside 10% significance level. There are also differences concerning the substances taken into account. According to model 3, the intensity of use effect in all countries is greater in the case of steel than in the case of cement or energy. This may reflect material substitution of steel for plastics as well as technological improvements in the molding and cutting of steel (McSweeney and Hirosako 1991). Furthermore, each of the models gives a better explanation for energy than for steel and cement. This can be explained by looking at the measurement of the data. Energy consumption data reflect *actual* consumption while the data on steel and cement consumption reflect *apparent consumption* including changes in stocks. Hence, an increase in stocks will be incorrectly counted as an increase in consumption and vice versa. Such developments may be reflected via a larger residual term.

The outcome of the time-series estimations indicates that the increase observed in the consumption of materials and energy during the 1980s (see Table 10.3) is simply the result of an increase in economic growth. Since the estimators for the growth effect range between 0.97 and 4.8, a higher economic growth leads to a more than proportional increase in the consumption of energy, steel, and cement given the reductions due to the intensity of use effect. Chow-tests

Table 10.4. Results from Estimation of Alternative Models for Materials and Energy Consumption

		Model 1: $\Delta\ln(I) = \beta_1\Delta\ln(Y) + \beta_3\Delta\ln(P)$			Model 2: $\Delta\ln(I) = \beta_1\Delta\ln(Y) + \beta_2\{\ln(Y_t)^2 - \ln(Y_{t-1})^2\} + \beta_3\Delta\ln(P)$				Model 3: $\Delta\ln(I) = \beta_1\Delta\ln(Y) + \beta_2\{\ln(Y_t) + \beta_3\ln(P)$			
Country	Material	β_1	β_3	R^2/DW	β_1	β_2	β_3	R^2/DW	β_1	β_2	β_3	R^2/DW
Japan	Steel	1.457	-0.055	24.7%	3.370	-0.503	-0.103	26.4%	2.354	-0.029	-0.016	32.7%
		(3.28)	(-0.12)	2.224	(1.25)	(-0.72)	(-0.22)	2.337	(3.36)	(-1.62)	(-0.04)	2.318
	Energy	0.939	-0.330	62.4%	2.249	-0.343	-0.300	67.6%	1.379	-0.013	-0.056	70.3%
		(7.40)	(-1.47)	1.979	(3.18)	(-1.88)	(-1.41)	2.323	(6.41)	(-2.42)	(-0.24)	2.442
	Cement	0.916	-0.142	33.6%	2.722	-0.482	-0.274	41.6%	1.578	-0.021	-0.101	51.1%
		(3.94)	(-0.42)	1.167	(2.05)	(-1.38)	(-0.79)	1.262	(4.74)	(-2.56)	(-0.33)	1.425
Nether-lands	Steel	1.807	-0.479	24.5%	-4.792	1.643	-0.468	26.0%	3.396	-0.030	-0.517	37.9%
		(2.48)	(-1.21)	2.303	(-0.49)	(0.67)	(-1.16)	2.294	(3.43)	(-2.19)	(-1.41)	2.547
	Energy	1.464	-0.692	60.7%	5.422	-0.984	-0.668	62.9%	2.032	-0.010	-0.553	66.4%
		(5.49)	(-2.84)	1.691	(1.54)	(-1.13)	(-2.75)	1.913	(5.26)	(-1.94)	(-2.30)	1.942
	Cement	0.779	-0.164	16.3%	-1.067	0.460	-0.152	16.7%	1.624	-0.016	-0.254	31.0%
		(1.95)	(-0.55)	2.284	(-0.20)	(0.34)	(-0.47)	2.282	(3.02)	(-2.16)	(-0.90)	2.806
United States	Steel	2.610	0.026	32.2%	-12.159	2.748	0.017	33.7%	4.786	-0.035	0.129	70.2%
		(3.35)	(0.06)	1.945	(-0.59)	(0.72)	(0.04)	1.897	(7.16)	(-5.29)	(0.47)	2.734
	Energy	0.773	-0.353	64.6%	5.488	-0.879	-0.368	67.0%	0.970	-0.003	-0.284	68.0%
		(5.19)	(-2.75)	1.111	(1.47)	(-1.27)	(-2.89)	1.360	(5.00)	(-1.53)	(-2.14)	1.230
	Cement	1.901	-0.201	41.2%	0.522	0.257	-0.196	41.2%	2.920	-0.017	-0.255	59.8%
		(3.68)	(-0.55)	1.261	(0.04)	(0.10)	(-0.52)	1.265	(5.40)	(-3.19)	(-0.82)	1.541

Notes:

1. Figures in parenthesis are t-statistics.
2. Income levels have been calculated in US$1,000 constant (1985) per capita.

were conducted to test for structural breaks in the data after 1985 and were all rejected so that we can assume that the estimated relationship is stable over time. This leads us to the conclusion that the intensity of use hypothesis is too static and deterministic for predicting the patterns of material and energy consumption, a fact that restates the criticism of others on this hypothesis (Auty 1985; Tilton 1986; Roberts 1990). The predictive power from translog models describing an inverted-U curve turned out to be rather poor, especially after 1980 (Valdes 1990; Vial 1992). The increase in consumption during the second half of the 1980s cannot be explained by dropping prices of metals and energy. In all the models, prices (parameter β_3) are insignificant at the 5% level, except the prices for energy in the Netherlands and the United States. This is in line with a number of other studies which found that in general prices of materials are not an important variable for determining materials consumption (Auty 1985; Holmes 1990a, 1990b; Tilton 1990; Mannaerts 1993). The demand for materials is typically a derived demand (i.e., dependent on the demand for final consumption goods). Material costs are only a small proportion of the total costs of producing these goods (Malenbaum 1978).

THE RATE OF SUSTAINABLE GROWTH

The existence of two counteractive effects, economic growth and intensity of use, implies that under conditions of low economic growth the consumption of materials and energy may decline. This provides a link with the literature on steady state economics and sustainable development. Daly (1991) proposes the installation of depletion quotas for exhaustible resources: "to be set near the existing extraction rates. The first task would be to stabilize, to get off the growth path. Later we could try to reduce quotas to a more sustainable level, if present flows proved too high." So in a first step towards a steady state, economic growth would only be acceptable if it does not result in a higher level of materials and energy consumption. We, rather informally, label this rate of acceptable growth here as *sustainable* growth, without claiming that economic growth that does not result in an increase in the consumption of materials and energy is sustainable in terms of all environmental aspects.[12]

From model 3 of the previous section, we can calculate which rate of economic growth does not result in a growth of materials consumption, but solely stems from improvements in the efficiency of transforming materials into valuable products and income. This rate of sustainable growth can be calculated by setting the right-hand side of equation (10.5) smaller than or equal to zero. This gives the following condition.

$$\beta_1 (\Delta \ln Y) + \beta_2 (\ln Y_t) \leq 0 \tag{10.6}$$

This equation can be expressed as:

$$\ln \left(\frac{Y_t}{Y_{t-1}}\right) \leq - \frac{\beta_2}{\beta_1} \cdot \ln Y_t \tag{10.7}$$

By solving the right-hand side of this equation for the parameters estimated from model 3 given a certain level of income, the maximum rate of sustainable growth is obtained. It indicates which rate of growth will not result in a higher consumption of the selected materials. Table 10.5 gives the application of condition (10.7) to the United States, Japan, and the Netherlands for energy, steel, and cement, where the prices have been assumed to remain constant over the whole period. The average level of GDP and the average growth between 1966–1990 have been chosen as explaining variables. In Table 10.5 the calculated rate of sustainable growth is compared with the actual rate of economic growth over the time period considered. In order to determine the accuracy of the calculations, Table 10.5 also gives a comparison of the actual annual growth rates of materials and energy consumption with the predicted growth rates according to the model. The predicted growth rates have been calculated using model 3 with the average level of GDP and the average economic growth rates between 1966–1990 with constant prices. In the last two columns of this model the actual and predicted growth rates of steel, cement, and energy have been given. This shows that the prediction of the model is rather good, except for energy consumption in the Netherlands and cement consumption in the United

Table 10.5. Annual Sustainable Growth Rates (1966–1990)

		GDP growth		Growth Consumption	
		Sustainable	Actual	Predicted*	Actual
Japan	Steel	2.68%	4.56%	4.19%	3.67%
	Energy	2.05%	4.56%	3.32%	3.39%
	Cement	2.92%	4.56%	2.43%	2.48%
Netherlands	Steel	1.83%	2.21%	1.19%	2.21%
	Energy	1.05%	2.21%	2.30%	0.94%
	Cement	2.08%	2.21%	0.16%	0.18%
United States	Steel	1.98%	1.56%	-2.11%	-1.71%
	Energy	0.80%	1.56%	0.72%	0.66%
	Cement	1.58%	1.56%	-0.10%	0.47%

*Calculated using average income level and average annual rate of growth between 1966–1990 and the parameters from OLS regression of model 3.

States. In column 3 the rate of sustainable growth, as has been calculated from equation (10.7), is presented.

In most cases, the rate of sustainable growth is below the actual annual growth rates over the period 1966–1990, indicating that the consumption of energy, steel, and cement has risen over the period considered. It is also interesting to notice that the rate of sustainable growth in all countries is lower for energy than for cement and steel, which indicates that economic growth will sooner result in a rise in energy consumption than in a rise in steel or cement consumption. Finally, it appears that the rate of sustainable growth in Japan for all materials is about 1% higher than in the Netherlands and the U.S. This indicates that the process of structural change may have played a more pronounced role in the Japanese economy than in the Dutch and U.S. economies.

The calculated rate of sustainable growth can also be placed in the light of sustainable development, as put forward in the Brundtland report (WCED, 1987). The Brundtland Commission estimated that an economic growth of 3% in developed economies would not be in conflict with the sustainable development. From our estimations we can see that an economic growth of 3% would in all cases lead to increasing consumption of steel, energy, and cement. This indicates that with present rates of technological and structural changes, an economic growth of 3% should not be termed sustainable.

DETERMINANTS OF THE DECLINE IN THE INTENSITY OF USE

The previous sections have shown that structural changes in material and energy consumption are determined by the intensity of use effect. It appeared that the intensity of use hypothesis underestimates the progressive influence of economic growth on material and energy demand. Yet, the intensity of use has declined in many developed countries. This phenomenon is poorly understood (Auty 1985). In order to predict future material demand, Tilton (1986) and Radetzki and Tilton (1990) have stressed the need for identifying the factors underlying the decline in the intensity of use. They emphasize two factors: changes in the structure of the economy and changes in technology. This distinction is also of importance for sustainable development. It has been stated that shifts in the structure of the economy are needed in order to transform our economies toward a sustainable economy (WCED 1987; Jänicke et al. 1989). Technological improvements, on the other hand, may come to an end once the opportunities for win-win situations have been exhausted. It will be clear that the prices of materials and energy should be an important determinant of technological changes.

In order to distinguish between both factors we propose a decomposition method of the intensity of use that employs sectoral data. We can then decompose the intensity of use into intrasectoral and intersectoral changes. Intrasectoral changes are changes of the intensity of use within a sector. The intensity of use within a sector may decrease as the result of increased process-efficiency (less inputs) or increased knowledge intensity of the products (more value added). These can be labeled as technological changes. Intersectoral changes refer to shifts in the composition of income: (i.e., changes in the economic structure). These can be labeled as structural changes.

The distinction between intrasectoral and intersectoral changes in the intensity of use can formally be specified and empirically applied. It requires that data are available on the input of material I, $I_{j,t}$, and value added $Y_{j,t}$ for a range of {j=1...n} sectors in the economy of a country. The intensity of use of material i by the total economy in year t (i_t) is then defined as:

$$i_t = \frac{I_{i,t}}{Y_t} \tag{10.8}$$

The sum of the value added of every sector gives (Y_t), the GDP of an economy.

$$Y_t = \sum_{j=1}^{n} Y_{j,t} \tag{10.9}$$

The sum of $I_{i,j,t}$, the consumption of material i by sector j gives the total demand for this material.

$$I_{i,t} = \sum_{j=1}^{n} I_{i,j,t} \tag{10.10}$$

Equation (10.8) can now be rewritten as the sum of the intensity of use of the individual sectors by weighting them with their market shares in outputs, skipping the material index i of the intensity of use variable I.

$$i_t = \sum_{j=1}^{n} \left(\frac{Y_{j,t}}{Y_t} \cdot \frac{I_{j,t}}{Y_{j,t}} \right) \tag{10.11}$$

The first ratio on the right hand side of (10.11) gives the market share of sector j, defined as the value added of a sector divided by the total value added of the economy. The second ratio gives the intensity of use of sector j. Taking the differences from equation (11) may result in the following expression for the change in the intensity of use of the whole economy, specified in changes in the intensity of use of the individual sectors.[13]

$$\frac{I_{t+1}}{Y_{t+1}} - \frac{I_t}{Y_t} = \sum_{j=1}^{n}\left(\frac{Y_{j,t}}{Y_t} \cdot \left(\frac{I_{j,t+1}}{Y_{j,t+1}} - \frac{I_{j,t}}{Y_{j,t}}\right)\right) +$$

$$\sum_{j=1}^{n}\left(\left(\frac{Y_{j,t+1}}{Y_{t+1}} - \frac{Y_{j,t}}{Y_t}\right) \cdot \frac{I_{j,t+1}}{Y_{j,t+1}}\right) \tag{10.12}$$

The first term on the left-hand side of equation (10.12) gives the intrasectoral changes as the sum of the changes in the intensities of use of the sectors individually, multiplied by their market shares. The second term on the left-hand side gives the intersectoral changes as the sum of each sector's change in market shares multiplied by the intensity of use. Simple rewriting will confirm that the sum of intrasectoral and intersectoral changes equals the total change in the intensity of use. It will be clear that the precise distinction of intersectoral and intrasectoral changes will depend on the number of sectors that will be used in the analysis. If data can be found that are sufficiently detailed, one could treat the lowest level of aggregation as the fabrication process itself. Intrasectoral changes would then truly represent actual improvements in physical efficiency.

This decomposition model will be used to determine which factors (structural or technological changes) have been most influential in the development of the energy intensity of the Dutch economy. Three years of reference have been chosen: 1970, 1980, and 1990. Data for these years have been collected on commercial energy consumption and value added of ten industrial and six nonindustrial sectors (Appendix 2). The industrial classification consists of the following manufacturing sectors: foodstuffs, textiles, paper, chemicals, fertilizers, refineries, building materials, base metals, metal products, and other industry. The nonindustrial classification consists of: agriculture, mining, power generation and distribution, other public utilities, construction and the service-sector, which includes commercial and governmental services. Energy consumption is measured as commercial energy input used for fuel combustion. This means that energy use by households and energy embodied in the final products (for example, naphtha as the output of refineries) has not been taken into account. A second consequence is that the consumption of electricity taken from public power plants is not counted as energy input for the industrial sectors. The energy input used for electricity generation is contributable to the power generation sector only. In general, the share of electricity in the total energy requirement is rather low in Dutch industry.

Table 10.6 shows that the energy intensities differ substantially between sectors. Especially the fertilizer industry and power generation have high intensities. Both typically have labor extensive and energy intensive production so that they add relatively little value compared to their energy inputs. Construction has the lowest energy intensity because of the labor intensive character of production.

Over the time period considered, energy intensities have changed considerably, as can be seen by the index numbers for 1980 and 1990. These indices show that the energy intensity of the whole Dutch economy has declined between 1970–1990. Considering sectoral developments, the decrease in the energy intensities of the public utilities and the mining sector is remarkable. The reason is the large-scale extraction of natural gas and the closing down of several coal mines. As the extraction of natural gas requires less energy than the

Table 10.6. Developments of Energy Intensities and Market Shares in the Netherlands for Selected Sectors (1970–1990)

Sector	Energy Intensity			Market Share		
	1970	1980	1990	1970	1980	1990
	Value	Index numbers (1970–100)				
Agriculture/Fisheries[1]	5.01	217	128	5.7%	3.6%	4.3%
Mining	9.60	6	20	1.6%	5.9%	3.2%
Total Industry	9.52	102	76	26.8%	18.5%	20.1%
Foodstuffs	5.41	118	83	4.7%	3.3%	3.7%
Textiles	4.95	119	53	1.1%	0.5%	0.4%
Paper	15.78	84	65	0.8%	0.6%	0.6%
Fertilizers	107.44	45	83	0.3%	0.2%	0.1%
Other chemicals	13.63	226	116	3.3%	1.7%	2.8%
Building materials	15.64	101	67	1.1%	0.8%	0.7%
Base metals	14.32	119	99	1.3%	0.8%	0.7%
Metal products	1.19	147	68	8.6%	5.7%	6.3%
Oil industry[2]	43.29	87	59	1.8%	1.2%	1.5%
Other industry	2.03	64	33	3.7%	3.7%	3.2%
Electricity[3]	99.63	98	80	1.6%	1.5%	1.5%
Other utilities	11.34	2	19	0.6%	0.7%	0.3%
Construction	0.84	67	29	8.0%	7.3%	5.5%
Service-sector	1.56	51	44	55.6%	62.6%	65.2%
Total economy[4]	5.60	78	61	100.0%	100.0%	100.0%

Notes:

1. Only commercial energy meant for fuel combustion is taken into account.
2. Energy data only for agriculture.
3. Oil industry consists of refineries and coke fabrication.
4. Electricity includes production and distribution of electricity.
5. Total economy excludes the consumption of fuel types by households.
6. Energy counted as TJ input, Value added in millions Dutch constant (1980) guilders.

extraction of coal, this has led to a much lower energy intensity for the mining sector. For the public utilities, the distribution of natural gas instead of the gasification of coal has resulted in much less energy use. On the other hand, energy intensities in the agriculture sector and the chemical industry more than doubled in 1980 compared to 1970. Cheap deliveries to horticulture of natural gas for heating purposes has increased the energy intensity in the agricultural sector.

In 1980 industry as a whole had a slightly higher energy intensity than in 1970. However, in 1990 energy intensity also declined in total industry. The share of industry in total GDP (the last three columns of Table 10.6) declined in 1980, but increased afterwards to 20% in 1990. The market share for the mining sector almost quadrupled due to the large scale extraction of natural gas. The contributions of base metals and building materials industry to the GDP have declined over all the years.

Using equation (10.12) for these data, it is possible to calculate which factors have contributed to the decreases in the energy intensity in the Netherlands. The results are depicted in Table 10.7. The time period under study is divided into two periods: 1970–1980 and 1980–1990. It is remarkable that explanations for the observed decline in the intensity of use differ completely for both periods in the context of the present method. In the first period, the energy intensity of the whole economy declined by more than 24%. It appears that this is solely explained by intersectoral changes. The structure of the Dutch economy has become less energy intensive. Despite the increase in prices in this period, technological changes (measured as intrasectoral changes) have played a marginal role during this decade. It appears that energy saving considerations within sectors were not an important aspect of production (this is largely due to the increase in energy consumption in the chemical sector). The explanation for the declining intensity of use in the second period is completely different: the individual sectors have succeeded in an energy saving per unit output of more than 22% due to technological changes. The structure of

Table 10.7. Decomposition of Energy Intensities in the Netherlands: Intrasectoral Changes and Intersectoral Changes (Expressed as Percentage Change)

	1970–1980	1980–1990
Total economy	-24.14%	-19.80%
o.w. intrasectoral	-0.53%	-22.46%
o.w. intersectoral	-23.61%	2.67%
Total industry	1.97%	-25.57%
o.w. intrasectoral	12.83%	-28.56%
o.w. intersectoral	-10.87%	2.99%

the economy, however, has become more energy intensive. Changes in the structure of the Dutch economy have resulted in an increase in the energy intensity of more than 2.5%.

If decomposition is performed for the Dutch industry only (i.e., excluding the service sector, transport, and communication sector) the above-described developments are more pronounced. As has already been noted, the energy intensity in Dutch industry increased slightly in 1980 compared to 1970. The reason is that, despite the shifts towards a less energy intensive production structure, the individual sectors showed a rise in their energy intensities, leading to an intrasectoral effect of almost 13%. It appears that Dutch industry was unsuccessful in applying energy saving technology until 1980.

In the second period this picture is reversed. The total energy intensity declined by more than 25%. The explanation for this decrease is intrasectoral change, indicating an increase in efficiency at the sectoral level. The intersectoral effect now is positive, indicating that in the 1980s the structure of production has become more energy intensive. This case study shows that the method of decomposition can provide additional insights into issues of dematerialization and de-linking. As a main conclusion of the application of this method to the Dutch economy, one can notice that the effect of the first oil shock (1973) resulted in a change in the structure of the Dutch economy, while the second oil shock (1979–1981) resulted in technological changes in production towards energy saving. These latter technological changes may have been accelerated by policy programs since the Dutch government launched a number of projects to promote energy efficiency in the 1980s.

CONCLUSIONS

There are serious drawbacks in equating an observed decline in materials and energy demand with dematerialization and ecological structural change. Studies that do so find important evidence for dematerialization and structural change, especially after 1970. The cross-section analysis performed in this chapter confirms that pattern, but adds the insight that the consumption of several materials and energy has increased in most developed economies since the mid-1980s. A time-series analysis was presented which provides insight into this apparent pattern of re-linking. It appears that materials and energy consumption is not necessarily de-linked from economic growth in general. An increase in economic growth may result in a more than proportional increase in the consumption of materials and energy. Furthermore, the growth rates of the Dutch, Japanese, and U.S. economies between 1966 and 1990 have been well above the rate of sustainable growth (i.e., the rate of growth below which the increase in income does not lead to a higher consumption of materials).

A statistical decomposition method can provide more insight into the factors underlying the changes in the intensity of use. Application of the method to commercial energy consumption in the Netherlands makes clear why energy saving was higher in the 1970s than in the 1980s despite the launching of energy saving programs in the 1980s. The process of structural change turned out to be rather absent in the 1980s. Contrary to the 1970s, the structure of the Dutch economy has become more energy intensive: sectors with a high energy intensity have expanded their production more than the sectors with a low energy intensity. This all casts serious doubts on the stability of the processes of dematerialization and structural change.

ACKNOWLEDGMENTS

The authors would like to thank Jan van der Straaten and Roebijn Heintz for comments on earlier drafts. All remaining errors are ours.

FOOTNOTES

1. Throughout this chapter consumption and income refer to per capita consumption and income.
2. It may be noted that in 1990 in the United States cement consumption accounted for 18% and steel consumption for 22% of all industrial minerals and metals consumed in the U.S. economy (construction materials like crushed stone, sand, and gravel excluded) (Berry 1994). The weight of freight transport in tons can be seen as a general indicator of the volume aspect of production.
3. It should be noted, however, that with such a low R^2 the decreasing slope is difficult to interpret and may fall within the statistical confidence intervals.
4. For a critical review of these studies on methodological grounds, see Stern et al. (1995) and De Bruyn, van den Bergh, and Opschoor (1995).
5. A considerable amount of resource input is being "stored" in the products that have been produced which become wastes after they have been discarded.
6. The sum of the deviations from the sample mean is zero by definition. Therefore, the arrows of individual countries indicate the developments from the relative position in 1970 to the relative position in 1985.
7. The sample consists of Belgium–Luxembourg, Denmark, Finland, France, Greece, Hungary, Italy, Japan, Luxembourg, the Netherlands, Norway, Poland, Spain, Sweden, Switzerland, Turkey, United Kingdom, the United States, Western Germany, and (former) Yugoslavia.
8. Poland has not been given in this figure for reasons of clearness.
9. The error term in the models is calculated as the first difference of the logarithmic expression for the usual error term. It should be added that first difference models have been chosen for ease of interpretation and that it does not necessarily mean that the lag structure is properly specified in this way.
10. The inverted-U form of the intensity of use curve has been given by Suslick and Harris (1990):

    $$I_t / Y_t = a_1 Y_t^{(a2+ \beta 2 \ln Yt)} + \varepsilon_t$$

    From this the demand function for I_t can be derived as:

    $$I_t = a_1 \cdot Y_t^{(\beta^1 + \beta^2 \ln Y^t)} + \varepsilon_t$$

where $\beta_1 + \alpha_2$. This model implies an income elasticity of demand that varies with the income level, which is characteristic for the inverted-U shaped intensity of use curve. Estimation of model 2 will thus, in general, give a higher value for β_1 than of the other models.

11. This may give a bias for the other countries involved in the estimations, because their rate of price increase may differ from the U.S. market. Unfortunately we have not been able until now to collect all the price indices for each country individually.
12. Some authors have pointed out that to realize sustainable development on a global scale, the developed countries should even reduce their consumption of materials and energy in order to create "environmental space" for growth in the consumption of materials and energy in developing countries (Weterings and Opschoor 1992).
13. This decomposition is not unique; there are other decomposition methods possible that give a different division between intersectoral and intrasectoral changes.

REFERENCES

Al-Rawahi, K. and M. Rieber. 1991. Embodied copper: Trade, intensity of use and consumption forecasts. *Resources Policy* 17:2–12.

Auty, R. 1985. Materials intensity of GDP: Research issues on the measurement and explanation of change. *Resources Policy* 11:275–283.

Barnet, H. J. and Ch. Morse. 1963. *Scarcity and Growth: The Economics of Natural Resource Availability.* London: Johns Hopkins Press.

Berry, D. 1994. Information needs for sustainable resource policy. In *Sustainable Resource Management and Resource Use: Policy Questions and Research Needs,* eds. F. J. Duinhouwer, G. J. van der Meer, and H. Verbruggen. Raad voor Milieu en Natuur Onderzoek (RMNO), RMNO, nr. 97:175–90.

Bossanyi, E. 1979. UK primary energy consumption and the changing structure of final demand. *Energy Policy* 7:253–258.

Chesshire, J. 1986. An energy-efficient future: A strategy for the UK. *Energy Policy* 14:395–412.

Daly, H. E. 1991. *Steady State Economics.* San Francisco: Freeman.

De Bruyn, S. M., J. C. J. M. van den Bergh, and J. B. Opschoor. 1995. Empirical Investigations in Environmental–Economic Relationships: Reconsidering the Empirical Basis of the Environmental Kuznets Curves and the De-linking of Pollution from Economic Growth. Tinbergen Discussion Paper, TI 95–140. Amsterdam: Tinbergen Institute.

Dobozi, I. 1990. The centrally planned economies: Extravagant consumers. In *World Metal Demand: Trends and prospects, Resources for the Future,* ed. J. E. Tilton. Washington DC.

Goldemberg, J. 1992. Energy, technology, development. *Ambio* 21:14–17.

Holmes, M. J. 1990a. Material substitution in battery electrodes: The UK 1983–1987. *Resources Policy* 16:22–34.

Holmes, M. J. 1990b. Material substitution in the UK window industry, 1983–1987. *Resources Policy* 16:128–142.

Jänicke, M., H. Monch, and M. Binder. 1992. *Environmental Improvement by Industrial Structural Change?* (in German). Berlin: Edition Sigma.

Jänicke, M., H. Monch, and M. Binder. 1993. Ecological aspects of structural change. *Intereconomics Review of International Trade and Development* 28:159–169.

Jänicke, M., H. Monch, T. Ranneberg, and U. E. Simonis. 1989. Economic structure and environmental impacts: East-west comparisons. *The Environmentalist* 9:171–182.

Larson, E. D., M. H Ross, and R. H. Williams. 1986. Beyond the era of materials. *Scientific American* 254: 34–41.

Malenbaum, W. 1978. *World Demand for Raw Materials in 1985 and 2000.* New York: McGraw-Hill.

Mannaerts, H. J. B. M. 1993. *The use of non-energetic resources in the long run* (in Dutch). Onderzoeksmemorandum 109. Den Haag: Centraal Planbureau.

McSweeney, Ch. and M. Hirosako. 1991. Understanding crude steel consumption: The perils of ignoring the role of technological change. *Resources Policy* 17:258–270.

Moll, H. C. 1993. *Energy Counts and Materials Matter in Models for Sustainable Development: Dynamic Lifecycle Modelling as a Tool for Design and Evaluation of Long-term Environmental Strategies.* Groningen: Styx.

Opschoor, J. B. 1990. Ecologically sustainable economic development: A theoretical idea and a refractory practice (in Dutch). In *Het Nederlands Milieu in de Europese Ruimte: Preadviezen van de Koninklijke Vereniging voor Staathuishoudkunde,* pp. 77–126, eds. P. Nijkamp and H. Verbruggen. Stenfert Kroese, Leiden.

Opschoor, J. B. and L. Reijnders. 1991. Towards sustainable development indicators. In *Search of Indicators for Sustainable Development,* eds. H. Verbruggen and O. Kuik. Dordrecht: Kluwer Academic Press.

Pezzey, J. 1989. Economic Analysis of Sustainable Growth and Sustainable Development. Environment Department Working Paper No 15. Washington DC: The World Bank.

Radetzki, M. 1990. Developing countries: The new growth markets. In *World Metal Demand: Trends and prospects,* ed. J. E. Tilton. Washington DC: Resources for the Future, pp. 77–112.

Radetzki, M. and J. E. Tilton. 1990. Conceptual and methodological issues. In *World Metal Demand: Trends and prospects,* ed. J. E. Tilton. Washington DC: Resources for the Future, pp. 13–34.

Rijksinstituut voor Volksgezondheid en Milieuhygiene (RIVM). 1991. *National Environmental Outlook* 2 (in Dutch), Alphen aan de Rijn: Samson-Willink.

Roberts, M. C. 1990. Predicting metal consumption: The case of US steel. *Resources Policy* 16:56–73.

Schmidt-Bleek, F. 1993. MIPS Re-Visited. *Fresensius Environmental Bulletin* 2:2–7.

Selden, T. M. and D. S. Song. 1994. Environmental quality and development: Is there a Kuznets curve for air pollution emissions? *Journal of Environmental Economics and Management* 27:147–162.

Shafik, N. and S. Bandyopadhyay. 1992. Economic Growth and Environmental Quality: Time-Series and Cross-Country Evidence, World Bank Working Papers WPS 904, Washington DC.

Simonis, U. E. 1989. Industrial Restructuring for Sustainable Development: Three Points of Departure. Science Centre Berlin FS II 89-401, Berlin.

Stern, D. I., M. S. Common, and E. B. Barbier. 1996. Economic growth and environmental degradation: A critique of the Environmental Kuznets Curve. *World Development* 24:1151–1160.

Suslick, S. B., and D. P. Harris. 1990. Long-range metal consumption forecasts using innovative methods: The case of aluminium in Brazil to the year 2000. *Resources Policy* 16:184–199.

Tilton, J. E. 1986. Beyond intensity of use. *Materials and Society* 10:245–250.

Tilton, J. E. 1990. The OECD countries: Demand trend setters. In *World Metal Demand: Trends and prospects,* ed. J. E. Tilton. Washington DC: Resources for the Future, pp. 35–76.

Valdes, R. M. 1990. Modelling Australian steel consumption. *Resources Policy* 16:172–183.

Vial, J. 1992. Copper consumption in the USA: Main determinants and structural changes. *Resources Policy* 18:107–121.

von Weizsacker, E.U., and F. Schmidt-Bleek. 1994. Signs of hope for the 21st century? (In Dutch) Committee for Long-Term Environmental Policy. *The Environment: Towards a Sustainable Future.* Dordrecht: Kluwer Academic Publishers.

Wadell, L. M. and W. C. Labys. 1988. Transmaterialization: Technology and materials demand cycles. *Materials and Society* 12:59–85.

Weterings, R. A. P. M. and J. B. Opschoor. 1992. The Ecocapacity as a Challenge to Technological Development. Advisory Council for Research on Nature and Environment, Rijswijk.

World Commission on Environment and Development (WCED). 1987. *Our Common Future.* Oxford: Oxford University Press.

APPENDIX 1

A METHODOLOGY FOR THE CALCULATION OF THE INDEX OF ENVIRONMENTAL DETERIORATION (IED)

The calculation of the Index of Environmental Deterioration (IED) in this chapter involves a two-step procedure. First, sample means have been calculated for all countries and all years. For each indicator this was done using the following formula:

$$\overline{x_i} = \frac{\sum_{t=1966}^{1990} \sum_{c=1}^{20} C_{i,c,t}}{\sum_{t=1966}^{1990} \sum_{c=1}^{20} POP_{c,t}} \tag{10.13}$$

where i = indicator (1..4), c = country (1..20), t = year (1966..1990), Xi = per capita consumption of indicator (i), POP = population, Ci = measured yearly consumption of indicator i.

The environmental index (IED) for each country (c) and for each year (t) is calculated by adding the relative deviations from the mean per capita consumption of each indicator ($X_{i,c,t}$), dividing it by four which results in an equal weighting of each indicator. In formula this becomes:

$$IED_{c,t} = \sum_{i=1}^{4} \frac{1}{4} \cdot \left(\frac{x_{i,c,t} - \overline{x_i}}{\overline{x_i}} \right) \tag{10.14}$$

The formula used by Jänicke et al. (1989) differs from this since the sample mean was year-specific (i.e., the summation signs over the years 1966–1990 in equation (10.14) were not taken into account). Both for 1970 and 1985 they calculated the IED as the deviations from the mean for, respectively, 1970 and 1985.

APPENDIX 2

DATA SOURCES AND DATA QUALITY

Our data were taken from a number of international sources as listed below.

Cement Consumption

Economic Commission for Europe of the United Nations: Annual Report on Housing and Building. United States: CRB: Commodity Yearbook. For Italy, Japan, Spain, and Turkey cement production has been taken as a proxy for consumption. Production data from Statistical Office of the United Nations: Statistical Yearbook.

Energy Consumption

Measured as Total Primary Energy Supply (TPES). International Energy Agency (IEA): Energy Balances of OECD Countries, Energy Balances of Non-OECD countries, World Energy Statistics and Balances; For Hungary, Poland, and Yugoslavia before 1971 growth rates extracted from: United Nations: "World Energy Supplies" (Total Final Consumption).

Steel Consumption

Until 1987: United Nations: Statistical Yearbook. After 1987: Eurostat: Eisen undStahl for EC members, Economic Commission for Europe of the United Nations: Annual Steel Statistics for Europe for non-EC members. Series have been standardized on each other using three-year averages.

Freight Transport

Economic Commission for Europe of the United Nations: Annual Bulletin of Transport Statistics for Europe. International Road Federation: World Road Statistics. European Conference of Ministers of Transport (ECMT): Transport Statistics. Especially for road transport series show differences. Series have been standardized. Rail transport has been used for all countries, road transport for all countries except the United States, Spain, Italy, and Greece. For these countries the quantity of freight transport was measured by train only. This is the same routine as Jänicke et al. took in their investigations in 1989.

Population

Worldbank: Worldtables.

Gross Domestic Product

Worldbank: Worldtables, 1992. For Yugoslavia, Poland, and Hungary before 1971 growth rates have been extracted from: Summers, R. A. Heston: A new set of international comparisons of real product and price levels estimates for 130 countries, 1950–1985 In: Income and Wealth 1988.

Structure of Dutch Economy

Value added data have been extracted from the Dutch Statistical Office (CBS) (1985): Nationale rekeningen 1969–1981 met herziene reeksen voor de jaren 1969–1976. From 1977 and on: CBS (1978–1991): Samenvattend overzicht industriestatistieken" For the fertilizer industry data taken from: Statistical Yearbook of the Netherlands (1972–1993). For power generation by public utilities, data taken from: Statistiek van de elektriciteitsvoorziening in Nederland

(1970–1991). Value added has been calculated as gross value added using market prices.

Energy Input in Dutch Economy

Delivered from CBS databases.

Name Index